Theilheimer's

Synthetic Methods

of Organic Chemistry

Vol. 83

Vol. 83 2014

Theilheimer's

Synthetic Methods

of Organic Chemistry

Editor **Gillian Tozer-Hotchkiss,** Wirral, UK

Assistant Editors Alan F. Finch, Cambridge, UK
Chris Hardy, Leeds, UK
Julian Hayward, Leeds, UK

Technical Editor Jill Entwistle, Berkhamsted, UK

Basel • Freiburg
Paris • London • New York
New Delhi • Bangkok
Bejing • Tokyo
Kuala Lumpur
Singapore • Sydney

KARGER

digital chemistry

Deutsche Ausgaben

Vol. 1	1946	1. Auflage
	1948	2., unveränderte Auflage
	1950	3., unveränderte Auflage
Vol. 2	1948	
Vol. 3	1949	with English Index key
	1953	2., unveränderte Auflage
	1966	3., unveränderte Auflage
	1975	4., unveränderte Auflage
Vol. 4	1950	with English Index key
	1966	2., unveränderte Auflage

English Editions

Vol. 1	1948	Interscience Publishers
	1975	(Karger) Second Edition
Vol. 2	1949	Interscience Publishers
	1975	(Karger) Second Edition
Vol. 5	1951	with Cumulative Reaction Titles and Index
	1966	Second Edition
Vol. 6	1952	1975 Second Edition
Vol. 7	1953	1975 Second Edition
Vol. 8	1954	1975 Second Edition
Vol. 9	1955	
Vol. 10	1956	with Cumulative Reaction Titles and Index
	1975	Second Edition
Vol. 11	1957	1975 Second Edition
Vol. 12	1958	1975 Second Edition
Vol. 13	1959	1975 Second Edition
Vol. 14	1960	1975 Second Edition
Vol. 15	1961	with Cumulative Reaction Titles and Index
Vol. 16	1962	
Vol. 17	1963	
Vol. 18	1964	
Vol. 19	1965	
Vol. 20	1966	with Cumulative Reaction Titles and Index
Vol. 21	1967	
Vol. 22	1968	
Vol. 23	1969	
Vol. 24	1970	
Vol. 25	1971	with Cumulative Reaction Titles and Index
Vol. 26	1972	
Vol. 27	1973	
Vol. 28	1974	
Vol. 29	1975	
Vol. 30	1976	with Cumulative Reaction Titles and Index
Vol. 31	1977	
Vol. 32	1978	
Vol. 33	1979	
Vol. 34	1980	
Vol. 35	1981	with Cumulative Reaction Titles and Index
Vol. 36	1982	
Vol. 37	1983	
Vol. 38	1984	
Vol. 39	1985	
Vol. 40	1986	with Cumulative Reaction Titles and Index
Vol. 41	1987	
Vol. 42	1988	
Vol. 43	1989	
Vol. 44	1990	
Vol. 45	1991	with Cumulative Reaction Titles and Index
Vol. 46	1992	
Vol. 47	1993	
Vol. 48	1994	
Vol. 49	1995	
Vol. 50	1996	with Cumulative Reaction Titles
Vol. 51	1997	
Vol. 52	1997	
Vol. 53	1998	
Vol. 54	1998	
Vol. 55	1999	
Vol. 56	1999	
Vol. 57	2000	
Vol. 58	2000	
Vol. 59	2001	
Vol. 60	2001	
Vol. 61	2002	
Vol. 62	2002	
Vol. 63	2003	
Vol. 64	2003	
Vol. 65	2004	
Vol. 66	2004	
Vol. 67	2005	
Vol. 68	2005	
Vol. 69	2006	
Vol. 70	2006	
Vol. 71	2007	
Vol. 72	2008	
Vol. 73	2008	
Vol. 74	2009	
Vol. 75	2009	
Vol. 76	2010	
Vol. 77	2011	
Vol. 78	2011	
Vol. 79	2012	
Vol. 80	2012	
Vol. 81	2013	
Vol. 82	2013	

Library of Congress, Cataloging-in-Publication Data

Theilheimer's synthetic methods of organic chemistry = Synthetische Methoden der organischen Chemie. –
Vol. 83 (2014) - Basel; New York: Karger, © 1982 -
v.
Continues: Synthetic methods of organic chemistry.
Editor: Gillian Tozer-Hotchkiss.
1. Chemistry, Organic – yearbooks I. Tozer-Hotchkiss, Gillian II. Finch, Alan F. III. Theilheimer, William, 1914–2005
ISBN: 978-3-318-02654-2
e-ISBN: 978-3-318-02655-9

Distributed by S. Karger AG, Allschwilerstrasse 10, P.O. Box, CH-4009 Basel (Switzerland)
ISBN: 978-3-318-02654-2
e-ISBN: 978-3-318-02655-9

Contents

Preface

This volume of *Theilheimer* contains abstracts of new synthetic methods and supplementary data mainly from papers published in the literature up to June 2013.

For browsing purposes, abstracts are displayed according to the Systematic Classification (symbol notation: summary p. VIII) so that reactions of the same type and associated data appear together. For example, all deprotections appear in the early symbols (under HO↓↑, HN↓↑, HS↓↑); reduction of oxo compds., imines and carbon-carbon multiple bonds under the HC⇓ sections; C-defunctionalization under the HC sections; oxy-functionalization under the OC sections; aminations, nitrations, peptide coupling etc. under the NC sections; halogenation under the HalC sections; sulfurations under the SC sections; selenation, stannylation, phosphorylation, etc. under the RemC sections; syntheses involving C-C bond formation in the latter half of the book under the CC sections; and data on resolutions (Res) at the end. A list of reaction symbols and references thereto is given in the Systematic Survey (p. XXII).

The displayed data are supported by the customary in-depth Subject Index (p. 466) and access to supplementary data can be made in the usual manner via the Supplementary Reference section, e.g. the reader interested in updates on the Biginelli synthesis (Synth. Meth. *55*, 337) will note from p. 523 that additional references can be found on p. 331 of this volume.

As usual, the volume contains a 'Reviews' section (p. 446), covering reviews published up to and including December 2013, and a 'Trends' section (p. XI) incorporating key developments in synthetic chemistry up to and including January 2014.

I would like to express my gratitude to Alan Finch for his continuing help and enthusiasm during the preparation of these volumes, as well as to Julian Hayward and Chris Hardy. We are also very grateful for the assistance and support of Jill Entwistle, Eliot Cartwright-Finch, Chloë Cyrus-Kent, Alan Hayward and, last but not least, Andrew Hotchkiss.

January 2014 *G. Tozer-Hotchkiss*, Editor

Advice to the User

General Remarks

New methods for the synthesis of organic compounds and improvements of known methods are being recorded continuously in this series.

Reactions are classified on a simple though purely formal basis by symbols, which can be arranged systematically. Thus searches can be performed without knowledge of the current trivial or author names (e.g. 'Oxidation' and 'Friedel-Crafts reaction').

Users accustomed to the common notations will find these in the subject index (see page 466). By consulting this index, use of the classification system may be avoided. It is thought that the volumes should be kept close at hand. The books should provide a quick survey, and obviate the immediate need for an elaborate library search. Syntheses are therefore recorded in the index by starting materials and end products, along with the systematic arrangement for the methods. This makes possible a sub-classification within the reaction symbols by reagents, a further methodical criterion. Complex compounds are indexed with cross reference under the related simpler compounds. General terms, such as synthesis, replacement, heterocyclics, may also be brought to the attention of the reader.

A brief review, *Trends and Developments in Synthetic Organic Chemistry* (see page XI), stresses highlights of general interest and calls attention to key methods too recent to be included in the body of the text.

The abstracts are limited to the information needed for an appraisal of the applicability of a desired synthesis. In order to carry out a particular synthesis it is therefore advisable to have recourse to the original papers or, at least, to an abstract journal. In order to avoid repetition, selections are made on the basis of most detailed description and best yields whenever the same method is used in similar cases. Continuations of papers already included will not be abstracted, unless they contain essentially new information. They may, however, be quoted at the place corresponding to the abstracted papers. These supplementary references (see page 519) make it possible to keep abstracts of previous volumes up-to-date.

Syntheses that are divided into their various steps and recorded in different places can be followed with the help of the notations such as *startg. m. f.* (starting material for the preparation of ...).

Method of Classification

Reaction Symbols. As summarized in the Systematic Survey (page XXII), reactions are classified firstly according to the bond formed in the synthesis, secondly according to the reaction type, and thirdly according to the bond broken or the element eliminated. This classification is summarized in the reaction symbol, e.g

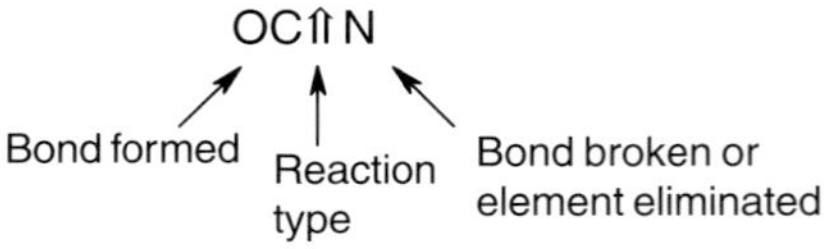

The first part of the symbol refers to the chemical bond formed during the reaction, expressed as a combination of the symbols for the two elements bonded together, e.g. HN, NC, CC. The order of the elements is as follows:

H, O, N, Hal (Halogen), S, Rem (Remaining elements), and C.

Thus, for the formation of a hydrogen-nitrogen bond, the notation is HN, not NH.

If two or more bonds are formed in a reaction, the 'principle of the latest position' applies. Thus, for the reduction

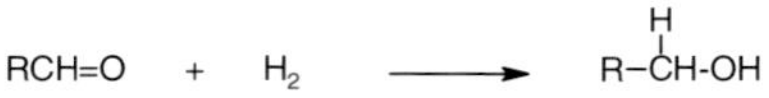

in which both hydrogen-oxygen and hydrogen-carbon bonds are formed, the symbol is HC⇓OC and not HO⇓OC.

The second part of the symbol refers to the reaction type. Four types are distinguished: addition (⇓), rearrangement (∩), exchange (⇅), and elimination (⇑), e.g.

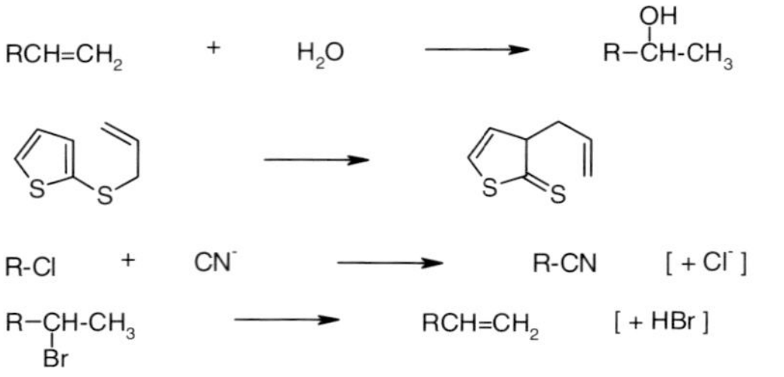

OC⇓CC

CC∩SC

CC⇅Hal

CC⇑Hal

Monomolecular reactions are either rearrangements (∩), where the molecular weight of the starting material and product are the same, or eliminations (⇑), where an organic or inorganic fragment is lost; bimolecular and multicomponent reactions are either additions (⇓), such as intermolecular Diels-Alder reactions, Michael addition and 1,4-addition of organometallics, or exchanges (↓↑), such as substitutions and condensations, where an organic or inorganic fragment is lost.

The last part of the symbol refers to the essential bond broken or, in the case of exchange reactions and eliminations, to a characteristic fragment which is lost. While the addition symbol is normally followed by the two elements denoting the bond broken, in the case of valency expansion, where no bonds are broken, the last part of the symbol indicates the atom at which the addition occurs, e.g.

$R_2S \longrightarrow R_2SO$ OS⇓S

$RONO \longrightarrow RONO_2$ ON⇓N

For addition, exchanges, and eliminations, the 'principle of the latest position' again applies if more than one bond is broken. However, for rearrangements, the most descriptive bond-breakage is used instead. Thus, for the thio-Claisen rearrangement depicted above, the symbol is CC∩SC, and not CC∩CC.

Deoxygenations, quaternizations, stable radical formations, and certain rare reaction types are included as the last few methods in the yearbook. The reaction symbols for these incorporate the special symbols El (electron pair), Het (heteropolar bond), Rad (radical), Res (resolutions), and Oth (other reaction types), e.g.

$R_2S{=}O \longrightarrow R_2S$ ElS⇑O

$R_3N + R'Cl \longrightarrow R_3N^+R'\ Cl^-$ Het⇓N

The following rules simplify the use of the reaction symbols:

1. The chemical bond is rigidly classified according to the structural formula without taking the reaction mechanism into consideration.

2. Double or triple bonds are treated as being equivalent to two or three single bonds, respectively.

3. Only stable organic compounds are usually considered: intermediates such as Grignard compounds and sodiomalonic esters, and inorganic reactants, such as nitric acid, are therefore not expressed in the reaction symbols.

Reagents. A further subdivision, not included in the reaction symbols, is based on the reagents used. The sequence of the reagents usually follows that of the periodic system. Reagents made up of several components are arranged according to the element significant for the reaction (e.g. $KMnO_4$ under Mn, NaClO under Cl). When a constituent of the reagent forms part of the product, the remainder of the reagent, which acts as a 'carrier' of this constituent, is the criterion for the classification; for example, phosphorus is the carrier in a chlorination with PCl_5 and sodium in a nitrosation with $NaNO_2$.

Further Trends and Developments in Synthetic Organic Chemistry 2013

There are many foundations to synthetic organic chemistry, but from a catalytic perspective the central pillars are transition metal catalysis, organocatalysis, and Brønsted/Lewis acid/base catalysis, with biocatalysis, and the relative newcomers on the block - carbene catalysis and the much heralded visible-light photoredox catalysis - as the capstones[1]. Providing the cement is a diverse subdivision, some elements clearly definable, some less so, some overlapping, some rather unnecessary, and in total a veritable nightmare from an indexing perspective. These include: dual and multiple catalysis, cooperative catalysis, synergistic catalysis, sequential catalysis, relay catalysis, bifunctional catalysis, bimetal catalysis and a host of others. *Merging* the pillars together is in the gift of *tandem catalysis*, let us say providing the lintels, with the facility of linking two of the central elements in one step through a dual catalytic cycle, either with the same catalyst or with two mutually compatible catalysts. This isn't new in so far as transition metal-catalyzed tandem conversions, as illustrated in a recent regioselective palladium-catalyzed Heck-Negishi coupling in the presence of $Pd(PPh_3)_4$[2]. Nor is merging transition metal catalysis with organocatalysis: it's as old as metalloenzymic catalysis itself, no doubt inspiring the seminal dynamic kinetic resolution of secondary alcohols via ruthenium-catalyzed inversion and enzymatic O-acylation (*53,* 500)[3]. Nowadays, such merging also embraces organocatalysis, notably in the sphere of asymmetric synthesis, and highlighted in the *Trends* of Volume *82* (s. *83,* 381)[4]. Note, as a more recent example, the waste-free asymmetric conversion of allyl alcohols to γ-functionalized alcohols by merging an iron-catalyzed 'borrowing hydrogen' step with regioselective asymmetric amine-catalyzed nucleophilic addition[5]. Merging acid catalysis with transition metal catalysis is also relatively commonplace, as illustrated by the rapid hydrogenolysis of alcohols and ethers to alkanes, wherein the substrates undergo initial Lewis acid-catalyzed dehydration/dealcoholation to the alkene (using a high-valent metal triflate) with subsequent hydrogenation with a supported palladium catalyst[6]; in the same vein, a Prins cyclization of an allyloxyindole is effected by merging iridium-catalyzed isomerization to an oxonium ion with subsequent Brønsted acid-catalyzed ring closure[7]. Merging visible-light photoredox catalysis with organocatalysis, however, is of more recent origin, but developing fast. Note, for example, application of the principle to the direct β-functionalization of ketones with aldehydes via an amine-catalyzed 'homo-aldol-type' condensation[8], and to an asymmetric synthesis of cyclic 2-aminoalcohols via an intramolecular pinacol-type coupling, wherein a chiral phosphoric acid monoester serves as organo-Brønsted acid[9]. Quite where a recent *synergistic* catalysis, merging gold catalysis with amine catalysis[10], and *relay* catalysis[11] fit into the equation is open to conjecture, while *cooperative* catalysis is arguably too loose a descriptor to merit a detailed analysis. Not so, however, the aspect of *multicatalytic multicomponent* synthesis, adding a complication but at the same time having enormous potential. A new 3-component dibenzazepine synthesis, for example, features a dual catalytic domino sequence with rhodium(I) and palladium(0)[12], while a recent 3-component synthesis of carboxylic acid N-(diarylmethyl)amides is reliant on $Yb(OTf)_3$ as Lewis acid and palladium(II)[13]. The future may not bring another central pillar to match transition metal catalysis and organocatalysis, at least in the foreseeable future, but more cement will doubtless be added from time to time, with perhaps *supramolecular* catalysis leading the field[14] ... and there is always the prospect of another element, on a par with tandem catalysis, to merge the pillars closer: bioinspired syntheses via supramolecular-catalyzed *self-organization* of simple components, effectively *'on an assembly line'* could well be it![15]

While the precious metal triads of the 'old' Group VIII - Ru,Rh,Pd and Os,Ir,Pt - have taken center stage in transition metal catalysis over the last three decades, comparisons with the less expensive and more available Fe,Co,Ni triad are always welcome[16]. A recent site-selective, [non-heme] iron-catalyzed *C(sp^3)-H* oxidation of complex molecules is of particular note, wherein a tetrakis(trifluoromethyl)ated cationic iron(II) [N,N′-bis(2-pyridylmethyl)]-2,2′-bipyrrolidinyl complex activates a normally unreactive C-H site by blocking the adjacent, conventionally reactive one[17]; (E)-selective partial hydrogenation of alkynes has also emerged courtesy of a benign acridine-based iron PNP-type pincer complex[18], as have: asymmetric transfer-hydrogenation of ketones and imines with a chiral iron amine(imino)di-(phosphine) complex[19]; a highly selective hydrogenation of diverse nitroarenes with iron(III) oxide particles surrounded by a nitrogen-doped carbon layer[20]; and a novel iron-catalyzed hydrogenolytic cleavage of phenolethers to provide arenes without ring hydrogenation[21]; note, too, a low-cost photochemical aromatic borylation under cooperative catalysis with a heterobimetallic Fe-Cu or Fe-Zn catalyst[22], and iron-catalyzed N-directed *ortho* functionalization of N-(8-quinolyl)benzamides[23].

C-H bond activation, as touched upon above, is arguably the most rapidly expanding and diverse aspect of transition metal catalysis[24]. Of particular note is the recent upsurge of *dehydrogenative cross-coupling of C(sp^3)-H bonds*, notably associated with α-functionalization of amines in the presence of a sacrificial oxidant[25] – and recently adapted in an iron-catalyzed dehydrogenative α-substitution of α-substituted α-aminocarboxylic acid amides[26]. Copper-catalyzed, *aerobic* cross-coupling is also developing, as illustrated in a synthesis of α-cyanoacetophenones from benzyl alcohols and acetonitrile involving loss of *four* hydrogen atoms[27]; and even more welcome are visible-light *photocatalytic* [literally dehydrogenative!] cross-couplings *without* any oxidant, exemplified by C_1-functionalization of 1,2,3,4-tetrahydroisoquinolines on graphene-supported RuO_2 nanocomposite with eosin Y as sensitizer[28]. Transition metal-catalyzed *dehydrogenative cross-coupling of C(sp*^{2}*)-H bonds* is rather more diversely represented, thanks largely to the prevalence of *o*-directed dehydrogenative Heck-type reactions. Here, rhodium-catalyzed *o*-vinylation of aryl N-tosylimines[29] and aroyl-silanes[30] are notable, as is a palladium-catalyzed *o*-vinylation of β-aryl-sulfoxides in which the sulfinyl group serves as a *remote* directing group with Selectfluor as oxidant[31]; rhodium-catalyzed N-deoxygenative *o*-vinylation of dialkyl(aryl)amine N-oxides, however, requires no oxidant, the traceless N-oxide group serving the purpose as well as directing substitution[32]. Dehydrogenative directed *o*-functionalization with allenes has also been reported[33], together with an interesting synthesis of 3,3-dialkyl-oxindoles from cinnamic acid anilides involving copper-catalyzed intramolecular hydroarylation and a rare dehydrogenative coupling *with a hydrocarbon*[34]. In recent years, transition metal-catalyzed directed *o*-substitution by carbon has developed side by side with *ortho* heteroatom substitution. In connection with C-N bond formation, newcomers include rhodium-catalyzed N-directed *o*-acylamination with hydroxamic acids[35], and a mild directed *o*-hydroxylamination via rhodium-catalyzed 1,2-addition to nitroso compounds[36]. Rather more notably, however, is the norbornene-mediated palladium-catalyzed *o*-amination *of aryl bromides or iodides* with O-benzoylhydroxylamines, being energetically more favorable than classical palladium-catalyzed N-arylation[37]. Phosphinyl-mediated *o*-hydroxylation is equally notable[38], alongside iridium-catalyzed *o*-borylation of anilines and N-heteroarenes based on *in situ* incorporation of a traceless N-boryl directing group[39]. Ruthenium-catalyzed amide-directed *o*-halogenation has also been reported[40], together with a mild palladium-catalyzed *o*-chlorination based on the weakly coordinating carbamyloxy directing group[41]; on the other hand, *o*-fluorination of N-heteroarenes with AgF_2 [*i.e.* at the C_2-site] likely proceeds via the directing influence of Ag-N coordination[42]; note, too, the emergence of a palladium-catalyzed *o*-deuteriation of arylacetic and benzoic acids, facilitated by the weakly coordinating carboxyl group as directing handle[43]. A suitably positioned cyano group, however, is somewhat anomalous in directing *meta* substitution, as in palladium-catalyzed *m*-arylation of β-arylcarboxylic acid N,N-bis(*o*-cyanophenyl)amides[44]; in a similar vein, palladium-catalyzed dehydrogenative *m*-vinylation of benzyl alcohols can be effected via substitution of their readily appended and cleavable O-(*o*-cyanophenylsiloxy)-derivatives[45].

Developments in dehydrogenative cross-coupling and arene coupling outlined above can indeed provide eco-friendly and waste-free alternatives to classical transition metal-catalyzed coupling of organometallics, organoboron, organosilicon and organotin compounds with halides and other electrophiles. Regioselectivity, functional group sensitivity to oxidizing agents, and the reliance on directing groups do however present problems and limitations, which are largely obviated in classical approaches. Hence, diversification of the latter is as topical as ever: note, for example, a green Suzuki coupling *in water* courtesy of amphiphilic α-tocopherol methoxy-polyethylene glycol succinate as surface active agent[46]; cross-coupling of boronic acids with chiral 2-ethylenealcohols has also been effected with inversion of configuration[47], while Hiyama cross-coupling of vinyl- and aryl-silanes with a wide range of saturated halides can now be conducted under copper catalysis via intramolecular activation of the C-Si bond through coordination with an appended alkoxide group[48]. Particularly noteworthy is the development of 'ring-opening' cross-coupling wherein the leaving group is present as a cyclic C-O or C-N function. This is exemplified in the regioselective synthesis of 2-aryl-N-sulfonylamines by palladium-catalyzed ring opening of N-sulfonylaziridines with arylboronic acids[49]; by the same token, aryl and vinyl halides (or triflates) participate in nickel-catalyzed ring-opening cross-coupling with epoxides under reducing conditions[50], while the synthesis of stereo-defined 2,4-dienecarboxylic acids is now achievable by ring-opening cross-coupling of 2-pyrones with Grignard reagents under iron catalysis[51]; the regioselective generation of quaternary carbon centers by classical asymmetric Grignard cross-coupling with allyl halides is also in the news, courtesy of a chiral bifunctional imidazolidin-2-ylidene as ligand which, notably, obviates the need for a transition metal as catalyst[52]. A transition metal-free Negishi coupling of arylzinc halides with unsatd. iodides has also been reported with LiCl as key additive, inducing an SET manifold[53].

The many and varied developments in heterogeneous catalysis can be overviewed quickly from the reviews literature[54]. Add to this a key publication on the prevention of base metal (e.g. copper) deactivation by *atomic layer deposition* (ALD) of an alumina protective coating, which minimizes both leaching in liquid-phase reactions and sintering[55]. The stabilization of palladium nanoparticles with a silicon nanowire array is a further advance, thereby generating a highly active, recyclable catalyst for performing a variety of transformations at extremely low catalytic levels of the metal (e.g. 0.000049 mol% for Heck arylation!)[56]. A well-defined, highly active silica-supported alkylidene(oxo)(phenolato)tungsten(VI) is also available for alkene metathesis[57], while a porous silica-bonded N-methyldisilazane serves as a highly selective nucleophilic catalyst for the Baylis-Hillman reaction[58]. In the sphere of asymmetric organocatalysis, a stable *textile*-immobilized cinchona derivative has been grafted for asymmetric organocatalysis, illustrated by a highly enantioselective desymmetrization of dicarboxylic acid anhydrides by alcoholysis, with the added advantage of recyclability (remarkably, up to 250 times!)[59] – so don′t throw your old coat away!

In asymmetric synthesis[60] a variety of new chiral ligands and organocatalysts has emerged in a variety of contexts: a chiral biphenyl-2,2′-diyl phospholan-1-ylmethyl phosphite, for example, offers high enantioselectivity and 'branched' regioselectivity for *solvent-free* rhodium-catalyzed hydroformylation of styrenes at very low catalyst loading[61], while asymmetric intermolecular Heck arylation with aryl halides has been effected efficiently with a chiral 1,1′-spirobiindan-based di(phosphine)[62]. In asymmetric organocatalysis, note also: the design of a highly active, helically chiral 1,4-diphosphinyl-1,3-diene as Lewis base for the asymmetric Hosomi-Sakurai reaction[63]; a bifunctional arylcarboxylic acid, conjugate base-stabilized by an appended chiral 2-acylaminothiourea residue, for Brønsted acid-catalyzed asymmetric Povarov reactions[64]; a catalytic peptide bifunctionalized with appended piperidine nitroxyl and imidazole residues, exhibiting improved enantioselectivity over the dual catalytic alternative for desymmetrization of glycols by oxidative esterification[65]; a sterically demanding (S)-proline-based 6-azabicyclo[3.2.1]octane-4-carboxylic acid for the α-benzylation of α-substituted aldehydes as a dynamic kinetic asymmetric transformation[66]; and the first oxidative asymmetric intramolecular enamine catalysis, linked cascade-fashion in a new asymmetric 1,2,3,4-tetrahydroquinoline synthesis[67]. More notable, perhaps, are asymmetric *organocatalyzed* multicomponent syntheses wherein multiple bond formation and introduction of multiple stereocenters are potentially achievable in one operation. This is illustrated in an organocatalyzed asymmetric 3-component synthesis of N-condensed 2,6-diazabicyclo[2.2.2]octan-3-ones by triple ring closure using a chiral 2-*tert*-aminothiourea as catalyst[68]. The multicomponent bi- and polyfunctionalization of carbon chains is a valuable subset, whether in one step or by multistep conversion, exemplified by the generation of 1,4-difunctionalized compounds with remote chirality control via Mannich reaction and subsequent [2.3]-sigmatropic rearrangement[69]. 1,2,3-Trifunctionalized chains possessing O-, S- and N-functions can also be crafted simply from chiral aziridine-2-carboxaldehydes with simultaneous C-C bond formation[70]; but the prize should arguably go to the formation of differentially 1,5-di-functionalized chains from alkenylcyclopropanes, based on selective C-H allylic activation and simultaneous ring opening to generate a 1,5-bis-nucleophile quenchable by two different electophiles[71].

Visible-light photoredox catalysis, highlighted in the *Trends* of Volume 81, has been duly reviewed[72] and developed in recent months. Note (in addition to the 'merging' element mentioned above) an application to the direct arylation of benzyl ethers via C-H bond activation[73], an intramolecular dehydrogenative carbazole synthesis *under continuous flow*[74], and metal-free radical conversions with electron-transfer from neutral organic donors[75].

Turning to specific, commonly encountered conversions, mention might be given to two *metal-free* procedures for α-hydroxylation of ketones[76]; to a simple copper-catalyzed deoxygenation of primary and secondary alcohols via triflates or iodides[77]; to an alternative, one-pot α-amination of ketones via aerobic copper-catalyzed α-bromination with $CuBr_2$[78]; to an eco-friendly, large-scale uncatalyzed α-alkylation of ketones with alcohols [with or without simultaneous reduction of the keto group][79], along with a rhodium-catalyzed α-methylation of ketones with DMF as methylating agent[80], and an extra-ordinary ruthenium-catalyzed deaminative-decarboxylative α-alkylation of ketones *with α- or β-aminocarboxylic acids*![81]. Note also: the elaboration of palladium-catalyzed β-arylation[82], and less familiar palladium-catalyzed β-oxylation of carboxylic acid amides[83]; a diverse array of carbonylative[84], decarboxylative[85] and N-dealkylative[86] syntheses of carboxylic acid amides, along with an efficient and generalized peptide synthesis in a microflow reactor[87] and a novel solution-phase procedure via group-assisted purification (GAP) to avoid chromatographic clean-up or recrystallization[88]. Tosyl-hydrazones are also in the news again as Michael-type sulfonyl transfer agents[89], while a synthesis of di- and tri-substituted vinyl triflones (and heterocyclic analogs) from readily available substrates is notable for involving six transformations in one pot[90]; and bringing up the rear is a kinetic resolution of allyl fluorides *via C-F bond activation,* based on organocatalyzed asymmetric cross-coupling with trifluoromethyl(trimethyl)silane[91].

And, lastly, we have an apparent challenge to conventional wisdom with the revelation that tertiary alcohols, in spite of steric hindrance, *can* undergo nucleophilic substitution *with inversion of configuration* - or so it seems with the isolation of inverted tertiary isonitriles by reaction of tertiary alcohols with trimethylsilyl cyanide in the presence of $Sc(OTf)_3$. But, alas, there is only 90% inversion, so assigning a precise mechanism must await further experimentation. In the meantime, let's call it $S_N1.8$ substitution![92]

1 For recent reviews s. *Reviews,* sections 1-5 (p. 446).

2 D.-C. Wang, H.-Y. Niu, M.-S. Xie, G.-R. Qu, H.-X. Wang, H.-M. Guo, Org. Lett. *2014*, *16* (1), 262-5 [DOI: 10.1021/ol4032683].

3 For a heterogeneous dynamic kinetic resolution of prim. amines with a biomimetic hybrid catalyst formed by co-immobilization of palladium and *Candida antarctica* lipase B in the compartments of mesoporous silica s. K. Engström, E.V. Johnston, O. Verho, K.P.J. Gustafson, M. Shakeri, C.-W. Tai, J.-E. Bäckvall, Angew. Chem., Int. Ed. *2013*, *52* (52), 14006-10 [DOI: 10.1002/anie.201306487]; for tandem catalysis with a metathesis catalyst and a metallo-enzyme to produce epoxides through a dynamic equilibrium of alkenes s. C.A. Denard, H. Huang, M.J. Bartlett, L. Lu, Y. Tan, H. Zhao, J.F. Hartwig, ibid. *2014*, *53* (2), 465-9 [DOI: 10.1002/anie.201305778].

4 Highlight s. M.T. Oliveira, M. Luparia, D. Audisio, N. Maulide, Angew. Chem., Int. Ed. *2013*, *52* (50), 13149-52 [DOI: 10.1002/anie.201305933].

5 A. Quintard, T. Constantieux, J. Rodriguez, Angew. Chem., Int. Ed. *2013*, *52* (49), 12883-7 [DOI: 10.1002/anie.201307295].

6 Z. Li, R.S. Assary, A.C. Atesin, L.A. Curtiss, T.J. Marks, J. Am. Chem. Soc. *2014*, *136* (1), 104-7 [DOI: 10.1021/ja411546r]; for an example of the merging of Brønsted acid catalysis with palladium catalysis s. C.-B. Yu, Y.-G. Zhou, Angew. Chem., Int. Ed. *2013*, *52* (50), 13365-8 [DOI: 10.1002/anie.201307036].

7 V.M. Lombardo, C.D. Thomas, K.A. Scheidt, Angew. Chem., Int. Ed. *2013*, *52* (50), 12910-4 [DOI: 10.1002/anie.201306462].

8 F.R. Petronijevic, M. Nappi, D.W.C. MacMillan, J. Am. Chem. Soc. *2013*, *135* (49), 18323-6 [DOI: 10.1021/ja410478a].

9 L.J. Rono, H.G. Yayla, D.Y. Wang, M.F. Armstrong, R.R. Knowles, J. Am. Chem. Soc. *2013*, *135* (47), 17735-8 [DOI: 10.1021/ja4100595]; for a further example of merging organocatalysis with photoredox catalysis in an asym. synthesis of highly functionalized decalins s. B.-C. Hong, C.-W. Lin, W.-K. Liao, G.-H. Lee, Org. Lett. *2013*, *15* (24), 6258-61 [DOI: 10.1021/ol403113c].

10 Z. Wang, X. Li, Y. Huang, Angew. Chem., Int. Ed. *2013*, *52* (52), 14219-23 [DOI: 10.1002/anie.201308835].

11 For a diaryl sulfoxide synthesis by relay catalysis s. T. Jia, A. Bellomo, S. Montel, M. Zhang, K. EL Baina, B. Zheng, P.J. Walsh, Angew. Chem., Int. Ed. *2014*, *53* (1), 260-4 [DOI: 10.1002/anie.201307172].

12 J. Tsoung, J. Panteleev, M. Tesch, M. Lautens, Org. Lett. *2014*, *16* (1), 110-3 [DOI: 10.1021/ol4030925].

13 T. Beisel, G. Manolikakes, Org. Lett. *2013*, *15* (23), 6046-9 [DOI: 10.1021/ol402949t].

14 For chiral amide-directed assembly of a diastereo- and enantio-pure supramolecular host for asym. catalysis with neutral substrates s. C. Zhao, Q.-F. Sun, W.M. Hart-Cooper, A.G. DiPasquale, F.D. Toste, R.G. Bergman, K.N. Raymond, J. Am. Chem. Soc. *2013*, *135* (50), 18802-5 [DOI: 10.1021/ja411631v].

15 A.G. Salles, Jr., S. Zarra, R.M. Turner, J.R. Nitschke, J. Am. Chem. Soc. *2013*, *135* (51), 19143-6 [DOI: 10.1021/ja412235e].

16 Recent overview s. S. Borman, Chem. Eng. News *2013*, *91* (50), 27.

17 P.E. Gormisky, M.C. White, J. Am. Chem. Soc. *2013*, *135* (38), 14052-5 [DOI: 10.1021/ja407388y]; cf. *72*, 65.

18 D. Srimani, Y. Diskin-Posner, Y. Ben-David, D. Milstein, Angew. Chem., Int. Ed. *2013*, *52* (52), 14131-4 [DOI: 10.1002/anie.201306629].

19 W. Zuo, A.J. Lough, Y.F. Li, R.H. Morris, Science *2013*, *342* (6162), 1080-3 [DOI: 10.1126/science.1244466].

20 R.V. Jagadeesh, A.-E. Surkus, H. Junge, M.-M. Pohl, J. Radnik, J. Rabeah, H. Huan, V. Schünemann, A. Brückner, M. Beller, Science *2013*, *342* (6162), 1073-6 [DOI: 10.1126/science.1242005].

21 Y. Ren, M. Yan, J. Wang, Z.C. Zhang, K. Yao, Angew. Chem., Int. Ed. *2013*, *52* (48), 12674-8 [DOI: 10.1002/anie.201305342].

22 T.J. Mazzacano, N.P. Mankad, J. Am. Chem. Soc. *2013*, *135* (46), 17258-61 [DOI: 10.1021/ja408861p].

23 S. Asako, L. Ilies, E. Nakamura, J. Am. Chem. Soc. *2013*, *135* (47), 17755-7 [DOI: 10.1021/ja4106368]; s.a. T. Matsubara, S. Asako, L. Ilies, E. Nakamura, ibid. *2014*, *136* (2), 646-9 [DOI: 10.1021/ja412521k].

24 For reviews on C-H bond activation see *Reviews*, section 7 (p. 454).

25 s. *72*, 491s*83*; for a recent review s. S.A. Girard, T. Knauber, C.-J. Li, Angew. Chem., Int. Ed. *2014*, *53* (1), 74-100 [DOI: 10.1002/anie.201304268], and highlight on dehydrogenative cross-coupling s. Q. Liu, R. Jackstell, M. Beller, ibid. *2013*, *52* (52), 13871-3 [DOI: 10.1002/anie.201307865]; for a recent dehydrogenative cross-coupling of benzylic ethers with aldehydes s. Z. Meng, S. Sun, H. Yuan, H. Lou, L. Liu, ibid. *2014*, *53* (2), 543-7 [DOI: 10.1002/anie.201308701].

26 K. Li, G. Tan, J. Huang, F. Song, J. You, Angew. Chem., Int. Ed. *2013*, *52* (49), 12942-5 [DOI: 10.1002/anie.201306181].

27 J. Shen, D. Yang, Y. Liu, S. Qin, J. Zhang, J. Sun, C. Liu, C. Liu, X. Zhao, C. Chu, R. Liu, Org. Lett. *2014*, *16* (2), 350-3 [DOI: 10.1021/ol403555n].

28 Q.-Y. Meng, J.-J. Zhong, Q. Liu, X.-W. Gao, H.-H. Zhang, T. Lei, Z.-J. Li, K. Feng, B. Chen, C.-H. Tung, L.-Z. Wu, J. Am. Chem. Soc. *2013*, *135* (51), 19052-5 [DOI: 10.1021/ja408486v].

29 T. Zhang, L. Wu, X. Li, Org. Lett. *2013*, *15* (24), 6294-7 [DOI: 10.1021/ol403178a].

30 P. Becker, D.L. Priebbenow, R. Pirwerdjan, C. Bolm, Angew. Chem., Int. Ed. *2014*, *53* (1), 269-71 [DOI: 10.1002/anie.201307446].

31 B. Wang, C. Shen, J. Yao, H. Yin, Y. Zhang, Org. Lett. *2014*, *16* (1), 46-9 [DOI: 10.1021/ol402921w].

32 X. Huang, J. Huang, C. Du, X. Zhang, F. Song, J. You, Angew. Chem., Int. Ed. *2013*, *52* (49), 12970-4 [DOI: 10.1002/anie.201307174].

33 T.-J. Gong, W. Su, Z.-J. Liu, W.-M. Cheng, B. Xiao, Y. Fu, Org. Lett. *2014*, *16* (2), 330-3 [DOI: 10.1021/ol403407q]; for an N-deoxygenative alternative s. R. Zeng, S. Wu, C. Fu, S. Ma, J. Am. Chem. Soc. *2013*, *135* (49), 18284-7 [DOI: 10.1021/ja409861s].

34 Z. Li, Y. Zhang, L. Zhang, Z.-Q. Liu, Org. Lett. *2014*, *16* (2), 382-5 [DOI: 10.1021/ol4032478]; for a related intramolecular carbonitration s. T. Shen, Y. Yuan, N. Jiao, Chem. Commun. *2014*, *50* (5), 554-6 [DOI: 10.1039/c3cc47336h].

35 B. Zhou, J. Du, Y. Yang, H. Feng, Y. Li, Org. Lett. *2014*, *16* (2), 592-5 [DOI: 10.1021/ol403477w]; for an alternative *o*-amination with O-benzoylhydroxylamines s. K. Wu, Z. Fan, Y. Xue, Q. Yao, A. Zhang, ibid. *2014*, *16* (1), 42-5 [DOI: 10.1021/ol402965d].

36 B. Zhou, J. Du, Y. Yang, H. Feng, Y. Li, Org. Lett. *2013*, *15* (24), 6302-5 [DOI: 10.1021/ol403187t].

37 Z. Dong, G. Dong, J. Am. Chem. Soc. *2013*, *135* (49), 18350-3 [DOI: 10.1021/ja410823e].

38 H.-Y. Zhang, H.-M. Yi, G.-W. Wang, B. Yang, S.-D. Yang, Org. Lett. *2013*, *15* (24), 6186-9 [DOI: 10.1021/ol403028a].

[39] S.M. Preshlock, D.L. Plattner, P.E. Maligres, S.W. Krska, R.E. Maleczka, Jr., M.R. Smith III, Angew. Chem., Int. Ed. *2013*, *52* (49), 12915-9 [DOI: 10.1002/anie.201306511].

[40] L. Wang, L. Ackermann, Chem. Commun. *2014*, *50* (9), 1083-5 [DOI: 10.1039/c3cc47852a].

[41] X. Sun, Y. Sun, C. Zhang, Y. Rao, Chem. Commun. *2014*, *50* (10), 1262-4 [DOI: 10.1039/c3cc47431c]; for a classical electrophilic *o*-chlorination of phenols with SO_2Cl_2/2,2,6,6-tetramethylpiperidine at elevated temperature cf. N.I. Saper, B.B. Snider, J. Org. Chem. *2014*, *79* (2), 809-13 [DOI: 10.1021/jo402424h].

[42] P.S. Fier, J.F. Hartwig, Science *2013*, *342* (6161), 956-60 [DOI: 10.1126/science.1243759].

[43] S. Ma, G. Villa, P.S. Thuy-Boun, A. Homs, J.-Q. Yu, Angew. Chem., Int. Ed. *2014*, *53* (3), 734-7 [DOI: 10.1002/anie.201305388].

[44] L. Wan, N. Dastbaravardeh, G. Li, J.-Q. Yu, J. Am. Chem. Soc. *2013*, *135* (48), 18056-9 [DOI: 10.1021/ja410760f].

[45] S. Lee, H. Lee, K.L. Tan, J. Am. Chem. Soc. *2013*, *135* (50), 18778-81 [DOI: 10.1021/ja4107034].

[46] N.A. Isley, F. Gallou, B.H. Lipshutz, J. Am. Chem. Soc. *2013*, *135* (47), 17707-10 [DOI: 10.1021/ja409663q].

[47] H.-B. Wu, X.-T. Ma, S.-K. Tian, Chem. Commun. *2014*, *50* (2), 219-21 [DOI: 10.1039/c3cc45772a].

[48] A. Tsubouchi, D. Muramatsu, T. Takeda, Angew. Chem., Int. Ed. *2013*, *52* (48), 12719-22 [DOI: 10.1002/anie.201306882].

[49] M.L. Duda, F.E. Michael, J. Am. Chem. Soc. *2013*, *135* (49), 18347-9 [DOI: 10.1021/ja410686v].

[50] Y. Zhao, D.J. Weix, J. Am. Chem. Soc. *2014*, *136* (1), 48-51 [DOI: 10.1021/ja410704d].

[51] C.-L. Sun, A. Fürstner, Angew. Chem., Int. Ed. *2013*, *52* (49), 13071-5 [DOI: 10.1002/anie.201307028]; for Ni-catalyzed ring opening of cyclic enolethers with arylmagnesium halides s. J. Cornella, R. Martin, Org. Lett. *2013*, *15* (24), 6298-301 [DOI: 10.1021/ol4031815].

[52] D. Grassi, A. Alexakis, Angew. Chem., Int. Ed. *2013*, *52* (51), 13642-6 [DOI: 10.1002/anie.201307591]; for metal-free phenylhydrazine-initiated direct radical arylation of arenes with KOBu-*t* as base s. A. Dewanji, S. Murarka, D.P. Curran, A. Studer, Org. Lett. *2013*, *15* (23), 6102-5 [DOI: 10.1021/ol402995e].

[53] E. Shirakawa, F. Tamakuni, E. Kusano, N. Uchiyama, W. Konagaya, R. Watabe, T. Hayashi, Angew. Chem., Int. Ed. *2014*, *53* (2), 521-5 [DOI: 10.1002/anie.201308200].

[54] For reviews on heterogeneous catalysis s. *Reviews,* section 3 (p. 448).

[55] B.J. O'Neill, D.H.K. Jackson, A.J. Crisci, C.A. Farberow, F. Shi, A.C. Alba-Rubio, J. Lu, P.J. Dietrich, X. Gu, C.L. Marshall, P.C. Stair, J.W. Elam, J.T. Miller, F.H. Ribeiro, P.M. Voyles, J. Greeley, M. Mavrikakis, S.L. Scott, T.F. Kuech, J.A. Dumesic, Angew. Chem., Int. Ed. *2013*, *52* (51), 13808-12 [DOI: 10.1002/anie.201308245].

[56] Y.M.A. Yamada, Y. Yuyama, T. Sato, S. Fujikawa, Y. Uozumi, Angew. Chem., Int. Ed. *2014*, *53* (1), 127-31 [DOI: 10.1002/anie.201308541]; N-formylation with palladium deposited on a shape-controllable Al_2O_3 nano-rod support s. X. Cui, Y. Zhang, Y. Deng, F. Shi, Chem. Commun. *2014*, *50* (2), 189-91 [DOI: 10.1039/c3cc46427j].

[57] M.P. Conley, V. Mougel, D.V. Peryshkov, W.P. Forrest, Jr., D. Gajan, A. Lesage, L. Emsley, C. Copéret, R.R. Schrock, J. Am. Chem. Soc. *2013*, *135* (51), 19068-70 [DOI: 10.1021/ja410052u].

[58] Y. Furukawa, M. Ogura, J. Am. Chem. Soc. *2014*, *136* (1), 119-21 [DOI: 10.1021/ja410781z].

[59] J.-W. Lee, T. Mayer-Gall, K. Opwis, C.E. Song, J.S. Gutmann, B. List, Science *2013*, *341* (6151), 1225-9 [DOI: 10.1126/science.1242196].

[60] For reviews on asym. synthesis s. *Reviews,* section 1 (p. 446).

[61] G.M. Noonan, C.J. Cobley, T. Mahoney, M.L. Clarke, Chem. Commun. *2014*, *50* (12), 1475-7 [DOI: 10.1039/c3cc48823c].

[62] C. Wu, J. Zhou, J. Am. Chem. Soc. *2014*, *136* (2), 650-2 [DOI: 10.1021/ja412277z]; asym. hydrogenation of 2-subst. allyl alcohols with a chiral aminoferrocenyldi(phosphine) [ChenPhos] as ligand s. Q. Wang, X. Liu, X. Liu, B. Li, H. Nie, S. Zhang, W. Chen, Chem. Commun. *2014*, *50* (8), 978-80 [DOI: 10.1039/c3cc47727d].

[63] M. Ogasawara, S. Kotani, H. Nakajima, H. Furusho, M. Miyasaka, Y. Shimoda, W.-Y. Wu, M. Sugiura, T. Takahashi, M. Nakajima, Angew. Chem., Int. Ed. *2013*, *52* (51), 13798-802 [DOI: 10.1002/anie.201308112].

[64] C. Min, N. Mittal, D.X. Sun, D. Seidel, Angew. Chem., Int. Ed. *2013*, *52* (52), 14084-8 [DOI: 10.1002/anie.201308196]; for asym. hetero-Diels-Alder reaction of Brassard's dienes with isatins using a chiral magnesium(II) cyclic bis(N-oxide) complex as catalyst s. J. Zheng, L. Lin, Y. Kuang, J. Zhao, X. Liu, X. Feng, Chem. Commun. *2014*, *50* (8), 994-6 [DOI: 10.1039/c3cc47800a]; for asym. [2+2]-cycloaddition with a chiral 1,3,2-oxazaborolidine·$AlBr_3$ complex s. R. Brimioulle, T. Bach, Science *2013*, *342* (6160), 840-3 [DOI: 10.1126/science.1244809].

[65] C. Hofmann, S.M.M. Schuler, R.C. Wende, P.R. Schreiner, Chem. Commun. *2014*, *50* (10), 1221-3 [DOI: 10.1039/c3cc48584f].

[66] B. List, I. Coric, O.O. Grygorenko, P.S.J. Kaib, I. Komarov, A. Lee, M. Leutzsch, S.C. Pan, A.V. Tymtsunik, M. van Gemmeren, Angew. Chem., Int. Ed. *2014*, *53* (1), 282-5 [DOI: 10.1002/anie.201306037].

[67] Y.K. Kang, D.Y. Kim, Chem. Commun. *2014*, *50* (2), 222-4 [DOI: 10.1039/c3cc46710d].

[68] M.d.M.S. Duque, O. Baslé, Y. Génisson, J.-C. Plaquevent, X. Bugaut, T. Constantieux, J. Rodriguez, Angew. Chem., Int. Ed. *2013*, *52* (52), 14143-6 [DOI: 10.1002/anie.201306656].

[69] T. Kano, R. Sakamoto, K. Maruoka, Chem. Commun. *2014*, *50* (8), 942-4 [DOI: 10.1039/c3cc47827k].

[70] S.K. Liew, Z. He, J.D. St. Denis, A.K. Yudin, J. Org. Chem. *2013*, *78* (23), 11637-45 [DOI: 10.1021/jo401489q].

[71] A. Masarwa, D. Didier, T. Zabrodski, M. Schinkel, L. Ackermann, I. Marek, Nature *2014*, *505* (7482), 199-203 [DOI: 10.1038/nature12761].

[72] For reviews s. *Reviews*, section 2 (p. 447).

[73] K. Qvortrup, D.A. Rankic, D.W.C. MacMillan, J. Am. Chem. Soc. *2014*, *136* (2), 626-9 [DOI: 10.1021/ja411596q].

[74] A.C. Hernandez-Perez, S.K. Collins, Angew. Chem., Int. Ed. *2013*, *52* (48), 12696-700 [DOI: 10.1002/anie.201306920].

[75] S. O'Sullivan, E. Doni, T. Tuttle, J.A. Murphy, Angew. Chem., Int. Ed. *2014*, *53* (2), 474-8 [DOI: 10.1002/anie.201306543].

[76] Y.-F. Liang, N. Jiao, Angew. Chem., Int. Ed. *2014*, *53* (2), 548-52 [DOI: 10.1002/anie.201308698]; s.a. C.-B. Miao, Y.-H. Wang, M.-L. Xing, X.-W. Lu, X.-Q. Sun, H.-T. Yang, J. Org. Chem. *2013*, *78* (22), 11584-9 [DOI: 10.1021/jo401866p].

[77] H. Dang, N. Cox, G. Lalic, Angew. Chem., Int. Ed. *2014*, *53* (3), 752-6 [DOI: 10.1002/anie.201307697].

[78] R.W. Evans, J.R. Zbieg, S. Zhu, W. Li, D.W.C. MacMillan, J. Am. Chem. Soc. *2013*, *135* (43), 16074-7 [DOI: 10.1021/ja4096472]; for an alternative NHC-catalyzed redox asym. α-amination [of α-aryloxyaldehydes] s. J.E. Taylor, D.S.B. Daniels, A.D. Smith, Org. Lett. *2013*, *15* (23), 6058-61 [DOI: 10.1021/ol402955f].

79 Q. Xu, J. Chen, H. Tian, X. Yuan, S. Li, C. Zhou, J. Liu, Angew. Chem., Int. Ed. *2014*, *53* (1), 225-9 [DOI: 10.1002/anie.201308642]; for a rhodium-catalyzed α-methylation of ketones with methanol s. L.K.M. Chan, D.L. Poole, D. Shen, M.P. Healy, T.J. Donohoe, Angew. Chem., Int. Ed. *2014*, *53* (3), 761-5 [DOI: 10.1002/anie.201307950].

80 Y. Li, D. Xue, W. Lu, C. Wang, Z.-T. Liu, J. Xiao, Org. Lett. *2014*, *16* (1), 66-9 [DOI: 10.1021/ol403040g].

81 N. Kalutharage, C.S. Yi, Angew. Chem., Int. Ed. *2013*, *52* (51), 13651-5 [DOI: 10.1002/anie.201307766].

82 For β-arylation of simple ketones with aryl iodides s. Z. Huang, G. Dong, J. Am. Chem. Soc. *2013*, *135* (47), 17747-50 [DOI: 10.1021/ja410389a]; of chiral N-protected α-aminocarboxylic acid amides with asym. induction s. Q. Zhang, K. Chen, W. Rao, Y. Zhang, F.-J. Chen, B.-F. Shi, Angew. Chem., Int. Ed. *2013*, *52* (51), 13588-92 [DOI: 10.1002/anie.201306625]; of chiral N-protected cyclopropaneacetamides s. N. Hoshiya, T. Kobayashi, M. Arisawa, S. Shuto, Org. Lett. *2013*, *15* (24), 6202-5 [DOI: 10.1021/ol4030452].

83 β-Alkoxylation of N-protected carboxylic acid amides s. G. Shan, X. Yang, Y. Zong, Y. Rao, Angew. Chem., Int. Ed. *2013*, *52* (51), 13606-10 [DOI: 10.1002/anie.201307090]; β-acoxylation s. L. Zhou, W. Lu, Org. Lett. *2014*, *16* (2), 508-11 [DOI: 10.1021/ol403393w].

84 By carbonylation of ar. amines or nitroarenes s. X. Fang, R. Jackstell, M. Beller, Angew. Chem., Int. Ed. *2013*, *52* (52), 14089-93 [DOI: 10.1002/anie.201308455]; by carbonylative coupling of amines with methyl groups s. H. Liu, G. Laurenczy, N. Yan, P.J. Dyson, Chem. Commun. *2014*, *50* (3), 341-3 [DOI: 10.1039/c3cc47015f].

85 By aerobic decarboxylative ammoxidation of phenylacetic and α-hydroxyphenylacetic acids with ammonia in water s. Q. Song, Q. Feng, K. Yang, Org. Lett. *2014*, *16* (2), 624-7 [DOI: 10.1021/ol403586c]; by aerobic decarboxylative amination of α-ketocarboxylic acids under visible-light photoredox catalysis s. J. Liu, Q. Liu, H. Yi, C. Qin, R. Bai, X. Qi, Y. Lan, A. Lei, Angew. Chem., Int. Ed. *2014*, *53* (2), 502-6 [DOI: 10.1002/anie.201308614].

86 By N-dealkylative coupling with carboxylic acid esters s. Y.-S. Bao, B. Zhaorigetu, B. Agula, M. Baiyin, M. Jia, J. Org. Chem. *2014*, *79* (2), 803-8 [DOI: 10.1021/jo4023974].

87 S. Fuse, Y. Mifune, T. Takahashi, Angew. Chem., Int. Ed. *2014*, *53* (3), 851-5 [DOI: 10.1002/anie.201307987].

88 J. Wu, G. An, S. Lin, J. Xie, W. Zhou, H. Sun, Y. Pana, G. Li, Chem. Commun. *2014*, *50* (10), 1259-61 [DOI: 10.1039/c3cc48509a].

89 M. Fernández, U. Uria, L. Orbe, J.L. Vicario, E. Reyes, L. Carrillo, J. Org. Chem. *2014*, *79* (1), 441-5 [DOI: 10.1021/jo402518q].

90 X.-H. Xu, M. Taniguchi, X. Wang, E. Tokunaga, T. Ozawa, H. Masuda, N. Shibata, Angew. Chem., Int. Ed. *2013*, *52* (48), 12628-31 [DOI: 10.1002/anie.201307535].

91 T. Nishimine, K. Fukushi, N. Shibata, H. Taira, E. Tokunaga, A. Yamano, M. Shiro, N. Shibata, Angew. Chem., Int. Ed. *2014*, *53* (2), 517-20 [DOI: 10.1002/anie.201308071]; generation of nucleophilic (Z)- or (E)-α-fluoroalkenylchromium reagents via carbon-fluorine bond activation s. T. Nihei, S. Yokotani, T. Ishihara, T. Konno, Chem. Commun. *2014*, *50* (13), 1543-5 [DOI: 10.1039/c3cc47219a]; for reviews on introducing trifluoromethyl groups s. *Reviews,* section 6 (p. 451); explosion warning regarding 1-(trifluoromethyl)-1,2-benziodoxol-3(1*H*)-one [Togni's reagent II] s. N. Fiederling, J. Haller, H. Schramm, Org. Process Res. Dev. *2013*, *17* (3), 318-9 [DOI: 10.1021/op400035b].

92 S.V. Pronin, C.A. Reiher, R.A. Shenvi, Nature *2013*, *501* (7466), 195-9 [DOI: 10.1038/nature12472].

Systematic Survey

Abbreviations and Symbols

abs.	absolute
alc.	alcoholic
aq.	aqueous
ar.	aromatic
atm.	atmosphere(s)
compd(s).	compound(s)
deriv(s).	derivative(s)
e.e.	enantiomeric excess
eq(s).	equivalent(s)
E.	Example
F.e.s.	Further example(s) see
M	molar
prepn.	preparation
prim.	primary
s*83*	supplementary reference in Volume *83*
sec.	secondary
startg. m.f.	starting material for (the preparation of ...)
subst.	substituted
sym.	symmetrical
tert.	tertiary
v.i.	via intermediates
w.a.r.	without additional reagents
Y *	Yield
ϟ	Electrolysis
///	Irradiation
[\\\\]	Microwave irradiation
○	Ring closure
◌	Ring contraction
◌	Ring expansion
C	Ring opening
Ⓗ	Ring hydrogenation
←	'see title or reagent on the left half of the page'

* Yields in parentheses refer to the immediately preceding step of a multi-step reaction

Formation of H-O Bond

Exchange ↓↑

Nitrogen ↑ HO ↓↑ N

Palladium-carbon *Pd-C*
3-Hydroxy- from 3-(arylideneaminooxy)-oxindoles ON=CHAr → OH
Chiral 3-subst. 3-hydroxyoxindoles s. *83*, 65

Remaining Elements ↑ HO ↓↑ Rem

Silica gel-sulfonic acid ←
O-Desilylation OSi≤ → OH
O-de-*tert*-butyldiphenylsilylation with TfOH-on-silica gel cf. *40*, 1s*82*; O-desilylation of various aryl and alkyl silyl ethers with silica gel-sulfonic acid s. H. Fujii, T. Yamada, K. Hayashida, M. Kuwada, A. Hamasaki, K. Nobuhara, S. Ozeki, H. Nagase, Heterocycles *2012*, *85* (11), 2685-91 [DOI: 10.3987/com-12-12577].

Carbon ↑ HO ↓↑ C

Microwaves s. under 1-Butyl-3-methylimidazolium bromide [\\\\]

Sodium ethoxide *NaOEt*
O-Deformylation OCHO → OH
en route to 2-alkoxy-3-methylene-1,2-oxaphospholane 2-oxides s. *83*, 202

Lithium 2,2,6,6-tetramethylpiperidide *$LiNR_2$*
Cleavage of 1,3-dioxolanes s. *8*, 22s*83* $C(OR)_2$ → CO

Silver hexafluoroantimonate/1,3,5-trimethoxybenzene *$AgSbF_6$/ArH*
O-De-*p*-methoxybenzylation s. *8*, 22s*83* OCH_2Ar → OH

Isopropanol s. under Pd-C and $Pd(OH)_2$ *i-PrOH*

1-Butyl-3-methylimidazolium bromide/microwaves *[bmim]Br/*[\\\\]
Cleavage of phenolethers by O-dealkylation OR → OH
of methyl phenolethers, update s. *35*, 7s*78*; cleavage of ethyl, propyl and benzyl phenolethers in the absence of solvent with 1-*n*-butyl-3-methylimidazolium bromide (3 eq.) under microwave irradiation s. S.K. Park, O. Battsengel, J. Chae, Bull. Korean Chem. Soc. *2013*, *34* (1), 174-8 [DOI: 10.5012/bkcs.2013.34.1.174]; **partial cleavage** of linear or branched *p*-quinol ethers with BBr_3 (cf. *19*, 8s*37*) s. C. Pasquini, A. Coniglio, M. Bassetti, Tetrahedron Lett. *2012*, *53* (46), 6191-4 [DOI: 10.1016/j.tetlet.2012.08.132]; **heterogeneous O-demethylation** with a robust, ferrocene-tagged Merrifield resin s. R. Kurane, V. Gaikwad, J. Jadhav, R. Salunkhe, G. Rashinkar, ibid. *53* (47), 6361-6 [DOI: 10.1016/j.tetlet.2012.09.021].

2,3-Dichloro-5,6-dicyanoquinone s. under t-BuONO *DDQ*
Triethylsilane s. under Pd-C *Et_3SiH*

tert-Butyl nitrite/2,3-dichloro-5,6-dicyanoquinone/oxygen *t-BuONO/DDQ/O_2*
Oxidative O-debenzylation s. *8*, 22s*83* OBn → OH

Ferrocene-tagged Merrifield resin ←
Heterogeneous O-demethylation of methyl phenolethers s. *35*, 7s*83* OMe → OH

Palladium-carbon/ethanolamine *Pd-C/$NH_2CH_2CH_2OH$*
Palladium-carbon/triethylsilane *Pd-C/Et_3SiH*
Palladium-carbon/palladium(II) hydroxide-carbon/isopropanol *Pd-C/$Pd(OH)_2$-C/i-PrOH*
Palladium(II) hydroxide-carbon/isopropanol *$Pd(OH)_2$-C/i-PrOH*
Cleavage of O-protective groups ←
update s. *8,* 22s*82*; **selective O-debenzylation** of protected polyethyleneglycols with $Pd(OH)_2$-C in isopropanol, and **selective O-deallylation** with a 1:1 combination of $Pd(OH)_2$-C and Pd-C in isopropanol s. A. Mart, M.S. Shashidhar, Tetrahedron *2012, 68* (47), 9769-76 [DOI: 10.1016/j.tet.2012.09.016]; O-debenzylation of protected carbohydrates and **cleavage of O,O-benzylidene derivs.** with Pd-C and triethylsilane s. A. Santra, T. Ghosh, A.K. Misra, Beilstein J. Org. Chem. *2013, 9,* 74-8 [DOI: 10.3762/bjoc.9.9]; oxidative O-debenzylation, **O-de-*p*-methoxybenzylation and O-de-*p*-phenylbenzylation** with *tert*-butyl nitrite under O_2 (1 atm.) with DDQ as catalyst s. Z. Shen, L. Sheng, X. Zhang, W. Mo, B. Hu, N. Sun, X. Hu, Tetrahedron Lett. *2013, 54* (12), 1579-83 [DOI: 10.1016/j.tetlet.2013.01.045]; mild and chemoselective O-de-*p*-methoxybenzylation with $AgSbF_6$ (5 mol%) and 1,3,5-trimethoxybenzene (0.5 eq.) s. N. Kern, T. Dombray, A. Blanc, J.-M. Weibel, P. Pale, J. Org. Chem. *2012, 77* (20), 9227-35 [DOI: 10.1021/jo301787v]; **O-depropargylation** of propargyl phenolethers and N-depropargylation of protected arylamines (cf. *5,* 32s*83*) with Pd-C in water with added ethanolamine s. D. Rambabu, S. Bhavani, N.K. Swamy, M.V. Basaveswara Rao, M. Pal, Tetrahedron Lett. *2013, 54* (9), 1169-73 [DOI: 10.1016/j.tetlet.2012.12.093]; base-mediated liberation of ketones by selective **cleavage of 1,3-dioxolanes** with lithium 2,2,6,6-tetramethylpiperidide s. C. Yuan, L. Yang, G. Yue, T. Yu, W. Zhong, B. Liu, Tetrahedron Lett. *2012, 53* (51), 6972-6 [DOI: 10.1016/j.tetlet.2012.10.037]; removal of protective groups after Fmoc solid-phase peptide synthesis with dilute HCl in hexafluoroisopropanol or trifluoroethanol, also **cleavage of the resin support**, s. P. Palladino, D.A. Stetsenko, Org. Lett. *2012, 14* (24), 6346-9 [DOI: 10.1021/ol303124r]; **photocleavage of phenacyl esters** bearing heteroaryloxy residues s. A.S.C. Fonseca, M.S.T. Gonçalves, S.P.G. Costa, Tetrahedron *2012, 68* (38), 8024-32 [DOI: 10.1016/j.tet.2012.06.100].

Formation of H-N Bond

Exchange ↓↑

Oxygen ↑ HN ↓↑ O

Irradiation s. under Alcohols ///
Sodium azide *NaN_3*
Preferential reduction of *o*-nitro groups $NO_2 \rightarrow NH_2$

1.

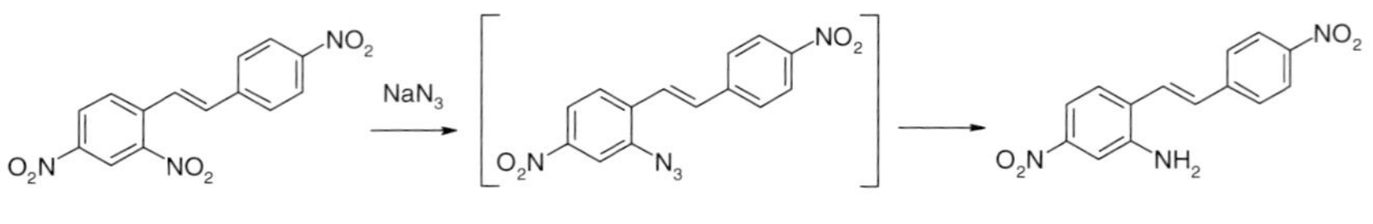

(E)-*o*-Amino-*p*-nitrostilbenes. A mixture of 2,4,4′-trinitrostilbene (1.7 mmol) and NaN_3 (1.74 eq.) in DMF (15 ml) stirred at 120° for 20 h, concentrated *in vacuo,* and purified chromatographically → 2-amino-4,4′-dinitrostilbene. Y 89%. This reaction was successful (in DMF or DMSO) for a series of 2,4-dinitrostilbenes (five examples; Y 75-89%), with an expected azidonitrostilbene observed as a transient intermediate in one case. It was argued that the more sterically-hindered *ortho* nitro group is not coplanar with the benzene ring and therefore more prone to reaction. The source of H-atoms is unclear, but is likely to be adventitious water (although deuterated amine products are observed using d_6-DMSO). F.e.s. K. Janowska, R. Matczak, J. Zakrzewski, H. Krawczyk, Tetrahedron Lett. *2012, 53* (48), 6504-7 [DOI: 10.1016/j.tetlet.2012.09.066].

Borane·ammonia/gold nanoparticles-on-titanium dioxide — *$BH_3 \cdot NH_3/Au\text{-}TiO_2$*
Borane·tert-butylamine/gold nanoparticles — *$BH_3 \cdot t\text{-}BuNH_2/Au$*
Ar. amines from nitro compds. — $NO_2 \rightarrow NH_2$
Also hydroxylamines from aliphatic nitro compds. s. *75*, 7s*83*

Alcohols/titanium dioxide/irradiation — *ROH/TiO_2////*
Glycerol/Fe_3O_4-supported nickel nanoparticles — ←
Ar. amines from nitro compds. s. *75*, 7s*83*

Titanium dioxide s. under Alcohols — *TiO_2*
Hydrazine s.a. under Magnetically-recyclable, hollow nanocomposite ruthenium catalyst — *N_2H_4*

Hydrazine/graphene-Fe_3O_4 nanocomposite — ←
Ar. amines from nitro compds. s. *75*, 7s*83*

(R)-1-[(S_P)-2-(Dicyclohexylphosphino)ferrocenyl]ethyl(di-tert-butyl)phosphine — *JosiPhos*
s. under Bis(1,5-cyclooctadiene)rhodium(I) hexafluoroantimonate

Nitrogen-enriched carbon-immobilized cobalt(II,III) oxide — ←
Prim. ar. amines from nitro compds.
by hydrogenation using a heterogenized cobalt oxide catalyst

2.

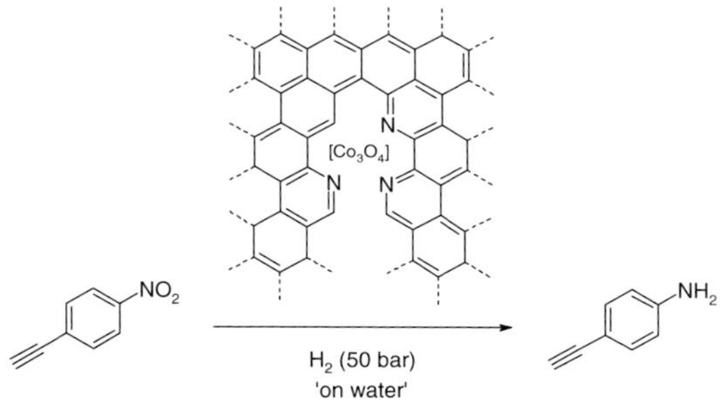

A novel, inexpensive and environmentally benign cobalt-based hydrogenation catalyst has been developed, that is highly selective for the reduction of nitroarenes to anilines and has applicability on an industrial scale. The catalyst is prepared by adsorption of a solution of $Co(phen)_2(OAc)_2$ onto commercially available carbon (Vulcan XC72R), followed by pyrolysis at 800°; it is readily recycled and can be reused several times without loss of activity. **E:** Cobalt catalyst (2 mol%), startg. nitroarene (10 mmol), THF (5 ml) and water (0.5 ml) added sequentially to a glass vial, the vial placed into a 300 ml steel Parr autoclave, the mixture hydrogenated at 110° under 50 bar H_2 for 4 h, cooled to room temp., diluted with ethyl acetate, filtered through a silica plug, and solvent removed *in vacuo* → product. Y 88%. High yields (generally 80-99%) and >99% conversion were recorded for the reduction of nitroarenes in the presence of a wide variety of functionality, notably including thioether, halo, carbonyl, cyano, alkene and alkyne moieties. Yields were somewhat depressed (53-75%) for N-heterocyclic analogs, however. Catalyst activity was significantly enhanced in the presence of water and the reaction could be successfully performed even in the absence of organic solvents. F.e. (thirty-four in all), optimization, and catalyst structure determination s. F.A. Westerhaus, R.V. Jagadeesh, G. Wienhöfer, M.-M. Pohl, J. Radnik, A.-E. Surkus, J. Rabeah, K. Junge, H. Junge, M. Nielsen, A. Brückner, M. Beller, Nat. Chem. *2013*, *5* (6), 537-43 [DOI: 10.1038/nchem.1645].

Fe_3O_4-supported nickel nanoparticles s. under Glycerol — ←

Magnetically-recyclable, hollow nanocomposite ruthenium catalyst/hydrazine — ←
Ar. amines from nitro compds. s. *83*, 458

Rhodium/nickel nanoparticles *Rh_3Ni*
Ar. amines from nitro compds. s. *75,* 7s*83* $NO_2 \rightarrow NH_2$

Bis(1,5-cyclooctadiene)rhodium(I) hexafluoroantimonate/(R)-1-[(S_P)-2-(dicyclohexylphosphino)ferrocenyl]ethyl(di-tert-butyl)phosphine *[Rh(I)]**
Acetylamines from O-acetyloximes C=N-OAc → CHNHAc
Asym. homogeneous hydrogenation

3. 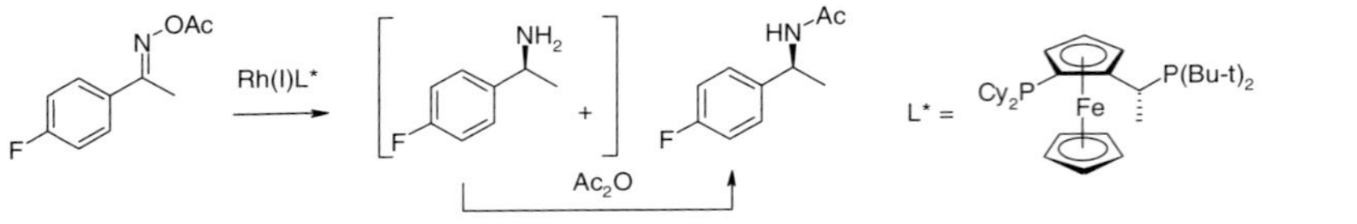

The first enantioselective hydrogenation of ketoxime acetates is described, affording chiral acetamides rather than the expected O-acetylhydroxylamines. **E:** A soln. of $[Rh(cod)_2]SbF_6$ (1 mol%) and JosiPhos SL-J009-1 ligand (1.1 mol%) in 1,4-dioxane (0.5 ml) stirred at room temp. under N_2 for 15 min in a glove box, a soln. of (E)-4-fluoroacetophenone oxime acetate (0.1 mmol) in the same solvent (1 ml) added, the mixture stirred under H_2 (50 atm.) in an autoclave at 40° for 24 h, pressure released cautiously, acetic anhydride (0.3 ml) added, the mixture stirred for 30 min, concentrated, and filtered through silica → (S)-N-[1-(4-fluorophenyl)ethyl]acetamide. Y 80% (e.e. 90%). The reaction appears general for electron-diverse aryl methyl ketoxime acetates and required acetylation of the crude product due to the presence of prim. amine by-product (thirteen examples; Y 52-80%; e.e. 74-90%). Ketoxime acetates containing 4-methoxyphenyl, 2-subst. phenyl or 2-naphthyl, or with ethyl/propyl in place of methyl were poorer substrates (seven examples; Y 37-68%; e.e. 39-60%). Optimum results were obtained using THF or dioxane as solvents, with use of protic solvents affording only ketoximes under these conditions. The mechanism was not established. F.e. and optimization s. K. Huang, S. Li, M. Chang, X. Zhang, Org. Lett. *2013, 15* (3), 484-7 [DOI: 10.1021/ol303282u].

Porous palladium nanostructures *Pd*
Ionic liquid-supported platinum nanoparticles ←
Ar. amines from nitro compds. $NO_2 \rightarrow NH_2$
under transition metal catalysis, update, s. *75,* 7s*82*; hydrogenation of nitrobenzene and styrene with highly porous palladium nanostructures [pPdNs] with perpendicular pore channels, also application in Suzuki coupling (cf. *37,* 902s*82*), s. X. Huang, Y. Li, Y. Chen, E. Zhou, Y. Xu, H. Zhou, X. Duan, Y. Huang, Angew. Chem., Int. Ed. *2013, 52* (9), 2520-4 [DOI: 10.1002/anie.201208901]; with highly active, magnetically recyclable hollow nanocomposite catalysts with a permeable carbon surface s. M. Shokouhimehr, J.E. Lee, S.I. Han, T. Hyeon, Chem. Commun. *2013, 49* (42), 4779-81 [DOI: 10.1039/c3cc41034j]; multi-gram selective reduction of ar. nitro compds. and nitroalkanes [to hydroxylamines] with $BH_3{\cdot}NH_3$ activated by Au nanoparticles-on-TiO_2 (even at the ppm level) s. E. Vasilikogiannaki, C. Gryparis, V. Kotzabasaki, I.N. Lykakis, M. Stratakis, Adv. Synth. Catal. *2013, 355* (5), 907-11 [DOI: 10.1002/adsc.201200983]; with BH_3·*tert*-butylamine activated with monodisperse Au nanoparticles in water or oil systems, selective reduction with retention of keto or aldehyde groups, s. J. Zhou, Q. Yang, Chem. Asian J. *2012, 7* (9), 2045-50 [DOI: 10.1002/asia.201200337]; via hydrosilylation (e.g. with TMDS) in the presence of magnetically separable, Fe_3O_4-supported Au nanoparticles s. S. Park, I.S. Lee, J. Park, Org. Biomol. Chem. *2013, 11* (3), 395-9 [DOI: 10.1039/c2ob27025k]; chemoselective hydrogenation of nitrostyrenes with basic ionic liquid-supported platinum nanoparticles, notably supported on carbon nanotubes, also selective hydrogenation of the alkene group under acidic conditions, s. M.J. Beier, J.-M. Andanson, A. Baiker, ACS Catal. *2012, 2* (12), 2587-95 [DOI: 10.1021/cs300529y]; reduction of nitroarenes with glycerol as hydrogen source (cf. *83,* 140) in the presence of magnetically recyclable, Fe_3O_4-supported nickel nanoparticles, also reduction of aryloxo compds., s. M.B. Gawande, A.K. Rathi, P.S. Branco, I.D. Nogueira, A. Velhinho, J.J. Shrikhande, U.U. Indulkar, R.V. Jayaram, C.A.A. Ghumman, N. Bundaleski, O.M.N.D. Teodoro, Chem. Eur. J. *2012, 18* (40), 12628-32 [DOI: 10.1002/chem.201202380]; by selective hydrogenation with highly active bimetallic Rh_3Ni_1 nanoparticles s. S. Cai, H. Duan, H. Rong, D. Wang, L. Li, W. He, Y. Li, ACS

Catal. *2013, 3* (4), 608-12 [DOI: 10.1021/cs300689w]; *metal-free* reduction with hydrazine and readily recoverable superparamagnetic graphene-Fe_3O_4 nanocomposite as catalyst s. C. Feng, H.-Y. Zhang, N.-Z. Shang, S.-T. Gao, C. Wang, Chin. Chem. Lett. *2013, 24* (6), 539-41 [DOI: 10.1016/j.cclet.2013.03.036]; by selective hydrogenation on photoactivated TiO_2 [rutile] with an alcohol as hydrogen source s. Y. Shiraishi, Y. Togawa, D. Tsukamoto, S. Tanaka, T. Hirai, ACS Catal. *2012, 2* (12), 2475-81 [DOI: 10.1021/cs300500p].

Sulfur ↑ HN ⇃↾ S

Magnesium/methanol *Mg/MeOH*
N-De(2-pyridylsulfonyl)ation of indolines s. *83*, 187 $NSO_2Py \rightarrow NH$

Zinc/ammonium chloride *Zn/NH_4Cl*
N-De(2-pyridylsulfonyl)ation of indoles or pyrroles s. *83*, 350

Potassium diphenylphosphide/hydrogen chloride *$KPPh_2/HCl$*
N-Desulf(i,o)nylation with phosphides $NSO_2Ar \rightarrow NH$

4. 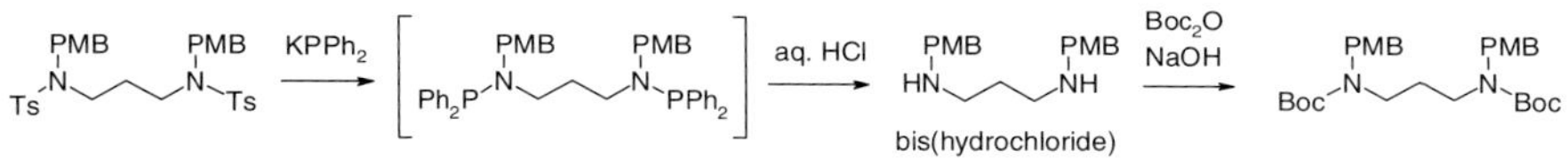

PMB = $4\text{-}MeOC_6H_4CH_2$

Interchange of N-protective groups under mild conditions. A soln. of K-diphenylphosphide (1.3 eq.) in THF (1.3 ml) added to a soln. of N,N′-bis(*p*-toluenesulfonyl)-N,N′-(*p*-methoxybenzyl)-1,3-propanediamine (0.5 mmol) in the same solvent (5 ml) at -78°, the mixture stirred for 2 h, quenched with 1 *M* aq. HCl (2 eq.), stirred at room temp. for 30 min, 1 *M* aq. NaOH (4 eq.) and di-*tert*-butyl dicarbonate (2 eq.) added, the mixture stirred for 12 h, extracted with ethyl acetate, concentrated *in vacuo,* and purified chromatographically → N,N′-di(*tert*-butoxycarbonyl)-N,N′-(*p*-methoxybenzyl)-1,3-propanediamine. Y 70% overall. The sulfonyl moiety activates the nitrogen towards nucleophilic attack by phosphide (other nucleophiles were unreactive), affording an unstable aminophosphine which may be easily hydrolyzed to the free amine (aq. HCl) or converted *in situ* to a carbamate under basic conditions. The method was applied to tert. *p*-toluenesulfonamides, benzenesulfonamides and *p*-toluenesulfinamides (seventeen examples; Y 70-100%; bulky N,N-diisopropyl-*p*-toluenesulfonamide gave 58%). Methanesulfonamide, sec. *p*-toluenesulfonamide, carbamate, ester, *p*-methoxybenzylamine, benzylamine and ar. halide functionality were inert to the conditions but an *o*-nitrobenzenesulfonamide gave a complex mixture of products. F.e.s. S. Yoshida, K. Igawa, K. Tomooka, J. Am. Chem. Soc. *2012, 134* (47), 19358-61 [DOI: 10.1021/ja309642r].

Carbon ↑ HN ⇃↾ C

Irradiation s. under 2,2′-Bipyridylbis(2-phenylpyridinato-N,C2′)iridium(III) hexafluorophosphate *///*

Piperidine *$(CH_2)_5NH$*
Prim. ar. amines from N-arylpyridinium salts s. *83*, 123 ←

Ammonium cerium(IV) nitrate *CAN*
N-De(4-hydroxy-3,5-di-methylphenyl)ation of indazoles s. *83*, 372 $NAr \rightarrow NH$

Triethylsilane/trifluoroacetic acid *Et_3SiH/CF_3COOH*
Cleavage of N-protective groups ←
update s. *5*, 32s*82*; **N-de(diphenylmethyl)ation** of protected isoxazolidines with triethylsilane/trifluoroacetic acid s. W. Wang, K.S. Rein, Tetrahedron Lett. *2013, 54* (14), 1866-8 [DOI: 10.1016/j.tetlet.2013.01.105]; regioselective protection of the amino group of lysine-containing substrates by a supramolecular enzyme-mimic approach s. J. Su, X. Sheng, S. Li, T. Sun, G. Liu, A. Hao, Org. Biomol. Chem. *2012, 10* (47), 9319-24 [DOI: 10.1039/c2ob27021h].

Air/ferrocenylacetic acid *O_2/$FcCH_2COOH$*

N-Demethylation NMe → NH

by modified Polonovski-type reaction s. *75,* 14; one-pot N-demethylation of N-methylated alkaloids under aerobic conditions with ferrocenylacetic acid as redox catalyst s. G.B. Kok, P.J. Scammells, Synthesis *2012, 44* (16), 2587-94 [DOI: 10.1055/s-0032-1316590]; *under continuous flow* with *m*-chloroperoxybenzoic acid as oxidant and $FeCl_3$ as catalyst s. Y. Nakano, G.P. Savage, S. Saubern, P.J. Scammells, A. Polyzos, Austr. J. Chem. *2013, 66* (2), 178-82 [DOI: 10.1071/ch12463].

2,2′-Bipyridylbis(2-phenylpyridinato-N,C2′)iridium(III) hexafluorophosphate/ triethylenediamine/water/irradiation ←

Mono-N-demethylation of ar. dimethylamines under visible-light photoredox catalysis

5.

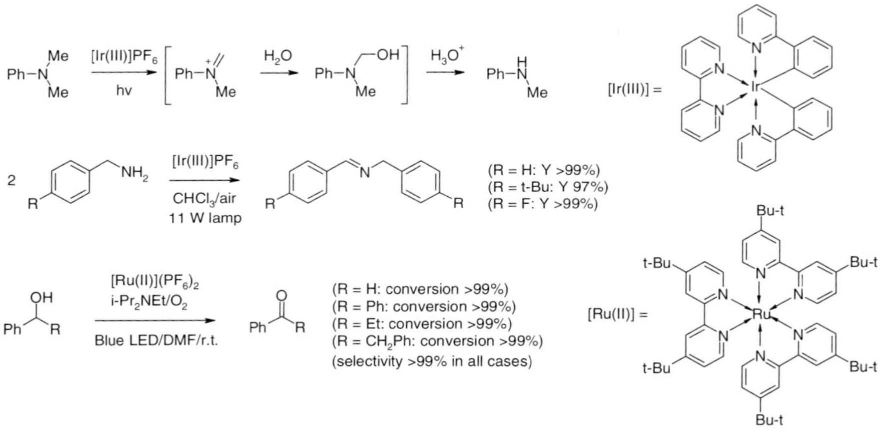

The photoredox catalyst [(2,2′-bpy)(ppy)$_2$Ir]PF_6 (1 mol%), DABCO (10 mol%), water (4.5 µl; 1 eq.), and N,N-dimethylaniline (0.25 mmol) in acetonitrile (2 ml) placed in a vial, the mixture stirred under irradiation with a 5 W fluorescent bulb (distance ~5 cm) for 48 h (TLC), 3 *M* aq. HCl (1 ml) added, the mixture stirred at 80° for 90 min, treated with 1 *M* NaOH (10 ml), extracted with methylene chloride, the extracts dried (Na_2SO_4), filtered, solvent removed under reduced pressure, and the residue purified by column chromatography → N-methylaniline. Y 82%. This photoredox N-demethylation method, which mimics cytochrome P-450 oxidation of tert. amines, proceeds under mild conditions and is applicable to a variety of N,N-dimethylanilines bearing *meta* and/or *para* substituents (nine examples; Y 35%, 55-80%). F.e., **also ar. N-benzylaldimines** (twelve examples; Y 65 to >99% by NMR) from two prim. benzylamine molecules (cf. *82*, 30; *79*, 160), and **oxo compds.** (fourteen examples; conversion 70 to >99%; selectivity >99% for all) from alcohols using a ruthenium catalyst, s. M. Rueping, C. Vila, A. Szadkowska, R.M. Koenigs, J. Fronert, ACS Catal. *2012*, *2* (12), 2810-5 [DOI: 10.1021/cs300604k].

Elimination ⇑

Oxygen ↑ HN ⇑ O

Indium/zinc/ammonium chloride *In/Zn/NH_4Cl*

N-Dehydroxylation NOH → NH

Chiral 2-aryl- or 2,4-diarylpiperidines s. *83*, 382

Formation of H-C Bond

Uptake ⇓

Addition to Oxygen and Carbon HC ⇓ OC

Irradiation s. under Dehydrogenase ///
Microwaves s. under Chiral N-condensed 1,3,2-oxazaborolidines [\\\\]

Copper(II) acetate/chiral 2,2′,6,6′-tetramethoxy-4,4′-bis[bis(3,5-dimethylphenyl)-phosphino]-3,3′-bipyridyl/phenylsilane ←
Asym. reduction of ketones via catalytic hydrosilylation CO → CHOH
under copper catalysis, update s. *62,* 18s*82*; of [hetero]aryl cycloalkyl ketones with $Cu(OAc)_2$/chiral 2,2′,6,6′-tetramethoxy-4,4′-bis[bis(3,5-dimethylphenyl)phosphino]-3,3′-bipyridyl and phenylsilane as reducing agent in air s. S.-B. Qi, M. Li, S. Li, J.-N. Zhou, J.-W. Wu, F. Yu, X.-C. Zhang, A.S.C. Chan, J. Wu, Org. Biomol. Chem. *2013, 11* (6), 929-37 [DOI: 10.1039/c2ob27040d]; of acylophenones *under nickel(II) catalysis* with the same chiral ligand and reducing agent cf. F.-F. Wu, J.-N. Zhou, Q. Fang, Y.-H. Hu, S. Li, X.-C. Zhang, A.S.C. Chan, J. Wu, Chem. Asian J. *2012, 7* (11), 2527-30 [DOI: 10.1002/asia.201200512].

Dendrimer-encapsulated copper(II) cluster complexes ←
Homogeneous hydrogenation of oxo compds. s. *32,* 29s*83*; *83,* 10

Borane-tetrahydrofuran s. under Chiral N-condensed 1,3,2-oxazaborolidines *BH_3-THF*

Borane·1,3-dimethylimidazol-2-ylidene/acetic acid *BH_3·NHC/AcOH*
Borane·carbene-based reduction of oxo compds.
update s. *15,* 39s*82*; with borane·1,3-dimethylimidazol-2-ylidene/acetic acid in ethyl acetate or methylene chloride, also selective reduction of aldehydes in the presence of ketones, s. V. Lamm, X. Pan, T. Taniguchi, D.P. Curran, Beilstein J. Org. Chem. *2013, 9,* 675-80 [DOI: 10.3762/bjoc.9.76].

Lithium tetrahydrido-borate or -aluminate *$LiBH_4$ or $LiAlH_4$*
β-Hydroxyketones from β-alkylidene-β-lactones
Stereoselective reductive ring opening

6. 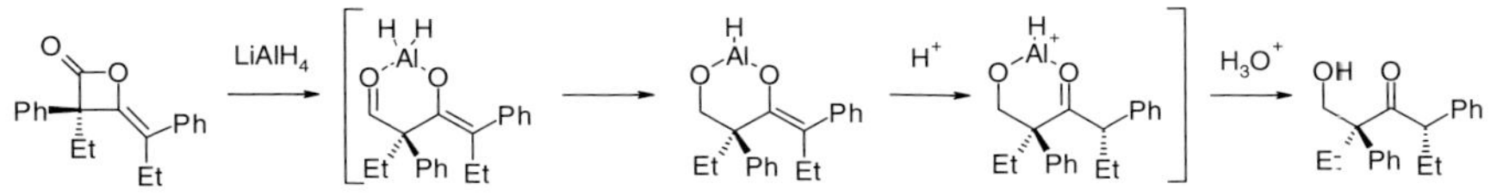

A soln. of $LiAlH_4$ (0.21 mmol) in ether (0.21 ml) reacted with enantioenriched ethyl(phenyl)ketene dimer (0.21 mmol; e.e. 90%) at -78°, quenched with satd. aq. NH_4Cl, and worked up with chromatographic purification → (2S,4S)-2,4-diethyl-1-hydroxy-2,4-diphenylbutan-3-one. Y 84% (d.r. >99:1; e.e. 91%). A series of ketene homo-/hetero-dimers underwent reductive ring opening with $LiAlH_4$ or $LiBH_4$ to afford the title products (eight examples; Y 35%, 62-90%; d.r. 2:1, 5:1 to >99:1), and with enantioenriched substrates proceeded with near complete **retention of chirality** (four examples; Y 66-92%; d.r. 3:1 to >99:1; e.e. 91-96%). A transition state model suggests that *syn*-diastereoselectivity is favored via protonation from the least hindered face of an intermediate cyclic aluminum enolate. F.e.s. P.-H. Wei, M.A. Gary, D. Nalla, G.D. Harzmann, A.A. Ibrahim, K.R. Dayak, N.J. Kerrigan, Tetrahedron Lett. *2013, 54* (8), 932-5 [DOI: 10.1016/j.tetlet.2012.12.011].

Chiral N-condensed 1,3,2-oxazaborolidines/borane-tetrahydrofuran/microwaves ← C
o-Hydroxy-o′-(hydroxymethyl)biaryls from dibenzo[b,d]pyran-6-ones
Corey-Bakshi-Shibata reductive ring opening with dynamic kinetic resolution under microwave irradiation

7. 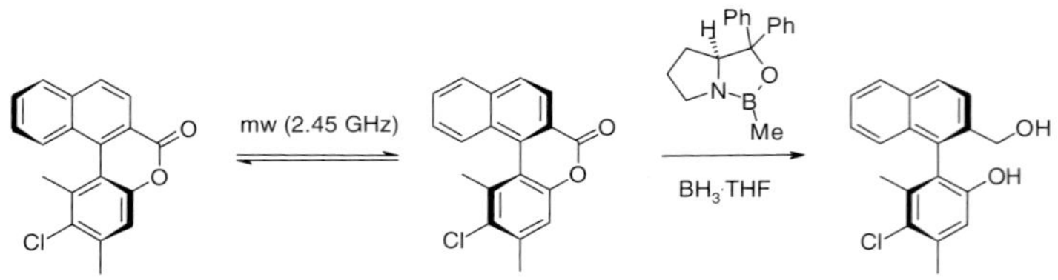

A soln. of startg. biaryl lactone (0.25 mmol) in THF (2.5 ml) added slowly to a soln. of chiral oxazaborolidine catalyst (10 mol%) and BH_3·THF (1.1 eq.) in the same solvent (3 ml), the mixture stirred under microwave irradiation (2.45 GHz) at 31.6° until lactone consumed (15 min), quenched with 1 *M* aq. HCl, extracted with ethyl acetate, and purified chromatographically → (*P*)-1-(3-chloro-6-hydroxy-2,4-dimethylphenyl)-2-hydroxymethylnaphthalene. Y 91% (e.e. 91%). Reductive cleavage of a series of biaryl lactones under microwave irradiation showed significant rate enhancement (up to ca. 550-fold) compared with conventional heating at the same temp., while maintaining a comparable level of enantioselectivity (e.e. 85-91%; four examples; Y 82-96%). Similar rates of reaction could be achieved by conventional heating at higher temperatures, but enantioselectivity was reduced. A direct effect of microwaves on the dynamic equilibrium between the *atropo*-isomers (*M* and *P*) of the substrate was presumed. F.e.s. K. Nushiro, S. Kikuchi, T. Yamada, Chem. Lett. *2013*, *42* (2), 165-7 [DOI: 10.1246/cl.2013.165].

Borane·lithium amide complex *BH_3·$LiNH_2$*
Borane·amine-based reduction of oxo compds. CO → CHOH
update s. *15*, 39s*82*; mild reduction of ketones with borane·lithium amide and related metal amide·borane complexes, also sec. amines from azomethines, s. W. Xu, G. Wu, W. Yao, H. Fan, J. Wu, P. Chen, Chem. Eur. J. *2012*, *18* (29), 13885-92 [DOI: 10.1002/chem.201201301].

Cerium dioxide s. under Ni *CeO_2*
Water s. under Tricarbonyl[1,3-bis(trimethylsilyl)-4,5,6,7-tetrahydro-2-indenone]iron(0) and Ni-Al *H_2O*
Isopropanol s. under Ferrocenylimidazolium salts, Ni and Polymer-based dichloro-(cyclopentadienyl)iridium(III) *i-PrOH*
1,4-Butanediol s. under Horse liver alcohol dehydrogenase *$HO(CH_2)_4OH$*
Carbon monoxide s. under Tricarbonyl[1,3-bis(trimethylsilyl)-4,5,6,7-tetrahydro-2-indenone]iron(0) *CO*
Formic acid/triethylamine s. under Dichloro(p-cymene)ruthenium(II) dimer *HCOOH·Et_3N*

Dehydrogenase ←
Dehydrogenase/graphene-based multi-anthraquinone-subst. porphyrin/2,2′-bipyridyl-(chloro)(pentamethylcyclopentadienyl)rhodium(III) chloride/irradiation ←
Asym. bioreduction of ketones
update s. *29*, 36s*82*; of sparingly water-soluble ketones *under continuous flow* in a dual dehydrogenase membrane reactor with recycling of the aq. phase containing an ionic liquid as detergent s. S. Leuchs, S. Na′amnieh, L. Greiner, Green Chem. *2013*, *15* (1), 167-76 [DOI: 10.1039/c2gc36558h]; asym. reduction of 3-aryl-2-oxopropanoic acids with lactate dehydrogenase, also comparison with classical asym. hydrogenation using a ruthenium(II) di(phosphine) complex, s. S. Lüttenberg, T.D. Ta, J. von der Heyden, J. Scherkenbeck, Eur. J. Org. Chem. *2013* (9), 1824-30 [DOI: 10.1002/ejoc.201201506]; asym. reduction of acetophenones *under solar energy* with a graphene-based multi-anthraquinone-subst. porphyrin as photocatalyst and [Cp*Rh(bpy)Cl]Cl as mediator coupled with an enzyme [LKADH] to mimic photosynthesis s. S. Choudhury, J.-O. Baeg, N.-J. Park, R.K. Yadav, Angew. Chem., Int. Ed. *2012*, *51* (46), 11624-8 [DOI: 10.1002/anie.201206019].

Horse liver alcohol dehydrogenase/nicotinamide adenine dinucleotide/1,4-butanediol ←
Improved NAD(P)H-dependent redox biocatalysis ←
using 1,4-butanediol as 'smart cosubstrate'

8. 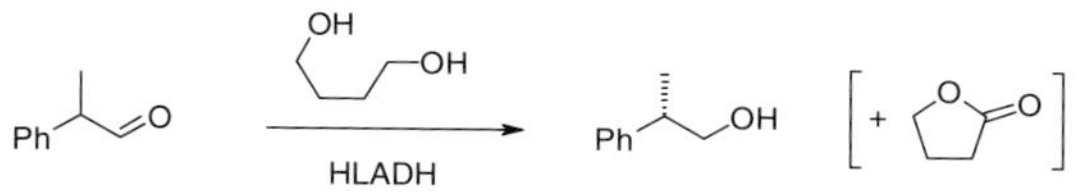

A significant reduction in the amount of H-donor required to achieve quantitative conversion has been demonstrated by employing 1,4-butanediol, which undergoes oxidation twice, producing γ-butyrolactone as a thermodynamically stable and kinetically inert coproduct, resulting in faster reaction rates. **E: 2-Aryl-*prim*-alcohols from α-arylaldehydes with dynamic kinetic resolution.** A mixture of 2-phenyl-1-propanal stock soln. in acetonitrile (0.5 *M*; 15 μl), 1,4-butanediol stock soln. in buffer (30 *mM*; 125 μl), NAD^+ stock soln. in buffer (15 *mM*; 10 μl) and Tris-HCl buffer (50 *mM*; pH 7.5; 1% v/v acetonitrile; 1200 μl) incubated at 30° for 3 min, treated with horse liver alcohol dehydrogenase [HLADH] soln. (150 μl; 0.1 g/l), and incubated at 30° in a screw-capped glass vial with shaking at 700 rpm for 24 h with monitoring by HPLC → product. Conversion 98% (e.e. 95%). In comparison, use of 0.5 eq. of ethanol or isopropanol resulted in conversions of 24% or 14%, respectively (confirming why they are usually used in huge molar excess). If ten times more biocatalyst was used the e.e. dropped to 56%, the enzymatic activity for the reduction outperforming the racemization rate under these conditions. The cosubstrate was also used for the enoate-reductase-catalyzed asym. reduction of a cyclic enedione (ketoisophorone) to the chiral dione ((R)-levodione) and 3-hydroxybenzoate-6-hydroxylase-catalyzed regioselective hydroxylation of 3-hydroxybenzoate to 2,5-dihydroxybenzoate. F. details s. S. Kara, D. Spickermann, J.H. Schrittwieser, C. Leggewie, W.J.H. van Berkel, I.W.C.E. Arends, F. Hollmann, Green Chem. *2013*, *15* (2), 330-5 [DOI: 10.1039/c2gc36797a]; ketones from sec. alcohols via alcohol dehydrogenase-catalyzed hydrogen transfer employing selected ketones (which can only be reduced but not oxidized) as hydrogen acceptors cf. I. Lavandera, A. Kern, V. Resch, B. Ferreira-Silva, A. Glieder, W.M.F. Fabian, S. de Wildeman, W. Kroutil, Org. Lett. *2008*, *10* (11), 2155-8 [DOI: 10.1021/ol800549f].

Organosilicon hydrides s. under (2-Pyrid-2-ylferrocenyl)borenium tetraalkoxyaluminates *⩾SiH*
Phenylsilane s. under Copper(II) acetate and Nickel(II) salts *$PhSiH_3$*
Chiral 2,2′,6,6′-tetramethoxy-4,4′-bis[bis(3,5-dimethylphenyl)phosphino]-3,3′-bipyridyl ←
s. under Copper(II) acetate and Nickel(II) salts
Chiral N,N′-bis[6-(diphenylphosphino)pyrazin-2-ylcarbonyl]-1,2-diaminocyclohexane ←
s. under Chloro(norbornadiene)rhodium(I) dimer
Chiral 7-(2-pyridylmethylamino)-7′-(diarylphosphino)-1,1′-biindanyl s. under ←
Chloro(1,5-cyclooctadiene)iridium(I) dimer
(S,S)-1,2-Bis(1-naphthyl)-N-(2,6-diphenylbenzenesulfonyl)-1,2-ethylenediamine s. under ←
Dichloro(p-cymene)ruthenium(II) dimer

(2-Pyrid-2-ylferrocenyl)borenium tetraalkoxyaluminates/organosilicon hydrides ←
Sec. alcohols from ketones via ferrocenylborenium-catalyzed hydrosilylation CO → CHOH

9. 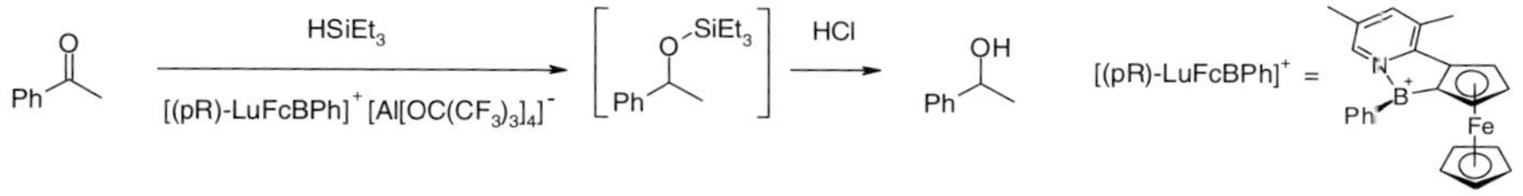

The first synthesis of an isolable planar-chiral ferrocenylborenium cation and proof-of-principle investigations into its utility for hydrosilylation of ketones are reported. **E:** A soln. of triethylsilane (1 eq.) in methylene chloride (1 *M*) added to $[(pR)\text{-LuFcBPh}]^+[Al[OC(CF_3)_3]_4]^-$ (5 mol%) dissolved in a minimum amount of methylene chloride, the mixture cooled to -37°, a soln. of acetophenone in methylene chloride (1 *M*) added, the mixture allowed to react at room temp. for ca. 12 h (until completion of the reaction by ^{1}H NMR), treated with dil. aq. HCl, and stirred vigorously overnight

→ (R)-(+)-product. Conversion 65% (e.e. 20%). The conversion was 92% (e.e. 6%) with dimethyl-(phenyl)silane in place of triethylsilane, while triphenylsilane was ineffective. It is expected that further modification of the steric demand of the pyridyl ligand and the exocyclic boron substituent may provide significantly enhanced selectivities, while the activity may be optimized through tuning of the electronic effect of the exocyclic B-aryl group. F.e. (four; conversion 29 to >95%; e.e. 0-5%) s. J. Chen, R.A. Lalancette, F. Jäkle, Chem. Commun. *2013*, *49* (43), 4893-5 [DOI: 10.1039/c3cc41556b].

Tricarbonyl[1,3-bis(trimethylsilyl)-4,5,6,7-tetrahydro-2-indenone]iron(0)/ carbon monoxide/water ←
Ferrocenylimidazolium salts/isopropanol ←
Transfer-hydrogenation of oxo compds. s. *33*, 47s*83* CO → CHOH

Tricarbonyl(1,3-disilyl-4,5,6,7-tetrahydro-2-indenone)iron(0) complex/potassium carbonate ←
[Bis[2-(dicyclohexylphosphino)ethyl]amine](trimethylsilylmethyl)cobalt(II) barfate *[Co(II)]*
Homogeneous hydrogenation of oxo compds. under transition metal catalysis

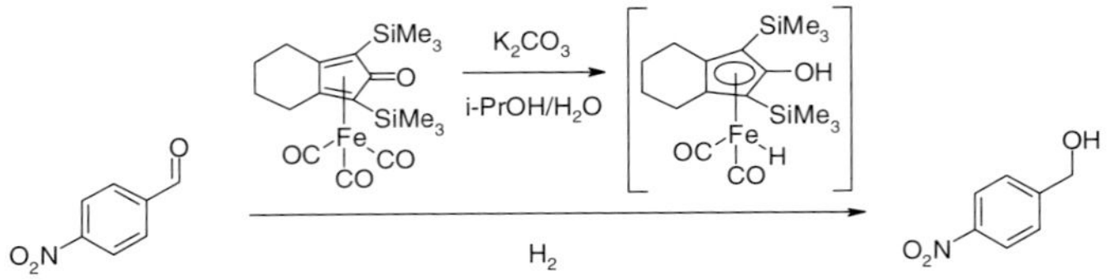

10.

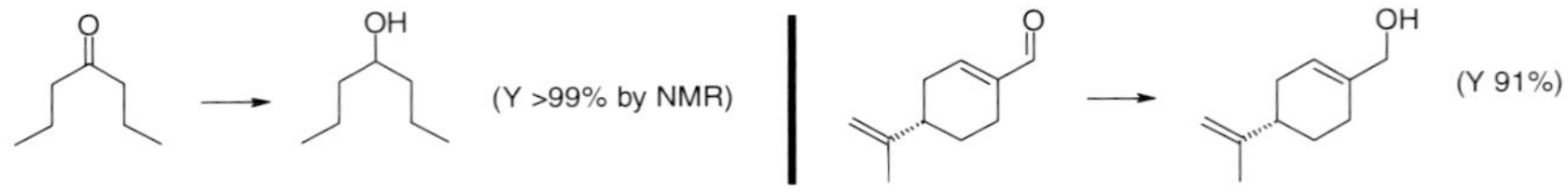

Chemoselective iron catalysis. A mixture of *p*-nitrobenzaldehyde (1 mmol), the tricarbonyl-(1,3-disilyl-4,5,6,7-tetrahydro-2-indenone)iron(0) complex (0.1 mol%) and K_2CO_3 (0.5 mol%) in isopropanol (0.5 ml) and water (0.2 ml) hydrogenated under 30 bar H_2 at 100° for 17 h → *p*-nitrobenzyl alcohol. Y 96%. Unlike previously described iron complexes (cf. *23*, 51s*72*), this complex and analogs are easily handled, being stable to air and water, may be prepared easily [from terminal diynes, chlorosilanes and $Fe(CO)_5$], and may be purified by column chromatography. Thirteen further examples of electron-diverse (het)aryloxo compds. underwent reduction in 76 to >99% GC yields (89-97% isolated yields for four examples), notably tolerating trifluoromethyl, ar. trifluoromethoxy, halo, amino, ester and amide groups, while benzil gave hydrobenzoin in >99% GC yield. TONs of up to 3800 may be achieved. Aliphatic oxo compds., including citronellal and cyclic ketones, similarly gave yields of 85 to >99% (seven examples). Furthermore, eight examples of **2-ethylene-*prim*-alcohols** were obtained selectively **from α,β-ethylenealdehydes** (Y 80% to >99%). F.e.s. S. Fleischer, S. Zhou, K. Junge, M. Beller, Angew. Chem., Int. Ed. *2013*, *52* (19), 5120-4 [DOI: 10.1002/anie.201301239]; homogeneous hydrogenation of ketones catalyzed by an iron pincer complex (s.a. *32*, 29s*82*) s. R. Langer, G. Leitus, Y. Ben-David, D. Milstein, ibid. *2011*, *50* (9), 2120-4 [DOI: 10.1002/anie.201007406]; mild procedure *under cobalt catalysis* with [bis[2-(dicyclohexylphosphino)ethyl]amine](trimethylsilylmethyl)cobalt(II) barfate, also hydrogenation of ethylene derivs. and azomethines, s. G. Zhang, B.L. Scott, S.K. Hanson, ibid. *2012*, *51* (48), 12102-6 [DOI: 10.1002/anie.201206051]; *under copper catalysis* in water with a regenerable copper(II) cluster encapsulated within a poly(amidoamine) dendrimer having hydroxyl surface groups [$(Cu(II))_n$@PAMAM-OH], also hydrogenation of alkenes, s. P. Maity, S. Yamazoe, T. Tsukuda, ACS Catal. *2013*, *3* (2), 182-5 [DOI: 10.1021/cs3007318].

Nickel-aluminum/water *Ni-Al/H$_2$O*
Sec. alcohols from ketones CO → CHOH
using water as hydrogen source under high hydrostatic pressure

11. 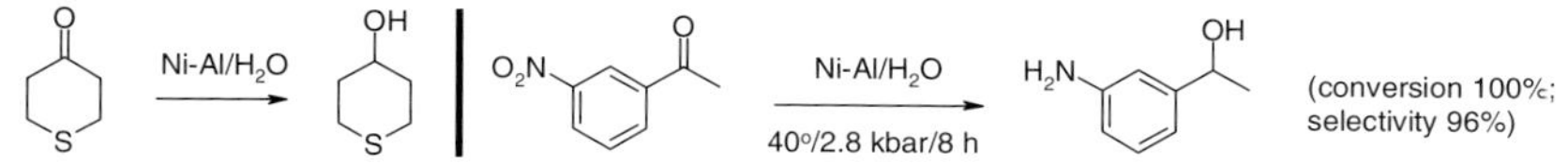

Startg. ketone (0.024 mmol), Ni-Al alloy (0.27 mmol) and water (110 μl) added to a 150 μl fluoropolymer MicroTube then placed in a stainless steel cartridge, the bench-top high pressure generator (Pressure BioSciences Barocycler HUB440) set to 80° and 2.8 kbar, after 6 h the cartridge removed from the barocycler, the sample filtered, the filtrate extracted with methylene chloride, and directly analyzed by GC-MS → product. Conversion 100%; selectivity 100%. The method is highly selective for formation of alcohols from acyclic, cyclic or aryl ketones, avoiding problems such as overhydrogenation or multiple product formation commonly associated with the use of Ni-Al alloy. It uses water as benign, readily available and inexpensive solvent as well as source of hydrogen, without the requirement for strong base, the high pressure ensuring that the formed hydrogen remains in solution so that carbonyl reduction occurs readily. In the case of *m*-nitroacetophenone, reduction of the nitro group occurred as well (conversion 100%; selectivity 96%). It is notable that the presence of sulfur in the substrate is tolerated, as well as N-carbo-*tert*-butoxy, spiroketal and ester groups. Conversions were low in the presence of *p*-halo substituents, however. F.e.s. A. Tomin, A. Lazarev, M.P. Bere, H. Redjeb, B. Török, Org. Biomol. Chem. *2012*, *10* (36), 7321-6 [DOI: 10.1039/c2ob25941a]; [N-unsubst.] **indolines from indoles or 1,2,3,4-tetrahydroquinolines from quinolines** under similar conditions s. H. Cho, F. Török, B. Török, ibid. *2013*, *11* (7), 1209-15 [DOI: 10.1039/c2ob26903a]; **alkylarenes from acylophenones** with Ni-Al alloy in water under microwave irradiation, incl. deutero-derivs. with D_2O, s. A. Miyazawa, M. Tashiro, G.K.S. Prakash, G.A. Olah, Bull. Chem. Soc. Jpn. *2006*, *79* (5), 791-2 [DOI: 10.1246/bcsj.79.791].

Nickel/cerium dioxide/isopropanol *Ni/CeO$_2$/i-PrOH*
Heterogeneous transfer-hydrogenation of ketones s. *33*, 47s*83*

Nickel(II) salts/chiral 2,2′,6,6′-tetramethoxy-4,4′-bis[bis(3,5-dimethylphenyl)phosphino]-3,3′-bipyridyl/phenylsilane ←
Asym. reduction of ketones via catalytic hydrosilylation s. *62*, 18s*83*

Ruthenium(II) complexes/H-donor/base ←
Transfer-hydrogenation of oxo compds.
under ruthenium catalysis, update s. *33*, 47s*82*; with chlorobis(triphenylphosphine)[3-methyl-6-[6-(3,5-dimethyl-1-pyrazolyl)-2-pyridyl]phenyl]ruthenium(II) as pre-catalyst in isopropanol containing KOPr-*i*, also reverse [Oppenauer] oxidation, s. W. Du, L. Wang, P. Wu, Z. Yu, Chem. Eur. J. *2012*, *18* (37), 11550-4 [DOI: 10.1002/chem.201201938]; transfer-hydrogenation of aryl ketones with ruthenium(II) complexes bearing pyridine-based tridentate or bidentate ligands (and KOH as base) s. O. Dayan, S. Dayan, I. Kani, B. Çetinkaya, Appl. Organomet. Chem. *2012*, *26* (12), 663-70 [DOI: 10.1002/aoc.2901]; with [2-(aminomethyl)piperidine](dichloro)di(phosphine)-ruthenium(II) complexes cf. H. Türkmen, ibid. 731-5 [DOI: 10.1002/aoc.2924]; *in water* with [N-alkyl-2,2′-bipyridylamine]-ruthenium(II) or -rhodium(III) complexes or their papain conjugates with formate as H-donor s. N. Madern, B. Talbi, M. Salmain, ibid. *2013*, *27* (1), 6-12 [DOI: 10.1002/aoc.2929]; with $[RuCl_2(p\text{-cymene})]_2$ and 2-(2-methyl-2,3-dihydro-1*H*-perimidin-2-yl)-benzene-1,3-diol as ligand s. M.E. Cucciolito, B. Panunzi, F. Ruffo, A. Tuzi, Tetrahedron Lett. *2013*, *54* (12), 1503-6 [DOI: 10.1016/j.tetlet.2012.12.088]; *under iron catalysis* with ferrocenyl-imidazolium salts and KOH as base in isopropanol cf. M.I. Ikhile, M.D. Bala, V.O. Nyamori, J.C. Ngila, Appl. Organomet. Chem. *2013*, *27* (2), 98-108 [DOI: 10.1002/aoc.2947]; selective reduction of aromatic, aliphatic and α,β-unsatd. aldehydes under water-gas shift conditions with tricarbonyl[1,3-bis(trimethylsilyl)-4,5,6,7-tetrahydro-2-indenone]iron(0) in water/DMSO with K_2CO_3 as base s. A. Tlili, J. Schranck, H. Neumann, M. Beller, Chem. Eur. J. *2012*, *18* (50), 15935-9 [DOI: 10.1002/chem.201203059]; *heterogeneous* transfer-hydrogenation of aromatic and aliphatic ketones *under nickel catalysis* with Ni/CeO_2 in isopropanol without base s. K. Shimura, K.-i. Shimizu, Green Chem. *2012*, *14* (11), 2983-5 [DOI: 10.1039/c2gc35836k]; **transfer-**

deuterogenation with cationic aqua(pentamethylcyclopentadienyl)(4,4′-dihydroxy-2,2′-bipyridyl)rhodium(III) or -iridium(III) complexes in D_2O/HCO_2H for an *efficient generation of deuterium* s. W.-H. Wang, J.F. Hull, J.T. Muckerman, E. Fujita, T. Hirose, Y. Himeda, Chem. Eur. J. *2012, 18* (30), 9397-404 [DOI: 10.1002/chem.201200576].

Chiral ruthenium(II) complexes *[Ru(II)]**
Asym. homogeneous hydrogenation of ketones CO → CHOH
under ruthenium catalysis, update s. *67,* 22s*82; 43,* 51s*82*; of 3-oxoglutaric acid derivs. with [RuCl(benzene)(S)-SunPhos]Cl [where SunPhos = (2,2,2′,2′-tetramethyl[4,4′-bibenzo[*d*][1,3]-dioxole]-5,5′-diyl)bis(diphenylphosphine); cf. *80,* 18], solvent effect on enantioselectivity, s. W. Li, X. Tao, X. Ma, W. Fan, X. Li, M. Zhao, X. Xie, Z. Zhang, Chem. Eur. J. *2012, 18* (51), 16531-9 [DOI: 10.1002/chem.201202614]; diastereo- and enantio-selective hydrogenation of α-acylamino-β-ketophosphonic acid esters *via dynamic kinetic resolution* with chiral (η^6-benzene) chloro[2,2′-bis(diarylphosphino)-5,6;5′,6′-bis(methylenedioxy)biphenyl]ruthenium(II) chloride complexes and added $CeCl_3{\cdot}7H_2O$ s. X. Tao, W. Li, X. Li, X. Xie, Z. Zhang, Org. Lett. *2013, 15* (1), 72-5 [DOI: 10.1021/ol303105d]; of methyl propionylacetate *under continuous flow* with a ruthenium(II) (S)-BINAP complex immobilized in a ionic liquid/supercritical CO_2 biphase medium with an added acid s. J. Theuerkauf, G. Franciò, W. Leitner, Adv. Synth. Catal. *2013, 355* (1), 209-19 [DOI: 10.1002/adsc.201200724]; of benzils to hydrobenzoins with Ru(OTf)(TsDPEN)(η^6-cymene) as pre-catalyst in methanol s. X. Huang, N. Li, Z. Geng, F. Pan, X. Wang, Chin. J. Chem. *2012, 30* (11), 2657-63 [DOI: 10.1002/cjoc.201200393]; of aliphatic α-subst. enones to chiral allyl alcohols *under iridium catalysis* with $[Ir(cod)Cl]_2$ and a chiral 7-(2-pyridylmethylamino)-7′-(diarylphosphino)-1,1′-biindanyl [(S)-SpiroPAP] as ligand with KOBu-*t* as base (cf. *50,* 19s*82*) s. Q.-Q. Zhang, J.-H. Xie, X.-H. Yang, J.-B. Xie, Q.-L. Zhou, Org. Lett. *2012, 14* (24), 6158-61 [DOI: 10.1021/ol302842h]; of acetophenone *under rhodium catalysis* in water with $[Rh(norbornadiene)Cl]_2$ and a chiral N,N′-bis[6-(diphenylphosphino)pyrazin-2-ylcarbonyl]-1,2-diaminocyclohexane [(R,R)-DAMPYPHOS] as ligand (cf. *23,* 51s*46-9*) s. N.I. Nikishkin, J. Huskens, W. Verboom, Tetrahedron Lett. *2013, 54* (14), 1857-61 [DOI: 10.1016/j.tetlet.2013.01.108].

Chiral ruthenium(II) complexes/H-donor/base *[Ru(II)]*/H-donor/base*
Asym. transfer-hydrogenation of ketones
update s. *46,* 42s*82*; with a chiral dichloro[2-(Δ^2-oxazolin-2-yl)-6-(pyrazol-5-yl)pyridine]-(triphenylphosphine)ruthenium(II) complex at low loading in isopropanol with KOPr-*i* as base s. W. Ye, M. Zhao, Z. Yu, Chem. Eur. J. *2012, 18* (35), 10843-6 [DOI: 10.1002/chem.201201703]; of 1-naphthyl ketones with a chiral, *ansa*-type (η^6-arene-functionalized) hydrido[N-aminosulfonyl-1,2-diphenylethylenediamine]ruthenium(II) complex in formic acid/Et_3N s. A. Kišic, M. Stephan, B. Mohar, Org. Lett. *2013, 15* (7), 1614-7 [DOI: 10.1021/ol400393j]; of N-protected α-amino-β-ketocarboxylic acid esters *in water* with dynamic kinetic resolution using $[RuCl_2(benzene)]_2$ and (1S,2S)-2-benzylamino-1,2-diphenylethanol as ligand and Na-formate as H-donor with added Tween 20 s. B. Seashore-Ludlow, F. Saint-Dizier, P. Somfai, Org. Lett. *2012, 14* (24), 6334-7 [DOI: 10.1021/ol303115v].

[1-[o-(Aminomethyl)phenyl]-3-methylimidazol-2-ylidene](pentamethylcyclopentadienyl)-(pyridine)ruthenium(II) hexafluorophosphate/potassium tert-butoxide ←
Ruthenium-catalyzed homogeneous hydrogenation of carboxylic acid esters or lactones under mild conditions ←

12.

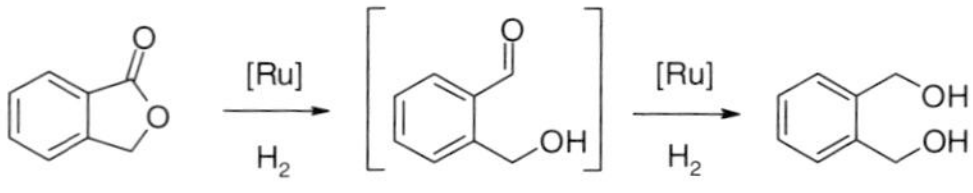

A ruthenium complex bearing an 1-[*o*-(*aminomethyl*)phenyl]-3-methylimidazol-2-ylidene ligand is highly effective for the homogeneous hydrogenation of carboxylic acid esters or lactones under lower (25 bar) hydrogen pressures and temperatures (50°) than those of existing routes. **E: Diols from lactones**. A soln. of [1-[*o*-(aminomethyl)phenyl]-3-methylimidazol-2-ylidene](pentamethylcyclopentadienyl)(pyridine)ruthenium(II) hexafluorophosphate (4.6 μmol), phthalide (6.9 mmol)

and *n*-tridecane (internal standard) in THF (4 ml), and a soln. of K-*tert*-butoxide (0.036 mmol) in the same solvent (2 ml), injected separately via syringe under argon into a reactor against a flow of hydrogen gas, the pressure of the latter adjusted to 25 bar, hydrogenated for 4 h, and the mixture worked up → 1,2-bis(hydroxymethyl)benzene. Y 96%.

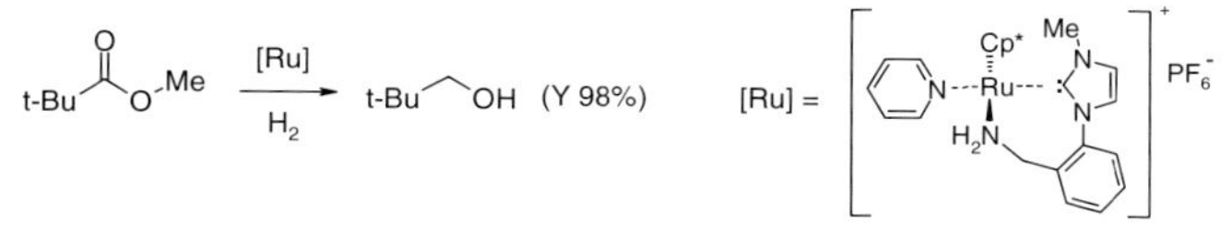

The procedure is applicable in high yield to aliphatic carboxylic acid esters, e.g. methyl pivalate and ethyl isobutyrate, or methyl benzoate to afford **prim. alcohols** and to a number of 5-membered lactones (six examples; Y 32-98%) with TOF numbers up to 1510 h^{-1}, but substituted benzoic acid esters and the hindered *tert*-butyl acetate were noticeably less reactive. DFT calculations suggest that hydrogenation proceeds via a concerted, asynchronous bifunctional mechanism. Kinetically, after splitting of hydrogen to generate an intermediate ruthenium hydride complex, the rate-limiting step of the cycle appears to be the transfer of a Ru-H/N-H pair to the carbonyl group through a 6-membered cyclic transition state involving hydrogen bonding of carbonyl oxygen with the N-H group of the appended aminomethyl residue. F.e.s. W.W.N. O, R.H. Morris, ACS Catal. *2013*, *3* (1), 32-40 [DOI: 10.1021/cs300619q].

Dichloro(p-cymene)ruthenium(II) dimer/(S,S)-1,2-bis(1-naphthyl)-N-(2,6-diphenylbenzenesulfonyl)-1,2-ethylenediamine/formic acid/triethylamine ←

α-Hydroxy- from acyl-phosphonic acid esters by ruthenium(II)-catalyzed asym. transfer-hydrogenation with dynamic kinetic resolution $C(O)PO_3R_2 \rightarrow CH(OH)PO_3R_2$

13. 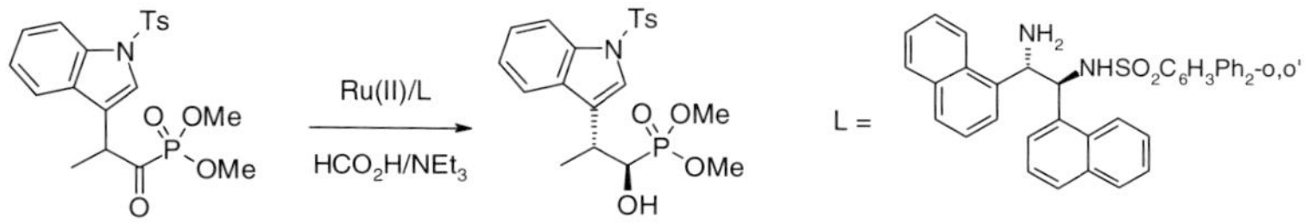

Asym. reduction of (het)arylated acylphosphonates has been shown to proceed from the opposite diastereotopic face to that reported for analogous reduction of β-aryl-α-keto-esters (*81*, 25). **E: Chiral β-aryl-α-hydroxyphosphonic acid esters**. A mixture of $[RuCl_2(p\text{-cymene})]_2$ (2 mol%) and chiral ligand (8 mol%) in DMSO (0.5 ml) stirred at 60° under N_2 for 30 min, cooled to room temp., a soln. of startg. acylphosphonate (0.155 mmol) in the same solvent (1 ml) and formic acid/triethylamine (5:2; 5 eq.) added, the mixture stirred for 10 h, extracted with ethyl acetate, and purified chromatographically → dimethyl [(1R,2R)-1-hydroxy-2-(1-tosyl-1*H*-indol-3-yl)-propyl]phosphonate. Y 94% (d.r. >20:1; e.e. 99%). This unexpected diastereoselectivity provided access to novel β-stereogenic α-hydroxyphosphonic acids not previously available in stereoisomerically pure form. The reaction appears general for acylphosphonates carrying both electron-diverse α-aryl (incl. indol-3-yl) and α-linear alkyl substituents (thirteen examples; Y 86-94%; d.r. 14:1 to >20:1; e.e. 97-99%), with somewhat lower diastereoselectivity observed for α-2-methoxyphenyl (d.r. 6:1; Y 84%; e.e. 97%) and -cyclopropyl substituents (d.r. 5:1; Y 86%; e.e. 99%). Analogs lacking a β-chiral center were also available by this method (four examples; Y 91-95%; e.e. 84-99%) using an alternative ethylenediamine-derived ligand (phenyl in place of 1-naphthyl). F.e. and optimization s. M.T. Corbett, J.S. Johnson, J. Am. Chem. Soc. *2013*, *135* (2), 594-7 [DOI: 10.1021/ja310980q]; reduction of β-aryl- or β-chloro-α-keto-esters s.a. K.M. Steward, M.T. Corbett, C.G. Goodman, J.S. Johnson, ibid. *2012*, *134* (49), 20197-206 [DOI: 10.1021/ja3102709].

Tripodal peptidic titanium phosphonate-encapsulated dichlororuthenium 2-amino-phosphine complex ←

Sec. alcohols from ketones by asym. heterogeneous hydrogenation CO → CHOH

14. 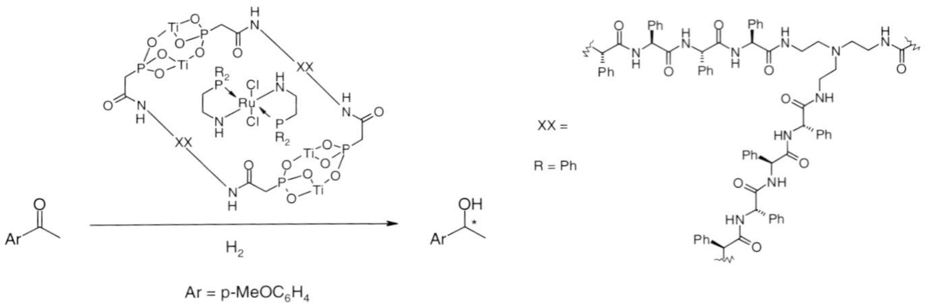

An inexpensive ‘off-the shelf’ achiral ruthenium-based hydrogenation catalyst, embedded within a homochiral amorphous matrix consisting of tripodal poly(phenylglycine) capped with phosphonate moieties and cross-linked with titanium oxide, provides an enantioselective, recyclable heterogeneous catalyst for ketone hydrogenation. **E:** The Ru-Ti-P catalyst (titanium/phosphorus 4.25:3; 10 mg) added to a 20 ml glass pressure tube, THF (500 µl; without stabilizer) added, followed by phosphate buffer (500 µl; pH 12.6, 0.15 *M* Na_3PO_4 and 0.03 *M* Na_2HPO_4), the tube purged with H_2 three times, then pressurized to 2 atm. with H_2, the contents stirred for 2 h, then startg. aryl ketone (10 µl) added, the suspension purged three times, pressurized to 2 atm. with H_2, and allowed to react for 60 h at room temp. (~22°) → product. Y 20% (e.e. 95%). Further optimization is required to improve yield and turnover numbers, but enantioselectivities for reduction of acetophenone (70% e.e.) or analogs bearing electron-donating groups (*p*-Me: 80% e.e.; *p*-Et: 88% e.e.) were good; *p*-chloroacetophenone, 1-acetylnaphthalene or ethyl pyruvate gave low e.e. (12-40%). Three variants of the encapsulated homochiral titanium phosphonate were prepared, slight variations in the preparation procedures leading to materials with different surface areas, pore structures, and morphologies; better results were obtained with a more ordered or defined pore network. F.e.s. A. Milo, R. Neumann, ACS Catal. *2012*, *2* (12), 2531-6 [DOI: 10.1021/cs3005715]; **heterogeneous asym. hydration of epoxides** using a tripodal peptidic titanium phosphonate cf. idem., Adv. Synth. Catal. *2010*, *352* (13), 2159-65 [DOI: 10.1002/adsc.201000373]; **asym. formation of glycols** by consecutive epoxidation and hydration of styrenes using an achiral manganese salen catalyst encapsulated in a peptidic phosphonate homochiral solid cf. idem., Chem. Commun. *2011*, *47* (9), 2535-7 [DOI: 10.1039/c0cc04205f].

Chloro(norbornadiene)rhodium(I) dimer/chiral N,N′-bis[6-(diphenylphosphino)-pyrazin-2-ylcarbonyl]-1,2-diaminocyclohexane *[Rh(I)]**

Asym. homogeneous hydrogenation of ketones s. *23,* 51s*83*

2,2′-Bipyridyl(chloro)(methyl)(pentamethylcyclopentadienyl)rhodium(III) chloride salt ←
s. under Dehydrogenase

Chloro(1,5-cyclooctadiene)iridium(I) dimer/chiral 7-(2-pyridylmethylamino)-7′-(diarylphosphino)-1,1′-biindanyl/potassium tert-butoxide *[Ir(I)]*/KOBu-t*

2-Ethylenealcohols from α,β-ethyleneketones s. *50,* 19s*83*

Polymer-based dichloro(cyclopentadienyl)iridium(III) complexes/ isopropanol/potassium tert-butoxide ←

Heterogeneous transfer-hydrogenation of ar. aldehydes using polymer-based iridium(III) or rhodium(III) complexes as catalysts CHO → CH_2OH

15. 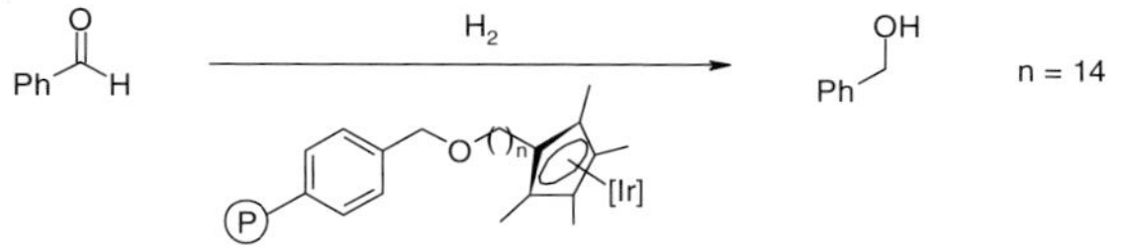

Immobilization of a series of hydroxyalkyl-substituted (tetramethylcyclopentadienyl)-iridium or -rhodium dichloride complexes onto a polymer [Wang] resin may be achieved by a robust and reproducible method, with advantages over tethering via a diamine ligand in terms of obviating leaching of the metal and/or cyclopentadienyl ligand and giving the opportunity to attach different ancillary ligands. **E:** Potassium *tert*-butoxide (0.004 mmol) and isopropanol (5 ml) added to the immobilized resin (n = 14; 57 mg; 5.3 mol%) in a carousel tube, the mixture stirred at 60° for 1 h, then benzaldehyde (0.5 mmol) added, and the mixture stirred at 60° for 24 h (GC monitoring) → benzyl alcohol. Y 100% by GC. A similar yield was obtained after the same time using an iridium catalyst with a 5-carbon tether but the initial rate was higher with the longer tether. Further runs were conducted by decanting the solution, and recharging the resin with potassium *tert*-butoxide (0.004 mmol), isopropanol (5 ml) and benzaldehyde (0.5 mmol) immediately, and it was found that the catalyst with a 5-carbon tether may be recycled 25 times without significant loss of activity over 48 h. Rhodium analogs also gave near-quantitative GC yields after 48 h. F. details s. S.J. Lucas, B.D. Crossley, A.J. Pettman, A.D. Vassileiou, T.E.O. Screen, A.J. Blacker, P.C. McGowan, Chem. Commun. *2013*, *49* (49), 5562-4 [DOI: 10.1039/c3cc42550a]; procedure for preparing organometallic nanoparticles s. S. Mavila, C.E. Diesendruck, S. Linde, L. Amir, R. Shikler, N.G. Lemcoff, Angew. Chem., Int. Ed. *2013*, *52* (22), 5767-70 [DOI: 10.1002/anie.201300362].

Addition to Nitrogen and Carbon HC ⇓ NC

Zinc triflate/chiral 3-tert-butyl-4,5-dihydro-3H-dinaphtho[2,1-c;2′,1′-e]phosphepine ←

Asym. homogeneous hydrogenation of azomethines s. *80,* 23s*83* C=NR → CHNHR

Bis(pentafluorophenyl)borane/chiral 3,3′-diaryl-2,2′-divinyl-1,1′-binaphthyls *$(C_6F_5)_2BH/L*$*

Asym. hydrogenation of ar. ketimines using frustrated Lewis pairs derived via *in situ*-hydroboration of a chiral 2,2′-divinyl-1,1′-binaphthyl

16. 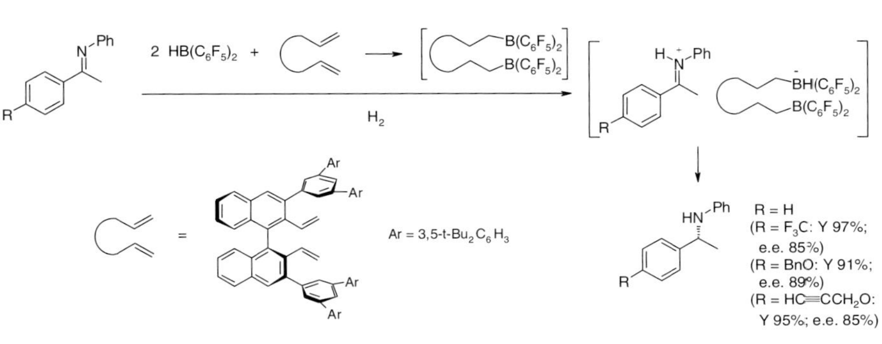

Direct hydroboration of chiral dienes bearing two terminal olefins with bis(pentafluorophenyl)-borane provides simple access to chiral borane catalysts for *metal-free* hydrogenation of imines (cf. *73*, 27) with promising (best to date) enantioselectivity. **E:** $(C_6F_5)_2BH$ (0.0125 mmol) added to a glass test tube (10 ml) in a glovebox under N_2, followed by the chiral 2,2′-divinyl-1,1′-binaphthyl (0.00625 mmol) and dry mesitylene (1 ml), the mixture stirred for 10 min at room temp., startg. imine (0.5 mmol) and dry mesitylene (1 ml) added, the tube moved to a stainless-steel autoclave, sealed, the autoclave purged three times with H_2 then the final H_2 pressure adjusted to 20 bar, the mixture stirred at room temp. for 15 h, the solvent removed under reduced pressure, and the crude residue purified by flash chromatography on silica gel → (R)-N-(1-phenylethyl)aniline. Y 98% (e.e. 78%). It was found that the steric bulkiness of the aryl substituent on the binaphthyl had a large impact on enantioselectivity, although all dienes tested resulted in a quantitative conversion of the imine. Eighteen further examples with electron-diverse [cyclic or acyclic] alkyl aryl ketimines gave enantiomeric excesses of 74-89% (Y 91-99%; 63% for a benzocycloheptene deriv.), notably tolerating ether, ar. halo, CF_3 and propargyloxy groups. Enantioselectivity was poor (<20% e.e.) with dialkyl or diaryl ketimines, however. F.e.s. Y. Liu, H. Du, J. Am. Chem. Soc. *2013*, *135* (18), 6810-3 [DOI: 10.1021/ja4025808].

Sodium trihydridocyanoborate *$NaBH_3CN$*
Pyrrolidines from Δ¹-pyrrolines C=NR → CHNHR
by diastereoselective reduction – 2,2-Difunctionalized pyrrolidines s. *83*, 189

Catecholborane/chiral 1,1′-binaphthyl-2,2′-diyl N-triflylphosphoramidates ←
Sec. amines from azomethines
via organocatalyzed asym. hydroboration s. *77,* 27s*83*

Chiral 2,2′-divinyl-1,1′-binaphthyls s. under $(C_6F_5)_2BH$ ←
Formic acid/triethylamine s. under Chiral ruthenium(II) complexes *$HCOOH/Et_3N$*
Formic acid s. under 1,4-Dihydropyridinium-functionalized rhodium(I) complexes *HCOOH*
Triethylsilane s. under $Pd(OAc)_2$ *Et_3SiH*

Trichlorosilane/(S)-N-acylprolines or chiral N-formylpipecolinic acid amides or N-nicotinoyl-4-pivaloyloxy-2(S)-[diaryl(hydroxy)methyl]pyrrolidine ←
Sec. amines from azomethines
via organocatalyzed asym. hydrosilylation
update s. *77,* 27s*81*; asym. reduction of N-arylazomethines with a chiral N-formylpipecolinic acid amide as Lewis base and $HSiCl_3$ as hydrosilylating agent s. Z. Wang, C. Wang, L. Zhou, J. Sun, Org. Biomol. Chem. *2013, 11* (5), 787-97 [DOI: 10.1039/c2ob26772a]; with (S)-N-acylprolines as Lewis base cf. M. Bonsignore, M. Benaglia, L. Raimondi, M. Orlandi, G. Celentano, Beilstein J. Org. Chem. *2013, 9*, 633-40 [DOI: 10.3762/bjoc.9.71]; asym. hydrosilylation of N-protected 1,2-diarylethylideneimines with N-nicotinoyl-4-pivaloyloxy-2(S)-[diaryl(hydroxy)-methyl]pyrrolidine as Lewis base s. Y. Zheng, Z. Xue, L. Liu, C. Shu, W. Yuan, X. Zhang, Org. Biomol. Chem. *2013, 11* (3), 412-5 [DOI: 10.1039/c2ob26672e]; **via organocatalyzed asym. hydroboration** with catecholborane and a chiral 1,1′-binaphthyl-2,2′-diyl N-triflylphosphoramidate as Brønsted acid cf. D. Enders, A. Rembiak, M. Seppelt, Tetrahedron Lett. *2013, 54* (6), 470-3 [DOI: 10.1016/j.tetlet.2012.11.055].

Chiral 3-tert-butyl-4,5-dihydro-3H-dinaphtho[2,1-c;2′,1′-e]phosphepine s. under Zinc triflate ←
Chiral 1,1′-binaphthyl-2,2′-diyl N-triflylphosphoramidates s. under Catecholborane ←
Chiral 1,1′-binaphthyl-2,2′-diyl hydrogen phosphates s. under Hydrido(pentamethylcyclopentadienyl)(N-tosylethylenediamine)iridium(II) complexes ←

Chiral ruthenium(II) complexes/formic acid/triethylamine *[Ru(II)]*/$HCOOH/Et_3N$*
Asym. transfer-hydrogenation of carbon-nitrogen double bonds C=N- → CHNH-
under ruthenium catalysis, update s. *46,* 42s*82*; parametric study with RuCl(η^6-*p*-cymene)((1S,2S)-N-*p*-toluenesulfonyl-1,2-diphenylethylenediamine) and formic acid/Et_3N as H-donor s. J. Pechácek, J. Václavík, J. Prech, P. Šot, J. Janušcák, B. Vilhanová, J. Vavrík, M. Kuzma, P. Kacer, Tetrahedron: Asym. *2013, 24* (4), 233-9 [DOI: 10.1016/j.tetasy.2013.01.010]; asym. transfer-hydrogenation of 1-aryl-3,4-dihydroisoquinolines s. Z. Wu, M. Perez, M. Scalone, T. Ayad, V. Ratovelomanana-Vidal, Angew. Chem., Int. Ed. *2013, 52* (18), 4925-8 [DOI: 10.1002/anie.201301134]; of 3,4-disubst. 1,2,5-thiadiazole 1,1-dioxides as precursors of chiral 1,2-diamines s. C. Schüttler, Z. Li-Böhmer, K. Harms, P. von Zezschwitz, Org. Lett. *2013, 15* (4), 800-3 [DOI: 10.1021/ol3034753].

Chloro[[bis(o-diphenylphosphinophenyl)boryl]-η^6-benzene]ruthenium(II) tetrakis(pentafluorophenyl)borate ←
Homogeneous hydrogenation of aldimines via frustrated Lewis pairs activated by an (η^6-arene)ruthenium(II) complex serving as C-based Lewis acid CH=NR → CH_2NHR

17. 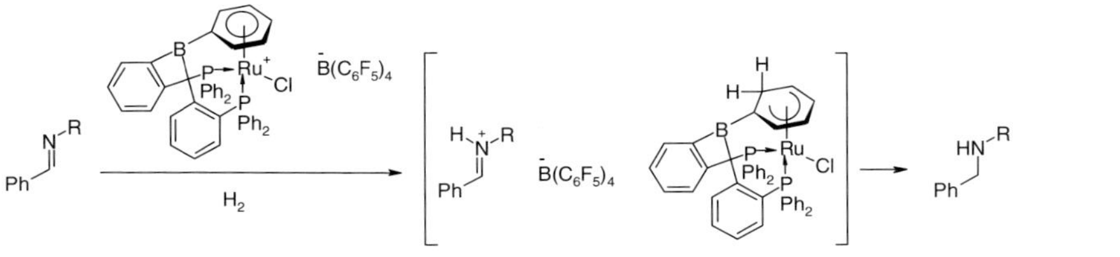

An ancillary metal center has been exploited to enhance the Lewis acidity of aromatic carbons, generating a C-based Lewis acid that can participate in frustrated Lewis pair [FLP] activation of H_2 and therefore act as a C-based hydrogenation catalyst. **E:** N-Benzylidene-*tert*-butylamine treated with [[(*o*-$Ph_2PC_6H_4)_2B(\eta^6$-Ph)]RuCl][$B(C_6F_5)_4$] (5 mol%) in methylene chloride under 102 atm. H_2 at 25° for 2 h → product. Y 99% by NMR. The same yield was obtained after 4 h with 2.5 mol% catalyst or after 8 h with 1 mol% catalyst or after 144 h using 4 atm. H_2 at 45° with 10 mol% catalyst. The [less basic] N-phenyl aldimine also gave a 99% NMR yield after 4 h with 5 mol% catalyst (Y 97% after 8 h with 2.5 mol% and 83% after 8 h with 1 mol%). F. details, also preparation of the catalyst, and formation and characterization of frustrated Lewis pair systems with tert. phosphines or trityl cation, s. M.P. Boone, D.W. Stephan, J. Am. Chem. Soc. *2013*, *135* (23), 8508-11 [DOI: 10.1021/ja403912n]; reductive cleavage of diaryl disulfides using a bisfluorenyl-substituted allene and [bulky] 1,3-bis(2,6-diisopropylphenyl)imidazol-2-ylidene to afford 9-[arylthio(fluoren-9-yl)methylene]fluorenes and 2-(arylthio)imidazolium salts s. B. Inés, S. Holle, R. Goddard, M. Alcarazo, Angew. Chem., Int. Ed. *2010*, *49* (45), 8389-91 [DOI: 10.1002/anie.201004149].

Dichloro[biphenyl-2,2′-diyl o-(diarylphosphino)aryl phosphite]((S,S)-1,2-diphenylethylenediamine)ruthenium(II) complexes/potassium tert-butoxide ←
Asym. homogeneous hydrogenation of azomethines C=NR → CHNHR
under ruthenium catalysis, update s. *80*, 23s*82*; of N-arylazomethines with a dichloro[(S,S)-DPEN]-[biphenyl-2,2′-diyl *o*-(diarylphosphino)aryl phosphite]ruthenium(II) complex and KOBu-*t* as base s. M. Vaquero, A. Suárez, S. Vargas, G. Bottari, E. Álvarez, A. Pizzano, Chem. Eur. J. *2012*, *18* (49), 15586-91 [DOI: 10.1002/chem.201203193]; of ketimines *under zinc catalysis* with $Zn(OTf)_2$ and chiral 3-*tert*-butyl-4,5-dihydro-3*H*-dinaphtho[2,1-*c*;2′,1′-*e*]phosphepine as ligand s. S. Werkmeister, S. Fleischer, K. Junge, M. Beller, Chem. Asian J. *2012*, *7* (11), 2562-8 [DOI: 10.1002/asia.201200309].

1,4-Dihydropyridinium-functionalized rhodium(I) complexes/formic acid *[Rh(I)]/HCOOH*
Sec. amines from azomethines by transfer-hydrogenation s. *48*, 40s*83*

Palladium-carbon *Pd-C*
Pyrrolidines from Δ^1-pyrrolines
by diastereoselective reduction – 2,2-Difunctionalized pyrrolidines s. *83*, 189

Palladium(II) acetate/triethylsilane *$Pd(OAc)_2/Et_3SiH$*
Sec. amines from azomethines
via Pd-catalyzed hydrosilylation s. *69*, 30s*76*; with $Pd(OAc)_2$/triethylsilane in ethanol s. M. Mirza-Aghayan, R. Boukherroub, M. Rahimifard, Appl. Organomet. Chem. *2013*, *27* (3), 174-6 [DOI: 10.1002/aoc.2961]; *via bioinspired transfer-hydrogenation* with a 1,4-dihydropyridinium-functionalized rhodium(I) complex and formic acid as H-donor (cf. *48*, 40) s. A. McSkimming, M.M. Bhadbhade, S.B. Colbran, Angew. Chem., Int. Ed. *2013*, *52* (12), 3411-6 [DOI: 10.1002/anie.201210086].

Chiral iridium(I) complexes *[Ir(I)]**
Hydrido(pentamethylcyclopentadienyl)(N-tosylethylenediamine)iridium(II) complexes/ chiral 1,1′-binaphthyl-2,2′-diyl hydrogen phosphates ←
Asym. homogeneous hydrogenation of cyclic imines C=NR → CHNHR
under iridium catalysis, update s. *46,* 47s*82*; *80,* 25s*82*; of 2*H*-1,4-benzoxazines with $[Ir(cod)Cl]_2$ and chiral 8-(diphenylphosphino)-1-(diphenylphosphinoamino)tetralin as ligand s. J. Hu, D. Wang, Z. Zheng, X. Hu, Chin. J. Chem. *2012, 30* (11), 2664-8 [DOI: 10.1002/cjoc.201200944]; also of 1,4-benzoxazin-2-ones, 1,4-benzothiazin-2-ones and 2(1*H*)-quinoxalinones with a chiral iridium(I) phosphine-phosphite complex at low loading cf. J.L. Núñez-Rico, A. Vidal-Ferran, Org. Lett. *2013, 15* (8), 2066-9 [DOI: 10.1021/ol400854a]; of isoquinoline *hydrochlorides* to chiral 1,2,3,4-tetrahydroisoquinolines with a chiral cationic trichlorodihydridobis[di(phosphine)]-diiridium(III) complex s. A. Iimuro, K. Yamaji, S. Kandula, T. Nagano, Y. Kita, K. Mashima, Angew. Chem., Int. Ed. *2013, 52* (7), 2046-50 [DOI: 10.1002/anie.201207748]; of isoquinolinium salts to N-subst. 1,2,3,4-tetrahydroisoquinolines s. Z.-S. Ye, R.-N. Guo, X.-F. Cai, M.-W. Chen, L. Shi, Y.-G. Zhou, ibid. *52* (13), 3685-9 [DOI: 10.1002/anie.201208300]; *under cooperative catalysis* through hydrogen bonding using an achiral hydrido(pentamethylcyclopentadienyl)-(N-tosylethylenediamine)iridium(II) complex and a chiral, Brønsted acidic 1,1′-binaphthyl-2,2′-diyl hydrogen phosphate s. W. Tang, S. Johnston, J.A. Iggo, N.G. Berry, M. Phelan, L. Lian, J. Bacsa, J. Xiao, ibid. *52* (6), 1668-72 [DOI: 10.1002/anie.201208774].

Platinum-carbon *Pt-C*
Piperidine-2-acetic from 2,3,4,5-tetrahydropyridine-2-acetic acid esters ←
by diastereoselective hydrogenation s. *83*, 125

Addition to Carbon-Carbon Bonds HC ⇓ CC

Triethylenediamine/tris(pentafluorophenyl)borane *DABCO/$(C_6F_5)_3B$*
Hydrogenation with frustrated Lewis pairs C=C → CHCH
s. *78,* 14; of electron-deficient allenes and alkenes with DABCO/tris(pentafluorophenyl)borane s. B. Inés, D. Palomas, S. Holle, S. Steinberg, J.A. Nicasio, M. Alcarazo, Angew. Chem., Int. Ed. *2012, 51* (49), 12367-9 [DOI: 10.1002/anie.201205348]; **of polyarenes**, incl. anthracenes, tetracene and tetraphene, with pentafluorophenyl(diphenyl)phosphine/tris(pentafluorophenyl)borane s. Y. Segawa, D.W. Stephan, Chem. Commun. *2012, 48* (98), 11963-5 [DOI: 10.1039/c2cc37190a].

Silver s. under Palladium *Ag*
Silver triflate s. under (Cp)Ru(cod)Cl* *AgOTf*

Nanoporous gold/dimethyl(phenyl)silane/pyridine/water *$Au/Me_2PhSiH/Py/H_2O$*
(Z)-Ethylene from acetylene derivs. s. *45,* 24s*83* C≡C → CH=CH

Nanoporous gold/dimethyl(phenyl)silane/water *$Au/Me_2PhSiH/H_2O$*
1,2,3,4-Tetrahydroquinolines from quinolines s. *23,* 66s*83* Ⓗ

Tris(pentafluorophenyl)borane s. under Triethylenediamine and $(F_5C_6)PPh_2$ *$(C_6F_5)_3B$*
Borane·ammonia s. under Tricyclohexylphosphine[N,N′-bis(2,6-diisopropylphenyl)-imidazolidin-2-ylidene]palladium(0) *$BH_3{\cdot}NH_3$*
Water s. under Au *H_2O*
Formic acid/triethylamine s. under Anionic hydrido(diiodo)(pentamethylcyclo-pentadienyl)rhodium(III) complexes *$HCOOH/Et_3N$*

Reductase ←
Asym. enzymatic reduction of carbon-carbon double bonds C=C → CHCH
s. *80,* 29; of electron-deficient ethylene derivs. (cyclic enones, diesters and maleimides) using *Shewanella* yellow enzyme [SYE-4] as novel recombinant enoate reductase s. N. Iqbal, F. Rudroff, A. Brigé, J. Van Beeumen, M.D. Mihovilovic, Tetrahedron *2012, 68* (37), 7619-23 [DOI: 10.1016/ j.tet.2012.05.092]; of α,β-ethylene-α-halogenocarboxylic acid esters with an ene-reductase s. G. Tasnádi, C.K. Winkler, D. Clay, N. Sultana, W.M.F. Fabian, M. Hall, K. Ditrich, K. Faber, Chem. Eur. J. *2012, 18* (33), 10362-7 [DOI: 10.1002/chem.201200990]; with an enoate reductase in the presence of *synthetic* nicotinamide cofactors s. C.E. Paul, S. Gargiulo, D.J. Opperman, I. Lavandera, V. Gotor-Fernández, V. Gotor, A. Taglieber, I.W.C.E. Arends, F. Hollmann, Org.

Lett. *2013, 15* (1), 180-3 [DOI: 10.1021/ol303240a]; with an enoate reductase in the presence of 1,4-butanediol s. *83*, 8.

Formic acid s. under (Ene or diene)palladium(0) diaminocarbene complexes *HCOOH*
Dimethyl(phenyl)silane s. under Au *Me_2PhSiH*
Organic/inorganic silica gel s. under Perfluoro-tagged rhodium or ruthenium nanoparticles ←

Hydrazine hydrate/rhodium-cobalt nanoparticles *N_2H_4/[Rh_4Co]*
Reduction of carbon-carbon multiple bonds s. *45*, 24s*83* ←

N,O-Bis(trifluoroacetyl)hydroxylamine/hydroxylamine *$CF_3CONHOCOCF_3/NH_2OH$*
Hydrogenation of carbon-carbon multiple bonds using *in situ*-generated diimide under continuous flow ←

18.

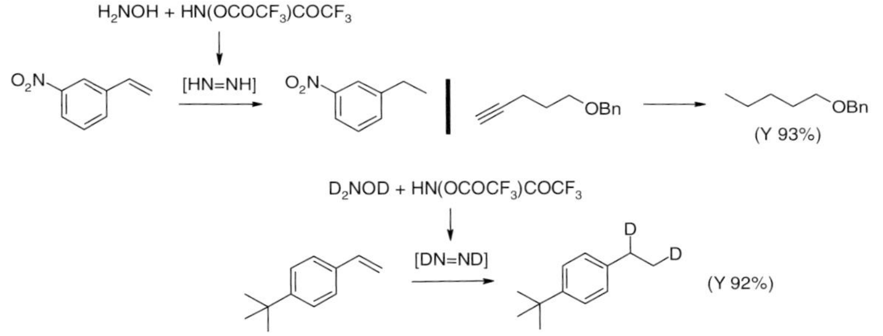

The first hydrogenation of alkenes/alkynes under continuous flow using diimide as hydrogen source is described. **E:** Dioxane solns. of 3-nitrostyrene (0.3 *M*) plus N,O-bis(trifluoroacetyl)-hydroxylamine (0.45 *M*) and hydroxylamine (1.5 *M*) pumped via a T-mixer into a reaction loop (4 ml) submerged in oil at 100° (20 min residence time), washed with satd. aq. $NaHCO_3$, extracted with ethyl acetate or hexanes, and concentrated → 1-ethyl-3-nitrobenzene. Y 97%. This novel method required no metal catalyst and was presumed to provide more efficient mixing and heating of reaction mixtures, with only limited amounts of reactive chemicals in contact at any time. In general, solns. of substrate plus hydroxylamine and N,O-bis(trifluoroacetyl)hydroxylamine were mixed in a T-mixer and then heated at 100° (with an optimum residence time of 20 min); however for hydroxylamine-sensitive substrates, the substrate was mixed with N,O-bis(trifluoroacetyl)-hydroxylamine instead. Products were generally >95% pure after simple aq. work up. The method was compatible with benzyl ether, silyl ether, alcohol, ester, amide, nitro and acid-sensitive Boc-amine and styrene functionality (twelve examples; Y 68-97%). Site-specific deuteration of 4-*tert*-butylstyrene (using d_3-hydroxylamine) afforded the corresponding dideutero deriv. (Y 92%; 90% D incorporation). F.e. and optimization s. A.S. Kleinke, T.F. Jamison, Org. Lett. *2013, 15* (3), 710-3 [DOI: 10.1021/ol400051n].

Pentafluorophenyl(diphenyl)phosphine/tris(pentafluorophenyl)borane *$(C_6F_5)PPh_2/(C_6F_5)_3B$*
Hydrogenation with frustrated Lewis pairs of polyarenes s. *78*, 14s*83* Ⓗ

Iron nanoparticles *Fe*
Heterogeneous hydrogenation of carbon-carbon multiple bonds s. *3*, 46s*83* ←

Nickel-aluminum/water *Ni-Al/H_2O*
Indolines from indoles or 1,2,3,4-tetrahydroquinolines from quinolines using water as hydrogen source under high hydrostatic pressure s. *83*, 11 Ⓗ

Rhodium-cobalt nanoparticles s. under N_2H_4 *[Rh_4Co]*

Chiral carbonyl(hydrido)[di(phosphine)]ruthenium cluster complexes *[Ru]**
Asym. homogeneous hydrogenation of α,β-ethylenecarboxylic acids s. *42*, 45s*83* C=C → CHCH

Chloro(1,5-cyclooctadiene)(pentamethylcyclopentadienyl)ruthenium(II)/silver triflate ←
(E)-Ethylene from acetylene derivs. C≡C → CH=CH
by ruthenium-catalyzed homogeneous hydrogenation

19.
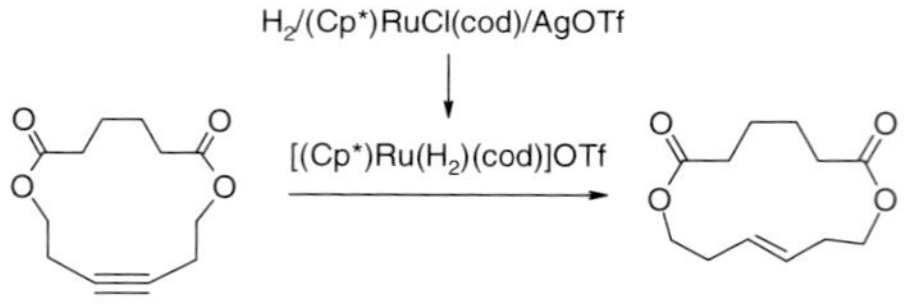

The first efficient, generally applicable, functional group-tolerant, *direct* transition metal-catalyzed hydrogenation of acetylene to (E)-ethylene derivs. is reported, being stereocomplementary with classical routes, e.g. hydrogenation with Lindlar's catalyst, which are uniquely (Z)-selective. **E:** A suspension of (Cp*)RuCl(cod) (5.5 mol%) and AgOTf (5 mol%) in methylene chloride (3.3 ml) stirred in a Schlenk tube for ca. 1 min under argon, the precipitate allowed to settle, the supernatant soln. transferred via syringe into a pre-dried autoclave under argon, pressurized with H_2 (10 bar), the mixture stirred at ambient temp. for 10-15 min, the pressure carefully released, a soln. of the startg. alkyne (1.29 mmol) in methylene chloride (3.2 ml) added under argon, the autoclave again set under 10 bar H_2 pressure, stirred for 30 min, the pressure carefully released, and worked up with passage of the crude material through a short column of silica → (E)-product. Y 89% (E/Z 98:2). The procedure is applicable to a wide range of acyclic and cyclic alkynes, being compatible with appended ester, lactone, amide, lactam, carboxyl, keto, *prim*-hydroxyl, siloxy, tosyloxy, nitro, alkyl bromide, hydroxamate, cyano, thioether and terminal alkene groups (eighteen examples; Y 60-96%; E/Z 87:13 to 98:2). The active catalyst is believed to be a cationic cyclooctadiene(pentamethylcyclopentadienyl)ruthenium(II) species which effectively splits the hydrogen molecule to initiate reaction, the ligated diene being critical in tempering the Lewis acidity of the central metal and in so doing ensuring high productivity while assisting *trans*-selectivity. A mechanism based on classical *cis*-hydrogenation and subsequent *cis*→*trans*-isomerization was experimentally disproved. Interestingly, 1,3-dienes and 1,3-enynes were not hydrogenated by the system, presumably because of their tight coordination to the active ruthenium species. Over-reduction was typically in the range 5-15%. F.e. and comparison of catalysts s. K. Radkowski, B. Sundararaju, A. Fürstner, Angew. Chem., Int. Ed. *2013, 52* (1), 355-60 [DOI: 10.1002/anie.201205946].

Perfluoro-tagged rhodium or ruthenium nanoparticles-on-organic/inorganic silica gel ←
Benzene ring hydrogenation Ⓗ
update s. *23,* 66s*82*; **of arenes and heteroarenes** with robust, highly active and recoverable perfluoro-tagged rhodium or ruthenium nanoparticles immobilized on highly fluorinated hybrid organic-inorganic silica gel under mild conditions s. S. Niembro, S. Donnici, A. Shafir, A. Vallribera, M.L. Buil, M.A. Esteruelas, C. Larramona, New J. Chem. *2013, 37* (2), 278-82 [DOI: 10.1039/c2nj40859g]; 1,2,3,4-tetrahydroquinolines from quinolines with recoverable, unsupported nanoporous gold and dimethyl(phenyl)silane with water as hydrogen source s. M. Yan, T. Jin, Q. Chen, H.E. Ho, T. Fujita, L.-Y. Chen, M. Bao, M.-W. Chen, N. Asao, Y. Yamamoto, Org. Lett. *2013, 15* (7), 1484-7 [DOI: 10.1021/ol400229z]; **transfer-hydrogenation** of heteroarenes [quinolines, isoquinolines and quinoxalines] with anionic hydrido(diiodo)(pentamethylcyclopentadienyl)rhodium(III) at low loading and formic acid/Et_3N as H-donor s. J. Wu, C. Wang, W. Tang, A. Pettman, J. Xiao, Chem. Eur. J. *2012, 18* (31), 9525-9 [DOI: 10.1002/chem.201201517].

Chiral rhodium(I) phosphine, di(phosphine), phosphite, bis(phosphite), *[Rh(I)]**
2,2′-bi(2,3-dihydro-1,3-benzoxaphosphole) or phosphoromonoamidite complexes
Asym. homogeneous hydrogenation C=C → CHCH
under rhodium catalysis, update s. *27,* 57s*78; 71,* 26s*82*; of 2-(aryl)enacylamines with a highly active (norbornadiene)[4,4′-di(anthracen-9-yl)-2,2′-bi(2,3-dihydro-1,3-benzoxaphosphole)]-rhodium(I) fluoroborate as catalyst having deep chiral pockets s. G. Liu, X. Liu, Z. Cai, G. Jiao,

G. Xu, W. Tang, Angew. Chem., Int. Ed. *2013, 52* (15), 4235-8 [DOI: 10.1002/anie.201300646]; with $[Rh(cod)_2]BF_4$ and chiral 1,1′-biindan-7,7′-diyl-based cyclic phosphines or phosphites as ligand cf. S.-F. Zhu, T. Liu, S. Yang, S. Song, Q.-L. Zhou, Tetrahedron *2012, 68* (37), 7685-90 [DOI: 10.1016/j.tet.2012.06.032]; asym. hydrogenation of α-acylamino-styrenes and -acrylates with a chiral cationic [1,2-bis[(*o-tert*-butoxyphenyl)(phenyl)phosphino]ethane]rhodium(I) complex s. B. Mohar, M. Stephan, Adv. Synth. Catal. *2013, 355* (2-3), 594-600 [DOI: 10.1002/adsc.201200780]; with a library of chiral, aroylamino-functionalized 1,1′-binaphthyl-2,2′-diyl phosphites [BenzaPhos] as ligand s. L. Pignataro, C. Bovio, M. Civera, U. Piarulli, C. Gennari, Chem. Eur. J. *2012, 18* (33), 10368-81 [DOI: 10.1002/chem.201201032]; with finely-tuned chiral 1,1′-binaphthyl-2,2′-diyl N-isopropyl-N-[3-(diphenylphosphino)alkyl]phosphoramidites as ligand in green media, e.g. propylene carbonate, s. S. Balogh, G. Farkas, Á. Szöllösy, F. Darvas, L. Ürge, J. Bakos, Tetrahedron: Asym. *2013, 24* (1), 66-74 [DOI: 10.1016/j.tetasy.2012.11.013]; of functionalized ethylene derivs. with $[Rh(cod)_2]BF_4$ and chiral 2-anilinoethyl, 2-pyridylmethyl or 2-(2-pyridyl)ethyl 1,1′-binaphthyl-2,2′-diyl phosphites as ligand s. K.A. Vallianatou, D.J. Frank, G. Antonopoulou, S. Georgakopoulos, E. Siapi, M. Zervou, I.D. Kostas, Tetrahedron Lett. *2013, 54* (5), 397-401 [DOI: 10.1016/j.tetlet.2012.11.023]; of enolesters or potassium (E)-3-cyano-5-methylhex-3-enoate with the chiral 2-imidazolidone-bridged diaryl phosphoramidite, DpenPhos, as ligand s. Y. Liu, Z. Wang, K. Ding, Tetrahedron *2012, 68* (37), 7581-5 [DOI: 10.1016/j.tet.2012.05.096]; of α,β-ethylenecarboxylic acids or esters with carbohydrate-bridged bis(1,1′-binaphthyl-2,2′-diyl phosphites) as ligand s. Q.-L. Zhao, X. Miao, L.-L. Wang, Tetrahedron: Asym. *2013, 24* (2-3), 104-7 [DOI: 10.1016/j.tetasy.2012.12.001]; of α,β-ethylenecarboxylic acids *under ruthenium catalysis* with chiral carbonyl(hydrido)[di(phosphine)]ruthenium cluster complexes of the type $[H_4Ru_4(CO)_{10}(\mu\text{-}1,2\text{-di(phosphine)})]$, $[H_4Ru_4(CO)_{10}(1,1\text{-di(phosphine)})]$ or $[H_4Ru_4(CO)_{11}(\text{di(phosphine)})]$, based on the JosiPhos or WalPhos ligand (cf. *42,* 45s*81*), s. V. Moberg, R. Duquesne, S. Contaldi, O. Röhrs, J. Nachtigall, L. Damoense, A.T. Hutton, M. Green, M. Monari, D. Santelia, M. Haukka, E. Nordlander, Chem. Eur. J. *2012, 18* (39), 12458-78 [DOI: 10.1002/chem.201200630]; design and synthesis of Janus-type chiral dendritic di(phosphines) and applications in asym. hydrogenation s. J. Liu, Y. Feng, B. Ma, Y.-M. He, Q.-H. Fan, Eur. J. Org. Chem. *2012* (34), 6737-44 [DOI: 10.1002/ejoc.201200999].

Anionic hydrido(diiodo)(pentamethylcyclopentadienyl)rhodium(III) complexes/ formic acid/triethylamine ←

Transfer-hydrogenation of 6-membered N-heteroarenes s. *23,* 66s*83* Ⓗ

Organosilica-encapsulated palladium nanoparticles ←
Palladium-boron nitride/diethylenetriamine or acetic acid ←
(Ene or diene)palladium(0) diaminocarbene complexes/formic acid ←

Ethylene from acetylene derivs. C≡C → CH=CH

update s. *45,* 24s*82*; (Z)-ethylene derivs. by hydrogenation with palladium nanoparticles encapsulated by sulfinyl-functionalized organosilica core shells s. T. Mitsudome, Y. Takahashi, S. Ichikawa, T. Mizugaki, K. Jitsukawa, K. Kaneda, Angew. Chem., Int. Ed. *2013, 52* (5), 1481-5 [DOI: 10.1002/anie.201207845]; semi-hydrogenation of *o*-nitrobenzenesulfonyl-functionalized alkynes with palladium/boron nitride and diethylenetriamine or acetic acid as additive, also hydrogenation of similarly functionalized alkenes, s. A. Kawanishi, C. Miyamoto, Y. Yabe, M. Inai, T. Asakawa, Y. Hamashima, H. Sajiki, T. Kan, Org. Lett. *2013, 15* (6), 1306-9 [DOI: 10.1021/ol4002448]; (Z)-ethylene derivs. *by transfer-hydrogenation* with (alkene)- or (diene)-palladium(0) cyclic diaminocarbene complexes and formic acid as H-donor s. P. Hauwert, J.J. Dunsford, D.S. Tromp, J.J. Weigand, M. Lutz, K.J. Cavell, C.J. Elsevier, Organometallics *2013, 32* (1), 131-40 [DOI: 10.1021/om300930w]; by reduction with hydrazine hydrate and rhodium-cobalt $[Rh_4Co]$ nanoparticles, also reduction of alkenes, s. J. Lin, J. Chen, W. Su, Adv. Synth. Catal. *2013, 355* (1), 41-6 [DOI: 10.1002/adsc.201200576]; (Z)-ethylene derivs. *under gold catalysis* using recoverable, unsupported nanoporous gold and dimethyl(phenyl)silane as reductant with water as hydrogen source and an amine (e.g. pyridine) as additive s. M. Yan, T. Jin, Y. Ishikawa, T. Minato, T. Fujita, L.-Y. Chen, M. Bao, N. Asao, M.-W. Chen, Y. Yamamoto, J. Am. Chem. Soc. *2012, 134* (42), 17536-42 [DOI: 10.1021/ja3087592].

Stabilized palladium or palladium/silver nanoparticles ←
Palladium/molecular sieves ←
Palladium(II)-on-polymer nanoparticles ←
Encapsulated platinum nanoparticles ←
Heterogeneous hydrogenation C=C → CHCH

of carbon-carbon double bonds, update s. *3*, 46s*82*; with palladium(II) adsorbed on polymer nanoparticles with basic functional groups, also application to Suzuki coupling (cf. *37,* 902s*83*), s. H. Seto, T. Morii, T. Yoneda, T. Murakami, Y. Hoshino, Y. Miura, Chem. Lett. *2013, 42* (3), 301-3 [DOI: 10.1246/cl.2013.301]; hydrogenation (and isomerization) of allyl alcohol with alkylamine- or alkanethiolate-stabilized Pd or $Pd_{91}Ag_9$ nanoparticles s. M. Moreno, L.N. Kissell, J.B. Jasinski, F.P. Zamborini, ACS Catal. *2012, 2* (12), 2602-13 [DOI: 10.1021/cs300361y]; selective hydrogenation of functionalized alkenes, alkynes and azides with palladium(0) supported on molecular sieves (MS3A or MS5A) s. T. Takahashi, M. Yoshimura, H. Suzuka, T. Maegawa, Y. Sawama, Y. Monguchi, H. Sajiki, Tetrahedron *2012, 68* (39), 8293-9 [DOI: 10.1016/j.tet.2012.07.044]; hydrogenation of alkenylaziridines with 2:1 palladium-carbon/Lindlar′s catalyst as a 'mixed palladium' catalyst s. M.-K. Ji, D.-H. Yoon, H.-J. Ha, P. Kang, W.K. Lee, Bull. Korean Chem. Soc. *2012, 33* (9), 2853-4 [DOI: 10.5012/bkcs.2012.33.9.2853]; of alkenes with platinum nanoparticles encapsulated in zeolitic imidazolate frameworks, Pt_n@ZIF-8, s. P. Wang, J. Zhao, X. Li, Y. Yang, Q. Yang, C. Li, Chem. Commun. *2013, 49* (32), 3330-2 [DOI: 10.1039/c3cc39275a]; of various alkenes, alkynes and weakly active carbonyl compds. under mild conditions with ultrasmall iron(0) nanoparticles as catalyst s. V. Kelsen, B. Wendt, S. Werkmeister, K. Junge, M. Beller, B. Chaudret, ibid. *49* (33), 3416-8 [DOI: 10.1039/c3cc00152k].

Tricyclohexylphosphine[N,N′-bis(2,6-diisopropylphenyl)imidazolidin-2-ylidene]-palladium(0)/borane·ammonia ←
Palladium-catalyzed generation of hydrogen from borane·amine complexes in tandem with homogeneous hydrogenation of carbon-carbon multiple bonds ←

20.

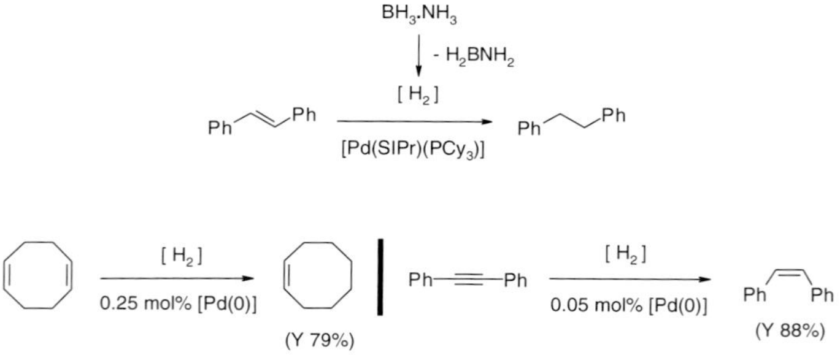

The novel palladium(0)-NHC catalyst, tricyclohexylphosphine[N,N′-bis(2,6-diisopropylphenyl)-imidazolidin-2-ylidene]palladium(0), effectively dehydrogenates borane-ammonia to liberate hydrogen gas for *in situ*-hydrogenation of carbon-carbon multiple bonds under catalysis with the same palladium(0) complex at very low catalyst loading. **E:** A mixture of stilbene (0.5 mmol), $[Pd(SIPr)(PCy_3)]$ (0.05 mol% as a 0.0017 *M* soln. in isopropanol), borane·ammonia (0.25 mmol) and isopropanol (1 ml) heated with stirring in a Schlenk flask at 50° for 16 h under argon, the solvent removed *in vacuo*, and the residue worked up with purification by flash chromatography → 1,2-diphenylethane. Y >99%. The procedure is highly efficient for the hydrogenation of a wide range of ethylene derivs., incl. diarylalkenes, allylbenzenes, methyl cinnamate and 1,4-cyclooctadiene (to cyclooctene or cyclooctane), but α,β-ethyleneketones were hydrogenated to the corresponding saturated alcohols (twelve examples; Y 69 to >99%). Alkynes were semi-hydrogenated to the corresponding (Z)-alkenes (two examples; Y 75-88%). With a thermally stable, solid hydrogen-storage material on hand, the method is a distinct improvement on classical hydrogenation in terms of safety and logistics. F.e.s. C.E. Hartmann, V. Jurcík, O. Songis, C.S.J. Cazin, Chem. Commun. *2013, 49* (10), 1005-7 [DOI: 10.1039/c2cc38145a].

Chiral iridium(I) phosphine, aminophosphine, phosphino-imidate, phosphino-Δ²-oxazoline, phosphinite or phosphoramidite complexes *[Ir(I)]**

Asym. homogeneous hydrogenation C=C → CHCH

of alkenes under iridium(I) catalysis, update s. *62,* 39s*82*; of minimally functionalized alkenes with a furanoside thioether-based phosphinite, phosphine or, preferably, phosphite as ligand s. M. Coll, O. Pàmies, M. Diéguez, Adv. Synth. Catal. *2013, 355* (1), 143-60 [DOI: 10.1002/adsc.201200711]; of unfunctionalized and minimally functionalized alkenes with a chiral cationic (cyclooctadiene)iridium(I) 2-[1-(1,3-dihydroisobenzofuran-1-ylideneamino)ethyl]ferrocenyl-phosphine complex cf. K. Bert, T. Noël, W. Kimpe, J.L. Goeman, J. Van der Eycken, Org. Biomol. Chem. *2012, 10* (42), 8539-50 [DOI: 10.1039/c2ob25871d]; of 1-aryl-1-(3,5-dimethoxyphenyl)-alkenes with a chiral cationic (cyclooctadiene)iridium(I) phosphoramidite complex based on PhosPrOx s. E.N. Bess, M.S. Sigman, Org. Lett. *2013, 15* (3), 646-9 [DOI: 10.1021/ol303465c]; of 1-aryl-1-(*o*-carboxyaryl)alkenes and aliphatic γ-methylenecarboxylic acids by *carboxyl-directed* asym. hydrogenation with a chiral iridium(I) spirophosphine-Δ²-oxazoline complex s. S. Song, S.-F. Zhu, Y.-B. Yu, Q.-L. Zhou, Angew. Chem., Int. Ed. *2013, 52* (5), 1556-9 [DOI: 10.1002/anie.201208606]; of α,β-ethylenecarboxylic acid esters with a chiral cationic (cyclooctadiene)-iridium(I) aminophosphine complex s. J.-Q. Li, X. Quan, P.G. Andersson, Chem. Eur. J. *2012, 18* (34), 10609-16 [DOI: 10.1002/chem.201200907]; with further P,N-ligands cf. D.H. Woodmansee, M.-A. Müller, L. Tröndlin, E. Hörmann, A. Pfaltz, ibid. *18* (43), 13780-6 [DOI: 10.1002/chem.201202397]; of α-alkylidenesuccinimides with a chiral cationic iridium(I) 2-diphenyl-phosphino-2′-(Δ²-oxazolin-2-yl)biphenyl complex at very low (0.05 mol%) loading s. Y. Liu, W. Zhang, Angew. Chem., Int. Ed. *2013, 52* (8), 2203-6 [DOI: 10.1002/anie.201209126]; *double* asym. hydrogenation of 2,6-di(1-phenylethenyl)-4-methylaniline s. E. Spahn, A. Albright, M. Shevlin, L. Pauli, A. Pfaltz, R.E. Gawley, J. Org. Chem. *2013, 78* (6), 2731-5 [DOI: 10.1021/jo3026548].

Platinum(IV) oxide *PtO_2*

Stereoselective reduction of 6-aryl-1-hydroxy-1,2,3,6-tetrahydropyridines s. *83*, 382 ←

Exchange ↓↑

Oxygen ↑ HC ↓↑ O

Microwaves s. under Ni-Al [\\\\]

(-)-Isopinocampheylborane·triethylenediamine/boron fluoride *(-)-$IpcBH_2$·TMEDA/BF_3*

Terminal 1,3-diols from β,γ-ethylenecarboxylic acid esters via asym. hydroboration C(OH)CHCH$_2$OH

21.

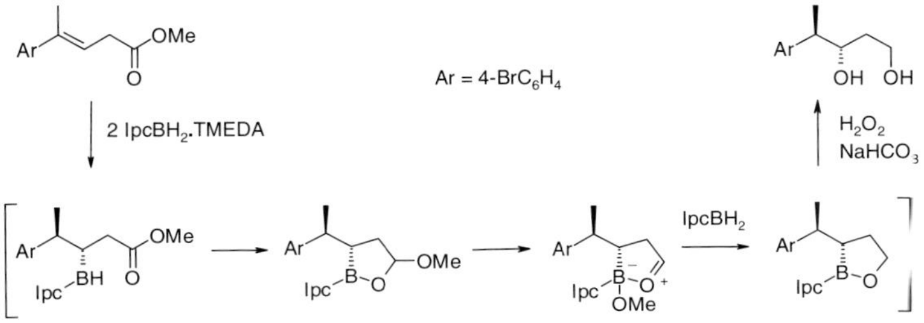

in one pot. BF_3-etherate (2 eq.) added dropwise to a soln. of chiral 2IpcBH_2·TMEDA (1.05 eq.) in dry THF (3.8 ml), the soln. stirred at room temp. for 3 h, cooled to -17°, a soln. of methyl (E)-4-(4-bromophenyl)pent-3-enoate (2.63 mmol) in the same solvent (1 ml) added dropwise, the mixture warmed to 0° overnight, 30% aq. H_2O_2 and satd. aq. $NaHCO_3$ added, stirred at room temp. for 6 h, extracted with ether, and purified chromatographically → (3S,4R)-4-(4-bromophenyl)pentane-

1,3-diol. Y 61% (e.e. 85%). Treatment of a series of 4-arylpent-3-enoates with the borane resulted in hydroboration and ester reduction to afford chiral 1,3-diols following oxidative work up (six examples; Y 41%, 54-77%; e.e. 67%, 75-85%). Interestingly, conducting the reaction of the 4-phenylpentenoate at lower temp. (-25°/48 h) and quenching with methanol before oxidative work-up afforded a β-hydroxy-ester (e.e. 75%), demonstrating that hydroboration was significantly faster than ester reduction. F.e. and substrate prepn. s. P.S. Fordred, S.D. Bull, Tetrahedron Lett. *2013*, *54* (1), 27-31 [DOI: 10.1016/j.tetlet.2012.10.047].

1,1,3,3-Tetramethyldisiloxane s. under $Ru_3(CO)_{12}$ — *$(Me_2SiH)_2O$*
(S)-2,2′-[(R,R)-Pentane-2,4-dioxy]-6,6′-bis[di(3,5-di-tert-butyl-4-methoxyphenyl)-phosphino]biphenyl s. under Chloro(1,5-cyclooctadiene)iridium(I) dimer ←

Nickel-aluminum alloy/water/microwaves — *Ni-Al/H_2O/[\\\\]*
Alkylarenes from acylophenones — CO → CH_2
Also deutero-derivs. with D_2O s. *83*, 11

[1-[o-(Aminomethyl)phenyl]-3-methylimidazol-2-ylidene](pentamethylcyclopentadienyl)-(pyridine)ruthenium(II) hexafluorophosphate/potassium tert-butoxide ←
Ruthenium-catalyzed homogeneous hydrogenation — COOR → CH_2OH
of carboxylic acid esters *s. 83*, 12

Dodecacarbonyltriruthenium/1,1,3,3-tetramethyldisiloxane — *$Ru_3(CO)_{12}$/$(Me_2SiH)_2O$*
Palladium/rhenium-on-graphite — *Pd/Re-C*
Amines from carboxylic acid amides — CON< → CH_2N<
under bimetal catalysis s. *46*, 34s*80*; from N-subst. and N,N-disubst. carboxylic acid amides by hydrogenation with a bimetallic Pd/Re catalyst on graphite under mild conditions s. M. Stein, B. Breit, Angew. Chem., Int. Ed. *2013*, *52* (8), 2231-4 [DOI: 10.1002/anie.201207803]; also of N-unsubst. carboxylic acid amides by catalytic reduction with 1,1,3,3-tetramethyldisiloxane using $Ru_3(CO)_{12}$ as catalyst s. J.T. Reeves, Z. Tan, M.A. Marsini, Z.S. Han, Y. Xu, D.C. Reeves, H. Lee, B.Z. Lu, C.H. Senanayake, Adv. Synth. Catal. *2013*, *355* (1), 47-52 [DOI: 10.1002/adsc.201200835].

Chloro(1,5-cyclooctadiene)iridium(I) dimer/(S)-2,2′-[(R,R)-pentane-2,4-dioxy]-6,6′-bis[di(3,5-di-tert-butyl-4-methoxyphenyl)phosphino]biphenyl — *[Ir(I)]**
Lactones from *meso*-dicarboxylic acid anhydrides — CO → CH_2
via iridium(I)-catalyzed homogeneous hydrogenation with desymmetrization

22.

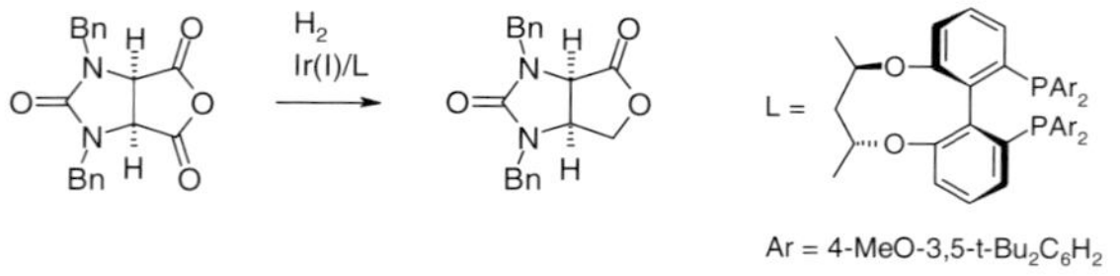

A soln. of $[Ir(cod)Cl]_2$ (0.1 mol%) and C_3*-DTBM-TunePhos (2.2 mol%) in ethyl acetate (0.1 ml) stirred under N_2 at room temp. for 30 min, added via syringe to a soln. of startg. anhydride (0.1 mmol) in the same solvent (1 ml), the mixture stirred in a steel autoclave under H_2 (80 bar) at 125° for 20 h, the pressure released, concentrated, and the residue purified chromatographically → (3aR,6aS)-1,3-dibenzyltetrahydro-1*H*-furo[3,4-*d*]imidazole-2,4-dione. Y 96% (e.e. 96%). Use of the rigid and sterically-hindered biaryl phosphine ligand promoted efficient and enantioselective reaction in this challenging anhydride → lactone conversion. The reaction was successful at low catalyst loading (0.1-0.01 mol%) at temperatures of 80-125° (eight examples; Y 89-96%; e.e. 80-99%). The use of methanol as solvent resulted in methanolysis to the corresponding hemi-esters. F.e. and optimization s. T.-L. Liu, W. Li, H. Geng, C.-J. Wang, X. Zhang, Org. Lett. *2013*, *15* (7), 1740-3 [DOI: 10.1021/ol400533g].

Via intermediates — *v.i.*
Methylene from keto groups via photoreduction of 1,3-dithianes — CO → $C(SR)_2$ → CH_2
s. *83*, 24

Nitrogen ↑ HC ↓↑ N

Lithium hydrido(diisobutyl)morpholinoaluminate *LiAlH(Bu-i)$_2$N(CH$_2$CH$_2$)$_2$O*
Aldehydes from N,N-disubst. carboxylic acid amides C(O)N< → CHO

23.

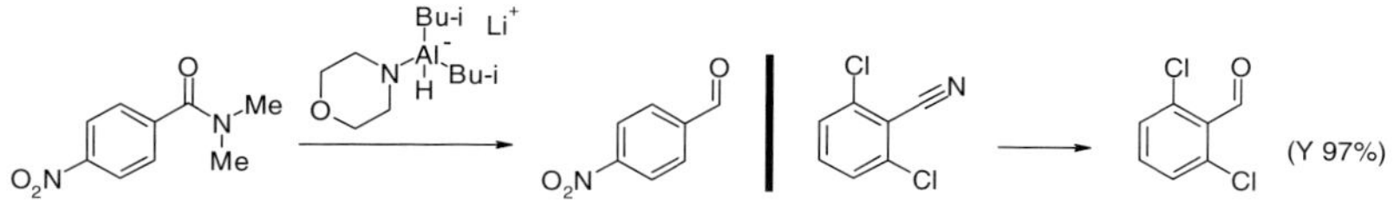

A soln. of Li-hydrido(diisobutyl)morpholinoaluminate (1.2 eq.) in THF (2.4 ml) added to a soln. of N,N-dimethyl-4-nitrobenzamide (1 mmol) in the same solvent (10 ml) at 0°, the mixture stirred for 30 min, quenched with 1 *M* aq. HCl, extracted with ether, and purified chromatographically → 4-nitrobenzaldehyde. Y 91%. The novel reducing agent (readily prepared from *i*-Bu$_2$AlH and lithium morpholide) provided general partial reduction of electron-diverse ar. and aliphatic tert. amides to aldehydes (fifteen examples; Y 80%, 89-99%), with a single hetaryl example, N,N-dimethyl-2-furoamide, affording 75%. **Ar. aldehydes** were similarly prepared **from** electron-diverse ar. **nitriles** (cf. *48*, 65, 65s*77*) in excellent yields (96-99%; thirteen examples), but aliphatic nitriles were unreactive, providing the possibility of *chemoselective reduction* of ar. nitriles in the presence of aliphatic ones. F.e.s. Y.R. Kim, D.K. An, Bull. Korean Chem. Soc. *2012, 33* (12), 4194-6 [DOI: 10.5012/bkcs.2012.33.12.4194].

Halogen ↑ HC ↓↑ Hal

Disodium N-methylindole-2,3-diolate ←
Replacement of ar. iodine by hydrogen s. *11,* 633s*83* I → H

Copper(I) chloride/1,2-bis(diphenylphosphino)benzene/dimethyl(phenyl)silane/ potassium tert-butoxide ←

(β-Diimine)aluminum dihydride complexes/zirconocene dichloride ←
Replacement of ar. or hetar. fluorine by hydrogen s. *76,* 40s*83* F → H

Samarium(II) bis(trimethylsilyl)amide *Sm[N(SiMe$_3$)$_2$]$_2$*
Replacement of fluorine by hydrogen
at saturated sites with Et$_3$SiH/[Ph$_3$C][B(C$_6$F$_5$)$_4$] cf. *69,* 37; solvent-dependent defluorination of aliphatic C-F bonds using Sm(HMDS)$_2$ in *n*-hexane s. M. Janjetovic, A.M. Träff, T. Ankner, J. Wettergren, G. Hilmersson, Chem. Commun. *2013, 49* (18), 1826-8 [DOI: 10.1039/c3cc37828d]; hydroxide-promoted hydrogenolysis of fluorocarbons with a ruthenium(II) 2,6-bis(imidazol-2-yl)-pyridine complex in aq. KOH, or under heterogeneous conditions with RuCl$_3$·3H$_2$O, s. M.M. Konnick, S.M. Bischof, R.A. Periana, B.G. Hashiguchi, Adv. Synth. Catal. *2013, 355* (4), 632-6 [DOI: 10.1002/adsc.201200870]; reductive cleavage of allylic or vinylic C-F bonds with titanocene difluoride and an organosilicon hydride as reductant s. M.F. Kuehnel, P. Holstein, M. Kliche, J. Krüger, S. Matthies, D. Nitsch, J. Schutt, M. Sparenberg, D. Lentz, Chem. Eur. J. *2012, 18* (34), 10701-14 [DOI: 10.1002/chem.201201125]; **methyl from trifluoromethyl groups** in CF$_3$-subst. anilines with a cationic (triarylphosphine)ruthenium(II) thiolate complex and methyl(diphenyl)-silane in *n*-hexane containing NaOMe [generating a sulfur-stabilized silylium ion as the effective reductant] s. T. Stahl, H.F.T. Klare, M. Oestreich, J. Am. Chem. Soc. *2013, 135* (4), 1248-51 [DOI: 10.1021/ja311398j].

Ammonium formate s. under Pt,Pd,Fe *HCOONH$_4$*
Dimethyl(phenyl)silane s. under Copper(I) chloride *Me$_2$PhSiH*
Methyl(diphenyl)silane s. under Cationic (triarylphosphine)ruthenium(II) thiolate complexes *MePh$_2$SiH*

Titanocene difluoride/organosilicon hydrides *Cp$_2$TiF$_2$/R$_3$SiH*
Replacement of allylic and vinylic fluorine by hydrogen s. *69,* 37s*83*

Zirconocene dichloride s. under (β-Diimine)aluminum dihydride complexes *Cp_2ZrCl_2*
1,2-Bis(diphenylphosphino)benzene s. under Copper(I) chloride *dppbz*

Cationic (triarylphosphine)ruthenium(II) thiolate complexes/methyl(diphenyl)silane/ sodium methoxide ←
Methyl from trifluoromethyl groups s. *69*, 37s*83* $CF_3 \rightarrow CH_3$

Platinum-palladium-iron nanoparticles/ammonium formate *$Pt,Pd,Fe/HCOONH_4$*
Replacement of ar. halogen by hydrogen Hal → H
update s. *11,* 633s*82*; with Pt/Pd/Fe trimetallic nanoparticles by transfer-hydrogenation, e.g. of chlorobenzene and electron-diverse ar. halides, with ammonium formate as H-donor, also reductive cleavage of aliphatic alkyl halides, s. R. Abazari, F. Heshmatpour, S. Balalaie, ACS Catal. *2013, 3* (2), 139-49 [DOI: 10.1021/cs300507a]; *metal-free* cleavage of ar. iodides (and reduction of other functionalized substrates) with disodium N-methylindole-2,3-diolate as electron donor s. R. Sword, S. O'Sullivan, J.A. Murphy, Austr. J. Chem. *2013, 66* (3), 314-22 [DOI: 10.1071/ch12480]; **replacement of ar. and hetar. fluorine by hydrogen** with a 4-coordinate (β-diimine)-aluminum dihydride complex as reductant and zirconocene dichloride as catalyst (cf. *76,* 40) s. S. Yow, S.J. Gates, A.J.P. White, M.R. Crimmin, Angew. Chem., Int. Ed. *2012, 51* (50), 12559-63 [DOI: 10.1002/anie.201207036]; with copper hydride complexes generated from CuCl/1,2-bis-(diphenylphosphino)benzene/dimethyl(phenyl)silane and KOBu-*t* as base s. H. Lv, Y.-B. Cai, J.-L. Zhang, ibid. *2013, 52* (11), 3203-7 [DOI: 10.1002/anie.201208364].

Sulfur ↑ HC ↓↑ S

tert-Butyllithium *t-B uLi*
C-Desulfinylation S(O)R → H
Allenes from *o*-allenylsulfoxides with retention of chirality s. *83*, 436

1,4-Cyclohexadiene/potassium tert-butoxide/irradiation ←
Methylene from keto groups via photoreduction of 1,3-dithianes $C(SR)_2 \rightarrow CH_2$

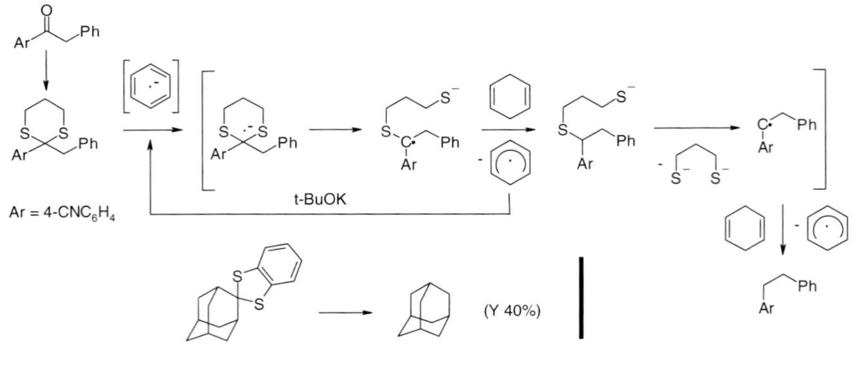

24.

K-*tert*-butoxide (3 eq.), 2-benzyl-2-(4-cyanophenyl)-1,3-dithiane (1 mmol) and 1,4-cyclohexadiene (3 eq.) added to dry DMSO (10 ml) under N_2, the mixture irradiated (>350 nm) at room temp. for 3 h, quenched with satd. aq. NH_4NO_3, extracted with methylene chloride, and purified chromatographically → 1-(4-cyanophenyl)-2-phenylethane. Y 83%. The reaction requires the presence of a good hydrogen acceptor (cyclohexadiene) to minimize the formation of dimerization products, and was shown to proceed via a photoinduced electron-transfer process between 1,3-dithiane and base. Moderate to good yields of products were obtained for 1,3-dithianes derived from electron-poor/neutral aryl ketones (three examples; Y 54-92%), with 1-(2,2-dimethylpropyl)-1-phenyl-1,3-dithiane affording a mixture of the required alkane (Y 77%) and the corresponding alkene (Y 23%), while the 4,4′-dimethoxybenzophenone-derived substrate afforded only 7% yield of the diarylmethane. The aliphatic ketones, 2-oxoadamantane and adamant-1-yl methyl ketone, gave moderate yields (40-44%; two examples) of the methylene compds. when derivatized as benzo-1,3-dithiolanes. F.e., optimization and substrate prepn. s. G. Oksdath-Mansilla, J.E. Argüello, A.B. Peñéñory, Tetrahedron Lett. *2013*, *54* (12), 1515-8 [DOI: 10.1016/j.tetlet.2012.12.118].

Remaining Elements ↑ HC ↓↑ Rem

(S)-3,3′-Bis(2,4,6-triisopropylphenyl)-1,1′-binaphthyl-2,2′-diyl hydrogen phosphate/ methanol ←

Deracemization of nitriles C=C=NSi≤ → CHCN

via organo-Brønsted acid-catalyzed asym. protonation of N-silylketenimines

25.

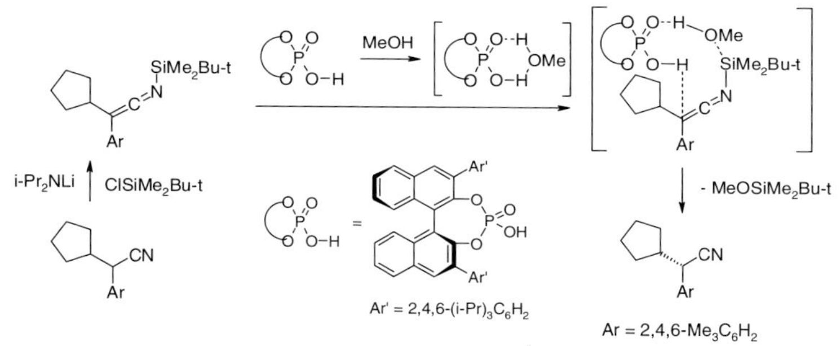

A mixture of chiral phosphoric acid diester (5 mol%), dry toluene (0.5 ml) and pentane (7.5 ml) stirred at room temp. for 15 min, anh. methanol (1.2 eq.) added, the mixture cooled to -78°, stirred for 15 min, a soln. of startg. N-silylketenimine (0.2 mmol) in dry pentane (0.5 ml) added dropwise, stirred until colorless (12 h), concentrated *in vacuo,* and purified chromatographically → (S)-2-cyclopentyl-2-mesitylacetonitrile. Y 90% (e.e. 96%). The illustrated chiral phosphate diester ((S)-TRIP), and a spirocyclic analog ((R)-STRIP), catalyzed the proto-desilylation of alkyl (incl. branched) aryl (incl. 2-thienyl) N-silylketenimines to the corresponding chiral nitriles using methanol as proton source (twenty-five examples; Y 76-97%; e.e. 80 to >98%). Lowest enantioselectivities were obtained with ester- or ketone-functionalized substrates, while dialkyl N-silylketenimines were inferior substrates (two examples; Y 91-94%; e.e. 6-70%). Use of stoichiometric catalyst as the only proton source also gave low enantioselectivity. F.e., optimization, substrate prepn. and use of the method in a gram-scale synthesis of (S)-cicloprofen s. J. Guin, G. Varseev, B. List, J. Am. Chem. Soc. *2013*, *135* (6), 2100-3 [DOI: 10.1021/ja312141b].

Carbon ↑ HC ↓↑ C

Potassium hydroxide/N-(9-anthracenylmethyl)cinchonidinium chloride/2-chloroethanol ←

C-Deacylative asym. protonation s. *83*, 87 C(O)R → H

1,8-Diazabicyclo[5.4.0]undec-7-ene *DBU*

Trifluoromethyl group as removable steering group CF_3 → H

en route to 3-subst. cyclopentenones s. *83*, 328

Lithium tetrahydridoaluminate *$LiAlH_4$*

Terminal glycols from N-tosyl-4-oxazolidones C

with retention of chirality s. *83*, 66

Nickel(II) acetoacetonate/tricyclohexylphosphine/1,1,3,3-tetramethyldisiloxane/ trimethylaluminum ←

Nickel-catalyzed replacement of cyano groups by hydrogen CN → H

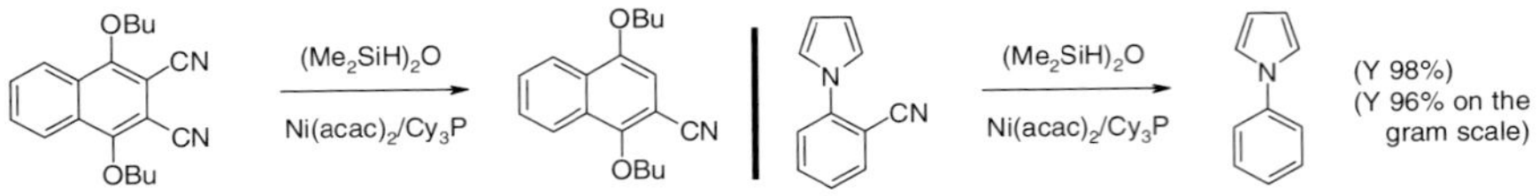

26.

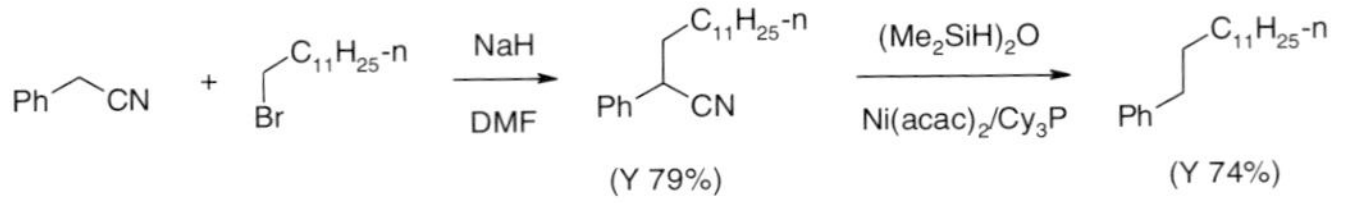

A general protocol for nickel-mediated decyanation of aryl or alkyl cyanides has been reported. It is applicable to a wide range of nitriles with varying electronic and steric substituents, including substrates possessing a β-hydrogen. **E: Partial and chemoselective replacement of ar. cyano groups.** An oven-dried resealable screw cap reaction tube charged with startg. ar. cyanide (0.5 mmol), $Ni(acac)_2$ (30 mol%) and Cy_3P (90 mol%), the tube evacuated and back-filled with N_2, three times, 1,1,3,3-tetramethyldisiloxane (0.5 mmol), Me_3Al in toluene soln. (1.5 mmol) and toluene (1 ml) added by syringe under a counter-flow of N_2, the tube placed in a preheated oil bath at 130°, the mixture stirred vigorously for 24 h, cooled to room temp., diluted with ethyl acetate, filtered through Celite, eluting with ethyl acetate, the filtrate concentrated, and the residue purified by column chromatography → 1,4-dibutoxy-2-naphthonitrile. Y 77%. Twenty further examples underwent (het)ar. decyanation in yields of 49-98%. The order of relative ease of defunctionalization was ar. SMe > CN > OAr > OMe > benzyl OMe. Seven examples of alkyl cyanides reacted in yields of 61-82%. Me_3Al was not essential but facilitated oxidative addition into the C-CN bond, and therefore was beneficial in terms of substrate scope and yields. The method was extended to α-alkylation of benzyl cyanide (Y 79%) followed by decyanation (Y 74%) and to biaryl synthesis via *o*-arylation of benzonitrile (Y unspecified) followed by decyanation (Y 80%). F.e.s. T. Patra, S. Agasti, Akanksha, D. Maiti, Chem. Commun. *2013*, *49* (1), 69-71 [DOI: 10.1039/c2cc36883h].

Elimination ⇑

Carbon ↑ HC ⇑ C

Piperidine/pyridine *$(CH_2)_5NH/Py$*
Copper(I) oxide/N,N,N′,N′-tetramethylethylenediamine *$Cu_2O/TMEDA$*
Decarboxylation COOH → H

update s. *1*, 112s*77*; of polyunsatd. α-cyano-α,β-ethylenecarboxylic acids with piperidine/pyridine s. L. Dufossé, D. Cartier, B. Valla, M. Fouillaud, R. Labia, A. Valla, Helv. Chim. Acta *2013*, *96* (2), 259-65 [DOI: 10.1002/hlca.201200162]; of arylcarboxylic acids with Cu_2O/TMEDA (or hexamethylenetetramine) as catalyst in N-methyl-2-pyrrolidone under mild conditions s. G. Cahiez, A. Moyeux, O. Gager, M. Poizat, Adv. Synth. Catal. *2013*, *355* (4), 790-6 [DOI: 10.1002/adsc.201201018].

Palladium(II) acetate/sodium carbonate/air *$Pd(OAc)_2/Na_2CO_3/O_2$*
C-Dehydroxymethylation $CH_2OH \rightarrow H$

27.
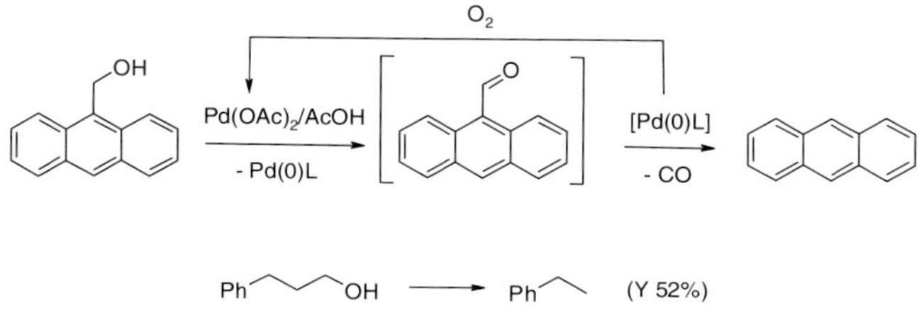

Cyclohexane (1.5 ml) added via syringe to a Schlenk tube containing 4 Å molecular sieves (150 mg), 9-hydroxymethylanthracene (0.5 mmol), Na_2CO_3 (1.5 eq.) and $Pd(OAc)_2$ (12 mol%), the tube tightly closed by a screw cap, placed in a preheated oil bath at 130°, stirred vigorously for 24 h under air, the mixture cooled to room temp., filtered through Celite, and worked up with purification by chromatography on silica gel → anthracene. Y 74%. The same catalyst effects the initial oxidation to the aldehyde and, in the same pot, the subsequent decarbonylation. The procedure is simple, scalable to the gram level, and generally applicable to electron-diverse arenes (notably possessing CF_3, OH, COOR, CN, NO_2 and even NH_2 at *ortho* sites) as well as heteroaromatic analogs, aliphatic prim. alcohols and complex hydroxymethylated natural products (thirty-seven examples; Y 42-93%). No scavenger for the liberated CO was required. F.e. and mechanistic considerations s. A. Modak, T. Naveen, D. Maiti, Chem. Commun. *2013*, *49* (3), 252-4 [DOI: 10.1039/c2cc36951f].

Formation of O-N Bond

Uptake ⇓

Addition to Nitrogen ON ⇓ N

Dimethyldioxirane ←
Mono-N-oxidation of N,N′-methylene-bridged N,N-heterocyclics ⩾N→O
en route to bicyclic hydrazines s. *83*, 94

Elimination ⇑

Hydrogen ↑ ON ⇑ H

Phenyl iodosoacetate/zinc oxide *$PhI(OAc)_2/ZnO$*
Δ^4-3-Isoxazolones from β-ketocarboxylic acid amides ○

28.
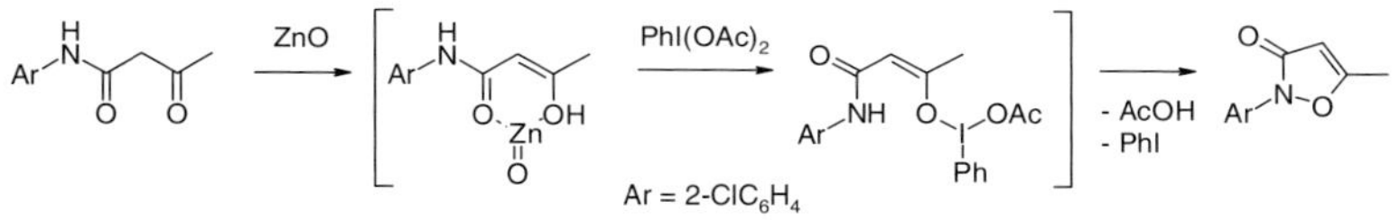

A mixture of 3-oxo-N-(2-chlorophenyl)butanamide (1 mmol), ZnO (20 mol%), phenyl iodosoacetate (1.3 eq.) and dioxane (2 ml) stirred at 100° for 5 h, and purified chromatographically → 2-(2-chlorophenyl)-5-methylisoxazol-3(2*H*)-one. Y 73%. This efficient and experimentally simple zinc(II)-catalyzed hypervalent iodine-promoted oxidative ring closure was effective for electron-diverse 3-oxo-N-arylbutanamides (ten examples; Y 73-86%). No reaction occurred in the absence of ZnO. F.e. and optimization s. W. Liu, P. Zhou, C. Chen, Q. Zhang, Z. Zhu, Org. Biomol. Chem. *2013, 11* (4), 542-4 [DOI: 10.1039/c2ob27145a].

Formation of O-S Bond

Uptake ⇓

Addition to Sulfur OS ⇓ S

Irradiation s. under Rose Bengal ///

Copper(II) phthalocyanine-on-silica magnetite nanoparticles/tetra-n-butylammonium peroxymonosulfate ←
Flavinium perchlorates/ascorbic acid ←
Rose Bengal/irradiation ←
Sulfoxides from thioethers >S → >SO
update s. *5*, 101s*82*; simple, *metal-free, aerobic* procedure under visible-light irradiation with Rose Bengal or solid-supported Rose Bengal as sensitizer s. X. Gu, X. Li, Y. Chai, Q. Yang, P. Li, Y. Yao, Green Chem. *2013, 15* (2), 357-61 [DOI: 10.1039/c2gc36683e]; *in phthalate-buffered water* under O_2 (1 atm.) with a flavinium perchlorate [5-ethyl-3,7,8,10-tetramethylisoalloxazinium perchlorate] as organocatalyst and ascorbic acid as reductant cf. Y. Imada, T. Kitagawa, H.-K. Wang, N. Komiya, T. Naota, Tetrahedron Lett. *2013, 54* (7), 621-4 [DOI: 10.1016/j.tetlet.2012.11.133]; selective oxidation with H_2O_2 and sulfamic acid as catalyst, also sulfones (cf. *78,* 40s*83*) with excess of the oxidant and an increased catalyst loading, s. H. Jafari, A. Rostami, F. Ahmad-Jangi, A. Ghorbani-Choghamarani, Synth. Commun. *2012, 42* (21), 3150-6 [DOI: 10.1080/00397911.2011.577924]; selective oxidation of aliphatic or aromatic thioethers with H_2O_2 and bis(3,5-dimethylpyrazole)-vanadyl chloride [$VO_2F(dmpz)_2$] as catalyst s. S. Hussain, D. Talukdar, S.K. Bharadwaj, M.K. Chaudhuri, Tetrahedron Lett. *2012, 53* (48), 6512-5 [DOI: 10.1016/j.tetlet.2012.09.067]; mild procedure *in a 3-phase microemulsion* containing water, a non-chlorinated solvent and dimethyl-dioctylammonium salts (hydrogen sulfate and dihydrogen phosphate) with H_2O_2 and recyclable didecyldimethylammonium tungstate as catalyst, also epoxidation of olefins (cf. *34,* 252s*83*), s. M. Fressancourt-Collinet, B. Hong, L. Leclercq, P.L. Alsters, J.-M. Aubry, V. Nardello-Rataj, Adv. Synth. Catal. *2013, 355* (2-3), 409-20 [DOI: 10.1002/adsc.201200514]; under *heterogeneous* conditions in water/ethanol with readily separable silica-coated magnetite nanoparticles coated with copper(II) phthalocyanine [via amino group-immobilization] using tetra-*n*-butylammonium peroxymonosulfate as oxidant, also epoxidation and oxidation of hydrocarbons, s. A. Rezaeifard, M. Jafarpour, A. Naeimi, R. Haddad, Green Chem. *2012, 14* (12), 3386-94 [DOI: 10.1039/c2gc35837a]; with purely inorganic porous frameworks having $[MnV_{13}O_{38}]^{7-}$ polyoxometalate clusters as nodes and rare earth ions as linkers s. D. Liu, Y. Lu, H.-Q. Tan, W.-L. Chen, Z.-M. Zhang, Y.-G. Li, E.-B. Wang, Chem. Commun. *2013, 49* (35), 3673-5 [DOI: 10.1039/c3cc40990b].

Flavinium-cyclodextrin conjugates/hydrogen peroxide ←
Chiral titanium(IV) salan complexes/hydrogen peroxide *[Ti(IV)]*/H_2O_2*
Sulfoxides from thioethers by asym. oxidation
update s. *39,* 83s*82*; of aliphatic or aromatic thioethers with H_2O_2 and flavinium-α- or β-cyclo-dextrin conjugates as catalyst s. T. Hartman, V. Herzig, M. Budešínský, J. Jindrich, R. Cibulka, T. Kraus, Tetrahedron: Asym. *2012, 23* (22-23), 1571-83 [DOI: 10.1016/j.tetasy.2012.10.017]; of bulky aryl benzyl sulfides and small alkyl phenyl sulfides with H_2O_2 and a chiral titanium(IV) salan complex s. E.P. Talsi, K.P. Bryliakov, Appl. Organomet. Chem. *2013, 27* (4), 239-44 [DOI: 10.1002/aoc.2968]; oxidation of thioethers using *m*-chloroperoxybenzoic acid with asym. induction by combining an experimental and computational study of diastereoselectivity s. A. Bottoni, M. Calvaresi, A. Ciogli, B. Cosimelli, G. Mazzeo, L. Pisani, E. Severi, D. Spinelli, S. Superchi, Adv. Synth. Catal. *2013, 355* (1), 191-202 [DOI: 10.1002/adsc.201200459].

Bis(3,5-dimethylpyrazole)vanadyl chloride/hydrogen peroxide $V(V)/H_2O_2$
Sulfoxides from thioethers s. *5*, 101s*83* $\geqslant S \rightarrow \geqslant SO$

Hydrogen peroxide [s.a. under Flavinium-cyclodextrin conjugates, Chiral titanium(IV) salan complexes, Bis(3,5-dimethylpyrazole)vanadyl chloride, Sulfamic acid and Didecyldimethylammonium tungstate] H_2O_2
Sulfones from thioethers $\geqslant S \rightarrow \geqslant SO_2$
update s. *78*, 40s*82*; heterogeneous, atom-economical, solvent- and catalyst-free oxidation of a wide range of thioethers using 30% aq. H_2O_2 with a simplified work-up s. M. Jereb, Green Chem. *2012, 14* (11), 3047-52 [DOI: 10.1039/c2gc36073j].

Sulfamic acid/hydrogen peroxide H_2NSO_3H/H_2O_2
Sulfoxides from thioethers s. *5*, 101s*83* $\geqslant S \rightarrow \geqslant SO$

Tetra-n-butylammonium peroxymonosulfate s. under Copper(II) phthalocyanine-on-silica magnetite nanoparticles ←

Didecyldimethylammonium tungstate/dimethyldioctylammonium salts/hydrogen peroxide ←
Sulfoxides from thioethers s. *5*, 101s*83*

Silica magnetite nanoparticles s. under Copper(II) phthalocyanine ←

Exchange ↓↑

Oxygen ↑ OS ↓↑ O

2,3-Diphenylcyclopropenone/triisobutylamine ←
Mesylates from sec. alcohols via alkoxycyclopropenium mesylates with inversion of configuration OH → OSO_2Me

29.

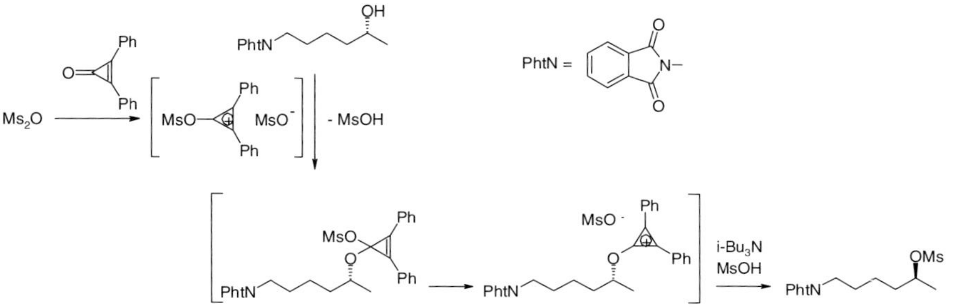

A soln. of 2,3-diphenylcyclopropenone (15 mol%) and methanesulfonic anhydride (1.5 eq.) in chloroform (0.64 ml) stirred at 55° for 10-15 min, a soln. of (R)-N-(5-hydroxyhexyl)phthalimide (0.1 mmol) and triisobutylamine (0.95 eq.) in the same solvent (0.36 ml) added via syringe pump over 18 h, additional triisobutylamine (0.25 eq.) added in one portion, stirring continued for 2 h (TLC), concentrated, dissolved in toluene (1 ml), tetraethylenepentamine (30 mol%) added, the mixture stirred until diphenylcyclopropenone consumed (TLC), diluted with ethyl acetate, washed with 1 *M* aq. HCl, satd. aq. $NaHCO_3$ and brine, dried, concentrated, and purified chromatographically → (S)-2-mesyloxy-6-phthalimidohexane. Y 74% (e.e. 92%). Careful optimization of the reaction conditions was required to exclude direct substitution of the alcohol with *retention of stereochemistry*. The reaction was successful on a gram scale and could use the activating agent in catalytic (seven examples; Y 72-83%; e.e. 92-94%) or stoichiometric amounts (fourteen examples; Y 64-98%; e.e. 88-94% or d.e. 76-96%). Sterically-hindered 2-methyl-3-pentanol was an inferior substrate (Y 45-68%; e.e. 70-82%), while a sec. homoallylic alcohol afforded a complex mixture. In cases where purification of products was difficult due to co-elution with the cyclopropenone, a scavenging procedure was used during work-up to remove the reagent. F.e., optimization and substrate/reagent prepn. s. E.D. Nacsa, T.H. Lambert, Org. Lett. *2013*, *15* (1), 38-41 [DOI: 10.1021/ol302970c]; **cyclic ethers from diols** (OC⇑O) with 2,3-diphenylcyclopropenone/Ms_2O s. B.D. Kelly, T.H. Lambert, ibid. *2011*, *13* (4), 740-3 [DOI: 10.1021/ol102980t].

Formation of O-Rem Bond

Uptake ⇓

Addition to Hydrogen and Oxygen ORem ⇓ HO

Galactokinase ←

Trichloroacetonitrile/tetra-n-butylammonium hydrogen phosphate $Cl_3CCN/(Bu_4NO)_2P(O)OH$

Phosphoric acid monoesters from alcohols OH → $OPO(OH)_2$

update s. *7,* 166s*76*; isolation of ammonium hydrogen phosphates by using tetra-*n*-butylammonium hydrogen phosphate in trichloroacetonitrile, followed by cation exchange chromatography with Dowex 50WX8/NH_4HCO_3, s. L.M. Lira, D. Vasilev, R.A. Pilli, L.A. Wessjohann, Tetrahedron Lett. *2013, 54* (13), 1690-2 [DOI: 10.1016/j.tetlet.2013.01.059]; **glycosyl dihydrogen phosphates** from aldoses with a highly efficient galactokinase from *Bifidobacterium infantis* having a broad substrate specificity s. L. Li, Y. Liu, W. Wang, J. Cheng, W. Zhao, P. Wang, Carbohydr. Res. *2012, 355*, 35-9 [DOI: 10.1016/j.carres.2012.04.022]; long-chain **phosphoric acid diesters from alcohols** with $POCl_3/Et_3N$ s. R.A. Aitken, C.J. Collett, S.T.E. Mesher, Synthesis *2012, 42* (14), 2515-8 [DOI: 10.1055/s-0031-1290823]; **phosphoric acid esters from alcohols** (primary or secondary) by phosphorylation with a wide range of pyrophosphates in the presence of Ti(OBu-*t*)$_4$ as Lewis acid s. O.S. Fenton, E.E. Allen, K.P. Pedretty, S.D. Till, J.E. Todaro, B.R. Sculimbrene, Tetrahedron *2012, 68* (44), 9023-8 [DOI: 10.1016/j.tet.2012.08.070]; rapid phosphorylation of alcohols with diphenyl phosphorochloridate in the presence of tris(pentafluorophenyl)borane/ Et_3N (cf. *2,* 315) s. W. Xue, L. Zhang, Chin. J. Chem. *2012, 30* (8), 1797-800 [DOI: 10.1002/ cjoc.201200264].

Addition to Remaining Elements ORem ⇓ Rem

Iodine I_2

Nucleoside, nucleotide and oligonucleotide chemistry ←

update s. *17,* 169s*82*; efficient microwave-assisted solid-phase coupling of nucleosides for generating small libraries s. H. Paritala, Y. Suzuki, K.S. Carroll, Tetrahedron Lett. *2013, 54* (14), 1869-72 [DOI: 10.1016/j.tetlet.2013.01.109]; solution-phase oligonucleotide synthesis based on a BH_3-masked H-phosphonate s. J. Huang, W. Lu, Z. Xi, ibid. *54* (17), 2183-6 [DOI: 10.1016/ j.tetlet.2013.02.054]; preparation of nucleoside phosphoromonoamidites by phosphitylating with an ionic liquid-stabilized phosphorodiamidite on ball-milling in the presence of pyridine·CF_3COOH s. K. Crossey, C. Hardacre, M.E. Migaud, Chem. Commun. *2012, 48* (98), 11969-71 [DOI: 10.1039/ c2cc36367d]; oxidation of nucleoside H-phosphonates with iodine via thermolabile, O-protected phosphoromonoiodates s. T. Ratajczak, M.K. Chmielewski, J. Org. Chem. *2012, 77* (18), 7866-72 [DOI: 10.1021/jo300937k]; one-pot, protection-free synthesis of **2′-deoxynucleoside 5′-tetraphosphates** based on monophosphorylation and subsequent reaction with tris(tri-*n*-butylammonium) triphosphate s. A.R. Kore, A. Senthilvelan, M. Shanmugasundaram, Tetrahedron Lett. *2012, 53* (44), 5868-70 [DOI: 10.1016/j.tetlet.2012.08.071]; **sugar nucleoside diphosphates** by coupling glycosyl dihydrogen phosphates with nucleotides in water in the presence of 2-(imidazol-1-yl)-1,3-dimethyl-Δ^2-imidazolium chloride s. H. Tanaka, Y. Yoshimura, M.R. Jørgensen, J.A. Cuesta-Seijo, O. Hindsgaul, Angew. Chem., Int. Ed. *2012, 51* (46), 11531-4 [DOI: 10.1002/ anie.201205433]; photocaging methodology for functionalization of oligonucleotides and application to labelling and cyclization s. M. Su, J. Wang, X.J. Tang, Chem. Eur. J. *2012, 18* (31), 9628-37 [DOI: 10.1002/chem.201103833].

Exchange ↓↑

Oxygen ↑ ORem ↓↑ O

Microwaves [\\\\]
Phosphorus(V) acid esters from acids ≥P(O)OH → ≥P(O)OR
update s. *33,* 101s*82*; **phosphinic acid esters** from long-chain alcohols under microwave irradiation, also phosphinic acid amides from chlorides, s. G. Keglevich, N.Z. Kiss, T. Körtvélyesi, Z. Mucsi, Phosphorus, Sulfur Silicon Relat. Elem. *2013, 188* (1-3), 29-32 [DOI: 10.1080/10426507.2012.743542].

Titanium tetra-tert-butoxide *Ti(OBu-t)$_4$*
Mixed phosphoric acid esters from alcohols ROH → ROP(O)(OR′)$_2$
by phosphorylation with pyrophosphates s. *7,* 166s*83*

Carbon ↑ ORem ↓↑ C

Via intermediates *v.i.*
Silanols from enesilanes with retention of chirality s. *83*, 246 ⩾SiOH

Elimination ⥣

Carbon ↑ ORem ⥣ C

Via intermediates *v.i.*
2-Alkoxy-3-methylene-1,2-oxaphospholane 2-oxides 
from β-phosphonyl-δ-lactolides via iodinative radical β-fragmentation s. *83*, 202

Formation of O-C Bond

Uptake ⇓

Addition to Hydrogen and Carbon OC ⇓ HC

Irradiation s. under Tetraphenylporphyrin *///*
(R)-1-(tert-Butylamino)-3-(3,4,5-trimethoxyphenoxy)propan-2-ol s. under tert-Butyl hydroperoxide ←
Hydroquinine s. under tert-Butyl hydroperoxide ←
Ammonium cerium(IV) nitrate s. under [1-(2-Pyridylmethyl)-4,7-dialkyl-1,4,7-triaza-cyclononane]iron(II) bis(triflate) *CAN*

Tetraphenylporphyrin/N-[3,5-bis(trifluoromethyl)benzyl]cinchoninium bromide/irradiation ←
tert-Butyl hydroperoxide/(R)-1-(tert-butylamino)-3-(3,4,5-trimethoxyphenoxy)propan-2-ol ←
tert-Butyl hydroperoxide/hydroquinine ←
Asym. α-hydroxylation of β-dicarbonyl compds. s. *83*, 30 H → OH

Hydroxylase/nicotinamide adenine dinucleotide/1,4-butanediol ←
Regioselective ar. hydroxylation s. *83*, 8

Phenyl iodosoacetate s. under Dichloro(p-cymene)ruthenium(II) dimer *PhI(OAc)$_2$*
Phenyl iodosotrifluoroacetate s. under Palladium(II) acetate or trifluoroacetate *PhI(OCOCF$_3$)$_2$*

3-Phenyl-N-tosyloxaziridine/chiral TADDOL-based cyclic guanidines ←
Organocatalyzed asym. α-hydroxylation of β-dicarbonyl compds. H → OH

30. 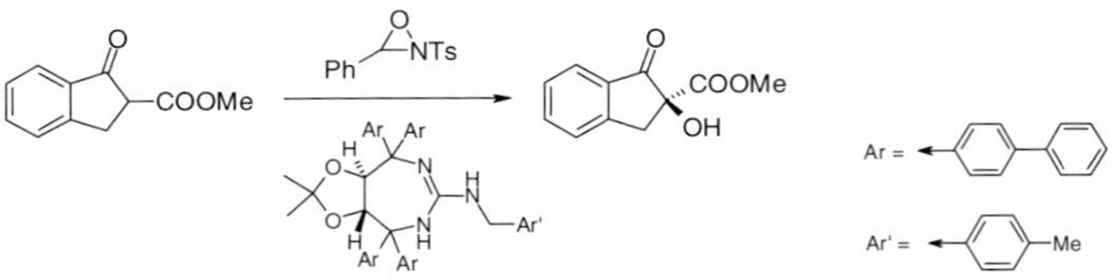

A library of chiral, TADDOL-based cyclic guanidines with tunable steric and electronic characteristics has been simply prepared from diethyl L-tartrate and assessed as catalysts for the asym. α-hydroxylation of β-dicarbonyl compds. **E:** The chiral guanidine (10 mol%) added under N_2 to a Schlenk tube containing the startg. β-dicarbonyl compd. (0.2 mmol) in toluene (1.5 ml), the soln. cooled to -78°, 3-phenyl-N-tosyloxaziridine (1.2 eq.) in toluene (0.5 ml) added via syringe, stirred at -78° for 6 h, warmed to room temp., the mixture concentrated *in vacuo,* and the residue purified by chromatography on silica gel → product. Y 99% (e.e. 93%). The procedure was applied successfully to 1-indanone-2-carboxylic acid esters and various 2-acyl-1-indanones, optimum enantioselectivity being recorded with cyclic ring-substituted N-benzylguanidine derivs. (eighteen examples; Y 90-99%; e.e. 79-93%). However, the choice of an N-tosyloxaziridine as oxidant was critical, reaction with H_2O_2 or *t*-BuOOH being sluggish with low enantioselectivity, while *m*-chloroperoxybenzoic acid gave racemic products. The ester group of the substrate had little effect on the reaction, but enantioselectivity was much lower with a 1-tetralone-2-carboxylic acid ester (e.e. 55%). F.e.s. L. Zou, B. Wang, H. Mu, H. Zhang, Y. Song, J. Qu, Org. Lett. *2013, 15* (12), 3106-9 [DOI: 10.1021/ol401306h]; asym. α-hydroxylation of α-subst. β-ketoamides, especially benzo-condensed cyclic β-ketoamides, with *t*-BuOOH/hydroquinine (in up to 83% e.e.) s. C. De Fusco, S. Meninno, C. Tedesco, A. Lattanzi, Org. Biomol. Chem. *2013, 11* (6), 896-9 [DOI: 10.1039/c2ob27283k]; of α-subst. β-ketoesters (in up to 88% e.e.) with *t*-BuOOH/chiral S-timolol derivs., especially (R)-1-(*tert*-butylamino)-3-(3,4,5-trimethoxyphenoxy)propan-2-ol, s. Y. Cai, M. Lian, Z. Li, Q. Meng, Tetrahedron *2012, 68* (38), 7973-7 [DOI: 10.1016/j.tet.2012.07.003]; by photooxygenation with molecular oxygen in the presence of 1 mol% TPP and 5 mol% N-[3,5-bis(trifluoromethyl)benzyl]cinchoninium bromide as chiral phase transfer catalyst (in up to 75% e.e.) s. M. Lian, Z. Li, Y. Cai, Q. Meng, Z. Gao, Chem. Asian J. *2012, 7* (9), 2019-23 [DOI: 10.1002/asia.201200358].

Tetra-n-butylammonium fluoride *Bu_4NF*
Aerobic hydroxylation
autoxidation with K/*t*-BuOH-DMSO s. *33,* 108; aerobic 3-hydroxylation of 3-subst. oxindoles under mild conditions with Bu_4NF as catalyst s. B.R. Buckley, B. Fernández D.-R., Tetrahedron Lett. *2013, 54* (8), 843-6 [DOI: 10.1016/j.tetlet.2012.11.083].

N-[3,5-Bis(trifluoromethyl)benzyl]cinchoninium bromide s. under Tetraphenylporphyrin ←

[1-(2-Pyridylmethyl)-4,7-dialkyl-1,4,7-triazacyclononane]iron(II) bis(triflate)/ ammonium cerium(IV) nitrate *[Fe(II)]/CAN*
Catalytic hydroxylation of hydrocarbon groups
update s. *50,* 53s*82*; *in water* with a [1-(2-pyridylmethyl)-4,7-dialkyl-1,4,7-triazacyclononane]-iron(II) bis(triflate) as catalyst and CAN as single electron oxidant, also *cis*-dihydroxylation (cf. *1,* 147s*83*) and epoxidation (cf. *34,* 252s*83*) of olefins, s. I. Garcia-Bosch, Z. Codolà, I. Prat, X. Ribas, J. Lloret-Fillol, M. Costas, Chem. Eur. J. *2012, 18* (42), 13269-73 [DOI: 10.1002/chem.201202147]; oxidation of satd. hydrocarbons with molecular oxygen activated with a metal-organic framework possessing open 2,2′-bipyridyl sites s. J. Long, L. Wang, X. Gao, C. Bai, H. Jiang, Y. Li, Chem. Commun. *2012, 48* (99), 12109-11 [DOI: 10.1039/c2cc36733e].

Dichloro(p-cymene)ruthenium(II) dimer/phenyl iodosoacetate *[Ru(II)]/PhI(OAc)$_2$*
Palladium(II) acetate or trifluoroacetate/phenyl iodosotrifluoroacetate *Pd(II)/PhI(OCOCF$_3$)$_2$*
Ar. hydroxylation H → OH
***o*-directed hydroxylation** *under ruthenium catalysis*, update s. *9*, 198s*82*; of arylhydroxamic acid esters [Weinreb esters] with [RuCl$_2$(*p*-cymene)]$_2$ and phenyl iodosoacetate as oxidant in trifluoroacetic acid/anhydride, also conversion to *o*-hydroxyaldehydes, s. F. Yang, L. Ackermann, Org. Lett. *2013, 15* (3), 718-20 [DOI: 10.1021/ol303520h]; of weakly coordinating aryl ketones s. V.S. Thirunavukkarasu, L. Ackermann, ibid. *2012, 14* (24), 6206-9 [DOI: 10.1021/ol302956s]; of aryl ketones (incl. diaryl ketones) *under palladium catalysis* with Pd(OAc)$_2$/phenyl iodosotrifluoroacetate s. P.Y. Choy, F.Y. Kwong, ibid. *2013, 15* (2), 270-3 [DOI: 10.1021/ol303088z]; of aryl ketones, arylcarboxylic acid esters and amides (incl. acetanilides) and arylsulfonamides with Pd(OAc)$_2$ in trifluoroacetic acid/anhydride as oxygen source s. G. Shan, X. Yang, L. Ma, Y. Rao, Angew. Chem., Int. Ed. *2012, 51* (52), 13070-4 [DOI: 10.1002/anie.201207458]; of aryl ketones with Pd(OCOCF$_3$)$_2$/phenyl iodosotrifluoroacetate s. F. Mo, L.J. Trzepkowski, G. Dong, ibid. 13075-9 [DOI: 10.1002/anie.201207479].

Addition to Oxygen and Carbon OC ⇓ OC

Potassium iodide s. under Cp$_2$TiCl$_2$ *KI*

Aluminum aminotris(aroxides)/tetra-n-butylammonium iodide ←
Cyclic carbonic acid esters from cyclic ethers ◯

31.

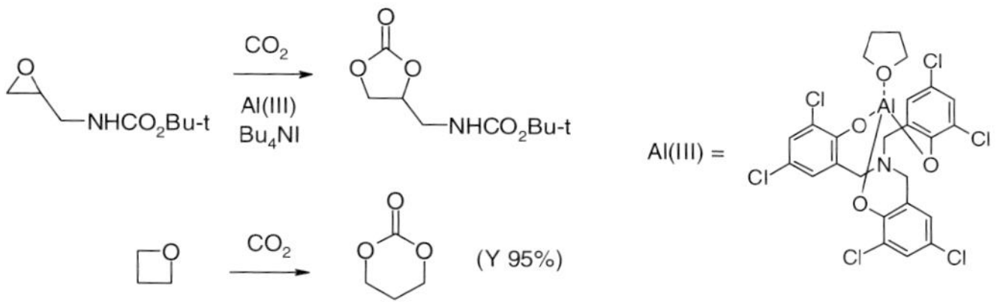

A mixture of *tert*-butyl oxiran-2-ylmethylcarbamate (1 g), [tris(3,5-dichlorobenzyl-2-olato)amine]-(tetrahydrofuran)aluminum (0.1 mol%) and tetra-*n*-butylammonium iodide (0.5 mol%) stirred in an autoclave under CO$_2$ (10 bar) at 70° for 18 h, and purified chromatographically → *tert*-butyl (2-oxo-1,3-dioxolan-4-ylmethyl)carbamate. Y 84%. The strong inductive effects of the aminotriphenoxide ligand gave rise to an oxophilic aluminum catalyst, which was used at low loading (0.05-0.1 mol%; initial TOF up to 36,000/h) for the effective conversion of terminal and internal epoxides to 1,3-dioxolan-2-ones using CO$_2$ (seventeen examples; Y 40-84%), with substantial loss of product observed in some cases during chromatography. The method was tolerant of ester, carbamate, alkene, alkyne, alcohol, ether, halide, sulfonate and silyl ether functionality. The catalyst was also applied to the more challenging oxetane system, at 0.5 mol% loading, affording 1,3-dioxan-2-one from the parent oxetane in high yield (95%) after 4 h at 70°. Sterically challenging 3,3-dimethyloxetane required longer reaction (66 h), resulting in increased polycarbonate formation and low conversion to the cyclic carbonate (59% conversion; Y 26%). F.e. and catalyst prepn. s. C.J. Whiteoak, N. Kielland, V. Laserna, E.C. Escudero-Adán, E. Martin, A.W. Kleij, J. Am. Chem. Soc. *2013*, *135* (4), 1228-31 [DOI: 10.1021/ja311053h].

Microorganisms ←
Glycols from epoxides by asym. ring opening ▽O▽ → C(OH)C(OH)
with kinetic resolution under cobalt catalysis, update s. *53*, 59s*82*; *gram-scale*, highly efficient asym. hydrolysis of racemic and *meso*-epoxides at high concentration with highly active recombinant *Escherichia coli* expressing epoxide hydrolase from *Sphingomonas* sp. HXN-200 s. S. Wu, A. Li, Y.S. Chin, Z. Li, ACS Catal. *2013, 3* (4), 752-9 [DOI: 10.1021/cs300804v]; racemic conversion with a Co(III)-salen complex encapsulated in silica-based nanoreactors (cf. *42*, 451) s. B. Li, S. Bai, X. Wang, M. Zhong, Q. Yang, C. Li, Angew. Chem., Int. Ed. *2012, 51* (46), 11517-21 [DOI: 10.1002/anie.201203774].

Titanocene dichloride/potassium iodide Cp_2TiCl_2/KI
1,3-Dioxolan-2-ones from epoxides by fixation of carbon dioxide s. *23,* 139s*83*

Tin(II) triflate $Sn(OTf)_2$
1,3-Dioxolane-4,4-dicarboxylic from α,β-epoxy-α-dicarboxylic acid esters and aldehydes
Regioselective Lewis acid-catalyzed [3+2]-cycloaddition

32. 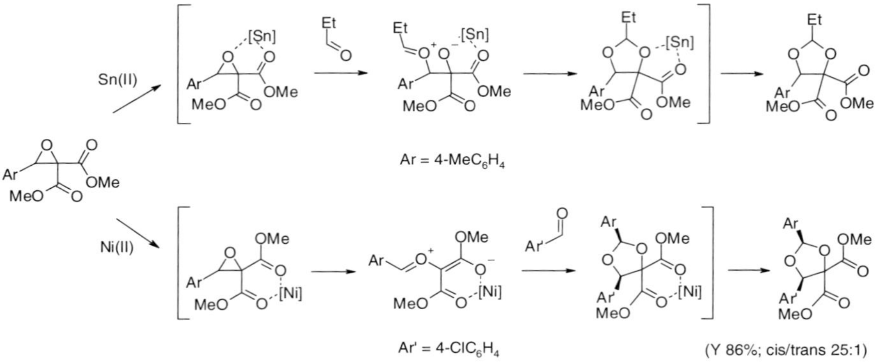

A chemodivergent cycloaddition of 3-aryloxirane-2,2-dicarboxylates with aldehydes is described, wherein C-O (cf. *24*, 128s*71,73*) or C-C cleavage of the oxirane may be controlled selectively by an appropriate Lewis acid. **E:** Propanal (1.1 eq.), dimethyl 3-(4-tolyl)oxirane-2,2-dicarboxylate (0.5 mmol) and methylene chloride (4 ml) added sequentially to a mixture of $Sn(OTf)_2$ and 4 Å molecular sieves in the same solvent (1 ml) at room temp. under N_2, the mixture stirred for 1 h, filtered through silica, concentrated *in vacuo*, and purified chromatographically → dimethyl 2-ethyl-5-(4-tolyl)-1,3-dioxolane-4,4-dicarboxylate. Y 83% (*cis/trans* 1:1.6). Extensive screening determined that oxirane C-O bond cleavage was optimally promoted by $Sn(OTf)_2$, whereas C-C cleavage was favored by use of $Ni(ClO_4)_2$. The reactions were successful for electron-diverse (het)ar. and aliphatic aldehydes (incl. cinnamal), with adducts arising from C-O cleavage obtained exclusively under tin catalysis with modest diastereoselectivity, with reactions generally complete within 1 h (seventeen examples; Y 59-86%; *cis/trans* 4.5:1 to 1:1.7). Reactions involving Ni-catalyzed C-C cleavage were significantly slower (6-72 h) and afforded isomeric adducts, generally with higher yields and diastereoselectivity than under tin catalysis (seventeen examples; Y 73-96%; *cis/trans* 2:1 to >99:1). F.e., optimization, substrate prepn. and reactions of the products s. Z. Chen, Z. Tian, J. Zhang, J. Ma, J. Zhang, Chem. Eur. J. *2012, 18* (28), 8591-5 [DOI: 10.1002/chem.201201453].

Azaphosphatranes ←
Fluorous polymer-based phosphonium chlorides ←
1,3-Dioxolan-2-ones from epoxides by fixation of carbon dioxide
update s. *23,* 139s*82*; *metal-free* procedure with tunable azaphosphatranes in the presence of chloride ion s. B. Chatelet, L. Joucla, J.-P. Dutasta, A. Martinez, K.C. Szeto, V. Dufaud, J. Am. Chem. Soc. *2013, 135* (14), 5348-51 [DOI: 10.1021/ja402053d]; with recyclable phosphonium chlorides covalently immobilized on a fluorous polymer in supercritical CO_2 s. Q.-W. Song, L.-N. He, J.-Q. Wang, H. Yasuda, T. Sakakura, Green Chem. *2013, 15* (1), 110-5 [DOI: 10.1039/c2gc36210d]; with Cp_2TiCl_2/KI as catalyst in THF under CO_2 (12 atm.) s. D. Bai, G. Nian, G. Wang, Z. Wang, Appl. Organomet. Chem. *2013, 27* (3), 184-7 [DOI: 10.1002/aoc.2967]; rapid procedure using an ionic liquid as catalyst in a microreactor [with a residence time of 14 s and TOF up to 14,000 h^{-1}] s. Y. Zhao, C. Yao, G. Chen, Q. Yuan, Green Chem. *2013, 15* (2), 446-52 [DOI: 10.1039/c2gc36612f].

Silica-supported N-(4-sulfobutyl)imidazolium hexafluorophosphate ←
Acylals from aldehydes CHO → $CH(OAc)_2$
update s. *51,* 55s*82*; metal- and solvent-free procedure with a reusable [covalently] silica-supported N-(4-sulfobutyl)imidazolium hexafluorophosphate as catalyst s. L.Q. Kang, Y.Q. Cai, L. Cheng, Monatsh. Chem. *2013, 144* (2), 247-9 [DOI: 10.1007/s00706-012-0777-6]; with inexpensive $Fe(OTs)_3 \cdot 6H_2O$ as catalyst, also O-acylation of a wide range of alcohols and phenols (cf. *29,* 184s*83*), s. N.J. Baldwin, A.N. Nord, B.D. O'Donnell, R.S. Mohan, Tetrahedron Lett. *2012, 53* (51), 6946-9 [DOI: 10.1016/j.tetlet.2012.10.033].

Tetra-n-butylammonium iodide s. under Aluminum aminotris(aroxides) *Bu_4NI*

Iron(III) tosylate *$Fe(OTs)_3$*
Acylals from aldehydes s. *51,* 55s*83*

Addition to Nitrogen and Carbon OC ⇓ NC

Cesium carbonate *Cs_2CO_3*
4-α-Sulfonylamino-1,3-dioxolan-2-ones from α-(N-methyl-N-sulfonylamino)ketones via 3-hydroxy-N-sulfonylazetidines

33. 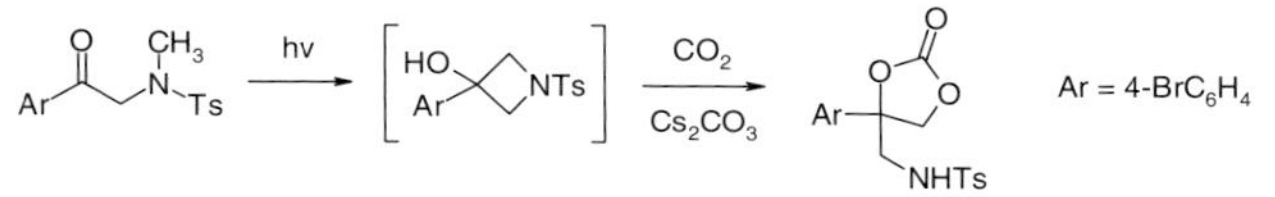

in one pot. A soln. of N-(4-bromophenacyl)-N-(toluenesulfonyl)methylamine (0.1 mmol) in DMA (1 ml) under CO_2 (1 atm.) in a Pyrex tube exposed to solar light (outside) for 8 h, Cs_2CO_3 (4 eq.) added to the formed 3-azetidinol, the mixture heated at 60° in a fume hood for 10 h, quenched with 2 *M* aq. HCl, extracted with ether, and purified chromatographically → 4-(4-bromophenyl)-4-(*p*-toluenesulfonylaminomethyl)-1,3-dioxolan-2-one. Y 85%. Irradiation of an *α-methylamino-ketone* under sunlight was shown to effect rearrangement to the corresponding azetidinol after eight hours exposure on a 'sunny' day (where a 'sunny' day provided 0.4 kW m^{-2} irradiation per hour). A series of azetidinols were generated in this way from electron-diverse α-N-arenesulfonyl-N-methylaminoacetophenones and trapped efficiently by CO_2 in the presence of base (eight examples; Y 57%, 73-87%). The corresponding pyrrolidin-3-ol was recovered unchanged from reaction with CO_2 under these conditions. F.e.s. N. Ishida, Y. Shimamoto, M. Murakami, Angew. Chem., Int. Ed. *2012, 51* (47), 11750-2 [DOI: 10.1002/anie.201206166].

Silver(I) oxide *Ag_2O*
2-(*o*-Carbalkoxyaryl)oxazoles from N-(alk-1-ynyl)phthalimides and alcohols

34. 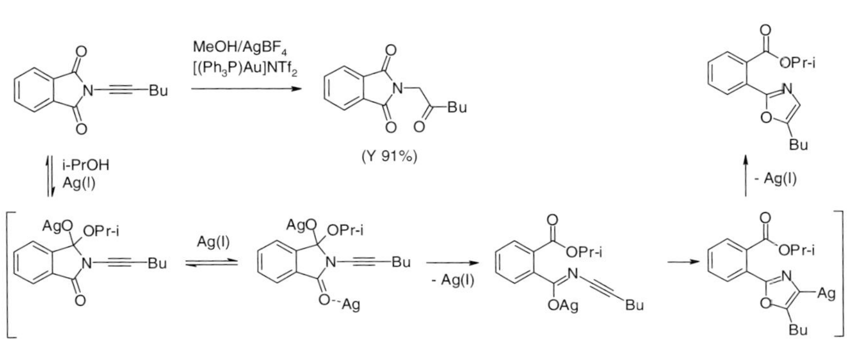

The first Au- or Ag-catalyzed reaction of internal ynimides with alcohols is described, affording N-β-keto-imides or oxazoles dependent on the catalyst system employed. **E:** A mixture of N-hex-

1-yn-1-ylphthalimide (0.1 mmol) and Ag_2O (30 mol%) in isopropanol (2 ml) stirred at 80° for 1 h (TLC), cooled to room temp., filtered through silica, concentrated *in vacuo*, and purified chromatographically → isopropyl 2-(5-butyloxazol-2-yl)benzoate. Y 92%. Use of $AgBF_4$/Et_3N or (optimally) Ag_2O afforded the oxazole derivs. as major products from phthalimide or succinimide derivs. with aliphatic alcohols (incl. *tert*-butanol and 2,2,2-trifluoroethanol) but failed with phenol (thirteen examples; Y 65-98%). Interestingly, use of $AgBF_4$ alone or (optimally) with $[(Ph_3P)Au]NTf_2$ as co-catalyst afforded **α-phthalimidoketones** as major/exclusive products (twelve examples; Y 69-91%). The products are presumed to arise from common intermediates formed by initial *nucleophilic addition of the alcohol at the imide carbonyl.* Use of aniline in place of alcohols produced an analogous 2-(*o*-carboxamido)oxazole (Y 80%) under Ag_2O catalysis, and an imino deriv. of the phthalimidoketone (Y 92%) with Ag(I)/Au(I) catalysis.

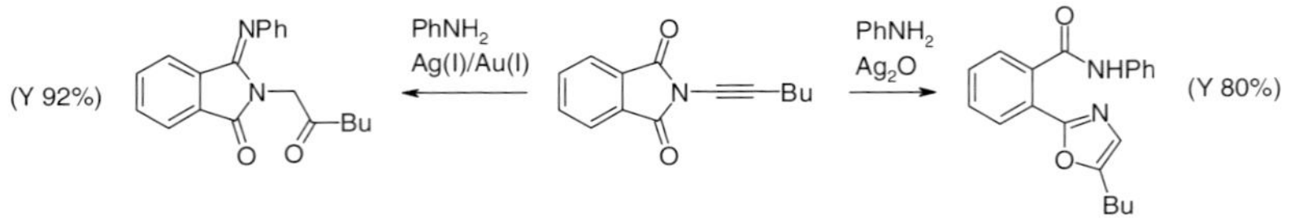

F.e., optimization and substrate prepn. s. T. Sueda, A. Kawada, Y. Urashi, N. Teno, Org. Lett. *2013*, *15* (7), 1560-3 [DOI: 10.1021/ol400338x].

(Trifluoromethyl)methyldioxirane
Nitrocarboxylic acids from lactams
Oxidative ring opening in water

35. 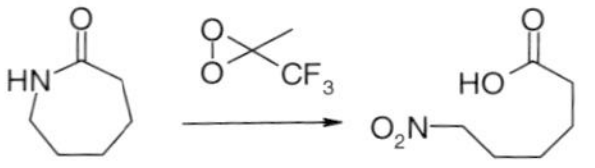

under mild conditions. A soln. of (trifluoromethyl)methyldioxirane (6.5 eq.) in 1,1,1-trifluoroacetone (4.4 ml) added rapidly to a soln. of caprolactam (0.327 mmol) in water (6 ml) at 0°, the mixture stirred vigorously until oxidant consumed (iodometry; 2 h), 1,1,1-trifluoroacetone distilled off (for regeneration of the oxidant), the residue extracted with ethyl acetate, and purified chromatographically → 6-nitrohexanoic acid. Y >99%. The dioxirane was sufficiently stable in water to effect unprecedented oxidative cleavage of the amide bond in small to large (4- to 12-membered) lactams (seven examples; Y 95 to >99%; 2-piperidone gave a 42:45:8 mixture of 5-nitropentanoic acid, glutaric acid and glutarimide). The largest lactam was poorly soluble in water and, of various modifications tested, the addition of a surfactant gave a 95% yield (based on 54% recovered substrate). Reaction of the bis-amide, 1,4-diketopiperazine, under these conditions gave no significant oxidation products. F.e.s. C. Annese, L. D'Accolti, R. Filardi, I. Tommasi, C. Fusco, Tetrahedron Lett. *2013*, *54* (6), 515-7 [DOI: 10.1016/j.tetlet.2012.17.074].

Addition to Carbon-Carbon Bonds OC ⇓ CC

Irradiation s. under 9-Mesityl-10-methylacridinium perchlorate

2(S)-[Fluoro(diphenyl)methyl)]pyrrolidine [or 2(S)-[bis[p-(hexylthio)phenyl]-(trimethylsiloxy)methyl]pyrrolidine]/hydrogen peroxide

Organocatalyzed asym. epoxidation of α,β-ethylenecarbonyl compds. C=C →

update s. *78*, 48s*82*; of challenging cyclic α,β-disubst., β,β-disubst. and α,β,β-trisubst. enals with H_2O_2 and 2(S)-[fluoro(diphenyl)methyl]pyrrolidine as catalyst s. E.-M. Tanzer, L.E. Zimmer, W.B. Schweizer, R. Gilmour, Chem. Eur. J. *2012*, *18* (36), 11334-42 [DOI: 10.1002/chem.201201316]; with H_2O_2 and eco-friendly 2(S)-[bis[*p*-(hexylthio)phenyl](trimethylsiloxy)-methyl]pyrrolidine and related catalysts, also epoxidation-Passerini reaction, s. A.M. Deobald, A.G. Corrêa, D.G. Rivera, M.W. Paixão, Org. Biomol. Chem. *2012*, *10* (38), 7681-4 [DOI: 10.1039/c2ob26247a]; asym. epoxidation of tetrasubst. β-(trifluoromethyl)-α,β-ethyleneketones by aerobic oxidation in the presence of methylhydrazine and a cinchona-based quaternary ammonium bromide as catalyst with Cs_2CO_3 as base s. H. Kawai, S. Okusu, Z. Yuan, E. Tokunaga, A. Yamano, M. Shiro, N. Shibata, Angew. Chem., Int. Ed. *2013, 52* (8), 2221-5 [DOI: 10.1002/anie.201209355]; of α,β-ethyleneketones *under Lewis acid catalysis* with a chiral ytterbium(III) 6-[3-(dimethylamino)prop-1-yn-1-yl]-1,1′-bi-2-naphthoxide complex, or related complexes, (alone or silica gel-supported) s. M.Y. El Kadiri, E. Framery, B. Andrioletti, Tetrahedron Lett. *2012, 53* (47), 6335-8 [DOI: 10.1016/j.tetlet.2012.09.004].

(R,R)-N,N′-Bis(diphenylacetyl)-N,N′-dihydroxycyclohexanediamine s. under Cumene hydroperoxide ←

6-tert-Butyl-2-chloropyridine N-oxide s. under Bis(2,4-di-tert-butylphenyl) 2′,4′,6′-trimethoxybiphenyl-2-ylphosphonite]gold(I) triflimide ←

2,6-Dichloropyridine N-oxide/methanesulfonic acid ←

2,3-Dihydro-4-quinolones from *o*-acetylene-*tert*-amines ○

Metal-free oxidative ring closure s. *83*, 286

3,5-Dibromopyridine N-oxide s. under (Triphenylphosphine)gold(I) chloride

8-Methylquinoline N-oxide s. under [Diadamant-1-yl[o-(cis-3,5-dimethylpiperidino)-phenyl]phosphine]gold(I) chloride ←

Silver hexafluoroantimonate s. under Dichloro[2-(Δ^2-oxazolin-2-yl)quinoline]palladium(II) ←

Silver triflimide s. under $FeCl_3$ *$AgNTf_2$*

Silver triflate *AgOTf*

Methyl ketones from terminal acetylene derivs. s. *83*, 42 C≡CH → $C(O)CH_3$

[Bis(2,4-di-tert-butylphenyl) 2′,4′,6′-trimethoxybiphenyl-2-ylphosphonite]gold(I) triflimide/ 6-tert-butyl-2-chloropyridine N-oxide ←

2-Indanones from propargylarenes ○

Improved gold(I)-catalyzed oxidative ring closure s. *83*, 293

Tetrachloroauric acid *$HAuCl_4$*

β-(*o*-Hydroxyaryl)ketones from 3-chromenes C

36.

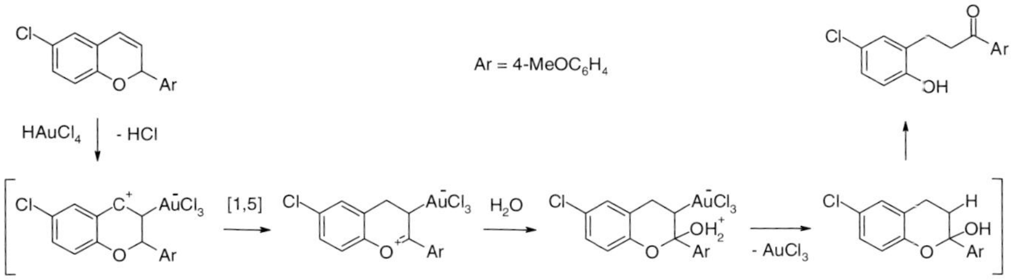

$HAuCl_4$ (2.5 mol%) added to a mixture of 6-chloro-2-(4-methoxyphenyl)-2*H*-chromene (0.5 mmol) in dry methylene chloride (5 ml) containing 1 drop of water, the mixture stirred at room temp. for 12 h, additional catalyst (2.5 mol%) added, stirring continued for 12 h (TLC), concentrated *in vacuo*, and purified chromatographically → 3-(5-chloro-2-hydroxyphenyl)-1-(4-methoxyphenyl)-propan-1-one. Y 87%. This novel and experimentally simple catalytic hydrative ring opening gave optimum results using $HAuCl_4$ (ammonium ceric nitrate was also effective) for electron-diverse 2-(het)aryl-3-chromenes (twenty examples; Y 52-87%), but failed for a sterically-hindered 2-(2,6-dichloro-phenyl) deriv. The proposed mechanism was based, in part, on deuterium-labelling experiments. F.e., optimization and substrate prepn. s. G. Maiti, U. Kayal, R. Karmakar, R.N. Bhattacharya, Tetrahedron Lett. *2012*, *53* (47), 6321-5 [DOI: 10.1016/j.tetlet.2012.08.117].

[Imidazol-2-ylidene]gold(I) chlorides *[NHC]AuCl*

Oxo compds. from acetylene derivs. C≡C → $CH_2C(O)$

by hydration, update s. *75*, 67s*82*; with air-stable [1-isopropyl-3-methyl-2,4-diphenylimidazol-2-ylidene]- or [1,4-diisopropyl-3-methyl-2-phenylimidazol-2-ylidene]-gold(I) chloride s. X. Xu,

S.H. Kim, X. Zhang, A.K. Das, H. Hirao, S.H. Hong, Organometallics *2013, 32* (1), 164-71 [DOI: 10.1021/om3009603].

(Triphenylphosphine)gold(I) chloride/silver(I) hexafluoroantimonate/3,5-dibromopyridine N-oxide ←
2-Alkylidenecyclobutanones from vinylidenecyclopropanes s. *83*, 296

[Diadamant-1-yl[o-(cis-3,5-dimethylpiperidino)phenyl]phosphine]gold(I) chloride/ sodium tetrakis[3,5-bis(trifluoromethyl)phenyl]borate/8-methylquinoline N-oxide ←
Acoxymethyl ketones C≡CH → $C(O)CH_2OC(O)R$
from terminal acetylene derivs. and carboxylic acids
via α-ketogold carbenoids

37. 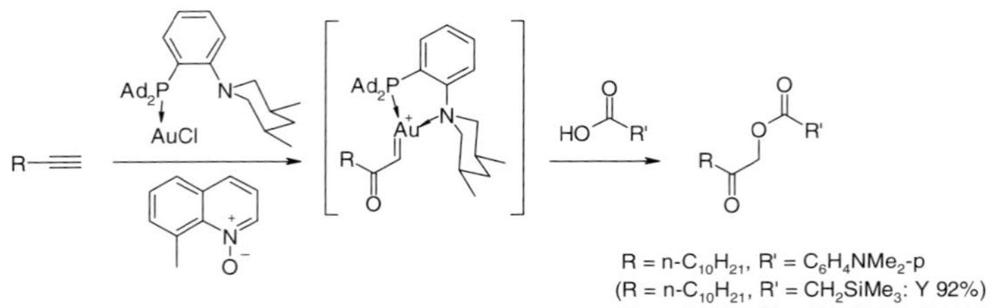

R = n-$C_{10}H_{21}$, R' = $C_6H_4NMe_2$-p
(R = n-$C_{10}H_{21}$, R' = CH_2SiMe_3: Y 92%)
(R = n-$C_{10}H_{21}$, R' = CH=CHPh: Y 94%)
(R = R' = Ph: Y 93%)

Formation of α-ketogold carbenoids as α-diazoketone equivalents (cf. *77*, 65) has been optimized for intermolecular trapping by carboxylic acids by using a gold(I) chloride complex ligated to configurationally fixed diadamant-1-yl[*o*-(3,5-dimethylpiperidino)phenyl]phosphine as catalyst, which allows formation of a *tri*coordinated gold carbenoid intermediate, the steric bulk shielding the carbene center from larger nucleophiles. **E:** *p*-Dimethylaminobenzoic acid (0.2 mmol), 1-dodecyne (1.3 eq.), [diadamant-1-yl[*o*-(*cis*-3,5-dimethylpiperidino)phenyl]phosphine]AuCl (5 mol%) and $NaBAr^F_4$ (10 mol%) added sequentially to chlorobenzene (2 ml) in a vial, the mixture stirred at room temp., a soln. of 8-methylquinoline N-oxide (1.5 eq.) in chlorobenzene (4 ml) then added via syringe pump over 12 h, the mixture concentrated *in vacuo*, and the residue purified by chromatography on silica gel → product. Y 96%. Twenty-nine further examples with (cyclo)alkyl-, (het)aryl- or cycloalkenyl-terminated ethynes and aliphatic, α,β-unsatd. or aromatic carboxylic acids gave yields of 54-97%, tolerating functional groups such as ar. bromine, trifluoromethyl, boronyl or nitro and aliphatic chlorine, silyl, ester or tosylate. F.e. and optimization, also conversion to an oxazole (with acetamide in the presence of BF_3-etherate), a β-hydroxy-γ-lactone or 2(5*H*)-furanones, s. K. Ji, Y. Zhao, L. Zhang, Angew. Chem., Int. Ed. *2013, 52* (25), 6508-12 [DOI: 10.1002/anie.201301601]; 2,4-disubst. **oxazoles from terminal acetylene derivs. and N-unsubst. carboxylic acid amides** with [Mor-DalPhos]AuCl/$NaBAr^F_4$/8-methylquinoline N-oxide [where Mor-DalPhos = diadamant-1-yl(*o*-morpholinophenyl)phosphine] s. Y. Luo, K. Ji, Y. Li, L. Zhang, J. Am. Chem. Soc. *2012, 134* (42), 17412-5 [DOI: 10.1021/ja307948m].

Magnesium oxide s. under Cumene hydroperoxide *MgO*

(-)-Isopinocampheylborane·triethylenediamine/boron fluoride *(-)-$IpcBH_2$·TMEDA/BF_3*
Terminal 1,3-diols from β,γ-ethylenecarboxylic acid esters $C(OH)CHCH_2OH$
via asym. hydroboration s. *83*, 21

Sodium tetrakis[3,5-bis(trifluoromethyl)phenyl]borate s. under [Diadamant-1-yl-[o-(cis-3,5-dimethylpiperidino)phenyl]phosphine]gold(I) chloride *$NaBAr^F_4$*
Atropoisomeric 7H-azepinium tetraphenylborates s. under Potassium peroxymonosulfate ←
Fluoroboric acid s. under $Pd(OAc)_2$ *HBF_4*

Chiral ytterbium(III) 6-[3-(dimethylamino)prop-1-yn-1-yl]-1,1′-bi-2-naphthoxide/oxidant ←
Lewis acid-catalyzed asym. epoxidation C=C → 
of α,β-ethylenecarbonyl compds. s. *78,* 48s*83*

Ammonium cerium(IV) nitrate *CAN*
β-(*o*-Hydroxyaryl)ketones from 3-chromenes s. *83,* 36 

2(S)-[Bis[p-(hexylthio)phenyl](trimethylsiloxy]methyl]pyrrolidine s. under 2(S)-[Fluoro-(diphenyl)methyl)]pyrrolidine ←

N-Condensed 1,2,4-triazol-3-ylidene/potassium acetate *NHC/KOAc*
O-Acyloximes from α,β-ethylenealdehydes and oximes under N-heterocyclic carbene catalysis C=NOH → C=NOCOR

38. 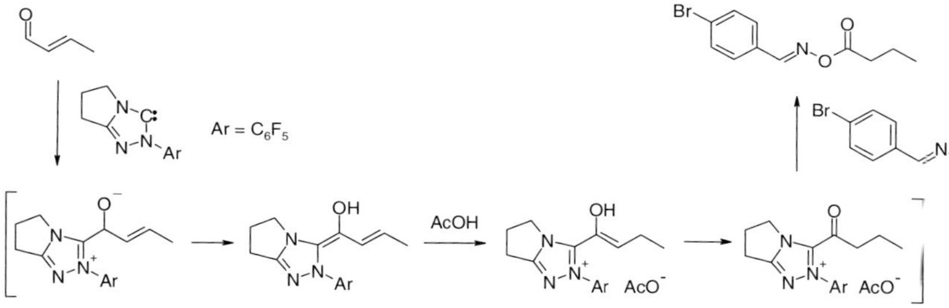

N-Condensed triazolium fluoroborate (NHC precursor; 10 mol%) and KOAc (0.5 eq.) added to a soln. of 4-bromobenzaldoxime (2 mmol) and 2-butenal (1.1 eq.) in chloroform (8 ml), the heterogeneous mixture stirred at room temp. for 24 h, and purified chromatographically → 4-bromobenzaldoxime O-butanoate. Y 90%. This experimentally simple and efficient redox esterification was applicable to (het)ar. and aliphatic aldoximes/ketoximes and aldehydes (twenty-three examples; Y 73-93%) with some methyl ketoximes giving lower yields (56-68%; two examples). F.e. and optimization s. D. Enders, A. Grossmann, D. Van Craen, Org. Biomol. Chem. *2013, 11* (1), 138-41 [DOI: 10.1039/c2ob26974k].

Perfluoro-tert-butanol s. under Cyclopropanomalonoyl peroxide *$(CF_3)_3COH$*
Benzoquinone s. under $Pd(OAc)_2$ *BQ*
2-Phenylmalononitrile s. under 9-Mesityl-10-methylacridinium perchlorate *$PhCH(CN)_2$*

Enzymes or Microorganisms ←
Metal-free asym. epoxidation C=C → 
by enzymatic catalysis s. *47,* 114s*80*; efficient, specifically (S)-selective epoxidation of a wide range of olefins in a 2-phase medium using recombinant *Escherichia coli* cells expressing *Rhodococcus* sp. ST-10 styrene monooxygenase and *Leifsonia* sp. S749 alcohol dehydrogenase genes s. H. Toda, R. Imae, N. Itoh, Tetrahedron: Asym. *2012, 23* (22-23), 1542-9 [DOI: 10.1016/j.tetasy.2012.09.017]; effect of added carboxylic acids on the face selectivity of styrene epoxidation using H_2O_2-dependent Cytochrome $P450_{SP\alpha}$ as catalyst s. T. Fujishiro, O. Shoji, N. Kawakami, T. Watanabe, H. Sugimoto, Y. Shiro, Y. Watanabe, Chem. Asian J. *2012, 7* (8), 2286-93 [DOI: 10.1002/asia.201200250]; epoxidation of unfunctionalized alkenes *under asym. organocatalysis* with an atropoisomeric cyclic [biphenylyl- or 1,1′-binaphthylyl-condensed] 7*H*-azepinium tetraphenylborate as catalyst and Oxone as oxidant with $NaHCO_3$ as base s. P.C.B. Page, C.J. Bartlett, Y. Chan, D. Day, P. Parker, B.R. Buckley, G.A. Rassias, A.M.Z. Slawin, S.M. Allin, J. Lacour, A. Pinto, J. Org. Chem. *2012, 77* (14), 6128-38 [DOI: 10.1021/jo300915u].

Monooxygenase ←
Baeyer-Villiger oxidation ←
under enzymatic catalysis s. *44*,114s*82*; stereoselective oxidation of a 4-alkylidenecyclohexanone to the corresponding *trans*- or *cis*-2-oxepanone with a wild-type and mutant-type cyclohexanone monooxygenase, respectively, s. Z.-G. Zhang, G.-D. Roiban, J.P. Acevedo, I. Polyak, M.T. Reetz, Adv. Synth. Catal. *2013, 355* (1), 99-106 [DOI: 10.1002/adsc.201200759]; spectrophotometric high-throughput quantification of Baeyer-Villiger monooxygenase activity (in cells or as the free enzyme) s. S. Saß, M. Kadow, K. Geitner, M.L. Thompson, L. Talmann, D. Böttcher, M. Schmidt, U.T. Bornscheuer, Tetrahedron *2012, 68* (37), 7575-80 [DOI: 10.1016/j.tet.2012.05.098]; simple, scalable, *heterogeneous* procedure with Sn-β-zeolite as a highly active, solid Lewis acid (cf. *46,* 128s*82*; *45,* 255s*82*), s. C. Hammond, S. Conrad, I. Hermans, Angew. Chem., Int. Ed. *2012, 51* (47), 11736-9 [DOI: 10.1002/anie.201206193]; Baeyer-Villiger oxidation of bulky cyclic ketones with stannosilicate zeolites possessing nanosheet morphology and MFI topology [Sn-MFI-ns] s. H.Y. Luo, L. Bui, W.R. Gunther, E. Min, Y. Román-Leshkov, ACS Catal. *2012, 2* (12), 2695-9 [DOI: 10.1021/cs300543z]; 2-phosphoryloxy-4,5-dihydrooxepins from 2-cyclohexenones by **Baeyer-Villiger oxidation-O-phosphorylation** s. K.C. Nicolaou, R. Yu, L. Shi, Q. Cai, M. Lu, P. Heretsch, Org. Lett. *2013, 15* (8), 1994-7 [DOI: 10.1021/ol4006689].

Acinetobacter sp. RS1/Escherichia coli [CHMO-GDH] ←
δ-Lactones from 2-alkylidenecyclopentanones ◌
via asym. enzymatic reduction-Baeyer-Villiger oxidation

39.

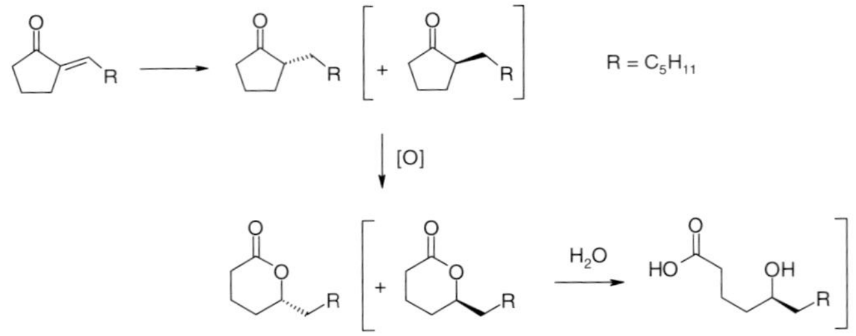

A novel one-pot two step sequence using whole cells for preparation of pure (R)-δ-alkyl-δ-lactones (not available by kinetic resolution of the racemic 2-alkylcyclopentanones using Baeyer-Villiger monooxygenases) is described. **E:** A mixture of 2-hexylidenecyclopentanone (25 µmol), glucose (100 mg) and *Acinetobacter sp.* RS1 (60 mg) in 50 m*M* Tris buffer (pH 7.4; 5 ml) incubated at 30° and 250 rpm for 3 h, a suspension of *E. coli* [CHMO-GDH] (150 mg) and glucose (150 mg) in 100 m*M* Tris buffer (15 ml) added, incubation continued for 3 h, extracted with ethyl acetate, and purified chromatographically → (R)-6-*n*-hexyltetrahydro-2*H*-pyran-2-one. Y 56% (e.e. 98%; GC yield 83%). Screening identified an *Acinetobacter* species containing a novel reductase for initial asym. reduction to 2-alkylcyclopentanones, and genetic engineering produced an *E. coli* expressing a cyclohexanone monooxygenase and glucose dehydrogenase for the Baeyer-Villiger oxidation step. Final specific hydrolysis of the minor enantiomer catalyzed by the *Acinetobacter sp.* produced pure (R)-lactones with high enantioselectivity (e.e. 97-98%; Y 41-56%; two examples). The process was more efficient than existing two-step methods. F.e. and optimization s. J. Liu, Z. Li, ACS Catal. *2013, 3* (5), 908-11 [DOI: 10.1021/cs400101v].

tert-Butyl hydroperoxide s. under Dichloro[2-(Δ^2-oxazolin-2-yl)quinoline]palladium(II) ←

Cumene hydroperoxide/hafnium tetra-tert-butoxide/(R,R)-N,N′-bis(diphenylacetyl)-N,N′-dihydroxycyclohexanediamine/magnesium oxide ←

Hafnium(IV)-catalyzed asym. epoxidation of 2- or 3-ethylene-*tert*-alcohols

40.

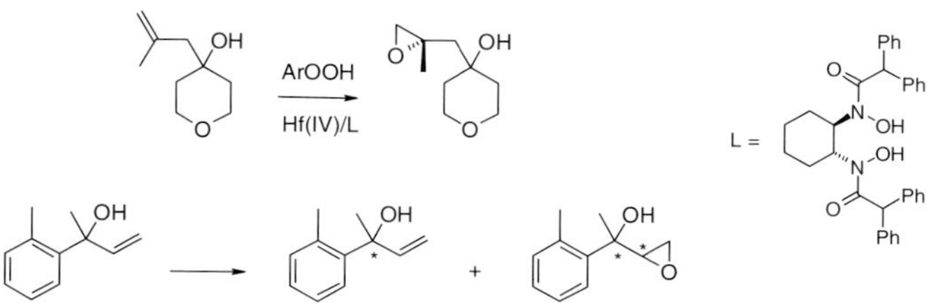

(Y 47%; e.e. 90%) (Y 46%; e.e. 80%)

$Hf(OBu\text{-}t)_4$ (10 mol%) added to a mixture of chiral hydroxamic acid ligand (1.1 eq.) and MgO (20 mol%) in toluene (3 ml), the mixture stirred at room temp. for 3 h, cooled to 0-5°, 4-(2-methyl-prop-2-enyl)tetrahydropyran-4-ol (0.4 mmol) and 80% cumene hydroperoxide (2 eq.) added sequentially, the mixture stirred efficiently for 48 h, and purified chromatographically → 4-(2(S)-methyloxiran-2-ylmethyl)tetrahydropyran-4-ol. Y 84% (e.e. 96%). The success of this novel asym. epoxidation of sterically-hindered (homo)allylic tert. alcohols was attributed to alleviation of steric interaction due to the long metal-O bonds in the hafnium-hydroxamic acid catalyst. The illustrated chiral ligand, based on diphenylacetohydroxamic acid, generally gave best results with less hindered allylic/homoallylic substrates lacking a chiral center, while an analogous ligand based on 3,3,3-triphenylpropanohydroxamic acid (assumed to provide a larger reaction pocket) was significantly more enantioselective for bulkier substrates (twenty-three examples; Y 74-99%; e.e. 53-99%). The catalyst system also proved effective for the **kinetic resolution** of racemic α-aryl-*tert*-allylic alcohols, giving good enantioselectivity for both unreacted allylic alcohol and epoxy-alcohol products (e.e. 80-90%; two examples), and for **desymmetrization** of prochiral bis(allylic)-*tert*-alcohols (no details). F.e.s. J.L. Olivares-Romero, Z. Li, H. Yamamoto, J. Am. Chem. Soc. *2013, 135* (9), 3411-3 [DOI: 10.1021/ja401182a].

Cyclopropanomalonoyl peroxide/perfluoro-tert-butanol ←

***cis*-Glycols from ethylene derivs.** s. *78,* 56s*83* C=C → C(OH)C(OH)

m-Chloroperoxybenzoic acid *mCPBA*

Dendritic fluoroalcohols/hydrogen peroxide ←

Metal-free epoxidation C=C →

with *m*-chloroperoxybenzoic acid s. *20,* 112; of the exocyclic double bond of 4-alkylidene-2-cyclopentenones s. M. Ahmar, S. Thomé, B. Cazes, Eur. J. Org. Chem. *2012* (36), 7093-105 [DOI: 10.1002/ejoc.201201030]; heterogeneous procedure with anhydrous [2-percarboxyethyl]-functionalized silica s. R. Mello, A. Alcalde-Aragonés, M.E.G. Núñez, G. Asensio, J. Org. Chem. *2012, 77* (15), 6409-13 [DOI: 10.1021/jo300533b]; epoxidation of enones *with H_2O_2 generated in situ* by aerobic photooxidation of isopropanol in the presence of anthraquinone-2-carboxylic acid and molecular O_2 under visible-light irradiation (with added KOH) s. L. Cui, S. Furuhashi, Y. Tachikawa, N. Tada, T. Miura, A. Itoh, Tetrahedron Lett. *2013, 54* (2), 162-5 [DOI: 10.1016/j.tetlet.2012.10.119]; epoxidation with H_2O_2 using recyclable 'dendritic fluoroalcohols' as catalyst [fluoroalcohols immobilized on dendritic polyglycerol] s. A. Berkessel, J. Krämer, F. Mummy, J.-M. Neudörfl, R. Haag, Angew. Chem., Int. Ed. *2013, 52* (2), 739-43 [DOI: 10.1002/anie.201206003].

Hydroxy(phenyl)(2-pyridyl)acetic acids s. under Iron(II) acetate ←

Tetrahydrothiophene *$(CH_2)_4S$*

2-α-Functionalized furans from 2,4-enynones s. *83,* 105

Phenyl iodosobenzoate s. under $PdCl_2(PhCN)_2$ *$PhI(OCOPh)_2$*

Hafnium tetra-tert-butoxide s. under Cumene hydroperoxide *$Hf(OBu\text{-}t)_4$*

Trimethylsilyl iodide/triphenylphosphine *Me_3SiI/Ph_3P*
Lactolides from cyclic enolethers ○
protected 2-deoxy-α-D-galactosides from galactals with a thiourea as catalyst s. *48,* 120s*82*; further 2-deoxy-α-D-glycosides with trimethylsilyl iodide/triphenylphosphine s. X.-K. Cui, M. Zhong, X.-B. Meng, Z.-J. Li, Carbohydr. Res. *2012, 358*, 19-22 [DOI: 10.1016/j.carres.2012.06.004].

Ammonia-modified titanosilicates/hydrogen peroxide ←
Epoxidation s. *34,* 252s*83* C=C → 

Tin-silicates or Tin-β-zeolite ←
Heterogeneous Baeyer-Villiger oxidation s. *46,* 128s*83* ←

Methylhydrazine/cinchona-based quaternary ammonium bromide/cesium carbonate ←
Organocatalyzed asym. epoxidation
of α,β-ethylenecarbonyl compds. s. *78,* 48s*83*

Triphenylphosphine s. under Me_3SiI *Ph_3P*
Tantalum nitride s. under Molybdenum sulfide ←
Hydrogen peroxide s. under 2(S)-[Fluoro(diphenyl)methyl)]pyrrolidine [or 2(S)-[bis[p-(hexylthio)phenyl](trimethylsiloxy)methyl]pyrrolidine], Dendritic fluoroalcohols, Ammonia-modified titanosilicates, WO_3, 14,28-[1,3-Diiminoisoindolinato]phthalocyaninatoiron(III), Iron(II) acetate, [1-(2-Pyridylmethyl)-4,7-dialkyl-1,4,7-triazacyclononane]iron(II) bis(triflate) and [N,N′-Bis-(2-pyridylmethyl)-1,2-diaminoethane]iron(II) bis(triflate) *H_2O_2*

o-Benzenedisulfonimide ←
Michael addition of O-nucleophiles C=C → CHC(OR)
update s. *46,* 976s*82*; to a wide range of electron-deficient alkenes with *o*-benzenedisulfonimide as a reusable Brønsted acid catalyst, also aza- and thia-Michael addition, s. M. Barbero, S. Cadamuro, S. Dughera, Synth. Commun. *2013, 43* (5), 758-67 [DOI: 10.1080/00397911.2011.614834]; (Z)- or (E)-selective Michael addition of phenols to α,β-acetylenenitriles with Na_2CO_3 or DABCO as base s. W. Zhou, Y. Zhang, P. Li, L. Wang, Org. Biomol. Chem. *2012, 10* (35), 7184-96 [DOI: 10.1039/c2ob25969a].

Methanesulfonic acid s. under 2,6-Dichloropyridine N-oxide *$MeSO_3H$*
2-Iodo-5-methylbenzenesulfonic acid s. under Potassium peroxymonosulfate *$ArSO_3H$*

Potassium peroxymonosulfate/atropoisomeric 7H-azepinium tetraphenylborates/ sodium hydrogen carbonate ←
Metal-free asym. epoxidation s. *47,* 114s*83* C=C → 

Potassium peroxymonosulfate/2-iodo-5-methylbenzenesulfonic acid ←
α-Arylketones from styrenes with 1,2-aryl migration s. *19,* 200s*83* ←

Molybdenum sulfide/tantalum nitride nanocomposite ←
Tungsten trioxide nanoparticles/hydrogen peroxide *WO_3/H_2O_2*
Epoxidation s. *34,* 252s*83* C=C → 

9-Mesityl-10-methylacridinium perchlorate/2-phenylmalononitrile/irradiation ←
Ethers from ethylene derivs. s *83*, 46 C=C → CHC(OR)

Cinchona-based quaternary ammonium bromide s. under Methylhydrazine ←

Transition metal complexes/oxidant ←
Transition metal-catalyzed asym. epoxidation C=C → 
update s. *46,* 106s*82*; of 3-vinyl-carbostyrils and -2-pyridones (at the vinylic site) with a chiral octahydro-1*H*-4,7-methanoisoindol-1-one-tethered carbonyl(porphyrinato)ruthenium(II) complex at low catalyst loading with 2,6-dichloropyridine N-oxide as oxidant s. P. Fackler, S.M. Huber, T. Bach, J. Am. Chem. Soc. *2012, 134* (30), 12869-78 [DOI: 10.1021/ja305890c]; of terminal and internal alkenes with a chiral vanadium(IV) or iron(III) salicylaldimine complex and aq. H_2O_2 as oxidant, also asym. sulfoxidation (cf. *39,* 83s*83*), s. K.A. Stingl, K.M. Weiß, S.B. Tsogoeva, Tetrahedron *2012, 68* (40), 8493-501 [DOI: 10.1016/j.tet.2012.07.052]; of allyl alcohols with chiral vanadium α-aminocarboxylic acid complexes modified by naturally occurring or synthetic

inorganic *nanosheets* as huge, rigid planar substituents (cf. *36*, 117) s. L.-W. Zhao, H.-M. Shi, J.-Z. Wang, J. He, Chem. Eur. J. *2012, 18* (32), 9911-8 [DOI: 10.1002/chem.201201659]; of styrenes in an aq. biphase medium with a manganese(III) salen complex embedded within cross-linked chiral polyethylenimine s. N. Levi, R. Neumann, Chem. Plus. Chem. *2012, 77* (11), 977-81 [DOI: 10.1002/cplu.201200203].

[14,28-(1,3-Diiminoisoindolinato)phthalocyaninato]iron(III)/hydrogen peroxide *[Fe(III)]/H_2O_2*

Epoxidation with hydrogen peroxide under mild conditions using a 'helmet-type' iron(III) phthalocyanine as catalyst C=C →

41.

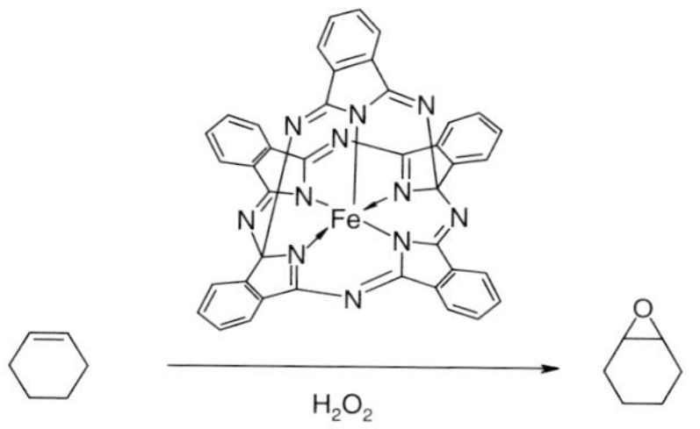

A soln. of cyclohexene in acetonitrile (0.3 *M*) treated with H_2O_2 (*2 eq.*) in the presence of [14,28-(1,3-diiminoisoindolinato)phthalocyaninato]iron(III) [hPcFe] (1 mol%) at 25° for 2 h under air → cyclohexene oxide. Conversion 97%; selectivity 92% (with 4% 2-cyclohexenol and 1% 2-cyclo-hexenone). 2-Cyclohexenone was obtained predominantly using *tert*-butyl hydroperoxide as oxidant (conversion 83%; selectivity 65%). The method was very efficient for epoxidation of cyclic and linear olefins, except [terminal] 1-octene (conversion 13%) and [sterically demanding] norbornene (conversion 14%), and for electron-diverse styrenes (notably without ar. aldehyde formation). High stereoselectivity was observed with *cis*-stilbene (60% conversion; 95% selectivity after 2 h; *cis/trans* 96:4). Similar results were obtained under air or argon, suggesting O_2 is not involved in the reaction. F.e. (eleven; conversion 72-100%; selectivity 60%, 71%, 90-96%), also evidence for the involvement of high-valent iron oxo species using cryospray mass spectrometry, s. I.Y. Skobelev, E.V. Kudrik, O.V. Zalomaeva, F. Albrieux, P. Afanasiev, O.A. Kholdeeva, A.B. Sorokin, Chem. Commun. *2013, 49* (49), 5577-9 [DOI: 10.1039/c3cc42301h]; formation of racemic iron(III) and cobalt(III) complexes containing a pentadentate 'helmet' phthalocyaninato ligand s. H.M. Kieler, M.J. Bierman, I.A. Guzei, P.J. Liska, R.W. McGaff, ibid. *2006* (31), 3326-8 [DOI: 10.1039/b606537f]; **hydroxylation** of cycloalkanes with the iron complex/H_2O_2 s. E.S. Brown, J.R. Robinson, A.M. McCoy, R.W. McGaff, Dalton Trans. *2011, 40* (22), 5921-5 [DOI: 10.1039/c1dt10147a].

Iron(II) acetate/hydroxy(phenyl)(2-pyridyl)acetic acids/hydrogen peroxide ←
[1-(2-Pyridylmethyl)-4,7-dialkyl-1,4,7-triazacyclononane]iron(II) bis(triflate)/ hydrogen peroxide *[Fe(II)]/H_2O_2*

***cis*-Glycols from ethylene derivs.** s. *1*, 147s*83* C=C → C(OH)C(OH)

[N,N'-Bis(2-pyridylmethyl)-1,2-diaminoethane]iron(II) bis(triflate)/hydrogen peroxide ←
(Hypochlorito)iron(III) porphyrin complexes *[Fe(III)]*

Transition metal-catalyzed epoxidation C=C →

update s. *34*, 252s*82*; with hypochlorito- and bis(hypochlorito)-iron(III) porphyrin complexes, also ar. chlorination and sulfoxidation (cf. *5*, 101s*83*), s. Z. Cong, S. Yanagisawa, T. Kurahashi, T. Ogura, S. Nakashima, H. Fujii, J. Am. Chem. Soc. *2012, 134* (51), 20617-20 [DOI: 10.1021/ja3108774]; epoxidation of unsatd. steroidal ketones with H_2O_2 and [N,N'-bis(2-pyridylmethyl)-1,2-diaminoethane]iron(II) bis(triflate) as catalyst, chemo- and stereo-selectivity, s. D. Clemente-Tejeda, A. López-Moreno, F.A. Bermejo, Tetrahedron *2012, 68* (45), 9249-55 [DOI: 10.1016/j.tet.2012.08.079]; with H_2O_2 and robust, gaseous ammonia-modified titanosilicates for improved epoxidation of alkenes s. L. Zhang, L. Xu, J. Sun, J. Jiang, Y. Liu, H. Wu, P. Wu, Chin. J. Chem. *2012, 30* (9), 2205-11 [DOI: 10.1002/cjoc.201200637]; biomimetic aerobic epoxidation with MoS_2/Ta_3N_5 nanocomposite, also f. oxidations, s. Q. Gao, C. Giordano, M. Antonietti, Angew.

Chem., Int. Ed. *2012, 51* (47), 11740-4 [DOI: 10.1002/anie.201206542]; with WO_3 nanoparticles and H_2O_2 as oxidant s. C. Hammond, J. Straus, M. Righettoni, S.E. Pratsinis, I. Hermans, ACS Catal. *2013, 3* (3), 321-7 [DOI: 10.1021/cs300826c].

Iron(III) chloride/silver triflimide *$FeCl_3/AgNTf_2$*
5,10,15,20-Tetrakis(4-sulfophenyl)porphyrinatocobalt(III) trisodium salt/methanol ←
Methyl ketones from terminal acetylene derivs. C≡CH → C(O)CH_3
via regioselective Markovnikov hydration
using water-soluble cobalt(III) porphyrin complexes

42. 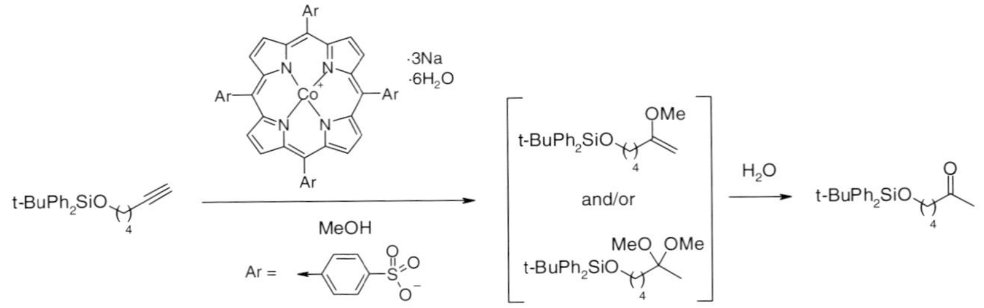

under mild conditions. A mixture of 6-*tert*-butyldiphenylsiloxyhex-1-yne (5 mmol), water (4.4 eq.) and cobalt porphyrin catalyst (2 mol%) in methanol (3.76 ml) heated at 80° under air in a sealed tube for 5 h (TLC/GC), cooled to 25°, diluted with water and hexane, extracted with hexane, and purified by distillation → 6-*tert*-butyldiphenylsiloxy-2-hexanone. Y 93%. The use of the preformed cobalt porphyrin complex under air enabled effective and scalable (to 5.1 g) hydration of aliphatic and electron-diverse aromatic terminal alkynes, incl. acid-sensitive substrates not easily hydrated by other methods (thirteen examples; Y 75-96%), with the addition of 0.3 mol% *triflimide* improving results in many cases. The method was compatible with silyl ether, trityl ether, allyl ether, benzyl ether, boronate, ester, halo, nitrile, nitro and amide functionality, and its use in late-stage reaction on carbohydrate (Y 97%) and amino acid (Y 91%) scaffolds was demonstrated. The method gave lower conversions under an inert atmosphere. The catalyst could be recovered from the aq. layer and was reused three times with no reduction in yield (although the reaction rate reduced somewhat in each run (5 → 14 h)). Initial tests indicated the reaction involves mono- and/or bis-hydroalkoxylation of the alkyne (with methanol) and subsequent hydrolysis of the enol ether/acetal intermediates. F.e., optimization and catalyst/substrate prepn. s. T. Tachinami, T. Nishimura, R. Ushimaru, R. Noyori, H. Naka, J. Am. Chem. Soc. *2013*, *135* (1), 50-3 [DOI: 10.1021/ja310282t]; with AgOTf s. R. Das, D. Chakraborty, Appl. Organomet. Chem. *2012*, *26* (12), 722-6 [DOI: 10.1002/aoc.2918]; ketones from internal or terminal alkynes **under iron(III) Lewis/Brønsted catalysis** with $FeCl_3/AgNTf_2$ and 3 eq. water in dioxane s. J.R. Cabrero-Antonino, A. Leyva-Pérez, A. Corma, Chem. Eur. J. *2012*, *18* (28), 11107-14 [DOI: 10.1002/chem.201200580].

Dinuclear ruthenium complexes *[Ru]-[Ru]*
Enolesters from acetylene derivs. C≡C → CH=C(OAc)
update s. *40,* 81s*82*; anti-Markovnikov addition of aliphatic carboxylic acids to phenylacetylene with high E/Z selectivity using a *dinuclear* ruthenium catalyst with a rigid anthracene spacer s. K.-C. Cheung, W.-L. Wong, M.-H. So, Z.-Y. Zhou, S.-C. Yan, K.-Y. Wong, Chem. Commun. *2013, 49* (7), 710-2 [DOI: 10.1039/c2cc38454j].

Palladium(II) acetate/benzoquinone/fluoroboric acid *$Pd(OAc)_2/BQ/HBF_4$*
Dichloro[2-(Δ^2-oxazolin-2-yl)quinoline]palladium(II)/silver hexafluoroantimonate/ tert-butyl hydroperoxide ←
Palladium(II) chloride *$PdCl_2$*
Wacker oxidation C=CH → CHC(O)
update s. *19,* 200s*82*; generally applicable route to ketones from internal olefins at room temperature with $Pd(OAc)_2$/benzoquinone (or aerobically) and added HBF_4 (0.27 *M*) in a mixed aq. organic

medium s. B. Morandi, Z.K. Wickens, R.H. Grubbs, Angew. Chem., Int. Ed. *2013, 52* (10), 2944-8 [DOI: 10.1002/anie.201209541]; with dichloro[2-(Δ^2-oxazolin-2-yl)quinoline]-palladium(II)/ $AgSbF_6$/*tert*-butyl hydroperoxide cf. R.J. DeLuca, J.L. Edwards, L.D. Steffens, B.W. Michel, X. Qiao, C. Zhu, S.P. Cook, M.S. Sigman, J. Org. Chem. *2013, 78* (4), 1682-6 [DOI: 10.1021/jo302638v]; ketones by regioselective Wacker oxidation of allylic functionalized ethylene derivs. with $PdCl_2$ under O_2 (1 atm.) in N,N-dimethylacetamide s. T. Mitsudome, S. Yoshida, Y. Tsubomoto, T. Mizugaki, K. Jitsukawa, K. Kaneda, Tetrahedron Lett. *2013, 54* (12), 1596-8 [DOI: 10.1016/j.tetlet.2013.01.049]; alternative route to α-arylketones from α-subst. styrenes with aryl shift using Oxone and 2-iodo-5-methylbenzenesulfonic acid as pre-catalyst s. V.C. Purohit, S.P. Allwein, R.P. Bakale, Org. Lett. *2013, 15* (7), 1650-3 [DOI: 10.1021/ol400432x].

Bis(benzonitrile)dichloropalladium(II)/phenyl iodosobenzoate *$PdCl_2(PhCN)_2/PhI(OCOPh)_2$*
Glycol esters from ethylene derivs. C=C → C(OAc)C(OAc)
update s. *29,* 152s*81*; di(benzoyloxyl)ation of chiral O-allylketoximes *with asym. induction* using $PdCl_2(PhCN)_2$ as catalyst and phenyl iodosobenzoate as oxidant/reactant s. S.R. Neufeldt, M.S. Sanford, Org. Lett. *2013, 15* (1), 46-9 [DOI: 10.1021/ol303003g]; rapid, green and safe di(acetoxyl)ation of alkenes in a microreactor under palladium catalysis with H_2O_2 (or with TfOH/peroxyacetic acid) s. J.H. Park, C.Y. Park, H.S. Song, Y.S. Huh, G.H. Kim, C.P. Park, ibid. *15* (4), 752-5 [DOI: 10.1021/ol3026445].

Hydroxo(oxo)[tris(2-pyridylmethyl)amine]osmium(V) complexes *[Os(V)]*
Glycols from ethylene derivs. C=C → C(OH)C(OH)
cis-dihydroxylation with OsO_4 s. *1,* 147s*80*; with a benign hydroxo(oxo)[tris(2-pyridylmethyl)-amine]osmium(V) complex as an alternative to toxic OsO_4 s. H. Sugimoto, K. Kitayama, S. Mori, S. Itoh, J. Am. Chem. Soc. *2012, 134* (46), 19270-80 [DOI: 10.1021/ja309566c]; mild and rapid *cis*-dihydroxylation with [1-(2-pyridylmethyl)-4,7-dialkyl-1,4,7-triazacyclononane]iron(II) bis(triflate) as catalyst and H_2O_2 as oxidant s. I. Prat, D. Font, A. Company, K. Junge, X. Ribas, M. Beller, M. Costas, Adv. Synth. Catal. *2013, 355* (5), 947-56 [DOI: 10.1002/adsc.201200938]; with $Fe(OAc)_2$ and a chiral hydroxy(phenyl)(2-pyridyl)acetic acid as ligand with H_2O_2 as oxidant cf. S.M. Barry, H. Mueller-Bunz, P.J. Rutledge, Org. Biomol. Chem. *2012, 10* (36), 7372-81 [DOI: 10.1039/c2ob25834j]; toluene dioxygenase-catalyzed *cis*-dihydroxylation of benzo[*b*]-thiophenes and benzo[*b*]furans under O_2 s. D.R. Boyd, N.D. Sharma, I.N. Brannigan, T.A. Evans, S.A. Haughey, B.T. McMurray, J.F. Malone, P.B.A. McIntyre, P.J. Stevenson, C.C.R. Allen, ibid. *10* (36), 7292-304 [DOI: 10.1039/c2ob26120k]; *cis*-cyclohexa-3,5-diene-1,2-diols from *p*-subst. arenes with the same enzyme s. J.F. Trant, J. Froese, T. Hudlicky, Tetrahedron: Asym. *2013, 24* (4), 184-90 [DOI: 10.1016/j.tetasy.2019.01.002]; stereoselective *selenium-catalyzed* dihydroxylation with H_2O_2 s. C. Santi, R. Di Lorenzo, C. Tidei, L. Bagnoli, T. Wirth, Tetrahedron *2012, 68* (51), 10530-5 [DOI: 10.1016/j.tet.2012.08.078]; *cis*-dihydroxylation of *trans*-olefins with cyclopropanomalonoyl peroxide as oxidant in the presence of perfluoro-*tert*-butanol as catalyst (cf. *78,* 56) s. S. Picon, M. Rawling, M. Campbell, N.C.O. Tomkinson, Org. Lett. *2012, 14* (24), 6250-3 [DOI: 10.1021/ol3030154].

Via intermediates *v.i.*
3-Oxetanones from allenes via 1,4-dioxaspiro[2.2]pentanes ○

43. 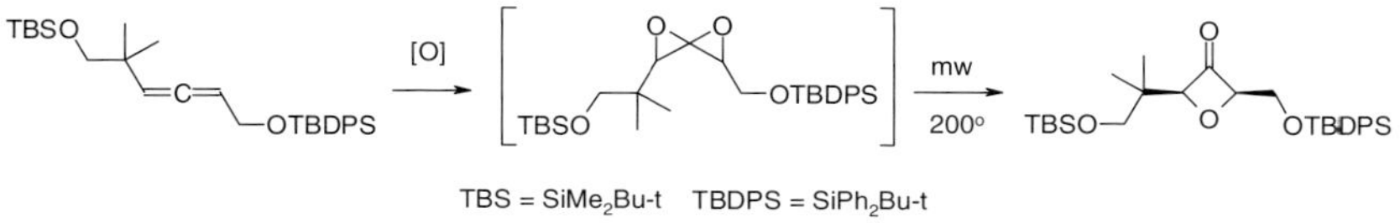

Two complementary methods are described for the preparation of the title compounds, involving thermal or bromide-induced rearrangement of spirodioxetanes, formed by oxidation of readily available allenes. **E:** A freshly prepared soln. of dimethyldioxirane (2.5 eq.) in chloroform (~0.2 *M*) added dropwise to a soln. of startg. allene (1 eq.) in the same solvent (0.1 *M*) at -20° under N_2, the mixture stirred for 1-2 h (TLC), concentrated *in vacuo*, toluene added, the mixture heated by microwaves at 200° for 1-1.5 h, cooled to room temp., and purified chromatographically → *cis*-2-[1-(*tert*-butyldimethylsilyloxy)-2-methylpropan-2-yl]-4-(*tert*-butyldiphenylsilyloxymethyl)-

oxetan-3-one. Y 73% (d.r. 8:1). Thermal rearrangement of the intermediate spirodioxetanes gave best results under microwave heating at 200° and appears general for tetra-, tri- and 1,1- or 1,3-di-subst. allenes (twelve examples; Y 65-98%; 51% in the case of tetramethyloxetanone, attributed to product and intermediate volatility), but diastereoselectivity was variable (1.4:1 to 8:1).

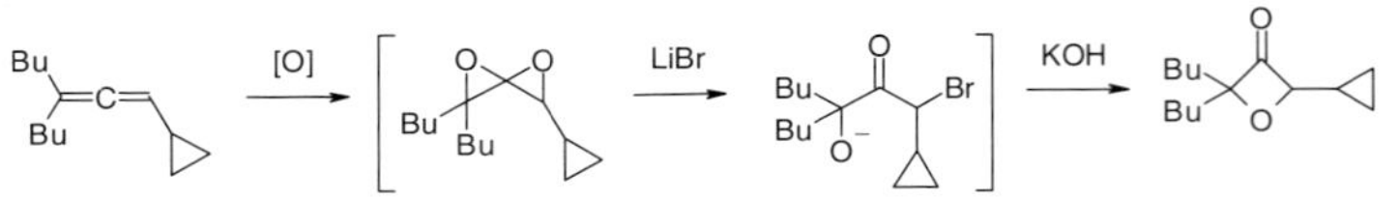

Oxidation and subsequent bromide-induced rearrangement was effective for simple trisubst. allenes (four examples; Y 86-93%), with more complex derivs. (three different substituents) giving moderate yields and diastereoselectivity (Y 41-45%; d.r. 1.1:1 to 1.4:1; three examples), while use of iodide was less general and often favored formation of hydroxyketones. F.e. and a mechanistic discussion s. R. Sharma, L.J. Williams, Org. Lett. *2013*, *15* (9), 2202-5 [DOI: 10.1021/ol400749e].

Rearrangement ⋂

Hydrogen/Carbon Type OC ⋂ HC

Glass wool ←
Chiral 6,6′-bis(2,4,6-triisopropylphenyl)-1,1′-biindan-7,7′-diyl N-triflylphosphoramidate ←
Aldehydes from epoxides (epoxide) → CHCHO
with zeolites cf. *65*, 83s*69*; α-arylaldehydes from styrenes, incl. challenging chlorinated derivs., by regioselective ring opening with glass wool in the vapor phase s. G.K. Ramaswamy, A. Somasundaram, B.K. Kuppuswamy, M. Velayudham, J. Chin. Chem. Soc. *2013, 60* (1), 97-102 [DOI: 10.1002/jccs.201200269]; **asym. ring opening** with chiral 6,6′-bis(2,4,6-triisopropylphenyl)-1,1′-biindan-7,7′-diyl N-triflylphosphoramidate as Brønsted acid (e.e. 19-50%) s. M. Zhuang, H. Du, Org. Biomol. Chem. *2013, 11* (9), 1460-2 [DOI: 10.1039/c3ob27285k].

[(R,R)-N-Tosyl-1,2-diphenylethylenediamine](cymene)ruthenium(II)/ sodium tert-butoxide/acetone/isopropanol ←
Asym. formation of lactones from ketoaldehydes s. *83,* 80 ○

Oxygen/Nitrogen Type OC ⋂ ON

Sodum hydroxide *NaOH*
N-Protected β-carbobenzoxyamino-α,β-ethyleneketones ⋂
from N-protected 2-acetylenehydroxylamines *en route* to pyrimidines *s. 83*, 175

N-Methyl-N-[2-(1-methyl-3-imidazolio)ethyl]chloroformiminium triflimide ←
Triphenylphosphine/iodine *Ph_3P/I_2*
Thionyl chloride/β-cyclodextrin *$SOCl_2$/β-CD*
Beckmann rearrangement ⋂
update s. *64,* 83s*80*; rapid procedure for rearrangement of ketoximes with triphenylphosphine/iodine s. F. Xu, N.-G. Wang, Y.-P. Tian, Y.-M. Chen, W.-C. Liu, Synth. Commun. *2012, 42* (19), 3532-9 [DOI: 10.1080/00397911.2011.585270]; with $SOCl_2$/β-cyclodextrin in aq. media, also dehydration of aldoximes (NC⇑O; cf. *39*, 701s*81,82*), s. D. Patil, D. Dalal, ibid. *2013, 43* (1), 118-28 [DOI: 10.1080/00397911.2011.592747]; with the *ionic liquid-based Vilsmeier reagent*, N-methyl-N-[2-(1-methyl-3-imidazolio)ethyl]chloroformiminium triflimide, also f. applications [e.g. converting prim. alcohols to chlorides, HalC⇃↑O], s. A.A. Hullio, G.M. Mastoi, Chin. J. Chem. *2012, 30* (7), 1647-57 [DOI: 10.1002/cjoc.201280028]; **carboxylic acid amides** (incl. peptides) **from nitrones** *under continuous flow* on irradiation (cf. *42,* 135) s. Y. Zhang, M.L. Blackman, A.B. Leduc, T.F. Jamison, Angew. Chem., Int. Ed. *2013, 52* (15), 4251-5 [DOI: 10.1002/anie.201300504].

Oxygen/Carbon Type OC ∩ OC

Microwaves or Bromide ion [\\\\] *or Br⁻*
3-Oxetanones from 1,4-dioxaspiro[2.2]pentanes s. *83*, 43

Carbon/Carbon Type OC ∩ CC

Irradiation s. under 9-Mesityl-10-methylacridinium perchlorate

Sodium hydride/N-(9-anthracenylmethyl)cinchonidinium chloride ←
Flavanones from *o′*-hydroxychalcones
Asym. intramolecular oxa-Michael addition under ion-pairing catalysis

44.

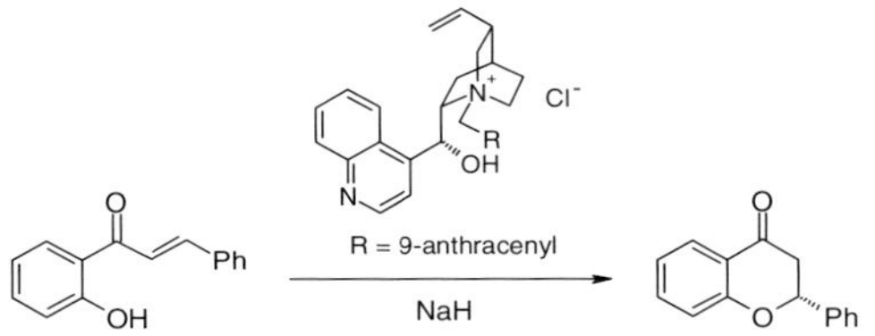

The first enantioselective ring closure of *o′*-hydroxychalcone to flavanone, which is problematic due to its reversibility in the absence of an additional activating group, is reported by asymmetric ion-pairing catalysis in homogeneous solution, which furthermore provides a chemical model of the chalcone isomerase reaction. **E:** *o′*-Hydroxychalcone (4.46 mmol) and NaH (60% in oil; 6 mol%) stirred in degassed toluene (20 ml) in a Schlenk vessel under argon, a soln. of N-(9-anthracenylmethyl)cinchonidinium chloride (4.7 mol%) in chloroform (7 ml; filtered through Al_2O_3) added to the resultant yellow soln./suspension, the mixture stirred for 4 h at room temp. (22°), quenched by addition of aq. HCl (2.4 *M*; 4 ml), the organic phase washed with water, and, after evaporation, the residue purified by column chromatography on silica → (+)-(2R)-product. Y 71% (e.e. 80%; e.e. 85% on the 0.4 mmol scale). Twelve further examples gave e.e. of 65-87% on the 0.4 mmol scale, although certain 2-polyaryl or benzannelated analogs gave e.e. of 38-52% (three examples). F.e.s. L. Hintermann, C. Dittmer, Eur. J. Org. Chem. *2012* (28), 5573-84 [DOI: 10.1002/ejoc.201200838]; asym. ring closure of activated *o′*-hydroxychalcones using cinchona alkaloids as chiral Brønsted bases s. C. Dittmer, G. Raabe, L. Hintermann, ibid. *2007* (35), 5886-98 [DOI: 10.1002/ejoc.200700682]; 4-chromanones (and flavanones) via asym. intramolecular oxa-Michael addition of α-(*o*-hydroxyaroyl)-α,β-ethylenecarboxylic acid *tert*-butyl esters using bifunctional thioureas as catalysts followed by acid-catalyzed decarbalkoxylation in one pot (cf. *74*, 135; *80*, 117) s. M.M. Biddle, M. Lin, K.A. Scheidt, J. Am. Chem. Soc. *2007*, *129* (13), 3830-1 [DOI: 10.1021/ja070394v]; application to a concise synthesis of the abyssinones and their evaluation as inhibitors of prostate cancer and MMP-2 expression s. R.L. Farmer, M.M. Biddle, A.E. Nibbs, X. Huang, R.C. Bergan, K.A. Scheidt, ACS Med. Chem. Lett. *2010*, *1* (8), 400-5 [DOI: 10.1021/ml100110x].

Polyoxometalate-supported (alkyldiarylphosphine)gold(I) complexes ←
Calcium, strontium or barium bis(trimethylsilyl)amide ←
Calcium triflimide/tetra-n-butylammonium hexafluorophosphate *$Ca(NTf_2)_2/Bu_4NPF_6$*
Catalytic cycloisomerization of unsatd. alcohols
update s. *36*, 148s*82*; mild, inexpensive procedure for preparing tetrahydro-furans or -pyrans from 4-ethylenealcohols with $Ca(NTf_2)_2/Bu_4NPF_6$ s. A.K. Diba, J.-M. Begouin, M. Niggemann, Tetrahedron Lett. *2012, 53* (49), 6629-32 [DOI: 10.1016/j.tetlet.2012.08.129]; cycloisomerization of acetylene- or allene-alcohols with Ca-, Sr- or Ba-bis(trimethylsilyl)amide, regioselectivity, s. C. Brinkmann, A.G.M. Barrett, M.S. Hill, P.A. Procopiou, S. Reid, Organometallics *2012, 31*

(20), 7287-97 [DOI: 10.1021/om3008663]; 2-vinyl- or 2-allyl-O-heterocyclics from 5-allene-alcohols with Fe(III)-trifluoroacetate or -tosylate, also N-heterocyclic analogs by intramolecular hydroamination (NC∩CC; cf. *70*, 147s*81*), s. M.S. Jung, W.S. Kim, Y.H. Shin, H.J. Jin, Y.S. Kim, E.J. Kang, Org. Lett. *2012, 14* (24), 6262-5 [DOI: 10.1021/ol303021d]; C-vinyl glycosides by gold(I)-catalyzed cycloisomerization of carbohydrate-based acetylenealcohols, also C-glycosides by intramolecular Michael addition (cf. *36*, 148s*82*), s. S. Redon, M. Wierzbicki, J. Prunet, Tetrahedron Lett. *2013, 54* (16), 2089-92 [DOI: 10.1016/j.tetlet.2013.02.022]; cycloisomerization of highly acid-sensitive, α-aryl-subst. tert. allenealcohols with *self-buffering* hybrid gold-polyoxometalate catalysts [prepared by covalent grafting - through tin - of (alkyldiarylphosphine)gold(I) complexes to organic polyoxometalates] s. N. Dupré, C. Brazel, L. Fensterbank, M. Malacria, S. Thorimbert, B. Hasenknopf, E. Lacôte, Chem. Eur. J. *2012, 18* (28), 12962-5 [DOI: 10.1002/chem.201201007]; 2-vinyl-O-heterocyclics from allenealcohols by gold(I)-catalyzed, *endo*-selective **asym. cycloisomerization** (cf. *36*, 148s*79*) at low (0.02 mol%) catalyst loading s. N. Cox, M.R. Uehling, K.T. Haelsig, G. Lalic, Angew. Chem., Int. Ed. *2013, 52* (18), 4878-82 [DOI: 10.1002/anie.201300174]; **organocatalyzed asym. intramolecular Michael addition** of 4-(2-hydroxyethoxy)-2,5-cyclohexadienones with a chiral 1,2-diamine and N-Boc-L-Pro or N-Boc-L-Phe s. W. Wu, X. Li, H. Huang, X. Yuan, J. Lu, K. Zhu, J. Ye, ibid. *52* (6), 1743-7 [DOI: 10.1002/anie.201206977].

2-Phenylmalononitrile s. under 9-Mesityl-10-methylacridinium perchlorate *$PhCH(CN)_2$*
Carbon tetrabromide s. under Ph_3P *CBr_4*

Silica gel *SiO_2*
4-α-Acoxypyrimidines from 5-acoxy-4,6-di(ene)-1,4,5,6-tetrahydropyrimidines ∩
s. *83*, 99

Triphenylphosphine/carbon tetrabromide *Ph_3P/CBr_4*
α-Aryloxo compds. from 2-aryl-2-ethylenealcohols via metal-free isomerization C=C(Ar)CHOH → CHC(Ar)CO

45.

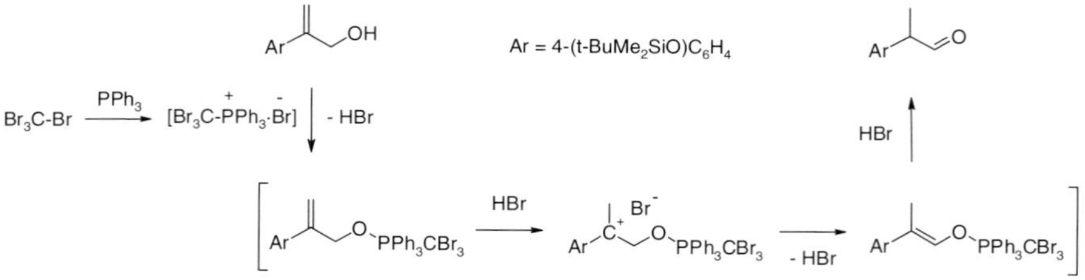

in one pot. Carbon tetrabromide (*20 mol%*) and triphenylphosphine (*20 mol%*) added to a soln. of 2-[4-(*tert*-butyldimethylsiloxy)phenyl]prop-2-en-1-ol (0.2 mmol) in dry methylene chloride (10 ml) at room temp., the mixture stirred for 4 h (TLC), quenched with ice-water, extracted with methylene chloride, and purified chromatographically → 2-[4-(*tert*-butyldimethylsilyloxy)phenyl]-propanal. Y 96%. Using stoichiometric CBr_4/Ph_3P the expected allylic bromides were formed as major products but use of *catalytic* reagents provided an efficient and atom-economical conversion of 2-aryl-allylic prim. and sec. alcohols to saturated aldehydes/ketones via internal redox reaction (twelve examples; Y 78-97%) in the presence of silyl and 4-methoxybenzyl ethers. Aliphatic allylic alcohols (geraniol, methallyl alcohol, 2-methylbut-1-en-3-ol) gave only minor conversion to the corresponding allyl bromides, with the majority of the substrates recovered unchanged. The method was used in a short synthesis of ibuprofen. F.e., optimization and substrate prepn. s. W. Gong, Y. Liu, J. Xue, Z. Xie, Y. Li, Chem. Lett. *2012, 41* (12), 1597-9 [DOI: 10.1246/cl.2012.1597].

9-Mesityl-10-methylacridinium perchlorate/2-phenylmalononitrile/irradiation ←
Cyclic ethers from ethylenealcohols ○

46.
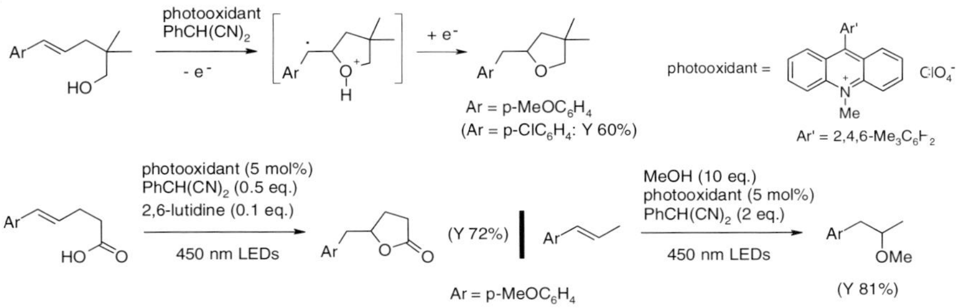

Employment of catalytic amounts of [commercially-available] 9-mesityl-10-methylacridinium perchlorate and 2-phenylmalononitrile as a redox-cycling source of H-atoms has been shown to mediate **anti-Markovnikov intramolecular hydroetherification** with complete (>20:1) regioselectivity, complementing traditional Markovnikov-based Brønsted acid catalyzed reactions. **E:** (E)-5-(4-Methoxyphenyl)-2,2-dimethylpent-4-en-1-ol (100 mg), 2-phenylmalononitrile (0.5 eq.) and 9-mesityl-10-methylacridinium perchlorate (0.05 eq.) added to a flame-dried vial, anh. 1,2-dichloroethane added (to 0.5 *M*), the vial sealed, the mixture degassed by the freeze pump thaw method, irradiated, with stirring, using a 450 nm bulb for 48 h (TLC), eluted through a short plug of silica (with methylene chloride), the eluent washed with 0.1 *M* NaOH, the aq. layer back-extracted with methylene chloride, the combined extracts dried (Na_2SO_4), concentrated, and purified by flash chromatography → 2-(4-methoxybenzyl)-4,4-dimethyltetrahydrofuran. Y 80%. Use of the acridinium salt renders the method more general than the prior art utilizing cyanoarene photooxidants, which was restricted to 1,1-diarylethylene derivs. This method is applicable to substrates possessing alkene groups that bear electron-diverse aryl groups or may be dialkylated, six further tetrahydrofurans being obtained in yields of 41-82% (d.r. 1.1:1 to 5:1 where applicable), a tetrahydropyran in 68% yield (d.r. 2.5:1) and an oxepane in 46% yield (d.r. 1.2:1). The method was extended to γ-lactone formation from an ethylenecarboxylic acid in the presence of 2,6-lutidine (Y 72%) and to intermolecular addition of methanol to anethole (Y 81%), both with complete anti-Markovnikov selectivity. F.e.s. D.S. Hamilton, D.A. Nicewicz, J. Am. Chem. Soc. *2012*, *134* (45), 18577-80 [DOI: 10.1021/ja309635w].

N-(9-Anthracenylmethyl)cinchonidinium chloride s. under Sodium hydride ←

Iron(III) trifluoroacetate or tosylate *$Fe(OCOCF_3)_3$ or $Fe(OTs)_3$*
Cycloisomerization of 5-allenealcohols s. *36*, 148s*83*

Dichloro[2,2′-bis(diphenylphosphino)-1,1′-binaphthyl]nickel(II)/ lithium triethylhydridoborate *$NiCl_2$(binap)/$LiBHEt_3$*
Acylsilanes from 2-ethylene-1,1-hydroxysilanes s. *83*, 270 CHCHC(O)Si≤

(p-Cymene)ruthenium(II) guanidinate complexes *[Ru(II)]*
Ketones from 2-ethylenealcohols C=C-CHOH → CHCHCO
Ruthenium-catalyzed isomerization in the absence of base

47.
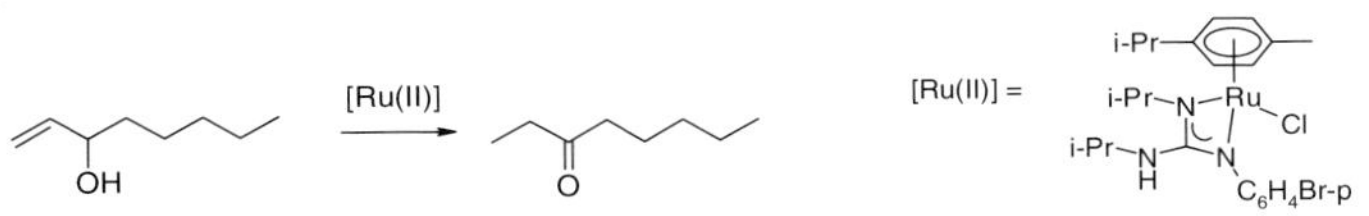

The first examples of ruthenium complexes containing asymmetrical monoanionic guanidinate ligands are reported and notably shown to be effective for redox isomerizations of allylic alcohols without the assistance of base (cf. *25*, 125s*69,70*). **E:** The ruthenium guanidinate complex [RuCl[κ²N,N′-C(N-C_6H_4Br-*p*)(NPr-*i*)NHPr-*i*](η⁶-*p*-cymene)] (1 mol%) added to a soln. of oct-1-en-

3-ol (4 mmol) in THF (20 ml) in a sealed tube under N_2, and the mixture stirred at 80° for 10 min → 3-octanone. Y >99% by GC. The same yield was obtained with 0.1 mol% catalyst after 1 h. An analogous amidinate complex showed much lower catalytic activity, as did reaction in the presence of an acid (1 mol% HCl), suggesting the pendant NHPr-*i* group facilitates the reaction by deprotonation of the allyl alcohol. Five further aliphatic allyl alcohols gave yields of 96 to >99%, generally within 10-20 min (60 min for 3-penten-2-ol bearing a disubst. C═C bond), while four examples of 1-aryl-2-propenols gave yields of 80-99% within 3-24 h, an electron-donating group (OMe) leading to increased reactivity whereas F or Cl lowered it. F.e., also prepn. and characterization of the catalysts, s. R. García-Álvarez, F.J. Suárez, J. Díez, P. Crochet, V. Cadierno, A. Antiñolo, R. Fernández-Galán, F. Carrillo-Hermosilla, Organometallics *2012, 31* (23), 8301-11 [DOI: 10.1021/om3009124].

Polyamine-type dendrimer-encapsulated palladium(II) complexes ←
Lactones by catalytic cycloisomerization of unsatd. carboxylic acids ○
update s. *35,* 85s*82*; γ-alkylidene-γ-lactones from γ,δ-acetylenecarboxylic acids with a polyamine-type dendrimer-encapsulated Pd(II) complex (prepared by complexation of $PdCl_2$ with internal *tert*-amino groups of a polyamino-type dendrimer) s. Z. Maeno, T. Mitsudome, T. Mizugaki, K. Jitsukawa, K. Kaneda, Heterocycles *2012, 86* (2), 947-54 [DOI: 10.3987/com-12-s(n)96]; *exo*-selective **asym. cycloisomerization of allene-carboxylic acids or -alcohols** (cf. *35,* 85s*48*) using a silver(I) catalyst possessing an anionic chiral ligand derived from an oxophosphorus(V) acid, also asym. intramolecular hydroamination of N-protected alleneamines (NC∩CC; cf. *72,* 185s*80*), s. J.L. Arbour, H.S. Rzepa, J. Contreras-García, L.A. Adrio, E.M. Barreiro, K.K. Hii, Chem. Eur. J. *2012, 18* (36), 11317-24 [DOI: 10.1002/chem.201200547].

Exchange ↓↑

Hydrogen ↑ OC ↓↑ H

Irradiation s. under Bi(O)Br and $Ru(bpy)_3Cl_2$ ///
Microwaves s. under MnO_2 [\\\\]
Lithium chloride s. under $PdBr_2$ *LiCl*

Chiral amines/aroyl peroxides ←
Syntheses via organocatalyzed asym. α-benzoyloxylation s. *83*, 59 H → OBz

Silver carbonate s. under $Pd(OAc)_2$ *Ag_2CO_3*

Copper(I) bromide/tert-butyl hydroperoxide *CuBr/t-BuOOH*
3-Acylindoles from *o*-acetylene-*tert*-amines ○

48.

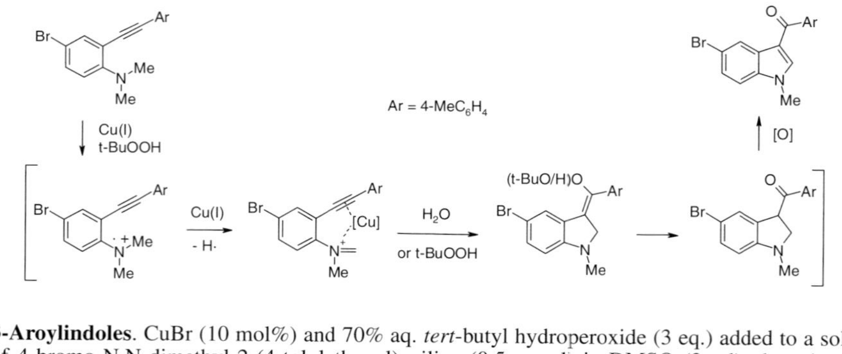

3-Aroylindoles. CuBr (10 mol%) and 70% aq. *tert*-butyl hydroperoxide (3 eq.) added to a soln. of 4-bromo-N,N-dimethyl-2-(4-tolylethynyl)aniline (0.5 mmol) in DMSO (2 ml), the mixture heated at 80° for 4 h, diluted with water, extracted with ethyl acetate, concentrated *in vacuo*, and purified chromatographically → 5-bromo-1-methyl-3-(4-toluoyl)indole. Y 70%. This **copper(I)-**

catalyzed oxidative ring closure was effective for generation of both C-C and C-O bonds without requirement for expensive palladium catalysts, apparently via a radical mechanism. The method was successful for electron-diverse N,N-dimethyl-*o*-(arylethynyl)anilines (twelve examples; Y 60-80%) with lowest yields obtained for electron-poor substrates. F.e. and optimization s. A. Gogoi, S. Guin, S.K. Rout, B.K. Patel, Org. Lett. *2013, 15* (8), 1802-5 [DOI: 10.1021/ol400692b]; **by palladium-copper cocatalyzed oxidative ring closure** with $PdBr_2$/CuI/LiCl/*t*-BuOOH s. X.-F. Xia, L.-L. Zhang, X.-R. Song, Y.-N. Niu, X.-Y. Liu, Y.-M. Liang, Chem. Commun. *2013, 49* (14), 1410-2 [DOI: 10.1039/c2cc37805a].

Copper(I) iodide s. under $PdBr_2$ *CuI*
Magnesium triflate s. under Chiral tetracyclic 1,2,4-triazolylidenes *$Mg(OTf)_2$*

Zinc bromide/hydrogen peroxide *$ZnBr_2/H_2O_2$*

Keto from methylene groups $CH_2 \rightarrow CO$

aryl ketones by benzylic oxidation, update s. *79,* 69s*82*; *under zinc catalysis* with $ZnBr_2/H_2O_2$ in trifluoroacetic acid/dioxane, also heteroaromatic analogs, s. X.-F. Wu, Tetrahedron Lett. *2012, 53* (45), 6123-6 [DOI: 10.1016/j.tetlet.2012.08.149]; with recyclable MnO_2 nanoparticles under microwave irradiation, also arylcarboxylic acid esters from benzyl ethers, s. B. Nammalwar, C. Fortenberry, R.A. Bunce, S.K. Lageshetty, K.D. Ausman, ibid. *2013, 54* (15), 2010-3 [DOI: 10.1016/j.tetlet.2013.02.009]; benzylic and allylic oxidation with $Rh_2(OAc)_4$(NHC) complexes as catalyst, K_2CO_3 as base and *t*-BuOOH (aerobically or anaerobically) as oxidant s. J.A.S. Coelho, A.F. Trindade, R. Wanke, B.G.M. Rocha, L.F. Veiros, P.M.P. Gois, A.J.L. Pombeiro, C.A.M. Afonso, Eur. J. Org. Chem. *2013* (8), 1471-8 [DOI: 10.1002/ejoc.201201300]; aerobic benzylic oxidation of 2-benzylpyridines with $FeCl_3$ as catalyst *under gas-liquid continuous flow* at high temperature and air pressure with propylene carbonate as solvent s. B. Pieber, C.O. Kappe, Green Chem. *2013, 15* (2), 320-4 [DOI: 10.1039/c2gc36896j].

Sodium fluoroborate s. under Chiral tetracyclic 1,2,4-triazolylidenes *$NaBF_4$*
N-(Pyrid-3-yl)-2,4,6-tris(4-tert-butylphenyl)pyridinium fluoroborate s. under Palladium(II) acetate ←
Acetone s. under $Pd(OAc)_2$ *Me_2CO*
p-Benzoquinone s. under Tetrakis(acetonitrile)palladium(II) bis(fluoroborate) *BQ*

1,3-Dimesitylimidazolidin-2-ylidene/oxygen *NHC/O_2*
Bicyclic 1,2,4-triazolylidenes/tris(2,2′-bipyrazinyl)ruthenium(II) bis(hexafluorophosphate)/oxygen ←

Carboxylic acid esters from aldehydes $CHO \rightarrow COOR$

update s. *44,* 149s*82*; [hetero]arylcarboxylic acid esters by aerobic oxidation *with carbene catalysis* using a bicyclic 1,2,4-triazolylidene *under cooperative catalysis* with tris(2,2′-bipyrazinyl)-ruthenium(II) bis(hexafluorophosphate), also arylcarboxylic acids without the cocatalyst, s. J. Zhao, C. Mück-Lichtenfeld, A. Studer, Adv. Synth. Catal. *2013, 355* (6), 1098-106 [DOI: 10.1002/adsc.201300034]; general procedure under oxygen with 1,3-dimesitylimidazolidin-2-ylidene as catalyst s. L. Möhlmann, S. Ludwig, S. Blechert, Beilstein J. Org. Chem. *2013, 9*, 602-7 [DOI: 10.3762/bjoc.9.65]; (E)-α,β-ethylenecarboxylic acid esters from α,β-ethylene-α-halogen-aldehydes with 1,3-dimesitylimidazol-2-ylidene as catalyst s. X.-B. Wang, X.-L. Zou, G.-F. Du, Z.-Y. Liu, B. Dai, Tetrahedron *2012, 68* (32), 6498-503 [DOI: 10.1016/j.tet.2012.05.109]; *organocatalyzed, non-carbenoid* procedure with *tert*-butyl hydroperoxide and Bu_4NI or Bu_4NBr as catalyst, also N-acyldisulfonylamines by oxidative amidation, s. B. Tan, N. Toda, C.F. Barbas III, Angew. Chem., Int. Ed. *2012, 51* (50), 12538-41 [DOI: 10.1002/anie.201205921]; with recyclable 4-acylamino-2,2,6,6-tetramethyl-N-oxopiperidinium fluoroborate as oxidant and pyridine as base cf. C.B. Kelly, M.A. Mercadante, R.J. Wiles, N.E. Leadbeater, Org. Lett. *2013, 15* (9), 2222-5 [DOI: 10.1021/ol400785d]; oxidative esterification of benzaldehyde and deactivated derivs. with NBS/pyridine or K_2CO_3 as base s. M.K. Agrawal, S. Adimurthy, P.K. Ghosh, Synth. Commun. *2012, 42* (19), 2931-6 [DOI: 10.1080/00397911.2011.572219]; **kinetic resolution of tert. alcohols**, notably 3-subst. 3-hydroxyoxindoles, by asym. esterification with cinnamaldehyde in the presence of a chiral tetracyclic 1,2,4-triazolylidene as catalyst in the presence of $MnO_2/Mg(OTf)_2/NaBF_4$ s. S. Lu, S.B. Poh, W.-Y. Siau, Y. Zhao, Angew. Chem., Int. Ed. *2013, 52* (6), 1731-4 [DOI: 10.1002/anie.201209043].

Chiral tetracyclic 1,2,4-triazolylidenes/manganese dioxide/magnesium triflate/ sodium fluoroborate ←

Kinetic resolution of tert. alcohols by oxidative esterification s. *44,* 149s*83* CHO → COOR

Mutant cytochrome P_{450} monooxygenase/alcohol dehydrogenase ←

Oxidation of hydrocarbons ←

s. *79,* 69s*81*; *50,* 53s83; **cycloalkanones** from cycloalkanes *under dual enzymatic catalysis* with a hydroxylating P_{450} BM-3 mutant monooxygenase and an alcohol dehydrogenase from *Lactobacillus kefir* under O_2 in water with $NAPD^+$ as cofactor and isopropanol as reductant s. S. Staudt, E. Burda, C. Giese, C.A. Müller, J. Marienhagen, U. Schwaneberg, W. Hummel, K. Drauz, H. Gröger, Angew. Chem., Int. Ed. *2013, 52* (8), 2359-63 [DOI: 10.1002/anie.201204464]; oxidation of alkanes to ketones (and C-C coupling) by visible light-induced activation of C-H bonds with chlorine radical [generated from Bi(O)Br/TiO_2 and HCl] s. R. Yuan, S. Fan, H. Zhou, Z. Ding, S. Lin, Z. Li, Z. Zhang, C. Xu, L. Wu, X. Wang, X. Fu, ibid. *52* (3), 1035-9 [DOI: 10.1002/anie.201207904].

4-Acylamino-2,2,6,6-tetramethyl-N-oxopiperidinium fluoroborate/pyridine ←

Carboxylic acid esters from aldehydes s. *44,* 149s*83* CHO → COOR

Trifluoroacetic anhydride s. under Dimethyl sulfoxide *$(CF_3CO)_2O$*

tert-Butyl hydroperoxide s.a. under CuBr, Rhodium(II) N-heterocyclic carbene complexes and Palladium(II) bromide *t-BuOOH*

tert-Butyl hydroperoxide/tetra-n-butylammonium halide *t-BuOOH/Bu_4NHal*

Carboxylic acid esters from aldehydes s. *44,* 149s*83*

tert-Butyl hydroperoxide/tetra-n-butylammonium iodide *t-BuOOH/Bu_4NI*

Arylcarboxylic acid benzyl esters from methylarenes and alkylarenes $ArCH_3 + RCH_2Ar' \rightarrow ArCOOCH(R)Ar'$

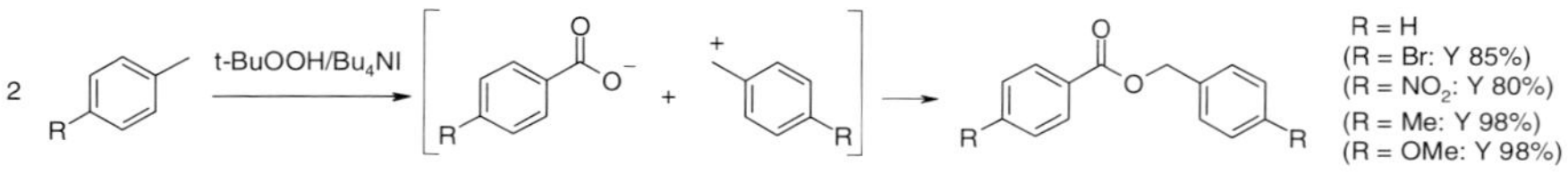

49.

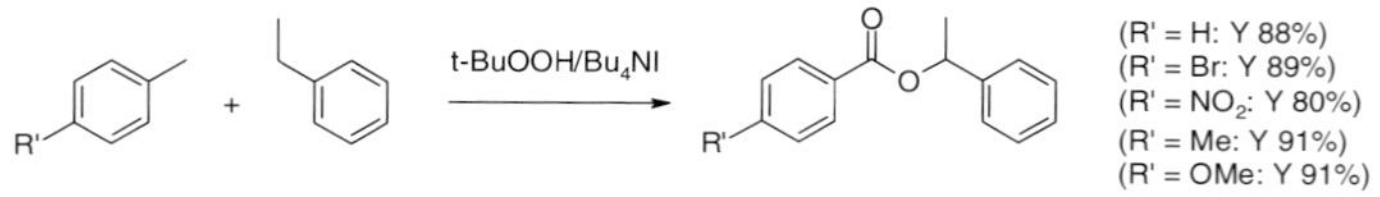

A novel **metal-free dehydrogenative coupling** is reported, involving *four* sp^3 C-H bond activations. **E:** An aq. soln. of *tert*-butyl hydroperoxide (70%; 6 eq.) added to a mixture of Bu_4NI (10 mol%) and toluene (1 ml), the mixture heated at 80° for 6 h (TLC), cooled to room temp., admixed with ethyl acetate, the ethyl acetate layer washed successively with 5% aq. $NaHCO_3$ and 5% aq. $Na_2S_2O_3$, dried (Na_2SO_4), solvent evaporated under reduced pressure, and the residue purified on a silica gel column → benzyl benzoate. Y 95%. Nine further examples of self-coupled products were obtained in yields of 80-98% from a variety of *p*-functionalized toluenes, *m*-methoxytoluene, *o*-, *m*- or *p*-xylene or mesitylene, with monoesters obtained exclusively for the latter examples. Selective oxidation of methylarenes in the presence of ethylarenes was also achieved, the former being more susceptible to oxidation to give the acids and the latter forming more stable carbocations, giving cross-coupled products in yields of 80-91% (ten examples); reaction of isopropylbenzene with toluene or its *p*-chloro-deriv. gave yields of 75% and 71%, the lower yield possibly being due to steric factors. F.e. and optimization s. G. Majji, S. Guin, A. Gogoi, S.K. Rout, B.K. Patel, Chem. Commun. *2013*, *49* (29), 3031-3 [DOI: 10.1039/c3cc40832a]; **from aldehydes** and alkylarenes under similar conditions (cf. *80*, 74) s. J. Huang, L.-T. Li, H.-Y. Li, E. Husan, P. Wang, B. Wang, ibid. *2012*, *48* (82), 10204-6 [DOI: 10.1039/c2cc35450k]; carboxylic acid benzyl esters **from carboxylic acids** (cf. *80*, 110) and alkylarenes similarly, notably incl. O-benzylation of N-Boc amino acids, s. J. Feng, S. Liang, S.-Y. Chen, J. Zhang, S.-S. Fu, X.-Q. Yu, Adv. Synth. Catal. *2012*, *354* (7), 1287-92 [DOI: 10.1002/adsc.201100920].

Dimethyl sulfoxide/trifluoroacetic anhydride *DMSO/($CF_3CO)_2O$*
2-α-Functionalization of indoles ←

50. 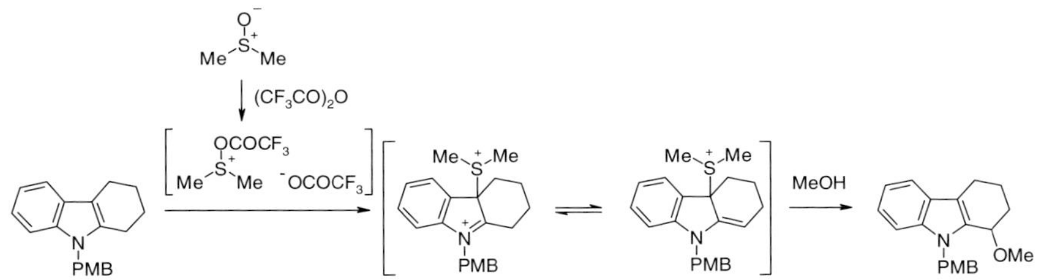

Trifluoroacetic anhydride (1 eq.) added under argon to a soln. of startg. tetrahydrocarbazole (0.4 mmol) and dimethyl sulfoxide (1 eq.) in methylene chloride (2 ml), stirred at -40° for 30 min, methanol (5 eq.) added, stirring continued for 10 min at the same temp., neutralized with satd. aq. $NaHCO_3$, extracted with methylene chloride, and purified by column chromatography → 1-methoxy-9-(4-methoxybenzyl)-1,2,3,4-tetrahydro-9*H*-carbazole. Y 95%. Better results were obtained with DMSO than with a variety of diaryl sulfoxides or methyl phenyl sulfoxide. Six further examples gave 2-α-methoxy-derivs. in yields of 61-90%, while seven examples with *p*-cresol, Me_3SiN_3, $BnNH_2$, [methyl, vinyl or allyl] Grignards or N-methylindole as nucleophiles gave yields of 70-100%. F.e.s. M. Tayu, K. Higuchi, M. Inaba, T. Kawasaki, Org. Biomol. Chem. *2013*, *11* (3), 496-502 [DOI: 10.1039/c2ob26944a]; s.a. K. Higuchi, M. Tayu, T. Kawasaki, Chem. Commun. *2011*, *47* (23), 6728-30 [DOI: 10.1039/c1cc11645b].

Phenyl iodosoacetate s. under $Pd(OAc)_2$ and Tetrakis(acetonitrile)palladium(II) bis(fluoroborate) *$PhI(OAc)_2$*

1,1,1-Triacetoxy-1,2-benziodoxol-3-one ←
5-(*o*-Iodobenzoyloxy)-Δ^3-2-pyrrolones from pyrroles ←

51. 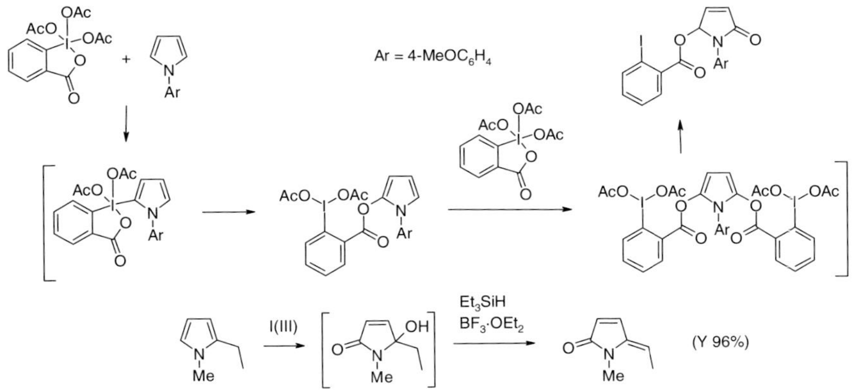

A novel method is described for the controlled oxidation of pyrroles that minimizes decomposition. **E:** 1-(4-Methoxyphenyl)pyrrole (1 eq.) added slowly to a soln. of Dess-Martin periodinane (2.5 eq.) in methylene chloride (0.1 *M*) at 0°, the mixture stirred for 2 h, diluted with water, quenched with $Na_2S_2O_5$, extracted with methylene chloride, washed with satd. aq. $NaHCO_3$, concentrated *in vacuo*, and purified chromatographically → 5-(*o*-iodobenzoyloxy)-1-(4-methoxyphenyl)-Δ^3-2-pyrrolone. Y 93%. Slow addition of the pyrrole to 2.5 eq. of the oxidant minimized the formation of decomposition products and unexpectedly afforded the iodobenzoyloxy deriv. (traces of the expected acetoxy derivs. were also observed). Interestingly, use of only 1 eq. of oxidant simply gave a reduced yield of the product (cf. monooxidation product). The method gave good yields for N-aryl- or N-alkyl-pyrroles (Y 81-96%; seven examples), with pyrrole itself giving a 56% yield. Oxidation of N-(2-hydroxyethyl)pyrrole proceeded with retention of the prim. alcohol (Y 61%), while N-methyl-2-alkylpyrroles gave unstable 5-hydroxy-5-alkylpyrrolones, attempted *in situ* reduction of which (Et_3SiH/BF_3) afforded **5-alkylidene-Δ^3-2-pyrrolones** (two examples; Y 89-96%). F.e., optimization and reactions of the products s. J.K. Howard, C.J.T. Hyland, J. Just, J.A. Smith, Org. Lett. *2013, 15* (7), 1714-7 [DOI: 10.1021/ol400491p].

N-Bromosuccinimide/pyridine or potassium carbonate — *NBS/py or K_2CO_3*
Carboxylic acid esters from aldehydes s. *44*, 149s*83* — CHO → COOR

Titanium dioxide s. under Bi(O)Br — *TiO_2*
2-Dicyclohexylphosphino-2′,4′,6′-triisopropylbiphenyl s. under $Pd(OAc)_2$ — *XPhos*

Bismuth(III) oxybromide/titanium dioxide/hydrogen chloride/irradiation — *$Bi(O)Br/TiO_2/HCl/$///*
Ketones from hydrocarbons s. *79*, 69s*83* — CH_2 → CO

Oxygen s. under 1,3-Dimesitylimidazolidin-2-ylidene, Bicyclic 1,2,4-triazolylidenes, $Pd(OAc)_2$, $PdBr_2$ and Dichloro(dipyrido[3,2-a:2′,3′-c]phenazine)platinum(II) — *O_2*
Hydrogen peroxide s. under $ZnBr_2$ — *H_2O_2*
Sulfuric acid s. under Dichloro(dipyrido[3,2-a:2′,3′-c]phenazine)platinum(II) — *H_2SO_4*
Potassium peroxymonosulfate s. under I_2 — *$KHSO_5$*
Potassium persulfate s. under Palladium(II) acetate — *$K_2S_2O_8$*

Pyridinium chlorochromate/periodic acid — *$[C_5H_5NH][CrO_3Cl]/H_5IO_6$*
1,(n+1)-Diacoxy-2,n-diketones from dienes
via oxidative ring opening of 2-α,2′-α-diacoxy-O-heterocyclics

52.

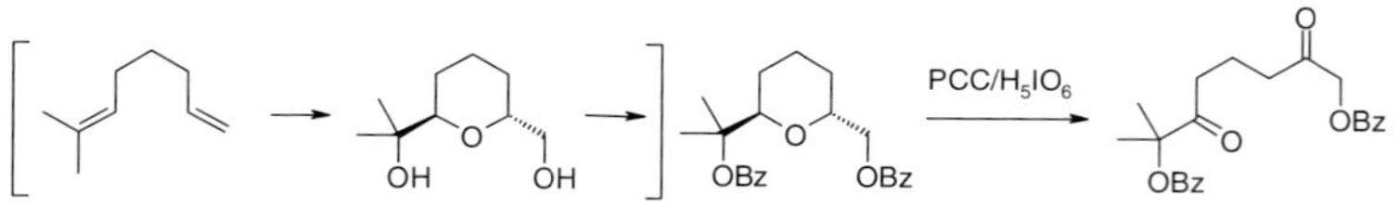

A range of tetrahydro-furan or -pyran diols have been prepared from 1,5- or 1,6-dienes, typically by well-established ruthenium- or osmium-catalyzed oxidative ring closures (cf. *61*, 71, 71s*62*), and, after O-acylation, subjected to oxidation with chlorochromatoperiodate to afford the title compounds, which are difficult to prepare directly. **E:** Pyridinium chlorochromate (0.01 *M* stock soln. in acetonitrile; 4 mol%) added to a suspension of H_5IO_6 (4 eq.) in acetonitrile at room temp. with vigorous stirring, after 5 min, a soln. of startg. *trans*-tetrahydropyran (0.068 mmol) in acetonitrile added (to give a final concentration of 0.05 *M*), after complete consumption of the starting material (TLC; typically 6-8 h), ethanol (excess) added, stirring continued until the color of the soln. turned from yellow to green (ca. 5 min), silica (excess) added, the solvent evaporated under reduced pressure, the resulting powder loaded onto a silica gel column, and subjected to flash chromatography → 7-methyl-2,6-dioxooctane-1,7-diyl dibenzoate. Y 95% (Y 83% from the *cis*-isomer). Eight further examples gave yields of 61-97%, incl. an acetoxy and a tosyloxy-deriv. and a bridged bicyclic ether. Reaction of an N-benzoylmorpholine gave a low yield (26%), but it is anticipated that this could be improved by tuning of the N-protective group. Two examples of acylated *cis*-reticulatacins gave yields of 30-40%. F.e. and conversion to 2-α,6-α-dioxypyridines s. V. Piccialli, S. D'Errico, N. Borbone, G. Oliviero, R. Centore, S. Zaccaria, Eur. J. Org. Chem. *2013* (9), 1781-9 [DOI: 10.1002/ejoc.201201554].

Iodine/potassium peroxymonosulfate *$I_2/KHSO_5$*
1,3-Dioxolanes from terminal ethylene derivs. s. *58,* 78s*83* ○

Periodic acid s. under Pyridinium chlorochromate *H_5IO_6*
Tetra-n-butylammonium halide s. under t-BuOOH *Bu_4NHal*
Hydrogen chloride s. under Bi(O)Br *HCl*
Manganese dioxide s.a. under Chiral tetracyclic 1,2,4-triazolylidenes *MnO_2*

Manganese dioxide nanoparticles/microwaves *MnO_2/*
Aryl ketones from alkylarenes s. *79,* 69s*83* CH_2 → CO

[6-(Di-tert-butylphosphinomethyl)-2,2′-bipyridyl]carbonyl(chloro)(hydrido)-ruthenium(II)/sodium hydroxide/water ←
Carboxylic acids from prim. alcohols using water as oxygen source under inert atmosphere CH_2OH → COOH

53.

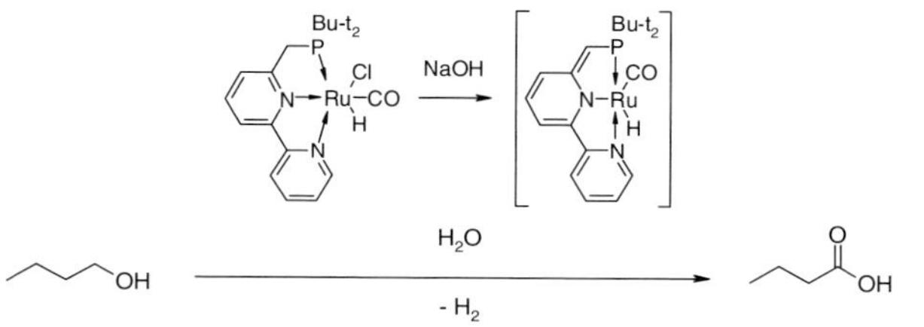

The first example of a homogeneously catalyzed conversion of prim. alcohols to carboxylic acids without oxidant or hydrogen acceptor, with liberation of hydrogen gas, is reported, as an extension of previous work on ruthenium-catalyzed dehydrogenative coupling of prim. alcohols with amines to afford amides (s. *73,* 160) and dehydrogenative homocoupling of prim. alcohols to afford esters [J. Zhang, D. Milstein et al., J. Am. Chem. Soc. *2005, 127* (31), 10840-1 [DOI: 10.1021/ja052862b]; cf. *75,* 99s*82*]. **E:** 1-Butanol (10 mmol), (6-*t*-Bu_2PCH_2-bipy)Ru(H)Cl(CO) (0.2 mol%) and NaOH (1.1 eq.) combined in a flame-dried Schlenk flask in a glove-box under N_2, the flask taken out of the glove-box, degassed, deionized water (2 ml) added, the mixture kept under a heavy flow of argon with occasional shaking for ca. 10 min, then refluxed under argon for 18 h, water (10 ml) added, the mixture extracted with ether, the aq. phase acidified with 5 *N* HCl, extracted with ethyl acetate, the combined extracts washed with brine, dried (Na_2SO_4), and evaporated under reduced pressure → butanoic acid. Y 84%. The pre-catalyst is air-stable, gives high turnovers and yields (61-100%; twelve examples) from aliphatic or benzylic alcohols, and reaction proceeds under mild conditions. Cinnamyl alcohol gave a 5:7:2 mixture of cinnamic acid, 3-phenylpropanol and 3-phenylpropanoic acid (68% conversion), while 4-aminobutanol gave 2-pyrrolidone in 85% yield by NMR. The presence of O_2 is detrimental to the reaction. F.e., also labelling studies (indicating that the incorporated oxygen is derived from water) and DFT calculations, s. E. Balaraman, E. Khaskin, G. Leitus, D. Milstein, Nat. Chem. *2013, 5* (2), 122-5 [DOI: 10.1038/nchem.1536].

Tris(2,2′-bipyrazinyl)ruthenium(II) bis(hexafluorophosphate) s. under Bicyclic 1,2,4-triazolylidenes ←

Dichlorotris(2,2′-bipyridyl)ruthenium(II)/irradiation *$Ru(bpy)_3Cl_2$/*
Directed *o*- or *p*-benzoyloxylation s. *77,* 94s*83* H → OBz

Rhodium(II) N-heterocyclic carbene complexes/tert-butyl hydroperoxide/potassium carbonate ←
Aryl ketones from alkylarenes s. *79,* 69s*83* CH_2 → CO

Palladium(II) acetate/silver carbonate/oxygen *$Pd(OAc)_2/Ag_2CO_3/O_2$*
(E)-Enolesters from terminal ethylene derivs. and carboxylic acids CH=C → C(OAc)=C
Regio- and stereo-selective palladium-catalyzed dehydrogenative O-vinylation

54.

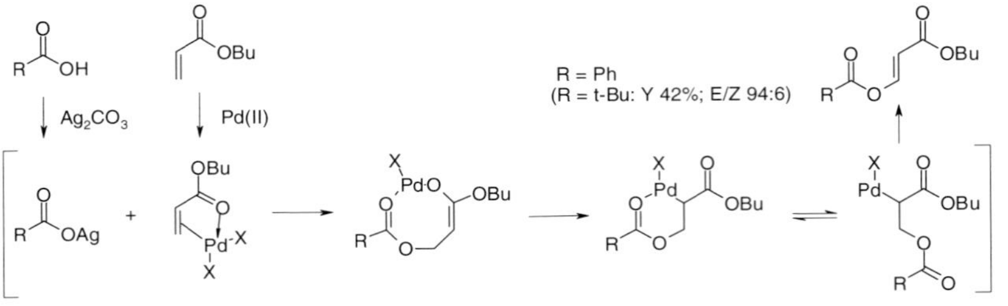

A soln. of benzoic acid (0.3 mmol), *n*-butyl acrylate (5 eq.), $Pd(OAc)_2$ (10 mol%) and Ag_2CO_3 (2 eq.) in acetonitrile (1.5 ml) refluxed at 80° under O_2 (1 atm.) for 24 h (TLC), the mixture cooled to room temp., diluted with ethyl acetate, filtered, dried (Na_2SO_4), solvent removed under reduced pressure, and the residue purified by flash chromatography on silica gel → 3-butoxy-3-oxoprop-1-enyl benzoate. Y 91% (E/Z 92:8). This provides an alternative to hydrocarboxylation of 1-alkynes for preparation of enolesters, offering good yields and excellent chemo- and regio-selectivity. Twelve further examples with *n*-butyl acrylate and a variety of (het)arylcarboxylic acids gave yields of 42-93% (E/Z 91:9 to >98:2), albeit using 20 mol% $Pd(OAc)_2$ for benzoic acids bearing electron-withdrawing groups, while reaction with cinnamic, cyclopropanecarboxylic or pivalic acid gave yields of 41-64% (E/Z 94:6 to >98:2). The method was extended to reaction with N,N-diethylacrylamide (Y 76%; E/Z >98:2), phenyl vinyl ketone (Y 33% due to polymerization; E/Z >98:2) and styrene (Y 58%; E/Z 67:33), while 4-benzyloxy- and 4-phenyl-1-butene gave the corresponding benzoyloxy-2-ethylenes (Y 65%, 72%). F.e.s. D. Yang, S. Ding, J. Huang, K. Zhao, Chem. Commun. *2013, 49* (12), 1211-3 [DOI: 10.1039/c2cc37779a].

Palladium(II) acetate/phenyl iodosoacetate *$Pd(OAc)_2/PhI(OAc)_2$*
Directed *o*-acoxylation H → OAc
update s. *77,* 94s*82*; of 1-aryl-3,4-dihydroisoquinolines with phenyl iodosoacetate as reactant/ oxidant and $Pd(OAc)_2$ as catalyst with added acetic anhydride, also ***o*-alkoxylation**, s. B.V.S. Reddy, N. Umadevi, G. Narasimhulu, J.S. Yadav, Tetrahedron Lett. *2012, 53* (45), 6091-4 [DOI: 10.1016/ j.tetlet.2012.08.124]; ***o*- or *p*-benzoyloxylation** of alkylarenes or phenolethers with dibenzoyl peroxide and $Ru(bpy)_3Cl_2{\cdot}6H_2O$ *under visible-light photoredox catalysis* at room temperature s. H. Rao, P. Wang, C.-J. Li, Eur. J. Org. Chem. *2012* (33), 6503-7 [DOI: 10.1002/ejoc.201201093]; benzylic-type acoxylation of N-protected indoles with phenyl iodosoacetate and tetra-*n*-butylammonium iodide as catalyst in methylene chloride/acetic acid s. H. Zaimoku, T. Hatta, T. Taniguchi, H. Ishibashi, Org. Lett. *2012, 14* (23), 6088-91 [DOI: 10.1021/ol302983t].

Palladium(II) acetate/2-dicyclohexylphosphino-2′,4′,6′-triisopropylbiphenyl/acetone/ potassium carbonate ←
Carboxylic acid esters from aldehydes and alcohols CHO → COOR

55.

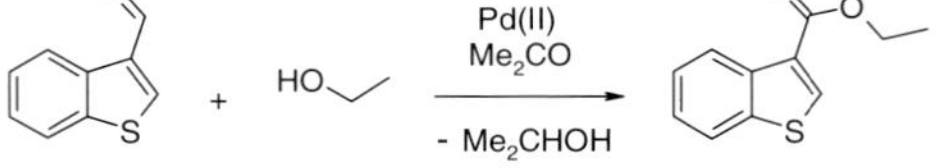

Anh. acetone (2.8 ml) added to a mixture of $Pd(OAc)_2$ (2.5 mol%) and XPhos (5 mol%) under N_2 at room temp., the mixture stirred in a sealed vial for 20 min, degassed ethanol (2.2 ml) added to a second vial containing benzothiophene-3-carboxaldehyde (1 mmol) and oven dried K_2CO_3 (18 mol%), the mixture stirred at room temp. for 20 min, the catalyst soln. added to the second vial via syringe, the mixture heated at 50° overnight, filtered through silica, concentrated *in vacuo*, and purified chromatographically → ethyl benzo[*b*]thiophene-3-carboxylate. Y 73%. Detailed investigation (incl. a screen of 95 additives!) identified optimal conditions for the direct conversion of (het)ar. and aliphatic aldehydes to esters, with the use of acetone as H-acceptor minimizing disproportionation of the aldehyde. The method was scalable (to 7 mmol) and gave good results with prim. alcohols at room temp. (twenty examples; Y 47-99%), but sec. alcohols and 2,2,2-trifluoroethanol reacted sluggishly, generally requiring prolonged reaction at 50° (nine examples; Y 7-60%), and *tert*-butanol was unreactive. Low yields were obtained with indole and imidazole containing substrates (two examples; 12-38%; thiophene and oxazole rings were tolerated), and with bulky diphenylacetaldehyde (Y 33%), while 4-chlorobenzaldehyde gave a complex mixture of products with the required ester obtained in only 5% yield. Poor results were obtained with alkene containing substrates, presumed due to complexation with the catalyst. F.e.s. B.A. Tschaen, J.R. Schmink, G.A. Molander, Org. Lett. *2013*, *15* (3), 500-3 [DOI: 10.1021/ol303298g].

Palladium(II) acetate/N-(pyrid-3-yl)-2,4,6-tris(4-tert-butylphenyl)pyridinium fluoroborate/potassium persulfate ←

Improved palladium(II)-catalyzed ar. acoxylation with an N-(pyrid-3-yl)pyridinium salt as phase transfer catalyst H → OAc

56.

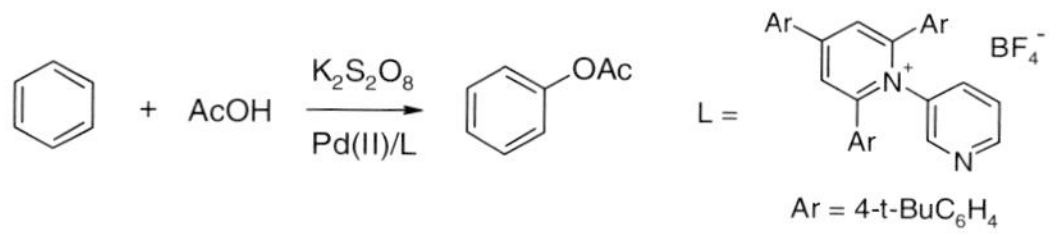

The development of pyridinium-based ligands to facilitate palladium-catalyzed C-H oxygenation of benzene with potassium persulfate is described. **E:** A mixture of $K_2S_2O_8$ (1.12 mmol), $Pd(OAc)_2$ (2 mol%), pyridinium ligand (2 mol%), glacial acetic acid (0.9 ml), acetic anhydride (0.1 ml) and benzene (10 eq.) stirred in a sealed vial at 80° for 24 h, the cooled mixture diluted with ethyl acetate, filtered through Celite, washed with satd. aq. K_2CO_3, and subjected to GC analysis → phenyl acetate. Y 71% (GC). $Pd(OAc)_2$ alone was a poor catalyst (Y 0.8%) for this reaction, with addition of pyridine (Y 37%) or pyridine and a phase transfer catalyst (Et_4NBF_4) (Y 52%) giving improved results, but optimum results were obtained using the illustrated cationic pyridinium salt as ligand. Experimental evidence showed the ligand's phase transfer effect to be key to the success of the reaction (cf. equilibrium or electronic effects). Biphenyl (Y 0.6-2%) was obtained as by-product in all cases. F.e. and ligand prepn. s. J.B. Gary, A.K. Cook, M.S. Sanford, ACS Catal. *2013*, *3* (4), 700-3 [DOI: 10.1021/cs300786j].

Tetrakis(acetonitrile)palladium(II) bis(fluoroborate)/phenyl iodosoacetate/p-benzoquinone ←

Methyl (E)-α,β-ethyleneketones from terminal ethylene derivs. CHCHCH=CH_2 → C=C-C(O)CH_3

Dehydrogenative Wacker oxidation under palladium(II)/iodine(III) co-catalysis

57.

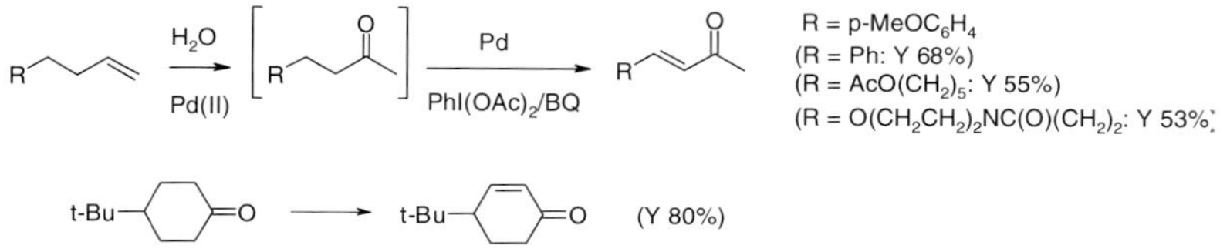

Startg. 1-alkene (0.3 mmol), *p*-benzoquinone (2 eq.), and $PhI(OAc)_2$ (0.25 eq.) added sequentially to a borosilicate vial without exclusion of air or moisture, deionized water (1 eq.) added via micropipette, $Pd(MeCN)_4(BF_4)_2$ (0.1 eq.) then carefully transferred from another vial, rinsing in with DMSO (3 x 0.15 ml), the vial sealed with a Teflon cap, placed in an aluminum block, stirred at 35° for 48 h, and the crude mixture purified directly by flash column chromatography → product. Y 69% (E/Z >20:1). The reaction proceeds **under mild conditions** to afford a wide range of β-aryl or -alkyl enones in reasonable yields (51-69%; sixteen examples) with high stereoselectivity (E/Z >20:1), tolerating functional groups such as acoxy/esters, enolesters, amides, phthalimido, benzyl ethers, internal olefin, trifluoromethyl, phenolethers or ar. bromides and *o*-substitution; it is notably chemoselective in that amides and esters did not undergo competitive dehydrogenation. It failed with a 4,4-disubst. 1-alkene (allylcyclohexane), however, the Wacker product being obtained instead. A preformed ketone, 4-*tert*-butylcyclohexanone, underwent dehydrogenation under similar conditions (Y 80%). Mechanistic investigations showed that the Wacker oxidation occurred rapidly (completion within 3 h) and was not influenced by the presence of phenyl iodosoacetate, while the latter acts as dehydrogenation catalyst rather than terminal oxidant, and no iodobenzene formation was observed. F.e.s. M.A. Bigi, M.C. White, J. Am. Chem. Soc. *2013*, *135* (21), 7831-4 [DOI: 10.1021/ja402651q].

Palladium(II) bromide/copper(I) iodide/lithium chloride/tert-butyl hydroperoxide ←
3-Acylindoles from *o*-acetylene-*tert*-amines ○
by palladium-copper-cocatalyzed oxidative ring closure s. *83*, 48

Dichloro(dipyrido[3,2-a:2′,3′-c]phenazine)platinum(II)/sulfuric acid/oxygen *[Pt(II)]/H_2SO_4/O_2*
Perfluoroalkylmethyl perfluorocarboxylates $2\ R_fCH_2OH \rightarrow R_fCH_2OC(O)R_f$
from two perfluoroalkylmethanol molecules
Platinum(II)-catalyzed aerobic oxidation

58.

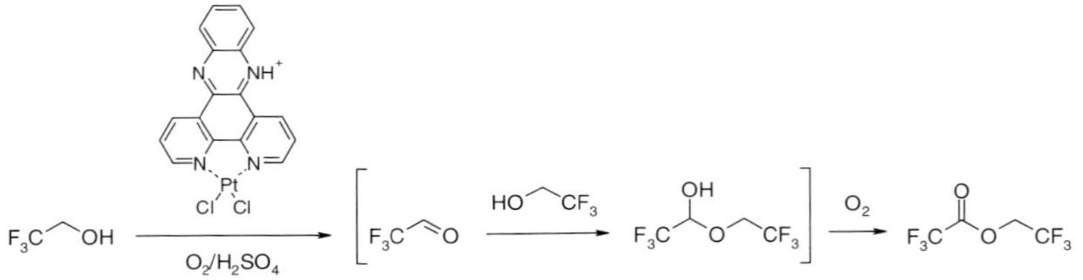

A mixture of 2,2,2-trifluoroethanol (2 ml), H_2SO_4 (98%; 0.25 ml) and Pt(dppz)Cl_2 (0.1 mol%) heated in an autoclave under a mixture of O_2 (10 bar) and methane (20 bar) for 24 h → 2,2,2-trifluoroethyl trifluoroacetate. Y 42%. Development of ligands for oxidation of methane using trifluoroethanol as an 'inert solvent' surprisingly effected efficient oxidation of the alcohol to the ester, with small amounts of trifluoroacetaldehyde and its acetal (2% combined yield) also observed. The presence of acid is essential to the catalytic activity, which was also applicable to pentafluoropropan-1-ol and heptafluorobutan-1-ol analogs, although turnover numbers were much lower. F.e.s. H. Ben-David, M.A. Iron, R. Neumann, Chem. Commun. *2013*, *49* (17), 1720-2 [DOI: 10.1039/c2cc37759d].

Via intermediates *v.i.*
1,(n+1)-Diacoxy-2,n-diketones from dienes ⊂
via oxidative ring opening of 2-α,2′-α-diacoxy-O-heterocyclics s. *83*, 52

Oxygen ↾ OC ⇃↾ O

Without additional reagents *w.a.r.*
5-(*o*-Iodobenzoyloxy)-Δ^3-2-pyrrolones from pyrroles s. *83*, 51 ←

Triisobutylamine ←
Mesylates from sec. alcohols via alkoxycyclopropenium mesylates OSO_2Me
with inversion of configuration s. *83*, 29

Chiral dibenzo-condensed N,N′-bis[3-(dimethylamino)-2-pyridyl]-2,3-diaminobicyclo-[2.2.2]octa-5,8-dienes/cesium carbonate ←
Organocatalyzed O-acylation of sec. alcohols with kinetic resolution OH → OAc
s. *52*, 497s*83*

N-(4-Pyridyl)pyrrolidine-2,5-dicarboxylic acid amides/collidine ←
Organocatalyzed O-acylation of *meso*-diols with desymmetrization OH → OAc
s. *52,* 497s*83*

Copper(II) trifluoroacetate/collidine ←
Regioselective O-benzoylation of glycopyranosides s. *29,* 184s*83* OH → OBz

Sodium tetrahydridoborate s. under Phenyl iodosoacetate *$NaBH_4$*

1,8-Diazabicyclo[5.4.0]undec-7-en-8-ium acetate ←
O-Acetylation s. *29,* 184s*83* OH → OAc

2(R)- or 2(S)-[Diphenyl(trimethylsiloxy)methyl]pyrrolidine s. under Phenyl iodosoacetate ←

Enzymes/hexadecyltrimethylammonium triflimide ←
Carboxylic acid esters from acids COOH → COOR
Flavor esters s. *48,* 169s*83*

Immobilized lipase/mesoporous silica-immobilized oxovanadium(V) complexes ←
Enzymatic dynamic kinetic resolution of sec. alcohols by O-acylation OH → OAc
s. *53,* 500s*83*

Ethyl 2-(tert-butoxycarbonyloxyimino)-2-cyanoacetate/ethyldiisopropylamine ←
Carboxylic acid esters from acids COOH → COOR
update on esterification s. *48,* 169s*82*; stoichiometric, racemization-free procedure with ethyl 2-(*tert*-butoxycarbonyloxyimino)-2-cyanoacetate [Boc-Oxyma] as coupling agent and ethyldiisopropylamine as base, also thioesterification (SC↓↑O), amidation (NC↓↑O) and soln. or solid-phase peptide synthesis (cf. *77,* 179s*83*), s. K. Thalluri, K.C. Nadimpally, M.P. Chakravarty, A. Paul, B. Mandal, Adv. Synth. Catal. *2013, 355* (2-3), 448-62 [DOI: 10.1002/adsc.201200645]; enzymatic preparation of flavor esters in the ionic liquid, hexadecyltrimethylammonium triflimide [being homogeneous at 50° but heterogeneous and separable by centrifugation on cooling], s. P. Lozano, J.M. Bernal, A. Navarro, Green Chem. *2012, 14* (11), 3026-33 [DOI: 10.1039/c2gc36081k]; study of esterification and transesterification (cf. *47,* 182s*83*) in Brønsted acidic ionic liquids s. C. Chiappe, S. Rajamani, F. D'Andrea, ibid. *2013, 15* (1), 137-43 [DOI: 10.1039/c2gc35941c]; stoichiometric esterification and transesterification without water removal using recyclable sulfonated polypyrene as acidic catalyst s. K. Tanemura, T. Suzuki, Tetrahedron Lett. *2013, 54* (15), 1972-5 [DOI: 10.1016/j.tetlet.2013.01.126].

2,2,6,6-Tetramethylpiperidine nitroxyl s. under Phenyl iodosoacetate *TEMPO*

(R)-Benzotetramisole/pivalic anhydride/ethyldiisopropylamine ←
Organocatalyzed asym. O-acylation of sec. alcohols OH → OAc
with kinetic resolution, update s. *52,* 497s*82*; of α-hydroxy-γ-lactones with diphenylacetic acid in the presence of pivalic anhydride/ethyldiisopropylamine and (R)-benzotetramisole as acyl transfer agent s. K. Nakata, K. Gotoh, K. Ono, K. Futami, I. Shiina, Org. Lett. *2013, 15* (6), 1170-3 [DOI: 10.1021/ol303453j]; of sec. alcohols with a C_2-symmetric dibenzo-condensed N,N′-bis-[3-(dimethylamino)-2-pyridyl]-2,3-diaminobicyclo[2.2.2]octa-5,8-diene as catalyst and Cs_2CO_3 as base s. E.Y. Yazicioglu, C. Tanyeli, Tetrahedron: Asym. *2012, 23* (24), 1694-9 [DOI: 10.1016/j.tetasy.2012.11.007]; **desymmetrization** of *meso*-diols with a C_2-symmetric N-(4-pyridyl)-pyrrolidine-2,5-dicarboxylic acid amide as catalyst and collidine as base s. H. Schedel, K. Kan, Y. Ueda, K. Mishiro, K. Yoshida, T. Furuta, T. Kawabata, Beilstein J. Org. Chem. *2012, 8*, 1778-87 [DOI: 10.3762/bjoc.8.203]; enzymatic **dynamic kinetic resolution** of sec. alcohols with immobilized lipase and a mesoporous silica-immobilized oxovanadium(V) complex as racemizing catalyst s. M. Egi, K. Sugiyama, M. Saneto, R. Hanada, K. Kato, S. Akai, Angew. Chem., Int. Ed. *2013, 52* (13), 3654-8 [DOI: 10.1002/anie.201208988]; **O-acetylative inversion of the hydroxyl group** of *L*-lactates by the immobilized lipase/ruthenium(II) couple (cf. *53,* 500s*79*) s. I.A. Shuklov, N.V. Dubrovina, J. Schulze, W. Tietz, K. Kühlein, A. Börner, Tetrahedron Lett. *2012, 53* (47), 6326-8 [DOI: 10.1016/j.tetlet.2012.08.126]; **non-enzymatic dynamic kinetic resolution** of sec. benzyl alcohols *with acetyl isopropyl carbonate* as acetylating agent using a ferrocene-condensed 4-dimethylaminopyridine as catalyst and carbonyl(chloro)(η^5-pentaphenylcyclopentadienyl)-ruthenium(II) as racemizing catalyst s. S.Y. Lee, J.M. Murphy, A. Ukai, G.C. Fu, J. Am. Chem. Soc. *2012, 134* (36), 15149-53 [DOI: 10.1021/ja307425g].

Phenyl iodosoacetate/2,2,6,6-tetramethylpiperidine nitroxyl/2(R)- or 2(S)-[diphenyl-(trimethylsiloxy)methyl]pyrrolidine/sodium tetrahydridoborate ←

Terminal glycol O²-monobenzoates from prim. alcohols $CHCH_2OH \rightarrow C(OBz)CH_2OH$
via TEMPO-mediated oxidation-organocatalyzed asym. α-benzoyloxylation-reduction

59. 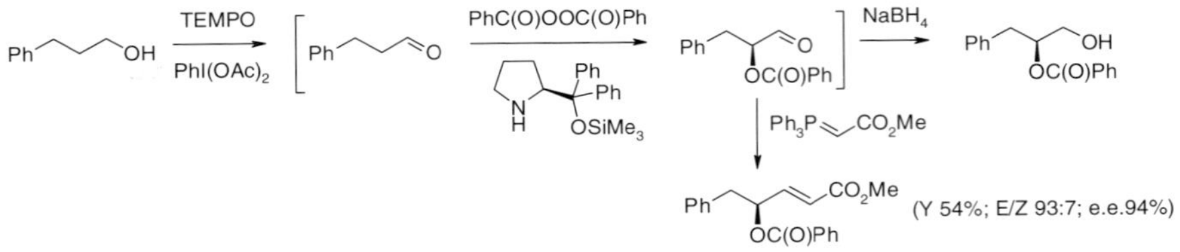

in one pot. Organocatalyzed asym. α-benzoyloxylation of aldehydes (cf. *76*, 90; *75*, 95) has been shown to be compatible with preceding TEMPO-mediated oxidation of prim. alcohols using phenyl iodosoacetate as co-oxidant, the co-produced acetic acid and phenyl iodide having minimal effect on subsequent steps. **E:** A mixture of phenyl iodosoacetate (0.25 mmol), TEMPO (10 mol%) and 3-phenylpropan-1-ol (1.1 eq.) in THF (1.5 ml) allowed to react at room temp. for 16 h, benzoyl peroxide (1.1 eq.) and (S)-diphenylprolinol trimethylsilyl ether (10 mol%) in THF (1 ml) added, allowed to react at room temp. for 5 h, then treated with $NaBH_4$ (4 eq.), and allowed to react at room temp. for 16 h → product. Y 52% (e.e. 94%). The (R)-catalyst gave the enantiomer in 93% e.e. (Y 49%). The method is simple, avoids handling [often unstable] aldehydes, and gives acceptable yields (20%, 36-50%; eight further examples) and high enantioselectivity (91-94% e.e.). F.e., **also chiral γ-benzoyloxy-α,β-ethylenecarboxylic acid esters** by reaction of the intermediate acoxyaldehyde with methyl (triphenylphosphoranylidene)acetate instead of $NaBH_4$ (two examples: Y 48%, 54%; E/Z 90:10, 93:7; e.e. 94-95%), and application to the synthesis of (S)-γ-octalactone, s. P. Hermange, F. Portalier, C. Thomassigny, C. Greck, Tetrahedron Lett. *2013*, *54* (9), 1052-5 [DOI: 10.1016/j.tetlet.2012.11.137]; **organocatalyzed asym. α-benzoyloxylation of** α-branched **aldehydes** using benzoyl peroxide and a cinchona alkaloid-derived prim. amine in the presence of (S)-TRIP as catalyst, **also chiral** [α-subst.] **α-benzoyloxy-β,γ-ethylenealdehydes** from enals with Cl_3CCOOH as the acid, s. N. Demoulin, O. Lifchits, B. List, Tetrahedron *2012*, *68* (37), 7568-74 [DOI: 10.1016/j.tet.2012.06.043]; direct asym. α-benzoyloxylation **of cyclic ketones** using a readily available cinchona-derived prim. amine/Cl_3CCOOH as organocatalyst cf. O. Lifchits, N. Demoulin, B. List, Angew. Chem., Int. Ed. *2011*, *50* (41), 9680-3 [DOI: 10.1002/anie.201104244]; of cyclohexanones and indanones with 9-amino-9-deoxy-*epi*-dihydroquinidine/salicylic acid cf. M.S. Jadhav, P. Righi, E. Marcantoni, G. Bencivenni, J. Org. Chem. *2012*, *77* (6), 2667-74 [DOI: 10.1021/jo2024976].

N-Chlorosuccinimide/fluorous hydrazodi(thionocarboxylic acid esters) ←

Acetalation $CO \rightarrow C(OR)_2$

update s. *58*, 78s*82*; of aldehydes with NCS and a readily recyclable *fluorous* hydrazodi(thionocarboxylic acid ester) as catalyst s. K. Tanemura, T. Suzuki, Tetrahedron Lett. *2013*, *54* (15), 1972-5 [DOI: 10.1016/j.tetlet.2013.01.126]; of acid-sensitive oxo compds. as 1,3-dioxolanes under mild conditions with N-hydroxybenzenesulfonamide (or its O-benzyl or O-tosyl derivs.) and triethylamine s. A. Hassner, C.R. Bandi, S. Panchgalle, Synlett *2012*, *23* (19), 2773-6 [DOI: 10.1055/s-0032-1317529]; one-pot formation of **1,3-dioxolanes** (or glycol monoesters) **from terminal ethylene derivs.** (cf. *26*, 143s*81*) and glycol with iodine as catalyst and subsequent oxidation with Oxone s. M.A. Kumar, P. Swamy, M. Naresh, M.M. Reddy, C.N. Rohitha, S. Prabhakar, A.V.S. Sarma, J.R.P. Kumar, N. Narender, Chem. Commun. *2013*, *49* (17), 1711-3 [DOI: 10.1039/c3cc38085h].

Ethyl 2-(3,4-dichlorophenyl)hydrazinecarboxylate s. under Triphenylphosphine ←
Fluorous hydrazodi(thionocarboxylic acid esters) s. under N-Chlorosuccinimide ←

Triphenylphosphine/ethyl 2-(3,4-dichlorophenyl)carbazate/iron phthalocyanine/air ←
Catalytic Mitsunobu reaction ←

60.

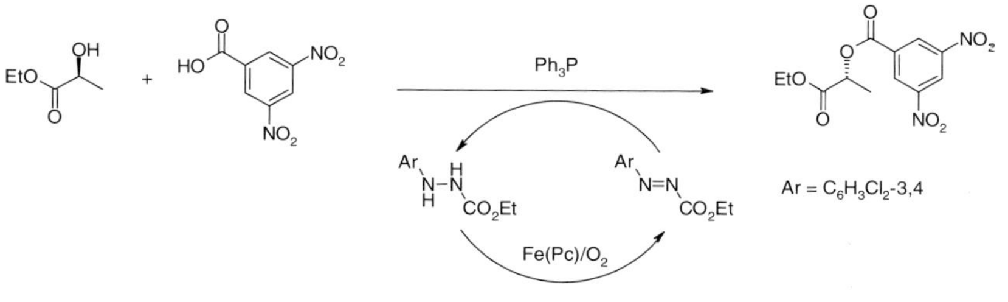

A rare example of a Mitsunobu reaction catalytic in the azo/hydrazo component [cf. T.Y.S. But, P.H. Toy, J. Am. Chem. Soc. *2006*, *128* (30), 9636-7 [DOI: 10.1021/ja063141v]] is reported, using air as terminal oxidant and an inexpensive, non-toxic iron catalyst. **E: Acoxy compds. from alcohols with inversion of configuration**. A mixture of (S)-ethyl lactate (0.85 mmol), 3,5-dinitrobenzoic acid (0.935 mmol), triphenylphosphine (2 eq.), ethyl 2-(3,4-dichlorophenyl)-carbazate (10 mol%), iron phthalocyanine (10 mol%) and activated 5 Å molecular sieves (400 mg) in THF (1.7 ml) heated at 65° under air (balloon) for 24 h, the mixture cooled to room temp., filtered, the solvent removed under reduced pressure, and the residue purified by silica gel chromatography → (R)-product. Y 79% (e.r. 98:2). Reaction on the 10 mmol scale gave 84% yield and 95:5 e.r. There was no reaction in the absence of molecular sieves. The hydrazine deriv. may be prepared readily in one step from commercially available 3,4-dichlorophenylhydrazine hydrochloride, in a highly pure form after recrystallization, and is stable under ambient conditions for more than a month; furthermore, ethyl 2-(3,4-dichlorophenyl)azocarboxylate may be recovered in 96% yield after the reaction. Low yield (28%) and inversion ratio (25:75 d.r.) were observed with sterically hindered (-)-menthol, however. F.e. (seven; Y 57-92%; e.r. 93:7 to 97:3 or d.r. 92:8, 97:3 where appropriate), also reaction of 3-phenylpropanol with phenol, N-hydroxyphthalimide, phthalimide or N-benzyl-*o*-nitrobenzenesulfonamide (Y 51-65%), s. D. Hirose, T. Taniguchi, H. Ishibashi, Angew. Chem., Int. Ed. *2013*, *52* (17), 4613-7 [DOI: 10.1002/anie.201300153].

Diacetoxytriphenylphosphorane *$Ph_3P(OAc)_2$*
O-Acetylation s. *29*, 184s*83* OH → OAc

Chiral C_2-symmetric cyclic phosphoryliminophosphoric acid diesters ←
Stereo-confined organo-Brønsted acid-catalyzed asym. acetalation CHO → $CH(OR)_2$

61.

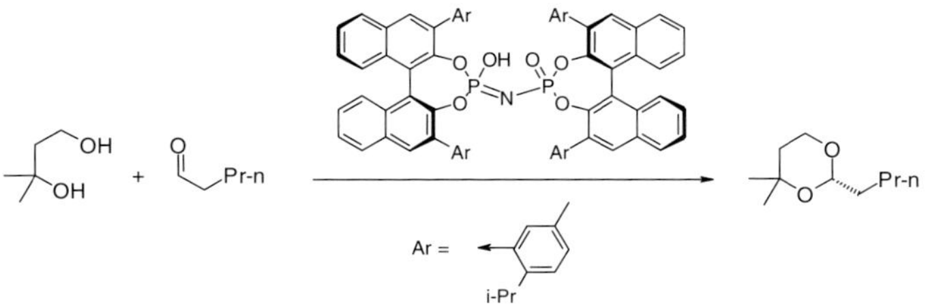

Chiral C_2-symmetric cyclic phosphoryliminophosphoric acid diesters, previously used for asym. spiroketalization of cyclic hydroxyenolethers (cf. *81*, 49), have now been applied in the first enantioselective acetalation of aldehydes. **E: Chiral 2-subst. 1,3-dioxanes**. Startg. aldehyde (2 eq.) added to a mixture of startg. diol (0.2 mmol), chiral N-phosphoryliminophosphoric acid diester catalyst (5 mol%), and 5 Å molecular sieves (40 mg) in toluene (2 ml), the mixture stirred vigorously at room temp. for 2 d, then treated with triethylamine (50 µl), and purified by column chromato-graphy on silica gel → product. Y 86% (e.r. 97.5:2.5). Highest enantioselectivities were obtained with catalysts bearing unsymmetrical aryl groups, the illustrated imidophosphate, derived from thymol, being optimal. Eleven further examples of chiral (benzo)dioxanes or 1,3-dioxolanes were obtained from aliphatic aldehydes in 72-93% yield and 91.5:8.5 to 99.5:0.5 e.r., while seven examples of chiral 2-aryl-1,3-dioxolanes were obtained from ar. aldehydes in 65-89% yield and 95.5:4.5 to 99.5:0.5 e.r.

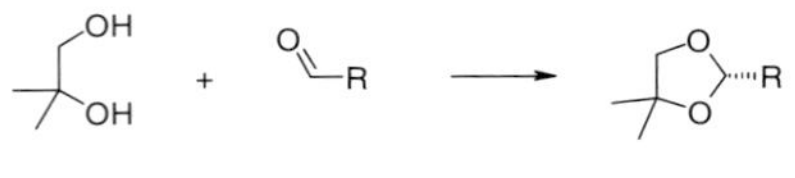

(R = CH_2CH_2Ph: Y 72%; e.r. 99:1)
(R = $CHEt_2$: Y 82%; e.r. 99.5:0.5)
(R = Ph: Y 89%; e.r. 95.5:4.5)
(R = $C_6H_3Br_2$-3,5: Y 83%; e.r. 99.5:0.5)

F.e.s. J.H. Kim, I. Coric, S. Vellalath, B. List, Angew. Chem., Int. Ed. *2013*, *52* (16), 4474-7 [DOI: 10.1002/anie.201300120]; **kinetic resolution of γ-hydroxyacetals** via organo-Brønsted acid-catalyzed asym. intramolecular transacetalation (cf. *78*, 123) using a chiral 1,1′-spirobiindan-based phosphate diester [STRIP] s. I. Coric, S. Müller, B. List, J. Am. Chem. Soc. *2010*, *132* (49), 17370-3 [DOI: 10.1021/ja108642s].

Mesoporous silica-immobilized oxovanadium(V) complexes s. under Immobilized lipase ←
Air s. under Triphenylphosphine *air*

Lithium bis(perfluoroalkanesulfonyl)imide *$LiN(SO_2R_f)_2$*
O-Acetylation s. *29,* 184s*83* OH → OAc

Hexadecyltrimethylammonium triflimide s. under Enzymes *$[C_{16}H_{33}NMe_3][NTf_2]$*

N-Hydroxybenzenesulfonamide/triethylamine *$PhSO_2NHOH/Et_3N$*
Acetalation s. *58,* 78s*83* CO → $C(OR)_2$

Sulfonated polypyrene ←
Carboxylic acid esters from acids s. *48,* 169s*83* COOH → COOR

Ferrocene-condensed 4-dimethylaminopyridine/carbonyl(chloro)(η^5-pentaphenyl-cyclopentadienyl)ruthenium(II) ←
Non-enzymatic dynamic kinetic resolution of sec. alcohols by O-acylation s. *53,* 500s*83* OH → OAc

Iron phthalocyanine s. under Triphenylphosphine *Fe(Pc)*

Iron(III) tosylate *$Fe(OTs)_3$*
O-Acylation
with carboxylic acid anhydrides, update s. *29,* 184s*82*; catalytic O-acylation of prim. or sec. alcohols, a few tert. alcohols, diols and phenols with $Fe(OTs)_3$, also acylals from aldehydes (cf. *51,* 55s*83*), s. N.J. Baldwin, A.N. Nord, B.D. O'Donnell, R.S. Mohan, Tetrahedron Lett. *2012, 53* (51), 6946-9 [DOI: 10.1016/j.tetlet.2012.10.033]; **O-acetylation** of alcohols and phenols with a recyclable lithium bis(perfluoroalkanesulfonyl)imide as catalyst s. G. Li, G. Zhao, Synth. Commun. *2013, 43* (1), 34-43 [DOI: 10.1080/00397911.2011.589563]; with diacetoxytriphenylphosphorane, prepared *in situ* from $Ph_3P/Br_2/NH_4OAc$, or under heterogeneous conditions with 1,3,2,4-diazadiphosphetidine, also S-acetylation, s. N. Iranpoor, H. Firouzabadi, E.E. Davan, Tetrahedron Lett. *2013, 54* (14), 1813-6 [DOI: 10.1016/j.tetlet.2013.01.068]; O-acetylation of alcohols and phenols, and N-acetylation of amines (NC⇃↾O), with the recyclable Lewis basic ionic liquid, 1,8-diazabicyclo[5.4.0]undec-7-en-8-ium acetate, as catalyst without solvent, s. L. Ji, C. Qian, X.-z. Chen, Monatsh. Chem. *2013, 144* (3), 369-74 [DOI: 10.1007/s00706-012-0820-7]; **regioselective O-benzoylation** of glycopyranosides in the presence of $Cu(OCOCF_3)_2$/collidine as promoter s. V. Evtushenko, Carbohydr. Res. *2012, 359,* 111-9 [DOI: 10.1016/j.carres.2012.06.020].

Carbonyl(chloro)(η^5-pentaphenylcyclopentadienyl)ruthenium(II) s. under Ferrocene-cyclocondensed 4-dimethylaminopyridine *[Ru(II)]*

Nitrogen ↑ OC ↓↑ N

Potassium carbonate K_2CO_3
Potassium carbonate/microwaves K_2CO_3/⁄\\\\\]
O-Functions from N-tosylhydrazones C=NNHTs → CH(O-)
phenolethers s. *78,* 88; **acoxy compds.** from carboxylic acids with excess of K_2CO_3 in dioxane s. A. Zhou, L. Wu, D. Li, Q. Chen, X. Zhang, W. Xia, Chin. J. Chem. *2012, 30* (8), 1862-6 [DOI: 10.1002/cjoc.201200387]; **benzyl alcohols** from ar. N-tosylhydrazones in water with K_2CO_3 under microwave enhancement, also α-deuteriobenzyl alcohols-OD, s. Á. García-Muñoz, A.I. Ortega-Arizmendi, M.A. García-Carrillo, E. Díaz, N. Gonzalez-Rivas, E. Cuevas-Yañez, Synthesis *2012, 42* (14), 2237-42 [DOI: 10.1055/s-0031-1290372].

Copper(II) acetate/cesium carbonate $Cu(OAc)_2/Cs_2CO_3$
Diaryl ethers from phenols and ar. nitro compds. ArOH + O_2NAr' → ArOAr'
Ligand-free copper-catalyzed O-arylation

62. 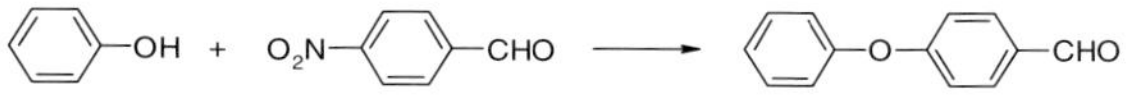

A mixture of *p*-nitrobenzaldehyde (0.5 mmol), phenol (2 eq.), $Cu(OAc)_2 \cdot H_2O$ (5 mol%) and Cs_2CO_3 (2 eq.) in DMF (3 ml) stirred in a Schlenk tube under N_2 at 100° (oil bath temp.) for 4 h, cooled to room temp., diluted with ethyl acetate, filtered, the filtrate concentrated *in vacuo*, and the residue worked up with purification by chromatography on silica gel → product. Y 92%. The procedure is facile, cost-effective, ligand-free, applicable on the gram scale and suitable for the coupling of electron-diverse phenols (bearing functional groups such as MeO, Me, halogen and CF_3) with ar. nitro compds. bearing CHO, COR, CN, COOMe, CF_3 or *p*-F groups. However, there was no reaction with nitrobenzene itself or with 3-nitrobenzaldehyde, while 1-fluoro-4-nitrobenzene underwent displacement of fluorine instead. F.e. and comparison of copper salts, bases and solvents s. J. Chen, X. Wang, X. Zheng, J. Ding, M. Liu, H. Wu, Tetrahedron *2012*, *68* (43), 8905-7 [DOI: 10.1016/j.tet.2012.08.032].

Silver trifluoroacetate s. under Bis(acetonitrile)dichloropalladium(II) $AgOCOCF_3$

Copper(I) chloride/2,2′-bis(diphenylphosphino)-1,1′-binaphthyl *CuCl/BINAP*
Hydroxyketones from oxaziridines C
Regioselective copper(I)-catalyzed oxidation

63. 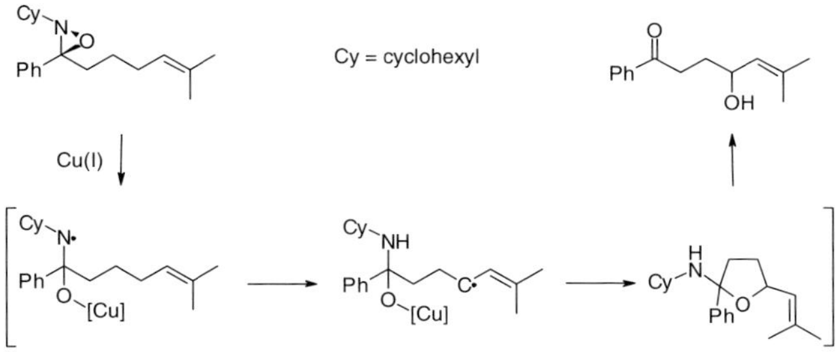

A soln. of CuCl (5 mol%) and *rac*-BINAP (5 mol%) in THF (12 ml) refluxed under argon for 30 min, a soln. of (2R,3S)-2-cyclohexyl-3-(5-methylhex-4-en-1-yl)-3-phenyl-1,2-oxaziridine (0.5 mmol) in the same solvent (3 ml) added slowly by syringe, heating continued for 1 h, the resulting mixture concentrated *in vacuo*, and purified chromatographically → 4-hydroxy-6-methyl-1-phenylhept-5-en-1-one. Y 75%. A series of 3-(het)aryl-3-pent-4-en-1-yloxaziridines underwent allylic hydroxylation in the presence of copper(I) [CuCl/BINAP or $[Cu(PPh_3)Cl]_4$], mediated by radical cleavage of the oxaziridine ring, to afford the title products in moderate yield (41-75%; eight examples). The method was applicable to oxidation of benzylic (Y 51%) and propargylic

C-H (Y 75%) using appropriate analogs, and even the unactivated *tert*-alkyl center of a 3-methylbut-1-yloxaziridine was oxidized to the hydroxyketone, albeit in low yield (16%). A homologous hex-5-enyloxaziridine gave only a low yield of the desired allylic alcohol (16%) in a mixture that included a hex-5-en-1-yl ketone as the major product (Y 35%). Interestingly, a *3-n-butyl*-3-pent-4-enyloxaziridine was cleaved to the expected hydroxyketone (Y 69%), whereas a *3-cyclohexyl* analog suffered only alternative cleavage of the oxaziridine to afford a 5-hexenamide in 74% yield. F.e. and optimization s. H.F. Motiwala, B. Gülgeze, J. Aubé, J. Org. Chem. *2012*, *77* (16), 7005-22 [DOI: 10.1021/jo3012336].

Lithium hydrido(diisobutyl)morpholinoaluminate *LiAlH(Bu-i)$_2$N(CH$_2$CH$_2$)$_2$O*
Ar. aldehydes from nitriles s. *83*, 23 CN → CHO

Carbon dioxide *CO$_2$*
Carbonic acid esters from formamide acetals HC(N<)(OR)$_2$ → C(O)(OR)$_2$
s. *83*, 73

Polyphosphoric, methanesulfonic or sulfuric acid ←
Isatins from prim. ar. amines via alkoximinoacetic acid anilides ○

64.

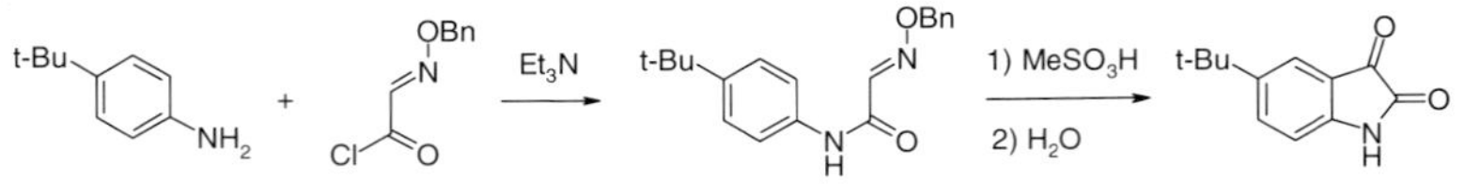

An improved stepwise method for preparation of isatins containing lipophilic groups is described. **E:** A mixture of 4-*tert*-butylaniline (1 eq.), benzyloximinoacetyl chloride (1.05 eq.) and triethylamine (1.15 eq.) stirred in methylene chloride at room temp. → N-(4-*tert*-butylphenyl)benzyloximinoacetamide (Y 88%), heated with methanesulfonic acid at 50-80°, and quenched with ice-water → 5-*tert*-butylisatin (Y 64%). In this modified and scalable (to 10g) version of the original Sandmeyer method (cf. *3*, 745), benzyl/methyloximinoacetanilides were prepared by coupling anilines with alkoximinoacetyl chlorides (using Et$_3$N) or free acids (using HOBt/EDAC) (sixteen examples; Y 46-93%), and cyclized by heating in sulfuric, methanesulfonic or polyphosphoric acids (seven examples; Y 40-85%). The success of the new method appears to be due to improved solubilization of the iminoacetamides in an appropriate acidic medium. F.e.s. L.L. Klein, M.D. Tufano, Tetrahedron Lett. *2013*, *54* (8), 1008-11 [DOI: 10.1016/j.tetlet.2012.12.035].

Palladium(II) nitrate/potassium peroxymonosulfate/lithium carbonate ←
4*H*-3,1-Benzoxazin-4-ones from *o*-acetyleneazides
via palladium-catalyzed oxidative rearrangement s. *83*, 153

Bis(acetonitrile)dichloropalladium(II)/silver trifluoroacetate/trifluoroacetic acid ←
9-Fluorenones from *o*-cyanobiaryls s. *83*, 426

Halogen ↑ OC ↓↑ Hal

Sodium hydride or hydroxide or Ethyldiisopropylamine *NaH or NaOH or i-Pr$_2$NEt*
O-Alkylation OH → OR
Hakomori permethylation s. *40*, 109; **regioselective polymethylation** of α-(1→4)-linked mannopyranosyl oligosaccharides with NaH via activation of *cis*-glycol residues by O,O-stannylidenation with Bu$_2$SnCl$_2$/TsCl/*i*-Pr$_2$NEt s. L. Xia, T.L. Lowary, J. Org. Chem. *2013*, *78* (7), 2863-80 [DOI: 10.1021/jo4001299]; **regioselective O-benzylation** of propargylic hydroxyl groups with benzyl bromide in acetylenic polyols with NaOH in DMF s. J.H. Lee, C.H. Oh, Eur. J. Org. Chem. *2012* (30), 5913-7 [DOI: 10.1002/ejoc.201201086]; regioselective O-benzylation of *prim*-hydroxyl groups in polyols (incl. carbohydrate derivs.; cf. *19*, 260) with *i*-Pr$_2$NEt (at 90°) s. M. Giordano, A. Iadonisi, Tetrahedron Lett. *2013*, *54* (12), 1550-2 [DOI: 10.1016/j.tetlet.2013.01.023].

Potassium tert-butoxide *KOBu-t*
Metal-free and ligand-free Ullmann diaryl ether synthesis OH → OAr
with diaryliodonium salts s. *54*, 90s*83*; *79*, 81s*83*

O⁹-(1-Naphthylmethyl)desmethylquinidine/potassium carbonate ←
Organocatalyzed asym. synthesis of 3-(arylideneaminooxy)- Hal → ON=CHAr
from 3-halogeno-oxindoles

65.

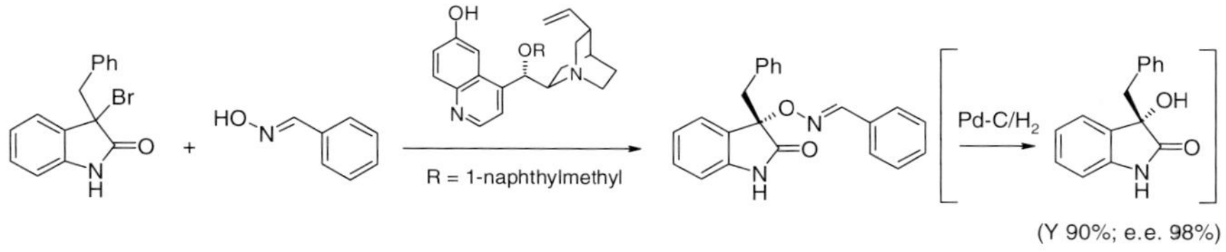

Electrophilic oxindoles (3-subst. 3-bromo- or 3-chloro-derivs.) have been used for the first time for the efficient and highly enantioselective formation of 3-subst. 3-oxyoxindoles by displacement of the halogen with (E)-benzaldoximes as nucleophile. **E:** A soln. of 3-benzyl-3-bromooxindole (0.1 mmol), (E)-benzaldoxime (1.2 eq.), O⁹-(1-naphthylmethyl)desmethylquinidine (20 mol%) and K_2CO_3 (0.8 eq.) in THF (1 ml) stirred at room temp. for 12 h, and worked up with purification by flash chromatography on silica gel → benzaldehyde (S,E)-O-(3-benzyl-2-oxoindolin-3-yl)-oxime. Y 89% (e.e. 99%). This mild, metal-free procedure is generally applicable to the coupling of electron-diverse 3-arylmethyl-, 3-naphthylmethyl-, 3-heteroarylmethyl-, 3-allyl- and [functionalized] 3-alkyl-3-bromo- or 3-chloro-oxindoles with electron-diverse (E)-benzaldoximes (and heteroaromatic analogs), notably bearing Br, Cl, Me, MeO or OH on the aromatic ring (thirty-one examples; Y 60-93%; e.e. 88-99%). A carbalkoxyamino group was tolerated in the alkyl chain of 3-subst. 3-bromooxindoles. Reaction with quinidine itself as catalyst gave racemic products, suggesting that the free hydroxyl group at $C_{6'}$ of the catalyst is critically important. It is thus proposed that, through hydrogen bonding, this hydroxyl group stabilizes the carbonyl group of an intermediate 2-indolone (formed by initial dehydrohalogenation), while the tertiary amino group of the catalyst synergistically activates the hydroxyl group of the attacking oxime, the developing (S)-configuration arising by attack of the oxime on the *re*-face of the 2-indolone. F.e., also conversion of the products to **chiral 3-subst. 3-hydroxyoxindoles** (by hydrogenation with Pd-C in methanol at room temp. over 20 h), s. Y.-H. Liao, Z.-J. Wu, W.-Y. Han, X.-M. Zhang, W.-C. Yuan, Chem. Eur. J. *2012*, *18* (29), 8916-20 [DOI: 10.1002/chem.201201395].

O-(Trimethylsilyl)quinidine/ethyldiisopropylamine/lithium tetrahydridoaluminate ←
Terminal glycols from carboxylic acid chlorides ○
via N-tosyl-4-oxazolidones
Asym. formation by [3+2]-cycloaddition with subsequent reduction

66.

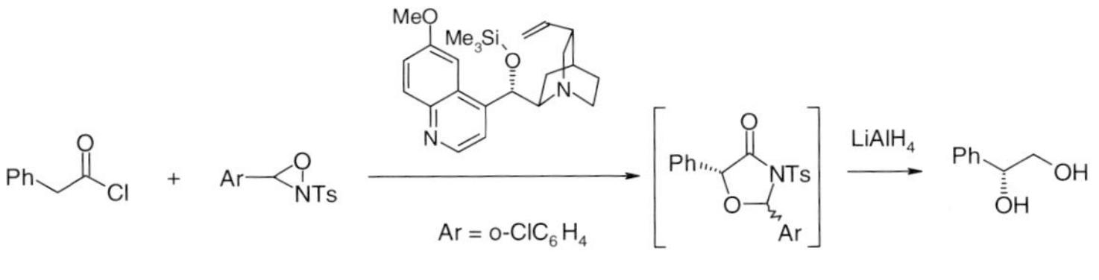

Phenylacetyl chloride (4 eq.) in THF (2 ml) added via syringe pump over 5 h to a soln. of 3-(*o*-chlorophenyl)-N-tosyloxaziridine (0.5 mmol), O-(trimethylsilyl)quinidine (10 mol%) and ethyldiisopropylamine (4 eq.) in the same solvent (3 ml) at -78°, stirred at the same temp. for 19 h, quenched with ethyl acetate, filtered through a plug of silica gel, worked up with purification by flash chromatography, the intermediate mixture of diastereoisomeric adducts dissolved in THF (5 ml), treated with $LiAlH_4$ (2 eq.) at 0°, stirred overnight at this temp., carefully quenched with methanolic HCl, and worked up with purification by flash chromatography on silica gel → (R)-1-phenylethane-1,2-diol. Y 72% (e.e. 99%). The procedure is generally high-yielding (except for reaction with *o*-chlorophenylacetyl chloride), highly enantioselective and applicable to electron-diverse aryl- or naphthyl-acetyl chlorides as well as aliphatic acyl chlorides (eleven examples; Y 34-73%;

e.e. 75-99%). The intermediate mixture of chiral *cis*- and *trans*-adducts was isolated in the case of phenylacetyl chloride and each diastereoisomer was shown to be virtually (99%) homochiral. F.e.s. P. Shao, L. Shen, S. Ye, Chin. J. Chem. *2012, 30,* (11), 2688-92 [DOI: 10.1002/cjoc.201200697].

Copper-ferrite nanoparticles/2,2,6,6-tetramethylheptane-3,5-dione ←
Copper(II) oxide nanoparticles/potassium hydroxide *CuO/KOH*
Copper(I) iodide/2-carbomethoxy-3-hydroxyquinoxaline di-N-oxide/cesium carbonate ←
Ullmann diaryl ether synthesis Hal → OAr
update s. *54,* 90s*82*; mild procedure with an excellent functional group compatibility using CuI/Cs_2CO_3 and 2-carbomethoxy-3-hydroxyquinoxaline di-N-oxide as ligand s. Y. Qiu, W. Jia, Z. Yao, F. Wu, S. Jiang, Org. Biomol. Chem. *2013, 11* (9), 1502-10 [DOI: 10.1039/c2ob26556g]; *heterogeneous* method with readily recyclable, magnetic copper ferrite nanoparticles [$CuFe_2O_4$ nps] and added 2,2,6,6-tetramethylheptane-3,5-dione s. S. Yang, C. Wu, H. Zhou, Y. Yang, Y. Zhao, C. Wang, W. Yang, J. Xu, Adv. Synth. Catal. *2013, 355* (1), 53-8 [DOI: 10.1002/adsc.201200600]; with reusable CuO nanoparticles and KOH in dimethylacetamide at room temperature without ligand, also diaryl sulfides from thiophenols and ar. halides (SC↓↑Hal), s. S.G. Babu, R. Karvembu, Tetrahedron Lett. *2013, 54* (13), 1677-80 [DOI: 10.1016/j.tetlet.2013.01.063]; inexpensive, mild, generally applicable, *metal- and ligand-free* Ullmann diaryl ether synthesis **with diaryliodonium salts** using K-*tert*-butoxide as base (cf. *79*, 81), also O-arylation of carboxylic (cf. *80*, 91) and sulfonic acids, s. N. Jalalian, T.B. Petersen, B. Olofsson, Chem. Eur. J. *2012, 18* (44), 14140-9 [DOI: 10.1002/chem.201201645]; **1-chloro-2,2,2-trifluoroethyl phenolethers** and thiophenolethers by copper-mediated radical cross-coupling with 2,2-dichloro-1,1,1-trifluoroethane [HCFC-123] in the presence of diethylamine as both activator and ligand s. X.-J. Tang, Q.-Y. Chen, Org. Lett. *2012, 14* (24), 6214-7 [DOI: 10.1021/ol302962p].

Zinc s. under $NiCl_2(DME)$ *Zn*

Hafnium(IV) triflate *$Hf(OTf)_4$*
Glycosides from glycosyl fluorides F → OR

67.

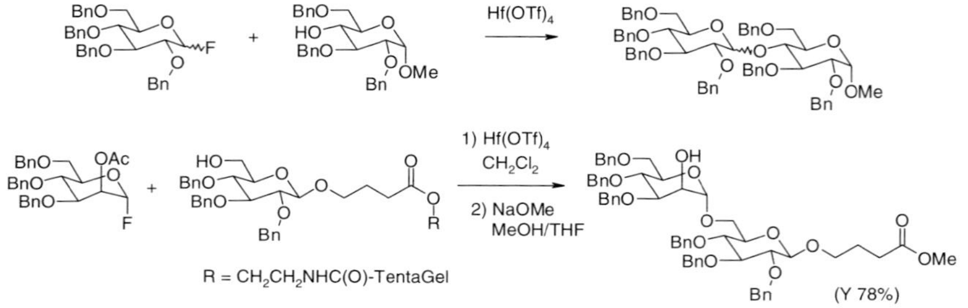

Hafnium(IV) triflate, a commercially available, stable, easily handled solid, is reported to be widely applicable *as sole activator* in operationally simple solution-phase or polymer-based glycosylations with armed or disarmed glycosyl fluorides. **E:** A soln. of startg. glycosyl fluoride (1.5 eq.; α/β 1:4) and the glycosyl acceptor (1 eq.) in methylene chloride (acceptor concentration 0.1 *M*) treated at -20° with $Hf(OTf)_4$ (1.5 eq.), after 0.5 h, the reaction quenched with satd. $NaHCO_3$, the aq. layer extracted several times with ethyl acetate, the combined extracts washed with brine, dried (Na_2SO_4), concentrated, and purified by prep. TLC → product. Y 80% (α/β 45:55). The familiar solvent effect was observed, β-glycoside selectivity increasing to 80% in acetonitrile (Y 94%) and α-glycoside selectivity increasing to 85% in 1:1 dioxane/toluene (Y 80%) under otherwise similar conditions. Seven further examples gave yields of 58-99%, with cleavage of an O,O-benzylidene group being suppressed in the presence of 2,6-di-*tert*-butyl-4-methylpyridine. Polymer-based reactions with a TentaGel-immobilized acceptor or a methyl poly(ethylene glycol)-based acceptor gave 78% or quantitative yields, respectively. F.e.s. S. Manabe, Y. Ito, J. Org. Chem. *2013, 78* (9), 4568-72 [DOI: 10.1021/jo400282x].

Nickel(II) chloride-glyme/1,3-bis(diphenylphosphino)propane/zinc *$NiCl_2$(DME)/dppp/Zn*
Aryl ketones from nitriles and ar. iodides s. *83*, 426 CN → C(O)Ar

Carbonyl(dihydrido)tris(triphenylphosphine)ruthenium(II)/bis[o-(diphenylphosphino)-phenyl] ether/potassium carbonate ←
***p*-Aryloxyacylophenones from 1-(*p*-fluoroaryl)-2-ethylenealcohols** ←
and phenols s. *83*, 164

Tris(dibenzylideneacetone)dipalladium/5-(di-tert-butylphosphino)-1′,3′,5′-triphenyl-1′H-1,4′-bipyrazole/cesium hydroxide *[Pd(0)]/CsOH*
Phenols from ar. halides ArHal → ArOH
under palladium catalysis, update s. *78*, 94s*82*; general procedure with robust $Pd_2(dba)_3$/5-(di-*tert*-butylphosphino)-1′,3′,5′-triphenyl-1′*H*-1,4′-bipyrazole [BippyPhos] and $CsOH \cdot H_2O$ as base under air in unpurified solvents, also from hydroxylated heteroaromatic analogs, s. C.B. Lavery, N.L. Rotta-Loria, R. McDonald, M. Stradiotto, Adv. Synth. Catal. *2013, 355* (5), 981-7 [DOI: 10.1002/adsc.201300088].

Sulfur ↑ OC ↓↑ S

Irradiation s. under Bis[3,5-difluoro-2-(5-trifluoromethylpyrid-2-yl)phenyl]-(4,4′-di-tert-butyl-2,2′-bipyridyl)iridium(III) hexafluorophosphate ///

Potassium tert-butoxide *KOBu-t*
***tert*-Butyl ynolethers** C≡C-SO_2NR_2 → C≡C-OBu-*t*
from N,N-disubst. α,β-acetylenesulfonic acid amides

68.

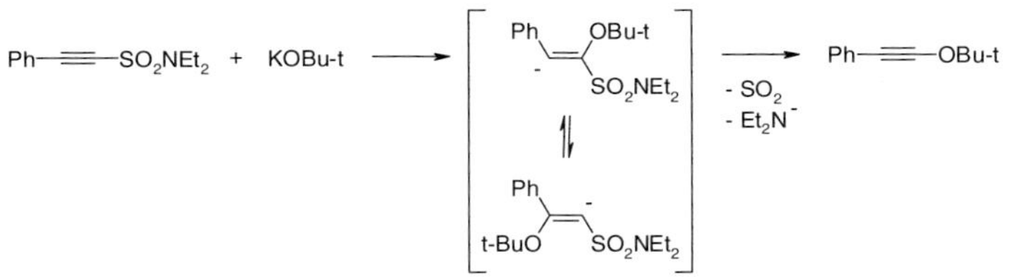

Arylethynyl *tert*-butyl ethers. A soln. of N,N-diethyl-2-phenylethynylsulfonamide (0.3 mmol) in *anhydrous* DMF (0.31 *M*) stirred under argon for 1 min at room temp., KOBu-*t* (1.2 eq.) added in one portion, the resulting brown soln. stirred for 20 min, diluted with methylene chloride, and worked up after passing the crude product through a plug of silica → (*tert*-butoxyethynyl)benzene. Y 70%. The procedure is applicable to both electron-rich and -deficient arylethynylsulfonamides (ten examples; Y 59-93%) and was also successful with K-neopentyl oxide in place of K-*tert*-butoxide (one example; Y 63%), but reaction failed with aliphatic substrates and with lithium, sodium, aluminum, magnesium or barium *tert*-butoxides. Based on isotopic labelling and DFT calculations, it is proposed that reaction involves a nucleophilic addition-elimination mechanism: an equilibrated mixture of vinylic anions is first formed by addition of *tert*-butoxide anion, after which the β-adduct eliminates SO_2 and amide anion to give the product. For this, however, the necessity of a rigorously dry medium and the absence of proton donors are essential. A radical mechanism was discounted, but it is possible that the less stable [non-hydrogen bonded] α-adduct eliminates to a carbenoid prior to product-forming Fritsch-Buttenberg-Wiechell rearrangement. F.e.s. V.J. Gray, B. Slater, J.D. Wilden, Chem. Eur. J. *2012*, *18* (49), 15582-5 [DOI: 10.1002/chem.201203015].

Silver triflate s. under [Tris(p-trifluoromethylphenyl)phosphine]gold(I) chloride and p-Nitrobenzenesulfenyl chloride *AgOTf*

[Tris(p-trifluoromethylphenyl)phosphine]gold(I) chloride/silver triflate ←

Glycosides from but-3-ynyl thioglycosides $SCH_2CH_2C{\equiv}CH \rightarrow OR$

69. 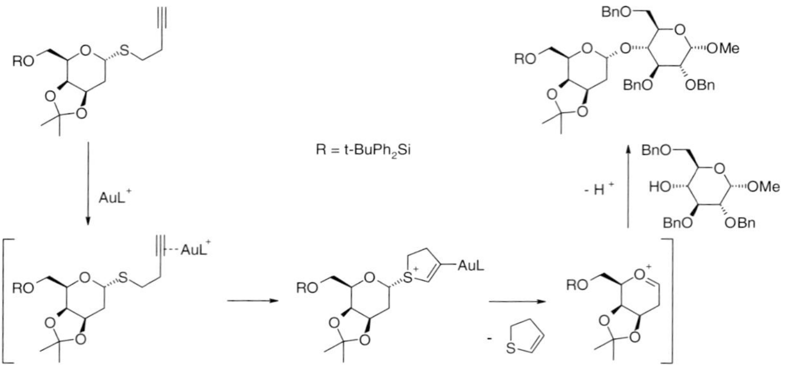

2-Deoxy- or 2,6-dideoxy-glycosides. 4 Å Molecular sieves (50 mg), AgOTf (10 mol%) and a freshly prepared soln. of gold(I) catalyst (5 mol%) in methylene chloride (1 ml) added to an azeotropically dried mixture of but-3-ynyl 6-O-*tert*-butyldiphenylsilyl-3,4-O-isopropylidene-2-deoxy-1-thio-α-D-*lyxo*-hexopyranoside (1.2 eq.) and methyl 2,3,6-tri-O-benzyl-α-D-glucopyranoside (0.1 mmol), the mixture stirred at room temp. for 0.5 h, quenched with solid $NaHCO_3$, filtered through Celite, concentrated, and purified chromatographically → methyl 6-O-*tert*-butyldiphenylsilyl-3,4-O-isopropylidene-2-deoxy-α-D-*lyxo*-hexopyranosyl-(1→4)-2,3,6-tri-O-benzyl-α-D-glucopyranoside. Y 80% (α only; Y 90% based on recovered acceptor). This atom-economical gold(I)-catalyzed glycosylation was successful **under mild conditions** for 2-deoxy- and 2,6-dideoxy-1-but-3-ynylthioglycoside donors with prim., sec. and tert. alcohol acceptors (incl. steroid and adamantane alcohols), generally with good α-selectivity (α/β 1.1:1 to 100:0; twenty examples; Y 69-99%). The method was compatible with acetate, benzoate, benzyl ether, silyl ether and acetonide functionality. The proposed mechanism was based on the isolation of 2,3-dihydrothiophene as by-product. F.e., optimization and substrate prepn. s. S. Adhikari, K.N. Baryal, D. Zhu, X. Li, J. Zhu, ACS Catal. *2013, 3* (1), 57-60 [DOI: 10.1021/cs300670k].

N-(p-Tolylthio)-ε-caprolactam/trimethylsilyl triflate ←

Diphenyl sulfoxide/trifluoromethanesulfonic anhydride *Ph_2SO/Tf_2O*

p-Nitrobenzenesulfenyl chloride/silver triflate *ArSCl/AgOTf*

Glycosides from thioglycosides SR → OR′

update s. *39,* 189s*82*; from 'armed' or 'unarmed' thioglycosides with N-(*p*-tolylthio)-ε-caprolactam/trimethylsilyl triflate as activator, also application to reactivity-based one-pot **oligosaccharide synthesis**, s. S.K. Maity, N. Basu, R. Ghosh, Carbohydr. Res. *2012, 354*, 40-8 [DOI: 10.1016/j.carres.2012.03.024]; from 2,3-O-carbonate-protected thio-glucosides and -galactosides, effect of added Lewis acid (BF_3-etherate cf. $SnCl_4$) on α- cf. β-selectivity on glycosidation of 4,6-O,O-diacetyl- cf. 4,6-O,O-dibenzyl-derivs., s. Y. Geng, Q. Qin, X.-S. Ye, J. Org. Chem. *2012, 77* (12), 5255-70 [DOI: 10.1021/jo3002084]; stereoselective glycosidation of substrates possessing a hydrogen-bonding O-picolinyl or O-picoloyl group at remote positions (C-3, C-4, and C-6) s. J.P. Yasomanee, A.V. Demchenko, J. Am. Chem. Soc. *2012, 134* (49), 20097-102 [DOI: 10.1021/ja307355n]; of *p*-tolyl thiogalacturonoside 3,6-lactones affording α-products by preactivation with *p*-nitrobenzenesulfenyl chloride/AgOTf, or β-products without preactivation, using diphenyl sulfoxide/Tf_2O, s. A.E. Christina, J.A. Muns, J.Q.A. Olivier, L. Visser, B. Hagen, L.J. van den Bos, H.S. Overkleeft, J.D.C. Codée, G.A. van der Marel, Eur. J. Org. Chem. *2012* (29), 5729-37 [DOI: 10.1002/ejoc.201200717]; automated solid-phase, linear or branched, oligoarabino-furanoside synthesis s. J. Kandasamy, M. Hurevich, P.H. Seeberger, Chem. Commun. *2013, 49* (40), 4453-5 [DOI: 10.1039/c3cc00042g].

Carbon tetrabromide s. under Bis[3,5-difluoro-2-(5-trifluoromethylpyrid-2-yl)phenyl]-(4,4′-di-tert-butyl-2,2′-bipyridyl)iridium(III) hexafluorophosphate *CBr_4*

Tin(II) chloride *$SnCl_2$*

β-Ketothiono- from β-ketodithio-carboxylic acid esters and alcohols C(S)SR → C(S)OR′

70. 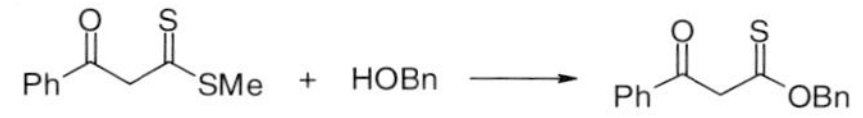

Solvent-free Lewis acid-catalyzed transesterification. A mixture of methyl 3-oxo-3-phenylpropanedithiocarboxylate (1 mmol), benzyl alcohol (1 eq.) and $SnCl_2 \cdot H_2O$ (5 mol%) heated at 110° for 1-2 h in a round bottom flask fitted with a distillation condenser to remove liberated methanethiol, the catalyst filtered off, the filtrate concentrated, and worked up with purification by column chromatography on silica gel → benzyl 3-oxo-3-phenylpropanethionate. Y 80%. The procedure is short, good-to-excellent yielding, inexpensive and generally applicable to the transesterification of a wide range of methyl alkyl-, aryl- or heteroaryl-β-oxodithiocarboxylates with alcohols such as benzyl alcohol, *n*- and *tert*-butanol, and isopropanol, reactivity decreasing in the order primary > secondary > tertiary (nineteen examples; Y 50-85%). Phenol, *p*-cresol, and *m*-chlorophenol, however, failed to react, while yields were lower with other Lewis acids ($CuCl_2$, $ZnCl_2$, $AlCl_3$ or $FeCl_3$). F.e.s. N.S. Devi, S.J. Singh, O.M. Singh, Tetrahedron Lett. *2013*, *54* (11), 1432-5 [DOI: 10.1016/j.tetlet.2013.01.001].

Trifluoromethanesulfonic anhydride s. under Diphenyl sulfoxide *Tf_2O*
Trimethylsilyl triflate s. under N-(p-Tolylthio)-ε-caprolactam *Me_3SiOTf*

Hypochlorous acid *HOCl*
Potassium dichloroiodate *$KICl_2$*

Replacement of carbonyl sulfur by oxygen C=S → CO

ureas from thioureas with chlorine cf. *24*, 231; with aq. K-dichloroiodate at room temperature, with simultaneous ar. iodination at elevated temperature, s. G.M. Viana, L.C.d.S. Aguiar, J.d.A. Ferrão, A.B.C. Simas, M.G. Vasconcelos, Tetrahedron Lett. *2013*, *54* (8), 936-40 [DOI: 10.1016/j.tetlet.2012.12.045]; thiocoumarins from coumarins with HOCl in a pH 4.8 buffered medium, and potential application to HOCl detection at low levels based on selective and efficient colorimetric and fluorogenic signaling behavior s. J.O. Moon, J.W. Lee, M.G. Choi, S. Ahn, S.-K. Chang, ibid. *2012*, *53* (48), 6594-7 [DOI: 10.1016/j.tetlet.2012.09.110].

(4,4′-Di-tert-butyl-2,2′-bipyridyl)bis[3,5-difluoro-2-(5-trifluoromethylpyrid-2-yl)phenyl]-iridium(III) hexafluorophosphate/carbon tetrabromide/irradiation ←

Glycosides from aryl thioglycosides under visible-light photoredox catalysis SAr → OR

71. 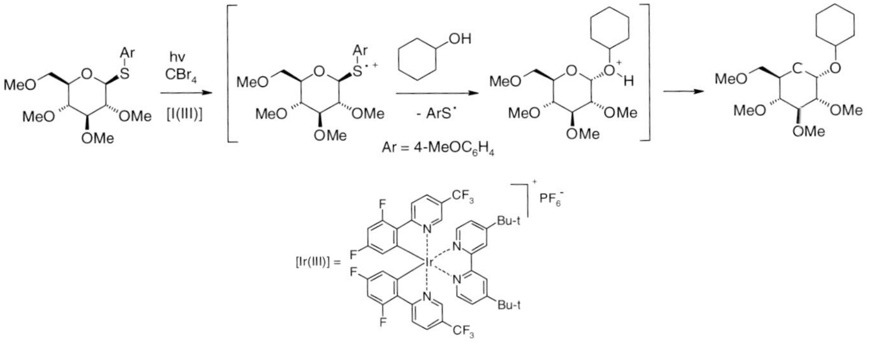

Cyclohexanol (2 eq.) added to a mixture of 2,3,4,6-tetra-O-methyl-D-glucose-4-methoxyphenyl-thioglycoside (0.223 mmol), anh. acetonitrile (1-2 ml), hexafluoro-2-propanol (10 eq.),

Ir[dF(CF_3)ppy]$_2$(dtbbpy)PF_6 (5 mol%) and carbon tetrabromide (0.25 eq.) under argon, the sealed mixture stirred and irradiated (blue LEDs; ~452 nm) for 6 h (NMR), concentrated, and purified chromatographically → cyclohexyl 2,3,4,6-tetra-O-methyl-*D*-glucoside. Y 87% (α/β 1.6:1). This novel visible light initiated glycosylation proceeded via a sulfur radical cation, with glycosyl donors of lower oxidation potential than the iridium catalyst (tetramethyl and tetrabenzyl ethers of 4-methoxyphenyl thioglycoside) being suitable substrates, while donors of higher oxidation potential (phenyl thioglycosides or tetraacetyl 4-methoxyphenyl thioglycosides) were unreactive (diacetylated 2-deoxy-*L*-4-methoxyphenyl thioglycoside was also a suitable substrate). The reaction was successful with relatively simple prim., sec. and tert. alcohol acceptors (nine examples; Y 61-87%; α/β 0.75:1 to 1.6:1), but a more highly-functionalized tribenzylglucoside-6-OH acceptor gave somewhat lower yields (23-42%; two examples; α/β 0.34:1 to 1.2:1). The use of the non-nucleophilic protic solvent and sub-stoichiometric amounts of co-oxidant (CBr_4 or $BrCCl_3$) were key to the success of the reaction. F.e.s. W.J. Wever, M.A. Cinelli, A.A. Bowers, Org. Lett. *2013, 15* (1), 30-3 [DOI: 10.1021/ol302941q].

Remaining Elements ↑ OC ↓↑ Rem

1,4-Bis(9-O-dihydroquin[id]ine)phthalazine s. under Osmium tetroxide ←

Copper(II) acetate/cesium carbonate *$Cu(OAc)_2/Cs_2CO_3$*
Copper(II) acetate/di-tert-butyl peroxide *$Cu(OAc)_2$/t-BuOOBu-t*
Copper(I) triflate/tert-butyl peroxybenzoate *CuOTf/PhC(O)OOBu-t*
Copper(II) triflate/triethylenediamine/triethylamine/potassium hydrogen phosphate ←
Chan-Lam-type O-arylation with arylboronic acids OH → OAr
of N-hydroxyphthalimides with $Cu(OAc)_2$/pyridine cf. *79,* 90; **of 6-subst. 2-hydroxypyridines** with $Cu(OTf)_2$/DABCO, Et_3N as base and K_2HPO_4 as additive s. T. Chen, Q. Huang, Y. Luo, Y. Hu, W. Lu, Tetrahedron Lett. *2013, 54* (11), 1401-4 [DOI: 10.1016/j.tetlet.2012.12.126]; **of ar. oximes** with $Cu(OAc)_2/Cs_2CO_3$ (cf. *61,* 100s*75*) s. M. Mondal, G. Sarmah, K. Gogoi, U. Bora, ibid. *2012, 53* (46), 6219-22 [DOI: 10.1016/j.tetlet.2012.09.003]; **aryl benzoates** with CuOTf and *tert*-butyl peroxybenzoic acid as oxidant/reactant, also N-arylation and oxidative Suzuki arylation of stilbenes, s. E. Shirakawa, R. Nishikawa, N. Uchiyama, I. Hata, T. Hayashi, Chem. Lett. *2013, 42* (3), 269-71 [DOI: 10.1246/cl.2013.269]; **O-alkylation of phenols** with B-alkylpinacolboranes using $Cu(OAc)_2$/*t*-BuOOBu-*t*, also N-alkylation, s. S. Sueki, Y. Kuninobu, Org. Lett. *2013, 15* (7), 1544-7 [DOI: 10.1021/ol400323z].

Aluminum oxide (acidic)/hydrogen peroxide *Al_2O_3/H_2O_2*
Phenols from arylboronic acids s. *80,* 97s*83* $B(OH)_2$ → OH

Di-tert-butyl peroxide s. under $Cu(OAc)_2$ *t-BuOOBu-t*
tert-Butyl peroxybenzoate s. under CuOTf *PhC(O)OOBu-t*

Amberlite IR-120/hydrogen peroxide ←
2-Aminothiophenol/potassium carbonate/air *$ArSH/K_2CO_3/O_2$*
Phenols from arylboronic acids
update s. *80,* 97s*82*; rapid, metal-, ligand-, solvent- and base-free method with recyclable Amberlite IR-120 resin as catalyst and aq. H_2O_2 as oxidant, also heteroaromatic analogs, s. N. Mulakayala, Ismail, K.M. Kumar, R.K. Rapolu, B. Kandagatla, P. Rao, S. Oruganti, M. Pal, Tetrahedron Lett. *2012, 53* (45), 6004-7 [DOI: 10.1016/j.tetlet.2012.08.087]; with reusable, acidic Al_2O_3 and 30% aq. H_2O_2 cf. A. Gogoi, U. Bora, ibid. *2013, 54* (14), 1821-3 [DOI: 10.1016/j.tetlet.2013.01.079]; aerobic, metal-free procedure with 2-aminothiophenol/K_2CO_3 as promoter in water [generating H_2O_2 *in situ*] s. P. Kaewmati, E. Somsook, R.N. Dhital, H. Sakurai, ibid. *2012, 53* (45), 6104-6 [DOI: 10.1016/j.tetlet.2012.08.142].

Hydrogen peroxide s. under Al_2O_3 and Amberlite IR-120 *H_2O_2*
Methanesulfonamide s. under Osmium tetroxide *$MeSO_2NH_2$*

Osmium tetroxide/1,4-bis(9-O-dihydroquin[id]ine)phthalazine/methanesulfonamide/ tert-butanol ←

α-Hydroxyketones from enol phosphates C=C-OPO(OR)$_2$ → C(OH)CO

72.

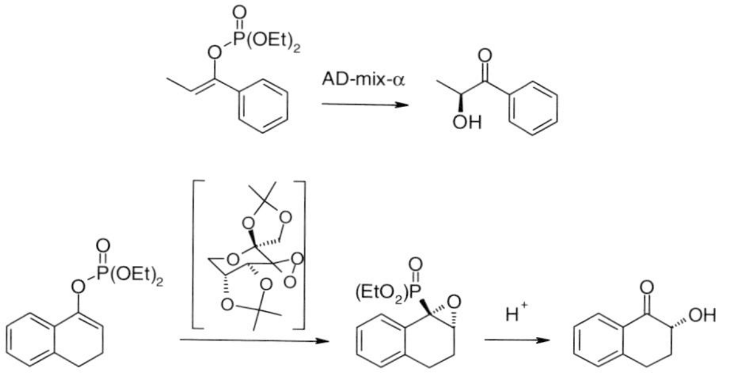

via asym. Sharpless oxidation. A mixture of Sharpless AD-mix-α (1.4 g) and methanesulfonamide (1 mmol) in 1:1 *tert*-butanol/water (10 ml) stirred at room temp. until nearly homogeneous, cooled to 0°, startg. enol phosphate (1 mmol) added in one portion, the resulting heterogeneous mixture stirred vigorously at 0° for 16 h (ice bath), reaction quenched at 0° by the addition of Na_2SO_3 (1 g), stirred for 1 h, and worked up with purification by flash chromatography on silica gel → (S)-2-hydroxy-1-phenylpropan-1-one. Y 47% (e.e. 95%). High yields and enantioselectivity were recorded for the oxidation of *acyclic* (Z)-enol phosphates derived from alkyl aryl ketones to give (S)-products (four examples; Y 40-67%; e.e. 92-100%), reaction with Sharpless AD-mix-β (based on dihydroquinidine) affording the (R)-enantiomers with the same efficiency. A 1,2-diarylenol phosphate, however, was very low yielding (Y 14%), although the enantioselectivity was nonetheless quantitative. Results with benzo-condensed *cyclic* enol phosphates were unsatisfactory, but success was achieved with an alternative procedure based on initial **asym. epoxidation** with an *in situ*-generated [fructose-based] dioxirane (four examples; Y 40-60%; e.e. 83-92%). Here, the degree of enantioselectivity and absolute configuration of the formed cyclic α-hydroxyketones were dependent on the geometry (E/Z) and substitution pattern of the startg. enol phosphates. For the Sharpless procedure, a mnemonic was devised to predict the facial selectivity. F.e.s. E. Krawczyk, G. Mielniczak, K. Owsianik, J. Luczak, Tetrahedron: Asym. *2012, 23* (20-21), 1480-9 [DOI: 10.1016/ j.tetasy.2012.09.012].

Carbon ↑ OC ↓↑ C

Without additional reagents *w.a.r.*

Carbonic acid esters from formamide acetals HC(N<)(OR)$_2$ + CO$_2$ → C(O)(OR)$_2$

73.

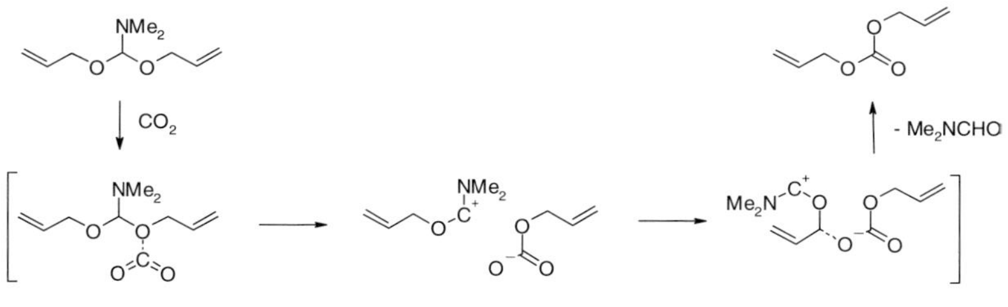

A soln. of N,N-dimethylformamide diallyl acetal (1 mmol) in DMSO-d_6 (2 ml) stirred under CO_2 (20 atm.) at 100° for 24 h, and purified chromatographically → diallyl carbonate. Y 73%. The use of DMSO as solvent was critical to this metal-free fixation of carbon dioxide using dialkyl/diallyl acetals of DMF, at CO_2 pressures of 1-20 atm. (six examples; Y 47-75%). Labelling experiments showed that both CO_2 oxygens were incorporated into the product, rationalized in terms of ambiphilic behavior of the substrates. F.e.s. Y. Takada, A. Matsuoka, Y. Du, H. Naka, S. Saito, Chem. Lett. *2013*, *42* (2), 146-7 [DOI: 10.1246/cl.2013.146].

Irradiation s. under 2-Chloroanthraquinone and Ru(bpy)$_3$Cl$_2$ ///

Microwaves s. under Manganese(II) carbonate or sulfate [\\\\\]

N,N,N',N'',N''-Pentamethyldiethylenetriamine *PMDETA*

Selective O-benzoylation of alcohols with α,α,α-trichloroacetophenone OH → OBz

74.

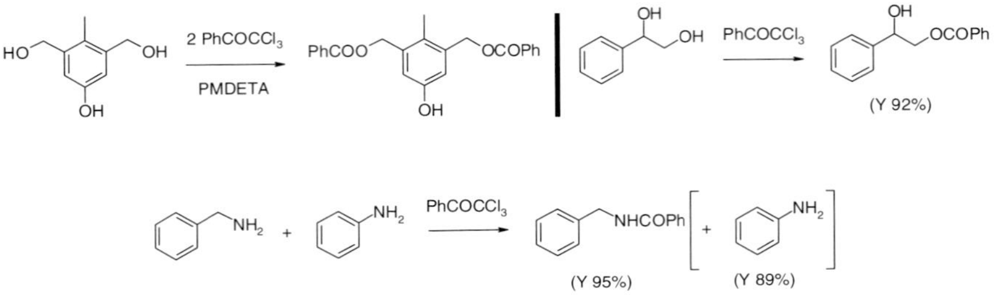

under mild conditions. A soln. of 2,2,2-trichloroacetophenone (2 eq.), 3,5-bis(hydroxymethyl)-4-cresol (2 mmol) and PMDETA (2 mol%) in ethyl acetate (5 ml) stirred at room temp. for 30 h (TLC), and purified chromatographically → 3,5-bis(benzoyloxymethyl)-4-cresol. Y 92%. The inexpensive *tert.* amine catalyst promoted benzoylation of prim. and sec. aliphatic alcohols (incl. benzylic, allylic and propargylic) in the presence of silyl ether, epoxide, ester and halo functionality (eighteen examples; Y 82-96%). Tert. alcohols and phenols were essentially inert under these conditions, and exclusive benzoylation of prim. and sec. alcohols in the presence of phenols was demonstrated (eighteen examples; Y 84-95%), and a single example, exploiting the lower reactivity of sec. alcohols, produced a prim. monobenzoate of 1-phenyl-1,2-ethanediol exclusively (Y 92%). In competition experiments, benzylamine was N-benzoylated (Y 95%) in the presence of aniline (89% recovery), but 3- and 4-aminobenzyl alcohols produced mixtures of O-benzoyl (Y 76-78%) and N,O-dibenzoyl derivs. (13-14%). F.e. and optimization s. R.N. Ram, V.K. Soni, D.K. Gupta, Tetrahedron *2012*, *68* (44), 9068-75 [DOI: 10.1016/j.tet.2012.08.051].

Copper/cesium carbonate *Cu/Cs$_2$CO$_3$*

***o-prim*-Aminodiaryl sulfides from benzothiazoles and ar. iodides** s. *83*, 238 C

Silver triflate s. under ICl *AgOTf*

Aluminum oxide *Al$_2$O$_3$*

N-Subst. carboxylic acid amides from α-iminonitriles C(CN)=NR → C(O)NHR

N-Subst. β-(organothio)carboxylic acid amides via β,γ-ethylene-α-iminonitriles in one-pot – Also N-subst. carboxylic acid amides from aldehydes or prim. alcohols and prim. amines s. *83*, 227

Indium(III) triflate *In(OTf)$_3$*

Transacetalation s. *29*, 187s*83* $C(OR)_2 \rightarrow C(OR')_2$

Boron fluoride *BF$_3$*

Glycosides ←

from acyl glycosides, update s. *44*, 206s*82*; peracetylated allyl β-glycosides from acyl β-glycosides with allyl alcohols and BF_3-etherate (with re-acetylation as necessary) s. J. Khamsi, R.A. Ashmus, N.S. Schocker, K. Michael, Carbohydr. Res. *2012*, *357*, 147-50 [DOI: 10.1016/j.carres.2012.05.008]; β-D-glycopyranosides from acyl D-glycopyranosides with trimethylsilyl triflate and 2-subst. pyridines via reaction of intermediate N-(D-glycopyranosyl)pyridinium triflates with alcohols, also α-selective formation of glycopyranosyl azides or C-glycopyranosides with trimethylsilyl azide or diorganozinc compds., respectively, as nucleophile s. H. Fujioka, Y. Minamitsuji, T. Moriya,

K. Okamoto, O. Kubo, T. Matsushita, K. Murai, Chem. Asian J. *2012, 7* (8), 1925-33 [DOI: 10.1002/asia.201200234]; glycosides **from *o*-(alk-1-ynyl)benzoyl glycosides** (cf. *73*, 89) with iodine as promoter, notably without affecting thioglycosidic bonds, s. S. Dutta, S. Sarkar, S.J. Gupta, A.K. Sen, Tetrahedron Lett. *2013, 54* (8), 865-70 [DOI: 10.1016/j.tetlet.2012.11.101]; glycosides **from 1-methyl- or 1-phenyl-cyclopropyl glycosides** with trimethylsilyl triflate, notably for coupling N-Boc-protected serinates, s. C. Scholl, T. Licisyn, C. Cummings, K. Hughes, D. Johnson, W. Boyko, R. Giuliano, Carbohydr. Res. *2012, 356*, 288-94 [DOI: 10.1016/j.carres.2012.03.008]; cyclic acetals by **transacetalation** of acyclic acetals or ketals with diols or triols without solvent using $In(OTf)_3$ as catalyst (cf. *29*, 187s*81*) s. B.M. Smith, T.M. Kubczyk, A.E. Graham, Tetrahedron *2012, 68* (38), 7775-81 [DOI: 10.1016/j.tet.2012.07.048].

1,3-Dialkylimidazol-2-ylidenes/1,3-dialkylimidazolium ionic liquids ←
Transesterification COOR → COOR′
update s. *47*, 182s*82*; with cathodically-generated 1,3-dialkylimidazol-2-ylidenes as catalyst in the corresponding 1,3-dialkylimidazolium ionic liquid as medium s. I. Chiarotto, M. Feroci, G. Sotgiu, A. Inesi, Eur. J. Org. Chem. *2013* (2), 326-31 [DOI: 10.1002/ejoc.201201023]; transesterification of ethyl β-ketocarboxylates with $MnSO_4$ or $MnCO_3$ as catalyst with acceleration under sonication or microwave irradiation s. G. Krishnaiah, B. Sandeep, D. Kondhare, K.C. Rajanna, J.N. Reddy, Y.R. Rao, P.K. Zhubaidha, Tetrahedron Lett. *2013, 54* (7), 703-6 [DOI: 10.1016/j.tetlet.2012.12.030].

Lipase ←
Regioselective enzymatic mono-O-acylation OH → OAc
update s. *43*, 200s*82*; conformational effects on regioselective 5′-O-acylation of nucleosides with *Candida antarctica* lipase B s. S. Martínez-Montero, S. Fernández, Y.S. Sanghvi, V. Gotor, M. Ferrero, Eur. J. Org. Chem. *2012* (28), 5483-90 [DOI: 10.1002/ejoc.201200609].

Lipase or immobilized lipase ←
Kinetic resolution of alcohols by asym. transesterification
with vinyl acetate, update s. *44*, 214s*82*; with (S)-selective *Candida antarctica* lipase A and (R)-selective *Pseudomonas stutzeri* lipase as a pair of stereocomplementary enzymes for the kinetic resolution of 1,2-diarylethanols, also of 1,2-diarylethanamines via asym. N-acylation, and **dynamic kinetic resolution** with a ruthenium complex as racemizing catalyst (cf. *53*, 500s*83*), s. S. Kim, Y.K. Choi, J. Hong, J. Park, M.-J. Kim, Tetrahedron Lett. *2013, 54* (10), 1185-8 [DOI: 10.1016/j.tetlet.2012.11.147]; enhancement of enzymatic activity in organic media with recyclable lipases immobilized on hydrophobic, room-temperature, solid-phase ionic liquids s. J.K. Lee, Bull. Korean Chem. Soc. *2012, 33* (10), 3458-60 [DOI: 10.5012/bkcs.2012.33.10.3458]; immobilized lipase-catalyzed kinetic resolution of sec. 2-furylcarbinols with Novozyme 435 in dry diisopropyl ether s. P. Hara, M.-C. Turcu, R. Sundell, M. Tosa, C. Paizs, F.-D. Irimie, L.T. Kanerva, Tetrahedron: Asym. *2013, 24* (2-3), 142-50 [DOI: 10.1016/j.tetasy.2012.11.016]; *continuous-flow* enzymatic resolution of tetrahydroisoquinoline-based N-protected 2-, 3- and 4-aminoalcohols s. L. Schönstein, E. Forró, F. Fülöp, ibid. *24* (4), 202-6 [DOI: 10.1016/j.tetasy.2013.01.006].

2-Chloroanthraquinone/oxygen/irradiation *2-Cl-AQN/O_2/hν*
Carboxylic acids from glycols CH(OH)C(OH) → COOH
via aerobic photooxidative C-cleavage

75.

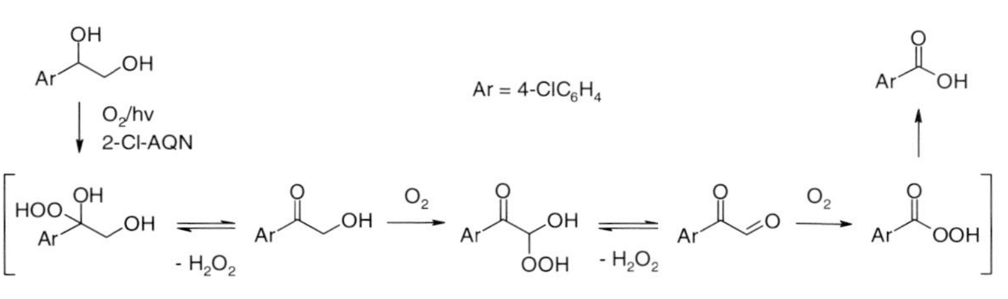

A soln. of 1-(4-chlorophenyl)-1,2-ethanediol (0.3 mmol) and 2-chloroanthraquinone (10 mol%) in dry ethyl acetate (3 ml) stirred and irradiated externally (400 W high pressure Hg lamp) in a

Pyrex tube under O_2 (balloon) for 20 h, concentrated *in vacuo*, dissolved in 1% aq. NaOH, acidified with 2 *M* aq. HCl, extracted with ether, and purified chromatographically → 4-chlorobenzoic acid. Y 89%. Photooxidation of terminal 1,2-diols using O_2 as sole oxidant was successful for electron-diverse 1-aryl-1,2-ethanediols (six examples; Y 80-89%; a 2-naphthyl analog gave only 30%). Linear alkyl 1,2-diols were less reactive (three examples; Y 52-78%; 0-10% recovered substrate), with cyclohexane-1,2-diol giving adipic acid (Y 59%; 18% recovered substrate), and tetraphenyl-1,2-ethanediol giving benzophenone (Y 70%).

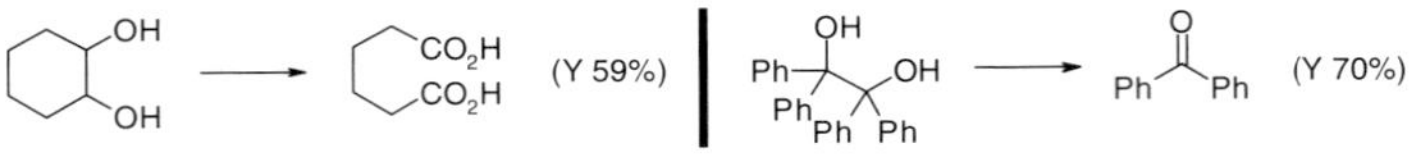

F.e. and optimization s. Y. Matsusaki, T. Yamaguchi, N. Tada, T. Miura, A. Itoh, Synlett *2012*, *23* (14), 2059-62 [DOI: 10.1055/s-0032-1316585].

Polymer-based triarylphosphine s. under Ozone ⓟ-PAr_2
Oxygen s. under 2-Chloroanthraquinone O_2

Ozone/polymer-based triarylphosphine O_3/ⓟ-PAr_2
Oxo compds. from ethylene derivs. by C-cleavage C=C → CO
under ozonolysis with a polymer-based triarylphosphine s. *41,* 231; formylacetic esters via ozonolysis of *trans*-β-hydromuconic esters s. S. Edvinsson, S. Johansson, A. Larsson, Tetrahedron Lett. *2012, 53* (50), 6819-21 [DOI: 10.1016/j.tetlet.2012.10.021]; ar. aldehydes from styrenes by a highly efficient oxidative cleavage with a (5,10,15,20-tetraphenylporphyrinato)cobalt(II) complex under molecular oxygen in benzotrifluoride s. X. Zhou, H. Ji, Chin. J. Chem. *2012, 30* (9), 2103-8 [DOI: 10.1002/cjoc.201200621].

Trimethylsilyl triflate Me_3SiOTf
Glycosides from cyclopropyl glycosides s. *82,* 106s*83* OR → OR′

Iodine I_2
Glycosides from *o*-(alk-1-ynyl)benzoyl glycosides s. *44,* 206s*83, 73,* 89s*83* OAc → OR

Perchloric acid-silica $HClO_4$-SiO_2
Glycosides from glycosyl trihalogenoacetimidates OC(=N-)$CHal_3$ → OR
update s. *60,* 103s*82*; α-ribosides from β-ribosyl N-phenyltrifluoroacetimidates with trimethylsilyl triflate or $HClO_4$-SiO_2, notably for coupling hydroxylated α-aminoesters, also formation of α-amino-ester-based α-acyl ribosides and acylamine analogs, s. H.A.V. Kistemaker, G.J. van der Heden van Noort, H.S. Overkleeft, G.A. van der Marel, D.V. Filippov, Org. Lett. *2013, 15* (9), 2306-9 [DOI: 10.1021/ol400929c]; synthesis of heparan sulfate **oligosaccharides** on a *fluorous* support for simplified purification s. C. Zong, A. Venot, O. Dhamale, G.-J. Boons, ibid. *15* (2), 342-5 [DOI: 10.1021/ol303270v].

(meso-Tetraphenylporphyrinato)manganese(III) chloride *TPPMnCl*
Ketones from acetylene derivs. R-C≡C-R′ → RC(O)R′
via oxidative 1,2-migration with loss of one C-atom

76. 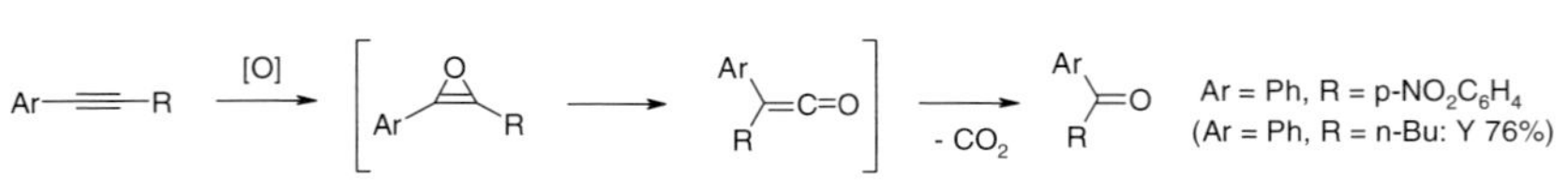

A mixture of 1-nitro-4-(2-phenylethynyl)benzene (1 mmol), TPPMnCl (0.1 mol%) and *n*-hexane (5 ml) in a pressure vessel stirred at 150° under O_2 (2 MPa) for 4 h, cooled to room temp., and purified directly by flash chromatography on silica gel → 4-nitrobenzophenone. Y 82% (conversion 95%). High conversions (89-95%) and good yields (70-85%) were obtained for a variety of diaryl- and arylalkyl-ethynes (seven examples), wherein aromatic rings may bear electron-donating or -withdrawing (MeO, Cl, NO_2) groups, and the alkyl groups are preferably ethyl, *n*-propyl or *n*-butyl (acetophenone afforded a 55% yield with 75% conversion). Both α- and β-naphthyl-(phenyl)-acetylenes afforded the same product (2-benzoylnaphthalene) in moderate yields (49-54%)

and conversions (58-62%), the former via rearrangement of the initially-formed 1-benzoyl-naphthalene. Dialkyl, terminal, and trimethylsilyl acetylenes all gave messy product mixtures, in which no more than trace amounts (<5%) of the desired carbonyl compounds could be detected. The illustrated mechanism was proposed following experiments that discounted a radical process and established that the rearrangement is strictly intramolecular, not involving either benzil or α-subst. acetic acid intermediates. F.e.s. W.-B. Sheng, Q. Jiang, W.-P. Luo, C.-C. Guo, J. Org. Chem. *2013*, *78* (11), 5691-3 [DOI: 10.1021/jo400209p].

Manganese(II) carbonate or sulfate/microwaves — *Mn(II)/[\\\\\]*
Transesterification s. *47*, 182s*83* — COOR → COOR′

(5,10,15,20-Tetraphenylporphyrinato)cobalt(II) complex — *[Co(II)]*
Ar. aldehydes from styrenes by C-cleavage s. *43*, 231s*83* — $CH{=}CH_2$ → CHO

Dichlorotris(2,2′-bipyridyl)ruthenium(II)/piperidine/irradiation — *$Ru(bpy)_3Cl_2/(CH_2)_5NH$/⫻*
Oxo compds. from aldehydes with loss of one C-atom under visible-light photoredox catalysis — CHCHO → CO

77. 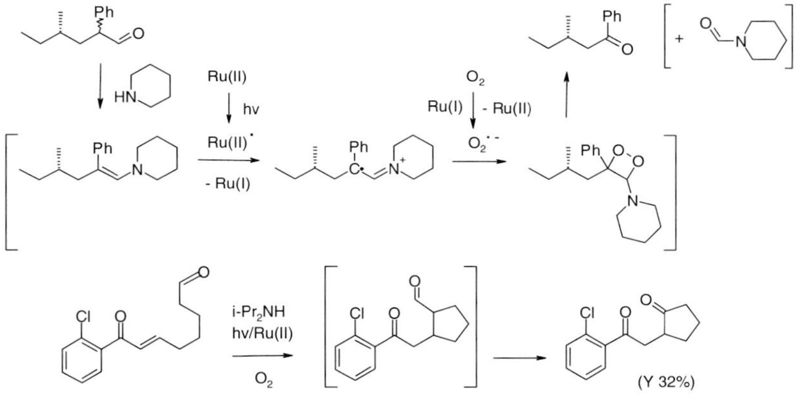

A mixture of 4(S)-methyl-2-phenylhexanal (0.1 mmol), piperidine (3 eq.) and $Ru(bpy)_3Cl_2$ (5 mol%) in acetonitrile (1 ml) irradiated (15 W fluorescent light bulb) under air at room temp. for 6 h, concentrated *in vacuo*, and purified chromatographically → (S)-3-methyl-1-phenylpentan-1-one. Y 76% (90% by GC). This novel and experimentally simple C-C cleavage using visible light was apparently general, affording ketones and aldehydes via oxidative deformylation (eleven examples; Y 60-83%), with concomitant isomerization of alkenes. The method was applied to a synthesis of **2-β-ketocyclopentanones** from ε,ζ-ethylene-η-ketoaldehydes via a one-pot intramolecular Michael addition-oxidative deformylation sequence, in moderate yields (20-33%; five examples). F.e., optimization and substrate prepn. s. H. Sun, C. Yang, F. Gao, Z. Li, W. Xia, Org. Lett. *2013*, *15* (3), 624-7 [DOI: 10.1021/ol303437m].

Elimination ⥣

Hydrogen ↾ OC ⥣ H

Irradiation s. under CaI_2, Trifluoroacetic anhydride, Pentaammine(chloro)cobalt(III) dichloride and Tris(4,4′-di-tert-butyl-2,2′-bipyridyl)ruthenium(II) bis(hexafluorophosphate) — ⫻
Silver acetate s. under $Pd(OAc)_2$ — *AgOAc*

Copper complexes or salts/oxygen (air) ←

Transition metal-catalyzed aerobic oxidation of alcohols CHOH → CO

update s. *26,* 463s*82*; *under copper catalysis* by a modification of Markó's method at room temperature with CuCl/4,7-diphenyl-1,10-phenanthroline and added di-*tert*-butyl azodicarboxylate/Cs_2CO_3 s. T. Nishii, T. Ouchi, A. Matsuda, Y. Matsubara, Y. Haraguchi, T. Kawano, H. Kaku, M. Horikawa, T. Tsunoda, Tetrahedron Lett. *2012, 53* (44), 5880-2 [DOI: 10.1016/j.tetlet. 2012.08.095]; aerobic oxidation of benzyl alcohols with $CuBr_2/HNO_3$ in 1-butyl-3-methylimidazolium fluoroborate s. C.M. Lim, S.J. Ha, J.C. Lee, Bull. Korean Chem. Soc. *2012, 33* (12), 4258-60 [DOI: 10.5012/ bkcs.2012.33.12.4258]; *heterogeneous* aerobic procedure *under ruthenium catalysis* with ruthenium ion-exchanged hydroxyapatites having different sizes and compositions s. S. Kim, J.-H. Jung, D.-H. Kim, D.K. Woo, J.B. Park, M.Y. Choi, K.-Y. Kwon, Bull. Korean Chem. Soc. *2013, 34* (1), 221-4 [DOI: 10.5012/bkcs.2013.34.1.221]; *under palladium catalysis* with recyclable, highly active palladium nanoparticles dispersed on amino-functionalized mesocellular foam under solvent-free conditions (TON up to 365,000 h^{-1}) s. E.V. Johnston, O. Verho, M.D. Kärkäs, M. Shakeri, C.-W. Tai, P. Palmgren, K. Eriksson, S. Oscarsson, J.-E. Bäckvall, Chem. Eur. J. *2012, 18* (39), 12202-6 [DOI: 10.1002/chem.201202157]; aerobic oxidation of aliphatic, allylic, heterocyclic, benzylic and heteroatom-containing alcohols 'on pure water' at low temperature with palladium nanoparticles supported on nano-fibrillated mesoporous carbon [prepared by the carbonization of 1-methyl-3-phenethylimidazolium hydrogen sulfate in the presence of SBA-15] s. B. Karimi, H. Behzadnia, M. Bostina, H. Vali, ibid. *2012, 18* (28), 8634-40 [DOI: 10.1002/ chem.201200373].

Copper complexes or salts/oxidant ←

Transition metal-catalyzed oxidation of alcohols

update s. *35,* 160s*82*; mild, gram-scale procedure *under neutral conditions* for oxidation of acid- or base-sensitive alcohols *under copper catalysis* with CuBr as catalyst and di-*tert*-butyldiaziridinone as oxidant s. Y. Zhu, B. Zhao, Y. Shi, Org. Lett. *2013, 15* (5), 992-5 [DOI: 10.1021/ol303431h]; selective oxidation of propargyl alcohols with a 2-hydroxyquinoxaline-condensed copper(II) salen complex and *tert*-butyl hydroperoxide as oxidant s. K.C. Weerasiri, A.E.V. Gorden, Eur. J. Org. Chem. *2013* (8), 1546-50 [DOI: 10.1002/ejoc.201201394]; oxidation of prim. and sec. alcohols *under zinc catalysis* with $ZnBr_2$ and Chloramine-T as oxidant s. P. Wang, J. Cai, J. Yang, C. Sun, L. Li, H. Hu, M. Ji, Tetrahedron Lett. *2013, 54* (6), 533-5 [DOI: 10.1016/j.tetlet.2012.11.076]; of benzyl alcohols to ar. aldehydes or arylcarboxylic acid esters (OC⇃↑H) with ligated $ZnBr_2$ in the presence of trifluoroacetic acid and H_2O_2 s. X.-F. Wu, Chem. Eur. J. *2012, 18* (29), 8912-5 [DOI: 10.1002/chem.201201105]; *under rhenium catalysis* with $HReO_4$ and di-*p*-tolyl sulfoxide as oxidant s. S.C.A. Sousa, J.R. Bernardo, A.C. Fernandes, Tetrahedron Lett. *2012, 53* (46), 6205-8 [DOI: 10.1016/j.tetlet.2012.08.145]; *heterogeneous* procedure *under cobalt catalysis* with magnetically recoverable cobalt nanoparticles-on-ferrite [$Co\text{-}Fe_3O_4$] and *tert*-butyl hydroperoxide as oxidant s. M.B. Gawande, A. Rathi, I.D. Nogueira, C.A.A. Ghumman, N. Bundaleski, O.M.N.D. Teodoro, P.S. Branco, ChemPlusChem. *2012, 77* (10), 865-71 [DOI: 10.1002/cplu.201200081]; oxidation of alcohols *in water* with $[Co(NH_3)_5Cl]Cl_2$ as a sacrificial oxidant *under visible-light photoredox catalysis* with a cationic dinuclear ruthenium(II) complex as sensitizer s. D. Chao, W.-F. Fu, Chem. Commun. *2013, 49* (37), 3872-4 [DOI: 10.1039/c3cc00305a]; selective oxidation of sec. alcohols *under ruthenium catalysis* with $[RuCl(PPh_3)_2(3\text{-phenylindenyl})]$ and K-bis(trimethylsilyl)amide as co-catalyst using acetone as oxidant s. S. Manzini, C.A. Urbina-Blanco, S.P. Nolan, Organometallics *2013, 32* (2), 660-4 [DOI: 10.1021/om301156v]; oxidation of prim. and sec. alcohols, incl. allyl alcohols, under heterogeneous conditions with a ruthenium trichloride-exchanged aluminum metal-organic framework and phenyl iodosoacetate as oxidant s. F. Carson, S. Agrawal, M. Gustafsson, A. Bartoszewicz, F. Moraga, X. Zou, B. Martín-Matute, Chem. Eur. J. *2012, 18* (48), 15337-44 [DOI: 10.1002/chem.201200885]; ketones from sec. alcohols *by dehydrogenation under iridium catalysis* with pincer-type iridium(I) bis(phosphinite) or di(phosphine) complexes *without* a sacrificial hydrogen acceptor s. A.V. Polukeev, P.V. Petrovskii, A.S. Peregudov, M.G. Ezernitskaya, A.A. Koridze, Organometallics *2013, 32* (4), 1000-15 [DOI: 10.1021/om300921q]; by heterogeneous dehydrogenation with cobalt/TiO_2 s. K.-i. Shimizu, K. Kon, M. Seto, K. Shimura, H. Yamazaki, J.N. Kondo, Green Chem. *2013, 15* (2), 418-24 [DOI: 10.1039/c2gc36555c].

Calcium iodide/oxygen/irradiation *CaI_2/O_2/⫻*
δ-Keto-γ-lactones from δ-ketocarboxylic acids ○
via aerobic photooxidative intramolecular α-acoxylation s. *83*, 78

Acetone/isopropanol s. under [(R,R)-N-Tosyl-1,2-diphenylethylenediamino]-(mesitylene)ruthenium(II) *Me_2CO/i-PrOH*

Piperidine nitroxyls/oxidant ←
Nitroxyl-mediated oxidation of alcohols CHOH → CO
update s. *39*, 225s*82*; *selective oxidation* of the benzylic *sec*-hydroxyl group of lignins with 4-acylamino-2,2,6,6-tetramethylpiperidine nitroxyl and added HCl/HNO_3 (10 mol% of each) s. A. Rahimi, A. Azarpira, H. Kim, J. Ralph, S.S. Stahl, J. Am. Chem. Soc. *2013*, *135* (17), 6415-8 [DOI: 10.1021/ja401793n]; oxidation of a wide range of functionalized prim. and sec. alcohols, α-hydroxyamides, and α-hydroxyesters with TEMPO using $Ca(OCl)_2$ as oxidant s. S.R. Reddy, S. Stella, A. Chadha, Synth. Commun. *2012*, *42* (23), 3493-503 [DOI: 10.1080/00397911.2011.584650]; of 2,2,2-trifluoroalcohols *with 4-acylamino-2,2,6,6-tetramethyl-1-oxopiperidinium fluoroborate*/pyridine [with regeneration of the oxidant by treatment of the by-product, 4-acylamino-2,2,6,6-tetramethylpiperidine nitroxyl, with aq. $NaOCl/HBF_4$] s. C.B. Kelly, M.A. Mercadante, T.A. Hamlin, M.H. Fletcher, N.E. Leadbeater, J. Org. Chem. *2012*, *77* (18), 8131-41 [DOI: 10.1021/jo301477s].

Trifluoroacetic anhydride/oxygen/irradiation *$(CF_3CO)_2O/O_2$/⫻*
δ-Keto-γ-lactones from δ-ketocarboxylic acids ○
via aerobic photooxidative intramolecular α-acoxylation

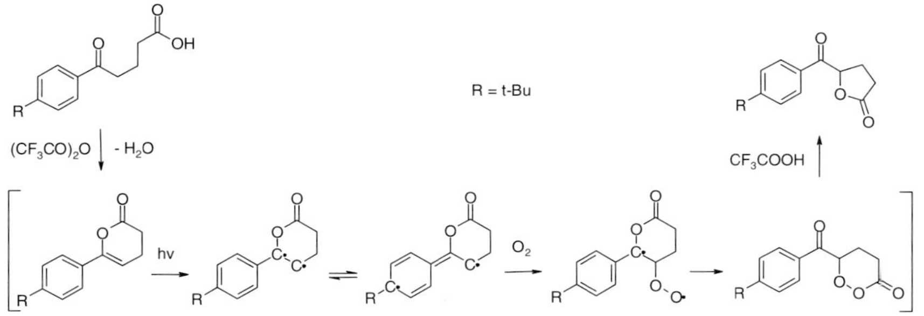

78.

in a fluorous solvent under mild conditions. A soln. of 4-(4-*tert*-butylbenzoyl)butyric acid (0.3 mmol) and trifluoroacetic anhydride (3 eq.) in FC-72 (5 ml) stirred and irradiated externally (2 x 22 W fluorescent tubes) in a Pyrex test tube under O_2 (balloon) for 15 h, quenched with satd. aq. $NaHCO_3$, extracted with ethyl acetate, and purified chromatographically → 5-(4-*tert*-butylbenzoyl)dihydrofuran-2(3*H*)-one. Y 80%. Use of the fluorous solvent was key to the success of the reaction (typical organic solvents gave low yields), attributed to high solubilization of oxygen (a low yield was obtained under air). Electron-diverse 4-aroylbutyric acids (incl. sterically-hindered *o*-methylbenzoyl derivs.) were converted to the corresponding lactones in moderate to good yields (53-80%; eight examples). Electron-rich 4-methoxybenzoyl- and 4-acetyl-butyric acids were poor substrates (Y 13-16%), while homologous 5-benzoylpentanoic and des-homologous 3-benzoylpropionic acids gave none of the corresponding lactones. Under argon atmosphere or in the absence of irradiation, 4-benzoylbutyric acid afforded dihydro-6-phenyl-2-pyrone (Y up to 89%). For an analogous conversion promoted by *in situ*-generated iodite/hypoiodite (from Bu_4NI/H_2O_2) s. *80*, 110. F.e. and optimization s. N. Tada, L. Cui, T. Ishigami, K. Ban, T. Miura, B. Uno, A. Itoh, Green Chem. *2012*, *14* (11), 3007-9 [DOI: 10.1039/c2gc36238d]; calcium iodide-catalyzed aerobic photo-oxidative oxylactonization (Y up to 74%) s. N. Tada, T. Ishigami, L. Cui, K. Ban, T. Miura, A. Itoh, Tetrahedron Lett. *2013*, *54* (3), 256-8 [DOI: 10.1016/j.tetlet.2012.11.014].

Acetic acid s. under N-Bromosuccinimide *AcOH*
N-Carbo-tert-butoxy-L-isoleucine s. under Pd(OAc)$_2$ *Boc-Ile-OH*

1-Methyl-3-[ω-(methylthio)alkyl]imidazolium tosylate/N-chlorosuccinimide/triethylamine ←
1-Methyl-3-[ω-(methylsulfinyl)alkyl]imidazolium tosylate/oxalyl chloride/triethylamine ←

Metal-free oxidation of alcohols CHOH → CO

update s. *21,* 285s*82*; Swern-type oxidation with an ionic liquid-tagged sulfoxide [e.g. 1-methyl-3-[ω-(methylsulfinyl)alkyl]imidazolium tosylate] in the presence of $(COCl)_2/Et_3N$, or by Corey-Kim-type oxidation with a 1-methyl-3-[ω-(methylthio)alkyl]imidazolium tosylate and added NCS/Et_3N, for a clean, odor-free oxidation of prim., sec. and allylic alcohols with a simplified purification s. D. Tsuchiya, M. Tabata, K. Moriyama, H. Togo, Tetrahedron *2012, 68* (34), 6849-55 [DOI: 10.1016/j.tet.2012.06.026]; direct oxidation of benzyl alcohols, benzyl halides or alkylbenzenes with Oxone in 1:1 acetonitrile/water, limitations, s. K.N. Parida, S. Jhulki, S. Mandal, J.N. Moorthy, ibid. *68* (47), 9763-8 [DOI: 10.1016/j.tet.2012.09.028]; α-diketones from α-hydroxyketones with mesoporous carbon nitride under visible-light irradiation (cf. *79,* 103) s. Z. Zheng, X. Zhou, Chin. J. Chem. *2012, 30* (8), 1683-6 [DOI: 10.1002/cjoc.201200284].

Phenyl iodosoacetate s. under Pd(OAc)$_2$ *PhI(OAc)$_2$*
N-Chlorosuccinimide s. under 1-Methyl-3-[ω-(methylthio)alkyl]imidazolium tosylate *NCS*

N-Bromosuccinimide/acetic acid *NBS/AcOH*

2-(Acylene)furans from 2-γ-ketofurans via oxidative recyclization C.O

79.

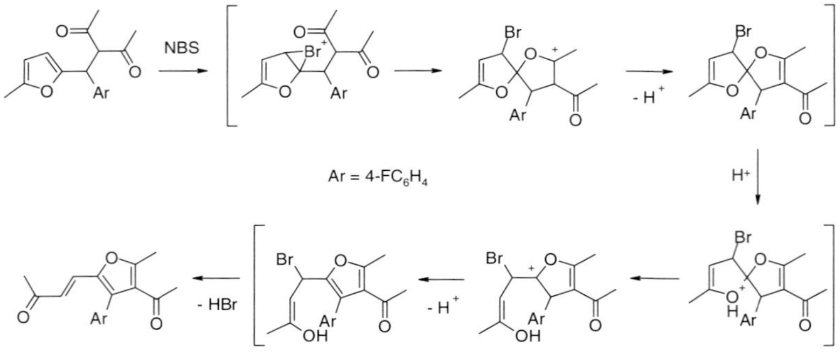

A mixture of 2-methyl-5-[2,2-diacetyl-1-(4-fluorophenyl)ethyl]furan (0.3 mmol), THF (3 ml), water (1 ml), acetic acid (10 mol%) and NBS (1.1 eq.) stirred at room temp. for 30 min (TLC), the mixture heated at 80° for 15 h, cooled to room temp., quenched with satd. aq. $Na_2S_2O_3$, extracted with ethyl acetate (with satd. aq. $NaHCO_3$ washing), and purified chromatographically → 3-acetyl-4-(4-fluorophenyl)-2-methyl-5-(3-oxobut-2-en-1-yl)furan. Y 74%. This experimentally simple procedure uses readily available substrates (from 2-furylcarbinols and active methylene compds.) to afford tri- and tetra-subst. furans via a dearomatization-spirocyclization-rearomatization sequence (fifteen examples; Y 28-75%). The choice of suitable acidic oxidizing conditions was essential for suppression of alternative reaction pathways. F.e. and optimization s. H. Yu, W. Zhong, T. He, W. Gu, B. Yin, Tetrahedron Lett. *2013, 54* (10), 1256-60 [DOI: 10.1016/j.tetlet.2012.12.085].

Oxalyl chloride s. under 1-Methyl-3-[ω-(methylsulfinyl)alkyl]imidazolium tosylate *(COCl)$_2$*
Oxygen s. under CaI$_2$, Tris(4,4′-di-tert-butyl-2,2′-bipyridyl)ruthenium(II) bis(hexafluorophosphate) and Transition metal complexes *O$_2$*

Ozone *O$_3$*

Metal-free oxidation of benzyl alcohols s. *21,* 285s*83* CHOH → CO

Transition metal complexes/oxygen (air) ←
Transition metal-catalyzed aerobic oxidation of alcohols s. *26,* 463s*83* CHOH → CO

Transition metal complexes/oxidant ←
Transition metal-catalyzed oxidation of alcohols s. *35,* 160s*83*

Pentaammine(chloro)cobalt(III) dichloride/cationic dinuclear ruthenium(II) complexes/ irradiation ←
Oxidation of alcohols in water
under visible-light photoredox catalysis s. *35,* 160s*83*

Tris(4,4′-di-tert-butyl-2,2′-bipyridyl)ruthenium(II) bis(hexafluorophosphate)/ triethylenediamine/oxygen/irradiation ←
Oxo compds. from alcohols
under visible-light photoredox catalysis s. *83,* 5

Cationic dinuclear ruthenium(II) complexes s. under Pentaammine(chloro)cobalt(III) dichloride *[Ru(II)]*

(Mesitylene)[(R,R)-N-tosyl-1,2-diphenylethylenediamine]ruthenium(II)/ sodium tert-butoxide/acetone/isopropanol ←
Lactones from *prim*-hydroxyketones ○
via ruthenium(II)-catalyzed asym. intramolecular ketone hydroacylation

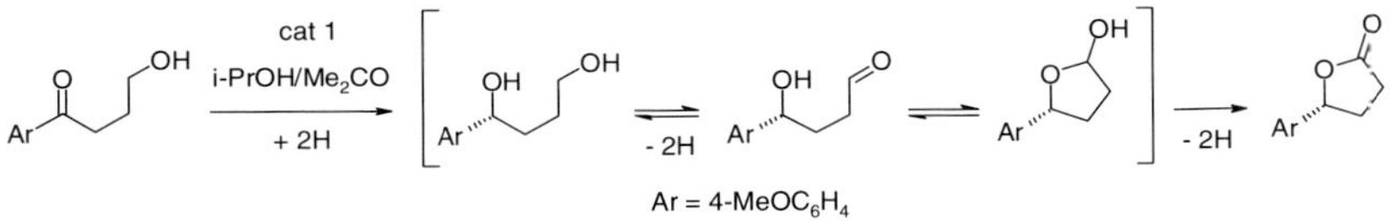

80. 

(Y 85%; e.e. 90%)

A novel strategy for the intramolecular asym. hydroacylation of 1,4- and 1,5-hydroxyketones is described, involving generation of sensitive ketoaldehyde substrates *in situ,* using Noyori's transfer-hydrogenation catalyst. **E: Chiral γ-aryl-γ-lactones**. Ethyl acetate (0.6 ml), acetone (1.2 eq.), isopropanol (3 eq.) and 4-(4-methoxyphenyl)-4-oxobutan-1-ol (0.3 mmol) added sequentially to a mixture of [(R,R)-TsDPEN](mesitylene)RuCl (catalyst precursor; 5 mol%) and Na-*tert*-butoxide (5 mol%), the sealed mixture stirred at 0° for 3 d, and purified chromatographically → (+)-(R)-5-(4-methoxyphenyl)-γ-butyrolactone. Y 70% (e.e. 92%). Using acetone as hydrogen acceptor and generating the required ruthenium catalyst and ketoaldehydes *in situ* provided clean and effective conversion to γ-lactones (ten examples; Y 65-91%; e.e. 87-93%), even for racemization-prone substrates (e.g. 4-methoxyphenyl ketones). The addition of isopropanol (also formed during the reaction) is thought to promote formation of an active ruthenium hydride species (larger quantities of isopropanol resulted in over-reduction). The method was also applied to the more strained δ-lactones and gave good results for electron-diverse aryl and furyl ketones (five examples; Y 65-81%; e.e. 90-96%). Interestingly, modest results obtained with simple 2-furyl and alkynyl ketones (three examples; Y <10 to 42%; e.e. 86%) were significantly improved for analogous substrates containing *gem*-dimethyl in the *pro*-4 position (three examples; Y 55-98%; e.e. 90%). A similar catalyst was also effective for direct cyclization of an *o*-ketobenzaldehyde to the corresponding phthalide (Y 85%; e.e. 90%). F.e. and substrate prepn. s. S.K. Murphy, V.M. Dong, J. Am. Chem. Soc. *2013,* *135* (15), 5553-6 [DOI: 10.1021/ja4021974].

Hydrido[1,3-bis(6-methylpyrid-2-ylimino)isoindolinato]bis(triphenylphosphine)-ruthenium(II) *[Ru(II)]*

Base-free, acceptorless ruthenium(II)-catalyzed dehydrogenation of alcohols CHOH → CO

81. 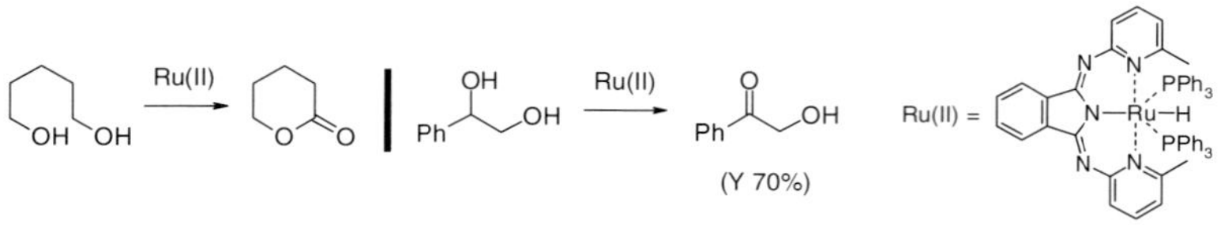

Lactones from diols (cf. *37*, 236s*72*). 1,5-Pentanediol (0.5 mmol) added to a mixture of ruthenium catalyst (5 mol%) and toluene (2 ml) under N_2, the mixture stirred under reflux in an open system for 24 h, and analyzed by GC-MS → tetrahydropyranone. Y 99%. In the absence of an H_2-acceptor, dehydrogenation of diols to lactones (two examples; Y 88-99%), prim. alcohols to esters (*n*-butanol: Y 99%; ethanol: Y 8%; benzyl alcohol: Y 50%), and a sec. alcohol to a ketone (one example; unspecified yield) was demonstrated. Not unsurprisingly, the reaction was significantly inhibited by H_2 pressure when carried out in a closed system. Selective dehydrogenation of a sec. alcohol in the presence of a prim. alcohol was demonstrated for 1-phenyl-1,2-ethanediol (Y 70%) and 1,3-butanediol (52%), with the corresponding hydroxyketones obtained as sole products. F.e. and catalyst prepn. s. K.-N.T. Tseng, J.W. Kampf, N.K. Szymczak, Organometallics *2013*, *32* (7), 2046-9 [DOI: 10.1021/om4000677].

Palladium(II) acetate/N-carbo-tert-butoxy-L-isoleucine/phenyl iodosoacetate/potassium acetate ←

Palladium(II) acetate/phenyl iodosoacetate/silver acetate/sodium acetate/cesium acetate ←

2(3*H*)-Benzofuranones from α-arylcarboxylic acids ○

Palladium(II)-catalyzed asym. dehydrogenative intramolecular O-arylation

82. 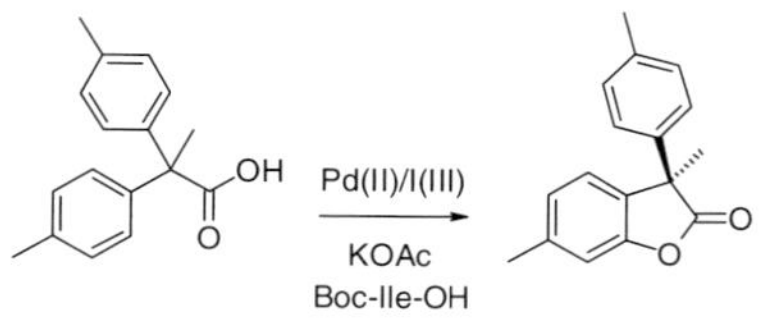

Chiral 3-aryl-2(3*H*)-benzofuranones. *tert*-Butanol (2 ml) added to a mixture of 4,4′,α-trimethyl-α,α-diphenylacetic acid (0.2 mmol), $Pd(OAc)_2$ (5 mol%), N-Boc-L-isoleucine (30 mol%), KOAc (2 eq.) and $PhI(OAc)_2$ (1.5 eq.) under air, the sealed mixture stirred at 100° for 12 h, cooled to room temp., concentrated *in vacuo*, and purified chromatographically → (3R)-6-methyl-3-(4-tolyl)-2(3*H*)-benzofuranone. Y 84% (e.e. 96%). This novel Pd(II)/Pd(IV)-catalyzed C-H activation/C-O bond formation was free of C-acetoxylation products and successful for a series of α,α-dialkyl-phenylacetic acids using Ac-Gly-OH as non-chiral additive (Boc-Val-OH at 80° preferred for spiro-alkyl derivs.) (twenty examples; Y 56-94%). Lowest yields were obtained from electron-poor substrates (thien-2-yl-2-methylpropanoic acid gave the corresponding thienofuranone in 50% yield) and the reaction was incompatible with *o*-subst. or α-H-phenylacetic acids. The presence of mono-protected amino acids was essential for optimal results (simple carboxylic acid additives were ineffective). Application of the conditions to α-aryl-α-alkylphenylacetic acids (using Boc-Ile-OH as optimum chiral additive) afforded the title products with good enantiocontrol (e.e. 89-96%; eight examples; Y 37-86%). F.e., optimization and substrate prepn. s. X.-F. Cheng, Y. Li, Y.-M. Su, F. Yin, J.-Y. Wang, J. Sheng, H.U. Vora, X.-S. Wang, J.-Q. Yu, J. Am. Chem. Soc. *2013*, *135* (4), 1236-9 [DOI: 10.1021/ja311259x]; achiral 3,3-disubst. 2(3*H*)-benzofuranones, incl. 3-spiro-derivs., with $Pd(OAc)_2/PhI(OAc)_2$/AgOAc/NaOAc/CsOAc s. M. Yang, X. Jiang, W.-J. Shi, Q.-L. Zhu, Z.-J. Shi, Org. Lett. *2013*, *15* (3), 690-3 [DOI: 10.1021/ol303569b].

Via intermediates *v.i.*
Diastereoselective formation of 2,7-dioxabicyclo[4.4.0]decan-4-ones from 2-α,5-γ-dihydroxyfurans via 1,6-dioxaspiro[4.5]dec-9-en-8-ones ○

83. 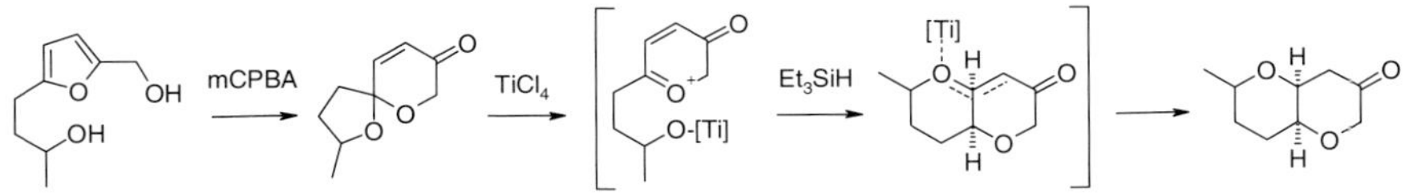

m-Chloroperoxybenzoic acid (*m*CPBA; 77%, 1.1 eq.) added portionwise to a soln. of startg. furfuryl alcohol (1 mmol) in methylene chloride (5 ml) at 0°, the mixture stirred for 2 h, camphorsulfonic acid (1 eq.) added, the mixture stirred for 15 min, quenched with satd. aq. $NaHCO_3$, extracted with methylene chloride, concentrated *in vacuo*, and purified chromatographically → 2-methyl-1,6-dioxaspiro[4.5]dec-9-en-8-one. (Y 78%; d.r. 3:1), a portion of which (0.1 mmol) dissolved in dry methylene chloride (1 ml) at -78°, triethylsilane (12 eq.) and $TiCl_4$ (12 eq.) added, the mixture stirred for 1 h, quenched with 1 *M* aq. HCl, extracted with methylene chloride, concentrated *in vacuo,* and purified chromatographically → 8-methyl-2,7-dioxabicyclo[4.4.0]decan-4-one (Y 53%; d.r. 3:1). Initial Achmatowicz rearrangement of the dihydroxyfurans (by treatment with *m*CPBA or N-bromosuccinimide) afforded spiroketals carrying H, methyl, allyl, propenyl, isopropyl, phenyl or siloxyalkyl substituents on the pyrone ring, with incorporation of α-methyl and α,α-dimethyl substituents in the tetrahydrofuran moiety being noteworthy (nine examples; Y 56-93%; d.r. 3:1 to 8:1). The analogous [6.6]-spiroketal was similarly obtained with high diastereoselectivity (d.r. 15:1; Y 84%), following treatment of the crude reaction mixture with Amberlyst 15. Subsequent unprecedented reductive intramolecular oxa-Michael addition of the spiroketals afforded the title products, as single diastereomers in some cases, (d.r. 4:1 to 100:0; eight examples; Y 32-65%; the [6.6]-spiroketal gave 35% of a single diastereomer via 7-*exo-trig* oxa-Michael cyclization, again accelerated by the presence of Amberlyst 15). The spiroketal incorporating the *gem*-dimethylated furan gave only decomposition products, however. F.e., substrate prepn., and experimental evidence for the mechanism of the reductive cyclization step s. L. Zhu, L. Song, R. Tong, Org. Lett. *2012*, *14* (23), 5892-5 [DOI: 10.1021/ol302813e].

Oxygen ↑ OC ⇑ O

2,3-Diphenylcyclopropenone/methanesulfonic acid ←
Cyclic ethers from diols s. *83*, 29 ○

Di-n-butyltin oxide/triethylamine/p-toluenesulfonyl chloride *$Bu_2SnO/Et_3N/p$-TsCl*
***trans*-5-Subst. 3-hydroxytetrahydrofurans from *syn*-1,2,4-triols via tin-catalyzed O[1]-tosylation**

84. 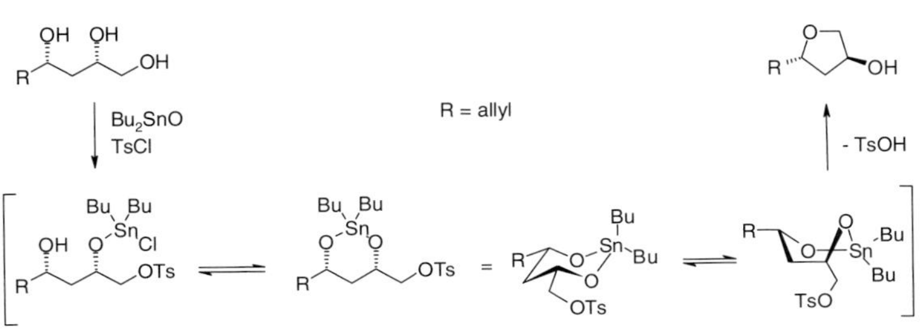

A mixture of *syn*-hept-6-ene-1,2,4-triol (0.175 mmol), *p*-toluenesulfonyl chloride (1.1 eq.), Bu_2SnO (5 mol%) and triethylamine (1.1 eq.) heated under reflux in methylene chloride (2 ml) for 3.5 h

(TLC), quenched with satd. aq. NH_4Cl, extracted with ethyl acetate, and purified chromatographically → *trans*-2-allyl-4-hydroxytetrahydrofuran. Y 82%. Initial regioselective O[1]-tosylation of *syn*-1,2,4-triols unexpectedly afforded tetrahydrofurans exclusively via subsequent intramolecular cyclization (three examples; Y 56-85%). Interestingly the corresponding *anti* isomers gave low yields of *cis*-tetrahydrofurans (Y 5-12%), the major products being the expected O[1]-tosylates (Y 66-83%). This difference in reactivity was rationalized in terms of formation of a 2,4-diol-tin chelate that would be significantly more stable for the *syn*-isomer due to decreased 1,3-diaxial steric repulsion compared to the *anti*-isomer. This difference in chelate stability was predicted by *ab initio* calculations. F.e.s. M.P. Gamedze, R.B. Maseko, F. Chigondo, C.M. Nkambule, Tetrahedron Lett. *2012*, *53* (44), 5929-32 [DOI: 10.1016/j.tetlet.2012.08.110].

Chiral 1,1′-binaphthyl-2,2′-diyl phosphoramidite s. under Chloro(1,5-cyclooctadiene)-iridium dimer ←

Chiral 1,1′-spirobiindan-based phosphate diester ←
Organo-Brønsted acid-catalyzed asym. intramolecular transacetalation ○
of γ-hydroxyacetals with kinetic resolution s. *83*, 61; *78*, 123s*83*

Methanesulfonic acid s. under 2,3-Diphenylcyclopropenone *Ms_2O*
p-Toluenesulfonyl chloride s. under Bu_2SnO *p-TsCl*

Chloro(1,5-cyclooctadiene)iridium dimer/chiral 1,1′-binaphthyl-2,2′-diyl phosphoramidite/ 1,8-diazabicyclo[5.4.0]undec-7-ene ←
Aryloxy-2-ethylenes from 2-ethylenecarbonic acid aryl esters ←
Iridium(I)-catalyzed asym. intramolecular decarboxylative O-allylation

85. 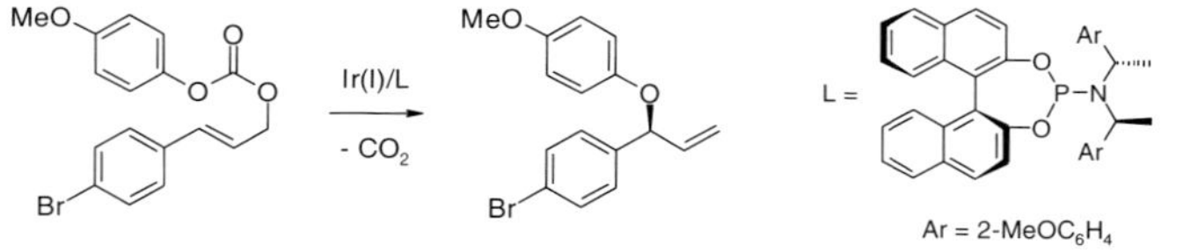

under mild conditions. A mixture of $[Ir(cod)Cl]_2$ (2 mol%), chiral phosphoramidite ligand (4 mol%) and THF (0.5 ml) stirred at 25° under N_2 until homogeneous, DBU (1 eq.) added (orange → yellow soln.), a soln. of 3-(4-bromophenyl)prop-2-en-1-yl 4-methoxyphenyl carbonate (0.2 mmol) in the same solvent (1 ml) added, the homogeneous soln. stirred until substrate consumed (TLC), concentrated, and purified chromatographically → (S)-1-bromo-4-[1-(4-methoxyphenoxy)allyl]-benzene. Y 85% (branched/linear 99:1; e.e. 96%). Use of the Ir(I) catalyst provided a general method for this intramolecular process, affording branched allylic ethers with high regio- and enantio-selectivity. The method was tolerant of a range of alkyl and aryl substituents at the allyl terminus and gave highest selectivity for electron-rich aryl carbonates (thirty examples; Y 70-91%; regioselectivity 87:13 to >99:1; e.e. 86 to >99%; a series of four 4-haloaryl carbonates gave e.e. 74-80%). A 2-bromophenyl terminated allyl carbonate was somewhat unreactive, requiring reaction at 60° (Y 56%; regioselectivity 99:1; e.e. 75%). Use of a more robust iridium catalyst [chloro(dibenzocyclooctatetraene)iridium dimer], extended the scope of the etherification to sterically-hindered substrates (incl. neopentyl), although higher temps. (60-85°) were required (two examples; Y 78-90%; regioselectivity 99:1; e.e. 98-99%); an intermolecular variant demonstrated similar selectivity but yields were poor (20-32%). For analogous ruthenium(I)-catalyzed reaction of cinnamyl derivs. cf. *74*, 136. F.e., optimization, substrate prepn. and use of the method in a synthesis of mono-protected chiral 1,3-diols s. D. Kim, S. Reddy, O.V. Singh, J.S. Lee, S.B. Kong, H. Han, Org. Lett. *2013*, *15* (3), 512-5 [DOI: 10.1021/ol3033237]; extension to the formation of chiral differentially-protected *syn*- or *anti*-3-ene-1,2-diols and 1,2,3,4-tetraols and prepn. of the startg. 4-oxy-2-ethylenecarbonates by cross-metathesis s. D. Kim, J.S. Lee, S.B. Kong, H. Han, Angew. Chem., Int. Ed. *2013*, *52* (15), 4203-6 [DOI: 10.1002/anie.201209112]; for iridium(I)-catalyzed stereoselective decarboxylative allylic amidation of chiral branched benzyl allyl imidodicarboxylates s. O.V. Singh, H. Han, Org. Lett. *2007*, *9* (23), 4801-4 [DOI: 10.1021/ol702115h].

Nitrogen ↑ OC ⇑ N

Boron fluoride *BF_3*

2,3-Dihydrobenzofurans from *o*-β-hydroxytriazenes
via Lewis acid-promoted denitrogenative intramolecular O-arylation ○

86.

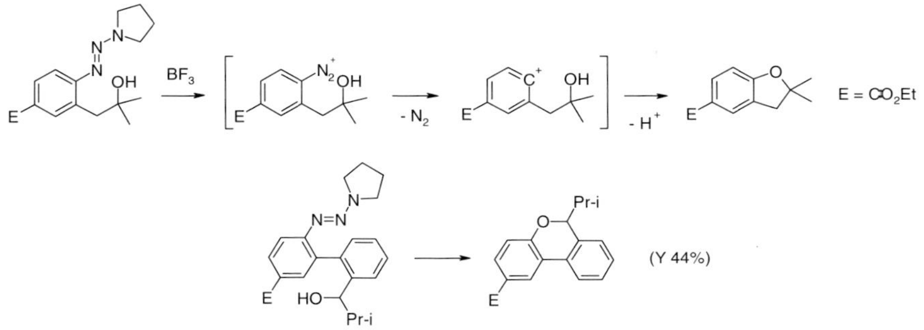

under mild conditions. BF_3-etherate (2 eq.) added slowly to a soln. of ethyl 3-(2-hydroxy-2-methylpropyl)-4-(pyrrolidin-1-yldiazenyl)benzoate (0.5 mmol) in methylene chloride (20 ml) at room temp., the mixture stirred at 40° for 4 h, cooled, quenched with 5% aq. $NaHCO_3$, extracted with methylene chloride, and purified chromatographically → ethyl 2,2-dimethyl-2,3-dihydrobenzofuran-5-carboxylate. Y 78%. This experimentally simple cyclization was unaffected by steric congestion at the reaction center and occurred in both polar and non-polar solvents (optimally methylene chloride) for a series of prim., sec., and tert. triazenyl-alcohols, in moderate to good yields (35-79%; seventeen examples; a 4-fluoro-triazenyl-alcohol gave 27%). The method was compatible with halo, nitrile, ester, azide and alkene functionality, and occurred with good retention of chirality (two examples; Y 59-62%; e.e. 96-98% from substrates with e.e. 99%). The method was extended to the synthesis of **6*H*-dibenzo[*b,d*]pyrans** (three examples; Y 44-57%). The reaction was unaffected by the addition of radical inhibitors, and a mechanism involving an aryl cation was suggested. F.e., optimization and substrate prepn. s. G. Zhao, B. Wang, W. Yang, H. Ren, Eur. J. Org. Chem. *2012* (31), 6236-47 [DOI: 10.1002/ejoc.201200902].

Carbon ↑ OC ⇑ C

Irradiation s. under K_2CO_3 ///

Potassium hydroxide/N-(9-anthracenylmethyl)cinchonidinium chloride/2-chloroethanol ←

α,β-Ethyleneketones from 2-acoxy-1,3-dienes C=C-OC(O)R → CHCO
via phase transfer-catalyzed asym. protonation

87.

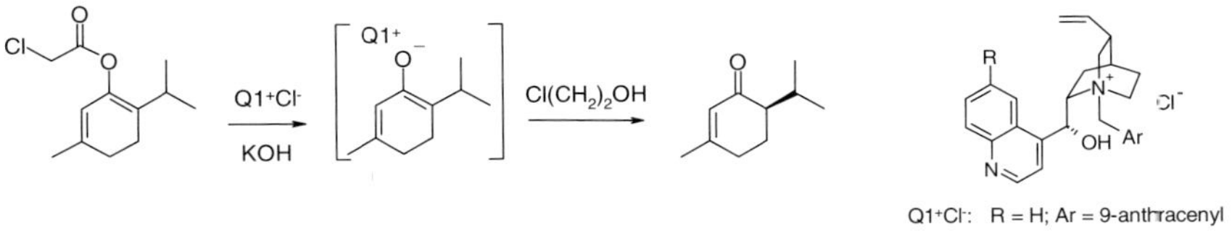

Chiral 6-subst. 2-cyclohexenones. A mixture of chiral phase transfer catalyst (2 mol%), 2-chloroethanol (0.5 eq.), 50% aq. KOH (0.1 ml) and chloroform/mesitylene (2:1; 0.4 ml) stirred at -40° under air for 10 min, 2-chloroacetoxy-1-isopropyl-4-methylcyclohexa-1,3-diene (0.1 mmol) added, the mixture stirred efficiently for 24 h, filtered through silica, and purified chromatographically → (R)-6-isopropyl-3-methylcyclohex-1-enone. Y 94% (e.e. 93%). A series of 2-chloroacetoxycyclohexa-1,3-dienes were hydrolyzed to the corresponding cyclohex-2-enones, with moderate to good enantioselectivity, in the presence of the chiral catalyst (fifteen examples; Y 83-99%; e.e. 59-94%). Moderately bulky 6-substituents (ethyl, allyl, isopropyl) and 5-methyl/aryl in the cyclohexenone products showed highest enantiodiscrimination, with lowest enantioselectivities observed for 4,4-dimethyl- (e.e. 59%), 6-benzyl- (e.e. 76%) and 5-ethoxy-cyclohexenones (e.e. 78%). A cycloheptenone homolog was similarly obtained with moderate enantioselectivity (e.e. 73%; Y 88%). Initial investigations using analogous catalysts identified an unprecedented alternative route to the products via **C-deacylative asym. protonation** (retro-Claisen condensation) of a γ,δ-ethylene-β-diketone (Y 86%; e.e. 89.5%).

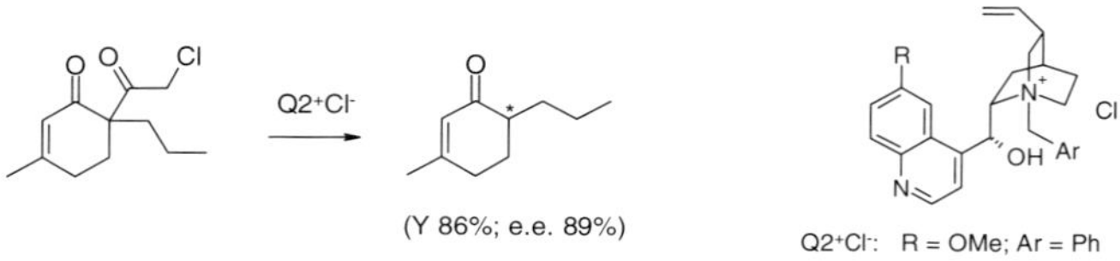

F.e. and substrate prepn. s. E. Yamamoto, D. Gokuden, A. Nagai, T. Kamachi, K. Yoshizawa, A. Hamasaki, T. Ishida, M. Tokunaga, Org. Lett. *2012*, *14* (24), 6178-81 [DOI: 10.1021/ol3027363].

Potassium carbonate/irradiation *K_2CO_3/⫻*

3,4-Dihydro-2,4-benzoxazepin-5-ones
from α-[N-(*o*-ketoaroyl)amino]carboxylic acids
by stereoselective decarboxylative photocyclization with memory of chirality

88. 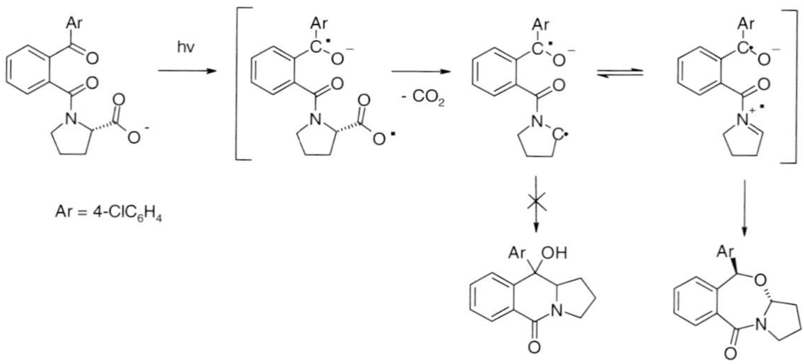

Novel one-pot photocyclization of N-(*o*-acylbenzoyl)-α-amino-acids affords bi- and tri-cyclic benzoxazepinones exclusively via C-O rather than C-C bond formation. **E:** A mixture of (S)-N-[2-(4-chlorobenzoyl)benzoyl]proline (0.9 mmol), K_2CO_3 (5.6 eq.) and water (8 ml) stirred for a few min, methanol (72 ml) added, the soln. irradiated by medium-pressure Hg lamp (500 W) for 8 h (TLC), solvent evaporated, and the residue worked up with chromatographic purification → (3R,5R)-5-(4-chlorophenyl)-1,2,3,3a-tetrahydrobenzo[*e*]pyrrolo[2,1-*b*][1,3]oxazepin-10(5*H*)-one. Y 41% (d.e. 70%; e.e. 96%). This experimentally simple photodecarboxylation afforded (3R,5R)-tricyclic products from (S)-proline-based substrates in moderate yields (35-48%; five examples; d.e. 70-90%; e.e. 87 to >99%), with racemic proline analogs affording mixtures of (3R,5R/3S,5S) products (three examples; Y 30-41%; d.e. 82-98%). The reaction was also applied to glycine and (S)-alanine-based substrates affording racemic and (3R,5R) products respectively in somewhat better yields, but stereoselectivity was not discussed (eight examples; Y 58-82%; 38-42% for 2-acetylbenzoyl derivs.). F.e. and substrate prepn. s. C.-E. Liu, Q. Han, N. Ma, Z.-S. Geng, R.-H. Zhang, Z.-Q. Jiang, Tetrahedron Lett. *2013*, *54* (6), 541-4 [DOI: 10.1016/j.tetlet.2012.11.077].

[Tris[tris(3,5-di-tert-butyl-4-methoxyphenyl)silylethynyl]phosphine]gold(I) triflimide *[Au(I)]*

2-β-Keto-O-heterocyclics from 2′-acetylenediol 1-monobenzoates

89.

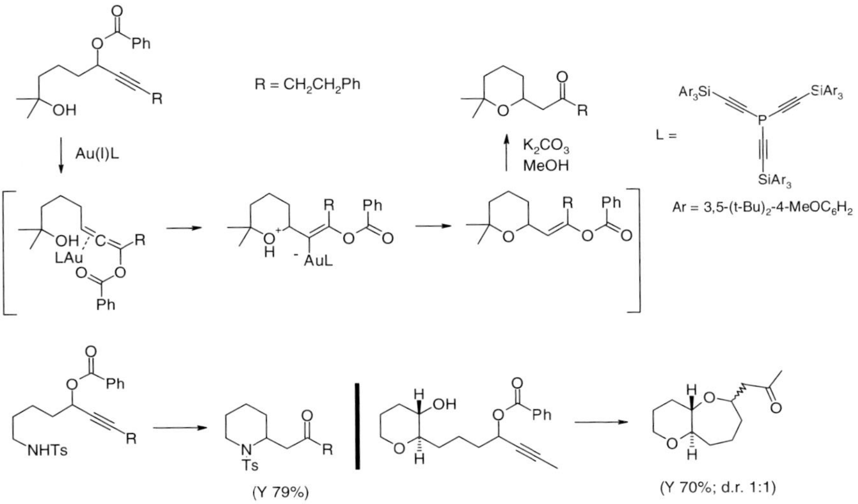

Application of a 'semihollow' trialkynylphosphine ligand to the gold(I)-catalyzed cyclization of α-(ω-hydroxy)propargylic esters has been shown to allow clean synthesis of 6- and 7-membered cyclic ethers under mild conditions. **E:** A soln. of gold(I) catalyst (0.5 mol%) in dry methylene chloride (1 ml) added dropwise to a stirred soln. of 9-hydroxy-9-methyl-1-phenyl-3-decyn-5-yl benzoate (0.198 mmol) in the same solvent (3 ml) at -40°, the mixture stirred for 1.5 h (TLC), filtered through silica, concentrated to dryness in an open vial, K_2CO_3 (5 eq.) and methanol (4 ml) added to the residue under air at room temp., the mixture stirred for 2 h (TLC), concentrated, diluted with water, extracted with ether, and purified chromatographically → 1-(6,6-dimethyltetrahydro-2*H*-2-pyranyl)-4-phenyl-2-butanone. Y 80%. Use of low temp. generally resulted in exclusive formation of the title compds. (incl. a sterically-congested 2,2,6,6-tetrasubst. tetrahydropyran) for a series of tethered prim., sec. and tert. alcohols (twelve examples; Y 49%, 65-96%) with lowest yields observed for tert. benzoate derivs. (two examples) due to formation of 2-alkynyltetrahydropyran by-products (Y 7-14%). An ω-tosylamino analog afforded a piperidine in good yield (79%; illustrated). F.e. and optimization s. H. Ito, A. Harada, H. Ohmiya, M. Sawamura, Adv. Synth. Catal. *2013*, *355* (4), 647-52 [DOI: 10.1002/adsc.201200949].

(Triphenylphosphine)gold(I) chloride/silver triflate/p-toluenesulfonic acid/water

2,4-Bridged 1,3-dioxolanes from acetylene-1,2-diol acetonides via gold(I)-catalyzed cleavage of the protective group with double ring closure

90.

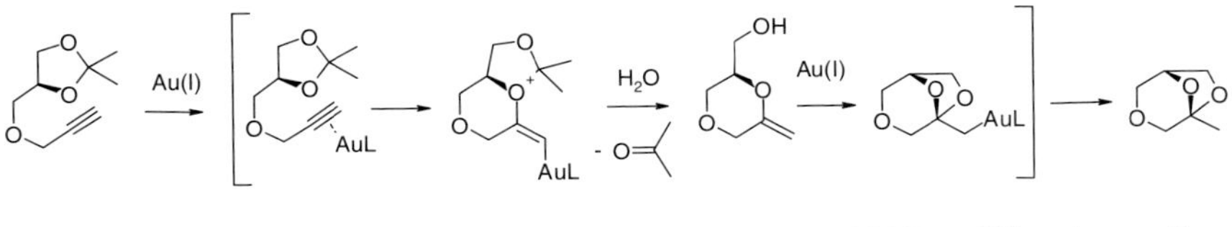

$(Ph_3P)AuCl$ (2.5 mol%), AgOTf (2.5 mol%), *p*-toluenesulfonic acid (10 mol%) and water (1 eq.) added to a stirred soln. of (S)-2,2-dimethyl-4-prop-2-ynyloxymethyl-1,3-dioxolane (0.37 mmol)

in methylene chloride (0.37 ml), the mixture heated in a sealed tube at 80° for 3 h, cooled to room temp., filtered through Celite, extracted with methylene chloride, and purified chromatographically → (1S,5R)-5-methyl-3,6,8-trioxabicyclo[3.2.1]octane. Y 100%. The presence of water was essential in this regioselective and convenient access to 2,4-bridged 1,3-dioxolanes, with the use of 1 eq. water affording superior yields to a 4:1 mixture of water/methylene chloride in the presence of a phase transfer catalyst. Internal/terminal prop-2-ynyloxymethyl-tethered dioxolanes afforded the corresponding 3,6,8-trioxabicyclo[3.2.1]octanes (seven examples; Y 52-100%) via 6-*exo*/5-*exo*-double cyclization [a pent-2-ynyloxymethyl deriv. gave the expected product (Y 34%) together with a 7-*endo*/5-*exo* regioisomer, a 4,7,9-trioxabicyclo[4.2.1]nonane, in 66% yield]. 3,6,9-Trioxabicyclo[4.2.1]nonanes were similarly obtained from the homologous prop-2-ynyl-oxy*ethyl*-tethered substrates (five examples; Y 42-59%), and 3-aza-6,8-dioxabicyclo[3.2.1]octanes from prop-2-ynylaminomethyl derivs. [three examples; Y 55-80%; a phenyl-terminated alkyne gave bicyclooctane (Y 60%) and bicyclononane isomers (Y 35%)].

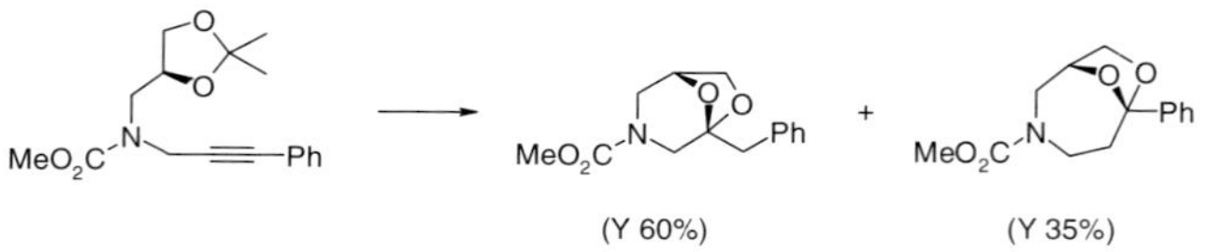

F.e., optimization and substrate prepn. s. B. Alcaide, P. Almendros, R. Carrascosa, Tetrahedron *2012*, *68* (46), 9391-6 [DOI: 10.1016/j.tet.2012.09.030].

Alumina *Al_2O_3*

1,3,4-Oxadiazol-2(3*H*)-ones from N-carbo-*tert*-butoxy-1,1-chlorohydrazones
Solvent-free ring closure on alumina

91.

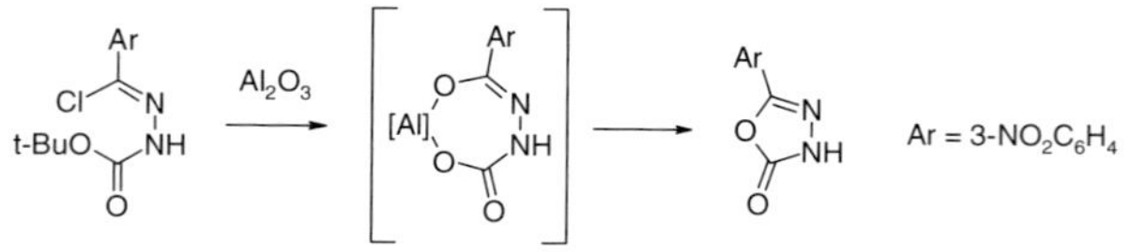

The first green synthesis of the title products on a solid support is reported. **E:** A soln. of *tert*-butyl α-chloro-N′-3-nitrobenzylidenecarbazate (0.5 mmol) in the minimum quantity of chloroform stirred with basic alumina (500 mg), dried *in vacuo*, heated at 85° for 12 min (TLC), cooled to room temp., extracted with ethyl acetate, and purified by recrystallization → 5-(3-nitrophenyl)-1,3,4-oxadiazol-2(3*H*)-one. Y 87%. This efficient method gave optimum results with basic alumina (neutral alumina and silica were less effective) for prepn. of a series of electron-diverse 5-aryl-1,3,4-oxadiazol-2(3*H*)-ones (thirteen examples; Y 80-99%) in the presence of nitro, nitrile, halo and ether functionality. The substrates were available in two steps from the corresponding aromatic aldehydes and the alumina catalyst was recycled six times without significant loss in activity. F.e., optimization and substrate prepn. s. K. Debnath, S. Pathak, A. Pramanik, Tetrahedron Lett. *2013*, *54* (8), 896-9 [DOI: 10.1016/j.tetlet.2012.11.128].

2-Chloroethanol s. under Potassium hydroxide *$Cl(CH_2)_2OH$*

Phenyl iodosotrifluoroacetate/trifluoroacetic acid *$PhI(OCOCF_3)_2/CF_3COOH$*
3-Methylene-1-oxa-4-azaspiro[5.5]undeca-7,10-diene-2,5,9-triones from α-(*p*-hydroxyaroylamino)acrylic acid esters
Iodine(III)-mediated oxidative spirocyclization

92.

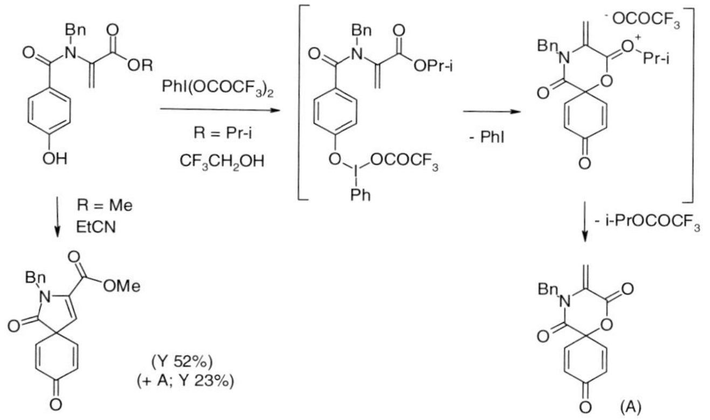

Trifluoroacetic acid (2 eq.) and phenyl iodosotrifluoroacetate (2 eq.) added sequentially to a stirred soln. of isopropyl N-benzyl-2-(4-hydroxybenzamido)acrylate (0.1 mmol) and 3 Å molecular sieves in 2,2,2-trifluoroethanol (2 ml) at 0°, the mixture stirred for 1 h, concentrated *in vacuo*, and purified chromatographically → 4-benzyl-3-methylene-1-oxa-4-azaspiro[5.5]undeca-7,10-diene-2,5,9-trione. Y 70%. The use of trifluoroethanol as solvent was critical to the success of this rare spirolactonization, with the benzyl ester giving a similar yield (68%), and methyl (48%), ethyl (54%) and *tert*-butyl esters (53%) being less effective. Interestingly, use of propionitrile as solvent afforded the spirolactone as minor product (four examples; Y 20-30%), the major products being spirolactam-esters, alkyl 2-azaspiro[4.5]deca-3,6,9-triene-1,8-dione-3-carboxylates (Y 35-52%), with the methyl ester giving the highest yield. F.e., optimization and substrate prepn. s. C. Hempel, N.M. Weckenmann, C. Maichle-Moessmer, B.J. Nachtsheim, Org. Biomol. Chem. *2012, 10* (47), 9325-9 [DOI: 10.1039/c2ob26815a].

Chiral 1,1′-binaphthyl-2,2′-diyl phosphoramidite s. under Chloro(1,5-cyclooctadiene)-iridium dimer ←

p-Toluenesulfonic acid [s.a. under $(Ph_3P)AuCl$] *TsOH*
Hydroxyspiroketals from lactones and 4-ω-sulfonyl-1,3-dioxolanes via Julia olefination

93.

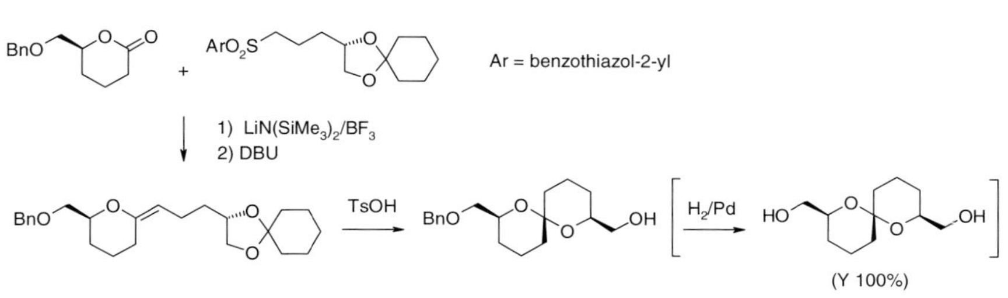

The scope of a two-step preparation of chiral spiroketals from lactones via enolethers [s. M. Corbet et al., Tetrahedron Lett. *2008, 49* (5), 750-4 [DOI: 10.1016/j.tetlet.2007.11.207]] has been extended to include minimally-subst. lactones and applied to the synthesis of some naturally occurring and biologically interesting products. **E:** 2(S)-[3-(Benzothiazol-2-ylsulfonyl)propyl]-1,4-dioxaspiro[4.5]decane (1.2 eq.) added to a soln. of (S)-6-(benzyloxymethyl)tetrahydro-2*H*-pyran-2-one (0.54 mmol) and BF_3-etherate (1 eq.) in THF (1 ml) at room temp., cooled to -78°, a soln. of Li-bis(trimethylsilyl)amide (2 eq.) in the same solvent (1 ml) added dropwise, the mixture stirred for 30 min, quenched with water, diluted with ethyl acetate, stirred at room temp. for 15 min, extracted with ethyl acetate, concentrated *in vacuo*, dissolved in THF (5 ml) at room temp., 1,8-diazabicyclo[5.4.0]undec-7-ene (2 eq.) added over 10 min, the mixture stirred for 30 min, concentrated *in vacuo*, and purified chromatographically → 2-[3-[(S)-6-(benzyloxymethyl)tetrahydro-2*H*-pyran-2-ylidene]propyl]-1,4-dioxaspiro[4.5]decane (Y 55%), a portion of which (0.22 mmol) dissolved in methylene chloride (3 ml), *p*-toluenesulfonic acid (0.5 eq.) added at room temp., the mixture stirred for 25 min, concentrated *in vacuo*, and purified chromatographically → ((2S,6S,8S)-8-benzyloxymethyl-1,7-dioxaspiro[5.5]undecan-2-yl)methanol (Y 94%). The use of acid-labile protection for the 1,2-diol moiety of the enol ether allowed spiroketalization to be completed in a single step. A *tert*-butyldiphenylsilyl protected product was prepared similarly (Y 31% for enol ether prepn. and Y 92% for spiroketalization), and both products were deprotected using standard methods to afford the required spirolactone (SPIKET-P) in quantitative yield. A further example reacting valerolactone with the benzothiazolyl sulfone afforded an enol ether in 60% yield which was cyclized to ((2S,6S)-1,7-dioxaspiro[5.5]undecan-2-yl)methanol in quantitative yield. F.e. and lactone prepn. s. L. Tomas, B. Bourdon, J.C. Caille, D. Gueyrard, P.G. Goekjian, Eur. J. Org. Chem. *2013* (5), 915-20 [DOI: 10.1002/ejoc.201201199].

N-(9-Anthracenylmethyl)cinchonidinium chloride s. under Potassium hydroxide ←

Chloro(1,5-cyclooctadiene)iridium dimer/chiral 1,1′-binaphthyl-2,2′-diyl phosphoramidite/ 1,8-diazabicyclo[5.4.0]undec-7-ene ←
Aryloxy-2-ethylenes from 2-ethylenecarbonic acid aryl esters ←
Iridium(I)-catalyzed asym. intramolecular decarboxylative O-allylation s. *83*, 84

Formation of N-N Bond

Exchange ⇂↾

Hydrogen ↾ NN ⇂↾ H

Sodium nitrite/acetic acid *$NaNO_2$/AcOH*
Azides from prim. amines $NH_2 \rightarrow N_3$
with $CF_3SO_2N_3$ cf. *24,* 286s*47*; rapid, mild and inexpensive method with hydrazine hydrate and $NaNO_2$/AcOH in methylene chloride, ar. azides, s. A.A. Siddiki, B.S. Takale, V.N. Telvekar, Tetrahedron Lett. *2013, 54* (10), 1294-7 [DOI: 10.1016/j.tetlet.2012.12.112].

Copper(I) bromide/N,N′-di-tert-butyldiaziridinone ←
Silver, gold or other metal nanoparticles/oxygen ←
Sym. azo compds. from amines 2 $NH_2 \rightarrow$ N=N
with CuBr/pyridine/O_2 cf. *78,* 128; ar. azo compds. on the gram scale with CuBr and N,N′-di-*tert*-butyldiaziridinone as oxidant, also sym. N,N′-disubst. hydrazo compds. from sec. ar. amines, s. Y. Zhu, Y. Shi, Org. Lett. *2013, 15* (8), 1942-5 [DOI: 10.1021/ol4005917]; sym. **and unsym. ar.**

azo compds. (from two different aniline molecules) under mild, aerobic oxidation with monodispersed metal nanoparticles (Ag, Pt, Co, Cu, Ni, Pd, or Au) as catalyst (cf. *82,* 116) s. S. Cai, H. Rong, X. Yu, X. Liu, D. Wang, W. He, Y. Li, ACS Catal. *2013, 3* (4), 478-86 [DOI: 10.1021/cs300707y].

Elimination ⇑

Carbon ↑ NN ⇑ C

Via intermediates *v.i.*

Bicyclic hydrazines from N,N′-methylene-bridged N,N-heterocyclics via Polonovski-type reaction

94.

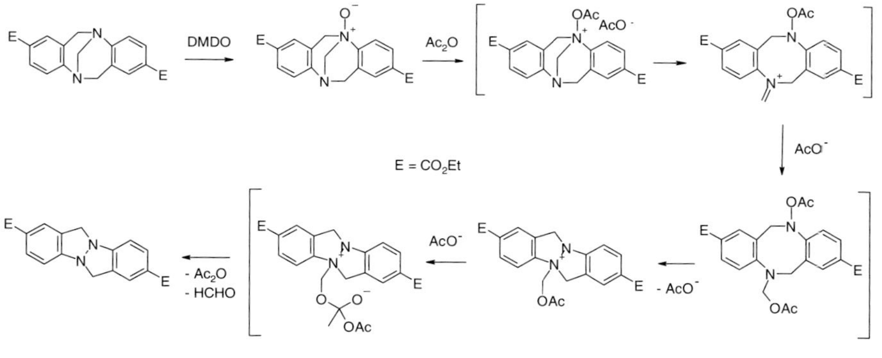

6,12-Dihydroindazolo[2,1-*a*]indazoles. A soln. of dimethyldioxirane (4 eq.) in acetone (8 ml) added to a soln. of diethyl 6,12-dihydro-5*H*-5,11-methanodibenzo[*b,f*][1,5]diazocine-2,8-dicarboxylate (0.2 mmol) in the same solvent (10 ml) at room temp., the mixture stirred for 1 h, and concentrated *in vacuo* → diethyl 6,12-dihydro-5*H*-5,11-methanodibenzo[*b,f*][1,5]diazocine-2,8-dicarboxylate 5-oxide (Y 100%), a portion of which (0.2 mmol) dissolved in acetone (2 ml) at room temp., acetic anhydride (2 eq.) added, the mixture stirred for 30 min, and purified chromatographically → diethyl 6,12-dihydroindazolo[2,1-*a*]indazole-2,8-dicarboxylate. Y 100%. This novel method was successful for a series of 6,12-dihydro-5*H*-5,11-methanodibenzo[*b,f*][1,5]-diazocines, affording 5-oxides (ten examples; Y 86-100%) which were converted to the ring contracted products via suggested elimination of formaldehyde (eight examples; Y 62-100%). A 2,8-diido deriv. gave an insoluble product on oxidation which was converted to the expected tetracycle in 83% overall yield, while a tetramethoxy 5-oxide deriv. gave mainly recovered substrate. The N-oxidation generally failed (with several oxidizing agents) where a substituent was *ortho* to a bridgehead nitrogen, but an N-oxide obtained from the 4,10-dimethyl substrate (using Oxone) gave a complex mixture on treatment with Ac_2O. Ring contraction also failed where the diazocine ring contained a substituent (e.g. 12-benzyl). F.e.s. X. Gao, C.S. Hampton, M. Harmata, Eur. J. Org. Chem. *2012* (36), 7053-6 [DOI: 10.1002/ejoc.201201341].

Formation of N-Rem Bond

Exchange ⇂↾

Hydrogen ↾ NRem ⇂↾ H

Copper(II) acetate/potassium carbonate/air *$Cu(OAc)_2/K_2CO_3/O_2$*
Copper(II)-catalyzed aerobic N-phosphorylation with phosphorous acid diesters $(RO)_2P(O)H \rightarrow (RO)_2P(O)N\lesssim$

95. 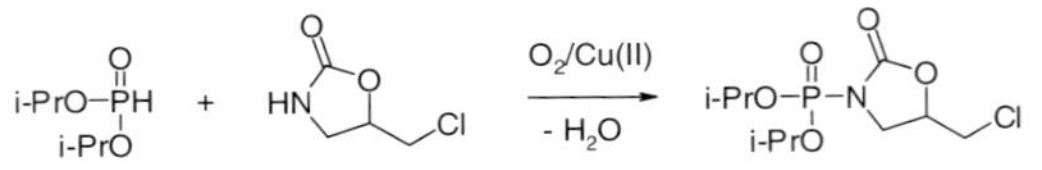

The first formation of P-N bonds by copper-catalyzed coupling of H-phosphonates and nitrogen nucleophiles using oxygen as the sole oxidant is described (for N-phosphorylation via the Todd-Atherton reaction cf. *2*, 314s*73*). **E:** A mixture of $Cu(OAc)_2$ (10 mol%), K_2CO_3 (1 eq.), 5-chloromethyloxazolid-2-one (3 eq.) and 4 Å molecular sieves (100 mg) in toluene (1 ml) stirred for 5 min, a soln. of diisopropyl H-phosphonate (0.2 mmol) in the same solvent (1 ml) added via syringe pump over 30 min at 80° under air, stirring continued for 10 min, filtered, concentrated, and purified chromatographically → 5-chloromethyl-1-diisopropylphosponyloxazolid-2-one. Y 99%. This experimentally simple, environmentally benign (water is the only by-product) and efficient procedure conveniently uses dialkyl H-phosphonates for reaction with a range of nitrogen nucleophiles (lactam, oxazolidone, urea, indole and sulfonamide) to afford the corresponding N-acylphosphoramidates (eighteen examples; Y 52-99%), with lowest yields obtained for indole derivs. F.e. and optimization s. X. Jin, K. Yamaguchi, N. Mizuno, Org. Lett. *2013*, *15* (2), 418-21 [DOI: 10.1021/ol303420g].

Zinc triflate/pyridine *$Zn(OTf)_2/py$*
Ruthenium(II) thiolate complexes *[Ru(II)]*
Dehydrogenative N-silylation $\gtrsim NH \rightarrow \gtrsim NSi\lesseqgtr$
update s. *80*, 121; base-free N-silylation of pyrroles, indoles, carbazoles and anilines without a hydrogen acceptor *under ruthenium catalysis* using a ruthenium(II) arylthiolate complex ligated to a phosphine and a sec. amine s. C.D.F. Königs, M.F. Müller, N. Aiguabella, H.F.T. Klare, M. Oestreich, Chem. Commun. *2013*, *49* (15), 1506-8 [DOI: 10.1039/c3cc38900f]; of indoles *under zinc catalysis* using $Zn(OTf)_2$/pyridine s. T. Tsuchimoto, Y. Iketani, M. Sekine, Chem. Eur. J. *2012*, *18* (31), 9500-4 [DOI: 10.1002/chem.201200651].

Sulfur ↾ NRem ⇂↾ S

Potassium phosphides *KP≤*
Aminophosphines from sulfonylamines s. *83*, 4 $NSO_2R \rightarrow NP\lesssim$

Formation of N-C Bond

Uptake ⇓

Addition to Oxygen and Nitrogen NC ⇓ ON

Copper(II) triflate/copper(I) chloride *$Cu(OTf)_2/CuCl$*
α-(N-Carbalkoxyhydroxylamino)-β-ketocarboxylic acid esters H → N(OH)COOR
from β-ketocarboxylic acid esters and *in situ*-generated C-nitrosoformic acid esters s. *83*, 113

Addition to Oxygen and Carbon NC ⇓ OC

Bis(1,5-cyclooctadiene)rhodium(I) triflate/(R)-1-[(S)-2-(diphenylphosphino)ferrocenyl]-ethyl(di-tert-butyl)phosphine/triethylamine hydrochloride/sodium cyanate ←

4-Vinyl-2-oxazolidone ring from 2,5-bridged 2,5-dihydrofurans via rhodium-catalyzed asym. ring opening-ring closure

96. 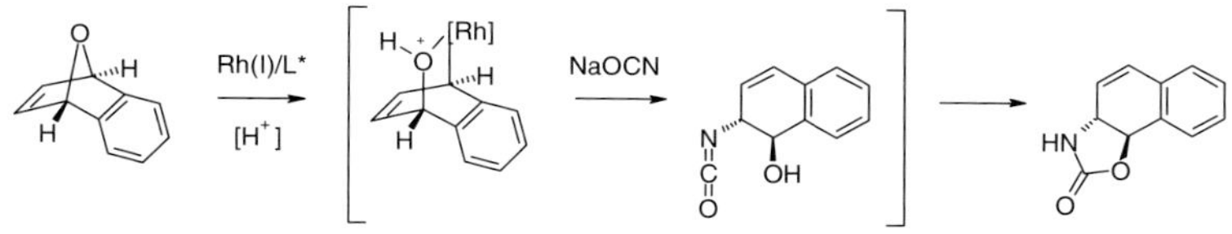

The novel use of cyanate as nucleophile in rhodium-catalyzed ring-opening reactions has led to a general procedure for the preparation of chiral oxazolidones, which are themselves widely used as chiral auxiliaries in asymmetric transformations. **E:** 1,2-Dichloroethane (2 ml) and distilled water (0.2 ml) added sequentially by syringe to a mixture of [Rh(cod)$_2$]OTf (8 mol%), (R,S)-PPF-P^{t}Bu$_2$ (8 mol%), startg. oxabicyclic alkene (0.194 mmol), sodium cyanate (2 eq.) and triethylamine hydrochloride (5 eq.) in a sealed vial under argon, the resulting biphasic mixture stirred vigorously at room temp. for 17 h, solvent removed *in vacuo*, and the crude mixture purified by flash chromatography on silica gel → (3aR,9bR)-3,3a-dihydronaphtho[2,1-*d*]oxazol-2(9b*H*)-one. Y 69% (e.e. 98%). Five further symmetrical examples, bearing electron-donating (Me, MeO, OCH$_2$O) or mildly electron-withdrawing (Br) groups, afforded yields of 52-73% (e.e. 94-98%). A more-highly electron-deficient (difluoro subst.) analog gave a much reduced yield (35%) with diminished e.e. (90%). A mixture of regioisomers was obtained from a single unsymmetrical (methoxy subst.) example (Y 34% and 26%; e.e. 92 and 99%). Optimization studies showed the reaction to be highly dependent on the nature of the solvent used, with the inclusion of water promoting higher yields at the expense of enantioselectivity, the combination of aq. DCE, JosiPhos ligand, and triethylamine hydrochloride as proton source proving optimal. Less reactive oxabicyclic alkenes, including non-benzo-fused and bridgehead-substituted examples, failed to afford the desired products, as did azabicyclic alkenes. F.e.s. G.C. Tsui, N.M. Ninnemann, A. Hosotani, M. Lautens, Org. Lett. *2013*, *15* (5), 1064-7 [DOI: 10.1021/ol4000668].

Addition to Nitrogen-Nitrogen Bonds NC ⇓ NN

Irradiation s. under (4,4′-Di-tert-butyl-2,2′-bipyridyl)bis[2-(pyrid-2-yl)phenyl]-iridium(III) fluoroborate

β-Isocupreidine ←

Organocatalyzed asym. C$_3$-amination of 3-(organothio)- or 3-alkoxy-oxindoles CH → C-N-NH

97. 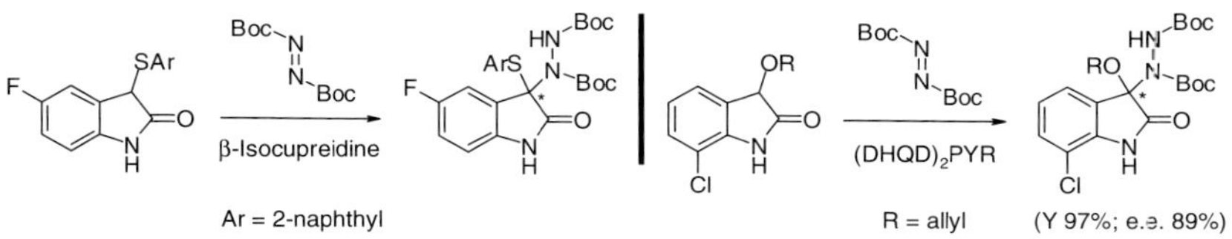

The first organocatalyzed formation of chiral 3,3-disubst. oxindoles bearing two heteroatoms at C3 is reported, optimally using cinchona alkaloid-derived catalysts. **E:** 5-Fluoro-3-naphth-2-yl-thiooxindole (0.25 mmol) and anh. methylene chloride (5 ml) added to β-isocupreidine (10 mol%), the mixture stirred vigorously at room temp. until dissolution of the substrate complete, cooled to -40°, stirred for ca. 30 min, di-*tert*-butyl azodicarboxylate (1.2 eq.) added, stirring continued for

9-12 h (TLC), and the mixture subjected directly to column chromatography on silica → 5-fluoro-3-[N,N′-bis(*tert*-butoxycarbonyl)hydrazine]-3-(naphth-2-ylthio)oxindole. Y 95% (e.e. 94%). A series of 3-naphth-2-yl- or -phenyl-thio-NH-oxindoles gave products of unspecified absolute configuration with high enantioselectivity (e.e. 83-94%; fourteen examples; Y 68-98%). Lower enantioselectivity was obtained for N-benzyl/-methyl (e.e. 79-82%; two examples; Y 92-96%) and for 3-allyl-/benzyl-thio derivs. (e.e. 47-56%; two examples; Y 82-88%). The reaction was extended to the less reactive 3-alkoxyoxindole analogs, and gave good enantioselectivity using $(DHQD)_2PYR$ as catalyst (e.e. 81-93%; sixteen examples; Y 50-97%). F.e., optimization and elaboration of products s. F. Zhou, X.-P. Zeng, C. Wang, X.-L. Zhao, J. Zhou, Chem. Commun. *2013*, *49* (20), 2022-4 [DOI: 10.1039/c3cc38819k].

Copper(II) oxide nanoparticles-on-silica *CuO-SiO_2*
Copper(II) triflate/1,3-bis(diphenylphosphino)propane *$Cu(OTf)_2$/dppp*
N-Hydroxyphthalimide ←
Substituting addition with azodicarboxylic acid esters CH → C-N-NH
update s. *33*, 286s*82*; **α-lateral amination** of *o*-alkyl-N-heteroarenes with $Cu(OTf)_2$/1,3-bis(diphenylphosphino)propane as catalyst s. B. Qian, L. Yang, H. Huang, Tetrahedron Lett. *2013, 54* (7), 711-4 [DOI: 10.1016/j.tetlet.2012.12.055]; ligand-free **ar. amination** of electron-diverse arenes with recyclable $RuCl_3$ as catalyst s. S.M. Inamdar, V.K. More, S.K. Mandal, ibid. *54* (6), 530-2 [DOI: 10.1016/j.tetlet.2012.11.075]; **α-amination** of benzo-condensed cyclic sec. amines *under visible-light photoredox catalysis* with (4,4′-di-*tert*-butyl-2,2′-bipyridyl)bis[2-(pyrid-2-yl)phenyl]-iridium(III) fluoroborate as sensitizer s. Y. Miyake, K. Nakajima, Y. Nishibayashi, Chem. Eur. J. *2012, 18* (51), 16473-7 [DOI: 10.1002/chem.201203066]; *organocatalyzed* radical amination of $C(sp^3)$-H bonds, incl. benzylic and propargylic amination, in the presence of N-hydroxyphthalimide as catalyst s. Y. Amaoka, S. Kamijo, T. Hoshikawa, M. Inoue, J. Org. Chem. *2012, 77* (22), 9959-69 [DOI: 10.1021/jo301840e]; **hydroacylation** of azodicarboxylates with aromatic or aliphatic aldehydes using recyclable CuO nanoparticles-on-silica (cf. *28*, 271) s. S.M. Inamdar, V.K. More, S.K. Mandal, Chem. Lett. *2012, 14* (11), 1484-6 [DOI: 10.1246/cl.2012.1484].

Peptides ←
Organocatalyzed asym. α-amination with azodicarboxylic acid esters
under organocatalysis, update s. *75*, 132s*82*; of aldehydes with a recyclable polymer-based tri- or tetra-peptide possessing an N-terminal D-Pro-Aib residue s. T. Tanaka, K. Akagawa, M. Mitsuda, K. Kudo, Adv. Synth. Catal. *2013, 355* (2-3), 294-6 [DOI: 10.1002/adsc.201200611].

Ruthenium trichloride *$RuCl_3$*
Ar. amination with azodicarboxylic acid esters s. *38,* 286s*83*

(4,4′-Di-tert-butyl-2,2′-bipyridyl)bis[2-(pyrid-2-yl)phenyl]iridium(III) fluoroborate/ irradiation ←
α-Amination of benzo-condensed cyclic sec. amines
under visible-light photoredox catalysis with azodicarboxylic acid esters s. *38,* 286s*83*

Addition to Nitrogen and Carbon NC ⇓ NC

Without additional reagents *w.a.r.*
Addition of amines to heterocumulenes ←
ureas from isocyanates s. *8,* 391; solvent-free formation of chiral α-ureido- from α-amino-nitriles and isocyanates without epimerization s. P. Ventosa-Andrés, J.A. González-Vera, M.T. García-López, R. Herranz, Org. Lett. *2013, 15* (3), 632-5 [DOI: 10.1021/ol303452r]; chiral and achiral bis(thioureas) from diamines (incl. sterically and electronically hindered derivs.) and isothiocyanates (cf. *1,* 284) by automated neat- or liquid-assisted ball milling (affording pure products without chromatographic purification or recrystallization) s. V. Štrukil, M.D. Igrc, M. Eckert-Maksic, T. Frišcic, Chem. Eur. J. *2012, 18* (27), 8464-73 [DOI: 10.1002/chem.201200632].

Sodium azide/Amberlyst-15 ←
Tetrazoles from nitriles ○
update s. *70,* 141s*82*; heterogeneous procedure with mild, recoverable Amberlyst-15 as catalyst s. R. Shelkar, A. Singh, J. Nagarkar, Tetrahedron Lett. *2013, 54* (1), 106-9 [DOI: 10.1016/j.tetlet.2012.10.116]; α-(5-aminotetrazol-1-yl)carboxylic acid esters from cyanogen azide and

α-*prim*-aminocarboxylic acid esters (as their hydrochlorides), limitations, s. Y.-H. Joo, S.G. Cho, E.M. Goh, D.A. Parrish, J.M. Shreeve, Eur. J. Org. Chem. *2013* (4), 688-92 [DOI: 10.1002/ejoc.201201153].

1,3-Dimesitylimidazolidin-2-ylidene/oxygen *NHC/O_2*
N-Acylureas from carbodiimides and aldehydes CHO → C(O)NRCONHR

98.

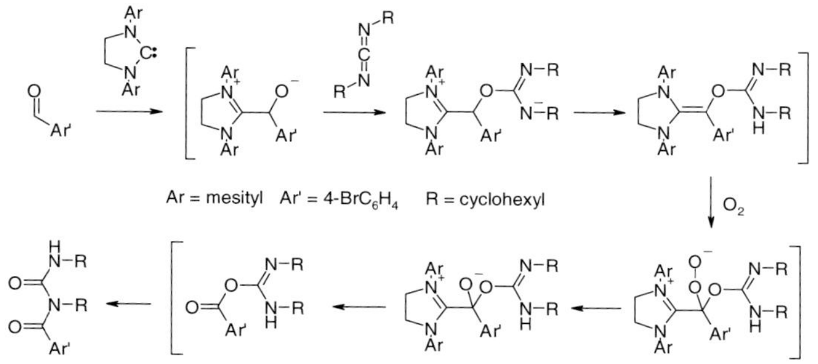

The first report of N-heterocyclic carbene [NHC] promoted coupling of aldehydes with N,N′-disubst. carbodiimides under aerobic conditions is described. **E:** K_2CO_3 (20 mol%) added to a soln. of 4-bromobenzaldehyde (1.2 eq.), N,N′-dicyclohexylcarbodiimide (0.25 mmol) and 1,3-dimesitylimidazolidinium chloride (NHC precursor; 20 mol%) in acetonitrile (0.25 ml) at room temp. under air (balloon), the mixture stirred for 18 h, diluted with water, extracted with chloroform, concentrated, and purified chromatographically → 4-bromo-N-cyclohexyl-N-(cyclohexylcarbamoyl)benzamide. Y 90%. The catalyst generally gave good yields for electron-diverse (het)ar. and aliphatic aldehydes (incl. cinnamal) reacting with symmetrical bis-prim./sec. alkyl-carbodiimides (nineteen examples; Y 70-97%), with lower yields obtained from pivalaldehyde (38%), di-*tert*-butylcarbodiimide (43%) and bis(4-methoxyphenyl)carbodiimide (21%), while 3-phenylpropynal gave a complex mixture. F.e. and optimization s. T. Soeta, Y. Tabatake, S. Fujinami, Y. Ukaji, Org. Lett. *2013*, *15* (9), 2088-91 [DOI: 10.1021/ol400672c].

Ammonium chloride-^{15}N *NH_4Cl-^{15}N*
1,2-Diamines from aziridines C
update s. *23*, 351s*82*; ^{15}N-labelled chiral 1,2-diamines with NH_4Cl-^{15}N s. G. Berger, M. Gelbcke, E. Cauët, M. Luhmer, J. Nève, F. Dufrasne, Tetrahedron Lett. *2013*, *54* (6), 545-8 [DOI: 10.1016/j.tetlet.2012.11.079].

Addition to Carbon-Carbon Bonds NC ⇓ CC

Without additional reagents *w.a.r.*
4-α-Acoxypyrimidines from electron-deficient 3-acoxy-1,4-diynes and amidines via 5-acoxy-4,6-di(ene)-1,4,5,6-tetrahydropyrimidines ○

99.

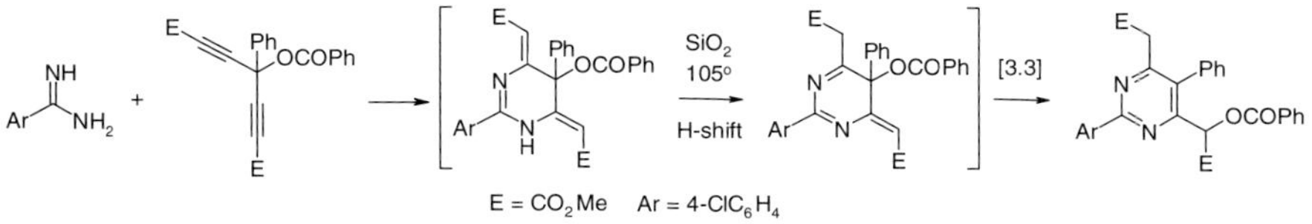

4-Chlorobenzamidine (1.2 eq.) added to a soln. of dimethyl 4-benzoyloxy-4-phenylhepta-2,5-diyne-1,7-dioate (1 mmol) in 1,2-dichloroethane (10 ml), the mixture heated under reflux for 1 h, cooled, concentrated *in vacuo*, the residue dissolved in methylene chloride, absorbed onto silica gel (2 g), solvent removed under reduced pressure, the residue heated at 105° for 3 d, extracted with ethyl acetate, and purified chromatographically → 1-[2-(4-chlorophenyl)-6-(2-methoxy-2-oxoethyl)-5-phenylpyrimidin-4-yl]-2-methoxy-2-oxoethyl benzoate. Y 53%. This novel method for synthesis of polysubstituted pyrimidines gave moderate yields of the title products from the skipped diyne derivs., 4-aroyloxy-4-arylhepta-2,5-diyne dioates, (ten examples; Y 37-53%) via double aza-Michael addition and aromatization of the cycloadducts under prolonged heating of the intermediates absorbed onto silica gel (heating with microwaves was less useful, leading either to incomplete conversion or partial decarboxylation of products). Cycloadduct formation from aliphatic analogs, 4-alkyl-4-acoxy-derivs., was only successful for branched alkyl chains (*i*-Pr or *t*-Bu), with aromatization only achieved using microwave heating (900 W/40 min), and then at the expense of decarboxylation of both unsaturated ester groups (two examples; Y 35-40%); interestingly, for the 4-*tert*-butyl-4-pivaloyloxy deriv., the acoxy group did not migrate, with aromatization occurring via expulsion of the *tert*-butyl group (as isobutylene).

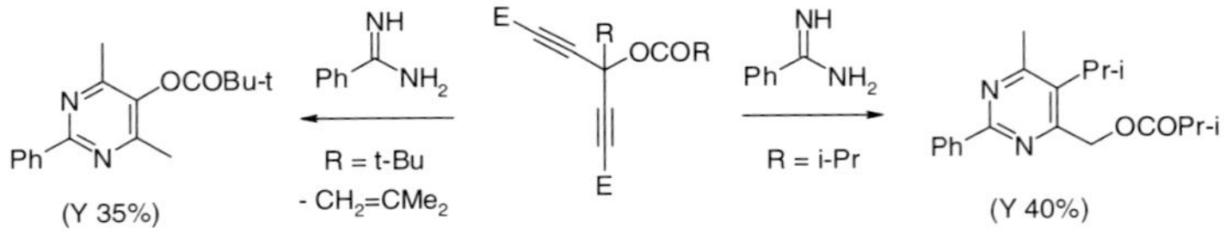

F.e.s. D. Tejedor, S. López-Tosco, F. García-Tellado, J. Org. Chem. *2013*, *78* (7), 3457-63 [DOI: 10.1021/jo400090w].

(Z,Z)-2-Acyl-2′-formamidodivinylamines from imidazoles and α,β-acetyleneketones via stereoselective 1,4-addition-hydrative ring opening C

100.

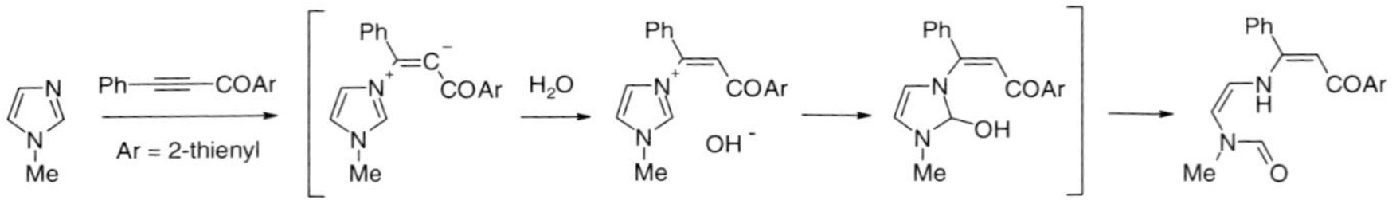

A soln. of 1-methylimidazole (0.5 mmol) in acetonitrile (0.2 ml) added to phenyl(thien-2-yl-carbonyl)acetylene (1 eq.) and water (1 eq.) in the same solvent (0.3 ml), the mixture stirred at 50-55° for 30 h, concentrated, and purified chromatographically → N-methyl-N-[(Z)-2-((Z)-3-oxo-1-phenyl-3-thien-2-yl-1-propenylamino)ethenyl]formamide. Y 80%. This unexpected cleavage of 1-alkylimidazoles (incl. benzyl and allyl) required no catalyst and occurred under relatively mild conditions (45-60°) for (het)aroyl(phenyl)acetylenes, affording 1,4-diaza-2,5-dienes, some of which were previously unknown (six examples; Y 60-80%). Lower yields were obtained with sterically hindered 1-isobutylimidazole (50%) or benzoyl(1-benzyl-4,5,6,7-tetrahydroindol-2-yl)-acetylene (15%), and a terminal benzoylacetylene gave only 5% yield of the diazadiene, the major product being the alkyne trimerization product, tribenzoylbenzene (Y 20%). F.e.s. B.A. Trofimov, L.V. Andriyankova, L.P. Nikitina, K.V. Belyaeva, A.G. Mal′kina, L.N. Sobenina, A.V. Afonin, I.A. Ushakov, Org. Lett. *2013*, *15* (9), 2322-4 [DOI: 10.1021/ol400844x].

Microwaves s. under Copper(II) acetate [\\\\]

Silver/carbon nanotubes *Ag/CNT*
Hydroamination of acetylene derivs. C≡C → CH=C(N<) or CH_2C=NR
s. *76*, 139s*83*

Nanoparticulate copper(I) oxide-on-melamine-formaldehyde resin ←
1,2,3-Triazoles from terminal acetylene derivs. and azides s. *64*, 141s*83* ○

Copper(II) acetate/boron fluoride/oxygen *$Cu(OAc)_2/BF_3/O_2$*
Oxazoles from acetylene derivs. and nitriles
Regioselective copper-catalyzed oxidative [2+2+1]-cycloaddition

101. 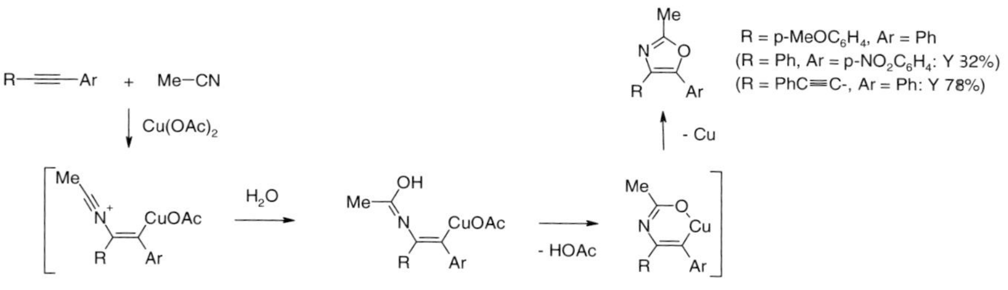

Previous work on alkene-alkyne coupling in nitrile-containing solvents (cf. *82*, 480), which afforded oxazole by-products, has been optimized to form them as the major products. **E:** A mixture of startg. alkyne (0.5 mmol), acetonitrile (3 eq.), water (5 eq.), $Cu(OAc)_2$ (10 mol%), BF_3-etherate (1 eq.) and nitromethane (2 ml) stirred in a dried Schlenk tube at 80° under 1 atm. O_2 for 12 h, then purified by prep. TLC → 4-(4-methoxyphenyl)-2-methyl-5-phenyloxazole. Y 86%. The method is applicable to symmetrical or unsymmetrical alkynes bearing electron-diverse aryl or alkyl groups and to aliphatic, α,β-unsatd. or aryl nitriles, especially electron-rich ones, thirty-two further examples giving yields of 43-89%, while an alkynoate gave a yield of 32%. The reaction failed with terminal acetylene derivs., however. An enacylamine gave the corresponding oxazole under the standard reaction conditions, suggesting it could be an intermediate, while ^{18}O labelling studies showed the oxygen atom to be derived from water. F.e. and application to the synthesis of a COX-2 inhibitor s. X. Li, L. Huang, H. Chen, W. Wu, H. Huang, H. Jiang, Chem. Sci. *2012, 3* (12), 3463-7 [DOI: 10.1039/c2sc21041j]; palladium-catalyzed [$PdCl_2$] cross-trimerization of acetylene derivs. and two ethylene deriv. molecules leading to **1,3,5-trienes** (with LiCl/K_2CO_3 in acetonitrile) **or 1,2,4,5-tetrasubst. benzenes** (with K_2CO_3 in acetonitrile/acetic acid) under O_2 (1 atm.) s. P. Zhou, L. Huang, H. Jiang, A. Wang, X. Li, J. Org. Chem. *2010, 75* (23), 8279-82 [DOI: 10.1021/jo101554r].

Copper(II) acetate/sodium ascorbate/microwaves ←
Copper(I) N-heterocyclic carbene complexes *[Cu(I)]*
Dithiocarboxylatobis(triphenylphosphine)copper(I) complexes *[Cu(I)]*
Copper(I) silica-sulfate nanoparticles *SiO_2-OSO_3Cu*
Copper ferrite nanoparticles/2,2′-bipyridyl/triethylamine *$CuFe_2O_4$/bpy/Et_3N*
Silver(I) complexes *[Ag(I)]*
1,2,3-Triazoles from terminal acetylene derivs. and azides
update on the CuAAC reaction s. *64*, 141s*82*; in aq. alcoholic media with a *mesoionic* copper(I) N-heterocyclic carbene complex (without added reducing agent) s. S. Hohloch, B. Sarkar, L. Nauton, F. Cisnetti, A. Gautier, Tetrahedron Lett. *2013, 54* (14), 1808-12 [DOI: 10.1016/j.tetlet.2013.01.054]; with heteroleptic copper(I) bis(N-heterocyclic carbene) complexes or (tri-*tert*-butylphosphine)[1,3-bis(2,6-diisopropylphenyl)imidazol-2-ylidene]copper(I) fluoroborate at very low (down to 5 ppm!) catalyst loading s. F. Lazreg, A.M.Z. Slawin, C.S.J. Cazin, Organometallics *2012, 31* (22), 7969-75 [DOI: 10.1021/om3006195]; with a dithiocarboxylatobis(triphenylphosphine)copper(I) complex (0.5 mol%) s. B.E. Velasco, G. López-Téllez, N. González-Rivas, I. García-Orozco, E. Cuevas-Yañez, Can. J. Chem. *2013, 91* (4), 292-9 [DOI: 10.1139/cjc-2012-0325]; exclusive formation of 1,4-disubst. 1,2,3-triazoles under copper(I) catalysis (0.5 mol%) *in liq. ammonia* at room temp. s. P. Ji, J.H. Atherton, M.I. Page, Org. Biomol. Chem. *2012, 10* (39), 7965-9 [DOI: 10.1039/c2ob26213d]; 4-subst. 1-(arylselenomethyl)-1,2,3-triazoles from aryl azidomethyl selenides with $Cu(OAc)_2 \cdot H_2O$/Na-ascorbate under microwave enhancement s. N. Seus, M.T. Saraiva, E.E. Alberto, L. Savegnago, D. Alves, Tetrahedron *2012, 68* (51), 10419-25 [DOI: 10.1016/j.tet.2012.07.019]; under *heterogeneous* conditions with nanoparticulate copper(I) silica-sulfate [SiO_2-OSO_3Cu], notably for 'click' cycloaddition with 2-azidoalcohols, s. M.N.S. Rad,

S. Behrouz, M.M. Doroodmand, A. Movahediyan, ibid. *68* (38), 7812-21 [DOI: 10.1016/j.tet.2012.07. 032]; with nanoparticulate copper(I) oxide-on-melamine-formaldehyde resin cf. M.N.S. Rad, S. Behrouz, A. Movahedian, M.M. Doroodmand, Y. Ghasemi, S. Rasoul-Amini, A.-R.A. Gandomani, R. Rezaie, Helv. Chim. Acta *2013, 96* (4), 688-701 [DOI: 10.1002/hlca.201200224]; with magnetically recoverable copper ferrite nanoparticles [$CuFe_2O_4$ nps] modified by 2,2′-bipyridyl using Et_3N as base s. S. Ishikawa, R. Hudson, A. Moores, C.-J. Li, Heterocycles *2013, 86* (2), 1023-30 [DOI: 10.3987/com-12-s(n)125]; with copper(I) immobilized by chelation to a hydrophilic [PEG-based], solid-supported tris(1,2,3-triazol-4-ylmethyl)amine for 'click' conjugation with deprotected oligonucleotides s. L. Eltepu, M. Jayaraman, K.G. Rajeev, M. Manoharan, Chem. Commun. *2013, 49* (2), 184-6 [DOI: 10.1039/c2cc36811k]; preparation of bis(1,2,3-triazole)-modified peptidomimetics s. T.-f. Niu, L. Gu, L. Wang, W.-b. Yi, C. Cai, Eur. J. Org. Chem. *2012* (34), 6767-76 [DOI: 10.1002/ejoc.201201096]; ^{11}C-labelled nucleosides and oligonucleotides by 'click' cycloaddition with methyl azide-^{11}C s. T. Bordenave, P.P. Hazari, D. James, A.K. Mishra, M. Szlosek-Pinaud, E. Fouquet, Eur. J. Org. Chem. *2013* (7), 1214-7 [DOI: 10.1002/ejoc.201201379]; *under silver catalysis* with a robust and efficient ligated silver(I) catalyst s. J. McNulty, K. Keskar, Eur. J. Org. Chem. *2012* (28), 5462-70 [DOI: 10.1002/ejoc.201200930]; high pressure-accelerated, *copper-free* cycloaddition to alkyne groups pre-organized in the crystalline state supramolecularly at room temperature s. H. Chen, B.-B. Ni, F. Gao, Y. Ma, Green Chem. *2012, 14* (10), 2703-5 [DOI: 10.1039/c2gc36069a]; alternative procedure with aryl azides generated *in situ* **from diazonium silica sulfates** in water using the classical $CuSO_4$/Na-ascorbate as catalyst/reductant for cycloaddition under microwaves s. A. Zarei, L. Khazdooz, A.R. Hajipour, H. Aghaei, G. Azizi, Synthesis *2012, 44* (21), 3353-60 [DOI: 10.1055/s-0032-1316783]; 1-β-alkylthio-1,2,3-triazoles from ethylene and acetylene derivs. via 1,2-azido-sulfenylation-1,3-dipolar cycloaddition, also N-β-hydroxy-1,2,3-triazoles from epoxides in water, s. *83*, 237.

Silver nitrate/1,8-diazabicyclo[5.4.0]undec-7-ene *$AgNO_3$/DBU*
5-Alkylidene-2-oxazolidones from 2-acetyleneamines ○
and carbon dioxide s. *83*, 104; *52*, 137s*83*

Silver triflate *AgOTf*
5-Alkylidene-1-tosyl-2-imidazolidones from 2-acetyleneamines
and tosyl isocyanate s. *83*, 104; *80*, 128s*83*

Bis[1-aryl-3-imidoylimidazol-2-ylidene]gold(I) dichloroaurates or Cationic gold(I) imidazole-2-thione complexes *[Au]*
Hydroamination of acetylene derivs. C≡C → CH=C(N<) or CH_2C=NR
update s. *76*, 139s*82*; Markovnikov addition of anilines to phenylacetylene with bis[1-aryl-3-imidoylimidazol-2-ylidene]gold(I) dichloroaurates or cationic gold(I) imidazole-2-thione complexes as catalyst s. E. Alvarado, A.C. Badaj, T.G. Larocque, G.G. Lavoie, Chem. Eur. J. *2012, 18* (38), 12112-21 [DOI: 10.1002/chem.201201448]; with bis(η^5-pentamethylcyclopentadienyl)dizinc as catalyst in the presence of stoichiometric [$PhNMe_2H$][$B(C_6F_5)_4$] [and comparison with [Cp^*_2Zn] and Et_2Zn] s. A. Lühl, L. Hartenstein, S. Blechert, P.W. Roesky, Organometallics *2012, 31* (20), 7109-16 [DOI: 10.1021/om300649q]; solvent-free Markovnikov hydroamination of alkynes with ar. amines using $GaCl_3$ as catalyst s. L. Li, G. Huang, Z. Chen, W. Liu, X. Wang, Y. Chen, L. Yang, W. Li, Y. Li, Eur. J. Org. Chem. *2012* (28), 5564-72 [DOI: 10.1002/ejoc.201200829]; catalytic Michael addition of anilines to activated alkynes with recyclable Ag/carbon nanotubes s. Y. Liu, G. Wu, Y. Cui, Appl. Organomet. Chem. *2013, 27* (4), 206-8 [DOI: 10.1002/aoc.2950]; **aldehyde hydrazones** from terminal acetylene derivs. by anti-Markovnikov hydrohydrazination with (diphenylhydrazono)(pyridine)[1,7-diisopropyl-4-methyl-1,6,7-triazaheptane-1,7-diyl]titanium(IV) as pre-catalyst s. A.D. Schwarz, C.S. Onn, P. Mountford, Angew. Chem., Int. Ed. *2012, 51* (49), 12298-302 [DOI: 10.1002/anie.201206249]; regioselective **hydroamination of allenes** (cf. *76*, 140) under kinetic control with an allyl(α,β-ethylene-β-phosphinoaldimine)palladium(II) triflate as catalyst s. N.C. Zingales, A.R. Shaffer, J.A.R. Schmidt, Organometallics *2013, 32* (2), 578-86 [DOI: 10.1021/om301102w].

Zwitterionic gold(I) phosphinocarba-closo-dodecaborate complexes *[Au(I)]*
Azomethines from acetylene derivs. C≡C → CH_2C=N
using a zwitterionic gold catalyst
bearing a perchlorinated carba-*closo*-dodecaborate anion

102. 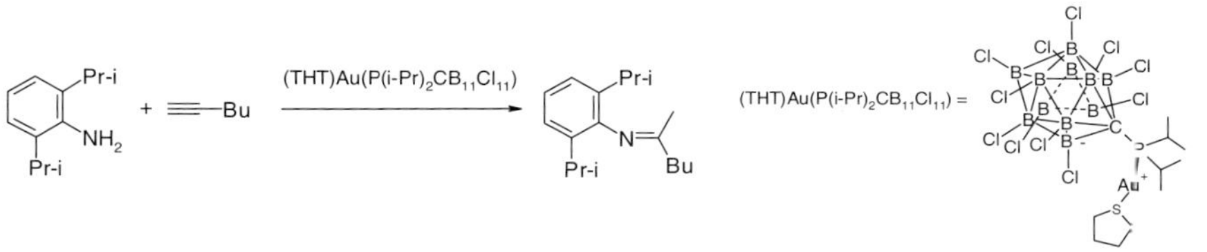

2,6-Diisopropylaniline and 1-hexyne (1 eq.) added to (THT)Au(P(i-Pr)$_2$CB$_{11}$Cl$_{11}$) [where THT = tetrahydrothiophene] (0.2 mol%), and the mixture stirred at 80° for 24 h → N-hex-2-ylidene-2,6-diisopropylaniline. Y >95% (NMR). The novel single-component gold catalyst promoted the addition of anilines to terminal and internal ar. and aliphatic alkynes (eighteen examples, NMR Y 54 to >95%). Aromatic terminal alkynes were hydroaminated at 50° and low catalyst loadings (0.001 mol%), with highest catalyst activity observed with bulky amines, and while diphenylacetylene and 1-hexyne required 80° and catalyst loadings of 0.2 mol%, their relatively lower activity remained comparable to that for typical Au catalysts. F.e. and catalyst prepn. s. V. Lavallo, J.H. Wright II, F.S. Tham, S. Quinlivan, Angew. Chem., Int. Ed. *2013*, *52* (11), 3172-6 [DOI: 10.1002/anie.201209107].

[3-(2,6-Diisopropylphenyl)-1,3-diazabicyclo[3.3.1]nonan-2-ylidene]gold(I) chloride/ potassium barfate *[Au(I)]*
Hydrazones from acetylene derivs. C≡C → CH_2C=NNH_2
Regioselective gold-catalyzed hydroamination with hydrazine under mild conditions

103. 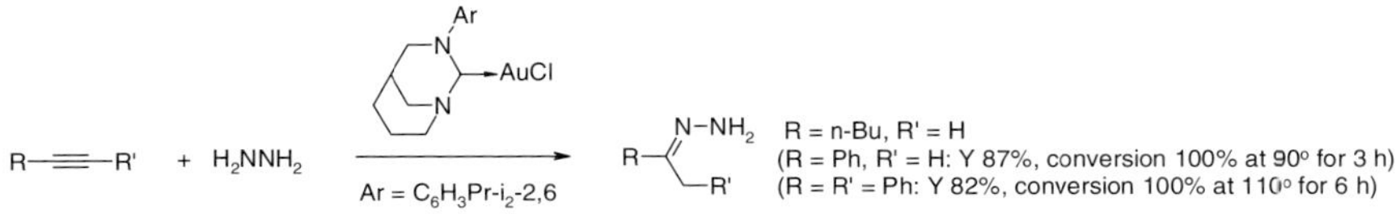

Gold-catalyzed hydroamination of acetylene derivs. with hydrazine (cf. *80*, 129) may now be carried out under milder conditions (room temperature in several cases) by using a gold(I) chloride complex ligated to a non-hindered anti-Bredt 3-aryl-1,3-diazabicyclo[3.3.1]nonan-2-ylidene, in which one of the two nitrogen atoms is placed in a strained bridgehead position; this restricts donation of its lone pair, thereby enhancing the π-accepting character of the carbene, while retaining its σ-donor properties. **E:** Benzene-d$_6$ (0.4 ml), 1,4-di-*tert*-butylbenzene (internal standard; 0.05 mmol), startg. alkyne (0.5 mmol) and anhydrous hydrazine (1 eq.) added sequentially to the gold(I) NHC complex (5 mol%) and KBAr$^F_4$ (5 mol%) in a J Young NMR tube under argon, after 3 h (monitored by ^{1}H and ^{13}C NMR) at room temp., the mixture promptly passed under a stream of argon through a short pad of well-dried Florisil® → product. Y 83% (conversion 91%). For alkyl-subst. terminal alkynes, the efficiency of the reaction decreases with increasing steric hindrance of the substrates; thus for *tert*-butylacetylene, only a 29% conversion was observed after 1.5 days at room temperature, completion of the reaction requiring heating at 90° for 6 h. For electron-poor arylethynes or 1-cyclohexenylacetylene or an internal alkyne [diphenyl-acetylene], reaction was performed at 75-110° (in chloroform for the latter example). F.e. (seven; Y 71-95%; conversion 78-100%) s. M.J. López-Gómez, D. Martin, G. Bertrand, Chem. Commun. *2013*, *49* (40), 4483-5 [DOI: 10.1039/c3cc41279b]; prepn. of the carbene s. D. Martin, N. Lassauque, B. Donnadieu, G. Bertrand, Angew. Chem., Int. Ed. *2012*, *51* (25), 6172-5 [DOI: 10.1002/anie.201202137].

(Triphenylphosphine)gold(I) chloride/silver triflate *$(Ph_3P)AuCl/AgOTf$*
Chemo- and regio-selective gold(I)- or silver(I)-catalyzed ring closure of 2-acetyleneamines and tosyl isocyanate ○

104.

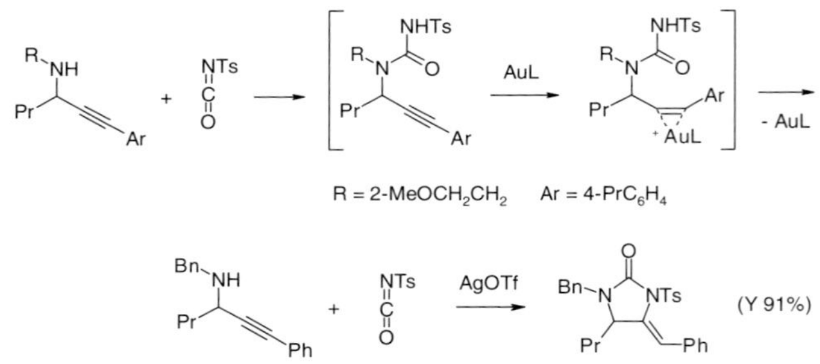

Two protocols have been developed for cyclization of propargylic ureas generated *in situ* from sec. propargylic amines and tosyl isocyanate. A protocol based on dual Au(I)/Ag(I) catalysis affords **5-alkylidene-2-(tosylimino)oxazolidines** as major/exclusive products via O-cyclization, while N-cyclized regioisomers, **5-alkylidene-1-tosyl-2-imidazolidones**, predominate using Ag(I) alone (cf. *80*, 128). **E:** Tosyl isocyanate (1.1 eq.) added to a soln. of N-(2-methoxyethyl)-1-(4-propylphenyl)hex-1-yn-3-amine (0.4 mmol) in dry methylene chloride (2 ml) at room temp., the mixture stirred for 5 min, AgOTf (5 mol%) and $(Ph_3P)AuCl$ (5 mol%) added, stirring continued at room temp. for 2 h in a sealed screw-cap vial under air, and the resulting mixture purified directly by chromatography on silica → 3-(2-methoxyethyl)-4-propyl-5-(4-propylbenzylidene)-2-(tosylimino)-oxazolidine. Y 90% (sole product). The O-cyclized products were formed at room temp. or 50° via 5-*endo-dig* cyclization of the intermediate propargylic ureas (incl. a terminal alkyne) and predominated in most cases (eleven examples; Y 63-100%; in two examples regioisomeric products predominated for reasons not understood). Isomeric N-cyclization products (6-28%) and traces (2-12%) of 6-*endo-dig* isomers were obtained as by-products in some cases. Use of 20 mol% AgOTf in toluene at higher temp. (80°) generated isomeric 5-alkylidene-1-tosyl-2-imidazolid-2-ones (thirteen examples; Y 42-91%) via preferential N-cyclization. A rationale for the observed chemoselectivity based on hard acid-hard base (O-cyclization) and soft acid-soft base (N-cyclization) is provided. F.e. and optimization s. O.P. Pereshivko, V.A. Peshkov, J. Jacobs, L. Van Meervelt, E.V. Van der Eycken, Adv. Synth. Catal. *2013*, *355* (4), 781-9 [DOI: 10.1002/adsc.201200905]; **5-alkylidene-2-oxazolidones** from 2-acetyleneamines and CO_2 (cf. *52*, 137s*68*) with $AgNO_3$ (cat)/DBU in DMSO under air s. M. Yoshida, T. Mizuguchi, K. Shishido, Chem. Eur. J. *2012*, *18* (49), 15578-81 [DOI: 10.1002/chem.201203366].

Bis(η^5-pentamethylcyclopentadienyl)dizinc/N,N-dimethylanilinium tetrakis(pentafluorophenyl)borate ←
Hydroamination of acetylene derivs. s. *76*, 139s*83* C≡C → CH=C(N<) or CH_2C=NR

Phenylboronic acid *$PhB(OH)_2$*
Differentially-protected 1,2-diamines from ethylene derivs. s. *83*, 291 ←

Boron fluoride s. under $Cu(OAc)_2$ *BF_3*

Gallium trichloride *$GaCl_3$*
Hydroamination of acetylene derivs. C≡C → CH=C(N<) or CH_2C=NR
in the absence of solvent s. *76*, 139s*83*

2(S)-[Bis[3,5-bis(trifluoromethyl)phenyl](trimethylsiloxy)methyl]pyrrolidine/benzoic acid ←
Organocatalyzed asym. aza-Michael addition C=C → CHC(N<)
update s. *56*, 129s*82*; of anilines to α,β-ethylene-γ,γ-difluoroaldehydes with 2(S)-[bis[3,5-bis(trifluoromethyl)phenyl](trimethylsiloxy)methyl]pyrrolidine/benzoic acid as catalyst and water (10 mol%), also asym. thia-Michael addition with mercaptans (SC⇓CC), s. A. Khalaf, D. Grée,

H. Abdallah, N. Jaber, A. Hachem, R. Grée, Eur. J. Org. Chem. *2013* (4), 653-7 [DOI: 10.1002/ejoc.201201427].

Tetrahydrothiophene ←

2-α-Functionalized furans from 2,4-enynones ○

Organocatalyzed ring closure under neutral conditions

105. 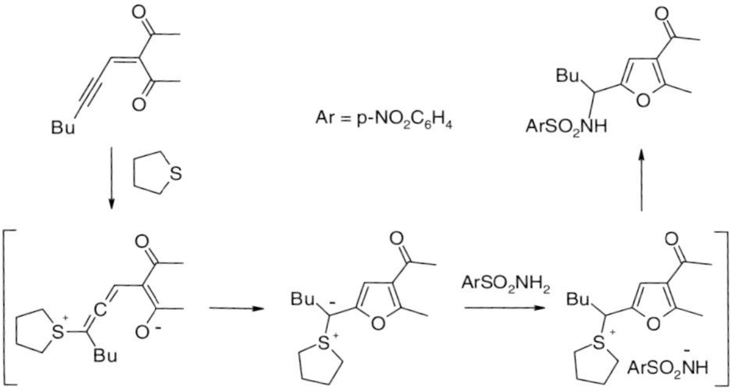

Tetrahydrothiophene serves as a simple nucleophilic organocatalyst for the ring closure of 2,4-enynones, whereby formation of an intermediate sulfonium ylid facilitates the trapping of various nucleophiles to provide a new route to 2-α-functionalized furans under *neutral* conditions. **E: 2-α-(Sulfonylamino)furans.** A soln. of 3-(hept-2-yn-1-ylidene)pentane-2,4-dione (0.52 mmol), *p*-nitrobenzenesulfonamide (1.1 eq.) and tetrahydrothiophene (10 mol%) in methylene chloride (1 *M*) heated under reflux for 18-48 h (TLC), and worked up with purification by flash chromatography → 5-[1-(*p*-nitrophenylsulfonamido)pentyl]-3-acetyl-2-methylfuran. Y 45%. The procedure is successful for the cyclization of enynones substituted at the alkyne terminus by alkyl, aryl or silyl groups, which couple with a variety of O- or N-nucleophiles, incl. alcohols, phenols, fluoroalcohols, electron-diverse arylcarboxylic acids (at 50 mol% catalyst loading) and sulfonylamines (twenty examples; Y 35-99%). The lowest yield was with *tert*-butanol, presumably for steric reasons. An electron-withdrawing group at the C_2 position of the enynone, however, is critical, there being no reaction with dec-3-en-5-yn-2-one. Reaction also failed with other organocatalysts. Mechanistically, it is proposed that ring closure is initiated by nucleophilic attack of the thioether on the triple bond, followed by cyclization with formation of a sulfonium ylid; the latter then deprotonates the nucleophile prior to displacement of the formed sulfonium ion, possibly via an S_N1 pathway. F.e. and comparison of solvents s. J.S. Clark, A. Boyer, A. Aimon, P.E. Garcia, D.M. Lindsay, A.D.F. Symington, Y. Danoy, Angew. Chem., Int. Ed. *2012*, *51* (48), 12128-31 [DOI: 10.1002/anie.201207300].

Cinchona-based 2-tert-aminothioureas ←

N-Tosyloxazolidines from electron-deficient 2-ethylenealcohols and N-tosylaldimines

Organocatalyzed asym. [formal] [3+2]-cycloaddition

106. 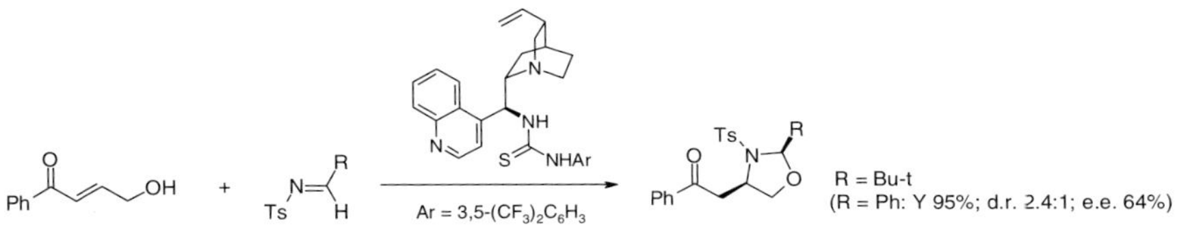

Organocatalyzed asym. [formal] [3+2]-cycloaddition via intramolecular Michael addition with electron-deficient 2-ethylenealcohols has been adapted by substituting aldimines in place of the previously-used aldehydes (cf. *81*, 44). **E: Chiral 4-β-keto-N-tosyloxazolidines**. A mixture of (E)-4-hydroxy-1-phenylbut-2-en-1-one (0.25 mmol) and startg. N-tosylaldimine (1 eq.) in toluene (0.5 ml) stirred for 10 min at 0°, treated with the cinchonidine-derived thiourea catalyst (10 mol%), stirred at 0° for 24 h, sequentially diluted with hexane/ethyl acetate (1:1), passed through a short silica gel pad, concentrated *in vacuo*, and purified by flash chromatography on silica gel → product. Y 62% (d.r. 2.6:1; e.e. 87%). The method is applicable to (het)aryl or alkyl N-tosylaldimines in reaction with (het)aryl or alkyl γ-hydroxyenones (twenty-two further examples; Y 45-99%; d.r. 1.9:1 to 11:1; e.e. 19-24%, 49-74%) or a phenyl ester analog (Y 77%; single diastereomer; e.e. 37%). F.e., gram-scale reaction and formation of an enantiomeric product (Y 84%; d.r. 9.1:1; e.e. 63%) s. Y. Fukata, K. Asano, S. Matsubara, Chem. Lett. *2013*, *42* (4), 355-7 [DOI: 10.1246/cl.121245].

Biphenyl-2,2′-bis[iodine(III) bis(ditosylamide)] *$ArI(NTs_2)_2$*
1,2-Diamination of ethylene derivs. C=C → C(N-)C(N-)
under metal-free conditions s. *79*, 128s*82*; N,N,N′,N′-tetratosyl-1,2-diamines with pre-formed or *in situ*-generated biphenyl-2,2′-bis[iodine(III) bis(ditosylamide)] as diaminating agent s. C. Röben, J.A. Souto, E.C. Escudero-Adán, K. Muñiz, Org. Lett. *2013, 15* (5), 1008-11 [DOI: 10.1021/ol3034884]; 3-vinyl-1,2,5-thiadiazolidine 1,1-dioxides by **regioselective asym. 1,2-diamination** of 1,3-dienes with di-*tert*-butylthiadiaziridine S,S-dioxide (cf. *74*, 166) using $Pd_2(dba)_3$ at catalyst and a chiral 1,1′-binaphthyl-2,2′-diyl phosphoramidite as ligand s. R.G. Cornwall, B. Zhao, Y. Shi, Org. Lett. *2013, 15* (4), 796-9 [DOI: 10.1021/ol303469a].

N-Bromosuccinimide/1,8-diazabicyclo[5.4.0]undec-7-ene *NBS/DBU*
Indirect Michael addition of succinimide to chalcones s. *56*, 129s*83* C=C → CHC(N<)

(Diphenylhydrazono)(pyridine)[1,7-diisopropyl-4-methyl-1,6,7-triazaheptane-1,7-diyl]titanium(IV) *[Ti(IV)]*
Aldehyde hydrazones from terminal acetylene derivs. C≡CH → CH_2CH=NN<
s. *76*, 139s*83*

Chiral 1,1′-binaphthyl-2,2′-diyl phosphoramidites s. under Tris(dibenzylideneacetone)-dipalladium *L**
Oxygen s. under $Cu(OAc)_2$ *O_2*

Tris(dibenzylideneacetone)dipalladium/chiral 1,1′-binaphthyl-2,2′-diyl phosphoramidites *[Pd(0)]**
Regioselective asym. 1,2-diamination of 1,3-dienes s. *74*, 166s*83* C=C → C(N-)C(N-)

Allyl(α,β-ethylene-β-phosphinoaldimine)palladium(II) triflate *[Pd(II)]*
Regioselective hydroamination of allenes s. *76*, 140s*83* C=C=C → CHC(N<)=C

Potassium osmate *$K_2OsO_2(OH)_4$*
1,2-Oxyamination of ethylene derivs. C=C → C(O-)C(N<)
update s. *31*, 312s*82*; 2-(carbalkoxyamino)alcohols from mono-, di- and tri-subst. ethylene derivs. and N-tosyloxyurethans with $K_2OsO_2(OH)_4$, regioselectivity, s. Masruri, A.C. Willis, M.D. McLeod, J. Org. Chem. *2012, 77* (19), 8480-91 [DOI: 10.1021/jo301372y]; 2-(tosylamino)alcohols with N,N′-dibromo-N,N′-1,2-ethanediylbis(*p*-toluenesulfonamide) in water containing β-cyclodextrin s. F. Kazemi, M.A. Kakroudi, Bull. Korean Chem. Soc. *2013, 34* (2), 500-4 [DOI: 10.5012/bkcs.2013.34.2.500].

Rearrangement ∩

Oxygen/Nitrogen Type NC ∩ ON

Potassium hexacyanoferrate(II)/1,10-phenanthroline *[Fe(II)]/phen*
4-α-Acoxy-2-oxazolidones from N-acoxycarbamyloxy-2-ethylenes via stereoselective iron(II)-catalyzed intramolecular 1,2-oxyamination ○

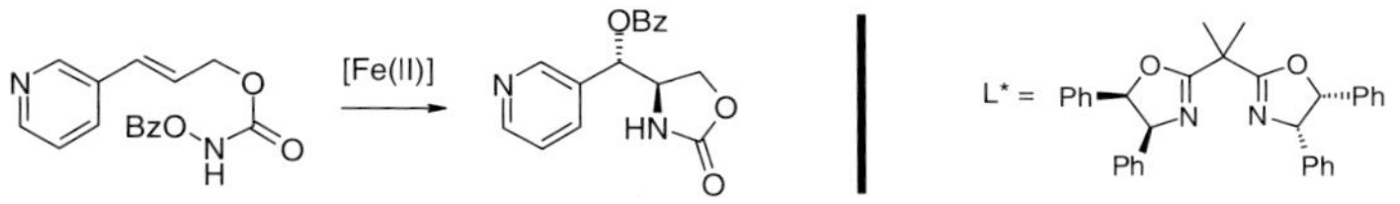

107. 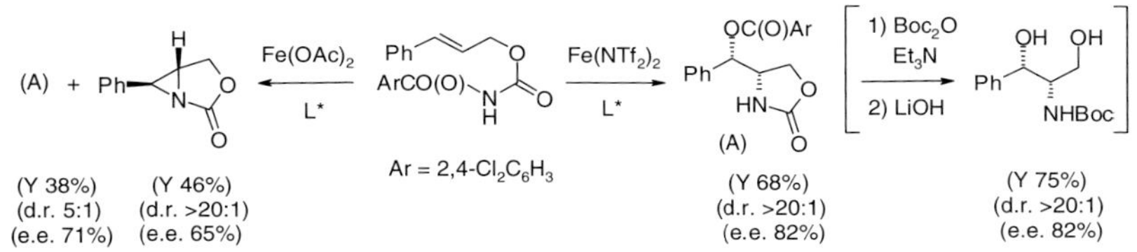

A mixture of $K_4Fe(CN)_6$ (10 mol%) and 1,10-phenanthroline (20 mol%) in anh. acetonitrile (3 ml) stirred vigorously at room temp. for 30 min, (E)-3-(pyridin-3-yl)allyl benzoyloxycarbamate (0.2 mmol) added, the mixture heated at 70° for 12 h, cooled, concentrated, and purified chromatographically → *anti*-(2-oxooxazolidin-4-yl)(pyridin-3-yl)methyl benzoate. Y 80% (d.r. >20:1). This novel intramolecular cyclization appears general for allyl N-acoxycarbamates carrying mono-/di-(het)aryl, methyl or alkynyl substituents on the allyl double bond (incl. a cyclic analog) affording rare *anti*-products generally with high diastereoselectivity (d.r. 1:1 to >20:1; seventeen examples; Y 71-92%; a substrate lacking a substituent on the double bond gave 40%). An asymmetric variant based on a chiral bis(oxazoline) ligand and $Fe(NTf_2)_2$ catalyst gave the same *syn*-product from E/Z isomers of a single substrate with good enantioselectivity (e.e. 82%; d.r. >20:1; Y 68%) and the product was cleaved to a 2-carbo-*tert*-butoxyamino-1,3-diol using a standard method (Boc_2O/Et_3N then LiOH) in 75% overall yield with no loss of chirality. Interestingly, use of $Fe(OAc)_2$ with the chiral ligand gave the product with lower selectivity (e.e. 71%; d.r. 5:1; Y 38%) as a mixture with an aziridinooxazolidinone (Y 46%; d.r. >20:1; e.e. 65%). F.e., optimization and substrate prepn. s. G.-S. Liu, Y.-Q. Zhang, Y.-A. Yuan, H. Xu, J. Am. Chem. Soc. *2013*, *135* (9), 3343-6 [DOI: 10.1021/ja311923z].

Carbon/Carbon Type NC ∩ CC

Microwaves s. under Platinum(II) chloride [\\\\]

Ethyldiisopropylamine *i-Pr$_2$NEt*
Intramolecular aza-Michael addition ○
update s. *38*, 301s*82*; 6-subst. *cis*-4-piperidone-2-carboxylic acid esters by 6-*endo-trig* cyclization with ethyldiisopropylamine s. M. Daly, A.A. Cant, L.S. Fowler, G.L. Simpson, H.M. Senn, A. Sutherland, J. Org. Chem. *2012*, *77* (22), 10001-9 [DOI: 10.1021/jo3022583]; effect of a Brønsted acid (TfOH or trifluoroacetic acid) on the diastereoselectivity of ring closure to N-protected 3-β-ketomorpholines s. C. Zhong, Y. Wang, C. O'Herin, D.W. Young, ACS Catal. *2013*, *3* (4), 643-6 [DOI: 10.1021/cs400031p].

Copper(II) triflate/chiral bis(Δ²-oxazolines)/1,4-cyclohexadiene/manganese dioxide/ 2,6-di-tert-butyl-4-methylpyridine ←
N-Sulfonylindolines from N-(*o*-allylaryl)sulfonamides
by asym. intramolecular hydroamination s. *72*, 185s*83*

Silver nitrate or triflate *$AgNO_3$ or AgOTf*
Gold(I) complexes *[Au(I)]*
(1,9-Diaminophenalene)zinc amide/N,N-dimethylanilinium tetrakis-(pentafluorophenyl)borate ←

Intramolecular hydroamination of carbon-carbon multiple bonds ○

update s. *70*, 147s*82*; **cycloisomerization of ethyleneamines** *under zinc catalysis* with a 1,9-diaminophenalene-based zinc amide in the presence of $[PhNMe_2H][B(C_6F_5)_4]$ s. A. Mukherjee, T.K. Sen, P.K. Ghorai, P.P. Samuel, C. Schulzke, S.K. Mandal, Chem. Eur. J. *2012, 18* (34), 10530-45 [DOI: 10.1002/chem.201200868]; *under iridium catalysis* with the bifunctional complex, (pentamethylcyclopentadienyl)[2-(pyrazolato-2-yl)phenyl]iridium(III), diastereoselectivity and functional group compatibility, s. Y. Kashiwame, S. Kuwata, T. Ikariya, Organometallics *2012, 31* (23), 8444-55 [DOI: 10.1021/om301063n]; effect of donor ligands (ethers, amines, ureas, alkoxides or ammonium salts) on Group 3 metal-catalyzed intramolecular hydroamination s. A.R. Smith, H.M. Lovick, T. Livinghouse, Tetrahedron Lett. *2012, 53* (47), 6358-60 [DOI: 10.1016/j.tetlet.2012.09.019]; N-sulfonylindolines from N-(*o*-allylaryl)sulfonamides by **asym. intramolecular hydroamination** with $Cu(OTf)_2$/(R,R)-Ph-box and excess of 1,4-cyclohexadiene in the presence of MnO_2 and 2,6-di-*tert*-butyl-4-methylpyridine as base (cf. *72*, 185s*82*) s. B.W. Turnpenny, K.L. Hyman, S.R. Chemler, Organometallics *2012, 31* (22), 7819-22 [DOI: 10.1021/om300744m]; 3-methylimidazo[1,2-*a*]pyridines by *silver-catalyzed* **intramolecular hydroamination of acetyleneamines** [N-(prop-2-yn-1-yl)pyridin-2-amines] with AgOTf s. M. Chioua, E. Soriano, L. Infantes, M.L. Jimeno, J. Marco-Contelles, A. Samadi, Eur. J. Org. Chem. *2013* (1), 35-9 [DOI: 10.1002/ejoc.201201258]; 3-subst. 1,2-benzothiazine 1,1-dioxides by intramolecular hydroamination with $AgNO_3$ as catalyst s. D. Rambabu, P.V.N.S. Murthy, K.R.S. Prasad, A. Kandale, G.S. Deora, M.V.B. Rao, M. Pal, Tetrahedron Lett. *2012, 53* (48), 6577-83 [DOI: 10.1016/j.tetlet.2012.09.102]; N-protected 3-perfluoroalkyl-1-alkylideneisoindolines or regioisomeric 1-perfluoroalkyl-1,2-dihydroisoquinolines under gold(I) catalysis s. S. Fustero, I. Ibáñez, P. Barrio, M.A. Maestro, S. Catalán, Org. Lett. *2013, 15* (4), 832-5 [DOI: 10.1021/ol3035142]; N-protected Δ^2-4-pyrrolones with $PtCl_2/K_2CO_3$ in PEG-3400 under microwave enhancement s. R. Spina, E. Colacino, B. Gabriele, G. Salerno, J. Martinez, F. Lamaty, J. Org. Chem. *2013, 78* (6), 2698-702 [DOI: 10.1021/jo302474r]; chiral morpholines and piperazines by **asym. intramolecular hydroamination of acetyleneamines** with chloro(*p*-cymene)[(S,S)-N-tosyl-1,2-diphenylethylenediamine]-ruthenium(II), making comparisons with bis(amidato)titanium(IV) and (dimethylamine)-bis(ureato)zirconium(IV) diamides, (cf. *77*, 158) s. H. Zhai, A. Borzenko, Y.Y. Lau, S.H. Ahn, L.L. Schafer, Angew. Chem., Int. Ed. *2012, 51* (49), 12219-23 [DOI: 10.1002/anie.201206826]; unprotected 4-vinyl-2,3,4,5-tetrahydro-1,5-benzoxazepines by **intramolecular hydroamination of allenes** under gold(I) catalysis with [LAuCl]/[AgX] s. D. Pflästerer, P. Dolbundalchok, S. Rafique, M. Rudolph, F. Rominger, A.S.K. Hashmi, Adv. Synth. Catal. *2013, 355* (7), 1383-93 [DOI: 10.1002/adsc.201300154].

1,4-Cyclohexadiene s. under Copper(II) triflate ←
Polyethylene glycol s. under Platinum(II) chloride *PEG*
Chiral bis(Δ^2-oxazolines) s. under Copper(II) triflate *box*

Carbonic acid esters/1,5,7-triazabicyclo[4.4.0]dec-5-ene *$(RO)_2CO$/TBD*

Catalytic Lossen rearrangement ∩

108. 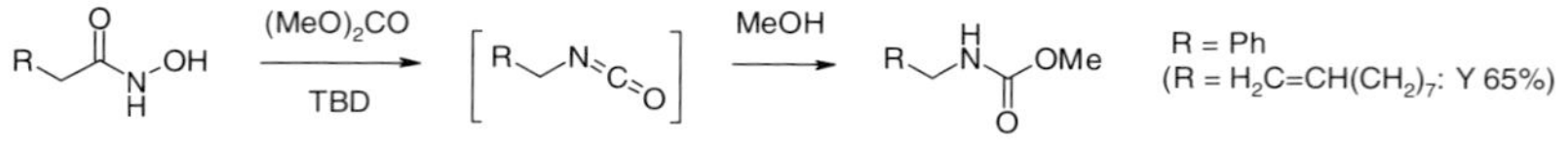

Urethans from aliphatic hydroxamic acids. Startg. hydroxamic acid (100 mmol) dissolved in dimethyl carbonate (20 eq.), methanol (2 eq.) added, the mixture heated to reflux, treated with 1,5,7-triazabicyclo[4.4.0]dec-5-ene (0.2 eq.), evaporated to dryness after 20 h (TLC and GCMS), and the residue purified by column chromatography → product. Y 73%. Four further examples gave methyl carbamates in yields of 62-65%, while a phenol-containing substrate gave a yield of ~76% with concomitant O-methylation of half of the product, and a bis(hydroxamic acid) gave a yield of 66% with DMSO as co-solvent to aid solubility. Diethyl, diallyl, dibenzyl and diphenyl carbonates gave the corresponding carbamates in the presence of the appropriate alcohol/phenol (four examples; Y 62-77%), while ethylene carbonate gave a mixture of glycol monocarbamate (Y 21%) and monocarbamate monocarbonate (Y 33%). **Prim. ar. amines** were obtained from arylhydroxamic acids (eight examples; Y 65-83%), typically with a reaction time of 3-5 h.

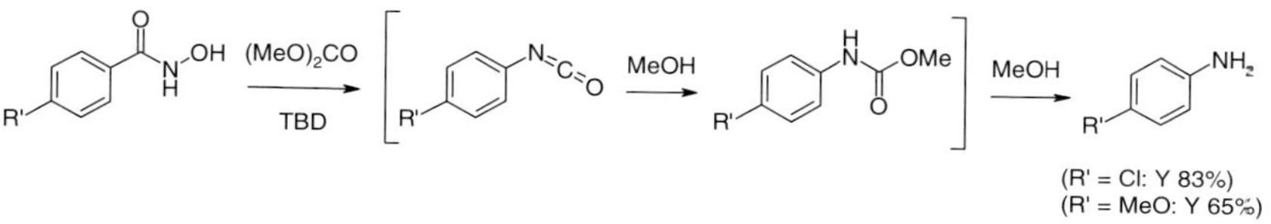

The dimethyl carbonate/methanol mixture may be recycled several times by simple distillation without decreased yields being observed. F.e. and alternative bases s. O. Kreye, S. Wald, M.A.R. Meier, Adv. Synth. Catal. *2013*, *355* (1), 81-6 [DOI: 10.1002/adsc.201200760]; **N-carballyloxyl-amines** with *p*-NsCl/*i*-Pr$_2$NEt and N-methylimidazole as catalyst in THF containing allyl alcohol, **also isocyanates** in the absence of the alcohol and N-methylimidazole (cf. *2*, 478s*77*), s. S. Yoganathan, S.J. Miller, Org. Lett. *2013*, *15* (3), 602-5 [DOI: 10.1021/ol303424b]; **unsym. ureas** by bromo-dimethylsulfonium bromide-mediated Lossen rearrangement using N-methylmorpholine as base s. D.K. Yadav, A.K. Yadav, V.P. Srivastava, G. Watal, L.D.S. Yadav, Tetrahedron Lett. *2012*, *53* (23), 2890-3 [DOI: 10.1016/j.tetlet.2012.03.129].

1-((S)-3,3-Dimethyl-1-((R)-2-(naphth-1-yl)pyrrolidin-1-yl)-1-oxobutan-2-yl)-3-[(1R,2R)-2-(2,5-diphenyl-1H-pyrrol-1-yl)cyclohexyl]thiourea/potassium carbonate ←
N-Acoxy- or -hydroxy-pyrrolidines from 4-ethylenehydroxylamines via organocatalyzed asym. intramolecular Cope-type hydroamination ○

109.

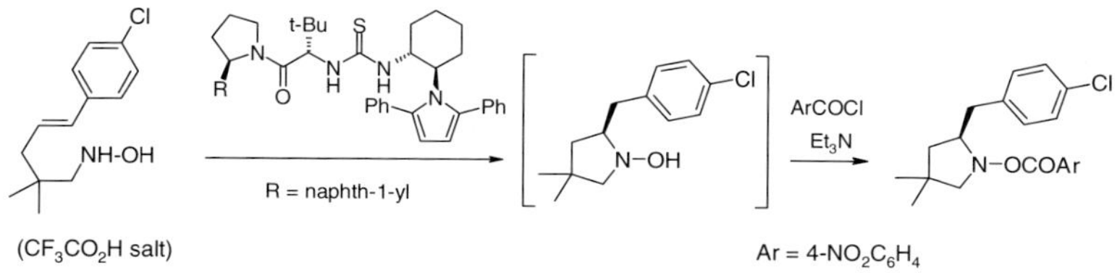

An enantioselective variant of *49*, 271 is reported. **E:** Hexanes (2.5 ml) and chiral thiourea catalyst (10 mol%) added to (E)-5-(4-chlorophenyl)-2,2-dimethylpent-4-ylhydroxylamine (0.2 mmol; as its trifluoroacetic acid salt, freshly generated from the bis-Boc protected hydroxylamine and trifluoroacetic acid), the mixture cooled to 0°, 20% aq. K_2CO_3 (5 eq.) added, the mixture stirred vigorously at 3° for 5 h, extracted with methylene chloride, 4-nitrobenzoyl chloride (4 eq.) and triethylamine (8 eq.) added, the mixture stirred for 2 h, quenched with 20% aq. K_2CO_3, extracted with methylene chloride, and purified chromatographically → (S)-2-(4-chlorobenzyl)-4,4-dimethyl-pyrrolidin-1-yl 4-nitrobenzoate. Y 87% (e.e. 95%). A series of thermally unstable 5-aryl-2,2-di-methylpent-4-enyl-hydroxylamines were freshly generated by the reduction of the corresponding oximes, or used as their freshly prepared trifluoroacetic acid salts, to afford the title products, isolated as their 4-nitrobenzoates, generally with high enantioselectivity (e.e. 87%, 91-96%; eight examples; Y 79-96%). Lower enantioselectivity was observed for analogs lacking *gem*-dimethyl groups (e.e. 81-83%; two examples; Y 68-93%), or terminal aryl groups (e.e. 78-85%; two examples; Y 86-91%). F.e., optimization, catalyst screening, and catalyst and substrate prepn. s. A.R. Brown, C. Uyeda, C.A. Brotherton, E.N. Jacobsen, J. Am. Chem. Soc. *2013*, *135* (18), 6747-9 [DOI: 10.1021/ja402893z].

Trifluoroacetic or trifluoromethanesulfonic acid *CF_3COOH or TfOH*
Intramolecular aza-Michael addition s. *38*, 301s*83*

Manganese dioxide s. under Copper(II) triflate *MnO_2*

Chloro(p-cymene)[(S,S)-N-tosyl-1,2-diphenylethylenediamine]ruthenium(II) *[Ru(II)]**
Asym. intramolecular hydroamination of acetyleneamines s. *72*, 185s*83*

(Pentamethylcyclopentadienyl)[2-(pyrazolato-2-yl)phenyl]iridium(III) *[Ir(III)]*
Platinum(II) chloride/potassium carbonate/polyethylene glycol-3400/microwaves ←
Intramolecular hydroamination of carbon-carbon multiple bonds s. *70*, 147s*83* ○

Exchange ⇃↾

Hydrogen ↾ NC ⇃↾ H

Sodium 5-methyl-2-pyridinolate s. under Copper(I) iodide ←

Potassium carbonate/air *K_2CO_3/O_2*
Pyrazole-4-carbonyl compds. from 1,3-enyne-2-carbonyl compds. and hydrazines via base-mediated aerobic ring closure ○

110. 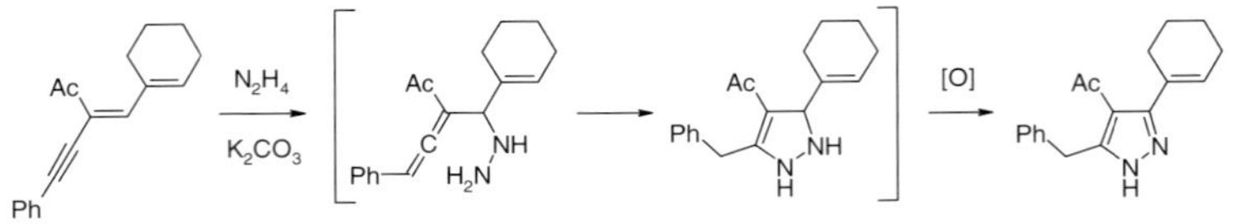

under mild conditions. K_2CO_3 (1.4 eq.) added in one portion to a soln. of 3-cyclohex-1-en-1-yl-methylidene-5-phenylpent-4-yn-2-one (0.2 mmol) and hydrazine hydrate (1.2 eq.) in N,N-dimethyl-acetamide (DMA; 2.5 ml) under air at room temp., the mixture stirred for 3 h (TLC), quenched with water, extracted with ether, and purified chromatographically → 4-acetyl-5-benzyl-3-cyclohex-1-en-1-ylpyrazole. Y 76%. This novel, experimentally simple and scalable (to 12.2 mmol) reaction occurred via tandem cyclization and aerobic aromatization of 1,3-enynes carrying an electron-withdrawing group (COR, CHO, CN, CO_2Me, $COCO_2Me$) at the 2-position, electron-diverse aryl at the alkyne terminus and aryl/alkene at the ene terminus. Cyclization, in DMA or acetonitrile, with hydrazine hydrate or monosubst. (Me, Ph, *t*-Bu) hydrazines afforded the title products (generally as single isomers for subst. hydrazines) in yields of 53-97% (thirty examples). The survival of carbonyl, cyano and styryl groups under these conditions is notable. A cyclic enyne analog, 2-(phenylethynyl)cyclohex-2-en-1-one, gave a bicyclic pyrazole deriv. in 75% yield, while tosylhydrazine cyclized with *concomitant desulfonation* to afford an N-H pyrazole in 72% yield. F.e. and optimization s. X. Yu, J. Zhang, Chem. Eur. J. *2012, 18* (41), 12945-9 [DOI: 10.1002/chem.201202583].

Potassium iodide/tert-butyl hydroperoxide *KI/t-BuOOH*
N-α-Acylaminodicarboxylic acid imides from dicarboxylic acid imides and acylamines NH → N-C-N(R)C(O)R′

111. 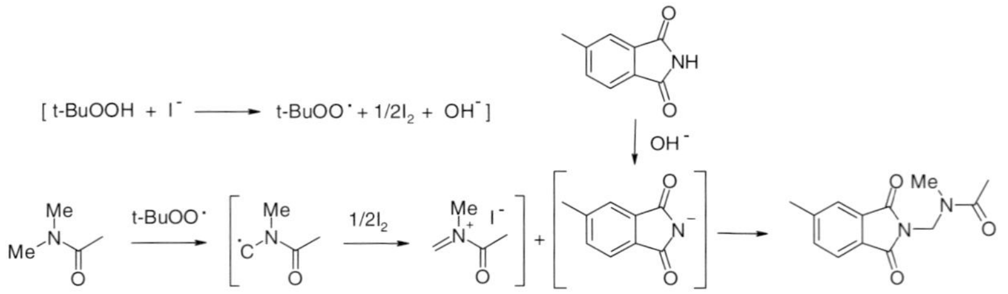

The first imidation of sp^3 C-H adjacent to an amide nitrogen is reported. **E:** 70% Aq. *t*-BuOOH (4 eq.) added rapidly to a mixture of 4-methylphthalimide (1 mmol), KI (20 mol%) and N,N-di-methylacetamide (1 ml; ca. 9 eq.), the mixture stirred at 90° for 1 h, quenched with aq. Na_2SO_3, extracted with methylene chloride, and purified chromatographically → N-(5-methyl-1,3-dihydro-1,3-dioxo-2*H*-isoindol-2-ylmethyl)-N-methylacetamide. Y 91%. The imidation of aliphatic N,N-dimethylamides and of N,N-dimethylaniline with electron-diverse (benzo)phthalimides was promoted successfully by KI (twelve examples; Y 85-95%; imidation of DMF gave 73% using 2 eq. of oxidant). NaI and Bu_4NI were also effective catalysts, but other halides gave rise to little or no imidation. Succinimide was an effective imide source (two examples; Y 87-89%), but lower yields were obtained with glutarimide (64%) and its β,β-dimethyl deriv. (46%), while use of N-tosyl-2-methylpropylamine in place of an imide gave only 44% yield of the coupled product after prolonged reaction (24 h). Interestingly, reaction of phthalimide with N-methylpyrrolidone occurred with high regioselectivity for the methylene C-H (Y 89%).

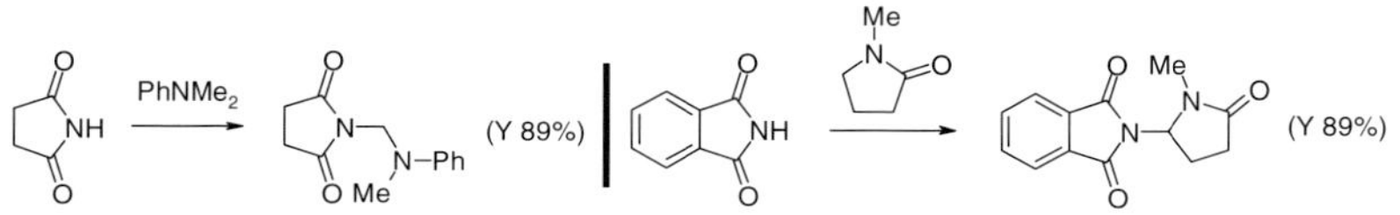

F.e. and optimization s. Z.-Q. Lao, W.-H. Zhong, Q.-H. Lou, Z.-J. Li, X.-B. Meng, Org. Biomol. Chem. *2012, 10* (39), 7869-71 [DOI: 10.1039/c2ob26430g].

Potassium iodide/tert-butyl peroxybenzoate/pivalic acid *KI/$PhCOO_2Bu$-t/t-BuCOOH*
N-Aminomethylation of indoles with simultaneous C_3-formylation NH → N-C-NMe_2
s. *83*, 385

N-Methylmorpholine N-oxide s. under Copper(II) acetate *NMMO*

Copper(I) oxide/tert-butyl hydroperoxide/potassium carbonate *Cu_2O/t-BuOOH/K_2CO_3*
Carboxylic acid amides from aldehydes s. *55,* 146s*83* CHO → CON<

Copper(II) acetoacetonate/oxygen *$Cu(acac)_2/O_2$*
Regioselective oxidative amination of azoles H → N<
update s. *75,* 149s*82*; C_2-amination of azoles with aliphatic or aromatic amines under base-free conditions with $Cu(acac)_2$ under O_2 s. Y.S. Wagh, B.M. Bhanage, Tetrahedron Lett. *2012, 53* (48), 6500-3 [DOI: 10.1016/j.tetlet.2012.09.064]; mild conversion with $CuBr_2/AcOH/O_2$ and an aldehyde as synergistic catalyst involving catalytic generation of aminals *in situ* s. Y. Xie, B. Qian, P. Xie, H. Huang, Adv. Synth. Catal. *2013, 355* (7), 1315-22 [DOI: 10.1002/adsc.201200944]; regioselective *prim*-amination of electron-rich azoles and other electron-rich heteroarenes with phenyl iodosoacetate/TsOH and aq. NaN_3 via fragmentation of the intermediate heteroaryl(phenyl)-iodonium azides with CuCl, and subsequent reduction of the generated azides with aq. ammonium sulfide, also *in situ*-'click' trapping as 1,2,3-triazol-1-ylheteroarenes, s. D. Lubriks, I. Sokolovs, E. Suna, J. Am. Chem. Soc. *2012, 134* (37), 15436-42 [DOI: 10.1021/ja305574k]; metal-free C_2-(azol-1-yl)ation of indoles with iodine in satd. aq. ammonium formate s. W.-B. Wu, J.-M. Huang, Org. Lett. *2012, 14* (23), 5832-5 [DOI: 10.1021/ol302609m]; 2-step C_2-amination of benzoxazoles via oxidative ring closure of intermediate N-(*o*-hydroxyaryl)formamidines with *o*-iodoxybenzoic acid s. Y.S. Wagh, N.J. Tiwari, B.M. Bhanage, Tetrahedron Lett. *2013, 54* (10), 1290-3 [DOI: 10.1016/j.tetlet.2012.12.127]; nucleophilic *o-prim*-amination of the pyrrole ring of 2-nitro-5,10,15,20-tetraarylporphyrins with N,N,N-trimethylhydrazinium iodide in the presence of KOH/DMSO s. S. Ostrowski, S. Grzyb, ibid. *2012, 53* (47), 6355-7 [DOI: 10.1016/j.tetlet.2012.09.024].

Copper(II) acetate/silver carbonate/N-methylmorpholine N-oxide ←
Copper-catalyzed N-directed dehydrogenative *o*-amination of arylcarboxylic acid N-(8-quinolyl)amides

112.

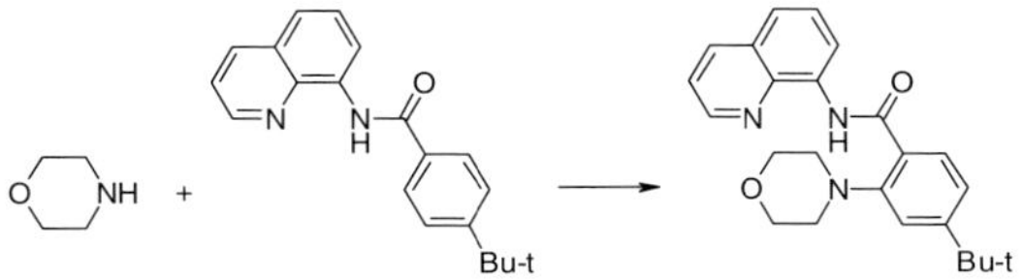

***o-tert*-Amination.** N-Methylmorpholine N-oxide (1 mmol) added to a vial containing N-4-*tert*-butyl-benzoyl-8-aminoquinoline (0.5 mmol), $Cu(OAc)_2$ (12 mol%) and Ag_2CO_3 (12 mol%) inside a glovebox, the vial removed from the latter, N-methyl-2-pyrrolidone (2 ml) and morpholine (2.06 eq.) added, the vial wrapped with aluminum foil, stirred at 110° for 11.25 h, the mixture cooled to room temp., dry absorbed on silica gel, and purified by column chromatography → N-(2-morpholino-4-*tert*-butylbenzoyl)-8-aminoquinoline. Y 81%. The procedure, based on a *readily removable* directing group, is inexpensive and generally applicable in good yield to the *o*-amination of electron-diverse arylcarboxylic acid N-(8-quinolyl)amides or 5- or 6-membered heteroaromatic analogs with sec. amines, notably possessing ester, cyano, ketal or BocNH groups; prim. amines (especially straight-chain examples) afforded the lowest yields (twenty examples in all; Y 20-87%). Significantly, mono-amination takes place exclusively at the least sterically-demanding position, and alkoxy, fluorine and ester groups were tolerated on the aromatic ring of the benzamide. Reactions can be scaled up at least ten-fold without significant loss of yield, and the aminoquinoline directing group can be readily removed by heating with NaOH in ethanol for 3 days. α,α-Disubst. N-benzylpicolinamides were also *o*-aminated under the same conditions in low yield (40-43%; two examples). The mechanism of the amination is as yet unclear. F.e.s. L.D. Tran, J. Roane, O. Daugulis, Angew. Chem., Int. Ed. *2013*, *52* (23), 6043-6 [DOI: 10.1002/anie.201300135]; **copper-catalyzed oxidative *o*-azidation** of prim. ar. amines with trimethylsilyl azide, CuBr (10 mol%) and *t*-BuOOH (2 eq.) in acetonitrile at room temp. s. C. Tang, N. Jiao, J. Am. Chem. Soc. *2012*, *134* (46), 18924-7 [DOI: 10.1021/ja3089907].

Silver p-nitrobenzoate s. under Chiral pincer-type bis(Δ^2-oxazoline)iron(II) complex *AgOCOAr*

Copper(II) triflate/copper(I) chloride/air *$Cu(OTf)_2/CuCl/O_2$*

α-(N-Carbalkoxyhydroxylamino)-β-ketocarboxylic acid esters from β-ketocarboxylic acid esters and N-hydroxyurethans via copper-catalyzed aerobic generation of C-nitrosoformic acid esters H → N(OH)COOR

113. 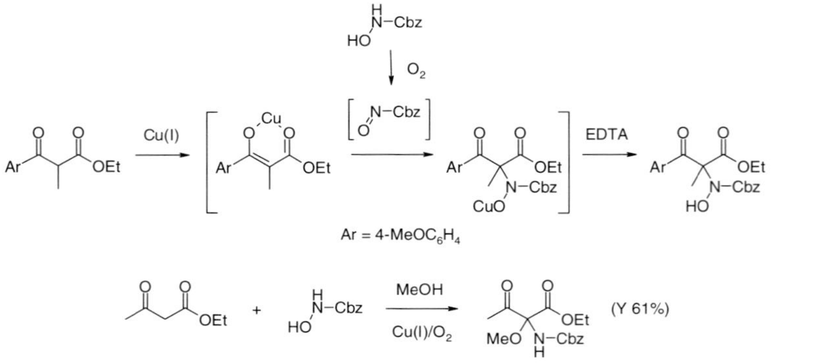

The first α-amination of carbonyl compds. using nitrosoformates generated *in situ* (via aerobic oxidation of N-hydroxycarbamates) and Lewis acid activated β-keto-esters is described (cf. *80*, 189 for use of nitrosoformates in intermolecular acylnitroso-ene reactions). **E:** A soln. of ethyl 2-methyl-2-(4-methoxybenzoyl)acetate (1.2 eq.) in methanol (2 ml) added to a mixture of benzyl N-hydroxy-carbamate (0.28 mmol), $Cu(OTf)_2$ (5 mol%) and CuCl (5 mol%) at room temp., the mixture stirred under air for 96 h (TLC), concentrated *in vacuo*, diluted with ethyl acetate, stirred with aq. EDTA (0.5 *M*;, pH 7) until color no longer persisted in the organic layer (ca. 30 min), extracted with ethyl acetate, and purified chromatographically → ethyl 2-[(benzyloxycarbonyl)(hydroxy)-amino]-3-(4-methoxyphenyl)-2-methyl-3-oxopropanoate. Y 90%. Careful optimization minimized the formation of O-coupled products. The method was scalable (to 1 g) and successful for acyclic α-alkyl(methyl, ethyl, benzyl)-β-keto-esters and N-hydroxycarbamates incorporating common N-protecting groups (twenty-two examples; Y 63-96%), and appears unaffected by the steric bulk of the ester moiety. Cyclic β-keto-esters gave somewhat lower yields (43-70%; two examples) due to significant formation of O-coupled isomers (Y 25-40%), and a low yield (42%) was also obtained with a bulky α-isopropyl-β-keto-ester. Interestingly, an *α-unsubstituted* ester analog underwent N-O bond cleavage with trapping of the presumed imino-keto-ester intermediate by methanol solvent to afford an α-Cbz-amino-α-methoxy coupled product (Y 61%). F.e.s. D. Sandoval, C.P. Frazier, A. Bugarin, J. Read de Alaniz, J. Am. Chem. Soc. *2012*, *134* (46), 18948-51 [DOI: 10.1021/ja310784f].

Copper(II) sulfate/tert-butyl hydroperoxide/calcium carbonate — *$CuSO_4$/t-BuOOH/$CaCO_3$*
Copper(I) iodide/sodium 5-methyl-2-pyridinolate — ←
Carboxylic acid amides from aldehydes s. *55*, 146s*83* — CHO → CON<

Copper(I) bromide s. under Me_3SiN_3 — *CuBr*

Copper(II) chloride/tert-butyl hydroperoxide — *$CuCl_2$/t-BuOOH*
N-Acyl-N-alkoxyureas — CONH(OR) → CON(OR)CON<
from hydroxamic acid esters and formamides
via copper(II)-catalyzed dehydrogenative coupling

114. 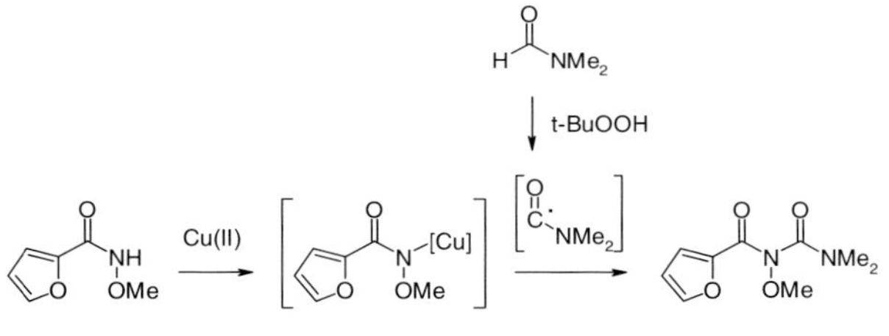

The first synthesis of N-acylureas via efficient coupling of two different amides under mild conditions is described. **E:** A mixture of N-methoxyfuran-2-carboxamide (0.5 mmol), DMF (52 eq.), $CuCl_2 \cdot 2H_2O$ (3 mol%) and 70% aq. *t*-BuOOH (0.2 ml; 3 eq.) stirred at room temp. under air for 5 h, quenched with satd. aq. Na_2SO_3, extracted with ethyl acetate, concentrated *in vacuo*, and purified chromatographically → N-(dimethylcarbamoyl)-N-methoxyfuran-2-carboxamide. Y 80%. This general method was applied to the coupling of prim./sec. N-alkoxyamides of (het)ar., cinnamic and phenylacetic acids with cyclic/acyclic tert. amides of formamide (twenty-four examples; Y 52-80%; N-formylpiperidine gave 21%, improved to 43% using DMSO as co-solvent), with lowest yields obtained from electron-poor N-alkoxyarylamides. The mechanism was presumed to involve C-H activation of the formamide via a radical process (in the presence of TEMPO, an adduct with DMF was obtained in 80% yield). Addition of ligands to the reaction did not improve the yields. F.e., optimization and substrate prepn. s. H. Jiang, A. Lin, C. Zhu, Y. Cheng, Chem. Commun. *2013*, *49* (8), 819-21 [DOI: 10.1039/c2cc36417d].

Copper(II) bromide/acetic acid/oxygen/aldehydes — *$CuBr_2$/AcOH/O_2/RCHO*
Regioselective oxidative amination of azoles s. *75*, 149s*83* — H → N<

Silver nitrite/2,2,6,6-tetramethylpiperidine nitroxyl — *$AgNO_2$/TEMPO*
Regio- and stereo-selective vinylic nitration — C=CH → C=C-NO_2
under mild conditions

115. 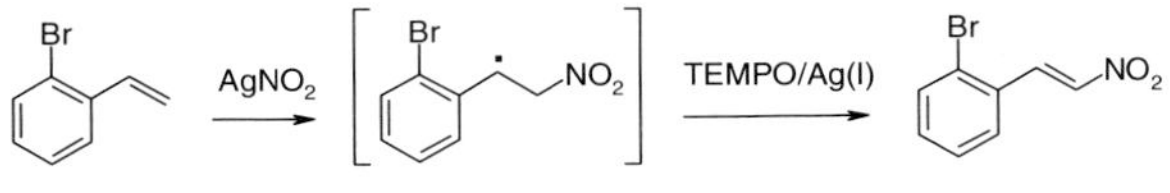

The first mild, practical, highly efficient, direct, regio- and stereo-selective vinylic nitration of a wide range of ethylene derivs. has been reported. **E:** *o*-Bromostyrene (0.5 mmol) and dichloroethane (2 ml) added via syringe to an oven-dried screw cap test-tube charged with $AgNO_2$ (3 eq.), TEMPO (0.4 eq.) and oven-dried 4 Å molecular sieves, the tube placed in a pre-heated oil bath at 70°, stirred vigorously for 12 h, cooled to room temp., the mixture filtered through Celite, and the organic extract concentrated before purification by chromatography on silica gel → (E)-*o*-bromo-β-nitrostyrene. Y 92%. The procedure is applicable to the nitration of an array of unactivated or functionalized aromatic, aliphatic or heteroaromatic terminal, 1,1- and 1,2-disubst. olefins with exclusive or high (E)-selectivity and in high yield (generally 50-96%; >forty examples). Monosubst. olefins were more reactive than disubst. ones, and both were significantly more reactive than trisubst. olefins so that **preferential nitration** takes place in substrates possessing a combination of differently substituted alkene groups. Notable was the (E)-selective mononitration of electron-diverse and sterically demanding styrenes (incl. naphthyl analogs), a variety of functional groups being supported on the benzene ring, even at the *ortho* site, e.g. alkyl, methoxy, formyl, carbalkoxy, cyano, chlorine, bromine, fluorine, formyl, acylamino. Selective mononitration of carboxylic acid esters possessing two terminal alkene groups is also reported, the double bond nearer to the carbalkoxy residue being left untouched.

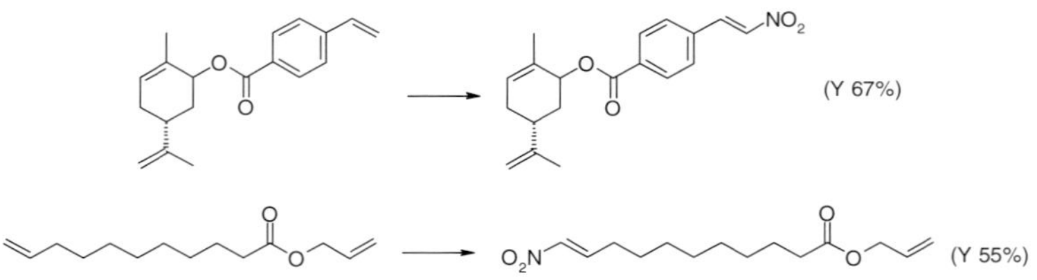

It is possible that reaction involves generation of a β-nitroalkyl radical by initial nitration with the silver salt, followed by abstraction of a hydrogen atom by TEMPO to give the product, with reformation of the nitroxyl by the Ag(I)→Ag(0) manifold in a catalytic cycle; however, trapping of the β-nitroalkyl radical, followed by quenching with TEMPO and subsequent elimination, could not be ruled out. F.e., optimization, and gram-scale procedure s. S. Maity, S. Manna, S. Rana, T. Naveen, A. Mallick, D. Maiti, J. Am. Chem. Soc. *2013, 135* (9), 3355-8 [DOI: 10.1021/ja311942e].

Dichloro(2,2′-bipyridyl)gold(III) chloride/N-bromosuccinimide *[Au(III)]/NBS*
Oxidative benzylic amidation s. *77,* 161s*83* H → N<

Zinc bromide/tert-butyl hydroperoxide *$ZnBr_2$/t-BuOOH*
Carboxylic acid amides from aldehydes CHO → CON<

update s. *55,* 146s*82*; arylcarboxylic acid N-alkylamides from ar. aldehydes with $ZnBr_2$ as catalyst and *tert*-butyl hydroperoxide as oxidant without solvent s. M. Zhang, X.-F. Wu, Tetrahedron Lett. *2013, 54* (9), 1059-62 [DOI: 10.1016/j.tetlet.2012.12.010]; N-unsubst. amides with NH_4Cl/Cu_2O/ *tert*-butyl hydroperoxide/K_2CO_3, and N-subst. or N,N-disubst. amides with $CuSO_4{\cdot}5H_2O$/*tert*-butyl hydroperoxide/$CaCO_3$, incl. racemization-free coupling with chiral amines, s. S.C. Ghosh, J.S.Y. Ngiam, A.M. Seayad, D.T. Tuan, C.L.L. Chai, A. Chen, J. Org. Chem. *2012, 77* (18), 8007-15 [DOI: 10.1021/jo301252c]; aerobic amidation of ar. aldehydes and cinnamaldehydes with sec. amines using CuI/sodium 5-methyl-2-pyridinolate s M. Zhu, K.-i. Fujita, R. Yamaguchi, ibid. *77* (20), 9102-9 [DOI: 10.1021/jo301553v]; amidation with $Co(OAc)_2$/N-hydroxysuccinimide under O_2, notably with chiral amines, s. H. Yao, K. Yamamoto, Chem. Asian J. *2012, 7* (7), 1542-5 [DOI: 10.1002/asia.201200017]; 2-step amidation with *p*-nitroanilines via coupling of intermediate *p*-azidonitroarenes with aldehydes using Na-*tert*-butoxide as electron-transfer agent in the presence of 3-benzyl-5-(2-hydroxyethyl)-4-methylthiazolium chloride as catalyst s. G. Carbone, J. Burnley, J.E. Moses, Chem. Commun. *2013, 49* (27), 2759-61 [DOI: 10.1039/c3cc40452h].

Aldehydes s. under $CuBr_2$ *RCHO*
Ammonium formate s. under I_2 *$HCOONH_4$*
1-Butyl-3-methylimidazolium hexafluorophosphate s. under $Bi(NO_3)_3$ *[bmim]PF_6*
N-Hydroxysuccinimide s. under Cobalt(II) acetate ←
2,2,6,6-Tetramethylpiperidine nitroxyl s. under $AgNO_2$ *TEMPO*
Trifluoroacetic anhydride s. under Dimethyl sulfoxide *$(CF_3CO)_2O$*
tert-Butyl hydroperoxide s. under KI, Cu_2O, $CuSO_4$, $CuCl_2$, $ZnBr_2$, N-Iodo-succinimide, Me_3SiN_3 and $FeCl_3$ *t-BuOOH*
tert-Butyl peroxybenzoate s. under KI *$PhCOO_2Bu$-t*

Acetic acid s. under $CuBr_2$ *AcOH*
Pivalic acid s. under KI and $Pd(OAc)_2$ *t-BuCOOH*
Carbon dioxide/hydrogen s. under $RuCl_2(dppe)_2$ *CO_2/H_2*
Dimethyl sulfoxide/trifluoroacetic anhydride *$DMSO/(CF_3CO)_2O$*
2-α-Functionalization of indoles s. *83*, 50 ←

Phenyl iodosoacetate s. under $Pd(OAc)_2$ *$PhI(OAc)_2$*
Bromobenzene s. under Dichlorobis(trimethylphosphine)palladium(II) *PhBr*
N-Bromosuccinimide s. under Dichloro(2,2′-bipyridyl)gold(III) chloride *NBS*
N-Iodosuccinimide/tert-butyl hydroperoxide *NIS/t-BuOOH*
α-Amination of propiophenones s. *77*, 161s*83* H → N<

Trimethylsilyl azide/copper(I) bromide/tert-butyl hydroperoxide *$Me_3SiN_3/CuBr/t$-BuOOH*
Copper-catalyzed oxidative *o*-azidation of prim. ar. amines s. *83*, 112 H → N_3

1-Azido-3,3-dimethyl-1,2-benziodoxole s. under Chiral pincer-type bis(Δ^2-oxazoline)-iron(II) complex ←

Imidazole-1-sulfonyl azide ←
Diazo group transfer $CH_2 \rightarrow C{=}N_2$
update s. *20*, 271s*82*; safe procedure with imidazole-1-sulfonyl azide as diazo transfer agent s. H. Ye, R. Liu, D. Li, Y. Liu, H. Yuan, W. Guo, L. Zhou, X. Cao, H. Tian, J. Shen, P.G. Wang, Org. Lett. *2013*, *15* (1), 18-21 [DOI: 10.1021/ol3028708].

tert-Butyl nitrite/oxygen *t-BuONO/O_2*
β- or δ-Nitro-γ-lactols from ethylene derivs. ○
via metal-free radical aerobic oxynitration and sp³ C-H oxidation

116.

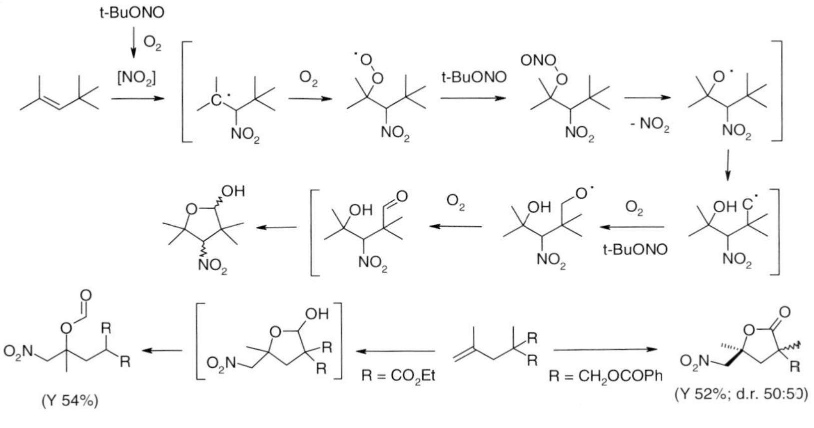

t-BuONO (5 eq.) added to a stirred soln. of 2,4,4-trimethylpent-2-ene (0.6 mmol) in dry DMSO (3 ml), the mixture stirred at room temp. under O_2 (balloon) for 6 h, diluted with water, extracted with ether, concentrated *in vacuo*, and purified chromatographically → 3,3,5,5-tetramethyl-4-nitro-tetrahydro-2-furanol. Y 44% (d.r. 75:25). A series of radical reactions involving oxynitration of the double bond, C-H oxidation of an adjacent sp³ carbon, and cyclization, afforded the title lactols with modest diastereoselectivity (d.r. 50:50 to 75:25 (as the derived lactones in some cases); eight examples; Y 45-55%). C-H oxidation was favored by carbons adjacent to sec. or tert. sites (linear 2-methylpent-1-ene gave only 15% yield of lactol), and included cyclization of 1-cyano-1-(2-methylprop-2-enyl)cyclohexane to a bicyclic lactol. 4,4-Bis(benzoyloxymethyl)-2-methylbut-2-enes cyclized preferentially at the carbon adjacent to the OCOPh moiety, affording the corresponding **β′-benzoyloxy-δ-nitro-γ-lactones** (two examples; Y 47-50%; d.r. 50:50 to 65:35). An α-methylmalonate-derived alkene apparently suffered isomerization (C-C cleavage) of the final lactol to afford a formate ester in 54% yield. F.e., optimization and further reactions of the lactols s. T. Taniguchi, Y. Sugiura, T. Hatta, A. Yajima, H. Ishibashi, Chem. Commun. *2013*, *49* (22), 2198-200 [DOI: 10.1039/c3cc00130j].

Tri-tert-butylphosphine s. under Pd(OAc)$_2$ *t-Bu*$_3$*P*

Bismuth(III) nitrate/1-butyl-3-methylimidazolium hexafluorophosphate *Bi(NO*$_3$*)*$_3$*/[bmim]PF*$_6$
Ar. nitration H → NO_2
update s. *1*, 343s*82*; of activated arenes with $Bi(NO_3)_3{\cdot}5H_2O$ in 1-butyl-3-methylimidazolium hexafluorophosphate or 1,2-dichloroethane s. J. Jacoway, G.G.K.S.N. Kumar, K.K. Laali, Tetrahedron Lett. *2012, 53* (50), 6782-5 [DOI: 10.1016/j.tetlet.2012.09.137]; *p*-nitration of mono-functionalized arenes with HNO_3 and reusable polymer-based 1-methyl-3-(3-sulfopropyl)imidazolium hydrogen sulfate as Brønsted acidic ionic liquid s. L.X. Li, Q.L. Ling, Z.L. Liu, X.D. Xing, X.Q. Zhu, X. Meng, Bull. Korean Chem. Soc. *2012, 33* (10), 3373-7 [DOI: 10.5012/bkcs.2012.33.10.3373].

Oxygen or Air s. under K$_2$*CO*$_3$*, Cu(OAc)*$_2$*, Cu(OTf)*$_2$*, CuBr*$_2$*, t-BuONO, Cobalt(II) acetate and Pd(OAc)*$_2$ *O*$_2$

Polymer-based 1-methyl-3-(3-sulfopropyl)imidazolium hydrogen sulfate ←
Ar. nitration s. *1*, 343s*83*

Iodine/ammonium formate *I*$_2$*/HCOONH*$_4$
C$_2$**-(Azol-1-yl)ation of indoles** s. *75*, 149s*83* H → N<

Chiral pincer-type bis(Δ^2*-oxazoline)iron(II) complex/silver p-nitrobenzoate/ 1-azido-3,3-dimethyl-1,2-benziodoxole* ←
Iron(II)-catalyzed asym. α-azidation H → N_3

117.

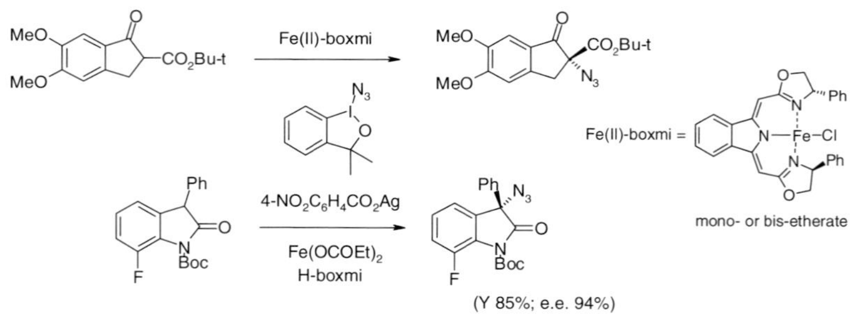

The first examples of iron-catalyzed enantioselective azidation **of cyclic β-ketocarboxylic acid esters or N-(carbo-*tert*-butoxy)oxindoles** are described, utilizing a strategy similar to that reported for the copper-catalyzed trifluoromethylation of β-keto-esters with Togni's reagent (*82*, 400). **E:** A mixture of chiral iron(II) catalyst (10 mol%), Ag-*p*-nitrobenzoate (10 mol%) and dry ether (2 ml) stirred at room temp. under argon with exclusion of light for 10 h, *tert*-butyl 5,6-dimethoxy-indan-1-one-2-carboxylate (0.1 mmol) added, the mixture stirred for 0.5 h, 1-azido-3,3-dimethyl-1,2-benziodoxole (1.5 eq.) added, the mixture stirred for 48 h (TLC), filtered through Celite, concentrated *in vacuo*, and purified chromatographically → *tert*-butyl (R)-2-azido-5,6-dimethoxy-indan-1-one-2-carboxylate. Y 87% (e.e. 93%). Use of the pre-formed iron(II)-boxmi catalyst and the readily available azide transfer agent provided efficient and enantioselective azidation of 2-*tert*-butoxycarbonyl-1-indanones (e.e. 81-93%; eight examples; Y 84-90%; a methyl ester gave e.e. 70%, Y 90%). Analogous 5-carbalkoxycyclopent-2-en-1-ones proved good substrates (three examples; Y 84-89%; e.e. 86-92%), but carbalkoxy-cyclopentanone and -tetralone analogs were less enantioselective (e.e. 67-76%; Y 87-88%; three examples). Application of the method to 3-aryloxindoles optimally required generation of the chiral iron catalysts *in situ* from the H-ligand and iron(II) propionate (nine examples; Y 85-90%; e.e. 87-94%). F.e., optimization and reactions of the azide products s. Q.-H. Deng, T. Bleith, H. Wadepohl, L.H. Gade, J. Am. Chem. Soc. *2013, 135* (14), 5356-9 [DOI: 10.1021/ja402082p].

Iron(III) chloride/tert-butyl hydroperoxide *$FeCl_3$/t-BuOOH*
Oxidative amination of *sp*^{3}-carbon-hydrogen bonds H → N⋜
update s. *77*, 161s*82*; regioselective, solvent-free oxidative N-alkylation of benzotriazoles and benzimidazoles with $FeCl_3$/*tert*-butyl hydroperoxide s. X. Liu, Y. Chen, K. Li, D. Wang, B. Chen, Chin. J. Chem. *2012, 30* (10), 2285-91 [DOI: 10.1002/cjoc.201200690]; benzylic amination with acylamines and sulfonylamines using dichloro(2,2′-bipyridyl)gold(III) chloride/N-bromosuccinimide s. Y. Zhang, B. Feng, C. Zhu, Org. Biomol. Chem. *2012, 10* (46), 9137-41 [DOI: 10.1039/c2ob26857d]; α-amination of propiophenones with NIS as catalyst and *tert*-butyl hydroperoxide as oxidant, also metal-free conversion of acetophenones to arylglyoxylic acid amides, s. M. Lamani, K.R. Prabhu, Chem. Eur. J. *2012, 18* (46), 14638-42 [DOI: 10.1002/chem.201202703].

Cobalt(II) acetate/N-hydroxysuccinimide/oxygen ←
Carboxylic acid amides from aldehydes s. *55*, 146s*83* CHO → CON⋜

Ruthenium(II) bis(N-heterocyclic carbene) complexes *[Ru(II)]*
N-Formylation with methanol s. *13*, 442s*83* NH → NCHO

Dichlorobis[1,2-bis(diphenylphosphino)ethane]ruthenium(II)/carbon dioxide/hydrogen ←
2-Unsubst. benzimidazoles from *o*-diamines ○
Ruthenium(II)-catalyzed dehydrative ring closure

118. 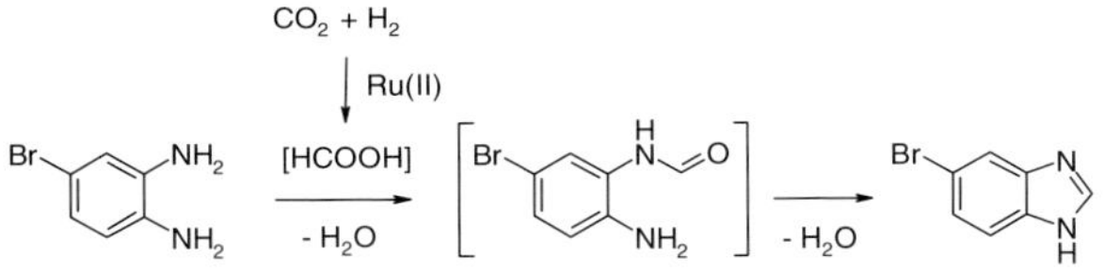

under solvent-free conditions. A mixture of 4-bromo-1,2-phenylenediamine (5 mmol) and $RuCl_2(dppe)_2$ stirred at 120° in an autoclave under CO_2/H_2 (3:2; 20 MPa) for 40 h, cooled in ice water, gas vented slowly, and the mixture purified chromatographically → 5-bromo-1*H*-benzimidazole. Y 90%. This novel and direct synthesis of benzimidazoles was successful with electron-diverse *o*-phenylenediamines (incl. N-phenyl-1,2-phenylenediamine) in the presence of halo, nitro, ester and ketone functionality (ten examples; Y 87-95%). Direct reaction of *o*-phenylenediamine with CO_2 to afford 2-hydroxybenzimidazole was the sole outcome in the absence of H_2, and the somewhat lower yield (73%) obtained from 4-trifluoromethylphenylenediamine was attributed to competing formation of the corresponding 2-hydroxybenzimidazole by-product (note the mechanism was shown not to proceed via reduction of the 2-hydroxybenzimidazole). Cyclization of N-acetyl-1,2-phenylenediamine gave only 2-methylbenzimidazole (Y 98%) under these conditions. F.e. and optimization s. B. Yu, H. Zhang, Y. Zhao, S. Chen, J. Xu, C. Huang, Z. Liu, Green Chem. *2013, 15* (1), 95-9 [DOI: 10.1039/c2gc36517k].

Palladium(II) acetate/pivalic acid/air *$Pd(OAc)_2$/t-BuCOOH/O_2*
Sec. ar. (Z)-enamines C=CH → C=C(NHAr)
from electron-deficient ethylene derivs. and [*o*-subst.] prim. ar. amines
Regio- and stereo-selective palladium-catalyzed dehydrogenative vinylic arylamination

119. 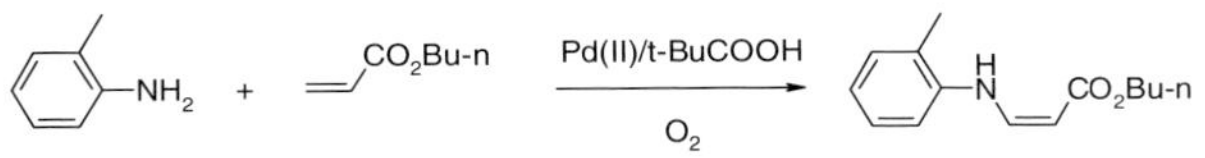

The first palladium-catalyzed oxidative amination of olefins using molecular oxygen as the sole oxidant has been reported, the use of *o*-substituted primary anilines being a key factor for success. **E:** A mixture of butyl acrylate (1 mmol) and *o*-toluidine (3 eq.) allowed to react in the presence of $Pd(OAc)_2$ (3 mol%) and pivalic acid (0.25 eq.) in NMP (0.5 ml) at 60° under open air for 15 h, the crude mixture cooled to room temp., ethyl acetate added, and the residue purified by chromatography on silica gel → (Z)-butyl 3-(*o*-tolylamino)acrylate. Y 90%. It is believed the *ortho* substituent hampers excessive coordination of the aniline nitrogen to the palladium center, preventing its deactivation. Seven further examples with acrylates gave yields of 58-89%, the lowest obtained with a bulky *tert*-butyl ester group. A low yield and poor stereoselectivity were observed with acrylonitrile, a poor yield with acrylamide (due to deactivation of the catalyst) and low conversion with styrene, while simple alkenes such as 1-octene were totally inactive. Phthalimide and cyclooctylamine were also inactive. F.e.s. Y. Mizuta, K. Yasuda, Y. Obora, J. Org. Chem. *2013*, *78* (12), 6332-7 [DOI: 10.1021/jo4010734].

Palladium(II) acetate/tri-tert-butylphosphine/phenyl iodosoacetate *$Pd(OAc)_2/t\text{-}Bu_3P/PhI(OAc)_2$*
Sterically-controlled, regioselective, dehydrogenative ar. amination NH → NAr

120. 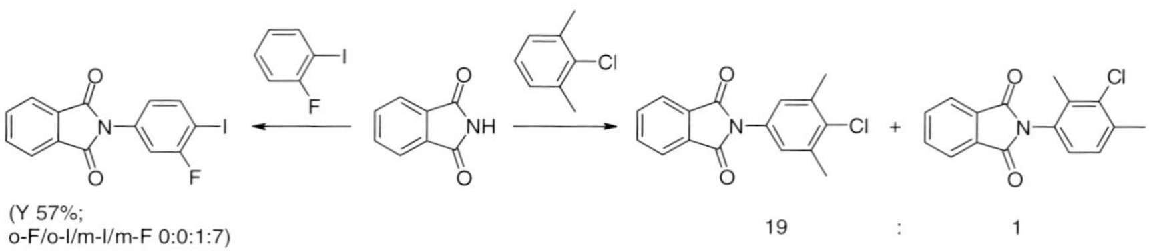

with phthalimide. $Pd(OAc)_2$ (10 mol%), *t*-Bu_3P (10 mol%), phthalimide (0.5 mmol) and $PhI(OAc)_2$ (2 eq.) weighed into an oven-dried scintillation vial in a nitrogen-filled glovebox, 2,6-dimethylchlorobenzene (5 ml) and dodecane (50 µl; internal standard) added via an automatic pipette, the vial capped with a septum, removed from the glovebox, stirred at 1200 rpm while maintaining the temp. at 100°, after 9 h and 24 h, the mixture cooled to room temp., taken back into the glovebox, treated with additional $PhI(OAc)_2$ (2 eq. each time) and, after 33 h total reaction time, purified by silica gel column chromatography → product. Y 69% (*o*/*m* 1:19). In this palladium-catalyzed intermolecular oxidative amination of mono-, di- or tri-subst. arenes, regioselectivity is guided by steric rather than electronic effects, and is in contrast to that for related palladium-catalyzed ar. acetoxylations. Yields were highest with electron-rich arenes while 4-methylphthalimide may also be used, but reactions with 4-chloro- or 3,4,5,6-tetrachloro-phthalimide, maleimide, succinimide or saccharin resulted in low conversions. F.e. (twenty-five; Y 46-90%) and optimization s. R. Shrestha, P. Mukherjee, Y. Tan, Z.C. Litman, J.F. Hartwig, J. Am. Chem. Soc. *2013*, *135* (23), 8480-3 [DOI: 10.1021/ja4032677].

Palladium(II) acetate/oxygen *$Pd(OAc)_2/O_2$*
2-Aminobenzimidazoles from *o*-diamines and isonitriles via palladium(II)-catalyzed aerobic ring closure ○

121. 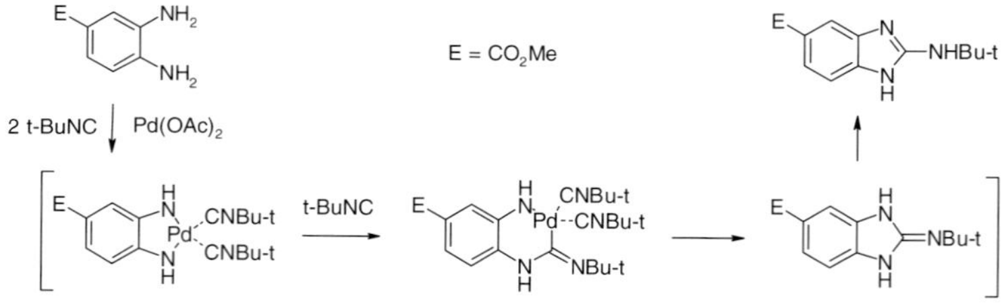

A soln. of $Pd(OAc)_2$ (1 mol%) in 2-methyltetrahydrofuran (5 ml) and *tert*-butyl isocyanide (1.2 eq.) added to a mixture of methyl 3,4-diaminobenzoate (1 mmol) and 4 Å molecular sieves (300 mg), the mixture stirred under O_2 (balloon) at 75° for 20 h, filtered through Celite, concentrated, and purified chromatographically → methyl 2-*tert*-butylaminobenzimidazole-5-carboxylate. Y 99%. This novel oxidative cyclization was applied to a series of electron-diverse *o*-phenylenediamines and *tert*-butyl isocyanide, generally giving excellent yields, with reactions somewhat slower (66-72 h) for highly electron-poor substrates (eleven examples; Y 83%, 90-99%), while 3,4-diaminopyridine (Y 65%) and a diaminouracil (Y 43%) afforded the corresponding hetero-analogs. Cyclization with isopropyl isocyanide required use of 10 mol% catalyst and slow addition of isocyanide (Y 54%) to minimize formation of a tricyclic by-product, but N-methyl/4-methoxyphenyl-*o*-phenylenediamines were cyclized with prim., sec. and tert. alkyl isocyanides without problem (five examples; Y 87-97%). 2-Aminobenzimidazole was prepared in a one-pot procedure by refluxing the crude 2-*tert*-butylaminobenzimidazole product with concd. HCl (Y 93%). The reaction was extended to the prepn. of benzoxazole (Y 99%), benzothiazole (Y 69%), quinazol-4-one (Y 79%) and 4-thiaquinazoline S,S-dioxide (Y 69%) analogs via cyclization of the appropriate *o*-functionalized anilines. 2-Aminomethyl- and 2-(N-methylaminomethyl)-anilines were cyclized with concomitant oxidation to the aromatic quinazoline (Y 85%) and quinazol-4-one (Y 77%), respectively, while N-phenylethylenediamine resisted aromatization, affording a 2-amino-2-imidazoline (Y 94%).

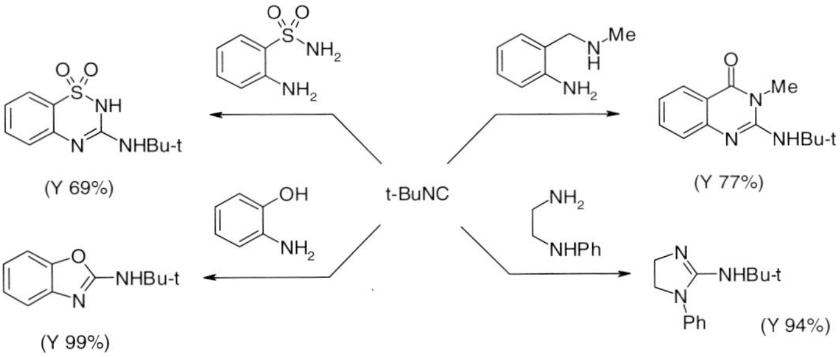

F.e., optimization and use of the method in the synthesis of astemizole and norastemizole s. T. Vlaar, R.C. Cioc, P. Mampuys, B.U.W. Maes, R.V.A. Orru, E. Ruijter, Angew. Chem., Int. Ed. *2012*, *51* (52), 13058-61 [DOI: 10.1002/anie.201207410].

Dichlorobis(trimethylphosphine)palladium(II)/bromobenzene/cesium carbonate ←
β-Amino-α,β-ethylenenitriles from nitriles $CH_2CHCN \rightarrow C(N{<})=C\text{-}CN$
Palladium-catalyzed dehydrogenative N-(2-cyano)vinylation

122. 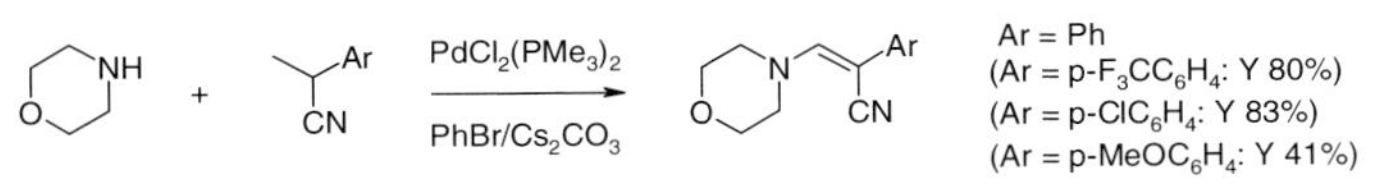

Cyano, which usually does not coordinate strongly to a metal catalyst in a side-on manner, has been used as a directing group for sp^3-hybridized C-H activation for the first time. **E: (Z)-β-Amino-α-arylacrylonitriles**. $PdCl_2(PMe_3)_2$ (10 mol%) and Cs_2CO_3 (4 eq.) placed in a screw-capped vial in a nitrogen-filled drybox, dry cyclopentyl methyl ether (2 ml) added, the vial sealed with a screw cap having a PTFE/silicone septum, then removed from the drybox, morpholine (2 eq.), bromobenzene (4 eq.) and 2-phenylpropionitrile (0.4 mmol) added by a micro-syringe at room temp., the mixture stirred at room temp. for 30 min, then heated at 100° for 40 h (GC), cooled to room temp., filtered, evaporated under reduced pressure, and the residue purified by flash column chromatography → (Z)-3-morpholino-2-phenylacrylonitrile. Y 82%. Use of a nickel catalyst was ineffective, and palladium with a small monophosphine ligand (to avoid α-arylation as a side-reaction) was found to be optimal. Ten further examples with cyclic or acyclic sec. amines and α-aryl-propionitriles gave yields of 41-92%, electron-withdrawing groups on the aromatic ring hardly affecting the yield but the lowest yield being for the *p*-methoxy-deriv. A prim. amine [*n*-octylamine] may also be used (Y 63%). There was no reaction with propionitrile or α-alkylated propionitriles. The illustrated product could be hydrolyzed to α-cyano(phenyl)acetaldehyde *in situ* (7% aq. HCl/40°/2 h; Y 80%) or converted to 3-amino-4-phenylpyrazole with hydrazine (AcOH/ethanol/ microwave irradiation (500 W)/10 min; Y 68%). F.e.s. S. Ueno, R. Maeda, S. Yasuoka, R. Kuwano, Chem. Lett. *2013*, *42* (1), 40-2 [DOI: 10.1246/cl.2013.40]; **3-*tert*-amino-*sec*-alcohols** from sec. alcohols and sec. amines by *in situ* reduction of β-*tert*-amino-α,β-ethyleneketones formed by nickel-catalyzed oxidative coupling (cf. *76*, 159) s. S. Ueno, K. Usui, R. Kuwano, Synlett *2011* (9), 1303-7 [DOI: 10.1055/s-0030-1260536].

Via intermediates *v.i.*

Anodic ar. *prim*-amination via N-arylpyridinium salts H → NH_2

123. 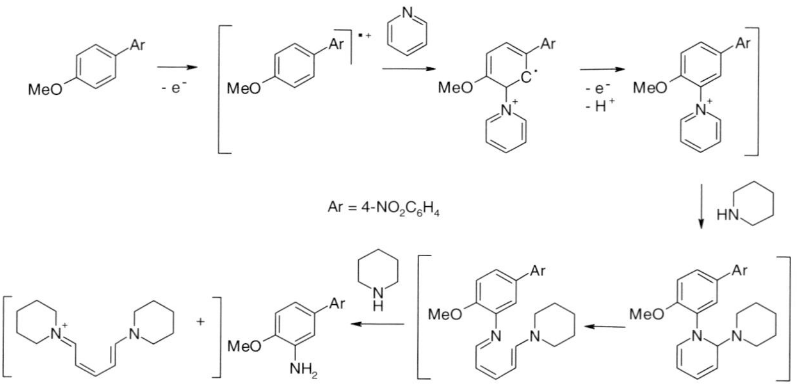

A soln. of 4-methoxy-4′-nitrobiphenyl (0.2 mmol), pyridine (0.5 ml) and electrolyte (0.3 *M* Bu_4NBF_4 in acetonitrile; 10 ml) placed in the anodic chamber of a H-type divided cell (4G glass filter; carbon fiber anode), triflic acid (0.2 ml) and additional electrolyte (10 ml) placed in the cathodic chamber, the mixture stirred and electrolyzed at constant current (8 mA) at 25° until 3.5 F/mol of electricity consumed, concentrated *in vacuo*, piperidine (10 eq.) and acetonitrile (10 ml) added, stirred at 80° for 12 h, concentrated *in vacuo,* and purified chromatographically → 3-amino-4-methoxy-4′-nitrobiphenyl. Y 100%. This novel and efficient method for prepn. of prim. aromatic amines apparently proceeds via one-electron oxidation of the substrate to a radical cation which is trapped with pyridine, as an N-arylpyridinium species, and subsequently cleaved by heating with piperidine. The method was applied to benzene, biphenyl and naphthalene derivs. (thirteen examples; Y 66-100%), was compatible with *iodo*, *nitro*, ester, ether, amide and ketone functionality, and gave mixtures of regioisomers in some cases. F.e. and use of the method in the synthesis of a key VLA-4 antagonist intermediate s. T. Morofuji, A. Shimizu, J.-i. Yoshida, J. Am. Chem. Soc. *2013*, *135* (13), 5000-3 [DOI: 10.1021/ja402083e].

C_2-Amination of benzoxazoles via N-(*o*-hydroxyaryl)formamidines s. *75*, 149s*83* H → N<

Oxygen ↾ NC ⇃↾ O

Without additional reagents *w.a.r.*

Uncatalyzed N-allylation with 2-ethylenealcohols s. *25*, 575s*83* NH → N-C-C=C

Aldimines from aldehydes and prim. amines
via solid-state grinding of alkylammonium carbamates

$NH_2 \rightarrow N{=}CHR$

124.

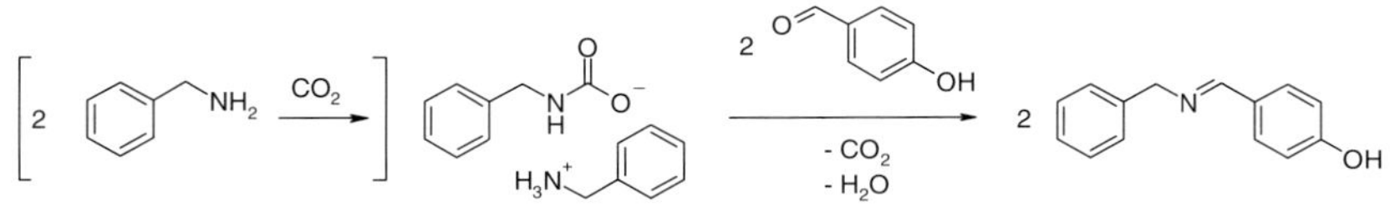

Ammonium carbamates produced from prim. amines (or 1,n-diamines) in liquid CO_2 have been found to be stable and effective alternatives to liquid amines and may be used in an environmentally benign solid-state reaction with ar. aldehydes to cleanly generate imines. **E:** A mixture of benzylammonium N-benzylcarbamate (5.03 mmol) and 4-hydroxybenzaldehyde (1.99 eq.) ground at room temp. using a pestle and mortar for 5 min, the mixture placed in a vial, and heated in an oven at 80° for up to 3 h → N-benzyl-4-hydroxybenzaldimine. Y 98%. Grinding a 1:2 mixture of carbamate salts and ar. aldehydes at room temp. was sufficient to generate imines within 24 h, with grinding for 5 min and simple heating (50-80°) of the ground powder in an oven providing a more rapid (3 h) alternative (nine examples; Y >97%). In some cases simple mixing of the powdered substrates before heating was equally effective. Comparable reactions using liquid prim. amines, either neat or as solutions in methanol, gave lower yields and generated by-products. F.e.s. B. Lee, K.H. Lee, B.W. Lim, J. Cho, W. Nam, N.H. Hur, Adv. Synth. Catal. *2013*, *355* (2-3), 389-94 [DOI: 10.1002/adsc.201200907].

Piperidine-2-acetic acid esters from 6-acyl-2-cyclohexenones
via diastereoselective hydrogenation of 2,3,4,5-tetrahydropyridine-2-acetic acid esters

125.

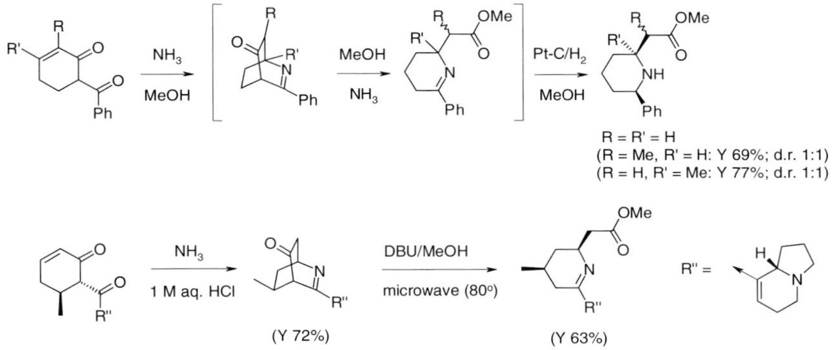

A telescoped route to ***syn*-6-subst. piperidine-2-acetic acid esters** is reported via Dieckmann-like ring opening followed by heterogeneous hydrogenation, inspired by speculation over grandisine alkaloid biosynthesis. **E:** 35% aq. NH_3 (2 ml) added to a stirred soln. of startg. diketone (0.5 mmol) in methanol (4 ml) at 0° under argon, the resulting yellow soln. held at 0° for 30 min, then warmed to room temp. for 3 h, the soln. concentrated *in vacuo*, the crude intermediate tetrahydropyridine (Y 97%; >90% purity by NMR) taken up in methanol (5 ml), treated at room temp. with Pt-C (10 wt%; 2.5 mol%), the resulting suspension stirred at room temp. for 1 h under H_2, filtered through Celite, washing with methanol, and the filtrate concentrated *in vacuo* → methyl [(2RS,6SR)-6-phenylpiperidin-2-yl]acetate. Y 90% overall. Five further examples with phenyl, pyrid-3-yl or cyclohexyl cyclohex-2-enon-6-yl ketones, optionally bearing methyl groups in the 2-, 3- or 4,4-positions of the cyclohex-2-enone ring, gave yields of 55-92% overall. For 5-methylated cyclohex-2-enone derivs., however, the intermediate **2-azabicyclo[2.2.2]oct-2-en-5-ones** were isolated on treatment with methanolic ammonia, subsequent ring opening requiring treatment with DBU in methanol at reflux or, preferably, under microwave irradiation (three examples; Y 81-93%). This procedure was adapted for the first total synthesis of (-)-grandisine G via formation of (-)-grandisine B from (+)-grandisine D (Y 72% and 63% for the two steps). *cis*-2-(Carbomethoxymethyl)-6-pentylpiperidine, a natural product from ladybirds of the *Calvia guttata* species, was also prepared in 90% yield over 2 steps. F.e.s. J.D. Cuthbertson, R.J.K. Taylor, Angew. Chem., Int. Ed. *2013, 52* (5), 1490-3 [DOI: 10.1002/anie.201208118].

Benzimidazoles from *o*-diamines and carboxylic acids ○
update s. *30,* 265s*82*; uncatalyzed synthesis of 2-(fluoroalkyl)benzimidazoles and aza-analogs, also benzothiazole analogs from *o*-aminomercaptans, s. O. René, A. Souverneva, S.R. Magnuson, B.P. Fauber, Tetrahedron Lett. *2013, 54* (3), 201-4 [DOI: 10.1016/j.tetlet.2012.09.069].

2,6,9-Triazabicyclo[3.3.1]nonanes
from two α,β-ethylenealdehyde molecules and three prim. amine molecules

126. 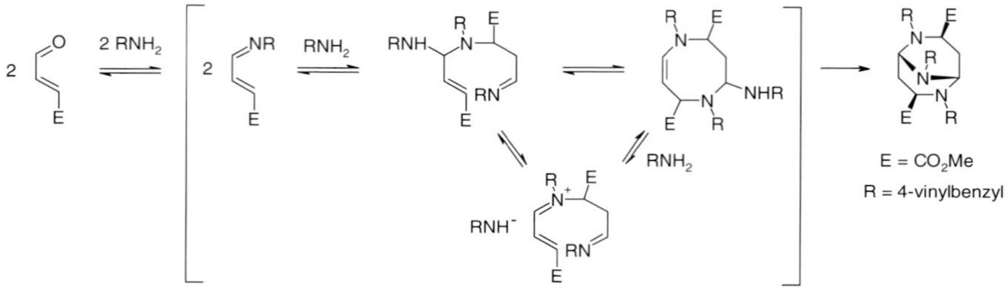

4-Vinylbenzylamine (*1.5 eq.*) added to a soln. of methyl 3-formylacrylate (87.6 μmol) in chloroform (0.1 ml) at room temp., the soln. stirred for 5 h, and purified chromatographically → dimethyl 2,6,9-tris(4-vinylbenzyl)-2,6,9-triazabicyclo[3.3.1]nonane-3,7-dicarboxylate. Y 100%. In the presence of at least 1.5 eq. of an electron-diverse prim. benzyl- or allyl-amine the formylacrylate cyclized via formal [4+4]-cycloaddition of the intermediate imines and subsequent bridging of the 8-membered ring with the prim. amine in quantitative yields (seven examples) in the presence of ether, alkene and Boc-amine functionality. The reaction failed with propyl-, propargyl-, phenethyl- and a bulky benzyl-amine (but treating the pre-formed benzylimine with propylamine incorporated the amine at the bridgehead position in 17% yield). Use of equimolar substrates gave only the imine which was unchanged after 5 days reaction, while in the absence of light or in the presence of a radical scavenger the reaction was slow and afforded lower yields. The procedure was extended to acrolein and its 2- and 3-methyl derivs. (three examples; Y 51-62%), but failed for 2,3-dimethyl or monophenyl derivs. Interestingly, mixing solns. from reactions of benzylamine with crotonaldehyde and the formylacrylate after 10 min, gave the cross-product, methyl 2,6,9-tribenzyl-7-methyl-2,6,9-triazabicyclo[3.3.1]nonane-3-carboxylate, in 48% yield. F.e.s. K. Tanaka, E.R.O. Siwu, S. Hirosaki, T. Iwata, R. Matsumoto, Y. Kitagawa, A.R. Pradipta, M. Okumura, K. Fukase, Tetrahedron Lett. *2012, 53* (44), 5899-902 [DOI: 10.1016/j.tetlet.2012.08.081]; f. reactions of prim. amines with acrolein, also relevance in biological systems with respect to the oxidative stress response to acrolein, s. A. Tsutsui, K. Tanaka, Org. Biomol. Chem. *2013, 11* (41), 7208-11 [DOI: 10.1039/c3ob41285g].

Microwaves [\\\\]
2(1*H*)-Quinoxalones from *o*-diamines
and 4-alkylidene-Δ^2-5-oxazolones cf. *43*, 319; **and α-ketocarboxylic acid esters**, uncatalyzed synthesis of 3-(6-phenylimidazo[2,1-*b*]thiazol-5-yl)quinoxalin-2(1*H*)-ones under microwave enhancement at high temperature (200°; 250 psi) s. C. Mukherjee, K.T. Watanabe, E.R. Biehl, Tetrahedron Lett. *2012, 53* (45), 6008-14 [DOI: 10.1016/j.tetlet.2012.08.093]; **from *o*-nitros-amines and carboxylic acid esters** with a weak base, functional group tolerance, s. Z. Wróbel, K. Stachowska, A. Kwast, A. Goscik, M. Królikiewicz, R. Pawlowski, I. Turska, Helv. Chim. Acta *2013, 96* (5), 956-68 [DOI: 10.1002/hlca.201200304].

Potassium hydroxide/air *KOH/O_2*
Aldimines from prim. alcohols and prim. amines $CH_2OH \rightarrow CH{=}NR$
update s. *77,* 188s*82*; simple, metal-free procedure with KOH in air s. J. Xu, R. Zhuang, L. Bao, G. Tang, Y. Zhao, Green Chem. *2012, 14* (9), 2384-7 [DOI: 10.1039/c2gc35714c]; with recyclable, carbon black-stabilized, polymer-incarcerated gold/palladium nanoparticles under O_2 s. J.-F. Soulé, H. Miyamura, S. Kobayashi, Chem. Commun. *2013, 49* (4), 355-7 [DOI: 10.1039/c2cc36213a]; **also ketimines from sec. alcohols** via acceptor-less dehydrogenation of the substrates with [2,2′-bis[2-(dicyclohexylphosphino)ethyl]amine](trimethylsilylmethyl)cobalt(II) barfate s. G. Zhang, S.K. Hanson, Org. Lett. *2013, 15* (3), 650-3 [DOI: 10.1021/ol303479f].

Potassium formate s. under Laccase *KOCHO*

Sodium azide/trifluoromethanesulfonic acid *NaN_3/TfOH*
Nitriles from aldehydes s. *55,* 146s*83* CHO → CN

Potassium phosphate/N-9-anthracenylmethyl-O-diphenylmethylcinchoninium bromide ←
1-Tosyldiaziridines from N-tosylaldimines and N,O-diprotected hydroxylamines under asym. phase transfer catalysis ○

127. 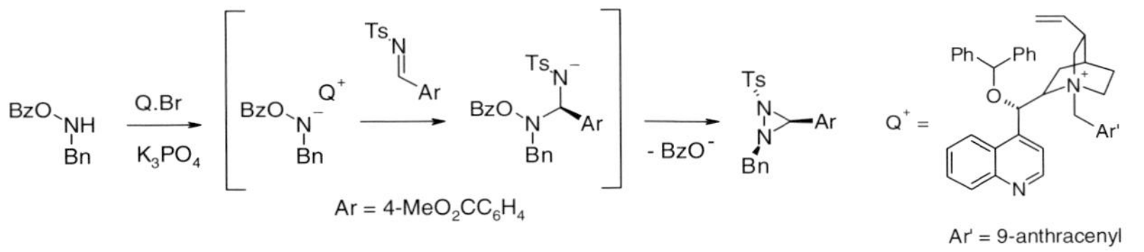

The first catalytic chiral diaziridine synthesis is reported. **E:** A mixture of chiral phase transfer catalyst (10 mol%), anh. K_3PO_4 (2 eq.) and anh. toluene (0.5 ml) stirred briefly at room temp., N-benzyl-O-benzoylhydroxylamine (1.5 eq.) added, the mixture cooled to -30°, N-tosyl-4-methoxycarbonylbenzaldimine (0.05 mmol) added, the mixture stirred at -12° for 72 h, cooled to -30°, diluted with pentane/ether (1:1; 4 ml), filtered through silica, concentrated *in vacuo*, and purified chromatographically → (1R,2R,3S)-1-benzyl-3-(4-methoxycarbonylphenyl)-2-tosyldiaziridine. Y 48% (e.e. 96%). Performing the reaction under phase transfer conditions, thereby avoiding the need for a stoichiometric amount of chiral base, and use of a controlled reaction temp., were critical to the success of this asym. insertion of an N-benzylhydroxylamine into an imine C=N (N-tosyl- and N-*tert*-butoxycarbonyl-protected hydroxylamines were not sufficiently nucleophilic). The method was successful with electron-diverse N-tosylaraldimines, affording chiral diaziridines as single diastereomers (ten examples; Y 42-73%; e.e. 75-96%). A sterically-hindered 2-fluorobenzaldimine gave low enantioselectivity (e.e. 41%; Y 45%) as did an aliphatic imine (cyclohexanal-derived) (e.e. 33%; Y 43%), and no products were obtained from 2-furfuraldimine or 4-methoxybenzaldimine (a 3-methoxybenzaldimine gave Y 42%; e.e. 92%). Orthogonal deprotection of the products without loss of enantiopurity was demonstrated in one case. F.e. and optimization s. L. Lykke, K.S. Halskov, B.D. Carlsen, V.X. Chen, K.A. Jørgensen, J. Am. Chem. Soc. *2013, 135* (12), 4692-5 [DOI: 10.1021/ja401150m].

Potassium iodide s. under H_5IO_6 *KI*

1,4-Dihydropyridines/S-benzylisothiouronium chloride ←
Organocatalyzed reductive N-alkylation with oxo compds. s. *17,* 436s*83* CO → CHN<

2,2′-Bipyridyl s. under CuI *bpy*

Cinchona derivs./benzylamines ←
Prim. amines from ketones CO → $CHNH_2$
by organocatalyzed asym. transamination s. *43,* 299s*83*

Copper(II) oxide nanoparticles-on-silica *CuO-SiO_2*
Benzimidazoles from *o*-diamines and aldehydes s. *69,* 171s*83* ○

Copper(II) triflate *Cu(OTf)$_2$*

N-Protected 4-(benzazol-1-yl)indoles
from 4-alkoxy-2-(alk-l-ynyl)-2,5-cyclohexadienimines

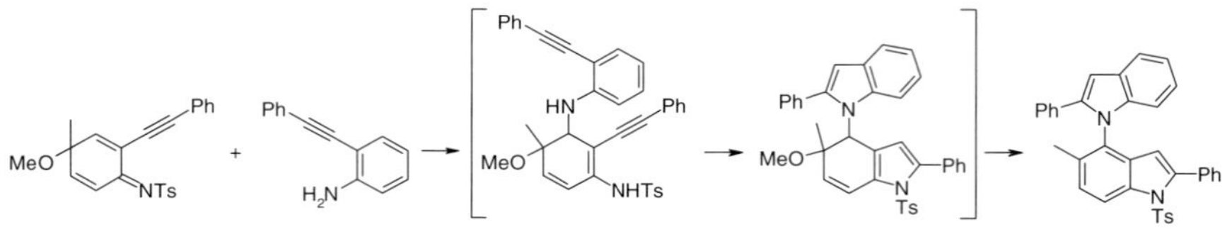

128.

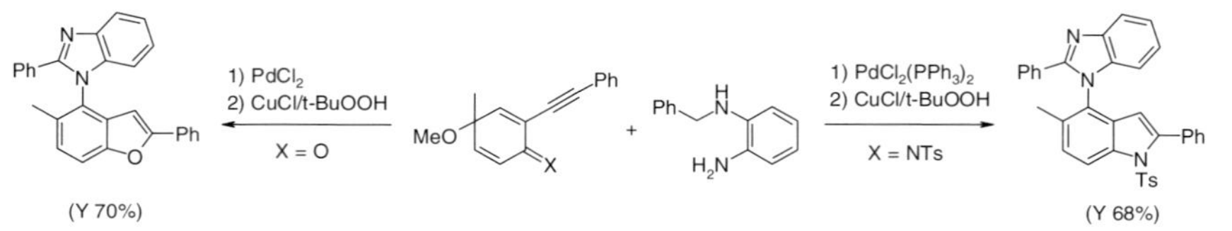

1′-Tosyl-1,4′-biindolyls. Cu(OTf)$_2$ (10 mol%) added to a Schlenk tube containing a soln. of the startg. 2,5-cyclohexadienimine (0.1 mmol) and *o*-aminoarylacetylene (1.2 eq.) in dichloroethane (2 ml) under N_2 at room temp., allowed to react to completion (TLC), quenched with satd. $NaHCO_3$, and worked up with purification by flash chromatography on silica gel → 1-(5-methyl-2-phenyl-1-tosyl-1*H*-indol-4-yl)-2-phenyl-1*H*-indole. Y 95%. This one-pot copper-catalyzed double ring closure is presumed to involve **1,4-addition-intramolecular hydroamination-aromatization**, each phase being catalyzed by the Lewis acid. The procedure is broadly applicable to the coupling of electron-diverse *o*-acetyleneamines with 4-alkoxy-2-(alk-l-ynyl)-N-sulfonyl-2,5-cyclohexadienimines possessing an alkyl, aryl or cyclopropyl group at the alkyne terminus and a methyl group at C_4 (fourteen examples; Y 31-99%). There was no reaction, however, with N-acetyl- or N-benzoyl-protected 2,5-cyclohexadienimines, nor with substrates possessing a phenyl, chloro or methoxy group at C_4 in place of methyl. Yields were lowest with *o*-ethynylanilines and *tert*-butylethynyl analogs. **4-(Indol-1-yl)benzofurans** were obtained similarly **from 4-alkoxy-2-(alk-l-ynyl)-2,5-cyclohexadienones** (fourteen examples; Y 30-97%), while **4-(benzimidazol-1-yl)-N-tosylindoles and -benzofurans** were obtained **from *o*-diamines** in place of the *o*-acetyleneamines using $PdCl_2(PPh_3)_2$ or $PdCl_2$, respectively, as catalyst, and subsequent catalytic oxidation with CuCl and *tert*-butyl hydroperoxide (nine examples in all; Y 44-72%). F.e. and comparison of metal salts and solvents s. M. Yang, J. Tang, R. Fan, Chem. Commun. *2012*, *48* (96), 11775-7 [DOI: 10.1039/c2cc36113b].

Copper(I) iodide/2,2′-bipyridyl/2,2,6,6-tetramethylpiperidine nitroxyl *CuI/bpy/TEMPO*

Nitriles from prim. alcohols $CH_2OH \rightarrow CN$

Copper(I)-catalyzed aerobic conversion under mild aq. conditions

129.

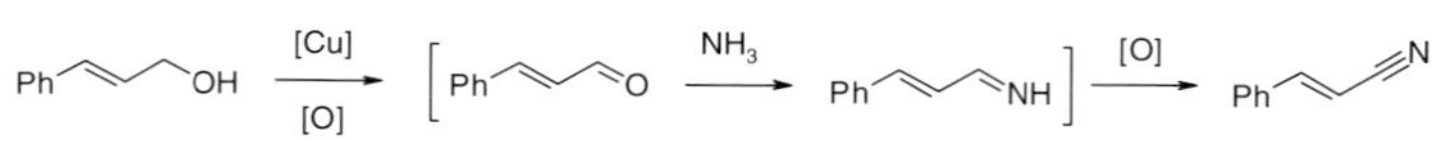

Cinnamyl alcohol (1 mmol), acetonitrile (2 ml) and aq. NH_3 (25-28% w/w; 2 eq.) added to a mixture of CuI (5 mol%), 2,2′-bipyridyl (5 mol%) and TEMPO (5 mol%) under O_2 (balloon), the mixture stirred at 50° for 24 h, concentrated, and purified chromatographically → cinnamonitrile. Y 98%. This general and experimentally simple procedure was successful for electron-diverse benzylic, allylic and simple aliphatic alcohols (incl. pyridine-derived alcohols) (twenty-seven examples; Y 68 to >99%). Strongly electron-deficient 4-nitrobenzyl alcohol gave a 60% yield of the nitrile due to competing oxidation to the benzoic acid. The method was compatible with amide, tert. amine, halo, ether, acetal and alkene functionality. Initial oxidation of the alcohol to an aldehyde was the rate-limiting step. F.e. and optimization, also use of the method in illustrative one-pot syntheses of triazolopyridine (two examples; Y 42%), thiazoline (Y 75%), oxazoline (Y 71%), tetrazole (Y 67%) and imidazoline derivs. (Y 75%) from startg. prim. alcohols, s. W. Yin, C. Wang, Y. Huang, Org. Lett. *2013, 15* (8), 1850-3 [DOI: 10.1021/ol400459y].

Gold/silver-molybdenum nano-rods/glycerol *Au/Ag-Mo/CH(OH)($CH_2OH)_2$*
Sec. or tert. amines from nitro compds. and prim. alcohols $NO_2 \rightarrow$ NHR or NR_2
via reductive mono- or di-N-alkylation with glycerol as reductant s. *83*, 140

Carbon black-stabilized, polymer-incarcerated gold/palladium nanoparticles/oxygen ←
Aldimines from prim. alcohols and prim. amines s. *77,* 188s*83* $CH_2OH \rightarrow CH{=}NR$

Magnesium *Mg*
***tert*-Amino(aryl)acetic acid esters** $C{=}NOSO_2R \rightarrow$ CHNR′R″
from aryl(sulfoximino)acetic acid esters and organomagnesium halides s. *83*, 362

Lithium amidoborane/aluminum chloride *$LiNH_2BH_3/AlCl_3$*
Sodium hydridotriacetoxoborate *Na[BH(OAc)$_3$]*
Reductive N-alkylation with oxo compds. CO → CHN<
update s. *17*, 436s*82*; rapid procedure with prim. amines using $AlCl_3$ as Lewis acid and Li-amidoborane [$LiNH_2BH_3$] as reductant s. W. Xu, X. Zheng, G. Wu, P. Chen, Chin. J. Chem. *2012, 30* (8), 1775-80 [DOI: 10.1002/cjoc.201200132]; one-pot **reductive N-alkylation-N-carbo-*tert*-butoxylation** of prim. amines with Na-hydridotriacetoxoborate/di-*tert*-butyl dicarbonate s. R. Neelarapu, P.A. Petukhov, Tetrahedron *2012, 68* (35), 7056-62 [DOI: 10.1016/j.tet.2012.06.055]; reductive N-alkylation of anilines with aromatic or aliphatic ketones by hydrogenation using $Cu(OAc)_2$ as catalyst s. S. Werkmeister, K. Junge, M. Beller, Green Chem. *2012, 14* (9), 2371-4 [DOI: 10.1039/c2gc35565e]; enhancement of selectivity under CO_2 using Pt/C (5%) poisoned by CO (generated by reverse water-gas shift reaction) s. S. Ichikawa, T. Seki, T. Ikariya, Chem. Lett. *2012, 41* (12), 1628-9 [DOI: 10.1246/cl.2012.1628]; *o*-carboxy-benzylamines from -benzaldehydes with cobalt(II) phthalocyanine/diphenylsilane, and subsequent phthalimidine ring closure, s. V. Kumar, U. Sharma, B. Singh, N. Kumar, Austr. J. Chem. *2012, 65* (12), 1594-8 [DOI: 10.1071/ch12321]; reductive N-alkylation of prim. acylamines with aldehydes by hydrogenation with $[Rh(cod)Cl]_2$/XantPhos and TsOH as acidic co-catalyst s. S. Raoufmoghaddam, E. Drent, E. Bouwman, Adv. Synth. Catal. *2013, 355* (4), 717-33 [DOI: 10.1002/adsc.201200837]; *organocatalyzed* reductive N-alkylation of prim. amines with ketones using Hantzsch ester and S-benzylisothiouronium chloride as recoverable catalyst s. Q.P.B. Nguyen, T.H. Kim, Synthesis *2012, 44* (13), 1977-82 [DOI: 10.1055/s-0031-1291125]; sec. methylamines from N-methylformamide by the Leuckart reaction under microwave enhancement (cf. *1,* 634), also subsequent N-formylation under microwaves (cf. *13,* 442s*83*), s. F. Barba, J. Recio, B. Batanero, Tetrahedron Lett. *2013, 54* (14), 1835-8 [DOI: 10.1016/j.tetlet.2013.01.097].

Boron fluoride *BF_3*
N-Protected 3,4-unsatd. 2-keto-N-glycosides from O^2,O^3,O^4-triacylglycals via stereoselective Ferrier-type N-glycosylation ←

130.

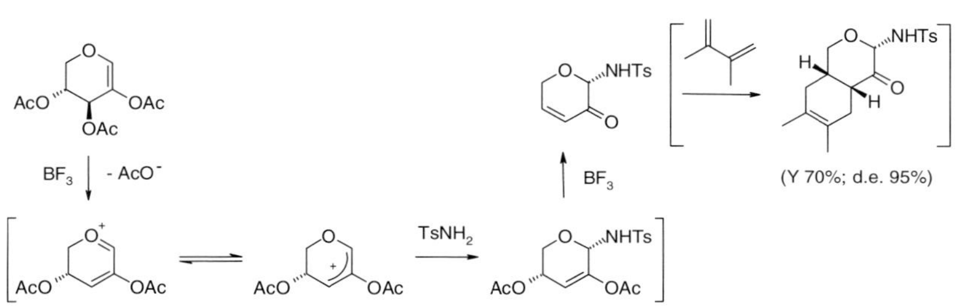

BF_3-etherate (4.4 eq.) added to a soln. of 2,3,4-tri-O-acetyl-2-hydroxy-D-xylal (0.25 mmol) and *p*-toluenesulfonamide (1.1 eq.) in methylene chloride (4 ml) at room temp. under N_2, the mixture stirred for 20 min, quenched with satd. aq. $NaHCO_3$, extracted with methylene chloride, and purified chromatographically → (S)-4-methyl-N-(3-oxo-3,6-dihydro-2*H*-pyran-2-yl)benzene-sulfonamide. Y 78% (α/β 92:8). Use of 0.5-2.2 eq. of BF_3 promoted formation of Ferrier rearranged N-glycosides of synthetically useful enone sugars (incl. a 6-deoxy enone), with 4.4 eq. not only favoring the illustrated products, but also showing good α-selectivity (other Lewis acids were inferior). The method was successful with electron-diverse ar. and aliphatic sulfonamides (incl. the sec. sulfonamide, N-tosyl-N-allylamine) and carbamates (incl. Fmoc-NH_2 and Cbz-NH_2) (fifteen examples; Y 42-86%; α/β 68:32 to 92:8). F.e., optimization and reactions of the products, s. F. Ding, R. William, X.-W. Liu, J. Org. Chem. *2013*, *78* (3), 1293-9 [DOI: 10.1021/jo302619b].

Aluminum chloride s. under $LiNH_2BH_3$ *$AlCl_3$*

Lanthanum(III) triflate *$La(OTf)_3$*

Benzo[1,4]heterocycle-condensed cyclopentanones from 2-furylcarbinols and *o*-difunctionalized arenes ←

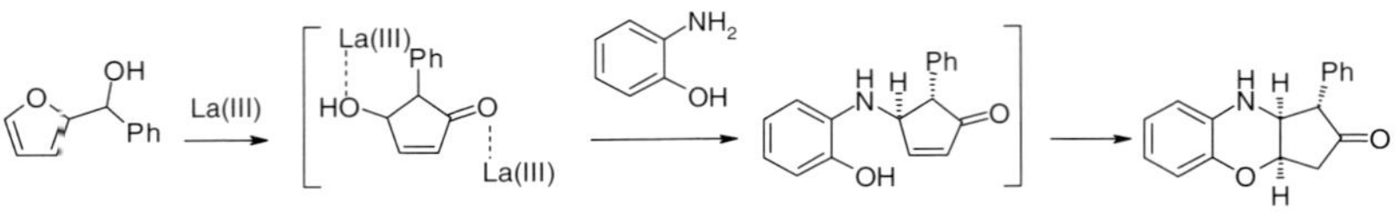

131.

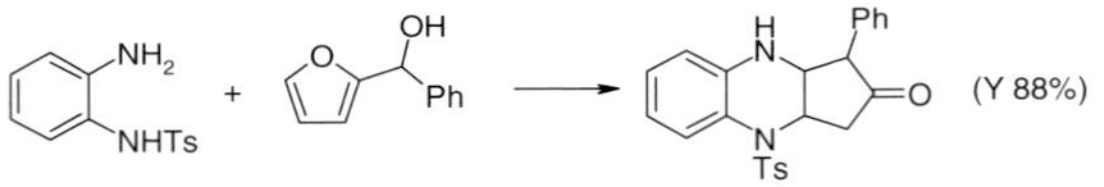

o-Aminophenols and *o*-diamines react with 2-furylcarbinols **via Piancatelli rearrangement** to give benzo[1,4]heterocycle-condensed cyclopentanones condensed through nitrogen or oxygen at the 3 and 4 positions. **E: 3,3a,9,9a-Tetrahydrocyclopenta[*b*][1,4]benzoxazin-2(1*H*)-ones.** 2-Furyl(phenyl)carbinol (0.6 mmol), *o*-aminophenol (0.5 mmol) and $La(OTf)_3$ (5 mol%) in acetonitrile (2 ml) stirred at 80° for 4 h, the mixture concentrated under reduced pressure, and worked up with purification by chromatography on silica gel → product. Y 81%. The procedure is inexpensive, cost- and energy-efficient, rapid and broadly applicable to the coupling of *o*-aminophenol (or analogs substituted by chlorine, methyl, *tert*-butyl or nitro groups) with electron-diverse aryl(2-furyl)carbinols (sixteen examples; Y 10%, 41-92%). Furthermore, a single diastereoisomer (with a *cis*-fusion) was obtained in each instance. Alkyl(2-furyl)carbinols, however, gave much lower yields (19-53%; three examples). *o*-Phenylenediamine, 2-aminobenzenethiol, pyrocatechol and 2-aminopyridin-3-ol as dinucleophile in place of the *o*-aminophenol gave messy results or failed to react at all. However, electron-diverse N-tosyl- or N-mesyl-*o*-diamines did participate, affording the corresponding cyclopentanone-condensed 1,2,3,4-quinoxalines, but yields were generally moderate and there was no diastereoselectivity (four examples; Y 45-88%). After the initial Piancatelli rearrangement, reaction is presumed to involve C-N coupling prior to intramolecular Michael addition of the phenolic hydroxyl group onto the formed 2-cyclopentenone ring. F.e., limitations, and comparison of Lewis/Brønsted acids and solvents s. J. Liu, Q. Shen, J. Yu, M. Zhu, J. Han, L. Wang, Eur. J. Org. Chem. *2012* (35), 6933-9 [DOI: 10.1002/ejoc.201200962]; ***trans*-4-arylamino-5-aryl-2-cyclopentenones** by aza-Piancatelli rearrangement (cf. *79*, 145) using 0.03 mol% phosphomolybdic acid in acetonitrile at reflux s. B.V.S. Reddy, G. Narasimhulu, P.S. Lakshumma, Y.V. Reddy, J.S. Yadav, Tetrahedron Lett. *2012*, *53* (14), 1776-9 [DOI: 10.1016/j.tetlet.2012.01.115].

Graphite *C*

Quinoxalines from *o*-diamines and α-diketones ○

update s. *7*, 446s*82*; eco-friendly, heterogeneous procedure with graphite in ethanol, also 2,3-dihydropyrazines from 1,2-diamines, s. H.K. Kadam, S. Khan, R.A. Kunkalkar, S.G. Tilve, Tetrahedron Lett. *2013, 54* (8), 1003-7 [DOI: 10.1016/j.tetlet.2012.12.041]; quinoxalines **from ketones** (cf. *7*, 446s*81*) via iodination-Kornblum oxidation-cyclization with I_2/CuO/DMSO or

$K_3PO_4 \cdot 3H_2O$ s. M. Lian, Q. Li, Y. Zhu, G. Yin, A. Wu, Tetrahedron *2012, 68* (47), 9598-605 [DOI: 10.1016/j.tet.2012.09.056].

Glycerol s. under $RuCl_3$ $CH(OH)(CH_2OH)_2$

β-Cyclodextrin β-*CD*

2,3-Dihydro-4(1*H*)-quinazolones from *o*-aminocarboxylic acid amides and oxo compds. ○

from aldehydes with HCl cf. *19,* 481; 2-aryl-derivs. from ar. aldehydes with β-cyclodextrin in water s. K. Ramesh, K. Karnakar, G. Satish, B.S.P.A. Kumar, Y.V.D. Nageswar, Tetrahedron Lett. *2012, 53* (51), 6936-9 [DOI: 10.1016/j.tetlet.2012.10.029]; *heterogeneous* synthesis of 2-spiro-derivs. from cyclic ketones with recyclable Amberlyst-15 under ultrasonication s. D. Rambabu, S.K. Kumar, B.Y. Sreenivas, S. Sandra, A. Kandale, P. Misra, M.V.B. Rao, M. Pal, ibid. *2013, 54* (6), 495-501 [DOI: 10.1016/j.tetlet.2012.11.057]; from oxo compds. with recoverable N-propyl-sulfamic acid-on-magnetic nanoparticles *in water* s. A. Rostami, B. Tahmasbi, H. Gholami, H. Taymorian, Chin. Chem. Lett. *2013, 24* (3), 211-4 [DOI: 10.1016/j.cclet.2013.01.032].

Ammonium acetate/trichloroisocyanuric or p-dodecylbenzenesulfonic acid ←

Multicomponent synthesis of imidazoles from α-diketones and aldehydes

2,4,5-trisubst. imidazoles by **3-component synthesis**, update s. *23,* 423s*82*; with trichloroisocyanuric acid as catalyst in recoverable 2,2,2-trifluoroethanol s. S.F. Hojati, S.A. Nezhadhoseiny, Z. Beykzadeh, Monatsh. Chem. *2013, 144* (3), 387-90 [DOI: 10.1007/s00706-012-0830-5]; in the same solvent *without catalyst*, also 1,2,4,5-tetrasubst. imidazoles **by 4-component synthesis**, s. S. Khaksar, M. Alipour, ibid. *144* (3), 395-8 [DOI: 10.1007/s00706-012-0834-1]; *in water* with *p*-dodecylbenzenesulfonic acid as catalyst s. B. Das, J. Kashanna, R.A. Kumar, P. Jangili, ibid. *144* (2), 223-6 [DOI: 10.1007/s00706-012-0770-0].

1-Butylimidazolium trifluoroacetate $[Hbim]OCOCF_3$

1-Ethyl-3-(3-dimethylaminopropyl)carbodiimide/(2,2-dimethyl-1,3-dioxolan-4-yl)methyl-2-cyano-2-(hydroxyimino)acetate/sodium hydrogen carbonate ←

N-Formylation NH → NCHO

update s. *13,* 442s*82*; with 85% formic acid under solvent-free conditions with 1-butylimidazole·trifluoroacetic acid as Brønsted acidic ionic liquid catalyst s. S. Majumdar, J. De, J. Hossain, A. Basak, Tetrahedron Lett. *2013, 54* (3), 262-6 [DOI: 10.1016/j.tetlet.2012.11.017]; in water or aq. DMF with EDC/$NaHCO_3$ and the Oxyma deriv., (2,2-dimethyl-1,3-dioxolan-4-yl)methyl-2-cyano-2-(hydroxyimino)acetate, as additive for the mild N-formylation of prim. amines, incl. α-amino-acids, s. B.A. Aleiwi, K. Mitachi, M. Kurosu, Tetrahedron Lett. *2013, 54* (16), 2077-81 [DOI: 10.1016/j.tetlet.2013.01.095]; of prim. or sec. amines, incl. α-amino-acids, *with formamide* and NaOMe s. S. Joseph, P. Das, B. Srivastava, H. Nizar, M. Prasad, ibid. *54* (8), 929-31 [DOI: 10.1016/j.tetlet.2012.12.009]; of prim. or sec. amines *with methanol* as formylating agent in the presence of a ruthenium(II) bis(N-heterocyclic carbene) complex s. N. Ortega, C. Richter, F. Glorius, Org. Lett. *2013, 15* (7), 1776-9 [DOI: 10.1021/ol400639m].

Peptide coupling agents ←

Peptide synthesis COOH → CON<

update s. *77,* 179s*82*; rapid, racemization-free, **soln.-phase peptide synthesis** by classical coupling with DCCI/HONB or HATU/HOAt/DIEA *under microwave enhancement* s. A. Mahindra, K.K. Sharma, R. Jain, Tetrahedron Lett. *2012, 53* (51), 6931-5 [DOI: 10.1016/j.tetlet.2012.10.028]; soln.-phase synthesis of **cyclic and bicyclic peptides** s. X. Elduque, E. Pedroso, A. Grandas, Org. Lett. *2013, 15* (8), 2038-41 [DOI: 10.1021/ol400726y]; **enzymatic peptide synthesis** of alternating oligopeptides under papain and protease catalysis (cf. *18,* 442s*46-8*) s. X. Qin, A.C. Khuong, Z. Yu, W. Du, J. Decatur, R.A. Gross, Chem. Commun. *2013, 49* (4), 385-7 [DOI: 10.1039/c2cc36381j]; of side chain-protected peptides using subtilisin A in anhydrous organic solvents s. T. Nuijens, A.H.M. Schepers, C. Cusan, J.A.W. Kruijtzer, D.T.S. Rijkers, R.M.J. Liskamp, P.J.L.M. Quaedflieg, Adv. Synth. Catal. *2013, 355* (2-3), 287-93 [DOI: 10.1002/adsc.201200694]; use of the *tert*-butyl side chain as a powerful means to lock peptoid amide bonds in the *cis*-conformation s. O. Roy, C. Caumes, Y. Esvan, C. Didierjean, S. Faure, C. Taillefumier, Org. Lett. *2013, 15* (9), 2246-9 [DOI: 10.1021/ol400820y]; mild and robust **solid-phase peptide synthesis** of challenging peptides on a controllable core-shell-type aminomethyl polystyrene resin s. H.-J. Cho, T.-K. Lee, J.W. Kim, S.-M. Lee, Y.-S. Lee, J. Org. Chem. *2012, 77* (20), 9156-62 [DOI: 10.1021/jo301716d];

of peptide acids on a 'click'-modified Wang resin with a dimethoxyphenyl linker s. V. Castro, H. Rodriguez, F. Albericio, Org. Lett. *2013, 15* (2), 246-9 [DOI: 10.1021/ol303367s]; of N-methylated peptides under microwave enhancement with COMU/ethyldiisopropylamine as coupling agent s. R. Rodbeen, S.L. Pedersen, M. Hosseini, K.J. Jensen, Eur. J. Org. Chem. *2012* (36), 7106-11 [DOI: 10.1002/ejoc.201201050]; of pseudopeptides, cyclic peptides and amino acids on a solid support with a pipecolic linker through the lysine side-chain s. N. Masurier, P. Zajdel, P. Verdié, M. Pawlowski, M. Amblard, J. Martinez, G. Subra, Chem. Eur. J. *2012, 18* (37), 11536-40 [DOI: 10.1002/chem.201201452]; of peptides containing backbone-fluorinated amino acids using the Fmoc-strategy s. L. Hunter, S. Butler, S.B. Ludbrook, Org. Biomol. Chem. *2012, 10* (44), 8911-8 [DOI: 10.1039/c2ob26596f]; of peptides phosphocholinated at serine, threonine, and tyrosine s. M.F. Albers, C. Hedberg, J. Org. Chem. *2013, 78* (6), 2715-9 [DOI: 10.1021/jo302587g]; **of depsipeptides** s. M.M. Nguyen, N. Ong, L. Suggs, Org. Biomol. Chem. *2013, 11* (7), 1167-70 [DOI: 10.1039/c2ob26893k]; of peptides possessing a C-terminal methyl ester via trityl side-chain anchoring s. V. Diaz-Rodriguez, D.G. Mullen, E. Ganusova, J.M. Becker, M.D. Distefano, Org. Lett. *2012, 14* (22), 5648-51 [DOI: 10.1021/ol302592v]; convergent solid-phase synthesis of **N-glycopeptides** based on a pseudoproline linker s. V. Ullmann, M. Rädisch, I. Boos, J. Freund, C. Pöhner, S. Schwarzinger, C. Unverzagt, Angew. Chem., Int. Ed. *2012, 51* (46), 11566-70 [DOI: 10.1002/anie.201204272]; synthesis of homogeneous N-linked glycopolypeptides by convergent aspartylation s. P. Wang, B. Aussedat, Y. Vohra, S.J. Danishefsky, ibid. 11571-5 [DOI: 10.1002/anie.201205038].

1-Butyl-3-methylimidazolium hexafluorophosphate s. under Trimethylsilyl azide *[bmim]PF$_6$*

Laccase/2,2,6,6-tetramethylpiperidine nitroxyl/(4,4′-dimethoxy-2,2′-bipyridyl)(pentamethylcyclopentadienyl)iridium(III) sulfate/potassium formate ←

N-Benzylation with benzyl alcohols via enzymatic oxidation-iridium(III)-catalyzed reduction with potassium formate as small molecule switch NH → NCH$_2$Ar

132. 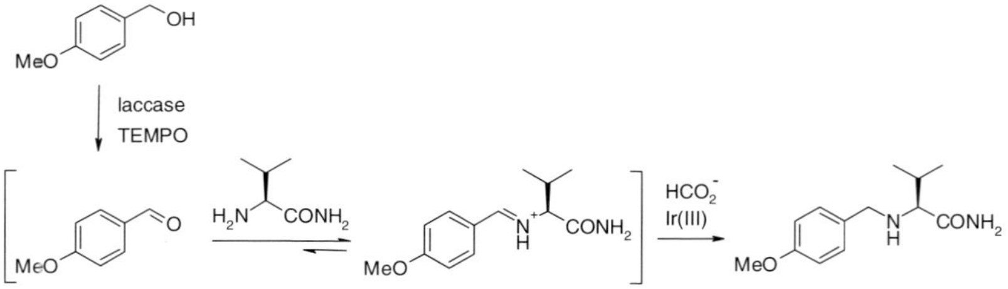

A model one-pot reaction has been described wherein the prim. amine of L-valinamide is N-*p*-methoxybenzylated via enzymatic oxidation of the benzyl alcohol, condensation with the amino acid, and reduction of the resultant iminum ion. Key to the reaction is the novel addition of formate to the sequence to effectively 'switch off' the oxidant thereby minimizing undesired reverse reaction. **E:** Anisyl alcohol (1.5 eq.) added to a soln. of laccase (5 U), TEMPO (0.3 eq.), Ir(III) complex (1 mol%) and L-valinamide·HCl (85.4 μmol) in 25 m*M* aq. oxygenated citric acid/Na-citrate buffer (pH 5; 10 ml), the mixture shaken at 40° for 40 h, K-formate (10 eq.) added, shaking continued for a further 25 h, quenched with satd. aq. NaHCO$_3$, extracted with methylene chloride, and purified chromatographically → N-4-methoxybenzyl-L-valinamide. Y 50% (conversion 91%). This model one-pot reaction was developed via optimization of the individual steps and gave a slightly higher yield (62%) from a 78% conversion after a total reaction time of 41 h. The authors note that application to other benzylic alcohols requires extensive experimentation in each case. F. details s. J.S. Willemsen, J.C.M. van Hest, F.P.J.T. Rutjes, Chem. Commun. *2013, 49* (30), 3143-5 [DOI: 10.1039/c3cc00126a].

Transaminase ←

Prim. amines from ketones by asym. enzymatic transamination CO → $CHNH_2$

update s. *43,* 299s*82*; of 1- or 2-tetralones and 3- or 4-chromanones with ω-transaminase and various NH_2-donors for R- or S-selectivity s. D. Pressnitz, C.S. Fuchs, J.H. Sattler, T. Knaus, P. Macheroux, F.G. Mutti, W. Kroutil, ACS Catal. *2013, 3* (4), 555-9 [DOI: 10.1021/cs400002d]; **organocatalyzed asym. transamination** of ar. ketones with *o*-hydroxybenzylamine as NH_2-donor and 9-O-pentylhydroquinine as chiral base (cf. *43,* 299s*82*) s. Y. Xie, H. Pan, X. Xiao, S. Li, Y. Shi, Org. Biomol. Chem. *2012, 10* (45), 8960-2 [DOI: 10.1039/c2ob26782a]; of trifluoromethyl ketones with a benzylamine as NH_2-donor and a cinchona deriv. as base s. M. Liu, J. Li, X. Xiao, Y. Xie, Y. Shi, Chem. Commun. *2013, 49* (14), 1404-6 [DOI: 10.1039/c2cc37423d].

Prim. amines from ketones by asym. enzymatic transamination using 3-aminocyclohexa-1,5-dienecarboxylic acid as prim. amino group donor

133. 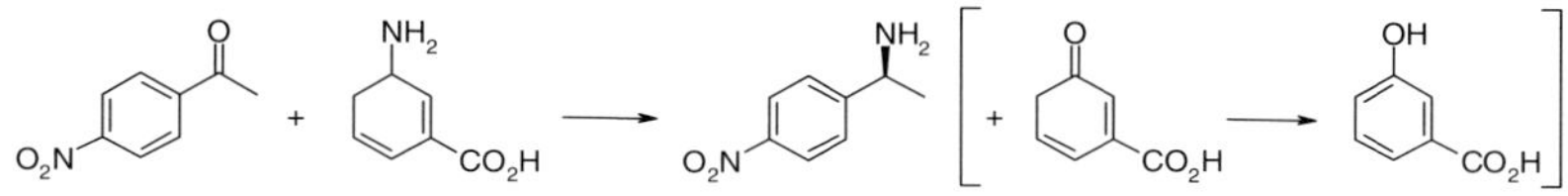

A novel single-enzyme conversion of acetophenones to chiral prim. benzylamines using an aminocyclohexadiene deriv. as amine source is described (cf. *43,* 299s*82,83*), conversion of the amine donor to the corresponding ketone, and its spontaneous tautomerization to 3-hydroxybenzoic acid, serving to obviate inhibition of the enzyme by the ketone by-product and also to drive the equilibrium forwards. The reaction may therefore be accomplished with only a slight excess of the amine donor with no additional enzymes or expensive cofactors required. **E:** 4-Nitroacetophenone (5 m*M*), 3-aminocyclohexa-1,5-dienecarboxylic acid (5.25 m*M*) and a preparation of *Chromobacterium violaceum* ω-transaminase (0.8 U) containing a balance amount of pyridoxal 5′-phosphate dissolved in 50 m*M* aq. HEPES buffer (1 ml), pH adjusted to 8.2, and the mixture stirred at 37° for 12 h → (S)-1-(4-nitrophenyl)ethylamine. Y >99% by GC (e.e. >99%). Experiments using transaminases from wild-type and mutant strains were individually optimized for a series of electron-diverse acetophenones (five examples) and gave complete conversion and high enantioselectivity (e.e. >99%). F.e., optimization and enzyme isolation/purification s. B. Wang, H. Land, P. Berglund, Chem. Commun. *2013, 50* (2), 161-3 [DOI: 10.1039/c2cc37232k].

2,2,6,6-Tetramethylpiperidine nitroxyl s. under CuI and Laccase *TEMPO*

Amberlyst-15 ←

2-Spiro-2,3-dihydro-4(1*H*)-quinazolones from *o*-aminocarboxylic acid amides ◌

and cyclic ketones – Heterogeneous conversion under ultrasonication s. *19,* 481s*83*

tert-Butyl methyl carbonate/potassium tert-butoxide *t-BuOC(O)OMe/KOBu-t*

3,4,5,6-Tetrahydro-1,3-oxazin-2-ones from 1,3-diols and N-nucleophiles

134. 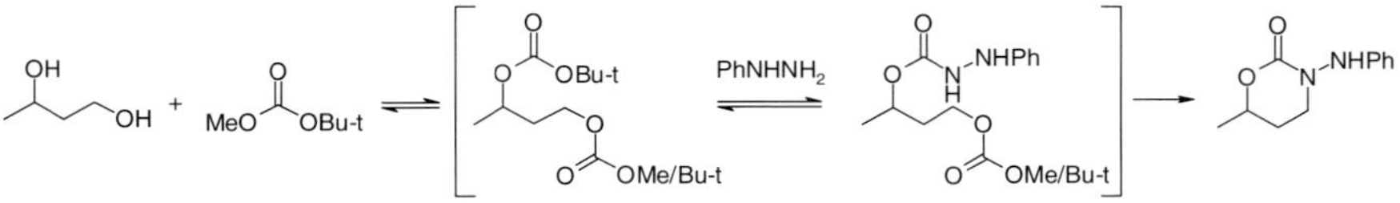

3-Amino-3,4,5,6-tetrahydro-1,3-oxazin-2-ones. A mixture of phenylhydrazine (10 mmol), butane-1,3-diol (1 eq.), *tert*-butyl methyl carbonate (30 eq.) and K-*tert*-butoxide (20 eq.) stirred at 90° for 2 h, quenched with water, extracted with methylene chloride, concentrated *in vacuo*, solids precipitated with ether, and purified by recrystallization → 6-methyl-3-phenylamino-1,3-oxazinan-2-one. Y 53%. Carbonate derivs., generated *in situ* by reaction of 1,3-diols with dialkyl carbonates, were cyclized with four N-nucleophiles (benzylamine, aniline, octylamine or phenylhydrazine) to the title products in one pot. Product formation was favored by bulkier dialkyl carbonates (attributed to easier formation of the pre-product carbamate with a bulky alkoxy leaving group), but gave reduced yields with increasing bulk of the 1,3-diol component. Highest (unoptimized) yields were obtained for the four N-nucleophiles with *tert*-butyl methyl carbonate (di-*tert*-butyl carbonate was too sterically hindered to generate the carbonate intermediate), and although generally moderate, were apparently higher than current literature values for propane-1,3-diol (25-57%), butane-1,3-diol (25-51%) and 3-methylbutane-1,3-diol (17-32%), with bulky 2-methylpentane-2,4-diol affording only 0.3% yield of 3-benzyl-4,6,6-trimethyloxazinanone. F.e.s. C.R. McElroy, F. Aricò, P. Tundo, Synlett *2012*, *23* (12), 1809-15 [DOI: 10.1055/s-0031-1290368].

Tri-n-butylmethylammonium methyl carbonate *[MeOCOO][MeNBu₃]*
N-Methylation NH → NMe
with dimethyl carbonate s. *41,* 325; of indole using a basic ionic liquid as catalyst, e.g. tri-*n*-butylmethylammonium methyl carbonate, *under continuous flow* at high pressure and temperature with a short residence time, also methylation of phenols, thiophenols and carboxylic acids, s. T.N. Glasnov, J.D. Holbrey, C.O. Kappe, K.R. Seddon, T. Yan, Green Chem. *2012, 14* (11), 3071-6 [DOI: 10.1039/c2gc36226k]; regioselective N-methylation of indazoles with methyl 2,2,2-trichloroacetimidate and added trifluoromethanesulfonic acid s. S.R. Baddam, N.U. Kumar, A.P. Reddy, R. Bandichhor, Tetrahedron Lett. *2013*, *54* (13), 1661-3 [DOI: 10.1016/j.tetlet.2013.01.030].

S-Benzylisothiouronium chloride s. under 1,4-Dihydropyridines ←

Benzotetramisoles or 7-Chloro-2,3-dihydroimidazo[1,2-a]quinolines ←
Organocatalyzed asym. N-acylation NH → NAc
with kinetic resolution, update s. *55,* 149s*82*; of lactams and thiolactams with a chiral amidine as catalyst [7-chloro-2,3-dihydroimidazo[1,2-*a*]quinoline or benzotetramisole derivs.] s. X. Yang, V.D. Bumbu, P. Liu, X. Li, H. Jiang, E.W. Uffman, L. Guo, W. Zhang, X. Jiang, K.N. Houk, V.B. Birman, J. Am. Chem. Soc. *2012, 134* (42), 17605-12 [DOI: 10.1021/ja306766n]; application to the abs. configuration determination of lactams and 2-oxazolidones with chiral [dihydropyrimidobenzothiazole-type] benzotetramisole analogs [(S)- or (R)-HBTM] as catalyst s. M.A. Perry, J.V. Trinidad, S.D. Rychnovsky, Org. Lett. *2013, 15* (3), 472-5 [DOI: 10.1021/ol303239t].

N,N-Diethyl-S,S-difluorosulfiliminium fluoroborate *$[F_2S{=}NEt_2]BF_4$*
N-Subst. carboxylic acid amides from acids COOH → CON<
update s. *23,* 415s*82*; from challenging carboxylic acids with N,N-diethyl-S,S-difluorosulfiliminium fluoroborate [XtalFluor-E] as coupling agent (without racemization/epimerization of chiral amines) s. A. Orliac, D.G. Pardo, A. Bombrun, J. Cossy, Org. Lett. *2013, 15* (4), 902-5 [DOI: 10.1021/ol400045d]; from unactivated carboxylic acids and prim. or sec. amines with Ti(OPr-*i*)$_4$ as catalyst s. H. Lundberg, F. Tinnis, H. Adolfsson, Synlett *2012, 23* (15), 2201-4 [DOI: 10.1055/s-0032-1316993]; preparation of radio-labelled, dendritic polymer-based carboxylic acid amides in water with EDC/N-hydroxysulfosuccinic acid as coupling agent s. G. Aljayyoussi, M. Gumbleton, Tetrahedron Lett. *2013, 54* (9), 1045-8 [DOI: 10.1016/j.tetlet.2012.11.088].

Trichloroisocyanuric acid s. under Ammonium acetate ←
Diphenylsilane s. under Cobalt(II) phthalocyanine *Ph_2SiH_2*
Silica s. under CuO and 18-Tungstophosphoric acid *SiO_2*

Sulfated titanium dioxide *TiO_2-SO_4^{2-}*
Pyrroles ○
by Paal-Knorr synthesis, update s. *1,* 335s*82*; *heterogeneous*, solvent-free route to N-aryl- and N-[cyclo]alkyl-pyrroles by grinding with [recyclable] sulfated anatase-titania [TiO_2-SO_4^{2-}] as solid acid catalyst s. K. Ravi, B. Krishnakumar, M. Swaminathan, Bull. Chem. Soc. Jpn. *2013*, *86* (3), 370-5 [DOI: 10.1246/bcsj.20120221]; N-arylpyrroles **from 2,5-dimethoxytetrahydrofuran** (cf. *76,* 162s*82*) *in an eco-friendly medium* [water, a deep eutectic solvent or polyethylene glycol] under ultrasound irradiation or thermal conditions s. N. Azizi, A. Davoudpour, F. Eskandari, E. Batebi, Monatsh. Chem. *2013*, *144* (3), 405-9 [DOI: 10.1007/s00706-012-0841-2].

Nanocrystalline titanium dioxide-perchloric acid *TiO_2-$HClO_4$*
N-Carbo-*tert*-butoxylation NH → NCOOBu-*t*
update s. *60,* 135s*82*; of amines with di-*tert*-butyl dicarbonate [*tert*-butoxyformic anhydride] under heterogeneous, solvent-free conditions with recyclable, mildly acidic nanocrystalline TiO_2-$HClO_4$ at room temperature s. F. Shirini, S.V. Atghia, M.G. Jirdehi, Chin. Chem. Lett. *2013, 24* (1), 34-6 [DOI: 10.1016/j.cclet.2012.12.005].

Trimethylsilyl azide/1-methyl-3-(4-sulfobutyl)imidazolium triflate/1-butyl-3-methyl-imidazolium hexafluorophosphate ←

Nitriles from aldehydes CHO → CN

update s. *55*, 146s*82*; mild Schmidt reaction with [hetero]aromatic aldehydes using trimethylsilyl azide and 1-methyl-3-(4-sulfobutyl)imidazolium triflate as catalyst in 1-butyl-3-methyl-imidazolium hexafluorophosphate s. G.C. Nandi, K.K. Laali, Tetrahedron Lett. *2013, 54* (17), 2177-9 [DOI: 10.1016/j.tetlet.2013.02.051]; rapid, scalable, chemoselective conversion of electron-diverse ar. aldehydes with NaN_3/TfOH s. B.V. Rokade, K.R. Prabhu, J. Org. Chem. *2012, 77* (12), 5364-70 [DOI: 10.1021/jo3008258]; nitriles **from aldehydes or prim. alcohols** with aq. NH_3 and H_5IO_6/KI s. A. Ghorbani-Choghamarani, M.A. Zolfigol, M. Hajjami, S. Sardari, Synth. Commun. *2013, 43* (1), 52-8 [DOI: 10.1080/00397911.2011.591035].

Titanium tetraisopropoxide *Ti(OPr-i)$_4$*

N-Subst. carboxylic acid amides from acids s. *23*, 415s*83* COOH → CON<

Tin(II) chloride *$SnCl_2$*

6,6a-Dihydroisoindolo[2,1-*a*]quinazoline-5,11-diones from *o*-nitrocarboxylic acid amides and *o*-aldehydocarboxylic acids via reductive double ring closure ○

135.

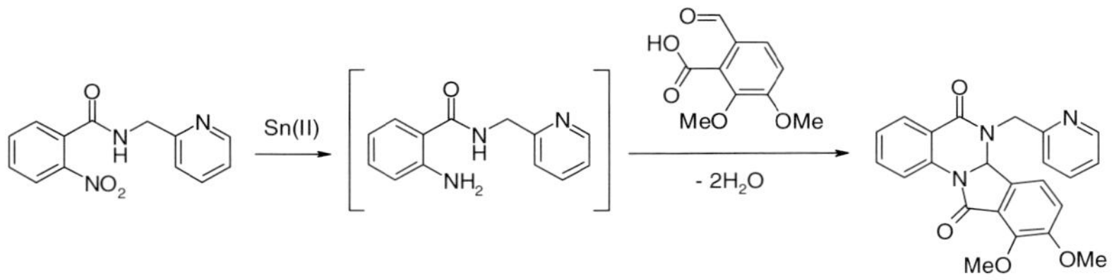

in one pot. A soln. of N-(pyrid-2-ylmethyl)-2-nitrobenzamide (1 mmol), 5,6-dimethoxy-2-formyl-benzoic acid (1 eq.) and $SnCl_2 \cdot H_2O$ (4 eq.) in ethanol (8 ml) stirred under reflux for 8 h, quenched with 3% aq. HCl, filtered, and purified by recrystallization → 6,6a-dihydro-9,10-dimethoxy-6-(pyrid-2-ylmethyl)isoindolo[2,1-*a*]quinazoline-5,11-dione. Y 80%. Reduction of the nitro group to amine and subsequent double ring closure with 2-formylbenzoic acids afforded the title tetracycles in good yields (80-90%; eight examples). F.e. and optimization s. M. Mahdavi, R. Najafi, M. Saeedi, E. Alipour, A. Shafiee, A. Foroumadi, Helv. Chim. Acta *2013, 96* (3), 419-23 [DOI: 10.1002/hlca.201200199].

Triphenylphosphine s.a. under $RuCl_3$ *Ph_3P*

Triphenylphosphine/ethyl 2-(3,4-dichlorophenyl)carbazate/ iron phthalocyanine/air ←

Catalytic Mitsunobu reaction s. *83*, 60 ←

Sodium diphenylphosphinobenzene-3-sulfonate s. under Palladium(II) acetate ←

Vanadyl acetoacetonate *$VO(acac)_2$*

Niobium pentachloride *$NbCl_5$*

18-Tungstophosphoric acid-silica *$H_6P_2W_{18}O_{62}$-SiO_2*

β-Amino-α,β-ethylene- from β-keto-carbonyl compds. COCH → C(NHR)=C

update s. *26*, 331s*78*; solvent-free (Z)-selective formation of β-amino-α,β-ethylene-carboxylic acid esters and -ketones from aromatic or aliphatic amines with $VO(acac)_2$ s. R.A. Laskar, N.A. Begum, M.H. Mir, S. Ali, A.T. Khan, Tetrahedron Lett. *2013, 54* (5), 436-40 [DOI: 10.1016/j.tetlet.2012.11.051]; with $NbCl_5$ cf. Y.-H. Liu, P. Wang, G.-T. Cheng, Monatsh. Chem. *2013, 144* (2), 191-6 [DOI: 10.1007/s00706-012-0783-8]; eco-friendly, *heterogeneous*, solvent-free formation of β-aminocrotonates with recyclable bulk- or silica gel-supported $H_6P_2W_{18}O_{62} \cdot 24H_2O$ s. L.M. Sanchez, Á.G. Sathicq, G.T. Baronetti, H.J. Thomas, G.P. Romanelli, Can. J. Chem. *2013, 91* (2), 137-42 [DOI: 10.1139/cjc-2012-0157].

Oxygen or Air s.a. under KOH, Polymer-incarcerated gold/palladium nanoparticles, Triphenylphosphine and I_2 *O_2*

Air O_2

Benzimidazoles from *o*-diamines and aldehydes ○

update s. *69,* 171s*82*; *uncatalyzed* procedure *in toluene* under air at room temperature s. C. Zhang, L. Zhang, N. Jiao, Green Chem. *2012, 14* (12), 3273-6 [DOI: 10.1039/c2gc36416f]; *in ethanol* cf. G.-F. Chen, H.-D. Shen, H.-M. Jia, L.-Y. Zhang, H.-Y. Kang, Q.-Q. Qi, B.-H. Chen, J.-L. Cao, J.-T. Li, Austr. J. Chem. *2013, 66* (2), 262-6 [DOI: 10.1071/ch12458]; *heterogeneous* conversion with recyclable CuO nanoparticles-on-silica, also benzoxazoles from *o*-aminophenols, and benzothiazoles from *o*-aminomercaptans, s. S.M. Inamdar, V.K. More, S.K. Mandal, Tetrahedron Lett. *2013, 54* (6), 579-83 [DOI: 10.1016/j.tetlet.2012.11.091]; with nanoparticulate organic-inorganic hybrid porous iron phosphonate [prepared from hexamethylenediamine-N,N,N′,N′-tetrakis-(methylphosphonic acid)] under temperature-induced morphological control s. A. Dutta, J. Mondal, A.K. Patra, A. Bhaumik, Chem. Eur. J. *2012, 18* (42), 13372-8 [DOI: 10.1002/chem.201201350].

p-Toluenesulfonyl chloride/ethyldiisopropylamine/hydrogen chloride *TsCl/i-Pr₂NEt/HCl*

2-*prim*-Aminopyridines from pyridine N-oxides ←

Regioselective conversion with saccharin as ammonia equivalent

136. 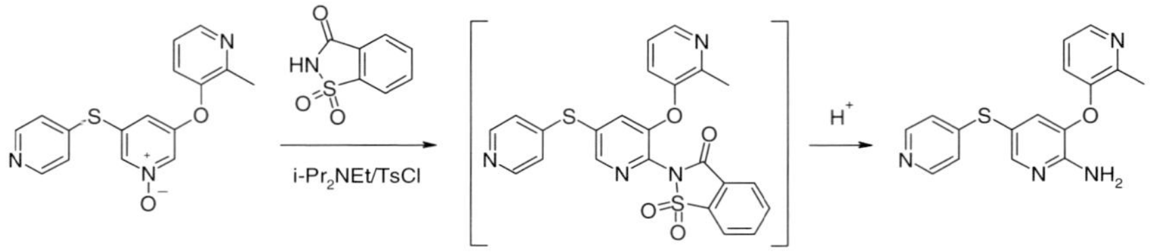

Ethyldiisopropylamine (2 eq.) added to a slurry of the startg. pyridine N-oxide (0.149 mmol) and saccharin (1.1 eq.) in methylene chloride (0.5 ml), the resulting soln. cooled to 0-5°, tosyl chloride (1.2 eq.) in the same solvent (0.1 ml) added via syringe over 10 min, stirring continued at 0-5° for 2 h, allowed to warm to room temp., the mixture treated with 5 *N* HCl (0.25 ml) with stirring, the lower methylene chloride layer removed, the resulting aq. product layer heated to 80°, treated in sequence with ethanol (0.5 ml), 2 *N* aq. K_2HPO_4 (1 ml) and 10 *N* NaOH (to pH 6), the aq. layer extracted then concentrated to an oil, and purified by crystallization → 3-[(2-methylpyridin-3-yl)oxy]-5-(pyridin-4-ylsulfenyl)pyridin-2-amine. Y 97% (conversion 100%; regioisomeric ratio of aminated products: 27.8:1). The procedure is mild, inexpensive, based on a readily available ammonia surrogate, high-yielding, and is applicable to various 3,5-difunctionalized pyridine N-oxides with regioselectivity for 2-amination varying from 2:1 to 84:1 depending on the electronic nature of the ring substituents. For 3-[het]aryloxy-5-aryl(thio- or sulfinyl)-derivs., the *o*-amination ratio [relative to the 3-substituent] ranged from 2.9:1 to 27.8:1, while for 3-alkoxy-5-bromo- and 3-aryloxy-5-bromo-derivs. the ratio ranged from 2:1 to 84:1. For 3-arylthio-5-bromopyridine N-oxide, however, amination took place preferentially at the *ortho* site relative to bromine. The initially formed *o*-saccharinyl derivs. were isolated (thirteen examples; Y 21-97%), of which the water-soluble products could be directly deprotected *in situ* on treatment with acid to give the desired *o*-amino-derivs. (two examples; Y 65-97%). Saccharin afforded higher yields and was generally cleaner than other aminating agents, e.g. *tert*-butylamine, hexamethyldisilazane, phthalimide, N-carbo-*tert*-butoxycarbonyl-*p*-toluenesulfonamide and N-protected phosphoromonoamidates, while tosyl chloride proved the most efficient and economical activating agent, at the same time interfering least with the aminating agent. F.e. and comparison of solvents s. R.P. Farrell, M.V. Silva Elipe, M.D. Bartberger, J.S. Tedrow, F. Vounatsos, Org. Lett. *2013, 15* (1), 168-71 [DOI: 10.1021/ol303218p].

Trifluoromethanesulfonic acid (s.a. under NaN_3) *TfOH*

N-Methylation with methyl 2,2,2-trichloroacetimidate s. *41,* 325s*83* NH → NMe

p-Dodecylbenzenesulfonic acid s. under Ammonium acetate *$ArSO_3H$*

1-Methyl-3-(4-sulfobutyl)imidazolium triflate s. under Trimethylsilyl azide ←

N-Propylsulfamic acid-on-magnetic nanoparticles ←
2,3-Dihydro-4(1*H*)-quinazolones ○
from *o*-aminocarboxylic acid amides and oxo compds. s.*19*, 481s*83*

Iodine/hydrogen chloride/oxygen $I_2/HCl/O_2$
2,4-Diaryl-1-benzylimidazoles from aryl ketones and two benzylamine molecules
Regioselective oxidative ring closure

137. 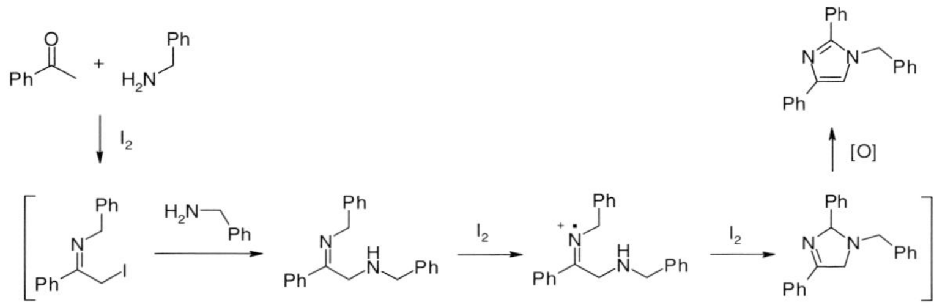

A mixture of acetophenone (0.5 mmol), benzylamine (5 eq.), iodine (20 mol%) and concd. HCl (10 ml; 20 mol%) stirred in DMSO (2 ml) at 90° under O_2 for 5 h, quenched with water, and worked up with purification by flash chromatography on silica gel → 1-benzyl-2,4-diphenyl-1*H*-imidazole. Y 85%. The procedure is applicable to the coupling of electron-diverse aryl ketones (incl. benzophenone and naphthyl analogs) with benzylamine and ring-functionalized analogs (with Me, OMe, Cl or F at the *m*- or *p*-site) (twenty-two examples; Y 35-90%). The yield was lowest with a propiophenone analog. Reaction takes place via iodine-catalyzed aerobic oxidation involving initial α-iodination of the ketone, nucleophilic displacement of iodide by the amine, two single electron transfers via an amine radical cation, and an aromatizing oxidative ring closure. The added Brønsted acid serves as promoter while at the same time preventing undesirable oxidative oxazole ring closure. F.e. and comparison of additives, oxidants and solvents, also subsequent elaboration of the products by 5-bromination and via N-debenzylation, s. H. Huang, X. Ji, W. Wu, H. Jiang, Adv. Synth. Catal. *2013*, *355* (1), 170-80 [DOI: 10.1002/adsc.201200582].

Perchloric acid s. under TiO_2 $HClO_4$

Periodic acid/potassium iodide H_5IO_6/KI
Nitriles from aldehydes or prim. alcohols s. *55*, 146s*83* CHO or CH_2OH → CN

N-9-Anthracenylmethyl-O-diphenylmethylcinchoninium bromide s. under Potassium phosphate ←

Iron/acetic acid *Fe/AcOH*
Oxazoles from α-nitroketones and orthocarboxylic acid esters ○

138. 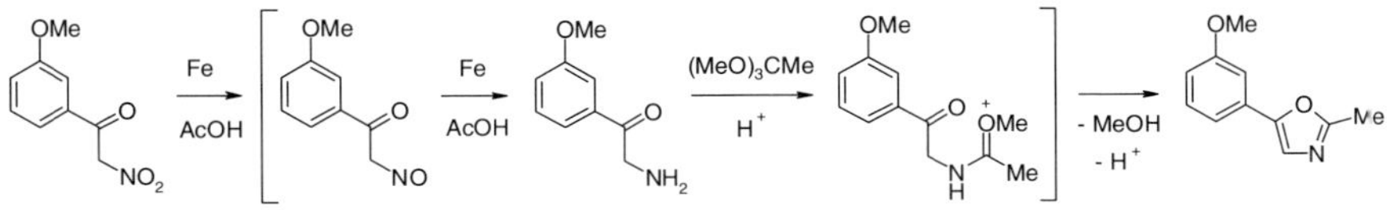

2-Subst. 5-aryloxazoles. A suspension of iron powder (2 mmol), the startg. 1-aryl-2-nitroethanone (0.5 mmol) and acetic acid (5 mmol) in acetonitrile (5 ml) stirred under N_2 at room temp. for 5 min, trimethyl orthoacetate (4 eq.) in acetonitrile (10 ml) added, the mixture refluxed for 8 h, filtered, and the filtrate worked up with purification by chromatography on silica gel → 5-(*m*-methoxyphenyl)-2-methyloxazole. Y 75%. The procedure is simple, inexpensive and applicable in good yield to the coupling of trimethyl ortho-acetate or -benzoate with electron-diverse 1-aryl-2-nitroethanones, notably leaving aromatic bromine, chlorine, fluorine, methoxy and methyl groups unaffected and with no obvious steric hindrance to reaction. Yields were poor with samarium or zinc as reducing agent, and the method is a significant improvement on a previously reported procedure with indium. Reaction is thought to involve initial formation of an α-nitrosoketone. F.e. and optimization s. S. Chen, D. Bai, F. Shi, J. Li, C. Li, X. Jia, Chin. J. Chem. *2012, 13* (7), 1464-8 [DOI: 10.1002/cjoc.201100683].

Iron/sulfur *Fe/S*

Benzimidazoles or benzoxazoles ○

from *o*-nitramines or *o*-nitrophenols and active methyl groups

139.

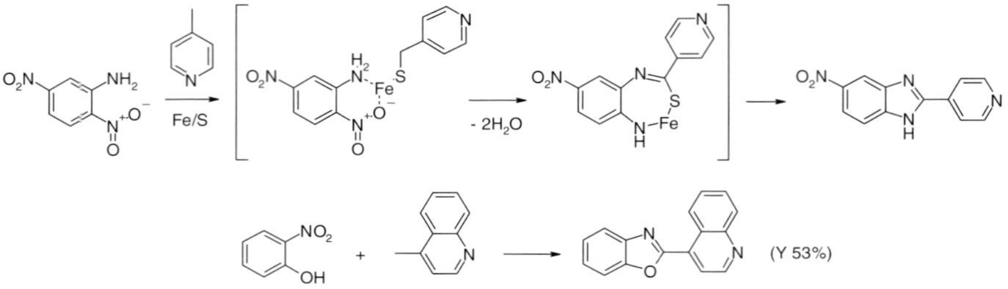

A mixture of 2,5-dinitroaniline (5 mmol), 4-methylpyridine (2 eq.), elemental sulfur (10 mol%) and iron powder (10 mol%) stirred under argon at 150° for 24 h, cooled to room temp., and purified chromatographically → 5-nitro-2-pyrid-4-yl-1*H*-benzo[*d*]imidazole. Y 57%. In this iron sulfide promoted dehydrative coupling, *in situ* generation of iron sulfide clusters from elemental iron and sulfur gave significantly better results than using pre-formed FeS (or other Fe catalysts), and provided a novel and experimentally simple and inexpensive route to the title products. Electron-diverse 2-nitroanilines (and mono-N-methyl derivs.) were successfully cyclized with 2- and 4-methylpyridines/quinolines (a 3-methylpyridine was unreactive) or 2-methylbenzimidazole (and its N-methyl deriv.) to the corresponding **2-(hetaryl)benzimidazoles** (eighteen examples; Y 56-90%). 2,4,6-Trimethylpyridine gave a mixture of 2-pyrid-2-yl- (Y 40%) and -4-yl-benzimidazoles (Y 22%). The method was also applicable to 2-nitrophenols, affording moderate yields of the corresponding **2-(hetaryl)benzoxazoles** (Y 51-53%; two examples). F.e. and optimization s. T.B. Nguyen, L. Ermolenko, A. Al-Mourabit, J. Am. Chem. Soc. *2013, 135* (1), 118-21 [DOI: 10.1021/ja311780a].

Transition metal complexes ←

Catalytic N-alkylation with alcohols NH + ROH → NR

update s. *76*, 173s*82*; of prim. amines with a wide range of alcohols under *iridium* catalysis using chloro(pentamethylcyclopentadienyl)((S)-prolinato)iridium(III) as catalyst in toluene or water without additive s. A. Wetzel, S. Wöckel, M. Schelwies, M.K. Brinks, F. Rominger, P. Hofmann, M. Limbach, Org. Lett. *2013, 15* (2), 266-9 [DOI: 10.1021/ol303075h]; of prim. or sec. amines with the iridium(I) bis(phosphaalkene) complex, chloro[2,6-bis(2-mesityl-2-phosphavinyl)-pyridine]iridium(I), and CsOH or KH_2PO_4 as base s. Y.-H. Chang, Y. Nakajima, F. Ozawa, Organometallics *2013, 32* (7), 2210-5 [DOI: 10.1021/om4000743]; of anilines with prim. or sec. alcohols using an acetonitrile[1-butyl-3-(2-hydroxyalkyl)imidazol-2-ylidene](pentamethylcyclopentadienyl)iridium(III) bis(fluoroborate) or alkoxy-tethered analogs as highly active catalysts (without base) s. A. Bartoszewicz, R. Marcos, S. Sahoo, A.K. Inge, X. Zou, B. Martín-Matute, Chem. Eur. J. *2012, 18* (45), 14510-9 [DOI: 10.1002/chem.201201845]; N,N′-dialkylation of phenylenediamines with tetracarbonyl(dichloro)[N′,N′-*m*-xylylbis(N-benzylimidazolidin-2-ylidene)]diiridium(I) as catalyst and CsOH as base without solvent s. H.-Y. Kuo, Y.-H. Liu, S.-M. Peng, S.-T. Liu, Organometallics *2012, 31* (20), 7248-55 [DOI: 10.1021/om300800c]; N-alkylation of prim. amines, and 2-alkylation of sec. alcohols (CC⇃↾O), with prim. alcohols using a recyclable mesoporous silica-supported chloro(pentamethylcyclopentadienyl)[1-(3-silylpropyl)-3-(pyrimidin-2-yl)imidazol-2-ylidene]iridium(III) chloride as catalyst s. D. Wang, X.-Q. Guo, C.-X. Wang, Y.-N. Wang, R. Zhong, X.-H. Zhu, L.-H. Cai, Z.-W. Gao, X.-F. Hou, Adv. Synth. Catal. *2013, 355*

(6), 1117-25 [DOI: 10.1002/adsc.201200732]; piperazines from 1,2-diamines and glycol with $[Cp^*IrCl_2]_2/NaHCO_3$ in water or toluene s. L.L.R. Lorentz-Petersen, L.U. Nordstrøm, R. Madsen, Eur. J. Org. Chem. *2012* (34), 6752-9 [DOI: 10.1002/ejoc.201201099]; N-alkylation of prim. amines with prim. alcohols under *palladium* catalysis using TiO_2 loaded with palladium nanoparticles under irradiation (>300 nm) [involving photocatalytic alcohol oxidation-imination-hydrogenation in sequence] s. Y. Shiraishi, K. Fujiwara, Y. Sugano, S. Ichikawa, T. Hirai, ACS Catal. *2013, 3* (3), 312-20 [DOI: 10.1021/cs300756f]; also of prim. acylamines under homogeneous conditions via initial palladium-catalyzed aerobic oxidation of the alcohols cf. X. Yu, L. Jiang, Q. Li, Y. Xie, Q. Xu, Chin. J. Chem. *2012, 30* (10), 2322-32 [DOI: 10.1002/cjoc.201200462]; general heterogeneous procedure under *nickel* catalysis with recyclable nickel nanoparticles on various oxides having acid-base sites (e.g. *theta*-Al_2O_3) s. K.-i. Shimizu, N. Imaiida, K. Kon, S.M.A.H. Siddiki, A. Satsuma, ACS Catal. *2013, 3* (5), 998-1005 [DOI: 10.1021/cs4001267]; **prim. amines from alcohols** and ammonia under the same conditions s. K.-i. Shimizu, K. Kon, W. Onodera, H. Yamazaki, J.N. Kondo, ibid. *3* (1), 112-7 [DOI: 10.1021/cs3007473]; N-protected allyl- or propargyl-amines from the corresponding unsatd. alcohols and protected amines with Re_2O_7 under mild conditions s. B.G. Das, R. Nallagonda, P. Ghorai, J. Org. Chem. *2012, 77* (13), 5577-83 [DOI: 10.1021/jo300706b]; *metal-free*, green and scalable N-alkylation of prim. amines and acylamines with prim. alcohols using *an aldehyde as catalyst* [based on an aerobic oxidation-imination-transfer-hydrogenation] s. Q. Xu, Q. Li, X. Zhu, J. Chen, Adv. Synth. Catal. *2013, 355* (1), 73-80 [DOI: 10.1002/adsc.201200881].

Transition metal salts or complexes ←
Transition metal-catalyzed hydrogenative N-alkylation CO → CHN<
with oxo compds. s. *17,* 436s*83*

Iron phthalocyanine s. under Triphenylphosphine *Fe(Pc)*
Chiral ferrocenyldi(phosphine) s. under Bis(1,5-cyclooctadiene)rhodium(I) hexafluoroantimonate ←

Nanoparticulate hybrid porous iron phosphonate ←
Benzimidazoles from *o*-diamines and aldehydes s. *69,* 171s*83* ⌒

Cobalt(II) phthalocyanine/diphenylsilane *Co(II)/Ph₂SiH₂*
***o*-Carboxy-benzylamines from -benzaldehydes** s. *17,* 436s*83* CO → CHN<

[2,2′-Bis[2-(dicyclohexylphosphino)ethyl]amine](trimethylsilylmethyl)cobalt(II) barfate —
Azomethines from alcohols and prim. amines s. *77,* 188s*83* CHOH → C=NR

Ruthenium trichloride/triphenylphosphine/potassium carbonate/glycerol ←
Sec. or tert. amines from nitro compds. and prim. alcohols via reductive mono- or di-N-alkylation with glycerol as reductant NO_2 → NHR or NR_2

140.

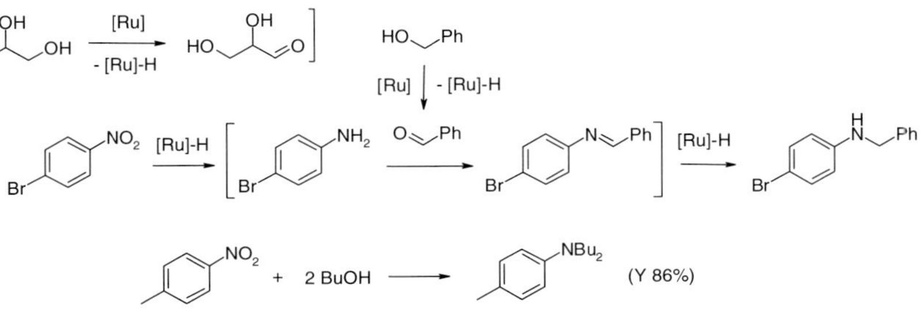

4-Bromo-1-nitrobenzene (1 mmol) and benzyl alcohol (1 eq.) added to a mixture of $RuCl_3{\cdot}xH_2O$ (2.5 mol%), triphenylphosphine (10 mol%), K_2CO_3 (10 mol%), glycerol (1.5 g) and trifluoromethylbenzene (0.5 ml) under argon, the mixture stirred at 130° for 24 h, cooled to room temp., extracted with ether, and purified chromatographically → N-benzyl-4-bromoaniline. Y 83%. This efficient and clean procedure demonstrates novel controlled synthesis of sec. or tert. anilines from electron-diverse nitrobenzenes, by use of 1 eq. (at 130°) or 2 eq. (at 150°) of a prim. aliphatic or benzylic alcohol (thirty-six examples; Y 69-93%). Glycerol appears to function both as reductant and co-solvent. F.e. and optimization s. X. Cui, Y. Deng, F. Shi, ACS Catal. *2013, 3* (5), 808-11 [DOI: 10.1021/cs400049b]; with Au/Ag-Mo nano-rods as heterogeneous catalyst and glycerol cf. X. Cui, C. Zhang, F. Shi, Y. Deng, Chem. Commun. *2012, 48* (75), 9391-3 [DOI: 10.1039/c2cc34178f].

Dichlorobis[1,2-bis(diphenylphosphino)ethane]ruthenium(II)/carbon dioxide/hydrogen ←
2-Unsubst. benzimidazoles from *o*-diamines ○
Ruthenium(II)-catalyzed dehydrative ring closure in the absence of solvent s. *83*, 118

*Bis(1,5-cyclooctadiene)rhodium(I) hexafluoroantimonate/chiral ferrocenyldi(phosphine) [Rh(I)]**
Acetylamines from O-acetyloximes C=N-OAc → CHNHAc
via asym. homogeneous hydrogenation s. *83*, 3

Dichloro(pentamethylcyclopentadienyl)rhodium(III) dimer/silver hexafluoroantimonate/ potassium acetate ←
Rhodium-catalyzed N-directed *o*-(carbalkoxyamin)ation under mild conditions H → NHCOOR

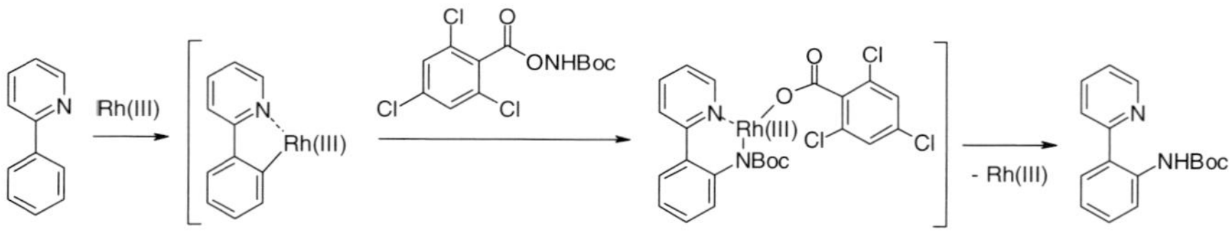

141. 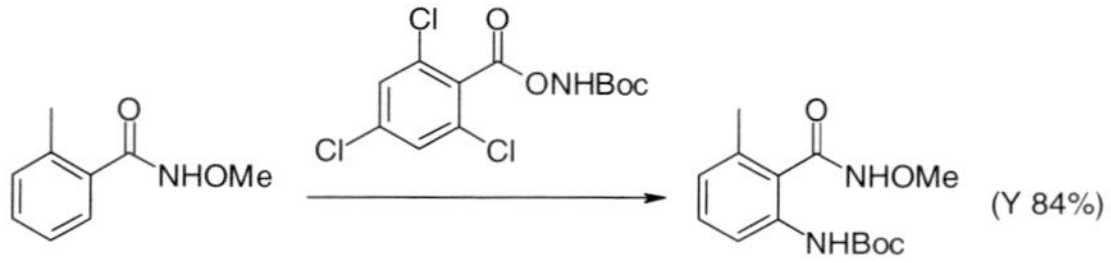

with N-(2,4,6-trichlorobenzoyloxy)carbamic acid esters. 2-Phenylpyridine (0.25 mmol), *tert*-butyl N-(2,4,6-trichlorobenzoyloxy)carbamate (1.2 eq.) and dry acetonitrile (2.5 ml) added under a stream of argon to a sealable tube containing $[Cp^*RhCl_2]_2$ (2.5 mol%), $AgSbF_6$ (10 mol%) and KOAc (1.2 eq.), the tube sealed, the mixture stirred at 60° for 16 h, cooled, and worked up with purification by flash chromatography on silica gel → product. Y 90%. The procedure is applicable to 2-phenylpyridine as well as 2-arylpyridines possessing an *o*- or *m*-methyl group or a *p*-methoxy group (but not a *p*-cyano-deriv.), delivering the NHBoc, NHCbz or NHFmoc function regioselectively at the *ortho* site (six examples; Y 55-98%). N-Directed *o*-(carbalkoxyamin)ation of N-methoxybenzamides, heteroaromatic analogs (incl. furan, thiophene, pyrrole, benzo[*b*]-thiophene and indole derivs. at a higher catalyst loading) and 1-naphthoic acid N-methoxyamide was also achieved under slightly modified conditions (without Ag(I)), although *m*- and *p*-subst. N-methoxybenzamides gave a mixture of diamidation products (ca. twenty-five examples in all; Y 20-98%). N-(*p*-Nitrobenzoyloxy)carbamic acid esters were also efficient *o*-(carbalkoxyamin)-ating agents. Reaction is thought to involve initial coordination of pyridyl or hydroxamate nitrogen to rhodium followed by intramolecular *o*-rhodation to form a 5-membered rhodacycle; the N-aroyloxy-carbamate then coordinates to the metal through nitrogen prior to delivery of the amido function and protodemetalation to complete the cycle. F.e., also N-directed β-(carbalkoxyamin)ation of 2-vinylpyridine and an α,β-ethylenecarboxylic acid N-methoxyamide (two examples; Y 31-70%), s. C. Grohmann, H. Wang, F. Glorius, Org. Lett. *2013, 15* (12), 3014-7 [DOI: 10.1021/ol401209f].

Palladium nanoparticles s. under Polymer-incarcerated gold/palladium nanoparticles *Pd*

Palladium complexes *[Pd]*

N-Allylation NH → N-C-C=C

with acoxy-2-ethylenes under palladium catalysis, update s. *42,* 354s*82*; α-(arylaminomethyl)-α,β-ethylenecarboxylic acid esters from Baylis-Hillman acetates with $Pd_2(dba)_3$ and a ferrocenyldi-(phosphine) as ligand with K_2CO_3 as base s. Y. Wang, L. Liu, D. Wang, Y.-J. Chen, Org. Biomol. Chem. *2012, 10* (34), 6908-13 [DOI: 10.1039/c2ob25976a]; *heterogeneous* N-allylation of prim. and sec. amines with a biodegradable and easily recyclable cellulose-palladium catalyst and the same base s. B.R. Vaddula, A. Saha, R.S. Varma, J. Leazer, Eur. J. Org. Chem. *2012* (34), 6707-9 [DOI: 10.1002/ejoc.201201241]; regioselective N-allylation of N-heterocyclics, incl. uridine and thymidine derivs., **with** linear or branched **2-ethylenealcohols** using $Pd(OAc)_2$/1,2-bis(diphenylphosphinomethyl)benzene (cf. *35,* 257s*80*; *50,* 195s*81*) s. D. Banerjee, R.V. Jagadeesh, K. Junge, H. Junge, M. Beller, Angew. Chem., Int. Ed. *2012, 51* (46), 11556-60 [DOI: 10.1002/anie.201206319]; mild, *uncatalyzed* N-allylation with 2-ethylenealcohols in 1,1,1,3,3,3-hexafluoroisopropanol or 2,2,2-trifluoroethanol (cf. *25,* 575), also allylation of Si- and C-nucleophiles, s. P. Trillo, A. Baeza, C. Nájera, J. Org. Chem. *2012, 77* (17), 7344-54 [DOI: 10.1021/jo301049w]; *metal-free* procedure with iodine as catalyst for preparing C_2-(N-arylamidomethyl)glycals from C_2-(acetoxymethyl)-glycals s. J.S. Yadav, G. Narasimhulu, N. Umadevi, Y.V. Reddy, B.V.S. Reddy, Tetrahedron Lett. *2013, 54* (8), 871-3 [DOI: 10.1016/j.tetlet.2012.11.106].

Palladium(II) acetate/sodium diphenylphosphinobenzene-3-sulfonate ←

4(3*H*)-Quinazolones from *o*-aminocarboxylic acid amides and prim. alcohols ○

under iridium catalysis cf. *66,* 178s*82*; 2-aryl derivs. with $Pd(OAc)_2$/Na-diphenylphosphinobenzene-3-sulfonate [TPPMS] *in water* s. H. Hikawa, Y. Ino, H. Suzuki, Y. Yokoyama, J. Org. Chem. *2012, 77* (16), 7046-51 [DOI: 10.1021/jo301282n].

Chiral palladium or iridium complexes *[Pd]* or [Ir]**

Asym. N-allylation NH → N-C-C=C

under *palladium* catalysis s. *50,* 195s*82*; chiral β-arylamino-α-methylenecarboxylic acid esters from Baylis-Hillman acetates with $Pd_2(dba)_3$ and a planar-chiral di(phosphino)[2,2′]cyclophane as ligand s. Y. Wang, T. Zhang, L. Liu, Chin. J. Chem. *2012, 30* (11), 2641-6 [DOI: 10.1002/cjoc.201200771]; regioselective asym. N-allylation of Na-benzotriazolide under *iridium* catalysis with $[Ir(cod)Cl]_2$ and a chiral phosphoromonoamidite as ligand with K_3PO_4 as base s. M. Zhang, X.-W. Guo, S.-C. Zheng, X.-M. Zhao, Tetrahedron Lett. *2012, 53* (51), 6995-8 [DOI: 10.1016/j.tetlet.2012.10.064].

Iridium-containing metal-organic frameworks *Ir-MOF*

Heterogeneous N-allylation with diallyl carbonate using iridium-containing metal-organic frameworks as catalysts

142.

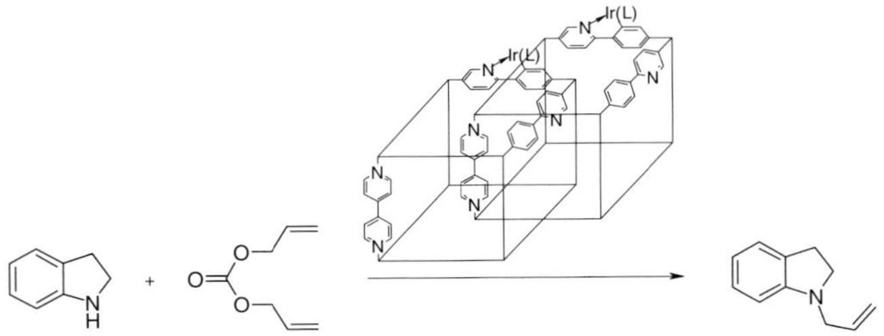

Heterogeneous N-allylation with stable, storable, recyclable, crystalline MOFs functionalized via iridium(I) cyclometalation is reported, reaction notably taking place cleanly under mild conditions, without rigorous exclusion of air or moisture. **E:** BMOF-1-Irdcppy (BMOF = 4,4′-bipyridyl MOF, dcppy = 2-phenylpyridine-5,4′-dicarboxylate; ca. 70 mg; ca. 5 mol% of Ir), indoline (0.46 mmol) and diallyl carbonate (10 eq.) added to $CDCl_3$ (3 ml) in a scintillation vial, then incubated in an isothermal oven at 55°, and allowed to react for ca. 8 h (with monitoring every 2 h by ^{1}H NMR) → N-allylindoline. Y ca. 99% by ^{1}H NMR. A similar yield was obtained after ca. 30 h using DMOF-1-Irdcppy (DMOF = dabco MOF). Yields of ca. 95% (by NMR) were also obtained for N-allylation of 4-phenylpiperidine and di-*n*-hexylamine. Although reactions may be carried out under air, the recyclability of the catalyst was improved by exclusion of air and moisture. The catalysts were prepared by treatment of BMOF-1-dcppy or DMOF-1-dcppy with $[Ir(cod)(OMe)]_2$, resulting in cyclometalation of ca. 32% and 20% of the ligand struts, respectively. F.e.s. P.V. Dau, S.M. Cohen, Chem. Commun. *2013*, *49* (55), 6128-30 [DOI: 10.1039/c3cc42119h].

(4,4′-Dimethoxy-2,2′-bipyridyl)(pentamethylcyclopentadienyl)iridium(III) sulfate *[Ir(III)]*
s. under Laccase

Via intermediates *v.i.*
1-Thiocarbamylindoles from α-(*o*-isocyanoaryl)ketones ○
via *o*-isothiocyanato-β-styryl carbonates

143. 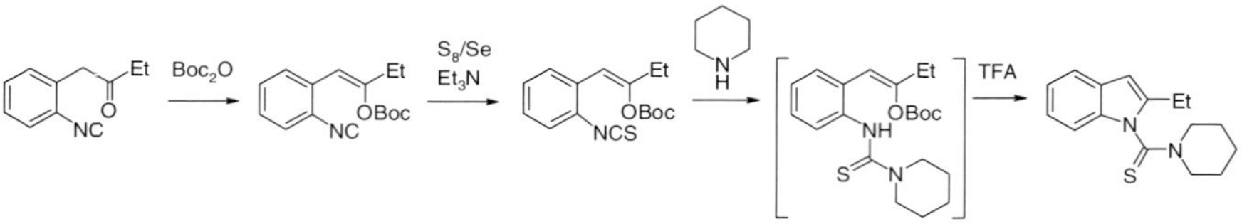

under mild conditions. A soln. of 1-(2-isocyanophenyl)butan-2-one (2.8 mmol), $(Boc)_2O$ (1 eq.) and DMAP (0.3 eq.) in pyridine (1 ml) stirred at room temp. for ca. 15 min (TLC), quenched with water, extracted with ethyl acetate, and purified chromatographically → *tert*-butyl (1Z)-1-(2-isocyanophenyl)but-1-en-2-yl carbonate (Y 63%), a portion of which dissolved in THF and stirred with triethylamine, S_8 and catalytic Se at room temp. (method cf. *47*, 483, 483s*48*) → *tert*-butyl (1Z)-1-(2-isothiocyanatophenyl)but-1-en-2-yl carbonate (Y 91%), a portion of which (0.81 mmol) dissolved in methylene chloride (1 ml) at room temp., piperidine (1 eq.) added, the mixture stirred for ca. 5 min (TLC), concentrated *in vacuo*, the resulting crude thiourea stirred with trifluoroacetic acid (1 ml) at room temp. for ca. 10 min (TLC), quenched with satd. aq. $NaHCO_3$, extracted with methylene chloride, and purified chromatographically → 2-ethyl-1*H*-indole-1-carbothiopiperidinamide (Y 90%). In this experimentally simple procedure, a series of *o*-isothiocyanato-β-styryl carbonates (five examples; Y 80-91%), prepared from the corresponding *o*-isocyanobenzyl ketones (Y 59-75%), were converted to thiourea derivs. with cyclic/acyclic sec. amines, and cyclized, under acidic conditions, in a one-pot procedure to afford the title products (ten examples; Y 55-90%). F.e.s. K. Kobayashi, K. Yamane, S. Fukamachi, Helv. Chim. Acta *2013*, *96* (1), 93-8 [DOI: 10.1002/hlca.201200236].

Nitrogen ↾ NC ⇃↾ N

Irradiation s. under Au, TiO_2 and 2,2′-Bipyridylbis(2-phenylpyridinato-N,C2′)iridium(III) hexafluorophosphate ⫻

Sodium methoxide *NaOMe*
N-Formylation with formamide s. *13*, 442s*83* NH → NCHO

Potassium tert-butoxide *KOBu-t*
N-Subst. carboxylic acid amides from aldehydes and azides s. *48*, 346s*83* CHO → CONHR

Cinchona derivs. ←
Organocatalyzed asym. N-alkylation with diazo compds. s. *52*, 164s*83* NH → NR

Silver oxide *Ag_2O*

β-Cyanamido-α,β-ethylenecarboxylic acid esters
from propiolic acid esters and tetrazoles
via stereoselective silver(I)-catalyzed addition of cyanamides

144.

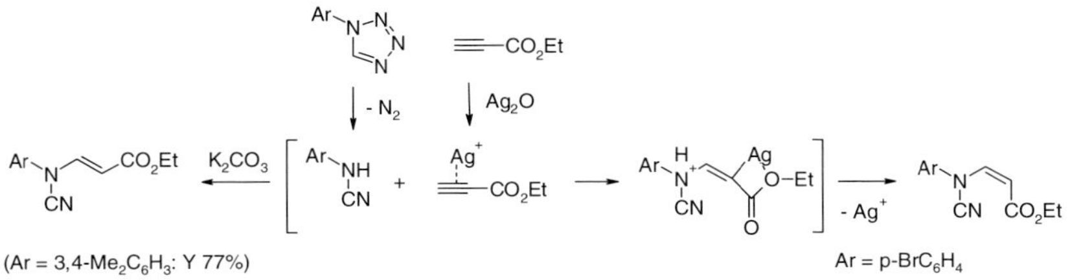

Ar. cyanamides, generated *in situ* from 1-aryltetrazoles, have been shown to undergo stereoselective 1,2-addition to propiolate esters in the presence of Ag_2O to afford (Z)-enamines exclusively. In the presence of added K_2CO_3, however, the reaction affords mainly (E)-enamines via allenoate isomerization of the initially-formed (Z)-isomers. **E:** A mixture of 1-(4-bromophenyl)tetrazole (0.5 mmol), ethyl propiolate (1.2 eq.), Ag_2O (10 mol%) and THF (2 ml) stirred at 100° in a screw-capped vial for 12 h, cooled to room temp., diluted with ethyl acetate, worked up, and purified chromatographically → (Z)-ethyl 3-[N-(4-bromophenyl)cyanamido]acrylate. Y 83%. This experimentally simple method uses readily available electron-diverse 1-aryltetrazoles as a relatively non-toxic source of cyanamide, affording (Z)- (twenty-four examples; Y 70-95%) or, with added K_2CO_3, (E)-enamines (twenty-four examples; Y 63-93%). A sample of N-phenylcyanamide (prepared by treatment of 1-phenyltetrazole with NaOH) afforded the same products on treatment with Ag_2O and Ag_2O/K_2CO_3, respectively. F.e., optimization and substrate prepn. s. K. Luo, L. Meng, Y. Zhang, X. Zhang, L. Wang, Adv. Synth. Catal. *2013*, *355* (4), 765-80 [DOI: 10.1002/adsc.201200868].

Copper(II) acetate s. under Dichloro(p-cymene)ruthenium(II) dimer *$Cu(OAc)_2$*

Copper(I) bromide/6,6′-di-tert-butyl-2,2′-bipyridyl/9-azabicyclo[3.3.1]nonan-3-one N-oxyl/ oxygen ←

Ar. N-benzylaldimines from two prim. benzylamine molecules s. *83*, 340 ArCH═NCH$_2$Ar

Copper(I) iodide/triethylamine/cesium carbonate *$CuI/Et_3N/Cs_2CO_3$*

4-α-Sulfonylimidazoles
from 2-acetyleneamines, terminal acetylene derivs. and sulfonic acid azides
via one-pot two-step 3-component ring closure

145.

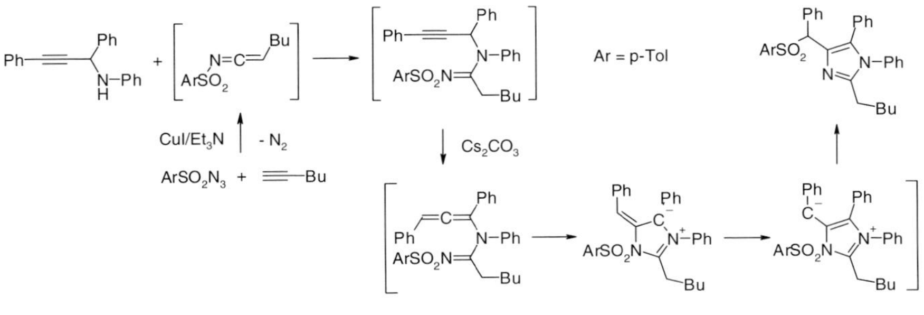

Triethylamine (1.2 eq.) added to a mixture of CuI (10 mol%), N,1,3-triphenylprop-3-ynylamine (1 eq.), 1-hexyne (1.1 eq.) and *p*-toluenesulfonyl azide [*caution*: potentially shock sensitive] (1.2 eq.) in acetonitrile (10 ml) under N_2, the mixture stirred at 25° for 6 h, Cs_2CO_3 (0.5 eq.) added, stirring continued at 80° for 4 h, concentrated *in vacuo,* and purified chromatographically

→ 2-pentyl-1,5-diphenyl-4-(α-tosylbenzyl)-1*H*-imidazole. Y 66%. In this efficient and scalable (to 3 mmol) approach to tetrasubst. imidazoles, formation of the initial 3-component product, an N-α-iminopropargylamine, and its subsequent base-induced cyclization, were optimized separately. The reaction generated N-sulfonylketenimines from electron-diverse terminal aryl-or alkyl-acetylenes and aryl- or alkyl-sulfonyl azides which were trapped with sec. N-alkyl- or -aryl-1,3-diaryl-propargylamines; treatment with base then effected 6π-electrocyclization and 1,3-sulfonyl shift to afford the products (twenty examples; Y 49-83%; electron-rich 4-methylphenylacetylene gave 39%). F.e.s. Z. Jiang, P. Lu, Y. Wang, Org. Lett. *2012, 14* (24), 6266-9 [DOI: 10.1021/ol303023y].

Copper(II) bromide *$CuBr_2$*
Quinoxalines from *o*-diamines and 1-nitroethylene derivs. ○

146.

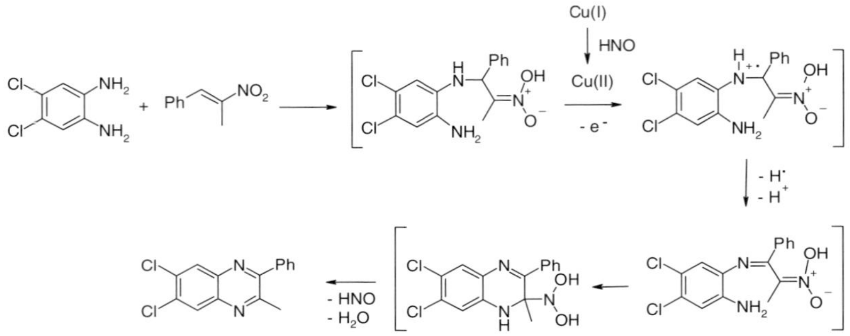

via copper(II)-catalyzed ring closure in one pot. A mixture of 2-nitro-1-phenylprop-1-ene (0.25 mmol), 4,5-dichloro-1,2-phenylenediamine (1 eq.), $CuBr_2$ (10 mol%) and ethanol (2 ml) stirred at 110° for 4 h, cooled to room temp., diluted with ethyl acetate, filtered, and purified chromatographically → 6,7-dichloro-2-methyl-3-phenylquinoxaline. Y 93%. This novel cyclization required no additional base and was successful for reaction of electron-diverse 1-aryl-2-nitroprop-1-enes with *o*-phenylenediamines (seventeen examples; Y 35-93%), with lowest yields obtained for a series of *o*-subst. aryl derivs., and the des-methyl substrate, β-nitrostyrene, only reacting in DMSO (Y 40%). *p*-Subst. *o*-phenylenediamines afforded mixtures of regioisomers with the 2-aryl-6-subst. quinoxalines predominating [just!] (1.1:1 to 2.1:1; four examples; Y 53-86%). The reaction failed with a 1-furylnitropropene and also with cyclohexane-1,2-diamine and 1,2-diaminoanthracene-9,10-dione. F.e. and optimization s. Y. Chen, K. Li, M. Zhao, Y. Li, B. Chen, Tetrahedron Lett. *2013, 54* (13), 1627-30 [DOI: 10.1016/j.tetlet.2012.11.127]; **imidazo[1,2-*a*]pyridines from 2-aminopyridines** and 1-nitroethylene derivs. with $FeCl_3$ (20 mol%) in DMF, incl. application to the synthesis of zolimidine, a useful drug for the treatment of peptic ulcer, s. S. Santra, A.K. Bagdi, A. Majee, A. Hajra, Adv. Synth. Catal. *2013, 355* (6), 1065-70 [DOI: 10.1002/adsc.201201112]; 3-nitro-derivs. with CuBr (10 mol%) and air as oxidant (NC⇃↾H) s. R.-L. Yan, H. Yan, C. Ma, Z.-Y. Ren, X.-A. Gao, G.-S. Huang, Y.-M. Liang, J. Org. Chem. *2012, 77* (4), 2024-8 [DOI: 10.1021/jo202447p].

Silver hexafluoroantimonate s. under Dichloro(p-cymene)ruthenium(II) dimer *$AgSbF_6$*

Gold nanoparticles/titanium dioxide/oxygen/irradiation *$Au/TiO_2/O_2$/⫻*
Aldimines from two prim. amine molecules s. *38,* 375s*83* $2\ RCH_2NH_2 \rightarrow RCH{=}NCH_2R$

Methanol s. under HCl *MeOH*
2,6-Dimethoxybenzoquinone s. under Shvo's ruthenium hydride complex *$2,6\text{-}(MeO)_2BQ$*

Azodiisobutyronitrile/oxygen *$AIBN/O_2$*
Aldimines from two prim. amine molecules
update s. *38,* 375s*82*; *metal-free* conversion with AIBN as initiator under aerobic oxidation, also α-cyanation of tert. dimethylamines with trimethylsilyl cyanide (cf. *47,* 715s*81*), s. L. Liu, Z. Wang, X. Fu, C.-H. Yan, Org. Lett. *2012, 14* (22), 5692-5 [DOI: 10.1021/ol302708r]; gram-scale procedure

under aerobic photocatalysis in water with TiO_2 nanoparticles under UV-irradiation s. N. Li, X. Lang, W. Ma, H. Ji, C. Chen, J. Zhao, Chem. Commun. *2013, 49* (44), 5034-6 [DOI: 10.1039/c3cc41407h]; *under visible-light photoredox catalysis* with rutile TiO_2-supported Au nanoparticles cf. S.-i. Naya, K. Kimura, H. Tada, ACS Catal. *2013, 3* (1), 10-3 [DOI: 10.1021/cs300682d].

1,3-Diisopropylimidazolium bromide s. under Dihydridotetrakis(triphenylphosphine)-ruthenium(II) ←

9-Azabicyclo[3.3.1]nonan-3-one N-oxyl s. under Copper(I) bromide ←

(S)-Proline *(S)-Pro-OH*
Transamidation CONR → CONR′
update s. *28,* 424s*82*; general, solvent-free conversion with (S)-proline as catalyst s. S.N. Rao, D.C. Mohan, S. Adimurthy, Org. Lett. *2013, 15* (7), 1496-9 [DOI: 10.1021/ol4002625].

Potassium ethyl xanthate *$EtOCS_2K$*
2-Acylimidazoles from two azidomethyl ketone molecules ○

147.
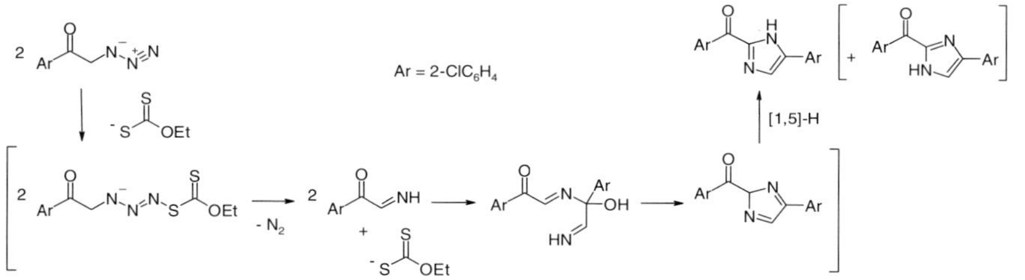

A mixture of α-azido-2-chloroacetophenone (1 mmol) and K-ethyl xanthate (0.5 eq.) in isopropanol (8 ml) stirred at 70° for 6 h → 2-(2-chlorobenzoyl)-5-(2-chlorophenyl)imidazole. Y 75% (as a 3.7:1 mixture with its 4-aryl tautomer). This novel, efficient and experimentally simple procedure generated ar. glyoxal imine intermediates which cyclized to afford mixtures of 2-acyl-5(4)-subst. imidazoles (fifteen examples; Y 55-82%; 5:4-subst. 1:1 to 6.7:1). The reaction failed with α-azido derivs. of electron-poor 4-nitroacetophenone or ethyl acetoacetate. The proposed mechanism was supported by cross-over experiments which produced four pairs of isomers from two α-azido-acetophenones. F.e. and optimization s. J. Chen, W. Chen, Y. Yu, G. Zhang, Tetrahedron Lett. *2013, 54* (12), 1572-5 [DOI: 10.1016/j.tetlet.2013.01.042].

Phenylsilane s. under Ph_3P *$PhSiH_3$*
Titanium dioxide s.a. under Au *TiO_2*

Titanium dioxide/oxygen/irradiation *$TiO_2/O_2/h\nu$*
Aldimines from two prim. amine molecules s. *38,* 375s*83* $2\,RCH_2NH_2 \rightarrow RCH{=}NCH_2R$

Triphenylphosphine/phenylsilane *$Ph_3P/PhSiH_3$*
N-Subst. carboxylic acid amides from azides $RN_3 \rightarrow RNHCOR'$
and carboxylic acids with Bu_3P/PhSeSePh cf. *48,* 346s*50*; under P(III)/P(V) redox catalysis by a traceless Staudinger ligation using triphenylphosphine as catalyst and phenylsilane as reductant for preparing arylcarboxamides, imides, lactams and dipeptides s. A.D. Kosal, E.E. Wilson, B.L. Ashfeld, Angew. Chem., Int. Ed. *2012, 51* (48), 12036-40 [DOI: 10.1002/anie.201206533]; with a diaryl or dialkyl chlorophosphite as dual acid/azide activator in the presence of base cf. A.D. Kosal, E.E. Wilson, B.L. Ashfeld, Chem. Eur. J. *2012, 18* (45), 14444-53 [DOI: 10.1002/chem.201201773]; **from aldehydes and azides** with K-*tert*-butoxide in DMF for a simple preparation of N-subst. arylcarboxamides s. S.S. Kulkarni, X. Hu, R. Manetsch, Chem. Commun. *2013, 49* (12), 1193-5 [DOI: 10.1039/c2cc37289d]; **from prim. alcohols and azides** under ruthenium catalysis with $RuH_2(PPh_3)_4$ and added 1,3-diisopropylimidazolium bromide/NaH for a clean procedure with elimination of N_2 and H_2 s. Z. Fu, J. Lee, B. Kang, S.H. Hong, Org. Lett.

2012, 14 (23), 6028-31 [DOI: 10.1021/ol302915g]; **N,N-disubst. formamides** from aldehydes and azides via Schmidt reaction with TfOH as promoter and aq. KOH as base (cf. *11*, 160), notably for preparing tosylamino-functionalized N-arylformamides and N,N-dialkylformamides, s. P. Gu, J. Sun, X.-Y. Kang, M. Yi, X.-Q. Li, P. Xue, R. Li, Org. Lett. *2013, 15* (5), 1124-7 [DOI: 10.1021/ol400213f].

Bismuth(III) triflate *$Bi(OTf)_3$*
1-Aryl-4-(arylamino)indoles ○
from 4-alkoxy-2-(alk-1-ynyl)-N-sulfonylcyclohexa-2,5-dienimines and two prim. ar. amine molecules s. *83*, 168

Oxygen (or air) s. under Copper(I) bromide, Au, Azodiisobutyronitrile, TiO_2 and Shvo's ruthenium hydride complex *O_2*

Trifluoromethanesulfonic acid/potassium hydroxide *TfOH/KOH*
N,N-Disubst. formamides from aldehydes and azides RCHO → R(R′)NCHO
via Schmidt reaction s. *11*, 160s*83*

Potassium peroxymonosulfate s. under Palladium(II) nitrate *$KHSO_5$*

Hydrogen chloride/methanol *HCl/MeOH*
Pyrimidines from nitriles and vinylogous amidinium salts ○

148. 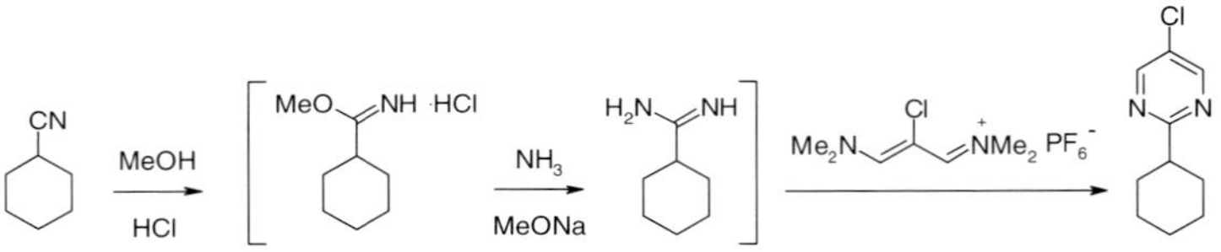

A soln. of cyclohexanecarbonitrile (25 mmol) in methanol (3.04 ml; 3 eq.) added over 30 min to HCl (3 eq.) in dioxane (4 *M*) pre-cooled to 0°, while maintaining the temp. below 10°, the mixture warmed to 22°, stirred for 6-8 h, cooled to 5°, treated with NaOMe in methanol (25 wt%; 11.4 ml; 2 eq.) while maintaining the temp. below 15°, stirred for 10-15 min at 15°, ammonia in methanol (7 *N*; 5.36 ml; 1.5 eq.) added, the mixture stirred for 2 h at room temp., then concentrated under reduced pressure at 50° (~130 Torr) to a volume of ca. 25 ml (to remove excess ammonia), the resulting heterogeneous mixture (containing the crude amidine) cooled to 20°, treated with NaOMe in methanol (25 wt%; 14.3 ml; 2.5 eq.), stirred for 30 min, (Z)-N-[2-chloro-3-(dimethylamino)-allylidene]-N-methylmethanaminium hexafluorophosphate (90% w/w; 0.95 eq.) added in two portions at room temp. over ca. 30 min, the mixture stirred for 3 h at room temp., then concentrated under reduced pressure at 60° to a volume of ca. 20 ml, diluted with 2-methyltetrahydrofuran (40.7 ml), concentrated again to ca. 20 ml under reduced pressure at 60°, diluted with 2-methyl-tetrahydrofuran (41 ml), re-cooled to <20°, water (20.4 ml) added, the mixture stirred for 5 min, the layers separated, the aq. layer extracted with 2-methyl-tetrahydrofuran (20.4 ml), the combined organic layers washed with 30% aq. NaOH (3 eq.), the organic layer dried ($MgSO_4$), concentrated under reduced pressure, and the residue purified by flash chromatography → 5-chloro-2-cyclo-hexylpyrimidine. Y 86%. This is a practical, robust, one-step process, avoiding the need to isolate intermediate iminoesters and amidines. It is applicable to the formation of a variety of **2,5-disubst. pyrimidines** bearing an assortment of functional groups in good yields (66-85%; eight further examples) using inexpensive reagents. F.e.s. R.P. Frutos, X. Wei, N.D. Patel, T.G. Tampone, J.A. Mulder, C.A. Busacca, C.H. Senanayake, J. Org. Chem. *2013*, *78* (11), 5800-3 [DOI: 10.1021/jo400720p].

Chloro(dipyrrinato)iron(II) complex *[Fe(II)]*
N-Protected cyclic sec. amines from azides s. *83*, 192

Iron(III) chloride *$FeCl_3$*
Imidazo[1,2-*a*]pyridines from 2-aminopyridines and 1-nitroethylene derivs. s. *83*, 146

Bis(salicylideniminato-3-propyl)methylaminocobalt(II) s. under Shvo's ruthenium hydride complex *[Co(II)]*

[2-(Di-tert-butylphosphinomethylene)-6-(2-pyridyl)-1,2-dihydropyridinato]carbonyl-(hydrido)ruthenium(II) *[Ru(II)]*
Aldimines from two nitrile molecules $2 RCN \rightarrow RCH{=}N\text{-}CH_2R$
via ruthenium(II)-catalyzed hydrogenation under mild, neutral conditions

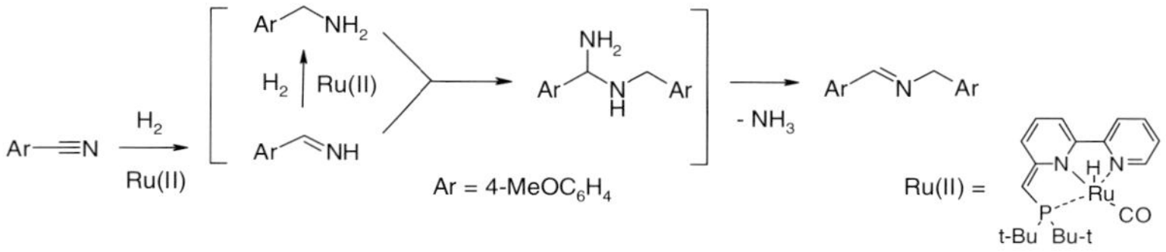

149.

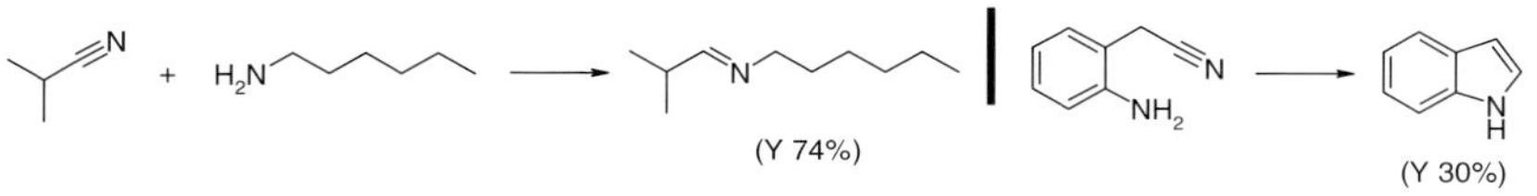

A mixture of ruthenium(II) PNN catalyst (0.6 mol%), 4-methoxybenzonitrile (2 mmol) and THF (2 ml) prepared under N_2 in a Fischer-Porter tube, stirred at 70° under H_2 (4 bar) for 4 h, cooled to room temp., and excess H_2 vented carefully → N-(4-methoxybenzylidene)-1-(4-methoxyphenyl)-methanamine. Y 92% (GC). The use of the PNN 'pincer complex' and relatively low temp. and pressure were key to the selectivity of this reductive coupling, in which hydrogenation of ar. nitriles afforded imine and amine intermediates, which combined with elimination of NH_3 to afford aldimines (three examples; Y 81-92%); at 110° and 10 bar H_2 benzonitrile gave a mixture of benzylamine and dibenzylamine. In the presence of an added prim. aliphatic amine, hydrogenation of (het)ar. or alkyl nitriles afforded cross-imines as major products (ten examples; Y 50%, 69-91%), with 2-cyanopyridine affording only 15% yield presumed due to chelation with the catalyst, while hydrogenation of 2-aminobenzonitrile resulted in intramolecular reaction to afford indole in modest yield (30%). F.e.s. D. Srimani, M. Feller, Y. Ben-David, D. Milstein, Chem. Commun. *2012, 48* (97), 11853-5 [DOI: 10.1039/c2cc36639h].

Dihydridotetrakis(triphenylphosphine)ruthenium(II)/1,3-diisopropylimidazolium bromide/ sodium hydride ←
N-Subst. carboxylic acid amides from prim. alcohols and azides $CH_2OH \rightarrow CONHR$
s. *48*, 346s*83*

Rhodium(II) acetate *$Rh_2(OAc)_4$*
N-Alkylation with diazo compds. NH → NR
cf. *52*, 164s*60*; selective N-alkylation of DNA and RNA in an aq. MES [2-(N-morpholino)-ethanesulfonic acid] buffer s. K. Tishinov, K. Schmidt, D. Häussinger, D.G. Gillingham, Angew. Chem., Int. Ed. *2012, 51* (48), 12000-4 [DOI: 10.1002/anie.201205201]; **asym. N-alkylation** of anilines with ethyl phenyldiazoacetate *under organocatalysis* with a cinchona alkaloid s. H. Saito, D. Morita, T. Uchiyama, M. Miyake, S. Miyairi, Tetrahedron Lett. *2012, 53* (49), 6662-4 [DOI: 10.1016/j.tetlet.2012.09.087].

Shvo's ruthenium hydride complex/bis(salicylideniminato-3-propyl)methylaminocobalt(II)/ 2,6-dimethoxybenzoquinone/air ←

2-Arylbenzoxazoles from *o*-aminophenols and benzylamines via biomimetic aerobic ring closure ○

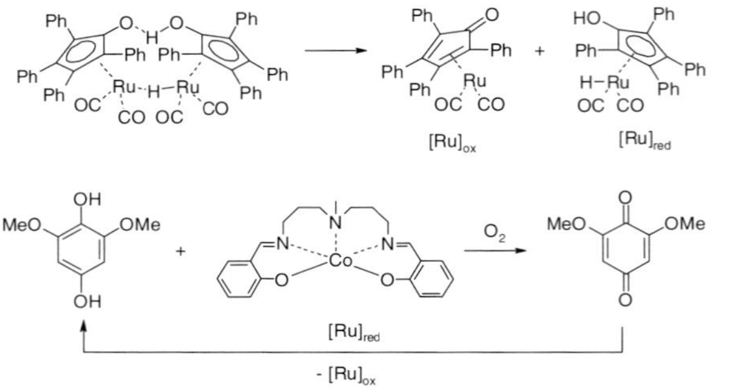

150.

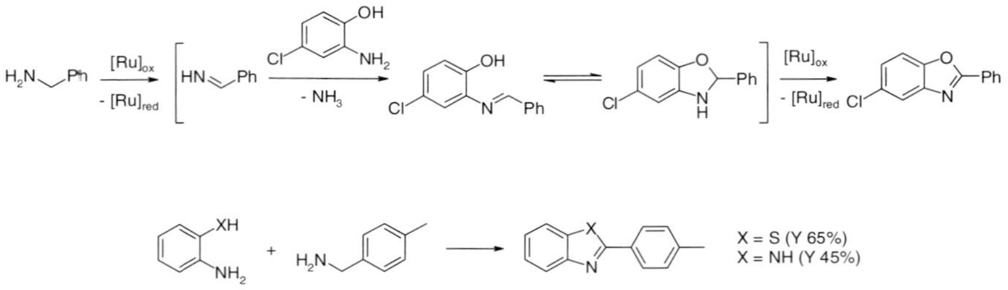

under mild conditions. 1-Hydroxytetraphenylcyclopentadienyl(tetraphenyl-2,4-cyclopentadien-1-one)-μ-hydridotetracarbonyldiruthenium(II) [Shvo's complex] (1 mol%), cobalt(II) salen complex (2 mol%) and 2,6-dimethoxy-1,4-benzoquinone (20 mol%) added to a soln. of 2-amino-4-chlorophenol (0.728 mmol) and benzylamine (1 eq.) in chlorobenzene (1 ml), the mixture stirred under air (balloon) at 120° for 24 h, the resulting suspension concentrated, and purified chromatographically → 5-chloro-2-phenylbenzoxazole. Y 90%. Electron-diverse *o*-aminophenols cyclized with benzylamine and hetero analogs in this biomimetic oxidation system, via elimination of ammonia (sixteen examples; Y 56-92%), with electron-deficient 4-fluorobenzylamine giving 40% yield, and 2-amino-4-fluorophenol only a trace of product. Aliphatic amines (*n*-octylamine and 2-phenylethylamine) gave low yields (trace to 6%) presumed due to rapid isomerization of the imine intermediates to enamines. Interestingly, benzyl alcohol was a poor substrate in this oxidation system (Y 11% after 24 h) and, in competitive tests, benzylamine was shown to oxidize more rapidly than benzyl alcohol. More highly electron-deficient tetrafluorobenzoquinone was an effective oxidant in a stoichiometric reaction (non-aerobic conditions) but was inefficient in the catalytic system due to slow re-oxidation of the corresponding hydroquinone. Mechanistically, the reaction is presumed to be initiated by dissociation of the ruthenium catalyst to afford an active oxidation catalyst that promotes two oxidation steps and is regenerated by the quinone (which is in turn regenerated by the cobalt catalyst in the presence of air). The reaction was extended to the prepn. of 2-(*p*-tolyl) derivs. of benzimidazole (Y 45%) and benzothiazole (Y 65%), but better yields (77% and 89% respectively) were obtained using 4-methylbenzaldehyde in place of the benzylamine. F.e. and optimization s. Y. Endo, J.-E. Bäckvall, Chem. Eur. J. *2012*, *18* (43), 13609-13 [DOI: 10.1002/chem.201202187].

Dichloro(p-cymene)ruthenium(II) dimer/silver hexafluoroantimonate/copper(II) acetate ←
Ruthenium-catalyzed *o*-sulfonylamination of aryl ketones with sulfonic acid azides H → $NHSO_2R$

151. 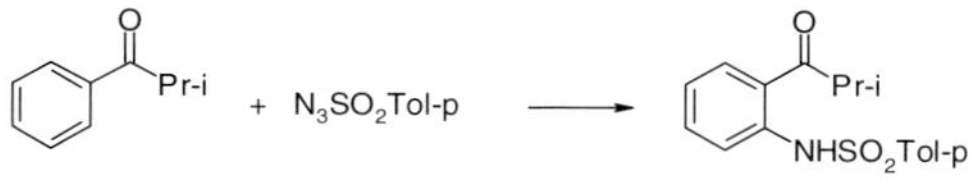

The first ruthenium(II)-catalyzed directed C-H amidation of weakly coordinating aryl ketones with azides is reported, thus forming *o*-aminoketones while avoiding the need for an external oxidant. **E:** A mixture of isobutyrophenone (0.2 mmol), *p*-toluenesulfonyl azide (2 eq.), $[Ru(p\text{-cymene})Cl_2]_2$ (2.5 mol%), $AgSbF_6$ (10 mol%) and $Cu(OAc)_2$ (30 mol%) in DCE (1 ml) allowed to react at 80° under argon for 21 h (TLC), the mixture concentrated, and purified by flash chromatography on silica gel → N-(2-isobutyrylphenyl)-*p*-toluenesulfonamide. Y 74% [48% with NaOAc in place of $Cu(OAc)_2$]. Twenty-three further examples gave yields of 30%, 32%, 46-88% from electron-diverse (cyclo)alkyl aryl or sym. diaryl ketones and *p*-toluene-, benzene-, *p*-nitrobenzene-, benzyl- or *n*-butyl-sulfonyl azides. The products may be used for quinoline or oxindole synthesis or for a novel route to the antiepileptic drug, oxcarbazepine [Trileptal], from acetophenone. F.e. and optimization s. Q.-Z. Zheng, Y.-F. Liang, C. Qin, N. Jiao, Chem. Commun. *2013*, *49* (50), 5654-6 [DOI: 10.1039/c3cc42613k]; rhodium-catalyzed 2-sulfonylamination (cf. *82*, 160) of N-(pyrimid-2-yl)indoles with sulfonic acid azides using $[(MeCN)_3Cp^*Rh](SbF_6)_2$ in DCE containing 10 eq. water at 80° under air s. J. Shi, B. Zhou, Y. Yang, Y. Li, Org. Biomol. Chem. *2012*, *10* (45), 8953-5 [DOI: 10.1039/c2ob26767e].

Dirhodium tetrakis(N-1,8-naphthoyl-(S)-tert-leucinate) *[Rh(II)]**
1-Sulfonyl-Δ⁴-2-imidazolones or 2-imino-3-sulfonyl-Δ⁴-thiazolines from 1-sulfonyl-1,2,3-triazoles and isocyanates or isothiocyanates via rhodium(II)-catalyzed denitrogenation to aza-vinylcarbenes ○

152. 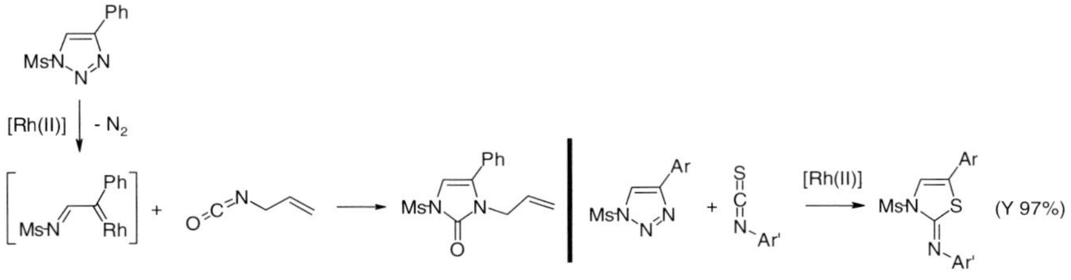

Allyl isocyanate (1.2 eq.) and anh. chloroform (5 ml) added to a mixture of 1-methanesulfonyl-4-phenyl-1*H*-1,2,3-triazole (1 mmol) and $Rh_2(S\text{-}NTTL)_4$ under an inert atmosphere, the mixture sealed, stirred at room temp. for 1-6 h (TLC), and purified chromatographically → 3-allyl-1-methanesulfonyl-4-phenyl-1*H*-imidazol-2(3*H*)-one. Y 82%. This novel method generated azavinyl rhodium-carbenes from electron-diverse 4-(het)aryl-N-mesyl-1,2,3-triazoles (N-tosyl-triazoles were poorer substrates) which were trapped with electron-diverse ar. or *prim*- or *sec*-alkyl isocyanates to afford imidazolones (eighteen examples; Y 80-97%; lower yields with 4-trifluoromethyl- (69%) or 4-nitrophenyl isocyanates (53%)). Notably, allyl isocyanate favored imidazolone formation over cyclopropanation of the double bond. The reactive intermediates could also be trapped with isothiocyanates to afford the iminothiazoline derivs. (fifteen examples; Y 45-97%). F.e., optimization, and methods for removal of the mesyl protecting group s. S. Chuprakov, S.W. Kwok, V.V. Fokin, J. Am. Chem. Soc. *2013*, *135* (12), 4652-5 [DOI: 10.1021/ja400350c]; **3-mesyl-Δ⁴-oxazolines** from aldehydes **or imidazoles** from aldimines (NC⇃↾S) s. M. Zibinsky, V.V. Fokin, Angew. Chem., Int. Ed. *2013*, *52* (5), 1507-10 [DOI: 10.1002/anie.201206388].

Tris(acetonitrile)(pentamethylcyclopentadienyl)rhodium(III) bis(hexafluoroantimonate) $[(MeCN)_3Cp^*Rh](SbF_6)_2$
Rhodium-catalyzed 2-sulfonylamination of N-(pyrimid-2-yl)indoles H → $NHSO_2R$
with sulfonic acid azides s. *83*, 151; *82*, 160s*83*

Palladium(II) nitrate/potassium peroxymonosulfate/lithium carbonate ←
4*H*-3,1-Benzoxazin-4-ones from *o*-acetyleneazides ○
via palladium-catalyzed oxidative rearrangement

153. 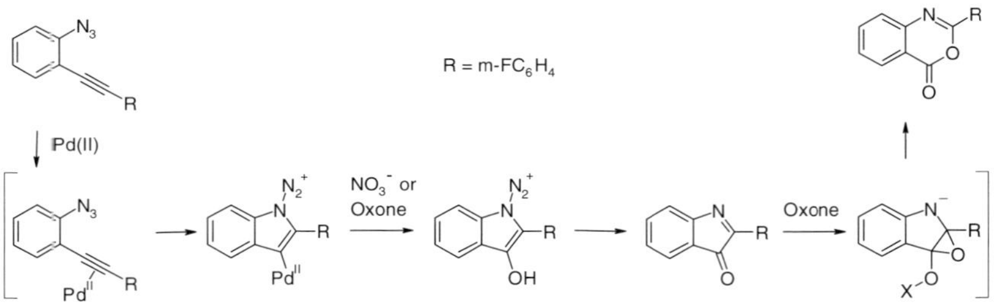

A novel route to 4*H*-3,1-benzoxazin-4-ones via an unusual C-C triple bond cleavage, believed to involve a palladium-catalyzed aminopalladation-oxidation-Baeyer-Villiger sequence, has been uncovered while investigating 3-fluoroindole synthesis. **E:** Startg. *o*-acetyleneazide (0.2 mmol), $Pd(NO_3)_2 \cdot 2H_2O$ (10 mol%), Oxone (1 eq.), Li_2CO_3 (2 eq.) and dry acetonitrile (2 ml) added to a brown vial, the mixture stirred in the dark at 60° for 68 h (TLC), after cooling, the mixture filtered through a pad of Celite, the filtrate evaporated, and the residue subjected to column chromatography → product. Y 76%. The method is applicable to substrates bearing electron-rich or -poor aromatic groups on the terminal triple bond (sixteen further examples; Y 35%, 46-73%, lowest being for the 6-nitro-2-phenyl-deriv.) or *tert*-butyl (Y 74%) or cyclopropyl (Y 35%); it failed with a linear alkyl group at the terminus, however. Functional groups such as ether, ester, halogen or amide are tolerated. NFSI could be used as oxidant in place of Oxone with slightly lower yields (60-69%; four examples). F.e. and mechanistic studies s. Q. Liu, P. Chen, G. Liu, ACS Catal. *2013*, *3* (2), 178-81 [DOI: 10.1021/cs300733s].

2,2′-Bipyridylbis(2-phenylpyridinato-N,C2′)iridium(III) hexafluorophosphate/ air/irradiation ←
Ar. N-benzylaldimines from two prim. benzylamine molecules s. *83*, 5 $ArCH{=}NCH_2Ar$

Halogen ↑ NC ↓↑ Hal

Without additional reagents *w.a.r.*
N-Alkylation NH → NR
update s. *20,* 340s*82*; hydrogen bond-assisted N-alkylation **with alkyl fluorides** in water s. P.A. Champagne, J. Pomarole, M.-È. Thérien, Y. Benhassine, S. Beaulieu, C.Y. Legault, J.-F. Paquin, Org. Lett. *2013, 15* (9), 2210-3 [DOI: 10.1021/ol400765a].

Irradiation s. under CuI ///
Microwaves s. under Copper(I) iodide [\\\\]

Sodium hydride *NaH*
5-α-Fluoro-2-oxazolidones from 2-hydroxy-1,3-difluorides and isocyanates s. *83*, 193 ○

Sodium hydride or n-Butyllithium *NaH or BuLi*

3-Fluoropyrazoles from 2-methylene-1,1,1-trifluorides and hydrazines via 2-(difluoromethylene)-N′-tosylhydrazines

154.

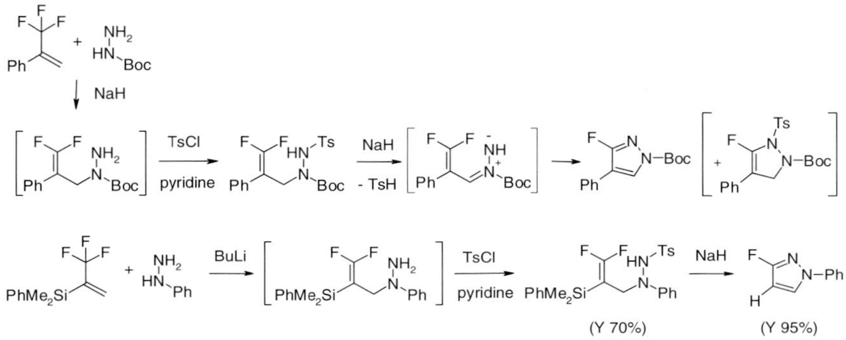

A novel ring closure with 2-trifluoromethyl-1-alkenes via sequential regioselective S_N2'-type reaction and intramolecular N-substitution is reported, using hydrazines as dinucleophiles. **E: 4-Aryl-1-carbo-*tert*-butoxy-3-fluoropyrazoles**. NaH (1.8 eq.) added to a THF (2 ml) soln. of α-trifluoromethylstyrene (1 mmol) and Boc-hydrazine (1.8 eq.) at 0°, stirred for 1 h at 0°, quenched with phosphate buffer (pH 7), extracted with ethyl acetate, the extracts washed with brine, dried (Na_2SO_4), solvent removed under reduced pressure, the residue transferred to another flask with ethyl acetate, solvent removed *in vacuo*, the residue azeotropically dried with pyridine under reduced pressure, the crude ethylenehydrazine taken up in pyridine (2 ml), tosyl chloride (2.7 mmol) added at room temp., stirred for 2.5 h, quenched with phosphate buffer (pH 7), filtered, extracted with ethyl acetate, the extracts washed with brine, dried (Na_2SO_4), solvent removed under reduced pressure, the residue purified by chromatography on silica gel, solvent removed under reduced pressure, hexane added to the viscous residue, and the obtained solid washed with hexane three times → intermediate tosylhydrazide (Y 89% overall for both steps), a portion of which (0.456 mmol) in DMF (0.9 ml) treated at room temp. with NaH (2.2 eq.), the mixture stirred for 5 h, quenched with phosphate buffer (pH 7), extracted with ethyl acetate, the extracts washed with brine, dried (Na_2SO_4), solvent removed under reduced pressure, and the residue purified by chromatography on silica gel → product (Y 86%, plus 12% fluoropyrazoline). Three further examples gave the intermediate N-Boc-N′-tosylhydrazines in 69-88% yield and the fluoropyrazoles in 55-85% yield. Reaction with *aryl*hydrazines using *n*-BuLi as base was also successful, giving the corresponding ethylenehydrazines in 30-73% yield, which underwent ring closure to **1,4-diaryl-3-fluoropyrazoles** in 96-98% yield. There was no reaction with 3,3,3-trifluoropropene but 4-unsubst. 3-fluoropyrazoles could be accessed using a 2-silylated propene (two examples; 1st step Y 70%, 73%; 2nd step Y 70%, 95%). Since the 1-Boc-2-tosylpyrazoline by-products could not be converted to the pyrazoles with NaH it is believed cyclization proceeds via azomethine imines rather than by the initially postulated intramolecular S_NV reaction, nitrene intermediates also being ruled out by experimental evidence. F.e.s. K. Fuchibe, M. Takahashi, J. Ichikawa, Angew. Chem., Int. Ed. *2012*, *51* (48), 12059-62 [DOI: 10.1002/anie.201206946].

Sodium azide/copper(II) sulfate-on-aluminum oxide *$NaN_3/CuSO_4$-Al_2O_3*

1,4-Disubst. 1,2,3-triazoles from terminal acetylene derivs.

and halides via *in situ*-generated alkyl azides, update s. *68*, 184s*82*; solvent-free, eco-friendly and safe formation of N-alkyl- and N-benzyl derivs. over recyclable copper(II) sulfate-on-alumina *by ball-milling* with a simple, chromatography-free work-up, also N-aryl derivs. with arylboronic acids, s. N. Mukherjee, S. Ahammed, S. Bhadra, B.C. Ranu, Green Chem. *2013*, *15* (2), 389-97 [DOI: 10.1039/c2gc36521a]; N-aryl derivs. from arylboronic acids with recyclable montmorillonite KSF-supported CuO nanoparticles cf. S. Mohammed, A.K. Padala, B.A. Dar, B. Singh, B. Sreedhar,

R.A. Vishwakarma, S.B. Bharate, Tetrahedron *2012, 68* (39), 8156-62 [DOI: 10.1016/j.tet.2012.07.080].

Sodium nitrite s. under (2-Dicyclohexylphosphino-2′,4′,6′-triisopropylbiphenyl)-o′-aminobiphenylylpalladium(II) mesylate *$NaNO_2$*

Lithium bromide s. under $FeCl_2$ *LiBr*

Triethylamine *Et_3N*

Alkoximinoacetic acid anilides from prim. ar. amines RON=CHCONHAr

en route to isatins s. *83*, 64

O-(Trimethylsilyl)quinidine/ethyldiisopropylamine ←

N-Tosyl-4-oxazolidones from carboxylic acid chlorides ○

via asym. [3+2]-cycloaddition *en route* to terminal glycols s. *83*, 66

2-Carbomethoxy-3-hydroxyquinoxaline di-N-oxide s. under Copper(I) iodide ←

Copper/lithium hydroxide *Cu/LiOH*

Copper-catalyzed N-arylation s. *62*, 171s*83* NH → NAr

Copper(I) oxide/N,N′-dimethylethylenediamine/cesium carbonate/oxygen ←

2-Arylquinazolines from *o*-bromobenzyl bromides and ar. amidines ○

Copper-catalyzed ring closure in water

155.

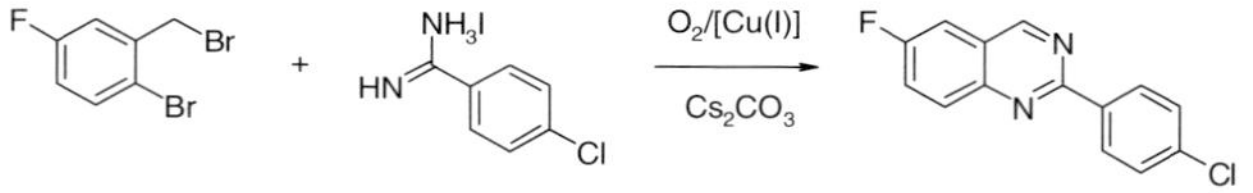

in one pot. Water (2 ml) added by syringe to a mixture of Cu_2O (10 mol%), N,N′-dimethylethylenediamine (20 mol%), Cs_2CO_3 (4 eq.), 2-bromo-5-fluorobenzyl bromide (0.5 mmol) and 4-chlorobenzamidinium iodide (2 eq.) under air, the mixture heated at 100° for 40 h, cooled to room temp., diluted with brine, extracted with methylene chloride, and purified chromatographically → 2-(4-chlorophenyl)-6-fluoroquinazoline. Y 80%. This novel and experimentally simple procedure was successful with electron-diverse substrates, affording 2-arylquinazolines as sole products (thirteen examples; Y 57-85%). Other solvents were found to be less effective. F.e. and optimization s. C.C. Malakar, A. Baskakova, J. Conrad, U. Beifuss, Chem. Eur. J. *2012, 18* (29), 8882-5 [DOI: 10.1002/chem.201200583].

Copper(II) oxide nanoflakes-on-polystyrene/potassium tert-butoxide ←

Copper(II) acetoacetonate/potassium phosphate *$Cu(acac)_2/K_3PO_4$*

Copper-catalyzed N-arylation s. *62*, 171s*83* NH → NAr

Tetrakis(acetonitrile)copper(I) fluoroborate *$[Cu(MeCN)_4][BF_4]$*

1-Sulfonyl-Δ^2-azetines

from α,β-ethylenediazo compds. and N-sulfonyliminoiodinanes

via regioselective copper(I)-catalyzed formal [3+1]-cycloaddition

156.

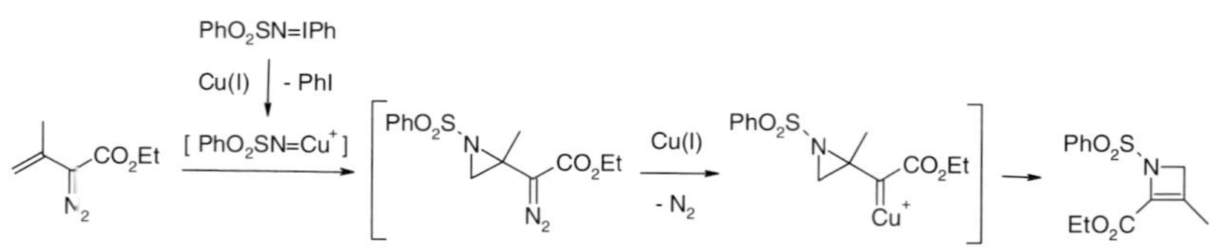

1-Sulfonyl-Δ^2-azetine-2-carboxylic acid esters. $[Cu(MeCN)_4]BF_4$ (5 mol%) added to a soln. of ethyl 2-diazo-3-methylbut-3-enoate (0.5 mmol) and phenyl(phenylsulfonylimino)iodinane (1 eq.) in acetonitrile (5 ml) at room temp., the mixture stirred for 2-3 h (TLC), solvent removed *in vacuo*, and the residue purified by flash chromatography on silica → ethyl 1-benzenesulfonyl-3-methyl-Δ^2-azetine-2-carboxylate. Y 66%. This experimentally simple prepn. of functionalized azetines was successful for 2-diazobut-3-enoates carrying phenyl or alkyl substituents on the double bond and also for a keto-stabilized substrate, 3-diazo-2-oxo-4-methylpent-4-ene (six examples; Y 44-66%), but failed for an ethoxycarbonyl terminated butenoate. From studies directed towards identifying proposed aziridine intermediates a 3-siloxybutenoate substrate afforded an N-silyl-2-diazo-3-oxo-4-sulfonamidobutanoate (Y 75%) under the reaction conditions, while methanol, as solvent, displaced the sulfonamide to afford a 2-diazo-4-methoxy-3-oxobutanoate (Y 73%).

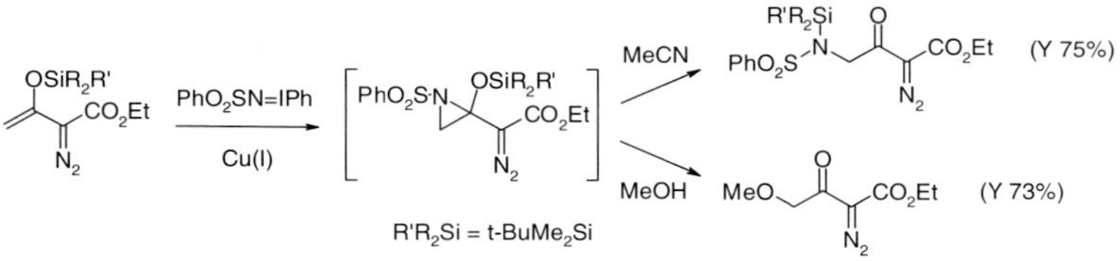

Interestingly, in methanol, ethyl 2-diazobut-3-enoate afforded a 2-methoxy-4-sulfonamidobut-2-enoate (Y 62%) via reaction of an aziridine intermediate with methanol. F.e., also conversion of products to **N-sulfonylazetidine-2-carboxylic acid esters** (Pd-C/H_2; Y 91%, 93%), s. J. Barluenga, L. Riesgo, G. Lonzi, M. Tomás, L.A. López, Chem. Eur. J. *2012*, *18* (30), 9221-4 [DOI: 10.1002/chem.201200998].

Silver acetate s. under Palladium(II) acetate *AgOAc*
Copper(II) sulfate s. under NaN_3 *$CuSO_4$*

Copper(I) bromide/N,N′-dimethylethylenediamine/cesium carbonate *CuBr/DMEDA/Cs_2CO_3*
Copper(I) bromide/4,7-dipyrrolidino-1,10-phenanthroline/polyethylene glycol-400/ potassium hydroxide ←
Copper(I) bromide/manganese(II) fluoride/trans-1,2-diaminocyclohexane/ potassium hydroxide ←
Copper-catalyzed N-arylation s. *62*, 171s*83* NH → NAr

Copper(I) iodide/lithium tert-butoxide/irradiation *CuI/LiOBu-t/hν*
Photoinduced copper-catalyzed N-alkylation NH → NR

157.

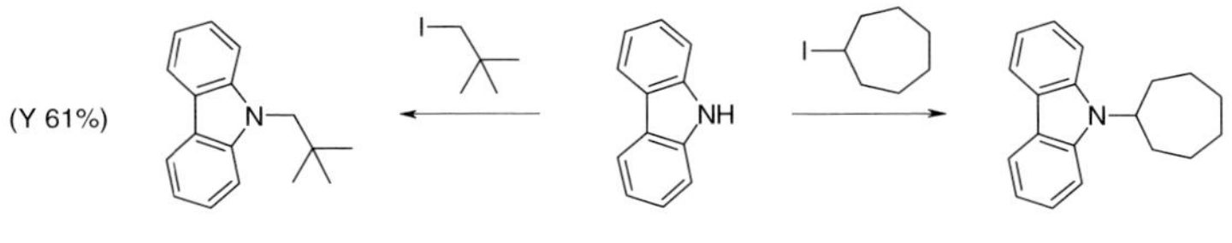

It is now reported that upon photolysis, a range of N-alkylations **of carbazoles** may be achieved with secondary and hindered primary alkyl iodides (or bromides) **under mild conditions** using a simple precatalyst (CuI) in the presence of a Brønsted base, complementing S_N2 reactions of unhindered primary electrophiles. **E:** Carbazole (1 mmol), LiOBu-*t* (1.9 eq.) and CuI (10 mol%) added to an oven-dried quartz test tube fitted with a rubber septum, the test tube evacuated and backfilled with N_2 (3 cycles), acetonitrile (4 ml) and cycloheptyl iodide (1.9 eq.) added in turn via syringe, the mixture stirred for 5 min, cooled in ice, irradiated with a 100 W Hg lamp for 10 h, passed through a long plug of silica gel with hexanes as eluant, the solvent removed, and the residue purified by chromatography on silica gel → product. Y 70% (Y 60% with cycloheptyl bromide at 30°). Ten further examples with cyclic or acyclic sec. iodides and carbazole or its fluoro-, ethyl- or methoxy-subst. derivs. gave yields of 45-80%, while reaction with *trans*-2-ethoxycyclohexyl iodide gave the *trans*-product with excellent diastereoselectivity (>20:1; Y 56%). Reaction with [hindered, prim.] neopentyl iodide gave a yield of 61%. For photoinduced Ullmann N-arylation cf. *82*, 166. F.e.s. A.C. Bissember, R.J. Lundgren, S.E. Creutz, J.C. Peters, G.C. Fu, Angew. Chem., Int. Ed. *2013*, *52* (19), 5129-33 [DOI: 10.1002/anie.201301202]; for an overview of light-mediated copper-catalyzed C-C and C-N bond formation s. M. Majek, A.J. von Wangelin, ibid. *52* (23), 5919-21 [DOI: 10.1002/anie.201301843].

Copper(I) iodide/sodium hydride or potassium hydroxide or potassium phosphate ←
Copper(I) iodide/poly(ethylene glycol)-bridged dicationic ionic liquid/potassium carbonate ←
Copper(I) iodide/bis(imidazolium)-functionalized calix[4]arene bis(α-aminocarboxylates)/ potassium carbonate/microwaves ←
Copper(I) iodide/8-hydroxyquinoline/potassium phosphate ←
Copper(I) iodide/2-carbomethoxy-3-hydroxyquinoxaline di-N-oxide/cesium carbonate ←

Copper-catalyzed N-arylation NH → NAr

update s. *62*, 171s*82*; *in water* under microwave enhancement with CuI and a hydrophobic/ hydrophilic bis(imidazolium)-functionalized calix[4]arene bis(α-aminocarboxylate) as both ligand and phase transfer catalyst using K_2CO_3 as base s. L. Huang, C. Jin, W. Su, Chin. J. Chem. *2012, 30* (10), 2394-400 [DOI: 10.1002/cjoc.201200623]; selective N-arylation of indoles, imidazoles and purines in the presence of aliphatic amines in water with CuBr/4,7-dipyrrolidino-1,10-phenanthroline and added PEG-400 using KOH as base s. J. Engel-Andreasen, B. Shimpukade, T. Ulven, Green Chem. *2013, 15* (2), 336-40 [DOI: 10.1039/c2gc36589h]; of N-heterocyclics in pure water with activated copper powder as catalyst and LiOH as base s. Q. Yang, Y. Wang, D. Lin, M. Zhang, Tetrahedron Lett. *2013, 54* (15), 1994-7 [DOI: 10.1016/j.tetlet.2013.02.004]; **of carboxylic and sulfonic acid amides** in water with CuI/MnF_2 *under bimetal catalysis* with *trans*-1,2-diaminocyclohexane as ligand and KOH as base under mild conditions s. Y.-C. Teo, F.-F. Yong, I.K. Ithnin, S.-H.T. Yio, Z. Lin, Eur. J. Org. Chem. *2013* (28), 515-24 [DOI: 10.1002/ejoc.201201218]; N-arylation in DMF with CuI/2-carbomethoxy-3-hydroxyquinoxaline di-N-oxide and Cs_2CO_3 as base s. D. Chen, K. Yang, H. Xiang, S. Jiang, Tetrahedron Lett. *2012, 53* (52), 7121-4 [DOI: 10.1016/j.tetlet.2012.10.083]; in air under copper catalysis *at the ppm level* with air-stable N,N″-dimethyldiethylenetriamine [DMDETA] as ligand s. P.-F. Larsson, P. Astvik, P.-O. Norrby, Beilstein J. Org. Chem. *2012, 8*, 1909-15 [DOI: 10.3762/bjoc.8.221]; N-arylation of aryl- and diaryl-amines with CuI/K_3PO_4 in DMF and 8-hydroxyquinoline as ligand s. C. Qian, S. Xu, Q. Zong, D. Fang, Chin. J. Chem. *2012, 30* (8), 1881-5 [DOI: 10.1002/cjoc.201200247]; *ligand-free* N-arylation of N-heterocyclics in DMSO with reusable CuI/NaH as an air-stable system s. Q. Yang, Y. Wang, B. Zhang, M. Zhang, Chin. J. Chem. *2012, 30* (10), 2389-93 [DOI: 10.1002/cjoc.201100715]; *heterogeneous* N-arylation with ar. bromides using CuO nanoflakes-on-polystyrene beads in DMSO with KOBu-*t* as base s. Y. Heo, D. Hyun, M.R. Kumar, H.M. Jung, S. Lee, Tetrahedron Lett. *2012, 53* (49), 6657-61 [DOI: 10.1016/j.tetlet.2012.09.097]; **of N-subst. carboxylic acid amides** with CuBr/DMEDA in toluene and Cs_2CO_3 as base s. M. Wang, H. Yu, X. You, J. Wu, Z. Shang, Chin. J. Chem. *2012, 30* (10), 2356-62 [DOI: 10.1002/cjoc.201200701]; racemization-free **selective N-arylation** of the amide nitrogen of chiral α-aminocarboxylic acid anilides in toluene with CuI/K_2CO_3 s. J. Dong, Y. Wang, Q. Xiang, X. Lv, W. Weng, Q. Zeng, Adv. Synth. Catal. *2013, 355* (4), 682-6 [DOI: 10.1002/adsc.201200752]; N-arylation **of hydroxamic acid esters** with ar. iodides using Cu_2O/DMEDA/K_2CO_3 s. T. Kukosha, N. Trufilkina, S. Belyakov, M. Katkevics, Synthesis *2012, 44* (15), 2413-23 [DOI: 10.1055/s-0031-1290384]; **of** pyridine-2-carbaldehyde **hydrazones** with CuI/K_3PO_4 in dioxane (without ligand), also conversion to N-arylhydrazines, s. W. Wu, X.-L. Li, X.-H. Fan, L.-M. Yang, Eur. J. Org. Chem. *2013* (5), 862-7 [DOI: 10.1002/ejoc.201201445]; N-arylation of D-glucosamines with CuI as catalyst s. C. Tao, F. Liu, W. Liu, Y. Zhu, Y. Li, X. Liu, J. Zhao, Tetrahedron Lett. *2012, 53* (52), 7093-6 [DOI: 10.1016/j.tetlet.2012.10.069]; **prim. ar. amines from ar. halides** (cf. *62*, 171s*82*) *in liq. ammonia* with a Cu(I) salt and ascorbic acid at 1 mol% loading s. P. Ji, J.H. Atherton, M.I. Page, J. Org. Chem. *2012, 77* (17), 7471-8 [DOI: 10.1021/jo301204t]; from ar. bromides or iodides with *aq. ammonia* in DMSO using CuI and the novel diol-functionalized poly(ethylene glycol)-bridged dicationic ionic liquid, [diol-PEG_{1000}-DIL][PF_6], as reusable ligand with K_2CO_3 as base, also from ar. chlorides possessing strongly electron-withdrawing groups with added Bu_4NBr, s. Y. Wang, J. Luo, T. Hou, Z. Liu, Austr. J. Chem. *2013, 66* (5), 586-93 [DOI: 10.1071/ch12487]; ligand-less procedure from N- or S-heteroaryl bromides *with gaseous ammonia* using $Cu(acac)_2/K_3PO_4$ s. S. Fantasia, J. Windisch, M. Scalone, Adv. Synth. Catal. *2013, 355* (4), 627-31 [DOI: 10.1002/adsc.201201010]; *metal-free* procedure from heteroaryl chlorides via Staudinger reaction with NaN_3/triphenylphosphine and subsequent treatment with aq. HCl s. S.R. Kandalkar, R.D. Kaduskar, P.A. Ramaiah, D.A. Barawkar, D. Bhuniya, A.M. Deshpande, Tetrahedron Lett. *2013, 54* (5), 414-8 [DOI: 10.1016/j.tetlet.2012.11.027].

Copper(I) iodide/N,N′-dimethylethylenediamine/base ←
Copper(I)-catalyzed reductive coupling of dibromomethylene compds. ←
with N-nucleophiles – N-Functionalized 1-bromoenamines, ketene aminals or ynamines s. *76*, 195s*83*; *79*, 225s*83*

Magnesium s. under $FeCl_2$ *Mg*

Indium/hydrogen chloride/water *In/HCl/*H_2O
3-Component synthesis of 2-subst. N-benzylbenzimidazoles from *o*-nitramines, benzyl bromides and aldehydes ○

158. 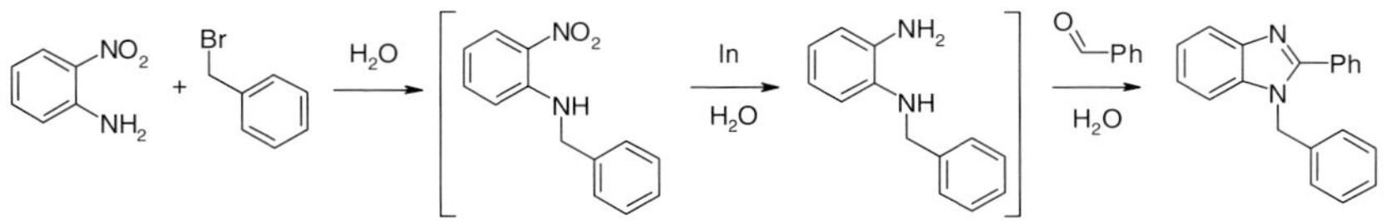

in water. A new *transition metal- and base-free*, one-pot 3-component synthesis of 2-subst. N-benzylbenzimidazoles has been effected *solely in water* [tap or pure] via N-monobenzylation, reduction and ring closure, wherein both the initial and terminating steps are presumed to involve *hydrogen bond-mediated activation of both electrophilic and nucleophilic partners by water*, while the latter improves the solvation of In^{3+} in the intermediate reductive phase. In this way substituents on the benzimidazole ring are disposed in a regio-defined manner. **E:** A mixture of *o*-nitroaniline (2 mmol) and benzyl bromide (1.2 eq.) in water (4 ml) stirred at 100° for 1.5 h, In (2.5 eq.) and 2 *N* aq. HCl (5 eq.) added, stirring continued for a further 30 min, benzaldehyde (1 eq.) added, stirred for a further 30 min, cooled to room temp., treated with $NaHCO_3$, and worked up with purification by flash chromatography → 1-benzyl-2-phenyl-1*H*-benzimidazole. Y 90%. The procedure is broadly applicable to the coupling of unfunctionalized substrates as well as those bearing methyl-, alkoxy- and chloro-substitution on the aromatic ring (thirty-four examples; Y 82-90%). There was no reaction without solvent and poor results were obtained in a wide range of conventional organic media, supporting the notion of an activating role of water through hydrogen bonding. The same mode of activation is presumed to hold sway in methanol or ethanol, although, here, yields were only moderate. Each of the three stages in this one-pot conversion was studied individually in support of the presumed reaction cascade and yields quoted accordingly. Interestingly, the first step appears to be exclusive to benzyl bromides as there was no condensation with non-benzylic alkyl bromides. F.e. and consideration of possible transition states s. D.N. Kommi, D. Kumar, R. Bansal, R. Chebolu, A.K. Chakraborti, Green Chem. *2012*, *14* (12), 3329-35 [DOI: 10.1039/c2gc36377a].

Aluminum oxide s. under NaN_3 Al_2O_3

Ethylaluminum chloride or Boron fluoride $EtAlCl_2$ *or* BF_3
N-Aryllactams from azidocarboxylic acid chlorides and arenes

159. 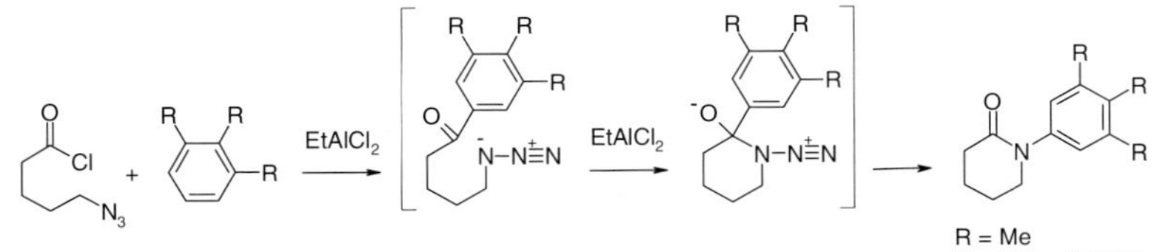

R = Me
(R = OMe: Y 97%)
(R = CO_2Me: Y 68%)

A quick, efficient one pot method for the formation of N-aryllactams *by Friedel-Crafts reaction-intramolecular Schmidt rearrangement* has been disclosed. **E:** An equimolar amount of startg. arene and azidoalkanoic acid chloride taken up in dry methylene chloride, stirred for 10 min at room temp., treated slowly in 2-3 small portions with $EtAlCl_2$ (10 mol%), allowed to react for 25 min (TLC), poured into distilled water, extracted with methylene chloride, the organic layer separated, dried (Na_2SO_4), and concentrated → product. Y 95%. Twenty-six further examples with 1,2- or 1,3-disubst. or 1,2,3-trisubst. arenes bearing electron-donating (Me, MeO) or -withdrawing (CO_2Me) groups gave N-aryl-γ-, -δ- or -ε-lactams in yields of 65-98%, yields being lower and reaction times slightly longer (40-50 min compared to 20-35 min) for electron-poor derivs. Of a variety of Lewis acids tested (incl. BF_3-etherate, $AlCl_3$, CF_3SO_3H, Me_3B), $EtAlCl_2$ was the most effective. The starting azides were prepared from cyclic ketones under standard conditions. F.e.s. D. Chaturvedi, A.K. Chaturvedi, N. Mishra, V. Mishra, Org. Biomol. Chem. *2012*, *10* (46), 9148-51 [DOI: 10.1039/c2ob26230d]; with BF_3-etherate cf. idem. Synlett *2012*, *23* (18), 2627-30 [DOI: 10.1055/s-0032-1317326].

Polyethylene glycol-400 s. under Copper(I) bromide *PEG-400*
Poly(ethylene glycol)-bridged dicationic ionic liquid s. under Copper(I) iodide ←
Bis(imidazolium)-functionalized calix[4]arene bis(α-aminocarboxylates) s. under Copper(I) iodide ←
Pivalic acid s. under Palladium(II) acetate *t-BuCOOH*

N-Chloro-N-sodiosulfonamides/benzyltriethylammonium chloride ←
4-Sulfonylimino-2-imidazolidones or 2-sulfonylimino-4-imidazolidones from α-isocyanocarboxylic acid amides and N-chloro-N-sodiosulfonic acid amides ○

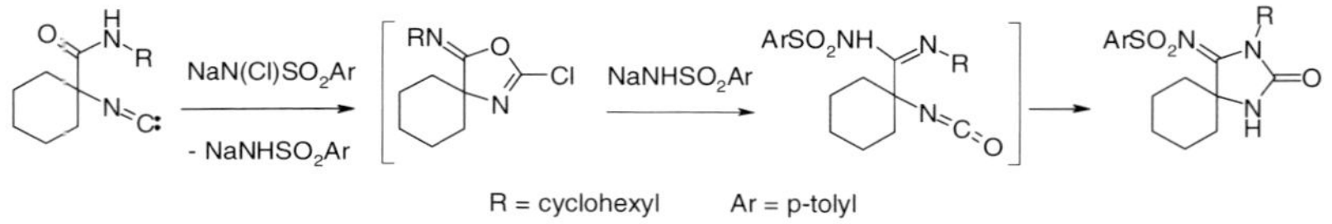

160.

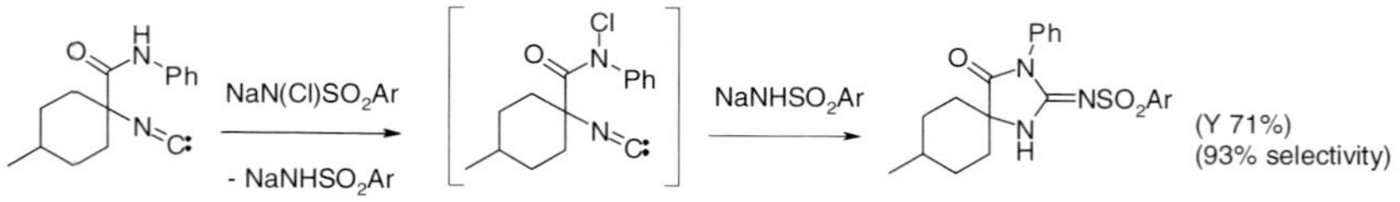

A remarkable **dependence of regioselectivity on amide substituent** is reported in the reaction of α-isocyanoamides with chloramine-B or -T, N-alkylated isocyanoamides affording 4-iminohydantoins as the major products and N-arylated isocyanoamides affording 2-iminohydantoins. **E:** A stirred suspension of startg. isocyanoamide (2.5 mmol) and dry chloramine-T (1 eq.) in dry chloroform (10 ml) treated with a little (30-35 mg) benzyltriethylammonium chloride, water added after 4 d, the organic layer separated, dried (Na_2SO_4 or $MgSO_4$), evaporated to dryness, and the residue recrystallized → 3-cyclohexyl-4-[(4-tolylsulfonyl)imino]-1,3-diazaspiro[4.5]decan-2-one. Y 76% (regioselectivity >95:5). Seven further examples with N-alkyl-derivs. gave yields of 61-70% (regioselectivity typically >95:5; 85:15 for the N-benzyl-spirocycloheptano-deriv.). In contrast, N-phenyl-derivs. gave 3-phenyl-2-sulfonylimino-4-imidazolidones in yields of 61-71% (three examples; regioselectivity 9:91 to 6:94), while an N-*p*-anisyl-deriv. gave a 39% yield of the corresponding 2-iminohydantoin and 23% of the 4-iminohydantoin. The regioselectivity is believed to reflect the different abilities of the substrates to undergo N-chlorination. For N-alkyl-isocyanoacetamides the rate of N-chlorination is relatively slow, allowing competitive α-addition on the isocyano group from the chloronium cation and the amide oxygen atom to give intermediate 2-chlorooxazolines, followed by ring opening with sulfonamides to give α-isocyano-N-sulfonyl-amidines which undergo intramolecular trapping by the more nucleophilic sulfonamide nitrogen. In contrast, N-aryl-derivs. undergo N-chlorination followed by ring closure with the resulting sulfonamides. F.e.s. M. García-Valverde, S. Marcaccini, A. González-Ortega, F.J. Rodríguez, J. Rojo, T. Torroba, Org. Biomol. Chem. *2013*, *11* (5), 721-5 [DOI: 10.1039/c2ob27098f].

Triphenylphosphine/2,2,6,6-tetramethylpiperidine *Ph_3P/TMP*

Regioselective 3-component synthesis of hydantoins from α-halogenocarboxylic acids, azides and isocyanates ○

161. 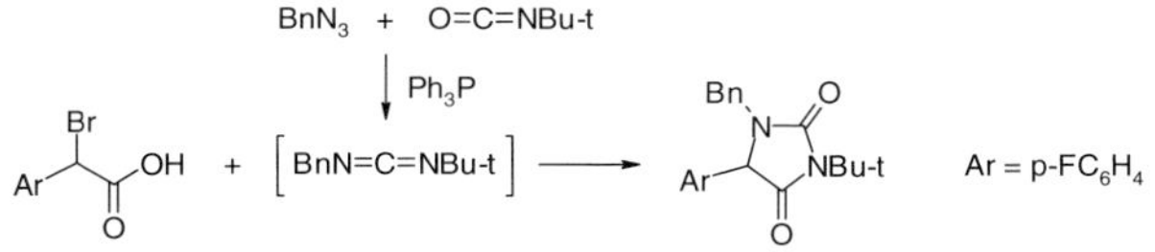

Solid triphenylphosphine (1.3 eq.) added to a stirred 0.1 *M* soln. of benzyl azide (1.3 eq.) and *tert*-butyl isocyanate (1.3 eq.) in methylene chloride, the mixture stirred until carbodiimide formation complete (TLC), 2,2,6,6-tetramethylpiperidine (1 eq.) added, followed by bromo-(*p*-fluorophenyl)acetic acid (1 eq.), stirring continued overnight, and the mixture worked up with purification by flash chromatography → 1-benzyl-3-*tert*-butyl-5-(4-fluorophenyl)imidazolidine-2,4-dione. Y 51%. With certain substrate combinations, where TLC monitoring indicated formation of the hydantoin in combination with the precursor N-(α-halogenoacyl)urea, the reaction mixture was converted solely to the hydantoin by subsequent treatment *in situ* with 2 *N* aq. NaOH (10% by volume) or K-*tert*-butoxide (1.3 eq.) at 0°. The procedure is simple, mild, atom-economical, eco-friendly, high-yielding and generally applicable to the condensation of [cyclo]alkyl, benzylic or aromatic azides and isocyanates with a wide range of mono- or di-subst. bromo- or chloro-acetic acids to give 1,3,5-trisubst. and 1,3,5,5-tetrasubst. hydantoins, respectively, incl. spirocyclic analogs (twenty-two examples; Y 44-95%). Where the initial Staudinger coupling provides an intermediate *unsym.* carbodiimide, coupling with the α-halogenocarboxylic acid progresses to give a *single* regioisomer, reaction with *tert*-butyl isocyanate being particularly notable in that the N-*tert*-butyl group of the formed N^1-subst. N^3-*tert*-butylhydantoin can be cleaved easily with methanesulfonic acid to give the corresponding N^3-unsubst. hydantoin for further elaboration. In a preliminary study, hydantoins were also prepared by direct coupling of the α-halogenocarboxylic acids with pre-formed carbodiimides (seventeen examples; Y 33 to >98%). F.e. and optimization, also with isothiocyanates in place of the isocyanates (one example; Y 68% as a mixture of regio-isomers), s. F. Olimpieri, M.C. Bellucci, T. Marcelli, A. Volonterio, Org. Biomol. Chem. *2012*, *10* (48), 9538-55 [DOI: 10.1039/c2ob26498f].

Bis[o-(diphenylphosphino)phenyl] ether s. under Ru(CO)(H)$_2$(PPh$_3$)$_3$ *DPEphos*

1,2,3,4-Tetraethyl-5-[o-(diphenylphosphino)phenyl]cyclopentadiene s. under Palladium(II) acetate ←

Oxygen s. under Cu$_2$O and Palladium(II) acetate *O_2*

p-Toluenesulfonic acid *TsOH*

Metal-free N-arylation with 2-halogeno-2-cyclohexenones NH → NAr

162. 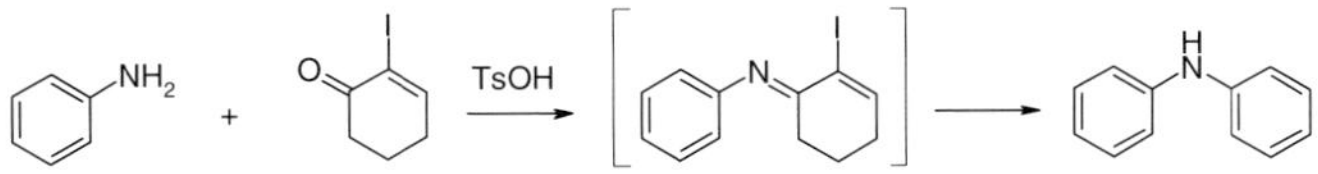

A soln. of 2-iodocyclohex-2-en-1-one (1.2 eq.), aniline (0.5 mmol) and anh. *p*-TsOH (20 mol%) in dry ethanol (0.5 ml) stirred at 75° for 1.5 h in a sealed screw-capped vessel under argon, concentrated, diluted with ethyl acetate, washed with 20% $NaHCO_3$ soln. and brine, and worked up with purification by flash chromatography on silica gel → diphenylamine. Y 87% (cf. Y 83% with 2-bromocyclohex-2-en-1-one after 1.5 h, or 85% with the 2-chloro-deriv. after 2 h). This method is simple, high yielding and regioselective, uses inexpensive materials and non-hazardous reaction conditions and has relatively short reaction times. Solvents such as acetonitrile or methanol may also be used but ethanol was chosen as being eco-friendly. It is applicable to a variety of prim. ar. amines (the presence of *o*-nitro or *p*-ester groups necessitated use of 30 mol% *p*-TsOH), prim. aliphatic amines or cyclic sec. amines (such as such as indoline, pyrrolidine or tetrahydro-quinoline, but not piperidine or acyclic sec. amines) and a wide range of 2-iodocyclohex-2-en-

1-ones. F.e. (thirty; Y 70-95%) s. M.T. Barros, S.S. Dey, C.D. Maycock, Eur. J. Org. Chem. *2013* (4), 742-7 [DOI: 10.1002/ejoc.201201263].

Sodium thiosulfate *$Na_2S_2O_3$*
N-*tert*-Butylanilines from ar. azides s. *83*, 217 NH_2 → NHBu-t

Benzyltriethylammonium chloride s. under N-Chloro-N-sodiosulfonamides *$BnEt_3NCl$*
Manganese(II) fluoride s. under Copper(I) bromide *MnF_2*

Iron(II) chloride/lithium bromide/magnesium *$FeCl_2$/LiBr/Mg*
Ar. amines from bromomagnesium amides and ar. bromides NH → NAr

163.

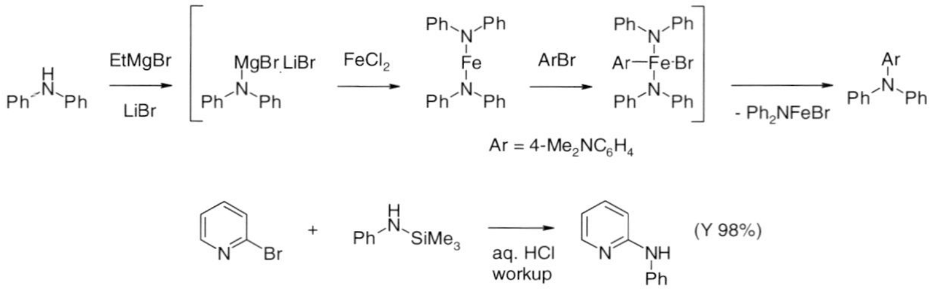

A soln. of EtMgBr (2 eq.) in ether (1.38 ml) added to a soln. of diphenylamine (2.02 eq.) in the same solvent (2 ml) at 0°, the mixture stirred at 40° for 2 h, solvent removed *in vacuo*, N,N-dimethyl-4-bromoaniline (1 mmol), $FeCl_2$ (5 mol%), LiBr (4 eq.) and xylene (4 ml) added, the mixture heated at 140° for 24 h, quenched with 1 *M* aq. HCl (2 ml) at room temp., extracted with ethyl acetate, the extracts filtered through Florisil, concentrated *in vacuo*, and purified chromatographically → N,N-diphenyl-4-dimethylaminoaniline. Y 88%. Iron(II) optimally catalyzed this prepn. of mixed triarylamines (tests showed the catalytic effect was not due to trace metal contaminants), with the addition of LiBr increasing reaction efficiency, presumed via disaggregation of the initial magnesium amide. Diarylamines were effectively arylated with electron-diverse bromo-benzene, -pyridine and -thiophene derivs. (incl. bis-amination of 1,4-dibromobenzene) (twelve examples; Y 59-96%). Prim. anilines were not arylated under these conditions, but temporary mono-protection (as N-trimethylsilylanilines) enabled synthesis of diarylaminosilanes, which were hydrolyzed during work-up to afford **diarylamines** (four examples; Y 76-98%). The proposed, non-conventional $Fe^{(II)}/Fe^{(IV)}$ mechanism was supported by DFT calculations. F.e. and optimization s. T. Hatakeyama, R. Imayoshi, Y. Yoshimoto, S.K. Ghorai, M. Jin, H. Takaya, K. Norisuye, Y. Sohrin, M. Nakamura, J. Am. Chem. Soc. *2012*, *134* (50), 20262-5 [DOI: 10.1021/ja309845k].

Carbonyl(dihydrido)tris(triphenylphosphine)ruthenium(II)/bis[o-(diphenylphosphino)-phenyl] ether ←
***p*-Aminoketones from 1-(*p*-fluoroaryl)-2-ethylenealcohols** ←

164.

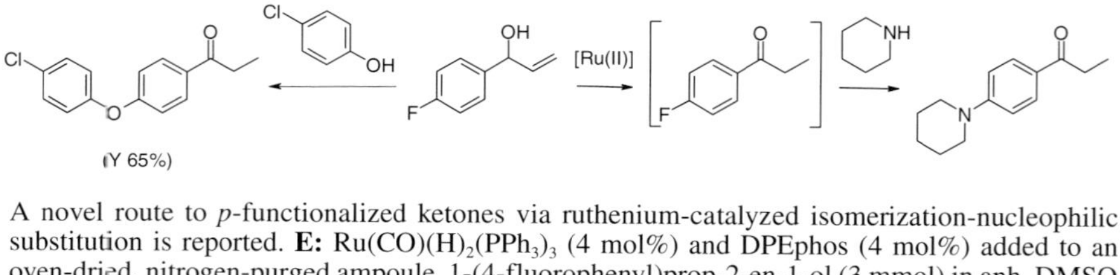

A novel route to *p*-functionalized ketones via ruthenium-catalyzed isomerization-nucleophilic substitution is reported. **E:** $Ru(CO)(H)_2(PPh_3)_3$ (4 mol%) and DPEphos (4 mol%) added to an oven-dried, nitrogen-purged ampoule, 1-(4-fluorophenyl)prop-2-en-1-ol (3 mmol) in anh. DMSO (3 ml) added, followed by piperidine (2.2 eq.), the vessel sealed with a Young's tap, the mixture agitated with heating to 115° for 24 h, allowed to cool to room temp., transferred to a separating funnel, washing with DMSO, water and brine added, extracted with ether, and purified by chromatography on silica gel → product. Y 87%. The method is applicable to a variety of 1-(*p*-fluoroaryl)-2-ethylenealcohols, which may bear additional F, Cl, CF_3 or Me groups on the aromatic ring and substitution on the alkene, in reaction with 5- to 7-membered cyclic sec. amines (fourteen further examples; Y 35-86%). An analogous homoallyl alcohol gave an 85% yield. The method was extended to formation of ***p*-aryloxyacylophenones from phenols** with added K_2CO_3 (eight examples; Y 55-65%). F.e.s. A.J.A. Watson, B.N. Atkinson, A.C. Maxwell, J.M.J. Williams, Adv. Synth. Catal. *2013*, *355* (4), 734-40 [DOI: 10.1002/adsc.201200879].

Palladium-carbon s. under (2-Dicyclohexylphosphino-2′,4′,6′-triisopropylbiphenyl)-o′-aminobiphenylylpalladium(II) mesylate *Pd-C*

Palladium(II) acetate/silver acetate/pivalic acid/oxygen ←
N-Subst. 2-pyridones from 2-chloropyridines and electron-deficient ethylene derivs. ←

165.
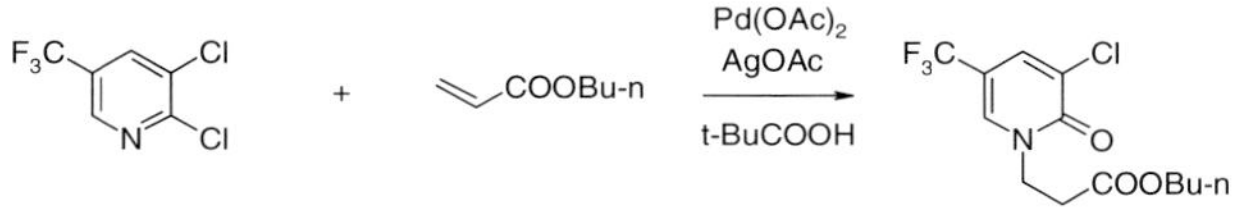

A simple and efficient route to pyridones from readily available starting materials is reported. **E: β-(2-Pyridon-1-yl)carboxylic acid derivs.** AgOAc (3 eq.), *n*-butyl acrylate (4 eq.), 2,3-dichloro-5-(trifluoromethyl)pyridine (0.5 mmol) and pivalic acid (2.5 eq.) in dry DMF (0.8 ml) added sequentially to $Pd(OAc)_2$ (10 mol%), the mixture stirred at 130° under O_2 (balloon) for 20 h, after cooling, the mixture diluted with methylene chloride, filtered, the filtrate concentrated, and the residue purified by flash column chromatography → *n*-butyl 3-[3-chloro-2-oxo-5-(trifluoromethyl)-pyridin-1(2*H*)-yl]propanoate. Y 80% (96% in the presence of 30 mol% 1,10-phenanthroline). The method was applicable to α,β-unsatd. esters, amides or nitriles but not vinyl acetate, while acrylic acid gave a low yield (18%); it was more efficient with 5-subst. pyridines bearing electron-withdrawing groups than with 3-subst. pyridines and, in particular, sterically hindered 3-subst. pyridines did not work well. The reaction failed with 2-chloropyrimidine. It is believed the source of the oxygen atom is pivalic acid and pyridinolates are important intermediates. F.e. (seventeen; Y 21%, 26%, 56-98%) s. K. Zhou, P. Wen, W. Chen, C. Ma, G. Huang, Eur. J. Org. Chem. *2013* (3), 448-52 [DOI: 10.1002/ejoc.201201151].

Palladium(II) acetate/1,2,3,4-tetraethyl-5-[o-(diphenylphosphino)phenyl]cyclopentadiene/lithium tert-butoxide ←
Pyrroles from 1,4-dibromo-1,3-dienes and sec. amines ○

166.
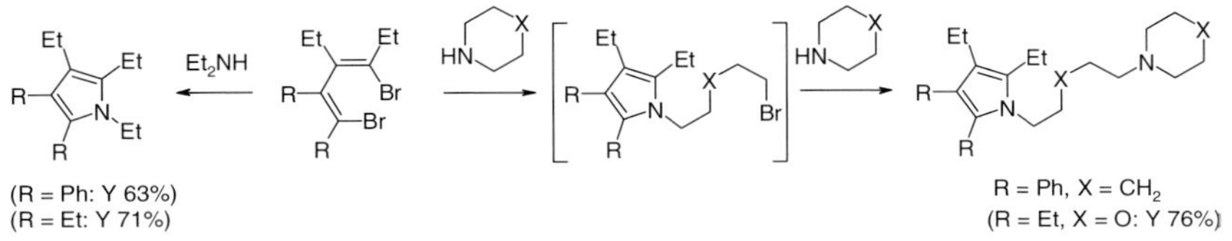

1-ω-Aminopyrroles from two cyclic sec. amine molecules. A mixture of $Pd(OAc)_2$ (2 mol%), 1,2,3,4-tetraethyl-5-[*o*-(diphenylphosphino)phenyl]cyclopentadiene (5 mol%) and LiOBu-*t* (3 eq.) in toluene (2 ml) stirred at room temp. for 5 min under N_2, startg. 1,4-dibromo-1,3-diene (0.5 mmol) and piperidine (0.2 ml) added, the mixture stirred at 110° for 12 h, quenched with water, extracted with ether, and purified by chromatography on silica gel → product. Y 84%. It appears that both the phosphino and the cyclopentadiene moieties are important for high efficiency in this palladium-catalyzed coupling. An excess (>5 eq.) of amine is also required for clean reaction. Six further examples with sym. or unsym. dibromodienes and cyclic sec. amines, including morpholine and pyrrolidine, gave yields of 62-90%. Reaction with acyclic sec. amines (diethyl-, dipropyl- or di-*n*-butyl-amine but not diisopropylamine) gave yields of 50-77% (six examples) with concomitant formation of the corresponding tert. amines. For unsym. acyclic sec. amines the bulkiness of the N-substituent and the C-N bond energy determine which C-N bond is more readily cleaved (three examples; Y 57-65%). F.e., also **indoles from *o*,β-dibromostyrenes** (six 1-ω-amino examples: Y 58-85%; three examples from acyclic amines: Y 63-83%), s. W. Geng, W.-X. Zhang, W. Hao, Z. Xi, J. Am. Chem. Soc. *2012, 134* (50), 20230-3 [DOI: 10.1021/ja308950d].

Palladium N-heterocyclic carbene, thiosemicarbazone or phosphine complexes/base [Pd]/base
Palladium-catalyzed N-arylation NH → NAr
update s. *51,* 171s*82*; *52,* 171s*82*; with highly active [Pd(IPr*^{OMe})(acac)Cl] [where IPr*^{OMe} = 1,3-bis-[4-methoxy-2,6-di(benzhydryl)phenyl]imidazol-2-ylidene] at 250-1000 ppm and K-*tert*-pentoxide as base s. S. Meiries, K. Speck, D.B. Cordes, A.M.Z. Slawin, S.P. Nolan, Organometallics *2013, 32* (1), 330-9 [DOI: 10.1021/om3011867]; *under bimetal catalysis* with a copper(II) iodide/(hydroxo)-methylpalladium(II) complex bridged with an N-[*o*-(diphenylphosphino)phenyl]-N′-8-quinolyl-formamidine ligand using Na-*tert*-butoxide as base s. N. Tsukada, N. Ohnishi, S. Aono, F. Takahashi, ibid. *2012, 31* (21), 7336-8 [DOI: 10.1021/om300902t]; with a palladium(II) thiosemicarbazone complex for N-arylation of various sec. amines with a wide range of aryl and heteroaryl halides (bromides or chlorides), incl. activated, neutral and deactivated substrates, s. R.N. Prabhu, R. Ramesh, Tetrahedron Lett. *2013, 54* (9), 1120-4 [DOI: 10.1016/j.tetlet.2012.12.070]; N-arylation of a wide range **of carboxylic acid amides** (incl. carbamates and lactams) with bulky ar. halides using electron-rich 2-dialkylphosphino-2′-alkoxy-1,1′-binaphthyl as ligand and Cs_2CO_3 or K_3PO_4 as base s. F. Ma, X. Xie, L. Zhang, Z. Peng, L. Ding, L. Fu, Z. Zhang, J. Org. Chem. *2012, 77* (12), 5279-85 [DOI: 10.1021/jo3005827]; of unprotected sulfonimidamides [ArS(O)(NH)NR_2] using RuPhos [2-dicyclohexylphosphino-2′,6′-diisopropoxybiphenyl] as ligand and NaOBu-*t* as base under microwave enhancement s. M.F. Maldonado, F. Sehgelmeble, F. Bjarnemark, M. Svensson, J. Åhman, P.I. Arvidsson, Tetrahedron *2012, 68* (36), 7456-62 [DOI: 10.1016/j.tet.2012.06.072]; safe N-arylation of hydrazine with ar. chlorides *under continuous flow*, also subsequent conversions of the resulting N-arylhydrazines, s. A. DeAngelis, D.-H. Wang, S.L. Buchwald, Angew. Chem., Int. Ed. *2013, 52* (12), 3434-7 [DOI: 10.1002/anie.201208544].

(2-Dicyclohexylphosphino-2′,4′,6′-triisopropylbiphenyl)-o′-aminobiphenylylpalladium(II) mesylate/potassium phosphate/palladium-carbon/sodium nitrite/hydrogen chloride ←
Benzotriazoles from *o*-nitrochlorides and prim. amines via diazotization under continuous flow ○

167.

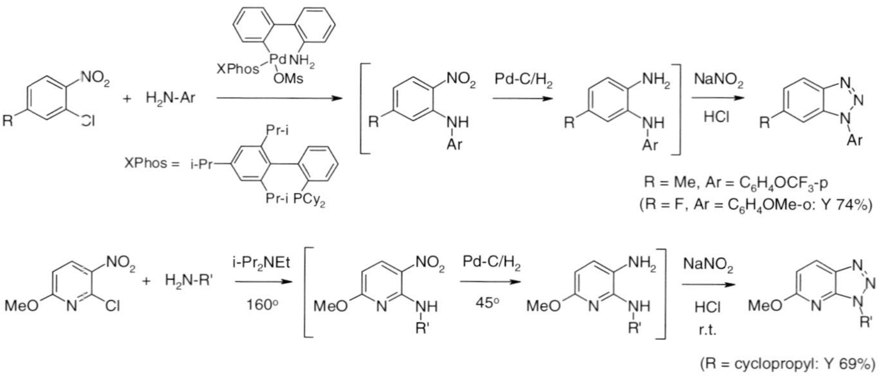

A novel route to 1-subst. benzotriazoles via a regioselective N-arylation-hydrogenation-diazotization-ring closure sequence under continuous flow is reported, providing a rare example of the performance of consecutive multiphase processes in a continuous line. **E: By ultrasound-promoted palladium-catalyzed N-arylation**. 2-Chloro-4-methyl-1-nitrobenzene (6 mmol) and 4-(trifluoromethoxy)aniline (1.3 eq.) treated sequentially with Pd-catalyst/K_3PO_4, Pd-C/H_2, and

$NaNO_2$/HCl under continuous flow conditions → 6-methyl-1-[4-(trifluoromethoxy)phenyl]-1*H*-benzo[*d*][1,2,3]triazole. Y 81%. Five further examples with *o*-nitrochlorides bearing F, Me or MeO groups and ar. or hetar. [pyrid-3-yl, pyrazol-5-yl] prim. amines gave yields of 51-89%. For *o*-nitrochlorides bearing an additional activating group, such as CF_3, CN or ester, or pyridine derivs. (affording **1,2,3-triazolo[4,5-*b* or -*c*]pyridines**), initial S_NAr substitution could be achieved with *i*-$PrNEt_2$ (and DMA/*n*-hexanol as solvent system at 160-180°) in the absence of Pd-catalyst using ar. or aliphatic prim. amines (nine examples; Y 63-93%). F.e., apparatus and experimental details s. M. Chen, S.L. Buchwald, Angew. Chem., Int. Ed. *2013*, *52* (15), 4247-50 [DOI: 10.1002/anie.201300615].

Sulfur ↑ NC ↓↑ S

Without additional reagents *w.a.r.*

N-Subst. carboxylic acid amides from thiolic acids and azides COSH → CONHR

s. *43*, 340; solvent-free and soln.-phase [with triphenylphosphine], thermal procedures s. S. Nagarajan, P. Shanmugavelan, M. Sathishkumar, N. Priyadharshini, P. Sudakar, A. Ponnuswamy, Synth. Commun. *2013*, *43* (5), 668-80 [DOI: 10.1080/00397911.2011.606043]; ligation of peptide thiolic acids with glycosylamines using $Cu(OAc)_2$/N-hydroxybenzotriazole for preparation of N-glycosylpeptides s. R. Joseph, F.B. Dyer, P. Garner, Org. Lett. *2013*, *15* (4), 732-5 [DOI: 10.1021/ol302961s]; decarboxylative amide formation **from α-ketocarboxylic acids and O-(*p*-nitrophenyl)hydroxylamines** in aq. DMSO s. S. Kumar, R. Sharma, M. Garcia, J. Kamel, C. McCarthy, A. Muth, O. Phanstiel, IV, J. Org. Chem. *2012*, *77* (23), 10835-45 [DOI: 10.1021/jo302175g].

Copper(I) cyanide s. under $PdCl_2$(bpy) *CuCN*

2-Picoline·borane ←

Reductive N-alkylation with aldehyde bisulfite adducts >NH → >NR

of sec. amines using $NaBH(OAc)_3/Et_3N$ cf. *77*, 208; of a variety of amines with 2-picoline·borane in methanol s. M. Faul, R. Larsen, A. Levinson, J. Tedrow, F. Vounatsos, J. Org. Chem. *2013*, *78* (4), 1655-9 [DOI: 10.1021/jo3023878].

Trimethylsilyl chloride *Me_3SiCl*

Formamidrazones from thiosemicarbazones RC=NNHC(S)N< → RC=NN=CHNHR′

and isonitriles s. *83*, 182

Bismuth(III) triflate *$Bi(OTf)_3$*

1-Aryl-4-(arylamino)indoles
from 4-alkoxy-2-(alk-l-ynyl)-N-sulfonylcyclohexa-2,5-dienimines
and two prim. ar. amine molecules ○

168.

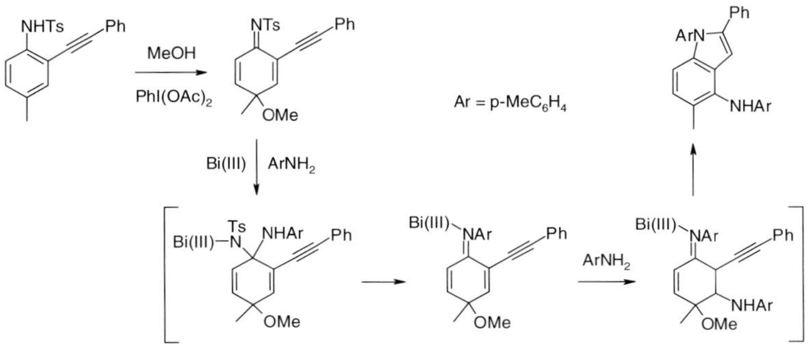

Startg. N-tosylcyclohexadienimine [prepared crude by oxidation of the corresponding *o*-(tosylamino)arylacetylene (0.1 mmol) with phenyl iodosoacetate (1.1 eq.) in methanol at 25° over 10 min] mixed with *p*-toluidine (3 eq.) in *tert*-butanol (2 ml), treated with $Bi(OTf)_3$ (10 mol%), stirred at 25° for 48 h, the mixture passed through a short silica gel column, concentrated under reduced pressure, and the residue purified by flash chromatography on silica gel → 5-methyl-2-phenyl-N,1-di-*p*-tolyl-1*H*-indol-4-amine. Y 90% [from the *o*-(tosylamino)arylacetylene]. The procedure is applicable to a wide range of electron-diverse prim. ar. amines (fourteen examples; Y 64-95%), but there was no reaction with terminal alkyne derivs. or heteroaromatic amines. Reaction is presumed to involve initial coordination of bismuth to the imine nitrogen, followed *by 1,2-addition-1,4-addition* of two amine molecules prior to pyrrole ring closure and bismuth-catalyzed elimination of methanol. The products react smoothly with a wide range of [hetero]-aromatic, α,β-ethylenic, and aliphatic aldehydes to give the corresponding **6,11-dihydro-2*H*-azepino[4,3,2-*cd*]indoles** (seventeen examples; Y 70-99%). F.e. and comparison of Lewis acids s. L. Zhang, Z. Li, R. Fan, Org. Lett. *2012*, *14* (23), 6076-9 [DOI: 10.1021/ol3029675].

Dichloro(2,2′-bipyridyl)palladium(II)/copper(I) cyanide/potassium carbonate ←
Palladium/copper-cocatalyzed desulfitative C_2-amination of benzoxazoles with sulfamyl chlorides H → N<

169. 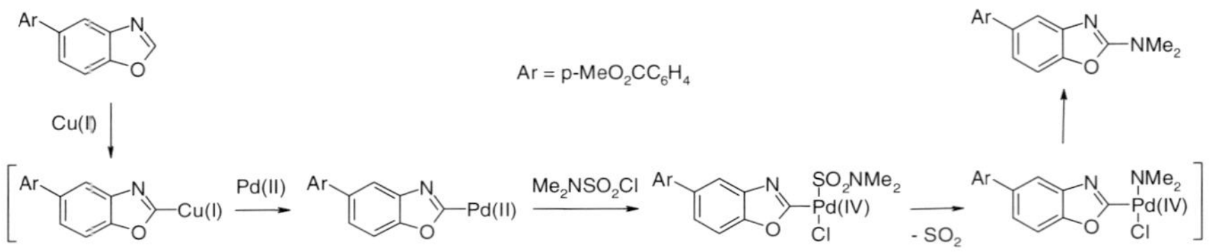

A mixture of 5-(4-methoxycarbonylphenyl)benzoxazole (0.2 mmol), dimethylsulfamyl chloride (1.5 eq.), $PdCl_2$(bpy) (10 mol%), CuCN (20 mol%), K_2CO_3 (1 eq.) and 4 Å molecular sieves (50 mg) in dry chlorobenzene (2 ml) heated under air at 150° in a sealed tube for 24 h → 2-dimethylamino-5-(4-methoxycarbonylphenyl)benzoxazole. Y 88%. Efficient and direct C-amination of benzoxazoles with dimethylsulfamyl chloride was successful with electron-diverse benzoxazoles (twelve examples; Y 48%, 60-91%) in the presence of halo, ester, keto, nitro and ether functionality. Moderate yields of products were obtained with N-sulfonyl chlorides of morpholine (61%), pyrrolidine (52%) and piperidine (33%) but reaction with chlorosulfonyl isocyanate was unsuccessful. F.e. and optimization s. S. Chen, K. Zheng, F. Chen, Tetrahedron Lett. *2012*, *53* (46), 6297-9 [DOI: 10.1016/j.tetlet.2012.09.044].

Via intermediates *v.i.*
Interchange of N-protective groups with phosphides ←
N-Carbo-*tert*-butoxyamines from tosylamines s. *83*, 4

Remaining Elements ↑ NC ↓↑ Rem

Sodium hydride *NaH*
Cyclic N-acyl-N′-(sulfonyl)ene-1,2-diamines from N-acyl-N′-sulfonyldiamines ○

170. 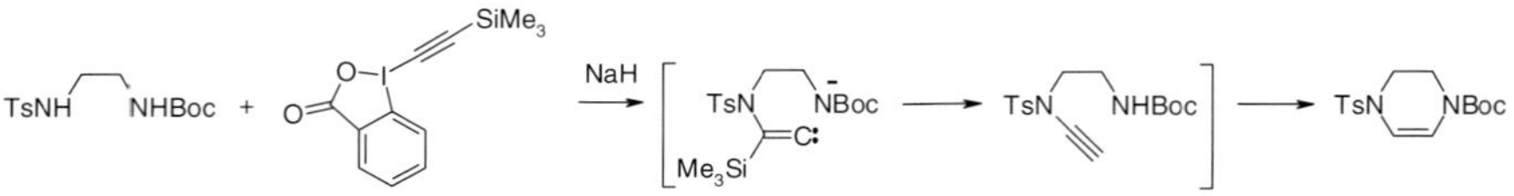

One-pot procedure via N-acyl-N′-(alk-l-ynyl)-N′-sulfonyldiamines. NaH (2 eq.; 60% dispersion in mineral oil), followed by 1-(trimethylsilylethynyl)-1,2-benziodoxol-3(1*H*)-one (1 eq.), added to a soln. of *tert*-butyl 2-(4-methylphenylsulfonamido)ethylcarbamate (0.2 mmol) in DMF (2 ml) at 0°, the ice-bath removed, the mixture stirred at room temp. for 2 h, and worked up with purification by flash chromatography on silica gel → *tert*-butyl 4-tosyl-3,4-dihydropyrazine-1(2*H*)-carboxylate. Y 56%. The procedure is applicable to the synthesis of **1-acyl- or 1-(carbo-*tert*-butoxy)-4-sulfonylpiperazines** (nine examples; Y 26-83%) and a 4,5,6,7-tetrahydro-1*H*-1,4-diazepine analog (Y 57%), reaction being dependent on the initial selective N-(alk-1-ynyl)ation at the N-sulfonyl rather than the N-acyl site of the starting material. Yields were markedly dependent on the choice of the N-sulfonyl group (being lowest with N-mesyl- and N-nosyl-derivs.), while reaction was favored by N-benzoyl groups bearing electron-donating substituents, and most efficient with N-alkanoyl-derivs. An intermediate carbene is thought to be involved in the process. F.e.s. T. Aubineau, J. Cossy, Chem. Commun. *2013, 49* (32), 3303-5 [DOI: 10.1039/c3cc40529j].

Sodium azide/copper(II) oxide/montmorillonite *NaN_3/CuO/clay*
Sodium azide/copper(II) sulfate-on-aluminum oxide/potassium carbonate ←
4-Subst. 1-aryl-1,2,3-triazoles from terminal acetylene derivs. and arylboronic acids ○
s. *68*, 184s*83*

Copper(I) oxide *Cu_2O*
Chan-Lam-Evans N-substitution with boronic acids NH → NR
N-arylation, update s. *55,* 166s*82*; mild N-heteroarylation of [benz]imidazoles with heteroarylboronic acids or 2-heteroaryl-4,4,5,5-tetramethyl-1,3,2-dioxaborolanes using Cu_2O in methanol s. A. Begouin, M.-J.R.P. Queiroz, Helv. Chim. Acta *2013, 96* (5), 853-63 [DOI: 10.1002/hlca.201200310]; N-arylation of imidazoles with arylboronic acids *in water* under copper catalysis with an amphiphilic surfactant under O_2 without base s. K. Inamoto, K. Nozawa, J. Kadokawa, Y. Kondo, Tetrahedron *2012, 68* (38), 7794-8 [DOI: 10.1016/j.tet.2012.07.042].

Copper(II) acetate/sodium pivalate/oxygen *$Cu(OAc)_2$/t-$BuCO_2Na$/O_2*
Benzimidazoles from amidines and arylboronic acids ○
via Chan-Lam-Evans N-arylation-dehydrogenative intramolecular N-arylation

171. 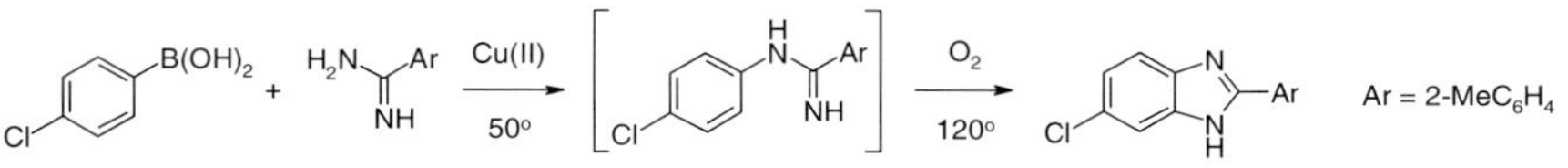

in one pot. A mixture of 2-methylbenzamidine (1.5 eq.), 4-chlorobenzeneboronic acid (0.2 mmol), $Cu(OAc)_2 \cdot xH_2O$ (20 mol%), Na-pivalate (40 mol%) and DMF (0.8 ml) stirred at 50° under air for 24 h (TLC), then heated at 120° under O_2 (balloon) for 4-24 h (TLC), cooled to room temp., quenched with aq. ammonia, extracted with ethyl acetate, and purified chromatographically → 6-chloro-2-(2-methylphenyl)benzimidazole. Y 90%. Mild conditions were established for the initial novel N-arylation of prim. and sec. benzamidines (seven examples; Y 56-81%), and one-pot conversion to **2-arylbenzimidazoles** was achieved by heating the crude reaction mixture under O_2. In this way, 2-aryl-NH-benzimidazoles were obtained from a variety of primary benzamidines and electron-diverse arylboronic acids (eleven examples; Y 54-90%; 4-acetylbenzeneboronic acid gave 24%). The reaction was extended to N-methylbenzamidines, affording the corresponding 1-methyl-2-arylbenzimidazoles (five examples; Y 45-68%), with lower yields obtained for N-propyl (34-38%; two examples) and N-isopropyl (6%) analogs. Yields were apparently compromised due to partial oxidative N-dealkylation of the products in each case. F.e. and optimization s. J. Li, S. Bénard, L. Neuville, J. Zhu, Org. Lett. *2012, 14* (23), 5980-3 [DOI: 10.1021/ol3028847].

Copper(II) acetate/1,5-bis(diphenylphosphino)pentane/sodium carbonate ←
α-Aminocarboxylic acid esters C=C(OSi⋜)OR → C(N⋜)COOR
from O-silyl O-alkyl keteneacetals and benzoyloxylamines

172. 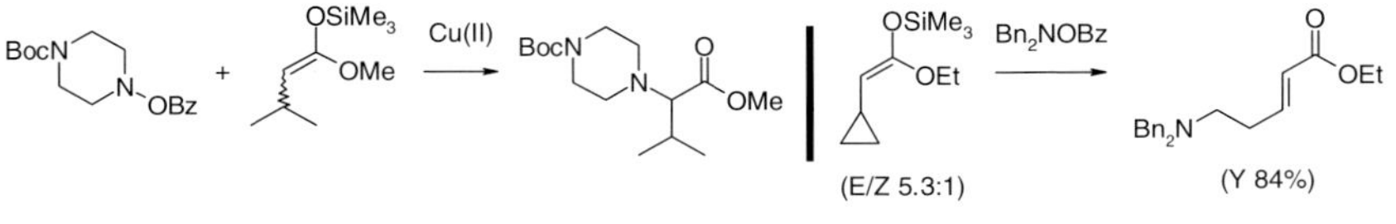

The copper(II)-promoted amination of ketene silyl acetals with tert. benzoyloxylamines (incl. cyclic examples) has been shown to provide a general route for the synthesis of unnatural α-amino acids. **E:** A suspension of N-benzoyloxy-N′-*tert*-butoxycarbonylpiperazine (0.25 mmol), $Cu(OAc)_2$ (10 mol%), 1,5-bis(diphenylphosphino)pentane (10 mol%) and Na_2CO_3 (2 eq.) in DMF (0.5 ml) stirred at room temp. for 15 min, a soln. of O-trimethylsilyl O-methyl isopropylketeneacetal (2 eq.) in the same solvent (1 ml) added dropwise, the soln. stirred for 4 h (GC), quenched with water, extracted with ethyl acetate, and purified chromatographically → *tert*-butyl 4-(1-methoxy-3-methyl-1-oxobut-2-yl)piperazine-1-carboxylate. Y 86%. The method was generally applicable, with bulkier substrates giving moderate yields in some cases (twenty-four examples; Y 37-90%). A single sec. amino acid was obtained in 41% yield. A general mechanism was not established but experimental evidence excluded a radical pathway. Interestingly, a cyclopropylketene silyl acetal gave none of the expected product, amination occurring via cleavage of the cyclopropane ring to afford a 5-aminopent-2-enoate in high yield (84%). F.e. and optimization s. N. Matsuda, K. Hirano, T. Satoh, M. Miura, Angew. Chem., Int. Ed. *2012*, *51* (47), 11827-31 [DOI: 10.1002/anie.201206755]; **tert. ar. amines** from arylstannanes [aryl-[*o*-(hydroxymethyl)phenyl]dimethylstannanes] and N,N-disubst. acoxylamines with CuI/LiOBu-*t* and *o*-(di-*tert*-butylphosphino)-biphenyl as ligand s. Y. Miki, K. Hirano, T. Satoh, M. Miura, Org. Lett. *2013*, *15* (1), 172-5 [DOI: 10.1021/ol303222s].

Copper(II) acetate/1,1′-bis(diphenylphosphino)ferrocene *Cu(OAc)₂/dppf*
N-Arylazomethines from hydrazonium salts and arylboronic acids C=NN⁺⋜ → C=NAr

173. 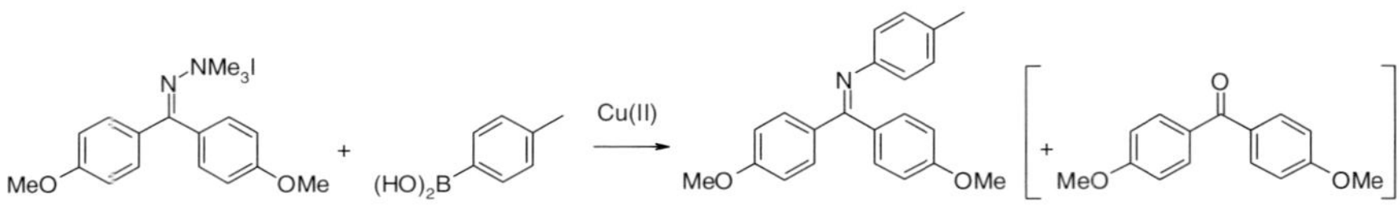

A mixture of 4,4′-dimethoxybenzophenone N,N,N-trimethylhydrazonium iodide (0.5 mmol), 4-tolylboronic acid (1.2 eq.), $Cu(OAc)_2$ (1.2 eq.) and dppf (10 mol%) in dry dioxane stirred under N_2 at 70° for 1 h, cooled to room temp., volatiles removed *in vacuo*, and the residue purified chromatographically → N-[bis(4-methoxyphenyl)methylene]-*p*-tolylamine. Y 60% (plus 17% 4,4′-dimethoxybenzophenone). Addition of the phosphine ligand improved the yields of imine products that at best were moderate (24-60%; seven examples; electron-deficient 4-cyanobenzeneboronic acid gave 10%), and accompanied by ketone (12-50%) and hydrazone (8-28%) by-products. Use of the hydrazonium hexafluorophosphate salt gave only a trace of imine, while copper halide catalysts were ineffective. F.e.s. M. Kitamura, Y. Tokuda, N. Tashiro, T. Okauchi, Austr. J. Chem. *2012*, *65* (12), 1687-90 [DOI: 10.1071/CH12350].

Copper(I) bromide/sodium trimethylsilanolate/di-tert-butyl peroxide ←
N-Alkylation of N-unsubst. carboxylic acid amides with alkylboronic acids NH → NR

174. 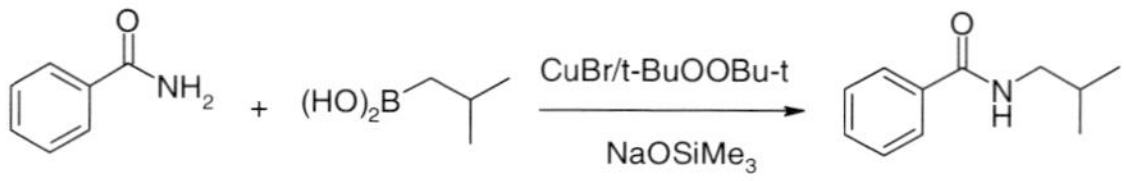

The first example of an efficient, catalytic **Chan-Lam-Evans alkylation** as an alternative route to N-monosubst. carboxylic acid amides is disclosed, key to which is the revelation of *Na-trimethylsilanolate as mild base* and the choice of di-*tert*-butyl peroxide as oxidant. **E:** CuBr (10 mol%), benzamide (1 eq.), *sec*-butylboronic acid (1.5 eq.) and Na-trimethylsilanolate (2.2 eq.) charged under N_2 into a flask, the latter evacuated and backfilled three times with N_2, anh. *tert*-butanol (2.5 ml) added, followed by di-*tert*-butyl peroxide (3 eq.) via syringe, the flask placed in an oil bath at 75°, stirred for 24 h, cooled to room temp., diluted with methylene chloride, filtered through Celite, the filtrate and washings treated with satd. aq. $Na_2S_2O_3$ and brine, and worked up with purification by flash chromatography on silica gel → benzoic acid N-*sec*-butylamide. Y 84%. The procedure is mild, inexpensive, and broadly applicable to a wide range of arylcarboxylic acid amides [possessing electron-donating or weakly electron-withdrawing (Cl, F) groups], as well as heteroaryl- and [cyclo]aliphatic carboxamides; the reactivity of alkylboronic acids was in the order *prim*- > *sec*- > *tert*-alkyl, tolerating a variety of substitutents, e.g. OH, CN, and siloxy, on the alkyl chain of *prim*-alkylboronic acids. Significantly, reaction is tolerant of β-hydrogen atoms and, in most instances, requires only a small excess of the alkylboronic acid; it also appears to be dependent on the choice of di-*tert*-butyl peroxide as oxidant, a whole range of other common oxidants and oxygen itself being ineffective. F.e. and comparison of copper salts and bases s. S.A. Rossi, K.W. Shimkin, Q. Xu, L.M. Mori-Quiroz, D.A. Watson, Org. Lett. *2013*, *15* (9), 2314-7 [DOI: 10.1021/ol401004r].

Copper(I) iodide/1,2-(di-tert-butylphosphino)biphenyl/lithium tert-butoxide ←
Tert. ar. amines from arylstannanes and N,N-disubst. acoxylamines ArSn≤ → ArN<
s. *83*, 172

Di-tert-butyl peroxide s. under CuBr *t-BuOOBu-t*
Sodium trimethylsilanolate s. under CuBr *$NaOSiMe_3$*
1,5-Bis(diphenylphosphino)pentane s. under $Cu(OAc)_2$ *dpppen*
1,2-(Di-tert-butylphosphino)biphenyl s. under Copper(I) iodide ←

Bismuth(III) nitrate *$Bi(NO_3)_3$*
Ar. nitro compds. from arylboronic acids $B(OH)_2$ → NO_2
with $Bi(NO_3)_3/K_2S_2O_8$ cf. *81,* 160; rapid, uncatalyzed, regioselective conversion with $Bi(NO_3)_3 \cdot 5H_2O$ for preparing [hetero]ar. nitro compds. s. R.R. Yadav, R.A. Vishwakarma, S.B. Bharate, Tetrahedron Lett. *2012, 53* (44), 5958-60 [DOI: 10.1016/j.tetlet.2012.08.121].

Oxygen s. under $Cu(OAc)_2$ *O_2*
1,1′-Bis(diphenylphosphino)ferrocene s. under $Cu(OAc)_2$ *dppf*

Carbon ↾ NC ⇃↾ C

Without additional reagents *w.a.r.*
N-Subst. carboxylic acid amides from α-ketocarboxylic acids COCOOH → CONHR
and O-(*p*-nitrophenyl)hydroxylamines s. *43,* 340s*83*

Potassium tert-butoxide *KOBu-t*
Pyrimidines from N-protected 2-acetylenehydroxylamines via N-protected β-amino-α,β-ethyleneketones ○

175.

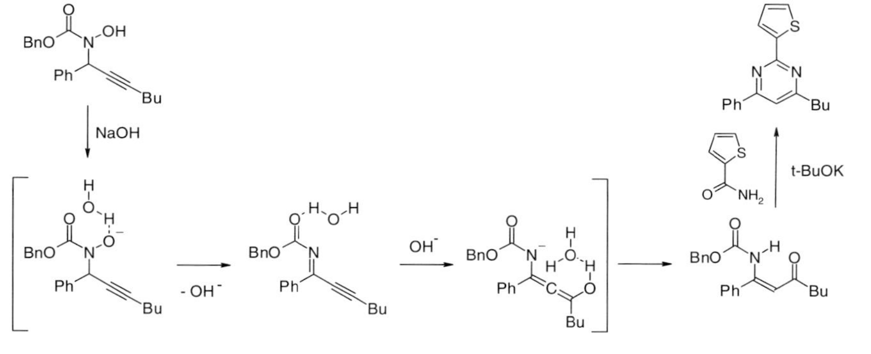

A mixture of N-benzyloxycarbonyl-N-(1-phenylhept-2-ynyl)hydroxylamine (0.2 mmol) and NaOH powder (10 mol%) in acetonitrile (2 ml) heated at 50° for 1 h, filtered through silica, concentrated, and purified chromatographically → (Z)-benzyl (3-oxo-1-phenylhept-1-en-1-yl)carbamate (Y 86%), a portion of which (0.2 mmol) dissolved in toluene (2 ml), thiophene-2-carboxamide (1.5 eq.), 4 Å molecular sieves (70 mg) and K-*tert*-butoxide (2 eq.) added sequentially, the mixture heated at 130° for 1 h, filtered through silica, concentrated, and purified chromatographically → 4-butyl-6-phenyl-2-thien-2-ylpyrimidine (Y 60%). Initial isomerization to a β-enaminone optimally required only catalytic amounts of NaOH (K_2CO_3 was less effective, and triethylamine was unreactive), and gave (Z)-products exclusively for electron-diverse aryl (incl. 2-thienyl) substituents at the propargylic position and alkyl (Me, *i*-Pr, *n*-Bu) at the acetylenic position (fourteen examples; Y 45-90%), with bulky propargylic substituents, 1-naphthyl (Y 29%) and 2-tolyl (Y 0%), giving poor results. Bulky *tert*-butyl at the acetylenic position gave only traces of the required product, the major product being an imine (via dehydration) in 41% yield (46% recovered startg. m.). Imines were also isolated (Y 14-18%; two examples) for aryl substituents at the acetylenic position (Y 51-53%). Interestingly, a trimethylsilyl-terminated deriv. suffered N-deprotection, desilylation and rearrangement to a β-aminoenal (Y 68%). The products cyclized on treatment with (het)ar. and aliphatic aldehydes (incl. propenal and formaldehyde) in the presence of *t*-BuOK to afford pyrimidines (twenty examples; Y 49-89%). Attempts to combine the two steps in one-pot were unsuccessful. F.e., optimization and substrate prepn. s. E. Gayon, M. Szymczyk, H. Gérard, E. Vrancken, J.-M. Campagne, J. Org. Chem. *2012*, *77* (20), 9205-20 [DOI: 10.1021/jo301675g].

Sodium azide s. under N-Iodosuccinimide *NaN_3*

Ammonia *NH_3*

N,N-Dimethylanilines from ar. halides Hal → NMe_2
Nucleophilic ar. substitution with dimethylamine generated from dimethylformamide under continuous flow

176. 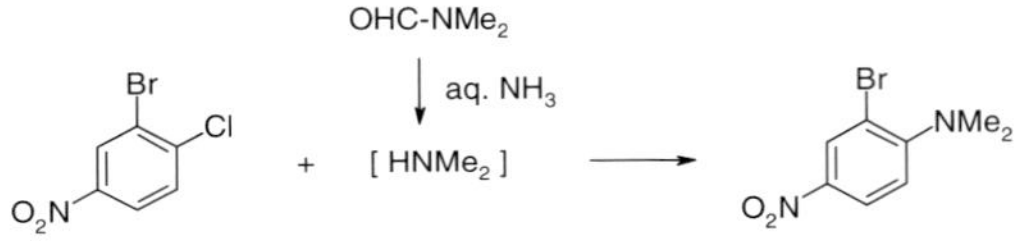

An experimentally simple flow protocol for the *in situ* formation of dimethylamine from DMF using aqueous ammonia has been reported for use in the substitution of activated aromatic and heteroaromatic halides. It overcomes difficulties such as pressure build up on heating dimethylamine or DMF in a sealed tube, while maintaining the advantages of sealed-tube microwave technology, such as fast, efficient heat transfer and safe heating above the boiling point of the solvent with no outgassing of the generated dimethylamine. **E:** A soln. of 2-bromo-1-chloro-4-nitrobenzene (2.998 mmol) in DMF (3 ml) loaded into a sample loop (1.043 ml), the soln. injected into a DMF stream pumped at 0.167 ml/min and mixed with a stream of aq. NH_3/DMF (12.4 *M* aq. NH_3 mixed with DMF (1:9); flow rate: 0.167 ml/min), the mixture passed through a high temperature stainless steel reactor (10 ml; 240°; residence time: 30 min) followed by a cooling element (1 m) and a back pressure regulator (25 bar), the collected product diluted with ethyl acetate (100 ml), washed with water (100 ml) and brine (100 ml), the organic phase dried ($MgSO_4$), concentrated, and the residue purified by column chromatography → 2-bromo-N,N-dimethyl-4-nitroaniline. Y 87%. For more sensitive substrates (e.g. those bearing nitriles or ethyl esters) a 2-step procedure was developed whereby dimethylamine generation and nucleophilic substitution were performed in separate reactors to allow independent variation of temperatures and reaction times. The method is applicable to 2- or 4-chloropyridines, chloropyrimidines, chloro- or bromoquinolines, electron-deficient chloro- or fluoro-benzenes and 2-chloro- or -bromo-azoles. It may be easily scaled up without further optimization by running the reaction for longer. F.e. (sixteen; Y 59-97%) s. T.P. Petersen, A.F. Larsen, A. Ritzén, T. Ulven, J. Org. Chem. *2013*, *78* (8), 4190-5 [DOI: 10.1021/jo400390t].

Silver carbonate/trimethylsilyl azide *Ag_2CO_3/Me_3SiN_3*
Nitriles from terminal acetylene derivs. with C-cleavage C≡CH → CN

177. 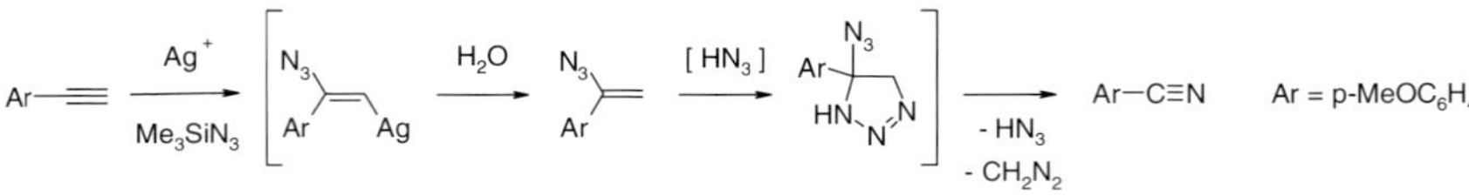

The first example is reported of a *direct* 'nitrogenative' conversion of terminal acetylene derivs. to nitriles by C-C cleavage (cf. *83*, 181). **E:** Ag_2CO_3 (10 mol%), DMSO (2 ml), *p*-methoxyphenyl-acetylene (0.5 mmol) and trimethylsilyl azide (2 eq.) added in turn to an oven-dried Schlenk tube, the mixture stirred at 80° under air for 12 h, cooled to room temp., water added, extracted with ethyl acetate, and worked up with purification by column chromatography → *p*-methoxybenzonitrile. Y 81%. The procedure is applicable to arylacetylenes possessing electron-donating (alkyl, alkoxy, protected amino) or weakly electron-withdrawing (Cl, Br) groups on the benzene ring, as well as to heteroaromatic (indolyl, thienyl) analogs, and both short- and long-chain aliphatic alkynes, incl. diynes (to dinitriles) and thioether-functionalized derivs. (thirty-four examples; Y 39-93%). There was no conversion with other azide sources or transition metal catalysts. Mechanistically, it is thought that reaction proceeds via 1,2-azidoargentation of the alkyne, followed by protonation to give an intermediate [isolable] enazide prior to a 'click' cycloaddition with the liberated HN_3 followed by direct, uncatalyzed cleavage to give the product with elimination of HN_3 and diazomethane. F.e. and solvent effect, also from pre-formed enazides, s. T. Shen, T. Wang, C. Qin, N. Jiao, Angew. Chem., Int. Ed. *2013*, *52* (26), 6677-80 [DOI: 10.1002/anie.201300193].

Copper(II) chloride/1,10-phenanthroline/sodium hydrogen carbonate/oxygen ←
Copper-catalyzed decarboxylative N-arylation with *o*-subst. arylcarboxylic acids NH → NAr

178. 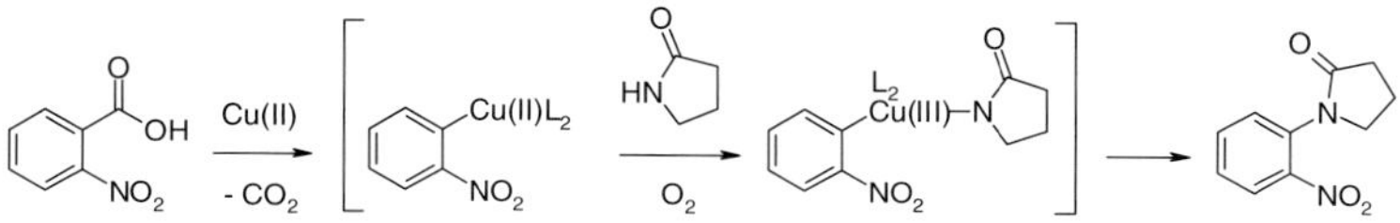

of amides. A soln. of *o*-nitrobenzoic acid (0.5 mmol), 2-pyrrolidone (5 eq.), $CuCl_2$ (20 mol%), 1,10-phenanthroline (20 mol%), $NaHCO_3$ (2 eq.) and 4 Å molecular sieves (50 mg) in anh. anisole (2 ml) purged with O_2 in an oven-dried pressure vessel, the latter sealed, the mixture heated at 170° for 14 h, cooled to ambient temp., diluted with ethyl acetate, acidified with 1 *N* HCl, filtered through a short plug of Celite, and the filtrate worked up with purification by flash chromatography on silica gel → 1-(2-nitrophenyl)pyrrolidin-2-one. Y 73%. The procedure employs a stable, inexpensive, relatively non-toxic (cf. aryl halides!) and widely available arylating agent and is generally applicable to the **N-arylation of carboxylic acid amides, lactams, cyclic ureas and sulfonamides** with arylcarboxylic acids possessing an electron-withdrawing function (nitro, acyl, carbalkoxy or sulfonyl) or a coordinating N-heteroarene group (2-pyridyl, 2-pyrimidinyl or 1-pyrazolyl) *at the ortho position* (twenty-seven examples; Y 34-82%). The yield was lowest with sterically demanding 2-methyl-6-nitrobenzoic acid. Although the mechanistic pathway is unclear, it is possible that reaction proceeds through an arylcopper(III) species with oxygen as oxidant of Cu(I) to Cu(II) after the reductive elimination. F.e. and comparison of copper salts, ligands and solvents s. Y. Zhang, S. Patel, N. Mainolfi, Chem. Sci. *2012*, *3* (11), 3196-9 [DOI: 10.1039/c2sc20606d].

Copper(II) perchlorate/tert-butyl hydroperoxide *$Cu(ClO_4)_2$/t-BuOOH*
N,N-Disubst. carboxylic acid amides from acids and N,N-disubst. formamides COOH + OHC-N< → CON<
Copper-catalyzed oxidative decarboxylative coupling

179. 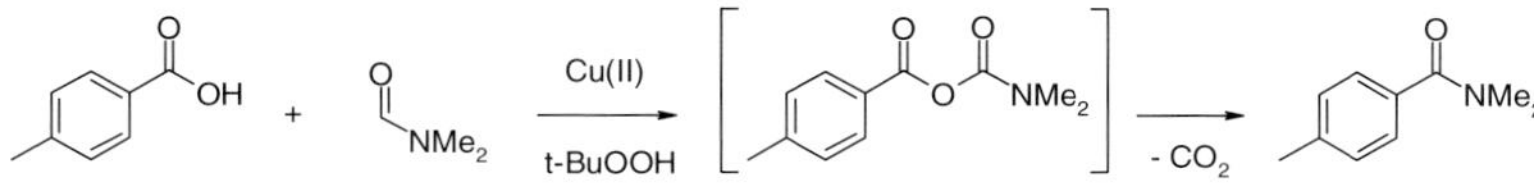

tert-Butyl hydroperoxide (1.5 eq.; 70 wt% in water) added slowly over 5 min to a mixture of *p*-methylbenzoic acid (1 mmol) and $Cu(ClO_4)_2{\cdot}6H_2O$ (10 mol%) in dimethylformamide (2 ml) at room temp., heated in an oil bath at 100° for 15 h with exposure to the atmosphere, cooled to room temp., and worked up with purification by chromatography on silica gel → N,N-dimethyl-*p*-methylbenzamide. Y 78%. The procedure is mild and generally applicable to the coupling of dimethyl- or diethyl-formamide with electron-diverse arylcarboxylic acids [electron-deficient substrates reacting fastest], aliphatic carboxylic acids, arylacetic acids [with electron-donating or weakly electron-withdrawing substituents] or N-protected chiral α-aminocarboxylic acids *with retention of configuration* (twenty-three examples; Y 45-90%). Reaction is thought to involve initial oxidation of the carboxylic acid to give the corresponding *tert*-butyl peroxycarboxylate, followed by copper-catalyzed coupling with the formamide to give an O-acyl carbamate prior to decarboxylation to give the product. F.e. and comparison of solvents, also from the pre-formed *tert*-butyl peroxycarboxylate, s. P.S. Kumar, G.S. Kumar, R.A. Kumar, N.V. Reddy, K.R. Reddy, Eur. J. Org. Chem. *2013* (7), 1218-22 [DOI: 10.1002/ejoc.201201544].

Silver triflate *AgOTf*

1-*sec*-Aminoisoquinolines from *o*-acetylenealdoximes and isothiocyanates ○

180. 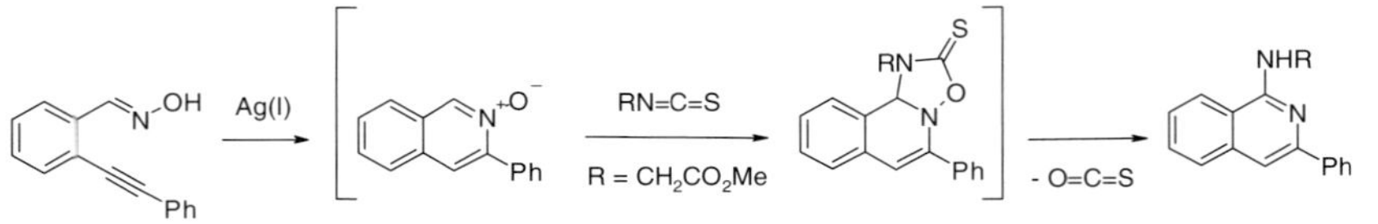

under neutral conditions. Startg. *o*-alkynylbenzaldoxime (0.3 mmol) and isothiocyanate (1.2 eq.) added to a soln. of AgOTf (10 mol%) in methylene chloride (2 ml), stirred for 24 h at room temp., and worked up with purification by chromatography on silica gel → methyl 2-(3-phenylisoquinolin-1-ylamino)acetate. Y 87%. The procedure is simple, mild, efficient, *base-free* and applicable to the coupling of a wide range of *o*-(alk-l-ynyl)aldoximes substituted by a [cyclo]alkyl or phenyl group at the alkyne terminus, and supporting functional groups such as fluorine, methoxy and methylenedioxy on the benzene ring of the aryl aldoxime; in addition, alkyl, benzyl and aryl isothiocyanates were willing partners, reaction being favored by electron-withdrawing groups on the aromatic ring of the latter (nineteen examples in all; Y 62-92%). *o*-(Silylethynyl)aldoximes, however, gave a complex mixture of products, while aryl aldoximes possessing a 5-nitro group failed to react. Reaction is presumed to involve initial 6-*endo* cyclization to give an intermediate isoquinoline N-oxide, followed by [3+2]-cycloaddition with the isothiocyanate and subsequent elimination of carbon oxide sulfide. F.e.s. W. Li, Y. Wang, T. Lu, Tetrahedron *2012*, *68* (34), 6843-8 [DOI: 10.1016/j.tet.2012.06.030].

Boron fluoride *BF_3*

N-Protected 3,4-unsatd. 2-keto-N-glycosides from O^2,O^3,O^4-triacylglycals ←

via stereoselective Ferrier-type N-glycosylation s. *83*, 130

2,2,6,6-Tetramethylpiperidine nitroxyl s. under t-BuONO *TEMPO*

tert-Butyl hydroperoxide s. under $Cu(ClO_4)_2$, I_2 and Bu_4NI *t-BuOOH*

N-Iodosuccinimide/sodium azide or trimethylsilyl azide *NIS/NaN_3 or Me_3SiN_3*

Nitriles from acetylene derivs. R-C≡C-R′ → RCN + R′CN

via direct C-C triple bond cleavage under metal-free conditions

181. 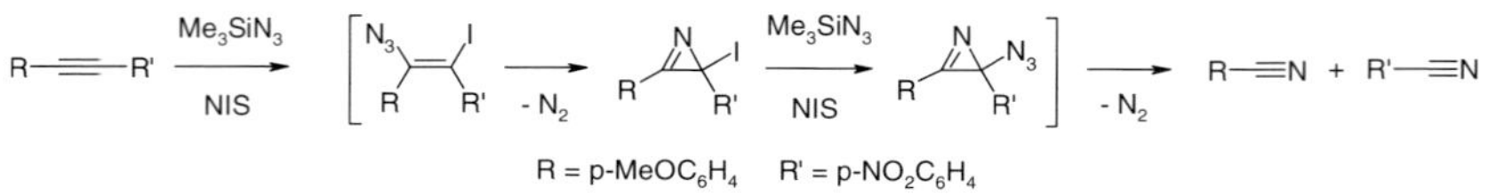

N-Iodosuccinimide (2.4 eq.) and trimethylsilyl azide (2.4 eq.) added to a soln. of 4-methoxyphenyl(4-nitrophenyl)ethyne (0.4 mmol) in 1,2-dichloroethane/acetonitrile (1:1; 2 ml), the mixture stirred at room temp. for 2 h then at 70° for 1 h (TLC), quenched with 10% aq. $Na_2S_2O_3$, extracted with ethyl acetate, and purified by column chromatography → 4-methoxybenzonitrile (Y 84%) and 4-nitrobenzonitrile (Y 84%). This rare, metal-free alkyne cleavage was applicable to diaryl, alkyl-aryl and terminal acetylene derivs. (twenty-seven examples; Y 13%, 43-84%), tolerating ether, carbamate, ester and amide functionality, with best results obtained from diarylacetylenes bearing at least one electron-donating group in the *para* position. NaN_3 could be used in place of

Me_3SiN_3, but NBS and NCS were poor substitutes for NIS. F.e. and support for the proposed mechanism s. N. Okamoto, M. Ishikura, R. Yanada, Org. Lett. *2013*, *15* (11), 2571-3 [DOI: 10.1021/ol401311h].

Trimethylsilyl azide s. under Ag_2CO_3 and N-Iodosuccinimide *Me_3SiN_3*

Trimethylsilyl chloride *Me_3SiCl*

Formamidrazones from thiosemicarbazones and isonitriles RC=NNHC(S)N< → RC=NN=CHNHR'

182. 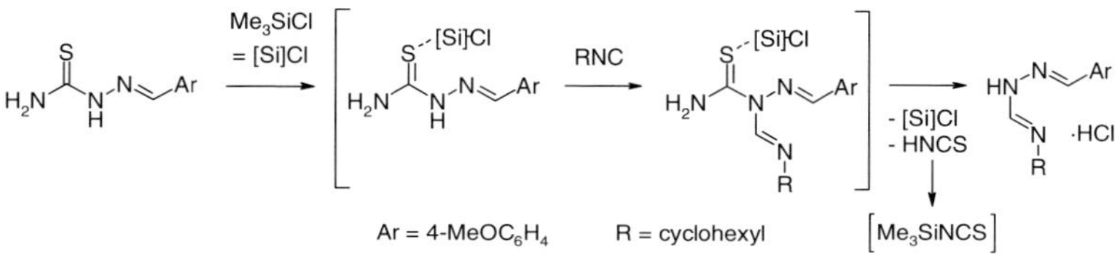

Me_3SiCl (1 eq.) added to a soln. of 4-methoxybenzaldehyde thiosemicarbazone (1 mmol) in dry acetonitrile (15 ml) at room temp., the mixture stirred for 10 min, cyclohexyl isocyanide (1.1 eq.) added under argon, the mixture stirred for 3-6 h (TLC), and the solids filtered off → 1-cyclohexyl-4-(4-methoxybenzylidene)formamidrazone hydrochloride. Y 94%. This novel and unexpected Me_3SiCl-promoted reaction appears general and afforded rare examples of formamidrazones in variable yields (21-94%; twenty-six examples) with lowest yields generally obtained from thiosemicarbazones of aliphatic aldehydes/ketones or from the only ar. isocyanide employed, 4-methoxyphenyl isocyanide. In some cases products conveniently precipitated from soln. as hydrochloride salts, but in general aqueous work up was required (with aq. $NaHCO_3$ wash), affording the products as free bases. The reaction was also applicable to thiosemicarbazide, affording a monosubst. formamidrazone with *t*-BuNC in 67% yield, but analogous conversion of thiourea to an amidine was low-yielding (6%). F.e.s. P. Sarnpitak, S. Tsirulnikov, M. Krasavin, Tetrahedron Lett. *2012*, *53* (48), 6540-3 [DOI: 10.1016/j.tetlet.2012.09.085].

tert-Butyl nitrite/2,2,6,6-tetramethylpiperidine nitroxyl *t-BuONO/TEMPO*

(E)-1-Nitroethylene derivs. from α,β-ethylenecarboxylic acids via stereoselective metal-free decarboxylative nitration COOH → NO_2

183. 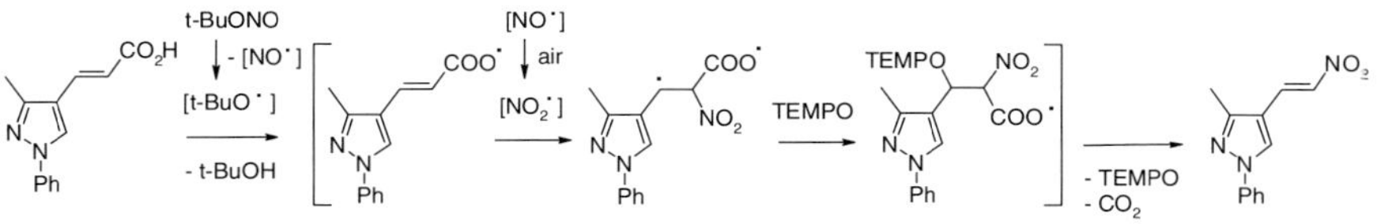

under mild conditions. *t*-BuONO (2 eq.) and acetonitrile (1 ml) added via syringe to a mixture of TEMPO (0.8 eq.) and (E)-3-(3-methyl-1-phenyl-1*H*-pyrazol-4-yl)acrylic acid (0.25 mmol) under air, the mixture stirred vigorously at 50° for 36 h, cooled to room temp., filtered through Celite, concentrated, and purified chromatographically → (E)-3-methyl-4-(2-nitrovinyl)-1-phenyl-1*H*-pyrazole. Y 74%. This acid- and metal-free alternative to the Henry reaction was scalable (6 mmol; 85%) and gave (E)-products exclusively (α-methylcinnamic acid gave E/Z 4:1). Good yields were generally obtained with electron-rich cinnamic acids and thiophene, furan or pyrazole analogs [fourteen examples; Y 70-95%; 3-(thien-2-yl)acrylic acid gave 54%], with yields somewhat lower for electron-poor cinnamic acids (six examples; Y 34-46%), while 1-carboxycyclohexene gave a 95% recovery of startg. m. The proposed radical mechanism involved *anti*-elimination from an energetically favorable conformation of the final intermediate to account for the observed stereoselectivity (in the absence of TEMPO a 1:1 mixture of E/Z products was obtained from 4-methoxycinnamic acid in 42% yield). F.e. and optimization s. S. Manna, S. Jana, T. Saboo, A. Maji, D. Maiti, Chem. Commun. *2013*, *49* (46), 5286-8 [DOI: 10.1039/c3cc41576g].

1,2,3,4-Tetraethyl-5-[o-(diphenylphosphino)phenyl]cyclopentadienes. under Palladium(II) acetate ←

Oxygen s. under $CuCl_2$ *O_2*

Iodine *I_2*

Nitriles from epoxides via oxidative C-cleavage in aq. medium → CN

with [potentially explosive] *o*-iodoxybenzoic acid cf. *80*, 184; with I_2 s. R.R. Jadhav, K.G. Akamanchi, Chem. Lett. *2013*, *42* (2), 162-4 [DOI: 10.1246/cl.2013.162].

Iodine/tert-butyl hydroperoxide/sodium hydroxide *I_2/t-BuOOH/NaOH*
Tetra-n-butylammonium iodide/tert-butyl hydroperoxide *Bu_4NI/t-BuOOH*
N-Subst. carboxylic acid amides CH_2OH + OHC-N< → CON<
from prim. alcohols and N-subst. formamides

184. 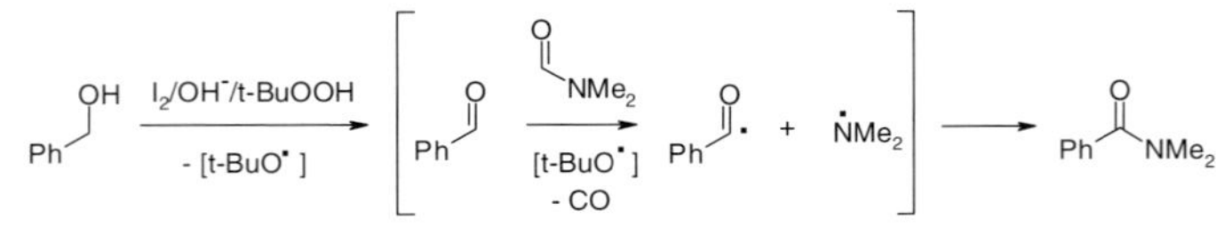

N,N-Disubst. arylcarboxylic acid amides. *tert*-Butyl hydroperoxide (4 eq.; 70 wt% aq. soln.) added to a reaction tube containing a mixture of benzyl alcohol (0.5 mmol), iodine (20 mol%), NaOH (20 mol%) and DMF (0.5 ml), stirred at 70° for 2 h, heated at 80° for 22 h, cooled to room temp., quenched with a satd. soln. of $Na_2S_2O_4$, and worked up with purification by flash column chromatography → N,N-dimethylbenzamide. Y 84%. The procedure is direct, efficient, low-cost, *metal-free* (cf. *80,* 139) and generally applicable to the coupling of electron-diverse benzyl alcohols (incl. *o*-subst. derivs., chlorobenzyl alcohols, and naphthyl analogs), as well as heteroaryl analogs and cinnamyl alcohols, with N,N-disubst. formamides (twenty examples; Y 62-89%) or N-methylformamide (one example; Y 41%). There was no reaction with aliphatic prim. alcohols, nor with hypervalent iodine compds., nor with oxidants other than *tert*-butyl hydroperoxide; sodium hydroxide also appears critical as base. A radical recombinant mechanism is proposed, involving initial hydroperoxy-mediated oxidation of the prim. alcohol to aldehyde through the I_2/I^- cycle, followed by *tert*-butoxyl radical-mediated decarbonylative generation and combination of acyl with aminyl radicals. F.e.s. K. Xu, Y. Hu, S. Zhang, Z. Zha, Z. Wang, Chem. Eur. J. *2012*, *18* (32), 9793-7 [DOI: 10.1002/chem.201201203]; with Bu_4NI (cat)/*t*-BuOOH (cf. *81*, 166 from aldehydes) s. H. Li, J. Xie, Q. Xue, Y. Cheng, C. Zhu, Tetrahedron Lett. *2012*, *53* (48), 6479-82 [DOI: 10.1016/j.tetlet.2012.09.039].

Iron(III) chloride *$FeCl_3$*

Iron-catalyzed syntheses with β-diketones as leaving groups ←

ar. benzylation, allylation or propargylation cf. *79*, 470; **N-benzylation** of prim. ar. amines or sulfonamide with 2-benzyl-1,3-diphenylpropane-1,3-diones s. W. Li, X. Zheng, Z. Li, Adv. Synth. Catal. *2013*, *355* (1), 181-90 [DOI: 10.1002/adsc.201200324].

Palladium(II) acetate/1,2,3,4-tetraethyl-5-[o-(diphenylphosphino)phenyl]cyclopentadiene/lithium tert-butoxide ←

Pyrroles from 1,4-dibromo-1,3-dienes and sec. amines ○

Also indoles from *o*,β-dibromostyrenes s. *83*, 166

Elimination ⇑

Hydrogen ↑ NC ⇑ H

Copper(II) acetate/sodium pivalate/oxygen *$Cu(OAc)_2$/t-$BuCO_2Na/O_2$*
Benzimidazoles from N-arylamidines s. *83,* 171 ○

Copper(II) acetate/potassium phosphate/phenyl iodosoacetate *$Cu(OAc)_2/K_3PO_4/PhI(OAc)_2$*
Cyclic amidines from N′-unsubst. N-alkylamidines s. *81,* 170s*83*

Magnesium oxide s. under Bis[rhodium (α,α,α′,α′-tetramethyl-1,3-benzene-dipropionic acid)] *MgO*

(S)-2-(6-Methyl-2-pyridyl)-4-phenyl-Δ²-oxazoline s. under Palladium(II) carboxylates ←

Laccase/2,2′-azinobis(3-ethylbenzothiazoline-6-sulfonic acid) diammonium salt/air ←

Pyridines from 1,4-dihydropyridines ←

185. 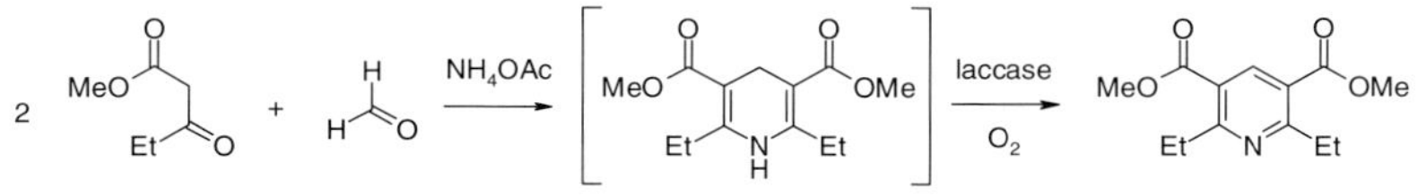

Laccase-catalyzed oxidation of 1,4-dihydropyridines using aerial oxygen as oxidant is reported to proceed highly selectively under mild conditions, and may be applied in a facile and environmentally benign **one pot enzymatic aromatizing Hantzsch pyridine synthesis**. **E:** A mixture of startg. aldehyde (1 mmol), β-dicarbonyl compd. (2 eq.) and ammonium acetate (1.5 eq.) stirred at 80° for 15 min (TLC), cooled to room temp., dissolved in methanol (3-6 ml), acetate buffer (0.2 *M*; pH 4.37; 30 ml), laccase from *Trametes versicolor* (168 U; 12 mg) and 2,2′-azinobis(3-ethylbenzothiazoline-6-sulfonic acid) [ABTS] diammonium salt (1.25 mol%) added, the mixture stirred at 50° for 26 h, extracted with methylene chloride, and purified by flash chromatography on silica gel → product. Y 74% (67% with 10 mol% ABTS after 20 h). Eight further examples of 4-unsubst. or 4-methylated pyridine diesters were obtained in yields of 30-61% (with 1.25 mol% ABTS) or 49-68% (with 10 mol% ABTS), while 3,5-diacetyl-2,6-dimethylpyridine was formed in 59% or 60% yield. Nine examples of laccase-catalyzed oxidation of preformed 4-unsubst. or 4-methylated dihydropyridine diesters gave yields of 60-94% (with 1.25 mol% ABTS) or 71-94% (with 10 mol% ABTS); a 4-phenyl-deriv. was obtained in 80% yield under the second set of conditions, and 3,5-diacetyl-2,6-dimethylpyridine was obtained in 86% or 90% yield under the two sets of conditions. F.e.s. H.T. Abdel-Mohsen, J. Conrad, U. Beifuss, Green Chem. *2012, 14* (10), 2686-90 [DOI: 10.1039/c2gc35950b].

Propionic acid s. under PhI(OCOBu-t)₂ *EtCOOH*

m-Chloroperoxybenzoic acid s. under PhI *mCPBA*

Phenyl iodosoacetate s.a. under Cu(OAc)₂, Chlorobis(α,α,α′,α′-tetramethyl-1,3-benzenedipropanecarboxamidate)dirhodium(II,III), Bis[rhodium-(α,α,α′,α′-tetramethyl-1,3-benzenedipropionic acid)] and Pd(OAc)₂ *PhI(OAc)₂*

Iodobenzene/m-chloroperoxybenzoic acid *PhI/mCPBA*

Phenyl iodosoacetate/cesium carbonate *$PhI(OAc)_2/Cs_2CO_3$*

Benzimidazoles from N-arylamidines s. *75,* 192s*83* ○

Phenyl iodosopivalate/propionic acid *$PhI(OCOBu\text{-}t)_2/EtCOOH$*

2,4-Pyrrolidinediones from β-ketocarboxylic acid amides s. *81,* 170s*83*

Oxygen (or air) s. under $Cu(OAc)_2$, $FeCl_3$, Shvo's ruthenium hydride complex and Palladium(II) carboxylates *O_2*

Iodine/n-butyllithium/zinc chloride *$I_2/BuLi/ZnCl_2$*

Benzo-fused cyclic amines from *o*-subst. ω-arylamines s. *83*, 194

Iodine/potassium carbonate *I_2/K_2CO_3*

2-Acyl-N-carbo-*tert*-butoxyindolines from β-[*o*-(carbo-*tert*-butoxyamino)aryl]ketones s. *83*, 194

Iron(III) bromide/oxygen *$FeBr_3/O_2$*

Oxidative intramolecular *o*-amination

under palladium catalysis, update s. *75,* 192s*82*; **indazoles** from ar. hydrazones *under iron catalysis* with $FeBr_3$ under O_2, also pyrazoles from α,β-ethylenehydrazones, s. T. Zhang, W. Bao, J. Org. Chem. *2013, 78* (3), 1317-22 [DOI: 10.1021/jo3026862]; *metal-free* conversion of N-arylamidines to **benzimidazoles** with *m*-chloroperoxybenzoic acid and iodobenzene as catalyst in hexafluoroisopropanol s. S.K. Alla, R.K. Kumar, P. Sadhu, T. Punniyamurthy, Org. Lett. *2013, 15* (6), 1334-7 [DOI: 10.1021/ol400274f]; with phenyl iodosoacetate/Cs_2CO_3 cf. J. Huang, Y. He, Y. Wang, Q. Zhu, Chem. Eur. J. *2012, 18* (44), 13964-7 [DOI: 10.1002/chem.201202271]; benzimidazoles from *o*-diamines and trialkylamines (NC⇃↑N) with Pd-C in toluene/water s. N.T. Tran, C.S. Cho, Bull. Korean Chem. Soc. *2012, 33* (12), 4188-90 [DOI: 10.5012/bkcs.2012.33.12.4188]; 5,6-dihydrophenanthridines from 2-alkyl-2′-azidobiphenyls by **denitrogenative intramolecular *o*-α-amination** with [Ru(TPP)CO] under O_2 with white-light irradiation (cf. *82,* 183) s. D. Intrieri, M. Mariani, A. Caselli, F. Ragaini, E. Gallo, Chem. Eur. J. *2012, 18* (34), 10487-90 [DOI: 10.1002/chem.201200888].

Shvo's ruthenium hydride complex/cobalt(II) complex/2,6-dimethoxybenzoquinone/air ←

Lactams from aminoalcohols ○

under rhodium catalysis cf. *73,* 160s*81*; N-protected 5-, 6- and 7-membered lactams in the presence of Shvo's ruthenium hydride complex and added cobalt(II) complex/2,6-dimethoxybenzoquinone under air s. B.P. Babu, Y. Endo, J.-E. Bäckvall, Chem. Eur. J. *2012, 18* (37), 11524-7 [DOI: 10.1002/chem.201202080].

Chlorobis(α,α,α′,α′-tetramethyl-1,3-benzenedipropanecarboxamidate)dirhodium(II,III)/ phenyl iodosoacetate ←

Regioselective oxidative intramolecular amination with sulfamic acid esters using a stable, mixed-valent dirhodium(II,III) catalyst at low loading

186. 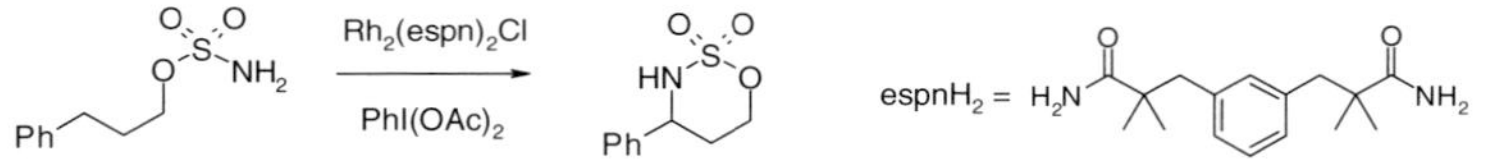

A new, stable, mixed-valent dirhodium(II,III) catalyst, $Rh_2(espn)_2Cl$ [where $espn^{2-}$ = α,α,α′,α′-tetramethyl-1,3-benzenedipropanecarboxamidate], that performs intramolecular C-H amination with TONs >1400, is reported. **E: 3,4,5,6-Tetrahydro-1,2,3-oxathiazine 2,2-dioxides**. Startg. sulfamate ester (20 mg) in freshly distilled methylene chloride (2 ml) containing 4 Å molecular sieves (~20 mg), treated with a stock soln. of $Rh_2(espn)_2Cl$ [*cis*-2,2-isomer] (0.05 mol%) by micropipette, solid $PhI(OAc)_2$ (1 eq.) added, the mixture allowed to react for 12 h, and the solvent evaporated → product. Y 72% by ^1H NMR (TON 1450). The catalyst may be prepared from $Rh_2(espn)_2$ and N-chlorosuccinimide and purified by chromatography. F.e. (one; Y 70% by NMR; TON 1400) and regioselectivity studies, also preliminary investigations into intermolecular reactions, s. K.P. Kornecki, J.F. Berry, Chem. Commun. *2012, 48* (99), 12097-9 [DOI: 10.1039/c2cc36614b].

Bis[rhodium (α,α,α′,α′-tetramethyl-1,3-benzenedipropionic acid)]/magnesium oxide/ phenyl iodosoacetate ←

Oxidative intramolecular amination at sp^3-carbon-hydrogen sites

intramolecular allylic and propargylic amination cf. *80,* 192; *81,* 170; N-protected N′-subst. 1,2-diamines via oxidative intramolecular amination of N-protected O-sulfamylhydroxylamines using $Rh_2(esp)_2$/MgO as catalyst and phenyl iodosoacetate as oxidant s. D.E. Olson, D.A. Roberts, J. Du Bois, Org. Lett. *2012, 14* (24), 6174-7 [DOI: 10.1021/ol302895f]; cyclic amidines from N-unsubst. N-alkylamidines with $Cu(OAc)_2/K_3PO_4$ and the same oxidant s. H. Chen, S. Sanjaya, Y.-F. Wang, S. Chiba, ibid. *2013, 15* (1), 212-5 [DOI: 10.1021/ol303302r]; 2,4-pyrrolidinediones from β-ketocarboxylic acid amides, incl. 3-spiro-derivs., by oxidative intramolecular α-acylamination with phenyl iodosopivalate and excess of propionic acid in nitromethane s. L. Mao, Y. Li, T. Xiong, K. Sun, Q. Zhang, J. Org. Chem. *2013, 78* (2), 733-7 [DOI: 10.1021/jo302144w].

Palladium(II) carboxylates/(S)-2-(6-methyl-2-pyridyl)-4-phenyl-Δ^2-oxazoline/oxygen ←

2-Vinyl-N-heterocyclics from ethyleneamines by Wacker-type oxidative asym. intramolecular N-allylation

s. *80,* 192; with $Pd(OAc)_2$ or $Pd(OCOCF_3)_2$ under O_2 with (S)-2-(6-methyl-2-pyridyl)-4-phenyl-Δ^2-oxazoline as ligand s. A.B. Weinstein, S.S. Stahl, Angew. Chem., Int. Ed. *2012, 51* (46), 11505-9 [DOI: 10.1002/anie.201206702].

Palladium(II) acetate/phenyl iodosoacetate *Pd(OAc)$_2$/PhI(OAc)$_2$*
N-(2-Pyridinesulfonyl)indolines from N-(2-pyridinesulfonyl)-2-arylamines ○
Palladium-catalyzed N-directed dehydrogenative intramolecular *o*-amination

187. 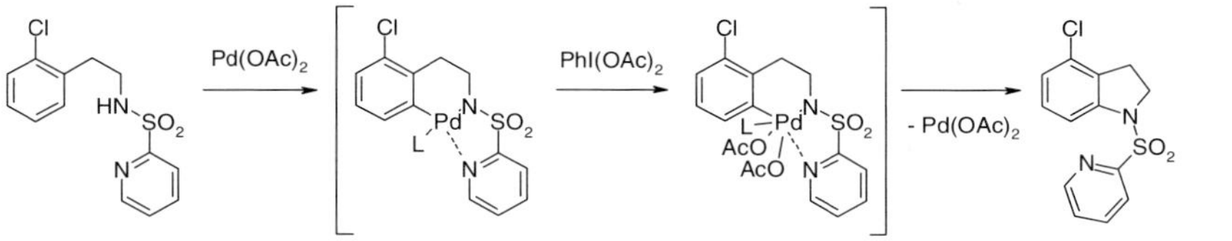

Indoline ring closure by dehydrogenative intramolecular amination has been significantly improved by using the 2-pyridinesulfonyl N-protective and -directing group in place of triflyl (cf. *74,* 232). **E:** A reaction tube containing the startg. 2-pyridinesulfonyl-protected phenethylamine (0.2 mmol), Pd(OAc)$_2$ (10 mol%) and PhI(OAc)$_2$ (2 eq.) in dry *toluene* (20 ml) sealed under air with a Teflon-lined cap, the mixture stirred at 130° for 4 h, cooled to room temp., concentrated *in vacuo*, and worked up with purification by chromatography on silica gel → 4-chloro-N-(2-pyridylsulfonyl)-indoline. Y 68%. The procedure is relatively inexpensive and faster than that with N-triflylated substrates, while removal the N-protective group can be effected under less harsh, racemization-free conditions by using magnesium in methanol at 0°. The method is applicable to substrates possessing electron-donating or -withdrawing groups on the aromatic ring (incl. Br and Cl), as well as substrates substituted at the benzylic and homobenzylic site, incl. carbomethoxy-derivs. (fifteen examples; Y 51-82%). Yields were lower with quinolyl or 2-methylpyridyl as directing group. Reaction is thought to involve initial coordination of palladium(II) to both nitrogens of the substrate, followed by intramolecular *o*-palladation prior to oxidation to the palladium(IV) state and subsequent C-N reductive elimination. F.e. and comparison of solvents s. T.-S. Mei, D. Leow, H. Xiao, B.N. Laforteza, J.-Q. Yu, Org. Lett. *2013*, *15* (12), 3058-61 [DOI: 10.1021/ol401246u].

Oxygen ↑ NC ⇑ O

Irradiation s. under Ethyldiisopropylamine ///

Sodium hydride *NaH*
2,2-Difunctionalized pyrrolidines from γ,γ-difunctionalized oxime O-derivs. ○
via diastereoselective reduction of 5,5-difunctionalized Δ^1-pyrrolines s. *56*, 172s*83*; *83*, 189

Ethyldiisopropylamine/magnesium perchlorate/dichlorotris(2,2′-bipyridyl)ruthenium(II)/ irradiation ←
Oximes from aliphatic nitro compds. CHNO$_2$ → C=NOH

188. 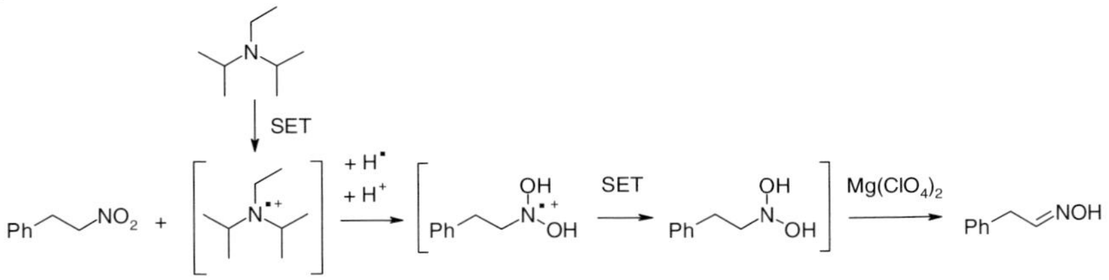

The reductive conversion of aliphatic nitro compds. to oximes has been effected synergistically with a combination of ethyldiisopropylamine as reductive quencher and Mg(ClO$_4$)$_2$ as Lewis acid **under visible-light photoredox catalysis. E:** A soln. of 2-phenylnitroethane (0.55 mmol), ethyl-diisopropylamine (5 eq.), dichlorotris(2,2′-bipyridyl)ruthenium(II) hexahydrate (5 mol%) and Mg(ClO$_4$)$_2$ (50 mol%) in acetonitrile (4 ml) irradiated with a household bulb (45 W) under argon at room temp. for 12 h (TLC), and worked up with purification by chromatography on silica gel → phenylacetaldehyde oxime. Y 79% (E/Z 2.3:1). The procedure is applicable to electron-diverse aryl-subst. nitro compds., heteroaryl analogs and aliphatic (including alicyclic) nitro compds. (twenty examples; Y 35-86%; E/Z 100:0 to 4.1:1).

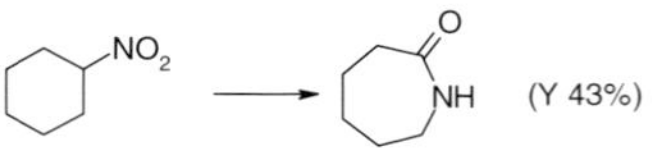

The nature of the tertiary amine, the Lewis acid and the sensitizer were critical, triethylamine and TMEDA having an inhibiting effect, while the organic sensitizer Eosin was practically ineffective. Aldoxime formation was generally more facile than ketoxime formation, and various functional groups were compatible, e.g. CN, Br, hydroxyl and alkynyl. Reaction is thought to involve generation of a dihydroxylamine radical cation via photoredox-mediated SET from the tertiary amine and subsequent hydrogen atom transfer, followed by a second SET to give the neutral dihydroxylamine which then undergoes Lewis acid-catalyzed dehydration to give the oxime. The procedure can be coupled in one pot for further elaboration of the oxime, e.g. by subsequent dehydration to the nitrile with tosyl chloride, or by subsequent Beckmann rearrangement with 85% H_2SO_4. F.e. and comparison of Lewis acids s. S. Cai, S. Zhang, Y. Zhao, D.Z. Wang, Org. Lett. *2013*, *15* (11), 2660-3 [DOI: 10.1021/ol4009443].

Trifluoroacetic acid *CF_3COOH*
1-Thiocarbamylindoles from *o*-thioureido-β-styryl carbonates s. *83*, 143 ○

Tris[3,5-bis(trifluoromethyl)phenyl]phosphine s. under Tris(dibenzylideneacetone)-dipalladium *Ar_3P*

Diethyl phosphorocyanidate/triethylamine or 1,8-diazabicyclo[5.4.0]undec-7-ene ←
Nitriles from aldoximes CH=NOH → CN
update s. *39*, 701s*82*; mild, general procedure with diethyl phosphorocyanidate/Et_3N or DBU s. K. Lee, H.-S. An, C.-Y. Hwang, Bull. Korean Chem. Soc. *2012*, *33* (10), 3173-4 [DOI: 10.5012/bkcs.2012.33.10.3173].

Magnesium perchlorate s. under Ethyldiisopropylamine *$Mg(ClO_4)_2$*

Iron/acetic acid *Fe/AcOH*
Pyrid[2,3-*b*]indoles from α-alkylidene-γ-cyano-γ-(*o*-nitroaryl)ketones s. *83*, 384 ○

Iron(III) chloride *$FeCl_3$*
Asym. intramolecular N-allylation
under palladium catalysis cf. *51*, 188s*67*; chiral N-tosyl-2-vinyl-N-heterocyclics from ω-tosyl-amino-2-ethylenecarbonic acid esters with various chiral ligands s. B. Olszewska, B. Kryczka, A. Zawisza, Tetrahedron Lett. *2012*, *53* (50), 6826-9 [DOI: 10.1016/j.tetlet.2012.10.014]; N-protected 1,2-dihydroquinolines from 1-(*o*-aminoaryl)-2-ethylenealcohols, racemic conversion under iron catalysis with $FeCl_3 \cdot 6H_2O$ (cf. *37*, 246), also **quinolines** by subsequent refluxing with NaOH, s. Z. Wang, S. Li, B. Yu, H. Wu, Y. Wang, X. Sun, J. Org. Chem. *2012*, *77* (19), 8615-20 [DOI: 10.1021/jo301560w].

Dichlorotris(2,2′-bipyridyl)ruthenium(II) s. under Ethyldiisopropylamine *$RuCl_2(bpy)_3$*

Tris(dibenzylideneacetone)dipalladium/tris[3,5-bis(trifluoromethyl)phenyl]phosphine/triethylamine ←
5-Vinyl-Δ^1-pyrrolines from O-pentafluorobenzoyl-γ,δ-ethyleneoximes

189. 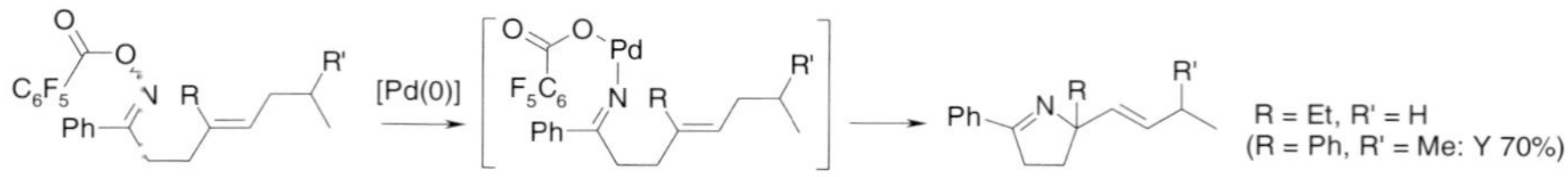

Narasaka-Heck-type intramolecular amination (cf. *56*, 177) has been optimized for reaction with 1,1-disubst. alkenes, providing an operationally simple route to synthetically challenging 5,5-disubst. Δ¹-pyrrolines. **E:** An oven-dried reaction tube charged with $Pd_2(dba)_3$ (3.75 mol%), $[3,5-(F_3C)_2C_6H_3]_3P$ (15 mol%) and (4E)-4-ethyl-1-phenyloct-4-en-1-one O-perfluorobenzoyloxime (0.14 mmol), the tube fitted with a rubber septum then purged with argon, anh. DMF (10 ml/mmol) and Et_3N (2 eq.; reagent grade) added via syringe, the mixture placed in a preheated oil bath (120°) for 2.25 h, cooled to room temp., concentrated *in vacuo*, and the residue purified by flash column chromatography → (E)-2-(but-1-en-1-yl)-2-ethyl-5-phenyl-3,4-dihydro-2*H*-pyrrole. Y 90%. The method is applicable to a variety of alkyl or (het)aryl homoallyl ketoxime esters, affording Δ¹-pyrrolines bearing subst. or unsubst. vinyl groups and 5-methyl, -ethyl or, notably, -phenyl groups or a 5-spiro-analog in high yields (61-85%; twelve examples); a substrate with an electron-deficient alkene group afforded a 5-carbethoxy-deriv. in 31% yield. Preliminary studies on *asym. ring closure* using an extensive range of commercially available electron-neutral/poor chiral ligands identified a [tunable] TADDOL-based phosphoromonoamidite as a promising starting point (three examples; Y 51-56%; e.e. 32-37%).

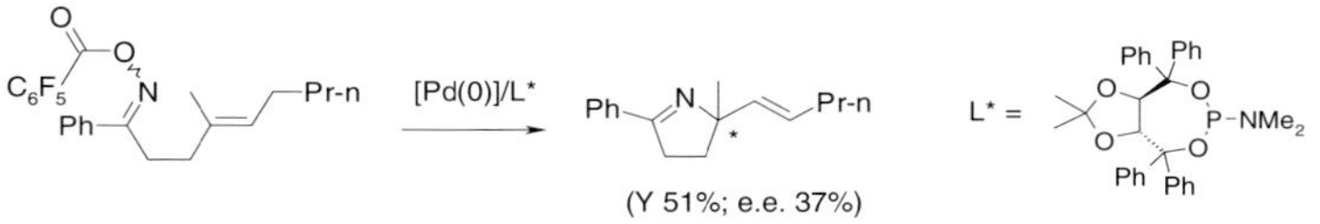

(Y 51%; e.e. 37%)

F.e.s. A. Faulkner, J.S. Scott, J.F. Bower, Chem. Commun. *2013*, *49* (15), 1521-3 [DOI: 10.1039/c2cc38944d]; ring closure with cyclic ethylene derivs. cf. A. Faulkner, J.F. Bower, Angew. Chem., Int. Ed. *2012*, *51* (7), 1675-9 [DOI: 10.1002/anie.201107511]; **4-difluoromethylene-Δ¹-pyrrolines** from O-pentafluorobenzoyl-β-trifluoromethyl-β,γ-ethyleneoximes with $Pd(PPh_3)_4/Ph_3P$ (NC⇑Hal) s. J. Ichikawa, R. Nadano, N. Ito, Chem. Commun. *2006* (42), 4425-7 [DOI: 10.1039/b610690k]; **2,2-difunctionalized pyrrolidines** from γ,γ-difunctionalized oxime O-derivs. (ether or O-tosyl) **via** ring closure with NaH (cf. *56*, 172) and **diastereoselective reduction of** the formed **5,5-difunctionalized Δ¹-pyrrolines** (with Pd-C/H_2 or $NaBH_3CN$) s. N. Chandan, A.L. Thompson, M.G. Moloney, Org. Biomol. Chem. *2012*, *10* (39), 7863-8 [DOI: 10.1039/c2ob26423d].

Nitrogen ↑ NC ⇑ N

Irradiation s. under Carbonyl(5,10,15,20-tetraphenylporphyrinato)ruthenium(II) ℋ

Copper(II) triflate/tert-butyl hydroperoxide *$Cu(OTf)_2$/t-BuOOH*

Nitriles from prim. azides $CH_2N_3 \rightarrow CN$

under ruthenium catalysis cf. *82*, 189; arylethynyl cyanides from arylpropargyl azides by oxidative denitrogenation with $Cu(OTf)_2 \cdot PhMe$ as catalyst and *tert*-butyl hydroperoxide as oxidant via Schmidt rearrangement s. T. Wang, H. Yin, N. Jiao, Adv. Synth. Catal. *2013, 355* (6), 1207-10 [DOI: 10.1002/adsc.201201056].

Oxygen s. under Carbonyl(5,10,15,20-tetraphenylporphyrinato)ruthenium(II) *O_2*

Trimethylsilyl triflate *Me_3SiOTf*

Cyclic α,β-ethyleneazomethines from ω-azido-2-ethylenealcohols ○

190.

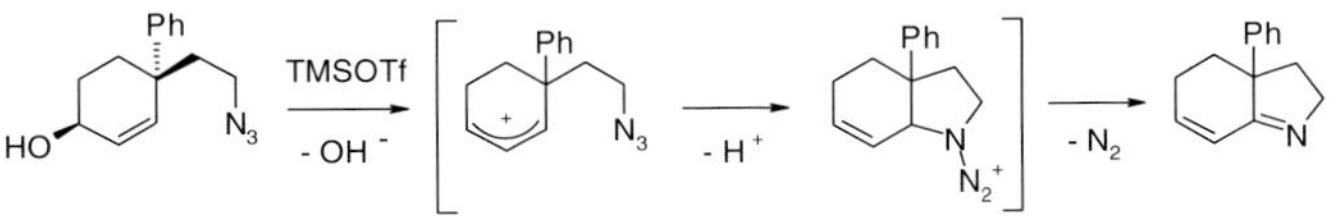

Trimethylsilyl triflate (1.2 eq.) added to a soln. of *trans*-4-(2-azidoethyl)-4-phenylcyclohex-2-en-1-ol in methylene chloride (0.1 *M*) at 0°, the mixture stirred for 10 min (TLC), quenched with satd. aq. $NaHCO_3$, extracted with ether or ethyl acetate, and purified chromatographically → 3a-phenyl-3,3a,4,5-tetrahydro-2*H*-indole. Y 83%. A series of sec. or tert. allylic alcohols incorporating an azidopropyl unit (typically 6-azidohex-2-en-1-ol derivs.) cyclized to the corresponding **2-vinyl-Δ¹-pyrrolines** under the conditions (seven examples; Y 60-98%, a single tert. alcohol requiring trifluoroacetic acid as catalyst), with an isomeric 6-azidohex-1-en-3-ol deriv. giving lower (35%)

and non-reproducible yields. The reaction appeared unaffected by double bond geometry, and yields were not improved by the use of better leaving groups (OAc, OSiMe$_2$Bu-*t*, OMOM). The reaction failed with prim. allylic alcohol analogs and also with a simple ω-azidoalkene, but an epoxy-azidoalkene was converted to the imine with concomitant cleavage of the epoxy ring (Y 37% using TsOH as catalyst).

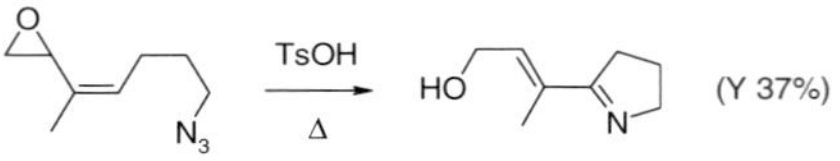

Experimental evidence suggested the reaction occurs via carbocation-mediated S$_N$2′-type Schmidt reaction. F.e., optimization, and use of the method in the synthesis of a Costa Rican ant venom alkaloid s. K. Hayashi, H. Tanimoto, H. Zhang, T. Morimoto, Y. Nishiyama, K. Kakiuchi, Org. Lett. *2012*, *14* (22), 5728-31 [DOI: 10.1021/ol302608q].

Iron(II) bromide *FeBr$_2$*

Indoles from *o*-β-alkoxyazides
via iron(II)-catalyzed intramolecular amination with 1,2-migration

191.

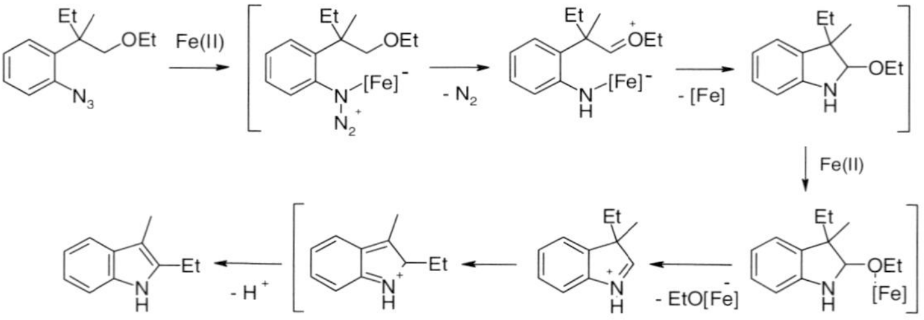

Toluene (1.2 ml) added to a mixture of 2-(1-ethoxy-2-methylbut-2-yl)phenyl azide (0.1 mmol) and FeBr$_2$ (20 mol%) in a Schlenk tube, the mixture heated at 140° for 16 h, cooled to room temp., and purified chromatographically → 2-ethyl-3-methylindole. Y 83%. Iron(II) appears to uniquely catalyze an intramolecular amination-elimination-1,2-migration sequence to afford 2,3-disubst. indoles (eight examples; Y 60-95%), with migration demonstrated for methyl, ethyl, isopropyl, spiroalkyl (ring expansion), phenyl and 4-methoxyphenyl (a 4-trifluoromethylphenyl deriv. gave decomposition products). For unsymmetrically subst. pro-3,3-indoline derivs. the order of migratory aptitude was shown to be Ph > sec. alkyl > prim. alkyl > methyl. The method was also applicable to the synthesis of 2-H,3-alkylindoles but the yields were somewhat lower (42-50%; two examples). Crossover experiments indicate the migration is likely a concerted process. F.e., optimization and substrate prepn. s. Q. Nguyen, T. Nguyen, T.G. Driver, J. Am. Chem. Soc. *2013*, *135* (2), 620-3 [DOI: 10.1021/ja3113565].

Chloro(dipyrrinato)iron(II) complex *[Fe(II)]*

N-Protected cyclic sec. amines from azides

192.

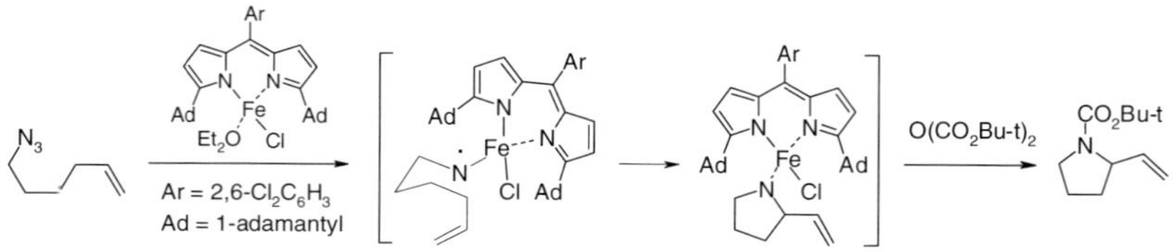

An enzyme-like, iron-mediated intramolecular amination with alkyl azides at remote sp^3 sites is reported, that is rendered catalytic in the presence of an N-protecting agent. **E: N-Carbo-*tert*-**

butoxypyrrolidines. ($^{Ad}L_{Cl2}$)FeCl(OEt$_2$) [where ($^{Ad}L_{Cl2}$)H = 1,9-di(1-adamantyl)-5-(2,6-dichlorophenyl)dipyrromethene] (10 mol%) and *tert*-butoxyformic anhydride (1 eq.) added to a stirred soln. of startg. azide (1 eq.) in benzene in a scintillation vial under N_2, the resultant inky, dark red soln. transferred to a pressure vessel, heated to 65° for 12 h, the mixture flash chromatographed through a short pipette of deactivated silica gel using methylene chloride/methanol (20:1) as eluent (to remove paramagnetic materials), the clear bright orange soln. concentrated by rotary evaporation, and the residue purified by silica gel chromatography → *tert*-butyl 2-vinylpyrrolidine-1-carboxylate. Y 72%. Reaction takes place at a variety of sp^3 sites (allylic, benzylic, sec., tert.) and notably proceeds with retention of configuration with a chiral substrate.

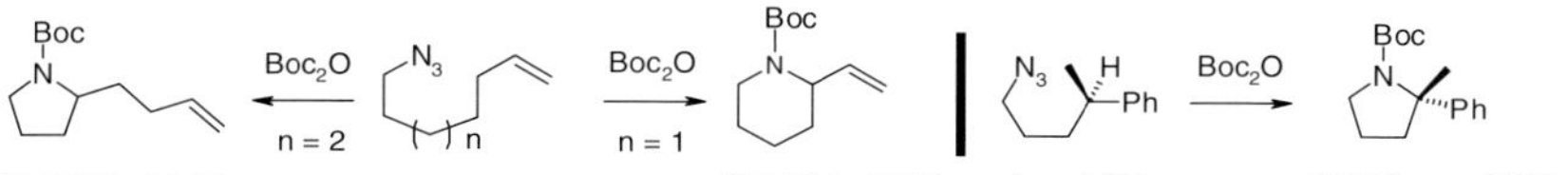

F.e. (ca. twenty; Y 11-19%, 47-98%) and regioselectivity s. E.T. Hennessy, T.A. Betley, Science *2013*, *340* (6132), 591-5 [DOI: 10.1126/science.1233701]; intermolecular reactions with alkyl or aryl azides, affording sec. benzylamines from toluene and 2-phenylaziridines from styrene, cf. E.R. King, E.T. Hennessy, T.A. Betley, J. Am. Chem. Soc. *2011*, *133* (13), 4917-23 [DOI: 10.1021/ja110066j].

Carbonyl(5,10,15,20-tetraphenylporphyrinato)ruthenium(II)/oxygen/irradiation *[Ru(II)]/O_2/↯*
5,6-Dihydrophenanthridines from 2-alkyl-2′-azidobiphenyls s. *82*, 183s*83* ○

Halogen ↑ NC ⇑ Hal

Cesium fluoride s. under N,O-Bis(trimethylsilyl)acetamide *CsF*

Copper/sodium tert-butoxide/polyethylene glycol-400 *Cu/NaOBu-t/PEG*
Copper(I) oxide/o-phenylenediamine/sodium tert-butoxide ←
Catalytic intramolecular N-arylation ○
by Ullmann coupling, update s. *63*, 191s*82*; **oxindoles** from α-(*o*-bromoaryl)carboxylic acid amides with Cu_2O/*o*-phenylenediamine and NaOBu-*t* as base under aerobic conditions s. Y.-H. Jhan, T.-W. Kang, J.-C. Hsieh, Tetrahedron Lett. *2013, 54* (9), 1155-9 [DOI: 10.1016/j.tetlet.2012.12.082]; **indazoles** from *o*-bromoaryl hydrazones with Cu-powder/NaOBu-*t* in PEG-400 s. H.K. Lee, C.S. Cho, Synth. Commun. *2013, 43* (6), 915-21 [DOI: 10.1080/00397911.2011.614714]; N-protected **indolines** from *o*-(2-aminoalkyl)diaryliodonium tosylates with Cs_2CO_3/TEMPO s. K.P. Landge, K.S. Jang, S.Y. Lee, D.Y. Chi, J. Org. Chem. *2012, 77* (13), 5705-13 [DOI: 10.1021/jo300874m].

2,2,6,6-Tetramethylpiperidine nitroxyl/cesium carbonate *TEMPO/Cs_2CO_3*
N-Protected indolines from *o*-(2-aminoalkyl)diaryliodonium tosylates s. *63,* 191s*83*

N,O-Bis(trimethylsilyl)acetamide/cesium fluoride *MeC(OSiMe$_3$)=NSiMe$_3$/CsF*
5-α-Fluoro-2-oxazolidones from 2-carbamyloxy-1,3-difluorides via intramolecular nucleophilic substitution of unactivated fluorine

193.

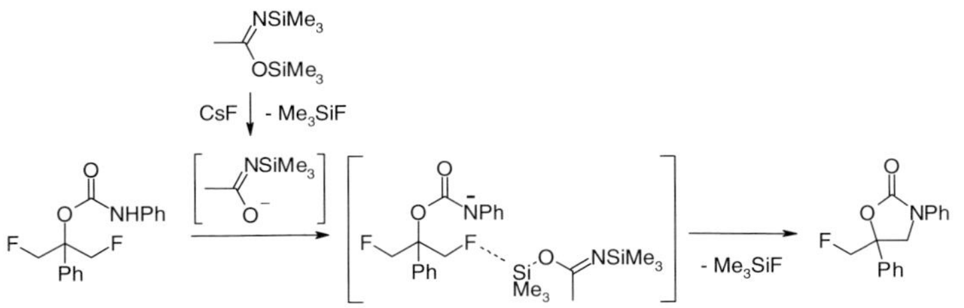

N,O-Bis(trimethylsilyl)acetamide [BSA] (1.2 eq.) added to a soln. of the startg. 2-carbamyloxy-1,3-difluoride (0.1 mmol) and CsF (10 mol%) in acetonitrile (1 ml) under N_2, stirred at room temp. for 1 h, quenched with satd. aq. NH_4Cl, and worked up with purification by chromatography on silica gel → 5-fluoromethyl-3,5-diphenyloxazolidin-2-one. Y 94%. High yields were recorded for the ring closure of substrates possessing electron-diverse aryl or alkyl substitution at the C_2-position as well as electron-diverse N-aryl groups (69-96%; eleven examples). The two reagents function synergistically by initial fluoride ion-generation of a catalytic amido anion from BSA for N-deprotonation of the substrate; this is then followed by BSA-mediated cleavage of the C-F bond prior to ring closure with formation of trimethylsilyl fluoride as the driving force of the reaction. Yields were lower with strong inorganic bases (NaOH, NaH), while organic bases were ineffective. Interestingly, a one-pot procedure **from 2-hydroxy-1,3-difluorides** and electron-diverse N-arylisocyanates was equally facile (sixteen examples; Y 45-94%); however, in this case, NaH alone proved to be the reagent of choice with elimination of NaF. F.e. and comparison of bases and solvents s. G. Haufe, S. Suzuki, H. Yasui, C. Terada, T. Kitayama, M. Shiro, N. Shibata, Angew. Chem., Int. Ed. *2012*, *51* (49), 12275-9 [DOI: 10.1002/anie.201207304].

Tetrakis(triphenylphosphine)palladium(0)/triphenylphosphine *$Pd(PPh_3)_4/Ph_3P$*
4-Difluoromethylene-Δ¹-pyrrolines ○
from O-pentafluorobenzoyl-β-trifluoromethyl-β,γ-ethyleneoximes s. *83*, 189

Sulfur ↑ NC ⇑ S

Sodium hydride or n-Butyllithium *NaH or BuLi*
3-Fluoropyrazoles from 2-(difluoromethylene)-N′-tosylhydrazines s. *83*, 154 ○

Iodine/potassium carbonate *I_2/K_2CO_3*
2-Acylindoles from β-[*o*-(tosylamino)aryl]ketones

194. 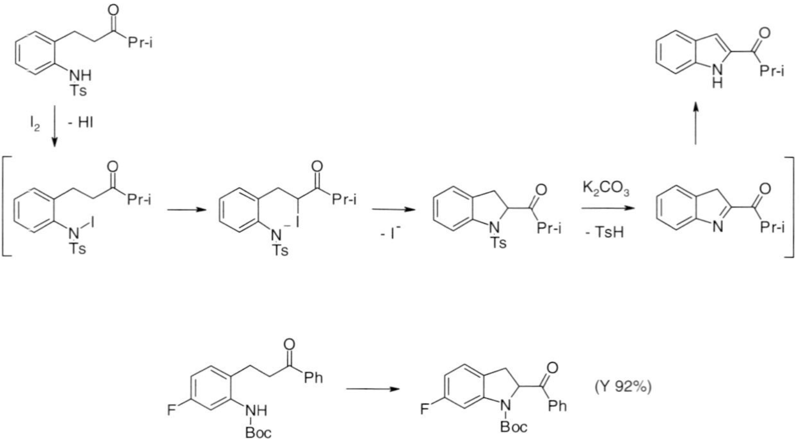

I_2 (1.1 eq.) added to a mixture of N-tosyl-2-(4-methyl-3-oxopentyl)aniline (0.2 mmol) and K_2CO_3 (3 eq.) in methanol (3 ml) at room temp., the mixture stirred at 60° for 1.5 h (TLC), concentrated *in vacuo*, quenched with satd. aq. $Na_2S_2O_3$, extracted with ethyl acetate, and purified chromatographically → 2-isobutanoylindole. Y 84%. This novel **iodine-mediated dehydrogenative intramolecular N-alkylation** was successful for a series of N-tosyl-β-2′-aminophenyl (het)aryl/alkyl ketones at 60°, with loss of the N-protecting group affording the corresponding 2-acyl/

(het)aroyl-NH-indoles [incl. bis(indol-2-yl) ketone] (fifteen examples; Y 71-93%; 63% for 2-pivaloylindole). A methyl group in the pro-3 position was also tolerated. Interestingly, when Boc was used to protect N, the reaction proceeded, generally at room temp., with *retention of the protecting group,* to afford the corresponding **2-acyl-N-carbo-*tert*-butoxyindolines** (twelve examples, Y 75-92%), with an α-iodoketone intermediate isolated in one case (which cyclized on treatment with K_2CO_3/MeOH). F.e. and optimization s. W.-C. Gao, S. Jiang, R.-L. Wang, C. Zhang, Chem Commun. *2013*, *49* (43), 4890-2 [DOI: 10.1039/c3cc40797g]; **benzo-fused cyclic amines from *o*-subst. ω-arylamines** with BuLi then $ZnCl_2$ then I_2 s. J.L. Jeffrey, E.S. Bartlett, R. Sarpong, Angew. Chem., Int. Ed. *2013*, *52* (8), 2194-7 [DOI: 10.1002/anie.201209591].

Remaining Elements ↑ NC ⇑ Rem

Sodium hydride *NaH*

3-Fluoropyrazoles from 2′-(N′-tosylhydrazino)-2,2-difluoroenesilanes s. *83*, 154 ○

Carbon ↑ NC ⇑ C

[Tris[tris(3,5-di-tert-butyl-4-methoxyphenyl)silylethynyl]phosphine]gold(I) triflimide *[Au(I)]*

2-β-Keto-N-tosylpiperidines from 5-benzoyloxy-6-acetylene-N-tosylamines s. *83*, 89 ○

Carbonic acid esters/1,5,7-triazabicyclo[4.4.0]dec-5-ene *$(RO)_2CO$/TBD*

Catalytic Lossen rearrangement ⌒

Prim. ar. amines from arylhydroxamic acids s. *83*, 108

Bis(dibenzylideneacetone)palladium(0)/chiral 2,3:2′,3′-bis(methylenedioxy)-6,6′-bis[bis(3,5-di-tert-butyl-4-methoxyphenyl)phosphino]biphenyl *[Pd(0)]**

2-Allene-N-tosylamines from N-tosylcarbamyloxy-2-allenes OC(O)N(R)Ts → N(R)Ts

195. 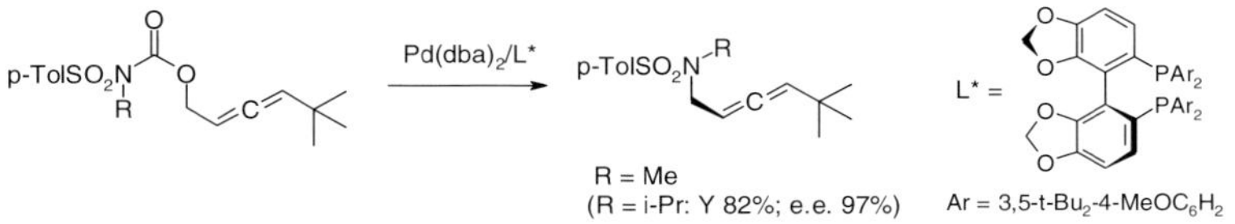

Axially-chiral 2-allene-N-tosylamines. $Pd(dba)_2$ (5 mol%), (S)-DTBM-SegPhos (7.5 mol%) and a 1:1 mixture of DMF/DME (1.5 ml) added sequentially under argon to a flame-dried Schlenk tube, the mixture stirred at 25° for 30 min, startg. allenyl N-tosylcarbamate (0.5 mmol) and 1:1 DMF/DME (0.5 ml) added, the mixture stirred at 25° for 84 h (TLC), diluted with ether, quenched with water, separated, extracted with ether, dried (Na_2SO_4), filtered, evaporated, and the residue purified by chromatography on silica gel → (S)-N-(5,5-dimethylhexa-2,3-dien-1-yl)-N,4-dimethylbenzenesulfonamide. Y 88% (e.e. 92%). Initially the formation of racemic allenyl amines was studied using $Pd(PPh_3)_4$ (5 mol%) in methylene chloride at 40° (fourteen examples; Y 73-93%). The method was then optimized for *palladium-catalyzed asym. decarboxylative amination*, and found to be compatible with additional functionality such as benzyl ether, *tert*-butyldimethylsilyl ether, aromatic bromine and N-but-2-ynyl groups; thus sixteen further examples gave yields of 76-93% (e.e. 90-97%). Formation of allenes having axial chirality and central chirality from substrates bearing a sec. siloxy group was also studied. F.e.s. B. Wan, S. Ma, Angew. Chem., Int. Ed. *2013*, *52* (1), 441-5 [DOI: 10.1002/anie.201204796].

Formation of Hal-C Bond

Uptake ⇓

Addition to Oxygen and Carbon HalC ⇓ OC

Triethylamine tris(hydrogen fluoride)/chloro(1,5-cyclooctadiene)rhodium(I) dimer/chiral ferrocenyldi(phosphines) *$Et_3N{\cdot}3HF$/[Rh]**

2-Fluoro-1-hydroxy-1,2-dihydronaphthalenes from 1,4-oxido-1,4-dihydronaphthalenes
Rhodium-catalyzed asym. fluorinative ring opening

196. 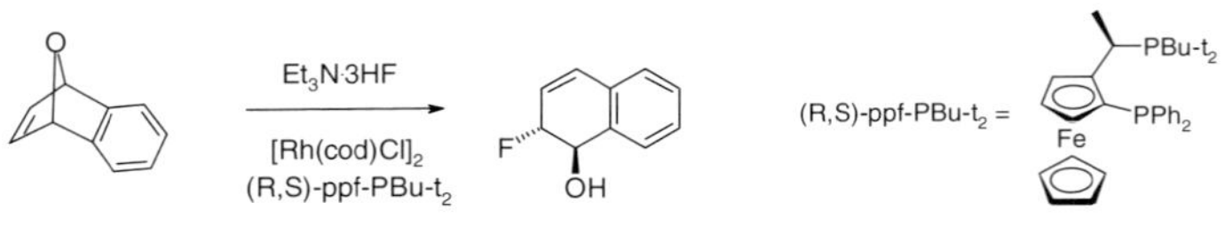

The first example of halide as nucleophile in the enantioselective ring opening of oxabicyclic alkenes is reported (cf. *82*, 41 for water as nucleophile). **E:** $[Rh(cod)Cl]_2$ (4 mol%), the chiral JosiPhos ligand (8 mol%) and startg. oxabenzonorbornadiene (0.2 mmol) added to a glass vial, the vial capped with a septum, flushed with argon, THF (2 ml) followed by triethylamine tris-(hydrogen fluoride) (3 eq.) added by a syringe, the mixture stirred at 50° for 30 min under argon, quenched with satd. aq. $NaHCO_3$, extracted with methylene chloride, and purified by flash column chromatography → product. Y 88% (e.e. 98%). Three further examples from substrates bearing difluoro, dibromo or dimethyl groups gave e.e. of 96% and yields of 53-79% within 1 h. For substrates bearing electron-donating groups such as (di)alkoxy or methylenedioxy, reaction was complete within 15 min at room temp. but the initial products decomposed to 1-naphthols on chromatography or after prolonged reaction times, so were subjected to hydrogenation over Pd-C after a rapid aqueous work-up, as was a triphenylene-deriv., to afford **chiral 2-fluoro-1-tetralols** (four examples; Y 36-67%; e.e. 94-99%). Non-benzo-fused substrates, bridgehead substituted substrates, and azabicyclic alkenes were unreactive under these conditions. Reaction of the illustrated substrate on the 4 mmol scale followed by hydrogenation afforded the corresponding tetralol in 72% yield (98% e.e.) which was subjected to further substitution and oxidation reactions. F.e. and optimization s. J. Zhu, G.C. Tsui, M. Lautens, Angew. Chem., Int. Ed. *2012, 51* (49), 12353-6 [DOI: 10.1002/anie.201207356].

Addition to Carbon-Carbon Bonds HalC ⇓ CC

Lithium bromide/phenyl iodosoacetate *$LiBr/PhI(OAc)_2$*

1,2-Alkoxybromides from ethylene derivs. C=C → C(OR)C(Br)

with N-bromosulfoximines/H_2SO_4 cf. *32*, 472; N-protected 1-alkoxy-2-bromamines from enamines with LiBr/phenyl iodosoacetate, diastereoselectivity, s. S. Nocquet-Thibault, P. Retailleau, K. Cariou, R.H. Dodd, Org. Lett. *2013, 15* (8), 1842-5 [DOI: 10.1021/ol400453b].

9-(p-Nitrobenzenesulfonylamino)-9-deoxyquinine s. under N-Bromosuccinimide ←

Diphenyl selenide s. under N-Chlorosuccinimide *Ph_2Se*

Phenyl iodosoacetate s. under LiBr *$PhI(OAc)_2$*

N-Chlorosuccinimide/diphenyl selenide *NCS/Ph_2Se*

N-Protected 1,2-halogenamines from ethylene derivs. C=C → C(N<)C(Hal)

update s. *76*, 132s*82*; 2-(chlor)acylamines with acetonitrile as reactant, N-chlorosuccinimide as chlorinating agent and diphenyl selenide as catalyst, diastereoselectivity, s. D.W. Tay, I.T. Tsoi, J.C. Er, G.Y.C. Leung, Y.-Y. Yeung, Org. Lett. *2013, 15* (6), 1310-3 [DOI: 10.1021/ol400249x].

N-Bromosuccinimide/9-(p-nitrobenzenesulfonylamino)-9-deoxyquinine ←
Regioselective asym. 1,2-oxybromination of 2-arylprop-2-enols C=C → C(OH)C(Br)
via *in situ* tethering with phenylboronic acid

197.

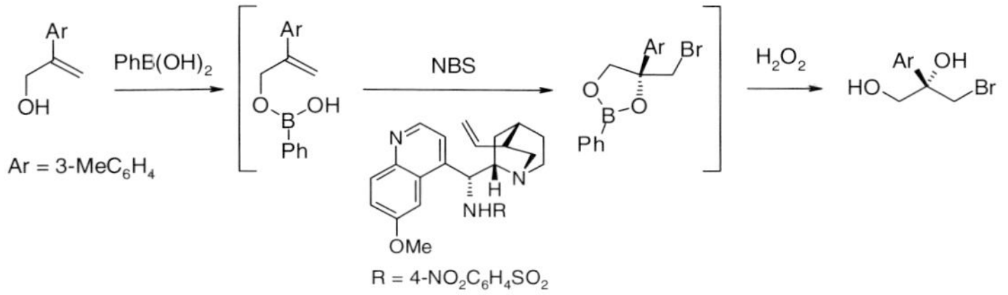

A soln. of N-bromosuccinimide [NBS] (1.2 eq.) in chloroform (2 ml) added dropwise to a suspension of 2-(3-tolyl)prop-2-enol (0.2 mmol), phenylboronic acid (1.2 eq.), quinine catalyst (5 mol%) and 4 Å molecular sieves (300 mg) in the same solvent (2 ml) at -60° under argon, the mixture stirred for 1 h, diluted with methylene chloride, filtered through Celite, concentrated, the residue dissolved in ethyl acetate/acetone (1:1; 2 ml), aq. H_2O_2 (41 µl; 2 eq.) added, the mixture stirred until reaction complete (TLC), solvent removed *in vacuo*, and the residue purified by chromatography on silica gel → 3-bromo-2-(3-tolyl)propane-1,2-diol. Y 71% (e.e. 96%). Control of regioselectivity via formation of a boronate ester *in situ*, and use of a bifunctional quinine-based catalyst for activation of both boronate and NBS were key to the success of this asym. bromohydroxylation of 2-arylprop-2-enols. Enantioselectivity was strongly dependent on the electronic properties of the aryl group, with moderately electron-rich/-poor substrates being most effective (e.e. 79-96%; nine examples; Y 55-90%), and yields/enantioselectivity significantly reduced for highly electron-rich/-poor substrates (e.e. 30-64%; five examples; Y 33-80%), and also for 2-benzyl-2-propenol (e.e. 61%; Y 37%). In some examples, particularly for electron-rich substrates, best results were obtained with the 4-methoxybenzenesulfonamide analog of the illustrated catalyst. F.e., optimization and reactions of the products s. Y. Zhang, H. Xing, W. Xie, X. Wan, Y. Lai, D. Ma, Adv. Synth. Catal. *2013*, *355* (1), 68-72 [DOI: 10.1002/adsc.201200782].

p-Xyleneiodonium difluoride *$ArIF_2$*
1,2-Aminofluorination of styrenes C=C → C(NHSO$_2$R)C(F)
2-Aryl-2-fluoro-N-sulfonylamines s. *83*, 203

Trimethylsilyl chloride/zinc triflate *$Me_3SiCl/Zn(OTf)_2$*
2-(1-Halogenovinyl)cyclopent-3-ene-1,1-dicarboxylic acid esters
from 2-(1,2,4-trien-4-yl)cyclopropane-1,1-dicarboxylic acid esters s. *83*, 307

Trimethylsilyl halides *Me_3SiHal*
Diphenylphosphine oxide *$Ph_2P(O)H$*
α,β-Ethyleneiodides from acetylene derivs. C≡C → CH=C(I)
with diphenylphosphine oxide s. *48*, 426s*77*; Markovnikov hydroiodination of terminal acetylenes with $I_2/Ph_2P(O)H$ or $I_2/(PhO)_2P(O)H/Ph_2P(O)OH$, also from silylacetylenes, s. S.-i. Kawaguchi, H. Masuno, M. Sonoda, A. Nomoto, A. Ogawa, Tetrahedron *2012*, *68* (47), 9818-25 [DOI: 10.1016/j.tet.2012.08.096]; (E)-1-(iodo or bromo)-N-tosylenamines from ynetosylamines with trimethylsilyl iodide or bromide s. A.H. Sato, K. Ohashi, T. Iwasawa, Tetrahedron Lett. *2013*, *54* (10), 1309-11 [DOI: 10.1016/j.tetlet.2012.12.101].

Exchange ↓↑

Hydrogen ↑ HalC ↓↑ H

Without additional reagents *w.a.r.*

4-Chloromethyl-3-iodofurans from 2-α-hydroxy-1,3-enynes ○

198. 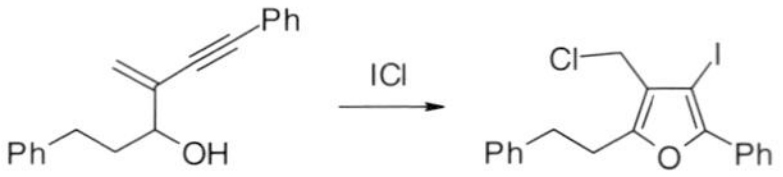

A soln. of ICl (2.2 eq.) in nitromethane (1 ml) added to a soln. of 1,6-diphenyl-4-hydroxy-3-methylenehex-1-yne (0.25 mmol) in the same solvent (5 ml) at room temp., the mixture stirred for 5 min, concentrated *in vacuo*, and purified chromatographically → 4-chloromethyl-3-iodo-2-phenyl-5-(2-phenylethyl)furan. Y 81%. This experimentally simple procedure made efficient use of ICl (I_2 was unreactive), giving good yields of 2-phenyl-4-chloromethyl-3-iodofurans (4-iodomethyl analogs were the expected products) from the corresponding sec. enynyl alcohols (Y 67-81%; four examples). Lower yields were reported for electron-diverse 2-aryl (45-62%; six examples) or 2-butyl (32%) derivs., while a terminal alkyne gave a complex mixture, and a prim. alcohol analog gave a mixture of three 5-subst. (H, Cl, I) derivs. in low (unspecified) yield. F.e. and optimization s. C. Thongsornkleeb, W. Rabten, A. Bunrit, J. Tummatorn, S. Ruchirawat, Tetrahedron Lett. *2012*, *53* (48), 6615-9 [DOI: 10.1016/j.tetlet.2012.09.124].

Sodium hydrogen carbonate *$NaHCO_3$*

N-Protected 4-iodo-Δ⁴-3-pyrrolones from 2,5-oxazolidiones via iodocyclization of α,β-acetylene-α′-aminoketones with retention of chirality

199. 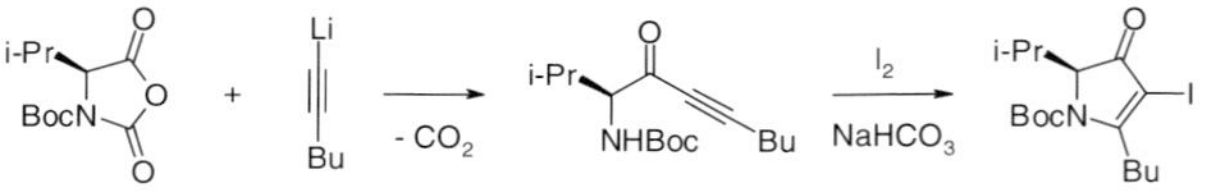

An improved general procedure for direct generation of α-amino-ynones from carboxyanhydride derivs. of N-protected α-amino acids, and their novel iodocyclization to chiral 4-iodo-Δ^4-3-pyrrolones, is described. **E:** (4S)-N-*tert*-Butoxycarbonyl-4-isopropyl-2,5-oxazolidione (2.32 mmol) in anh. THF added slowly to a soln. of hex-1-yn-1-yllithium [freshly prepd. from hexyne (1.33 eq.), butyllithium (1.33 eq.) and LiBr (0.94 eq.) at -78° for 30 min] in THF/hexanes under N_2, the mixture stirred for 4 h, warmed to room temp., quenched with satd. aq. NH_4Cl, extracted with ethyl acetate, and concentrated → *tert*-butyl ((2S)-2-methyl-4-oxodec-5-yn-3-yl)-carbamate (Y 81%; e.e. 96%), a portion of which (0.1 mmol) dissolved in acetonitrile (0.5 ml), $NaHCO_3$ (3 eq.) and a soln. of I_2 (3 eq.) in the same solvent (0.5 ml) added at room temp., stirred for 2 h, concentrated *in vacuo*, the residue dissolved in ethyl acetate, washed with satd. aq. $Na_2S_2O_3$, and concentrated → N-*tert*-butoxycarbonyl-5-butyl-4-iodo-2-isopropyl-Δ^4-3-pyrrolone (Y 96%; e.e. 93%). Initial generation of the amino-ynones was optimally achieved using freshly generated lithium alkynylides (the corresponding Grignard reagents gave low yields), generally without appreciable racemization (e.e. 65%, 81-100%; eight examples; Y 35-95%), although some phenylalanine-derived products gave poor results (e.e. 10-18%; two examples; Y 60-64%). Iodocyclization in acetonitrile at room temp. occurred with only minimal loss of chirality (seven examples; Y 67-98%; e.e. 60%, 78-99%), but use of microwave irradiation at 50° in PEG 3400 was less effective, as was the use of methylene chloride as solvent. F.e., optimization and illustrative palladium-catalyzed cross-coupling reactions of the products s. R. Spina, E. Colacino, B. Gabriele, G. Salerno, J. Martinez, F. Lamaty, Org. Biomol. Chem. *2012*, *10* (45), 9085-9 [DOI: 10.1039/c2ob26427g].

Potassium carbonate/iodine *K_2CO_3/I_2*

7-Iodo-1,3-dioxolo[4,5-*c*]pyridines from 3-alkoxy-4-hydroxypyridines

200.

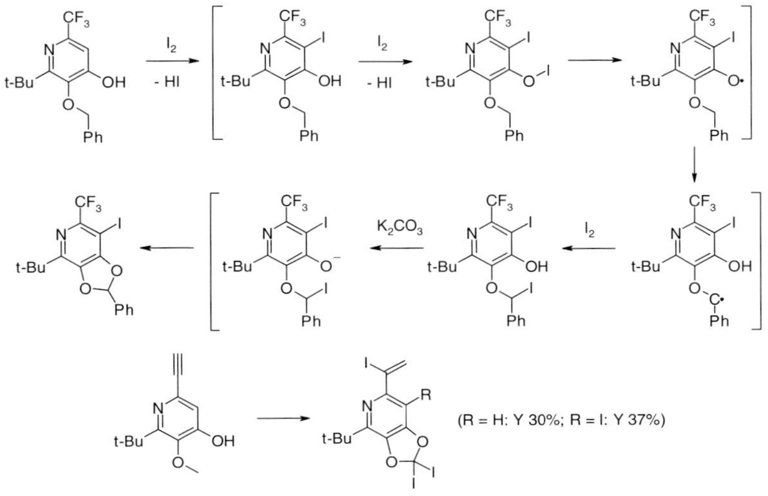

A mixture of 3-benzyloxy-2-*tert*-butyl-4-hydroxy-6-trifluoromethylpyridine (1.2 mmol), I_2 (3 eq.) and K_2CO_3 (3 eq.) in methylene chloride (4 ml) stirred in a sealed tube at 70° for 72 h, diluted with water, extracted with methylene chloride, and purified chromatographically → 4-*tert*-butyl-7-iodo-2-phenyl-6-trifluoromethyl-[1,3]dioxolo[4,5-*c*]pyridine. Y 87%. Attempted 5-iodination of 4-hydroxypyridines in the presence of base unexpectedly resulted in concomitant formation of the dioxolane ring, optimally using chlorinated solvents (chloroform, methylene chloride and 1,2-dichlorobenzene) for the illustrated substrate and its analogous methyl ether (Y 58-87%). Interestingly, use of 1,2-dichloroethane as solvent with the 3-methoxy analog afforded a 5-iodo-4-(2-chloroethoxy)pyridine (Y 50%) while acetonitrile gave only the expected 5-iodo derivs. (three examples; Y 81-90%). A further substrate, containing a 6-ethynyl group, gave a mixture of 7-H/-iodo dioxolopyridines with concomitant hydroiodination of the ethyne and *gem*-diiodination of the dioxolane carbon (Y 30% and 37% respectively). The dioxolane ring was assumed to be formed by a radical process (Hofmann-Löffler-Freytag reaction). F.e. and substrate prepn. s. T. Lechel, G. Podolan, B. Brusilowskij, C.A. Schalley, H.-U. Reissig, Eur. J. Org. Chem. *2012* (29), 5685-92 [DOI: 10.1002/ejoc.201200564].

Sodium chloride (or bromide) s. under Tryptophan-7-halogenase *NaCl or NaBr*

Lithium bromide s. under Pd(OAc)$_2$ *LiBr*

Sodium iodide/vanadyl acetoacetonate/hydrogen peroxide *NaI/VO(acac)$_2$/H_2O_2*

Copper(II) oxide *CuO*

α-Halogenation H → Hal

update s. *38,* 473s*82*; direct, one-pot formation of iodomethyl ketals from methyl ketones **via α-iodination** with I_2/CuO prior to I_2-catalyzed ketalization s. Y. Yang, M. Gao, W.-M. Shu, L.-M. Wu, D.-X. Zhang, A.-X. Wu, Org. Biomol. Chem. *2013, 11* (7), 1226-33 [DOI: 10.1039/c2ob27028e]; α-iodination of β-dicarbonyl compds. with VO(acac)$_2$/H_2O_2/NaI at 0-5°, also ar. iodination (cf. *6*, 565s*83*) of anilines and pyrazoles, s. A.T. Khan, S. Ali, Bull. Chem. Soc. Jpn. *2012, 85* (11), 1239-43 [DOI: 10.1246/bcsj.20120155]; **α-fluorination** and α,α-difluorination of functionalized thioethers with air- and moisture-stable, solid IF_5/pyridine/HF, also 1,1-difluorides from mercaptals (cf. *46*, 454), and trifluoromethylarenes from aryltrithioorthocarboxylic acid esters, s. S. Hara, M. Monoi, R. Umemura, C. Fuse, Tetrahedron *2012, 68* (49), 10145-50 [DOI: 10.1016/j.tet.2012.09.104]; trihalogenomethyl from methyl sulfones, α,α-dihalogenation of nitromethyl sulfones, and preparation of mixed trihalogenomethyl sulfones with Na-hypohalites s. K.M. Borys, M.D. Korzyński, Z. Ochal, Tetrahedron Lett. *2012, 53* (48), 6606-10 [DOI: 10.1016/j.tetlet.2012.09.121].

Magnesium bis(2,2,6,6-tetramethylpiperidide)·bis(lithium chloride)/boron fluoride ←
Chlorozinc 2,2,6,6-tetramethylpiperidide·lithium chloride/boron fluoride *TMPZnCl·LiCl/BF₃*

Regioselective metalation of N-heteroarenes ←

using boron fluoride-etherate and hindered magnesium- or zinc-amides as frustrated Lewis pairs s. *83*, 410

3-Hexenol s. under N′-Chloromethyl-N-fluoro-1,4-diazoniabicyclo[2.2.2]octane bis(fluoroborate) *ROH*

Chiral N-(2-tert-aminocyclohexyl)-N′-arylsquaramides s. under N-Iodosuccinimide ←

Methyl N-carbo-tert-butoxy-L-dimethylaminoalaninyl-D-prolinyl-α-aminoisobutyryl-L-phenylalaninate s. under 1,3-Dibromo-5,5-dimethylhydantoin ←

Chiral N-(2′-hydroxy-1,1′-binaphthyl-2-ylmethyl)amidines s. under N-Iodosuccinimide ←

Tryptophan-7-halogenase/flavin reductase/glucose dehydrogenase/sodium chloride (or bromide) ←

Regioselective multi-enzymatic ar. halogenation H → Hal

201.

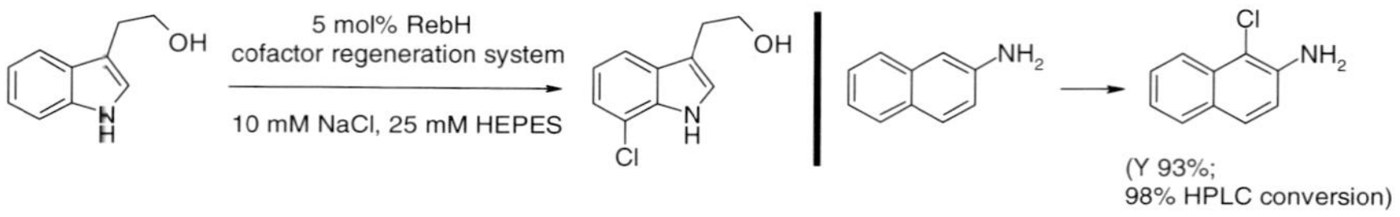

Co-expression of the tryptophan-7-halogenase, RebH, with GroEL/ES and fusion of its cognate flavin reductase, RebF, to maltose-binding protein, MBP, enabled production of both enzymes on scales sufficient for preparative biocatalysis without the need for specialized fermentation equipment or bioreactors. The system has been found to be applicable to regioselective ar. halogenations, in particular to **7-chlorination of 3-subst. indoles**, in contrast to the use of PrnA, which lead to halogenations at the more activated 2-position. **E:** A soln. of substrate (10 mg; 0.5 m*M* final concentration) in HEPES buffer (5 ml; 25 m*M*; pH 7.4) or isopropanol added to a crystallization dish, solns. in HEPES buffer (25 m*M*; pH 7.4) of NAD (0.2 eq.), FAD (0.2 eq.), NaCl (20 eq.), and a glucose dehydrogenase (50 U/ml final concentration CDX-901) added, solns. in HEPES/glycerol buffer (25 m*M*; pH 7.5; 10% glycerol v/v) of RebH (0.05 eq.) and MBP-RebF (0.005 eq.) added, the mixture diluted with HEPES buffer (25 m*M*; pH 7.4), an aq. soln. of glucose (40 eq.) added to initiate the cofactor regeneration cycle, the dish covered with perforated aluminum foil, stirred with a magnetic stir bar at 60 rpm, reactions being monitored by HPLC, upon completion quenched with aq. HCl (5 *M*, until pH < 2), NaCl added to saturation, precipitated protein filtered out through a pad of Celite, washing with water and/or methylene chloride, the filtrate extracted into methylene chloride, and the crude material purified by flash chromatography on silica gel → 7-chlorotryptophol. Y 95% (95% conversion by HPLC). 7-Chlorination of L- or D-tryptophan, tryptamine or its N-Ω- or 2-methyl-analogs proceeded in 60-78% yields (65%, 94-99% conversions) and 7-bromination of L-tryptophan in 85% yield (97% conversion). 5-Methyltryptamine underwent 6-chlorination (Y 58%) while tryptoline gave a ca. 1:1 mixture of 5- and 6-chloro-derivs. (Y 54%). Indole itself underwent 3-chlorination (Y 53%), 2-aminonaphthalene underwent 1-chlorination (Y 93%) and 1-naphthol underwent 2-chlorination (Y 59%). F.e.s. J.T. Payne, M.C. Andorfer, J.C. Lewis, Angew. Chem., Int. Ed. *2013*, *52* (20), 5271-4 [DOI: 10.1002/anie.201300762].

Iodosobenzene s. under Manganese(III) salen complex *PhIO*

Phenyl iodosoacetate *PhI(OAc)₂*

2-Alkoxy-3-methylene-1,2-oxaphospholane 2-oxides from β-phosphonyl-δ-lactols via iodinative radical β-fragmentation

202. 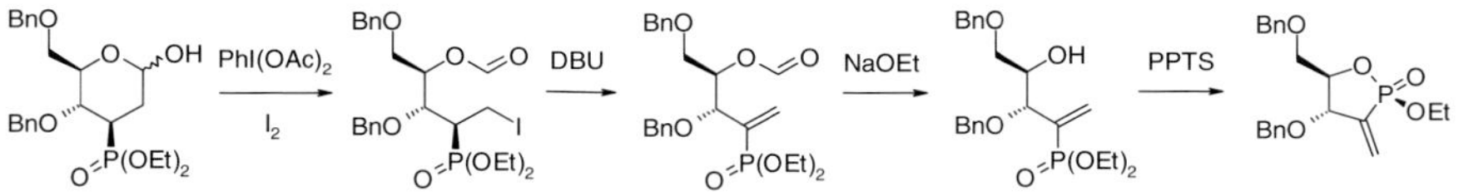

Phenyl iodosoacetate (1.1 eq.) and I_2 (1.1 eq.) added to a soln. of 4,6-di-O-benzyl-2,3-dideoxy-3-diethoxyphosphoryl-D-*arabino*-hexopyranose (0.398 mmol) in methylene chloride (10 ml), the mixture stirred at room temp. for 1 h, quenched with aq. $Na_2S_2O_3$, extracted with methylene chloride, and purified chromatographically → 3,5-di-O-benzyl-1,2-dideoxy-2-diethoxyphosphoryl-4-O-formyl-1-iodo-D-arabinitol (Y 82%), a portion of which (0.297 mmol) dissolved in dry benzene (7.4 ml), cooled to 9-10°, DBU (1.98 eq.) added, the mixture stirred for 15 min, quenched with satd. aq. $NaHSO_4$, extracted with ethyl acetate, and purified chromatographically → 3,5-di-O-benzyl-1,2-dideoxy-2-diethoxyphosphoryl-4-O-formyl-D-*erythro*-pent-1-enitol (Y 99%), a portion of which (0.432 mmol) stirred with Na-ethoxide (8.8 mol%) in ethanol (3.8 ml) under N_2 at room temp. for 3 h, concentrated *in vacuo*, and purified chromatographically → 3,5-di-O-benzyl-1,2-dideoxy-2-diethoxyphosphoryl-D-*erythro*-pent-1-enitol (Y 95%), a portion of which (0.41 mmol) dissolved in toluene (2 ml), pyridinium tosylate (2 eq.) added, the mixture stirred at 100° under N_2 for 29 h, quenched with water, extracted with ethyl acetate, and purified chromatographically → (S_P)-3,5-di-O-benzyl-1,2-dideoxy-1-ethoxy-2-methylene-1-phospha-1-oxo-D-*erythro*-pento-furanose (Y 31% with 16% of the R_P enantiomer). Pent-2-enono-1,4- and hex-2-enono-1,5-lactones were β-phosphonylated [$(EtO)_3P$] and reduced [*i*-Bu_2AlH] to afford the corresponding lactol substrates, which were subjected to alkoxy radical β-fragmentation [ARF; (cf. *61*, 195)] to afford β-iodophosphonates (three examples; Y 73-95%) and dehydroiodinated to the key vinylphosphonates (Y 90-99%), one of which was deformylated and cyclized to the illustrated mixture of oxaphospholane 2-oxides. A similar sequence of reactions using lactols generated via phosphinylation ($Ph_2PCl/AlCl_3$) and hydration ($Ph_3P{\cdot}HBr$) of perbenzylated D-glucal, -galactal and -lactal afforded the corresponding **β-iodophosphine oxides** (three examples; Y 67-79%) and **vinylphosphine oxides** (Y 83-91%) which were not further elaborated. F.e. and reactions of the oxaphospholane 2-oxide product s. D. Hernández-Guerra, M.S. Rodríguez, E. Suárez, Org. Lett. *2013*, *15* (2), 250-3 [DOI: 10.1021/ol302939z].

Chiral o-α,o′-α-difunctionalized iodobenzene difluoride

Asym. intramolecular 1,2-fluoroamination of unactivated ethyleneamines

203. 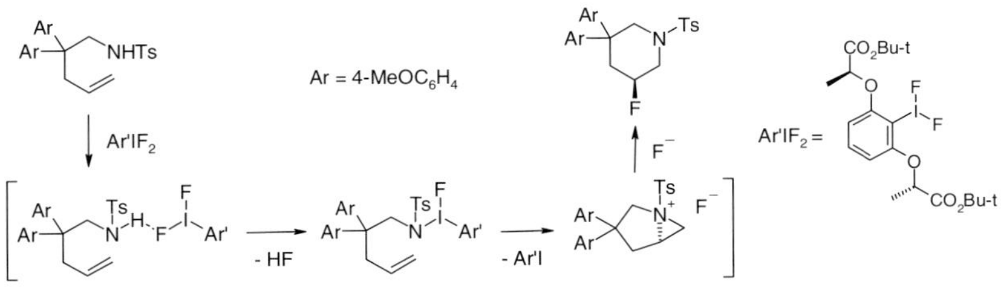

The first intramolecular, regioselective, metal-free aminofluorination of N-protected pent-4-enylamines, affording rare **chiral N-protected 3-fluoropiperidines**, has been reported. **E:** A soln. of 2,6-dihydroxyiodobenzene difluoride O²,O⁶-bis(*tert*-butyl(R)-lactate) (2.5 eq.) in toluene (1 ml) added to a soln. of N-tosyl-2,2-bis(4-methoxyphenyl)pent-4-enamine (0.1 mmol) in the same solvent (1 ml), the mixture stirred at 25° until reaction complete (TLC), concentrated *in vacuo*, and the residue purified chromatographically → (S)-5-fluoro-3,3-bis(4-methoxyphenyl)-1-tosyl-piperidine. Y 63% (e.e. 72% improved to 99% by recrystallization). Pent-4-enylamines, N-protected as sulfonamide or carbamate, and carrying 2,2-diaryl or 1,1-dimethyl substituents, underwent exclusive 6-*endo*-cyclization in the presence of (R,R)- or (S,S)-bis-lactate derived iodonium difluorides (the by-product iodoarenes were recovered for regeneration) to afford (R)/(S)-3-fluoro-piperidines with moderate enantioselectivity (e.e. 61-88%; eleven examples; Y 63-79%), generally increased by a single recrystallization (90 to >99%). Cyclizations of 2,2-di- and 2,2,4-tri-methyl-pent-4-enamines were not enantioselective (three examples; Y 64-90%), while a 2-phenyl deriv. cyclized with high *cis*-diastereoselectivity (20:1; Y 71%). Extension of the reaction to hex-5-enyl-amine analogs required the presence of a gold(III) Lewis acid for effective 7-*endo*-trig cyclization, affording **chiral N-protected 2-fluoroazepans** (six examples; Y 50-62%; e.e. 72-77% improved to 91-99% by recrystallization). Intermolecular aminofluorination of styrene derivs. is also described using a non-chiral fluorinating agent, *p*-xyleneiodonium difluoride, and N-methylarene-sulfonamides, to afford novel **2-aryl-2-fluoro-N-sulfonylamines** with 2 eq. of the styrenes required (recoverable) to minimize formation of diamination products (eleven examples; Y 52-80%; a single sec. sulfonamide gave 53%).

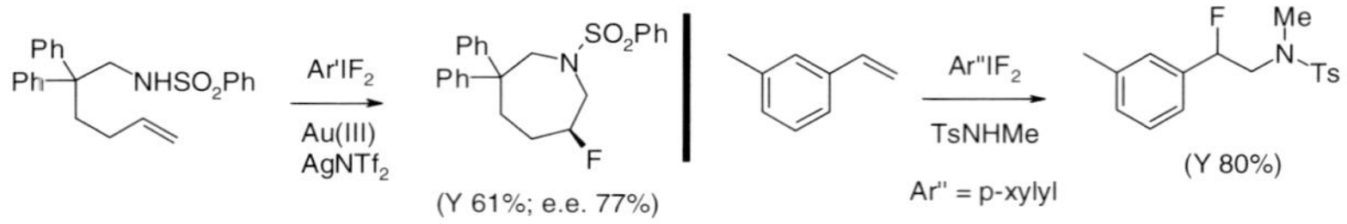

F.e. and optimization s. W. Kong, P. Feige, T. de Haro, C. Nevado, Angew. Chem., Int. Ed. *2013, 52* (9), 2469-73 [DOI: 10.1002/anie.201208471].

Electrophilic halogen sources ←

Ar. halogenation H → Hal

update s. *6,* 565s*82*; aerobic ar. chlorination and bromination with recyclable multifunctional ionic liquids generated by addition of HCl or HBr to alkylammonium nitrates or 3-methyl-1-(4-sulfobutyl)imidazolium nitrate as both medium and halogenating agent s. R. Prebil, K.K. Laali, S. Stavber, Org. Lett. *2013, 15* (9), 2108-11 [DOI: 10.1021/ol4001476]; *heterogeneous* ar. chlorination, bromination and iodination with the appropriate halogen source generated *in situ* by aerobic oxidation of organic or inorganic halides over nanocrystalline ceria s. A. Leyva-Pérez, D. Cómbita-Merchán, J.R. Cabrero-Antonino, S.I. Al-Resayes, A. Corma, ACS Catal. *2013, 3* (2), 250-8 [DOI: 10.1021/cs300644s]; regioselective **ar. chlorination** (notably of deactivated arenes) with satd. NH_4Cl/acetonitrile or [for ar. dichlorination] with brine/acetonitrile and a suitable excess of K-persulfate s. L. Gu, T. Lu, M. Zhang, L. Tou, Y. Zhang, Adv. Synth. Catal. *2013, 355* (6), 1077-82 [DOI: 10.1002/adsc.201300062]; ar. chlorination with $NaCl/NaIO_4$ in acetic acid containing Na-lauryl sulfate as surfactant s. T. Mahajan, L. Kumar, K. Dwivedi, D.D. Agarwal, Synth. Commun. *2012, 42* (24), 3655-63 [DOI: 10.1080/00397911.2011.589560]; ar. mono-chlorination of electron-rich arenes with LiCl/TsOH·H_2O/*m*-chloroperoxybenzoic acid and iodo-benzene as catalyst s. T.-T. Li, C. Xu, C.-B. Xiang, J. Yan, Chin. Chem. Lett. *2013, 24* (6), 535-8 [DOI: 10.1016/j.cclet.2013.03.049]; **ar. bromination** with $NaBr/H_2O_2$ in aq. medium containing a *water-soluble* bis(3-methylimidazoliomethyl)phenyl diselenide dibromide as catalyst s. E.E. Alberto, A.L. Braga, M.R. Detty, Tetrahedron *2012, 68* (51), 10476-81 [DOI: 10.1016/j.tet.2012.08.004]; regioselective ar. bromination with poly(4-vinylpyridinium bromochromate) s. J. Albadi, H. Tajik, M. Keshavarz, M. Abedini, Monatsh. Chem. *2013, 144* (2), 179-81 [DOI: 10.1007/s00706-012-0779-4]; *heterogeneous* ar. bromination of unactivated arenes with Br_2 and recyclable Fe_2O_3/zeolite as catalyst s. Y. Nishina, K. Takami, Green Chem. *2012, 14* (9), 2380-3 [DOI: 10.1039/c2gc35821b]; with readily recoverable $AuCl_3$ as catalyst supported on poly(N-vinyl-2-pyrrolidone)-grafted magnetic [covalently anchored] $Fe_3O_4@SiO_2$ cf. B. Li, L. Gao, F. Bian, W. Yu, Tetrahedron Lett. *2013, 54* (9), 1063-6 [DOI: 10.1016/j.tetlet.2012.12.017]; **ar. iodination** with dichloroiodic(I) acid [$HICl_2$] generated catalytically from I_2/H_2O_2 and a little HCl s. L. Bedrac, J. Iskra, Adv. Synth. Catal. *2013, 355* (7), 1243-8 [DOI: 10.1002/adsc.201300127]; with thermally stable, water-soluble, recyclable potassium 4-iodylbenzenesulfonate s. M.S. Yusubov, R.Y. Yusubova, V.N. Nemykin, A.V. Maskaev, M.R. Geraskina, A. Kirschning, V.V. Zhdankin, Eur. J. Org. Chem. *2012* (30), 5935-42 [DOI: 10.1002/ejoc.201201064]; rapid mono- or di-halogenation of N-heteroarenes with the appropriate N,N′-dihalogeno-5,5-dimethylhydantoin and

H_2SO_4 as activating catalyst s. A.H. Sandtorv, H.-R. Bjørsvik, Adv. Synth. Catal. *2013, 355* (2-3), 499-507 [DOI: 10.1002/adsc.201200723]; **directed *o*-halogenation** (cf. *6,* 565s*82*) of variously functionalized electron-deficient arenes (notably possessing acylamino, carbonyl or sulfamyl groups) with NCS or NBS/TfOH and a co-oxidant (e.g. $Na_2S_2O_8$) under palladium(II) catalysis with $Pd(OAc)_2$ s. X. Sun, G. Shan, Y. Sun, Y. Rao, Angew. Chem., Int. Ed. *2013, 52* (5), 4440-4 [DOI: 10.1002/anie.201300176]; *o*-bromination of aryl carbamates with NBS in the presence of $Pd(OAc)_2$ as catalyst and an acidic additive (trifluoroacetic acid, TfOH or TsOH) s. A. John, K.M. Nicholas, J. Org. Chem. *2012, 77* (13), 5600-5 [DOI: 10.1021/jo300713h]; regioselective *o*-bromination and *o*-iodination of azobenzenes with NBS or NIS and $Pd(OAc)_2$/TsOH s. X.-T. Ma, S.-K. Tian, Adv. Synth. Catal. *2013, 355* (2-3), 337-40 [DOI: 10.1002/adsc.201200902]; gram-scale *cyano-directed o*-halogenation of ar. nitriles with the appropriate N-halogenosuccinimide under palladium(II) catalysis with added TsOH s. B. Du, X. Jiang, P. Sun, J. Org. Chem. *2013, 78* (6), 2786-91 [DOI: 10.1021/jo302765g]; ***p*-fluorination** of N-pivaloylanilines with phenyl iodosopivalate/HF·pyridine s. T. Tian, W.-H. Zhong, S. Meng, X.-B. Meng, Z.-J. Li, ibid. *78* (2), 728-32 [DOI: 10.1021/jo302099d]; **asym. directed *o*-iodination** of N,N-diisopropylferrocenecarboxamide and methyl ferrocenecarboxylate with iodine via *o*-deprotonation with chiral Li-dialkylamide/$ZnCl_2$·TMEDA/MeLi s. G. Dayaker, D. Tilly, F. Chevallier, G. Hilmersson, P.C. Gros, F. Mongin, Eur. J. Org. Chem. *2012* (30), 6051-7 [DOI: 10.1002/ejoc.201200965]; *vinylic* α-iodination of acrylic acid esters with N-iodophthalimide/3-quinuclidinol on KF/Celite in the dark s. V. Pace, G. Vilkauskaite, A. Sackus, W. Holzer, Org. Biomol. Chem. *2013, 11* (7), 1085-8 [DOI: 10.1039/c2ob27341a].

Electrophilic halogenocyclization ○

update s. *35,* 351s*82*; *cis*-β,γ-disubst. (Z)-δ,ε-ethylene-δ-iodo-γ-lactones from γ-allenecarboxylic acids **by chlorolactonization** with asym. induction using N-iodonaphthalene-2,3-dicarboximide and Cs_2CO_3 as base s. X. Zhang, C. Fu, Y. Yu, S. Ma, Chem. Eur. J. *2012, 18* (42), 13501-9 [DOI: 10.1002/chem.201200229]; **asym. bromolactonization** (cf. *78,* 220s*82*) of internal ethylenecarboxylic acids with 1,3-dibromo-5,5-dimethylhydantoin and a chiral, C_3-symmetric 1,3,5-tris-(Δ^2-imidazolin-2-yl)benzene as catalyst, notably for preparing chiral δ,δ-disubst. δ-lactones, s. K. Murai, A. Nakamura, T. Matsushita, M. Shimura, H. Fujioka, ibid. *18* (27), 8448-53 [DOI: 10.1002/chem.201200647]; chiral δ-bromo-γ-lactones from γ,δ-ethylenecarboxylic acids with 2,4,4,6-tetrabromo-2,5-cyclohexadienone [TBCO] and chiral acetonitrile(aqua)[2,2′-bis(diphenylphosphino)-1,1′-binaphthyl]palladium(II) bis(hexafluoroantimonate) as catalyst s. H.J. Lee, D.Y. Kim, Tetrahedron Lett. *2012, 53* (51), 6984-6 [DOI: 10.1016/j.tetlet.2012.10.051]; **desymmetrization** of 1,4-cyclohexadien-3-ylcarboxylic acids by bromolactonization with N-bromosuccinimide and $(DHQD)_2$PHAL, also bromolactonization **with kinetic resolution**, s. K. Ikeuchi, S. Ido, S. Yoshimura, T. Asakawa, M. Inai, Y. Hamashima, T. Kan, Org. Lett. *2012, 14* (23), 6016-9 [DOI: 10.1021/ol302908a]; chiral 2-aryl-3-chlorotetrahydrofurans by **asym. intramolecular chloroetherification** of homoallyl alcohols with N-chlorosuccinimide and N-[*p*-(trifluoromethyl)benzyl]-cinchoninium bromide as catalyst s. X. Zeng, C. Miao, S. Wang, C. Xia, W. Sun, Chem. Commun. *2013, 49* (24), 2418-20 [DOI: 10.1039/c2cc38436a]; regioselective **intramolecular iodoetherification** of *o*-acetylenecarboxylic acid amides to give 3-(α-iodoalkylidene)phthalimidines or 4-iodoisocarbostyrils s. C. Schlemmer, L. Andernach, D. Schollmeyer, B.F. Straub, T. Opatz, J. Org. Chem. *2012, 77* (22), 10118-24 [DOI: 10.1021/jo3017378]; chiral 5-α-fluoro-2-pyrrolidones from γ,δ-ethylenecarboxylic acids by **intramolecular fluoroamination** (cf. *35,* 351s*82*) with asym. induction using Selectfluor/$AgNO_3$, also 2-oxazolidone and 2-imidazolidone analogs, s. Z. Li, L. Song, C. Li, J. Am. Chem. Soc. *2013, 135* (12), 4640-3 [DOI: 10.1021/ja400124t]; cyclic 2-fluoroamines from ethyleneamines with N-iodosuccinimide/Bu_4NF and AgOTf s. H.-T. Huang, T.C. Lacy, B. Blachut, G.X. Ortiz, Jr., Q. Wang, Org. Lett. *2013, 15* (8), 1818-21 [DOI: 10.1021/ol4003866]; cyclic 2-(fluoro)tosylamines from ethylenetosylamines with HF/pyridine and phenyl iodosopivalate/boron fluoride etherate, diastereoselectivity, s. Q. Wang, W. Zhong, X. Wei, M. Ning, X. Meng, Z. Li, Org. Biomol. Chem. *2012, 10* (43), 8566-9 [DOI: 10.1039/c2ob26664d]; **intramolecular chloramination** of ethyleneamines with $CuCl_2 \cdot 2H_2O$ as both promoter and chlorine source s. G.-Q. Liu, W. Li, Y.-M. Li, Adv. Synth. Catal. *2013, 355* (2-3), 395-402 [DOI: 10.1002/adsc.201200352]; diastereoselective formation of N-protected *syn*-3-fluoro-2-α-iodopyrrolidines and 2-pyrrolidone analogs by *5-exo-trig* **intramolecular iodoamination** with $I_2/NaHCO_3$ s. L.E. Combettes, M. Schuler, R. Patel, B. Bonillo, B. Odell,

A.L. Thompson, T.D.W. Claridge, V. Gouverneur, Chem. Eur. J. *2012, 18* (41), 13126-32 [DOI: 10.1002/chem.201201576]; chiral N-protected 2-α-bromopyrrolidines from 4-ethyleneamines by **asym. intramolecular bromamination** with N-bromophthalimide and a chiral C_2-symmetric, mannitol-based tetrahydroselenophene as catalyst, also silver(I)-catalyzed rearrangement of the products to chiral N-protected 3-piperidinols, s. F. Chen, C.K. Tan, Y.-Y. Yeung, J. Am. Chem. Soc. *2013, 135* (4), 1232-5 [DOI: 10.1021/ja311202e]; chiral N-protected 3-bromo-2,3-dihydro-4(1*H*)-quinolone-3-carboxylic acid esters with N-bromosuccinimide and a chiral bis(guanidinium salt) as catalyst, also asym. intramolecular aza-Michael addition, s. X. Xiao, X. Liu, S. Dong, Y. Cai, L. Lin, X. Feng, Chem. Eur. J. *2012, 18* (50), 15922-6 [DOI: 10.1002/chem.201203216]; 3-α-bromooxindoles from α,β-ethylenecarboxylic acid anilides by **intramolecular bromocarbocyclization** with N-bromosuccinimide and N-(*o*-iodobenzoyl)glycine as catalyst or with KBr/Oxone, stereoselectivity, s. D.C. Fabry, M. Stodulski, S. Hoerner, T. Gulder, ibid. *18* (35), 10834-8 [DOI: 10.1002/chem.201201232].

N-Halogenosuccinimides ←
Electrophile-induced rearrangements of 3-allenyl-3-hydroxyoxindoles ←
with selenium-based electrophiles cf. *81,* 222; ring expansion or oxycyclization with NCS or NBS in the presence or absence of precious metal salts s. B. Alcaide, P. Almendros, A. Luna, N. Prieto, Org. Biomol. Chem. *2013, 11* (7), 1216-25 [DOI: 10.1039/c2ob27359d].

N-Iodosuccinimide/chiral N-(2′-hydroxy-1,1′-binaphthyl-2-ylmethyl)amidines ←
N-Iodosuccinimide/chiral N-(2-tert-aminocyclohexyl)-N′-arylsquaramides ←
Organocatalyzed asym. iodolactonization ○

204.

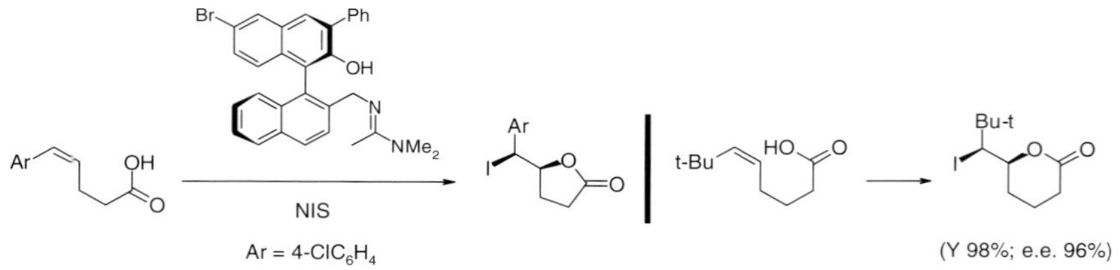

of (Z)-ethylenecarboxylic acids. NIS (1.2 eq.) added to a soln. of (Z)-5-(4-chlorophenyl)pent-4-enoic acid (0.5 mmol) and chiral catalyst (10 mol%) in toluene/methylene chloride (2:1; 15 ml) at -20°, the soln. stirred for 14 h, quenched with satd. aq. $Na_2S_2O_3$, warmed to room temp. with vigorous stirring, extracted with methylene chloride, and purified chromatographically → (S)-5-[(S)-iodo-(4-chlorophenyl)methyl]dihydrofuran-2(3*H*)-one. Y 89% (e.e. 96%). The chiral bifunctional catalyst (s. *81*, 185 for related *bromo*lactonizations), promoted regioselective iodo-lactonization of 5-alkyl/aryl-4(*Z*)-pentenoic acids to afford γ-lactones with good enantiocontrol (e.e. 94-98%; eight examples; Y 89-99%; electron-rich 5-(4-methoxyphenyl)pentenoic acid gave only decomposition products). The homologous 5(Z)-hexenoic acids were similarly cyclized to the corresponding δ-lactones (e.e. 96-98%; four examples; Y 88-98%), but iodolactonization of 5-alkyl/aryl-4(*E*)-pentenoic acids generally proceeded with poor regio- and enantio-selectivity. The method was also applied to kinetic resolution of the cyclic pentenoic acid, *rac*-2-cyclopentene-1-acetic acid, affording a 44% yield of a bicyclic iodolactone with moderate enantioselectivity (e.e. 66%). F.e., substrate prepn. and conversion (Cs_2CO_3/MeOH) of a γ-lactone product to a chiral epoxide with retention of enantiopurity (Y 87%) s. C. Fang, D.H. Paull, J.C. Hethcox, C.R. Shugrue, S.F. Martin, Org. Lett. *2012, 14* (24), 6290-3 [DOI: 10.1021/ol3030555]; asym. iodolactonization of terminal ethylenecarboxylic acids using chiral N-(2-*tert*-aminocyclohexyl)-N′-arylsquaramides and NIS (e.e. up to 96%) s. J.E. Tungen, J.M.J. Nolsøe, T.V. Hansen, ibid. *14* (23), 5884-7 [DOI: 10.1021/ol302798g].

1,3-Dibromo-5,5-dimethylhydantoin/methyl N-carbo-tert-butoxy-L-dimethylamino-alaninyl-D-prolinyl-α-aminoisobutyryl-L-phenylalaninate ←

Peptide-catalyzed asym. di- or tri-bromination of N,N-disubst. *m*-hydroxycarboxylic acid amides H → Br

205. 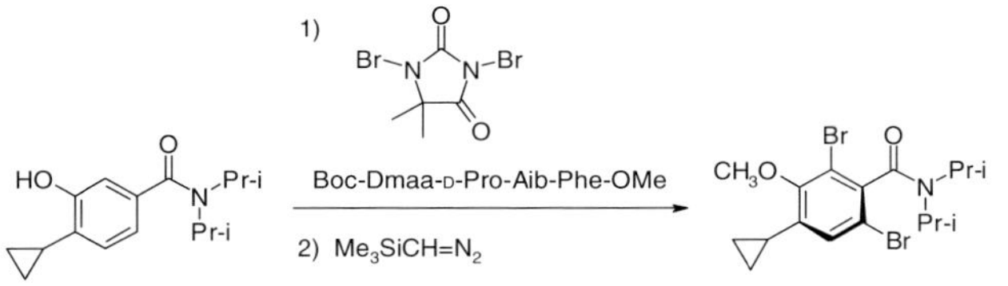

Tert. ar. amides exhibit hindered rotation about the C_{Ar}-C_{CO} bond, and organocatalyzed *o*-bromination, as described here, afforded atropisomers with a high barrier to racemization (a sample analyzed after 40 days showed no reduction in e.e.). **E:** 1,3-Dibromo-5,5-dimethylhydantoin [DBDMH] (1 eq.) added to a soln. of N,N-diisopropyl-3-hydroxy-4-cyclopropylbenzamide (0.2 mmol) and tetrapeptide catalyst (10 mol%) in chloroform (10 ml) at -40°, the mixture stirred until reaction complete (24-48 h), quenched with a soln. of butyl vinyl ether (3.75 eq.) in methanol (0.5 ml), warmed to room temp., methylene chloride/methanol (ca. 5:1; 2 ml) and a soln. of trimethylsilyldiazomethane (4 eq.) in hexanes (0.4 ml) added, the mixture stirred until methylation complete (TLC; 15-120 min), quenched with silica, filtered, and worked up with chromatographic purification → 2,6-dibromo-N,N-diisopropyl-3-methoxy-4-cyclopropylbenzamide. Y 79% (e.e. 92%). A series of N,N-diisopropyl-3-hydroxybenzamides optionally substituted (Br, Ph, Me, cyclopropyl) in the 4- or 5-positions afforded polybrominated products, with good enantio-selectivity, isolated as methyl ethers (e.e. 80-92%; nine examples; Y 69-80%), whereas substrates with *o*-bromo/methyl substitution gave poor enantioselectivity (e.e. 4-26%; three examples; Y 75-90%). Quenching experiments at low conversion showed that bromination occurs initially at the most hindered 2-position (initial *uncatalyzed bromination* occurred at the 6-position), suggesting this as the stereo-defining event. Lower enantioselectivity was obtained with analogous piperazin- (e.e. 50%; Y 86%) and piperidin-amides (e.e. 80%; 92%). F.e., substrate and catalyst prepn., and selected 2-lithiation/iodination of a 2,6-dibrominated product without loss of chirality (e.e. 86%; Y 69%), s. K.T. Barrett, S.J. Miller, J. Am. Chem. Soc. *2013*, *135* (8), 2963-6 [DOI: 10.1021/ja400082x].

N′-Chloromethyl-N-fluoro-1,4-diazoniabicyclo[2.2.2]octane bis(fluoroborate)/(S)-6,6′-di-n-octyl-3,3′-bis(2,4,6-triisopropylphenyl)-1,1′-binaphthyl-2,2′-diyl hydrogen phosphate/ sodium carbonate/3-hexanol ←

Cyclic α-fluoro-N-benzoylketimines from N-benzoylenamines C=C-NH → C(F)C=N

Asym. electrophilic fluorination under chiral anionic phase transfer catalysis

206. 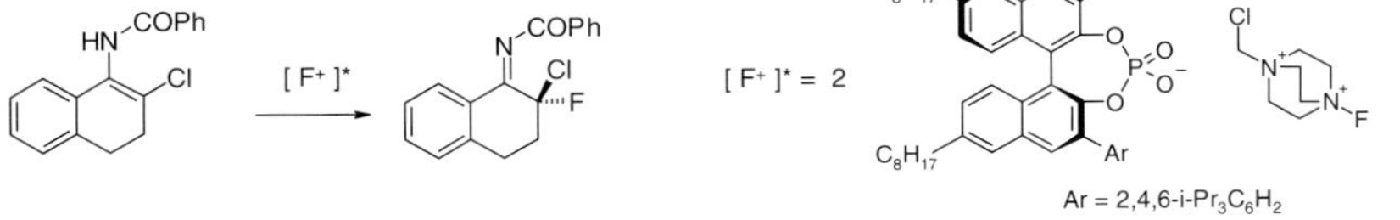

Anh. Na_2CO_3 (1.5 eq.), anh. Selectfluor (1.5 eq.), 3-hexanol (1 eq.) and (S)-C_8-TRIP (5 mol%) added to a soln. of N-(2-chloro-3,4-dihydronaphthalen-1-yl)benzamide (0.25 mmol) in hexanol (2.5 ml), the mixture capped, stirred rapidly and efficiently at room temp. for 24 h, diluted with ethyl acetate, quenched with satd. aq. $NaHCO_3$, extracted with ethyl acetate, concentrated *in vacuo*, and purified chromatographically → (S)-N-(2-chloro-2-fluoro-3,4-dihydronaphthalen-1(2*H*)-ylidene)benzamide. Y 84% (e.e. 95%). The phase transfer catalyst was key to the success of this fluorination of N-benzoylenamines derived from tetralone, indanone, benzocycloheptanone and 2-phenylcyclohexanone, generally with excellent enantioselectivity (nineteen examples;

Y 58-94%; e.e. 82-99%). Enamides carrying α-alkyl (methyl, benzyl, allyl), -phenyl and -halo (chloro, bromo) substituents were good substrates, while fluorination of α-unsubstituted derivs. was surprisingly unproblematic. Analogs derived from 2-phenylchromanone gave excellent enantioselectivity in low yields (20-41%; two examples; e.e. 93-99%), but the 2-methylcyclohexanone-derived enamide was unreactive. Other acyl groups in place of benzoyl were less effective. F.e., optimization, substrate prepn., and reduction ($LiAlH_4$) of one product to a 2-fluoroamine with full retention of chirality (Y 79%) s. R.J. Phipps, K. Hiramatsu, F.D. Toste, J. Am. Chem. Soc. *2012*, *134* (20), 8376-9 [DOI: 10.1021/ja303959p].

N′-Chloromethyl-N-fluoro-1,4-diazoniabicyclo[2.2.2]octane bis(fluoroborate)/(S)-3,3′-bis-(2,4,6-tricyclohexylphenyl)-1,1′-binaphthyl-2,2′-diyl hydrogen phosphate/sodium carbonate ←
6-Fluoro-2,4- or 4-fluoro-2,5-cyclohexadienones from phenols ←
via dearomatizing asym. fluorination under chiral anionic phase transfer catalysis

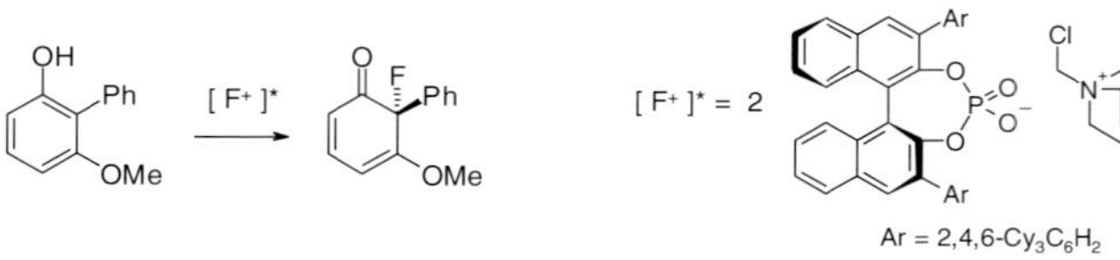

207.

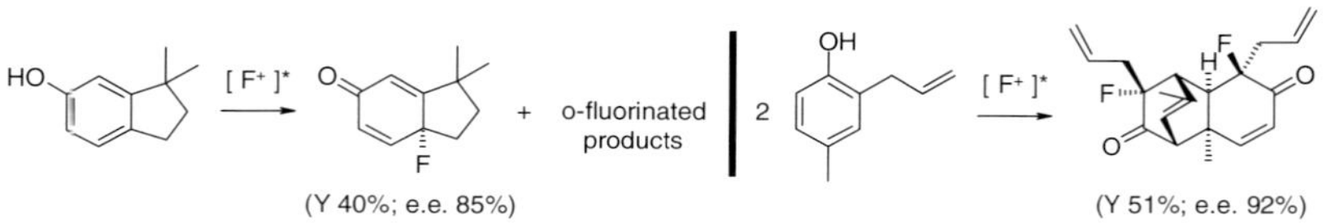

Anh. Na_2CO_3 (1.5 eq.), Selectfluor (1.5 eq.) and chiral phosphate diester (5 mol%) added sequentially to a soln. of 2-hydroxy-6-methoxybiphenyl (0.25 mmol) in toluene (1.25 ml), the capped mixture stirred rapidly at room temp. for 46 h, diluted with ethyl acetate, quenched with satd. aq. $NaHCO_3$, extracted, and purified chromatographically → (R)-1-fluoro-6-methoxybiphenyl-2(1*H*)-one. Y 70% (e.e. 91%). Combination of the cationic fluorinating agent and chiral anionic catalyst enabled use of relatively inactive phenols as nucleophiles in this enantioselective dearomative fluorination process. The position and selectivity of fluorination was largely controlled by steric factors, with simple 2,3-disubst. phenols (incl. 5,6,7,8-tetrahydro-1-naphthols) undergoing selective *o*-fluorination to afford 6-fluoro-2,4-cyclohexadienones (eight examples; Y 42-75%; e.e. 87-96%), with N-Boc-5-hydroxy-1,2,3,4-tetrahydroquinoline (wherein N is a *meta* substituent of the phenol moiety) affording only 28% yield (e.e. 87%; the isoquinoline analog gave Y 57%; e.e. 96%). Notable was the compatibility of an *iodophenol* under these conditions, since aryl iodides are usually oxidized to I(III) by Selectfluor and, indeed, analogous homogeneous reaction gave a complex mixture. Phenols lacking a 3-substituent suffered [4+2]-dimerization of the products (ten examples; Y 51-96%; e.e. 79-97%). 5,6,7,8-Tetrahydro-2-naphthol gave a mixture of S_EAr *o*-fluorinated products together with a *p*-fluoro dearomatized product, which was formed with moderate enantioselectivity (e.e. 63%; Y 41%). Geminal 8,8-dimethyl substitution improved the enantioselectivity (e.e. 86%; Y 29%), with the [6,5]-analog, 1,1-dimethyl-6-hydroxyindan, showing similar enantioselectivity (e.e. 85%; Y 40%). F.e., optimization, use of the method in a synthesis of a fluoro analog of (-)-grandifloracin, and derivatization of the [4+2]-cycloadducts, s. R.J. Phipps, F.D. Toste, J. Am. Chem. Soc. *2013*, *135* (4), 1268-71 [DOI: 10.1021/ja311798q].

N′-Chloromethyl-N-fluoro-1,4-diazoniabicyclo[2.2.2]octane bis(fluoroborate)/ iron(II) acetoacetonate or N,N-dihydroxypyromellitimide ←
Benzylic fluorination H → F
with $CsOSO_3F$ cf. *47*, 439; benzylic monofluorination with Selectfluor and $Fe(acac)_2$ as catalyst s. S. Bloom, C.R. Pitts, R. Woltornist, A. Griswold, M.G. Holl, T. Lectka, Org. Lett. *2013*, *15* (7),

1722-4 [DOI: 10.1021/ol400424s]; radical procedure with Selectfluor and N,N-dihydroxypyromellitimide as catalyst, also fluorination of aliphatic compds., s. Y. Amaoka, M. Nagatomo, M. Inoue, ibid. *15* (9), 2160-3 [DOI: 10.1021/ol4006757].

N-Fluorobenzenesulfonimide/lithium or sodium bis(trimethylsilyl)amide ←
Asym. α-fluorination H → F
update s. *60,* 186s*82*; α-fluorination of chiral 3-acyl-2-(trifluoromethyl)oxazolidines **with asym. induction** using N-fluorobenzenesulfonimide/Na-bis(trimethylsilyl)amide as base s. H. Lubin, C. Dupuis, J. Pytkowicz, T. Brigaud, J. Org. Chem. *2013, 78* (7), 3487-92 [DOI: 10.1021/jo400210j]; *syn*-selective α-fluorination of chiral, N-protected β-aminocarboxylic acid esters with asym. induction using N-fluorobenzenesulfonimide/*i*-Pr_2NLi s. V. Peddie, A.D. Abell, Helv. Chim. Acta *2013, 95* (12), 2460-73 [DOI: 10.1002/hlca.201200520].

Methoxy(trimethyl)silane s. under Bis(pentafluorobenzonitrile)((S)-xylyl-phanephos)platinum(II) bis(fluoroborate) *$MeOSiMe_3$*
Chiral 1,1′-binaphthyl-2,2′-diyl hydrogen phosphates s. under N′-Chloromethyl-N-fluoro-1,4-diazoniabicyclo[2.2.2]octane bis(fluoroborate) ←
Vanadyl acetoacetonate s. under NaI *$VO(acac)_2$*
Oxygen s. under $Pd(OAc)_2$ *O_2*
Hydrogen peroxide s. under NaI *H_2O_2*
Iodine s. under K_2CO_3 *I_2*

Iodine pentafluoride/pyridine·hydrogen fluoride *IF_5/py·HF*
α-Fluorination of thioethers s. *42,* 494s*83*

Sodium hypohalites *NaOHal*
α-Halogenation of sulfones s. *38,* 473s*83* H → Hal

Pyridinium dichlorobromate *$[Py][BrCl_2]$*
Regioselective ar. halogenation
in single-walled carbon nanotubes [SWNT] as nanoreactors

208.

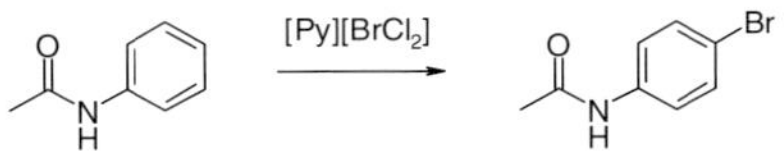

The use of single-walled carbon nanotubes, which exhibit high mechanical, chemical and thermal stability, and are highly polarizable, as substrate-confining nanoreactors for preparative chemical reactions, has been disclosed for the first time. Notably, use of those prepared by the high-pressure carbon monoxide disproportionation [HiPCO] process has been shown to affect the regioselectivity and kinetics of ar. halogenations. **E:** N-Phenylacetamide@SWNT [e.g. 5.5 mg, of which 5 mg is SWNT; prepared by adding purified HiPCO SWNT (5 mg) to molten N-phenylacetamide (100 mg; 120°), leaving the mixture for 5 min to allow complete immersion, then heating at 120° under reduced pressure (190 mbar) until the sample mass reduced to 5-10 mg (typically 5-12 h)] added to an aq. soln. of pyridinium dichlorobromate (0.6 eq.; 1.5 mmol dm^{-3}), the suspension stirred vigorously at room temp. for 24 h, quenched with excess Na-bisulfite, extracted with ether, the organic soln. washed with water, dried (Na_2SO_4), filtered, and dried *in vacuo* → product. Conversion ca. 100% (*o/p* 3:97; cf. 32:68 in the absence of the nanotube or 31:69 with a closed nanotube). For chlorination with an excess of Cl_2, the *o/p* ratio changed from 85:15 in solution to 58:42 using the SWNT. F. details s. S.A. Miners, G.A. Rance, A.N. Khlobystov, Chem. Commun. *2013, 49* (49), 5586-8 [DOI: 10.1039/c3cc42414f].

Manganese(III) salen complex/triethylamine tris(hydrogen fluoride)/iodosobenzene ←

Manganese-catalyzed oxidative benzylic fluorination H → F

209.

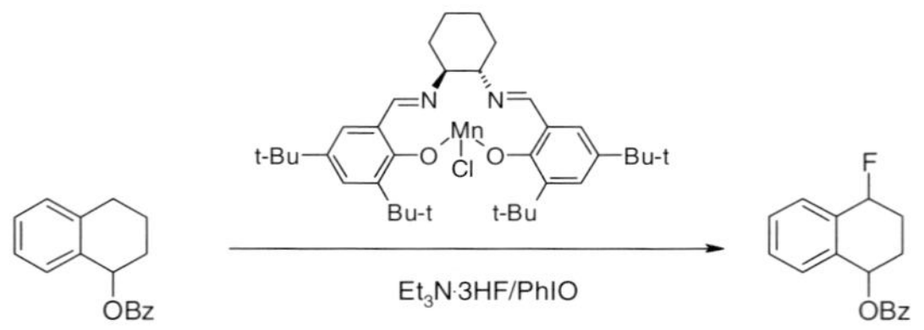

Mn-salen complex (20 mol%), 1-benzoyloxytetralin (0.8 mmol) and $Et_3N·3HF$ (0.2 ml; 1.5 eq.) added to an oven-dried Schlenk flask under N_2, followed by degassed acetonitrile (0.5 ml), the mixture heated to 50°, iodosobenzene (6-8 eq.) added slowly over 6-8 h with monitoring by GC/MS, the soln. allowed to cool to room temp., diluted with hexanes (2 ml), and worked up with purification by silica gel column chromatography → product. Y 70% (d.r. 3:1). The conditions and catalyst previously described for manganese-catalyzed monofluorination of unactivated aliphatic hydrocarbons (*82*, 203) could not be used since they gave benzylic alcohols and ketones as major by-products with alkylarenes as substrates. This method using the Mn-salen complex has broad scope and high functional group tolerance, being compatible with amide, ether, ester, keto, halide, imide, and aryl groups. For substrates containing electron-withdrawing groups, a combination of $Et_3N·3HF$ (0.5 eq.) and AgF (3 eq.) was found to be advantageous. The method may be used for late-stage fluorination of drug-like compounds, and ibuprofen methyl ester was fluorinated in 20% yield within 30 min using KF/18-crown-6/AgOTf in combination with the Mn-catalyst and PhIO, conditions that may be of interest for ^{18}F PET imaging. F.e. (seventeen; Y 44-67%) s. W. Liu, J.T. Groves, Angew. Chem., Int. Ed. *2013*, *52* (23), 6024-7 [DOI: 10.1002/anie.201301097].

Iron(II) acetoacetonate or N,N-dihydroxypyromellitimide s. under N′-Chloromethyl-N-fluoro-1,4-diazoniabicyclo[2.2.2]octane bis(fluoroborate) *Fe^{2+}*

Palladium(II) acetate/lithium bromide/oxygen *$Pd(OAc)_2/LiBr/O_2$*

N-Aryl-2-bromoenamines CH═CH → C(Br)═C(NHAr)

from terminal ethylene derivs. and prim. ar. amines

via stereoselective palladium(II)-catalyzed dehydrogenative bromoamination

210.

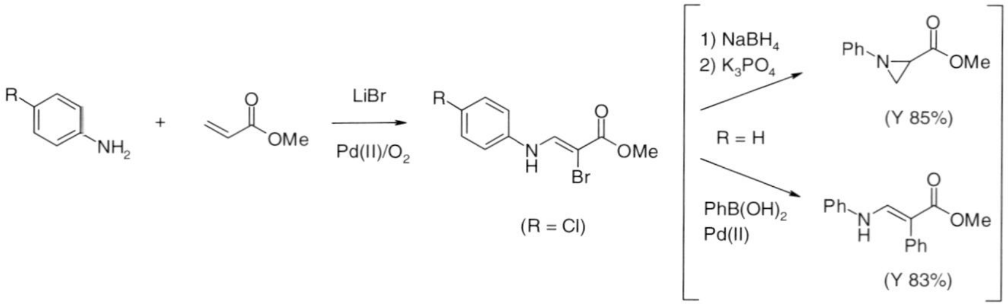

A mixture of $Pd(OAc)_2$ (10 mol%), 4-chloroaniline (0.5 mmol), methyl acrylate (1.6 eq.), LiBr (4 eq.) and THF (2 ml) stirred under O_2 (0.5 MPa) in an autoclave at 60° for 24 h, cooled to room temp., surplus O_2 vented, concentrated *in vacuo,* and purified chromatographically → (Z)-methyl 2-bromo-3-(4-chlorophenylamino)acrylate. Y 77% (Z/E >20:1). This novel, efficient and scalable (to 5 mmol) bromoamination utilized inexpensive LiBr as the bromine source and O_2 as the sole oxidant. The reaction was successful with electron-diverse anilines, affording the title products with high regio- and stereo-selectivity (twenty-six examples; Y 63-87%; Z/E 12:1 to >20:1) in the presence of ester, halo, sulfone, ether, nitrile, amide and *prim. alcohol* functionality. The reaction failed with benzylamine or N-methylaniline, and with LiCl gave a modest yield (13%; Z/E 5.2:1) of the 2-chloroenamine. The reaction mechanism was not established but does not appear to involve a radical process. F.e., optimization and f. transformations of the products s. X. Ji, H. Huang, W. Wu, H. Jiang, J. Am. Chem. Soc. *2013*, *135* (14), 5286-9 [DOI: 10.1021/ja401034g].

Bis(pentafluorobenzonitrile)((S)-xylyl-phanephos)platinum(II) bis(fluoroborate)/ xenon difluoride/methoxy(trimethyl)silane ←

2-Fluoro-1,2,3,4,4a,9a-hexahydroxanthenes from *o*-(2,6-dienyl)phenols via platinum(II)-catalyzed asym. fluorinative double ring closure ○

211.

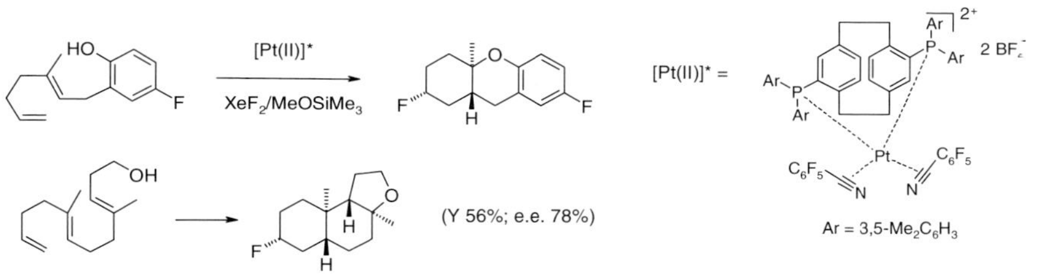

CD_3NO_2 (0.1 ml) added to a mixture of (S)-(xylyl-phanephos)PtI_2 (10 mol%), $AgBF_4$ (25 mol%) and pentafluorobenzonitrile (30 mol%) under foil at room temp. in a glovebox, the mixture stirred for 1 h, filtered through a PTFE syringe filter into a frozen (liquid N_2) soln. of (E)-2-(3-methylhepta-2,6-dien-1-yl)-4-fluorophenol (0.04 mmol) in the same solvent (0.1 ml), a soln. of XeF_2 (1.25 eq.) in the same solvent (0.1 ml) added, the mixture capped, methoxy(trimethyl)silane (1.25 eq.) added, stirred at 0° for 18 h, filtered through silica, and purified chromatographically → product. Y 80% (e.e. 81%; plus a trace of elimination product). The use of XeF_2 as fluorine source (other sources favored formation of elimination products) and addition of Me_3SiOMe, to minimize aromatic fluorination via removal of generated HF, were key features of this dicationic platinum catalyst promoted double ring closure. A series of 2,6-dienylphenols afforded the title products (six examples; Y 56-80%; e.e. 73-87%) together with elimination by-products (trace to 24%). A dienylnaphthol analog suffered subsequent aromatic fluorination (Y 49%; e.e. 86%; elimination 15%). An aliphatic 3,7-dienol required use of polystyrene-supported piperidine in place of Me_3SiOMe, to afford a volatile perhydrobenzofuran (GC Y 63%; elimination 22%), conditions which were **also** effective for **triple ring closure** of a 3,7,11-trienol (Y 56%; e.e. 78%; elimination 0%) and 2,6,10-trienylphenol (Y 69%; e.e. 10%; elimination 10%). F.e., optimization and substrate prepn. s. N.A. Cochrane, H. Nguyen, M.R. Gagne, J. Am. Chem. Soc. *2013*, *135* (2), 628-31 [DOI: 10.1021/ja3116795].

Via intermediates *v.i.*

Regioselective ar. iodination of functionalized arenes via borylation H → I

212.

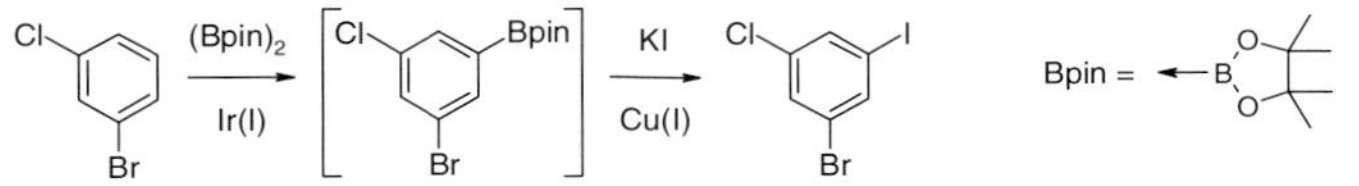

It is reported that the position of iodination maybe controlled by an initial borylation step, which occurs at the least hindered site (cf. *65*, 62), in a one pot two step process. The overall reaction is therefore controlled only by steric effects (cf. electronic or directing effects). **E: *m*-Iodination.** 3-Chloro-1-bromobenzene (1.25 mmol) added to soln. of $[Ir(cod)OMe]_2$ (0.1 mol%), 4,4′-di-*tert*-butyl-2,2′-bipyridyl (0.2 mol%) and bis(pinacolato)diborane (0.75 eq.) in THF (2.5 ml), the soln. sealed under argon, heated at 80° until borylation complete (GC/MS; 24-48 h), cooled to room temp., concentrated *in vacuo*, the residue dissolved in methanol (10 ml), the soln. added to a mixture of CuI (10 mol%), 1,10-phenanthroline (20 mol%) and KI (1.5 eq.) at room temp., water (2.5 ml) added, the flask sealed under air, the mixture heated at 80° until iodination complete (GC/MS; 1 h), cooled to room temp., diluted with water, extracted with ether, and purified chromatographically (using an aluminum foil wrapped column to minimize product decomposition) → 3-chloro-5-iodobromobenzene. Y 82%. This mild prepn. of (het)ar. iodides complements existing methods, affording single products (twenty-three examples; Y 33-82%), with hetarene derivs. (pyridine, pyrrole, indole, quinoline, benzofuran) generally giving lowest yields. The two-step method tolerates halo, ester, ether, tert. amine, nitrile, silane and tert. amide functionality, but a ketone was partially oxidized to ester. Alternative iodination of pre-formed boronates using limited iodide (i.e. a slight excess - 1.2 eq. - of boronate) allowed ar. iodination in the presence of aldehyde, carboxylic acid, nitro, sec. amide and alkene functionality (eight examples; Y 62-87%; 3-iodobenzoic acid was obtained in 30% yield), with a β-styrylboronate affording the corresponding vinyl iodide (Y 45%).

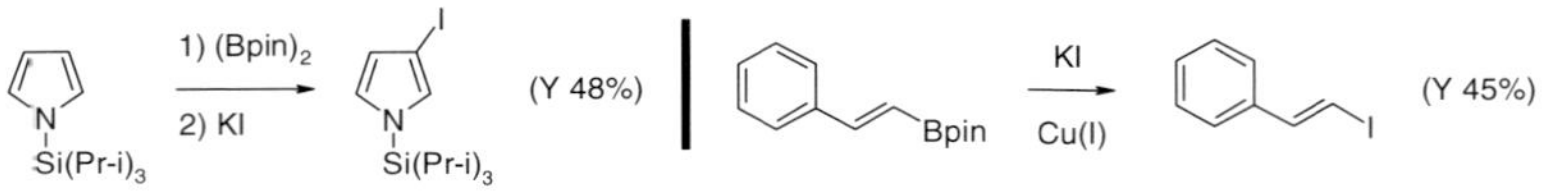

The authors suggest the method may be suitable for late-stage introduction of radiolabeled iodine. F.e., optimization and use of the method for synthesis of a SPECT imaging agent s. B.M. Partridge, J.F. Hartwig, Org. Lett. *2013*, *15* (1), 140-3 [DOI: 10.1021/ol303164h].

Oxygen ↾ HalC ⇃↾ O

Without additional reagents *w.a.r.*

o-Iodophenolethers or o-diiodides from 5-en-2-ynones via iodine-mediated carbocyclization

213.

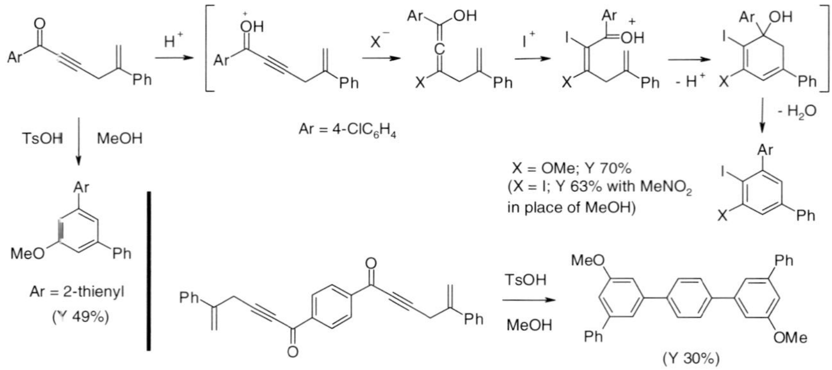

I_2 (1.5 eq.) added to a soln. of 1-(4-chlorophenyl)-5-phenylhex-5-en-2-yn-1-one (0.2 mmol) in methanol (3 ml), the mixture stirred at 65° until reaction complete (TLC), quenched with satd. aq. $Na_2S_2O_3$, extracted with ether, and purified chromatographically → 4-chloro-6′-iodo-5′-methoxy-1,1′:3′,1″-terphenyl. Y 70%. Electrocarbocyclization of hex-5-en-2-yn-1-ones carrying 1-(het)aryl/alkyl (incl. styryl) and 5-aryl substituents, with I_2/MeOH afforded the corresponding *o*-iodomethoxybenzene derivs. (fifteen examples; Y 35-70%; a 1,4-benzene-bisenynone gave 27% of a quinquephenyl deriv.). Bulky 2,5-dimethoxyphenyl or 1-naphthyl substituents gave only 1,2-diiodobenzene derivs. in low yield (20%), which were major products from similar reactions using nitromethane in place of methanol (eight examples; Y 43-70%). Use of ethanol in place of methanol gave the corresponding *o*-iodoethoxybenzene (Y 57%) but phenol gave only the 1,2-diiodobenzene deriv. (Y 62%). Cyclization was also effected using Brønsted acids (optimally TsOH) to afford the corresponding des-iodo derivs. (fourteen examples; Y 39-86%; ethanol gave a 25% yield of an ethoxybenzene; a quinquephenyl deriv. was obtained - Y 30% - from a bis-enynone), with the same bulky 1-substituents giving rise to low yields (trace to 20%), and cyclization of a 1-styryl-enyone affording only decomposition products. F.e., optimization and substrate prepn. s. F. Yang, Y.-F. Qiu, K.-G. Ji, Y.-N. Niu, S. Ali, Y.-M. Liang, J. Org. Chem. *2012*, *77* (20), 9029-37 [DOI: 10.1021/jo301567p].

9-(1-Iodovinyl)-3,4-dihydro-1(2*H*)-acridones ○
from *o*-[3-(cyclohex-2-enon-3-yloxy)prop-1-ynyl]amines

214.

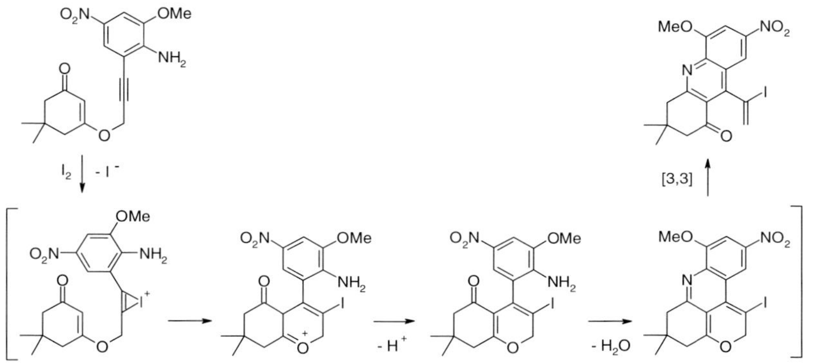

A mixture of 5,5-dimethyl-3-[3-(2-amino-3-methoxy-5-nitrophenyl)prop-2-yn-1-yloxy]cyclohex-2-en-1-one and iodine (2 eq.) in methylene chloride stirred at room temp. for 2 h → 3,3-dimethyl-9-(1-iodovinyl)-5-methoxy-7-nitro-3,4-dihydro-1(2*H*)-acridone. Y 59%. This novel and experimentally simple iodination/6-*endo*-dig cyclization appears general for a wide variety of 3-(3-arylprop-2-yn-1-yloxy)cyclohex-2-en-1-ones (fourteen examples; Y 50-68%) and gives best results in the absence of added base. F.e. and optimization s. M. Sivaraman, D. Muralidharan, P.T. Perumal, Tetrahedron Lett. *2012*, *53* (45), 6039-43 [DOI: 10.1016/j.tetlet.2012.08.107].

Microwaves s. under Tris(dibenzylideneacetone)dipalladium [\\\\\]
Cesium fluoride s. under Tris(dibenzylideneacetone)dipalladium *CsF*

Alkali metal fluoride-¹⁸F ←
Preparation of fluorides-¹⁸F ←
update s. *80,* 214s*82*; from mesylates with or without stereoretention using $K^{18}F$/Kryptofix 222 for preparing 1,4-disubst. 3-[^{18}F]fluoropiperidines s. R. Koudih, G. Gilbert, M. Dhilly, A. Abbas, L. Barré, D. Debruyne, F. Sobrio, Org. Biomol. Chem. *2012, 10* (42), 8493-500 [DOI: 10.1039/c2ob26378e]; 2-tosyloxy-1,1,1-trifluorides-^{18}F from 2-tosyloxy-1,2-ethylene-1,1-difluorides with fluoride ion-$^{18}F/K_2CO_3$ and Kryptofix 222 s. P.J. Riss, V. Ferrari, L. Brichard, P. Burke, R. Smith, F.I. Aigbirhio, ibid. *10* (34), 6980-6 [DOI: 10.1039/c2ob25802a]; aryl fluorides-^{18}F from arylboronic acid esters with NaOH/AgOTf and subsequent reaction of the intermediate arylsilver(I) compds. with Selectfluor s. I.S.R. Stenhagen, A.K. Kirjavainen, S.J. Forsback, C.G. Jørgensen, E.G. Robins, S.K. Luthra, O. Solin, V. Gouverneur, Chem. Commun. *2013, 49* (14), 1386-8 [DOI: 10.1039/c2cc38646a]; sulfonyl fluorides-^{18}F from sulfonyl chlorides with aq. CsF/$Cs^{18}F$ and Cs_2CO_3/pyridine in aq. *tert*-butanol s. J.A.H. Inkster, K. Liu, S. Ait-Mohand, P. Schaffer, B. Guérin, T.J. Ruth, T. Storr, Chem. Eur. J. *2012, 18* (35), 11079-87 [DOI: 10.1002/chem.201103450].

Copper(I) halides-dimethyl sulfide *CuHal·SMe₂* ←

2-(Diarylmethyleneamino)-3-halogenoindoles from indoles and benzophenone O-acyloximes via copper(I)-catalyzed haloamination

215.

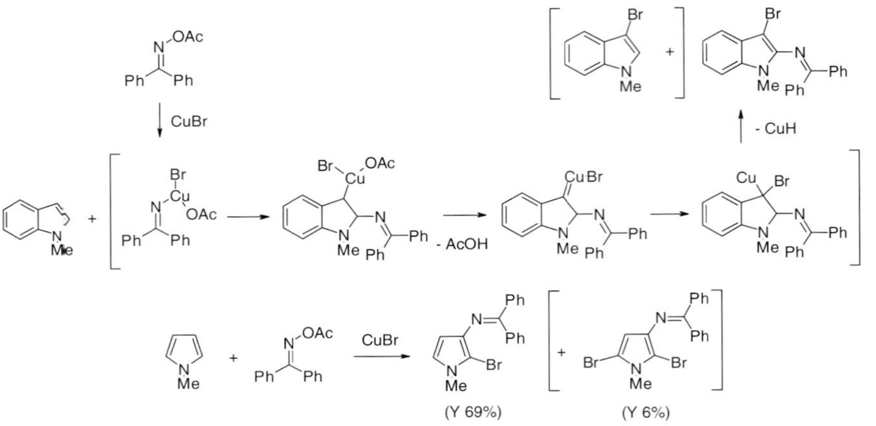

Benzophenone O-acetyloxime (2 eq.) and $CuBr \cdot SMe_2$ (1 eq.) added to a stirred soln. of 1-methylindole (0.013 mmol) in dry, degassed acetonitrile (ca. 2 ml), the mixture stirred at 70° for 16 h, cooled to room temp., filtered through silica, concentrated *in vacuo*, and purified chromatographically → 3-bromo-1-methyl-2-(diphenylmethylideneamino)indole. Y 58% (and 42% 3-bromo-1-methylindole). This novel C-H bifunctionalization is applicable to electron-diverse N-methyl/-benzyl-indoles using stoichiometric $CuHal \cdot SMe_2$ (Hal = Cl, Br, I), but only gave significant amounts of the halogenaminated products with benzophenone O-acetyl/-benzoyl- and 4,4′-dimethoxybenzophenone O-acetyl-oximes (nine examples; Y 29-58%), with attempts to produce catalytic turnover being unsuccessful. In all cases, significant amounts of 3-haloindole by-products were also obtained (Y 36-48% in three cases), but were major products (Y 40-85%; four examples) with other oxime esters. Experimental evidence suggested that the by-products were formed via a separate pathway. Applying the reaction to N-methylpyrrole gave a mixture of 2-bromo-3-(diphenylmethylideneamino)pyrrole (Y 14%) and its 2,5-dibromo-analog (Y 39%) using the same reagent stoichiometry, but with less oxime ester (1.25 eq.) gave a 69:6 mixture of mono- and dibromo products. F.e.s. A. John, K.M. Nicholas, Organometallics *2012*, *31* (22), 7914-20 [DOI: 10.1021/om300553b].

Fluoroboric acid s. under Bis(pyridine)iodine(I) fluoroborate *HBF_4*

1-Methyl-3-butylimidazolium halides *[bmim]Hal*

Halides from sulfonic acid esters OSO_2R → Hal

from tosylates s. *5,* 402; eco-friendly, solvent-free procedure from tosylates or mesylates with recyclable 1-methyl-3-butylimidazolium halides as halogenating agent, also nucleophilic displacement with other 1-methyl-3-butylimidazolium salts (acetate, thiocyanate) s. Y. Liu, Y. Xu, S.H. Jung, J. Chae, Synlett *2012, 23* (18), 2692-8 [DOI: 10.1055/s-0032-1317473]; nucleophilic displacement with polyethylene glycols, e.g. $HOCH_2CH_2O(CH_2CH_2O)_3CH_2CH_2OH$, as highly efficient multifunctional promoters s. V.H. Jadhav, S.H. Jang, H.-J. Jeong, S.T. Lim, M.-H. Sohn, J.-Y. Kim, S. Lee, J.W. Lee, C.E. Song, D.W. Kim, Chem. Eur. J. *2012, 18* (13), 3918-24 [DOI: 10.1002/chem.201102455].

N-[Chloro(phenylthio)methylene]dimethylammonium chloride *$[Cl(PhS)C{=}NMe_2]^+Cl^-$*
s. under Tetra-n-butylammonium bromide

Polymethylhydrosiloxane *PMHS*
Glycosyl halides from acyl glycosides OAc → I
cf. *19*, 523; **glycosyl iodides** with I_2/polymethylhydrosiloxane s. M. Giordano, A. Iadonisi, Eur. J. Org. Chem. *2013* (1), 125-31 [DOI: 10.1002/ejoc.201201084].

2′-Di-tert-butylphosphino-2,4,6-triisopropyl-3′,6′-dimethoxybiphenyl s. under Tris(dibenzylideneacetone)dipalladium ←

Perfluoro-1-butanesulfonyl fluoride s. under Tris(dibenzylideneacetone)dipalladium $C_4F_9SO_2F$

Tetra-n-butylammonium bromide/N-[chloro(phenylthio)methylene]dimethylammonium chloride ←
Halides from alcohols OH → Hal
update s. *81,* 188s*82*; prim. bromides from alcohols in unprotected carbohydrates or glycosides with tetra-*n*-butylammonium bromide and added N-[chloro(phenylthio)methylene]dimethyl-ammonium chloride s. W. Xue, L. Zhang, Chin. J. Chem. *2012, 30* (7), 1429-32 [DOI: 10.1002/cjoc.201200211].

Bis(pyridine)iodine(I) fluoroborate/hydrogen fluoride·pyridine or fluoroboric acid ←
Glycosyl fluorides from pent-4-enyl glycosides OR → F
or carbohydrate pent-4-enyl O^1,O^2-orthoesters s. *45*, 291s*83*

Tris(dibenzylideneacetone)dipalladium/2′-di-tert-butylphosphino-2,4,6-triisopropyl-3′,6′-dimethoxybiphenyl/cesium fluoride/perfluoro-1-butanesulfonyl fluoride/microwaves *[Pd(0)]/CsF/$C_4F_9SO_2F$/[\\\\\\]*
Ar. fluorides from phenols via aryl nonaflates OH → F
Palladium(0)-catalyzed nucleophilic fluorination under microwave irradiation

216.
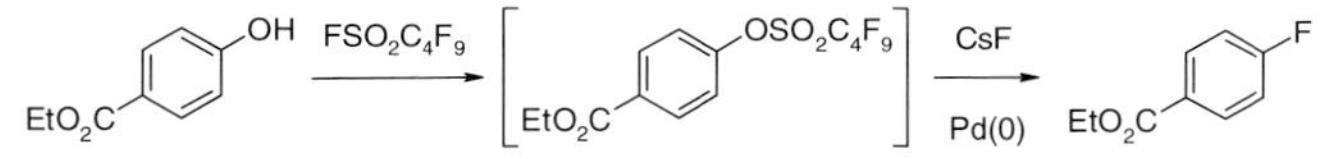

in one pot. A soln. of ethyl 4-hydroxybenzoate (1 mmol) and perfluoro-1-butanesulfonyl fluoride (1.5 eq.) in dry toluene (2 ml) added via syringe to CsF (2 eq.) in a capped vial under N_2, the mixture stirred at 50° for 24 h, a mixture of $Pd_2(dba)_3$ (2 mol%) and *t*-BuBrettPhos (6 mol%) in the same solvent (2 ml) stirred at room temp. for 5 min, added via syringe, the resulting mixture stirred under microwave irradiation at 180° for 30 min, cooled to room temp., and purified chromatographically → ethyl 4-fluorobenzoate. Y 71% (GC yield 83%). Initial development identified nonaflate as a cost-effective and convenient alternative to triflate, and its efficient generation *in situ* allowed direct conversion of phenols to ar. fluorides (eleven examples; Y 41-85%), with low isolated yields (cf. GC yields) attributed to product volatility. The method was compatible with nitrile, ketone, ester, ether, trifluoromethyl and tert. amine functionality. In competitive experiments between CsF, CsCl and CsBr, fluorination was the dominant process, a result that was supported by density functional theory [DFT] calculations. F.e.s. J. Wannberg, C. Wallinder, M. Ünlüsoy, C. Sköld, M. Larhed, J. Org. Chem. *2013*, *78* (8), 4184-9 [DOI: 10.1021/jo400255m]; **palladium-catalyzed cross-coupling with aryl nonaflates generated *in situ* from phenols**, incl. Suzuki-Miyaura, Sonogashira and Stille reactions and Buchwald-Hartwig N-arylation, s. T. Ikawa, K. Saito, S. Akai, Synlett *2012*, *23* (15), 2241-6 [DOI: 10.1055/s-0032-1317076].

Via intermediates *v.i.*
Fluorides from oxo compds. via tosylhydrazones s. *83*, 218 CO → CHF

Nitrogen ↑ HalC ↓↑ N

Sodium nitrite/fluoroboric acid *$NaNO_2/HBF_4$*
Ar. fluorides by the Balz-Schiemann reaction $ArNH_2$ → ArF
with $NaNO_2$/HF·pyridine cf. *17*, 622s*46*; *under continuous flow* in a two-reactor system for the initial diazotization and subsequent fluorodediazoniation of the intermediate diazonium fluoroborates s. Z.-q. Yu, Y.-w. Lv, C.-m. Yu, W.-k. Su, Tetrahedron Lett. *2013, 54* (10), 1261-3 [DOI: 10.1016/j.tetlet.2012.12.084].

tert-Butyl iodide *t-BuI*
Alkyl iodides from azides $N_3 \rightarrow I$

217.

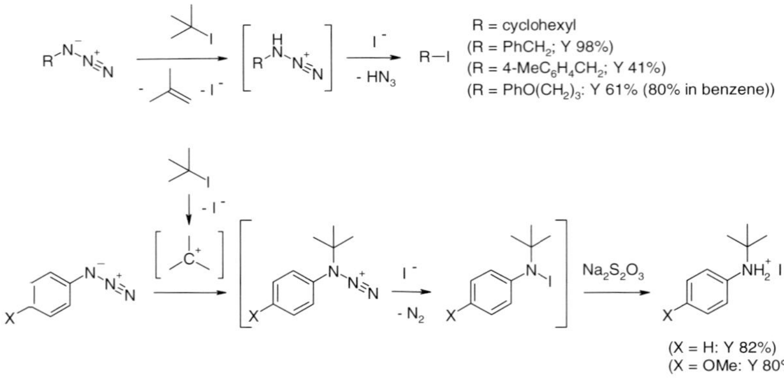

tert-Butyl iodide (4.6 eq.) added to a soln. of azidocyclohexane (0.4 mmol) in methylene chloride (2 ml) at room temp. under argon, stirred for 18 h, washed with satd. aq. $Na_2S_2O_3$, the organic phase dried ($MgSO_4$), filtered, concentrated *in vacuo*, and the residue purified by flash chromatography on silica gel → iodocyclohexane. Y 61%. This unexpected transformation afforded prim. and sec. alkyl iodides in moderate to high yield (41-98%; five examples) and relied on the alkyl azide moiety being sufficiently basic in apolar media to abstract a proton from *tert*-butyl iodide. Hydriodide salts of **N-*tert*-butylanilines** were obtained **from ar. azides** under the same conditions (two examples; Y 80-82%), via an alternative pathway, and possibly arising from reductive workup of initially-formed N-*tert*-butyl-N-iodoanilines. Bromination was not observed using *t*-BuBr, and sec. alkyl iodides were ineffective. F.e. and mechanistic experiments s. J. Maury, L. Feray, M.P. Bertrand, A. Kapat, P. Renaud, Tetrahedron *2012*, *68* (47), 9606-11 [DOI: 10.1016/j.tet.2012.09.066].

Triethylamine tris(hydrogen fluoride)/potassium carbonate *$Et_3N{\cdot}3HF/K_2CO_3$*
Fluorides from oxo compds. via tosylhydrazones C=NNHTs → CHF

218.

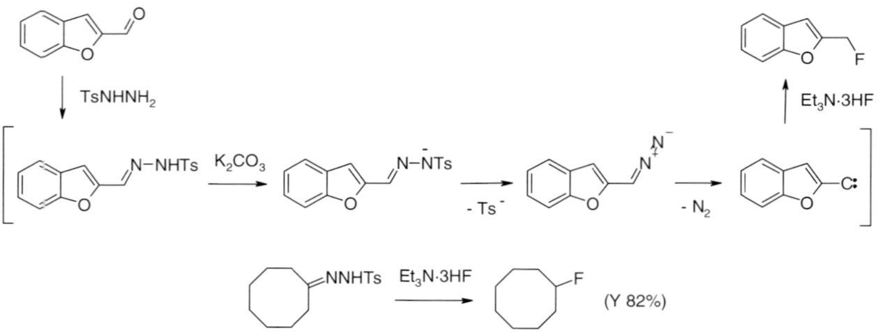

in one pot. Tosylhydrazine (1 eq.) added to a soln. of benzofuran-2-carbaldehyde (1 mmol) in toluene (4 ml), the mixture stirred at 80° for 1.5 h, K_2CO_3 (5 eq.) and $Et_3N{\cdot}3HF$ (1.5 eq.) added, the mixture stirred at reflux (110°) under N_2 for 2 h (TLC), quenched with satd. aq. $NaHCO_3$, extracted with ethyl acetate, and concentrated → 2-fluoromethylbenzofuran. Y 73%. N-Fluoro reagents (pyridine·nHF and $Et_3N\cdot 3HF$) were uniquely effective in converting tosylhydrazones of electron-diverse ar. and aliphatic aldehydes and ketones to the fluoroalkyl derivs. (BF_3 as fluoride source afforded alkenes via supposed dimerization of carbene intermediates, whereas KF and CsF were ineffective), with the presence of mild and insoluble K_2CO_3 thought to control the flow of diazo intermediate into soln., thereby reducing side-reactions. The reaction was successful with pre-formed tosylhydrazones (eighteen examples; Y 63-93%), or via generation of the tosylhydrazones *in situ* with only marginal reduction in yields (60-87%; six examples). F.e. and optimization s. A.K. Yadav, V.P. Srivastava, L.D.S. Yadav, Chem. Commun. *2013*, *49* (21), 2154-6 [DOI: 10.1039/c3cc00122a].

Halogen ↑ HalC ↓↑ Hal

Fluoride ion F^-

Halogen exchange Hal → Hal′

update s. *40,* 357s*82*; 5-fluoro- from 5-iodo-1,2,3-triazoles with fluoride ion s. B.T. Worrell, J.E. Hein, V.V. Fokin, Angew. Chem., Int. Ed. *2012, 51* (47), 11791-4 [DOI: 10.1002/anie.201204979].

Sulfur ↑ HalC ↓↑ S

^{18}F-Fluoride-Kryptofix 222 $^{18}F^-$

Ar. fluorides-^{18}F from diaryl sulfoxides ArS(O)Ar′ → ArF

219.

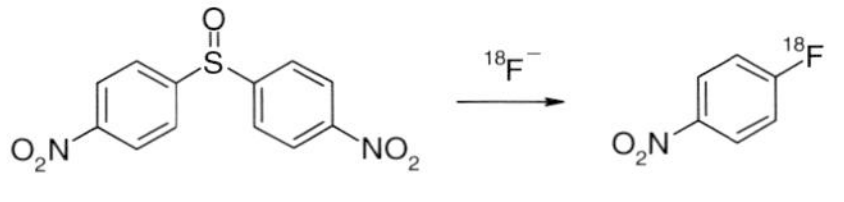

A soln. of dry ^{18}F anion (as its Kryptofix 222 complex) in DMSO (0.1 ml) added to a soln. of bis(4-nitrophenyl) sulfoxide (5 mg) in the same solvent (0.2 ml), the mixture heated at 200° for 15 or 30 min, and an aliquot (ca. 0.05 ml) quenched with aq. acetonitrile and analyzed by radio-HPLC → 4-^{18}F-fluoronitrobenzene. Radiochemical yield 90%. The radiofluorination of symmetrical and unsymmetrical bis(het)ar. sulfoxides was initially examined in a fused silica micro-reactor. Substrates (incl. electronically diverse unsymmetrical examples) containing electron-poor (het)ar. groups (4-nitrophenyl, 4-cyanophenyl, 4-trifluoromethylphenyl and 4-/2-pyridyl) gave good radiochemical yields (64-93%) of the corresponding (het)ar. fluorides, with electronically similar unsymmetrical sulfoxides affording mixtures of products (36:45 to 46:43). Performing the reactions under batch conditions using DMF/180° or DMSO/200° gave comparable radiochemical yields in most cases, but those for 2- and 4-pyridyl fluorides were somewhat lower. There was no evidence for nitro substitution in the radiofluorination of 4-nitrophenyl-containing substrates. F.e. and substrate prepn. s. J.-H. Chun, C.L. Morse, F.T. Chin, V.W. Pike, Chem. Commun. *2013*, *49* (21), 2151-3 [DOI: 10.1039/c3cc37795d].

S,S-Difluorosulfiliminium fluoroborates $[F_2S{=}NR_2]BF_4$

Glycosyl fluorides from thioglycosides SR → F

with *p*-methyliodobenzene difluoride cf. *46,* 454; with S,S-difluorosulfiliminium fluoroborates (with or without N-bromosuccinimde), also from glycosyl sulfoxides, selenides or tellurides, s. S. Tsegay, R.J. Williams, S.J. Williams, Carbohydr. Res. *2012, 352*, 16-22 [DOI: 10.1016/j.carres.2012.04.016]; with N,N-diethylaminosulfur trifluoride, also acceleration of the reaction in the presence of dimethyl(methylthio)sulfonium triflate, s. K. Suzuki, Y. Ito, O. Kanie, ibid. *2012*, 359, 81-91 [DOI: 10.1016/j.carres.2012.07.003]; glycosyl fluorides from pent-4-enyl glycosides with bis(pyridine)iodine(I) fluoroborate and added HF/pyridine (HalC↓↑O; cf. *45*, 291), also from carbohydrate pent-4-enyl O^1,O^2-orthoesters with added HBF_4 (or with HF·pyridine alone), s. B. Fraser-Reid, J.C. Lopez, P. Bernal-Albert, A.M. Gomez, C. Uriel, J. Ventura, Can. J. Chem. *2013*, *91* (1), 51-65 [DOI: 10.1139/cjc-2012-0285].

Iodine pentafluoride/pyridine·hydrogen fluoride $IF_5/py\cdot HF$

1,1-Difluorides from mercaptals $C(SR)_2 \rightarrow CF_2$

Also trifluoromethylarenes from aryltrithioorthocarboxylic acid esters s. *46,* 454s*83*

Remaining Elements ↑ HalC ↓↑ Rem

Bis(pivalonitrile)copper(I) triflate s. under N-Fluoro-2,4,6-trimethylpyridinium triflate ←

Copper(I) iodide/1,10-phenanthroline/potassium iodide *CuI/phen/KI*
Regioselective ar. iodination via arylboronic aid esters s. *83*, 212 H → I

m-Chloroperoxybenzoic acid *mCPBA*
α,β-Acetylene(aryl)iodonium salts ArI → ArI(OTs)C≡C-R
from ar. iodides and α,β-acetyleneboronic acid esters

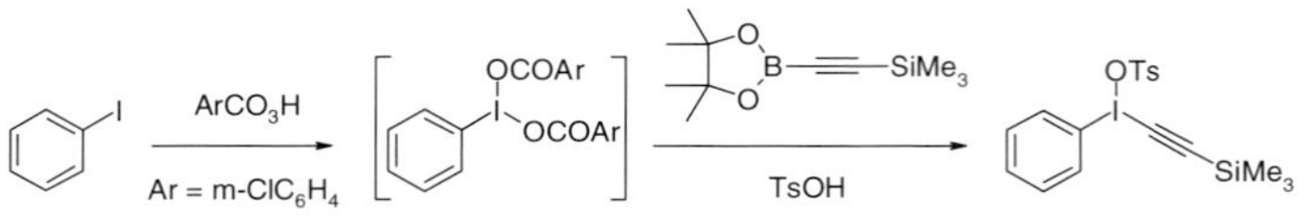

220.

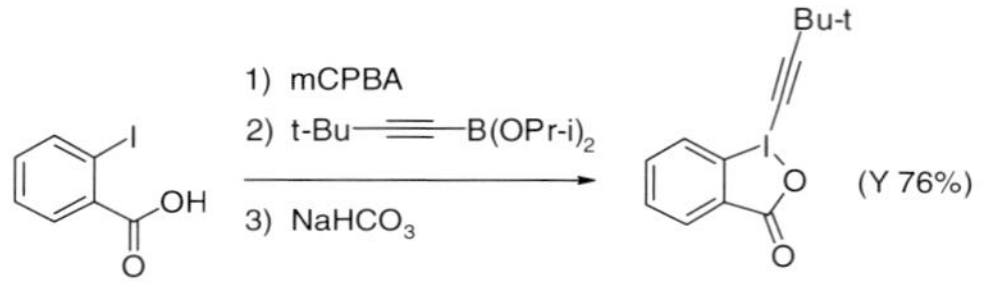

A novel and general route to alkynyl(aryl)iodonium salts (and alkynyl-1,2-benziodoxolones) is described that is unaffected by air or moisture. **E:** A soln. of iodobenzene (0.27 mmol), *m*CPBA (1.1 eq.) and *p*-toluenesulfonic acid monohydrate (1 eq.) in methylene chloride/2,2,2-trifluoroethanol (1:1; 1 ml) stirred at room temp. for 30 min, trimethylsilylethynyl-pinacolborane (1.4 eq.) added, stirring continued for 30 min, the resulting mixture filtered, the filtrate concentrated *in vacuo*, and the residue triturated with ether → trimethylsilylethynyl(phenyl)iodonium tosylate. Y 97%. Oxidation of iodoarenes with *m*-chloroperoxybenzoic acid gave aryl iodosocarboxylates, which reacted with sulfonic acids and alkynylboronates to afford the corresponding iodonium tosylates (eight examples; Y 94-100%) or triflates (six examples; Y 69-83%), with analogous iodonium fluoroborate salts (formed using BF_3 in place of RSO_3H) proving unstable during purification. Initial attempts using terminal alkynes were unsuccessful. Application of the method to *o*-iodobenzoic acid provided the alkynyl(*o*-carboxyphenyl)iodonium salts which cyclized to **1-(alk-1-ynyl)-1,2-benziodoxol-3-ones** on treatment with base in a one-pot procedure (six examples; Y 71-90%). F.e. and optimization s. M.J. Bouma, B. Olofsson, Chem. Eur. J. *2012*, *18* (45), 14242-5 [DOI: 10.1002/chem.201202977].

Electrophilic halogen sources ←
Replacement of silyl by halogen Si≤ → Hal
cf. *33*, 501s*38,39*; regioselective formation of β,γ-ethylene-α-halogenoboronic acid [pinacol] esters from α,β-ethylene-γ-silylboronic acid esters with various halonium sources, e.g. Selectfluor, N-halogenosuccinimides, also regioselective displacement of the silyl group with other electrophiles, s. A. Macé, F. Tripoteau, Q. Zhao, E. Gayon, E. Vrancken, J.-M. Campagne, B. Carboni, Org. Lett. *2013*, *15* (4), 906-9 [DOI: 10.1021/ol4000263].

N-Fluoro-2,4,6-trimethylpyridinium triflate/bis(pivalonitrile)copper(I) triflate ←
Ar. fluorides from arylstannanes Sn≤ → F
with Selectfluor under silver(I) catalysis cf. *78*, 229; mild conversion with N-fluoro-2,4,6-trimethylpyridinium triflate under copper(I) catalysis with bis(pivalonitrile)copper(I) triflate, also from potassium aryl(trifluoro)borates, s. Y. Ye, M.S. Sanford, J. Am. Chem. Soc. *2013*, *135* (12), 4648-51 [DOI: 10.1021/ja400300g].

S,S-Difluorosulfiliminium fluoroborates *$[F_2S{=}NR_2]BF_4$*
Glycosyl fluorides from glycosyl selenides or tellurides s. *46*, 454s*83* SeR or TeR → F

Carbon ↾ HalC ⇃↾ C

Oxalyl bromide *$(COBr)_2$*
Dealkylative bromination of 5-membered 2-acyl-5-alkoxyheteroarenes ←

221. 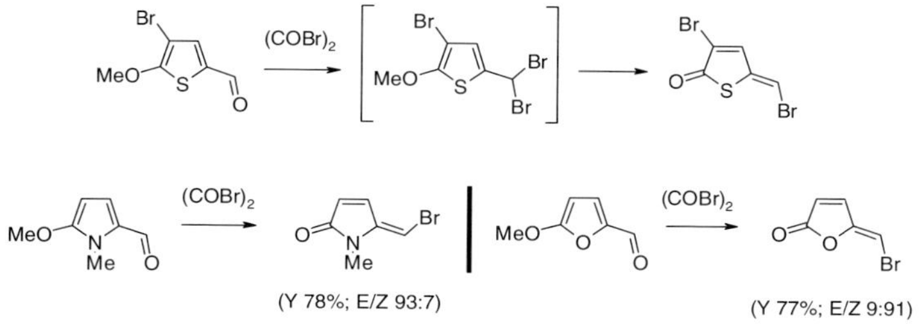

5-(α-Bromoalkylidene)-2(5*H*)-thiophenones. Oxalyl bromide (1.1 eq.) added dropwise to a soln. of 4-bromo-5-methoxythiophene-2-carbaldehyde (8.87 mmol) in methylene chloride (20 ml) at 0°, the mixture stirred at room temp. for 45 min, concentrated *in vacuo*, and purified chromatographically → (Z)-3-bromo-5-bromomethylene-2(5*H*)-thiophenone. Y 93% (E/Z 16:84). Oxalyl bromide was cleaner and generally higher yielding than acetyl bromide and was not required in excess. Demethylative bromination gave variable stereoselectivity for prepn. of the title compds. (E/Z 10:90 to 61:39; four examples; Y 44-93%), **5-(α-bromoalkylidene)-2(5*H*)-furanones** (E/Z 9:91 to 10:90; two examples; Y 40-77%) and **5-(α-bromoalkylidene)-Δ³-2-pyrrolones** (E/Z 85:15 to 93:7; two examples; Y 65-78%). F.e.s. T. Benneche, E.J. Chamgordani, I. Reimer, Tetrahedron Lett. *2012*, *53* (51), 6982-3 [DOI: 10.1016/j.tetlet.2012.10.048].

Elimination ⥮

Oxygen ↾ HalC ⥮ O

Dichloro(p-cymene)ruthenium(II) dimer/diphenylacetylene *$[RuCl_2(p\text{-}cymene)]_2/PhC{\equiv}CPh$*
***o*- or *m*-Halogenonitriles from ar. alkoximinohalides** ←
Regioselective ruthenium(II)-catalyzed intramolecular eliminative ar. halogenation

222. 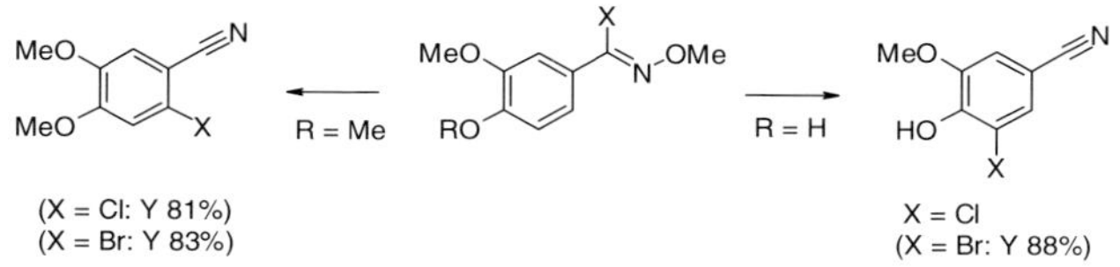

O-Methyl-4-hydroxy-3-methoxybenzohydroximoyl chloride (1 mmol) and isopropanol (3 ml) added via syringe to a mixture of $[RuCl_2(p\text{-}cymene)]_2$ (3 mol%) and diphenylacetylene (30 mol%) under N_2, the mixture sealed with a screw cap, stirred at 100° for 16 h, cooled to room temp., diluted with methylene chloride, filtered through Celite, concentrated, and purified chromatographically → 3-chloro-4-hydroxy-5-methoxybenzonitrile. Y 85%. A series of 4-subst. (alkoxy, OH, NHMe, NMe_2) and also 3,4-dihydroxy- or 4-hydroxy-3-methoxy-benzohydroximoyl chlorides/bromides afforded the corresponding *m*-halobenzonitriles exclusively (thirteen examples; Y 84-97%). However, similar treatment of 3-methoxy, 3,4/5-dimethoxy, 3,4-methylenedioxy derivs., and also O-methyl-1-naphthohydroximoyl chloride, afforded the corresponding *o*-halobenzonitriles (halogenation occurring at the least hindered *ortho* site for unsymmetrical substrates) (eight examples; Y 81-94%). The mechanism for this highly regioselective halogenation was not established. F.e., optimization, substrate prepn. and illustrative conversions of products to **5-aryl-tetrazoles**, s. R.K. Chinnagolla, S. Pimparkar, M. Jeganmohan, Chem. Commun. *2013*, *49* (30), 3146-8 [DOI: 10.1039/c3cc41124a].

Formation of S-S Bond

Exchange ⇂↾

Hydrogen ↾ SS ⇂↾ H

Electrolysis ϟ
Sym. disulfides from mercaptans 2 RSH → $(RS)_2$
update s. *47,* 468s*82*; with a soluble support-bound peptide allowing electron transfer at the surface of the electrodes and facilitating isolation of product s. S. Kitada, M. Takahashi, Y. Yamaguchi, Y. Okada, K. Chiba, Org. Lett. *2012, 14* (23), 5960-3 [DOI: 10.1021/ol302863r]; polymer-based cyclic disulfide formation in peptide synthesis using N-chlorosuccinimide, compatibility with oxidation-prone methionine and tryptophan, also application to the synthesis of oxytocin and a regioselective synthesis of an α-conotoxin, s. T.M. Postma, F. Albericio, ibid. *2013*, *15* (3), 616-9 [DOI: 10.1021/ol303428d].

Nitrogen ↾ SS ⇂↾ N

Scandium(III) triflate/1-butyl-3-methylimidazolium hexafluorophosphate/sodium salt ←
Thiolsulfonic acid esters $SO_2Na \rightarrow SO_2SR$
from sulfinic acid salts and N-(organothio)succinimides
Metal triflate-catalyzed conversion in ionic liquid/water

223. 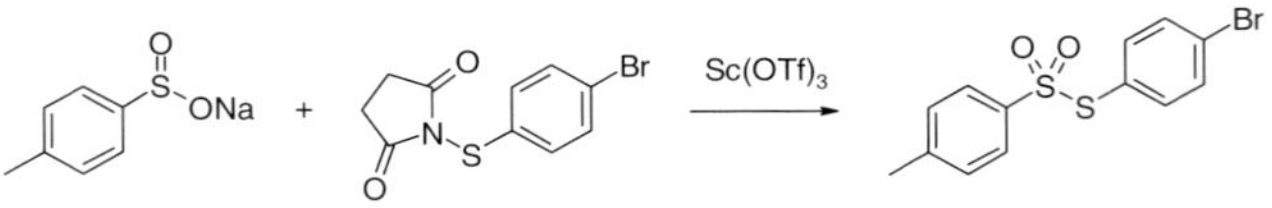

$Sc(OTf)_3$ (5 mol%) added to a soln. of N-(*p*-bromophenylthio)succinimide (0.3 mmol) and Na-*p*-toluenesulfinate (1.2 eq.) in [bmim]PF_6 (3 ml) and water (1 ml) at 30°, the mixture stirred for 2 h (TLC/GC-MS), washed with water, extracted with ether, and purified by chromatography on silica → S-*p*-bromophenyl *p*-toluenethiosulfinate. Y 81%. This novel and experimentally simple sulfenylation was catalyzed effectively by several metal triflates [e.g. Cu(II)-, Mg-, Zn- or Yb(III)-triflate in water] and appears general for N-(aryl- or benzyl-thio)succinimides and arene- or methane-sulfinates (thirty-four examples; Y 68-92%), with highest yields generally obtained with electron-rich substrates. The catalyst and ionic liquid may be recycled five times with only slight reduction in reactivity (Y 88% → 79%). F.e. and optimization s. G. Liang, J. Chen, J. Chen, W. Li, J. Chen, H. Wu, Tetrahedron Lett. *2012*, *53* (50), 6768-70 [DOI: 10.1016/j.tetlet.2012.09.132].

Elimination ⇑

Hydrogen ↾ SS ⇑ H

N-Chlorosuccinimide *NCS*
Cyclic disulfides from dithiols ○
in solid-phase peptide synthesis s. *47,* 468s*83*

Formation of S-Rem Bond

Uptake ⇓

Addition to Remaining Elements SRem ⇓ Rem

1,2,4-Dithia-zol-5-ones or -zole-5-thiones ←
Phosphorus(V) sulfides from phosphorus(III) compds. ⩾P → ⩾P=S
thionophosphoric acid esters with dimethylthiuram sulfide cf. *75,* 53; P-thionation with 3-subst. 1,2,4-dithia-zol-5-ones or -zole-5-thiones, e.g. for thionation of triphenyl phosphite, s. O. Ponomarov, A.P. Laws, J. Hanusek, Org. Biomol. Chem. *2012, 10* (44), 8868-76 [DOI: 10.1039/c2ob26460a].

Formation of S-C Bond

Uptake ⇓

Addition to Oxygen and Carbon SC ⇓ OC

Ammonium thiocyanate/2-chloro-1-methylpyridinium iodide ←
2-Hydroxythiocyanates from epoxides  → C(OH)C(SCN)
update s. *59,* 203s*80*; fast, mild, solvent-free procedure using the Mukaiyama reagent with perfect regioselectivity s. R. Azadi, B. Mokhtari, H. Oghabi, Phosphorus, Sulfur Silicon Relat. Elem. *2012, 187* (11), 1377-82 [DOI: 10.1080/10426507.2012.685670].

Addition to Nitrogen and Carbon SC ⇓ NC

Cinchona-based squaramides ←
N-Protected 1-aminothioethers from aldimines and mercaptans by organocatalyzed asym. 1,2-addition CH=N- → CH(SR)NH-
from N-acylaldimines s. *80,* 216; asym. addition to an N-protected trifluoroacetaldimine with a bifunctional cinchona-based squaramide as catalyst s. X. Fang, Q.-H. Li, H.-Y. Tao, C.-J. Wang, Adv. Synth. Catal. *2013, 355* (2-3), 327-31 [DOI: 10.1002/adsc.201200845].

Addition to Carbon SC ⇓ C

Selenium/triethylamine *Se/Et$_3$N*
Isothiocyanates from isonitriles NC: → NCS
o-Isothiocyanato-β-styryl carbonates *en route* to 1-thiocarbamylindoles s. *83,* 143

Addition to Carbon-Carbon Bonds SC ⇓ CC

Irradiation s. under 2-(o-Nitrophenyl)propyl N-hexylcarbamate ///

Sodium azide/copper nanoparticles-on-activated carbon *NaN$_3$/nano-Cu-C*
1-β-Alkylthio-1,2,3-triazoles from ethylene and acetylene derivs. ○
via 1,2-azidosulfenylation-1,3-dipolar cycloaddition s. *83,* 237

Tert. amines *R$_3$N*
Thia-Michael addition C=C → CHC(SR)
update s. *47,* 487s*82*; effect of base [triethylamine or DBU] and structural parameters on Michael addition of mercaptans to polycyclic enones s. C.J. Rosenker, E.H. Krenske, K.N. Houk, P. Wipf, Org. Lett. *2013, 15* (5), 1076-9 [DOI: 10.1021/ol400094k]; Michael addition of thioacetic acid to aliphatic or aromatic α,β-ethylenenitriles (excepting β,β-disubst. acrylonitriles) with tri-*n*-butyl-

amine, also conversion to 1-subst. homotaurines, s. Y. Ma, J. Xu, Synthesis *2012, 44* (14), 2225-30 [DOI: 10.1055/s-0031-1289773]; preparation of green organogelators with a sulfide linkage and free amino groups using N-benzylcinchonidinium bromides as phase transfer catalyst s. F. Delbecq, K. Tsujimoto, Y. Ogue, T. Kawai, Tetrahedron Lett. *2013, 54* (7), 651-6 [DOI: 10.1016/j.tetlet 2012.11.145]; Michael addition of thiophenols to enones *under nickel catalysis* with a cationic (acetonitrile)nickel(II) perchlorate complex possessing an N_3O_2 pentadentate ligand [BMPP or BPPP] s. W.-Z. Lee, C.-W. Chiang, G.M. Kulkarni, T.-S. Kuo, J. Chin. Chem. Soc. *2012, 60* (3), 245-50 [DOI: 10.1002/jccs.201200386]; base-catalyzed, solvent-dependent Michael addition (and double Michael addition) of mercaptans to propiolic acid esters s. V.X. Truong, A.P. Dove, Angew. Chem., Int. Ed. *2013, 52* (15), 4132-6 [DOI: 10.1002/anie.201209239]; one-pot synthesis of β-(organothio)carboxylic acid amides via Michael addition of mercaptans to intermediate β,γ-ethylene-α-iminonitriles s. *83*, 227.

Copper nanoparticles-on-activated carbon s. under NaN_3 *nano-Cu-C*

Scandium(III) triflate/chiral 6-α,6′-α-dihydroxy-2,2′-bipyridyls/pyridine ←
Asym. thia-Michael addition in water s. *47,* 487s*83* C=C → CHC(SR)

Chiral aminosquaramides ←
Asym. thia-Michael addition to 1-nitroethylene derivs.
of thioacetic acid s. *47,* 487s*83*

2-(o-Nitrophenyl)propyl N-hexylcarbamate/irradiation *$ROCONHC_6H_{13}$/⫻*
Photo-initiated Michael addition with mercaptans using an *in situ*-generated prim. amine as catalyst

224.

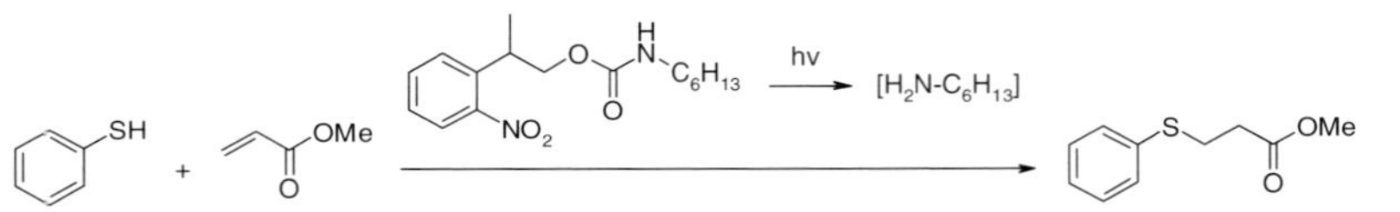

Methyl acrylate (1 eq.) added to a mixture of benzenethiol (2 mmol) and 2-(*o*-nitrophenyl)propyl N-hexylcarbamate (5 mol%), and the mixture stirred under irradiation (320-390 nm) for 90 min → methyl 3-(phenylthio)propionate. Y 93%. Photolysis of the carbamate released hexylamine, which is an efficient catalyst for thia-Michael addition reactions (experiments indicated the prim. amine was released completely during 30 min irradiation). The method was effective for the thiolation of methyl acrylate with aliphatic and phenyl thiols (four examples; Y 91-94%), and also initiated a Michael photopolymerization reaction of two tetrathiols and bisphenol A ethoxy diacrylate. No reaction occurred in the absence of catalyst or irradiation. F.e.s. W. Xi, M. Krieger, C.J. Kloxin, C.N. Bowman, Chem. Commun. *2013*, *49* (40), 4504-6 [DOI: 10.1039/c3cc41123k].

Cinchona-based 2-tert-aminothioureas ←
Asym. thia-Michael addition
to N-(α,β-ethyleneacyl)-2-oxazolidones s. *47,* 487s*83*

Chiral 1,1′-binaphthyl-based tetraaminophosphonium barfates/2,6-lutidine ←
Asym. thia-Michael addition
update s. *47,* 487s*82*; of mercaptans to nitroolefins *under synergistic catalysis* with a chiral 1,1′-binaphthyl-based tetraaminophosphonium barfate as Brønsted acidic ionic liquid coupled with 2,6-lutidine as achiral base s. D. Uraguchi, N. Kinoshita, D. Nakashima, T. Ooi, Chem. Sci. *2012, 3* (11), 3161-4 [DOI: 10.1039/c2sc20698f]; diastereoselective asym. Michael addition of thioacetic acid to trisubst. nitroolefins on a 10 gram scale with a bifunctional chiral aminosquaramide as catalyst s. W. Yang, D.-M. Du, Org. Biomol. Chem. *2012, 10* (34), 6876-84 [DOI: 10.1039/c2ob26068a]; to α-subst. N-(α,β-ethyleneacyl)-2-oxazolidones via asym. protonation with a cinchona-based 2-*tert*-aminothiourea as catalyst s. R.A. Unhale, N.K. Rana, V.K. Singh, Tetrahedron Lett. *2013, 54* (15), 1911-5 [DOI: 10.1016/j.tetlet.2013.01.005]; to enones *in water* under scandium(III) catalysis with $Sc(OTf)_3$ and a chiral 6-α,6′-α-dihydroxy-2,2′-bipyridyl as ligand and pyridine as base s. T. Kitanosono, M. Sakai, M. Ueno, S. Kobayashi, Org. Biomol. Chem. *2012, 10* (35), 7134-47 [DOI: 10.1039/c2ob26264a].

N-Benzylcinchonidinium bromides ←
Thia-Michael addition s. *47,* 487s*83* C=C → CHC(SR)

(Acetonitrile)nickel(II) perchlorate complexes *[Ni(II)]*
Nickel-catalyzed thia-Michael addition s. *47,* 487s*83*

Dichloro[N-(dicyclohexylphosphino)piperidine]palladium(II)/sodium hydroxide ←
***cis*-Thioenolethers from acetylene derivs. and mercaptans** C≡C → CH=C(SR)
Regio- and stereo-selective palladium(II)-catalyzed hydrothiolation

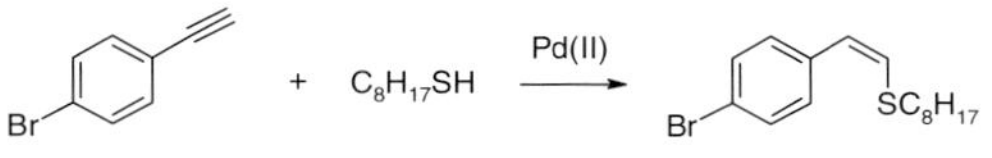

225.

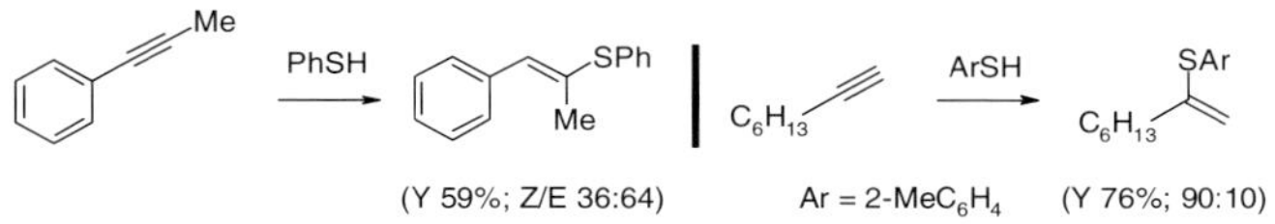

A mixture of 4-bromophenylacetylene (1.3 eq.), *n*-octanethiol (1 mmol), NaOH (1 eq.), Pd(II) catalyst (0.05 mol%), N-methylpyrrolidone (2 ml) and THF heated at 120° for a few min, and purified by flash chromatography on silica → (Z)-β-octylthio-4-bromostyrene. Y 92%. Palladium-assisted hydrothiolation of (het)ar. terminal alkynes with (het)ar. or benzylic thiols afforded the *anti*-Markovnikov adducts with high/exclusive *cis*-selectivity (nineteen examples; Y 73-95%; Z/E 86:14 to 100:0), with somewhat lower stereoselectivity reported for octanethiol (Z/E 70:30; Y 66%). Internal diarylethynes generally showed similar *cis*-selectivity (Z/E 87:13 to 100:0; four examples; Y 78-93%), but methyl(phenyl)acetylene was much less stereoselective (Z/E 29:71 to 51:49; three examples; Y 51-68%), and diphenylacetylene almost completely (E)-selective in reaction with *o*-methylthiophenol (Z/E 2:98; Y 71%). Interestingly, terminal *aliphatic alkynes* afforded *Markovnikov* adducts as major isomers (1,1/1,2-disubst. ethenes 75:25 to 90:10; six examples; Y 61-80%). F.e. and a mechanistic discussion s. R. Gerber, C.M. Frech, Chem. Eur. J. *2012, 18* (29), 8901-5 [DOI: 10.1002/chem.201200388].

Exchange ↓↑

Hydrogen ↑ SC ↓↑ H

Copper(II) oxide/iodine/cesium carbonate/1,10-phenanthroline/dimethyl sulfoxide/ sodium hydrogen sulfide ←
2-Aroylbenzothiazoles from prim. ar. amines and acetophenones ○

226.

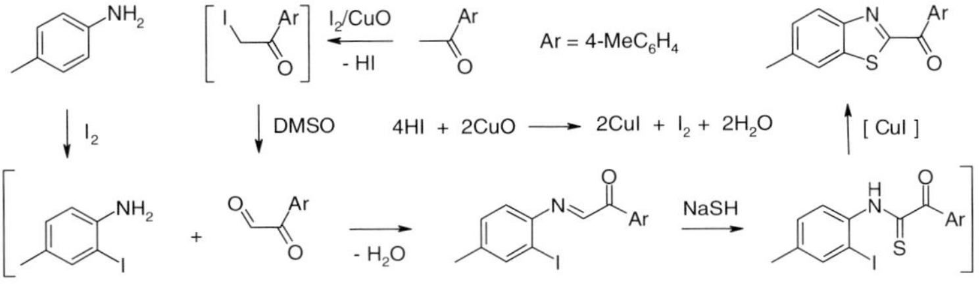

I_2 (2.5 eq.), CuO (1 eq.), Cs_2CO_3 (0.5 eq.) and 1,10-phenanthroline (25 mol%) added to a stirred soln. of 4-methylacetophenone (1.1 eq.), 4-toluidine (0.5 mmol) and NaSH·nH_2O (2 eq.) in DMSO (3 ml), the mixture stirred at 120° for 24 h, diluted with ethyl acetate and water, quenched with satd. aq. $Na_2S_2O_3$, the organic layer concentrated *in vacuo*, and purified chromatographically → 6-methyl-2-(4-methylbenzoyl)benzo[*d*]thiazole. Y 75%. This novel method for benzothiazole synthesis was developed within the concepts of self-sequencing networks, and involved the rational integration and optimization of six separate reactions. The method was successful for electron-diverse acetophenones and anilines, being most effective with electron-rich components (nineteen examples; Y 30-80%). Electron-poor 4-nitroacetophenone gave only 23% yield, while 4-nitroaniline was unreactive. F.e. and optimization s. W.-J. Xue, Y.-Q. Guo, F.-F. Gao, H.-Z. Li, A.-X. Wu, Org. Lett. *2013*, *15* (4), 890-3 [DOI: 10.1021/ol400029t]; **from 2-unsubst. benzothiazoles and acetophenones** (SC↓↑C) with I_2/KOH s. Q. Gao, X. Wu, F. Jia, M. Liu, Y. Zhu, Q. Cai, A. Wu, J. Org. Chem. *2013*, *78* (6), 2792-7 [DOI: 10.1021/jo302754c].

Via intermediates *v.i.*

N-Subst. β-(organothio)carboxylic acid amides from α,β-ethylenealdehydes and prim. amines C(SR)CHCON<

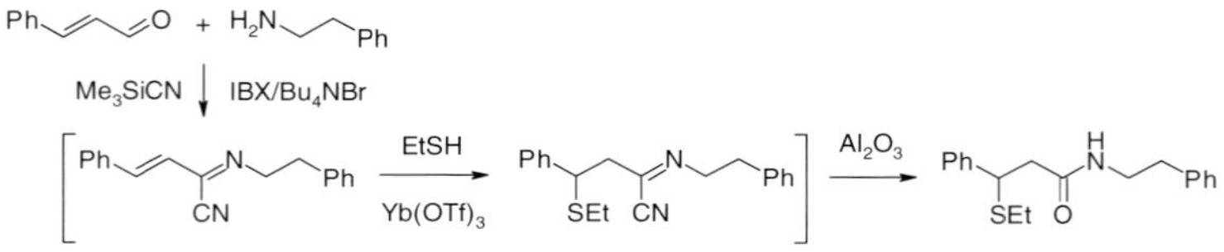

227. 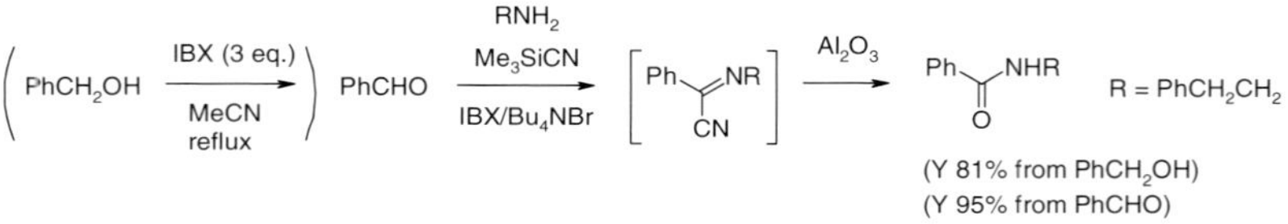

It is reported that carboxylic acid amides may be obtained by mild alumina-promoted hydrolysis of α-iminonitriles, the latter being obtainable via oxidative 3-component Strecker reaction between aldehydes, amines and trimethylsilyl cyanide (*73*, 374), and the procedure may be extended to a **one-pot synthesis** of the title compounds **via Michael addition** of mercaptans **to** intermediate **β,γ-ethylene-α-iminonitriles**. **E:** *o*-Iodoxybenzoic acid (1 eq.) and tetra-*n*-butylammonium bromide (1 eq.) added to a stirred soln. of cinnamaldehyde (0.1 mmol), phenethylamine (1.1 eq.) and trimethylsilyl cyanide (1.1 eq.) in toluene (0.25 ml) at room temp., stirring continued at room temp. until startg. ms. consumed, ethanethiol (5 eq.) added at 0°, stirred for 5 min, $Yb(OTf)_3$ (0.1 eq.) added, the resulting mixture stirred at 0° until reaction complete, diluted with methylene chloride (2 ml), Al_2O_3 (1.5 g) added, volatiles removed *in vacuo*, the resulting powder left to stand at 40° for 4 h, taken up in ethyl acetate, filtered through Celite, the filtrate evaporated under reduced pressure, and the residue purified by chromatography on silica gel → 3-(ethylthio)-N-phenethyl-3-phenylpropanamide. Y 98%. Thirteen further examples afforded yields of 58-96%, with lowest due to steric (e.g. *t*-BuSH) or electronic (e.g. 4-methoxycinnamal) factors, or to the presence of a free hydroxyl group. Substrates included both cinnamaldehyde and crotonaldehyde derivs., and aliphatic or aromatic amines and thiols, tolerating a wide variety of ring or chain substituents. The protocol was also used in one-pot syntheses of **N-subst. carboxylic acid amides from aldehydes and prim. amines** (five examples; Y 73-100%), **or from prim. alcohols** and prim. amines (nine examples; Y 76-84%), via tandem oxidation with IBX. F.e. and optimization s. J.-B. Gualtierotti, X. Schumacher, P. Fontaine, G. Masson, Q. Wang, J. Zhu, Chem. Eur. J. *2012*, *18* (46), 14812-9 [DOI: 10.1002/chem.201202291].

Oxygen ↑ SC ↓↑ O

Irradiation s. under CdS ///

Electrolysis ϟ

2,3-Unsatd. thioglycosides from O^3-acylglycals and mercaptans ←

thia-Ferrier rearrangement with $LiBF_4$ cf. *57*, 202, and with $NaHSO_4$ supported on silica gel cf. *24,* 240s*82*; under electrolysis using a *zirconium* anode, also thia- and aza-Michael addition to methyl vinyl ketone, s. D. Stevanovic, A. Pejovic, I. Damljanovic, M. Vukicevic, G.A. Bogdanovic, R.D. Vukicevic, Tetrahedron Lett. *2012, 53* (46), 6257-60 [DOI: 10.1016/j.tetlet.2012.09.023]; with I_2, stereoselectivity, s. B.V. Subba Reddy, Ch. Divyavani, J.S. Yadav, Synthesis *2010* (10), 1617-20 [DOI: 10.1055/s-0029-1218722].

Sodium sulfide/copper(I) iodide *Na_2S/CuI*

Sym. diaryl trithiocarbonates 2 $ArOSO_2R$ → ArSC(S)SAr

from aryl tosylates or triflates s. *83*, 234

Ammonium thiocyanate/triphenylphosphine/4,4′-azopyridine *NH_4SCN/Ph_3P/(4-Py)N═N(Py-4)*

Sym. disulfides from alcohols 2 ROH → $(RS)_2$

228. 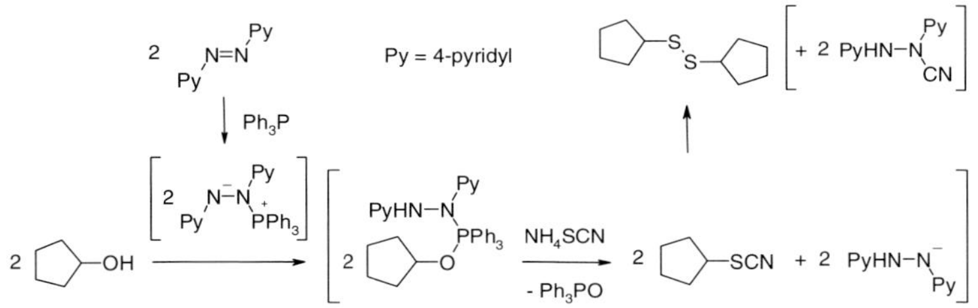

Cyclopentanol (1 mmol) added to a refluxing soln. of 4,4′-azopyridine (1.8 eq.) and triphenylphosphine (1.8 eq.) in acetonitrile (5 ml), the mixture stirred under reflux for 20 min, NH_4SCN (2.5 eq.) added in one portion, stirring continued until substrate consumed (TLC/GC; 5 h), filtered, solvent removed, the residue diluted with water, extracted with ether, and purified chromatographically → dicyclopentyl disulfide. Y 83%. Attempted conversion of aliphatic alcohols to thiocyanates under Mitsunobu conditions led instead to disulfides. This efficient and direct reaction was general for prim. and sec. (a)cyclic aliphatic alcohols (benzylic alcohols gave thiocyanates only), using 4,4′-azopyridine or 5,5′-dimethyl-3,3′-azoisoxazole and soluble NH_4SCN (ten examples; Y 73-91%). By-product N-cyanohydrazines were isolated for recycling by simple filtration. F.e. and optimization s. N. Iranpoor, H. Firouzabadi, D. Khalili, Tetrahedron Lett. *2012, 53* (51), 6913-5 [DOI: 10.1016/j.tetlet.2012.10.013]; sym. disulfides **from** [alkyl or aryl] **halides** (SC⇃↾Hal) with CuCl and potassium 5-methyl-1,3,4-oxadiazole-2-thiolate as the base, ligand and sulfur-transfer reagent s. M. Soleiman-Beigi, F. Mohammadi, ibid. *53* (52), 7028-30 [DOI: 10.1016/j.tetlet.2012.10.016].

Copper(I) iodide s. under Na_2S *CuI*

Mesoporous cadmium sulfide nanospheres/irradiation *CdS/⫻*

Benzothiazoles from *o*-aminomercaptans and aldehydes ○

update s. *19,* 674s*82*; under heterogeneous conditions with recyclable mesoporous CdS nanospheres under visible-light irradiation for coupling with aliphatic and electron-diverse aromatic (and heteroaromatic) aldehydes s. S. Das, S. Samanta, S.K. Maji, P.K. Samanta, A.K. Dutta, D.N. Srivastava, B. Adhikary, P. Biswas, Tetrahedron Lett. *2013, 54* (9), 1090-6 [DOI: 10.1016/j.tetlet.2012.12.044].

Boron fluoride *BF_3*

Thioglycosides from acyl glycosides OAc → SR

update s. *20,* 447s*82*; ethyl per-O-benzoyl-α-thioglycosides from per-O-benzoylated carbohydrates with BF_3-etherate s. D. Sail, P. Kovác, Carbohydr. Res. *2012, 357*, 47-52 [DOI: 10.1016/j.carres.2012.05.012].

Triphenylphosphine/4,4′-azopyridine s. under NH_4SCN *Ph_3P/(4-Py)N═N(Py-4)*

Triphenyl phosphite/tetra-n-butylammonium bromide *$(PhO)_3P/Bu_4NBr$*
Benzothiazoles from *o*-aminomercaptans and carboxylic acids ○
with $MeSO_3H/P_2O_5$ cf. *34*, 528; with triphenylphosphine as catalyst and tetra-*n*-butylammonium bromide as ionic medium, functional group tolerance, s. S. Meghdadi, M. Amirnasr, P.C. Ford, Tetrahedron Lett. *2012, 53* (51), 6950-3 [DOI: 10.1016/j.tetlet.2012.10.035].

Phosphorus pentasulfide *P_2S_5*
Carboxylic acid thioamides from ketoximes R(R′)C=NOH → RC(S)NHR′
via Beckmann rearrangement s. *83*, 229; *80*, 221s*83*

O,O-Diethyl dithiophosphoric acid s. under Hydroxylamine hydrochloride *$(EtO)_2PSSH$*

Sulfamic acid/silica gel *H_2NSO_3H/SiO_2*
Mercaptals from oxo compds. CO → $C(SR)_2$
with sulfuric acid/silica s. *8*, 667s*82*; with recyclable sulfamic acid/silica gel, also monothioacetals, s. T. Aoyama, T. Suzuki, T. Nagaoka, T. Takido, M. Kodomari, Synth. Commun. *2013, 43* (4), 553-66 [DOI: 10.1080/00397911.2011.604458].

Iodine *I_2*
2,3-Unsatd. thioglycosides from O³-acylglycals and mercaptans ←
Stereoselective thia-Ferrier rearrangement s. *57*, 202s*83*; *24*, 240s*83*

Hydroxylamine hydrochloride/sodium acetate/O,O-diethyl dithiophosphoric acid ←
Carboxylic acid thioamides from oxo compds. RC(O)R′ → RC(S)NHR′
via Beckmann rearrangement

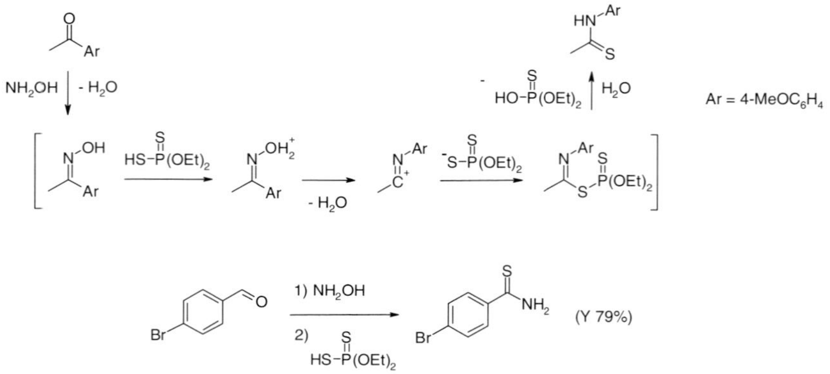

229.

in one pot. Hydroxylamine·HCl (1 eq.), Na-acetate (1 eq.) and O,O-diethyl dithiophosphoric acid (1 eq.) added to a soln. of 4-methoxyacetophenone (1 mmol) in dioxane (2 ml) under N_2, the mixture stirred at 80-90° for 2 h, cooled to room temp., neutralized with aq. $NaHCO_3$, extracted with ethyl acetate, and purified chromatographically → N-(4-methoxyphenyl)thioacetamide. Y 94%. The reagent functions both as source of S and as an acid in this efficient and experimentally simple transformation, which appears general for electron-diverse aryl and aliphatic ketones (incl. cyclic ones), affording the corresponding sec. thioamides (via preferential aryl migration for alkyl aryl ketones) from intermediate oximes (nine examples; Y 76-94%). Under similar conditions electron-diverse benzaldehydes gave N-unsubst. arylcarboxylic acid thioamides (six examples; Y 76-93%). F.e. and optimization s. A.K. Yadav, V.P. Srivastava, L.D.S. Yadav, Tetrahedron Lett. *2012*, *53* (52), 7113-6 [DOI: 10.1016/j.tetlet.2012.10.089]; **from ketoximes** via Beckmann rearrangement with P_2S_5 (cf. *80*, 221) s. J. Li, C. Cheng, X. Zhang, Z. Li, F. Cai, Y. Xue, W. Liu, Chin. J. Chem. *2012*, *30* (8), 1687-9 [DOI: 10.1002/cjoc.201200448].

Chiral iridium(I) complexes/base *[Ir(I)]*/base*
Regioselective asym. synthesis of 2-ethylenethioethers OCOOR → SR
from 2-ethylenecarbonic acid esters under iridium catalysis, update s. *79*, 206s*81*; with Na-2-aminobenzenethiolate as nucleophile, also conversion to benzo-condensed N,S-heterocyclics, s. N. Gao, X.-W. Guo, S.-C. Zheng, W.-K. Yang, X.-M. Zhao, Tetrahedron *2012, 68* (46), 9413-8 [DOI: 10.1016/j.tet.2012.09.010].

Nitrogen ↾ SC ⇃↾ N

Sodium sulfide/trichloroacetic acid *Na_2S/CCl_3COOH*
Replacement of ar. triazenyl groups ArN=N-N< → ArX
ar. halides cf. *36*, 492s*62*; **arylmercaptans** from ar. triazenes under mild conditions with Na_2S/trichloroacetic acid s. A. Khazaei, M. Kazem-Rostami, A.R. Moosavi-Zare, M. Bayat, S. Saednia, Synlett *2012, 23* (13), 1893-6 [DOI: 10.1055/s-0032-1316557].

Sodium iodide s. under $CeCl_3$ *NaI*

Copper(II) acetate/triethylamine *$Cu(OAc)_2/Et_3N$*
Benzothiazoles from *o*-aminomercaptans and nitriles ○

230.

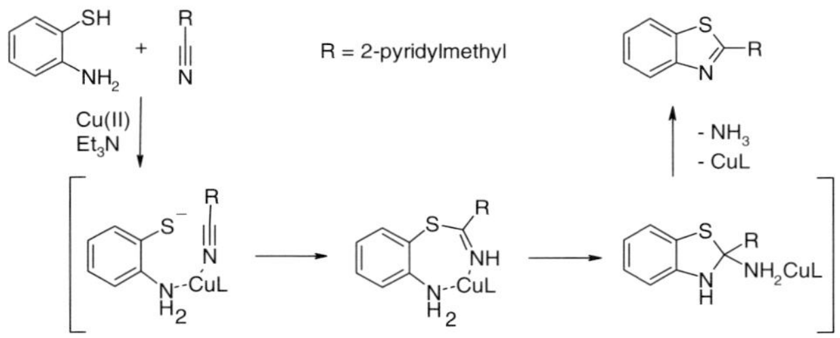

A mixture of 2-aminobenzenethiol (0.5 mmol), 2-pyrid-2-ylacetonitrile (1 eq.), $Cu(OAc)_2$ (10 mol%), Et_3N (1 eq.) and ethanol (2.5 ml) stirred at 70° for 6 h (TLC), cooled to room temp., quenched with aq. Na_2CO_3, extracted with ethyl acetate, and purified by chromatography on silica → 2-(2-pyridylmethyl)benzothiazole. Y 95%. This efficient, experimentally simple and scalable (up to 10 mmol) copper(II)-catalyzed ring closure was successful with electron-diverse (het)ar. and aliphatic nitriles (twenty-four examples; Y 75-97%) in the presence of ether, tert. amine, nitrile (1,4-dicyanobenzene gave the mono-product) and halo functionality. The reaction was, however, sensitive to *ortho* substituents and failed for 2-methyl- and 2-bromo-benzonitriles. Other copper and transition metal catalysts were less effective. F.e. and optimization s. Y. Sun, H. Jiang, W. Wu, W. Zeng, X. Wu, Org. Lett. *2013*, *15* (7), 1598-601 [DOI: 10.1021/ol400379z].

Magnesium methoxide *$Mg(OMe)_2$*
Replacement of nitro by organothio groups NO_2 → SR
with AIBN cf. *33*, 546s*41*; **ar. thioethers** from nitro compds. (possessing electron-withdrawing groups) and aliphatic or aromatic mercaptans using $Mg(OMe)_2$ as solid base s. H. Naeimi, M. Moradian, Synlett *2012, 23* (15), 2223-6 [DOI: 10.1055/s-0032-1317079].

Cerium(III) chloride/sodium iodide *$CeCl_3$/NaI*
3-Sulfenylation of indoles with N-(organothio)phthalimides H → SR
Also 2-sulfenylation of pyrroles s. *83*, 232

2,6-Dimethoxybenzoquinone s. under Shvo's ruthenium hydride complex *2,6-$(MeO)_2$BQ*

Chiral squaramides —
C_3-Sulfenylation of oxindoles s. *70*, 237s*83*

Trichloroacetic acid s. under Na_2S *CCl_3COOH*

Chiral 1,1′-binaphthyl-based cyclic selenophosphoric acid triamides/ethanesulfonic acid ←
***trans*-1-Subst. 2-(arylthio)tetralins from (E)-homoallylarenes** ○
Organocatalyzed asym. intramolecular carbosulfenylation

231. 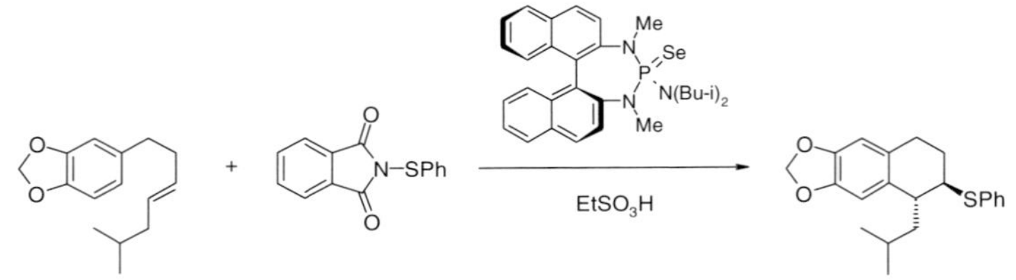

The first catalytic asym. carbosulfenylation of alkenes, providing access to enantioenriched *trans*-1,2-disubst. 1,2,3,4-tetrahydronaphthalenes, is described. **E:** A soln. of ethanesulfonic acid (0.75 eq.) in methylene chloride (0.75 ml) added over 20 s to a mixture of N-phenylthiophthalimide (1 eq.), chiral selenophosphoramide catalyst (10 mol%) and (E)-5-(6-methylhept-3-en-1-yl)benzo[*d*]-1,3-dioxole (1 mmol) in the same solvent (4.25 ml) at -20°, the mixture stirred for 3 d, quenched with triethylamine, diluted with water, extracted with ether, and purified chromatographically → (5R,6R)-5-isobutyl-6-phenylthio-5,6,7,8-tetrahydronaphtho[2,3-*d*]-1,3-dioxole. Y 77% (e.e. 92%). A series of terminally-subst. (E)-butenylarenes carrying a 3,4-methylenedioxy group cyclized exclusively, at the least hindered site, to the *trans*-products in the presence of the Brønsted acid, chiral Lewis base and N-phenylsulfenylphthalimide, with moderate to good enantiocontrol (e.e. 64-94%; fourteen examples; Y 45-92%). Terminally-disubst. analogs gave good yields of products containing a quaternary center, but enantioselectivity was somewhat lower (e.e. 42-60%; Y 82-90%; two examples), while similar cyclization of 3-methoxy-(E)-alkenylbenzenes afforded mixtures of regioisomers (45:45 to 60:20; e.e. 80-92%; two examples). A single (Z)-substrate gave a 91% yield of the *cis*-product with little enantioselectivity (e.e. 8%). F.e. and optimization s. S.E. Denmark, A. Jaunet, J. Am. Chem. Soc. *2013*, *135* (17), 6419-22 [DOI: 10.1021/ja401867b].

Iodine I_2
Iodine-catalyzed regioselective sulfenylation of indoles H → SR
with sulfonic acid hydrazides

232. 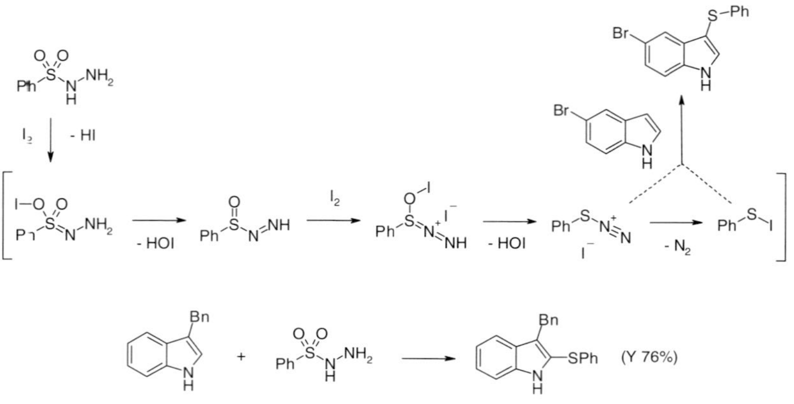

3-Sulfenylation. 5-Bromoindole (0.4 mmol) and I_2 (10 mol%) added to a soln. of benzenesulfonyl hydrazide (1.2 eq.) in ethanol (1 ml), the mixture stirred at 70° for 5 h, cooled to room temp., and purified chromatographically → 5-bromo-3-phenylsulfanylindole. Y 89%. This unprecedented C_3-sulfenylation of indoles with (het)aryl and aliphatic sulfonyl hydrazides was optimally catalyzed by I_2, with sulfenylation occurring at C2 for C3-subst. indoles, (twenty-eight examples; Y 56-90%). The method was compatible with halo, nitro, ether, ester and nitrile functionality. In cases where sulfenylation was slow (the reagent was slowly decomposed by I_2), the use of 2 eq. of reagent was required. F.e. and optimization s. F.-L. Yang, S.-K. Tian, Angew. Chem., Int. Ed. *2013*, *52* (18), 4929-32 [DOI: 10.1002/anie.201301437]; 3-sulfenylation of indoles **with N-(organothio)-phthalimides** using $CeCl_3 \cdot 7H_2O$/NaI, **also 2-sulfenylation of pyrroles**, s. E. Marcantoni, R. Cipolletti, L. Marsili, S. Menichetti, R. Properzi, C. Viglianisi, Eur. J. Org. Chem. *2013* (1), 132-40 [DOI: 10.1002/ejoc.201201100].

Shvo's ruthenium hydride complex/bis(salicylideniminato-3-propyl)methylaminocobalt(II)/ 2,6-dimethoxybenzoquinone/air ←

2-Arylbenzothiazoles from *o*-aminomercaptans and benzylamines s. *83*, 150 ○

Dirhodium tetrakis(N-1,8-naphthoyl-(S)-tert-leucinate) *[Rh(II)]**

2-Imino-3-sulfonyl-Δ⁴-thiazolines from 1-sulfonyl-1,2,3-triazoles ○C

and isothiocyanates s. *83*, 152

Halogen ↑ SC ⇃↾ Hal

Irradiation s. under Iridium(III) sensitizer ///

Microwaves s. under KSCN and Copper(I) iodide [\\\\]

Cesium carbonate *Cs_2CO_3*

2-Aminothiazole ring from *o*-halogenamines and isothiocyanates ○

2-aminobenzothiazoles, update s. *76,* 256s*82*; 6-aminothiazolo[4,5-*c*]pyridazines from 3-amino-4-bromopyridazines with Cs_2CO_3 s. A. Stella, S. De Jonghe, K. Segers, P. Herdewijn, Tetrahedron Lett. *2013, 54* (8), 830-3 [DOI: 10.1016/j.tetlet.2012.11.072].

Sodium thiocyanate/cyclohexapeptoids ←

Thiocyanates from halides Hal → SCN

with NH_4SCN cf. *5,* 448; with NaSCN and cyclohexapeptoids as phase transfer catalyst s. G.D. Sala, B. Nardone, F. De Riccardis, I. Izzo, Org. Biomol. Chem. *2013, 11* (5), 726-31 [DOI: 10.1039/c2ob26764k].

Potassium thiocyanate/microwaves *KSCN/*[\\\\]

Ar. hydroxythioethers from diols and diaryliodonium salts Ar_2I^+ + ROH → ArSR

233.

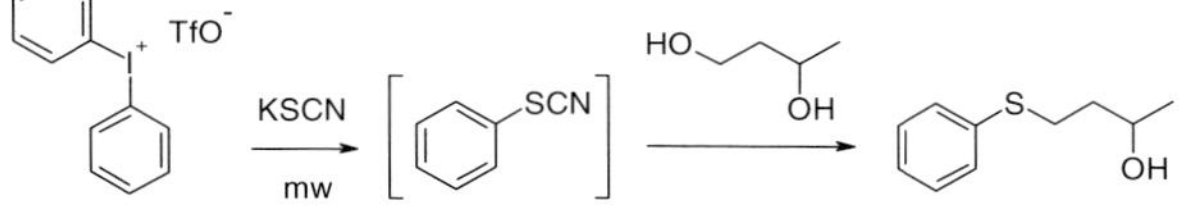

Regioselective conversion. A mixture of diphenyliodonium triflate (1 mmol), KSCN (2 eq.) and butane-1,3-diol (1 ml) heated by microwaves at 120° for 15 min, quenched with water, extracted with ethyl acetate, and purified by chromatography on silica → 4-phenylsulfanyl-2-butanol. Y 75%. This efficient and experimentally simple conversion was effective for electron-diverse diaryl-iodonium salts (triflate, tosylate, hexafluorophosphate; bromide also gave ar. bromide) reacting with ethane-1,2- and propane-1,3-diols (ten examples; Y 65-95%). The illustrated 1,3-diol reacted chemoselectively at the prim. center as did glycerol [affording 3-(phenylsulfanyl)propane-1,2-diol and diphenyl disulfide in unspecified yields]. The reaction failed, however, with simple monohydric alcohols (ethanol, methanol). In the absence of diol, formation of the intermediate ar. thiocyanate was observed. F.e.s. B.R. Vaddula, R.S. Varma, J. Leazer, Eur. J. Org. Chem. *2012* (35), 6852-5 [DOI: 10.1002/ejoc.201201313].

Potassium thioacetate s. under Copper(I) iodide *KSAc*

Sodium sulfide/copper(I) iodide *Na_2S/CuI*
Sym. diaryl trithiocarbonates 2 ArHal → ArSC(S)SAr
from ar. bromides, iodides, tosylates or triflates
Copper(I)-catalyzed ligand-free conversion

234. 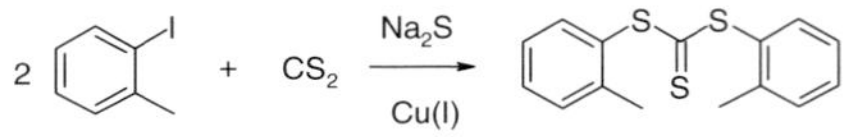

in one pot. Na_2S (1.1 eq) added to a stirred mixture of carbon disulfide (5 eq.) and DMF (2 ml) at room temp., stirring continued for 15 min, CuI (5 mol%) and 2-iodotoluene (2 eq.) added, the mixture stirred at 130° for 24 h, cooled to room temp., extracted with ethyl acetate, solvent evaporated, and the residue purified by chromatography on silica → di-2-tolyl trithiocarbonate. Y 78%. This novel, scalable (up to 20 mmol ArI) and experimentally simple method avoids the use of toxic thiophosgene or ar. thiols, and was apparently general for electron-diverse (het)arenes containing a suitable leaving group (I, Br, OTf or OTs) (fifteen examples; Y 62%, 70-87%), with the lowest yield obtained for 4-nitrophenyl bromide (other 4-nitrophenyl substrates gave 82-87%). Interestingly, heating reaction mixtures containing two different substrates gave only sym. trithiocarbonates. F.e. and optimization s. M. Gholinejad, Eur. J. Org. Chem. *2013* (2), 257-9 [DOI: 10.1002/ejoc.201201126].

Sodium sulfide/copper(I) iodide/potassium phosphate/polyethylene glycol ←
Sym. diaryl sulfides from ar. halides s. *79,* 211s*83* 2 ArHal → Ar_2S

Sodium sulfide/dichlorobis(triphenylphosphine)palladium(II)/1,4-bis(diphenylphosphino)butane/1,8-diazabicyclo[5.4.0]undec-7-ene/sulfuric acid ←
4-Thiochromones from ar. halides and propiolic acid ○
via (Z)-3-(arylthio)acrylic acids

235. 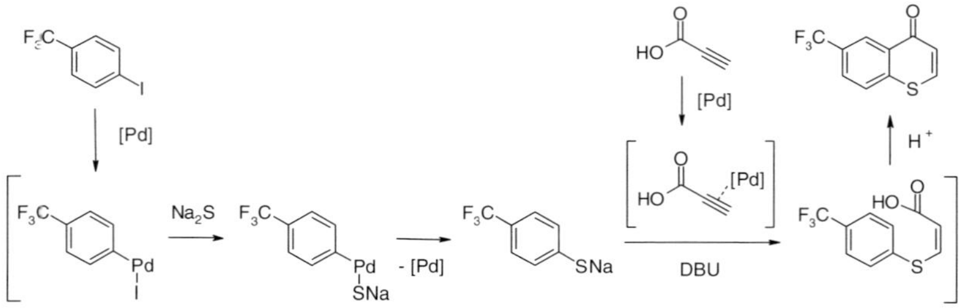

in one pot. A mixture of 4-iodo-1-trifluoromethylbenzene (3 mmol), $Na_2S{\cdot}5H_2O$ (2 eq.), $PdCl_2(PPh_3)_2$ (2.5 mol%) and dppb (5 mol%) stirred in DMSO (15 ml) for 5 min, DBU (2 eq.) and propiolic acid (1.2 eq.) added, the mixture stirred at 50° for 12 h, cooled, diluted with methylene chloride, washed with water, filtered through Celite, concentrated, the residue treated with concd. H_2SO_4 (10 mol%) at 100° for 1.5 h, cooled, quenched with ice, extracted with ethyl acetate, solvent evaporated, and the residue purified chromatographically → 6-trifluoromethyl-4*H*-thiochromen-4-one. Y 73%. Initial development identified general optimized conditions for stereoselective 3-component formation of (Z)-3-(arylthio)acrylic acids from propiolic acid (Z/E 91:9 to 100:0; nineteen examples; Y 47-85%). Thiolation gave lower yields with simple β-subst. propiolic acids (35-50%; three examples) and failed completely for bulkier propiolic acids. The method was subsequently extended to a synthesis of the title heterocycles from propiolic acid, via cyclization of the crude intermediates with acid (ten examples; Y 51-80%), with 2-butynoic acid affording a 2-methylthiochromone in lower yield (32%). The one-pot cyclization failed for iodoanisoles and 1-iodonaphthalene. F.e.s. T. Palani, K. Park, K.H. Song, S. Lee, Adv. Synth. Catal. *2013*, *355* (6), 1160-8 [DOI: 10.1002/adsc.201201106].

Sodium salt *Na^+*

Sulfones from sulfinic acids $SO_2Na \rightarrow SO_2Ar$

and halides s. *3,* 524; aryl [hetero]aryl sulfones from sodium [hetero]arylsulfinates and diaryliodonium salts under mild, metal-free conditions s. N. Umierski, G. Manolikakes, Org. Lett. *2013, 15* (1), 188-91 [DOI: 10.1021/ol303248h].

8-Hydroxyquinoline N-oxide s. under Copper(I) iodide ←

Copper(I) chloride/potassium 5-methyl-1,3,4-oxadiazole-2-thiolate ←

Sym. disulfides from halides s. *83*, 228 $2\ RHal \rightarrow (RS)_2$

Copper(I) iodide s.a. under Na_2S *CuI*

Copper(I) iodide/potassium carbonate *CuI/K_2CO_3*

Sulfonylation $H \rightarrow SO_2R$

ar. sulfonylation with $AlCl_3$ cf. *5,* 447; C_2-sulfonylation of quinoline N-oxides with arenesulfonyl chlorides using CuI/K_2CO_3 s. Z. Wu, H. Song, X. Cui, C. Pi, W. Du, Y. Wu, Org. Lett. *2013, 15* (6), 1270-3 [DOI: 10.1021/ol400178k]; stereoselective β-sulfonylation of enacylamines with aliphatic or aromatic sulfonyl chlorides *under visible-light photoredox catalysis* with an iridium(III) sensitizer, $Ir(ppy)_2(dtbbpy)[PF_6]$, and Na_2HPO_4 as base s. H. Jiang, X. Chen, Y. Zhang, S. Yu, Adv. Synth. Catal. *2013, 355* (4), 809-13 [DOI: 10.1002/adsc.201200874].

Copper(I) iodide/2,2′-bipyridyl/aluminum/magnesium chloride or sodium carbonate ←

Sym. diaryl sulfides from ar. halides $2\ ArHal \rightarrow Ar_2S$

update s. *79,* 211s*82*; from ar. iodides and sulfur with $CuI/bpy/MgCl_2$ and Al as reductant, also sym. diaryl disulfides with Na_2CO_3 in place of $MgCl_2$, and selenium analogs with Al or Mg as reductant (RemC↓↑Hal), s. N. Taniguchi, Tetrahedron *2012, 68* (51), 10510-5 [DOI: 10.1016/j.tet.2012.09.019]; from electron-diverse and *o*-subst. ar. bromides and iodides with $Na_2S{\cdot}9H_2O/CuI/K_3PO_4$ in PEG-400 s. J. Chen, Y. Zhang, L. Liu, T. Yuan, F. Yi, Phosphorus, Sulfur Silicon Relat. Elem. *2012, 187* (11), 1284-90 [DOI: 10.1080/10426507.2012.694003].

Copper(I) iodide/1,10-phenanthroline/potassium thioacetate/microwaves ←

Copper(I) iodide/8-hydroxyquinoline N-oxide/cesium carbonate ←

Ar. thioethers from ar. halides and mercaptans $Hal \rightarrow SR$

under copper(I) catalysis, update s. *31,* 522s*82*; also ar. selenides and arylphosphonic acid esters from ar. halides (RemC↓↑Hal), s. M. Font, T. Parella, M. Costas, X. Ribas, Organometallics *2012, 31* (22), 7976-82 [DOI: 10.1021/om3006323]; coupling of ar. halides (incl. ar. chlorides) with aliphatic or aromatic mercaptans *under nickel catalysis* with $Ni(OAc)_2$/1,3-bis(2,6-diisopropylphenyl)imidazol[idin]-2-ylidene and KOBu-*t* as base s. P. Guan, C. Cao, Y. Liu, Y. Li, P. He, Q. Chen, G. Liu, Y. Shi, Tetrahedron Lett. *2012, 53* (45), 5987-92 [DOI: 10.1016/j.tetlet.2012.08.055]; [hetero]aryl thioglycosides from glycosyl mercaptans [1-thiosugars] *under palladium catalysis* with exclusive α- or β-selectivity using $Pd(OAc)_2$/XantPhos/Et_3N s. E. Brachet, J.-D. Brion, S. Messaoudi, M. Alami, Adv. Synth. Catal. *2013, 355* (2-3), 477-90 [DOI: 10.1002/adsc.201200695]; ar. thioethers from electron-diverse ar. iodides with K-thioacetate/CuI/1,10-phenanthroline under microwave enhancement s. S.M. Soria-Castro, A.B. Peñéñory, Beilstein J. Org. Chem. *2013, 9,* 467-75 [DOI: 10.3762/bjoc.9.50]; **diaryl sulfides** from ar. iodides and aryl mercaptans with CuI/8-hydroxyquinoline N-oxide/Cs_2CO_3 s. K. Su, Y. Qiu, Y. Yao, D. Zhang, S. Jiang, Synlett *2012, 23* (19), 2853-7 [DOI: 10.1055/s-0032-1317518]; from pre-formed, dimeric (1,10-phenanthroline)(μ-thiophenolato)copper(I) complexes s. C. Chen, Z. Weng, J.F. Hartwig, Organometallics *2012, 31* (22), 8031-7 [DOI: 10.1021/om300711c]; **thioenolethers from α,β-ethylenehalides** and mercaptans under iron catalysis with $FeCl_3$/XantPhos/KOBu-*t* (cf. *68,* 230s*80*) s. Y.-Y. Lin, Y.-J. Wang, C.-H. Lin, J.-H. Cheng, C.-F. Lee, J. Org. Chem. *2012, 77* (14), 6100-6 [DOI: 10.1021/jo3008397]; clean, metal- and ligand-free formation of **β-styryl thioethers** from halides and K-thioacetate with retention of configuration s. A.A. Heredia, A.B. Peñéñory, Eur. J. Org. Chem. *2013* (5), 991-7 [DOI: 10.1002/ejoc.201201163].

Magnesium/N-chlorosuccinimide *Mg/NCS*
Ar. thioethers from mercaptans and arylmagnesium bromides via sulfenyl chlorides SH → SAr

236. 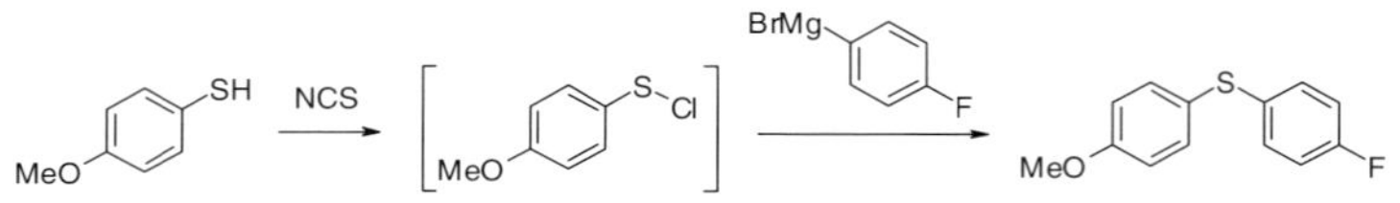

under mild conditions in one pot. A mixture of 4-methoxybenzenethiol (1 mmol), NCS (1.1 eq.) and toluene (1.5 ml) stirred at room temp. for 20 min, a soln. of 4-fluorophenylmagnesium bromide (1.5 eq.) in the same solvent (1 ml) added by syringe under N_2, stirring continued for 10 min, quenched with satd. aq. K_2CO_3, diluted with ethyl acetate, filtered through silica, the organic layer concentrated, and the residue purified chromatographically → 4-fluorophenyl 4-methoxyphenyl sulfide. Y 80%. This convenient method optimally uses NCS as the halogenating agent, with NBS giving a lower yield and NIS producing only a trace of the sulfide, presumed due to dimerization of the reactive sulfenyl iodide intermediate to a disulfide (no details). Electron-diverse (het)ar. and aliphatic thiols reacted with electron-diverse aryl Grignard reagents to afford the products within 30 min at room temp. (nineteen examples; Y 65-93%; ethyl 2-mercaptoacetate gave 46%). F.e. and optimization s. J.-H. Cheng, C. Ramesh, H.-L. Kao, Y.-J. Wang, C.-C. Chan, C.-F. Lee, J. Org. Chem. *2012*, *77* (22), 10369-74 [DOI: 10.1021/jo302088t].

Zinc s. under $NiCl_2$ *Zn*
Magnesium chloride s. under Copper(I) iodide *$MgCl_2$*
Aluminum s. under Copper(I) iodide *Al*
Polyethylene glycol s. under Sodium sulfide *PEG*
1,3-Bis(2,6-diisopropylphenyl)imidazol[idin]-2-ylidene s. under Nickel(II) acetate *NHC*
Cyclohexapeptoids s. under Sodium thiocyanate ←
Potassium 5-methyl-1,3,4-oxadiazole-2-thiolate s. under Copper(I) chloride ←
N-Chlorosuccinimide s. under Mg *NCS*
1,4-Bis(diphenylphosphino)butane s. under Na_2S *dppb*
4,5-Bis(diphenylphosphino)-9,9-dimethylxanthene s. under Iron(III) chloride and $Pd(OAc)_2$ *XantPhos*
Sulfuric acid s. under Na_2S *H_2SO_4*

Iron(III) chloride/4,5-bis(diphenylphosphino)-9,9-dimethylxanthene/potassium tert-butoxide ←
Thioenolethers from α,β-ethylenehalides and mercaptans s. *68*, 230s83 Hal → SR

Nickel(II) acetate/1,3-bis(2,6-diisopropylphenyl)imidazol[idin]-2-ylidene/potassium tert-butoxide ←
Ar. thioethers from ar. halides and mercaptans s. *31*, 522s*83*

Nickel(II) chloride/zinc/2,2'-bipyridyl *$NiCl_2$/Zn/bpy*
Ar. thioethers from ar. iodides and mercaptans s. *83*, 240 I → SR

Palladium(II) acetate/4,5-bis(diphenylphosphino)-9,9-dimethylxanthene/triethylamine ←
[Hetero]aryl thioglycosides from glycosyl mercaptans [1-thiosugars] Hal → SR
s. *31*, 522s*83*

Dichlorobis(triphenylphosphine)palladium(II) s. under Na_2S *$PdCl_2(PPh_3)_2$*

Iridium(III) sensitizer/sodium hydrogen phosphate/irradiation *[Ir(III)]/Na_2HPO_4/hν*
Stereoselective β-sulfonylation of enacylamines H → SO_2R
under visible-light photoredox catalysis s. *5*, 447s*83*

Sulfur ↑ SC ↓↑ S

Microwaves s. under I_2 [\\\\\]

Sodium azide/copper nanoparticles-on-activated carbon *NaN_3/nano-Cu-C*
1-β-Alkylthio-1,2,3-triazoles from ethylene and acetylene derivs. via 1,2-azidosulfenylation-1,3-dipolar cycloaddition ○

237.

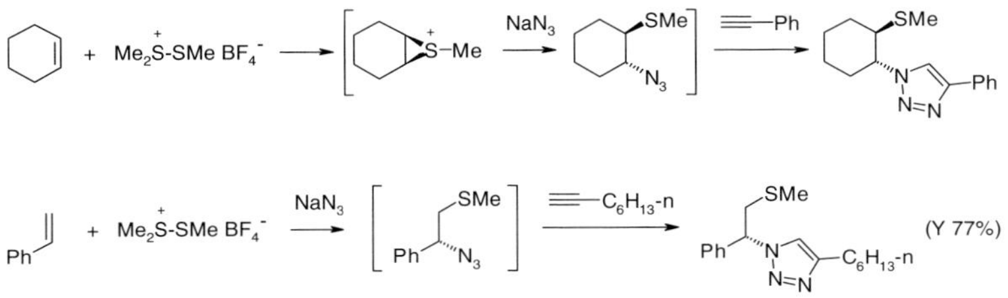

A one-pot protocol for 1,2,3-triazole synthesis starting from unactivated alkenes has been reported, based on two 'click' reactions: 1,2-azidosulfenylation of the carbon-carbon double bond (via ring opening of *in situ*-formed thiiranium salts by azide ion) and copper-catalyzed azide-alkyne cycloaddition using oxidized copper nanoparticles-on-activated carbon (cf. *68,* 184s*79-81*). **E: Diastereoselective conversion**. NaN_3 (1.2 eq.), freshly-prepared dimethyl(methylthio)-sulfonium tetrafluoroborate (1.2 eq.) and cyclohexene (0.5 mmol) added to a suspension of nano-Cu-C (10 mg; 0.5 mol% Cu) in acetonitrile (2 ml) at room temp. under argon, the mixture stirred for 1 h, phenylacetylene (1 eq.) added, the mixture warmed to 70° for 16 h (TLC), quenched with water, extracted with ethyl acetate, and purified by column chromatography → 1-[(1R*,2R*)-2-(methylthio)cyclohexyl]-4-phenyl-1*H*-1,2,3-triazole. Y 81%. The catalyst could not be recycled (possibly due to poisoning by sulfur), but is used at low loading, whereas a variety of commercially-available copper catalysts were inactive for this transformation. It is believed the catalyst is active for both steps. F.e. (eleven; Y 37-91%) and regioselectivity, also conversion to 1-vinyl- or 1-unsubst. derivs. (one example of each) s. F. Alonso, Y. Moglie, G. Radivoy, M. Yus, J. Org. Chem. *2013, 78* (10), 5031-7 [DOI: 10.1021/jo400110m]; N-β-hydroxy-1,2,3-triazoles from epoxides in water (cf. *72,* 171s*79*) s.a. ibid. *2011, 76* (20), 8394-405 [DOI: 10.1021/jo2016339].

Sodium hydrogen sulfide s. under I_2 *NaHS*

1,5,7-Triazabicyclo[4.4.0]dec-5-ene *TBD*
α-Sulfenylation H → SR
under rhodium catalysis cf. *33,* 513s*82*; of cyclic ketones with tetramethylthiuram disulfide using 1,5,7-triazabicyclo[4.4.0]dec-5-ene as base, **also** α-sulfenylation of 4-subst. cyclohexanones **with desymmetrization**, s. B. Poladura, Á. Martínez-Castañeda, H. Rodríguez-Solla, C. Concellón, V. del Amo, Tetrahedron *2012, 68* (32), 6438-46 [DOI: 10.1016/j.tet.2012.05.124]; **asym. α-sulfenylation** (cf. *70,* 237s*82*) of N-protected oxindoles (at the C_3 site) with N-(organothio)-phthalimides using a chiral squaramide as catalyst s. C. Wang, X. Yang, C.C.J. Loh, G. Raabe, D. Enders, Chem. Eur. J. *2012, 18* (37), 11531-5 [DOI: 10.1002/chem.201201262].

Copper nanoparticles-on-activated carbon s. under NaN_3 *nano-Cu-C*
Magnesium oxide s. under Trichloroisocyanuric acid *MgO*

Chiral squaramides ←
Asym. C_3-sulfenylation of N-protected oxindoles s. *70,* 237s*83*

Trichloroisocyanuric acid/magnesium oxide *TCICA/MgO*
Ar. sulfenylation
update s. *48,* 519s*82*; C_3-sulfenylation of indoles with diaryl sulfides using trichloroisocyanuric acid/MgO, also C_3-selenylation, s. C.C. Silveira, S.R. Mendes, L. Wolf, G.M. Martins, L. von Mühlen, Tetrahedron *2012, 68* (51), 10464-9 [DOI: 10.1016/j.tet.2012.09.022].

Phosphazene base [t-Bu-P4] ←
Trifluoromethyl thioethers from disulfides $(RS)_2 \rightarrow RSCF_3$
Metal-free generation of trifluoromethyl carbanion from fluoroform s. *83,* 269

Iodine/microwaves I_2/[\\\\]
Iodine/sodium hydrogen sulfide/microwaves I_2/NaHS/[\\\\]
Thioethers from alcohols and disulfides OH → SR
from prim. alcohols with Bu_3P/pyridine cf. *31,* 561; from dialkyl disulfides with I_2 in N-methyl-2-pyrrolidone under microwave enhancement, also sym. thioethers with I_2/NaHS in acetonitrile, s. B. Hu, H. Hu, L. Sun, R. Tang, Chin. J. Chem. *2012, 30* (10), 2556-62 [DOI: 10.1002/cjoc.201200723].

Remaining Elements ↑ SC ↓↑ Rem

Without additional reagents *w.a.r.*
Ar. trifluoromethyl thioethers from arylboronic acids $B(OH)_2 \rightarrow SCF_3$
and 3,3-dimethyl-1-(trifluoromethylthio)-1,2-benziodoxole s. *81,* 206s*83,* 211s*83*

Copper(II) triflate/4,4′-di-tert-butyl-2,2′-bipyridyl/cesium carbonate *Cu(OTf)$_2$/t-Bu$_2$bpy/Cs$_2$CO$_3$*
Ar. trifluoromethyl thioethers from arylboronic acids
with tetramethylammonium trifluoromethylthiolate s. *81,* 206s*83,* 211s*83*

Copper(II) fluoride/2,2′-bipyridyl *CuF$_2$/bpy*
Introduction of trifluoromethylthio groups SCF_3
ar. trifluoromethyl thioethers from ar. halides with tetramethylammonium trifluoromethylthiolate cf. *81,* 206; with trifluoromethyl(trimethyl)silane using air-stable CuF_2/bpy/S_8 s. Z. Weng, W. He, C. Chen, R. Lee, D. Tan, Z. Lai, D. Kong, Y. Yuan, K.-W. Huang, Angew. Chem., Int. Ed. *2013, 52* (5), 1548-52 [DOI: 10.1002/anie.201208432]; **from arylboronic acids** (cf. *81,* 211) and 3,3-dimethyl-1-(trifluoromethylthio)-1,2-benziodoxole, also α-trifluoromethylsulfenylation, and trifluoromethyl (E)-thioenolethers from trifluoromethyl 1-acetylenethioethers, s. X. Shao, X. Wang, T. Yang, L. Lu, Q. Shen, ibid. *52* (12), 3457-60 [DOI: 10.1002/anie. 201209817]; ar. and vinylic trifluoromethyl thioethers from unsatd. boronic acids with tetramethylammonium trifluoromethylthiolate using $Cu(OTf)_2$/4,4′-di-*tert*-butyl-2,2′-bipyridyl/Cs_2CO_3 in air s. C.-P. Zhang, D.A. Vicic, Chem. Asian J. *2012, 7* (8), 1756-8 [DOI: 10.1002/asia.201200347]; trifluoromethyl 2-ethylenethioethers from 2-ethylenesilanes, regioselective conversion with N-(trifluoromethylthio)aniline as electrophile and acetyl chloride as promoter (cf. *37,* 589) s. J. Liu, L. Chu, F.-L. Qing, Org. Lett. *2013, 15* (4), 894-7 [DOI: 10.1021/ol400032g].

Silver triflate *AgOTf*
Ar. thioethers from arylboronic acids and mercaptans ArSR
with $Cu(OAc)_2$/py cf. *60,* 201; diaryl sulfides from aryl mercaptans, general procedure with AgOTf, **also biaryls** from ar. iodides with AgOAc, s. R. Das, D. Chakraborty, Tetrahedron Lett. *2012, 53* (52), 7023-7 [DOI: 10.1016/j.tetlet.2012.09.127].

Acetyl chloride *AcCl*
Trifluoromethyl 2-ethylenethioethers C=C-C(Si≤) → C(SCF_3)C=C
from 2-ethylenesilanes s. *37,* 589s*83*

Carbon ↑ SC ↓↑ C

Potassium tert-butoxide/tetra-n-butylammonium chloride *KOBu-t/Bu$_4$NCl*
Trifluoromethyl thioethers from disulfides $(RS)_2 \rightarrow RSCF_3$
Nucleophilic trifluoromethylation with hexafluoroacetone hydrate-1,8-diazabicyclo[5.4.0]undec-7-ene salt s. *83,* 257

Copper/cesium carbonate *Cu/Cs₂CO₃*

o-prim-Aminodiaryl sulfides from benzothiazoles and ar. iodides via copper nanoparticle-catalyzed S-arylative ring opening

238. 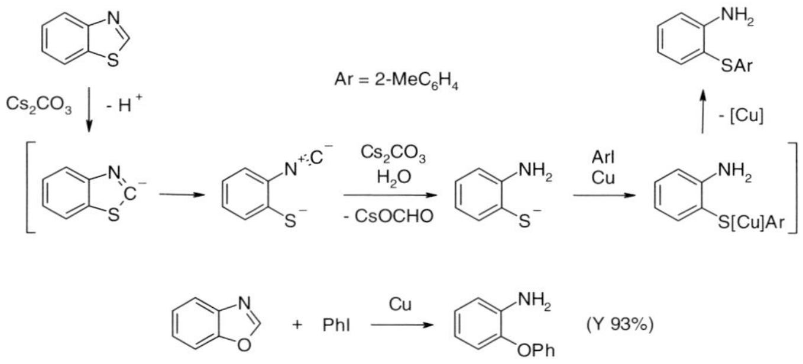

Benzothiazole (0.5 mmol) and 2-iodotoluene (2 eq.) added sequentially to a mixture of Cu powder (10 mol%), Cs_2CO_3 (2 eq.) and PEG-600 (2 g) under N_2, the mixture stirred at 140° for 7 h, quenched with brine, extracted with ethyl acetate, and purified chromatographically → 2-(2-methylphenylthio)aniline. Y 94%. A number of copper catalysts promoted this experimentally simple and environmentally benign S-arylation, with optimum results obtained using copper powder (shown to react as nanoparticles), which may be recycled up to six times. The reaction was successful for electron-diverse ar. iodides (incl. a bis-product from 1,4-diiodobenzene) (fifteen examples; Y 79-96%). Benzoxazole and iodobenzene gave the corresponding *o-prim*-aminodiaryl ether (Y 93%). Polyethylene glycols of higher molecular weight (PEG-2000 and PEG-6000) gave low yields. F.e. and optimization s. L. Yao, Q. Zhou, W. Han, S. Wei, Eur. J. Org. Chem. *2012* (35), 6856-60 [DOI: 10.1002/ejoc.201201166].

Dimethyl sulfoxide/oxygen *DMSO/O₂*

2-Arylbenzothiazoles from o-aminomercaptans and acylophenones via oxidative ring closure under metal- and iodine-free conditions

239. 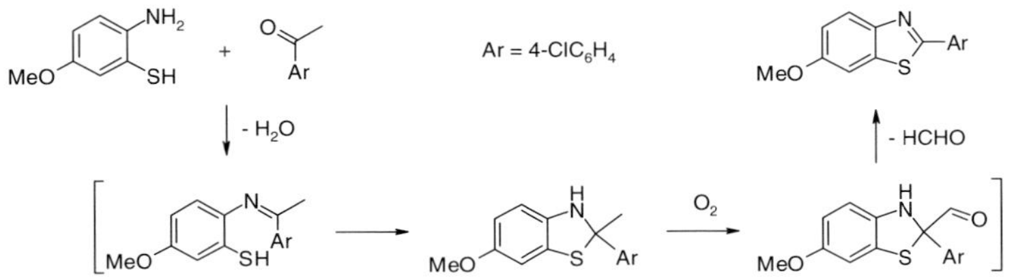

A mixture of 2-amino-5-methoxybenzenethiol (1.8 eq.), 4-chloroacetophenone (0.5 mmol), *DMSO (2 eq.)* and chlorobenzene (2 ml) stirred under O_2 at 140° for 16 h, cooled to room temp., concentrated *in vacuo*, and purified chromatographically → 2-(4-chlorophenyl)-6-methoxybenzo[*d*]thiazole. Y 68%. This efficient and experimentally simple procedure was successful with electron-diverse *o*-aminobenzenethiols reacting with propiophenone, α-carbethoxyacetophenone and benzaldehyde but generally gave best results with electron-diverse (het)ar. methyl ketones (twenty-four examples; Y 55-80%; 3-chloro-2-aminobenzenethiol gave only a trace of product). The use of 2 eq. DMSO was key to the success of the reaction, which was unsuccessful in other solvent systems and also failed with aliphatic ketones. F.e. and optimization s. Y. Liao, H. Qi, S. Chen, P. Jiang, W. Zhou, G.-J. Deng, Org. Lett. *2012, 14* (23), 6004-7 [DOI: 10.1021/ol302902e].

Iodine/potassium hydroxide *I_2/KOH*
2-Aroylbenzothiazoles from 2-unsubst. benzothiazoles and acetophenones s. *83*, 226 ←

Elimination ⇑

Nitrogen ↑ SC ⇑ N

(Triphenylphosphine)copper(I) iodide tetramer *[(Ph_3P)CuI]$_4$*
Cu(I)-catalyzed [3.3]-sigmatropic rearrangement-regioselective N→C-sulfonyl migration ∩
(E,E)-2-Sulfonyl-1,3-dienes from N-propargyl-N-sulfonylhydrazones s. *83*, 492

Halogen ↑ SC ⇑ Hal

Copper(II) oxide nanoparticles/potassium carbonate *CuO/K_2CO_3*
2-Aminobenzothiazoles from *o*-halogenothioureas ○
with CuI/1,10-phenanthroline/Cs_2CO_3 cf. *68*, 233; 'on water' under mild, eco-friendly conditions with recyclable CuO nanoparticles/K_2CO_3 s. S.K. Rout, S. Guin, J. Nath, B.K. Patel, Green Chem. *2012, 14* (9), 2491-8 [DOI: 10.1039/c2gc35575b].

Nickel(II) chloride/zinc/2,2′-bipyridyl *$NiCl_2$/Zn/bpy*
Nickel-mediated intramolecular S-arylation s. *83*, 240

Carbon ↑ SC ⇑ C

Nickel(II) chloride/zinc/2,2′-bipyridyl *$NiCl_2$/Zn/bpy*
Nickel-mediated intramolecular S-arylation ○

240.

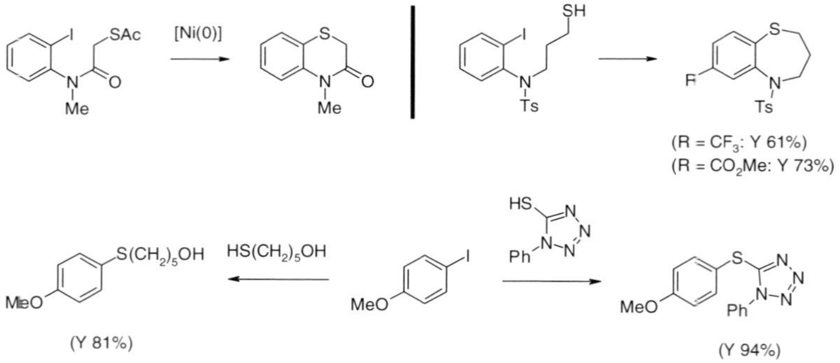

with S-deacylation. 2,2′-Bipyridyl (60 mol%) added to a stirred slurry of Zn (2 eq.) and $NiCl_2$ (50 mol%) in pyridine (0.5 ml) at room temp., vigorous stirring continued at 55° for 15 min, the resulting Ni(0) complex cooled to room temp., a soln. of substrate (1 mmol) in pyridine (1 ml) and methanol (10 ml) added dropwise over 1-1.5 h, the mixture stirred for a further 3.5 h, filtered through a short plug of silica (with ether), the filtrate washed with 1 *N* HCl, water and brine, dried ($MgSO_4$), concentrated, and the crude product purified by flash chromatography on silica gel → product. Y 88%. Three further examples afforded yields of 51-84%, the highest obtained on a 2.4 gram scale using only 5 mol% catalyst at 40° (increasing to 92% with 10 mol% catalyst at room temp.). Nine further intramolecular examples utilizing unprotected thiols (with 2,2′-bipyridyl replaced by ethyl crotonate) afforded cyclized products in yields of 42%, 54-88%, notably C-S coupling being more efficient for the construction of 7-membered rings than C-C coupling using the same reagent [s. C.-S. Yan, Y. Peng, X.-B. Xu, Y.-W. Wang, Chem. Eur. J. *2012*, *18* (19), 6039-48 [DOI: 10.1002/chem.201200190]]. The corresponding intermolecular process was also successful, enabling the preparation of **ar. thioethers from** electron-diverse **ar. iodides and** both ar. (incl. the illustrated heteroaromatic example) and prim., sec. or tert. aliphatic **mercaptans** (twenty-one examples; Y 39%, 50-94%), notably tolerating functionality such as acetal, silyl ether, phthalimido and free hydroxyl. F.e., a proposed mechanism, and application to the synthesis of chuangxinmycin, a medicinally important antibacterial, notably effective for the treatment of *E. coli* infection, s. X.-B. Xu, J. Liu, J.-J. Zhang, Y.-W. Wang, Y. Peng, Org. Lett. *2013, 15* (3), 550-3 [DOI: 10.1021/ol303366u].

Formation of Rem-C Bond

Uptake ⇓

Addition to Oxygen and Carbon RemC ⇓ OC

Alkaline earth bis(trimethylsilyl)amides ←
α-Hydroxyphosphorus(V) compds. from oxo compds. CO → C(OH)P(O)<
update s. *41,* 556s*82*; α-hydroxyphosphonic acid esters from aldehydes or non-activated ketones with heteroleptic or homoleptic Ca, Sr or Ba complexes, e.g. $Ca[N(SiMe_3)_2]_2\cdot(THF)_2$, at very low (down to 0.02 mol%) loading with perfect chemoselectivities and unprecedented TOFs under solvent-free conditions s. B. Liu, J.-F. Carpentier, Y. Sarazin, Chem. Eur. J. *2012, 18* (42), 13259-64 [DOI: 10.1002/chem.201201489].

(S)-Bis(acetonitrile)[5,6;5′,6′-bis(methylenedioxy)-3,3′-bis[bis(3,5-di-tert-butyl-4-methoxyphenyl)phosphine]biphenyl]copper(I) hydrogen bifluoride/methanol ←
1-Hydroxysilanes from aldehydes CHO → CH(OH)Si≤
via copper(I)-catalyzed asym. 1,2-addition

241.

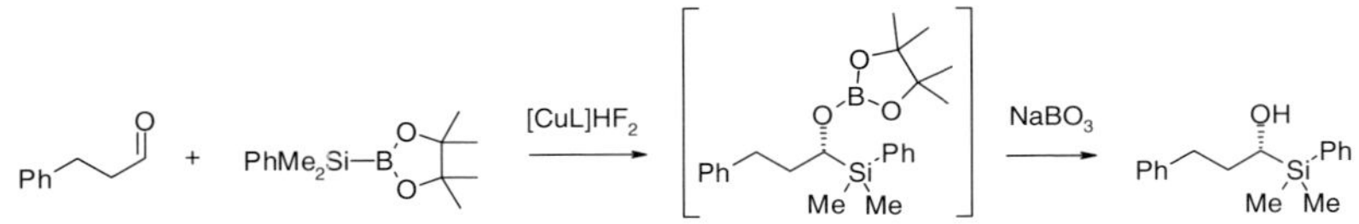

The first enantioselective copper-catalyzed 1,2-addition of Si-nucleophiles to aldehydes is reported. **E:** Freshly distilled 3-phenylpropanal (0.5 mmol), dimethyl(phenyl)(4,4,5,5-tetramethyl-1,3,2-dioxaborolan-2-yl)silane (1.2 eq.) and methanol (4 eq.) added sequentially to [(S)-DTBM-SegPhos]$(MeCN)_2CuHF_2$ (5 mol%) in THF (3 ml) under argon at room temp., the mixture stirred overnight, $NaBO_3$ (4 eq.), ether (1 ml) and water (3 ml) added, stirred for 3 h, extracted with ether (with satd. aq. NH_4Cl and satd. aq. $NaHCO_3$ washes), and purified by flash chromatography on silica → (R)-1-dimethylphenylsilyl-3-phenylpropan-1-ol. Y 88% (e.e. >99%). Development of novel chiral copper(I) phosphine complexes enabled general α-silylation of electron-diverse ar. and aliphatic aldehydes with excellent enantioselectivity (e.e. 93 to >99%; sixteen examples; Y 51-99%), with the somewhat lower selectivity observed for propanal (e.e. 87%; Y 60%) being attributed to lack of steric bulk. A single α,β-unsaturated aldehyde, citral, gave a modest result for the (R)-alcohol (e.e. 66%; Y 30%), using the (R)-enantiomer of the illustrated catalyst. The addition of methanol to the reaction was presumed to play a significant part in catalyst turnover. Dependent on the source of silver bifluoride used in its prepn., the catalyst was often contaminated with dihydrogen trifluoride anion, which was not detrimental to its activity. F.e., catalyst development and optimization s. V. Cirriez, C. Rasson, T. Hermant, J. Petrignet, J. Díaz Álvarez, K. Robeyns, O. Riant, Angew. Chem., Int. Ed. *2013, 52* (6), 1785-8 [DOI: 10.1002/anie.201209020].

Addition to Nitrogen and Carbon RemC ⇓ NC

Without additional reagents *w.a.r.*
α-Aminophosphorus(V) compds. from azomethines C=N- → C(NH-)PO<
update s. *77,* 267s*82*; *uncatalyzed* synthesis of N-protected α-aminophosphonic acid esters from aldimines and dimethyl trimethylsilyl phosphite s. Z. Cai, Y. Fan, G. Du, L. He, Chin. J. Chem. *2012, 30* (7), 1658-62 [DOI: 10.1002/cjoc.201200119].

Copper(I) benzimidazol-2-ylidene complexes *[Cu(I)]*
α-(Sulfinylamino)boronic acid esters from N-sulfinylimines with asym. induction ←
with a copper(I) imidazol-2-ylidene complex cf. *73,* 244; from S-chiral aldehyde- or ketone-derived N-sulfinylimines with an electronically tunable copper(I) benzimidazol-2-ylidene complex s. K. Wen, H. Wang, J. Chen, H. Zhang, X. Cui, C. Wei, E. Fan, Z. Sun, J. Org. Chem. *2013, 78* (7), 3405-9 [DOI: 10.1021/jo4000477].

Diethylzinc/9-picolinoylamino-9-deoxy-epi-cinchonine *Et_2Zn/L^**
N-Protected β-aminophosphonic acid esters ▽N → C(NH-)C-PO(OR)$_2$
from aziridines and phosphorous acid diesters with desymmetrization

242. 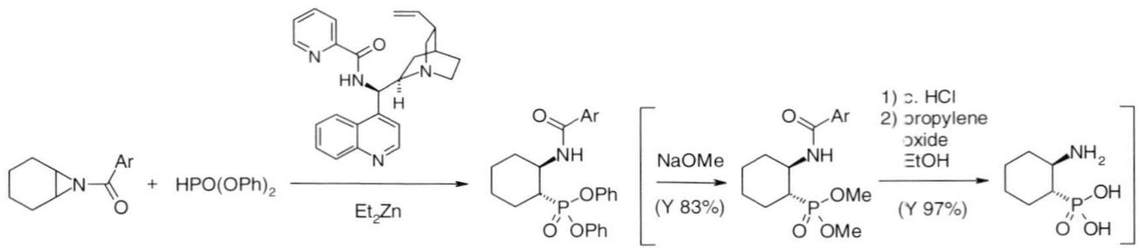

Ar = 2-pyridyl

A soln. of Et_2Zn (10 mol%) in toluene (0.02 ml) added to a suspension of *epi*-cinchona catalyst (10 mol%) and dry 4 Å molecular sieves (100 mg) in benzene (1 ml), the mixture stirred for 10 min, a soln. of 7-picolinoyl-7-azabicyclo[4.1.0]heptane (0.2 mmol) and diphenyl phosphite (1.5 eq.) in benzene (0.5 ml) added, stirred for 12 h, diluted with ethyl acetate, filtered through Celite, concentrated *in vacuo*, and purified chromatographically → diphenyl 2-(picolinamido)cyclohexylphosphonate. Y 84% (e.e. 99%). Chiral catalysts based on 9-amino-9-deoxy-*epi*-cinchona alkaloids promoted this novel and efficient desymmetrization of aziridines with phosphite, with selection of the nitrogen protecting group key to controlling the reactivity of the aziridine. Cyclohexano-aziridine and cyclohexene, tetralin and cyclopentene analogs as well as simple monocyclic 2,3-disubst. aziridines were effective substrates (twelve examples; Y 76-90%; e.e. 95-99%), but pyrrolidinoaziridines (two examples; Y 64-65%; e.e. 95-97%) and cycloheptanoaziridines (single example; Y 19%; e.e. 90%) reacted sluggishly due to competing phospha-Brook rearrangement. The illustrated product was converted to a β-aminophosphonic acid with no loss of chirality (Y 81%; two steps). F.e., optimization, substrate and catalyst prepn. s. M. Hayashi, N. Shiomi, Y. Funahashi, S. Nakamura, J. Am. Chem. Soc. *2012, 134* (47), 19366-9 [DOI: 10.1021/ja309963w].

Addition to Carbon-Carbon Bonds RemC ⇓ CC

Without additional reagents *w.a.r.*
Hydroboration of ethylene derivs. C=C → CHC-BR$_2$
with pinacolborane cf. *29,* 497s*49*; with odor-free and easily recoverable poly(propylene sulfide)-borane s. K. Smith, A.A. Balakit, G.A. El-Hiti, Tetrahedron *2012, 68* (38), 7834-9 [DOI: 10.1016/j.tet.2012.07.037]; regioselective hydroboration of the terminal alkene group of 1,n-dienes with pinacolborane using [6-(di-*tert*-butylphosphino)-2,2′-bipyridyl](dichloro)iron(II) and added $NaBHEt_3$ s. L. Zhang, D. Peng, X. Leng, Z. Huang, Angew. Chem., Int. Ed. *2013, 52* (13), 3676-80 [DOI: 10.1002/anie.201210347]; **1,4-hydroboration** of α,β-ethylenecarbonyl compds. with pinacolborane using tert. phosphines as catalyst (avoiding Brønsted bases!) s. C. Pubill-Ulldemolins,

A. Bonet, H. Gulyás, C. Bo, E. Fernández, Org. Biomol. Chem. *2012, 10* (48), 9677-82 [DOI: 10.1039/c2ob26899j]; regioselective **asym. hydroboration** (cf. *76,* 266; *78,* 251s*80*) of β-methylene-carboxylic acid esters and amides with a TADDOL-based chiral rhodium(I) cyclic phosphite complex s. S.M. Smith, G.L. Hoang, R. Pal, M.O.B. Khaled, L.S.W. Pelter, X.C. Zeng, J.M. Takacs, Chem. Commun. *2012, 48* (100), 12180-2 [DOI: 10.1039/c2cc36199j]; **of eneboranes** [notably 1,8-naphthalenediaminatoborylalkenes] with pinacolborane to give chiral 1,1-diboryl compds. using CuCl and a chiral 6,6′-bis(diarylphosphino)-2,3;2′,3′-bis(methylenedioxy)biphenyl or a ferrocenyldi(phosphine) as ligand and Na-*tert*-butoxide as base s. X. Feng, H. Jeon, J. Yun, Angew. Chem., Int. Ed. *2013, 52* (14), 3989-92 [DOI: 10.1002/anie.201208610]; **asym. 1,4-hydroboration** (cf. *73,* 249s*76*; *78,* 251s*80*) of α,β-ethyleneketones with bis(pinacolato)diboron using Cu_2O and a chiral N-condensed N[1]-([2.2]paracyclophanyl)-1,2,4-triazolylidene as ligand (having both planar and central chirality) s. L. Zhao, Y. Ma, F. He, W. Duan, J. Chen, C. Song, J. Org. Chem. *2013, 78* (4), 1677-81 [DOI: 10.1021/jo302694d]; of N-(α,β-ethyleneacyl)-2-oxazolidones with a diphenyl-(hydroxy)methyl-subst. chiral N-condensed N[1]-([2.2]paracyclophanyl)-1,2,4-triazolylidene as ligand s. L. Zhao, Y. Ma, W. Duan, F. He, J. Chen, C. Song, Org. Lett. *2012, 14* (22), 5780-3 [DOI: 10.1021/ol302839d]; of enones, α,β-ethylenenitriles and α,β-ethylenecarboxylic acid esters or amides *in water* with bis(pinacolato)diboron using $Cu(OH)_2$/AcOH and a chiral 6-α,6′-α-di-hydroxy-2,2′-bipyridyl as ligand s. S. Kobayashi, P. Xu, T. Endo, M. Ueno, T. Kitanosono, Angew. Chem., Int. Ed. *2012, 51* (51), 12763-6 [DOI: 10.1002/anie.201207343]; of α,β-ethyleneketimines with a chiral copper(I) 1,1′-binaphthyl-2,2′-diyl phosphoramidite complex as catalyst s. C. Sole, A. Bonet, A.H.M. de Vries, J.G. de Vries, L. Lefort, H. Gulyás, E. Fernández, Organometallics *2012, 31* (22), 7855-61 [DOI: 10.1021/om300194k].

Hydrophosphination C=C → CHC(P<)

of ethylene derivs. with AIBN initiation s. *18,* 705; solvent-free, *uncatalyzed,* regioselective hydrophosphination of alkenes s. F. Alonso, Y. Moglie, G. Radivoy, M. Yus, Green Chem. *2012, 14* (10), 2699-702 [DOI: 10.1039/c2gc35898k]; mild, regio- and stereo-selective hydrophosphination **of acetylene derivs.** (notably chiral propargylamines and propargyl alcohols) with air-stable di-*tert*-butylphosphine·BH_3 in the presence of NaH (cf. *23,* 639) s. C.A. Busacca, B. Qu, E. Farber, N. Haddad, N. Grět, A.K. Saha, M.C. Eriksson, J.-P. Wu, K.R. Fandrick, S. Han, N. Grinberg, S. Ma, H. Lee, Z. Li, M. Spinelli, A. Gold, G. Wang, P. Wipf, C.H. Senanayake, Org. Lett. *2013, 15* (5), 1132-5 [DOI: 10.1021/ol400309y].

Phospha-Michael addition

update s. *45,* 340s*78*; C- and O-phosphorylation of *p*-quinones by 1,4- and 1,6-addition, respectively, of >P(O)H compds. s. B. Xiong, R. Shen, M. Goto, S.-F. Yin, L.-B. Han, Chem. Eur. J. *2012, 18* (52), 16902-10 [DOI: 10.1002/chem.201202074]; **asym. phospha-Michael addition** (cf. *76,* 267s*82*) of diarylphosphine oxides to enones or enoates with chiral magnesium BINOLates as cooperative Brønsted/Lewis acid-base catalysts s. M. Hatano, T. Horibe, K. Ishihara, Angew. Chem., Int. Ed. *2013, 52* (17), 4549-53 [DOI: 10.1002/anie.201300938]; of diarylphosphines to enones using chiral benzylamine-type *o*-iodopallad[II]acyclic imidazol-2-ylidene complexes as catalyst s. S. Sabater, J.A. Mata, E. Peris, Organometallics *2013, 32* (4), 1112-20 [DOI: 10.1021/om400007a].

1,2-Ethylene-1,2-di(silanes) from acetylene derivs. C≡C → C(Si≤)=C(Si≤)

update s. *44,* 513s*50*; regio- and (Z)-selective *metal-free* 1,2-disilylation with N-aryl-N-chlorosilyl-N-(trimethylsilyl)amine·1,3-diisopropyl-4,5-dimethylimidazol-2-ylidene complexes s. H. Cui, B. Ma, C. Cui, Organometallics *2012, 31* (21), 7339-42 [DOI: 10.1021/om3009958]; cyclic derivs. by *dehydrogenative* coupling of alkynes with bis(organosilicon hydrides), e.g. hexamethyldisiloxane, over gold nanoparticles-on-TiO_2 (cf. *80,* 243) s. V. Kotzabasaki, I.N. Lykakis, C. Gryparis, A. Psyllaki, E. Vasilikogiannaki, M. Stratakis, ibid. *2013, 32* (2), 665-72 [DOI: 10.1021/om3011678].

Potassium fluoride/aluminum oxide/polyethylene glycol/microwaves ←

Ene-1,2-di(selenides) from acetylene derivs. C≡C → C(SeR)=C(SeR)

and diselenides under irradiation cf. *47,* 533s*48*; (Z)-1,2-bis(arylseleno)alkenes from terminal alkynes with reusable KF/Al_2O_3 in PEG-400 under microwave enhancement s. R.G. Lara, P.C. Rosa, L.K. Soares, M.S. Silva, R.G. Jacob, G. Perin, Tetrahedron *2012, 68* (51), 10414-8 [DOI: 10.1016/

j.tet.2012.08.055]; (E)-products from terminal or internal alkenes with $FeCl_3$ as catalyst s. G. Sartori, J.S.S. Neto, A.P. Pesarico, D.F. Back, C.W. Nogueira, G. Zeni, Org. Biomol. Chem. *2013, 11* (7), 1199-208 [DOI: 10.1039/c2ob27064a].

Chiral 6-α,6′-α-dihydroxy-2,2′-bipyridyls s. under Copper(II) hydroxide ←

Copper(I) oxide chiral N-condensed N¹-([2.2]paracyclophanyl)-1,2,4-triazolylidenes ←
Copper(II) hydroxide/acetic acid/chiral 6-α,6′-α-dihydroxy-2,2′-bipyridyls ←
Chiral copper(I) 1,1′-binaphthyl-2,2′-diyl phosphoramidite complexes *[Cu(I)]**
Asym. 1,4-hydroboration s. *73,* 249s*83*; *78*, 251s*83* C=C → CHC-BR_2

Copper(I) chloride/o-bis(diphenylphosphino)benzene/lithium tert-butoxide ←
β-Aminoboronic acid esters C=C → C(N<)C-$B(OR)_2$
from ethylene derivs. and aroyloxylamines
Regio- and stereo-selective copper(I)-catalyzed 1,2-aminoboration

243.

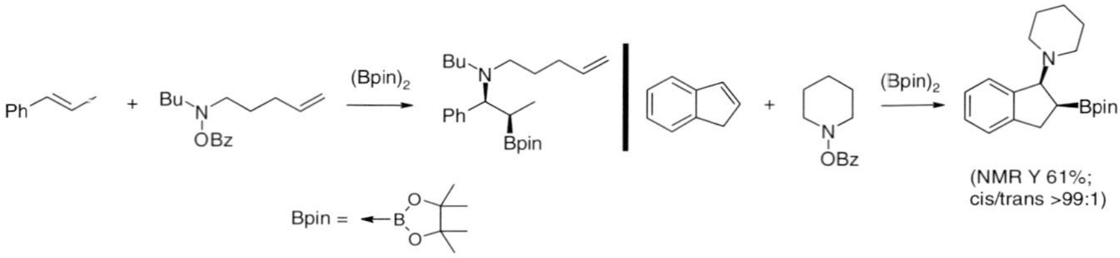

A mixture of CuCl (10 mol%), *o*-bis(diphenylphosphino)benzene (10 mol%), Li-*tert*-butoxide (3 eq.) and THF (0.5 ml) stirred at room temp. under N_2 for 15 min, a soln. of *trans*-β-methylstyrene (0.25 mmol), bis(pinacolato)diboron (1.5 eq.) and O-benzoyl-N-butyl-N-pent-4-en-1-ylhydroxylamine (1.5 eq.) in the same solvent (1 ml) added dropwise, the soln. stirred for 4 h, quenched with water, extracted with ethyl acetate, and purified chromatographically → *syn*-N-butyl-N-[1-phenyl-2-(4,4,5,5-tetramethyl-1,3,2-dioxaborolan-2-yl)propyl]pent-4-en-1-amine. Y 86% (*syn/anti* >99:1). This simultaneous formation of C-N and C-B bonds was apparently general and scalable (to 1 mmol) for internal and terminal alkenes, affording the *syn*-adducts from (E)-styrenes, generally with >99% selectivity. Isolated yields were often significantly lower than the NMR yields due to partial decomposition of products during chromatographic purification (twenty-one examples; Y 42-95%). Similar treatment of (Z)-β-methylstyrene afforded an *anti*-product (*anti/syn* Y >99:1; Y 52%) while the cyclic alkene, indene, afforded a *cis*-adduct (*cis/trans* >99:1; Y 61% by NMR). Initial investigations of the **asym. variant** with (S,S)-Me-Duphos as chiral ligand gave promising enantioselectivity (e.e. 80-86%; three examples; Y 51-83%). F.e., optimization, and reactions of the products s. N. Matsuda, K. Hirano, T. Satoh, M. Miura, J. Am. Chem. Soc. *2013*, *135* (13), 4934-7 [DOI: 10.1021/ja4007645].

Copper(I) iodide/chiral 1,3-bis[o-(2-naphthyl)phenyl]imidazolidin-2-ylidenes *[Cu(I)]**
Asym. 1,4-hydrosilylation of α,β-ethylenecarbonyl compds. C=C → CHC(Si≤)
under copper catalysis s. *77,* 269; of 5-membered α,β-ethylenelactones *with kinetic resolution* using CuI and a chiral 1,3-bis[*o*-(2-naphthyl)phenyl]imidazolidin-2-ylidene as ligand s. V. Pace, J.P. Rae, H.Y. Harb, D.J. Procter, Chem. Commun. *2013, 49* (45), 5150-2 [DOI: 10.1039/c3cc42150k]; of 2,4-dienones and 2,4-dienoates with chiral copper(I) NHC complexes, **also asym. 1,6-addition** (where there is trisubstitution of the carbonyl-conjugated ene group), s. K.-s. Lee, H. Wu, F. Haeffner, A.H. Hoveyda, Organometallics *2012, 31* (22), 7823-6 [DOI: 10.1021/om300790t].

Gold nanoparticles/titanium dioxide *Au/TiO_2*
1,2-Ethylene-1,2-di(silanes) from acetylene derivs. s. *44,* 513s*83* C≡C → C(Si≤)=C(Si≤)

Chiral magnesium 1,1′-bi-2-naphthoxides ←
Asym. phospha-Michael addition s. *76,* 267s*83* C=C → CHC(P<)

Sodium hydridotriethylborate s. under [6-(Di-tert-butylphosphino)-2,2′-bipyridyl]-(dichloro)iron(II) *$NaBHEt_3$*
Aluminum oxide s. under Potassium fluoride *Al_2O_3*
Polyethylene glycol s. under Potassium fluoride *PEG*
Chiral 1,3-bis[o-(2-naphthyl)phenyl]imidazolidin-2-ylidenes s. under Copper(I) iodide ←
Chiral N-condensed N^1-([2.2]paracyclophanyl)-1,2,4-triazolylidenes s. under Cu_2O *NHC**
Acetic acid s. under Copper(II) hydroxide *AcOH*
Titanium dioxide s. under Au *TiO_2*

Tert. phosphines *R_3P*
1,4-Hydroboration of α,β-ethylenecarbonyl compds. s. *29,* 497s*83* C=C → CHC-BR_2

o-Bis(diphenylphosphino)benzene s. under CuCl *dppbz*
Chiral TADDOL-derived phosphonites s. under Tris(dibenzylideneacetone)platinum(0) ←

Dinitrogen[2,6-bis(imidoyl)pyridine]iron(0) complexes *[Fe(0)]*
Iron-catalyzed anti-Markovnikov hydroboration of ethylene derivs.

244.

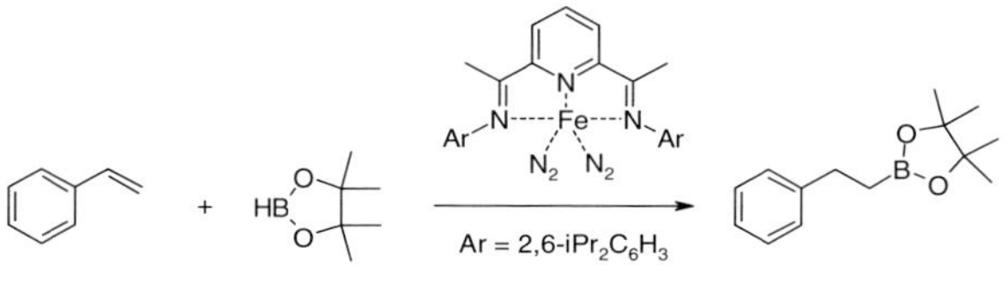

under mild, solvent-free conditions. A scintillation vial charged with styrene (0.96 mmol), pinacolborane (1 eq.) and (^{iPr}PDI)Fe(N_2)$_2$ (1 mol%), the vial capped, the mixture stirred until reaction complete (GC-FID; 15 min), and quenched by exposure to air → product. Conversion >98% (by NMR). A number of closely-related bis(imidoyl)pyridine iron dinitrogen complexes were shown to catalyze the anti-Markovnikov hydroboration of terminal and internal (incl. cyclic) olefins, including challenging styrene derivs., offering distinct advantages in substrate scope and selectivity over previously reported base- and precious-metal catalysts. The use of solvent-free conditions for liquid olefins and stoichiometric quantities of substrates are noteworthy. F.e. (ca. ten; conversion >98%, dependent on choice of catalyst) s. J.V. Obligacion, P.J. Chirik, Org. Lett. *2013, 15* (11), 2680-3 [DOI: 10.1021/ol400990u].

[6-(Di-tert-butylphosphino)-2,2′-bipyridyl](dichloro)iron(II)/sodium hydridotriethylborate ←
Regioselective hydroboration of terminal ethylene derivs. s. *29,* 497s*83*

Iron(III) chloride *$FeCl_3$*
(E)-Ene-1,2-di(selenides) from acetylene derivs. s. *47,* 533s*83* C≡C → C(SeR)=C(SeR)

Rhodium nanoparticles-in-graphitized carbon nanofibers *Rh-C*
Rhodium-platinum nanoparticles-in-graphitized carbon nanofibers *Rh/Pt-C*
Hydrosilylation of acetylene derivs. C≡C → CH=C(Si≤)
update s. *15,* 494s*82*; regioselective, (E)-selective hydrosilylation of terminal aliphatic alkynes with Rh- or Rh/Pt-nanoparticles encapsulated *inside* hollow graphitized carbon nanofibers, also (Z)-selective coupling of arylacetylenes with aryl(hydrido)silanes, s. W.A. Solomonsz, G.A. Rance, M. Suyetin, A. La Torre, E. Bichoutskaia, A.N. Khlobystov, Chem. Eur. J. *2012, 18* (41), 13180-7 [DOI: 10.1002/chem.201201542]; 2-(hydroxysilyl)-1(E),3-dienes from 1,3-enynes with a chloro-(hydrido)dimethylsilane under palladium catalysis with a palladium(0) phosphine complex s. H. Zhou, C. Moberg, Org. Lett. *2013, 15* (7), 1444-7 [DOI: 10.1021/ol4001334].

Chiral rhodium(I) cyclic phosphite complexes *[Rh(I)]**
Regioselective asym. hydroboration C=C → CHC-BR_2
of β-methylenecarboxylic acid esters and amides s. *76,* 266s*83*

Chiral palladacyclic imidazol-2-ylidene complexes *[Pd(II)]**
Asym. phospha-Michael addition to α,β-ethyleneketones s. *76,* 267s*83* C=C → CHC(P<)

Chiral bis(acetonitrile)[1-[α-(phosphino)alkyl]-2-naphthyl]palladium(II) perchlorate/triethylamine *[Pd(II)]*/Et$_3$N*

Tert. 2-ethylene-3-(N-tosylamino)phosphines from α,β-ethylene-N-tosylimines via palladium(II)-catalyzed asym. 1,4-hydrophosphination C(P<)C=C(NHTs)

245. 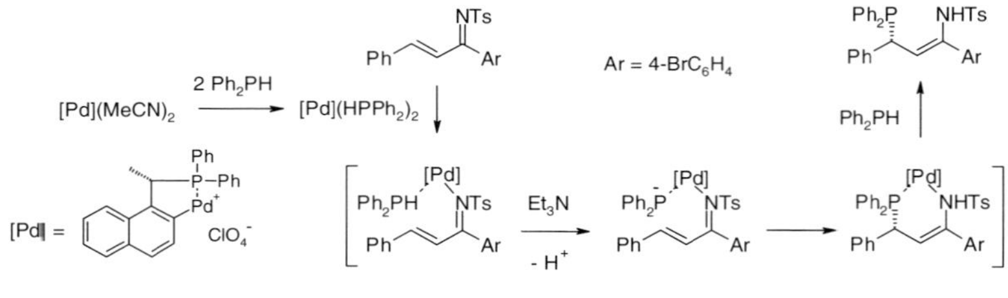

Chiral palladium(II) catalyst (6 mol%) added to a soln. of diphenylphosphine (0.27 mmol) in degassed THF (5 ml) at room temp., the soln. stirred for 5 min, cooled to -80°, startg. ketimine (1 eq.) added, a soln. of triethylamine (1 eq.) in the same solvent (0.5 ml) added dropwise, the soln. stirred until reaction complete (NMR), warmed to room temp., concentrated *in vacuo*, the residue dissolved in methylene chloride, filtered through silica in a stream of N_2, and concentrated → (R)-N-tosyl-1-(4-bromophenyl)-3-phenyl-3-(diphenylphosphino)prop-1-enamine. Y 95% (e.e. 96%). A series of α,β-ethylene-N-tosylimines carrying electron-diverse aryl at the alkene terminus and (het)aryl or styryl at the imine carbon afforded 1,4-adducts with high enantioselectivity (e.e. 91-99%; fifteen examples; Y 94-99%; a 2-chlorophenyl-terminated ketimine gave e.e. 70%). Reactivity was decreased for substrates bearing weaker electron-withdrawing groups (e.g. Ph rather than Ts) on nitrogen, while corresponding aldimines gave complex mixtures of products. F.e. and optimization s. Y. Huang, R.J. Chew, S.A. Pullarkat, Y. Li, P.-H. Leung, J. Org. Chem. *2012*, *77* (16), 6849-54 [DOI: 10.1021/jo300893s]; **asym. hydrophosphination of** 3-methyl-**4-nitro-5-vinylisoxazoles** with diphenylphosphine using bis(acetonitrile)[1-[1(S)-(diphenylphosphino)ethyl]-2-naphthyl]palladium(II) perchlorate/Et$_3$N s. R.J. Chew, Y. Huang, Y. Li, S.A. Pullarkat, P.-H. Leung, Adv. Synth. Catal. *2013, 355* (7), 1403-8 [DOI: 10.1002/adsc.201300164].

Tris(dibenzylideneacetone)platinum(0)/chiral TADDOL-derived phosphonites *[Pt(0)]**

Desymmetrization of diorganosilicon dihydrides by Si-vinylation with acetylene derivs. C≡C → CH=C(SiH<)

246. 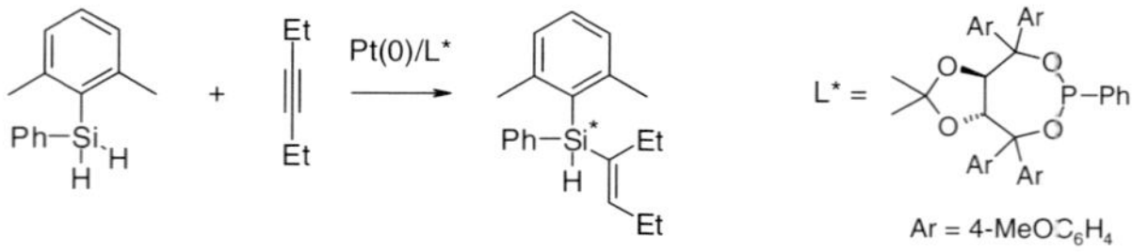

Chiral phosphonite (2 mol%) added to a soln. of Pt(dba)$_3$ (1 mol%) in toluene (2 ml) at 25°, the mixture stirred for 10 min, 3-hexyne (1 eq.) added, stirring continued for 5 min, 2,6-dimethylphenyl(phenyl)dihydrosilane (0.183 mmol) added, stirring continued for 3 h, concentrated *in vacuo*, and purified by chromatography on silica → (-)-2,6-dimethylphenyl(hex-3-en-3-yl)-(phenyl)hydrosilane. Y 83% (e.e. 67%; Y 58% based on recovered substrate, e.e. 86% at -30°). Initial development identified monodentate TADDOL-derived phosphonites as effective ligands for promotion of selective mono-vinylation with moderate enantiocontrol at 25° (e.e. 30-67%; five examples; Y 63-96%). Enantioselectivity was improved by reaction at lower temp. (0° or -30°) (e.e. 68-86%; five examples), usually with reduced conversion (Y 57-88% mostly based on recovered substrate). Further reactions of the product are described, incl. a three step conversion (Si-butylation, ozone oxidation and aminolysis) to **chiral silanols** (cf. *74*, 67). F.e., optimization, and substrate and ligand prepn. s. K. Igawa, D. Yoshihiro, N. Ichikawa, N. Kokan, K. Tomooka, Angew. Chem., Int. Ed. *2012*, *51* (51), 12745-8 [DOI: 10.1002/anie.201207361].

Exchange ⇃↾

Hydrogen ↾ RemC ⇃↾ H

Potassium tert-butoxide/dimethylformamide-d₇/microwaves *KOBu-t/DMF-d_7/[\\\\\\]*
Microwave-assisted base-catalyzed perdeuteration of polyarenes s. *83*, 249 H → D

Silver nitrate/magnesium nitrate *$AgNO_3/Mg(NO_3)_2$*
3-α-Phosphinyloxindoles from α,β-ethylenecarboxylic acid anilides
Silver-catalyzed oxidative intramolecular 1,2-carbaphosphinylation

247. 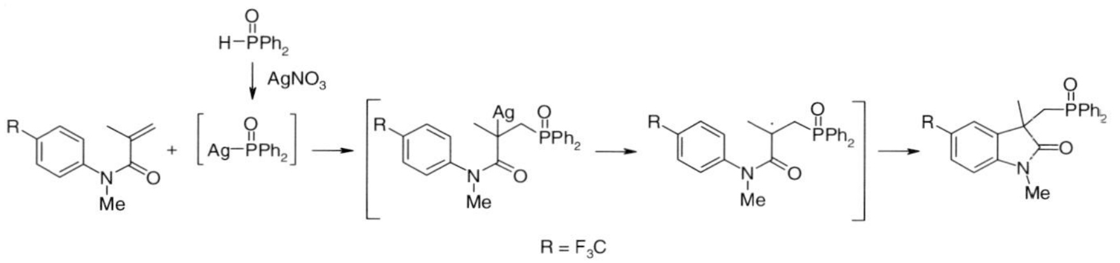

Startg. acrylanilide (0.3 mmol), $HP(O)Ph_2$ (2 eq.), $AgNO_3$ (5 mol%) and $Mg(NO_3)_2 \cdot 6H_2O$ (0.5 eq.) added to a Schlenk tube, the tube charged with argon three times, anh. acetonitrile (3 ml) added, the mixture stirred at 100° for 12 h (TLC), cooled to room temp., treated with NaOBu-*t* (4 eq.), stirred for an additional 10 min at room temp., solvent removed *in vacuo*, and the residue purified by column chromatography → product. Y 87%. The method uses a cheap, non-toxic silver reagent as catalyst, without the need for additional ligand or oxidant, and the starting materials are readily available. The yield was low with an N-tosyl-deriv. and the reaction did not work at all with an N-unsubst. anilide. The aromatic ring may bear an electron-donating or -withdrawing group in the *para* position (Me, MeO, CF_3, CN or halo); lower yields were obtained for *o*-subst. derivs., while *m*-subst. ones gave good yields but showed poor regioselectivity. The α-substituent on the olefin may be phenyl, benzyl, acetoxymethyl or phthalimidomethyl or be part of a cyclohexene ring, affording a spirocyclohexane-deriv., but there was no reaction in the absence of α-substitution. Preliminary studies support a radical mechanism. F.e. (twenty-four; Y 37-87%) and with diethyl phosphite (Y 83%) or ethyl phenylphosphinate (Y 43%) in place of $HP(O)Ph_2$, **also 4-α-phosphinyl-3,4-dihydrocarbostyrils** from β,γ-ethylenecarboxylic acid anilides (two examples; Y 63%, 69%), s. Y.-M. Li, M. Sun, H.-L. Wang, Q.-P. Tian, S.-D. Yang, Angew. Chem., Int. Ed. *2013*, *52* (14), 3972-6 [DOI: 10.1002/anie.201209475].

Gold-on-manganese oxide-based octahedral molecular sieves/oxygen —
Silylacetylenes C≡CH → C≡C-Si⋜
from terminal acetylene derivs. and organosilicon hydrides
Heterogeneous aerobic dehydrogenative coupling

248. 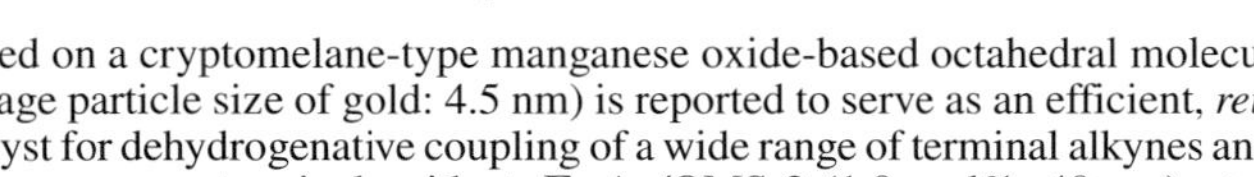

Gold supported on a cryptomelane-type manganese oxide-based octahedral molecular sieve (Au/OMS-2; average particle size of gold: 4.5 nm) is reported to serve as an efficient, *reusable* heterogeneous catalyst for dehydrogenative coupling of a wide range of terminal alkynes and monohydrosilanes using oxygen as terminal oxidant. **E:** Au/OMS-2 (1.9 mol%; 40 mg), startg. alk-1-yne (0.5 mmol) and toluene (1 ml) added to a Pyrex glass reactor, a toluene soln. of triethylsilane (0.55 *M*; 1 ml) added over 1 h by syringe pump at 80° under 1 atm. O_2, when reaction complete the catalyst removed by filtration (>95% recovery), the filtrate concentrated, and the residue subjected to column chromatography on silica gel → triethyl(4-fluorophenylethynyl)silane. Y 90% (Y 94% by GC). The method is applicable to aliphatic or aromatic alkynes, optionally bearing alkyl chloride or ar. Br, CF_3, NO_2 or (in lower yield) NH_2 groups, or to a 1,3-enyne, in reaction with $PhMe_2SiH$, $i\text{-}Pr_3SiH$, $t\text{-}BuMe_2SiH$, $t\text{-}BuPh_2SiH$ or $(EtO)_3SiH$ as well as Et_3SiH. No reaction took place in the presence of analogous platinum, ruthenium or rhodium catalysts, while only alkyne-alkyne homocoupling resulted using a copper catalyst. F.e. (eighteen; Y 54-98% by GC) s. K. Yamaguchi, Y. Wang, T. Oishi, Y. Kuroda, N. Mizuno, Angew. Chem., Int. Ed. *2013*, *52* (21), 5627-30 [DOI: 10.1002/anie.201300988].

Magnesium nitrate s. under $AgNO_3$ $Mg(NO_3)_2$

Mesitylenium tetrakis(pentafluorophenyl)borate/benzene-d$_6$ *[mesitylene-H]*$[B(C_6F_5)_4]/C_6D_6$
Arenium acid-catalyzed deuteration H → D

249. 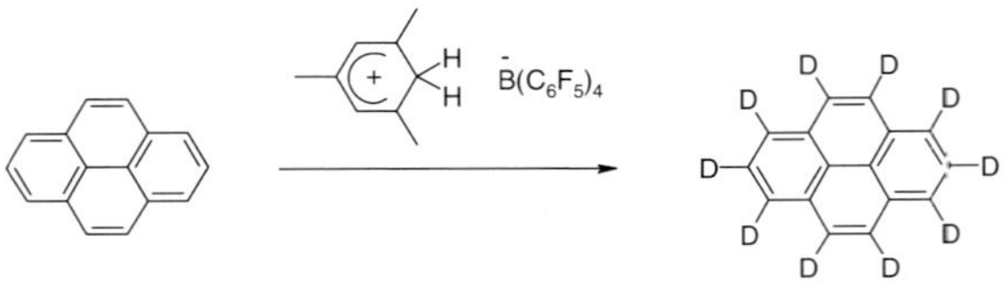

The arenium acid, [mesitylene-H]$^+$, is reported to be an extraordinarily active H/D exchange catalyst for the **perdeuteriation of polyarenes** using C_6D_6 as an inexpensive deuterium source **under mild conditions. E:** A vial (pre-dried at 150° for >12 h) charged with pyrene (56.7 mg; 280 µmol), [mesitylene-H]$[B(C_6F_5)_4]$ (6.7 mol%) and C_6D_6 (2 ml) in a glovebox under N_2, the vial taken out of the glovebox into a fumehood then heated to 70° for 3 d, water (1 ml) and concd. aq. NH_4Cl (1 ml) added, the mixture extracted with methylene chloride, the organic layers filtered through silica gel, and evaporated to dryness → pyrene-d$_{10}$. Y 96% (89-90% deuterium incorporation). Re-subjecting 26.7 mg of this material to the deuteration conditions for 1 d at 70° (7.1 mol% catalyst loading) afforded the product in 90% overall yield with 99% deuterium incorporation. Clean perdeuteriation was also observed for *p*-terphenyl (Y 96%; 94-96% D incorporation), fluoranthene (Y 93%; 94-95% D incorporation), triphenylene (Y 98%; 94-95% D incorporation) and corannulene (Y 76%; 96% D incorporation in toluene-d$_8$), all at room temperature. F.e.s. S. Duttwyler, A.M. Butterfield, J.S. Siegel, J. Org. Chem. *2013*, *78* (5), 2134-8 [DOI: 10.1021/jo302201a]; complementary, rapid, **microwave-assisted base-catalyzed perdeuteriation** of polyarenes with KOBu-*t*/DMF-d$_7$ s. A.K. Greene, L.T. Scott, ibid. 2139-43 [DOI: 10.1021/jo301903m].

Dimethylformamide-d$_7$ *s. under KOBu-t* *DMF-d*$_7$
Oxygen s. under Gold-on-manganese oxide-based octahedral molecular sieves O_2

Manganese(III) acetate $Mn(OAc)_3$
Regioselective oxidative ar. phosphonylation H → P(O)(OR)$_2$
with dialkyl phosphites, update s. *81,* 220s*82*; 5-phosphonylation of uracils in acetic acid with $Mn(OAc)_3$ as oxidant s. S.H. Kim, S.H. Kim, C.H. Lim, J.N. Kim, Tetrahedron Lett. *2013, 54* (13), 1697-9 [DOI: 10.1016/j.tetlet.2013.01.056].

Ruthenium-carbon *Ru-C*
Shvo's ruthenium hydride complex ←
Carbonyl(chloro)(hydrido)[6-(di-tert-butylphosphino)-2,2′-bipyridyl]ruthenium(II)/ sodium hydroxide ←
Deuteriation H → D
update s. *23,* 642s*81*; regio- and stereo-selective multiple deuteriation **of sugars** in water under hydrogen using Ru-C as catalyst s. Y. Sawama, Y. Yabe, H. Iwata, Y. Fujiwara, Y. Monguchi, H. Sajiki, Chem. Eur. J. *2012, 18* (51), 16436-42 [DOI: 10.1002/chem.201202852]; **α,β-di-deuteriation** of bioactive amines using Shvo's ruthenium hydride complex s. L. Neubert, D. Michalik, S. Bähn, S. Imm, H. Neumann, J. Atzrodt, V. Derdau, W. Holla, M. Beller, J. Am. Chem. Soc. *2012, 134* (29), 12239-44 [DOI: 10.1021/ja3041338]; α- and β-deuteriation of prim. and sec. alcohols with D_2O using carbonyl(chloro)(hydrido)[6-(di-*tert*-butylphosphino)-2,2′-bi-pyridyl]ruthenium(II) as catalyst at very low loading (down to 0.01 mol%) and NaOH as base s. E. Khaskin, D. Milstein, ACS Catal. *2013, 3* (3), 448-52 [DOI: 10.1021/cs400092p].

Palladium(II) acetate *Pd(OAc)$_2$*
N-Directed *o*-borylation H → $B(OR)_2$
with bis(pinacol)diboron, update s. *64,* 219s*82*; of 2-arylpyridines with 9-borabicyclo[3.3.1]nonane with $Pd(OAc)_2$ as catalyst at room temperature s. Y. Kuninobu, T. Iwanaga, T. Omura, K. Takai, Angew. Chem., Int. Ed. *2013, 52* (16), 4431-4 [DOI: 10.1002/anie.201210328]; potassium trifluoro(*o*-hydroxyaryl)borates from phenols via treatment of aryl diisopropylphosphinite-boranes with triflimide then KHF_2 s. C. Cazorla, T.S. De Vries, E. Vedejs, Org. Lett. *2013, 15* (5), 984-7 [DOI: 10.1021/ol303203m].

Oxygen ↑ RemC ↓↑ O

Microwaves [\\\\]
3-Component synthesis of α-aminophosphorus(V) compds. from oxo compds. CO → CH(NHR)P(O)<
Kabachnik-Fields reaction, update s. *33,* 593s*82*; under microwave enhancement, also microwave-enhanced phospha-Michael addition (cf. *45,* 340s*83*) and alcoholysis of dialkyl phosphites, s. E. Bálint, E. Fazekas, J. Takács, Á. Tajti, A. Juranovic, M. Kocevar, G. Keglevich, Phosphorus, Sulfur Silicon Relat. Elem. *2013, 188* (1-3), 48-50 [DOI: 10.1080/10426507.2012.743544]; synthesis of N-carbobenzoxy-protected α-aminophosphonic acid diaryl esters with HBF_4 in acetonitrile s. M. Psurski, M. Pigula, J. Ciekot, L. Winiarski, J. Wietrzyk, J. Oleksyszyn, Tetrahedron Lett. *2012, 53* (44), 5845-7 [DOI: 10.1016/j.tetlet.2012.08.037]; of N-carbobenzoxy-protected α-aminophosphonic acid fluoroalkyl esters with acetic acid s. M. Skorenski, J. Oleksyszyn, M. Sienczyk, ibid. *2013, 54* (12), 1566-8 [DOI: 10.1016/j.tetlet.2013.01.039].

1,8-Diazabicyclo[5.4.0]undec-7-ene *DBU*
Pyrid-2-yl-phosphinic or -thionophosphinic acid esters from N-methoxypyridinium salts and phosphonous or thiophosphonous acid O-monoesters P-H → P-Py

250.

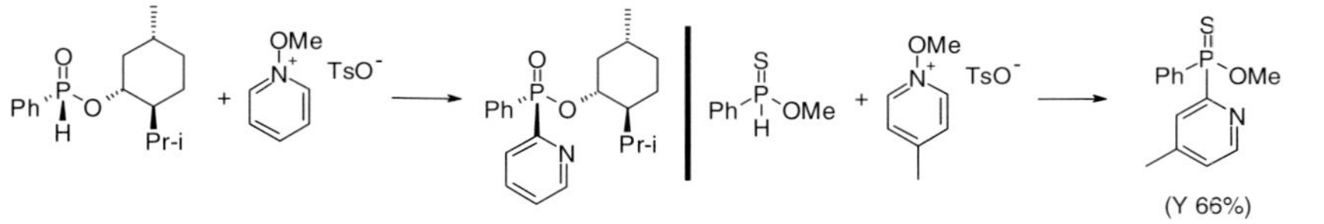

with retention of chirality. A soln. of N-methoxypyridinium tosylate (1.5 eq.) in dry acetonitrile (1 ml) added dropwise to a stirred soln. of (1R,2S,5R,R_P)-5-methyl-2-(1-methylethyl)cyclohexyl phenyl-H-phosphinate (1 mmol; R_p/S_p 95:5) and dry DBU (1.5 eq.) in the same solvent (1 ml) at -40°, the mixture stirred for 30 min, quenched with 0.2 *M* aq. NaH_2PO_4, extracted with ethyl acetate, and purified by chromatography on silica → (1R,2S,5R,S_P)-5-methyl-2-(1-methylethyl)-cyclohexyl phenyl(2-pyridyl)phosphinate. Y 71% (R_p/S_p 5:95). This experimentally simple method provides a cost-effective alternative to metal-catalyzed coupling and is optimally run at low temp. to avoid formation of hydroxymethylphosphinate by-products via reaction with HCHO generated from decomposition of the N-methoxypyridinium salt. Complete conversions were obtained for reaction of the unsubst. pyridinium salt with prim. and sec. alkyl H-phosphinates (Y 90-98% by NMR; three examples; 93% isolated yield for the methyl phosphinate), with ethyl and isopropyl phosphinate products contaminated with the methyl ester (Y 7-10% by NMR) formed by trans-esterification. Incomplete conversions were observed for reactions of the methyl phosphinate with 2-methyl- (68-75%; Y 33-37%) or 4-methyl-pyridinium salts (89-93%; Y 61-69%) under modified reaction conditions. Attempts to prepare thiophosphinate analogs directly from the products gave complex mixtures, but O-S exchange prior to 2-pyridylation (using Lawesson's reagent) enabled prepn. of methyl 2-pyridylthiophosphinate (Y 73%) and its 4-methylpyrid-2-yl analog (Y 66%). F.e., optimization and substrate prepn. s. N. Oka, K. Ito, F. Tomita, K. Ando, Chem. Lett. *2012*, *41* (12), 1630-2 [DOI: 10.1246/cl.2012.1630].

Nitrogen ↑ RemC ↓↑ N

Copper(II) chloride/potassium carbonate *$CuCl_2/K_2CO_3$*
Phosphorus(V) oxides C=NNHTs → CHP(O)<
from phosphorus(III) acids and tosylhydrazones
Copper-catalyzed ligand-free reductive coupling

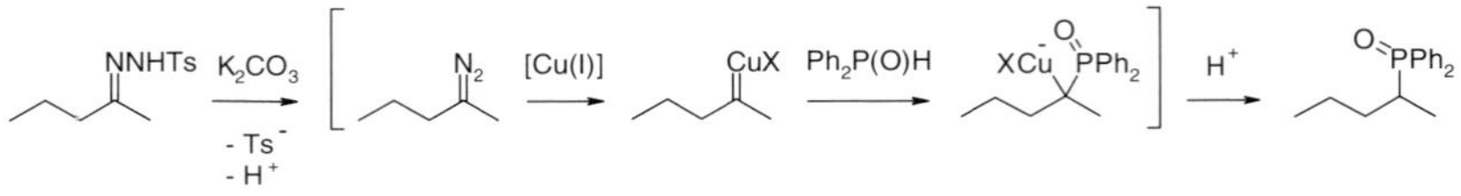

251.

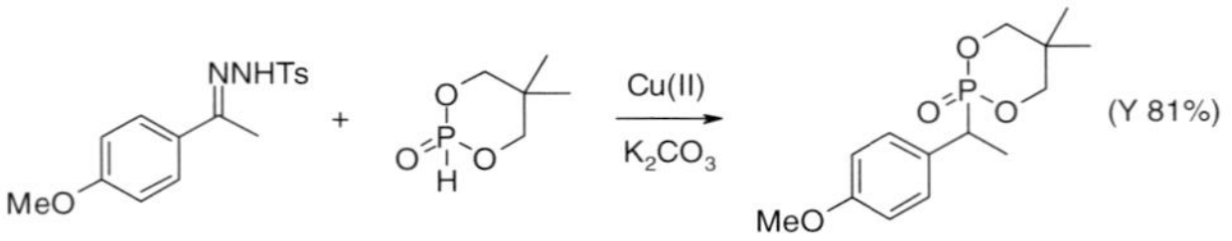

Novel C(sp[3])-P bond formation from reaction of H-phosphine oxides or -phosphonates with metal carbenes generated *in situ* from N-tosylhydrazones is reported. **E:** A mixture of diphenylphosphine oxide (1 eq.), pentan-2-one N-tosylhydrazone (1 mmol), K_2CO_3 (3 eq.) and $CuCl_2$ (20 mol%) in DMF (5 ml) heated under N_2 at 110° for 2 h (TLC), solvent evaporated, and the residue purified chromatographically → diphenyl(2-pentyl)phosphine oxide. Y 67%. The reaction was successful, using either dioxane or DMF as solvent, with tosylhydrazones of electron-diverse aryl and aliphatic aldehydes/ketones reacting with electron-diverse diarylphosphine oxides or diaryl/dialkyl (incl. cyclic) phosphonates (twenty examples; Y 50-81%). It is proposed the Cu(II) salt is reduced *in situ* to a Cu(I) species by the diazo compd. generated from the startg. N-tosylhydrazone under basic and thermal conditions; decomposition of the diazo compd. by the Cu(I) species then generates a copper carbene complex which undergoes nucleophilic attack by $Ph_2P(O)^-$, and the adduct undergoes protonation to give the final product. Other copper catalysts may be used but $CuCl_2$ was most effective. F.e. and optimization s. L. Wu, X. Zhang, Q.-Q. Chen, A.-K. Zhou, Org. Biomol. Chem. *2012, 10* (39), 7859-62 [DOI: 10.1039/c2ob26414e].

Ferrocene *Cp_2Fe*
Arylboronic acid derivs. from diazonium fluoroborates $ArN_2^+ BF_4^-$ → $ArB(OR)_2$
via sec. aryl(amino)boranes

252.

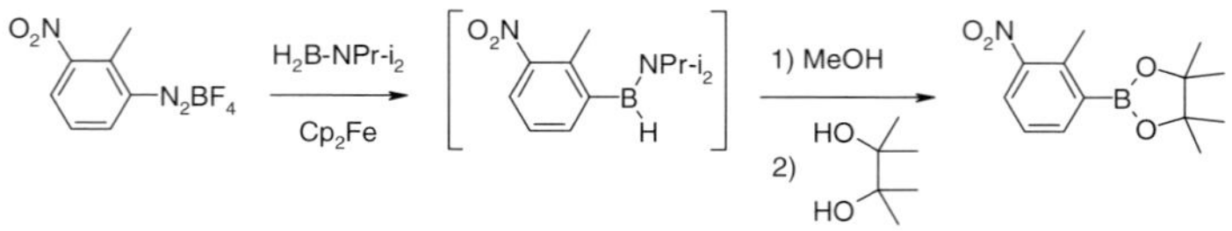

A straightforward **ferrocene-catalyzed denitrogenative ar. borylation** of diazonium salts is reported, the key to success being to balance generation of the intermediate aryl radical with a fast addition to the aminoborane and reduction of the formed ferricinium ion by the aminoborane(s). **E: Arylboronic acid esters**. Diisopropylaminoborane (2 eq.) added to a soln. of startg. diazonium salt (1 mmol) and ferrocene (*0.1 mol%*) in anh. acetonitrile (2 ml) under argon, the soln. stirred at room temp. for 2.5 h, quenched by slow addition of anh. methanol (2 ml) at 0°, stirred for 1 h at room temp., after removal of all volatiles, pinacol (1.3 eq.) in ether (2 ml) added, the mixture stirred at room temp. for 4 h, washed with a $CuCl_2$ soln. (50 g/l; 2 x 5 ml), the organic layer dried (Na_2SO_4), filtered, concentrated to dryness, the resultant oil dissolved in methylene chloride, and filtered through a pad of silica gel, eluting with methylene chloride → 2-(2-methyl-3-nitrophenyl)-4,4,5,5-tetramethyl-1,3,2-dioxaborolane. Y 91%. The method is compatible with a variety of functional groups (such as MeO, NO_2, CF_3, CN, F, Cl, and, notably, Br and I), giving yields of 47-87% for twenty-four further examples; a *p*-benzoyl-deriv. afforded only benzophenone, however. The intermediate amino(aryl)borane derived from *p*-tolyldiazonium fluoroborate was converted to a boronic acid (Y 91%), a naphthalene-1,8-diyldiaminoborane (Y 98%) or a potassium aryl(trifluoro)borate (Y 90%) as well as to the pinacol boronate (Y 97%). Because of safety concerns the reaction of *p*-methoxyphenyldiazonium fluoroborate was briefly optimized in a continuous flow system, affording an 80% yield on a 10 mmol scale. F.e.s. L.D. Marciasini, N. Richy, M. Vaultier, M. Pucheault, Adv. Synth. Catal. *2013*, *355* (6), 1083-8 [DOI: 10.1002/adsc.201200942].

Halogen ↑ RemC ↓↑ Hal

Potassium hydrogen fluoride s. under Palladium phosphine complexes *KHF_2*

Copper(I) iodide/N,N′-dimethylethylenediamine/base *CuI/DMEDA/base*

Copper(I)-catalyzed reductive coupling of dibromomethylene compds. with heteronucleophiles ←

(E)-α,β-ethylenephosphonic acid esters from dialkyl phosphites with CuI/DMEDA/K_3PO_4 cf. *79*, 225; also 1-bromoenolethers, ketene acetals or ynolethers from alcohols and N-functionalized 1-bromoenamines, ketene aminals or ynamines from N-nucleophiles (cf. *76*, 195) by fine-tuning of the reaction conditions, s. K. Jouvin, A. Coste, A. Bayle, F. Legrand, G. Karthikeyan, K. Tadiparthi, G. Evano, Organometallics *2012*, *31* (22), 7933-47 [DOI: 10.1021/om3005614].

Magnesium *Mg*

Arylboronic acids from arylmagnesium bromides s. *51*, 259s*83* MgBr → $B(OH)_2$

Nickel(II) complexes/base *[Ni(II)]/base*

B-Arylation >BH → >BAr

cf. *54*, 244s*82*; arylboronic acid esters by B-borylation of 4,4,6-trimethyl-1,3,2-dioxaborinane using $NiCl_2$(dppp or dppf)/Et_3N s. M. Murata, Y. Sogabe, T. Namikoshi, S. Watanabe, Heterocycles *2012*, *86* (1), 133-8 [DOI: 10.3987/com-12-s(n)28].

Nickel(II) or palladium(II) complexes/base *[Ni(II)] or [Pd(II)]/base*

Palladium-catalyzed P-arylation Hal → P(O)<

under palladium or nickel catalysis, update s. *38*, 584s*82*; arylphosphonic acid esters from ar. chlorides, bromides or iodides (incl. *unactivated electron-rich* ar. chlorides) with diisopropyl H-phosphonate using a ferrocenyl ketimine-based *o*-pallad^{II}acyclic catalyst at low loading and K_2CO_3 (or K-*tert*-butoxide) as base, also P-benzylation with K_2CO_3 alone, s. K. Xu, H. Hu, F. Yang, Y. Wu, Eur. J. Org. Chem. *2013* (2), 319-25 [DOI: 10.1002/ejoc.201201230]; triarylphosphine oxides from ar. chlorides with $NiCl_2(PPh_3)_2$ or $NiCl_2$(DME) and K_2CO_3 or Na-*tert*-butoxide s. H.-Y. Zhang, M. Sun, Y.-N. Ma, Q.-P. Tian, S.-D. Yang, Org. Biomol. Chem. *2012*, *10* (48), 9627-33 [DOI: 10.1039/c2ob26874d]; **P-vinylation** with vinyl halides or triflates under palladium catalysis with XantPhos or dppf/ethyldiisopropylamine and ethylene glycol or 1,2-dimethoxyethane as additive, also P-allylation and P-(alk-l-ynyl)ation, s. O. Berger, C. Petit, E.L. Deal, J.-L. Montchamp, Adv. Synth. Catal. *2013*, *355* (7), 1361-73 [DOI: 10.1002/adsc.201300069].

Palladium phosphine complexes/potassium acetate or potassium hydrogen fluoride ←

Boronic acid esters from halides Hal → $B(OR)_2$

update s. *51*, 259s*81*; **arylboronic acid** [*o*-phenylene or pinacol] **esters** from ar. bromides and 2,2′-bis(1,3,2-benzodioxaborole) or bis(pinacolato)diboron using $PdCl_2$(dppf) as catalyst and KOAc as base s. J. Takagi, T. Yamakawa, Tetrahedron Lett. *2013*, *54* (2), 166-9 [DOI: 10.1016/j.tetlet.2012.10.118]; arylboronic acid esters from ar. halides or triflates with tetrakis(dimethylamino)diboron and a diol under palladium catalysis in the presence of base, also α,β-ethyleneboronic acid esters from vinyl halides or triflates under the same conditions, s. C.S. Bello, J. Schmidt-Leithoff, ibid. *2012*, *53* (46), 6230-5 [DOI: 10.1016/j.tetlet.2012.09.009]; potassium aryl(trifluoro)borates from ar. halides or triflates with tetrahydroxydiboron using a palladium(0) phosphine complex (based on XantPhos) and added KHF_2 s. G.A. Molander, S.L.J. Trice, S.M. Kennedy, S.D. Dreher, M.T. Tudge, J. Am. Chem. Soc. *2012*, *134* (28), 11667-73 [DOI: 10.1021/ja303181m]; **arylboronic acids from arylmagnesium bromides** with diisopropylaminoborane and subsequent hydrolysis, also alkylboronic acids, s. C.L. Bailey, C.L. Murphy, J.W. Clary, S. Eagon, N. Gould, B. Singaram, Heterocycles *2012, 86* (1), 331-42 [DOI: 10.3987/com-12-s(n)14].

Cyclooctadiene(methoxo)iridium(I) dimer/2,9-dimethyl-1,10-phenanthroline ←
Borylation of sp^3 carbon-hydrogen bonds H → B(OR)$_2$
update s. *75,* 245s*82*; regio- and diastereo-selective borylation of monosubst. cyclopropanes with bis(pinacolato)diboron using [Ir(cod)(OMe)]$_2$ [or (η^6-mesityl)tris(pinacolatoboryl)iridium(III)] and 2,9-dimethyl-1,10-phenanthroline as ligand s. C.W. Liskey, J.F. Hartwig, J. Am. Chem. Soc. *2013, 135* (9), 3375-8 [DOI: 10.1021/ja400103p].

Remaining Elements ↑ RemC ↓↑ Rem

Copper(II) oxide nanoparticles/potassium carbonate *CuO/K$_2$CO$_3$*
Copper(II)/zinc or indium(I) bromide *Cu(II)/Zn or InBr*
1-Acetylene-1-selenides from diselenides C≡C(SeR)
and terminal acetylene derivs. s. *45,* 352s*46*; with recyclable CuO nanoparticles and K$_2$CO$_3$ as base under air, **also 1-acetylene-1-tellurides** from ditellurides, s. M. Godoi, E.W. Ricardo, T.E. Frizon, M.S.T. Rocha, D. Singh, M.W. Paixão, A.L. Braga, Tetrahedron *2012, 68* (51), 10426-30 [DOI: 10.1016/j.tet.2012.08.086]; **from α,β-acetylenehalides** and the appropriate dichalcogenide with alumina-supported copper(II) and Zn dust or InBr (for reductive cleavage of the dichalcogenide) s. S. Ahammed, S. Bhadra, D. Kundu, B. Sreedhar, B.C. Ranu, ibid. *68* (51), 10542-9 [DOI: 10.1016/j.tet.2012.08.046].

Copper(I) chloride/o-bis(diphenylphosphino)benzene/lithium tert-butoxide ←
β-Aminoboronic acid esters from ethylene derivs. C=C → C(N<)C-B(OR)$_2$
and aroyloxylamines – Regio- and stereo-selective copper-catalyzed 1,2-aminoboration s. *83*, 243

(S)-Bis(acetonitrile)[5,6;5′,6′-bis(methylenedioxy)-3,3′-bis[bis(3,5-di-tert-butyl-4-methoxyphenyl)phosphine]biphenyl]copper(I) hydrogen bifluoride/methanol ←
1-Hydroxysilanes from aldehydes CHO → CH(OH)Si≤
via copper(I)-catalyzed asym. 1,2-addition s. *83*, 241

Cyclooctadiene(methoxo)iridium(I) dimer/4,4′-di-tert-butyl-2,2′-bipyridyl *[Ir]*
Regioselective ar. iodination of functionalized arenes via borylation s. *83*, 212 H → B(OR)$_2$

Carbon ↑ RemC ↓↑ C

Magnesium bis(2,2,6,6-tetramethylpiperidide)·bis(lithium chloride)/boron fluoride ←
Regioselective metalation of N-heteroarenes ←
using boron fluoride-etherate and hindered magnesium-amides as frustrated Lewis pairs s. *83*, 410

Trifluoroacetic acid *CF$_3$COOH*
1,4-Dihydro-3(2*H*)-isoquinolone-1-phosphonic acid esters ○
from *o*-aldehydoarylacetic acid esters, prim. amines and trialkyl phosphites

253. 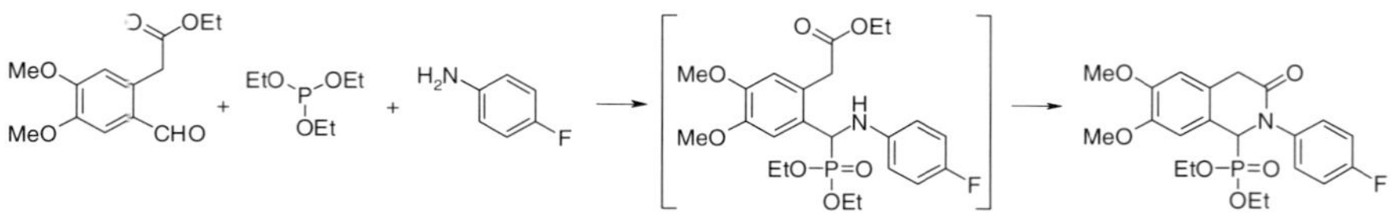

The first method for the formation of N-aryl-1,4-dihydro-3(2*H*)-isoquinolone-1-phosphonic acid esters is reported, using a convenient, 3-component one pot method with readily available substrates. **E:** 4 Å Molecular sieves (1 g) added to a soln. of ethyl 2-(2-formyl-4,5-dimethoxyphenyl)acetate (1 mmol), *p*-fluoroaniline (1.1 eq.) and triethyl phosphite (1.2 eq.) in dry acetonitrile (15 ml) under N$_2$, followed by trifluoroacetic acid (20 mol%) via syringe, the yellow mixture stirred at room temp. for 6 h, additional trifluoroacetic acid (6 eq.) added via syringe under N$_2$, the mixture refluxed for 17 h, acetonitrile removed under reduced pressure, the mixture poured into ice cold water, extracted with methylene chloride, the combined organic layers washed with aq. $NaHCO_3$ soln. (10%) and water, dried (Na_2SO_4), solvent evaporated, and the residue chromatographed on silica gel → product. Y 70%. Initially the intermediate α-aminophosphonate was isolated in 72% yield and converted to the lactam in 61% yield. Six further 1,4-dihydro-3(2*H*)-isoquinolones were obtained from ethyl 2-(2-formyl-4,5-dimethoxyphenyl)acetate and a variety of anilines (optionally bearing Cl, Br or alkyl groups in the *m*- or *p*-position) in 64-74% yield by the one pot method. F.e.s. A.U. Borse, N.L. Patil, M.N. Patil, R.S. Mali, Tetrahedron Lett. *2012*, *53* (51), 6940-2 [DOI: 10.1016/j.tetlet.2012.10.030].

Chloro(cyclooctadiene)rhodium(I) dimer/4,5-bis(diphenylphosphino)-9,9-dimethyl-xanthene/triethylenediamine ←
Arylboronic acid esters from ar. nitriles CN → $B(OR)_2$
s. *81*, 230; s.a. H. Kinuta, Y. Kita, E. Rémond, M. Tobisu, N. Chatani, Synthesis *2012*, *44* (19), 2999-3002 [DOI: 10.1055/s-0031-1290370].

Tris(dibenzylideneacetone)dipalladium/1,1′-bis(diphenylphosphino)ferrocene/ sodium ethoxide or cesium fluoride or 1,8-diazabicyclo[5.4.0]undec-7-ene ←
Ar. phosphine sulfides from ar. halides >P(S)CH_2O- → >P(S)Ar

254.
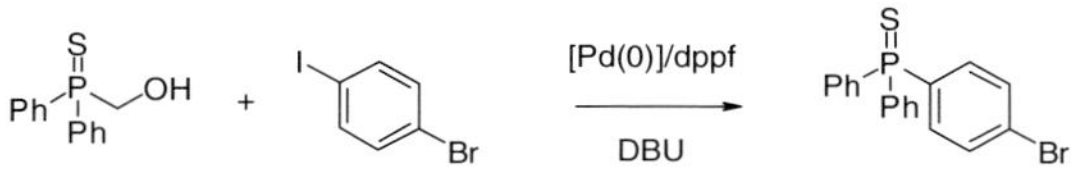

and hydroxymethylphosphine sulfides. A soln. of $Pd_2(dba)_3 \cdot CHCl_3$ (1 mol%) and dppf (2 mol%) in toluene (2 ml) stirred at room temp. for 15 min under argon, startg. hydroxymethylphosphine sulfide (0.5 mmol), ar. iodide (1.05 eq.), toluene (3 ml) and DBU (2 eq.) added successively, the mixture stirred at 40° for 3 h, quenched with aq. HCl (1 *N*), extracted with ethyl acetate, the combined extracts dried, evaporated to dryness, and the residue purified by column chromatography on silica gel → product. Y 99%. Fourteen further examples from ar. iodides bearing electron-donating or -withdrawing groups gave yields of 86-99%; hindered 2,6-dimethylphenyl iodide gave only 32% yield after 6 h at reflux, however. *tert*-Butyldimethylsilyl ethers and benzoate esters were stable under these basic conditions, such that (*tert*-butyldimethylsilyloxymethyl)-(hydroxymethyl)phenylphosphine sulfide underwent selective dehydroxymethylative P-arylation in 99% yield while (benzoyloxymethyl)(hydroxymethyl)phenylphosphine sulfide underwent selective reaction in 87% yield. With *CsF* in place of DBU (in THF at 70°), however, the title compounds were obtained **from *tert*-butyldimethylsiloxymethylphosphine sulfides** in 87-99% yields (eight examples), while with *NaOEt* in place of CsF they were obtained **from benzoyloxy-methylphosphine sulfides** in 32-99% yields (seven examples; lowest yields from ar. iodides bearing electron-withdrawing groups). (Benzoyloxymethyl)(*tert*-butyldimethylsilyloxymethyl)-(*p*-tolyl)phosphine sulfide also underwent the appropriate monocoupling in the presence of CsF or NaOEt (Y 99% in either case). The method was therefore extended to formation of **triarylphosphine sulfides from three different ar. halide molecules** and (benzoyloxymethyl)(*tert*-butyldimethyl-silyloxymethyl)(hydroxymethyl)phosphine sulfide by sequential reactions in the presence of DBU, NaOEt and CsF (four examples; Y 82-85% overall; cf. Y 80% for one example by reaction under DBU then CsF then NaOEt conditions).

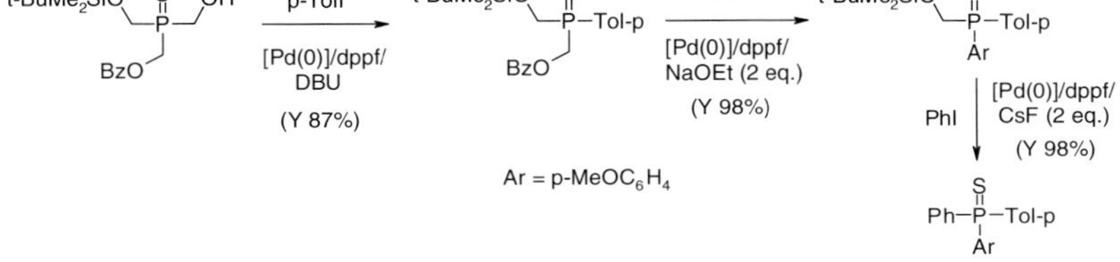

The products may be readily converted to triarylphosphines. F.e.s. M. Hayashi, T. Matsuura, I. Tanaka, H. Ohta, Y. Watanabe, Org. Lett. *2013*, *15* (3), 628-31 [DOI: 10.1021/ol303448j].

Elimination ⇑

Hydrogen ↑ RemC ⇑ H

Chloro(1,5-cyclooctadiene)rhodium(I) dimer/1,3-bis(diphenylphosphino)propane ←
Regiospecific rhodium(I)-catalyzed intramolecular dehydrogenative C-silylation ○

255. 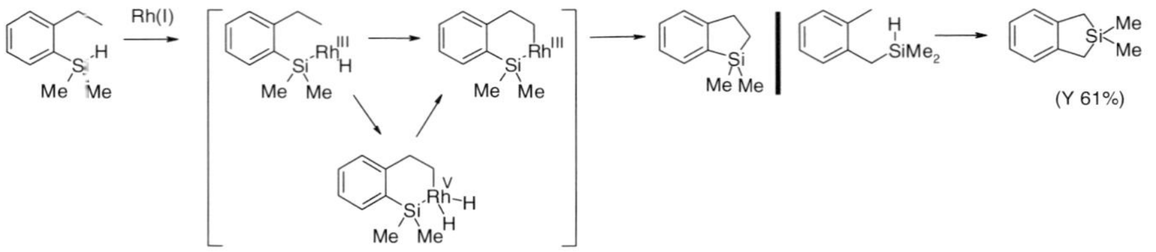

Novel intramolecular silylation of unactivated sp^3 C-H is described. **E:** A mixture of $[RhCl(cod)]_2$ (1.5 mol%), dppp (4.5 mol%) and 1,4-dioxane (0.25 ml) stirred at 25° for 30 min, (2-ethylphenyl)-dimethylhydrosilane (0.25 mmol) added, the mixture stirred at 180° for 24 h, concentrated *in vacuo*, and purified chromatographically → 1,1-dimethyl-2,3-dihydro-1*H*-benzo[*b*]silole. Y 66%. Rhodium-catalyzed cyclization of *o*-diphenyl/dimethyl-silyl ethyl/*tert*-butylarenes occurred at the terminal prim. carbon atoms to afford the corresponding **benzo- or naphtho-[*b*]siloles** in moderate to good yields (59-80%; six examples), (*o*-*n*-heptylphenyl)hydrosilane cyclized exclusively at an internal C to afford the analogous 2-pentylbenzo[*b*]silole in somewhat lower yield (37%), and tri-*n*-butylhydrosilane afforded the monocyclic silole in 43% yield. The method is also applicable to cyclization at the benzylic C in dimethyl(*o*-xylyl)hydrosilane, affording a benzo[*c*]silole analog (Y 61%). The mechanism was presumed to proceed via Rh^{III} rhodacycles, formed directly from the initial rhodium insertion product, or via reductive elimination from a Rh^{V} rhodacycle. F.e. and substrate prepn. s. Y. Kuninobu, T. Nakahara, H. Takeshima, K. Takai, Org. Lett. *2013*, *15* (2), 426-8 [DOI: 10.1021/ol303353m].

Formation of C-C Bond

Uptake ⇓

Addition to Hydrogen and Carbon CC ⇓ HC

Silver triflate *AgOTf*
Regioselective silver-promoted ar. formylation with dichloromethyl methyl ether under mild conditions H → CHO

256. 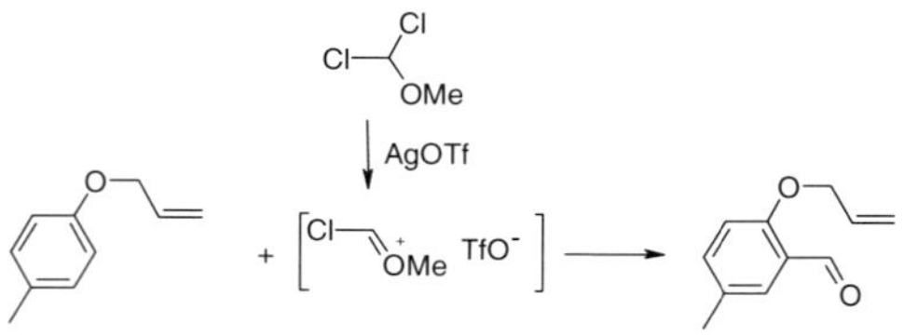

Formylation **of phenolethers** with dichloromethyl methyl ether under catalysis with traditional Lewis acids, e.g. $TiCl_4$, is invariably associated with O-dealkylation; however, *with silver triflate* as promoter *at low temperature*, no such ether cleavage takes place. **E:** A soln. of dichloromethyl

methyl ether (3 mmol) in dry methylene chloride (1.5 ml) added at -78° under argon to a suspension of the startg. phenolether (1 mmol) and AgOTf (3 mmol in the same solvent (4.5 ml), stirred for 10 min, the mixture quenched with satd. aq. $NaHCO_3$, stirred at room temp. for 30 min, filtered through a pad of Celite, and worked up with purification by flash chromatography on silica gel → product. Y 96%. Reaction is applicable in good yield to the *o*-formylation of methyl, allyl and benzyl phenolethers, incl. 1,3-dimethoxybenzene and anisole derivs. possessing an electron-withdrawing group. Monobrominated arenes (not 3,5-dibromotoluene!) and polyalkylbenzenes (except the hindered 1,3,5-tri-*tert*-butylbenzene) were also formylated, as were 1-methoxy-naphthalene (to the 4-formyl deriv). and N-tosylindole (to the 3-formyl deriv.), while 3,5-dimethyl-phenol gave a mixture of formyl derivs. (sixteen examples; Y 51-83%). There was no reaction at all, however, with O-acetyl- or O-tosyl-phenols in view of the presence of the electron-withdrawing group. Reaction is presumed to involve initial activation of dichloromethyl methyl ether by silver(I) to a highly reactive electrophilic species which readily formylates the arene in Friedel-Crafts fashion prior to hydrolysis. Reaction with other silver(I) salts failed. F.e.s. K. Ohsawa, M. Yoshida, T. Doi, J. Org. Chem. *2013*, *78* (7), 3438-44 [DOI: 10.1021/jo400056k].

Addition to Oxygen and Carbon CC ⇓ OC

Without additional reagents *w.a.r.*

Diastereoselective uncatalyzed aldol condensation CO → C(OH)C-CO
of 2,4-thiazolidinediones 'on water' s. *43*, 710s*83*

Passerini reaction CO → C(OAc)CONHR
with aldehydes cf. *47*, 720; synthesis of 3-(acoxy)oxindole-3-carboxylic acid amides from isatins in acetonitrile containing 4 Å molecular sieves s. A.A. Esmaeili, S.A. Ghalandarabad, S. Jannati, Tetrahedron Lett. *2013*, *54* (5), 406-8 [DOI: 10.1016/j.tetlet.2012.11.025]; solvent-free synthesis of α-acoxycarboxylic acid N-(azulenyl)amides *on grinding* the reactants s. K. Sato, T. Ozu, N. Takenaga, ibid. *54* (7), 661-4 [DOI: 10.1016/j.tetlet.2012.11.148].

Electrolysis s. under Sm

Potassium tert-butoxide/tetra-n-butylammonium chloride *KOBu-t/Bu$_4$NCl*

Nucleophilic trifluoromethylation CO → C(OH)CF_3
with hexafluoroacetone hydrate-1,8-diazabicyclo[5.4.0]undec-7-ene salt

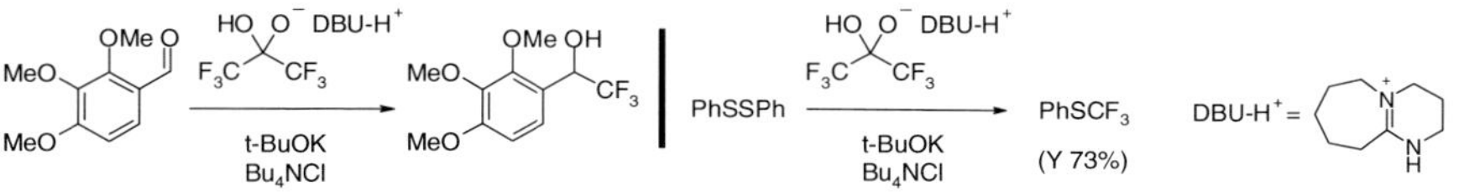

257.

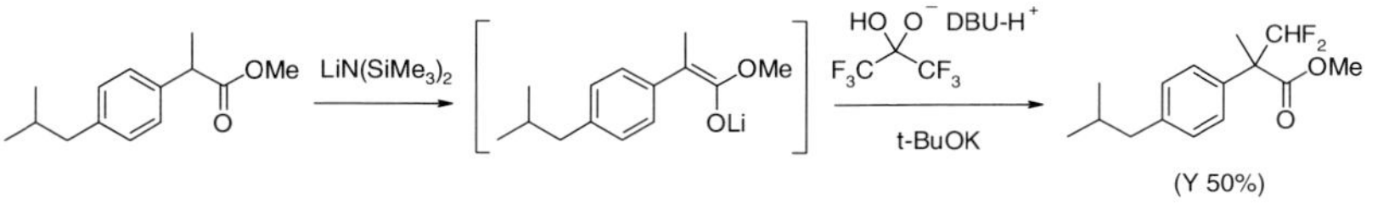

A novel moisture- and air-stable amidinate, prepared in one step from DBU and hexafluoroacetone hydrate, has been developed to generate trifluoromethyl carbanion, via base-induced fragmentation, in an anhydrous environment for trapping by added electrophiles. **E: 2,2,2-Trifluoroalcohols from oxo compds.** Hexafluoroacetone hydrate-1,8-diazabicyclo[5.4.0]undec-7-ene salt (1.2 eq.) added to a mixture of K-*tert*-butoxide (4.4 eq.) and tetra-*n*-butylammonium chloride (4.4 eq.) in anh. DMF (1 ml) at -30°, the mixture stirred for 30 min, 2,3,4-trimethoxybenzaldehyde (0.1 mmol) added, stirring continued for 1.5 h, the mixture warmed to room temp., quenched with satd. aq. NH_4Cl, extracted with ethyl acetate, and purified by flash chromatography on silica → 2,2,2-tri-fluoro-1-(2,3,4-trimethoxyphenyl)ethanol. Y 95%. The reagent was effective for gram-scale trifluoromethylation of electron-diverse ar. aldehydes and ketones (incl. cyclic) in good yields

(78-96%; ten examples). A single disulfide (PhSSPh) was converted to the corresponding trifluoromethyl thioether (Y 73%). Use of the reagent to generate fluoroform (either *in situ* or for bubbling through a reaction mixture) was demonstrated in the difluoromethylation of the lithium enolate of methyl ibuprofenate (Y 50%). F.e., optimization and reagent prepn. s. M.V. Riofski, A.D. Hart, D.A. Colby, Org. Lett. *2013*, *15* (1), 208-11 [DOI: 10.1021/ol303291x].

Organolithium compds. *RLi*

Synthesis of 5-alkoxy-4-arylidene-2-oxazolidinethiones from (Z)-α-isothiocyanatocinnamic acid esters and organolithium compds. ○

258.

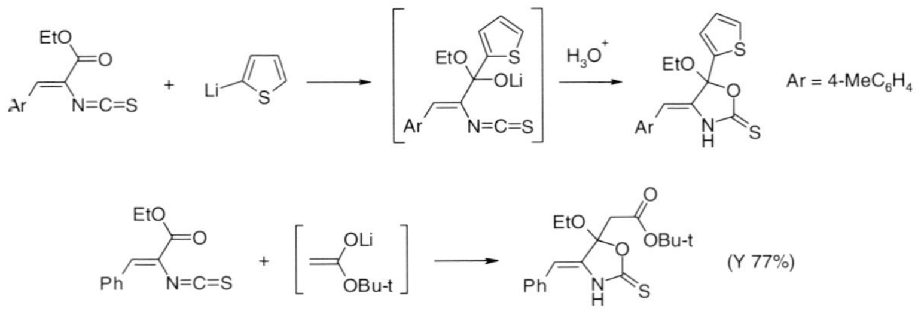

A soln. of 2-thienyllithium (1 eq.) in hexanes (0.32 ml) added dropwise to a soln. of ethyl (Z)-3-(4-tolyl)-2-isothiocyanatoprop-2-enoate (0.51 mmol) in THF (4 ml) at -78°, the mixture stirred for 15 min, quenched with satd. aq. NH_4Cl, warmed to room temp., extracted with ethyl acetate, and purified chromatographically → (4Z)-5-ethoxy-4-(4-methylbenzylidene)-5-thien-2-yl-1,3-oxazolidine-2-thione. Y 67%. This experimentally simple and mild method effected the cyclization of α-thiocyanato-β-(het)arylacrylates with alkyl- or (het)aryl-lithium compds. (eight examples; Y 63-79%) or with *i*-Pr_2NLi-generated lithium enolates of acetate esters (five examples; Y 62-77%) to afford the title products as single stereoisomers. F.e.s. K. Kobayashi, K. Ezaki, H. Hashimoto, Helv. Chim. Acta *2013*, *96* (3), 431-6 [DOI: 10.1002/hlca.201200479].

Lithium diisopropylamide *i-Pr₂NLi*

α-Diazo-β,β′-dihydroxyketones from two different aldehyde molecules via desilylative aldol condensation and aldol condensation with α-diazo-α-(triethylsilyl)acetone ←

259.

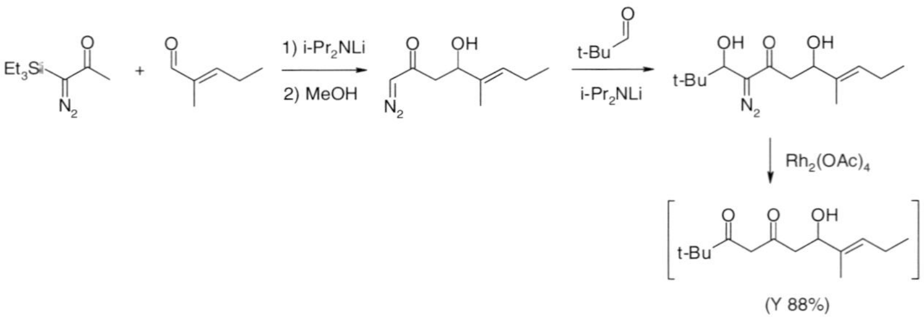

A soln. of α-diazo-α-(triethylsilyl)acetone (1 eq.) in anh. THF (0.25 *M*) added to a soln. of *i*-Pr_2NLi (freshly titrated; 1.82 eq.) in the same solvent (0.5 *M*) at -50°, the mixture stirred for 30 min, cooled to -100°, a soln. of (E)-2-methyl-2-pentenal (1 eq.) in the same solvent (0.25 *M*) added dropwise, the mixture stirred for 1.5 h, quenched with satd. aq. NH_4Cl, warmed slowly to room temp., extracted with ether, concentrated *in vacuo*, the residue stirred with methanol (10 ml) at room temp. for 24 h, concentrated *in vacuo*, and purified by chromatography on silica →

(E)-1-diazo-4-hydroxy-5-methyloct-5-en-2-one (Y 67%), a portion of which (1 eq.) dissolved in anh. THF (0.67 *M*), cooled to -100°, a soln. of *i*-Pr_2NLi (1.25 eq.) in the same solvent (0.5 *M*) added, the mixture stirred for 1.5 h, a soln. of acetic acid (1.5 eq.) in ether (500 µl) added slowly, the mixture warmed to room temp., extracted with ether (with aq. $NaHCO_3$ wash), and purified by chromatography on silica → (E)-4-diazo-3,7-dihydroxy-2,2,8-trimethylundec-8-en-5-one (Y 53%). This sequential double aldolization using α-diazo-α-(triethylsilyl)acetone provides an alternative to reaction of acetone with [unstable and difficult to prepare] acid chlorides then aldehydes. It was determined that methyl-side aldolization should be the first step due to instability of the diazo-side aldol under basic conditions, hence the azomethine proton required protection via silylation. The first aldolization, using electron-diverse ar. and aliphatic aldehydes was accompanied by partial desilylation, which was completed by methanolysis, to afford α-diazo-β′-hydroxyketones in moderate yields (42-68%; eight examples), compromised by chromatographic instability. Diazo-side aldolization of representative intermediates was accomplished with ar. and aliphatic aldehydes (five examples; Y 50-65%) to afford the title products which were dediazotized [$Rh_2(OAc)_4$] to the corresponding **δ-hydroxy-β-diketones** (five examples; Y 70-88%). F.e.s. A. Lancou, H. Haroun, U.K. Kundu, F. Legros, N. Zimmermann, M. Mathé-Allainmat, J. Lebreton, G. Dujardin, C. Gaulon-Nourry, P. Gosselin, Tetrahedron *2012, 68* (47), 9652-7 [DOI: 10.1016/j.tet.2012.09.063].

Cesium carbonate *Cs_2CO_3*
Friedel-Crafts α-hydroxyalkylation H → CH(OH)R
update s. *75,* 270s*80*; synthesis of aryl(3-indolyl)carbinols with Cs_2CO_3 in DMF s. R. Elayaraja, R.J. Karunakaran, Tetrahedron Lett. *2012, 53* (51), 6901-4 [DOI: 10.1016/j.tetlet.2012.10.002]; **asym. Friedel-Crafts α-hydroxyalkylation** (cf. *60,* 227s*81*) of indoles with alkyl trifluoropyruvates using a readily recoverable C_3-symmetric cinchonine-based squaramide as catalyst s. X. Han, B. Liu, H.-B. Zhou, C. Dong, Tetrahedron: Asym. *2012, 23* (18-19), 1332-7 [DOI: 10.1016/j.tetasy.2012.08.015].

Lithium chloride or sodium iodide s. under In *LiCl or NaI*

1-Butyl-3-methylimidazolium hydroxide/iodine *[bmim]OH/I_2*
α-Acylamino-γ-lactones from terminal epoxides and Δ^2-5-oxazolones ←
Regio- and stereo-selective conversion in a basic ionic liquid

260.

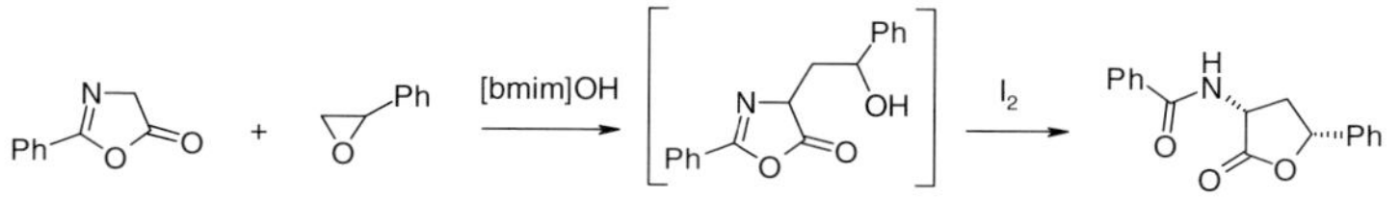

γ-Subst. *cis*-α-benzoylamino-γ-lactones. A mixture of 2-phenyl-Δ^2-5-oxazolone (2 mmol), styrene oxide (1 eq.) and iodine (10 mol%) in [bmim]OH (6 ml) stirred at room temp. for 4.5 h (TLC), and worked up with purification by recrystallization → α-benzoylamino-γ-phenyl-γ-lactone. Y 94% (*cis/trans* 95:5). The procedure is simple, mild, based on an inexpensive catalyst, free from by-products, and generally applicable to electron-diverse styrene oxides, naphthyl analogs, aliphatic alkene oxides, epichlorohydrin and aryloxy-functionalized epoxides (fourteen examples; Y 84-95%; *cis/trans* 93:7 to 98:2). The readily recyclable basic ionic liquid is presumed to deprotonate the 5-oxazolone, which then cleaves the epoxide regioselectively prior to iodine-catalyzed lactonization and oxazole ring opening. F.e. and comparison of ionic liquid and traditional solvents s. V.K. Rai, R. Sharma, A. Kumar, Tetrahedron Lett. *2013, 54* (9), 1071-5 [DOI: 10.1016/j.tetlet.2012.12.026].

Chiral diamines, cinchona alkaloids, chiral quaternary ammonium salts, (S)-prolinamides, enzymes, chiral supported α-aminocarboxylic acids, cinchona-based 2-tert-amino-thioureas or chiral polyfunctionalized thioureas ←
Organocatalyzed asym. aldol condensation CO → C(OH)C-CO
update s. *58,* 245s*82*; of cyclohexanone with aldehydes using (1R,2R)-cyclohexane-1,2-diamine *l*-tartrate salt s. S. Wu, L. Wang, J. Tang, D. Mao, X. Liu, Monatsh. Chem. *2013, 144* (3), 381-6 [DOI: 10.1007/s00706-012-0821-6]; *syn*-selective coupling of hindered pyruvate esters with sugar-based aldehydes using *cinchona* alkaloids as catalyst, also *anti*-isomers with chiral dinuclear metal complexes, s. O. El-Sepelgy, J. Mlynarski, Adv. Synth. Catal. *2013, 355* (2-3), 281-6 [DOI:

10.1002/adsc.201200914]; general procedure with N-tosyl-(S)-proline N-(*p*-fluorophenyl)amide as catalyst in aq. DMF s. P.B. Thorat, S.V. Goswami, B.C. Khade, S.R. Bhusare, Tetrahedron Lett. *2012, 53* (45), 6083-6 [DOI: 10.1016/j.tetlet.2012.08.123]; with an N-terminal prolyl dipeptide amide·CF_3COOH and added $Zn(OAc)_2$ or NaOAc, effect of dipeptide stereochemistry on absolute configuration and stereoselectivity, s. C. Andreu, M. del Olmo, G. Asensio, Tetrahedron *2012, 68* (38), 7966-72 [DOI: 10.1016/j.tet.2012.07.006]; with highly active sugar-based (S)-prolinamides under solvent-free conditions with high enantio- and diastereo-selectivity s. J. Agarwal, R.K. Peddinti, Eur. J. Org. Chem. *2012* (32), 6390-406 [DOI: 10.1002/ejoc.201200522]; with a chimeric D-glucosamine-based (S)-prolinamide cf. M. De Nisco, S. Pedatella, S. Bektas, A. Nucci, R. Caputo, Carbohydr. Res. *2012, 356*, 273-7 [DOI: 10.1016/j.carres.2012.03.041]; with a readily recoverable, reusable (15 times!) C_2-symmetric bis(imidazolium)-functionalized bis(4-acoxy-(S)-prolinamide) as catalyst in the presence of water (100 eq.) s. S.V. Kochetkov, A.S. Kucherenko, G.V. Kryshtal, G.M. Zhdankina, S.G. Zlotin, Eur. J. Org. Chem. *2012* (36), 7129-34 [DOI: 10.1002/ejoc.201201144]; with a reusable (5 times), water-tolerant, multifunctional, C_3-symmetric dendritic prolinamide possessing a hydrophobic aromatic core and hydrophilic 1,2,3-triazolium arms cf. S. Paladhi, J. Das, P.K. Mishra, J. Dash, Adv. Synth. Catal. *2013, 355* (2-3), 274-80 [DOI: 10.1002/adsc.201200856]; with chiral amino- or hydroxy-functionalized, (S)-proline-derived 2,2-disubst. pyrrolidines s. R.L. Sutar, N.N. Joshi, Tetrahedron: Asym. *2013, 24* (1), 43-9 [DOI: 10.1016/j.tetasy.2012.11.018]; with cinchona-based 2-*tert*-aminothioureas for regioselective aldol condensation of methyl β-diketones (at the methyl carbon!) s. H. Liu, H. Wu, Z. Luo, J. Shen, G. Kang, B. Liu, Z. Wan, J. Jiang, Chem. Eur. J. *2012, 18* (38), 11899-903 [DOI: 10.1002/chem.201201874]; with the same organocatalysts self-assembled with chiral α-amino-acids s. D. Sinha, T. Mandal, S. Gogoi, J.J. Goldman, J.C.-g. Zhao, Chin. J. Chem. *2012, 30* (11), 2624-30 [DOI: 10.1002/cjoc.201200639]; *under enzymatic catalysis* with bovine pancreatic lipase in acidic buffer (cf. *58*, 245s*81*) s. Z.-B. Xie, N. Wang, G.-F. Jiang, X.-Q. Yu, Tetrahedron Lett. *2013, 54* (8), 945-8 [DOI: 10.1016/j.tetlet.2012.12.022]; *under heterogeneous conditions* with α-amino-acids attached to rigid-layered double hydroxide nanosheets for enhancement of enantioselectivity, also application to zinc(II)-catalyzed asym. transformations, s. L.-W. Zhao, H.-M. Shi, J.-Z. Wang, J. He, Chem. Eur. J. *2012, 18* (48), 15323-9 [DOI: 10.1002/chem.201202604]; in aq. chitosan aerogels under heterogeneous enamine catalysis in the presence of surfactants and acid catalysts s. C. Gioia, A. Ricci, L. Bernardi, K. Bourahla, N. Tanchoux, M. Robitzer, F. Quignard, Eur. J. Org. Chem. *2013* (3), 588-94 [DOI: 10.1002/ejoc.201201187]; **asym. cross-aldol condensation** of aldehydes with acetaldehyde in brine using a novel 2(S)-[bis[*p*-(dioctylaminomethyl)phenyl]-(hydroxy)methyl]pyrrolidine as catalyst and an imidazolium-functionalized benzoic acid as co-catalyst (cf. *77*, 297) s. Y. Qiao, Q. Chen, S. Lin, B. Ni, A.D. Headley, J. Org. Chem. *2013, 78* (6), 2693-7 [DOI: 10.1021/jo302442g]; *anti*-selective asym. cross-coupling of aldehydes with 2-(2-chloroethoxy)acetaldehyde using 2(S)-[diphenyl(hydroxy)methyl]pyrrolidine as catalyst s. R.T. Sawant, J. Stevenson, L.R. Odell, P.I. Arvidsson, Tetrahedron: Asym. *2013, 24* (2-3), 134-41 [DOI: 10.1016/j.tetasy.2012.12.004]; *anti*-selective **asym. vinylogous aldol condensation** (cf. *78*, 285s*82*) with 2(5*H*)-furanones using N-(3,5-di-*tert*-butylbenzyl)quininium bromide/N,O-bis-(trimethylsilyl)acetamide and added sodium *p*-methoxyphenoxide as co-catalyst s. A. Claraz, S. Oudeyer, V. Levacher, Adv. Synth. Catal. *2013, 355* (5), 841-6 [DOI: 10.1002/adsc.201201041].

Chiral N,N′-bis[2-phenyl-2-(tosylamino)ethyl]-1,2-diphenylethylenediamine s. under Copper(II) acetate ←

Chiral 2-aminoalcohols or N,N′-bis(o-hydroxybenzyl)-cis-2,5-diaminobicyclo[2.2.2]octanes or N-(o-hydroxybenzylidene)-9(S)-amino-9-deoxy-epi-quinines or 2(S)-[diphenyl-(trimethylsiloxy)methyl]-N-(2-pyridylmethyl)pyrrolidines or chiral 2-[hetero]aryl-Δ²-oxazolines s. under Copper(II) acetate or triflate ←

Chiral 2-(morpholinomethyl)cyclopropyl(diaryl)carbinols or 2,6-bis[2(S)-[diphenyl-(hydroxy)methyl]pyrrolidin-1-ylmethyl]-p-cresol s. under Dimethylzinc ←

2(S)-[Diphenyl(hydroxy)methyl]pyrrolidine ←

2(S)-[Bis[p-(dioctylaminomethyl)phenyl](hydroxy)methyl]pyrrolidine/imidazolium-functionalized benzoic acids ←

Asym. cross-aldol condensation of aldehydes s. *77*, 297s*83* CHO → CH(OH)C-CHO

7-Hydroxypyrrolo[1,2-a]imidazole/sodium dodecyl sulfate ←
Isocupreidine/2,6-dimethoxybenzoic acid ←
Baylis-Hillman reaction CHO → CH(OH)C(=CH_2)CO
update s. *39,* 593s*82*; *in water* with bifunctional 7-hydroxypyrrolo[1,2-*a*]imidazole as catalyst and sodium dodecyl sulfate as surfactant s. J.C. Gomes, M.T. Rodrigues Jr., A. Moyano, F. Coelho, Eur. J. Org. Chem. *2012* (35), 6861-6 [DOI: 10.1002/ejoc.201200950]; **asym. Baylis-Hillman reaction** (cf. *58,* 233s*82*) of acrolein with ar. aldehydes using β-isocupreidine and 2,6-dimethoxybenzoic acid as co-catalyst s. X. Zeng, Y. Liu, C. Ji, J. Zhou, Chin. J. Chem. *2012, 30* (11), 2631-5 [DOI: 10.1002/cjoc.201200937]; of acrylates with aldehydes using a glucose-based 2-*tert*-phosphinothiourea as catalyst s. W. Yang, F. Sha, X. Zhang, K. Yuan, X. Wu, ibid. 2652-6 [DOI: 10.1002/cjoc.201200740]; **asym. Baylis-Hillman equivalent** via isomerizing asym. aldol condensation of N-protected Δ^3-2-pyrrolones with aryl α-ketocarboxylates using a chiral bifunctional thiourea as catalyst s. J. Zhang, X. Liu, X. Ma, R. Wang, Chem. Commun. *2013, 49* (32), 3300-2 [DOI: 10.1039/c3cc40717a].

Copper(II) acetate/chiral N,N′-bis[2-phenyl-2-(tosylamino)ethyl]-1,2-diphenylethylenediamine/camphorsulfonic acid ←
Asym. Henry reaction-intramolecular oxa-Michael addition C

261. 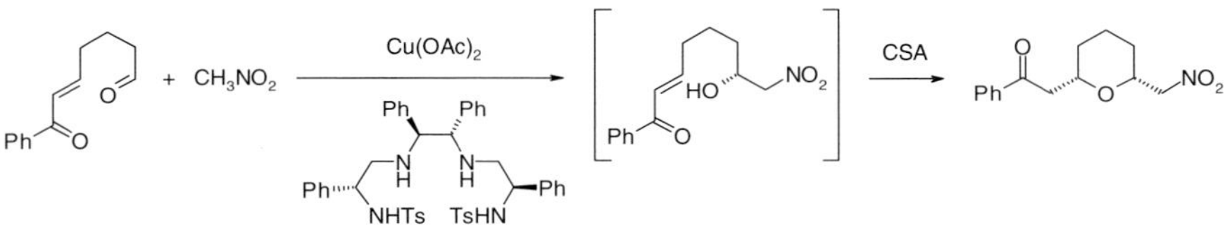

Chiral 2-β-keto-6-(nitromethyl)tetrahydropyrans. A soln. of the chiral N,N′-bis[2-(tosylamino)-ethyl]-1,2-diphenylethylenediamine (5 mol%) and $Cu(OAc)_2 \cdot H_2O$ (5 mol%) in ethanol (0.4 ml) stirred under argon at room temp. for 1 h, (E)-7-oxo-7-phenylhept-5-enal (0.1 mmol) and nitromethane (0.1 ml) added in one portion, stirring continued for 60 h (TLC), treated with camphorsulfonic acid (10 mol%), stirred for a further 12 h, volatiles removed under reduced pressure, and the residue purified by chromatography on silica gel → 2(R)-nitromethyl-6(S)-(2-phenyl-2-oxoethyl)tetrahydropyran. Y 86% (d.r. >99:1; e.e. 98%). The procedure is applicable to a wide range of (E)-7-oxo-7-arylhept-5-enals, irrespective of the electronic character and position of the aryl substituent, as well as aliphatic substrates (fourteen examples in all; Y 84-96%; d.r. >99:1; e.e. 98-99%). The high enantioselectivity originates from the initial Henry reaction, while the high diastereoselectivity is determined by the subsequent acid-catalyzed intramolecular oxa-Michael addition. F.e. and with Ooi's chiral tetraaminophosphonium chloride as precatalyst, also application to the formation of a chiral morpholine analog, s. Q. Dai, N.K. Rana, J.C.-G. Zhao, Org. Lett. *2013, 15* (12), 2922-5 [DOI: 10.1021/ol400900h].

Copper(II) acetate or triflate/chiral 2-aminoalcohols or N,N′-bis(o-hydroxybenzyl)-cis-2,5-diaminobicyclo[2.2.2]octanes or N-(o-hydroxybenzylidene)-9(S)-amino-9-deoxy-epi-quinines or 2(S)-[diphenyl(trimethylsiloxy)methyl]-N-(2-pyridylmethyl)pyrrolidines or chiral 2-[hetero]aryl-Δ^2-oxazolines ←
Chiral polymeric [N,N,N′,N′-tetrakis(Δ^2-oxazolin-2-yl)diamine]copper(II) coordination complexes *[Cu(II)]**
Catalytic asym. Henry [nitroaldol] reaction CO → C(OH)C(NO_2)
update s. *62,* 250s*82*; with ar. aldehydes using $Cu(OAc)_2 \cdot H_2O$ and a chiral 2-*tert*-aminoalcohol as catalyst for high (R)-selectivity s. D.-D. Qin, W.-H. Lai, D. Hu, Z. Chen, A.-A. Wu, Y.-P. Ruan, Z.-H. Zhou, H.-B. Chen, Chem. Eur. J. *2012, 18* (34), 10515-8 [DOI: 10.1002/chem.201201565]; asym. coupling of nitromethane with various aldehydes using $Cu(OTf)_2$ and a chiral, norephedrine-based N-(2-thienylmethyl)-2-aminoalcohol as ligand s. A.E. Aydin, Appl. Organomet. Chem. *2013, 27* (5), 283-9 [DOI: 10.1002/aoc.2969]; general procedure with $Cu(OTf)_2$·toluene and a chiral N,N′-bis(*o*-hydroxybenzyl)-*cis*-2,5-diaminobicyclo[2.2.2]octane as ligand s. J.D. White, S. Shaw, Org. Lett. *2012, 14* (24), 6270-3 [DOI: 10.1021/ol3030023]; with $Cu(OAc)_2$ and a chiral, bulky 2(S)-[diphenyl(trimethylsiloxy)methyl]-N-(2-pyridylmethyl)pyrrolidine as catalyst s. B. Ni, J. He,

Tetrahedron Lett. *2013, 54* (6), 462-5 [DOI: 10.1016/j.tetlet.2012.11.053]; with $Cu(OTf)_2/Na_2CO_3$ and an N-(*o*-hydroxybenzylidene)-9(S)-amino-9-deoxy-*epi*-quinine as catalyst s. L. Yao, Y. Wei, P. Wang, W. He, S. Zhang, Tetrahedron *2012, 68* (44), 9119-24 [DOI: 10.1016/j.tet.2012.08.029]; asym. coupling of *p*-nitrobenzaldehydes with nitromethane using $Cu(OTf)_2$ and a chiral, norephedrine-based 2-[hetero]aryl-Δ^2-oxazoline as ligand s. A.E. Aydin, S. Yuksekdanaci, Tetrahedron: Asym. *2013, 24* (1), 14-22 [DOI: 10.1016/j.tetasy.2012.11.022]; *anti*-selective asym. coupling *in water* or THF s. K. Xu, G. Lai, Z. Zha, S. Pan, H. Chen, Z. Wang, Chem. Eur. J. *2012, 18* (39), 12357-62 [DOI: 10.1002/chem.201201775]; *heterogeneous* coupling with a soluble copper(II) coordination polymer based on a chiral N,N,N′,N′-tetrakis(Δ^2-oxazolin-2-yl)diamine [or a chiral bis[1,1-bis(Δ^2-oxazolin-2-yl)alkane]], recyclable via precipitation at room temperature, s. B. Angulo, J.I. García, C.I. Herrerías, J.A. Mayoral, A.C. Miñana, J. Org. Chem. *2012, 77* (13), 5525-32 [DOI: 10.1021/jo3005974].

Magnesium/lanthanum(III) chloride·lithium chloride — *Mg/LaCl₃·LiCl*
Isopropylmagnesium chloride·lithium chloride — *i-PrMgCl·LiCl*
Chloromagnesium 2,2,6,6-tetramethylpiperidide·boron fluoride — *ClMgNR₂·BF₃*

Grignard synthesis of alcohols CO → C(OH)R

from oxo compds., update s. *1,* 680s*82*; diastereoselective synthesis of *tert-tert*-1,2-diol monosilyl ethers with $LaCl_3{\cdot}2LiCl$ as catalyst s. B.M. Loertscher, P.R. Young, P.R. Evans, S.L. Castle, Org. Lett. *2013, 15* (8), 1930-3 [DOI: 10.1021/ol4005799]; regioselective synthesis of 2-thienyl-, 2-furyl- and 2-pyridyl-carbinols from the corresponding heteroaromatic dibromides via regioselective Br→Mg-exchange with, for example, *i*-PrMgCl·LiCl, s. C. Sämann, B. Haag, P. Knochel, Chem. Eur. J. *2012, 18* (50), 16145-52 [DOI: 10.1002/chem.201202230]; alternative route to heteroarylcarbinols from the heteroarene via generation of chloromagnesium 2-heteroaryl(trifluoro)borates with the frustrated Lewis pair, chloromagnesium 2,2,6,6-tetramethylpiperidide·BF_3, cf. S.M. Manolikakes, M. Jaric, K. Karaghiosoff, P. Knochel, Chem. Commun. *2013, 49* (21), 2124-6 [DOI: 10.1039/c3cc39040c].

Magnesium/titanium tetraisopropoxide/chiral 2′-hydroxy-1,1′-binaphthyl-2-yl(4-pyridyl)-carbinols or 2′-hydroxy-1,1′-binaphthyl-2-yl(2-naphthyl)carbinols ←
Chiral dilithium dibutylmagnesium TADDOLates ←

Asym. Grignard synthesis s. *47,* 589s*83*

cf. *47,* 589s*81*; of sec. dialkylcarbinols from alkylmagnesium bromides (incl. MeMgBr) and aliphatic aldehydes with added $Ti(OPr\text{-}i)_4$ and (S_a,R)-2′-hydroxy-1,1′-binaphthyl-2-yl(4-pyridyl)-carbinol as ligand s. E. Fernández-Mateos, B. Maciá, M. Yus, Adv. Synth. Catal. *2013, 355* (7), 1249-54 [DOI: 10.1002/adsc.201201117]; with the related central-axially chiral 2′-hydroxy-1,1′-binaphthyl-2-yl(2-naphthyl)carbinol and analogous 'Ar-BINMOLs' s. L.-S. Zheng, K.-Z. Jiang, Y. Deng, X.-F. Bai, G. Gao, F.-L. Gu, L.-W. Xu, Eur. J. Org. Chem. *2013* (4), 748-55 [DOI: 10.1002/ejoc.201201301]; asym. synthesis of aryl(pyrazin-2-yl)carbinols from 2-iodopyrazines and ar. aldehydes via I→Mg exchange with ((R,R)-TADDOLate)Bu_2MgLi_2 s. O. Payen, F. Chevallier, F. Mongin, P.C. Gros, Tetrahedron: Asym. *2012, 23* (24), 1678-82 [DOI: 10.1016/j.tetasy.2012.11.003].

Dimethylzinc/chiral 2-(morpholinomethyl)cyclopropyl(diaryl)carbinols or 2,6-bis[2(S)-[diphenyl(hydroxy)methyl]pyrrolidin-1-ylmethyl]-p-cresol ←

Asym. synthesis of 2-acetylenealcohols from oxo compds. CO → C(OH)C≡C

update s. *58,* 236s*82*; general procedure with dimethylzinc and 2,6-bis[2(S)-[diphenyl(hydroxy)-methyl]pyrrolidin-1-ylmethyl]-*p*-cresol as ligand s. B.M. Trost, M.J. Bartlett, A.H. Weiss, A.J. von Wangelin, V.S. Chan, Chem. Eur. J. *2012, 18* (51), 16498-509 [DOI: 10.1002/chem.201202085]; chiral 2,4-diynols from terminal 1,3-diynes with dimethylzinc and a chiral 2-(morpholinomethyl)cyclopropyl(diaryl)carbinol as ligand s. B. Zheng, S.-N. Li, J.-Y. Mao, B. Wang, Q.-H. Bian, S.-Z. Liu, J.-C. Zhong, H.-C. Guo, M. Wang, ibid. *18* (30), 9208-11 [DOI: 10.1002/chem.201200728]; asym. synthesis of β,γ-acetylene-α-hydroxy-α-(trifluoromethyl)carboxylic acid esters with a chiral pincer-type diacetoxo(aqua)[2,6-bis(Δ^2-imidazolin-2-yl)phenyl]-rhodium(III) complex s. T. Wang, J.-L. Niu, S.-L. Liu, J.-J. Huang, J.-F. Gong, M.-P. Song, Adv. Synth. Catal. *2013, 355* (5), 927-37 [DOI: 10.1002/adsc.201200967].

Diethylzinc s. under Nickel(II) acetoacetonate — *Et_2Zn*
Magnesium bromide s. under $NiCl_2$(binap) — *$MgBr_2$*

Indium/lithium chloride or sodium iodide/ferrocene *In/LiCl or NaI/Cp_2Fe*
2,2,2-Trifluoroalcohols from aldehydes s. *83*, 269 CHO → $CH(OH)CF_3$

Lithium hydridotriethylborate s. under $NiCl_2$(binap) *$LiBHEt_3$*

Sodium triisopropoxy(phenyl)borate *Na[PhB(OPr-i)$_3$]*
Aldol condensation s. *43,* 710s*83* CO → C(OH)C-CO

(+)- or (-)-Diisopinocampheylchloroborane/triethylamine *Ipc_2BCl/Et_3N*
***anti*-1,3-Diols from aldehydes and ketones** CHO → CH(OH)C-CHOH
via stereoselective boron-mediated reductive aldol condensation

262. 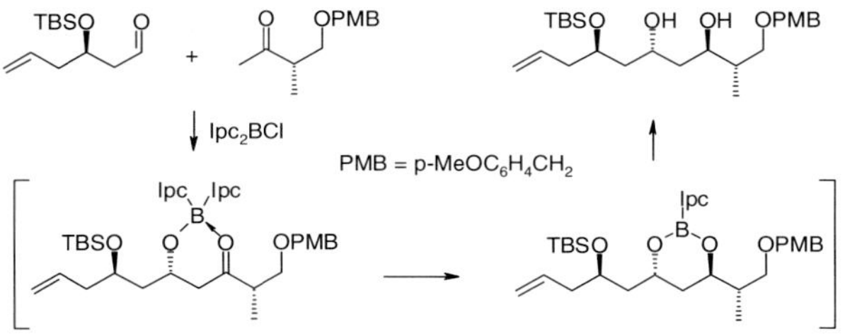

A soln. of startg. methyl ketone (1.3 eq.) in ether (0.6 ml) at 0° under argon treated with dry Et_3N (3 eq.) and a soln. of (+)-Ipc_2BCl (2.6 eq.) in ether (0.4 ml), the white mixture stirred for 2 h at 0° then cooled to -78°, a soln. of startg. aldehyde (127 µmol) in ether (0.6 ml) added slowly, the mixture stirred for 1 h at -78°, allowed to warm slowly to room temp. overnight (16-18 h, room temp. attained after ca. 4-5 h), quenched with aq. buffer (pH 7) and brine, the aq. layer extracted with methylene chloride, the combined organic layers concentrated *in vacuo*, the resulting crude cyclic boronate taken up in THF (1.5 ml), aq. NaOH (3 *M*; 2 ml) and aq. H_2O_2 (35%; 1 ml) added at 0°, the mixture stirred at room temp. for 30-120 min (TLC), aq. NH_4Cl soln. and brine added at 0°, the mixture extracted with ether, and purified by flash chromatography on silica gel → 1,3-*anti*-product. Y 85% (d.r. to other *anti*-isomer 10:1; *anti/syn* >19:1). The method is operationally simple and allows the rapid construction of complex polyketide fragments. In all cases 1,3-*anti*-diols were obtained with very high selectivities (*anti/syn* >19:1; Y 42-95%; eleven examples), independently of the Ipc-isomer used, even with sterically less hindered aldehydes. F.e.s. M. Dieckmann, D. Menche, Org. Lett. *2013*, *15* (1), 228-31 [DOI: 10.1021/ol3033303].

Samarium/tetra-n-butylammonium iodide/electrolysis *Sm/Bu_4NI/⚡*
Syntheses with electrochemically-generated samarium diiodide CO → C(OH)C(OH)

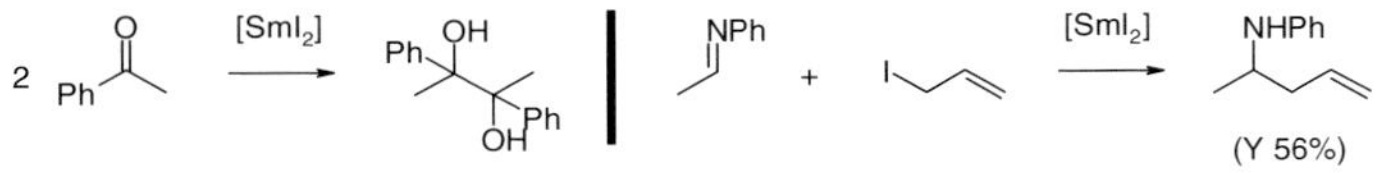

263. 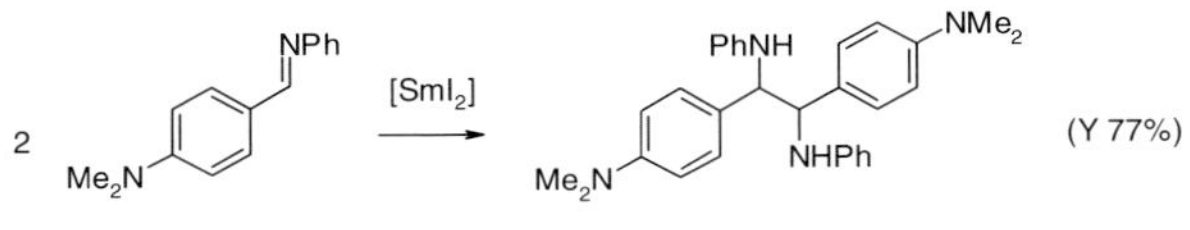

The first electrochemical preparation of SmI_2 and its use in homo-coupling and Barbier allylation reactions are reported. **E: Pinacols from oxo compds.** A soln. of acetophenone (4 mmol), *n*-Bu_4NI (2.5 eq.) and *n*-Bu_4NPF_6 (0.25 eq.) in THF (50 ml) introduced into a one-compartment cell fitted with a Sm anode and Pt grid cathode under argon, the soln. electrolyzed at 50 mA until substrate consumed (TLC/GC; 4.5 h (2 Fmol^{-1})), solvent evaporated, the residue quenched with 1 *M* aq. HCl, extracted with ether, and purified by flash chromatography → 2,3-diphenylbutane-2,3-diol.

Y 95% (DL/*meso* 80:20). Electrogeneration of SmI_2 (characterized by electrochemistry and UV/Vis spectroscopy) in the presence of aldehydes, ketones or imines afforded the corresponding sym. glycols or **1,2-diamines**, respectively, with moderate diastereoselectivity for aromatic substrates (DL/*meso* 52:48 to 80:20; eight examples; Y 42-96%) and poor diastereoselectivity (43:57 to 50:50; two examples; Y 82%) for more challenging aliphatic substrates (octan-2-one and its N-phenylimine). Barbier allylation of analogous substrates gave **3-ethyleneamines** only (five examples; Y 31-80%). Interestingly, the reactions are essentially self-indicating since on consumption of substrate a blue-coloration persists due to accumulation of an SmI_2 complex. F.e.s. K. Sahloul, L. Sun, A. Requet, Y. Chahine, M. Mellah, Chem. Eur. J. *2012*, *18* (36), 11205-9 [DOI: 10.1002/chem.201201390].

Carbon nanotube-confined neodymium(III) isopropoxide/sodium bis(trimethylsilyl)amide/ chiral α-[N-(m-hydroxybenzoyl)amino]carboxylic acid N-(o-hydroxyaryl)amides ←
Enhancement of catalytic activity of chiral, self-assembled heterobimetallic catalysts by confinement in carbon nanotubes ←

264. 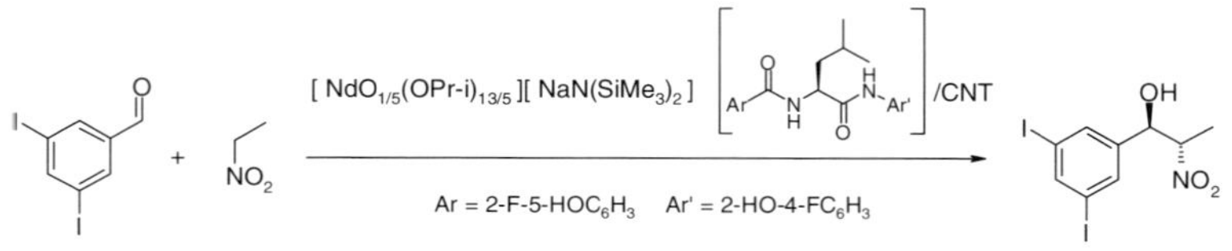

In the *anti*-selective **asym. Henry reaction**, Shibasaki-type self-assembled chiral heterobimetallic catalysts afford significantly higher yields on confinement within carbon nanotubes than not, and catalyst recovery for reuse is much simplified. **E:** THF (2.8 ml) and nitroethane (10 eq.) added via syringe under argon at room temp. to the heterobimetallic catalyst (0.5 mol%) [prepared as a suspension by mixing the multi-walled carbon nanotube, Baytubes, with a 2:1:2 catalytic mixture of chiral α-[N-(*m*-hydroxybenzoyl)amino]carboxylic acid N-(*o*-hydroxyaryl)amide, $NdO_{1/5}(OPr\text{-}i)_{13/5}$ and Na-bis(trimethylsilyl)amide in the presence of a little nitroethane] in THF (0.16 ml), the resulting black suspension cooled to -60°, a soln. of 3,5-diiodobenzaldehyde (0.4 mmol) in THF (1 ml) added dropwise via syringe over 1 min, stirring continued for 64 h under argon, quenched with a 0.2 *M* THF soln. of acetic acid, stirred at -60° for a further 1 h, warmed to room temp., 1 *N* aq. HCl added, the resulting biphasic mixture filtered through a Celite pad under reduced pressure, and the filtrate worked up with purification by chromatography on silica gel → product. Y 98% (*anti/syn* 96:4; e.e. of the *anti*-isomer 95%). At 1 mol% catalyst loading, the yield was 99% after 20 h (*anti/syn* 98:2; e.e. *anti*-isomer 99%), while the yield was only 24% with the unconfined catalyst under the same conditions (although the *anti/syn* selectivity and enantioselectivity were effectively the same). The high activity of the catalyst is presumed to be due to its self-assembly in the *fibrous* nanotube network which leads to the formation of small (20-200 nm), confined cluster complexes with a high specific surface area (five examples; Y 52-99%; *anti/syn* 83:17 to 98:2; e.e. 87-99%). The catalyst was simply recovered by filtration and retained its activity after six cycles. F.e.s. T. Ogawa, N. Kumagai, M. Shibasaki, Angew. Chem., Int. Ed. *2013, 52* (24), 6196-201 [DOI: 10.1002/anie.201302236].

Scandium(III) triflate *Sc(OTf)₃*
Carbonyl-ene reaction C(OH)C-C=C
with Et_2AlCl cf. *36*, 665; with $Sc(OTf)_3$ as catalyst for preparing 3-ethylene-*prim*-alcohols s. S. Sultana, S. Bondalapati, K. Indukuri, P. Gogoi, P. Saha, A.K. Saikia, Tetrahedron Lett. *2013, 54* (12), 1576-8 [DOI: 10.1016/j.tetlet.2013.01.046].

Ytterbium(III) triflate *$Yb(OTf)_3$*
1,2-Addition to oxo compds. CO → $C(OH)CH_2Ar$
of active methylene groups to perfluoroacetone with pyridine cf. *17*, 738; **of *o*-methyl-N-hetero-arenes** to trifluoromethyl ketones with $Yb(OTf)_3$ as Lewis acid s. V.B. Graves, A. Shaikh, Tetrahedron Lett. *2013*, *54* (7), 695-8 [DOI: 10.1016/j.tetlet.2012.12.013]; to aldehydes with acetic acid (1 eq.) as catalyst s. F.-F. Wang, C.-P. Luo, Y Wang, G. Deng, L. Yang, Org. Biomol. Chem. *2012, 10* (43), 8605-8 [DOI: 10.1039/c2ob26604k]; 1,2-addition of lutidines or picolines

to isatins with I_2 as catalyst s. S.V.N. Vuppalapati, Y.R. Lee, Tetrahedron *2012, 68* (39), 8286-92 [DOI: 10.1016/j.tet.2012.07.051].

Lanthanum(III) chloride·lithium chloride s. under Mg *$LaCl_3·LiCl$*
Chiral quaternary ammonium salts or (S)-prolinamides s. under Chiral diamines ←
cis-3,3′-Bis[hydroxy(diphenyl)methyl]azobenzenes s. under Tri-n-butylphosphine ←

Chiral N-condensed 1,2,4-triazolylidene/ethyldiisopropylamine *NHC*/i-Pr_2NEt*
α-Hydroxy-β-ketocarboxylic acid esters CHO → C(O)C(OH)COOR
from α-ketocarboxylic acid esters and [aliphatic] aldehydes
via organocatalyzed asym. cross-benzoin reaction

265. 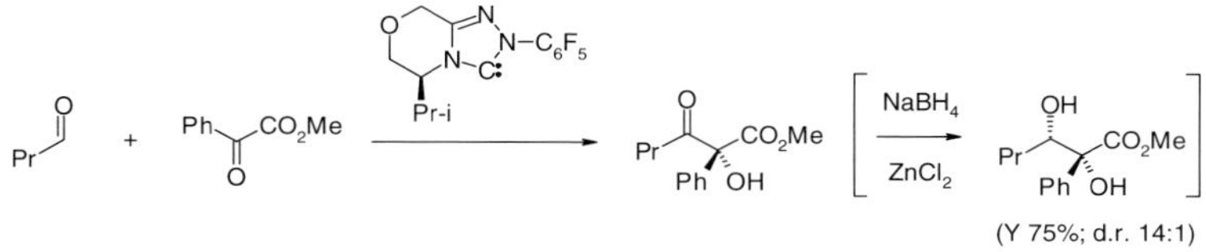

under mild conditions. Butanal (1.5 eq.) added to a suspension of methyl benzoylformate (1 eq.), (S)-5-isopropyl-2-pentafluorophenyl-6,8-dihydro-5*H*-[1,2,4]triazolo[3,4-*c*]oxazin-2-ium tetrafluoroborate (NHC precursor; 10 mol%), and powdered 4 Å molecular sieves (1:1 w/w with keto-ester) in anh. methylene chloride (0.2 *M*) under inert atmosphere at room temp., the mixture stirred for 10 min, ethyldiisopropylamine (1 eq.) added, the mixture stirred until reaction complete (TLC), quenched with acetic acid, filtered through silica, concentrated *in vacuo*, and purified chromatographically → (-)-(R)-methyl 2-hydroxy-3-oxo-2-phenylhexanoate. Y 92% (e.e. 91%). This challenging coupling of aliphatic aldehydes, promoted by NHC-catalysis, was optimal using methyl arylglyoxylate acceptors (cf. ethyl and *tert*-butyl esters) (eleven examples; Y 43-98%; e.e. 85-94%; an α-pyrid-3-yl-α-keto-ester gave 77%). Lowest yields were obtained with a β-branched aldehyde (α-branched analogs were unreactive) or sterically-hindered α-naphth-2-yl- or electron-rich -*p*-tolyl-keto-esters, while alkyl subst. keto-esters were unreactive. F.e., optimization, catalyst and substrate prepn., and diastereoselective reduction of products ($NaBH_4$/$ZnCl_2$) to **chiral *syn*-α,β-dihydroxycarboxylic acid esters** (two examples; Y 69-75%; d.r. 14:1 to >20:1) s. K. Thai, S.M. Langdon, F. Bilodeau, M. Gravel, Org. Lett. *2013*, *15* (9), 2214-7 [DOI: 10.1021/ol400769t].

Biguanides ←
Henry reaction CO → C(OH)C(NO_2)
update s. *34,* 610s*82*; rapid procedure with a readily recoverable, *superbasic* biguanide as catalyst, namely metformin, for the solvent-free reaction with aliphatic or aromatic aldehydes s. A. Alizadeh, M.M. Khodaei, G. Abdi, D. Kordestani, Bull. Korean Chem. Soc. *2012, 33* (11), 3640-4 [DOI: 10.5012/bkcs.2012.33.11.3640]; *heterogeneous* conversion with bifunctional mesoporous, inorganic/organic hybrid silica nanoparticles bearing acidic and basic sites s. A.T. Dickschat, F. Behrends, M. Bühner, J. Ren, M. Weiß, H. Eckert, A. Studer, Chem. Eur. J. *2012, 18* (52), 16689-97 [DOI: 10.1002/chem.201200499]; with recyclable, imidazolium-tethered functionalized mesoporous silica [SBA-15] s. B. Motos-Pérez, J. Roeser, A. Thomas, P. Hesemann, Appl. Organomet. Chem. *2013, 27* (5), 290-9 [DOI: 10.1002/aoc.2974]; with a recyclable, polymer-based ethyl 3-(triaryl-phosphonio)propionate betaine under solvent-free conditions s. L. Rokhum, G. Bez, Can. J. Chem. *2013, 91* (4), 300-6 [DOI: 10.1139/cjc-2012-0164]; with porous and robust zinc(II) metal-organic frameworks with coordinatively unsatd. metal sites s. X.-M. Lin, T.-T. Li, Y.-W. Wang, L. Zhang, C.-Y. Su, Chem. Asian J. *2012, 7* (12), 2796-804 [DOI: 10.1002/asia.201200601].

Cinchonine-based squaramides ←
Organocatalyzed asym. Friedel-Crafts α-hydroxyalkylation s. *60,* 227s*83* H → CH(OH)R

Enzymes s. under Chiral diamines ←
Chiral 2′-hydroxy-1,1′-binaphthyl-2-yl(4-pyridyl)carbinols or 2′-hydroxy-1,1′-binaphthyl-2-yl(2-naphthyl)carbinols s. under Mg ←
Chiral α-[N-(m-hydroxybenzoyl)amino]carboxylic acid N-(o-hydroxyaryl)amides s. under Carbon nanotube-confined neodymium(III) isopropoxide ←
Cinchona-based salicylaldimines/ethyl cyanoformate s. under Titanium tetraisopropoxide ←

Acetic acid *AcOH*
1,2-Addition of *o*-methyl-N-heteroarenes to aldehydes s. *17*, 738s*83* CO → C(OH)CH_2Ar

Benzoic acid s. under 1-((1R,2R)-2-Amino-1,2-diphenylethyl)-3-[3,5-bis(trifluoromethyl)phenyl]thiourea *PhCOOH*
2,6-Dimethoxybenzoic acid s. under Isocupreidine *ArCOOH*
Chiral supported α-aminocarboxylic acids s. under Chiral diamines ←

Helically chiral poly(α-aminocarboxylic acids) ←
Asym. synthesis of α-siloxynitriles from oxo compds. CO → C(OSi≤)CN
update s. *43*, 576s*82*; with *helically* chiral poly(amino-acids) s. K. Akagawa, K. Kudo, Tetrahedron Lett. *2012, 53* (45), 5981-3 [DOI: 10.1016/j.tetlet.2012.07.034]; **asym. synthesis of cyanohydrin carbonates from aldehydes** with ethyl cyanoformate using Ti(OPr-*i*)$_4$ and a bifunctional cinchona-based salicylaldimine as ligand (cf. *63*, 253s*78*) s. N. Ji, L. Yao, W. He, Y. Li, Appl. Organomet. Chem. *2013, 27* (4), 209-13 [DOI: 10.1002/aoc.2951].

Chiral bifunctional thioureas ←
Asym. Baylis-Hillman equivalent CHO → CH(OH)C(=C)CO
via isomerizing asym. aldol condensation s. *58*, 233s*83*

Chiral 2-tert-aminothioureas ←
(E)-α,β-Ethylene-γ-(3-hydroxyoxindol-3-yl)ketones from isatins and β,γ-ethyleneketones CO → C(OH)C-C=C-CO
Organocatalyzed asym. vinylogous aldol condensation

266. 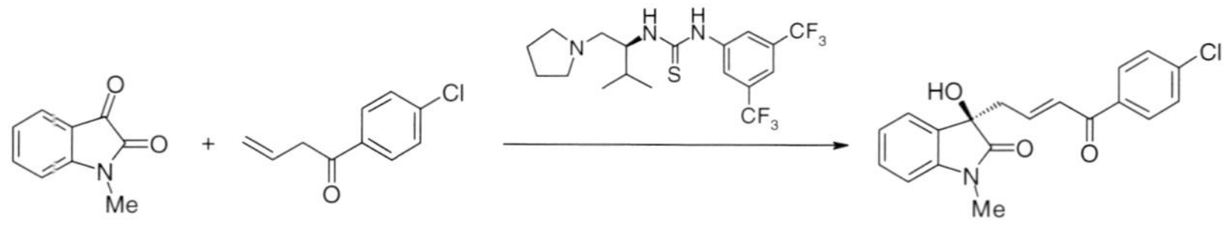

The first examples of *direct* vinylogous aldol condensation *with β,γ-ethyleneketones* is reported, showing advantages over use of the corresponding, traditional Mukaiyama-type (1,3-dien)oxysilanes which suffer from low atom-economy and efficiency. **E:** The startg. ketone (1.5 eq.) and chiral 2-*tert*-aminothiourea (10 mol%) in ether (1 ml) stirred at -20° for 5 min, N-methylisatin (0.1 mmol) added, stirring continued at the same temp. for 5 h, and the mixture directly loaded onto a short silica gel column → product. Y 93% (e.e. 96%; E/Z >99:1). The procedure is simple, direct, mild, applicable on the gram scale at low catalyst loading (down to 1 mol%), and suitable for coupling both N-subst. and [more efficiently] N-unsubst. isatins with a range of β,γ-ethyleneketones (twenty-six examples; Y 57-98%; e.e. 88 to >99%; E/Z 13:1 to >99:1). DFT calculations were in keeping with the observed (R)-selectivity and predominant E-selectivity, arising from favorable π-π*-interactions as well as (a) hydrogen bonding of the electrophile to the two N-H groups of the thiourea and (b) binding of the enolate to the pyrrolidinium arm. F.e. and comparison of organocatalysts and solvents s. B. Zhu, W. Zhang, R. Lee, Z. Han, W. Yang, D. Tan, K.-W. Huang, Z. Jiang, Angew. Chem., Int. Ed. *2013, 52* (26), 6666-70 [DOI: 10.1002/anie.201302274].

1-((1R,2R)-2-Amino-1,2-diphenylethyl)-3-[3,5-bis(trifluoromethyl)phenyl]thiourea/ benzoic acid ←

Organocatalyzed asym. vinylogous aldol reaction with α-ketocarboxylic acid esters H → C(OH)COOR

267. 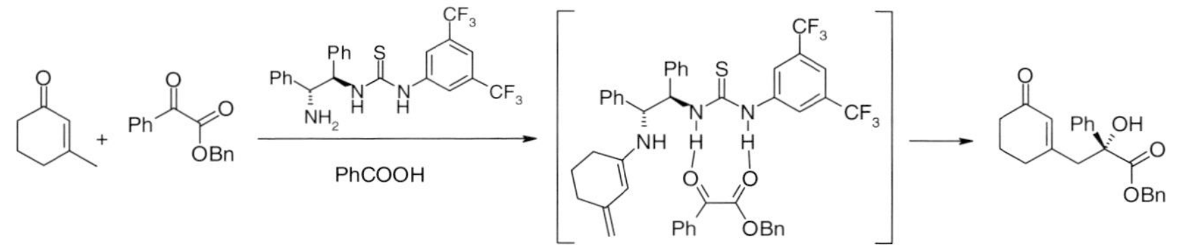

The direct enantioselective vinylogous aldol reaction of 3-methyl-2-cyclohexen-1-one with α-keto-esters is reported by using a chiral bifunctional *prim*-aminothiourea catalyst that can combine *hydrogen-bond-directing activation* of the electrophilic substrate *and dienamine activation* of the enone donor. **E:** A vial equipped with a screw cap charged with 1-((1R,2R)-2-amino-1,2-di-phenylethyl)-3-[3,5-bis(trifluoromethyl)phenyl]thiourea (20 mol%) without exclusion of air or moisture, benzoic acid (50 mol%) and toluene (synthesis grade, >99%; 0.2 ml) added, the soln. stirred at room temp. for 10 min, 3-methylcyclohex-2-enone (5 eq.) and startg. keto-ester (0.1 mmol) added sequentially, the vial sealed, the mixture stirred at 4° for 48 h, diluted with ether, quenched with satd. aq. $NaHCO_3$, the organic phase dried (Na_2SO_4), filtered, solvent removed under reduced pressure, and the residue purified by flash column chromatography → (R)-benzyl 2-hydroxy-3-(3-oxocyclohex-1-en-1-yl)-2-phenylpropanoate. Y 72% (e.e. 92%). Nineteen further examples gave yields of 32%, 65-89% and e.e. of 46%, 79-94%. The method failed with 3-methyl-2-cyclo-pentenone, however. It is believed, in addition to hydrogen bonding involving the urea moiety, the *ortho* proton of the 3,5-bis(trifluoromethyl)phenyl group participates as a hydrogen bond donor. F.e.s. D. Bastida, Y. Liu, X. Tian, E. Escudero-Adán, P. Melchiorre, Org. Lett. *2013, 15* (1), 220-3 [DOI: 10.1021/ol303312p].

Cinchona-based 2-tert-aminothioureas or chiral polyfunctionalized thioureas s. under Chiral diamines ←

Bifunctional mesoporous silica nanoparticles ←
Imidazolium-tethered mesoporous silica ←
Heterogeneous Henry reaction s. *34,* 610s*83* CO → C(OH)C(NO_2)

N,O-Bis(trimethylsilyl)acetamide s. under N-(3,5-Di-tert-butylbenzyl)quininium bromide *BSA*
Titanium tetraisopropoxide s.a. under Mg *Ti(OPr-i)₄*

Titanium tetraisopropoxide/cinchona-based salicylaldimines/ethyl cyanoformate ←
Asym. synthesis of cyanohydrin carbonates from aldehydes CHO → CH(OCOOR)CN
s. *63,* 253s*83*

Chlorobis(cyclopentadienyl)titanium(III) *Cp_2TiCl*
Barbier synthesis of 3-acetylenealcohols from oxo compds. CO → C(OH)C-C≡C
and propargyl halides with Cp_2TiCl s. *40,* 567s*76*; also 2-allenealcohols from aldehydes, s. J. Muñoz-Bascón, I. Sancho-Sanz, E. Álvarez-Manzaneda, A. Rosales, J.E. Oltra, Chem. Eur. J. *2012, 18* (45), 14479-86 [DOI: 10.1002/chem.201201720].

Trichlorotitanium isopropoxide/ethyldiisopropylamine *$Cl_3Ti(OPr-i)/i-Pr_2NEt$*
Silicon tetrachloride/(S)-2,2′-bis(diphenylphosphinyl)-1,1′-binaphthyl/N-methyl-dicyclohexylamine ←
Silicon tetrachloride/N-(diphenylphosphinyl)-2(S)-(diphenylphosphinylaminomethyl)-pyrrolidine ←
Titanium tetrachloride *$TiCl_4$*
Aldol condensation CO → C(OH)C-CO
update s. *43,* 710s*82*; with chiral methyl α-benzyloxyketones using trichlorotitanium isopropoxide/ ethyldiisopropylamine with *anti*-1,4-asym. induction s. M. Pellicena, J.G. Solsona, P. Romea,

F. Urpí, Tetrahedron *2012, 68* (50), 10338-50 [DOI: 10.1016/j.tet.2012.09.096]; with chiral isopropyl ketones using $TiCl_4$/ethyldiisopropylamine with *syn*-1,4-asym. induction s. J. Zambrana, P. Romea, F. Urpí, Chem. Commun. *2013, 49* (40), 4507-9 [DOI: 10.1039/c3cc41640b]; aldol condensation with chiral N-acyl-3,4,5,6-tetrahydro-2(1*H*)-pyrimidinones using $TiCl_4$ or Li-bis(trimethylsilyl)amide for 1,2-*syn*- or 1,2-*anti*-asym. induction, also base-mediated asym. α-alkylation, s. M. Chouhan, R. Sharma, V.A. Nair, Org. Lett. *2012, 14* (22), 5672-5 [DOI: 10.1021/ol302653g]; aldol condensation *in water* with (S)-proline-functionalized β-cyclodextrin s. E.G. Doyagüez, A. Fernández-Mayoralas, Tetrahedron *2012, 68* (36), 7345-54 [DOI: 10.1016/j.tet.2012.06.089]; enhancement of activity of supported prim. amines over sec. amines in water, and reversal of the effect in hexane, s. K. Kandel, S.M. Althaus, C. Peeraphatdit, T. Kobayashi, B.G. Trewyn, M. Pruski, I.I. Slowing, ACS Catal. *2013, 3* (2), 265-71 [DOI: 10.1021/cs300748g]; *uncatalyzed* diastereoselective aldol condensation of 2,4-thiazolidinediones with [hetero]aryl aldehydes 'on water' or in DMSO s. S. Paladhi, A. Chauhan, K. Dhara, A.K. Tiwari, J. Dash, Green Chem. *2012, 14* (11), 2990-5 [DOI: 10.1039/c2gc35819k]; *anti*-selective condensation of erythrulose derivs. with chiral aldehydes by the boron enolate procedure using dicyclohexylborinyl chloride/Et_3N (cf. *43,* 710s*81*), also with double diastereoselection, s. S. Díaz-Oltra, P. Ruiz, E. Falomir, J. Murga, M. Carda, J.A. Marco, Org. Biomol. Chem. *2012, 10* (34), 6937-44 [DOI: 10.1039/c2ob25803j]; *syn*- or *anti*-selective α-condensation with β,γ-ethylene- or α,β-ethylene-carboxylic acid esters via (Z)- or (E)-1-alkoxy-(1,3-dien)olborinates s. P.V. Ramachandran, D. Nicponski, B. Kim, Org. Lett. *2013, 15* (6), 1398-401 [DOI: 10.1021/ol400381q]; aldol condensation mediated by sodium triisopropoxy-(phenyl)borate to minimize eliminative side-reactions s. T. Müller, K. Djanashvili, I.W.C.E. Arends, J.A. Peters, U. Hanefeld, Chem. Commun. *2013, 49* (4), 361-3 [DOI: 10.1039/c2cc37047f]; *anti*- or *syn*-selective **asym. aldol condensation** with phenylacetic acid esters via the boron enolate procedure with diisopinocampheylborinyl triflate/triethylamine (cf. *43,* 568s*46*) s. P.V. Ramachandran, P.B. Chanda, ibid. *49* (30), 3152-4 [DOI: 10.1039/c3cc40860d]; **Lewis acid/base-catalyzed asym. aldol condensation** with 2,2,2-trifluoroethyl thiolocarboxylates using $SiCl_4$ as Lewis acid and N-(diphenylphosphinyl)-2(S)-(diphenylphosphinylaminomethyl)pyrrolidine as Lewis base (with moderate enantioselectivity) s. M. Bonsignore, M. Benaglia, F. Cozzi, A. Genoni, S. Rossi, L. Raimondi, Tetrahedron *2012, 68* (39), 8251-5 [DOI: 10.1016/j.tet. 2012.07.062]; β,β′-dihydroxyketones from alkyl methyl ketones and aldehydes by asym. *double* aldol condensation using $SiCl_4$/N-methyldicyclohexylamine and (S)-2,2′-bis(diphenylphosphinyl)-1,1′-binaphthyl as Lewis base s. Y. Shimoda, T. Kubo, M. Sugiura, S. Kotani, M. Nakajima, Angew. Chem., Int. Ed. *2013, 52* (12), 3461-4 [DOI: 10.1002/anie.201209848].

Polymer-based ethyl 3-(triarylphosphonio)propionate betaine ←
Heterogeneous Henry reaction s. *34,* 610s*83* CO → C(OH)C(NO_2)

Tri-n-butylphosphine/cis-3,3′-bis[hydroxy(diphenyl)methyl]azobenzenes ←
Baylis-Hillman reaction CHO → CH(OH)C(=C)CO
with photoswitchable cooperative catalysts

268.

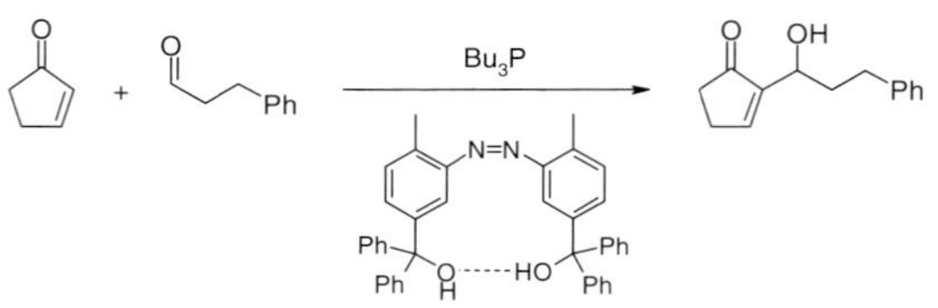

While *trans*-3,3′-bis[hydroxy(diphenyl)methyl]azobenzenes exert only a trivial accelerating effect as cooperative catalysts in the classical phosphine-mediated Baylis-Hillman reaction, the corresponding *cis*-isomers, obtained by UV irradiation of the *trans*-isomers, exhibit a pronounced enhancement with a significant increase in yield. **E:** 2-Cyclopenten-1-one (0.5 mmol), 3-phenylpropanal (1.5 eq.) and tri-*n*-butylphosphine (20 mol%) added at 20° under argon to a soln. of *cis*-3,3′-bis[hydroxy(diphenyl)methyl]-6,6′-dimethylazobenzene (5 mol%; *cis/trans* 100:0) in THF

(0.75 ml), the mixture stirred at the same temp. for 2 h, and worked up with purification by chromatography on silica gel → product. Y 52.3% (25.8% acceleration relative to the same reaction without the azobenzene). With *trans*-3,3′-bis[hydroxy(diphenyl)methyl]-6,6′-dimethylazobenzene, however, the yield was 26.8% under the same conditions with an acceleration of only 0.3%! The enhanced cooperative effect of the *cis*-isomer is attributed to the *closer proximity of the two hydroxyl groups* by which their acidity is increased through hydrogen bonding. The yield was also superior to that recorded with previously reported diol-type cooperative catalysts, e.g. (S)-BINOL. Interestingly, the *cis*-azobenzene can be switched back to the *trans*-isomer under visible light irradiation (>420 nm) or thermally. F.e. and comparison of related azobenzene-type cooperative catalysts s. T. Imahori, R. Yamaguchi, S. Kurihara, Chem. Eur. J. *2012*, *18* (35), 10802-7 [DOI: 10.1002/chem.201201383].

Dicyclohexyl(phenyl)phosphine s. under Nickel(II) acetoacetonate *Cy_2PPh*

Chiral 2-tert-phosphinothioureas ←
Asym. Baylis-Hillman reaction s. *58,* 233s*83* CHO → CH(OH)C(=CH_2)CO

(S)-2,2′-Bis(diphenylphosphinyl)-1,1′-binaphthyl s. under Silicon tetrachloride ←
N-(Diphenylphosphinyl)-2(S)-(diphenylphosphinylaminomethyl)pyrrolidine s. under Silicon tetrachloride ←

Phosphazene base [t-Bu-P4] ←
Metal-free generation of trifluoromethyl carbanion from fluoroform using a sterically-demanding organo-superbase ←

269.

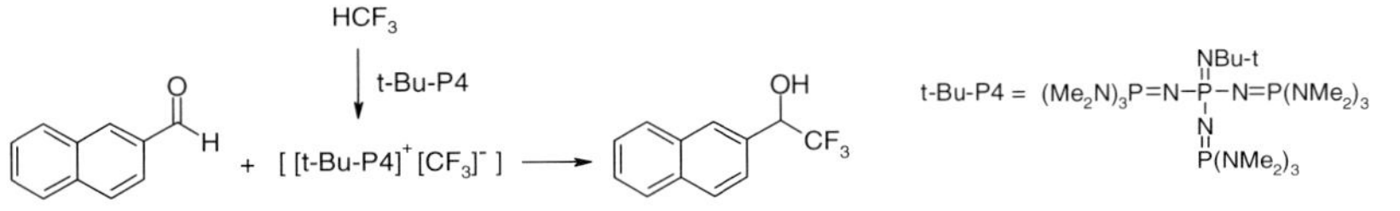

The facile decomposition of 'naked' trifluoromethyl carbanion to difluorocarbene has been avoided in the past by using copper(I) to generate a more manageable trifluoromethylcopper, or in the presence of DMF which converts the carbanion into stable 1-(dimethylamino)-2,2,2-trifluoroethoxide ion prior to reaction. An alternative metal- and DMF-free procedure has now been devised by using a sterically-demanding phosphazene as superbase *in THF* which effectively stabilizes the carbanion in the form of an ion pair. The methodology is illustrated in an alternative route to **2,2,2-trifluoroalcohols from oxo compds. E:** A Schlenk tube containing 2-naphthaldehyde (0.1 mmol) in THF (0.5 ml) charged with excess fluoroform by cooling in liquid nitrogen under vacuum, *t*-Bu-P4 phosphazene base (1.5 eq. as a 1 *M* soln. in hexane) added at -30°, fluoroform bubbled into the mixture for 1 min at the same temp., stirred for 2 h, quenched with satd. aq. NH_4Cl, and worked up with purification by chromatography on silica gel → 2,2,2-trifluoro-1-(2-naphthyl)ethanol. Y 88%. Other organobases, e.g. DBU and guanidine derivs., and other, less hindered phosphazene bases were ineffective. The procedure is applicable to electron-diverse aromatic aldehydes and ketones, incl. bulky anthracenyl derivs. and heteroaryl analogs, as well as α,β-unsatd. ketones (thirteen examples in all; Y 52-92%), but reaction failed with aliphatic analogs. The proportion of non-toxic, ozone-friendly and readily-available fluoroform can be reduced to 5 eq. without significant reduction in yield. The formation of a series of intermediate ion pairs during reaction is discussed. F.e., also trifluoromethyl thioethers from disulfides (one example; Y 74%), s. H. Kawai, Z. Yuan, E. Tokunaga, N. Shibata, Org. Biomol. Chem. *2013*, *11* (9), 1446-50 [DOI: 10.1039/c3ob27368g]; direct formation of *trifluoromethylindium* from trifluoromethyl iodide with In, LiCl or NaI and ferrocene and application in nucleophilic trifluoromethylation of aldehydes, also perfluoroalkyl analogs, s. M. Fujiu, Y. Nakamura, H. Serizawa, K. Aikawa, S. Ito, K. Mikami, Eur. J. Org. Chem. *2012* (36), 7043-7 [DOI: 10.1002/ejoc.201201116].

Camphorsulfonic acid s. under Copper(II) acetate *RSO_3H*
Sodium dodecyl sulfate s. under 7-Hydroxypyrrolo[1,2-a]imidazole *$NaOSO_2OC_{12}H_{25}$*

Iodine (s.a. under 1-Butyl-3-methylimidazolium hydroxide) *I_2*
1,2-Addition of *o*-methyl-N-heteroarenes to isatins CO → C(OH)CH_2Ar
s. *17*, 738s*83*

Tetra-n-butylammonium chloride s. under KOBu-t *Bu_4NCl*
Tetra-n-butylammonium iodide s. under Sm *Bu_4NI*

N-(3,5-Di-tert-butylbenzyl)quininium bromide/N,O-bis(trimethylsilyl)acetamide/ sodium p-methoxyphenoxide ←
Asym. vinylogous aldol condensation s. *78*, 285s*83* ←

Ferrocene s. under In *Cp_2Fe*

Nickel(II) acetoacetonate/dicyclohexyl(phenyl)phosphine/diethylzinc *[Ni(II)]/Et_2Zn*
3-Ethylene-3,1-fluorohydrins from aldehydes CH(OH)C-C(F)═C-CH
and 3-ethylene-2,2-difluorohalides – Regio- and stereo-selective synthesis s. *83*, 413

Dichloro[2,2′-bis(diphenylphosphino)-1,1′-binaphthyl]nickel(II)/ lithium triethylhydridoborate/magnesium bromide *$NiCl_2$(binap)/$LiBHEt_3$/$MgBr_2$*
β-Hydroxyacylsilanes CHO → CH(OH)C(CH)(O)Si⋞
from 2-ethylene-1,1-hydroxysilanes and aldehydes
via stereoselective nickel-catalyzed isomerizing aldol condensation

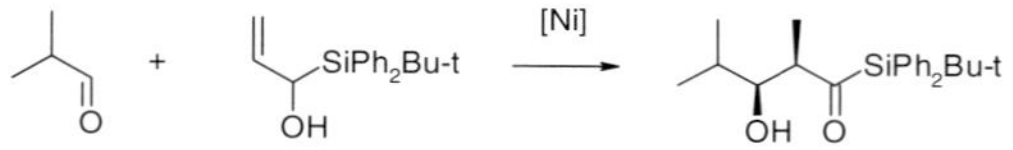

270. 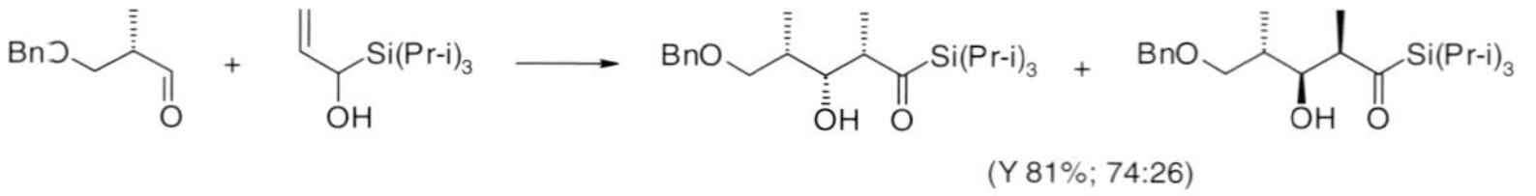

Superhydride (20 mol%) added to a stirred soln. of $NiCl_2$(binap) (20 mol%) in dry THF (1 ml) at 25° under argon, the mixture stirred for 15 min, added to dry $MgBr_2$ (0.6 eq.), stirring continued for 15 min, the mixture cooled to -50°, isobutanal (0.5 mmol) and a soln. of α-hydroxyallyl-*tert*-butyldiphenylsilane (1 eq.) in the same solvent (1 ml) added, stirred for 24 h (TLC), quenched with satd. aq. NH_4Cl, diluted with ether, filtered through Celite, extracted with ether, and purified by chromatography on silica → *syn*-1-(*tert*-butyldiphenylsilyl)-3-hydroxy-2,4-dimethylpentan-1-one. Y 65% (d.r. 91:9). Nickel hydride-catalyzed isomerization of terminal and internal 2-ethylene-1,1-hydroxysilanes afforded **acylsilanes** for both simple and bulky silanes (ten examples; Y 56-92%). In the presence of electron-diverse (het)ar. or aliphatic aldehydes, tandem isomerization and aldol reaction occurred with good *syn*-diastereoselectivity, particularly for bulky silanes (d.r. 84:16 to 98:2; eight examples; Y 62-90%), but with less selectivity in reactions with cinnamaldehydes (d.r. 60:40; two examples; Y 78-85%). Two chiral α-methyl-aliphatic aldehydes showed good face selectivity for the *syn-syn* diastereomer, particularly for a triisopropylsilane deriv., which gave a ca. 3:1 mixture of *syn-syn* and *anti-syn* derivs. as the only products (two examples; Y 81-91%). F.e. and substrate prepn. s. G.P.K. Reddy, J.S. Reddy, S. Das, T. Roisnel, J.S. Yadav, S. Chandrasekhar, R. Grée, Org. Lett. *2013*, *15* (7), 1524-7 [DOI: 10.1021/ol4003064].

Chiral diacetoxo(aqua)[2,6-bis(Δ²-imidazolin-2-yl)phenyl]rhodium(III) complexes *[Rh(III)]**
Asym. synthesis of 2-acetylenealcohols from oxo compds. s. *58*, 236s*83* CO → C(OH)C≡C

Addition to Nitrogen and Carbon

CC ⇓ NC

Without additional reagents *w.a.r.*

5-Acylamino-1,2-dihydro-3-benzazocin-4(3*H*)-ones from 3,4-dihydroisoquinolines and Δ^2-5-oxazolones

271.

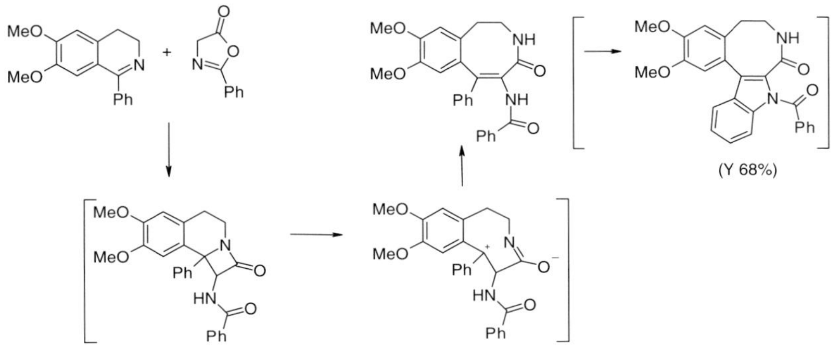

A soln. of the startg. azlactone (1 eq.) and 3,4-dihydroisoquinoline (1 eq.) in acetonitrile (6 ml/mmol of startg. m.) heated at reflux for 2 h, the mixture allowed cool to room temp., concentrated under reduced pressure, and the crude product purified by recrystallization → 5-benzoylamino-6-phenyl-8,9-dimethoxy-1,2-dihydrobenz[*d*]azocin-4-one. Y 74%. Ring expansion is thought to proceed via a carbocation stabilized by two aromatic rings (ten examples; Y 30-74%). The products undergo **oxidative intramolecular *o*-amination** to the indole-fused ring system (ten examples; Y 14-91%) with $Pd(OCOCF_3)_2/Cu(OAc)_2$ under O_2 (cf. *75*, 192s*83*), the N-benzoyl group being removable if required by subsequent treatment with NaOH in one pot. F.e.s. S. Boonya-udtayan, M. Eno, S. Ruchirawat, C. Mahidol, N. Thasana, Tetrahedron *2012*, *68* (50), 10293-301 [DOI: 10.1016/j.tet.2012.10.011].

Electrolysis s. under Sm ϟ

n-Butyllithium *BuLi*

β-Amino-α-oxycarboxylic acid amides from N-sulfinylaldimines via 1,2-addition with 2-phenyl-4-triflyloxyoxazole with asym. induction ←

272.

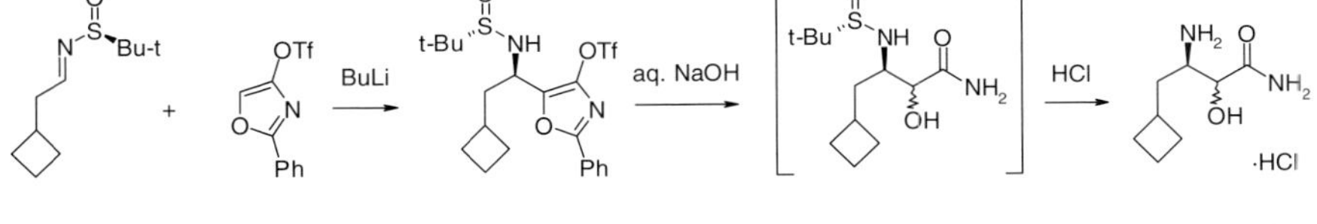

A soln. of 2-phenyloxazol-4-yl triflate (1 eq.) in THF (10 volumes) cooled to -78°, *n*-butyllithium (1.2 eq.) in hexanes (2.5 *M*) added at a rate so as to maintain a temp. of < -70°, stirring continued for 10 min, a soln. of chiral N-*tert*-butanesulfinyl-2-cyclobutylethanaldimine (1 eq.) in THF (2 volumes) added over 3-5 min, the mixture stirred for 10 min, quenched with 10% aq. NH_4Cl, extracted with methylene chloride, and purified by flash chromatography → 5-[1-(*tert*-butanesulfinylamino)-2-cyclobutylethyl]-2-phenyl-1,3-oxazol-4-yl triflate (Y 70%; d.r. 95:5), a portion of which (1.01 mmol) dissolved in ethanol (5 ml), the soln. cooled to 0°, 1 *M* aq. NaOH (3 eq.) added, the mixture stirred for 4 h, neutralized with 1 *M* aq. HCl, extracted with methylene chloride, concentrated, 5 *M* HCl (5 eq.) in isopropanol (1 ml) added at room temp., the mixture stirred for

3 h, *tert*-butyl methyl ether (10 ml) added, the mixture stirred for 30 min, filtered, and recrystallized → (2R)-4-amino-1-cyclobutyl-3-hydroxy-4-oxobutan-2-amine hydrochloride (Y 80%). The product was obtained as a mixture of diastereomeric alcohols (ultimately oxidized to a single enantiomeric ketone in a prepn. of the HCV protease inhibitor, boceprevir). Only the illustrated adduct was hydrolyzed to the amide, but the scope for asym. addition of the chiral sulfinimine to the oxazolyl triflate was examined. Modest yields and good diastereoselectivity were reported for α-aryl and -alkyl derivs. of the sulfinimine (d.r. 95:5; ten examples; Y 38-65%), but oxy-containing derivs. (benzyl or silyl ethers) gave 1:1 mixtures of diastereomers (two examples; Y 28-30%), presumed due to unfavorable coordination with the lithio intermediate, while an α-4-methoxyphenyl deriv. was unreactive. F.e.s. W.J. Morris, K.K. Muppalla, C. Cowden, R.G. Ball, J. Org. Chem. *2013*, *78* (2), 706-10 [DOI: 10.1021/jo301856f]; **2-α-(sulfinylamino)benzothiazoles** by 1,2-addition of benzothiazole with asym. induction using $KN(SiMe_3)_2$ (1.5 eq.) and Li-2-(dimethylamino)ethoxide (1.05 eq.), also addition of 4-methyl-5-vinylthiazole or isomerizable alkyl imines, s. J. Zhang, Y. Yang, M. Wang, L. Lin, R. Wang, Tetrahedron Lett. *2012*, *53* (51), 6893-6 [DOI: 10.1016/j.tetlet.2012.09.131].

Lithium diisopropylamide *i-Pr₂NLi*

α-(Sulfinylamino)carboxylic acid amides from N-sulfinylimines and formamides with asym. induction C=NSOR → C(NHSOR)CON<

273. 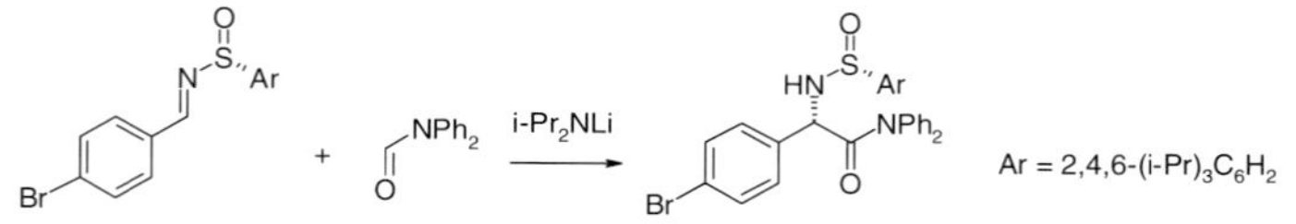

A soln. of i-Pr_2NLi (commercial 2 *M*; 3 eq.) added dropwise to a soln. of (R)-N-(2,4,6-triisopropylbenzenesulfinyl)-4-bromobenzaldimine (1.15 mmol), N,N-diphenylformamide (3.1 eq.) and THF (10 ml) at -78°, the mixture stirred for 10 min, quenched with water, warmed to room temp., extracted with ethyl acetate, and purified by flash chromatography → (S)-2-(4-bromophenyl)-N,N-diphenyl-2-((R)-2,4,6-triisopropylbenzenesulfinamido)acetamide. Y 78% (d.r. 98:2). This experimentally simple and atom economical addition allows direct introduction of a carbonyl group without requirement for additional masking/unmasking steps. Bulky chiral N-sulfinyl aldimines/ketimines underwent reaction with tert. formamides (and a single thioformamide analog) with high diastereoselectivity (d.r. 90:10 to 99.5:0.5; twenty-two examples; Y 67-90%). Removal of the sulfinyl group was readily achieved with HCl/MeOH (three examples; Y 91-98%) and the products converted to an α-amino ester deriv. (Y 75%; two steps) and a 1,2-diamine (Y 77%) and, after N-formylation (Y 86%) and N-methylation (Y 92%), reaction with a second sulfinimine molecule afforded a dipeptide analog (Y 79%; d.r. >97:3). F.e., optimization and substrate prepn. s. J.T. Reeves, Z. Tan, M.A. Herbage, Z.S. Han, M.A. Marsini, Z. Li, G. Li, Y. Xu, K.R. Fandrick, N.C. Gonnella, S. Campbell, S. Ma, N. Grinberg, H. Lee, B.Z. Lu, C.H. Senanayake, J. Am. Chem. Soc. *2013*, *135* (15), 5565-8 [DOI: 10.1021/ja402647m].

Sodium bis(trimethylsilyl)amide *$NaN(SiMe_3)_2$*

Synthesis of N-sulfinylamines from N-sulfinylimines with asym. induction C=N- → C(NH-)R

update s. *74*, 323s*82*; diastereoselective synthesis of chiral *syn*-β-(N-heteroaryl)-α-methyl-β-sulfinylaminocarboxylic acid 2-oxazolidonides with Na-bis(trimethylsilyl)amide as base s. J. Bian, D. Blakemore, J.S. Warmus, J. Sun, M. Corbett, C.R. Rose, B.M. Bechle, Org. Lett. *2013*, *15* (3), 562-5 [DOI: 10.1021/ol3033785]; of chiral 1,1-disubst. 3-ethylene- and 3-acetylenesulfinylamines in the presence of zinc s. T. Guo, R. Song, B.-H. Yuan, X.-Y. Chen, X.-W. Sun, G.-Q. Lin, Chem. Commun. *2013*, *49* (47), 5402-4 [DOI: 10.1039/c3cc42481b]; diastereoselective **radical synthesis** of sulfinylamines from a wide range of S-chiral N-sulfinylaldimines by 1,2-addition of *prim*-, *sec*- or *tert*-alkyl iodides in the presence of BF_3-etherate and tri-*n*-butyltin hydride or Et_3B/O_2 as alkyl radical generators s. J.A. Fernández-Salas, M.C. Maestro, M.M. Rodríguez-Fernández, J.L. García-Ruano, I. Alonso, Org. Lett. *2013*, *15* (7), 1658-61 [DOI: 10.1021/ol400439g].

Potassium carbonate/chiral 1,1′-binaphthyl-based spirocyclic quaternary ammonium bromides ←

1-α-(Sulfonylamino)allene-1,3-dicarboxylic acid esters ←
from allene-1,3-dicarboxylic acid esters and N-sulfonylaldimines
via regio- and diastereo-selective asym. Mannich-type reaction
under phase transfer catalysis

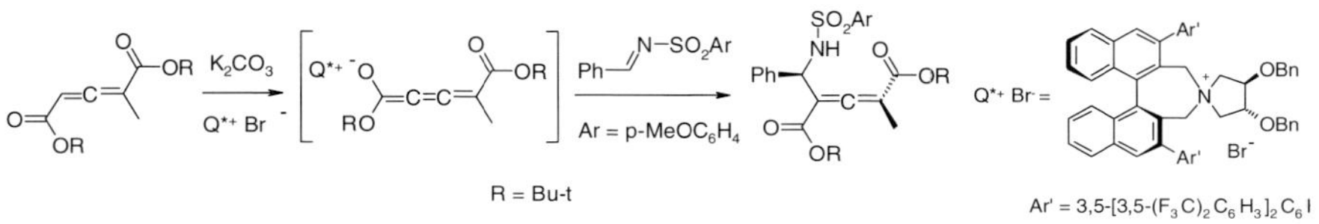

274. 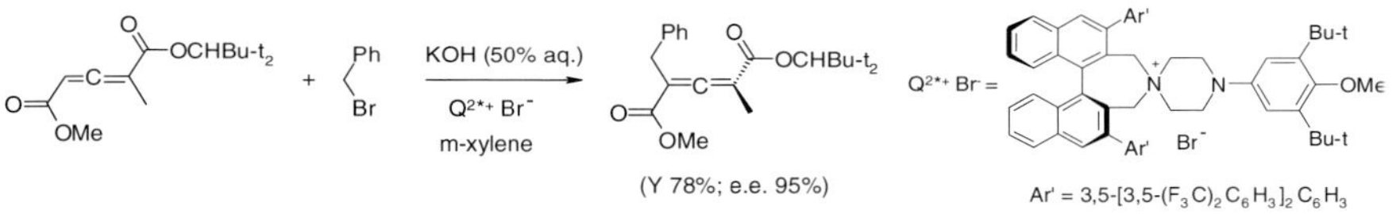

The first use of a cumulenolate, generated from the corresponding allenoate, as a prochiral nucleophilic template in concert with a chiral phase transfer catalyst is reported, allowing access to chiral *tetrasubst.* allenes in a catalytic manner. **E:** A mixture of the chiral phase transfer catalyst (2 mol%), cyclopentyl methyl ether (1 ml), di-*tert*-butyl 1-methylallene-1,3-dicarboxylate (0.1 mmol) and (E)-N-benzylidene-4-methoxybenzenesulfonamide (1.2 eq.) cooled to -20°, treated with powdered K_2CO_3 (5 eq.) in one portion, stirred vigorously at the same temp. for 36 h, the mixture diluted with hexane/ether (1:1), passed through a short pad of silica gel with hexane/ether (1:1), and, after evaporation of the solvent, the residue purified by preparative TLC on silica gel → product. Y 79% (single diastereomer; e.e. 94%). Reaction on a 1 mmol scale gave an 89% yield and 96% e.e. Sixteen further examples with a variety of ar. aldimines (not the *p*-nitro-deriv.) or a pivaldehyde deriv. gave tetrasubst. allenes (bearing alkyl, benzyl, allyl, prenyl or propargyl groups) in yields of 73-98% and e.e. of 85-96%. The concept was extended to **regioselective asym. α-alkylation** with benzyl, allyl or propargyl bromides, optimally using mixed allene-1,3-dicarboxylic acid esters to bias the regioselectivity, a spirocyclic ammonium salt containing N-arylpiperazine unit as phase transfer catalyst and 50% aq. KOH in *m*-xylene (seventeen examples; Y 50%, 65-97%; regioisomeric ratio 71:29 to >95:5, highest with *o*-subst. benzyl bromides; e.e. 90-96%). F.e., also examples of pyrroline, 2-eneazetidine and 2(5*H*)-furanone ring closures with chirality transfer, s. T. Hashimoto, K. Sakata, F. Tamakuni, M.J. Dutton, K. Maruoka, Nat. Chem. *2013, 5* (3), 240-4 [DOI: 10.1038/nchem.1567].

Cinchonine or 2(S)-[Dialkyl(triphenylsiloxy)methyl]pyrrolidine/p-nitrobenzoic acid ←
Organocatalyzed asym. Mannich reaction C=N- → C(NH-)C-CO
update s. *63,* 266s*82*; asym. synthesis of *anti*-γ,δ-acetylene-β-aminoaldehydes with 2(S)-[dialkyl-(triphenylsiloxy)methyl]pyrrolidine as organocatalyst *and p-nitrobenzoic acid* as Brønsted acid s. E. Gómez-Bengoa, J. Jiménez, I. Lapuerta, A. Mielgo, M. Oiarbide, I. Otazo, I. Velilla, S. Vera, C. Palomo, Chem. Sci. *2012, 3* (10), 2949-57 [DOI: 10.1039/c2sc20590d]; of N-protected β-aminomalononitriles by asym. addition of malononitrile to N-protected aldimines with cinchonine as catalyst (1 mol%) s. Z. Lu, C. Xie, W. Zhou, Z. Duan, J. Han, Y. Pan, Chin. J. Chem. *2012, 30* (10), 2333-7 [DOI: 10.1002/cjoc.201200586].

Cinchona-based squaramides or carbamates ←
Organocatalyzed asym. aza-Henry reaction C=N- → C(NH-)C-NO_2
update s. *67,* 290s*82*; with N-(benzothiazol-2-yl)aldimines using a cinchona-based squaramide as catalyst, also 3-component asym. Mannich reaction (cf. *68,* 361s*82,83*) from aldehydes, s. H.-X. He, W. Yang, D.-M. Du, Adv. Synth. Catal. *2013, 355* (6), 1137-48 [DOI: 10.1002/

adsc.201200 957]; diastereoselective synthesis of N-protected α-(4-nitropyrrolidin-2-yl)carboxylic from ω-nitro-α,β-ethylenecarboxylic acid esters and N-protected aldimines via asym. Henry reaction **with dynamic kinetic resolution** using a cinchona-based carbamate as catalyst, also piperidine analogs, s. T. Cheng, S. Meng, Y. Huang, Org. Lett. *2013, 15* (8), 1958-61 [DOI: 10.1021/ol4006129].

Mesitylcopper/1,2-bis[(2S,5S)-2,5-diphenylphospholano]ethane ←
Asym. synthesis of N-protected 2-acetyleneamines from terminal acetylene derivs. C=N- → C(NH-)C≡C
and N-diphenylphosphinyl-protected aldimines cf. *66*, 290s*82*; and N-diphenylthiophosphinyl-protected ketimines with mesitylcopper and 1,2-bis[(2S,5S)-2,5-diphenylphospholano]ethane as ligand s. L. Yin, Y. Otsuka, H. Takada, S. Mouri, R. Yazaki, N. Kumagai, M. Shibasaki, Org. Lett. *2013, 15* (3), 698-701 [DOI: 10.1021/ol3035609].

Copper(II) triflate/ethyldiisopropylamine/chiral 2,6-bis(Δ²-imidazolidin-2-yl)pyridines ←
Catalytic asym. Mannich reaction C=N- → C(NH-)C-CO
under zinc catalysis cf. *64*, 249s*82*; of N-sulfonylimines with N-benzylideneglycinates to give chiral *syn*-adducts using $Cu(OTf)_2$/ethyldiisopropylamine and a *tridentate* chiral 2,6-bis(Δ²-imidazolidin-2-yl)pyridine [PyBidine] as ligand s. T. Arai, A. Mishiro, E. Matsumura, A. Awata, M. Shirasugi, Chem. Eur. J. *2012, 18* (36), 11219-22 [DOI: 10.1002/chem.201201537].

Copper(I) iodide/2,2′-bipyridyl/sodium tert-butoxide *CuI/bpy/NaOBu-t*
β-Amino-α,β-ethyleneketones from ketones and nitriles COCH → COC=C(NH_2)
with sodium amide cf. *33*, 618; from ar. nitriles under copper catalysis with CuI/2,2′-bipyridyl/NaOBu-*t* s. X. Yu, L. Wang, X. Feng, M. Bao, Y. Yamamoto, Chem. Commun. *2013, 49* (28), 2885-7 [DOI: 10.1039/c3cc40466h].

Silver triflate/(S)-proline *AgOTf/(S)-Pro-OH*
Synthesis of 4-functionalized 1(Z)-alkylidene-1,2,3,4-tetrahydrophthalazine ring from cyclic *o*-acetyleneazomethine imides via regioselective silver-catalyzed ring closure-1,2-addition ○

275. 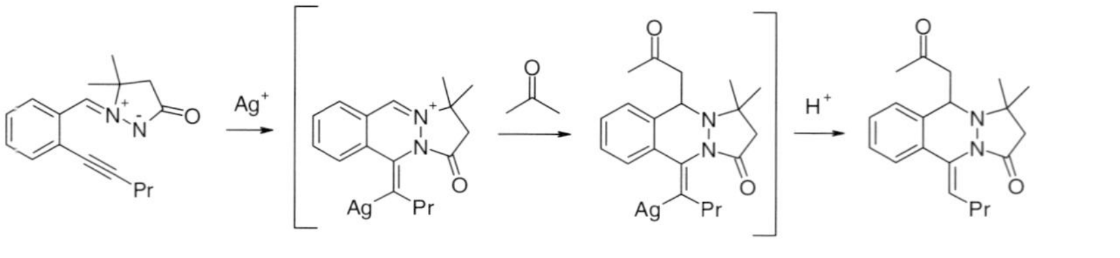

1(Z)-Alkylidene-4-β-keto-1,2,3,4-tetrahydrophthalazine ring. Acetone (3 ml) added to a Schlenk tube charged with the startg. azomethine imide (0.3 mmol), (S)-proline (20 mol%) and AgOTf (20 mol%), the tube wrapped with aluminum foil, the mixture stirred at room temp. in the dark for 12 h, diluted with methylene chloride, filtered through Celite, volatiles removed under reduced pressure, and the residue purified by chromatography on silica gel → product. Y 90%. The terminal substituent of the alkyne may be an electron-diverse aryl group (substituted at the *o*-, *m*- or *p*-position) or an alkyl or heteroaryl group, while both electron-donating and -withdrawing groups are tolerated on the benzene ring of the internal aryl residue of the substrate. The attacking nucleophile may be a methyl ketone (incl. methyl trifluoromethyl ketone), diethyl ketone or cyclopentanone, as well as nitromethane and water itself, but there was no reaction with activated carbonyl compds. of higher nucleophilicity, e.g. dimethyl malonate or ethyl acetoacetate. AgOTf proved the most efficient of several silver(I) and gold(I) salts tested, and (S)-proline enhanced reaction by increasing the nucleophilicity of the attacking molecule. It is proposed that silver(I) activates the alkyne residue towards *6-exo-dig* cyclization to give an intermediate alkenylsilver(I) compd. prior to attack by the nucleophile across the iminium bond. F.e.s. Y. Liu, W. Zhen, W. Dai, F. Wang, X. Li, Org. Lett. *2013, 15* (4), 874-7 [DOI: 10.1021/ol4000108].

(2,6-Diisopropylimidazol-2-ylidene)gold(I) triflimide *[Au(I)]*

Catalytic Mannich reaction C=N- → C(NH-)C-CO

β-aminoketones with BF_3 cf. *5,* 485s*33*; reaction of β-dicarbonyl compds. with N-protected aldimines *under gold catalysis* with (2,6-diisopropylimidazol-2-ylidene)gold(I) triflimide s. M. Delgado-Rebollo, R. Moreno, M.R. Fructos, A. Prieto, Eur. J. Org. Chem. *2013* (1), 31-4 [DOI: 10.1002/ejoc.201201252]; of N-protected aldimines with N-(9-fluorenylidene)amino-acetonitrile using 1,1,3,3-tetramethylguanidine as catalyst s. Y. Yamashita, M. Matsumoto, Y.-J. Chen, S. Kobayashi, Tetrahedron *2012, 68* (37), 7558-63 [DOI: 10.1016/j.tet.2012.06.044]; diastereoselective conversion with chiral N[1]-acylhydantoins using $TiCl_4$/ethyldiisopropylamine *with asym. induction* s. X.-R. Li, C.-F. Lu, Z.-X. Chen, Y. Li, G.-C. Yang, Tetrahedron: Asym. *2012, 23* (18-19), 1380-4 [DOI: 10.1016/j.tetasy.2012.09.006]; 3-component synthesis of cyclic β-(glycosylamino)acetals from N-glycosylaldimines, cyclic enolethers and alcohols via Mannich reaction with asym. induction using $Sc(OTf)_3$ s. G. Wang, B. Li, Q. Lou, Z. Li, X. Meng, Adv. Synth. Catal. *2013, 355* (2-3), 303-7 [DOI: 10.1002/adsc.201200761].

Magnesium/trimethylsilyl chloride/triethylamine *$Mg/Me_3SiCl/Et_3N$*

N-(Aryl)arylglycines from ar. N-arylaldimines by reductive carboxylation C=NR → C(NHR)COOH

276.

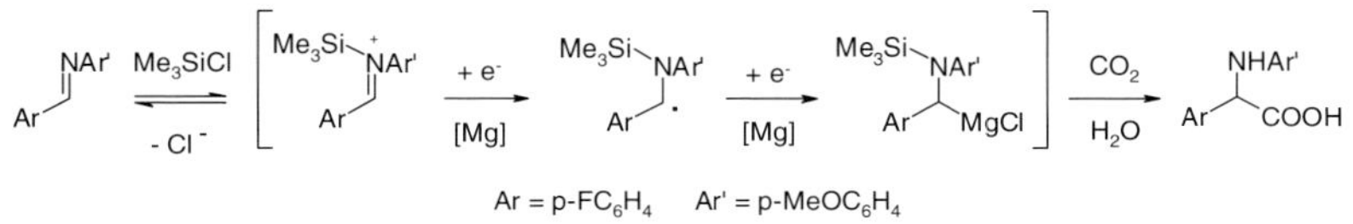

Trimethylsilyl chloride (3 eq.) added to a test-tube containing the startg. aldimine (1 mmol), magnesium (5 eq.), triethylamine (3 eq.) and DMF (5 ml), the tube quickly transferred to an autoclave charged with molecular sieves and Dry Ice, sealed tightly, heated to sublime the Dry Ice, the final pressure adjusted to 45 atm. after the vessel had warmed up to room temp., the vessel depressurized after 24 h, the mixture diluted with 10% aq. NaOH, and worked up after extraction and precipitation of the resulting amino acid at pH 4 (by adding concd. HCl), followed by trituration with 10% ether in hexanes → product. Y 72%. The procedure is simple, low-cost, applicable on the multigram scale, and a useful alternative to the Strecker reaction by avoiding toxic cyanide reagents. It is also applicable to a wide range of electron-diverse aryl aldimines, notably supporting pendant cyano, ester and amide functions, and tolerating various N-aryl substituents (fourteen examples; Y 48-73%). The yield was low, however, with a hindered 2,4,6-trimethylbenzaldimine (Y 20%), and there was no reaction with alkyl aldimines or ketimines. Reaction is presumed to involve generation of a nucleophilic α-metaloamine via two single electron transfers from the reducing magnesium. Chromatographic work-up is not required. F.e.s. A.A. Sathe, D.R. Hartline, A.T. Radosevich, Chem. Commun. *2013, 49* (44), 5040-2 [DOI: 10.1039/c3cc42057d].

Zinc *Zn*

Synthesis of 3-ethylene- or 3-acetylene-N-sulfinylamines C=N- → C(NH-)R

from N-sulfinylimines with asym. induction s. *74,* 323s*83*

Synthesis of siloximes from nitrones and organozinc halides C=N(O)R → C(R′)N(OSi≤)R

277.

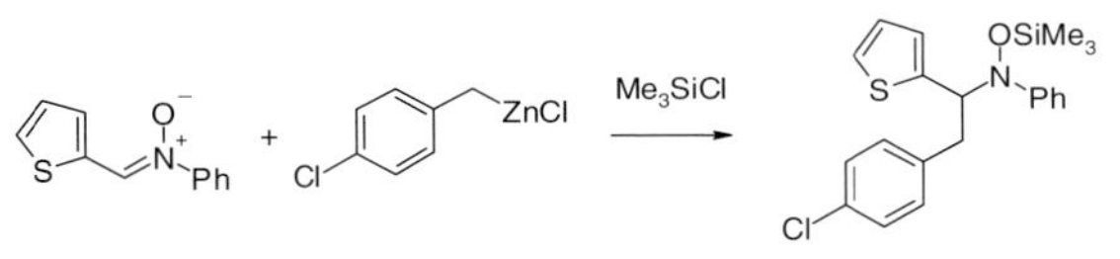

under mild conditions. A soln. of trimethylsilyl chloride (2 eq.) in THF (5 ml) added slowly to a freshly prepared soln. of 4-chlorobenzylzinc chloride (1.5 eq.) and N-phenyl-C-(2-thienyl)nitrone (3 mmol) in THF (10 ml) at room temp., the mixture stirred for 30 min, quenched with satd. aq.

NH_4Cl, extracted with ethyl acetate, and purified by chromatography on silica → N-[2-(4-chlorophenyl)-1-(thien-2-yl)ethyl]-N-phenyl-O-(trimethylsilyl)hydroxylamine. Y 79%. The use of Me_3SiX (X = Cl, Br, I or OTf) was key to the success of this novel nucleophilic addition of organozinc halides to nitrones, functioning both as Lewis acid catalyst and protecting group for the air unstable products (bulkier silyl halides were ineffective), with Me_3SiCl being the cost-effective reagent of choice. The reaction was general for addition of electron-diverse (het)ar. and aliphatic (incl. branched alkyl and benzylic) organozinc halides to α-(het)aryl-N-(het)aryl/benzyl nitrones (thirty-three examples; Y 62-92%), but failed for alkyl/aryl-zinc halides with hydroxybenzaldehyde-derived nitrones. Two examples of reduction [Zn/Cu$(OAc)_2$] of the products to **sec. amines** are also described (Y 72-77%). F.e., optimization and use of the procedure in a three-step conversion of a D-mannose-derived cyclic nitrone to an α-branched polyhydroxypyrrolidine s. Y. Fu, Y. Liu, Y. Chen, H.M. Hügel, M. Wang, D. Huang, Y. Hu, Org. Biomol. Chem. *2012*, *10* (38), 7669-72 [DOI: 10.1039/c2ob26202a].

Zinc/indium *Zn/In*
Reductive aza-Henry reaction C=N- → C(NH-)C-NO_2
with In cf. *46,* 592s*82*; N-(*p*-methoxyphenyl)-protected 2-nitramines from the corresponding aldimines and 1,1-nitrobromides with Zn as reductant and In in catalytic amount, also with asym. induction from carbohydrate-based aldimines, s. R.G. Soengas, A.M.S. Silva, Tetrahedron Lett. *2013, 54* (17), 2156-9 [DOI: 10.1016/j.tetlet.2013.02.041].

Triethylborane/oxygen *Et_3B/O_2*
Radical synthesis of N-sulfinylamines from N-sulfinylimines C=N- → C(NH-)R
with asym. induction s. *74,* 323s*83*

Indium(III) chloride *$InCl_3$*
1-Hydroxy-3-oxo-2,3-dihydro-2,3′-biindolyls from isatogens and indoles ←

278.

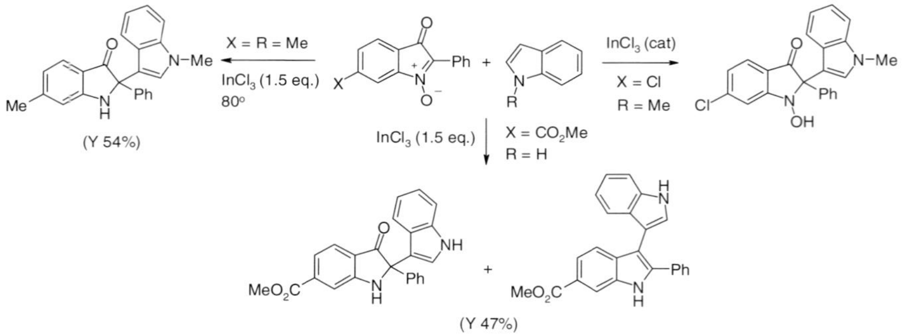

1-Methylindole (1 eq.) added to a soln. of 6-chloro-2-phenylisatogen (1 eq.) in acetonitrile (4 ml) under argon at room temp., anh. $InCl_3$ (20 mol%) added, the mixture stirred for 4-8 h (TLC), solvent evaporated, and the residue purified by chromatography on silica → 6-chloro-1-hydroxy-1′-methyl-2-phenyl-2,3′-biindolin-3-one. Y 67%. This $InCl_3$-catalyzed addition appears generally applicable for coupling of indole and its N-methyl deriv. with electron-diverse isatogens (ten examples; Y 49-74%), but failed with N-Boc-indole. The reaction was diastereoselective for an isatogen carrying a 2-(1-furanosyl) substituent, giving a single diastereomer with N-methylindole and a 2:1 mixture with NH-indole. Interestingly, use of stoichiometric $InCl_3$ (optimally in toluene at 80°) resulted in concomitant deoxygenation at the isatogen nitrogen to afford the corresponding **3-oxo-2,3-dihydro-2,3′-biindolyls** in moderate yield (41-56%; six examples), with a 6-(methoxycarbonyl)isatogen forming by-products (unspecified yields) by deoxygenative coupling at C3. F.e., optimization, and use of the method in a ten-step synthesis of 13-*deoxy*-isatisine A, s. C.V.S. Kumar, V.G. Puranik, C.V. Ramana, Chem. Eur. J. *2012*, *18* (31), 9601-11 [DOI: 10.1002/chem.201103604].

Samarium/tetra-n-butylammonium iodide/electrolysis *$Sm/Bu_4NI/$*
Syntheses with electrochemically-generated samarium diiodide ←
Sym. 1,2-diamines – 3-Ethyleneamines s. *83*, 263

Chiral quaternary ammonium phenoxide betaines ←
Organocatalyzed asym. aza-Henry-type reaction with 1-nitroethylene derivs. C=C-C(NO_2)C(NH-)

279. 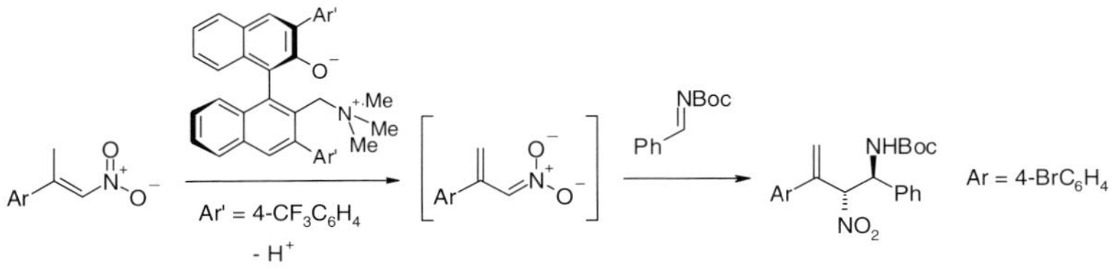

The use of an axially chiral betaine as a base catalyst for the aza-Henry-type reaction of nitroolefins with imines is reported (cf. *74*, 320 for use of an analogous catalyst in the asym. Mannich reaction of imines with α-nitrocarboxylates). **E: Chiral 3-ethylene-2-nitro-N-carbo-*tert*-butoxyamines**. 2-(4-Bromophenyl)-1-nitroprop-1-ene (1.1 eq.) and N-Boc-benzaldimine (0.1 mmol) added sequentially to a mixture of chiral ammonium phenoxide betaine (2 mol%) and toluene (0.5 ml) under argon at -30°, the mixture stirred for 15 h, quenched with 1 *M* aq. HCl at 0°, extracted with ethyl acetate, and purified by chromatography on silica → (1S,2R)-N-*tert*-butoxycarbonyl-3-(4-bromophenyl)-2-nitro-1-phenylbut-3-enylamine. Y 93% (*anti/syn* >20:1; e.e. 98%). Generally, substrates were reacted as mixtures of β-subst. 1-nitro-1/2-propenes, with proton abstraction occurring from the more reactive 1-nitro-2-propene to afford a vinylogous nitronate intermediate, which was trapped by electron-diverse benzaldimines and acetaldimine, to afford aza-Henry adducts with excellent diastereo- and enantio-selectivity (*anti/syn* >20:1; e.e. 95-99%; thirteen examples; Y 87-99%). The procedure tolerated substitution at the γ-position, affording similar levels of selectivity (*anti/syn* 18:1 to >20:1; e.e. 95-97%; two examples; Y 90%; >95% E-selectivity for a 1-nitrobut-1/2-ene). F.e., optimization, and further reactions of the adducts, s. D. Uraguchi, K. Oyaizu, T. Ooi, Chem. Eur. J. *2012, 18* (27), 8306-9 [DOI: 10.1002/chem.201201259].

Chiral 2,6-bis(Δ²-imidazolidin-2-yl)pyridines s. under Copper(II) triflate ←

1,1,3,3-Tetramethylguanidine *TMG*
Catalytic Mannich reaction s. *5*, 485s*83* C=N- → C(NH-)C-CO

Chiral 3,3′-di(9-phenanthryl)-1,1′-bi-2-naphthol s. under Chiral 2,2′-bis(adamantoyl-amino)-3-[o-(phospholidin-1-yl)phenyl]-5,5′,6,6′,7,7′,8,8′-octahydro-1,1′-binaphthyl ←

(S)-Proline s. under AgOTf *(S)-Pro-OH*

Cinchona-based 2-tert-aminothioureas/trimethylsilyl cyanide ←
Asym. Strecker reaction C=N- → C(NH-)CN
update s. *58*, 261s*82*; with N-Boc-protected 3-iminooxindoles using a quinine-based 2-*tert*-amino-thiourea as catalyst s. D. Wang, J. Liang, J. Feng, K. Wang, Q. Sun, L. Zhao, D. Li, W. Yan, R. Wang, Adv. Synth. Catal. *2013, 355* (2-3), 548-58 [DOI: 10.1002/adsc.201200630]; asym. synthesis of chiral N¹-protected 4-trifluoromethyl-4-cyano-3,4-dihydro-2(1*H*)-quinazolones with f. cinchona-based 2-*tert*-aminothioureas as catalyst and added ethanol (2 eq.) s. H. Xie, A. Song, X. Song, X. Zhang, W. Wang, Tetrahedron Lett. *2013, 54* (11), 1409-11 [DOI: 10.1016/j.tetlet.2012.12.119].

Trimethylsilyl cyanide (s.a. under Cinchona-based 2-tert-aminothioureas) *Me_3SiCN*
Strecker reaction
with diethylaluminum cyanide s. *31*, 616s*80*; *uncatalyzed* conversion *in ethylene glycol*, notably by reaction of α-difluoromethyl- or α-trifluoromethyl-subst. ketoimines with trimethylsilyl cyanide, also f. uncatalyzed procedures in the same solvent, e.g. Baylis-Hillman reaction and aldol-type condensation, s. Y.-L. Liu, X.-P. Zeng, J. Zhou, Chem. Asian J. *2012, 7* (8), 1759-63 [DOI: 10.1002/asia.201200289].

Trimethylsilyl chloride s. under Mg *Me_3SiCl*

Titanium tetrachloride/ethyldiisopropylamine *$TiCl_4$/i-Pr_2NEt*

Catalytic Mannich reaction with asym. induction s. *5,* 485s*83* C=N- → C(NH-)C-CO

1,2-Bis[(2S,5S)-2,5-diphenylphospholano]ethane s. under Mesitylcopper ←

Chiral [2.2]-paracyclophanol-based m-(di-m-xylylphosphino)biphenyls ←

Chiral 2,2′-bis(adamantoylamino)-3-[o-(phospholidin-1-yl)phenyl]-5,5′,6,6′,7,7′,8,8′-octahydro-1,1′-binaphthyl/chiral 3,3′-di(9-phenanthryl)-1,1′-bi-2-naphthol ←

Asym. aza-Baylis-Hillman reaction s. *68,* 272s*83* C=N- → C(NH-)C(=CH_2)CO

Oxygen s. under Et_3B *O_2*

Tetra-n-butylammonium iodide s. under Sm *Bu_4NI*

Chiral 1,1′-binaphthyl-based spirocyclic quaternary ammonium bromides s. under Potassium carbonate ←

(5,10,15,20-Tetraarylporphyrinato)ruthenium(III) bis(hexafluoroantimonate) *[Ru(III)]*

Friedel-Crafts α-aminoalkylation H → CH(NH-)

with Cu(OTf)$_2$/BINAP cf. *70,* 285; with a (5,10,15,20-tetraarylporphyrinato)ruthenium(III) bis(hexafluoroantimonate) as catalyst (1 mol%) s. T. Terada, T. Kurahashi, S. Matsubara, Heterocycles *2012, 85* (10), 2415-9 [DOI: 10.3987/com-12-12556].

Dichloro(pentamethylcyclopentadienyl)rhodium(III)/silver hexafluoroantimonate ←

Aza-Baylis-Hillman reaction C=N- → C(NH-)C(=CH_2)CO

with sulfonylimines using triethylenediamine cf. *40,* 456s*74*; N-protected 2-α-tosylaminoindoles from indoles and aliphatic or aromatic N-tosylaldimines with [Cp*$RhCl_2$]$_2$/$AgSbF_6$ s. B. Zhou, Y. Yang, S. Lin, Y. Li, Adv. Synth. Catal. *2013, 355* (2-3), 360-4 [DOI: 10.1002/adsc.201200909]; **asym. aza-Baylis-Hillman reaction** (cf. *68,* 272s*80*) for the synthesis of chiral 2-α-tosylamino-2-cyclopentanones with chiral 2,2′-bis(adamantoylamino)-3-[*o*-(phospholidin-1-yl)phenyl]-5,5′,6,6′,7,7′,8,8′-octahydro-1,1′-binaphthyl and chiral 3,3′-di(9-phenanthryl)-1,1′-bi-2-naphthol as co-catalyst s. Y.-L. Yang, Y. Wei, M. Shi, Org. Biomol. Chem. *2012, 10* (36), 7429-38 [DOI: 10.1039/c2ob25932j]; with a bifunctional chiral, [2.2]-paracyclophanol-based *m*-(di-*m*-xylylphosphino)biphenyl as catalyst s. S. Kitagaki, Y. Ohta, R. Takahashi, M. Komizu, C. Mukai, Tetrahedron Lett. *2013, 54* (5), 384-6 [DOI: 10.1016/j.tetlet.2012.11.021].

Dichloro(pentamethylcyclopentadienyl)rhodium(III)/silver hexafluoroantimonate ←

Regioselective Friedel-Crafts-type reaction with N-protected aziridines C

under indium(III) catalysis cf. *65,* 250; reaction with functionalized arenes via *o-directed* C-H activation using [Cp*$RhCl_2$]$_2$/$AgSbF_6$ s. X. Li, S. Yu, F. Wang, B. Wan, X. Yu, Angew. Chem., Int. Ed. *2013, 52* (9), 2577-80 [DOI: 10.1002/anie.201209887]; Lewis acid-catalyzed synthesis of N-protected (S)-tryptophans from chiral aziridine-2-carboxylic acid esters s. I. Tirotta, N.L. Fifer, J. Eakins, C.A. Hutton, Tetrahedron Lett. *2013, 54* (7), 618-20 [DOI: 10.1016/j.tetlet.2012.11.139].

Addition to Carbon-Carbon Bonds CC ⇓ CC

Without additional reagents *w.a.r.*

Michael addition C=C → CHC(R)

update s. *47,* 631s*82*; of active methylene compds. to nitroalkenes *in water* s. W. Dong, D. Xu, J. Xie, Chin. J. Org. Chem. *2012, 30* (8), 1771-4 [DOI: 10.1002/cjoc.201200228]; of α-nitroketones to various electron-deficient ethylene derivs. in water or *'on water'* under very dilute conditions with K_2CO_3 as base s. G. Giorgi, P. López-Alvarado, S. Miranda, J. Rodriguez, J.C. Menéndez, Eur. J. Org. Chem. *2013* (7), 1327-36 [DOI: 10.1002/ejoc.201201431]; under *neutral* conditions *in aq. KCl* with an added crown ether, also Michael addition of chiral N-acyl-2-oxazolidones *with asym. induction*, s. S. Shirakawa, L. Wang, A. Kasai, K. Maruoka, Chem. Eur. J. *2012, 18* (28), 8588-90 [DOI: 10.1002/chem.201201107]; Michael addition of diethyl 1-fluoro-1-phenylsulfonylmethanephosphonate with Cs_2CO_3 as base s. S. Opekar, R. Pohl, V. Eigner, P. Beier, J. Org. Chem. *2013, 78* (9), 4573-9 [DOI: 10.1021/jo400297f]; of S-chiral N-(*tert*-butanesulfinyl)-iminoesters to alkylidenemalonic acid esters with asym. induction using *i*-Pr_2NLi as base s. J. Wang, Y. Zhou, L. Zhang, Z. Li, X. Chen, H. Liu, Org. Lett. *2013, 15* (7), 1508-11 [DOI: 10.1021/ol400277h].

Friedel-Crafts reaction with electron-deficient ethylene derivs. C=C → CHC(R)
by 1,4-addition of indoles, update s. *11,* 770s*82*; *uncatalyzed* 1,4-addition to 3-nitro-1,2-dihydronaphthalenes *in water*, also addition of azaindoles and pyrroles, s. R. Malhotra, R. Ghosh, T.K. Dey, S. Chakrabarti, A. Ghosh, S. Dutta, S. Asijaa, S. Roy, S. Dutta, S. Basu, S. Hajra, Eur. J. Org. Chem. *2013* (4), 772-80 [DOI: 10.1002/ejoc.201201288]; 1,4-addition of indoles to alkylidenemalonic acid esters with an *o*-ureidophenylpalladium(II) complex as a *transition metal-assisted hydrogen-bonding catalyst* s. D.M. Nickerson, A.E. Mattson, Chem. Eur. J. *2012, 18* (27), 8310-4 [DOI: 10.1002/chem.201201206]; *Brønsted acid-catalyzed heterogeneous* 1,4-addition to nitroalkenes with magnetically retrievable nanoparticulate 3-sulfopropyl-functionalized γ-Fe_2O_3, also synthesis of 1,1-bis(indol-3-yl)alkanes, s. S. Sobhani, R. Jahanshahi, New J. Chem. *2013, 37* (4), 1009-15 [DOI: 10.1039/c3nj40899j]; Brønsted acid-enhanced rhodium-catalyzed **N-directed *o*-(3-ketoalkyl)ation** of aryl-N-heterocyclics by 1,4-addition to enones with $[Cp^*RhCl_2]_2/AgSbF_6/$ acetic acid s. L. Yang, B. Qian, H. Huang, Chem. Eur. J. *2012, 18* (31), 9511-5 [DOI: 10.1002/chem.201201348]; N-directed *o*-substitution of 2-phenylpyridines by 1,4-addition to enones and N-(α,β-ethyleneacyl)pyrroles using (η^6-benzene)(pentamethylcyclopentadienyl)cobalt(III) bis(hexafluorophosphate) as catalyst s. T. Yoshino, H. Ikemoto, S. Matsunaga, M. Kanai, Angew. Chem., Int. Ed. *2013, 52* (8), 2207-11 [DOI: 10.1002/anie.201209226].

Diels-Alder reaction ○
update s. *1,* 526-9s*82*; efficient, *solvent-free* cycloaddition of *volatile* dienes to various dienophiles *in a closed, batch system* s. D. Sun, F. Sato, Y. Yamada, S. Sato, Bull. Chem. Soc. Jpn. *2013, 86* (2), 276-82 [DOI: 10.1246/bcsj.20120247]; facile *endo*-cycloaddition and intramolecular cycloaddition (cf. *34,* 693s*81*) of carbocyclic α,β-ethylene-α-halogenoketones s. A.G. Ross, S.D. Townsend, S.J. Danishefsky, J. Org. Chem. *2013, 78* (1), 204-10 [DOI: 10.1021/jo302230m]; facile cycloaddition of Se-phenyl prop-2-eneselenoate (relative to acrylates) **as an ethylene equivalent** s. M.E. Jung, F. Perez, C.F. Regan, S.W. Yi, Q. Perron, Angew. Chem., Int. Ed. *2013, 52* (7), 2060-2 [DOI: 10.1002/anie.201208294]; Δ^5-isoquinuclidenes from highly subst. 1,2-dihydropyridines and electron-deficient ethylene derivs., regio- and stereo-selectivity, s. R.M. Martin, R.G. Bergman, J.A. Ellman, Org. Lett. *2013, 15* (3), 444-7 [DOI: 10.1021/ol303040r]; cycloaddition of S-chiral 2-hydroxy-3,5-dienesulfoxides to a condensed *p*-quinone with remote asym. induction and regioselectivity s. I. Almodovar, W. Cardona, T. Delgado-Castro, G. Zapata-Torres, M.C. Rezende, R. Araya-Maturana, M.C. Maestro, J.L. García Ruano, Tetrahedron: Asym. *2013, 24* (1), 56-61 [DOI: 10.1016/j.tetasy.2012.11.017]; *endo*-selective **Lewis acid-catalyzed Diels-Alder reaction** (cf. *28,* 655s*82*) of chiral α-phosphonylacrylates with asym. induction using $SnCl_4$ s. J-L. Zhu, P.-E. Chen, H.-W. Huang, ibid. 23-36 [DOI: 10.1016/j.tetasy.2012.11.021]; Povarov-type **hetero-Diels-Alder reaction** (cf. *31,* 623s*82*) of α,β-ethylenediazo compds. to N-arylaldimines under acid catalysis with TfOH (or under gold(I) catalysis) s. A.M. Jadhav, V.V. Pagar, R.-S. Liu, Angew. Chem., Int. Ed. *2012, 51* (47), 11809-13 [DOI: 10.1002/anie.201205692]; highly diastereoselective, *phosphine-catalyzed* Rauhut-Currier-type [4+2]-cycloaddition of vinyl ketones to α,β-ethyleneketimines with triphenylphosphine/4-methoxyphenol (cf. *80,* 284) s. Z. Shi, Q. Tong, W.W.Y. Leong, G. Zhong, Chem. Eur. J. *2012, 18* (32), 9802-6 [DOI: 10.1002/chem.201201318]; [4+2]-cycloaddition of allenoates to cyclic N-sulfonylketimines with tris(*p*-fluorophenyl)phosphine s. X.-Y. Chen, S. Ye, Eur. J. Org. Chem. *2012* (29), 5723-8 [DOI: 10.1002/ejoc.201200747].

Stereoselective synthesis of 5-imino-Δ^2-oxazolines from enacylamines and isonitriles

280.

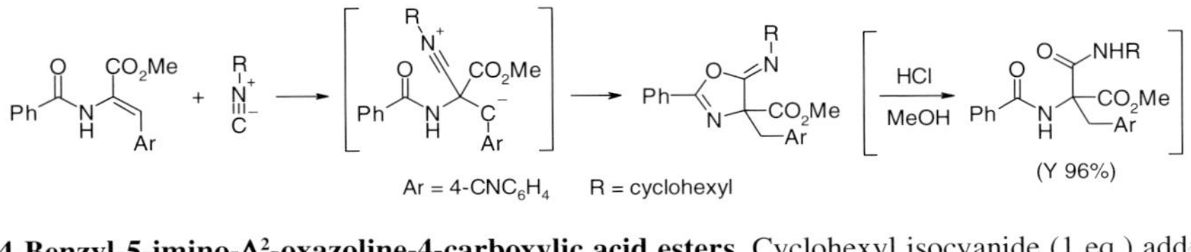

4-Benzyl-5-imino-Δ^2-oxazoline-4-carboxylic acid esters. Cyclohexyl isocyanide (1 eq.) added to molten methyl (Z)-2-(benzamido)-3-(4-cyanophenyl)acrylate (1 mmol), the mixture stirred for 5 min, then at 175° for 1 h, cooled to room temp., and purified chromatographically → methyl (5Z)-5-(cyclohexylimino)-4-(4-cyanobenzyl)-2-phenyl-4,5-dihydro-1,3-oxazole-4-carboxylate.

Y 69%. α-Benzamido-4-subst. (CN, NO_2) cinnamates were cyclized **under solvent-free conditions** in this novel and atom-economical synthesis of the title products (twelve examples; Y 40-93%). The presence of an electron-withdrawing substituent on the cinnamate ring was essential (the reaction failed for H or electron-releasing substituents), while the use of *tert*-butyl- or 1,1,3,3-tetramethylbutyl-isocyanides was unsuccessful. F.e., also hydrolysis of products (HCl/ROH) to the corresponding **acylamino(benzyl)malonic acid amide esters** (four examples; Y 55-96%) s. A.A. Esmaeili, S. Shahmansouri, A. Habibi, A. Reza Fakhari, Tetrahedron *2012*, *68* (38), 8046-51 [DOI: 10.1016/j.tet.2012.06.073].

Irradiation [s.a. under 2(S)-[Diphenyl(trimethylsiloxy)methyl]pyrrolidine, Acetylenedicobalt hexacarbonyl complexes, Cationic ruthenium(II) sensitizers and Tris(2,2′-bipyridyl)ruthenium(II) bis(hexafluorophosphate) ///

Regioselective radical hydroalkylation of ethylene derivs. with alkyl bromides by photoinduced reductive bromine atom transfer C=C → CHC(R)

281. 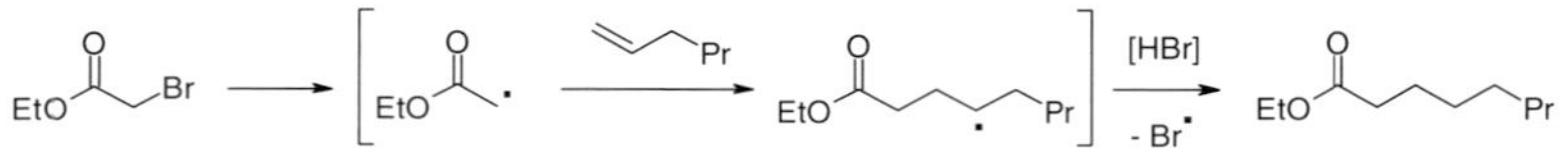

Whereas alkyl bromides undergo classical Kharasch addition to ethylene derivs. in the presence of radical initiators, reaction under irradiation *with a low-pressure mercury lamp* takes a different course, resulting in *hydroalkylation* via an unusual reductive bromine atom transfer. **E:** A stirred mixture of ethyl bromoacetate (0.25 mmol), 1-hexene (*10 eq.*), hexane (10 ml) and water (100 ml) irradiated in a quartz tube by a 6 W low-pressure mercury lamp for 6 h, concentrated *in vacuo*, and subjected to silica gel column chromatography → product. Y 77%. The procedure is applicable to the hydroalkylation of terminal and cyclic olefins, enisocyclics and vinyl ethers with various α-bromocarbonyl compds., although yields were low with α-bromoacetophenones unless a 100-fold excess of the alkene was used (fifteen examples; Y 45-99%). A non-chain radical mechanism is proposed, involving a dual cycle: heterolysis of the carbon-bromine bond initially generates a bromine atom and an alkyl radical which then adds to the alkene; the newly formed radical is then converted to the product by accepting a hydrogen atom from HBr, generated in a second cycle by the bromine atom accepting an allylic hydrogen atom from the alkene (in excess); hence no added reducing agent is required. F.e. and comparison of solvents s. S. Sumino, A. Fusano, I. Ryu, Org. Lett. *2013*, *15* (11), 2826-9 [DOI: 10.1021/ol4011536].

α,β-Ethylene-β-silylketones from electron-deficient acetylene derivs. and acylsilanes s. *83*, 337 C(Si≤)=C-COR

Microwaves (s.a. under 3,4,5-Trimethylthiazol-2-ylidene) [\\\\]

Regioselective uncatalyzed 1,4-hydroalkylation of electron-deficient ethylene derivs. s. *67*, 302s*83* C=C → CHC(R)

Potassium tert-butoxide/diphenyl phosphate/(triphenylphosphine)gold(I) chloride/silver hexafluoroantimonate/silica ←

N-Protected 4-nitro-2-vinylpyrrolidines from 1,3-nitroallenes and aldimines via diastereoselective aza-Henry [nitro-Mannich] reaction-gold(I)-catalyzed intramolecular hydroamination ○

282. 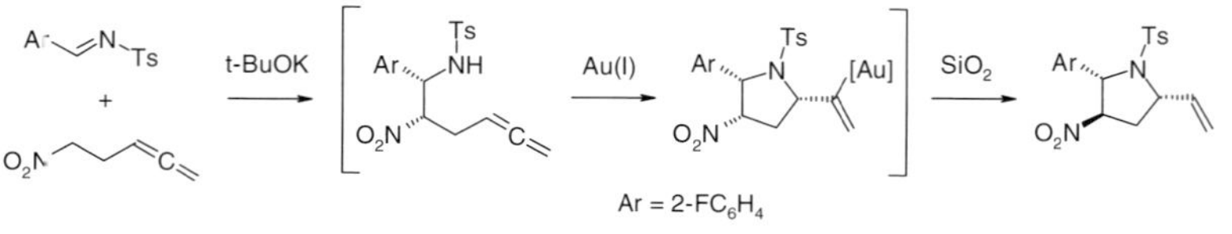

in one pot. A mixture of N-(*p*-toluenesulfonyl)-2-fluorobenzaldimine (0.2 mmol) and 5-nitropenta-1,2-diene (1.5 eq.) in toluene (2 ml) treated with KOBu-*t* (5 mol%) at room temp., stirred for 16 h, diluted with 1,2-dichloroethane (6 ml), diphenyl phosphate (20 mol%), $(Ph_3P)AuCl$ (10 mol%)

and $AgSbF_6$ (20 mol%) added, the mixture stirred at 70° for 12 h, silica gel (pre-treated with triethylamine) added, stirring continued for 2 h, filtered through Celite, concentrated under a stream of N_2, and purified by recrystallization → (2SR,3RS,5SR)-1-(*p*-toluenesulfonyl)-3-nitro-2-(*o*-fluorophenyl)-5-vinylpyrrolidine. Y 79% (d.r. 94:6). This method provides convenient access to pyrrolidines containing three stereocenters from readily available compounds, via a nitro-Mannich-hydroamination sequence (for a pyrrole synthesis from 4-nitrobut-1-yne cf. *82*, 277). A final heating step in the presence of silica gave a cleaner product mixture, via full epimerization at the nitro position (partial epimerization was observed during chromatographic purification of the crude mixture). 5-Nitropenta-1,2-diene reacting with electron-diverse benzaldimines afforded the title products as isomer pairs, diastereomeric at the vinyl carbon, in variable yields and diastereoselectivity (Y 31-81%; d.r. 62:38 to 94:6; fourteen examples), with a substrate incorporating pentamethylene at the allene terminus giving a lower yield (24%; d.r. 75:25). The reaction failed with phosphoryl, carbo-*tert*-butoxy or nitrobenzenesulfonyl protection for the aldimine nitrogen. F.e., optimization, substrate prepn., and further reactions of the products s. A. Duriš, D.M. Barber, H.J. Sanganee, D.J. Dixon, Chem. Commun. *2013*, *49* (27), 2777-9 [DOI: 10.1039/c3cc40729b].

Lithium diisopropylamide *i-Pr_2NLi*
Michael addition with S-chiral (N-*tert*-butanesulfinyl)iminoesters C=C → CHC(R)
with asym. induction s. *47*, 631s*83*

Potassium or cesium carbonate or potassium chloride/crown *K_2CO_3 or Cs_2CO_3 or KCl/crown*
Michael addition
Also with asym. induction s. *47*, 631s*83*

Cesium fluoride s. under Cyclooctadiene(8-quinolinato)rhodium(I) *CsF*

Triethylamine *Et_3N*
2-Amino-3-cyano-3a,6a-dihydrofuro[2,3-*b*]furans from 2-ene-1,4-diones by base-catalyzed double ring closure with malononitrile ○

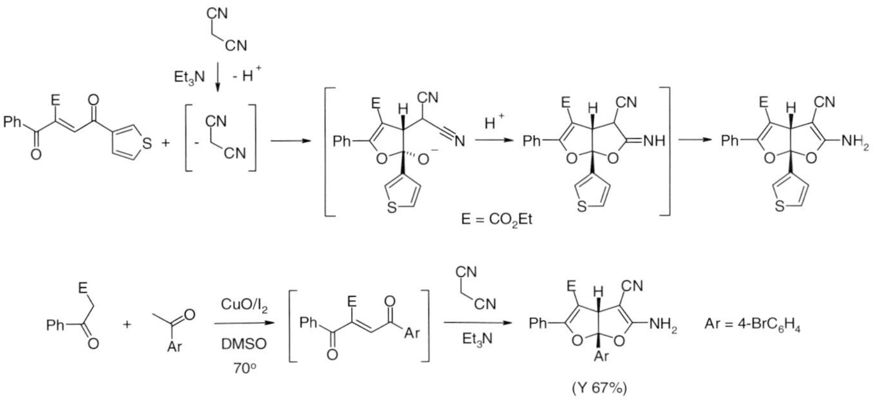

283.

under mild conditions. A mixture of ethyl 2-benzoyl-4-oxo-4-thien-3-ylbut-2-enoate (1 mmol), malononitrile (1 eq.) and triethylamine (1 eq.) in dry methanol (3 ml) stirred at room temp. for 2 min, diluted with water, extracted with ethyl acetate, and purified by chromatography on silica → ethyl 5-amino-4-cyano-2-phenyl-6a-(thien-3-yl)-3a,6a-dihydrofuro[2,3-*b*]furan-3-carboxylate. Y 86%. This novel, rapid and experimentally simple method was promoted by (in)organic bases (optimally triethylamine) and generated the title compds. in good yields for electronically diverse 1-(het)aryl-4-aryl/methyl-subst. 2-ene-1,4-diones carrying an activating substituent (COR, CN, CO_2Et) at C3 (64-96%; twenty-four examples). In a further development the enediones were generated *in situ* via an efficient domino strategy, affording the title products in one-pot **from**

(het)ar. **methyl ketones** and β-ketocarbonyl compds. via an iodination-Kornblum oxidation-Knoevenagel condensation-bicyclization sequence (nine examples; Y 58-75%). F.e. and optimization s. W.-M. Shu, Y. Yang, D.-X. Zhang, L.-M. Wu, Y.-P. Zhu, G.-D. Yin, A.-X. Wu, Org. Lett. *2013, 15* (3), 456-9 [DOI: 10.1021/ol400016c].

Chiral amines, diamines or triamines or helically chiral polyarene-condensed 2-amino-quinoline salts ←

Organocatalyzed asym. Diels-Alder reaction ○

update s. *46,* 662s*81*; of β-subst. cyclic enones with polyconjugated malononitriles with cinchona-based prim. amines as catalyst, regioselective and *endo/exo*-stereodivergent synthesis via dienamine catalysis, s. X. Feng, Z. Zhou, R. Zhou, Q.-Q. Zhou, L. Dong, Y.-C. Chen, J. Am. Chem. Soc. *2012, 134* (48), 19942-7 [DOI: 10.1021/ja3106219]; chiral amine-catalyzed cycloaddition of 2,4-dienals with alkylidene-rhodanine and -hydantoin derivs. via trienamine catalysis in the presence of *o*-fluorobenzoic acid s. K. Zhu, H. Huang, W. Wu, Y. Wei, J. Ye, Chem. Commun. *2013, 49* (21), 2157-9 [DOI: 10.1039/c3cc00023k]; asym. cycloaddition with α-acoxy- or α-acylamino-α,β-ethylenealdehydes as dienophile using a chiral *prim-sec-tert*-triamine as catalyst (as the TfOH salt) s. A. Sakakura, H. Yamada, K. Ishihara, Asian J. Org. Chem. *2012, 1* (2), 133-7 [DOI: 10.1002/ajoc.201200054]; *with nitroalkenes as dienophile* using a *helically chiral* polyarene-condensed 2-aminoquinoline salt as catalyst s. M.J. Narcis, D.J. Sprague, B. Captain, N. Takenaka, Org. Biomol. Chem. *2012, 10* (46), 9134-6 [DOI: 10.1039/c2ob26674a]; inverse electron-demand **organocatalyzed asym. hetero-Diels-Alder reaction** with enals as dienophile using a bifunctional, hydrogen-bonding (S)-proline-based amine as catalyst s. L. Albrecht, G. Dickmeiss, C.F. Weise, C. Rodríguez-Escrich, K.A. Jørgensen, Angew. Chem., Int. Ed. *2012, 51* (52), 13109-13 [DOI: 10.1002/anie.201207122]; with α-subst. cyclic ketones as dienophile under bifunctional organocatalysis cf. X. Jiang, L. Wang, M. Kai, L. Zhu, X. Yao, R. Wang, Chem. Eur. J. *2012, 18* (36), 11465-73 [DOI: 10.1002/chem.201201102]; asym. *vinylogous* Povarov reaction with 1-acylamino-1,3-butadienes using a chiral 5,5′,6,6′,7,7′,8,8′-octahydro-1,1′-binaphthyl-2,2′-diyl hydrogen phosphate as organo-Brønsted acid catalyst s. L. Caruana, M. Fochi, S. Ranieri, A. Mazzanti, L. Bernardi, Chem. Commun. *2013, 49* (9), 880-2 [DOI: 10.1039/c2cc38002a]; **N-heterocyclic carbene-catalyzed asym. [4+2]-cycloaddition of ketenes** with 3-aroylcoumarins using a hydroxylated chiral N-condensed 1,2,4-triazolylidene as catalyst s. T.-Y. Jian, X.-Y. Chen, L.-H. Sun, S. Ye, Org. Biomol. Chem. *2013, 11* (1), 158-63 [DOI: 10.1039/c2ob26804c]; asym. phospha-Diels-Alder reaction with phospholes as diene component s. T. Möller, M.B. Sárosi, E. Hey-Hawkins, ibid. *18* (52), 16604-7 [DOI: 10.1002/chem.201203671].

Chiral prim. amines/benzoic acids or chiral 1,1′-binaphthyl-2,2′-diyl hydrogen phosphate ←

Organocatalyzed asym. syntheses via the Michael-Michael addition pathway ←

update s. *76,* 303s*82*; chiral 3-spiro-2-benzofuranones by **asym. double Michael addition** of 2-benzofuranone to 1,4-dien-3-ones using a cinchona-based prim. amine and 1,1′-binaphthyl-2,2′-diyl hydrogen phosphate as co-catalysts s. X. Luo, L. Wang, L. Peng, J. Bai, L. Jia, F. Tian, X. Xu, L. Wang, Chin. J. Org. Chem. *2012, 30* (11), 2703-6 [DOI: 10.1002/cjoc.201200676]; chiral 4-spiro-Δ^2-5-pyrazolones from 4-alkylidene-Δ^2-5-pyrazolones and enones by **organocatalyzed asym. Michael addition-intramolecular Michael addition** with a cinchona-based prim. amine and *o*-fluorobenzoic acid as co-catalysts s. J.-X. Zhang, N.-K. Li, Z.-M. Liu, X.-F. Huang, Z.-C. Geng, X.-W. Wang, Adv. Synth. Catal. *2013, 355* (4), 797-808 [DOI: 10.1002/adsc.201200925]; with the same chiral organocatalyst and benzoic acid cf. J. Liang, Q. Chen, L. Liu, X. Jiang, R. Wang, Org. Biomol. Chem. *2013, 11* (9), 1441-5 [DOI: 10.1039/c2ob27095a]; chiral polyfunctionalized cyclohexylacetic acid esters from ε-nitro-α,β-ethylenecarboxylic acid esters and α,β-ethylene-α-sulfonylnitriles with a chiral bifunctional thiourea as catalyst s. S. Rajkumar, K. Shankland, J.M. Goodman, A.J.A. Cobb, Org. Lett. *2013, 15* (6), 1386-9 [DOI: 10.1021/ol400 356k]; chiral tetrahydro-furans and -pyrans by **asym. oxa-Michael addition-intramolecular Michael addition** with kinetic resolution or dynamic kinetic resolution using 2(S)-[diphenyl(trimethylsiloxy)-methyl]pyrrolidine as catalyst s. P.G. McGarraugh, R.C. Johnston, A. Martínez-Muñoz, P.H.-Y. Cheong, S.E. Brenner-Moyer, Chem. Eur. J. *2012, 18* (34), 10742-52 [DOI: 10.1002/chem.201104029]; chiral 4-β-keto-3-nitro-1,2,3,4-tetrahydroquinolines by **asym. aza-Michael addition-intramolecular Michael addition** with a chiral bifunctional thiourea as catalyst s. Z.-X. Jia, Y.-C. Luo, Y. Wang, L. Chen, P.-F. Xu, B. Wang, ibid. *18* (41), 12958-61 [DOI: 10.1002/chem.201201362].

Chiral diamines, tetramines, cinchona derivs., squaramides, (S)-proline derivs., 2(S)-[diaryl(siloxy)methyl]pyrrolidines, chiral 2-tert-aminoureas, 2-tert-aminothioureas, bis(thioureas), 2-prim-aminoguanidines, 2-(aroylamino)phosphines or N-phosphinyl-1,2-diamines ←

Organocatalyzed asym. Michael addition C=C → CHC(R)

update s. *62,* 282s*82*; of nitroalkanes to enones with chiral camphor-based *1,2-diamines* as catalyst s. Y. Zhou, Q. Liu, Y. Gong, Org. Biomol. Chem. *2012, 10* (37), 7618-27 [DOI: 10.1039/c2ob25922b]; of α-chloroaldehydes to vinyl sulfones with a chiral 9,10-dihydroanthracene-bridged *trans*-1,2-diamine combined with a trifluoromethanesulfonamide and a carboxylic acid to generate a bifunctional organocatalyst *in situ* s. S.A. Moteki, P.G. Kirira, S. Arimitsu, K. Maruoka, Asian J. Org. Chem. *2012, 1* (1), 26-9 [DOI: 10.1002/ajoc.201200040]; of β-dicarbonyl compds. to succinimide with a chiral bifunctional N,N′-bis(benzimidazol-2-yl)-*trans*-cyclohexane-1,2-diamine and added trifluoroacetic acid s. E. Gómez-Torres, D.A. Alonso, E. Gómez-Bengoa, C. Nájera, Eur. J. Org. Chem. *2013* (8), 1434-40 [DOI: 10.1002/ejoc.201201046]; of ketones to nitroalkenes and chalcones with an N,N′-bis(pyrrolidin-2(S)-ylmethyl)-*trans*-cyclohexane-1,2-diamine and added 3,5-dimethylphenol s. S. Ma, L. Wu, M. Liu, Y. Wang, Chin. J. Org. Chem. *2012, 30* (11), 2707-13 [DOI: 10.1002/cjoc.201200214]; of ketones or aldehydes to nitroalkenes with a chiral N,N′-bis(pyrrolidin-2(S)-ylmethyl)-2,2′-diaminobiphenyl or a 6,6′-diamino-2,2′-bipyridyl analog as catalyst cf. H.-W. Zhao, H.-L. Li, Y.-Y. Yue, Z.-H. Sheng, Eur. J. Org. Chem. *2013* (9), 1740-8 [DOI: 10.1002/ejoc.201201306]; solvent-free generation of adjacent quaternary and tertiary chiral centers by asym. Michael addition *on grinding* the reactants with the *cinchona* deriv., 9-O-benzylcupreine, s. P. Chauhan, S.S. Chimni, Asian J. Org. Chem. *2012, 1* (2), 138-41 [DOI: 10.1002/ajoc.201200048]; asym. Michael addition of nitroalkanes to nitroalkenes with O-desmethylquinidine 9-O-acetate as catalyst s. Y.-Q. Deng, Z.-W. Zhang, Y.-H. Feng, A.S.C. Chan, G. Lu, Tetrahedron: Asym. *2012, 23* (24), 1647-52 [DOI: 10.1016/j.tetasy.2012.11.008]; of α-nitrocarboxylic acid esters to nitroalkenes with 1,4-bis[9-O-(O-desmethyl)dihydroquinine]phthalazine as catalyst (at 1 mol%) s. Y.-Z. Li, F. Li, P. Tian, G.-Q. Lin, Eur. J. Org. Chem. *2013* (8), 1558-65 [DOI: 10.1002/ejoc.201201444]; of β-dicarbonyl compds. to nitroalkenes with a polyfunctionalized cinchona-based *squaramide* as catalyst s. Z. Dong, G. Qiu, H.-B. Zhou, C. Dong, Tetrahedron: Asym. *2012, 23* (22-23), 1550-6 [DOI: 10.1016/j.tetasy.2012.10.016]; of α-nitrophosphonic acid esters to vinyl sulfones with a quinine-based squaramide s. K. Bera, I.N.N. Namboothiri, Adv. Synth. Catal. *2013, 355* (7), 1265-70 [DOI: 10.1002/adsc.201300224]; of ketones to nitroalkenes with an *imidazolium 4(R)-sulfato-(S)-prolinate* as ionic catalyst in [bmim]NTf_2 s. V. Gauchot, J. Gravel, A.R. Schmitzer, Eur. J. Org. Chem. *2012* (31), 6280-4 [DOI: 10.1002/ejoc.201201068]; asym. Michael addition of dimethyl malonate to 6-subst. 2-cyclohexenones *with dynamic kinetic resolution* using 5-(pyrrolidin-2(S)-yl)tetrazole as catalyst with added piperidine s. G. Pandey, P.A. Adate, V.G. Puranik, Org. Biomol. Chem. *2012, 10* (41), 8260-7 [DOI: 10.1039/c2ob26597d]; of 2-(1-ethylpropoxy)acetaldehyde to N-[(1Z)-2-nitroethenyl]acetamide with *2(S)-[diphenyl-(trimethylsiloxy)methyl]pyrrolidine* as catalyst s. V. Hajzer, A. Latika, J. Durmis, R. Šebesta, Helv. Chim. Acta *2012, 95* (12), 2421-8 [DOI: 10.1002/hlca. 201200527]; of aldehydes to trisubst. nitroalkenes with the same catalyst and added benzoic acid for preparing chiral *syn,anti*-adducts s. L. Wang, X. Zhang, D. Ma, Tetrahedron *2012, 68* (37), 7675-9 [DOI: 10.1016/j.tet.2012.05.088]; of malonates to α,β-ethylenealdehydes *in aq. isopropanol* with readily recyclable 2(S)-[bis[bis-(*p*-dimethylaminomethyl)phenyl](trimethylsiloxy)methyl]pyrrolidine and added benzoic acid s. S.K. Ghosh, K. Dhungana, A.D. Headley, B. Ni, Org. Biomol. Chem. *2012, 10* (41), 8322-5 [DOI: 10.1039/c2ob26248g]; of aldehydes to nitroalkenes with 2(S)-[diaryl(siloxy)methyl]pyrrolidines, catalyst survey, s. D. Seebach, X. Sun, M.-O. Ebert, W.B. Schweizer, N. Purkayastha, A.K. Beck, J. Duschmalé, H. Wennemers, T. Mukaiyama, M. Benohoud, Y. Hayashi, M. Reiher, Helv. Chim. Acta *2013, 96* (5), 799-852 [DOI: 10.1002/hlca.201300079]; of cyclic ketones to 1-nitro-1,3-dienes with 3-(pyrrolidin-2(S)-ylmethyl)-2-thiouracils as catalyst and added *p*-nitrobenzoic acid s. M. Tsakos, M. Trifonidou, C.G. Kokotos, Tetrahedron *2012, 68* (41), 8630-5 [DOI: 10.1016/j.tet.2012.07.078]; of benzoylacetates to cyclic enones with *lithium 3(S)-amino-4-(tert-butyldiphenylsiloxy)butyrate* as catalyst s. M. Yoshida, A. Kubara, S. Hara, Chem. Lett. *2013, 42* (2), 180-2 [DOI: 10.1246/cl.2013.180]; of Δ^2-5-pyrazolones to nitroalkenes with a chiral *2-tert-aminothiourea* as catalyst s. F. Li, L. Sun, Y. Teng, P. Yu, J.C.-G. Zhao, J.-A. Ma, Chem. Eur. J. *2012, 18* (45), 14255-60 [DOI: 10.1002/chem.201202636]; of malonates to 3-cyano-2-imino-3-chromenes s. W. Li, J. Huang, J. Wang, Org. Biomol. Chem. *2013, 11* (3), 400-6 [DOI: 10.1039/

c2ob27102h]; of α-subst. cyanoacetates to maleimides with a chiral isosteviol-based 2-*tert*-aminothiourea as catalyst s. Z.-w. Ma, Y. Wu, B. Sun, H.-l. Du, W.-m. Shi, J.-c. Tao, Tetrahedron: Asym. *2013, 24* (1), 7-13 [DOI: 10.1016/j.tetasy.2012.11.009]; of activated 2-subst. N-protected 3-indolinones to nitroalkenes with a cinchona-based 2-*tert*-aminothiourea s. C.-Y. Jin, Y. Wang, Y.-Z. Liu, C. Shen, P.-F. Xu, J. Org. Chem. *2012, 77* (24), 11307-12 [DOI: 10.1021/jo301886j]; of 3-aryloxindoles to an ethylene-1,1-di(phosphonic acid ester) s. M.-X. Zhao, T.-L. Dai, R. Liu, D.-K. Wei, H. Zhou, F.-H. Ji, M. Shi, Org. Biomol. Chem. *2012, 10* (39), 7970-9 [DOI: 10.1039/c2ob25966d]; of Δ^2-5-thiazolones to nitroalkenes s. X. Liu, H. Song, Q. Chen, W. Li, W. Yin, M. Kai, R. Wang, Eur. J. Org. Chem. *2012* (33), 6647-55 [DOI: 10.1002/ejoc.201201094]; of oxo compds. to maleimides with a cinchona-based 2-*tert*-aminothiourea and added (S)-(*o*-chlorophenyl)glycine s. S. Muramulla, J.-A. Ma, J.C.-G. Zhao, Adv. Synth. Catal. *2013, 355* (7), 1260-4 [DOI: 10.1002/adsc.201300041]; of formaldehyde N-*tert*-butyl hydrazone to nitroalkenes (affording chiral 3-azonitro compds., effectively in a diaza-ene reaction) using a chiral 1,1′-binaphthyl-based *bis(thiourea)* as catalyst s. D. Monge, S. Daza, P. Bernal, R. Fernández, J.M. Lassaletta, Org. Biomol. Chem. *2013, 11* (2), 326-35 [DOI: 10.1039/c2ob26963e]; of nitromethane to β-subst. α,β-ethylenealdehydes in aq. methanol with a polymer-based peptide as catalyst s. K. Akagawa, K. Kudo, Angew. Chem., Int. Ed. *2012, 51* (51), 12786-9 [DOI: 10.1002/anie.201206916]; of α-subst. aldehydes to maleimides in aq. DMF with a chiral *trans*-cyclohexane-1,2-diamine-based *2-prim-aminoguanidine* as catalyst s. A.R. Chinchilla, C. Nájera, Tetrahedron: Asym. *2012, 23* (24), 1625-7 [DOI: 10.1016/j.tetasy.2012.11.002]; of ketones to 3-halogeno-1-nitro-1,3-dienes with 2(S)-(2-pyridylthiomethyl)pyrrolidine as catalyst and added *p*-(trifluoromethyl)benzoic acid as co-catalyst without solvent s. A.-B. Xia, C. Wu, D.-Q. Xu, Y.-F. Wang, X.-H. Du, Z.-B. Li, Z.-Y. Xu, J. Org. Chem. *2013, 78* (3), 1254-9 [DOI: 10.1021/jo302288z]; of N-protected oxindoles to Michael acceptors with a chiral *2-(aroylamino)phosphine* as catalyst s. F. Zhong, X. Dou, X. Han, W. Yao, Q. Zhu, Y. Meng, Y. Lu, Angew. Chem., Int. Ed. *2013, 52* (3), 943-7 [DOI: 10.1002/anie.201208285]; of 4-hydroxycoumarins to enones with a chiral *N-(diphenylphosphinyl)-1,2-diamine* as catalyst with added *p*-toluic acid s. J. Dong, D.-M. Du, Org. Biomol. Chem. *2012, 10* (40), 8125-31 [DOI: 10.1039/c2ob26334c]; of ketones to alkylidenemalonates *in water* with an N-(2(S)-pyrrolidinomethyl)-1,1′-binaphthyl-2,2′-diyl*disulfonimide* as catalyst s. S. Jia, C. Luo, D. Du, Chin. J. Org. Chem. *2012, 30* (11), 2676-80 [DOI: 10.1002/cjoc.201200910]; (E)-selective Michael addition of Δ^2-4-oxazolones **to α,β-acetyleneketones** with a chiral *8-α-hydroxy-1,5,7-triazabicyclo[4.3.0]non-5-ene* as catalyst s. T. Misaki, N. Jin, K. Kawano, T. Sugimura, Chem. Lett. *2012, 41* (12), 1675-7 [DOI: 10.1246/cl.2012.1675]; (Z)-selective addition of N-protected oxindoles to an acetylenedicarboxylate with a chiral *2-tert-aminourea* as catalyst s. G. Kang, Q. Wu, M. Liu, Q. Xu, Z. Chen, W. Chen, Y. Luo, W. Ye, J. Jiang, H. Wu, Adv. Synth. Catal. *2013, 355* (2-3), 315-20 [DOI: 10.1002/adsc.201200715].

1,5,7-Triazabicyclo[4.4.0]dec-5-ene *TBD*

O$^{\beta}$,O$^{\delta}$-Alkylidene-β,γ-ethylene-β,β′,δ-trihydroxycarboxylic acid esters from α-allenecarboxylic acid esters and three aldehyde molecules via diastereoselective sequential Baylis-Hillman reaction-acetalation-intramolecular Michael addition ○

284.

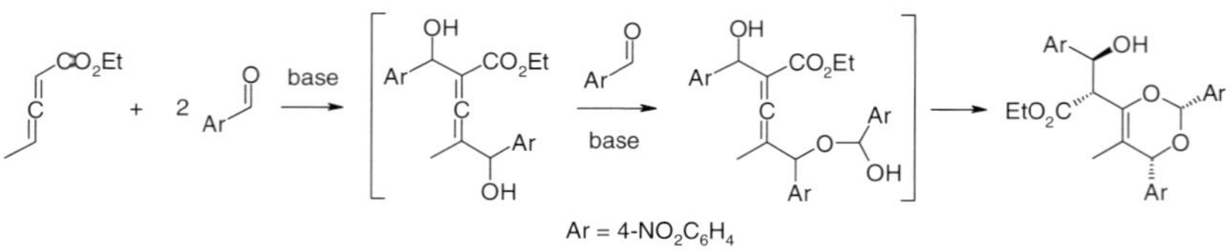

Four stereogenic carbon centers have been created in one pot with high diastereoselectivity from *γ-alkyl-substituted* α-allenoates and three aldehyde molecules, made possible by inverting the traditional electrophilic reactivity at both the α and [for the first time] the γ-site via sequential Baylis-Hillman reaction with 1,5,7-triazabicyclo[4.4.0]dec-5-ene as Lewis base; the latter then serves as a Brønsted base to effect acetalation, followed by intramolecular Michael addition to complete the reaction. **E:** 1,5,7-Triazabicyclo[4.4.0]dec-5-ene (15 mol%) added in one portion to a soln. of ethyl 2,3-pentadienoate (856 µmol) and 4-nitrobenzaldehyde (4 eq.) in acetonitrile (17 ml) at -25°, the mixture stirred at -20° to -25° for 1 h, quenched with solid NH_4Cl (excess), warmed to room temp., filtered, and the filtrate worked up with purification by flash chromatography on silica → product. Y 59% (d.r. 95/5). Given the complexity of the reaction, yields were good for the coupling of benzaldehyde itself and electron-deficient substrates (twenty-seven examples; Y 41-60%), but yields were significantly lower (even at 25 mol% catalyst loading) with electron-rich ar. aldehydes (21-50%; four examples) and with thiophene-2-carboxaldehyde (Y 15%) due to complicating retro-aldol cleavage. Aliphatic aldehydes were also low yielding (0-23%; two examples). The presumed intermediate 2,3-diene-1,5-diol was isolable. F.e. and comparison of Lewis bases and solvents, also application on the gram scale, s. P. Selig, A. Turockin, W. Raven, Adv. Synth. Catal. *2013*, *355* (2-3), 297-302 [DOI: 10.1002/adsc.201200807].

8(S)-[Diphenyl(hydroxy)methyl]-1,5,7-triazabicyclo[4.3.0]non-5-enes ←
Asym. Michael addition-intramolecular Henry reaction s. *67,* 313s*83* ○

Quinine ←
Asym. Michael addition-thiolactamization s. *67,* 313s*83*

O-Desmethyl-9-amino-9-deoxy-epi-quinidine/2,6-bis(trifluoromethyl)benzoic acid ←
Organocatalyzed asym. 1,6-addition-intramolecular aldol condensation

285. 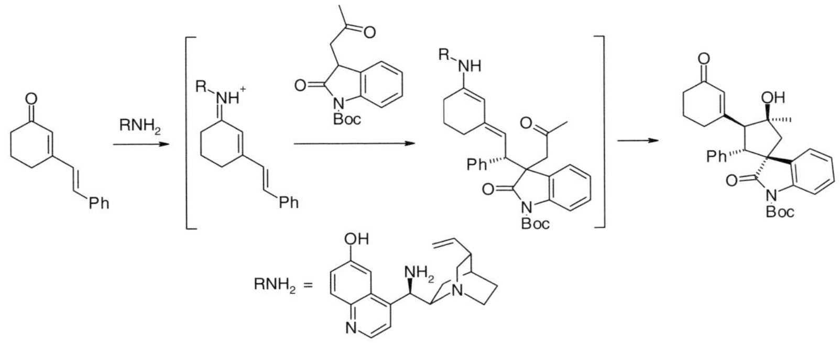

The first example of an efficient, highly enantioselective organocatalyzed 1,6-addition is reported wherein the stereochemical information is transferred *remotely through a conjugated π-system* from the chiral center of an intermediate iminium ion. This is illustrated in a novel asym. 1,6-addition-intramolecular aldol condensation to give **chiral spirocyclic β-(2-hydroxycyclopentyl)-α,β-ethyleneketones. E:** 2,6-Bis(trifluoromethyl)benzoic acid (15 mol%) and toluene (1 ml) added sequentially to a vial charged with O-desmethyl-9-amino-9-deoxy-*epi*-quinidine (10 mol%), the mixture stirred at room temp. for 10 min to allow the catalyst salt formation, the startg. dienone (1 mmol) and oxindole (1.5 eq.) added sequentially, the vial sealed, immersed in a silicon bath (at 40°), stirring continued for 48 h, the crude mixture flushed through a short plug of silica, solvent removed under reduced pressure, and the residue worked up with purification by column chromatography → product. Y 83% (d.r. 12.5:1; e.e. 97%). Reaction supports electron-diverse [hetero]aryl residues at the δ-position of the dienone (irrespective of ring-substitution pattern) as well as various functional groups on the benzene ring of the oxindole; an alkyl group at the dienone terminus is also acceptable (albeit with reduced diastereoselectivity), and the isocyclic ring may be 6-membered (fourteen examples in all; Y 50-86%; d.r. 3.2:1 to >20:1; e.e. 87-97%). There was no reaction, however, with acyclic dienones. Interestingly, ring closure generates four contiguous chiral centers, of which two are quaternary. Mechanistically, it is presumed that the organocatalyst activates both reactants simultaneously, with the acidic co-catalyst enhancing reactivity overall. F.e.s. X. Tian, P. Melchiorre, Angew. Chem., Int. Ed. *2013*, *52* (20), 5360-3 [DOI: 10.1002/anie.201301017].

Pyridine N-oxides s.a. under [Bis(2,4-di-tert-butylphenyl) 2′,4′,6′-trimethoxybiphenyl-2-ylphosphonite]gold(I) triflimide and (Triphenylphosphine)gold(I) chloride ←

2,6-Dichloropyridine N-oxide/methanesulfonic acid ←
2,3-Dihydro-4-quinolones from *o*-acetylene-*tert*-amines ○
Metal-free oxidative ring closure

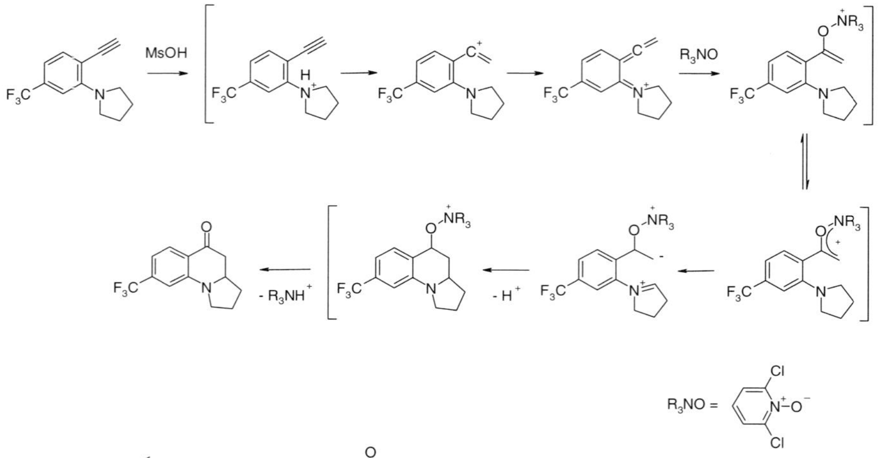

286.

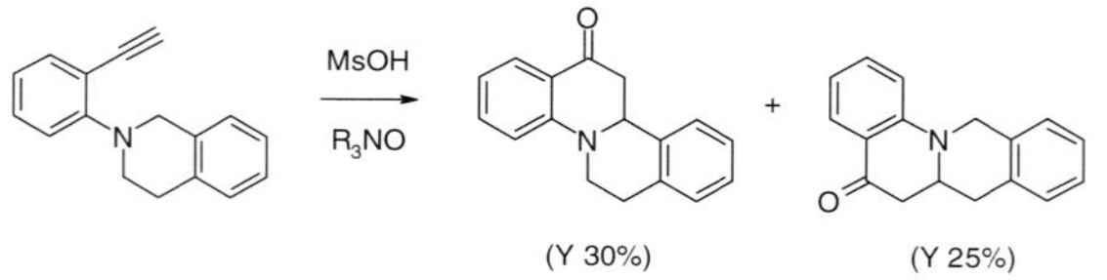

A soln. of methanesulfonic acid (4 eq.) in methylene chloride (1 ml) added to a soln. of 2-ethynyl-5-trifluoromethyl-1-pyrrolidin-1-ylbenzene (0.2 mmol) and 2,6-dichloropyridine N-oxide (2 eq.) in the same solvent (1 ml) at 30°, the mixture stirred for 12 h, quenched with triethylamine, solvent evaporated, and the residue purified by chromatography on silica → 8-trifluoromethyl-2,3,3a,4-tetrahydropyrrolo[1,2-*a*]quinolin-5(1*H*)-one. Y 76%. This metal-free cyclization was effected by pyridine N-oxide oxidants and crucially promoted by methanesulfonic acid. A series of electron-diverse 2-ethynylpyrrolidin-1-ylbenzenes were cyclized to the corresponding tricyclic dihydroquinolones (twelve examples; Y 48-84%), with the reaction giving a moderate yield with the N,N-dibenzylaniline analog (40%) and cyclizing regioselectively (Y 50%) at the benzylic C for an N-methyl-N-benzyl analog. Application to a piperidine homolog, N-(2-ethynylphenyl)tetrahydroisoquinoline, gave a mixture of the two regioisomeric tetracycles with little regioselectivity (25:30; Y 55%). A mechanism is proposed but alternative pathways could not be completely excluded. F.e., optimization and substrate prepn., s. D.-F. Chen, Z.-Y. Han, Y.-P. He, J. Yu, L.-Z. Gong, Angew. Chem., Int. Ed. *2012*, *51* (49), 12307-10 [DOI: 10.1002/anie.201205062].

Cyclic bis(N-oxides) s. under Chiral trans-perhydroindolic acid ←

Tetrakis(acetonitrile)copper(I) fluoroborate/(S)-2-amino-3,3′-dibromo-2′-diphenylphosphinamino-4,4′,6,6′-tetrakis(trifluoromethyl)biphenyl/cesium carbonate ←

2,3,4,7a-Tetrahydro-1*H*-cyclopenta[*c*]pyridine-3-carboxylic acid esters from fulvenes and (alkylideneamino)acetic acid esters via copper(I)-catalyzed asym. [6+3]-cycloaddition ○

287.

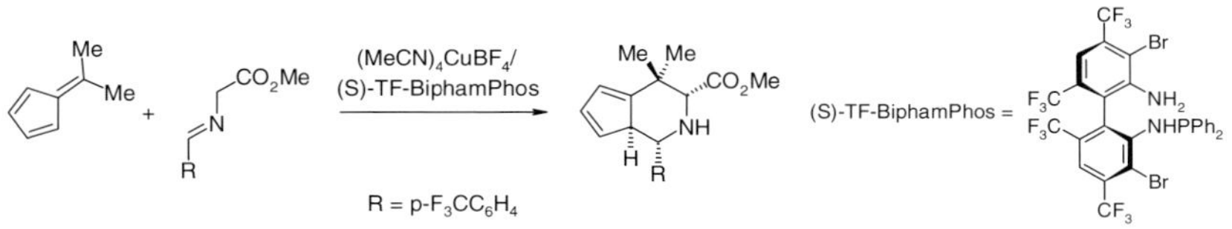

A soln. of (S)-TF-BiphamPhos (0.015 mmol) and $(MeCN)_4CuBF_4$ (0.013 mmol) in methylene chloride (1.5 ml) stirred at room temp. for ca. 30 min under argon, then startg. imine (0.25 mmol) and Cs_2CO_3 (0.25 mmol) added sequentially, the mixture cooled to -40°, startg. fulvene (0.5 mmol) added, when reaction complete by TLC, the organic solvent removed, and the residue purified by column chromatography → (1S,3R,7aS)-methyl 4,4-dimethyl-1-[4-(trifluoromethyl)phenyl]-2,3,4,7a-tetrahydro-1*H*-cyclopenta[*c*]pyridine-3-carboxylate. Y 87% (e.e. 99%). In this mild method, nineteen further examples with (alkylideneamino)acetic acid esters derived from electron-diverse ar., hetar. or aliphatic aldehydes and sym. fulvenes gave yields of 65-92% and 86-99% e.e., incl. formation of spiro compds., while five examples with unsym. (monosubst.) fulvenes gave yields of 55-73%, d.r. 4:1 to 6:1 and e.e. 90-93% for the major isomer (85-94% for the minor isomer). For copper(I)-catalyzed asym. [6+3]-cycloaddition-[4+2]-cycloaddition cf. *82*, 262. F.e., optimization and reactions of the products (reduction, Diels-Alder reactions) s. Z.-L. He, H.-L. Teng, C.-J. Wang, Angew. Chem., Int. Ed. *2013*, *52* (10), 2934-8 [DOI: 10.1002/anie.201208799].

Chiral copper(II) bis(Δ²-oxazoline) complexes *[Cu(II)]**

Ring expansion of cyclopropanes by [3+2]-cycloaddition ○

with ethylene derivs. under copper catalysis cf. *42*, 654s*77*; condensed 2-siloxycyclopentane-1,1-dicarboxylic from cyclopropane-1,1-dicarboxylic acid esters and cyclic enoxysilanes by **asym. ring expansion** with a chiral copper(II) bis(Δ²-oxazoline) complex as catalyst s. H. Xu, J.-P. Qu, S. Liao, H. Xiong, Y. Tang, Angew. Chem., Int. Ed. *2013, 52* (14), 4004-7 [DOI: 10.1002/anie.201300032]; electron-deficient vinyl-cyclopentanes from -cyclopropanes and electron-deficient ethylene derivs. under palladium catalysis with remote asym. induction s. B.M. Trost, P.J. Morris, S.J. Sprague, J. Am. Chem. Soc. *2012, 134* (42), 17823-31 [DOI: 10.1021/ja309003x].

Copper(I) bis(trimethylsilyl)amide/(S)-2,2′-bis[bis(3,5-diisopropylphenyl)phosphino]-1,1′-binaphthyl ←

N-Condensed Δ³-pyrazolines from cyclic azomethine imides and terminal acetylene derivs. via regioselective copper(I)-catalyzed asym. 1,3-dipolar cycloaddition ○

288.

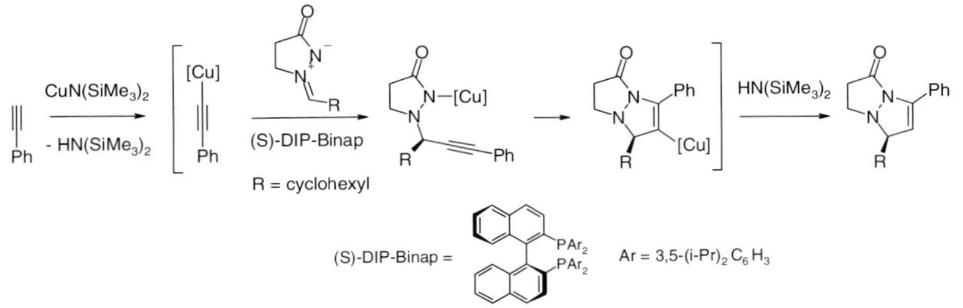

A soln. of $CuN(SiMe_3)_2$ (5 mol%) and (S)-DIP-Binap (5.5 mol%) in cyclopentyl methyl ether (2 ml) prepared in a glovebox stirred at room temp. for 30 min, 5 Å molecular sieves (50 mg) and startg. azomethine imide (0.4 mmol) added, phenylacetylene (1.1 eq.) added over 16 h to the stirred mixture at 40°, stirring continued for 8 h, quenched with water, extracted with methylene chloride, and purified chromatographically → (R)-5-cyclohexyl-7-phenyl-2,3-dihydro-5*H*-pyrazolo-[1,2-*a*]pyrazol-1-one. Y 94% (e.e. 95%). The reaction appears general for electron-diverse aryl- and alkyl-subst. substrates, affording novel chiral 5,7-disubst. 2,3-dihydro-5*H*-pyrazolo[1,2-*a*]-pyrazol-1-ones (seventeen examples; Y 83-96%; e.e. 82-95%), with the use of the strongly basic disilylamide essential for driving the cyclization to completion. Interestingly, use of bis(oxazoline) or 2,2′-bipyridyl in place of 1,1′-binaphthyl ligands, in one case, resulted in complete reversal of regioselectivity, affording a 5,6-disubst. deriv. (Y >99%; 5,6/5,7 >99:1), while 2,2′-disubst. biphenyl ligands afforded mixtures of regioisomers.

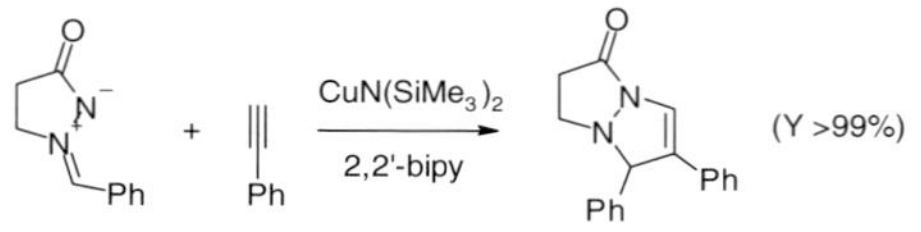

F.e. and optimization s. T. Imaizumi, Y. Yamashita, S. Kobayashi, J. Am. Chem. Soc. *2012*, *134* (49), 20049-52 [DOI: 10.1021/ja311150n].

Copper(II) acetate/chiral N,N′-bis[2-phenyl-2-(tosylamino)ethyl]-1,2-diphenyl-ethylenediamine/camphorsulfonic acid ←

Asym. Henry reaction-intramolecular oxa-Michael addition ○

Chiral 2-β-keto-6-(nitromethyl)tetrahydropyrans s. *83*, 261

Silver carbonate *Ag_2CO_3*

Pyrroles from acetylene derivs. and prim. isonitriles

289.

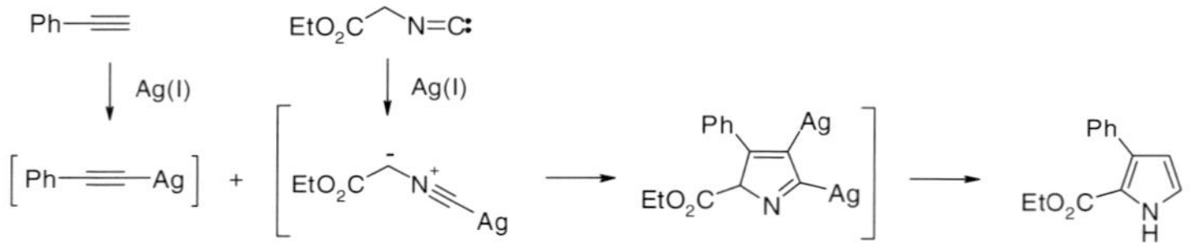

A simple, efficient, mild, rapid and atom-economical ['click'] pyrrole synthesis has been devised by silver-catalyzed [3+2]-cycloaddition of acetylene derivs. and isonitriles, inspired by the classical synthesis of 1,2,3-triazoles by copper-catalyzed [3+2]-cycloaddition of terminal acetylene derivs. with azides. **E: Pyrrole-2-carboxylic acid esters from terminal acetylene derivs.** A mixture of phenylacetylene (0.5 mmol), ethyl isocyanoacetate (1.5 eq.) and Ag_2CO_3 (10 mol%) in N-methyl-2-pyrrolidone (3 ml) stirred under N_2 at 80° for 1 h, the solid filtered off then washed with methylene chloride, the combined filtrate and washings concentrated, and worked up with purification by flash chromatography on silica gel → ethyl 3-phenylpyrrole-2-carboxylate. Y 89%. There was no undesirable homocoupling of the alkyne. The procedure is applicable to the coupling of a wide range of terminal aliphatic alkynes and electron-diverse ethynylarenes (irrespective of the position of the substituent on the benzene ring) with ethyl isocyanoacetate or diethyl isocyanomethyl-phosphonate (sixteen examples; Y 40-98%). Other silver and copper salts were ineffective. The silver salt is presumed to activate both partners in the cycloaddition to give a dimetalated pyrrolenine prior to protonation and tautomerization. F.e. and solvent effect, also from internal alkynes (two examples; Y 89-99%), and from the pre-formed silver complexes, s. M. Gao, C. He, H. Chen, R. Bai, B. Cheng, A. Lei, Angew. Chem., Int. Ed. *2013*, *52* (27), 6958-61 [DOI: 10.1002/anie.201302604].

Copper(I) or copper(II) triflate/chiral 2-(o-hydroxybenzylaminomethyl)-Δ²-imidazolines ←

Asym. Friedel-Crafts reaction with electron-deficient ethylene derivs. C=C → CHC(Ar) **by 1,4-addition of indoles**, update s. *67*, 336s*82*; asym. 1,4-addition to 3-(nitromethylene)oxindoles with $Cu(OTf)_2$ or CuOTf and a chiral 2-(*o*-hydroxybenzylaminomethyl)-Δ^2-imidazoline as ligand

s. T. Arai, Y. Yamamoto, A. Awata, K. Kamiya, M. Ishibashi, M.A. Arai, Angew. Chem., Int. Ed. *2013, 52* (9), 2486-90 [DOI: 10.1002/anie.201208918]; to 2-(trifluoromethyl)-1-nitroethylene derivs. with $Ni(ClO_4)_2 \cdot 6H_2O$ and a chiral bis(Δ^2-oxazoline) as ligand s. J.-R. Gao, H. Wu, B. Xiang, W.-B. Yu, L. Han, Y.-X. Jia, J. Am. Chem. Soc. *2013, 135* (8), 2983-6 [DOI: 10.1021/ja400650m]; of N-methylindoles to nitroalkenes with $Zn(OTf)_2$ and a chiral abietic acid-based 2-*prim*-amino-thiourea as ligand s. W.-g. Huang, H.-s. Wang, G.-b. Huang, Y.-m. Wu, Y.-m. Pan, Eur. J. Org. Chem. *2012* (29), 5839-43 [DOI: 10.1002/ejoc.201200584]; **organocatalyzed asym. Friedel-Crafts reaction** of indoles with oxindol-3-ylideneacetaldehydes using a chiral 2-[diaryl(trimethylsiloxy)-methyl]pyrrolidine as catalyst s. R. Liu, J. Zhang, Org. Lett. *2013, 15* (9), 2266-9 [DOI: 10.1021/ol400845c].

Copper(I) chloride/1,2-bis(diphenylphosphino)benzene/sodium tert-butoxide/ pinacolborane/potassium tert-butoxide ←

Generation of benzyl carbanion equivalents via regioselective copper(I)-catalyzed hydroboration ←

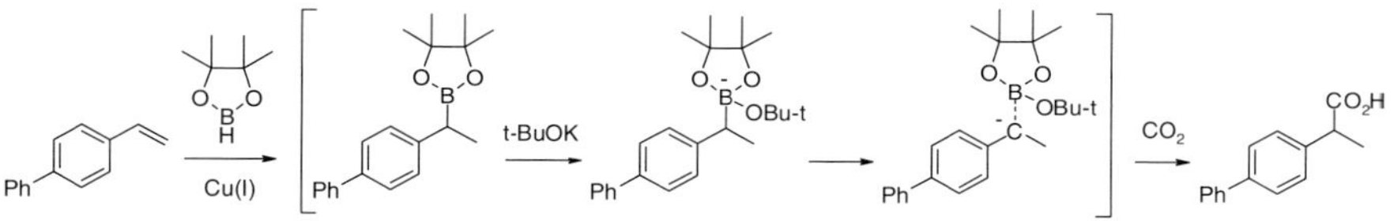

290.

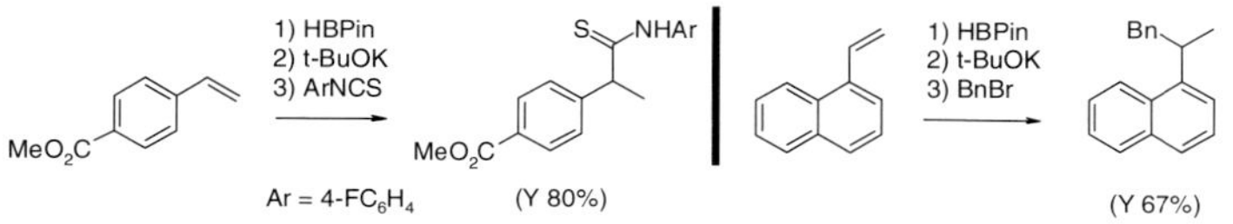

α-Arylcarboxylic acid derivs. from styrenes in one pot. A mixture of CuCl (3 mol%), dppbz (3.3 mol%) and Na-*tert*-butoxide (3 mol%) in dry toluene (0.33 ml) stirred at room temp. for 10 min, pinacolborane (1.1 eq.) added, the mixture stirred for 10 min, a soln. of 4-phenylstyrene (1 eq.) in the same solvent added by syringe, the mixture stirred at 65° overnight, diluted with THF (3:2 toluene/THF), a soln. of K-*tert*-butoxide (2.2 eq.) in THF (0.22 ml) added, the mixture stirred for 15 min, cooled to -78°, stirred under CO_2 (1 atm.) for 1 h, quenched with aq. HCl, extracted with methylene chloride, and purified by chromatography on silica → 2-biphenyl-4-yl-propanoic acid. Y 78%. Initial experiments identified optimum conditions for convenient generation of benzyl carbanions from bulky boronates, using a bulky base, and their carboxylation with CO_2 (ten examples; Y 69-85%). The conditions proved to be compatible with *in situ* generation of boronates from electron-diverse styrene precursors, and a one-pot procedure was developed wherein the generated benzyl carbanion was quenched with electrophiles (CO_2, CS_2, RNCS, RNCO) to afford α-aryl-(thio)carboxylic acids or amides (twenty-six examples; Y 62-92%); styrenes containing ester substituents gave lower yields with isocyanate electrophiles (26-47%; four examples). Alkylation of the benzylic carbanions with alkyl/benzyl halides was also demonstrated (three examples; Y 59-67%), but use of other electrophiles proved problematic, and the method was not applicable to non-benzylic boronates. F.e., optimization, and use of the method in a short synthesis of flurbiprofen s. R.D. Grigg, J.W. Rigoli, R. Van Hoveln, S. Neale, J.M. Schomaker, Chem. Eur. J. *2012*, *18* (30), 9391-6 [DOI: 10.1002/chem.201200642].

Copper(I) bromide/1,10-phenanthroline/trimethylsilyl cyanide *CuBr/phen/Me_3SiCN*

β-(Disulfonylamino)nitriles from ethylene derivs. C=C → $C(N(SO_2R)_2)C-CN$

by regio- and stereo-selective copper(I)-catalyzed 1,2-aminocyanation

291.

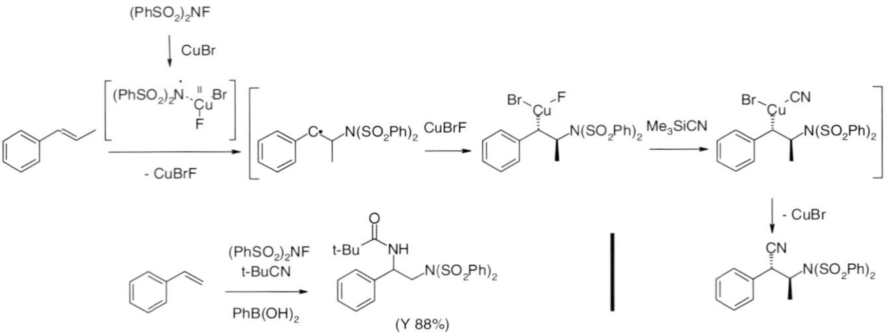

A novel copper-promoted method for the generation of a nitrogen-centered radical from N-fluorobenzenesulfonimide is reported. The radical may be employed in the presence of trimethylsilyl cyanide for regioselective aminocyanation of alkenes, or in the presence of organic nitriles and benzeneboronic acid for regioselective diamination, with the nitrile unusually serving as a second nitrogen source. **E:** Trimethylsilyl cyanide (1.4 eq.) and (E)-prop-1-enylbenzene (0.5 mmol) added to a mixture of CuBr (10 mol%), 1,10-phenanthroline (10 mol%) and N-fluorobenzenesulfonimide (1.4 eq.) in methylene chloride (2 ml), the mixture sealed, stirred at 70° for 1 h, quenched with water, extracted with methylene chloride, and purified by chromatography on silica → *trans*-N-(1-cyano-1-phenylpropan-2-yl)-N-(phenylsulfonyl)benzenesulfonamide. Y 85%. The (Z)-isomer gave the same yield of the same product. Successful cyanoamination was achieved for electron-diverse terminal styrenes (thirteen examples; Y 68-92%), with the illustrated β-methyl deriv. (and indene) affording *trans* adducts (Y 78-85%). Low yields were obtained with 4-methoxystyrene (25%), vinyl alkyl ethers (21-40%; two examples) and vinylcyclohexane (20%). The use of inorganic cyanides was ineffective. Alternative diamination with electron-diverse ar. or aliphatic nitriles in the presence of $PhB(OH)_2$ was successful for electron-diverse styrenes to afford **differentially-protected 1,2-diamines** (nineteen examples; Y 40-97%), with lowest yields generally obtained from electron-rich substrates (4-methoxystyrene, vinyl ethers and unactivated alkenes were unreactive), and β-methylstyrene affording 1:1 mixtures of *cis/trans* adducts (two examples; Y 80-92%). F.e. and optimization s. H. Zhang, W. Pu, T. Xiong, Y. Li, X. Zhou, K. Sun, Q. Liu, Q. Zhang, Angew. Chem., Int. Ed. *2013*, *52* (9), 2529-33 [DOI: 10.1002/anie.201209142].

Copper(I) iodide *CuI*

5-Alkylidene-2-oxazolidones ○

from two terminal acetylene deriv. molecules, prim. amines and carbon dioxide

292.

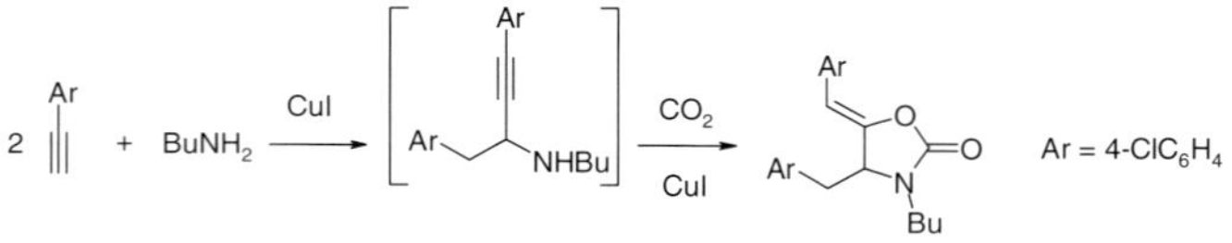

5-Arylidene-4-benzyl-2-oxazolidones. A mixture of CuI (5 mol%), 4-chlorophenylacetylene (2 mmol) and butylamine (1 eq.) heated at 90° under CO_2 (2 MPa) in a stainless steel autoclave for 24 h, cooled to 0°, pressure released slowly, extracted with ethyl acetate, and purified by chromatography on silica → (Z)-4-(4-chlorobenzyl)-5-(4-chlorobenzylidene)-3-butyloxazolidin-2-one. Y 81%. This **copper(I)-catalyzed multi-component ring closure** was effective **under solvent-**

free conditions for reaction of aliphatic prim. amines (incl. branched alkyl and benzyl) with electron-diverse terminal arylacetylenes (thirteen examples; Y 69-89%), with somewhat lower yields obtained for sterically hindered 2-methyl- (58%) and 4-propylphenyl-phenylacetylenes (52%), and no cyclization observed with 4-nitrophenylacetylene or 1-pentyne. Other copper catalysts were inferior, as were Au(I) and Ag(I) salts. F.e. and optimization s. J. Zhao, H. Jiang, Tetrahedron Lett. *2012*, *53* (51), 6999-7002 [DOI: 10.1016/j.tetlet.2012.10.073].

Copper(I) chloride *CuCl*
Regio- and stereo-selective synthesis of enol triflates C≡C → C(R)=C(OTf)
from acetylene derivs. and unsatd. iodonium triflates s. *83*, 405

Copper(I) salts s.a. under Magnesium and Triorganoaluminum compds. *Cu^+*

Copper(I) or silver(I) salts/chiral phosphines, di(phosphines) or N-phosphinodiamines ←
Asym. 1,3-dipolar cycloaddition of azomethinium ylids ○
for the synthesis of **chiral pyrrolidines**, update s. *67*, 301s*82*; with methyl α-fluoroacrylate using $[Cu(MeCN)_4]BF_4/Et_3N$ and chiral 3,3′-dibromo-4,4′,6,6′-tetrakis(trifluoromethyl)-N-(diphenylphosphino)-2,2′-diaminobiphenyl [TF-BiphamPhos] as ligand s. D. Yan, Q. Li, C. Wang, Chin. J. Org. Chem. *2012*, *30* (11), 2714-20 [DOI: 10.1002/cjoc.201200765]; chiral 4,7-dihydroxyisoindoline-1-carboxylic acid esters by one-pot aromatizing cycloaddition to *p*-quinones with $[Cu(MeCN)_4]BF_4$/triethylamine and (S,R_P)-2-(1-methoxyethyl)ferrocenyl(diphenyl)phosphine as ligand, followed by treatment with silica gel, s. Z. He, T. Liu, H. Tao, C.-J. Wang, Org. Lett. *2012*, *14* (24), 6230-3 [DOI: 10.1021/ol302987h]; **desymmetrization** by cycloaddition to 4-spiro-2,5-cyclohexadienones with $AgOAc/Et_3N$ and the same chiral ligand s. K. Liu, H.-L. Teng, L. Yao, H.-Y. Tao, C.-J. Wang, ibid. *2013*, *15* (9), 2250-3 [DOI: 10.1021/ol400821q]; asym. synthesis of 4-acylaminopyrrolidine-2,4-dicarboxylic acid esters via cycloaddition to alkylideneazlactones with AgOAc and (R)-DTBM-SegPhos as ligand s. M. González-Esguevillas, J. Adrio, J.C. Carretero, Chem. Commun. *2013*, *49* (41), 4649-51 [DOI: 10.1039/c3cc41663a]; asym. cycloaddition to maleimides and 1,2-bis(phenylsulfonyl)ethylene with $AgSbF_6$ or $AgOCOCF_3$ and (S)-BINAP as ligand s. J. Mancebo-Aracil, M. Martín-Rodríguez, C. Nájera, J.M. Sansano, P.R.R. Costa, E. Crizanto de Lima, A.G. Dias, Tetrahedron: Asym. *2012*, *23* (22-23), 1596-606 [DOI: 10.1016/j.tetasy.2012.10.015]; asym. *exo*-selective synthesis of spiro[pyrrolidine-3,3′-oxindole]s from 3-alkylideneoxindoles with $Ni(OAc)_2/Et_3N$ and a chiral 2-(*o*-hydroxybenzylaminomethyl)-1-tosyl-Δ^2-imidazoline as ligand s. A. Awata, T. Arai, Chem. Eur. J. *2012*, *18* (27), 8279-82 [DOI: 10.1002/chem.201201249].

Copper(II) salt s. under Magnesium, Dialkylzinc and Triorganoaluminum compds. *Cu^{2+}*
*Silver hexafluoroantimonate s. under Potassium tert-butoxide, [2-(Di-tert-butylphosphino)biphenyl]gold(I) chloride and [Cp*RhCl$_2$]$_2$* *$AgSbF_6$*

Silver triflate/(S)-proline *AgOTf/(S)-Pro-OH*
Synthesis of 4-functionalized 1(Z)-alkylidene-1,2,3,4-tetrahydrophthalazine ring ○
from cyclic *o*-acetyleneazomethine imides via regioselective silver-catalyzed ring closure-1,2-addition – 1(Z)-Alkylidene-4-β-keto-1,2,3,4-tetrahydrophthalazine ring s. *83*, 275

[Bis(2,4-di-tert-butylphenyl) 2′,4′,6′-trimethoxybiphenyl-2-ylphosphonite]gold(I) triflimide/ 6-tert-butyl-2-chloropyridine N-oxide ←
2-Indanones from propargylarenes ○
Improved gold(I)-catalyzed oxidative ring closure

293. 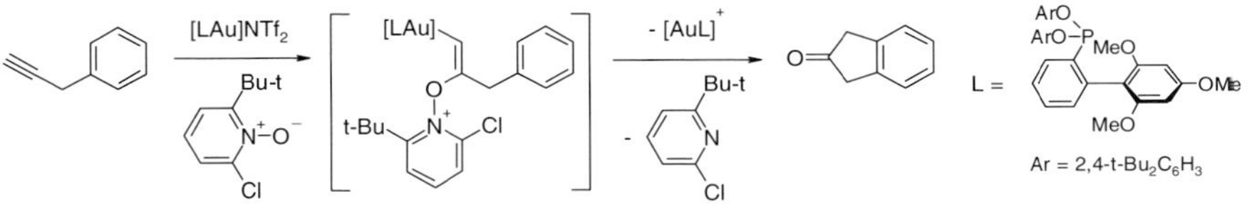

Novel cationic gold(I) *phosphonite* complexes and *hindered* pyridine N-oxides are more electronically and sterically in tune than more familiar gold(I) phosphine complexes and unhindered pyridine N-oxides for the oxidative ring closure of propargylarenes to 3-indanones. **E:** 6-*tert*-Butyl-2-chloropyridine N-oxide (1.2 eq.) added to a soln. of propargylbenzene (0.2 mmol) in chloroform (1 ml) at room temp., followed by [bis(2,4-di-*tert*-butylphenyl) 2′,4′,6′-trimethoxybiphenyl-2-yl-phosphonite]gold(I) triflimide (4 mol%), the mixture stirred at 60° for 2 h, cooled to room temp., and worked up with purification by flash chromatography → indan-2-one. Y 77%. With complexes possessing such *electron-poor* phosphonite ligands and with a hindered chloropyridine N-oxide as oxidant, reaction is not inhibited by coordination of the metal cation to the liberated pyridine, as generally takes place to varying degrees with gold(I) XPhos-type complexes and unhindered pyridine N-oxides. The procedure is applicable to a wide range of propargylarenes and heteroaromatic analogs with excellent functional group tolerance on the aromatic ring (e.g. alkoxy, amino, halogen, alkyl and aryl); with α,α-disubst. propargylarenes, however, the less bulky 3-bromo-pyridine N-oxide (2 eq.) and added methanesulfonic acid (1.2 eq.) afforded the highest yields with the same gold(I) phosphonite complex (twenty-seven examples in all; Y 47-77%). Although the precise mechanism of the conversion was not ascertained, it is thought likely that a *syn*-2-pyridyloxy-gold(I) species is implicated as the key reactive intermediate. F.e. and comparison of ligands and N-oxides s. G. Henrion, T.E.J. Chavas, X. Le Goff, F. Gagosz, Angew. Chem., Int. Ed. *2013*, *52* (24), 6277-82 [DOI: 10.1002/anie.201301015].

[1,3-Bis(2,6-diisopropylphenyl)imidazol-2-ylidene]gold(I) chloride/silver fluoroborate ←
γ,δ-Ethyleneketones from 2-ethylenealcohols and acetylene derivs. ←
via stereoselective gold(I)-catalyzed Claisen rearrangement

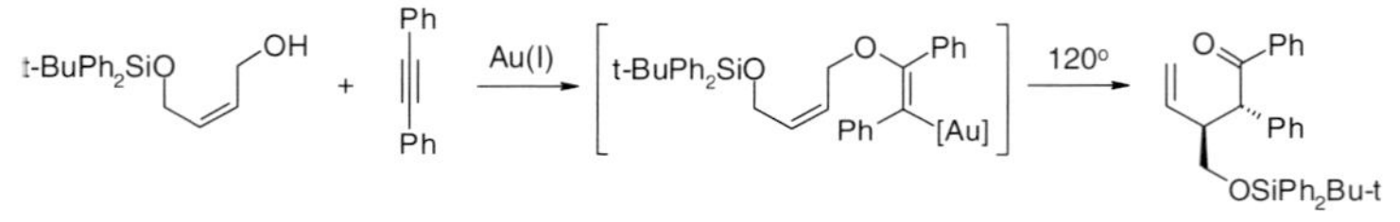

294.

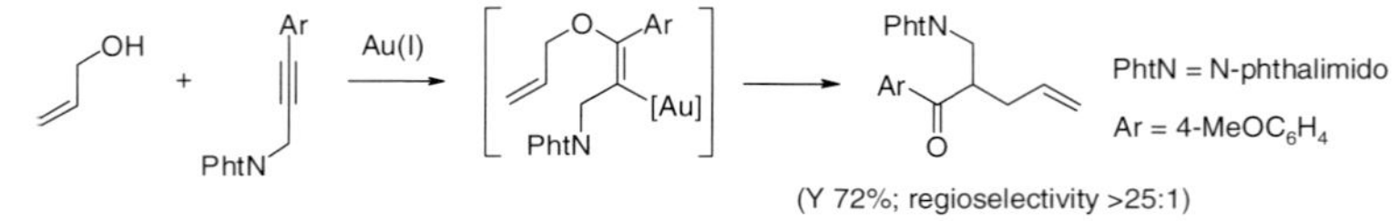

in one pot. A soln. of (IPr)AuCl (5 mol%) and $AgBF_4$ (5 mol%) in THF (0.5 ml) prepared in a glovebox, stirred for ca. 10 min, a soln. of diphenylacetylene (3 eq.) in the same solvent (0.5 ml) added, the mixture stirred at 65° during slow addition (12 h; syringe pump) of a soln. of (Z)-4-*tert*-butyl-diphenylsiloxybut-1-ene (0.2 mmol) in the same solvent (1 ml), stirring continued for 3 h, the reaction vessel sealed, heated at 120° for 6 h [precautionary blast shield used], cooled to room temp., filtered through silica, solvent evaporated, and the residue purified by flash chromatography → *syn*-1,2-diphenyl-3-(*tert*-butyldiphenylsiloxymethyl)pent-4-en-1-one. Y 70% (d.r. >25:1). Careful development was required to minimize undesired reactions (particularly ketalization via reaction of the intermediate allyl vinyl ether with a second molecule of alcohol). The optimized conditions, involving slow addition of allylic alcohol (incl. the parent allyl alcohol, and with disubstitution at C3 also tolerated) to an excess of alkyne, and a heating step to drive the rearrangement to completion, provided the title compds. in moderate to good yield (50-99%; sixteen examples). Diastereoselectivity was variable (d.r. 1:1 to >25:1) with *cis*-allylic alcohols giving better d.r. (in favor of *syn*-isomers) than *trans* ones (for *anti*-isomers). Alkyl(aryl)ethynes gave mixtures of regioisomers, with the preponderance of aryl (d.r. 8:1 to 25:1; six examples) or alkyl ketones (d.r. 3:1 to 8:1; three examples) largely influenced by electronic factors. F.e.s. J.M. Ketcham, B. Biannic, A. Aponick, Chem. Commun. *2013*, *49* (39), 4157-9 [DOI: 10.1039/c2cc37166a].

(Triphenylphosphine)gold(I) hexafluoroantimonate *$[(Ph_3P)Au]SbF_6$*
1-Alkoxy-1(Z),4-diene-2-carboxylic acid esters from alkoxy-2-ethylenes and propiolic acid esters ←

295.

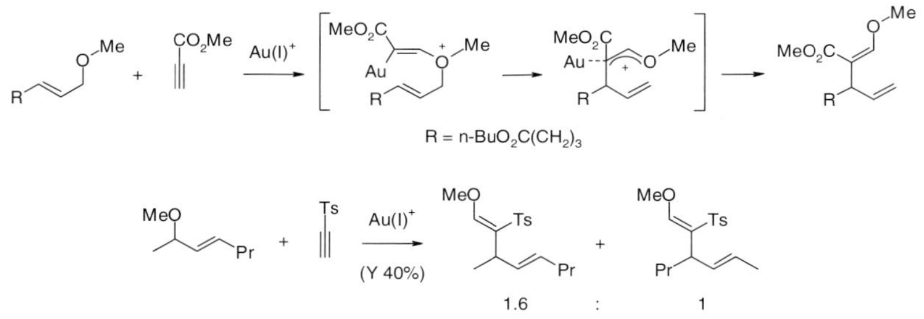

A soln. of butyl (E)-7-methoxyhept-5-enoate (0.4 mmol) and methyl propiolate (10 eq.) in dry nitromethane (1 ml) added to dry $[(Ph_3P)Au]SbF_6$ [freshly generated from $(Ph_3P)AuCl$ (8 mol%) and $AgSbF_6$ (8 mol%)], the mixture stirred at -15° for 40 h (TLC), quenched with triethylamine, stirred for 10 min, concentrated *in vacuo*, and purified chromatographically → (Z)-7-butyl 1-methyl 2-(methoxymethylene)-3-vinylheptanedioate. Y 72% (based on 92% conversion). This novel and highly efficient synthesis of (Z)-β-alkoxyacrylates **via gold(I)-catalyzed Claisen rearrangement** utilized pre-generated gold(I) cationic catalysts incorporating a phosphine ligand, and was free of by-products derived from alkyne-alkene coupling, affording (Z)-branched product isomers exclusively. Even with a stoichiometric excess of propiolate ester, yields were based on incomplete conversion for terminal/internal (incl. α-subst.) allylic ethers (60-84%; fifteen examples), tolerating ester, nitrile, silyl ether and benzyl ether functionality. Crossover experiments, using a 1:1 mixture of different allylic ethers, showed the reaction to occur via a concerted process. Interestingly, the more reactive *p*-toluenesulfonylethyne gave best results with a *stoichiometric excess of allylic ether* (three examples; Y 40-75%), with internal derivs. giving mixtures of branched/linear isomers (13:1 to 1:1.6). F.e. and optimization s. S.R. Park, C. Kim, D.-g. Kim, N. Thrimurtulu, H.-S. Yeom, J. Jun, S. Shin, Y.H. Rhee, Org. Lett. *2013*, *15* (6), 1166-9 [DOI: 10.1021/ol4001087].

(Triphenylphosphine)gold(I) chloride s.a. under Potassium tert-butoxide *$(Ph_3P)AuCl$*

(Triphenylphosphine)gold(I) chloride/silver(I) hexafluoroantimonate/ 3,5-dibromopyridine N-oxide ←
2-Alkylidenecyclobutanones from vinylidenecyclopropanes

296.

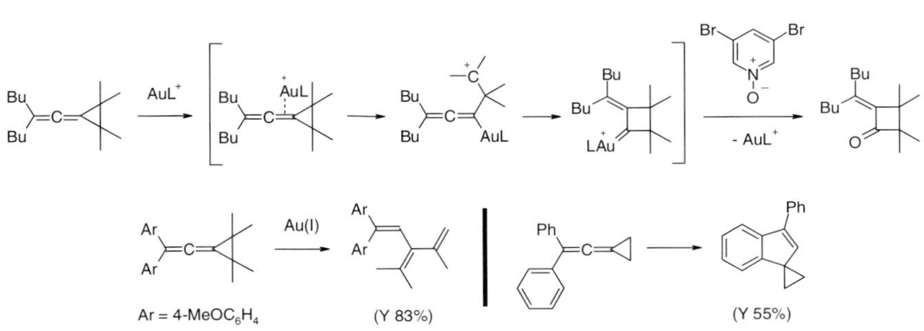

A mixture of (dibutylvinylidene)tetramethylcyclopropane (0.2 mmol), 3,5-dibromopyridine N-oxide (2 eq.), $(Ph_3P)AuCl$ (5 mol%), $AgSbF_6$ (10 mol%) and 1,2-dichloroethane stirred at room temp. under argon for 1 h, and purified directly by chromatography on silica → 2-(nonan-5-ylidene)-3,3,4,4-tetramethylcyclobutanone. Y 78%. This novel oxidative ring expansion appears general for electron-diverse (diaryl/dialkyl-vinylidene)tetramethylcyclopropanes, optimally using 3,5-dibromopyridine N-oxide as oxidant (other derivs. were less effective), affording the title products generally in high yield (78-95%; eleven examples), with unsymmetrical diarylvinylidene derivs. affording mixtures of E/Z derivs. (1:1; a 4-chlorophenyl-methyl deriv. gave 6:1). Lower yields were obtained with an unsubst. (23%) or bis(spirocyclohexyl)-subst. cyclopropane (39%), while an electron-rich bis(4-methoxyphenyl)vinylidene deriv. underwent ring opening to a branched triene (Y 83%), and diphenylvinylidenecyclopropane formed a spirocyclopropyl-indene via intramolecular cycloisomerization (Y 55%). An ethoxy-subst. cyclopropane gave a complex mixture. F.e., optimization and substrate prepn. s. W. Yuan, X. Dong, Y. Wei, M. Shi, Chem. Eur. J. *2012*, *18* (34), 10501-5 [DOI: 10.1002/chem.201201161].

[2-(Di-tert-butylphosphino)biphenyl]gold(I) chloride/silver hexafluoroantimonate ←

3-Oxa-2-azabicyclo[3.3.0]oct-6-en-8-ones from 1,3-dien-5-ynes and nitroso compds. Stereoselective gold(I)-catalyzed oxidative double ring closure ○

297.

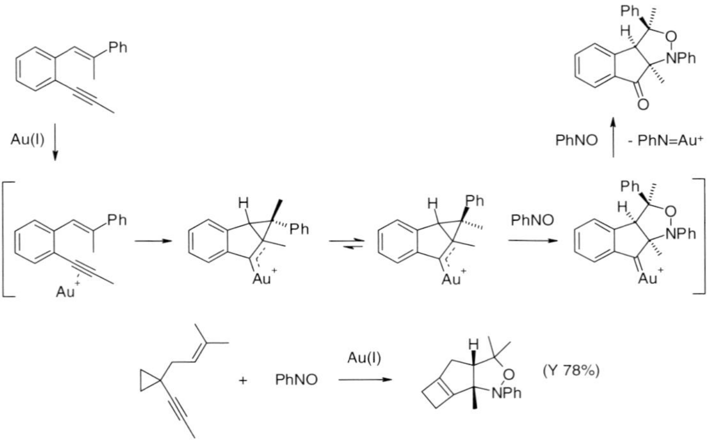

3,3a-Dihydro-1*H*-indeno[2,1-*c*]isoxazol-8(8a*H*)-ones. A soln. of gold(I) catalyst (5 mol%) and $AgSbF_6$ (5 mol%) in methylene chloride (0.7 ml) stirred at 25° for 10 min, a soln. of (E)-1-(2-phenylprop-1-en-1-yl)-2-(prop-1-yn-1-yl)benzene (0.6 mmol) and nitrosobenzene (3.8 eq.) in the same solvent (1.3 ml) added over 1 h, stirring continued for 2 h, filtered through Celite, solvent evaporated under reduced pressure, and the residue purified by flash chromatography → (3SR,3aRS,8aRS)-3,8a-dimethyl-1,3-diphenyl-3,3a-dihydro-1*H*-indeno[2,1-*c*]isoxazol-8(8a*H*)-one. Y 75% (plus 10% of an enyne cycloisomerization product). This unprecedented oxidative [3+2]-cycloaddition of electron-diverse 4-subst. nitrosobenzenes with cyclic 1,5-enynes or acyclic 1,3,5-dienynes proceeded with good stereocontrol, affording the required cycloadducts in good yield (63-91%; twenty-one examples). The products were generally accompanied by enyne cycloisomerization products (3-vinylindenes) in low yields (0-11%), with by-product formation apparently influenced by electronic factors, since a 4-methoxy-1-propynyl-2-propenylbenzene afforded the cycloisomerization product exclusively (Y 80%), while the 5-methoxy isomer gave the expected cycloadduct as the major product (Y 78%). Interestingly, isomeric E/Z enynes afforded the same products in similar yields, rationalized by equilibration of an initial cyclopropyl carbene intermediate to an energetically favorable conformation. A cyclopropane-tethered 1,5-enyne reacted with concomitant ring expansion via 1,2-migration to generate the first formal [2+2+1]-cycloadduct from enyne/nitroso functionalities in good yield (78%). F.e., optimization, and substrate prepn. s. C.-H. Chen, Y.-C. Tsai, R.-S. Liu, Angew. Chem., Int. Ed. *2013*, *52* (17), 4599-603 [DOI: 10.1002/anie.201209850].

Chiral (1,1′-binaphthyl-2,2′-diyl phosphoramidite)gold(I) chloride/silver triflimide ←
Asym. [3+2]-cycloaddition of allenes to nitrones s. *78,* 332s*83* C

Calcium triflimide/tetra-n-butylammonium hexafluorophosphate $Ca(NTf_2)_2/Bu_4NPF_6$
Ring expansion of cyclopropanes by [3+2]-cycloaddition C
cyclopentanes, update s. *42,* 654s*82*; *trans*-selective regioselective cycloaddition of 2-(alk-1-ynyl)-cyclopropane-1,1-dicarboxylic acid esters to aldehydes to give **5-(alk-1-ynyl)tetrahydrofuran-3,3-dicarboxylic acid esters** with $Ca(NTf_2)_2/Bu_4NPF_6$ s. S. Haubenreisser, P. Hensenne, S. Schröder, M. Niggemann, Org. Lett. *2013, 15* (9), 2262-5 [DOI: 10.1021/ol400809n]; **also pyrrolidine analogs** by cycloaddition to aldimines with a thiolate-bridged diruthenium complex via a ruthenium allenylidene complex s. Y. Miyake, S. Endo, T. Moriyama, K. Sakata, Y. Nishibayashi, Angew. Chem., Int. Ed. *2013, 52* (6), 1758-62 [DOI: 10.1002/anie.201207801]; *cis*-selective regioselective cycloaddition of electron-deficient vinylcyclopropanes to aldimines with $Ni(cod)_2$/1,2-bis(dimethyl-phosphino)ethane, also **asym. conversion** with a chiral di(phosphine) as ligand, s. R. Tombe, T. Kurahashi, S. Matsubara, Org. Lett. *2013, 15* (8), 1791-3 [DOI: 10.1021/ol4005068].

Chiral, polymer-based [2,6-bis(Δ²-oxazolin-2-yl)pyridine]calcium chloride complexes ←
Catalytic asym. Michael addition under continuous flow s. *49,* 657s*83* C=C → CHC(R)

Magnesium s.a. under Titanocene dichloride Mg

Magnesium/copper(I) or copper(II) salts/chiral 1-(2-hydroxyalkyl)-3-mesitylimidazolidin-2-ylidenes or 2-(Δ²-oxazolin-2-yl)ferrocenyl(diphenyl)phosphines or ferrocenyl-di(phosphines) ←
Asym. 1,4-addition of Grignards s. *37,* 657s*83*

Dialkylzinc/copper(II) salt/chiral N-heterocyclic carbenes $R_2Zn/Cu^{2+}/NHC^*$
Asym. 1,4-addition
of dialkylzincs, update s. *52,* 297s*82*; to acyclic enones under copper catalysis, e.g. with Cu(II)-hexa-fluoroacetoacetonate and a chiral, hydroxy- or carbalkoxy-functionalized 1-alkyl-3-(carbamyl-methyl)benzimidazol-2-ylidene as ligand (with reversal of face selectivity), s. K. Dohi, J. Kondo, H. Yamada, R. Arakawa, S. Sakaguchi, Eur. J. Org. Chem. *2012* (29), 7143-52 [DOI: 10.1002/ejoc.201201082]; to 2-cyclohexenone under copper catalysis with chiral imidazol-2-ylidenes or 1,2,4-triazolylidenes as ligand, also application of the N-heterocyclic carbenes to other asym. conversions, s. R.B. Strand, T. Helgerud, T. Solvang, A. Dolva, C.A. Sperger, A. Fiksdahl, Tetrahedron: Asym. *2012, 23* (18-19), 1350-9 [DOI: 10.1016/j.tetasy.2012.09.004]; **asym. 1,4-addition of Grignards** to linear 2,4-dienones with $[Cu(MeCN)_4]ClO_4$ as catalyst and a chiral 2-(Δ²-oxazolin-2-yl)ferrocenyl(diphenyl)phosphine as ligand (cf. *37,* 657s*81*) s. Z. Ma, F. Xie, H. Yu, Y. Zhang, X. Wu, W. Zhang, Chem. Commun. *2013, 49* (46), 5292-4 [DOI: 10.1039/c3cc42088d]; to cyclic 2,4-dienones and 2,4-enynones with $Cu(OTf)_2$ and a chiral 1-(2-hydroxyalkyl)-3-mesitylimida-zolidin-2-ylidene as ligand s. M. Tissot, D. Poggiali, H. Hénon, D. Müller, L. Guénée, M. Mauduit, A. Alexakis, Chem. Eur. J. *2012, 18* (28), 8731-47 [DOI: 10.1002/chem.201200502]; to 2-pyrones under copper catalysis with a chiral ferrocenyldi(phosphine) as ligand, e.g. (R,S)-Rev-Josiphos, also conversion to chiral δ-aldehydocarboxylic and γ-bromo-δ-hydroxycarboxylic acid esters, s. B. Mao, M. Fañanás-Mastral, B.L. Feringa, Org. Lett. *2013, 15* (2), 286-9 [DOI: 10.1021/ol303141x]; **asym. 1,4-addition of triorganoaluminum compds.** (cf. *52,* 297s*82*), notably trimethylaluminum, to acyclic β-subst. enones under copper catalysis with $CuCl_2$ and a chiral 3-(diphenylphosphino)-1,1′-bi-2-naphthol as ligand s. K. Endo, D. Hamada, S. Yakeishi, T. Shibata, Angew. Chem., Int. Ed. *2013, 52* (2), 606-10 [DOI: 10.1002/anie.201206297]; of trialkylaluminums and enalanes to α,β-ethylenelactams with Cu(II)-naphthenate and a chiral N,N-bis[2-(2-naphthyl)ethyl]amino-(diaryl)phosphine as ligand s. P. Cottet, D. Müller, A. Alexakis, Org. Lett. *2013, 15* (4), 828-31 [DOI: 10.1021/ol303505k]; racemic 1,4-addition of lithium (5-chloropenten-2-yl)(isobutyl)methyl-aluminate to cyclic β-subst. enones with Cu(II)-naphthenate as catalyst s. D. Müller, A. Alexakis, ibid. *15* (7), 1594-7 [DOI: 10.1021/ol400378v].

Dimethylzinc/boron fluoride/tert-butyl hydroperoxide $Me_2Zn/BF_3/t\text{-}BuOOH$
Radical 1,4-addition
update s. *39,* 646s*81*; of pivaloyloxymethyl iodide to alkylidenemalononitriles *under argon* using Me_2Zn/BF_3-etherate/*tert*-butyl hydroperoxide as radical generator, also 1,2-bis(oxymethylation)

in air, s. K.-i. Yamada, T. Konishi, M. Nakano, S. Fujii, R. Cadou, Y. Yamamoto, K. Tomioka, J. Org. Chem. *2012, 77* (13), 5775-80 [DOI: 10.1021/jo300944f]; 1,4-addition of perfluoroalkyl iodides to chiral N-(α,β-ethyleneacyl)-2-oxazolidones using Et_3B/O_2 and $Yb(OTf)_3$ as Lewis acid **with asym. induction** s. H. Erdbrink, I. Peuser, U.I.M. Gerling, D. Lentz, B. Koksch, C. Czekelius, Org. Biomol. Chem. *2012, 10* (43), 8583-6 [DOI: 10.1039/c2ob26810h]; of alkyl iodides to S-chiral (*o*-sulfinyl-α-styryl)carbonyl compds. with asym. induction using Et_3B/O_2 s. J.A. Fernández-Salas, M.C. Maestro, M.M. Rodríguez-Fernández, J.L. García Ruano, Eur. J. Org. Chem. *2013* (9), 1796-804 [DOI: 10.1002/ejoc.201201642].

Diethylzinc s. under Nickel(II) acetoacetonate *Et_2Zn*

Chiral magnesium salen complex *[Mg(II)]**

6-Nitro-2-cyclohexenols from α,β-ethyleneketones and 1-nitroethylene derivs. via asym. vinylogous Michael addition-intramolecular Henry reaction ○

298. 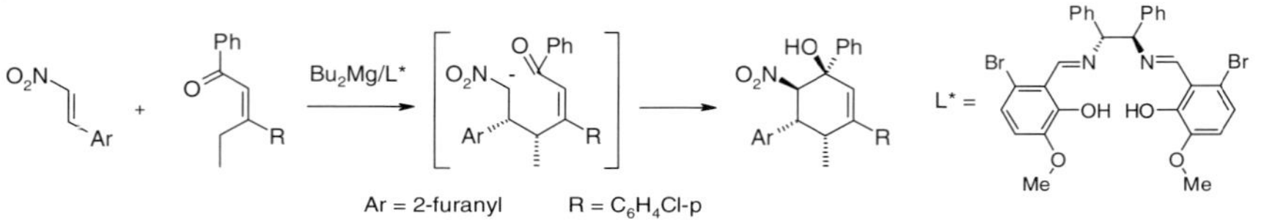

The use of a chiral magnesium Schiff base complex to direct γ-deprotonation of linear enones is reported, and applied in a novel cyclohexenol ring closure, affording products not readily accessible by other methodologies. **E: Chiral 1,3,4-trisubst. 5-aryl-6-nitro-2-cyclohexenols**. Bu_2Mg (40 µl; 1 *M* in heptanes; 20 mol%) added to a stirred soln. of the chiral ligand (20 mol%) and 4 Å molecular sieves (200 mg) in *p*-xylene (1 ml) under argon, the mixture stirred at room temp. for 0.5 h, mixed with startg. enone (2 eq.) and nitroalkene (0.2 mmol) in *p*-xylene (1 ml) by syringe, the mixture stirred at 60° for 24 h, quenched with satd. NH_4Cl, extracted with methylene chloride, and purified by chromatography on silica gel → product. Y 82% (d.r. 9:1; e.e. 98%). The method is applicable to nitroalkenes bearing electron-rich or -poor aryl or hetaryl groups but failed with aliphatic ones. The linear enones were mostly [aryl] β-alkyl-β-aryl-derivs., but β-aryl-β-benzyl- or sym. β,β-dialkyl-derivs. may also be used, or the substituents may be part of a ring. It is believed the methoxy groups on the ligand may act as Brønsted bases to assist deprotonation. F.e. (twenty-nine; Y 42-79%; d.r. 3:1 to 11:1; e.e. 86-99%) and optimization s. D. Yang, L. Wang, F. Han, D. Zhao, B. Zhang, R. Wang, Angew. Chem., Int. Ed. *2013, 52* (26), 6739-42 [DOI: 10.1002/anie.201301146].

Zinc triflate s.a. under Me_3SiCl *$Zn(OTf)_2$*

Zinc triflate/chiral 2-prim-aminothioureas ←

Asym. Friedel-Crafts reaction of indoles C=C → CHC(Ar)

with electron-deficient ethylene derivs. s. *67,* 336s*83*

Lithium hydridotriethylborate/magnesium bromide s. under $NiCl_2$(binap) *$LiBHEt_3/MgBr_2$*

Triethylborane/oxygen *Et_3B/O_2*

Radical 1,4-addition s. *39,* 646s*83* C=C → CHC(R)

Triorganoaluminum compds./copper(I) or copper(II) salts/chiral 3-(diphenylphosphino)-1,1′-bi-2-naphthol or N,N-bis[2-(2-naphthyl)ethyl]amino(diaryl)phosphines ←

Asym. 1,4-addition of triorganoaluminum compds. s. *52,* 297s*83*

Sodium tetrakis[3,5-bis(trifluoromethyl)phenyl]borate s. under $[IrCl(cod)]_2$ *$NaBAr^F_4$*

Pinacolborane s. under Copper(I) chloride *HBpin*

Boron fluoride (s.a. under Me_2Zn) *BF_3*
β-Acoxy-α-methyleneketones from acoxy-2-acetylenes and aldehydes $CH(OAc)C(=CH_2)C(O)R$

299.
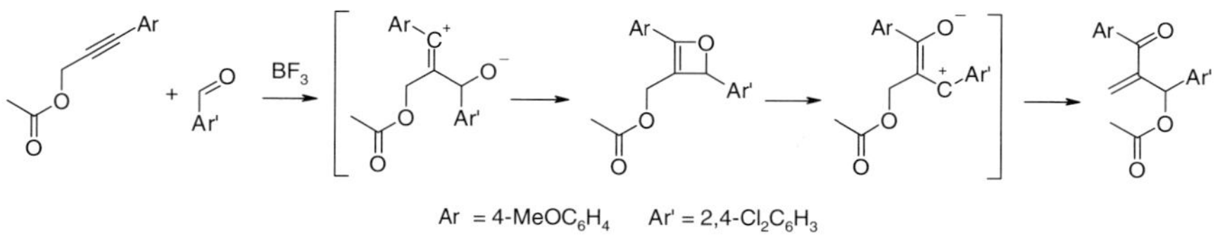

under mild conditions. BF_3-etherate (1 eq.) added to a soln. of 3-(4-methoxyphenyl)prop-2-yn-1-yl acetate (1 mmol) and 2,4-dichlorobenzaldehyde (1 eq.) in dry methylene chloride (5 ml) at room temp., the soln. stirred for 5 min (TLC), quenched with aq. $NaHCO_3$, and worked up with chromatographic purification → 1-(2,4-dichlorophenyl)-2-(4-methoxybenzoyl)allyl acetate. Y 86%. This experimentally simple and efficient approach to rare Morita-Baylis-Hillman adducts was available only for a combination of electron-neutral/rich 3-arylprop-2-yn-1-yl esters reacting with electron-poor ar. aldehydes, where the products were formed via [3.3]-migration of acoxy from the initial metathesis adducts (seventeen examples; Y 40-90%). Under these conditions aliphatic aldehydes generally afforded (E)-configured metathesis products, while electron-rich ar. aldehydes reduced the reaction rate allowing 2:1 adducts to predominate. F.e. and optimization s. I. Karpaviciene, I. Cikotiene, Org. Lett. *2013*, *15* (1), 224-7 [DOI: 10.1021/ol303319a].

Ethylaluminum dichloride or Tin(IV) chloride *$EtAlCl_2$ or $SnCl_4$*
3-Ethyleneacylamines from β,γ-ethyleneketones and aldimines via stereoselective 3-oxy-2-azonia-Cope rearrangement C(N-Ac)C-C=C-CH

300.
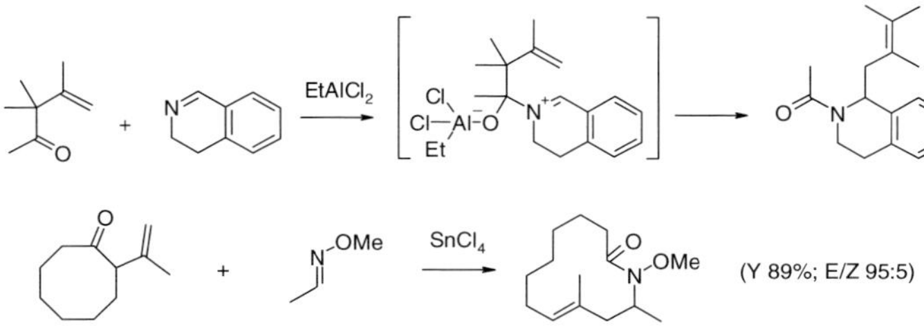

Novel and atom-economical production of homoallylic amides is reported via rearrangement of zwitterionic intermediates obtained from Lewis acid-catalyzed nucleophilic addition of aldimines [or O-alkylaldoximes] to β,γ-ethyleneketones. **E:** A soln. of ethylaluminum dichloride (1 eq.) in toluene (5.56 ml) added dropwise to a mixture of 2,3,3-trimethylpent-1-en-4-one (10 mmol) and 3,4-dihydroisoquinoline (1.2 eq.) in 1,2-dichloroethane (50 ml) under argon at room temp., the mixture stirred for 24 h, quenched with satd. aq. $NaHCO_3$, extracted with *tert*-butyl methyl ether, and purified by chromatography on silica → 1-[1-(2,3-dimethylbut-2-en-1-yl)-1,2,3,4-tetrahydroisoquinolin-2-yl]ethanone. Y 83%. Reaction of the trimethylpentenone or mono/di-desmethyl analogs containing 1/2-α-hydrogens with aldimines (incl. N-methylformaldimine generated *in situ* from N,N′,N″-trimethyltriazine) or O-alkylaldoximes (to afford **N-homoallylhydroxamic acid esters**) was catalyzed by $EtAlCl_2$ or $SnCl_4$ (eight examples; Y 67%, 83-97%). Lower yields were obtained from indole (58%; reacting as its 3H-tautomer), benzaldoxime O-methyl ether (39%) and 1-piperideine (29%). A single 1,2-disubst. alkene product was obtained exclusively as the (E)-stereoisomer. The reaction was extended to cyclic β,γ-ethyleneketones (2-propenylcycloalkanones), affording the corresponding [n+4] ring expanded **N-3-ethylenelactams or N-alkoxy-analogs**, generally with good yields/stereoselectivities for C6-C8 ketones (twelve examples;

Y 59-96%; E/Z 95:5 to >99:1). Lower yields were obtained with cyclopentanone (35%; E/Z <1:99) and cyclododecanone (35%; E/Z 25:75) analogs. F.e., optimization and use of the method in a short synthesis of the macrocyclic alkaloid motuporamine G, s. L. Zhou, Z. Li, Y. Zou, Q. Wang, I.A. Sanhueza, F. Schoenebeck, A. Goeke, J. Am. Chem. Soc. *2012, 134* (49), 20009-12 [DOI: 10.1021/ja310002m].

Dialkyl(cyclopentadienyl)yttrium(III) complexes/trityl tetrakis(pentafluorophenyl)borate ←
Regioselective hydroalkylation of ethylene derivs. C=C → CHC(R)
with active methylene groups using $AuCl_3/AgOTf$ cf. *67*, 302; with 2-alkylpyridines using dialkyl-(cyclopentadienyl)yttrium(III) complexes and trityl tetrakis(pentafluorophenyl)borate as co-catalyst s. B.-T. Guan, B. Wang, M. Nishiura, Z. Hou, Angew. Chem., Int. Ed. *2013, 52* (16), 4418-21 [DOI: 10.1002/anie.201208867]; of electron-deficient alkenes by thermal 1,4-hydroalkylation with *o*-alkyl-N-heteroarenes in DMSO s. H.-Y. Li, L.-J. Xing, T. Xu, P. Wang, R.-H. Liu, B. Wang, Tetrahedron Lett. *2013, 54* (8), 858-60 [DOI: 10.1016/j.tetlet.2012.11.100]; 1,4-hydroalkylation of β-nitrostyrenes with 2-methylquinolines *in water* under microwave enhancement s. N. Nageswara Rao, H.M. Meshram, ibid. *54* (10), 1315-7 [DOI: 10.1016/j.tetlet.2013.01.007].

(Pentamethylcyclopentadienyl)rare-earth(III) complexes/trityl tetrakis(pentafluorophenyl)borate ←
***o*-Alkylation of phenolethers with ethylene derivs.** s. *3,* 585s*83* H → R

Lanthanum(III) triflate/chiral 2,6-bis(Δ²-oxazolin-2-yl)pyridines *La(OTf)₃/PyBox*
Catalytic asym. Michael addition C=C → CHC(R)
update s. *49,* 657s*82*; of aliphatic nitro compds. to (E)-2-azachalcones with $La(OTf)_3$ and a chiral, bulky 2,6-bis(Δ^2-oxazolin-2-yl)pyridine as ligand s. G. Blay, C. Incerti, M.C. Muñoz, J.R. Pedro, Eur. J. Org. Chem. *2013* (9), 1696-705 [DOI: 10.1002/ejoc.201201579]; of 2-(carbamylmethyl)- or 2-(carbalkoxymethyl)-N-heteroarenes to nitroalkenes with a chiral bis(*trans*-cyclohexane-1,2-diamine)nickel(II) dibromide as catalyst s. C. Fallan, H.W. Lam, Chem. Eur. J. *2012, 18* (36), 11214-8 [DOI: 10.1002/chem.201202093]; *heterogeneous* Michael addition of β-dicarbonyl compds. to nitroalkenes with silyl-functionalized chiral bis(*trans*-cyclohexane-1,2-diamine)-nickel(II) dibromide incorporated in periodic mesoporous organosilicas s. K. Liu, R. Jin, T. Cheng, X. Xu, F. Gao, G. Liu, H. Li, ibid. *18* (48), 15546-53 [DOI: 10.1002/chem.201202407]; *under continuous flow* with a chiral, polymer-based [2,6-bis(Δ^2-oxazolin-2-yl)pyridine]calcium chloride complex as catalyst cf. T. Tsubogo, Y. Yamashita, S. Kobayashi, ibid. *18* (43), 13624-8 [DOI: 10.1002/chem.201202896]; Michael addition of *in situ*-generated azlactones to enones *en route* to chiral quaternary α-amino-acids and dipeptides with a bis(palladacyclic) complex in cooperation with an organo-Brønsted acid (e.g. acetic acid) and a Brønsted base, e.g. NaOAc, s. M. Weber, S. Jautze, W. Frey, R. Peters, ibid. *18* (46), 14792-804 [DOI: 10.1002/chem.201202455].

Trityl tetrakis(pentafluorophenyl)borate s. under Dialkyl(cyclopentadienyl)yttrium(III) complexes and (Pentamethylcyclopentadienyl)rare-earth(III) complexes ←

tert-Butanol s. under Titanocene dichloride *t-BuOH*

Crown s. under K_2CO_3 or Cs_2CO_3 or KCl *crown*

Carbon monoxide s. under PtI_4 *CO*

Chiral squaramides s.a. under Chiral diamines ←

Cinchona-based squaramides ←
Asym. Michael addition-intramolecular Henry reaction s. *67,* 313s*83*

Chiral 2-tert-aminoureas s. under Chiral diamines ←

2,2′-Azobis(4-methoxy-2,4-dimethylvaleronitrile)/sodium phosphate ←
Radical synthesis of 2-bromo-1,6-dienes C
from alkylidenecyclopropanes and β,γ-ethylenebromides

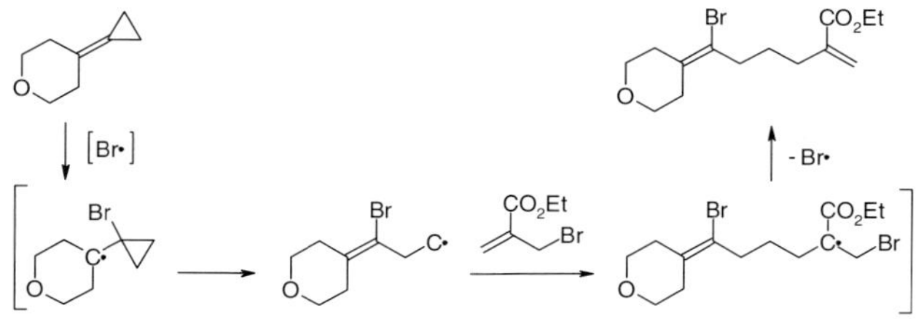

301.

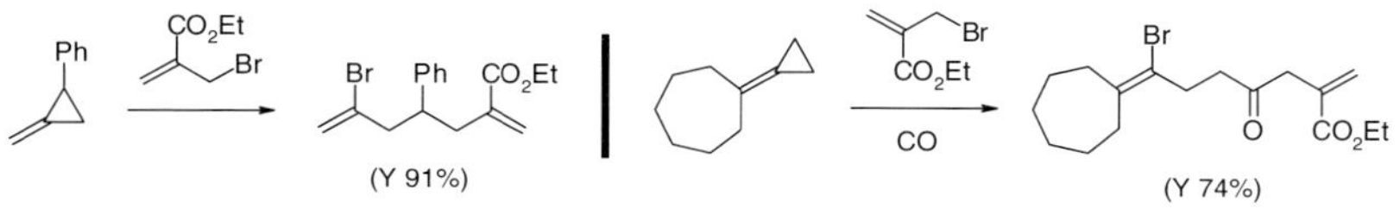

A novel synthesis of the title products is reported wherein bromination of alkylidenecyclopropanes, by generated bromine radicals, and subsequent cyclopropane cleavage affords homoallylic radicals which are trapped by activated allylic bromides. **E:** A mixture of tetrahydropyran-4-ylidene-cyclopropane (1 mmol), ethyl 2-(bromomethyl)acrylate (1.8 eq.), 2,2′-azobis(4-methoxy-2,4-dimethylvaleronitrile) [V-70] (10 mol%), Na_3PO_4 (20 mol%) and degassed benzene (1 ml) stirred under argon at 40° for 2 h, filtered through Celite, concentrated *in vacuo*, and purified chromatographically → ethyl 6-bromo-6-(tetrahydropyran-4-ylidene)-2-methylenehexanoate. Y 87%. The addition of an HBr scavenger and use of relatively moderate temps. (generally 40°), to minimize isomerization of the substrate, were key features of the method. The reaction was successful for a series of mono/di-alkylmethylenecyclopropanes (ten examples; Y 74-97%), with monosubst. substrates showing moderate (Z)-selectivity (E/Z 36:64 to 27:73). A phenyl-subst. cyclopropane gave a single product (Y 91%) via preferential cleavage of the cyclopropane to generate a benzyl radical. The intermediate homoallylic radicals were also trapped with CO to afford the corresponding **7-bromo-1,7-dien-4-ones** (three examples; Y 71-74%). F.e., optimization, substrate prepn. and further reactions of the products s. T. Kippo, K. Hamaoka, I. Ryu, J. Am. Chem. Soc. *2013, 135* (2), 632-5 [DOI: 10.1021/ja311821h].

Chiral 2-[diaryl(trimethylsiloxy)methyl]pyrrolidines (s.a. under Chiral diamines) ←
Organocatalyzed asym. Friedel-Crafts reaction of indoles C=C → CHC(Ar)
with electron-deficient ethylene derivs. s. *67,* 336s*83*

Organocatalyzed asym. ring closures via Michael addition ○
update s. *67,* 313s*82*; *78,* 319s*82*; **asym. Michael addition-intramolecular aldol condensation** of 3-alkylideneoxindoles with 1,5-pentanedial using 2(R)-[diphenyl(trimethylsiloxy)methyl]-pyrrolidine for preparing chiral 3-spirooxindoles s. X.-F. Huang, Z.-M. Liu, Z.-C. Geng, S.-Y. Zhang, Y. Wang, X.-W. Wang, Org. Biomol. Chem. *2012, 10* (44), 8794-9 [DOI: 10.1039/c2ob26205c]; chiral 4-halogeno-2-hydroxy-3-ketocyclopentanecarboxaldehydes from β-halogeno-α-diketones and enals with 2(S)-[diaryl(trimethylsiloxy)methyl]pyrrolidines s. A. Lefranc, L. Guénée, A. Alexakis, Org. Lett. *2013, 15* (9), 2172-5 [DOI: 10.1021/ol400697n]; asym. synthesis of partially-protected 5-[het]aryl-2,3,6-trihydroxy-4-nitrocyclohexanols with 5 stereocenters from β-[het]aryl-α-nitro-α,β-ethylenealdehydes s. F. Cagide-Fagín, O. Nieto-García, H. Lago-Santomé, R. Alonso, J. Org. Chem. *2012, 77* (24), 11377-82 [DOI: 10.1021/jo3022567]; asym. synthesis of 5-hydroxy-6-nitrobicyclo[3.2.1]octan-2-ones by **asym. Michael addition-intramolecular Henry reaction**

with a 3-(pyrrolidin-2(S)-ylmethyl)-2-thiouracil as catalyst s. M. Tsakos, M.R.J. Elsegood, C.G. Kokotos, Chem. Commun. *2013, 49* (22), 2219-21 [DOI: 10.1039/c3cc39165e]; of 3-β-keto-2-nitrocyclohexanols with a 8(S)-[diphenyl(hydroxy)methyl]-1,5,7-triazabicyclo[4.3.0]non-5-ene as catalyst s. Q. Dai, H. Huang, J.C.-G. Zhao, J. Org. Chem. *2013, 78* (8), 4153-7 [DOI: 10.1021/jo4001806]; of chiral 3-subst. 2,4-dinitrotetralols with a cinchona-based squaramide as catalyst s. D. Enders, R. Hahn, I. Atodiresei, Adv. Synth. Catal. *2013, 355* (6), 1126-36 [DOI: 10.1002/adsc.201300039]; of chiral 4-hydroxy-3-nitro-3,4-dihydro-2*H*-thiopyrano[2,3-*b*]quinolines **by asym. thia-Michael addition-intramolecular Henry reaction** using the same catalyst s. L. Wu, Y. Wang, H. Song, L. Tang, Z. Zhou, C. Tang, ibid. *355* (6), 1053-7 [DOI: 10.1002/adsc.201300086]; of chiral (3-hydroxy-2-nitrocyclohexyl)acetic acid esters by asym. Henry reaction-intramolecular Michael addition-retro-Henry reaction-Henry reaction with a cinchona-based catalyst s. A. Quintavalla, M. Lombardo, S.P. Sanap, C. Trombini, ibid. *355* (5), 938-46 [DOI: 10.1002/adsc.201201135]; of chiral 2-hydroxy-3-tosylaminomethyl-3,4-dihydro-2*H*-pyran-6-carboxylic from β,γ-ethylene-α-ketocarboxylic acid esters by **asym. Michael addition-lactolization** with chiral *trans*-perhydroindolic acid as catalyst and DABCO as co-catalyst s. J. Shen, Q. An, D. Liu, Y. Liu, W. Zhang, Chin. J. Org. Chem. *2012, 30* (11), 2681-7 [DOI: 10.1002/cjoc.201200839]; of chiral condensed 2-hydroxy-3,4-dihydro-2*H*-pyran-2-carboxylic acid esters under catalysis with a chiral 1,2,3,4-tetrahydroisoquinoline-based cyclic bis(N-oxide) as ligand cf. Z.E.D. Cele, S.C. Sosibo, P.G. Andersson, H.G. Kruger, G.E.M. Maguire, T. Govender, Tetrahedron: Asym. *2013, 24* (4), 191-5 [DOI: 10.1016/j.tetasy.2013.01.004]; chiral spiro(pyrrolidine[2,3′]oxindole-5-thiones) from 3-isothiocyanatooxindoles by **asym. Michael addition-thiolactamization** with quinine as catalyst s. X.-L. Liu, W.-Y. Han, X.-M. Zhang, W.-C. Yuan, Org. Lett. *2013, 15* (6), 1246-9 [DOI: 10.1021/ol400183k]; chiral 2-amino-3-cyano-4*H*-pyran ring by asym. Michael addition-cyclization with a cinchona-based squaramide as catalyst s. Y. Gao, D.-M. Du, Tetrahedron: Asym. *2012, 23* (18-19), 1343-9 [DOI: 10.1016/j.tetasy.2012.09.011].

2(S)-[Diphenyl(trimethylsiloxy)methyl]pyrrolidine ←
Asym. oxa-Michael addition-intramolecular Michael addition ○
with kinetic resolution or dynamic kinetic resolution s. *76*, 303s*83*

2(S)-[Diphenyl(trimethylsiloxy)methyl]pyrrolidine/irradiation ←
Organocatalyzed [2+2]-cycloaddition under visible-light photocatalysis s. *22*, 705s*83* □

2(S)-[Diphenyl(trimethylsiloxy)methyl]pyrrolidine/tris(dibenzylideneacetone)dipalladium/ 1,2-bis(diphenylphosphino)ethane ←
Asym. synthesis of 4-cyano-2-vinylcyclopentanecarboxaldehydes ○
from ε-acoxy-γ,δ-ethylenenitriles and α,β-ethylenealdehydes
Chiral amine/palladium-cocatalyzed asym. Michael addition-intramolecular α-allylation

302.

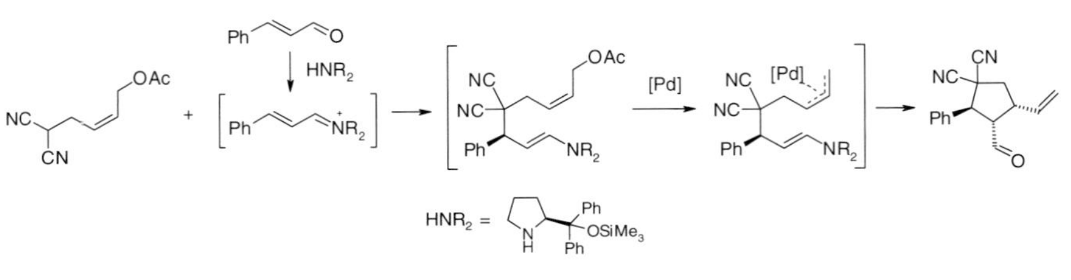

Acetonitrile (0.2 ml) added through a septum to an oven-dried vial charged with $Pd_2(dba)_3$ (5 mol%) and 1,2-bis(diphenylphosphino)ethane (10 mol%) under N_2, the mixture stirred at room temp. for 5 min, the contents of a second vial containing 2(S)-[diphenyl(trimethylsiloxy)methyl]pyrrolidine (20 mol%) and the startg. allyl acetate (1.5 eq.) and enal (0.1 mmol) (preliminarily stirred under N_2 at room temp. for 5 min) added via syringe, stirring continued at room temp. for 5 h, and worked up with purification by chromatography on silica gel → product. Conversion 90% (d.r. 96:4; e.e. >99%). Reaction proceeds via initial enamine-mediated Michael addition of the acoxynitrile to the enal to give a mixture of diastereoisomeric adducts, followed by Pd-catalyzed intramolecular α-allylation; the *ent*-Michael adduct (of that shown in the scheme), however, is relatively slow reacting in the ring closure, and under equilibration racemizes to boost the yield of the faster-reacting adduct leading almost exclusively to one diastereoisomeric product with high enantioselectivity – thus effecting an overall *dynamic asym. cascade transformation*. The procedure is applicable to the coupling of aliphatic or aromatic (E)- or (Z)-enals with the unsatd. acoxy-1,1-dinitrile as well as the analogous α-cyano-ester (fourteen examples; Y 66-88%; d.r. 80:20 to 95:5; e.e. 99%). One example is also given of cyclohexane ring closure from a ζ-acoxy-δ,ε-ethylenenitrile (Y 65%; d.r. >97:3; e.e. 99%). F.e. and comparison of catalysts and solvents s. G. Ma, S. Afewerki, L. Deiana, C. Palo-Nieto, L. Liu, J. Sun, I. Ibrahem, A. Córdova, Angew. Chem., Int. Ed. *2013*, *52* (23), 6050-4 [DOI: 10.1002/anie.201300559].

(S,E)-N-Benzhydryl-1-(N,N′-dicyclohexylcarbamimidoyl)pyrrolidine-2-carboxamide ←
Organocatalyzed asym. Michael addition with acylsilanes C=C → CHC-C-C(O)Si⩽

303. 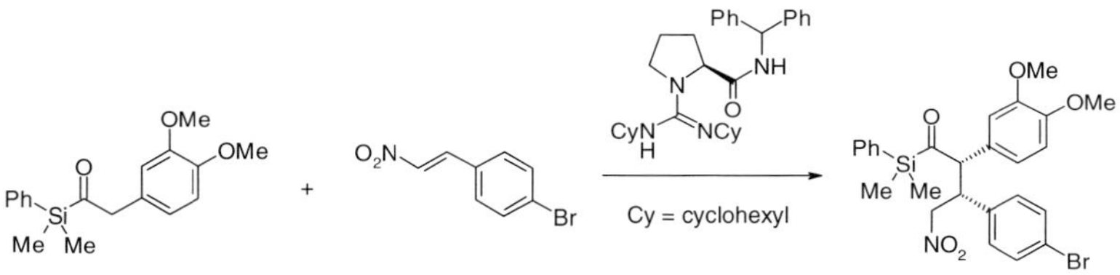

Novel Michael addition of acylsilanes **to 1-nitroethylene derivs.** is reported, providing rare examples of α-alkylation of acylsilanes. **E:** A mixture of dimethyl(phenyl)(3,4-dimethoxyphenacetyl)silane (0.1 mmol), 4 Å molecular sieves (60 mg) and chiral catalyst (15 mol%) in toluene (1.5 ml) stirred under dry N_2 at 0° for 12 h, a satd. soln. of 2-(4-bromophenyl)-1-nitroethylene (2.5 eq.) in the same solvent added in four portions over 8 h, and purified by flash chromatography → (2S,3R)-3-(4-bromophenyl)-2-(3,4-dimethoxyphenyl)-1-[dimethyl(phenyl)silyl]-4-nitrobutan-1-one. Y 77% (*syn/anti* 35:1; e.e. 95%). Initial optimization of both catalyst and the silane moiety enabled provision of Michael adducts from electron-diverse β-(het)aryl-nitroethylene derivs. and arylacetylsilanes with high *syn*- and enantio-selectivity (*syn/anti* 16:1 to >99: e.e. 89-97%; twenty-eight examples; Y 61-86%; a bulky 1-naphthylacetylsilane gave 51%). Aliphatic acylsilanes were typically unreactive, but β-cyclohexylnitroethylene was a good substrate (*syn/anti* 12:1; e.e. 90%; Y 85%). In initial tests, a trimethylsilane deriv. gave excellent selectivity (*syn/anti* >99:1; e.e. 91%) but the yield was moderate (43%). Experiments suggest the organocatalyst activates both components simultaneously. F.e., optimization, acylsilane and catalyst prepn., and use of the method for synthesis of γ-amino acids (see supplementary information) and a chiral 3,4-diaryl-2-pyrrolidone s. L. Wu, G. Li, Q. Fu, L. Yu, Z. Tang, Org. Biomol. Chem. *2013*, *11* (3), 443-7 [DOI: 10.1039/c2ob26950c].

Chiral 5-benzyl-4-imidazolidones ←
Organocatalyzed asym. 1,3-dipolar cycloaddition with nitrones s. *67,* 301s*83* ○

Chiral bis(Δ²-oxazolines) s. under Nickel(II) perchlorate *box*
Chiral 2,6-bis(Δ²-oxazolin-2-yl)pyridines s. under La(OTf)₃ *PyBox*
Chiral N-heterocyclic carbenes s.a. under Dialkylzinc *NHC**
Chiral 1-(2-hydroxyalkyl)-3-mesitylimidazolidin-2-ylidenes s. under Magnesium ←
Chiral 2-(o-hydroxybenzylaminomethyl)-Δ²-imidazolines s. under Copper(I) or copper(II) triflate and Nickel(II) acetate ←

Chiral N-condensed 1,2,4-triazolylidenes *NHC**
N-Heterocyclic carbene-catalyzed asym. [4+2]-cycloaddition of ketenes s. *46,* 662s*83*

Asym. Stetter hydroacylation of acrylic acid esters C=C → CHC-CCR
Synthesis of chiral α-subst. γ-ketocarboxylic acid esters s. *68,* 335s*83*

Chiral N-condensed 1,2,4-triazolylidenes/sodium acetate *NHC*/NaOAc*

Asym. synthesis of δ-nitrocarboxylic acid esters from α,β-ethylenealdehydes and 1-nitroethylene derivs. ←

304.

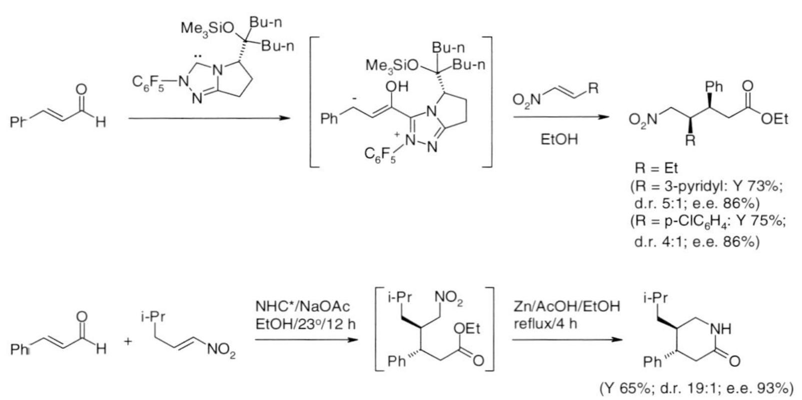

A novel chiral N-heterocyclic carbene catalyst is reported that, in the reaction of enals with nitroalkenes, favors δ-nitro-ester formation via the homoenolate pathway over Stetter reaction, and affords previously unreported ***syn*-isomers**. **E:** A screw cap vial charged with the chiral triazolium fluoroborate [NHC precursor] (10 mol%) and NaOAc (50 mol%), the vial fitted with a rubber septum, then evacuated and refilled with argon three times, ethanol (0.75 ml) added via syringe, (E)-1-nitrobut-1-ene (0.25 mmol) added to the soln. followed by *trans*-cinnamaldehyde (1.5 eq.), the septum quickly replaced with a screw cap, the mixture stirred at room temp. for 12 h, concentrated via rotary evaporation, and the residue purified by silica gel chromatography → (3S,4R)-ethyl 4-(nitromethyl)-3-phenylhexanoate. Y 70% (d.r. 17:1; e.e. 93%). Nineteen further examples with β-(het)aryl enals and aliphatic or aryl-subst. nitroalkenes gave yields of 35-95%, d.r. 3:1 to 20:1 and e.e. 79-96%, notably tolerating acetal and terminal alkene groups. A nitroolefin with a bulky *tert*-butyl group did not participate and *trans*-2-pentenal gave the Stetter product predominantly with the illustrated catalyst, while use of an alternative fluorinated catalyst gave only a 25% yield of nitro-ester (d.r. 19:1; e.e. 91%). The method was extended to a one pot synthesis of **chiral 2-piperidones** by addition of Zn dust and acetic acid to the crude reaction mixture after 12 h, following by heating for a further 4 h (five examples; Y 60-70%; d.r. 3:1 to 20:1; e.e. 82-93%); one example of further reduction to a chiral piperidine is also reported (Y 89%; e.e. 93%). F.e.s. N.A. White, D.A. DiRocco, T. Rovis, J. Am. Chem. Soc. *2013*, *135* (23), 8504-7 [DOI: 10.1021/ja403847e]; ***anti*-isomers** from β-alkyl or β-aryl enals and a broad range of nitroalkenes (such as nitrostyrenes or aryl-subst. 1-nitro-1,3-dienes or -enynes) using a chiral aminoindanol-derived 1,2,4-triazolylidene (non-asym. variant cf. *78*, 306) s. B. Maji, L. Ji, S. Wang, S. Vedachalam, R. Ganguly, X.-W. Liu, Angew. Chem., Int. Ed. *2012*, *51* (33), 8276-80 [DOI: 10.1002/anie.201203449].

Chiral 2-prim-aminoguanidines s. under Chiral diamines ←

tert-Butyl hydroperoxide s. under Me_2Zn *t-BuOOH*

Chiral 3,3′-dibromo-1,1′-bi-2-naphthol *3,3′-Br_2BINOL*

Asym. synthesis of *o*-hydroxy-1,1-diarylalkanes from *o*-quinone methods s. *83*, 449 ←

Acetic acid s. under $[Cp^*RhCl_2]_2$ *AcOH*

Benzoic acids s. under Chiral prim. amines and O-Desmethyl-9-amino-9-deoxy-quinidine *ArCOOH*

(S)-Proline derivs. s. under Chiral diamines ←

(S)-Proline s. under AgOTf *(S)-Pro-OH*

Chiral trans-perhydroindolic acid/triethylenediamine or cyclic bis(N-oxides) ←
Asym. Michael addition-lactolization s. *67,* 313s*83* ○

3-(Pyrrolidin-2(S)-ylmethyl)-2-thiouracils ←
Asym. Michael addition-intramolecular Henry reaction s. *67,* 313s*83*

3,4,5-Trimethylthiazol-2-ylidene/microwaves *NHC/*[\\\\]
Stetter-type reaction with a formyl carbanion equivalent C=C-CO → C(CHO)CHCO

305.

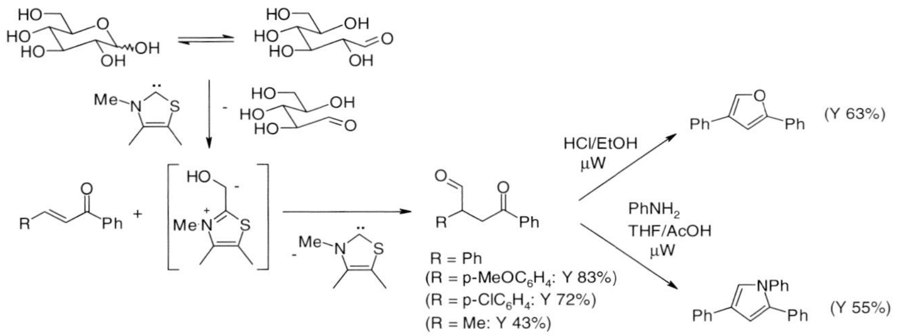

The first successful method for generating formaldehyde carbanion equivalents for a non-self-benzoin-type process is reported, such one-carbon nucleophile equivalents being **generated from** a variety of **aldoses**, *accessible from biomass*, **by** a novel **retro-benzoin** [retro-formose] **condensation**, thus minimizing the presence of free formaldehyde. **E: γ-Ketoaldehydes from α,β-ethyleneketones under microwave irradiation**. A microwave reaction tube charged with D-(+)-glucose (1 eq.), startg. enone (0.2 mmol), 3,4,5-trimethylthiazolium iodide [NHC precursor] (20 mol%), K_2CO_3 (20 mol%) and acetonitrile (2 ml), the tube sealed with a Teflon-lined snap cap, heated with stirring in a CEM Discover® SP System microwave reactor at 130° for 30 min, cooled by compressed air flow, the mixture transferred to a round bottom flask, rinsing in with methylene chloride, solvent evaporated *in vacuo*, and the residue purified by flash column chromatography over silica gel → product. Y 79% (Y 80% by ^{1}H NMR, or 58% after 10 min, or 53% with 0.2 eq. glucose, or 45% by oil bath heating for 5 h). The method gave NMR yields of 45-83% with a variety of C3-, C5- or C6-sugars but gave low yields with di- or poly-saccharides and was unsuccessful with imidazolium- or triazolium-based catalysts. A variety of chalcones, optionally bearing MeO, Me, Cl or Br on the aromatic rings and including furan-2-yl or 2-naphthyl analogs, gave yields of 57-86% with D-(-)-arabinose within 10 min (fourteen further examples; Y 81% for the illustrated example); a β-methylated enone gave 43%. F.e., also one pot conversion to the corresponding furan with HCl in ethanol (Y 63% overall) or pyrroles (overall Y 48% with NH_4OAc, 55% with $PhNH_2$), s. J. Zhang, C. Xing, B. Tiwari, Y.R. Chi, J. Am. Chem. Soc. *2013*, *135* (22), 8113-6 [DOI: 10.1021/ja401511r].

Chiral 2-prim-aminothioureas s. under Zinc triflate ←
Chiral 2-tert-aminothioureas or bis(thioureas) s. under Chiral diamines ←
Silica s. under Potassium tert-butoxide *SiO_2*
Trimethylsilyl vinyl ether s. under Benzylidene(dichloro)(tricyclohexylphosphine)(1,3-dimesitylimidazolidin-2-ylidene)ruthenium *$Me_3SiOCH{=}CH_2$*
Trimethylsilyl cyanide s. under CuBr and Bis(1,5-cyclooctadiene)nickel(0) *Me_3SiCN*

Trimethylsilyl chloride s.a. under Titanocene dichloride *Me_3SiCl*

Trimethylsilyl chloride/triethylenediamine/1,8-diazabicyclo[5.4.0]undec-7-ene ←

3-Alkylidenecyclohex-1-enecarboxylic acids
from acryloyloxy-2-acetylenes and electron-deficient ethylene derivs.
via Ireland-Claisen rearrangement-Diels-Alder reaction ○

306. 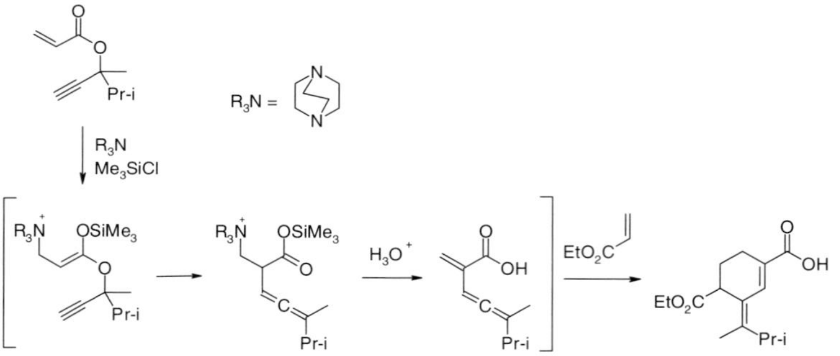

Novel Ireland-Claisen rearrangement of propargylic acetates, using conditions developed for rearrangement of *allylic acetates* [s. Y.X. Li et al., Synlett *2007* (2), 288-92 [DOI: 10.1055/s-2007-967985]], have been shown to afford unstable allenes (4-carboxy-1,2,4-trienes), which may be trapped with electron-poor dienophiles (acrylates, acrolein, acrylonitrile and maleic anhydride) to afford the title compds. in moderate yield. **E:** DABCO (20 mol%), trimethylsilyl chloride (3 eq.), DBU (2 eq.) and hydroquinone (1 mol%) added to a soln. of 2-isopropylbut-3-yn-2-yl acrylate (7.25 mmol) and ethyl acrylate (4 eq.) in acetonitrile (24 ml), the mixture stirred under reflux for 10-24 h, concentrated *in vacuo*, the residue dissolved in ether, washed with 3 *M* aq. HCl, and worked up with chromatographic purification → (E)-4-ethoxycarbonyl-3-(1,2-dimethylpropylidene)cyclohexene-1-carboxylic acid. Y 43% (E/Z >99:1). F.e. (nine in all; Y 35-45%) and substrate prepn. s. Y.-X. Li, Y.-L. Sheng, B.-S. Zhang, Chin. Chem. Lett. *2013*, *24* (2), 137-9 [DOI: 10.1016/j.cclet.2013.01.004].

Trimethylsilyl chloride/zinc triflate *$Me_3SiCl/Zn(OTf)_2$*

2-(1-Halogenovinyl)cyclopent-3-ene-1,1-dicarboxylic acid esters
from 2-(1,2,4-trien-4-yl)cyclopropane-1,1-dicarboxylic acid esters ○

307. 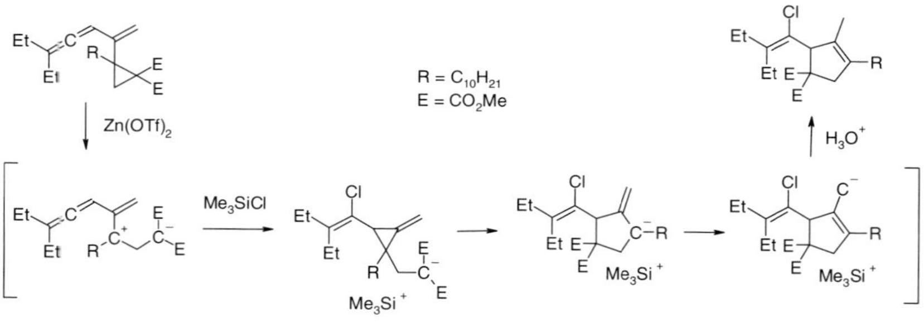

Me_3SiCl (1.2 eq.) added to a mixture of $Zn(OTf)_2$ (5 mol%) and methylene chloride (1 ml) at -10° under dry N_2, a soln. of dimethyl 2-decyl-2-(3-ethyl-3,4,6-heptatrien-6-yl)cyclopropane-1,1-dicarboxylate (0.2 mmol) in the same solvent (4 ml) added dropwise over 5 min, the mixture stirred at 30° for 6 h (TLC), quenched with water, extracted with ethyl acetate, and purified by flash chromatography → dimethyl 2-(1-chloro-2-ethylbut-1-enyl)-4-decyl-3-methylcyclopent-3-ene-1,1-dicarboxylate. Y 96%. This novel, unexpected (cyclohexene analogs were anticipated) and scalable (to 4.7 mmol) cleavage and halogenative (Cl, Br) recyclization of dialkyl cyclopropane-dicarboxylate derivs. was catalyzed by a variety of Lewis acids [optimally $Zn(OTf)_2$] in *methylene chloride* (a number of other common solvents were ineffective). The method tolerated aliphatic (incl. benzyl and allyl) substituents at C-2 and dialkyl substitution (incl. cyclic) at the allene terminus (fourteen examples; Y 44%, 50-96%), with the lowest yield obtained for a 2-allyl-cyclopropane deriv. F.e. and optimization s. Q. Yu, S. Ma, Chem. Commun. *2012*, *48* (96), 11784-6 [DOI: 10.1039/c2cc36672j].

Titanocene dichloride/magnesium/trimethylsilyl chloride/tert-butanol ←

N-Protected γ-aminocarbonyl compds. ←

from α,β-ethylenecarbonyl compds. and 1,1-aminoalcohols

via titanocene-catalyzed dehydroxylative radical 1,4-addition

308.

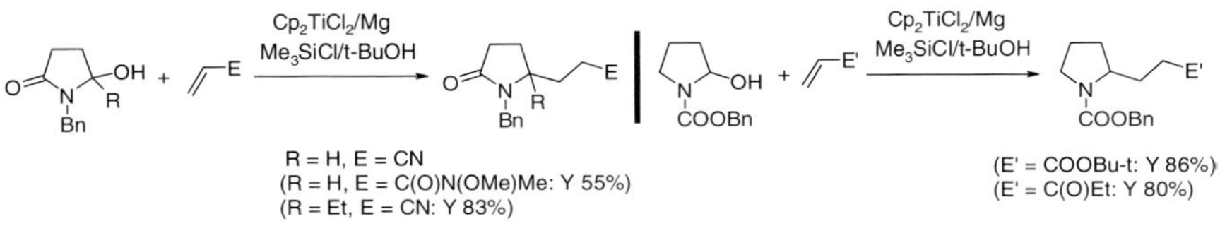

5-Subst. 2-pyrrolidones. Trimethylsilyl chloride (4 eq.) added dropwise at room temp. to a suspension of titanocene dichloride (0.0125 mmol) and Mg turnings (5 eq.) in anh. THF (1.5 ml) under N_2, the mixture stirred until it turned green (ca. 10 min), a soln. of startg. hemiaminal (0.5 mmol) and α,β-unsatd. nitrile (2 eq.) in anh. THF (1 ml) added, then *tert*-butanol (0.2 ml; 4 eq.), the orange mixture stirred for ca. 2-3 h until the color turned back to light green, filtered, washed with ethyl acetate (15 ml), the filtrate washed with brine, dried (Na_2SO_4), filtered, concentrated under reduced pressure, and the residue purified by flash chromatography on silica gel → 1-benzyl-5-(2-cyanoethyl)pyrrolidin-2-one. Y 94%. This is a milder and more chemoselective single-electron reducing system than SmI_2 (cf. Y 54% for the illustrated product with SmI_2/BF_3-etherate/*t*-BuOH at *-40°*). Mg-powder may also be used, and only 2 eq. was required, but Mg-turnings were preferred because they are cheap and readily available, convenient to handle and recycle (ca. 3.5 eq. were recovered after the reaction and their reuse gave the same result), and they do not form a suspension, allowing reaction progress to be assessed easily based on the color changes of the mixture. Ten further examples for the coupling of the 5-hydroxy-2-pyrrolidone with α-methylene esters, enones, an α-methylenehydroxamic acid ester, a 2,4-dienoate, an α-allene ester, α-methylene-γ-lactone or 2(5*H*)-furanone gave yields of 45-93%, while nine examples with 5-subst. 5-hydroxy-2-pyrrolidone gave products with a quaternary carbon center in yields of 68-84%. Five examples with N-carbobenzoxy-2-hydroxypyrrolidine and enoates, an enone, acrylonitrile or methyl propargylate gave yields of 71-86%. It is believed reaction proceeds via chlorosilane-mediated deoxychlorination. F.e. and large scale reaction, also application in an efficient six-step total synthesis of (±)-9,10-diepi-stemoamide, s. X. Zheng, X.-J. Dai, H.-Q. Yuan, C.-X. Ye, J. Ma, P.-Q. Huang, Angew. Chem., Int. Ed. *2013*, *52* (12), 3494-8 [DOI: 10.1002/anie.201210088]; coupling of N-protected 1,1-aminoethers with electron-deficient ethylene derivs. using SmI_2/BF_3-etherate/*t*-BuOH, incl. application to the stereoselective synthesis of pyrrolizidine alkaloid, (+)-xenovenine, s. Y.-G. Xiang, X.-W. Wang, X. Zheng, Y.-P. Ruan, P.-Q. Huang, Chem. Commun. *2009*, *45* (45), 7045-7 [DOI: 10.1039/b915488d]; SmI_2-mediated cross-coupling of N-protected 5-hydroxy-, 5-(2-pyridylthio)- or 5-sulfonyl-2-pyrrolidones with electron-deficient ethylene derivs. for the synthesis of 8-aza-analogs of PGE_2 s. K.-Z. Hu, J. Ma, S. Qiu, X. Zheng, P.-Q. Huang, J. Org. Chem. *2013*, *78* (5), 1790-801 [DOI: 10.1021/jo301277n].

Tri-n-butyltin hydride/azodiisobutyronitrile *$Bu_3SnH/AIBN$*

5-Amino-Δ³-2-pyrrolones from terminal acetylene derivs. and amidines
Radical [2+2+1]-cycloaddition

309. 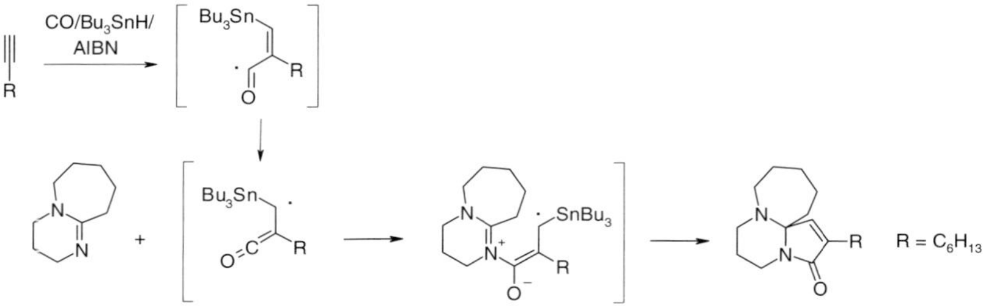

Radical cyclocarbonylation to form unsatd. lactams via the addition of vinyl radicals to carbon monoxide followed by cyclization onto C═N bonds, previously described in intramolecular fashion (*71*, 302; cf. *67*, 310), has now been extended to the intermolecular variant using amidines as substrates, providing the first example of a radical-mediated **aza-Pauson-Khand reaction**. **E:** AIBN (15 mol%), benzene (10 ml), Bu_3SnH (30 mol%), 1-octyne (0.51 mmol) and DBU (5 eq.) placed in a 50 ml stainless autoclave, the latter closed, purged three times with CO (10 atm.), pressurized to 80 atm. with CO, then heated at 80° for 6 h, excess CO discharged at room temp., solvent removed under reduced pressure, and the residue purified by flash chromatography on silica gel → 2-hexyl-5,6,8,9,10,11-hexahydro-4*H*,7*H*-3a,6a-diazacyclohepta[*d*]inden-3-one. Y 66%. The method is applicable to bicyclic, [6- but not 5-membered] monocyclic or acyclic amidines in reaction with alk-1-ynes terminated by aliphatic (optionally bearing acetoxy, silyl ether or tert. amino groups) or aryl (optionally bearing chloro or methoxy groups), yields of 55-69% being obtained for eleven further examples. Three examples with ar. aldimines gave 5-aryl-Δ³-2-pyrrolones in 15-51% yield, an electron-donating group enhancing the yield, while a cyclic iminoester, 5,6-dihydro-2-methyl-4*H*-1,3-oxazine, gave a 5-oxy-deriv. in 63% yield. F.e.s. T. Fukuyama, N. Nakashima, T. Okada, I. Ryu, J. Am. Chem. Soc. *2013*, *135* (3), 1006-8 [DOI: 10.1021/ja312654q].

Tetra-n-butylammonium hexafluorophosphate s. under $Ca(NTf_2)_2$ *Bu_4NPF_6*

Tert. phosphines [s.a. under Nickel(II) acetoacetonate, Cyclooctadiene-(8-quinolinato)rhodium(I) and Bis(dibenzylideneacetone)palladium(0)] *R_3P*

4-[Carbalkoxy(tosylamino)methylene]cyclopent-2-enecarboxylic acid esters
from β,γ-ethylene-α-(tosylimino)carboxylic acid esters and α-allenecarboxylic acid esters
Diastereoselective phosphine-mediated [3+2]-cycloaddition

310. 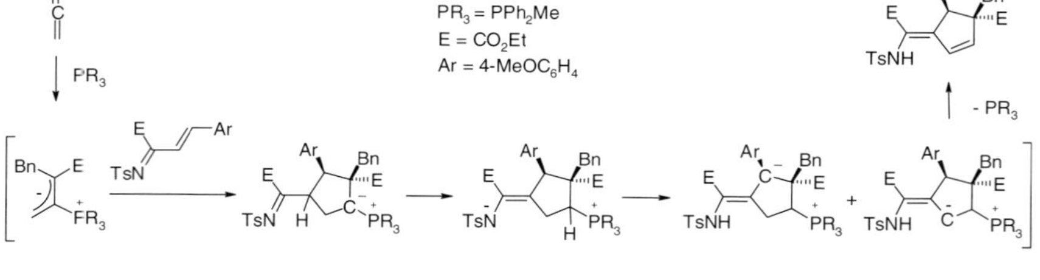

A novel variation of the Lu [3+2]-cycloaddition with allenes [s. Z. Xu, X. Lu, Tetrahedron Lett. *1997*, *38* (19), 3461-4 [DOI: 10.1016/S0040-4039(97)00656-4]; cf. *53*, 313] is reported, using ester-activated α,β-unsaturated imines as the 2-carbon component to afford highly functionalized cyclopentenes stereoselectively with an all-carbon quaternary center. **E:** A mixture of ethyl α-benzyl-allenoate (1.5 eq.), ethyl 4-(4-methoxyphenyl)-2-(*p*-toluenesulfonylimino)but-3-enoate (0.25 mmol) and methyl(diphenyl)phosphine (20 mol%) in toluene (2 ml) stirred under N_2 at room temp. for 12 h, concentrated *in vacuo*, and purified by flash chromatography → (*trans*,E)-ethyl 1-benzyl-4-[carbethoxy(tosylamino)methylene]-4-(4-methoxyphenyl)cyclopent-2-enecarboxylate. Y 78% (E/Z 91:9). Methyl(diphenyl)phosphine catalyzed the [3+2]-cycloaddition of α-alkyl-(incl. benzyl- or propargyl-) allenoates with electron-diverse ester-activated N-tosylcinnamald-imines (incl. 2-naphthyl and 2-thienyl analogs) to afford *trans*-4,5-disubst. cyclopent-2-ene-carboxylates, generally with good stereoselectivity (E/Z 77:23 to 94:6; twenty-five examples; Y49-78%). An ester-activated penta-2,4-dienaldimine homolog was a poorer substrate (Y 30%; E/Z 64:36). F.e.s. J. Tian, Z. He, Chem. Commun. *2013*, *49* (20), 2058-60 [DOI: 10.1039/c3cc38264h]; regioselective [3+2]-cycloaddition of allenes to 3-alkylidenebenzofuran-2-ones with triphenylphosphine to give spiro[cyclopent-2-ene-1,3′-benzofuran-2′-ones] s. X. Li, F. Wang, N. Dong, J.-P. Cheng, Org. Biomol. Chem. *2013, 11* (9), 1451-5 [DOI: 10.1039/c3ob27288e]; to 3-acylcoumarins as ene-component with tri-*n*-butylphosphine, also [4+2]-cycloaddition with DABCO, s. Y. Wang, Z.-H. Yu, H.-F. Zheng, D.-Q. Shi, ibid. *2012, 10* (38), 7739-46 [DOI: 10.1039/c2ob26300a]; phosphine-catalyzed synthesis of **5-vinyl-2,3-dihydrofuran-2-ylacetic acid esters** from α-allenoates and enones s. P. Xie, W. Lai, Z. Geng, Y. Huang, R. Chen, Chem. Asian J. *2012, 7* (7), 1533-7 [DOI: 10.1002/asia.201200140].

Phosphine-catalyzed Rauhut-Currier-type [4+2]-cycloaddition s. *31*, 623s*83* ○

Tris(2,6-dimethoxyphenyl)phosphine *Ar_3P*

5-Hydroxycyclopent-3-en-2-onecarboxylic acid esters
from α-allenecarboxylic acid esters and α-diketones
Diastereoselective phosphine-mediated ring closure

311. 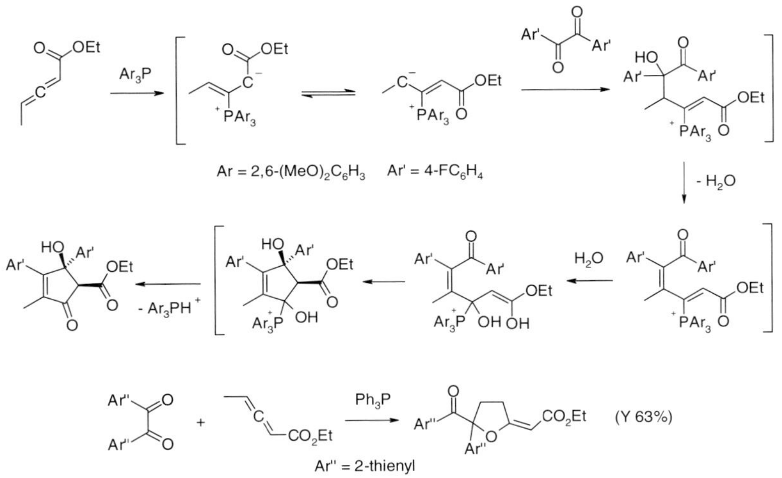

Tris(2,6-dimethoxyphenyl)phosphine (1.5 eq.) added to a soln. of 1,2-bis(4-fluorophenyl)ethane-1,2-dione (0.5 mmol) and ethyl penta-2,3-dienoate (1.5 eq.) in dry THF (5 ml) under argon at room temp., the mixture stirred for 3 h, solvent evaporated, and the residue purified by chromatography on silica → ethyl 2,3-bis(4-fluorophenyl)-2-hydroxy-4-methyl-5-oxocyclopent-3-enecarboxylate. Y 96%. This novel cyclization appears general for methyl/phenyl [or benzyl] terminated allenoates reacting with electron-diverse symmetrical di(het)aryl 1,2-diones (fifteen examples; Y 78-96%; a bis(2-furanyl)dione gave 65%), with a single unsym. diaryl 1,2-dione affording an inseparable mixture of two regioisomers (1:0.88; Y 95%). Interestingly, the use of triphenylphosphine as promoter afforded **5-(het)aroyl-5-(het)aryl-2-(carbalkoxymethylene)tetrahydrofurans** as major products from ethyl penta-2,3-dienoate under similar conditions (five examples; Y 35-77%, together with the cyclopentenones: 0-20%). F.e. and optimization s. A. Jose, K.C.S. Lakshmi, E. Suresh, V. Nair, Org. Lett. *2013, 15* (8), 1858-61 [DOI: 10.1021/ol400467v].

2-(Dicyclohexylphosphino)-1-methylimidazole s. under Dodecacarbonyltriruthenium ←
Chiral phosphines s. under Copper(I) or silver(I) salts ←
Chiral 3-(diphenylphosphino)-1,1′-bi-2-naphthol or N,N-bis[2-(2-naphthyl)ethyl]-amino(diaryl)phosphines s. under Triorganoaluminum compds. ←

Chiral 2-diphenylphosphino-2′-thioureido-1,1′-binaphthyl ←
Asym. [3+2]-cycloaddition with allenes ○
update s. *78,* 332s*82*; *79,* 294s*82*; cf. *53*, 313; asym. cycloaddition of α-allenecarbonyl compds. **to 4,4-dicyano-2-methylenebut-3-enoates** for the synthesis of chiral polysubst. vinylcyclopent-3-enes using a chiral 2-diphenylphosphino-2′-thioureido-1,1′-binaphthyl as Lewis base, also racemic conversion with tris(*p*-fluorophenyl)phosphine, s. X.-N. Zhang, M. Shi, ACS Catal. *2013, 3* (4), 507-12 [DOI: 10.1021/cs300751a]; regio- and stereo-selective asym. cycloaddition **to nitrones** with a chiral (1,1′-binaphthyl-2,2′-diyl phosphoramidite)gold(I) chloride as catalyst and added $AgNTf_2$ s. G.-H. Li, W. Zhou, X.-X. Li, Q.-W. Bi, Z. Wang, Z.-G. Zhao, W.-X. Hu, Z. Chen, Chem. Commun. *2013, 49* (42), 4770-2 [DOI: 10.1039/c3cc41769g].

1,2-Bis(diphenylphosphino)benzene s. under Copper(I) chloride *dppbz*
2-(Dicyclohexylphosphino)-1-[o-(dicyclohexylphosphino)phenyl]pyrrole s. under Palladium(II) acetoacetonate ←
Chiral di(phosphines) s. under $CuN(SiMe_3)_2$, Copper(I) or silver(I) salts, Carbonyl(dihydrido)-tris(triphenylphosphine)ruthenium(II) and Chloro(cyclooctene)iridium(I) dimer ←
Chiral N-phosphinodiamines s. under Copper(I) or silver(I) salts ←
(S)-2-Amino-3,3′-dibromo-2′-diphenylphosphinamino-4,4′,6,6′-tetrakis-(trifluoromethyl)biphenyl s. under $(MeCN)_4CuBF_4$ *TF-BiphamPhos*
Chiral 2-(aroylamino)phosphines or N-phosphinyl-1,2-diamines s. under Chiral diamines ←

Chiral P-spirocyclic iminophosphoramides ←
Regioselective organocatalyzed asym. 1,6-addition of Δ²-5-oxazolones to N-(2,4-dieneacyl)pyrroles ←

312.

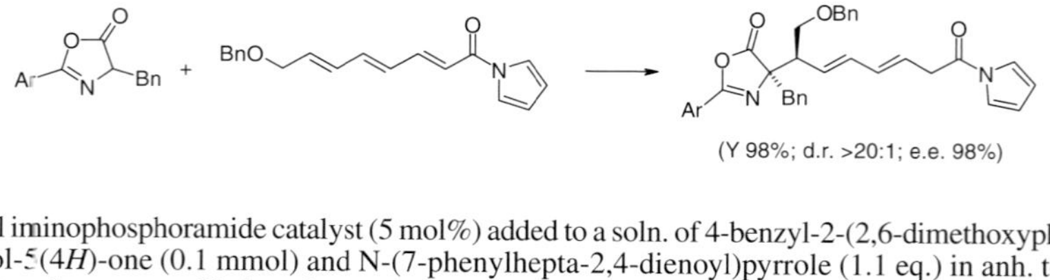

Chiral iminophosphoramide catalyst (5 mol%) added to a soln. of 4-benzyl-2-(2,6-dimethoxyphenyl)-oxazol-5(4*H*)-one (0.1 mmol) and N-(7-phenylhepta-2,4-dienoyl)pyrrole (1.1 eq.) in anh. toluene (1 ml) under argon at 0°, the mixture stirred for 3 h, quenched with a soln. of trifluoroacetic acid in toluene, solvent evaporated, and the residue purified by chromatography on silica → N-[7-phenyl-5-(4-benzyl-2-(2,6-dimethoxyphenyl)-5(4*H*)-oxazolon-4-yl)hept-3-enoyl]pyrrole. Y 99% (d.r. >20:1; e.e. 98%). Variation of substituents on the strongly basic chiral P-spiro-iminophosphoramide skeleton gave rise to remarkable regio-, diastereo- and enantio-selective reactions of terminally subst. N-(2,4-dieneacyl)pyrroles to afford 1,6-adducts exclusively (twelve examples; Y 84-99%; d.r. 12:1 to >20:1; e.e. 90-98%). The reaction was extended to near exclusive **asym. 1,8-addition** to N-(2,4,6-trieneacyl)pyrroles (five examples; Y 89-98%; 1,8/1,6/1,4 14:<1:1 to >20:1<1:1; d.r. 8:1 to >20:1; e.e. 98-99%). F.e., optimization, substrate and catalyst prepn., and further elaboration of azlactone and N-acylpyrrole moieties, s. D. Uraguchi, K. Yoshioka, Y. Ueki, T. Ooi, J. Am. Chem. Soc. *2012*, *134* (47), 19370-3 [DOI: 10.1021/ja310209g].

Chiral TADDOL-derived phosphite-phosphine s. under Bis(1,5-cyclooctadiene)nickel(0) ←
Diphenyl phosphate s. under Potassium tert-butoxide $(PhO)_2PO_2H$
Chiral TADDOL-based phosphoric acid diesters s. under Carbonyl(dihydrido)tris-(triphenylphosphine)ruthenium(II) ←
Chiral 1,1′-binaphthyl-2,2′-diyl hydrogen phosphate s. under Chiral prim. amines ←
Chiral 5,5′,6,6′,7,7′,8,8′-octahydro-1,1′-binaphthyl-2,2′-diyl hydrogen phosphates ←

Organocatalyzed asym. hetero-Diels-Alder reaction s. *46*, 662s*83* ○

Oxygen s. under Et_3B O_2
Methanesulfonic acid s. under 2,6-Dichloropyridine N-oxide $MeSO_3H$
Trifluoromethanesulfonic acid *TfOH*

Hetero-Diels-Alder reaction with α,β-ethylenediazo compds. s. *31*, 623s*83*

Camphorsulfonic acid s. under Copper(II) acetate RSO_3H
p-Toluenesulfonic acid s. under Palladium(II) acetoacetonate *TsOH*

Bromo(pentacarbonyl)manganese/dicyclohexylamine $MnBr(CO)_5/Cy_2NH$

Stereoselective manganese-catalyzed N-directed *o*-vinylation with terminal acetylene derivs. H → C=CH

313. 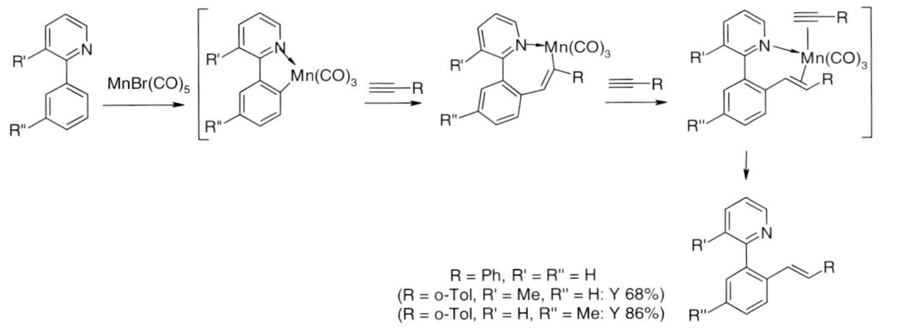

The first *manganese-catalyzed* aromatic C-H vinylation with alk-1-ynes is reported, the method featuring an operationally simple, commercially available catalyst system and displaying high chemo-, regio-, and stereo-selectivity. **E:** $MnBr(CO)_5$ (10 mol%), ether (1.2 ml), 2-phenylpyridine (2 eq.), phenylacetylene (0.5 mmol) and dicyclohexylamine (20 mol%) added sequentially under N_2 to a flame-dried Teflon-screw-capped tube, the mixture stirred at 80° for 6 h, cooled to room temp., diluted with methylene chloride (10 ml), filtered through a short pad of silica gel, washed with ethyl acetate (30 ml), the filtrate pre-absorbed onto silica gel, concentrated by rotary evaporation, and purified by silica gel column chromatography → (E)-product. Y 76%. The method is applicable to a variety of arylethynes (which may bear electron-donating or -withdrawing groups in the *ortho*, *meta* or *para* positions) or hetaryl analogs or [more sluggishly] to aliphatic alk-1-ynes, but there was no reaction with internal alkynes. The arylpyridines may also bear a variety of functional groups, with *m*-subst. arylpyridines reacting with complete regioselectivity; 2-thienylpyridine or 1-phenylpyrazole may also be used. Mechanistic investigations and DFT calculations show that the reaction is initiated by a C-H activation step involving cooperation between manganese and base, while a 5-membered manganacycle (which could be isolated) and alkynylmanganese species are the key reaction intermediates, and the ligand-to-ligand H-transfer and alkynyl-assisted C-H activation are the key steps rendering the reaction catalytic in manganese. F.e. (thirty-seven; Y 27%, 46-86%) s. B. Zhou, H. Chen, C. Wang, J. Am. Chem. Soc. *2013*, *135* (4), 1264-7 [DOI: 10.1021/ja311689k].

Transition metal complexes ←

Regioselective transition metal-catalyzed hydro[het]arylation of ethylene derivs. H → R

update s. *49*, 679s*82*; highly regioselective *cobalt-catalyzed* 4-alkylation of pyridines with ethylene derivs. using $CoBr_2$/$LiBHEt_3$/Et_3B s. T. Andou, Y. Saga, H. Komai, S. Matsunaga, M. Kanai, Angew. Chem., Int. Ed. *2013*, *52* (11), 3213-6 [DOI: 10.1002/anie.201208666]; *o*-(1-arylethyl)ation of ar. aldimines with styrenes using $CoBr_2$/triphenylphosphine/Me_3SiCH_2MgCl s. P.-S. Lee, N. Yoshikai, ibid. *52* (4), 1240-4 [DOI: 10.1002/anie.201207958]; regioselective *ruthenium-catalyzed* hydroarylation of unactivated alkenes, incl. highly fluorinated substrates, with an (η^6-arene)ruthenium(II) bis(carboxylate) s. M. Schinkel, I. Marek, L. Ackermann, ibid. *52* (14), 3977-80 [DOI: 10.1002/anie.201208446]; ruthenium-catalyzed N-directed *o*-substitution of 2-arylpyridines and other functionalized arenes with methylenecyclopropanes using $[RuCl_2(cod)]_n$ and a di(phosphine) as ligand, e.g. *rac*-BINAP, s. L. Ackermann, S.I. Kozhushkov, D.S. Yufit, Chem. Eur. J. *2012*, *18* (38), 12068-77 [DOI: 10.1002/chem.201200406]; regioselective [Markovnikov] *palladium-catalyzed* hydrohetarylation of α-methylated arylalkenes with $[Pd(cod)Cl\text{-}SnCl_3]$ as heterobimetal catalyst s. D. Das, S. Pratihar, S. Roy, Tetrahedron Lett. *2013*, *54* (4), 335-8 [DOI: 10.1016/j.tetlet.2012.11.038]; regioselective *iridium-catalyzed* **asym. hydrohetarylation** of bicycloalkenes with $[Ir(coe)Cl]_2$ and (S)-DTBM-SegPhos as ligand s. C.S. Sevov, J.F. Hartwig, J. Am. Chem. Soc. *2013*, *135* (6), 2116-9 [DOI: 10.1021/ja312360c].

Regioselective catalytic hydroarylation of acetylene derivs. C≡C → CH=C(Ar)

update s. *59*, 311s*82*; under *gold catalysis* with mono- or di-nuclear gold(I) N-heterocyclic carbene complexes, e.g. [IPrAu]Cl, with added $AgBF_4$ s. C. Tubaro, M. Baron, A. Biffis, M. Basato, Beilstein J. Org. Chem. *2013*, *9*, 246-53 [DOI: 10.3762/bjoc.9.29]; regio- and stereo-selective *ruthenium catalyzed o*-directed vinylation of aryl carbamates with $[RuCl_2(p\text{-cymene})]_2$/$AgSbF_6$/pivalic acid s. M.C. Reddy, M. Jeganmohan, Chem. Commun. *2013*, *49* (5), 481-3 [DOI: 10.1039/c2cc37758f]; *o*-hydroxy-1,1-diarylalkenes from phenols and ethynylarenes under *heterogeneous iron catalysis* with a recyclable iron-containing mesoporous aluminosilicate [Fe-Al-MCM-41] s. S. Haldar, S. Koner, Beilstein J. Org. Chem. *2013*, *9*, 49-55 [DOI: 10.3762/bjoc.9.6]; regioselective hydroarylation of trifluoromethylacetylenes *under superacid catalysis* with trifluoromethanesulfonic or fluorosulfonic acid s. H.M.H. Alkhafaji, D.S. Ryabukhin, V.M. Muzalevskiy, A.V. Vasilyev, G.K. Fukin, A.V. Shastin, V.G. Nenajdenko, Eur. J. Org. Chem. *2013* (6), 1132-43 [DOI: 10.1002/ejoc.201201375].

Regioselective asym. hydroalkynylation of ethylene derivs. with terminal acetylene derivs. C=C → CHC-C≡C

update s. *65*, 266s*82*; of 1,4-oxido-1,4-dihydronaphthalenes with terminal alkynes using $[Ir(cod)Cl]_2$ and (R)-2,3;2′,3′-bis(ethylenedioxy)-6,6′-bis(diphenylphosphino)biphenyl [(R)-SynPhos] as ligand s. J. Hu, Q. Yang, J. Xu, C. Huang, B. Fan, J. Wang, C. Lin, Z. Bian, A.S.C. Chan, Org. Biomol. Chem. *2013*, *11* (5), 814-20 [DOI: 10.1039/c2ob26775f]; of α-arylidene-β-diketones with ethynyl-[het]arenes **by asym. 1,4-addition** using Et_2Zn/(R)-3,3′-diphenyl-2,2′-bi-1-naphthol s. G. Blay, L. Cardona, J.R. Pedro, A. Sanz-Marco, Chem. Eur. J. *2012*, *18* (41), 12966-9 [DOI: 10.1002/chem.201201765]; **by asym. 1,6-addition** of triisopropylsilylacetylene to 2,4-dienecarboxylic acid derivs. with $Co(OAc)_2{\cdot}4H_2O$/Zn and a (S,S)-1,2-bis(2,5-dialkyl-phospholano)benzene as ligand s. T. Sawano, A. Ashouri, T. Nishimura, T. Hayashi, J. Am. Chem. Soc. *2012*, *134* (46), 18936-9 [DOI: 10.1021/ja309756k].

[2+2+2]-Cycloaddition ○

update s. *33*, 658s*82*; **condensed pyridines** with dimeric (1,3-diisopropylimidazol-2-ylidene)-(η^1-η^2-nitrile)nickel(II) complexes s. R.M. Stolley, H.A. Duong, D.R. Thomas, J. Louie, J. Am. Chem. Soc. *2012*, *134* (36), 15154-62 [DOI: 10.1021/ja3075924]; condensed 2-aminopyridine ring from diynes and cyanamides, and isolated 2,4,6-trisubst. analogs from two terminal acetylene molecules, with $FeCl_2$/N,N′-dimesitylpyridine-2,6-dicarboxaldimine/Zn as catalyst s. T.K. Lane, B.R. D'Souza, J. Louie, J. Org. Chem. *2012*, *77* (17), 7555-63 [DOI: 10.1021/jo3012418]; chiral condensed cyclohex-2-ene-1-carboxamides from enynes and acrylamide by regio- and diastereoselective **asym. [2+2+2]-cycloaddition** (cf. *64,* 287s*82*) with [Rh(cod)$_2$]BF$_4$/(R)-H$_8$-BINAP s. K. Masutomi, N. Sakiyama, K. Noguchi, K. Tanaka, Angew. Chem., Int. Ed. *2012, 51* (52), 13031-5 [DOI: 10.1002/anie.201206122]; chiral condensed 2,3-di(methylene)cyclohexanecarboxylic acid esters from 1,2,n-trienes and allenoates using [Rh(CH=CH$_2$)$_2$Cl]$_2$/AgOTf/(R)-H$_8$-BINAP s. A.T. Brusoe, R.V. Edwankar, E.J. Alexanian, Org. Lett. *2012, 14* (23), 6096-9 [DOI: 10.1021/ol303024q]; chiral condensed N-sulfonyl-1,2-dihydropyridines from diynes and N-sulfonylaldimines with [Rh(1,5-hexadiene)Cl]$_2$/AgSbF$_6$ and a chiral di(phosphine) as ligand, e.g. (R)-di-*tert*-Bu-MeO-Biphep, cf. M. Amatore, D. Lebœuf, M. Malacria, V. Gandon, C. Aubert, J. Am. Chem. Soc. *2013, 135* (12), 4576-9 [DOI: 10.1021/ja309301n].

Chiral transition metal complexes
Catalytic asym. Diels-Alder reaction
update s. *46,* 662s*82*; asym. cycloaddition of N^2-acryloyl-N^1-(1-naphthylmethyl)-5,5-dimethyl-3-pyrazolidone with N-carbalkoxy-1,2-dihydropyridines as diene component with a chiral copper(II) bis(Δ^2-oxazoline) as catalyst s. N.D.M.R. Hutabarat, C. Seki, T. Shimizu, M. Hirama, Y. Kohari, H. Nakano, K. Uwai, N. Takano, E. Kwon, H. Matsuyama, Heterocycles *2012, 86* (1), 203-17 [DOI: 10.3987/com-12-s(n)2]; with α-alkylidene-β-ketocarboxylic acid N-carbalkoxyamides as dienophile, also asym. Hosomi-Sakurai reaction, s. K. Orimoto, H. Oyama, Y. Namera, T. Niwa, M. Nakada, Org. Lett. *2013, 15* (4), 768-71 [DOI: 10.1021/ol303381c]; asym. cycloaddition with a chiral heterobimetallic lithium uranium(IV) BINOL-ate as catalyst s. J.R. Robinson, P.J. Carroll, P.J. Walsh, E.J. Schelter, Organometallics *2013, 32* (5), 1493-9 [DOI: 10.1021/om3011849]; spirocyclics by asym. cycloaddition of N-protected 3-alkylideneoxindoles with 1-alkoxy-3-siloxy-1,3-dienes using a chiral magnesium 3,3′-di(9-phenanthryl)-1,1′-binaphthyl-2,2′-diyl phosphate as catalyst s. G. Li, T. Liang, L. Wojtas, J.C. Antilla, Angew. Chem., Int. Ed. *2013, 52* (17), 4628-32 [DOI: 10.1002/anie.201209295]; **catalytic asym. hetero-Diels-Alder reaction** of cyclic enoxysilanes with 2-(α,β-ethyleneacyl)pyridine N-oxides as diene component, also asym. Michael-type addition, s. A. Livieri, M. Boiocchi, G. Desimoni, G. Faita, Chem. Eur. J. *2012, 18* (37), 11662-8 [DOI: 10.1002/chem.201201214]; with aldimines using a chiral *Brønsted-acidic* cobalt(III) salen complex s. V.I. Maleev, T.V. Skrupskaya, L.V. Yashkina, A.F. Mkrtchyan, A.S. Saghyan, M.M. Il′in, D.A. Chusov, Tetrahedron: Asym. *2013, 24* (4), 178-83 [DOI: 10.1016/j.tetasy.2013.01.011].

Nanoparticulate 3-sulfopropyl-functionalized γ-iron(III) oxide
Heterogeneous Friedel-Crafts reaction of indoles C=C → CHC(R)
with electron-deficient ethylene derivs. s. *11,* 770s*83*

Silica-supported magnetite nanoparticles *Fe_3O_4-SiO_2*
4,5,6,7-Tetrahydro-1*H*-1,4-diazepine-5-carboxylic acid amides
from 1,2-ethylene-1,2-diamines, two terminal acetylene deriv. molecules and isonitriles

314.

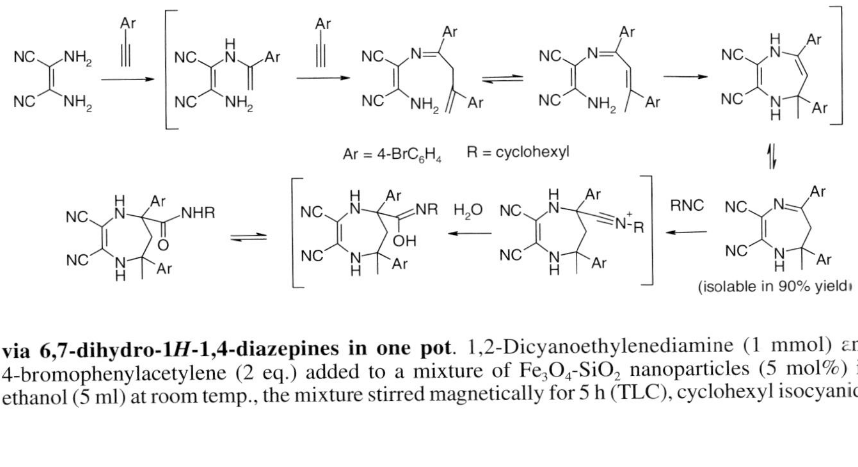

via 6,7-dihydro-1*H*-1,4-diazepines in one pot. 1,2-Dicyanoethylenediamine (1 mmol) and 4-bromophenylacetylene (2 eq.) added to a mixture of Fe_3O_4-SiO_2 nanoparticles (5 mol%) in ethanol (5 ml) at room temp., the mixture stirred magnetically for 5 h (TLC), cyclohexyl isocyanide

(1 eq.) added, the mixture stirred for 7 h (TLC), filtered (removing magnetic stirring bar with catalyst attached), and purified by recrystallization → N-cyclohexyl-5,7-bis(4-bromophenyl)-2,3-dicyano-7-methyl-4,5,6,7-tetrahydro-1*H*-1,4-diazepine-5-carboxamide. Y 83%. This novel multi-component reaction was catalyzed by non-toxic and environmentally benign superparamagnetic iron oxide nanoparticles, which were readily removed magnetically. Dihydrodiazepine formation was successful from 1,2-dicyanoethylenediamine (three examples; Y 90-95%) or electron-diverse *o*-phenylenediamines, in reaction with electron-diverse phenylacetylenes, the latter affording **2,3-dihydro-1*H*-1,5-benzodiazepines** (five examples; Y 70-92%), with the 3-subst. 7-methyl regioisomer predominating (88:12 to 95:5) for 4-subst. *o*-phenylenediamines. The monocyclic diazepines were reacted *in situ* with cyclohexyl isocyanide (illustrated) to afford the corresponding amides (two further examples; Y 90%, 92%). The catalyst was recycled five times without significant loss of activity. F.e. and catalyst prepn. s. A. Maleki, Tetrahedron Lett. *2013*, *54* (16), 2055-9 [DOI: 10.1016/j.tetlet.2013.01.123]; **from two oxo compd. molecules** in place of the acetylene derivs. (CC⇃↾O) s. A. Maleki, Tetrahedron *2012*, *68* (38), 7827-33 [DOI: 10.1016/j.tet.2012.07.034]; from 1,2-dicyanoethylenediamine with *p*-TsOH in water cf. A. Shaabani, A. Maleki, H. Mofakham, J. Moghimi-Rad, J. Org. Chem. *2008*, *73* (10), 3925-7 [DOI: 10.1021/jo8002512]; from *o*-diamines under similar conditions s. A. Shaabani, A. Maleki, H. Mofakham, J. Comb. Chem. *2008*, *10* (4), 595-8 [DOI: 10.1021/cc8000635].

Chiral ferrocenyldi(phosphines) or 2-(Δ²-oxazolin-2-yl)ferrocenyl(diphenyl)phosphines ←
s. under Magnesium

Iron(III) chloride *$FeCl_3$*
Regioselective ar. alkylation with ethylene derivs. H→R
Friedel-Crafts reaction, update s. *3*, 585s*82*; N-protected 3-[het]arylindolines from [het]arenes and indoles as the ene component, incl. *o*-substitution of phenols, with $FeCl_3$ s. R. Beaud, R. Guillot, C. Kouklovsky, G. Vincent, Angew. Chem., Int. Ed. *2012, 51* (50), 12546-50 [DOI: 10.1002/anie.201206611]; ***o*-alkylation** of phenolethers with 'half-sandwich' rare-earth(III) complexes, of the type Cp^*MR_2 (M = Sc, Y, Gd), with added trityl tetrakis(pentafluorophenyl)borate, also *o*-α-alkylation, s. J. Oyamada, Z. Hou, ibid. *51* (51), 12828-32 [DOI: 10.1002/anie.201206233].

Acetylenedicobalt hexacarbonyl complexes ←
Regioselective Pauson-Khand reaction ○
with dicobalt hexacarbonyl-complexed trifluoromethylacetylenes *en route* to 3-subst. 2-cyclopentenones s. *83*, 328

Acetylenedicobalt hexacarbonyl complexes/irradiation ←
Additive-free, photoinduced Pauson-Khand reaction under continuous flow

315.

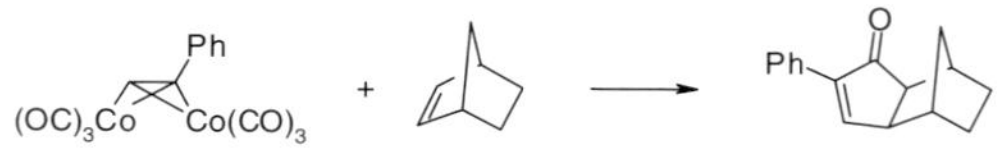

Pauson-Khand reactions may be performed at ambient temperature in the absence of additive using a photochemical flow microreactor, the activation of the alkyne-cobalt complexes being more efficient than in a conventional batch reactor, while scale-up is easy by extending the operation time. **E:** A soln. of startg. alkyne-dicobalt complex (0.05 mmol) and 2-norbornene (1 eq.) in toluene (0.05 *M*) passed by syringe pump through the fused quartz glass-covered microchannel (depth 200 μm; width 1000 μm; length 916 mm) of a photochemical flow microreactor (YMC, KeyChem-Lumino 2, fitted with an 80 W medium pressure Hg lamp) at a flow rate of 0.2 ml/min at 25° with a residence time of 55 s, the soln. collected, then stirred for 5 min at room temp., concentrated under reduced pressure, and the residue purified by flash chromatography → product. Y 88% by ¹H NMR (conversion 94%). The analogous reaction by the batch method gave only 32% yield. Continuous operation of the photochemical flow microreactor system for 1 h gave 85% yield (85% conversion) on a 0.6 mmol scale. Six further examples gave yields of 52-90% with electron-rich or -poor arylethyne-cobalt complexes while an aliphatic deriv. gave 92% with norbornene or 20% with cyclopentene. Four examples of **intramolecular Pauson-Khand reaction**

gave yields of 65-80% (88-99% conversion) with 1,2-dimethoxyethane as solvent. F.e.s. K. Asano, Y. Uesugi, J.-i. Yoshida, Org. Lett. *2013*, *15* (10), 2398-401 [DOI: 10.1021/ol4008519].

(η^6-Benzene)(pentamethylcyclopentadienyl)cobalt(III) bis(hexafluorophosphate) *[Co(III)]*
N-Directed *o*-substitution by 1,4-addition to enones s. *11*, 770s*83* C=C → CHC(R)

Bis(1,5-cyclooctadiene)nickel(0)/chiral TADDOL-derived phosphite-phosphine/trimethylsilyl cyanide *[Ni(0)]*/Me_3SiCN*
α-Arylnitriles from styrenes C=C → CHC-CN
Nickel(0)-catalyzed asym. hydrocyanation

316.

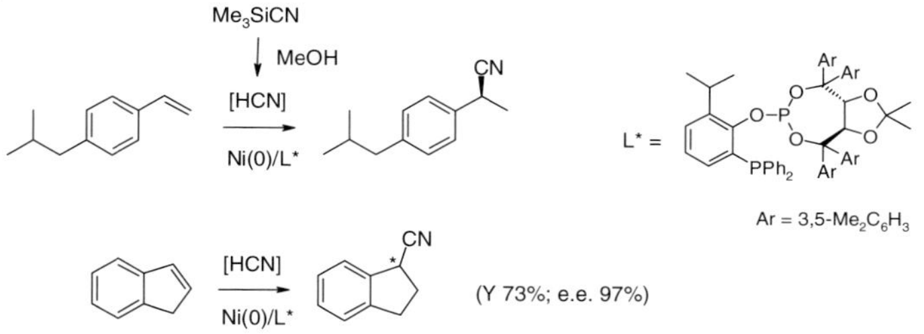

A mixture of chiral phosphite-phosphine ligand (5 mol%), [Ni(cod)$_2$] (5 mol%) and toluene (6.2 ml) stirred at room temp. for 5 min, solvent evaporated, the residue dried *in vacuo*, dissolved in THF (52 ml), 4-isobutylstyrene (6.24 mmol) added, a soln. of trimethylsilyl cyanide (1.5 eq.) in THF (69.9 ml) and methanol (5 ml) added via syringe pump over 2 h, the mixture concentrated *in vacuo*, and purified by chromatography on silica → (S)-(-)-2-(4-isobutylphenyl)propanenitrile. Y 100% (e.e. 92%). Development of novel chiral TADDOL-derived bidentate phosphite-phosphine ligands and extensive optimization identified conditions for clean, regioselective, enantioselective and scalable (for the illustrated synthesis of an ibuprofen precursor) hydrocyanation. Generation of HCN *in situ* at a controlled rate was key to the success of the reaction, and achieved by slow addition of Me_3SiCN in the presence of methanol. The method was successful for electron-diverse styrenes and indene (eight examples; Y 73-100%; e.e. 80-97%), but gave a lower yield with 1,2-dihydronaphthalene (34%; e.e. 90%). A β-methyl substituent was well tolerated in the electron-poor 4-cyanostyrene (Y 82%; e.e. 83%) but low yields were observed for electron-rich analogs (15-28%; two examples; e.e. 81-84%), and reduced enantioselectivity was observed for bulky 4-cyano-β-isopropylstyrene (e.e. 67%; Y 70%). F.e., optimization and ligand prepn. s. A. Falk, A.-L. Göderz, H.-G. Schmalz, Angew. Chem., Int. Ed. *2013*, *52* (5), 1576-80 [DOI: 10.1002/anie.201208082].

Nickel(II) acetate/chiral 2-(o-hydroxybenzylaminomethyl)-1-tosyl-Δ^2-imidazolines/triethylamine ←
Asym. 1,3-dipolar cycloaddition of azomethinium ylids s. *67*, 301s*83* ○

Nickel(II) acetoacetonate/dicyclohexyl(phenyl)phosphine/diethylzinc *[Ni(II)]/Et_2Zn*
Regio- and stereo-selective synthesis of 3-ethylene-3,1-fluorohydrins CH(OH)C-C(F)=C-CH
from aldehydes and *in situ*-generated 2-fluoro-1,3-dienes s. *83*, 413

Dichloro[2,2′-bis(diphenylphosphino)-1,1′-binaphthyl]nickel(II)/lithium hydridotriethylborate/magnesium bromide *$NiCl_2$(binap)/$LiBHEt_3$/$MgBr_2$*
β-Hydroxyacylsilanes from 2-ethylene-1,1-hydroxysilanes CHO → CH(OH)C(CH)(O)Si⋚
and aldehydes via stereoselective nickel-catalyzed isomerizing aldol condensation s. *83*, 270

Nickel(II) perchlorate *$Ni(ClO_4)_2$*
1,3-Dioxolane-4,4-dicarboxylic from oxirane-2,2-dicarboxylic acid esters and aldehydes
Regioselective Lewis acid catalyst-dependent [3+2]-cycloaddition s. *83*, 32

Nickel(II) perchlorate/chiral bis(Δ^2-oxazolines) *[Ni(II)]**
Asym. Friedel-Crafts reaction of indoles C=C → CHC(Ar)
with electron-deficient ethylene derivs. s. *67*, 336s*83*

Chiral bis(cyclohexane-1,2-diamine)nickel(II) dibromide or mesoporous organo-silica-supported variants *[Ni(II)]**
Catalytic asym. Michael addition s. *49*, 657s*83* C=C → CHC(R)

Dodecacarbonyltriruthenium/2-(dicyclohexylphosphino)-1-methylimidazole ←
Hydroformylation-hydrogenation s. *76*, 343s*83* C=C → CHC-CH_2OH

Carbonyl(dihydrido)tris(triphenylphosphine)ruthenium(II)/(S)-2,3;2′,3′-bis(methylene-dioxy)-6,6′-bis(diphenylphosphino)biphenyl/chiral TADDOL-based phosphoric acid diesters ←
3-Ethylenealcohols from [aliphatic] prim. alcohols $RCH(OH)CH(CH_3)CH{=}CH_2$
***syn*-Diastereoselective asym. synthesis with addition of four C-atoms**

317. 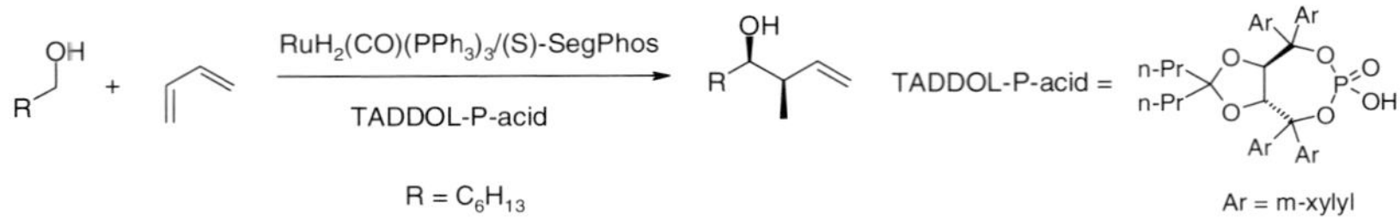

A notable **effect of the chiral anion** has been uncovered in hydrohydroxyalkylation of 1,3-butadiene with prim. aliphatic alcohols **under ruthenium catalysis**. Thus, with the ruthenium catalyst generated *in situ* from $RuH_2(CO)(PPh_3)_3$, (S)-SegPhos, and a TADDOL-based phosphate diester the observed diastereo- and enantio-selectivity are the opposite of those found using BINOL-derived phosphate counterions in combination with dppf (cf. *81*, 306) or even (S)-SegPhos. **E:** $RuH_2(CO)(PPh_3)_3$ (7 mol%), (S)-SegPhos (7 mol%) and TADDOL-phosphoric acid (14 mol%) added to a re-sealable pressure tube, the tube sealed with a rubber septum, purged with argon, heptanol (0.3 mmol) and acetone (0.3 ml) added, the soln. cooled to -78°, butadiene (0.1 ml; 4 eq.) quickly added, the septum quickly replaced with a screw cap, the mixture heated at 95° (oil bath temp.) for 3 d, allowed to cool to room temp., concentrated *in vacuo*, and purified by flash chromatography on silica gel → (3R,4R)-3-methyl-1-decen-4-ol. Y 82% (*syn/anti* 4.4:1; e.e. 95%). The enantio-selectivity was found to increase with increasing size of the TADDOL aryl substituent, with the 3,5-xylyl-substituted acid being most effective of those tested. It is believed the *syn*-diastereo-selectivity may arise as a consequence of kinetically preferred hydrometalation of the s-*cis* conformer of butadiene to furnish the *anti*-π-crotylruthenium isomer, the steric demand of the TADDOL-based phosphate counterion retarding isomerization to the *syn*-π-crotylruthenium isomer, and carbonyl addition occurring by way of the (Z)-σ-crotylruthenium haptomer via a closed Zimmerman-Traxler-type transition state. F.e. (nine; Y 62-89%; d.r. 3.8:1 to 5.5:1; e.e. 91-96%) and match/mismatch effects, also the first X-ray crystal structure of a ruthenium complex modified by a chiral phosphate counterion, s. E.L. McInturff, E. Yamaguchi, M.J. Krische, J. Am. Chem. Soc. *2012*, *134* (51), 20628-31 [DOI: 10.1021/ja311208a].

Cationic ruthenium(II) sensitizers/irradiation *Ru(II)/⫻*
[2+2]-Cycloaddition under visible-light photoredox catalysis s. *22*, 705s*83* □

Tris(2,2′-bipyridyl)ruthenium(II) bis(hexafluorophosphate)/irradiation *[Ru(II)]/⫻*
3-Acylamino-1,1,1-trifluorides from ethylene derivs. and nitriles C=C → $C(NHAc)C(CF_3)$
Regioselective radical synthesis with addition of one C-atom
under visible-light photoredox catalysis

318. 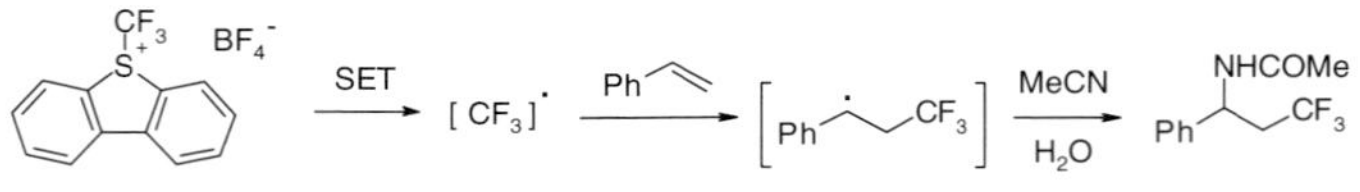

3-Aryl-3-acylamino-1,1,1-trifluorides. Umemoto′s trifluoromethylating agent (0.25 mmol), $[Ru(bpy)_3](PF_6)$ (0.5 mol%), styrene (1.1 eq.), acetonitrile (5 ml) and water (1 eq.) placed in a Schlenk tube under N_2, the mixture degassed three times by the freeze-pump-thaw method, the tube placed at a distance of 2-3 cm from a 3 W Blue LED lamp (425 nm), the resulting orange soln. stirred for 3 h at room temp. (water bath) under irradiation, and worked up with purification by chromatography on silica gel → N-(1-phenyl-3,3,3-trifluoropropyl)acetamide. Y 91%. The procedure is mild and applicable (even on the gram scale at 0.3 mol% catalyst loading and under sunlight) to the reaction of various nitriles with electron-diverse *terminal or internal* styrenes, incl. *o*-subst. and steroidal derivs., and is notably tolerant of aldehyde, ester, NHBoc, OAc and boronate groups (twenty-six examples; Y 53-91%). There was no reaction, however, with Togni's trifluoromethylating agents. Reaction is thought to involve initial generation of trifluoromethyl radical via photoredoxidative SET, followed by regioselective radical addition to the alkene; the formed radical adduct then undergoes a second SET to generate a β-trifluoromethylated carbocation prior to nucleophilic attack of the nitrile in the aq. medium. F.e.s. Y. Yasu, T. Koike, M. Akita, Org. Lett. *2013, 15* (9), 2136-9 [DOI: 10.1021/ol4006272].

Thiolate-bridged diruthenium complexes *[Ru-Ru]*
Pyrrolidines from cyclopropanes and aldimines s. *42,* 654s*83* ◌

Benzylidene(dichloro)(tricyclohexylphosphine)(1,3-dimesitylimidazolidin-2-ylidene)-ruthenium/trimethylsilyl vinyl ether ←
1,3-Dienes from terminal acetylene derivs. and ethylene derivs. via isomerizing ene-yne metathesis ←

319. 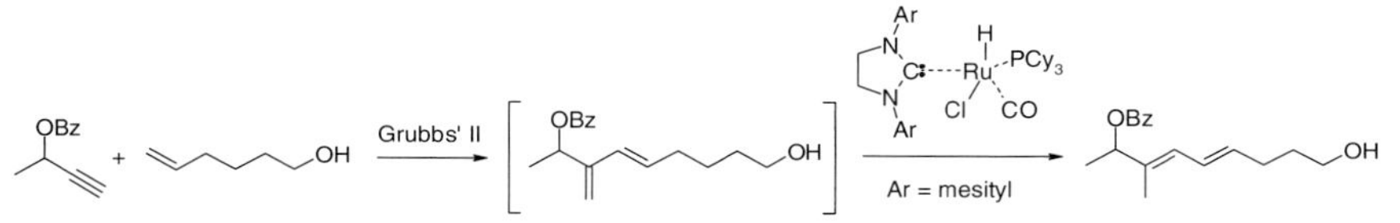

Internal 1,3-dienes (generally with E-configuration) may be obtained by novel ruthenium hydride promoted isomerization of the initial terminal 1,3-dienes formed from Grubbs' II catalyst-promoted ene-yne metathesis **in one pot**. **E:** Grubbs' 2nd generation catalyst (10 mol%), added to a soln. of 3-benzoyloxy-1-butyne (0.287 mmol) and hex-5-en-1-ol (6 eq.) in toluene (5 ml), the flask sealed under N_2, stirred at 60° for 45 min, trimethylsilyl vinyl ether (1 eq.) added, stirring continued for 1 h, then at 95° overnight, cooled to room temp., quenched with a soln. of K-cyanoacetate (0.5 eq.) in methanol (0.5 ml), solvent removed *in vacuo*, and the residue purified by chromatography on silica → 8-benzoyloxy-7-methylnona-4,6-dien-1-ol. Y 55%. Optimization of the isomerization step using the ruthenium hydride, shown to be formed by decomposition of the Grubbs' catalyst during metathesis [and accelerated by addition of trimethylsilyl vinyl ether in accordance with M. Arisawa et al., J. Org. Chem. *2006, 71* (11), 4255-61 [DOI: 10.1021/jo060308u]], was successful for a range of terminal 1,3-dienes, with rapid Z→E isomerization of 3(Z)-alkenes shown to precede isomerization. The addition of an alcohol additive (optimally 1-butanol) was essential for optimum yields (twelve examples; Y 55-87%), tolerating silyl ether, ester and sulfonamide functionality. The one pot metathesis-isomerization sequence gave moderate yields of internal 1,3-dienes (48-60%; four examples), with alcohol additive only required in one case. F.e. and optimization s. J.R. Clark, J.R. Griffiths, S.T. Diver, J. Am. Chem. Soc. *2013, 135* (9), 3327-30 [DOI: 10.1021/ja4011207].

Polymer-supported rhodium nanoparticles ←
Rhodium(I) phosphine, bis(diphenylphosphinite), bis(aryl phosphorodiamidite) or tris(phosphorodiamidite) complexes *[Rh(I)]*
Hydroformylation C=C → CHC-CHO
update s. *4,* 667s*82*; *branched* hydroformylation of various alkenes with $Rh(acac)(CO)_2$ and a novel 1,3-diol-based bis(diphenylphosphinous acid ester) as ligand s. S.R. Khan, B.M. Bhanage, Appl. Organomet. Chem. *2013, 27* (3), 313-7 [DOI: 10.1002/aoc.2983]; *linear* hydroformylation of terminal and internal alkenes with a 2,2′,6-trihydroxybiphenyl-based tris(phosphorodiamidite) as ligand s. C. Chen, Y. Qiao, H. Geng, X. Zhang, Org. Lett. *2013, 15* (5), 1048-51 [DOI: 10.1021/ol400033k]; with a 2,2′,6,6′-tetrahydroxybiphenyl-based tetrakis(phosphorodiamidite) as ligand s. C. Cai, S. Yu, B. Cao, X. Zhang, Chem. Eur. J. *2012, 18* (32), 9992-8 [DOI: 10.1002/chem.201201396]; with a spiroketal-based bis(aryl phosphorodiamidite) as ligand s. X. Jia, Z. Wang, C. Xia, K. Ding, ibid. *18* (48), 15288-95 [DOI: 10.1002/chem.201203042]; *heterogeneous* hydroformylation *in supercritical carbon dioxide* with highly active, recyclable rhodium nanoparticles in a hyper cross-linked polystyrene matrix s. S.E. Lyubimov, E.A. Rastorguev,

K.I. Lubentsova, A.A. Korlyukov, V.A. Davankov, Tetrahedron Lett. *2013, 54* (9), 1116-9 [DOI: 10.1016/j.tetlet.2012.12.063]; linear hydroformylation of styrenes *under supramolecular control* based on non-covalent ligand-substrate interactions for reversal of traditional regioselectivity s. P. Dydio, J.N.H. Reek, Angew. Chem., Int. Ed. *2013, 52* (14), 3878-82 [DOI: 10.1002/anie.201209582]; branched hydroformylation at high temperature with supramolecular encapsulated rhodium catalysts, e.g. tricarbonyl(hydrido)[tris(3-pyridyl)phosphine]rhodium(I), cf. T. Besset, D.W. Norman, J.N.H. Reek, Adv. Synth. Catal. *2013, 355* (2-3), 348-52 [DOI: 10.1002/adsc.201200790]; α,β-ethylenealdehydes by regio- and stereo-selective **hydroformylation of acetylene derivs.** with Pd(acac)$_2$/2-(dicyclohexylphosphino)-1-[*o*-(dicyclohexylphosphino)phenyl]pyrrole/TsOH s. X. Fang, M. Zhang, R. Jackstell, M. Beller, Angew. Chem., Int. Ed. *2013, 52* (19), 4645-9 [DOI: 10.1002/anie.201300759]; linear prim. alcohols from terminal alkenes by **hydroformylation-hydrogenation** (cf. *76,* 343s*82*) with [$Ru_3(CO)_{12}$] and 2-(dicyclohexylphosphino)-1-methylimidazole as ligand s. I. Fleischer, K.M. Dyballa, R. Jennerjahn, R. Jackstell, R. Franke, A. Spannenberg, M. Beller, ibid. *52* (10), 2949-53 [DOI: 10.1002/anie.201207133].

Cyclooctadiene(8-quinolinato)rhodium(I)/tris[p-(trifluoromethyl)phenyl]phosphine/ cesium fluoride ←

[2+2]-Cycloaddition □

with acetylene derivs. or allenes, update s. *22,* 705s*82*; **cyclobutenes** by regioselective rhodium-catalyzed cycloaddition of terminal alkynes with electron-deficient alkenes using cyclooctadiene-(8-quinolinato)rhodium(I)/tris[*p*-(trifluoromethyl)phenyl]phosphine/CsF s. K. Sakai, T. Kochi, F. Kakiuchi, Org. Lett. *2013, 15* (5), 1024-7 [DOI: 10.1021/ol303510k]; photochemical [2+2]-cycloaddition of *trans*-1,2-bis(4-pyridyl)ethylenes coordinated to a rhenium complex as a remote ancillary regulatory ligand s. Z.-Z. Lu, C.-C. Lee, M. Velayudham, L.-W. Lee, J.-Y. Wu, T.-S. Kuo, K.-L. Lu, Chem. Eur. J. *2012, 18* (49), 15714-21 [DOI: 10.1002/chem.201202301]; gram-scale cross-[2+2]-cycloaddition of electron-rich styrenes under visible-light photoredox catalysis with $[Ru(bmp)_3]^{2+}$ as sensitizer s. M.A. Ischay, M.S. Ament, T.P. Yoon, Chem. Sci. *2012, 3* (9), 2807-11 [DOI: 10.1039/c2sc20658g]; **organocatalyzed asym. [2+2]-cycloaddition** with enals using 2(S)-[diphenyl(trimethylsiloxy)methyl]pyrrolidine as catalyst via iminium-enamine activation initiated by a vinylogous Friedel-Crafts alkylation s. G.-J. Duan, J.-B. Ling, W.-P. Wang, Y.-C. Luo, P.-F. Xu, Chem. Commun. *2013, 49* (41), 4625-7 [DOI: 10.1039/c3cc41785a].

Chiral 1,1′-binaphthyl-based dibenzoato(cyclopentadienyl)rhodium(III) complex *[Rh(III)]**

Rhodium-catalyzed N-directed asym. *o*-allylation of arylhydroxamic acid esters with allenes H→CHCH=C

320.

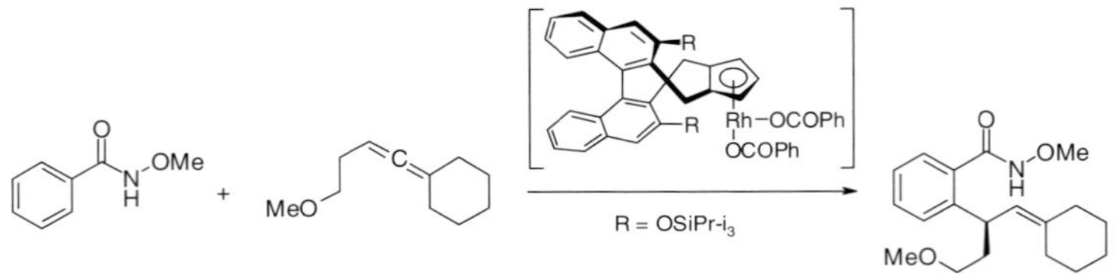

A class of chiral cyclopentadienyl ligands having a tunable biaryl backbone has been reported and applied in enantioselective rhodium(III)-catalyzed N-directed C-H allylations with allenes (an asym. variant of *82,* 301). It is believed the rhodium cyclopentadienyl complex creates a well-defined chiral pocket, and, while the ligands described previously for dihydroquinolone synthesis (*82,* 376) allowed adjustment of the back wall (forcing the approach of the incoming reactant from the unsubstituted side), this class of ligand allows adjustment of the size of the side wall (which is responsible for aligning the three-coordinated intermediate to minimize steric interactions). **E:** Chiral cyclopentadienylbis(ethylene)rhodium(I) complex (5 mol%), dibenzoyl peroxide (5 mol%) and N-methoxybenzamide (1.2 eq.) weighed into a vial without protection from O_2 and moisture, then sealed with a rubber septum, methylene chloride (500 µl) added, the mixture stirred at 23° for 15 min, cooled to -20°, startg. allene (0.1 mmol) added, the mixture kept at -20° for 18 h, quickly filtered through a pad of silica gel, eluting with ethyl acetate, concentrated under reduced pressure, and the residue purified further by silica gel chromatography → (R)-2-(1-cyclohexylidene-4-methoxybutan-2-yl)-N-methoxybenzamide. Y 80% (e.r. 96.5:3.5). ^{1}H NMR showed the crude product was a 9:1 mixture of *o*-allyl and *o,o'*-diallyl-deriv., the latter predominating using 2.2 eq. of allene at 0° (Y 70%; e.r. 99.5:0.5). The method is mild, selective and displays good functional group compatibility, with electron-donating or -withdrawing groups tolerated on the arylhydroxamate, and ether, silyl ether or free hydroxyl on the trisubst. allene. The rhodium(I) pre-catalyst was obtained by treating chiral 2,6-bis(triisopropylsilyloxy)-3,5-dihydrospiro[cyclohepta[1,2-*a*:7,6-*a'*]dinaphthalene-4,1'-cyclopenta-2,4-diene] with TlOEt then [Rh($H_2C{=}CH_2)_2$Cl]$_2$]. Use of dibenzoyl peroxide allows base-free generation of the active rhodium(III) species. F.e. (sixteen; Y 66-91%; e.r. 82:18 to 99:1) and prepn. of the chiral cyclopentadienyl ligands s. B. Ye, N. Cramer, J. Am. Chem. Soc. *2013*, *135* (2), 636-9 [DOI: 10.1021/ja311956k].

(1,5-Cyclooctadiene)(1,3-dimesitylimidazolidin-2-ylidene)rhodium(I) perchlorate ←
Cycloocta-2,5-dienones from γ-allenealdehydes and terminal acetylene derivs. Stereoselective ring closure ○

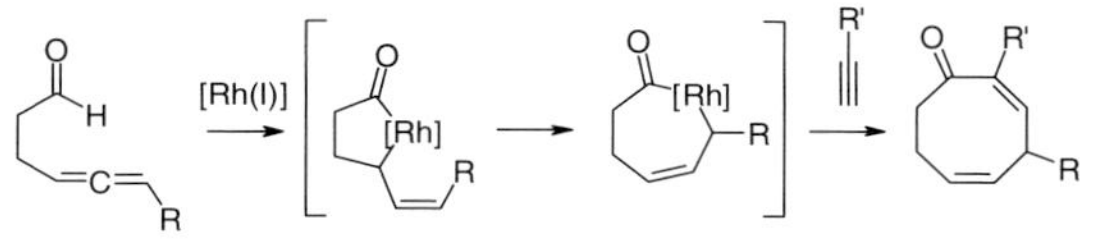

321.

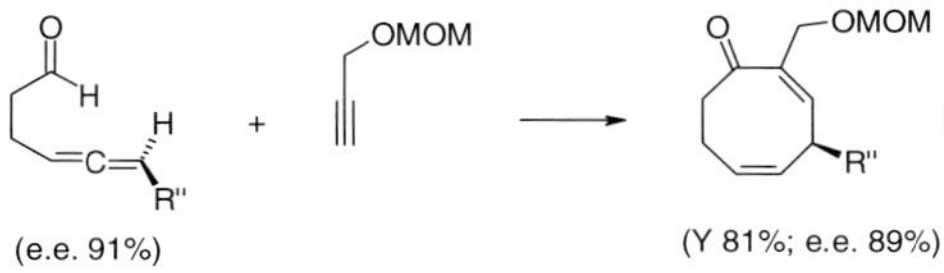

Cycloocta-2,5-dienone ring formation via rhodium(I)-catalyzed formal [6+2]-cycloaddition, previously described in intramolecular mode (s. *80*, 341), has now been extended to the intermolecular variant. **E:** A soln. of Rh(SIMes)(cod)Cl (10 mol%) and $AgClO_4$ (10 mol%) in degassed 1,2-dichloroethane (0.61 ml) stirred at room temp. for 10 min, a soln. of startg. allene-aldehyde (0.16 mmol) and alk-1-yne (3 eq.) in degassed 1,2-dichloroethane (0.99 ml) added at 0°, the mixture stirred for 18 h (TLC), and, after removal of the solvent, the residue purified by column chromatography on silica gel → product. Y 81% (single isomer with no vinylcyclohexenone formation). Nineteen further examples gave yields of 37-84% (plus 5-23% of the regioisomer in four cases and 11-28% of a 4-vinyl-2-cyclohexenone in three cases) with alk-1-ynes terminated by alkoxyalkyl, aroyloxymethyl, N-methyl-N-tosylaminomethyl, hydroxymethyl, alkyl or ester groups or, notably, gaseous acetylene itself in reaction with 4-allenals terminated by alkoxyalkyl, N-methyl-N-tosylaminoethyl, phenyl or, notably, trimethylsilyl. With an axially chiral allenal (e.e. 91%), reaction proceeded with high *transfer of chirality* (Y 81%; e.e. 89%). F.e.s. Y. Oonishi, A. Hosotani, Y. Sato, Angew. Chem., Int. Ed. *2012*, *51* (46), 11548-51 [DOI: 10.1002/anie.201206508].

(η^6-Fluorobenzene)[N,N-bis(diisopropylphosphino)methylamine]rhodium(I) barfate [Rh(I)]
Hydroacylation with aldehydes C=C → CHC-COR
under rhodium catalysis, update s. *52*, 319s*82*; linear hydroacylation of 1-octene and 1-octyne with *o*-(methylthio)benzaldehyde using (η^6-fluorobenzene)[N,N-bis(diisopropylphosphino)-methylamine]rhodium(I) barfate as catalyst, exploring ligands with small bite-angle s. I. Pernik, J.F. Hooper, A.B. Chaplin, A.S. Weller, M.C. Willis, ACS Catal. *2012*, *2* (12), 2779-86 [DOI: 10.1021/cs300541m]; **Stetter hydroacylation** (cf. *28*, 648s*82*) of styrenes with highly electron-rich N-heterocyclic carbenes as catalyst s. M. Schedler, D.-S. Wang, F. Glorius, Angew. Chem., Int.

Ed. *2013, 52* (9), 2585-9 [DOI: 10.1002/anie.201209291]; **asym. Stetter hydroacylation** of acrylates to give chiral α-subst. γ-ketocarboxylic acid esters with a chiral N-condensed N-(2,6-dimethoxyphenyl)-1,2,4-triazolylidene as catalyst (cf. *68,* 335s*83*) s. N.E. Wurz, C.G. Daniliuc, F. Glorius, Chem. Eur. J. *2012, 18* (51), 16297-301 [DOI: 10.1002/chem.201202432].

Chiral rhodium(I) di(phosphine) complexes *[Rh(I)]**
Chiral zinc-coordinated dicarbonyl(hydrido)rhodium(I) 3,3′-di(3- or 4-pyridyl)-5,5′,6,6′,7,7′,8,8′-octahydro-1,1′-binaphthyl-2,2′-diyl N-methyl-N-[o-(diphenylphosphino)phenyl]phosphoramidites ←
Asym. hydroformylation C=C → CHC-CHO
update s. *49,* 683s*82*; at the terminus of 1,1-disubst. N-allylphthalimides (*en route* to chiral γ-aminocarboxylic acids) with $Rh(acac)(CO)_2$ and (S,S)-1,2-bis(2,5-diphenylphospholano)ethane as ligand s. X. Zheng, B. Cao, T.-l. Liu, X. Zhang, Adv. Synth. Catal. *2013, 355* (4), 679-84 [DOI: 10.1002/adsc.201200960]; optimization of regioselective asym. hydroformylation based on libraries of chiral 1,2-bis(3,4-diazaphospholano)benzenes as ligand s. T.T. Adint, G.W. Wong, C.R. Landis, J. Org Chem. *2013, 78* (9), 4231-8 [DOI: 10.1021/jo400525w]; asym. hydroformylation of styrenes with chiral, zinc-coordinated dicarbonyl(hydrido)rhodium(I) 3,3′-di(3- or 4-pyridyl)-5,5′,6,6′,7,7′,8,8′-octahydro-1,1′-binaphthyl-2,2′-diyl N-methyl-N-[*o*-(diphenylphosphino)-phenyl]phosphoramidites as catalyst *under supramolecular control* s. R. Bellini, J.N.H. Reek, Chem. Eur. J. *2012, 18* (42), 13510-9 [DOI: 10.1002/chem.201202044].

Dicarbonyl(chloro)rhodium(I) dimer *$[Rh(CO)_2Cl]_2$*
3-Acoxy-2-alkylidene-3-cyclohexenones from 3-cyclopropyl-1-acoxy-2-acetylenes via rhodium(I)-catalyzed 1,3-acoxy migration-carbonylative ring expansion ○

322.

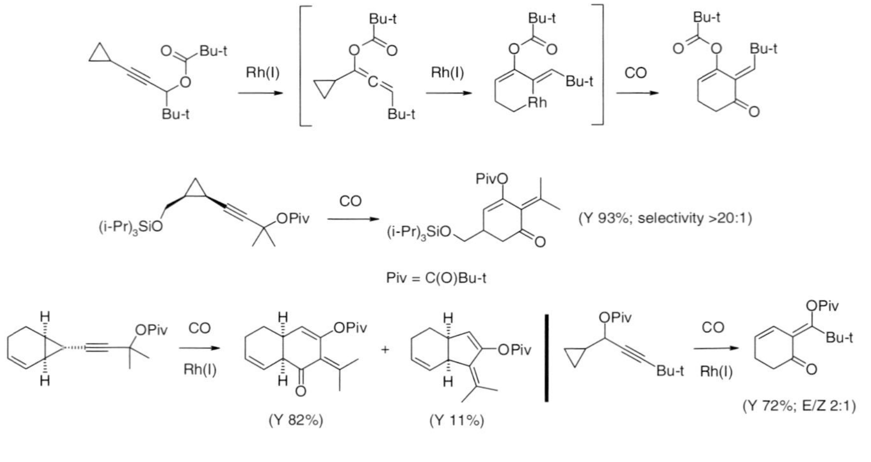

A mixture of $[Rh(CO)_2Cl]_2$ (5 mol%) and 1-cyclopropyl-4,4-dimethylpent-1-yn-3-yl pivalate (0.1 mmol) in anh. toluene (1 ml) stirred under CO (balloon) at 60° for 5 h (TLC), solvent evaporated, and the residue purified by chromatography on silica → (E)-6-(2,2-dimethylpropylidene)-5-oxocyclohex-1-enyl pivalate. Y 92% (E/Z 10:1). A series of cyclopropyl-terminated sec./tert. propargylic pivalates reacted via [Saucy-Marbet] 1,3-acoxy migration and trapping of the intermediate allenylcyclopropane by [5+1]-cycloaddition with CO to afford the title products, with bulkier propargylic substituents favoring (E)-alkylidene products (E/Z 1:1.3 to 10:1; seven examples; Y 81-95%). A single tert. propargylic *acetate* gave an 85% yield but a sec. analog was unreactive. Substrates containing additional substituents on the cyclopropane ring underwent cleavage preferentially at the least hindered C-C bond, affording mixtures of regioisomers, with selectivity generally more pronounced for *cis*-cyclopropanes and exclusive for cyclopropanes containing a quaternary center (selectivity 3.5:1 to 100:0; thirteen examples; Y 64-93%; a *trans*-phenylcyclopropyl deriv. showed reverse regioselectivity of 1:2.5). A chiral 2,2-disubst. cyclopropane reacted with full retention of chirality. Bicyclic cyclohex(a/e)nocyclopropanes gave the corresponding decalinone (two examples; Y 80-82%) as mixtures with indenes formed by cycloisomerization of the intermediate allenes (Y 11-19%; 82-88% in the absence of CO). The reaction was extended to isomeric 1-cyclopropyl-1-acoxy-2-acetylenes, affording **2-(α-acoxy-alkylidene)-3-cyclohexenones** (eight examples; Y 52-91%; E/Z 3.3:1 to 1:2), with a sec. acetate being unreactive and a phenyl-terminated alkyne affording a complex mixture of products. F.e. and optimization s. D. Shu, X. Li, M. Zhang, P.J. Robichaux, I.A. Guzei, W. Tang, J. Org. Chem. *2012*, *77* (15), 6463-72 [DOI: 10.1021/jo300973r].

Dichloro(pentamethylcyclopentadienyl)rhodium(III) dimer/cesium acetate *[$Cp*RhCl_2]_2/CsOAc$*
***o*-[β-(Acylamino)vinyl]phenols or benzofurans** ←
from acetylene derivs. and aryl hydroxamates

323.

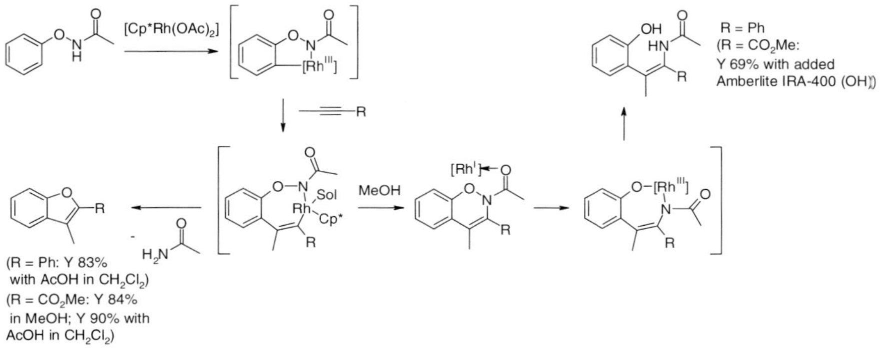

A regio- and stereo-selective rhodium-catalyzed redox-neutral synthesis of *o*-(β-acylaminovinyl)-phenols from alkynes is reported, based on an acetylaminooxy group as both *o*-directing group and acetamido transfer agent. **E:** Startg. N-aryloxyacetamide (0.2 mmol), the alkyne (1.2 eq.), $[Cp*RhCl_2]_2$ (2.5 mol%) and CsOAc (0.25 eq.) weighed into a vial without exclusion of oxygen or moisture, methanol (0.4 *M*) added, the mixture stirred at room temp. for 16 h, then diluted with ethyl acetate, transferred to a round bottom flask, silica gel added, volatiles evaporated under reduced pressure, and the residue purified by flash chromatography on silica gel → product. Y 83%. Ten further examples with sym. diaryl- or alkyl(aryl)-acetylenes or ynamides gave yields of 32-82%. Similar results were obtained with $Cp*Rh(OAc)_2$, suggesting this is the active catalyst. Three examples with α,β-acetylenecarboxylic acid esters gave yields of 58-72% with added Amberlite IRA-400 (hydroxide form). In the absence of Amberlite methyl but-2-ynoate gave methyl 3-methylbenzofuran-2-carboxylate in 84% yield in methanol, which was improved to 90% in methylene chloride as non-coordinating solvent containing acetic acid (illustrated); fourteen further examples gave benzofurans in yields of 48-90% under the latter conditions. F.e. and optimization s. G. Liu, Y. Shen, Z. Zhou, X. Lu, Angew. Chem., Int. Ed. *2013*, *52* (23), 6033-7 [DOI: 10.1002/anie.201300881].

Dichloro(pentamethylcyclopentadienyl)rhodium(III) dimer/silver hexafluoroantimonate/ acetic acid –
N-Directed *o*-substitution by 1,4-addition to enones s. *11*, 770s*83* C=C → CHC(R)

Palladium(II) acetoacetonate/2-(dicyclohexylphosphino)-1-[o-(dicyclohexylphosphino)phenyl]pyrrole/p-toluenesulfonic acid *[Pd(II)]/TsOH*
Hydroformylation of acetylene derivs. s. *4*, 667s*83* C≡C → CH=C-CHO

Bis(dibenzylideneacetone)palladium(0)/tri-p-tolylphosphine *[Pd(0)]*
2-(Arylselenomethyl)-1,3-dienes C=C=CH$_2$ → C=C(CH$_2$SeAr)C=C
from terminal allenes and ar. eneselenides

324. 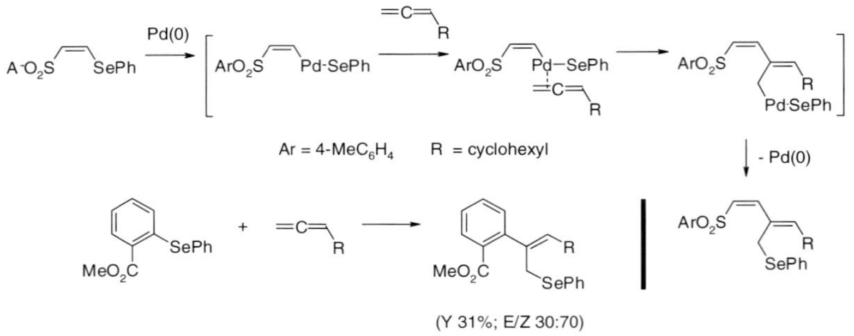

A mixture of Pd(dba)$_2$ (5 mol%), tri-*p*-tolylphosphine (10 mol%) and acetonitrile (0.3 ml) stirred under N$_2$ at room temp. for 10 min, (Z)-1-phenylseleno-2-(*p*-tolylsulfonyl)ethene (0.3 mmol) and cyclohexylallene (2 eq.) added, the brown mixture heated under reflux for 8 h, filtered through Celite, solvent removed *in vacuo*, and the residue purified chromatographically → 1-(4-cyclohexyl-3-phenylselenomethylbuta-1(Z),3(Z)-dienyl)sulfonyl-4-methylbenzene. Y 63% (3E/3Z 19:81). **Regioselective palladium-catalyzed vinylselenation** of terminal allenes with activated vinyl selenides occurred with full retention of the vinyl selenide stereochemistry and variable stereoselectivity at the retained allene double bond (E/Z 68:32 to 8:92; eight examples; Y 40-79%). Experiments suggested that the allene double bond was initially formed with (Z)-stereoselectivity but partially isomerized to the more thermodynamically stable (E)-stereoisomer. The reaction was also successful with methyl 2-phenylselenobenzoate (Y 31%; E/Z 30:70) but failed with phenyl styryl selenide. Analogous vinylthiolation with a β-phenylthioacrylamide gave a poor yield of the corresponding (Z)-phenylthio-1,3-diene (5%; E/Z 0:100). F.e. and optimization, also conversion of one product to a lactone (*m*CPBA then NaHCO$_3$) in 82% yield, s. S.-i. Fujiwara, M. Okuyama, S. Tsuda, T. Iwasaki, H. Kuniyasu, N. Kambe, Tetrahedron *2012*, *68* (51), 10523-9 [DOI: 10.1016/j.tet.2012.10.003].

o-Ureidophenylpalladium(II) complexes *[Pd(II)]*
Friedel-Crafts reaction of indoles C=C → CHC(R)
with electron-deficient ethylene derivs. s. *11*, 770s*83*

Palladium(II) chloride *PdCl$_2$*
(E)-γ,δ-Ethylenecarboxylic acid esters C=CH → C=C-CH$_2$CHCOOR
from ethylene derivs. and acrylic acid esters
via ligand-free palladium(II)-catalyzed 1,4-addition

325. 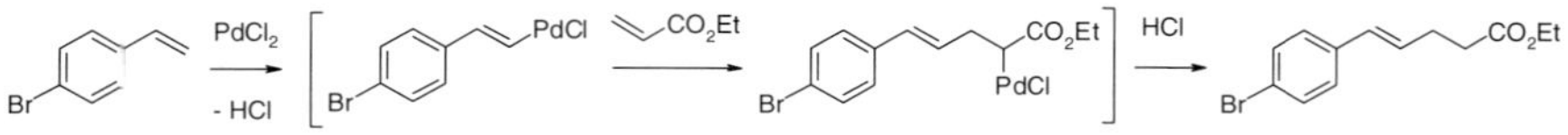

A mixture of 4-bromostyrene (0.5 mmol), ethyl acrylate (1.5 eq.), PdCl$_2$ (6 mol%) and chlorobenzene (2 ml) stirred in a sealed tube at 110° for 72 h (TLC), diluted with water, extracted with ethyl acetate, and purified chromatographically → (E)-ethyl 5-(4-bromophenyl)pent-4-enoate. Y 74%. This general and efficient addition was effective for styrenes and their α-methyl derivs. coupling with acrylate esters (no reaction occurred with bulkier crotonate esters), affording (E)-products exclusively (eleven examples; Y 68-85%). The reaction failed using 1-hexene in place of styrene. No reaction occurred in the presence of inorganic/organic base or ligand additives. F.e. and optimization s. P. Liu, H.-s. Wang, Y.-m. Pan, W.-l. Dai, H. Liang, Z.-F. Chen, Chem. Commun. *2013*, *49* (46), 5295-7 [DOI: 10.1039/c3cc41742e].

Chiral aqua(hydrido)[di(phosphine)](phosphino-Δ²-oxazoline)iridium(III) bis(hexafluoroantimonate) *[Ir(III)]**

Asym. 1,3-dipolar cycloaddition C

with nitrones, update s. *67,* 301s*82*; asym. cycloaddition to acrolein with a chiral aqua(hydrido)-[di(phosphine)](phosphino-Δ²-oxazoline)iridium(III) bis(hexafluoroantimonate) and related complexes as catalyst, also *endo*-selective iridium-catalyzed asym. Diels-Alder reaction (cf. *46,* 662s*83*), s. D. Carmona, J. Ferrer, N. García, P. Ramírez, F.J. Lahoz, P. García-Orduña, L.A. Oro, Organometallics *2013, 32* (6), 1609-19 [DOI: 10.1021/om301089c]; *exo*-selective cycloaddition of nitrones to chiral α,β-ethylenelactones *with asym. induction* using $Sc(OTf)_3$ as catalyst (compared with *endo*-selective thermal cycloaddition) s. M. Sniezek, S. Stecko, I. Panfil, B. Furman, Z. Urbanczyk-Lipkowska, M. Chmielewski, Tetrahedron: Asym. *2013, 24* (2-3), 89-103 [DOI: 10.1016/j.tetasy.2012.12.006]; *exo*-selective **organocatalyzed asym. 1,3-dipolar cycloaddition** of alanine-derived nitrones to crotonaldehyde with a chiral 5-benzyl-4-imidazolidone as catalyst s. K.B. Selim, A. Beauchard, J. Lhoste, A. Martel, M.Y. Laurent, G. Dujardin, ibid. *2012, 23* (24), 1670-7 [DOI: 10.1016/j.tetasy.2012.11.010].

Chloro(1,5-cyclooctadiene)iridium(I) dimer/triethylenediamine/ sodium tetrakis[3,5-bis(trifluoromethyl)phenyl]borate *$[IrCl(cod)]_2$/DABCO/$NaBAr^F_4$*

1-Sulfonylamino-3-vinylindans from aryl N-sulfonylketimines and 1,3-dienes
Regio- and stereo-selective iridium(I)-catalyzed ring closure

326.

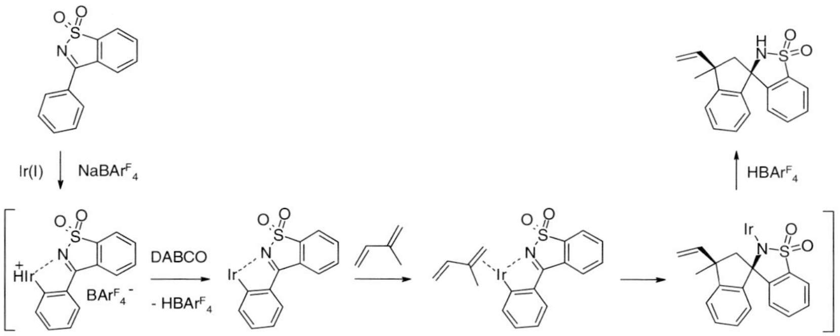

Toluene (0.8 ml), 2-methylbuta-1,3-diene (1.5 eq.), a soln. of $[IrCl(cod)]_2$ (2 mol%) in toluene (20 µl) and a soln. of DABCO (2 mol%) in toluene (16 µl) added sequentially to a mixture of 3-phenylbenzo[*d*]isothiazole S,S-dioxide (0.2 mmol), and $NaBAr^F_4 \cdot 2H_2O$ (4 mol%) under N_2, the sealed mixture stirred at 80° for 20 h, filtered through alumina, solvent evaporated, and the residue purified chromatographically → product. Y 96% (as a 96:4 mixture of regioisomers). This novel [3+2]-cycloaddition of cyclic N-sulfonylketimines (3-arylbenz[*d*]isothiazole, 3-phenyl-Δ²-isothiazoline and 4,5-diphenyl-[2,3,1]-thiadiazole S,S-dioxides) with 1,3-dienes afforded 3-vinyl-1-sulfonylamino-spiroindan derivs. as single diastereomers. Unsymmetrical dienes cyclized predominantly at the more electron-rich double bond (d.r. 91:9 to 100:0; nineteen examples; Y 79-99%). F.e., optimization and substrate prepn. s. T. Nishimura, Y. Ebe, T. Hayashi, J. Am. Chem. Soc. *2013*, *135* (6), 2092-5 [DOI: 10.1021/ja311968d].

Chloro(cyclooctene)iridium(I) dimer/(S)-2,2′-bis[bis(3,5-di-tert-butyl-4-methoxyphenyl)phosphino]-5,6;5′,6′-bis(methylenedioxy)biphenyl *[Ir(I)]**

Regioselective asym. hydrohetarylation of bicycloalkenes s. *49,* 679s*83* C=C → CHC(Ar)

Platinum(IV) iodide/carbon monoxide *PtI_4/CO*

5,6-Dihydro-4*H*-cyclopenta[*b*]furans
from 1,3-enyne-2-carboxaldehydes and ethylene derivs.
Regioselective platinum-catalyzed double ring closure via [3+2]-cycloaddition

327. 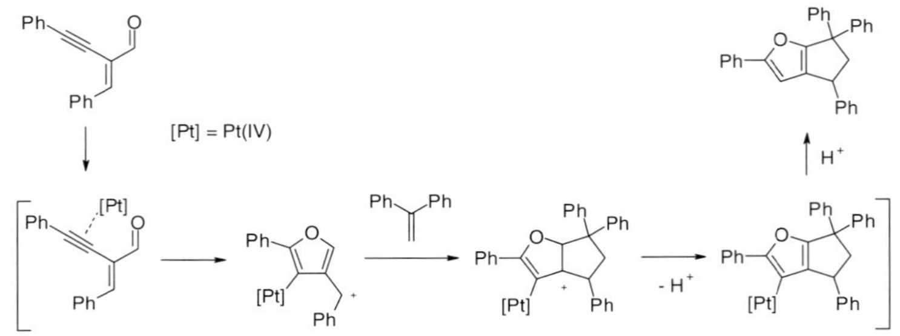

4,6-Diaryl-5,6-dihydro-4*H*-cyclopenta[*b*]furans. The startg. enynal (0.2 mmol) and ethylene deriv. (2 eq.) added to a soln. of PtI_4 (10 mol%) in anh. toluene (2 ml) under CO, the mixture stirred at 60° for 2 h (TLC), evaporated *in vacuo*, and the residue purified by chromatography on silica gel → product. Y 84%. The procedure is chemo- and regio-selective, and generally applicable to enynals substituted at the alkyne terminus by electron-donating or -withdrawing aryl groups as well as alkyl groups; the aryl substituent at the alkene terminus may also be electron-diverse, as may that of the ethylenic partner in the reaction, which may be a simple styrene or a 1,1-diarylethylene (nineteen examples; Y 54-98%). There was no reaction, however, with stilbene, enols or vinyl ethers. Reaction is thought to involve initial activation of the alkyne group by platinum(IV), followed by ring closure to give a furyl-substituted carbocation; this then undergoes [3+2]-cycloaddition with the alkene prior to aromatizing deprotonation and protonolysis of the C-Pt bond. F.e. and comparison of platinum catalysts and solvents s. T. He, P. Gao, S.-C. Zhao, Y.-D. Shi, X.-Y. Liu, Y.-M. Liang, Adv. Synth. Catal. *2013*, *355* (2-3), 365-9 [DOI: 10.1002/adsc.201200597].

Via intermediates *v.i.*

3-Subst. 2-cyclopentenones from terminal acetylene derivs.
via regioselective Pauson-Khand reaction
with dicobalt hexacarbonyl-complexed trifluoromethylacetylenes

328. 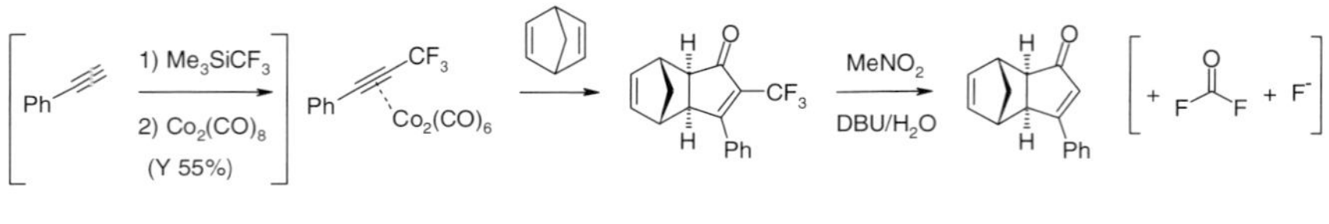

While terminal acetylenes ordinarily participate in the Pauson-Khand reaction to give *2-subst.* 2-cyclopentenones, the regioselectivity can be effectively reversed by initial conversion to dicobalt hexacarbonyl-complexed trifluoromethylacetylenes, which, after cycloaddition and a novel cleavage of the trifluoromethyl group, provide the regioisomeric *3-subst.* 2-cyclopentenones exclusively. The **trifluoromethyl group** in effect thus serves **as a removable steering group** for the cycloaddition. **E:** Norbornadiene (10 eq.) added under N_2 to a soln. of the startg. trifluoromethylacetylene-cobalt carbonyl complex (0.21 mmol) [prepared in 55% yield by coupling phenylacetylene with trifluoromethyl(trimethyl)silane in the presence of $CuCl/Ag_2CO_3$/1,10-phenanthroline and K-*tert*-butoxide as base in anh. DMF, followed by complexation with $Co_2(CO)_8$] in anh. toluene (7 ml), the mixture stirred at 70° for ca. 24 h (TLC), solvent removed under reduced pressure, and the residue worked up with purification by chromatography on silica gel → intermediate 2-(trifluoromethyl)-2-cyclopentenone (Y 77%), a portion of which (0.11 mmol) dissolved in nitromethane (7 ml), water (5 eq.) and DBU (1 eq.) added under N_2, the mixture heated under reflux for ca. 1.5 h (TLC), solvent removed *in vacuo*, and the residue worked up with purification by chromatography on silica gel → product (Y 55%). The acetylene derivs. may possess an alkyl, alkenyl or aryl residue, irrespective of electronic character (ten examples: cycloaddition step: Y 70-99%; CF_3 cleavage step: Y 40-68%). F.e. and application to the formal synthesis of cuparenone, also mechanistic consideration of the CF_3 cleavage (based on 1,4-addition of nitromethane, dehydrofluorination, hydration, and elimination of nitromethane with COF_2), s. N. Aiguabella, C. del Pozo, X. Verdaguer, S. Fustero, A. Riera, Angew. Chem., Int. Ed. *2013*, *52* (20), 5355-9 [DOI: 10.1002/anie.201300907].

Rearrangement ∩

Hydrogen/Oxygen Type CC ∩ HO

(Triphenylphosphine)gold(I) triflimide *[(Ph$_3$P)Au]NTf$_2$*

Gold(I)-catalyzed double ring closure of (*o*-diyne)carboxylic acids

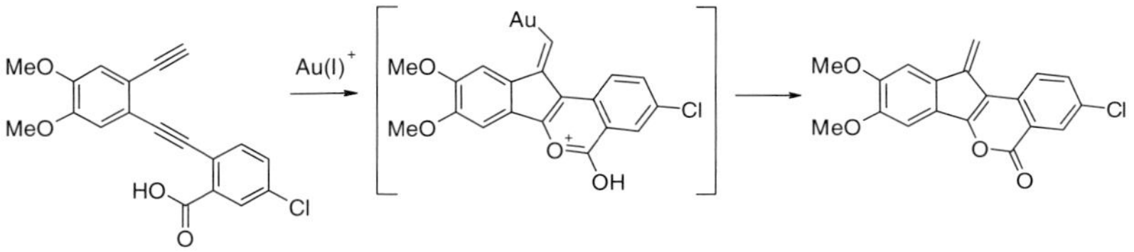

329.

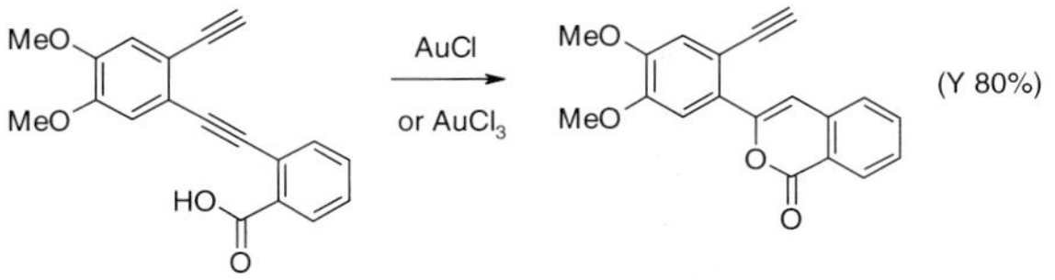

[(Ph$_3$P)Au]NTf$_2$ (5 mol%) added in one portion to a soln. of 5-chloro-2-[(2-ethynyl-4,5-dimethoxyphenyl)ethynyl]benzoic acid (0.5 mmol) in anh. methylene chloride (10 ml) at room temp., the mixture stirred for 15 min, concentrated *in vacuo*, and purified chromatographically → 3-chloro-8,9-dimethoxy-11-methyleneindeno[1,2-*c*]isochromen-5(11*H*)-one. Y 90%. The method appears general for electron-neutral/rich *o*-diynes carrying a terminal *o*-carboxyphenyl or 2-carboxyvinyl substituent on one ethyne, affording the corresponding **11-methyleneindeno[1,2-*c*]isochromen-5(11*H*)-ones** or **5-methyleneindeno[1,2-*b*]pyran-2(5*H*)-ones**, respectively (ten examples; Y 51-90%). A 4,5-dichlorophenyl-*o*-diyne gave only 23% yield, however. Experimental evidence indicates that the reaction involves an initial 5-*endo*-dig cyclization, forming both rings in a concerted process. Interestingly, use of simpler gold catalysts (AuCl, $AuCl_3$) promoted 6-*endo*-dig intramolecular hydrocarboxylation of the internal alkyne to afford a 3-(*o*-ethynylphenyl)isocoumarin (Y 80%). F.e., optimization and substrate prepn. s. Q. Hou, Z. Zhang, F. Kong, S. Wang, H. Wang, Z.-J. Yao, Chem. Commun. *2013*, *49* (7), 695-7 [DOI: 10.1039/c2cc36245g].

Hydrogen/Oxygen Type CC ∩ HC

Irradiation ///

(E)→(Z)-Rearrangement of 2-ene-1,4-diones s. *83*, 344 —

Lithium diisopropylamide *i-Pr$_2$NLi*

Allenes from acetylene derivs. CHC≡C → C=C=CH

Chiral *o*-allenylsulfoxides s. *83*, 436

Chiral prim. amines or 2-prim-Aminothioureas/p-nitrobenzoic acid ←
Organocatalyzed asym. intramolecular Michael addition ○
with cinchona-based 2-*tert*-aminothioureas s. *77,* 376; asym. synthesis of 4-nitrocyclohexanones using a cinchona-based prim. amine as catalyst with time-dependent diastereoselectivity s. K.L. Jensen, C.F. Weise, G. Dickmeiss, F. Morana, R.L. Davis, K.A. Jørgensen, Chem. Eur. J. *2012, 18* (38), 11913-8 [DOI: 10.1002/chem.201202353]; of chiral *trans*-2-acyl-3-nitromethyl-2,3-dihydrobenzofurans using a carbohydrate-based 2-*prim*-aminothiourea and *p*-nitrobenzoic acid as co-catalysts s. A. Lu, K. Hu, Y. Wang, H. Song, Z. Zhou, J. Fang, C. Tang, J. Org. Chem. *2012, 77* (14), 6208-14 [DOI: 10.1021/jo301006e]; N-protected 2-azetidinone-4-carboxylic from chiral N-(α,β-ethyleneacyl)-α-aminocarboxylic acid esters **with asym. induction** via axially chiral enolates using Cs_2CO_3 as base in ethanol as proton donor (cf. *42,* 682) s. T. Yoshimura, M. Takuwa, K. Tomohara, M. Uyama, K. Hayashi, P. Yang, R. Hyakutake, T. Sasamori, N. Tokitoh, T. Kawabata, Chem. Eur. J. *2012, 18* (48), 15330-6 [DOI: 10.1002/chem.201201339].

[1,3-Bis(2,6-diisopropylphenyl)imidazol-2-ylidene]gold(I) chloride/silver hexafluoroantimonate *(IPr)AuCl/AgSbF$_6$*
3-Tosylaminoindenes from 1-arylallene-N-tosylamines
via gold(I)-catalyzed imino-Nazarov cyclization

330. 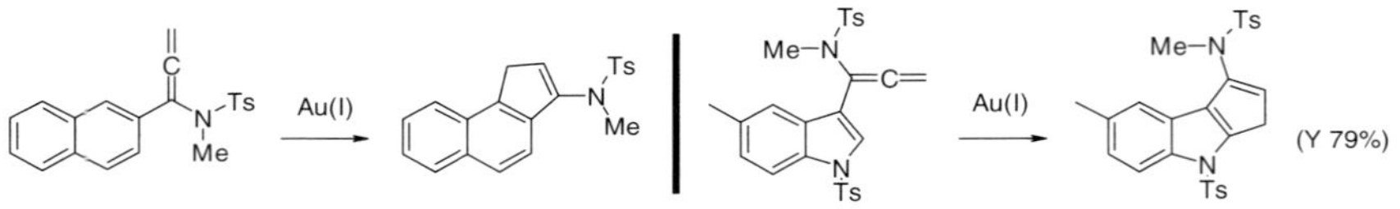

under mild conditions. A soln. of (IPr)AuCl (5 mol%) and $AgSbF_6$ (5 mol%) in methylene chloride (0.5 ml) added to a soln. of 1-naphth-2-ylallene-N-methyl-N-tosylamine (0.1 mmol) in the same solvent (1.5 ml) at room temp. under argon, the mixture stirred for 1 h, quenched with triethylamine, concentrated *in vacuo*, and purified by chromatography on silica → N-(1*H*-cyclopenta[*a*]naphthalen-3-yl)-4,N-dimethylbenzenesulfonamide. Y 97%. Initial development identified favorable conditions for the cyclization ($AuCl_3$ and AuCl effected dimerization of the allenamine prior to cyclization, to afford inseparable mixtures of products). The reaction was successful with electron-rich/neutral 1-arylallene-N-methyl-N-tosylamines (ten examples; Y 81-97%; a 2-methoxyphenyl deriv. gave 16% and a 4-bromophenyl deriv. was unreactive), and was also applicable to a 5-methylindol-3-ylallene (Y 79%) and its 5-H analog (Y 42%). Cyclization of unsymmetrical substrates to afford single regioisomers was apparently controlled by electronic (rather than steric) factors. F.e., optimization and substrate prepn. s. Z.-X. Ma, S. He, W. Song, R.P. Hsung, Org. Lett. *2012, 14* (22), 5736-9 [DOI: 10.1021/ol302743k].

Indium(III) triflate *In(OTf)$_3$*
9-(Arylseleno)phenanthrenes from *o*-(arylselenoethynyl)biaryls
Catalyst-dependent regioselective cycloisomerization s. *83*, 338

Aluminum chloride or Lanthanum(III) bis(trimethylsilyl)amide *$AlCl_3$ or $La[N(SiMe_3)_2]_3$*
Migration of carbon-carbon double bonds C=C-CH → CHC=C
update s. *48,* 704s*82*; *34,* 668s*82*; 3-aryl-3-ethylenealcohols from 3-aryl-2-ethylenealcohols with $La[N(SiMe_3)_2]_3$ s. S.Y. Seo, X. Yu, T.J. Marks, Tetrahedron Lett. *2013, 54* (14), 1828-31 [DOI: 10.1016/j.tetlet.2013.01.111]; migration of the styrene double bond in tetrahydro-1-xanthones with $AlCl_3$ s. A.M. Linsenmeier, S. Bräse, Eur. J. Org. Chem. *2012* (32), 6455-9 [DOI: 10.1002/ejoc.201201009]; 1,2,3,4-tetrahydro- from 1,2,3,6-tetrahydro-pyridinium ring of vinca alkaloids with DBU (cf. *5,* 512s*31*) s. Q.A. Ngo, F. Roussi, S. Thoret, F. Guéritte, Tetrahedron Lett. *2012, 53* (44), 5821-3 [DOI: 10.1016/j.tetlet.2012.07.137].

Chiral N-condensed 1,2,4-triazolylidenes *NHC**
Asym. intramolecular hydroacylation with aldehydes ○
by the Stetter reaction, update s. *79,* 319s*82*; with chiral N[4]-condensed N[1]-(2-pyridyl)-1,2,4-triazolylidenes as catalyst for preparing chiral chroman-4-on-3-ylacetic acid esters s. T. Soeta, Y. Tabatake,

Y. Ukaji, Tetrahedron *2012, 68* (49), 10188-93 [DOI: 10.1016/j.tet.2012.09.095]; diastereoselective **desymmetrization** of 2,5-cyclohexadienones using a chiral N^4-condensed N^1-(pentafluorophenyl)-1,2,4-triazolylidene as catalyst s. M.-Q. Jia, C. Liu, S.-L. You, J. Org. Chem. *2012, 77* (23), 10996-1001 [DOI: 10.1021/jo3022555].

2-prim-Aminothioureas/p-nitrobenzoic acid s. under Chiral prim. amines ←

N-Bromosuccinimide/triphenyl phosphite *NBS/(PhO)$_3$P*
Nazarov cyclization ○
update s. *29,* 688s*77*; dihydropyran-condensed cyclopentanones with a cationic bromonium complex (formed from NBS/triphenyl phosphite) by diastereoselective cyclization s. F. Guo, L. Wang, S. Mao, C. Zhang, J. Yu, J. Han, Tetrahedron *2012, 68* (39), 8367-70 [DOI: 10.1016/j.tet.2012.06.107]; **asym. Nazarov cyclization** of activated or unactivated cross-conjugated dienones with a chiral cationic chromium(III) salen complex as catalyst (cf. *67,* 339s*78*), notably providing chiral hydrindenones, s. G.E. Hutson, Y.E. Türkmen, V.H. Rawal, J. Am. Chem. Soc. *2013, 135* (13), 4988-91 [DOI: 10.1021/ja401908m].

Trimethylsilyl vinyl ether s. under Benzylidene(dichloro)(tricyclohexylphosphine)-(1,3-dimesitylimidazolidin-2-ylidene)ruthenium ←
Triphenyl phosphite s. under N-Bromosuccinimide *(PhO)$_3$P*

Chiral cationic chromium(III) salen complexes *[Cr(III)]**
Asym. Nazarov cyclization s. *67,* 339s*83*

Tetra-n-butylammonium iodide s. under Ru$_3$(CO)$_{12}$ *Bu$_4$NI*

Transition metal catalysts ←
Cycloisomerization of enynes
under transition metal catalysis cf. *41,* 671s*82*; cycloisomerization of N-propargyl-β-amino-α,β-ethyleneketones to 3-acylpyrroles or 5-acyl-1,2-dihydropyridines with $AgNO_3$ as catalyst s. M.A.P. Martins, M. Rossatto, C.P. Frizzo, E. Scapin, L. Buriol, N. Zanatta, H.G. Bonacorso, Tetrahedron Lett. *2013, 54* (8), 847-9 [DOI: 10.1016/j.tetlet.2012.11.086]; 1-alkylidene-2-vinyl-indans from *o*-(homoallyl)arylacetylenes with $Co_2(CO)_8$ (with or without a Lewis basic ligand) s. P. Xing, Z.-g. Huang, Y. Jin, B. Jiang, ibid. *54* (7), 699-702 [DOI: 10.1016/j.tetlet.2012.12.016]; chiral 1-acylamino-3-aza- and 3-oxa-bicyclo[4.1.0]hept-2-enes by **asym. cycloisomerization** with a chiral chloro(diene)rhodium(I) dimer as catalyst based on a bis(ferrocenyl)-subst. benzobicyclo-[2.2.2]octatriene as ligand (cf. *60,* 307s*80*) s. T. Nishimura, Y. Takiguchi, Y. Maeda, T. Hayashi, Adv. Synth. Catal. *2013, 355* (7), 1374-82 [DOI: 10.1002/adsc.201300148]; 4-oxabicyclo[4.1.0]-hept-2-enes from chiral 4-oxa-1,6-enynes **with asym. induction** using $PtCl_2$ s. E.T. Newcomb, E.M. Ferreira, Org. Lett. *2013, 15* (7), 1772-5 [DOI: 10.1021/ol400625f]; chiral *o*-methoxybiaryls by **asym. aromatizing cycloisomerization** with $[Pd(MeCN)_4][BF_4]_2$ and (R)-BINAP as ligand s. N. Kadoya, M. Murai, M. Ishiguro, J. Uenishi, M. Uemura, Tetrahedron Lett. *2013, 54* (6), 512-4 [DOI: 10.1016/j.tetlet.2012.11.071].

Intramolecular hydroarylation
of acetylene derivs., update s. *51,* 339s*82*; functionalized 6- and 7-membered benzo-condensed carbo- and hetero-cyclics by intramolecular hydroarylation of ynamines or 1-acetylene-1-thioethers with $FeCl_3$/AgOTf as catalyst s. D. Eom, J. Mo, P.H. Lee, Z. Gao, S. Kim, Eur. J. Org. Chem. *2013* (3), 533-40 [DOI: 10.1002/ejoc.201201270]; 4-aryl-3-chromenes from [(3-arylprop-2-ynyl)oxy]-benzenes under Ag(I) or Au(I) catalysis s. A. Arcadi, F. Blesi, S. Cacchi, G. Fabrizi, A. Goggiamani, F. Marinelli, Org. Biomol. Chem. *2012, 10* (48), 9700-8 [DOI: 10.1039/c2ob26763b]; naphthalene-2-carboxylic acid esters by **intramolecular aromatizing hydroarylation** under Pt catalysis s. D. Kang, J. Kim, S. Oh, P.H. Lee, Org. Lett. *2012, 14* (22), 5636-9 [DOI: 10.1021/ol302437v]; phenanthrenes from *o*-propargylbiphenyls under iridium(III) catalysis s. Y. Kwon, H. Cho, S. Kim, ibid. *2013, 15* (4), 920-3 [DOI: 10.1021/ol400073s].

Dodecacarbonyltriruthenium/tetra-n-butylammonium iodide *$Ru_3(CO)_{12}/Bu_4NI$*
2-Pyrrolidones from 2-, 3- or 4-ethyleneformylamines ○
Stereoselective ruthenium-catalyzed intramolecular hydrocarbamylation

331. 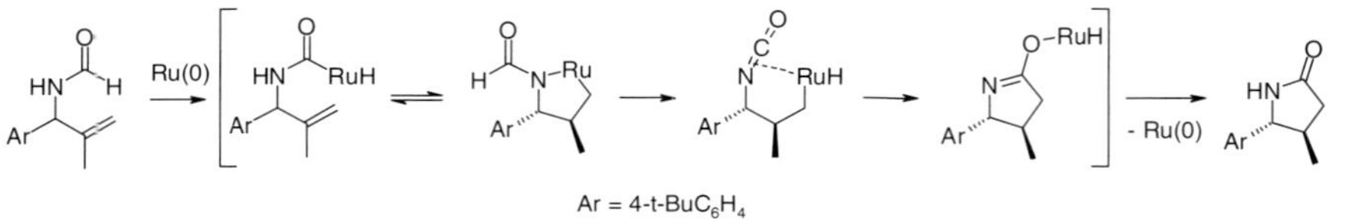

A suspension of N-[1-(4-*tert*-butylphenyl)-2-methylprop-2-en-1-yl]formamide (0.627 mmol), $Ru_3(CO)_{12}$ (5 mol%) and Bu_4NI (15 mol%) in DMF (1.1 ml) purged with CO (balloon) for 15 min, the sealed mixture heated at 150° for 4 h, cooled to room temp., and worked up with chromatographic purification → *trans*-5-[4-(*tert*-butyl)phenyl]-4-methylpyrrolidin-2-one. Y 82% (d.r. 20:1). This novel and atom-economical cyclization used NaI or Bu_4NI as iodide additive and gave cleaner products when the reaction mixture was purged with CO before heating. Cyclization was successful with N-(prop-2-en-1-yl)formamides carrying electron-diverse aryl or alkyl (incl. dialkyl) at C1, and H or one additional alkyl at C2 or C3 (fourteen examples; Y 70-90%), with 1,1-disubst. alkenes affording the more thermodyamically stable *trans*-products (d.r. 8:1 to 20:1; two examples). A single chiral allylic formamide (prepared in one step from the alcohol) cyclized with full retention of chirality (Y 77%). Interestingly, homoallylic (Y 81%) and bis-homoallylic formamides (Y 66%) cyclized to *pyrrolidones only*, suggesting alkene isomerization (which could not be observed experimentally). Deuterium-labelling experiments incorporated D at C2 and C3 in the product indicating that the hydrometalation step was reversible. F.e., optimization and substrate prepn. s. N. Armanino, E.M. Carreira, J. Am. Chem. Soc. *2013*, *135* (18), 6814-7 [DOI: 10.1021/ja4026787].

Benzylidene(dichloro)(tricyclohexylphosphine)(1,3-dimesitylimidazolidin-2-ylidene)-ruthenium/trimethylsilyl vinyl ether ←
Internal from terminal 1,3-dienes – Isomerizing ene-yne metathesis s. *83*, 319 ←

Oxygen/Carbon Type CC ⋂ OC

1,3-Dimesitylimidazolidin-2-ylidene *NHC*
4-Subst. coumarins from *o*-vinyloxycinnamaldehydes ○
via N-heterocyclic carbene-catalyzed intramolecular Michael addition of homoenolates

332. 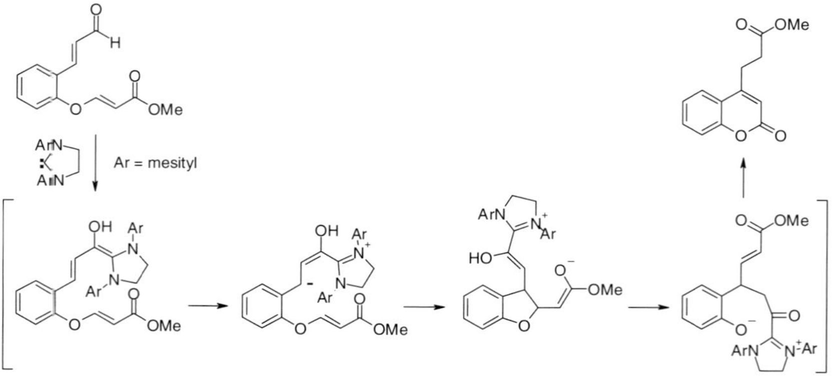

β-(Coumarin-4-yl)carboxylic acid esters. Dry toluene (7 ml) and DBU (20 mol%) added sequentially to a mixture of methyl 3-[2-((E)-3-oxoprop-1-enyl)phenoxy]acrylate (100 mg) and

1,3-dimesitylimidazolinium chloride (NHC precursor; 15 mol%), refluxed for 2 h (TLC) under argon, and worked up with purification by chromatography on silica gel → product. Y 60%. The procedure supports electron-diverse substituents (except nitro) on the benzene ring of the cinnamaldehyde, while the (E)- or (Z)-vinyloxy function may be substituted at the β-position or at both α- and β-positions by a carbalkoxy group (sixteen examples; Y 21-93%). Reaction is presumed to be initiated by formation of a Breslow intermediate which leads to the generation of a homoenolate by electron shift from an imidazole nitrogen atom; this then undergoes intramolecular Michael addition onto the enoate residue prior to carbalkoxy-activated Grob-type fragmentation and subsequent lactonization. F.e. and comparison of N-heterocyclic carbenes s. C.R. Sinu, D.V.M. Padmaja, U.P. Ranjini, K.C.S. Lakshmi, E. Suresh, V. Nair, Org. Lett. *2013*, *15* (1), 68-71 [DOI: 10.1021/ol303091m].

Titanium tetrachloride *$TiCl_4$*
Stereoselective synthesis of 2,6-*trans*-disubst. 4-acyltetrahydropyrans ◯C
from 2,4-dioxabicyclo[5.1.0]octanes s. *83*, 383

Bis(1,5-cyclooctadiene)rhodium(I) fluoroborate/1,1′-bis(diphenylphosphino)ferrocene ←
Cyclopentanones from diallyl ethers ◯
via rhodium-catalyzed isomerizing Claisen rearrangement-intramolecular hydroacylation

333. 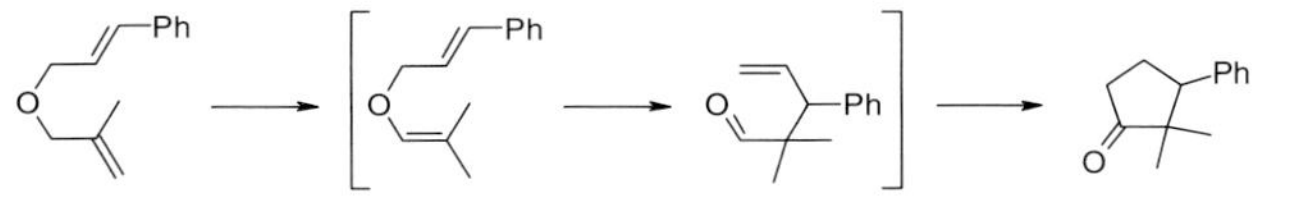

A mixture of $[Rh(cod)_2]BF_4$ (5 mol%) and dppf (5 mol%) in methylene chloride (2 ml) stirred at room temp. for 0.5 h, hydrogen introduced to the resulting soln. in a Schlenk tube, stirring continued at room temp. for 0.5 h, concentrated to dryness, the residue taken up in 1,2-dichloroethane (0.5 ml), a soln. of the startg. diallyl ether (0.3 mmol) in the same solvent (1 ml) added at room temp., stirred at 80° for 16 h, the mixture concentrated, and purified by preparative TLC → 2,2-dimethyl-3-phenylcyclopentanone. Y 83%. The procedure is milder than Eilbracht′s original methodology [P. Eilbracht et al., Synthesis *1995* (3), 330-4 [DOI: 10.1055/s-1995-3909]], and is generally applicable to diallyl ethers possessing a combination of 1,2- and 1,1-disubst. (or 1,1,2-trisubst.) ethylenic residues, which may be substituted by an electron-diverse aryl group at the terminus of the 1,2-subst. ethylenic residue (ten examples; Y 58-87%); diallyl ethers with a mono-subst. ethylenic residue and those with an alkyl group at the terminus of the 1,2-disubst. ethylenic residue, however, gave low yields (Y 30-37%), as did reaction with cycloalkene analogs (29-46%). Deuterium studies on the mechanism of the reaction suggest that the cascade involves initial chemoselective, rhodium-catalyzed O-allyl→O-vinyl isomerization, followed by thermal Claisen rearrangement, and terminating in classical rhodium-catalyzed intramolecular hydroacylation. F.e. and comparison of catalyst and ligands s. R. Okamoto, K. Tanaka, Org. Lett. *2013*, *15* (9), 2112-5 [DOI: 10.1021/ol400574s].

Nitrogen/Nitrogen Type CC ∩ NN

Chiral 3,3′-disubst. 1,1′-binaphthyl-2,2′-diyl hydrogen phosphates ←
Axially-chiral 2,2′-diamino-1,1′-binaphthyls from N,N′-di(2-naphthyl)hydrazines ∩
via asym. [3.3]-sigmatropic rearrangement under Brønsted acid catalysis

334. 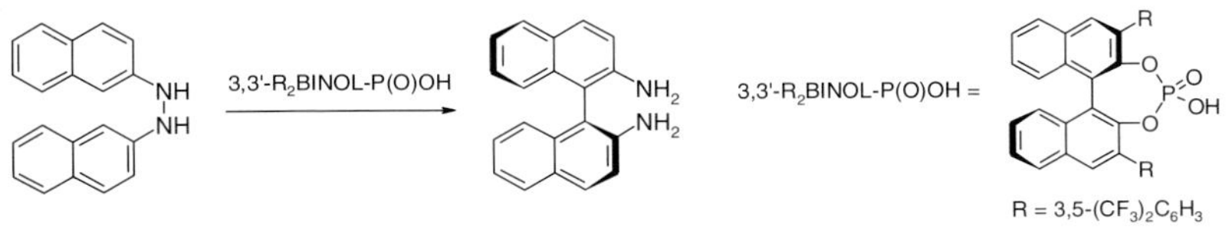

The first metal-free, catalytic C(sp²)-C(sp²) bond formation between two aromatic rings with *de novo* atroposelective generation of an axis of chirality is reported. **E:** An oven-dried flask charged with activated 4 Å molecular sieves (20 mg) and the chiral phosphate diester catalyst (0.2 eq.) under argon, the flask evacuated and refilled with argon three times, toluene (2 ml) added via syringe, the mixture cooled to -20°, stirred for 30 min, startg. N,N′-binaphthyl hydrazine (0.05 mmol) added in one portion, the mixture stirred at -20° for 24 h (monitored by HPLC), quenched with satd. aq. Na_2CO_3 soln., extracted with methylene chloride, the combined organic layers washed twice with brine, dried (Na_2SO_4), concentrated under reduced pressure, and the residue purified by column chromatography → (-)-product. Y 89% (e.r. 93:7). The corresponding 5,5′,6,6′,7,7′,8,8′-octahydrobinaphthyl hydrogen phosphate gave a similar result at 5 mol% loading with a reaction temperature of 0° (Y 86%; e.r. 92:8). Similar results were also obtained on a 2.8 g scale. Four further examples with substrates bearing methoxy groups in the 5,5′- or 7,7′-position or alkyl groups in the 3,3′-position gave yields of 69-85% and e.r. of 81.5:18.5 to 87.5:12.5 with the BINOL-derived catalyst or yields of 67-90% and e.r. of 54.5:45.5 to 91.5:8.5 with the H_8-BINOL-derived catalyst for this metal-free, non-coupling biaryl synthesis; substrates bearing methoxy groups in the 3,3′-positions or 3,3′,6,6′-positions showed only moderate enantioselectivity with either catalyst (e.r. 51:49 to 65.5:34.5; Y 80-85%). There was no enantioselectivity with an unsymmetrical substrate (N-naphthyl-N′-phenylhydrazine). DFT calculations indicate that in the transition state the phosphate diester proton is fully transferred to one of the nitrogen atoms of the hydrazine, and the phosphate counterion interacts with the hydrazinium ion in a bidentate manner involving two hydrogen bonding interactions. F.e.s. G.-Q. Li, H. Gao, C. Keene, M. Devonas, D.H. Ess, L. Kürti, J. Am. Chem. Soc. *2013*, *135* (20), 7414-7 [DOI: 10.1021/ja401709k].

Nitrogen/Carbon Type CC ∩ NC

n-Butyllithium or Lithium diisopropylamide *BuLi or i-Pr$_2$NLi*

1,5,6,7-Tetrahydro-1,2-diazepin-4-ones from N-aziridinyl-α,β-epoxyimines

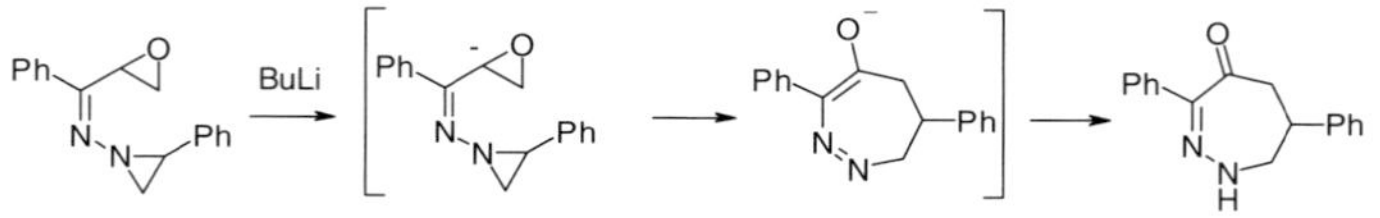

335.

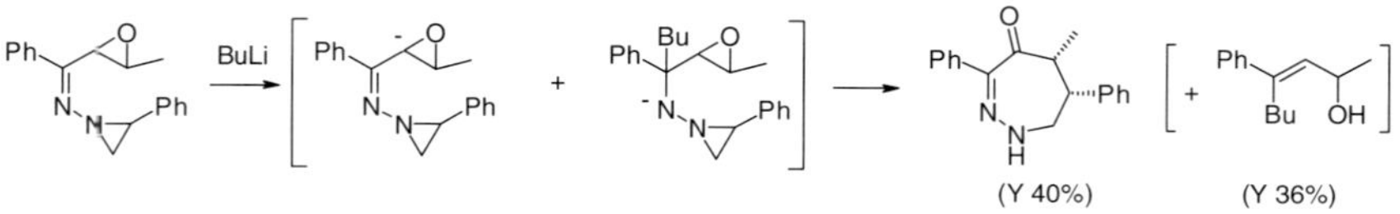

A soln. of N-[oxiran-2-yl(phenyl)methylene]-2-phenylaziridin-1-amine (0.11 mmol) in dry THF (3 ml) treated with *n*-BuLi (2.5 *M* in hexane; 1.5 eq.) at -78°, the mixture stirred at this temp. for 20 min, then quenched with satd. aq. $NaHCO_3$, warmed to room temp., extracted with ethyl acetate, the combined organic extracts dried (Na_2SO_4), filtered, concentrated, and the residue purified by chromatography on silica gel → 3,6-diphenyl-6,7-dihydro-1*H*-1,2-diazepin-4(5*H*)-one. Y 96%. Neither anionic addition nor Shapiro-type reaction were observed as side reactions. A similar yield was obtained with *i*-Pr_2NLi (94%), while MeLi gave 86%, but there was no reaction with K_2CO_3, KOBu-*t*, $LiN(SiMe_3)_2$ or $KN(SiMe_3)_2$, and NaOMe gave no isolable product. In this novel base-promoted rearrangement, seven further examples gave 3-aryl/*tert*-butyl-6-aryl-derivs. in yields of 57%, 73-93% using *n*-BuLi (lowest yield for a *p*-bromophenyl-deriv. with concomitant debromination) or 81-98% using *i*-Pr_2NLi (with retention of bromine). Treatment of a substrate having a 1,2-disubst. epoxy group with *n*-BuLi gave the corresponding cyclic ketone stereoselectively in 40% yield, along with 36% of a 2-ethylenealcohol obtained by fragmentation involving loss of styrene and molecular nitrogen; neither product was obtained with *i*-Pr_2NLi, however. The starting materials may be prepared readily from α,β-epoxyketones and 1-aminoaziridines by dehydrative condensation. F.e.s. F. Wu, M. Dong, F. Fang, Y. Sang, Y. Li, B. Cheng, L. Zhang, H. Zhai, Org. Lett. *2013*, *15* (4), 914-6 [DOI: 10.1021/ol400070j].

(Acetonitrile)[2-(dicyclohexylphosphino)biphenyl]gold(I) hexafluoroantimonate *[Au(I)]*
2,3-Dihydro-1*H*-pyrrolo[1,2-*a*]indoles from 2-(alk-1-ynyl)-N-arylazetidines

336.

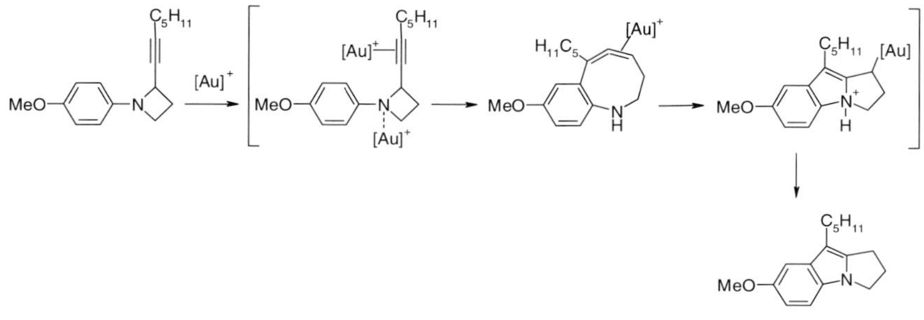

A novel, efficient and rapid *gold-catalyzed rearrangement* of 2-alk-1-ynyl-N-arylazetidines to afford 1*H*-pyrrolo[1,2-*a*]indole rings, found in a number of natural products, is reported. **E:** A soln. of 2-(hept-1-yn-1-yl)-1-(4-methoxyphenyl)azetidine (0.2 mmol) in methylene chloride (1 ml) treated with [(Cy_2)JohnPhosAu]SbF_6·MeCN (2.5 mol%) at 20°, stirred for 0.5 h (TLC), solvent removed *in vacuo*, and the residue purified by flash chromatography on silica or alumina → 7-methoxy-9-pentyl-2,3-dihydro-1*H*-pyrrolo[1,2-*a*]indole. Y 99%. The alkynyl group may be terminated by phenyl as well as (cyclo)alkyl, while the N-aryl group may bear electron-donating or -withdrawing groups and the azetidine may carry substituents in the 3-position (e.g. benzyloxy, dimethyl). β-Lactam analogs failed to give the corresponding pyrrolo[1,2-*a*]indolones and, although the mechanism has not been fully elucidated, this supports an azaphilic activation pathway. F.e. (sixteen; Y 47-98%) incl. applications to two formal syntheses of bioactive non-natural compounds (7-methoxymitosene and a 5-HT_{2C} receptor agonist) s. N. Kern, M. Hoffmann, A. Blanc, J.-M. Weibel, P. Pale, Org. Lett. *2013*, *15* (4), 836-9 [DOI: 10.1021/ol303518n].

Halogen/Carbon Type CC ∩ HalC

Palladium(II) acetate/1,1′-bis(diphenylphosphino)ferrocene *$Pd(OAc)_2$/dppf*
Radical ring closures via halogen atom transfer
update s. *38*, 718s*82*; ring closures of unactivated ethyleneiodides with $Pd(OAc)_2$/dppf s. H. Liu, Z. Qiao, X. Jiang, Org. Biomol. Chem. *2012*, *10* (36), 7274-7 [DOI: 10.1039/c2ob25990g].

Remaining Elements/Carbon Type CC ∩ RemC

Irradiation
Cyclic α,β-ethylene-β-silylketones from (acetylene)acylsilanes
Stereospecific visible-light-induced cycloisomerization

337.

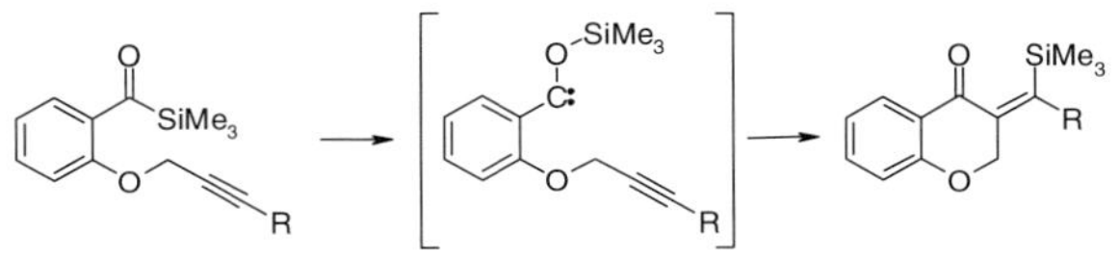

R = Ph
(R = C_6H_4F-p: Y 80%)
(R = Bu: Y 60%)

3-(1-Silylalkylidene)-4-chromanones. An oven-dried Schlenk tube charged with startg. acylsilane (0.2 mmol) and methylene chloride (1 ml), the soln. irradiated with an energy saving lamp (Philips Genie Long Life 23 W) with stirring overnight, solvent removed *in vacuo*, and the residue purified by flash chromatography on silica gel → product. Y 92%. This mild, catalyst-free method provides a facile and atom-economic route to chromones bearing a handle for further elaboration. Nine further examples with substrates in which the acetylene is terminated by aryl (optionally bearing *o*-OMe, *p*-Me or *p*-F) gave yields of 75-93%; a cyclohexenyl-terminated deriv. gave 89% but (cyclo)alkyl-terminated derivs. gave lower yields (38-60%; three examples) and no chromone was obtained from an acylsilane containing a terminal triple bond. No reaction was observed between benzoyl(trimethyl)silane and 3-hexyne or diphenylacetylene, but two examples with acylsilanes and dimethyl acetylenedicarboxylate gave yields of 82% and 87%, albeit with extended reaction times (3-13 days).

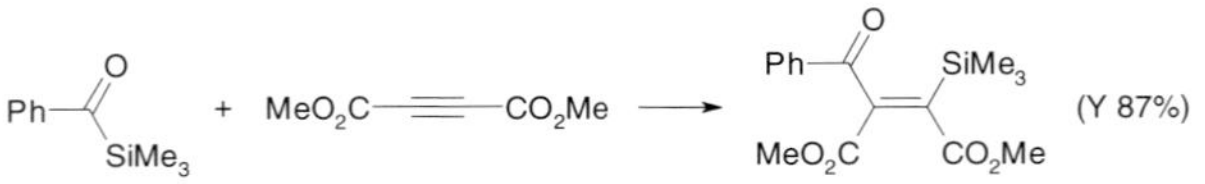

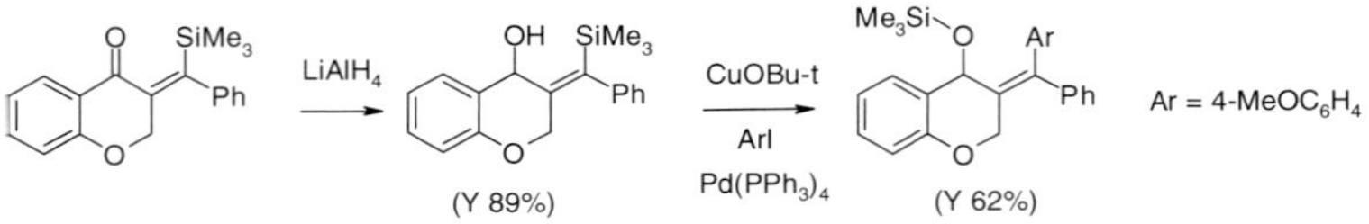

F.e. and evidence for a mechanism involving Brook rearrangement rather than acyl radicals, also Pd-catalyzed cross-coupling of the product after reduction ($LiAlH_4$) and Cu-promoted C→O-silyl migration, s. H.-J. Zhang, P. Becker, H. Huang, R. Pirwerdjan, F.-F. Pan, C. Bolm, Adv. Synth. Catal. *2012*, *354* (11-12), 2157-61 [DOI: 10.1002/adsc.201200510].

[1,3-Bis(2,6-diisopropylphenyl)imidazol-2-ylidene]gold(I) hexafluoroantimonate [(IPr)Au]SbF$_6$

9-(Arylseleno)phenanthrenes from *o*-(arylselenoethynyl)biaryls

Catalyst-dependent regioselective cycloisomerization

338. 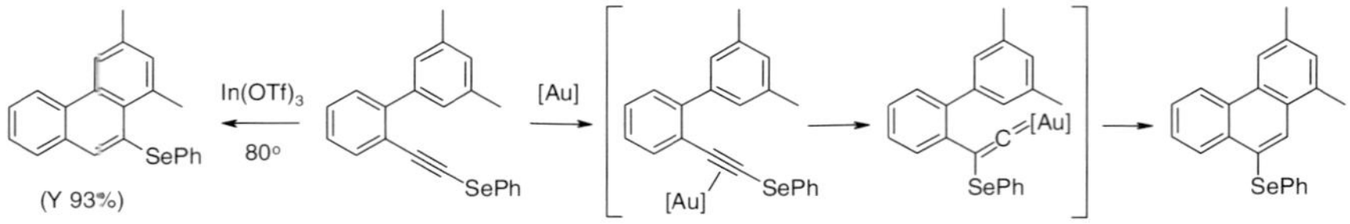

with 1,2-arylseleno group migration. A soln. of $AgSbF_6$ (0.01 mmol) in methylene chloride (1.5 ml) added to a soln. of (IPr)AuCl (0.01 mmol) in methylene chloride (1.5 ml), the mixture stirred for 10 min at room temp., the precipitate filtered through a pad of Celite, the solvent removed under reduced pressure, the residual oil dried under reduced pressure for 2 h, a soln. of [(3′,5′-dimethylbiphenyl-2-yl)ethynyl] phenyl selenide (0.2 mmol) in methylene chloride (4 ml) added to the residue, the mixture stirred for 15 min, the solvent removed under reduced pressure, the resulting yellow solid purified by flash chromatography on silica gel then further purified by recrystallization → 1,3-dimethylphenanthren-9-yl phenyl selenide. Y 97%. The selectivity of ring closure may be controlled perfectly by use of gold (I) or indium(III) catalysts. In the former case reaction proceeds via 1,2-migration with formation of vinylidene-gold(I) complexes (ten further examples; Y 36%, 86-97%). With $In(OTf)_3$ as catalyst (5 mol% in toluene at 80°) cycloisomerization occurs without migration, giving yields of 49%, 82-99% (eleven examples). F.e. and optimization s. W. Lim, Y.H. Rhee, Eur. J. Org. Chem. *2013* (3), 460-4 [DOI: 10.1002/ejoc.201201340].

Carbon/Carbon Type CC ∩ CC

Irradiation

Intramolecular [2+2]-cycloaddition

under irradiation, update s. *22,* 761s*82*; of N-tethered alkenyl-2(5*H*)-furanones to give conformationally restricted tricyclic pyrrolidines s. D.A. Fort, T.J. Woltering, M. Nettekoven, H. Knust, T. Bach, Chem. Commun. *2013, 49* (29), 2989-91 [DOI: 10.1039/c3cc40757h]; synthesis of the 2-oxatricyclo[4.2.1.$0^{4,9}$]nonan-3-one ring of lactiflorin s. P. Lu, E. Herdtweck, T. Bach, Chem. Asian J. *2012, 7* (8), 1947-58 [DOI: 10.1002/asia.201200295]; macrocyclic-condensed cyclobutenes from 1,n-enynes under gold(I) catalysis with (acetonitrile)[2-(di-*tert*-butylphosphino)-2′,4′,6′-triisopropylbiphenyl]gold(I) hexafluoroantimonate s. C. Obradors, D. Leboeuf, J. Aydin, A.M. Echavarren, Org. Lett. *2013, 15* (7), 1576-9 [DOI: 10.1021/ol400358f]; cyclobutanone ring of (-)-esermethole and Takayama′s intermediate for (+)-psychotrimine synthesis by intramolecular cycloaddition with carbalkoxyamino-functionalized ketenes s. T. Araki, T. Ozawa, H. Yokoe, M. Kanematsu, M. Yoshida, K. Shishido, ibid. *15* (1), 200-3 [DOI: 10.1021/ol303204v].

(Acetonitrile)[2-(di-tert-butylphosphino)-2′,4′,6′-triisopropylbiphenyl]gold(I) hexafluoroantimonate *[Au(I)]*

Macrocyclic-condensed cyclobutenes from 1,n-enynes s. *22,* 761s*83*

Chloro(1,5-cyclooctadiene)rhodium(I) dimer/(R)-2,3:2′,3′-bis(methylenedioxy)-6,6′-bis[bis(3,5-di-tert-butyl-4-methoxyphenyl)phosphino]biphenyl *[Rh(I)]*

Rhodium-catalyzed asym. intramolecular carboacylation with benzocyclobutenones

339. 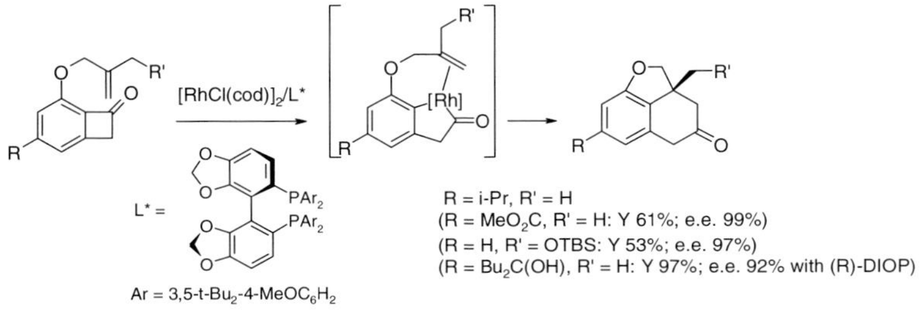

The first highly enantioselective rhodium-catalyzed carboacylation of olefins via C-C bond activation of benzocyclobutenones is reported (cf. *81*, 339 for nickel(0)-catalyzed asym. intramolecular carboacylation with cyclobutanones). **E: Chiral *peri*-fused 2-tetralones**. A vial charged with $[Rh(cod)Cl]_2$ (5 mol%) and (R)-DTBM-SegPhos (12 mol%) in a nitrogen-filled glovebox, a soln. of startg. alkenyloxybenzocyclobutenone in dioxane (0.1 *M*) added, the vial capped, the soln. kept at 133° for 48 h, then purified by flash chromatography → product. Y 74% (e.e. 98%). A range of allyloxybenzocyclobutenones bearing functional groups such as ester, ketone, ether, free tert. hydroxyl, silyl ether and CF_3 may be used and the disubst. ethylene moiety may bear alkyl or aryl groups (ten further examples with (R)-DTBM-SegPhos: Y 40-81%, e.e. 92-99%; two examples with (R)-DIOP: Y 76%, 97%, e.e. 92% for both); one example with an unsubst. homoallyloxy group gave the corresponding tetrahydropyran-fused product in 44% yield (e.e. 94%). A carbon-tethered substrate reacted with excellent diastereoselectivity but with low enantioselectivity, however. Benzene ring hydrogenation $[[RhCl(1,5\text{-hexadiene})]_2/Bu_4NHSO_4]$ followed by Ley oxidation of the partially-reduced keto moieties gave chiral saturated tricyclic ketones with four contiguous stereocenters (three examples; Y 41-70%). F.e.s. T. Xu, H.M. Ko, N.A. Savage, G. Dong, J. Am. Chem. Soc. *2012, 134* (49), 20005-8 [DOI: 10.1021/ja309978c]; racemic products with 1,4-bis(diphenylphosphino)butane as ligand s. T. Xu, G. Dong, Angew. Chem., Int. Ed. *2012, 51* (30), 7567-71 [DOI: 10.1002/anie.201202771].

Exchange ⇃↾

Hydrogen ↾ CC ⇃↾ H

Irradiation s. under Bis(2-phenylpyridinato-C2,N)(2,2′-bipyridyl)iridium(III) hexafluorophosphate ⫻

Electrolysis/palladium(II) acetate/copper(I) iodide/p-benzoquinone/triethylamine ←
Site-selective sequential coupling controlled by 'electrochemical reaction site switching' ←
Electro-oxidative homo-coupling of bromoarylethynes s. *83*, 440

Lithium chloride s. under $PdCl_2$ *KF*
Potassium fluoride-aluminum oxide s. under N-Bromosuccinimide *$KF\text{-}Al_2O_3$*

Potassium iodide/potassium peroxymonosulfate *$KI/KHSO_5$*
Oxidative 1,3-dipolar cycloaddition ○
s. *32,* 716; Δ^2-isoxazolines from aldoximes and ethylene derivs. with hypoiodite generated from KI as catalyst and Oxone as oxidant in aq. methanol s. A. Yoshimura, C. Zhu, K.R. Middleton, A.D. Todora, B.J. Kastern, A.V. Maskaev, V.V. Zhdankin, Chem. Commun. *2013, 49* (42), 4800-2 [DOI: 10.1039/c3cc41164h]; *heterogeneous* procedure with NBS and KF/Al_2O_3 under mild conditions s. M. Boruah, D. Konwar, Synth. Commun. *2012, 42* (22), 3261-8 [DOI: 10.1080/00397911.2011.558665].

Copper/2,3-dichloro-5,6-dicyanoquinone/chiral bis(Δ^2-oxazolines) *Cu/DDQ/box*
Oxidative asym. C_1-(alk-1-ynyl)ation of 1,2,3,4-tetrahydroisoquinolines H → C≡CR
with terminal acetylene derivs. by high-speed ball-milling s. *72,* 491s*83*

Copper(II) oxide s. under Iodine *CuO*
Silver carbonate s. under Palladium(II) acetate *Ag_2CO_3*
Copper(II) acetate s. under Dichloro(p-cymene)ruthenium(II) dimer, $Pd(OAc)_2$, Tris(dibenzylideneacetone)dipalladium and $PdCl_2(MeCN)_2$ *$Cu(OAc)_2$*
Silver acetate s. under $Pd(OAc)_2$ and Tris(dibenzylideneacetone)dipalladium *AgOAc*

Mesoporous silica-supported [1-aza-4-azoniabicyclo[2.2.2]octane]copper(I) chloride/piperidine/air ←
Sym. 1,3-diynes from terminal acetylene derivs. 2 RC≡CH → RC≡C-C≡CR
by Glaser coupling, update s. *16,* 780s*82*; *heterogeneous* coupling with mesoporous [covalently] silica-supported [1-aza-4-azoniabicyclo[2.2.2]octane]copper(I) chloride complexes at room temperature under air with added piperidine s. H. Li, M. Yang, X. Zhang, L. Yan, J. Li, Y. Qi, New J. Chem. *2013, 37* (5), 1343-9 [DOI: 10.1039/c3nj41006d]; with an amino-functionalized silica-supported copper complex cf. L. van Gelderen, G. Rothenberg, V.R. Calderone, K. Wilson, N.R. Shiju, Appl. Organomet. Chem. *2013, 27* (1), 23-7 [DOI: 10.1002/aoc.2933].

Copper(I) bromide/6,6′-di-tert-butyl-2,2′-bipyridyl/9-azabicyclo[3.3.1]nonan-3-one N-oxyl/oxygen ←
Copper-catalyzed aerobic dehydrogenative α-substitution of sec. amines via *in situ* formation of azomethines ←

340.

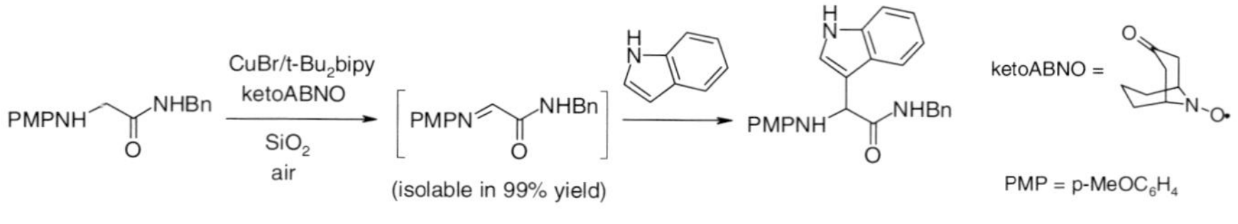

A mild, general and chemoselective catalytic aerobic oxidation of amines to imines is reported, the key to high catalytic activity being a combination of a novel, less sterically demanding, electron-deficient N-oxyl radical [ketoABNO] and a copper(I) salt. **E: Oxidative Friedel-Crafts α-amino-alkylation**. CuBr (10 μmol), 6,6′-*t*-Bu_2bipy (10 μmol) and ketoABNO (10 μmol) added to a flame-dried test tube containing THF (0.4 ml), the mixture stirred for ca. 1 h under argon, startg. glycinamide (0.2 mmol), indole (1.5 eq.) and silica gel (100 mg) added, the atmosphere replaced with air (1 atm.; balloon), the mixture stirred for 20 h at room temp., directly loaded onto a short pad of basic alumina, eluting with ethyl acetate, volatiles removed under reduced pressure, and the residue purified by flash chromatography on silica gel → product. Y 93%. Initial investigations involved optimization of the preparation of aldimines from a variety of [dibenzyl/aryl(benzyl)/alkyl(benzyl)/alkyl(aryl)] sec. amines (twelve examples; Y 72-99%) or of **ar. N-benzylimines from two prim. benzylamine molecules** (four examples; Y 86-95%). Tribenzylamine did not react at all. Due to the broad scope and mildness of the system the method was then extended to C-C bond formation with *in situ*-generated imines, establishing applicability to oxidative Friedel-Crafts α-aminoalkylation of pyrroles or indole (two further examples; Y 61%, 68%), stepwise addition of MeMgBr (Y 60%), oxidative aza-Diels-Alder reaction with Danishefsky's diene (Y 53%) and oxidative Strecker reaction using HCN generated from Me_3SiCN and *tert*-butanol (Y 90%). Furthermore the method was extended to *asym. coupling* with an aliphatic nitro compd. using 2(CuOTf)·toluene and (-)-Ph-Box as ligand (Y 78%; *syn/anti* 19.6:1; e.e. *syn* 95%).

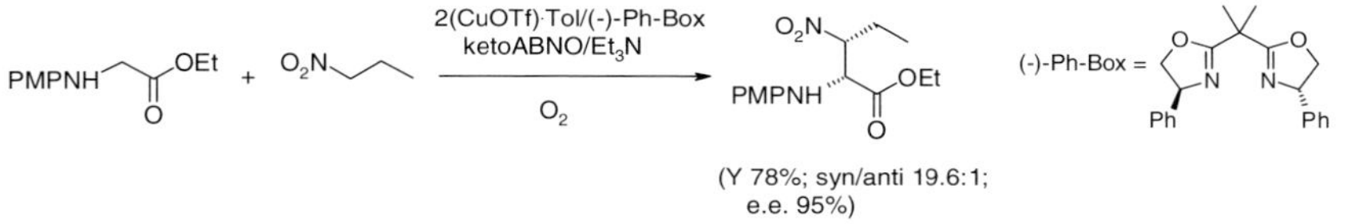

F.e.s. T. Sonobe, K. Oisaki, M. Kanai, Chem. Sci. *2012*, *3* (11), 3249-55 [DOI: 10.1039/c2sc20699d].

Copper(I) bromide/tert-butyl hydroperoxide *CuBr/t-BuOOH*
3-Acylindoles from *o*-acetylene-*tert*-amines ○
via copper(I)-catalyzed oxidative ring closure s. *83,* 48

Copper(I) iodide s.a. under Electrolysis *CuI*

Copper(I) iodide/benzyl cyanide *CuI/BnCN*
Ar. cyanation H → CN
update s. *3,* 600s*82*; C_3-cyanation of indoles and aza-analogs with CuI and benzyl cyanide as cyanating agent in open air s. O.Y. Yuen, P.Y. Choy, W.K. Chow, W.T. Wong, F.Y. Kwong, J. Org. Chem. *2013, 78* (7), 3374-8 [DOI: 10.1021/jo3028278].

Copper(II) chloride/tert-butyl hydroperoxide *$CuCl_2$/t-BuOOH*
α,β-Ethyleneketones from ethylene derivs. and aldehydes by regioselective dehydrogenative C-acylation C=CH → C=C-C(O)R

341. 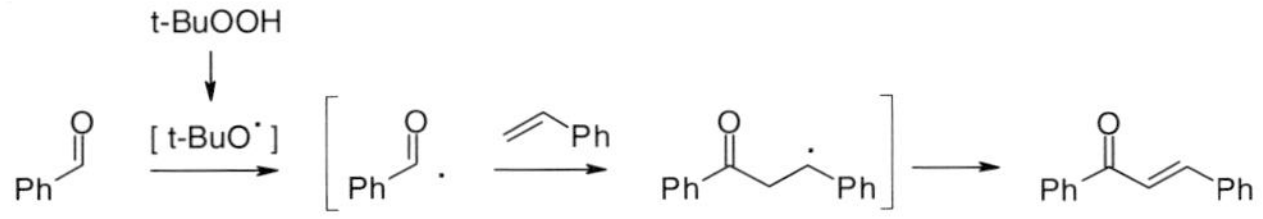

Chalcones. Benzaldehyde (5 eq.), styrene (0.5 mmol) and *tert*-butyl hydroperoxide (2.5 eq.) injected sequentially into a Schlenk tube containing $CuCl_2$ (20 mol%) in an argon-filled glove-box, the mixture stirred at 80° for 12 h, quenched with water, extracted with ether, and purified by flash chromatography on silica gel → (E)-chalcone. Y 63%. The method was applied to styrenes optionally bearing Me, *t*-Bu, F, Cl or OAc on the aromatic ring and to 1,1-diarylethylenes optionally bearing MeO, in reaction with (het)ar. aldehydes optionally bearing Me, MeO, MeS or Cl (eighteen examples; Y 30%, 34%, 50-80%). It failed with Ni-catalysts such as $Ni(PPh_3)_4$, $NiCl_2$ or $Ni(acac)_2$, or with $CoCl_2$, CuCl, $FeCl_2$, or $CuBr_2$. Reaction was inhibited by BHT, indicating a radical mechanism. F.e. and optimization s. J. Wang, C. Liu, J. Yuan, A. Lei, Angew. Chem., Int. Ed. *2013*, *52* (8), 2256-9 [DOI: 10.1002/anie.201208920].

Copper(II) bromide s. under I_2 *$CuBr_2$*

Silver nitrate/magnesium nitrate *$AgNO_3/Mg(NO_3)_2$*
3-α-Phosphinyloxindoles from α,β-ethylenecarboxylic acid anilides ○
Silver-catalyzed oxidative intramolecular 1,2-carbaphosphinylation s. *83*, 247

Silver hexafluoroantimonate s. under Dichloro(p-cymene)ruthenium(II) dimer *$AgSbF_6$*
Zinc acetate s. under Bis(2-phenylpyridinato-C2,N)(2,2′-bipyridyl)iridium(III) hexafluorophosphate *$Zn(OAc)_2$*
Magnesium nitrate s. under $AgNO_3$ *$Mg(NO_3)_2$*

Zinc triflate/air *$Zn(OTf)_2/O_2$*
3-Acoxy-4-acylaminopyridines from α-[acyl(vinyl)amino]aldehydes and isonitriles
via zinc(II)-mediated [5+1]-cycloaddition, aerobic oxidation, O-acylation with carboxylic acid chlorides and N→O-acyl migration s. *83*, 411

Benzoquinone s. under Electrolysis *BQ*
2,2′,6,6′-Tetra-tert-butyldiphenoquinone s. under Chiral N-condensed 1,2,4-triazolylidenes ←
4-Nitrophenyl isocyanate s. under Trifluoromethanesulfonic acid *ArNCO*
N-Hydroxyphthalimide s. under Sodium hexachloroantimonate ←
Chiral bis(Δ^2-oxazolines) s. under Copper *box*

Chiral N-condensed 1,2,4-triazolylidenes/2,2′,6,6′-tetra-tert-butyldiphenoquinone ←
Oxidative asym. α-functionalization of aldehydes ←
under N-heterocyclic carbene catalysis
Chiral ester enolate equivalents as intermediates

342.

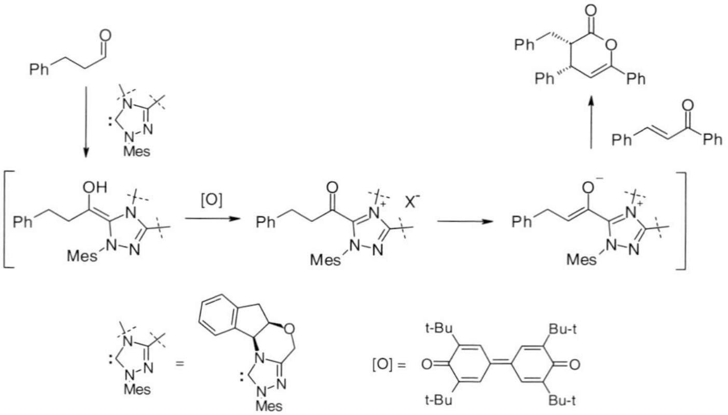

Asym. synthesis of 3,4-dihydro-2-pyrones from α,β-ethyleneketones. Freshly distilled anh. THF (1 ml) added under N_2 to a Schlenk tube containing 3-phenylpropanal (2.5 eq.), chalcone (0.1 mmol), the chiral 1,2,4-triazolium fluoroborate (NHC precursor; 10 mol%), 2,2′,6,6′-tetra-*tert*-butyldiphenoquinone (2.5 eq.) and Cs_2CO_3 (20 mol%), the tube closed with a septum, evacuated, refilled with N_2, stirred at room temp. for 3 h, the mixture concentrated under reduced pressure, and worked up with purification by chromatography on silica gel → product. Y 83% (d.r. 20:1; e.e. 99%). The procedure is complementary to reaction under direct enamine or indirect NHC catalysis. It is applicable to the coupling of electron-diverse chalcones (and heteroaryl analogs) with β-aryl- or β-alkyl-subst. aldehydes (fifteen examples; Y 70-95%; d.r. generally 20:1; e.e. uniformly 99%), but reaction failed with α-aryl- and α,α′-diaryl-aldehydes. Furthermore, electrophiles other than chalcones could be used to trap the intermediate ester enolate, e.g. trifluoromethyl ketones which provided **chiral β-trifluoromethyl-β-lactones** (two examples; Y 49-53%; e.e. 82-92%). F.e. and comparison of chiral reagents, bases, solvents and oxidants, also catalytic procedure with 10 mol% of the diphenoquinone and MnO_2 as re-oxidant, s. J. Mo, R. Yang, X. Chen, B. Tiwari, Y.R. Chi, Org. Lett. *2013*, *15* (1), 50-3 [DOI: 10.1021/ol303035r].

Laccase/oxygen ←
Oxidative α-arylation with catechols s. *30*, 542s*83* H → Ar

Benzyl cyanide s. under CuI *BnCN*
2,2,6,6-Tetramethylpiperidine nitroxyl s.a. under Palladium(II) acetate *TEMPO*

2,2,6,6-Tetramethylpiperidine nitroxyl-on-magnetic core-shell nanoparticles/ tert-butyl nitrite/oxygen ←
Oxidative Passerini reaction with prim. or sec. alcohols CHOH → C(OAc)CONHR

343. 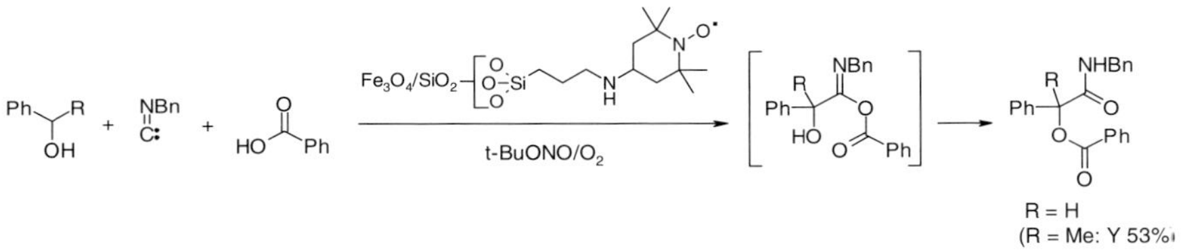

Benzyl isocyanide (*2 eq.*) added to a mixture of 2,2,6,6-tetramethylpiperidine N-oxyl-on-magnetic core-shell nanoparticles (1 mol%), *tert*-butyl nitrite *(10 mol%),* benzoic acid (1 mmol) and benzyl alcohol (1 eq.) in dry toluene (2 ml), the reaction flask purged with pure O_2 (filled balloon), vigorously stirred at room temp. for 24 h under O_2, diluted with methylene chloride, the catalyst separated using an external magnet, aq. NaOH (0.2 *M*) added to the organic layer, the latter extracted, and worked up with purification by chromatography on silica gel → product. Y 84%. With methyl(phenyl)carbinol, as an example of reaction with a sec. alcohol, the product yield was 53% after 36 h. This is the first example of an oxidative Passerini reaction with alcohols **under metal- and halogen-free conditions**, and the *first application of sec. alcohols* in the conversion. The procedure is generally applicable to the coupling of aliphatic or aromatic carboxylic acids and isonitriles *(2 eq.)* with aliphatic prim. alcohols and electron-diverse benzyl alcohols as well as aliphatic, benzylic and cyclic sec. alcohols, notably without undesirable over-oxidation (thirty-three examples; Y 46-94%). The catalyst is readily recoverable and was reused fourteen times with only a slight decrease in activity. F.e.s. B. Karimi, E. Farhangi, Adv. Synth. Catal. *2013*, *355* (2-3), 508-16 [DOI: 10.1002/adsc.201200449].

9-Azabicyclo[3.3.1]nonan-3-one N-oxyl s. under CuBr *ketoABNO*
Trifluoroacetic anhydride s. under Dimethyl sulfoxide *$(CF_3CO)_2O$*
tert-Butyl hydroperoxide s. under CuBr, $CuCl_2$ and $Pd(OAc)_2$ *t-BuOOH*
Di-tert-butyl peroxide s. under $FeCl_3$ *$(t\text{-}BuO)_2$*

Dimethyl sulfoxide/trifluoroacetic anhydride *$DMSO/(CF_3CO)_2O$*
2-α-Functionalization of indoles s. *83*, 50 ←

Phenyl iodosotrifluoroacetate/triethylamine *$PhI(OCOCF_3)_2/Et_3N$*
Oxidative α-arylation with arenes H → Ar
of malonic acid esters using CAN cf. *30*, 542s*41*; of β-ketocarbonyl compds. with the benzene ring of vindoline using phenyl iodosotrifluoroacetate/Et_3N s. T.C. Turner, K. Shibayama, D.L. Boger, Org. Lett. *2013*, *15* (5), 1100-3 [DOI: 10.1021/ol400135n]; regioselective α-arylation of cyclic β-dicarbonyl compds. with catechols using the laccase from *Pleurotus ostreatus* under oxygen s. J. Pietruszka, C. Wang, Green Chem. *2012*, *14* (9), 2402-9 [DOI: 10.1039/c2gc35662g].

Phenyl iodosotrifluoroacetate/trimethylsilyl azide *$PhI(OCOCF_3)_2/Me_3SiN_3$*
Metal-free directed *o*-acylation of N-heteroarenes s. *26*, 775s*83* H → COR

2,3-Dichloro-5,6-dicyanoquinone s. under Copper *DDQ*

N-Bromosuccinimide/potassium fluoride-aluminum oxide *NBS/KF-Al_2O_3*
Heterogeneous oxidative 1,3-dipolar cycloaddition ○
Δ^2-Isoxazolines from aldoximes and ethylene derivs. s. *32,* 716s*83*

Trimethylsilyl azide s. under $PhI(OCOCF_3)_2$ *Me_3SiN_3*
tert-Butyl nitrite s. under 2,2,6,6-Tetramethylpiperidine nitroxyl-on-magnetic core-shell nanoparticles *t-BuONO*
Tricyclohexylphosphine s. under Palladium(II) acetate *Cy_3P*
2,2′-Bis(diphenylphosphino)biphenyl s. under $Ru_3(CO)_{12}$ *BIPHEP*

Sodium hexachloroantimonate/N-hydroxyphthalimide/oxygen ←
Dehydrogenative α-substitution of amines CH(N<) → C(N<)R
update s. *72,* 491s*82*; with various C- and P-nucleophiles using $NaSbCl_6$/N-hydroxyphthalimide under O_2 (1 atm.) s. A. Tanoue, W.-J. Yoo, S. Kobayashi, Adv. Synth. Catal. *2013, 355* (2-3), 269-73 [DOI: 10.1002/adsc.201200999]; oxidative C_1-substitution of 1,2,3,4-tetrahydroisoquinolines with indoles using $NaAuCl_4$/*t*-BuOOH (cf. *72,* 491s*81*) s. Y. Zhang, S. Luo, B. Feng, C. Zhu, Chin. J. Org. Chem. *2012, 30* (12), 2741-6 [DOI: 10.1002/cjoc.201201154]; with other C-nucleophiles *in water* under *heterogeneous* conditions with poly(*o*-sulfoaniline)-supported gold nanoparticles under O_2 cf. T. Amaya, T. Ito, T. Hirao, Heterocycles *2012, 86* (2), 927-32 [DOI: 10.3987/com-12-s(n)71]; α-(indol-3-yl)ation of α-aminocarbonyl compds. (cf. *83,* 340) **under visible-light photoredox catalysis** with dichlorotris(2,2′-bipyridyl)ruthenium(II) as sensitizer under O_2 s. Z.-Q. Wang, M. Hu, X.-C. Huang, L.-B. Gong, Y.-X. Xie, J.-H. Li, J. Org. Chem. *2012, 77* (19), 8705-11 [DOI: 10.1021/jo301691h]; of glycine derivs. and peptides with an added Lewis acid cf. S. Zhu, M. Rueping, Chem. Commun. *2012, 48* (98), 11960-2 [DOI: 10.1039/c2cc36995h]; α-substitution of tert. amines with nitroalkanes in water with *water-soluble* graphenesulfonic acid-supported RuO_2 nanoparticles under O_2 s. Q.-Y. Meng, Q. Liu, J.-J. Zhong, H.-H. Zhang, Z.-J. Li, B. Chen, C.-H. Tung, L.-Z. Wu, Org. Lett. *2012, 14* (23), 5992-5 [DOI: 10.1021/ol3028785]; *metal-free* dehydrogenative α-substitution of tert. amines with C(sp^3)-nucleophiles using I_2 as catalyst and aq. H_2O_2 s. T. Nobuta, N. Tada, A. Fujiya, A. Kariya, T. Miura, A. Itoh, ibid. *2013, 15* (3), 574-7 [DOI: 10.1021/ol303389t]; aerobic nucleophilic C_1-substitution of 1,2,3,4-tetrahydroisoquinolines with I_2 as catalyst cf. J. Dhineshkumar, M. Lamani, K. Alagiri, K.R. Prabhu, ibid. *2013, 15* (5), 1092-5 [DOI: 10.1021/ol4001153]; solvent-free **oxidative asym. C_1-(alk-1-ynyl)ation** of 1,2,3,4-tetrahydroisoquinolines with terminal acetylene derivs. by high-speed ball-milling with copper balls in the presence of DDQ and a chiral PyBox-type bis(Δ^2-oxazoline) as ligand s. J. Yu, Z. Li, K. Jia, Z. Jiang, M. Liu, W. Su, Tetrahedron Lett. *2013, 54* (15), 2006-9 [DOI: 10.1016/j.tetlet.2013.02.007].

Oxygen or Air s. under Mesoporous silica-supported [1-aza-4-azoniabicyclo[2.2.2]-octane]copper(I) chloride, CuBr, $Zn(OTf)_2$, Laccase, 2,2,6,6-Tetramethylpiperidine nitroxyl, Sodium hexachloroantimonate, I_2, Chiral dimeric hydroxoiron(III) salan complexes, Ruthenium dioxide, Dichloro(p-cymene)ruthenium(II) dimer, Dichloro-tris(2 2′-bipyridyl)ruthenium(II), $Pd(OAc)_2$, $PdCl_2$, PdI_2 and Bis(2-phenyl-pyridinato-C2,N)-(2,2′-bipyridyl)iridium(III) hexafluorophosphate *O_2*
Hydrogen peroxide s. under I_2 *H_2O_2*

Trifluoromethanesulfonic acid/4-nitrophenyl isocyanate *TfOH/ArNCO*
Sym. benzophenones from arenes 2 ArH → Ar_2CO
4-Nitrophenyl isocyanate as phosgene equivalent s. *83*, 494

Potassium peroxymonosulfate s. under KI *$KHSO_5$*

Iodine/copper(II) oxide/triethylamine *I_2/CuO/Et_3N*
2-Amino-3-cyano-3a,6a-dihydrofuro[2,3-*b*]furans from 2-ene-1,4-diones ○
from methyl ketones via base-catalyzed double ring closure with malononitrile s. *83*, 283

Iodine/copper(II) bromide/oxygen *$I_2/CuBr_2/O_2$*
(E)-2-Ene-1,4-diones from two methyl ketone molecules RC(O)CH=CHC(O)R'

344.
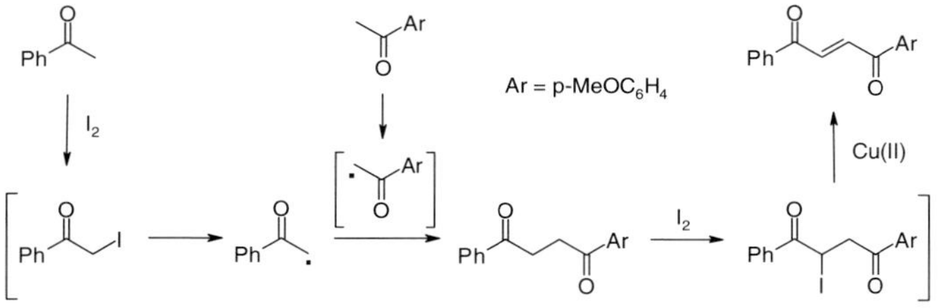

Unsym. (E)-1,4-diaryl-2-ene-1,4-diones. A mixture of acetophenone (*0.5 mmol*), *p*-methoxy-acetophenone (*0.25 mmol*), $CuBr_2$ (0.1 mmol), I_2 (1 mmol) and dry DMF (0.5 ml) heated in a sealed tube at 80° for 20 h, and worked up with purification by flash column chromatography → (E)-1-(4-methoxyphenyl)-4-phenylbut-2-ene-1,4-dione. Y 83%. Reaction is applicable to electron-diverse acetophenones with the proviso that one of the ketones is in 2-fold excess over the other (twelve examples; Y 58-83%). With either of the acetophenones on the 0.5 mmol scale (with all else the same), the corresponding **sym. (E)-1,4-diaryl-2-ene-1,4-diones** were isolated, generally in good yield (sixteen examples; 29%, 61-91%). Here, electron-donating groups on the aromatic ring were found to have a positive effect, with fluorinated acetophenones affording only moderate yields (Y 61-72%) while *p*-(trifluoromethyl)acetophenone was low-yielding (Y 29%). A radical mechanism is proposed, involving iodomethyl ketones and γ-diketones as key intermediates: iodine is thus implicated in both iodination and generation of a phenacyl radical, while in the last phase of the reaction copper(II) is inserted into the C-I bond prior to β-hydride elimination with re-oxidation of Cu(I) to Cu(II) with aerobic oxygen. Interestingly and very conveniently **(E)→(Z)-rearrangement** of the products was effected simply and in high yield by irradiation with white light (23 W) in 1:6 ethyl acetate/hexane at room temperature over 5 h (nine examples; Y 81-99%). F.e. and solvent effect s. K. Xu, Y. Fang, Z. Yan, Z. Zha, Z. Wang, Org. Lett. *2013, 15* (9), 2148-51 [DOI: 10.1021/ol4006344].

Iodine/oxygen or hydrogen peroxide *I_2/O_2 or H_2O_2*
Metal-free dehydrogenative α-substitution of amines s. *72,* 491s*83* CHN< → C(N<)R

Transition metal catalysts/oxidant ←
Regioselective transition metal-catalyzed dehydrogenative coupling of [hetero]arenes with ethylene derivs. H → C=C
dehydrogenative Heck arylation, update s. *69,* 369s*82*; dehydrogenative β-arylation of α-(phosphonylmethyl)cinnamic acid esters with $Pd(OCOCF_3)_2$/AgOAc/*t*-BuCOOH s. H.S. Lee, C.H. Lim, H.J. Lee, J.N. Kim, Bull. Korean Chem. Soc. *2012, 33* (11), 3817-22 [DOI: 10.5012/bkcs.2012.33.11.3817]; of 3-alkylideneoxindoles with $Pd(OAc)_2$ as catalyst s. H.J. Lee, K.H. Kim, S.H. Kim, J.N. Kim, Tetrahedron Lett. *2013, 54* (2), 170-5 [DOI: 10.1016/j.tetlet.2012.10.122]; regioselective chelation-free dehydrogenative *m*- and *p*-vinylation of arenes with $[RhCl(cod)]_2/Cu(OAc)_2 \cdot H_2O$/ trichloroacetic acid under air s. L. Zheng, J. Wang, Chem. Eur. J. *2012, 18* (31), 9699-704 [DOI: 10.1002/chem.201200657]; aerobic (E)-selective dehydrogenative β-hetarylation of styrenes under palladium(II) catalysis (without transition metal salts as co-oxidant) s. A. Vasseur, D. Harakat, J. Muzart, J. Le Bras, Adv. Synth. Catal. *2013, 355* (1), 59-67 [DOI: 10.1002/adsc.201200787]; dehydrogenative 3-vinylation of indoles with $Pd(OAc)_2/CF_3COOH$ under O_2 s. W.-L. Chen, Y.-R. Gao, S. Mao, Y.-L. Zhang, Y.-F. Wang, Y.-Q. Wang, Org. Lett. *2012, 14* (23), 5920-3 [DOI: 10.1021/ol302840b]; dehydrogenative C_3- or C_5-vinylation of N-protected 2-pyridones under palladium(II) catalysis with a copper(II) or Ag(I) salt as oxidant, also 3,5-divinylation and dehydrogenative arylation, s. Y. Chen, F. Wang, A. Jia, X. Li, Chem. Sci. *2012, 3* (11), 3231-6 [DOI: 10.1039/c2sc20869e]; dehydrogenative 5-vinylation of 4-chromones under rhodium(III) catalysis with $[Cp^*RhCl_2]_2$/ $AgSbF_6/Cu(OAc)_2$ s. R. Samanta, R. Narayan, A.P. Antonchick, Org. Lett. *2012, 14* (23), 6108-11 [DOI: 10.1021/ol303067f]; **N-directed *o*-vinylation** (cf. *69,* 369s*82*) of 2-aryl-Δ^2-oxazolines with

$[RuCl_2(p\text{-cymene})]_2/Cu(OAc)_2 \cdot H_2O$ under air with added 1,1′-binaphthyl-2,2′-diyl hydrogen phosphate s. B. Li, K. Devaraj, C. Darcel, P.H. Dixneuf, Green Chem. *2012, 14* (10), 2706-9 [DOI: 10.1039/c2gc36111f]; C_2,C_8-divinylation of N-(2-pyridylsulfonyl)carbazoles, and C_7-vinylation of indolines, with $Pd(OAc)_2$ and an N-fluoro compd. as oxidant, e.g. 1-fluoro-2,4,6-trimethylpyridinium triflate, s. B. Urones, R.G. Arrayás, J.C. Carretero, Org. Lett. *2013, 15* (5), 1120-3 [DOI: 10.1021/ol400206k]; *phosphonate-directed o*-vinylation of benzylphosphonic acid monoesters with $Pd(OAc)_2$/AgOAc s. X. Meng, S. Kim, ibid. *15* (8), 1910-3 [DOI: 10.1021/ol400565r]; *ether-directed o*-vinylation of benzyl ethers with $Pd(OAc)_2/Ag_2CO_3$ in the presence of an N-protected α-amino acid, e.g. N-acetylglycine, s. G. Li, D. Leow, L. Wan, J.-Q. Yu, Angew. Chem., Int. Ed. *2013, 52* (4), 1245-7 [DOI: 10.1002/anie.201207770]; C_2-vinylation of ferrocenyl ketones with acrylates using $[RuCl_2(p\text{-cymene})]_2/AgSbF_6$ and $Cu(OAc)_2 \cdot H_2O$ as oxidant under aerobic conditions s. K.S. Singh, P.H. Dixneuf, Organometallics *2012, 31* (21), 7320-3 [DOI: 10.1021/om3008162]; **N-directed *o*-vinylation of N-aryltriazenes** with $[Cp^*RhCl_2]_2/AgSbF_6$ and AgOAc as oxidant s. B. Liu, Y. Fan, Y. Gao, C. Sun, C. Xu, J. Zhu, J. Am. Chem. Soc. *2013, 135* (1), 468-73 [DOI: 10.1021/ja3099245]; **dehydrogenative coupling of terminal acetylene derivs.** with hetarenes (cf. *71*, 337s*80*) under palladium catalysis with $Pd_2(dba)_3$ s. X. Jie, Y. Shang, P. Hu, W. Su, Angew. Chem., Int. Ed. *2013, 52* (13), 3630-3 [DOI: 10.1002/anie.201210013]; dehydrogenative C_2-(alk-1-ynyl)ation of [benz]azoles using Ag_2CO_3 as oxidant and added acetic acid s. F. Shibahara, Y. Dohke, T. Murai, J. Org. Chem. *2012, 77* (12), 5381-8 [DOI: 10.1021/jo3008385].

Bi[hetero]aryls from two different [hetero]arene molecules Ar-Ar′
Regioselective transition metal-catalyzed dehydrogenative ar. cross-coupling
update s. *72,* 496s*82*; **of electron-rich and -deficient arenes** under redox control with $[C_6F_5Au(PPh_3)]Cl_2$ s. X.C. Cambeiro, T.C. Boorman, P. Lu, I. Larrosa, Angew. Chem., Int. Ed. *2013, 52* (6), 1781-4 [DOI: 10.1002/anie.201209007]; **1,1′-bi-2-naphthols** by aerobic cross-coupling of electron-rich and -deficient 2-naphthols with (1,3-dimesitylimidazolidin-2-ylidene)-copper(I) bromide/$AgNO_3$ and diethyl malonate under O_2 (1 atm.), also cross-coupling of naphthylamines, s. M. Holtz-Mulholland, M. de Léséleuc, S.K. Collins, Chem. Commun. *2013, 49* (18), 1835-7 [DOI: 10.1039/c3cc38675a]; 2-arylation of 4-chromones with unactivated arenes using $Pd(OCOCF_3)_2$/AgOAc/CsOCOBu-*t*/*t*-BuCOOH s. M. Min, H. Choe, S. Hong, Asian J. Org. Chem. *2012, 1* (1), 47-50 [DOI: 10.1002/ajoc.201200036]; **directed *o*-[het]arylation of functionalized arenes** with a (pentamethylcyclopentadienyl)rhodium(III) complex and $Cu(OAc)_2$/hexabromobenzene as co-oxidants s. J. Wencel-Delord, C. Nimphius, H. Wang, F. Glorius, Angew. Chem., Int. Ed. *2012, 51* (52), 13001-5 [DOI: 10.1002/anie.201205734]; ***o*-azolylation** of arylcarboxylic acid N-(8-quinolyl)amides with $Cu(OAc)_2 \cdot H_2O$ s. M. Nishino, K. Hirano, T. Satoh, M. Miura, ibid. *2013, 52* (16), 4457-61 [DOI: 10.1002/anie.201300587]; ***o*-hetarylation** of aryl-N-heteroarenes under rhodium or ruthenium catalysis s. J. Dong, Z. Long, F. Song, N. Wu, Q. Guo, J. Lan, J. You, ibid. *52* (2), 580-4 [DOI: 10.1002/anie.201207196]; ***o*-chalcogenophenylation** of functionalized arenes and heterocyclic analogs with $[Cp^*RhCl_2]_2/Cu(OAc)_2/AgSbF_6$ s. V.P. Reddy, R. Qiu, T. Iwasaki, N. Kambe, Org. Lett. *2013, 15* (6), 1290-3 [DOI: 10.1021/ol400230y]; C_2- or C_3-hetarylation of N-functionalized pyrroles and indoles with $Pd(OAc)_2$/dppb/$Cu(OAc)_2$, also tandem 4-fold C-H activation, s. Z. Wang, F. Song, Y. Zhao, Y. Huang, L. Yang, D. Zhao, J. Lan, J. You, Chem. Eur. J. *2012, 18* (52), 16616-20 [DOI: 10.1002/chem.201203004]; oxy-functionalized **terphenyls** by sequential *metal-free* arylation of phenols with two different arenes via oxidation-rearomatization with phenyl iodosoacetate s. T. Dohi, T. Kamitanaka, S. Watanabe, Y. Hu, N. Washimi, Y. Kita, Chem. Eur. J. *2012, 18* (43), 13614-8 [DOI: 10.1002/chem.201202086].

Regioselective transition metal-catalyzed dehydrogenative annelation with acetylene derivs. ○
update s. *79,* 346s*82*; **3-tosylpyridines** or 4-tosylaminocyclopent-2-enones from N-allyltosylamines with $[Cp^*RhCl_2]_2$ and AgOAc or AgOCOBu-*t* (in water) s. D. Wang, F. Wang, G. Song, X. Li, Angew. Chem., Int. Ed. *2012, 51* (49), 12348-52 [DOI: 10.1002/anie.201206918]; **azolopyridine ring** from N-vinyl- or N-aryl-azoles with the same rhodium catalyst and $Cu(OAc)_2 \cdot H_2O$ s. J.-R. Huang, Q.-R. Zhang, C.-H. Qu, X.-H. Sun, L. Dong, Y.-C. Chen, Org. Lett. *2013, 15* (8), 1878-81 [DOI: 10.1021/ol400537b]; **azoloisoquinoline ring** from N-arylazoles with $[Cp^*RhCl_2]_2/Cu(OAc)_2 \cdot H_2O$ cf. J.-R. Huang, L. Dong, B. Han, C. Peng, Y.-C. Chen, Chem. Eur. J. *2012, 18* (29), 8896-900 [DOI: 10.1002/chem.201201207]; pyrazolo[3,2-*a*]isoquinolines from 5-arylpyrazoles under ruthenium catalysis with added $AgSbF_6$ as co-catalyst and

$Cu(OAc)_2{\cdot}H_2O$ under air s. W. Ma, K. Graczyk, L. Ackermann, Org. Lett. *2012, 14* (24), 6318-21 [DOI: 10.1021/ol303083n]; isocarbostyril-condensed isoquinoline ring from 2-arylisocarbostyrils with $[(p\text{-cymene})RuCl_2]_2/Cu(OAc)_2{\cdot}H_2O/Na_2CO_3$ s. B. Li, H. Feng, N. Wang, J. Ma, H. Song, S. Xu, B. Wang, Chem. Eur. J. *2012, 18* (40), 12873-9 [DOI: 10.1002/chem.201201862]; **isocarbostyrils** from arylcarboxylic acid amides with $Pd(OAc)_2/Cu(OAc)_2/NaI{\cdot}2H_2O/K_2CO_3$ under air s. N. Zhang, B. Li, H. Zhong, J. Huang, Org. Biomol. Chem. *2012, 10* (47), 9429-39 [DOI: 10.1039/c2ob26668g]; N-alkoxyisocarbostyrils from arylhydroxamic acid esters (without the copper salt) cf. S. Lu, Y. Lin, H. Zhong, K. Zhao, J. Huang, Tetrahedron Lett. *2013, 54* (15), 2001-5 [DOI: 10.1016/j.tetlet.2013.02.003]; 4-cyclopropyl-isocarbostyrils and -isocoumarins under ruthenium catalysis with added $Cu(OAc)_2{\cdot}H_2O$ s. M. Deponti, S.I. Kozhushkov, D.S. Yufit, L. Ackermann, Org. Biomol. Chem. *2013, 11* (1), 142-8 [DOI: 10.1039/c2ob26250a]; **N-acyl-pyrroles** from enacylamines under the same conditions, also with N-acylation, s. B. Li, N. Wang, Y. Liang, S. Xu, B. Wang, Org. Lett. *2013, 15* (1), 136-9 [DOI: 10.1021/ol303159h]; **N-subst. pyrroles** from enamines s. L. Wang, L. Ackermann, ibid. 176-9 [DOI: 10.1021/ol303224e]; **naphthalenes** from allylarenes with $Pd(OAc)_2/Cu(OAc)_2/CF_3COOH$ s. P. Gandeepan, C.-H. Cheng, ibid. *15* (9), 2084-7 [DOI: 10.1021/ol400792y]; **benzo[*b*]thiophenes** from aryl mercaptans under manganese catalysis with added benzoic acid under O_2 s. K. Liu, F. Jia, H. Xi, Y. Li, X. Zheng, Q. Guo, B. Shen, Z. Li, ibid. *15* (8), 2026-9 [DOI: 10.1021/ol400719d]; regioselective dehydrogenative annelation **with ethylene derivs.** (cf. *79*, 345) for preparing **3-arylisocoumarins** or 3-arylidene-phthalides from benzoic acids and styrenes with $Pd(OAc)_2$ and Ag_2O as oxidant s. D. Nandi, D. Ghosh, S.-J. Chen, B.-C. Kuo, N.M. Wang, H.M. Lee, J. Org. Chem. *2013, 78* (7), 3445-51 [DOI: 10.1021/jo400174w]; **isochromene ring** from α-arylketones under palladium or ruthenium catalysis with $Cu(OAc)_2$ as oxidant s. S.R. Chidipudi, M.D. Wieczysty, I. Khan, H.W. Lam, Org. Lett. *2013, 15* (3), 570-3 [DOI: 10.1021/ol3033835]; **phenanthridines** from *o*-aminobiaryls with $PdCl_2$ and a copper(II) salt s. Z. Liang, L. Ju, Y. Xie, L. Huang, Y. Zhang, Chem. Eur. J. *2012, 18* (49), 15816-21 [DOI: 10.1002/chem.201202672].

Chiral dimeric hydroxoiron(III) salan complexes/air *[Fe(III)]*/air*

1-α-Nitro-2(1*H*)-naphthalenones from 2-naphthols and aliphatic nitro compds. ←
Iron(III)-catalyzed dearomatizing asym. dehydrogenative α-substitution

345.

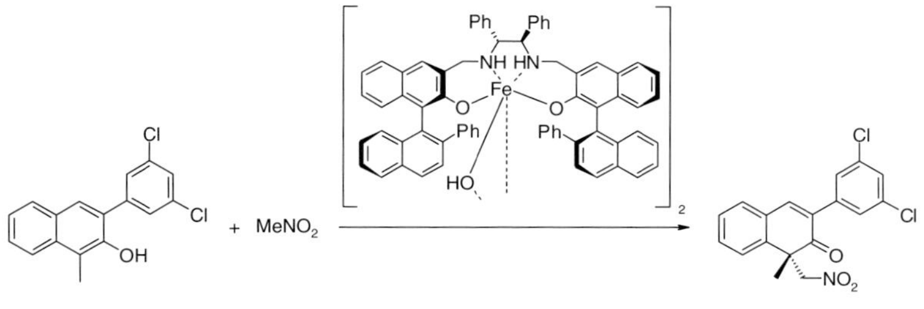

The first example of simultaneous oxidative dearomatization and formation of a chiral all-carbon quaternary center is reported, using an iron catalyst *under air*. **E:** Chiral iron(III) salan complex (4 mol%) added to a soln. of 3-(3,5-dichlorophenyl)-1-methyl-2-naphthol (0.5 mmol) in toluene/nitromethane (9:1; 5 ml) at 50° under air, the mixture stirred for 48 h, and purified directly by chromatography on silica → (S)-(+)-1-methyl-3-(3,5-dichlorophenyl)-1-(nitromethyl)naphthalen-2(1*H*)-one. Y 81% (e.e. 96%). The environmentally benign iron catalyst effected the atom-efficient oxidative dearomatization and alkylation of a series of 1-*prim*-alkyl- (Me, Et, hexyl) 3-subst. 2-naphthols with nitro(m)ethane in air with high enantioselectivity (e.e. 88-96%; twelve examples; Y 75-92%; a 1-cyclohexyl deriv. gave 43%; d.r. 9:1 to 10:1 for nitroethane). No reaction occurred under argon, or with 1-naphthol or phenol analogs (but 2,4,6-trimethylphenol was nitromethylated at the 4-methyl group in 15% yield). F.e., optimization and substrate prepn. s. T. Oguma, T. Katsuki, J. Am. Chem. Soc. *2012, 134* (49), 20017-20 [DOI: 10.1021/ja310203c].

Iron(II) chloride *$FeCl_2$*

Iron(II)-catalyzed 3-component synthesis of γ-keto-α-peroxy- from α,β-ethylene-carboxylic acid esters and aldehydes C=C → C(OOR)C-C(O)R′

346.

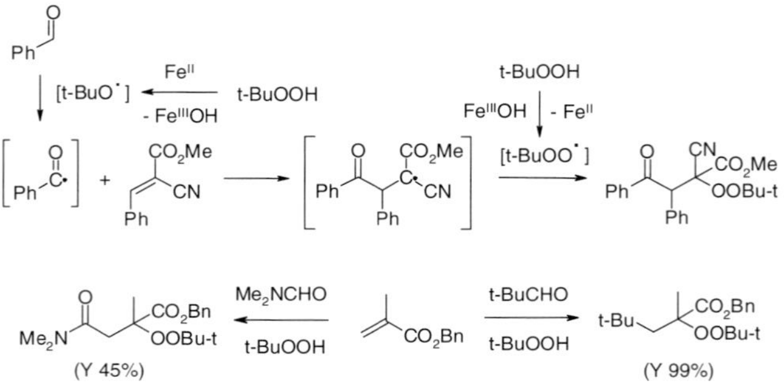

An extension to previous 3-component reaction of terminal alkenes, aldehydes and hydroperoxides (*80*, 363) is reported, using electron-deficient ethylene derivs. in place of styrenes. **E:** A soln. of *tert*-butyl hydroperoxide (4 eq.) in decane (5-6 *M*) added dropwise to a mixture of methyl α-cyano-cinnamate (0.5 mmol), benzaldehyde (5 eq.), $FeCl_2$ (2.5 mol%) and acetonitrile (3 ml) under N_2 at room temp., the mixture stirred at 85° for 1 h, cooled to room temp., filtered through silica, concentrated *in vacuo,* and purified by flash chromatography → methyl 2-(*tert*-butylperoxy)-2-cyano-4-oxo-3,4-diphenylbutanoate. Y 90% (d.r. 3:2). This general and practical method for synthesis of multi-functionalized peroxides was catalyzed by several metal salts (Co, Cu, Mn; optimally $FeCl_2$) and was successful with terminal/internal acrylate esters and electron-diverse (het)ar. and aliphatic aldehydes (also with formamides, used in excess in the absence of solvent, to afford **α-peroxysuccinic acid amide esters**) (forty-one examples; Y 45-94%), in the presence of carbonate, ester, nitrile, silyl ether, halo, amide, alcohol and unactivated alkene functionality. Lower yields were obtained with bulky 1-naphthaldehyde (13%), a tetrasubst. acrylate (18%) and dimethyl benzylidenemalonate with benzaldehyde (30%; aliphatic aldehydes gave 75-84%; two examples). Bulky pivalaldehyde reacted via decarbonylation (Y 99%), and simple methyl acrylate gave a dimeric by-product (Y 20%), formed from a 2:1:1 stoichiometry of acrylate/aldehyde/hydroperoxide, in addition to the expected product (Y 71%). Diastereoselectivity was moderate (1:1 to 2:1). F.e. and optimization s. K. Liu, Y. Li, X. Zheng, W. Liu, Z. Li, Tetrahedron *2012*, *68* (50), 10333-7 [DOI: 10.1016/j.tet.2012.09.110].

Iron(III) chloride/4-dimethylaminopyridine/di-tert-butyl peroxide *$FeCl_3$/DMAP/$(t\text{-}BuO)_2$*

2-Subst. tert. enamines from tert. amines via regioselective iron(III)-catalyzed oxidative coupling C(N<)=C(R)

347.

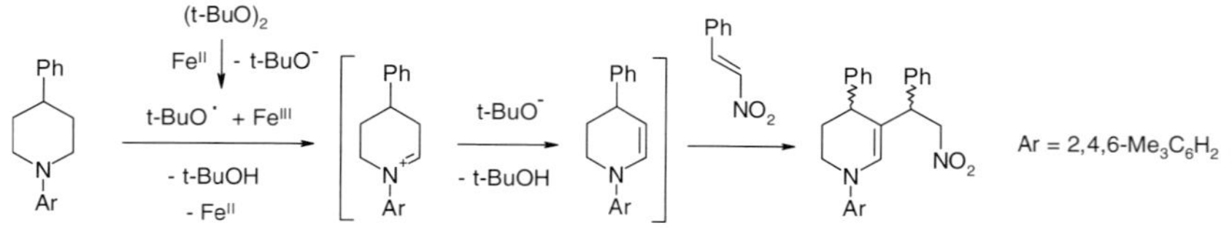

Tert. 4-nitroenamines. A mixture of $FeCl_3$ (5 mol%) and DMAP (7.5 mol%) in dry 1,2-dichloro-ethane (0.7 ml) stirred at room temp. for 20 min, a soln. of β-nitrostyrene (0.2 mmol) and N-mesityl-4-phenylpiperidine (4 eq.) in the same solvent (0.3 ml) added under argon, di-*tert*-butyl peroxide (2.5 eq.) added via syringe, the mixture stirred at 60° for 18 h, cooled to room temp., filtered through silica, concentrated *in vacuo*, and purified by flash chromatography → N-mesityl-3-(2-nitro-1-phenylethyl)-4-phenyl-2-piperideine. Y 75% (d.r. 4.1). Oxidation of cyclic amines, protected as their N-2,6-dimethyl-4-methoxyphenyl or N-mesityl derivs., generated enamines (via the corresponding iminium ions) which were trapped with electrophiles, β-(het)aryl-/alkyl-nitroethylenes (twenty-five examples; Y 44-91%; d.r. 1.1:1 to 4.1:1), with a single trisubst. nitroalkene, β-methyl-β-nitrostyrene giving 41% yield. The mildness of the conditions allowed the reaction to be extended to acyclic amines, with the unstable enamine products isolated following reduction ($NaBH_3CN$) to **1,4-nitramines** (three examples; Y 32-64%; d.r. 1.1:1 to 2:1).

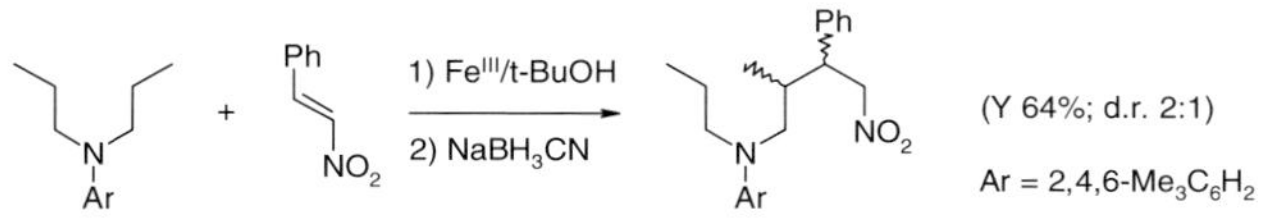

F.e., optimization and further reactions of the products s. N. Takasu, K. Oisaki, M. Kanai, Org. Lett. *2013*, *15* (8), 1918-21 [DOI: 10.1021/ol400568u].

Graphenesulfonic acid-supported ruthenium dioxide nanoparticles/oxygen ←
Dehydrogenative α-substitution of amines in water CHN< → C(N<)R
with nitroalkanes s. *72*, 491s*83*

Dodecacarbonyltriruthenium/tricyclohexylphosphine or 2,2′-bis(diphenylphosphino)biphenyl *$Ru_3(CO)_{12}/Cy_3P$ or BIPHEP*
Cyclohex-4-ene-1,2-diols from glycols and 1,3- or 1,4-dienes via ruthenium-catalyzed oxidative [4+2]-cycloaddition ○

348.

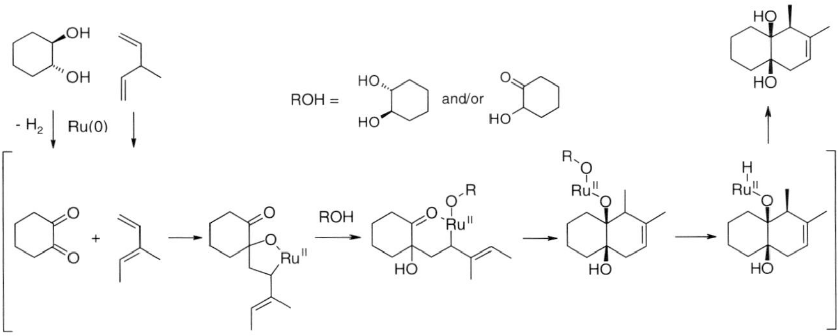

A novel strategy is reported for the formal oxidative [4+2]-cycloaddition of terminal and internal 1,2-diols with dienes, utilizing excess diene as hydrogen receptor. The method is also applicable to α-hydroxyketones under redox neutral, and internal 1,2-diones under reductive, conditions (using formic acid as hydrogen donor). **E:** Toluene (0.15 ml) and freshly distilled 3-methyl-1,4-pentadiene (5 eq.) added to a mixture of $Ru_3(CO)_{12}$ (2 mol%), BIPHEP (6 mol%) and *trans*-1,2-cyclohexanediol (0.3 mmol) under argon, the sealed mixture heated at 130° for 48 h, cooled to room temp., solvent removed *in vacuo*, and the residue purified by flash chromatography → (4aR*,5R*,8aR*)-5,6-dimethyl-1,2,3,4,4a,5,8,8a-octahydronaphthalene-4a,8a-diol. Y 91% (d.r. >20:1). This novel method was successful for reaction of mono/di-subst. dienes with terminal or internal 1,2-diols optimally using Cy_3P or the bidentate ligand, BIPHEP, respectively (seventeen examples; Y 60-93%; d.r. 1.4:1 to >20:1). The reaction was equally effective with *cis*- or *trans*-diols, and internal or non-conjugated dienes (2,4-hexadiene or 3-methyl-1,4-pentadiene) underwent prior isomerization to terminal conjugated dienes. Cyclization of α-hydroxyketones occurred under similar conditions (five examples; Y 61-98%), but use of 1,2-diones was restricted to internal diketones using HCOOH as hydrogen donor and $RuH_2(CO)(PPh_3)$ as catalyst (three examples; Y 35-70%). Experimental evidence suggests that substrates react initially as the derived 1,2-diones.

F.e. and optimization s. L.M. Geary, B.W. Glasspoole, M.M. Kim, M.J. Krische, J. Am. Chem. Soc. *2013*, *135* (10), 3796-9 [DOI: 10.1021/ja400691t].

Dichloro(p-cymene)ruthenium(II) dimer/copper(II) acetate *[Ru(II)]/Cu(OAc)$_2$*
1-α,1-α′-Diketo-1-spiroindenes
from cyclic α-aryl-β-dicarbonyl compds. and acetylene derivs.
Regioselective ruthenium-catalyzed oxidative annelation

349.

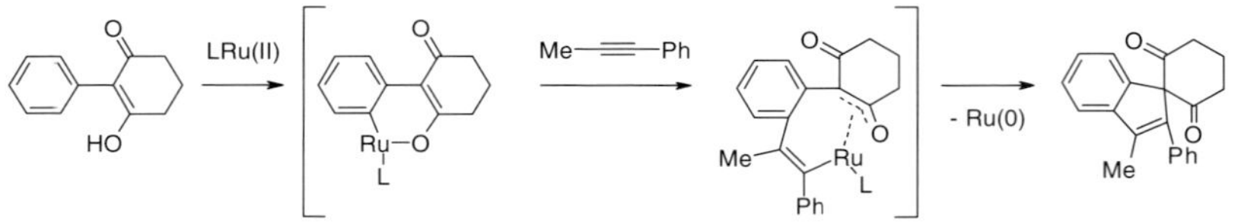

An *enolate-directed* oxidative annulation with alkynes has been achieved via an unusual ruthenium-catalyzed *dual activation of a C(sp^3)-H and a C(sp^2)-H bond* to give spirocyclic products bearing a *quaternary carbon center.* **E:** The startg. alkyne (1.5 eq.) in dioxane (0.5 ml) added to a mixture of 2-phenyl-1,3-cyclohexanedione (0.5 mmol), Cu(OAc)$_2$ (2.2 eq.) and [RuCl$_2$(*p*-cymene)]$_2$ (5 mol%) in dioxane (1 ml), stirred at 90° for 4 h, filtered through a pad of silica gel, the filtrate concentrated under reduced pressure, and the residue purified by column chromatography → 3′-methyl-2′-phenylspiro[cyclohexane-1,1′-indene]-2,6-dione. Y 80%. The procedure is generally applicable to di[hetero]aryl-, alkyl(aryl)- or dialkyl-acetylenes in moderate-to-high yield (twenty-two examples; Y 48-84%) and with high regioselectivity in the case of unsym. acetylenes; furthermore, the aryl residue on the triple bond may be electron-diverse, but there was no reaction with [hindered] *o*-subst. arylacetylenes nor with terminal alkynes. The cyclic β-dicarbonyl compd. may be a 1,3-cyclohexanedione, a Meldrum′s acid deriv. or a barbituric acid, but acyclic β-dicarbonyl compds. and 2-phenylcyclohexanone remained inactive. Reaction is thought to involve initial generation of an enolate which directs cyclometalation to form a 6-membered ruthenacycle into which the alkyne inserts regioselectively to give an oxa-π-allylruthenium complex prior to reductive elimination of ruthenium(0) to give the product and re-oxidation of the metal. F.e. and comparison of catalysts and solvents s. S.R. Chidipudi, I. Khan, H.W. Lam, Angew. Chem., Int. Ed. *2012*, *51* (48), 12115-9 [DOI: 10.1002/anie.201207170].

Dichloro(p-cymene)ruthenium(II) dimer/silver hexafluoroantimonate/copper(II) acetate/air ←
Ruthenium-catalyzed dehydrogenative C$_2$-vinylation of indoles H → C=C
with N-benzoyl as directing group

350.

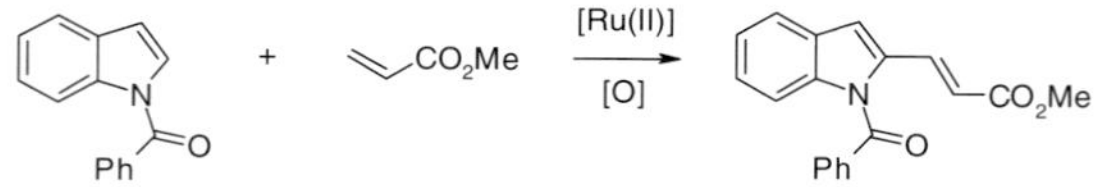

Challenging direct C$_2$-alkenylation of indoles has been achieved via ruthenium(II)-catalyzed oxidative coupling with acrylates, using N-benzoyl as directing group; notably, the reaction is highly regioselective in the presence of the more active indole C3- and C7-positions as well as the *ortho* position of the benzoyl moiety. **E:** Dichloroethane (2 ml) added to a mixture of N-benzoylindole (0.452 mmol), [Ru(*p*-cymene)Cl$_2$]$_2$ (5 mol%), AgSbF$_6$ (20 mol%) and Cu(OAc)$_2$ monohydrate (1 eq.), the mixture stirred for 1 min, methyl acrylate (1.2 eq.) added, stirring continued at 100° under air for 3 h (TLC), cooled to room temp., and worked up with purification by chromatography on silica gel → (E)-methyl 3-(1-benzoyl-1*H*-indol-2-yl)acrylate. Y 80% (97% by NMR). The reaction was successful for N-benzoylindoles, optionally substituted in the 5- or 6-positions with methoxy, chloro or bromo, in reaction with a variety of alkyl or aryl acrylates (thirteen examples; Y 64-95%). Diacrylates [containing a 1,4-benzene (Y 92%) or 1,2-ethane tether (Y 60%)] and *tert*-butyl acrylates were also suitable substrates, the latter suffering concomitant cleavage of the *tert*-butyl group to afford acrylic acid derivs. in moderate yield (50-62%; four examples). Attempted use of N-acetyl as directing group was less satisfactory (one example; Y 32%), and conducting the reaction at less than 100° (room temp. or 60°) gave rise to significant C7-alkenylation and recovered starting material. F.e., optimization, 1 g scale-up, and facile removal of the N-benzoyl groups using NaOH/aq. acetonitrile, s. V. Lanke, K.R. Prabhu, Org. Lett. *2013*, *15* (11), 2818-21 [DOI: 10.1021/ol4011486]; **palladium(II)-catalyzed C_2-vinylation of N-(2-pyridylsulfonyl)-indoles and -pyrroles** using $PdCl_2(MeCN)_2/Cu(OAc)_2$ (cf. *69*, 369), also cleavage of the N-protective group with Zn/NH_4Cl, s. A. García-Rubia, R.G. Arrayás, J.C. Carretero, Angew. Chem., Int. Ed. *2009*, *48* (35), 6511-5 [DOI: 10.1002/anie.200902802].

Dichlorotris(2,2′-bipyridyl)ruthenium(II)/oxygen — *$Ru(bpy)_3Cl_2/O_2$*

Oxidative α-arylation of α-aminocarboxylic acid esters with arenes — H → Ar

under visible-light photoredox catalysis s. *83*, 355

Palladium(II) acetate s.a. under Electrolysis — *$Pd(OAc)_2$*

Palladium(II) acetate/copper(II) acetate/air — *$Pd(OAc)_2/Cu(OAc)_2/O_2$*

Palladium(II) acetate/silver acetate — *$Pd(OAc)_2/AgOAc$*

Palladium(II) iodide/air — *PdI_2/O_2*

Ring closures via oxidative carbonylation — ○

oxidative cyclocarbonylation under palladium catalysis, update s. *69*, 370s*82*; phenanthridones from *o-sec*-aminobiaryls with $Pd(OAc)_2/AgOAc$ s. V. Rajeshkumar, T.-H. Lee, S.-C. Chuang, Org. Lett. *2013*, *15* (7), 1468-71 [DOI: 10.1021/ol4001922]; from N-unprotected *o*-aminobiaryls with $PdCl_2(MeCN)_2/Cu(OCOCF_3)_2/CF_3COOH$ cf. D. Liang, Z. Hu, J. Peng, J. Huang, Q. Zhu, Chem. Commun. *2013*, *49* (2), 173-5 [DOI: 10.1039/c2cc36817j]; 3-aryl-3,4-dihydroisocarbostyril-3-carboxylic acid esters from racemic α-arylalaninates under palladium(II) catalysis, also 3-benzylphthalimidine-3-carboxylic acid esters under stoichiometric conditions, s. J. Albert, X. Ariza, T. Calvet, M. Font-Bardia, J. Garcia, J. Granell, A. Lamela, B. López, M. Martinez, L. Ortega, A. Rodriguez, D. Santos, Organometallics *2013*, *32* (2), 649-59 [DOI: 10.1021/om301140t]; carbostyrils from *o*-aminostyrenes with $Pd(OAc)_2/Cu(OAc)_2$ under air s. J. Ferguson, F. Zeng, N. Alwis, H. Alper, Org. Lett. *2013*, *15* (8), 1998-2001 [DOI: 10.1021/ol4006739]; 3-(carbamylmethylene)phthalimidines from *o*-ethynylarylcarboxylic acid amides by **carbonylative ring closure** (cf. *44*, 683s*82*) with PdI_2 under air s. B. Gabriele, R. Mancuso, I. Ziccarelli, G. Salerno, Tetrahedron Lett. *2012*, *53* (49), 6694-6 [DOI: 10.1016/j.tetlet.2012.09.109].

Palladium(II) acetate/tert-butyl hydroperoxide — *$Pd(OAc)_2$/t-BuOOH*

Directed *o*-acylation with aldehydes — H → COR

update s. *26*, 775s*82*; **of diar. azo compds.** with $Pd(OAc)_2$/*tert*-butyl hydroperoxide, also subsequent conversion to 2-arylindazoles, s. H. Li, P. Li, L. Wang, Org. Lett. *2013*, *15* (3), 620-3 [DOI: 10.1021/ol303434n]; **of 2-aryl-benzothiazoles and -benzoxazoles** with ar. aldehydes under the same conditions s. A. Banerjee, S.K. Santra, S. Guin, S.K. Rout, B.K. Patel, Eur. J. Org. Chem. *2013* (7), 1367-9 [DOI: 10.1002/ejoc.201201503]; **of anilides** *in water* with $Pd(OAc)_2$/*tert*-butyl hydroperoxide and added trifluoroacetic acid and sodium dodecyl sulfate under air s. F. Szabó, J. Daru, D. Simkó, T.Z. Nagy, A. Stirling, Z. Novák, Adv. Synth. Catal. *2013*, *355* (4), 685-91 [DOI: 10.1002/adsc.201200948]; of N-benzyltriflamides in acetonitrile/DMF with added acetic acid s. S. Sharma, J. Park, E. Park, A. Kim, M. Kim, J.H. Kwak, Y.H. Jung, I.S. Kim, ibid. *355* (2-3), 332-6 [DOI: 10.1002/adsc.201200889]; *metal-free* *o*-acylation of N-heteroarenes with phenyl iodosotrifluoroacetate/trimethylsilyl azide s. K. Matcha, A.P. Antonchick, Angew. Chem., Int. Ed. *2013*, *52* (7), 2082-6 [DOI: 10.1002/anie.201208851].

o-Acylaminobenzophenones from ar. acylamines and methylarenes via regioselective palladium(II)-catalyzed dehydrogenative o-aroylation H → COAr

351.
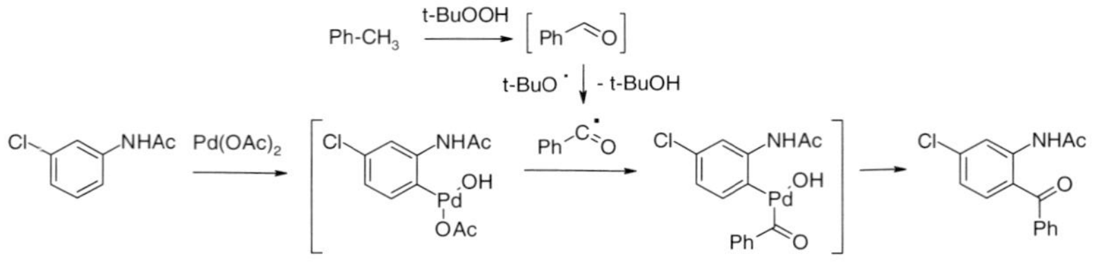

A mixture of 3-chloroacetanilide (0.5 mmol), toluene (2 eq.), $Pd(OAc)_2$ (5 mol%), 70% aq. *t*-BuOOH (4 eq.) and DMSO (1 ml) stirred at 100° in a sealed tube for 20 h, cooled to room temp., solvent removed *in vacuo*, and the residue purified by flash chromatography → N-(2-benzoyl-5-chlorophenyl)acetamide. Y 75% (82% conversion). A series of electron-diverse N-acylanilines were *o*-acylated with electron-diverse toluene derivs. (incl. 1- and 2-methylnaphthalene) via palladium(II)-catalyzed direct activation of sp^2 C-H, although none of the reactions went to completion (twenty-six examples; Y 43-85%; conversion 53-90%). Electron-poor 4-nitroacetanilide gave only a trace of product as did sterically-hindered 2-chloroacetanilide. F.e. and optimization s. Z. Yin, P. Sun, J. Org. Chem. *2012*, *77* (24), 11339-44 [DOI: 10.1021/jo302125h]; with 20 mol% $Pd(OAc)_2$ at 95° cf. J. Weng, Z. Yu, X. Liu, G. Zhang, Tetrahedron Lett. *2013*, *54* (10), 1205-7 [DOI: 10.1016/j.tetlet.2012.12.059]; with 10 mol% $Pd(OAc)_2$ at 80° cf. Y. Wu, P.Y. Choy, F. Mao, F.Y. Kwong, Chem. Commun. *2013*, *49* (7), 689-91 [DOI: 10.1039/c2cc37352a].

Palladium(II) acetate/tricyclohexylphosphine/2,2,6,6-tetramethylpiperidine nitroxyl/ silver carbonate/lithium acetate ←

Chalcones from propiophenones and arenes $C(O)CH_2CH_3 \rightarrow C(O)CH{=}CHAr$

352.
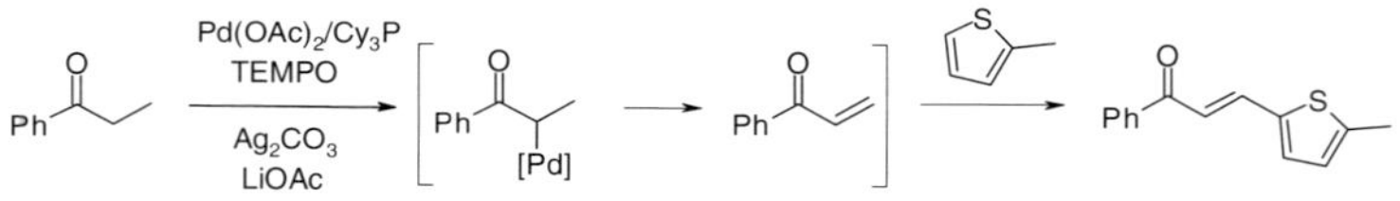

A versatile, facile and atom economical route to chalcones, or heterocyclic analogs, by **palladium-catalyzed dehydrogenative cross-coupling** of (hetero)aryl ethyl ketones and (hetero)arenes is reported. **E:** A Schlenk tube charged with $Pd(OAc)_2$ (0.1 eq.), Cy_3P (0.2 eq.), LiOAc (1.5 eq.), TEMPO (0.4 eq.) and Ag_2CO_3 (3 eq.) in a glovebox, the tube fitted with a rubber septum then removed from the glovebox, DME (2 ml), propiophenone (3 eq.) and 2-methylthiophene (0.3 mmol) added in turn via syringe, the septum replaced with a Teflon screwcap under N_2, the mixture stirred at 100° for 24 h, cooled, diluted with ether, filtered through a pad of silica gel, washing with ether, concentrated under reduced pressure, and the residue purified by flash chromatography on silica gel → (E)-3-(5-methyl-2-thienyl)-1-phenyl-2-propen-1-one. Y 85%. The use of [strongly electron-donating] Cy_3P as ligand was crucial to suppress homocoupling of thiophene, probably by reducing the electrophilicity of palladium and thus inhibiting further electrophilic attack of palladium to a second thiophene molecule. Twenty-one further examples of the reaction of propiophenone with subst. thiophenes, furans, benzothiophenes or indoles gave yields of 44-90% (tolerating groups such as formyl, acetyl, ester/acetoxy, ketal or Cl), while thiophene itself gave only 24%. Eighteen examples with 2-methylthiophene and a variety of propiophenones (bearing Me, F, Cl, CF_3, CN, aryl, NO_2, Me_2N, OMe or CO_2Me on the aromatic ring) gave yields of 44%, 61-88%, while 3-propionoylpyridine gave 53%. Three examples with thiophenes and dialkyl ketones gave yields of 44-69% with reaction exclusively at the less sterically hindered site. Eight examples with polyfluorobenzenes or 2,3,5,6-tetrafluoropyridine gave yields of 45-85% using tetramethylene sulfoxide as ligand in place of Cy_3P and KF as base; benzene itself gave 26% yield. Mechanistic investigations indicate that reaction proceeds via *in situ* dehydrogenation to the enone followed by C-H olefination of the (hetero)arene. F.e.s. Y. Shang, X. Jie, J. Zhou, P. Hu, S. Huang, W. Su, Angew. Chem., Int. Ed. *2013*, *52* (4), 1299-303 [DOI: 10.1002/anie.201208627].

Tris(dibenzylideneacetone)dipalladium/4,7-diphenyl-1,10-phenanthroline/silver acetate/ copper(II) acetate ←

Benzofurans from phenols and acetylene derivs. ○
by palladium-catalyzed oxidative annelation

353. 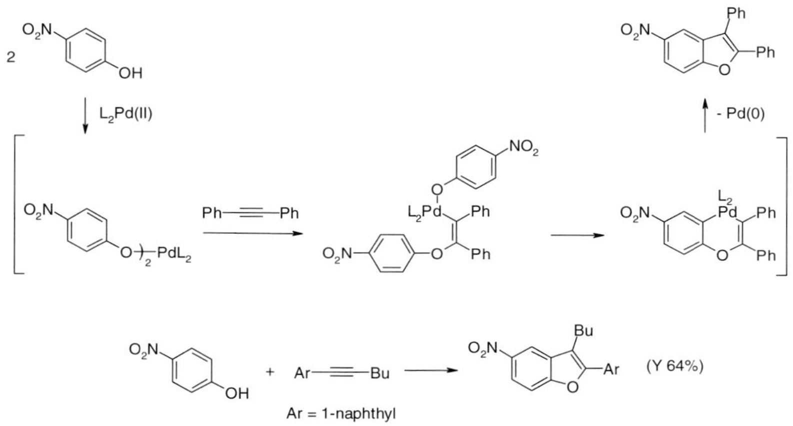

2,3-Diarylbenzofurans. A mixture of *p*-nitrophenol (2 eq.), tolan (1 mmol), $Pd_2(dba)_3$ (5 mol%), bathophenanthroline (10 mol%), AgOAc (2 eq.), $Cu(OAc)_2 \cdot H_2O$ (2 eq.) and 1,4-dioxane (8 ml) stirred in a dry Schlenk tube at 130° for 24 h, allowed to cool to room temp., diluted with methylene chloride, filtered through a small pad of Celite, solvent evaporated under reduced pressure, and the residue worked up with purification by chromatography on silica gel → 5-nitro-2,3-diphenyl-benzofuran. Y 92%. **Regioselective oxidative annelation** took place with unsym. internal alkynes; for example, *p*-nitrophenol with 1-(1-naphthyl)hex-1-yne gave 3-butyl-2-(1-naphthyl)-5-nitro-benzofuran exclusively after 48 h (Y 64%). Although the yield with phenol itself was low (22%), substrates with electron-withdrawing and -donating groups on the ring participated in the coupling efficiently, notably with retention of cyano, ketone, aldehyde, ester, fluorine and chlorine groups (reaction with *m*-subst. phenols being highly regioselective through formation of the C-C bond at the less hindered site). Bis(heteroaryl)acetylenes also took part in the conversion, but dialkyl-acetylenes, lacking conjugation, were less reactive, while terminal acetylene derivs., 4-aminophenol and hydroquinone were unreactive (twenty-eight examples in all; Y 34-92%). Of two possible mechanisms which the authors proposed, the more likely is centered on alkyne insertion into an intermediate (bathophenanthroline)palladium(II) bis(aroxide), followed by base-mediated intra-molecular *o*-palladation prior to reductive elimination of palladium(0) on ring closure to the product; palladium(0) is then re-oxidized to palladium(II) with copper(II) to complete the cycle. F.e. and comparison of catalysts, bases, solvents and ligands s. M.R. Kuram, M. Bhanuchandra, A.K. Sahoo, Angew. Chem., Int. Ed. *2013*, *52* (17), 4607-12 [DOI: 10.1002/anie.201210217].

Palladium(II) chloride/potassium carbonate/acetic acid/oxygen $PdCl_2/K_2CO_3/AcOH/O_2$
1,2,4,5-Tetrasubst. benzenes from acetylene derivs. and two ethylene deriv. molecules
by palladium-catalyzed cross-trimerization s. *83*, 101

Palladium(II) chloride/lithium chloride/potassium carbonate/oxygen $PdCl_2/LiCl/K_2CO_3/O_2$
1,3,5-Trienes from acetylene derivs. and two ethylene deriv. molecules ←
by palladium-catalyzed cross-trimerization s. *83*, 101

Bis(acetonitrile)dichloropalladium(II)/copper(II) acetate $PdCl_2(MeCN)_2/Cu(OAc)_2$
Palladium(II)-catalyzed C_2-vinylation of indoles or -pyrroles H → C=C
with N-(2-pyridylsulfonyl) as directing group s. *83*, 350

Bis(2-phenylpyridinato-C2,N)(2,2′-bipyridyl)iridium(III) hexafluorophosphate/ oxygen/irradiation [Ir(III)]/O_2/\\\\

Oxidative Ugi-type multicomponent reactions under visible-light photoredox catalysis ←

354.

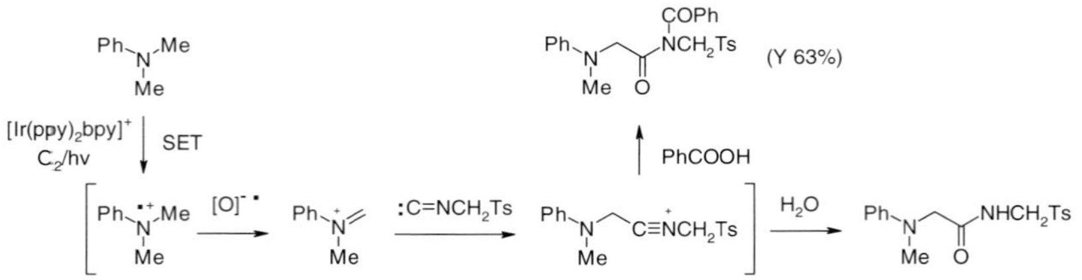

N-Arylglycinamides. A mixture of bis(2-phenylpyridinato-C2,N)(2,2′-bipyridyl)iridium(III) hexafluorophosphate (1 mol%), startg. isocyanide (1.5 eq.), N,N-dimethylaniline (0.15 mmol) and water (10 eq.) in acetonitrile (1 ml) stirred in a vial for 2 d under irradiation by a 1 W blue LED 30 cm strip (at a distance of ca. 5 cm), and worked up with purification by chromatography on silica gel → product. Y 75%. The procedure is applicable to the coupling of isonitriles with electron-diverse N,N-dimethylanilines substituted in the *m*- or *p*-position (twelve examples; Y 42-75%), as well as with N-ethyl-N-methylaniline, which reacted regioselectively at the methyl group (one example; Y 48%). Coupling was also effected with variously substituted aromatic, heteroaromatic and aliphatic carboxylic acids in place of water to give the corresponding **N-arylglycine N-acylamides** (fourteen examples; Y 32-63%). The catalytic cycle is presumed to involve two photoredoxidative SETs with generation of an intermediate N,N-dimethylaniline radical cation and oxygen radical anion which couple to generate a methyleneiminium ion; this then undergoes nucleophilic attack by the isonitrile to give a nitrilium ion prior to quenching with water or the carboxylic acid to give the product. F.e. and comparison of sensitizers and solvents s. M. Rueping, C. Vila, Org. Lett. *2013*, *15* (9), 2092-5 [DOI: 10.1021/ol400317v].

Bis(2-phenylpyridinato-C2,N)(2,2′-bipyridyl)iridium(III) hexafluorophosphate/ oxygen/zinc acetate/irradiation ←

Dehydrogenative α-functionalization of α-*sec*-aminocarboxylic acid esters combining visible-light photoredox catalysis and Lewis acid catalysis H → R

355.

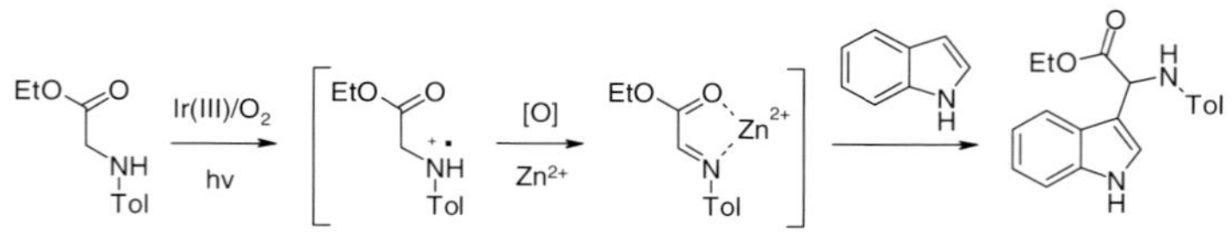

N-Subst. α-(3-indolyl)glycinates. A vial containing a mixture of [Ir(ppy)$_2$bpy]PF$_6$ (1 mol%), the startg. glycine ester (0.2 mmol), Zn(OAc)$_2$ (10 mol%) and indole (1.5 eq.) in acetonitrile (2 ml) stirred for 48 h under irradiation with an 11 W fluorescent bulb (Osram Dulux S), and worked up with purification by column chromatography → product. Y 72%. This is the first application of visible-light photoredox catalysis to glycine derivs., reaction taking place with initial generation of an amine radical cation under single electron transfer to photoactivated iridium(III), followed by oxidation to a Lewis acid-activated α-imino-ester prior to addition of the nucleophile (in this case, indole). The procedure is simple, efficient, mild, and applicable to the coupling of electron-diverse N-arylglycinates with variously substituted indoles, although N-acylindoles failed to react (twenty-six examples; Y 42-86%). Reaction was also successful with N-arylglycylglycinates without detectable peptide bond cleavage (ten examples; Y 50-76%), and electron-rich 1,3,5-trimethoxybenzene served as nucleophile in place of indoles (one example; Y 71%). Reaction failed, however, with α-*sec*-aminoketones. F.e. and comparison of Lewis acids, solvents and light sources s. S. Zhu, M. Rueping, Chem. Commun. *2012*, *48* (98), 11960-2 [DOI: 10.1039/c2cc36995h];

α-(indol-3-yl)ation of α-aminocarbonyl compds. under visible-light photoredox catalysis with dichlorotris(2,2′-bipyridyl)ruthenium(II) as sensitizer under O_2 s. Z.-Q. Wang, M. Hu, X.-C. Huang, L.-B. Gong, Y.-X. Xie, J.-H. Li, J. Org. Chem. *2012, 77* (19), 8705-11 [DOI: 10.1021/jo301691h].

Oxygen ↑ CC ↓↑ O

Without additional reagents *w.a.r.*

Knoevenagel condensation CO → C=C

update s. *46,* 713s*82*; mild *uncatalyzed* synthesis of isatinylidenerhodanines in DMSO s. F. Xue, A.D. MacKerell Jr., G. Heinzl, K. Hom, Tetrahedron Lett. *2013, 54* (13), 1700-3 [DOI: 10.1016/j.tetlet.2013.01.069]; of 1,1-bis(triflyl)alkenes via *self-promoting* condensation of Tf_2CH_2 with aldehydes s. H. Yanai, S. Egawa, T. Taguchi, ibid. *54* (17), 2160-3 [DOI: 10.1016/j.tetlet.2013.02.039]; mild, green and general (E)-selective conversion *in water* or without solvent with triethylenediamine as catalyst s. Y.-Q. Yu, Z.-L. Wang, J. Chin. Chem. Soc. *2013, 60* (3), 288-92 [DOI: 10.1002/jccs.201200391]; mild, rapid, green, solvent-free procedure with recyclable Na-benzoate, also under ultrasonication (and in ethanol), s. Q. Liu, H.-M. Ai, Synth. Commun. *2012, 42* (20), 3004-10 [DOI: 10.1080/00397911.2011.574245]; (E)-selective alkylidenation of sulfonylacetic acid esters *under Lewis acid catalysis* with $Ti(OPr\text{-}i)_4$ s. J. Yang, W. Yin, R. Liu, C. Chu, J. Chin. Chem. Soc. *2013, 60* (4), 412-7 [DOI: 10.1002/jccs.201200466]; under *heterogeneous* conditions with a highly cross-linked polycarbosilane-supported siloxytitanium(IV) triisopropoxide as catalyst s. K. Mangala, K. Sreekumar, Appl. Organomet. Chem. *2013, 27* (2), 73-8 [DOI: 10.1002/aoc.2941]; with a recyclable N-(2-pyridinioethyl)-*o*-hydroxyaldimine phosphotungstate as bifunctional (acid/base) ionic solid catalyst under solvent or solvent-free conditions s. M. Zhang, P. Zhao, Y. Leng, G. Chen, J. Wang, J. Huang, Chem. Eur. J. *2012, 18* (40), 12773-82 [DOI: 10.1002/chem.201201338]; with a reusable proline-functionalized polyacrylonitrile fiber s. G.-W. Li, J. Xiao, W.-Q. Zhang, Chin. Chem. Lett. *2013, 24* (1), 52-4 [DOI: 10.1016/j.cclet.2012.12.007]; in aq. media within the confined nanospace of a water-soluble self-assembled Pd_6 open cage with triimidazole walls (prepared from a palladium(II) acceptor and a triimidazole donor), also Diels-Alder reaction, s. D. Samanta, S. Mukherjee, Y.P. Patil, P.S. Mukherjee, Chem. Eur. J. *2012, 18* (39), 12322-9 [DOI: 10.1002/chem.201201679]; **1-nitroethylene derivs.** from aldehydes and nitroalkanes with ethylenediamine as catalyst (cf. *19,* 844s*82*) s. J. Yang, J. Dong, X. Lü, Q. Zhang, W. Ding, X. Shi, Chin. J. Org. Chem. *2012, 30* (12), 2827-33 [DOI: 10.1002/cjoc.201201094].

4-Component synthesis of α-(*o*-halogenoaroylamino)carboxylic acid amides ←

en route to phthalimidines s. *83*, 499

Irradiation s. under NH_4OH (or $HCOONH_4$) and Au

Microwaves (s.a. under Et_3N, Copper(II) 2-ethylhexanoate, CuCl and I_2) [\\\\\]

Hetero-Diels-Alder reaction with *in situ*-generated azomethines ◌

update s. *52,* 363s*82*; *uncatalyzed*, solvent-free synthesis of 5,6-dihydropyrido[4,3-*d*]pyrimidine-2,4-diones under microwave enhancement s. M.M. Sarmah, R. Sarma, D. Prajapati, W. Hu, Tetrahedron Lett. *2013, 54* (3), 267-71 [DOI: 10.1016/j.tetlet.2012.11.018]; synthesis of 2,3-diaryl-2-azabicyclo[2.2.2]octan-5-ones from 2-cyclohexenones *in water* with a layered α-zirconium hydrogen phosphate as catalyst and sodium calix[4]arenesulfonates as surfactant s. L. Huang, J.-P. Chen, C. Jin, W.-K. Su, Chin. Chem. Lett. *2013, 24* (4), 347-50 [DOI: 10.1016/j.cclet.2013.01.045]; 1,2,3,4-tetrahydro-3-spiroquinolines by uncatalyzed and solvent-free **Povarov reaction** under microwave enhancement s. D. Bhuyan, R. Sarma, D. Prajapati, Tetrahedron Lett. *2012, 53* (47), 6460-3 [DOI: 10.1016/j.tetlet.2012.09.081]; *trans*-2-aryl-4-arylamino-1,2,3,4-tetrahydroquinolines via Povarov reaction with ethyl vinyl ether and subsequent coupling with the aniline component (in excess) using CAN as catalyst s. P. Ribelles, V. Sridharan, M. Villacampa, M.T. Ramos, J.C. Menéndez, Org. Biomol. Chem. *2013, 11* (4), 569-79 [DOI: 10.1039/c2ob26754c]; **organocatalyzed asym. Povarov reaction** with isatin as the ketone component using 3,3′-bis(2,4,6-triisopropylphenyl)-1,1′-binaphthyl-2,2′-diyl hydrogen phosphate as catalyst (cf. *75,* 403) s. F. Shi, G.-J. Xing, R.-Y. Zhu, W. Tan, S. Tu, Org. Lett. *2013, 15* (1), 128-31 [DOI: 10.1021/ol303154k]; with *o*-hydroxystyrene as ene component using chiral 3,3′-bis(triphenylsilyl)-1,1′-binaphthyl-2,2′-diyl hydrogen phosphate as catalyst s. F. Shi, G.-J. Xing, Z.-L. Tao, S.-W. Luo, S.-J. Tu, L.-Z. Gong, J. Org. Chem. *2012, 77* (16), 6970-9 [DOI: 10.1021/jo301174g].

Uncatalyzed Biginelli synthesis under microwave irradiation s. *55,* 337s*83* ○

Organolithium compds./copper(I) cyanide *RLi/CuCN*
Stereoselective 3-component synthesis of 3-siloxyenesilanes RCH(OSi≤)C=C(Si≤)R′
from 3,3-disilylacroleins via copper(I)-mediated 1,4-C→O-silyl group migration – 6-Siloxy-4-silyl-1,2,4-trienes s. *83*, 402

Potassium carbonate *K_2CO_3*
4-Functionalized (Z)-3-alkylidenecyclopent-1-enecarboxylic acid esters from γ,δ-acetylene-β-acoxy-α-methylenecarboxylic acid esters and active methylene groups ○

356. 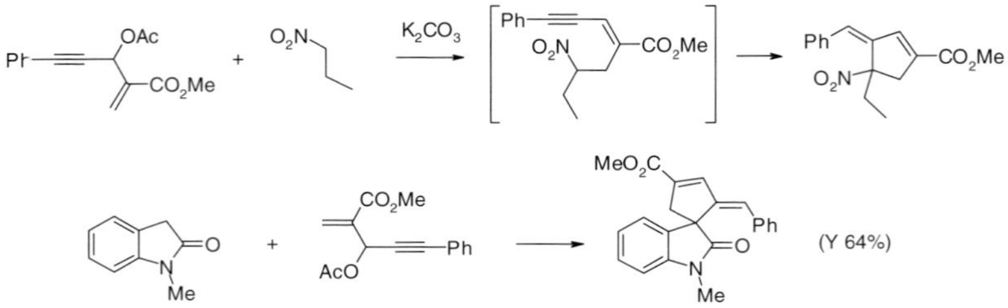

A novel method for synthesis of cyclopentene derivs. via formal [4+1]-annelation of Morita-Baylis-Hillman acetates with active methylene groups **under mild, metal-free conditions** is reported. **E:** K_2CO_3 (2.5 eq.) added to a soln. of methyl 3-acetoxy-2-methylene-5-phenylpent-4-ynoate (0.48 mmol) and nitropropane (1.1 eq.) in DMF (3 ml) at room temp., the mixture stirred for 8 h, diluted with water, extracted with ethyl acetate, and purified by chromatography on silica → (Z)-methyl 3-benzylidene-4-ethyl-4-nitrocyclopent-1-enecarboxylate. Y 73%. Formation of the title products was promoted by a number of bases (DABCO, Cs_2CO_3, optimally K_2CO_3) but Et_3N gave mainly the intermediate allylic substitution product (Y 62%; one example). The reaction was successful for common active methylene compds. (incl. nitropropane) reacting with electron-diverse (het)aryl-terminated ynoates, with electron-rich substrates sometimes reacting sluggishly (sixteen examples; Y 46-85%). N-Methyloxindole was also a suitable substrate, affording the spiro-oxindole in 64% yield, but an alkyl terminated ynoate gave only the dialkylated intermediate (Y 72%). The (Z)-selectivity suggests that the 5-*exo*-dig carbocyclization proceeds throughan *anti*-addition of nucleophile on to the alkyne. F.e. and optimization s. C.R. Reddy, P. Kumaraswamy, M.D. Reddy, Org. Biomol. Chem. *2012*, *10* (45), 9052-7 [DOI: 10.1039/c2ob26934a].

Sodium benzoate *PhCOONa*
Knoevenagel condensation s. *46,* 713s*83* CO → C=C

Potassium phosphate *K_3PO_4*
3-Component Mannich reaction s. *68,* 361s*83* CHO → CH(NHR)C-CO

Potassium iodide s. under PdI_2 *KI*

Ammonium hydroxide (or formate)/irradiation *NH_4OH (or $HCOONH_4$)/⫻*
Hantzsch synthesis of 1,4-dihydropyridines ○
N-unsubst. sym. 1,4-dihydropyridines s. *47,* 727s*82*; in aq. ethanol *under visible-light irradiation* with NH_4OH as ammonia source s. S. Ghosh, F. Saikh, J. Das, A.K. Pramanik, Tetrahedron Lett. *2013, 54* (1), 58-62 [DOI: 10.1016/j.tetlet.2012.10.079]; in 50% aq. ethyl lactate under visible-light irradiation with ammonium formate as ammonia source, also 4-component Hantzsch synthesis (cf. *68,* 368s*83*), s. P.P. Ghosh, S. Paul, A.R. Das, ibid. *54* (2), 138-42 [DOI: 10.1016/j.tetlet.2012.10.106].

Triethylamine/microwaves *Et_3N/[\\\\]*
3-Component [3+2]-cycloaddition with azomethinium ylids
s. *40,* 501; diastereoselective synthesis of pyrrolidine-2,5-dicarboxylic acid esters from electron-deficient ethylene derivs. and ylids derived from ethyl glyoxylate and α-aminocarboxylic acid

esters with Et_3N under microwave enhancement, also pyrroline analogs from acetylene derivs., s. J. Mancebo-Aracil, C. Nájera, J.M. Sansano, Org. Biomol. Chem. *2013, 11* (4), 662-75 [DOI: 10.1039/c2ob27072b]; of pyrrolizidines by **double [3+2]-cycloaddition** under the same conditions s. Q. Lu, G. Song, J.P. Jasinski, A.C. Keeley, W. Zhang, Green Chem. *2012, 14* (11), 3010-2 [DOI: 10.1039/c2gc36066g]; diastereoselective **3-component asym. [3+2]-cycloaddition** with aryl ethynyl ketones as dipolarophile using chiral 3,3′-di(9-anthracenyl)-1,1′-binaphthyl-2,2′-diyl hydrogen phosphate as catalyst s. F. Shi, G.-J. Xing, W. Tan, R.-Y. Zhu, S. Tu, Org. Biomol. Chem. *2013, 11* (9), 1482-9 [DOI: 10.1039/c2ob26566d].

Triethylenediamine *DABCO*
Knoevenagel condensation s. *46,* 713s*83* CO → C=C

3,5-Dicarbethoxy-2,6-dimethyl-1,4-dihydropyridine s. under (Acetonitrile)[2-(di-tert-butylphosphino)biphenyl]gold(I) hexafluoroantimonate ←

Cinchona-based prim. amines s. under Chloro(1,5-cyclooctadiene)iridium(I) dimer R^*NH_2

[Hydro]quinidine ←
Regioselective organocatalyzed asym. α-allylation s. *82,* 366s*83* H → C-C=C

β-Isocupreidine/trifluoroacetic acid ←
Organocatalyzed asym. synthesis of γ,γ-dicarbalkoxy-α-methylene-γ-lactones from β-carbalkoxyoxy-α-methylenecarboxylic acid esters and α-carbalkoxyoxymalonic acid esters ○

357.

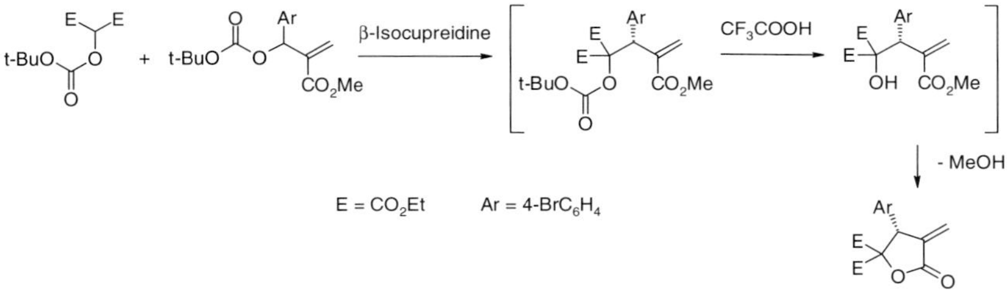

in one pot. A mixture of methyl 3-(4-bromophenyl)-3-*tert*-butoxycarbonyloxy-2-methylenepropanoate (2 eq.), diethyl 2-*tert*-butoxycarbonyloxymalonate (0.1 mmol) and β-isocupreidine (20 mol%) in toluene (1 ml) stirred at room temp. until reaction complete (NMR; 1-5 d), diluted with methylene chloride, trifluoroacetic acid (0.1 ml) added in one portion, the mixture stirred overnight, diluted with water, neutralized with Na_2CO_3, extracted with ethyl acetate, and purified by flash chromatography → diethyl 3-(4-bromophenyl)-4-methylene-5-oxodihydrofuran-2,2(3*H*)-dicarboxylate. Y 93% (e.e. 91%). This novel and mild synthesis generated a homoallylic alcohol deriv. via allylation of Boc-protected 2-hydroxymalonates with Morita-Baylis-Hillman carbonates, which underwent 5-*exo-trig* cyclization on removal of the protecting group *in situ* (alternative intramolecular Michael reaction of the intermediate via 5-*endo-trig* cyclization was disfavored). A series of electron-diverse 3-aryl-derivs. was prepared by this method (seven examples; Y 67-93%; e.e. 85-93%; a 3-naphth-1-yl deriv. gave e.e. 68%). The use of $(DHQD)_2PHAL$ in place of the isocupreidine catalyst gave enantiomeric products (six examples; Y 72-94%; e.e. 81-94%; a 3-naphth-1-yl deriv. gave e.e. 72%). F.e., optimization, substrate prepn. and testing of products against a human leukemia cell line, with all compds. tested exhibiting a consistent cytotoxic activity with IC_{50} in the low micromolar range, s. X. Companyó, A. Mazzanti, A. Moyano, A. Janecka, R. Rios, Chem. Commun. *2013*, *49* (12), 1184-6 [DOI: 10.1039/c2cc38659c].

Copper/silica Cu/SiO_2
Heterogeneous catalytic Biginelli synthesis s. *55,* 337s*83*

Copper(II) 2-ethylhexanoate/microwaves *$Cu(OCOR)_2$/⁄⁄\\\\]*
Copper(I) bis[2-(phenylthio)benzylidene]-1,2-ethylenediamine complexes-on-mesoporous aluminosilicate ←
3-Component synthesis of 2-acetyleneamines s. *66*, 353s*83* CO → C(N<)C≡C
Silver acetate s. under $[Cp^*RhCl_2]_2$ *AgOAc*
Copper(I) cyanide s. under Organolithium compds. *CuCN*
Copper(I) chloride/microwaves *CuCl/⁄⁄\\\\]*
Copper(I)-catalyzed *o*-(α-*tert*-aminoalkyl)ation of azoles with oxo compds. and sec. amines H → C-N<

358.

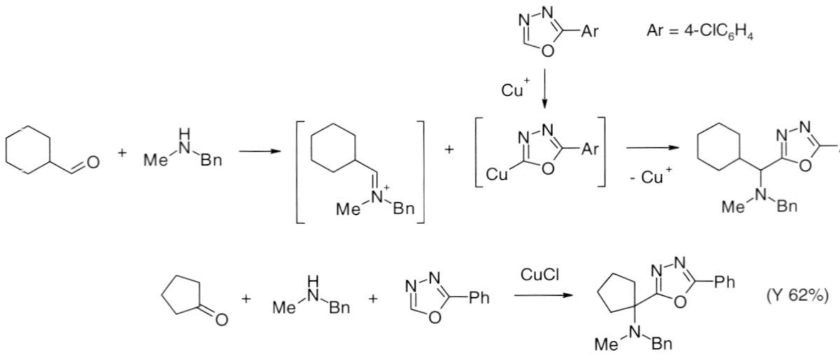

A mixture of N-methylbenzylamine (0.5 mmol), cyclohexanal (1.5 eq.) and toluene (0.2 ml) stirred at room temp. for 5 min, 2-(4-chlorophenyl)-1,3,4-oxadiazole (1.2 eq.) and CuCl (20 mol%) added, the sealed mixture heated by microwaves (200 W) at 140° for 50 min, diluted with ethyl acetate, washed with water, and worked up with purification by chromatography on silica → N-benzyl-N-methyl-5-[(α-cyclohexyl)aminomethyl]-2-(4-chlorophenyl)-1,3,4-oxadiazole. Y 72%. This novel, direct copper(I)-catalyzed *o*-(α-aminoalkyl)ation was successful for electron-diverse 2-aryl-1,3,4-oxadiazoles reacting with (a)cyclic aldehydes (incl. benzaldehyde) or cyclic ketones and sec. (a)cyclic amines (incl. N-methylaniline) affording the corresponding 5-aminoalkylated derivs. in variable yields (30-85%; seventeen examples). The reaction failed with benzylamine or acetophenone. A single 1,3,4-thiadiazole example gave a 39% yield (the analogous oxadiazole gave 30%). F.e. and optimization s. D.D. Vachhani, A. Sharma, E. Van der Eycken, Angew. Chem., Int. Ed. *2013*, *52* (9), 2547-50 [DOI: 10.1002/anie.201209312].

Copper(I) chloride/1,10-phenanthroline/lithium tert-butoxide *CuCl/phen/LiOBu-t*
Regioselective synthesis of 1(E),4-enynes from terminal acetylene derivs. and 2(Z)-ethylenephosphoric acid esters C≡CH → C≡C-C-C=C

359.

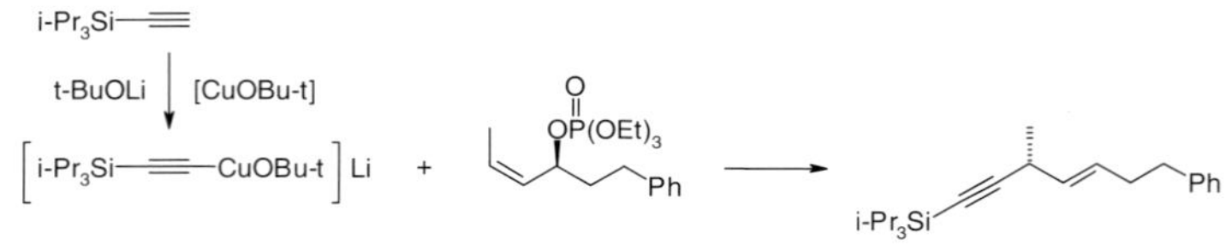

with asym. induction. THF (0.4 ml) added through a septum to a vial containing CuCl (10 mol%), 1,10-phenanthroline (12 mol%) and Li-*tert*-butoxide (1 eq.) under argon, the mixture stirred at room temp. to form a pale yellow-colored soln., triisopropylsilylacetylene (1.2 eq.) added, stirring continued for a few min, the startg. (S)-(Z)-allyl phosphate (0.2 mmol; e.e. 95%) added, stirred for 24 h at 40°, the mixture (containing a yellow precipitate) diluted with ether, filtered through a short plug of silica gel, the volatiles removed under reduced pressure, and the residue worked up with purification by flash chromatography on silica gel → product. Y 92% (γ/α selectivity >99:1; E/Z >99:1; e.e. 90%). The procedure is high-yielding (from both chiral and racemic 2-ethylenephosphoric acid esters) with very high (up to >99:1) γ-regioselectivity, even with more bulky substituents (e.g. *i*-Bu, cyclohexyl) in place of methyl, and various functional groups (e.g. esters and silyl ethers) were tolerated along the chain; the alkyne partner may be aromatic (possessing electron-donating or -withdrawing groups), heteroaromatic or aliphatic, notably supporting chloro, pivaloyl, dibenzylamino or phthalimido groups (thirty-seven examples in all; Y 44-99%; e.e. 79%, 88-94% for chiral substrates). The regioselectivity and yield were much lower with (E)-substrates, while there was no reaction at all with allyl chlorides or carbonates. Reaction is presumed to involve an intermediate allylcopper(III) species. F.e.s. Y. Makida, Y. Takayama, H. Ohmiya, M. Sawamura, Angew. Chem., Int. Ed. *2013*, *52* (20), 5350-4 [DOI: 10.1002/anie.201300785].

Copper(I) bromide (s.a. under Zinc bromide) *CuBr*
3-Component synthesis of 2-acetyleneamines s. *66*, 353s*83* CO → C(N<)C≡C

Copper(I) iodide/diisopropylamine *CuI/i-Pr$_2$NH*
Pyridines from O-acyloximes and α,β-ethylenealdehydes under cooperative copper/iminium catalysis ○

360.

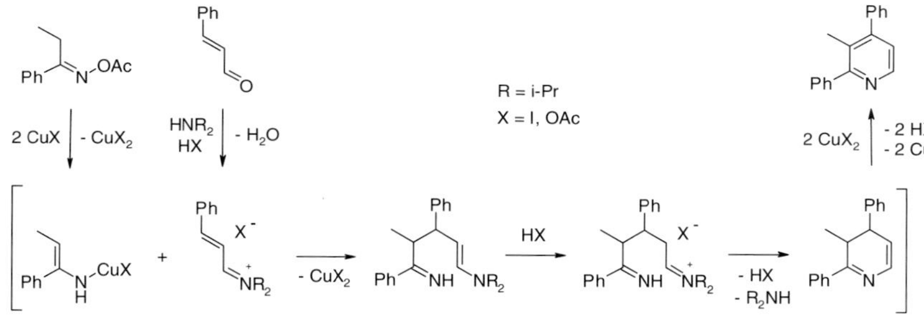

Cinnamaldehyde (1.5 eq.), diisopropylamine (2 eq.) and DMSO (1 ml) added to a mixture of (E)-propiophenone O-acetyloxime (0.2 mmol) and CuI (20 mol%) under N_2, the sealed mixture stirred at 60° for 16 h, cooled to room temp., diluted with ethyl acetate, filtered through silica, washed with water, and worked up with purification by flash chromatography → 3-methyl-2,4-di-phenylpyridine. Y 92%. This experimentally simple and scalable (to 10 mmol) method combined the redox activity of copper(I) with iminium activation [using either diisopropylamine (2 eq.) or pyrrolidinium perchlorate (20 mol%) as iminium ion precursor] to afford di- and tri-subst. pyridines from oxime acetates and enals, generally as single regioisomers (thirty-three examples; Y 36%, 50-92%; 2-methylacetophenone oxime acetate and cinnamaldehyde gave a 72/18 mixture of 2,4- and 2,6-disubst. pyridines using *i*-Pr$_2$NH). The method was compatible with ester, nitrile, nitro, ether and halo functionality. α,β-Unsaturated ketones (benzylideneacetone or chalcone) reacted sluggishly, affording the corresponding pyridines in <5% yields. F.e., optimization and substrate prepn. s. Y. Wei, N. Yoshikai, J. Am. Chem. Soc. *2013*, *135* (10), 3756-9 [DOI: 10.1021/ja312346s].

Copper(II) chloride *CuCl$_2$*
Copper(II) chloride/titanium tetraethoxide *CuCl$_2$/Ti(OEt)$_4$*
Gold nanoparticles/zinc oxide/irradiation *Au/ZnO/ↁ*
Gold nanoparticles/cerium dioxide *Au/CeO$_2$*
3-Component synthesis of 2-acetyleneamines CO → C(N<)C≡C
update s. *66*, 353s*82*; solvent-free conversion with cyclohexanone as ketone component with inexpensive CuCl$_2$ as catalyst s. C.J. Pierce, C.H. Larsen, Green Chem. *2012*, *14* (10), 2672-6 [DOI: 10.1039/c2gc35713e]; solvent-free coupling of unactivated *aliphatic* ketones *under dual catalysis* with CuCl$_2$/Ti(OEt)$_4$ s. C.J. Pierce, M. Nguyen, C.H. Larsen, Angew. Chem., Int. Ed.

2012, 51 (49), 12289-92 [DOI: 10.1002/anie.201206674]; *syn*-selective coupling of alkynes with chiral α-oxyaldehydes and dibenzylamine *with asym. induction* using CuBr s. S.C. Deshmukh, A. Roy, P. Talukdar, Org. Biomol. Chem. *2012, 10* (37), 7536-44 [DOI: 10.1039/c2ob26007g]; *heterogeneous* conversion with a reusable copper(I) bis[2-(phenylthio)benzylidene]-1,2-ethylenediamine [thiosalen] complex covalently anchored onto the nanopores of mesoporous MCM-41 s. H. Naeimi, M. Moradian, Appl. Organomet. Chem. *2013, 27* (2), 300-6 [DOI: 10.1002/aoc.2976]; coupling of *in situ*-generated sec. amines (from prim. amines and alcohols) in a *dual* catalytic cycle effected by Au nanoparticles-on-ceria s. A. Corma, J. Navas, M.J. Sabater, Chem. Eur. J. *2012, 18* (44), 14150-6 [DOI: 10.1002/chem.201201837]; rapid procedure in the presence of Au-nanoparticles and ZnO as activator under LED light irradiation with plasmons [oscillations of free electron density] s. M. González-Béjar, K. Peters, G.L. Hallett-Tapley, M. Grenier, J.C. Scaiano, Chem. Commun. *2013, 49* (17), 1732-4 [DOI: 10.1039/c3cc38287g]; alternative 3-component synthesis of 2-acetylene-*tert*-amines from terminal acetylene derivs., aldehydes and sec. amines **by N-alkylative α-(alk-1-ynyl)ation** of the latter with copper(II) 2-ethylhexanoate as catalyst under microwave enhancement cf. D. Das, A.X. Sun, D. Seidel, Angew. Chem., Int. Ed. *2013, 52* (13), 3765-9 [DOI: 10.1002/anie.201300021].

*Silver hexafluoroantimonate s. under [1,3-Bis(2,6-diisopropylphenyl)imidazol-2-ylidene]gold(I) chloride and [Cp*RhCl$_2$]$_2$* *$AgSbF_6$*

(Acetonitrile)[2-(di-tert-butylphosphino)biphenyl]gold(I) hexafluoroantimonate/ triethylamine/chiral 3,3′-diaryl-1,1′-binaphthyl-2,2′-diyl hydrogen phosphate/ 3,5-dicarbethoxy-2,6-dimethyl-1,4-dihydropyridine ←

Asym. synthesis of 1,2,3,4-tetrahydroquinolines ○
from *o-prim*-aminooxo compds. and acetylene derivs.
via pyridine ring annelation-Brønsted acid-catalyzed asym. transfer-hydrogenation

361. 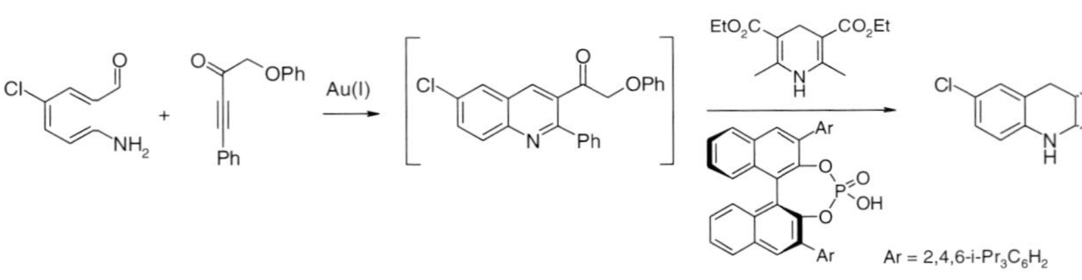

in one pot. 2-Amino-5-chlorobenzaldehyde (0.4 mmol) and phenoxyacetyl(phenyl)acetylene (1.2 eq.) added to a mixture of gold(I) catalyst (3 mol%) and 5 Å molecular sieves (1 g) in dry benzene (6 ml) at room temp., the mixture stirred at 60° until reaction complete (TLC), triethylamine (4 mol%), chiral cyclic phosphate diester (5 mol%) and Hantzsch ester (2.5 eq.) added successively, the mixture stirred until reaction complete (TLC), filtered through silica, concentrated *in vacuo*, and the residue purified by chromatography on silica → (2R,3S)-6-chloro-3-phenoxyacetyl-2-phenyl-1,2,3,4-tetrahydroquinoline. Y 96% (d.r. >20:1; e.e. >94%). This efficient and experimentally simple gold-catalyzed formation, and subsequent asym. reduction, of quinolines was successful for *o*-amino-benzaldehydes and -benzophenones reacting with electron-diverse arylacetylenes carrying H or acyl/aroyl termini, affording the title products with high regio-, diastereo- and enantio-selectivity (forty-one examples; Y 67-98%; d.r. 5:1 to >20:1, e.e. 85 to >99%). The method was compatible with halo, nitrile, ether, ketone and ester functionality. F.e. and optimization s. X.-Y. Liu, Y.-P. Xiao, F.-M. Siu, L.-C. Ni, Y. Chen, L. Wang, C.-M. Che, Org. Biomol. Chem. *2012, 10* (35), 7208-19 [DOI: 10.1039/c2ob25753j].

[1,3-Bis(2,6-diisopropylphenyl)imidazol-2-ylidene]gold(I) chloride/ silver hexafluoroantimonate *[Au(I)]/$AgSbF_6$*

Cycloisomerization-Pictet-Spengler cyclization s. *8,* 823s*83*

Magnesium/1,3-dibromo-5,5-dimethylhydantoin ←

Synthesis of α-subst. *tert*-amino(aryl)acetic acid esters C=NOSO$_2$R → C(R′)NR″$_2$
from aryl(O-sulfonyloximino)acetic acid esters and organomagnesium halides

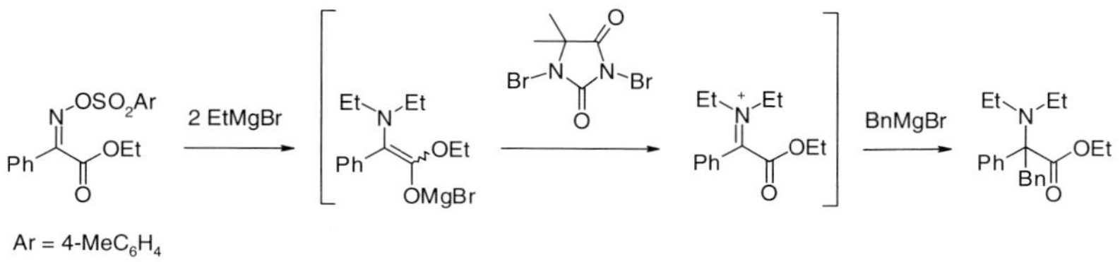

362.

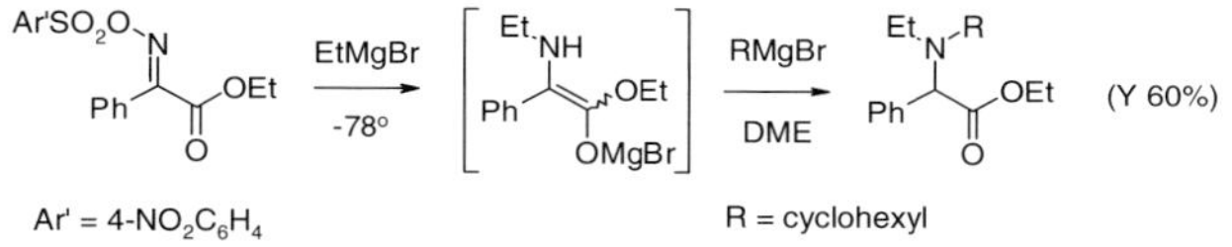

in one pot. A soln. of ethylmagnesium bromide (2.2 eq.) in ether added to a soln. of ethyl (Z)-α-(*p*-toluenesulfonyloximino)phenylacetate in toluene (0.1 *M*) at 30° under argon, the mixture stirred for 15 min, acetonitrile (0.1 *M*) and 1,3-dibromo-5,5-dimethylhydantoin (0.6 eq.) added, stirring continued for 15 min, a soln. of benzylmagnesium bromide (2 eq.) in ether added, stirring continued for a further 15 min, quenched with satd. aq. NH_4Cl, extracted with ethyl acetate, and purified chromatographically → ethyl 2-diethylamino-2,3-diphenylpropanoate. Y 68%. Initial development identified the (Z)-*p*-toluenesulfonyloximino deriv. of ethyl phenylacetate as the best activating group ((E)-analogs were less reactive), which was N,N-dialkylated in moderate to good yields with ethyl- (85%), *n*-propyl- (65%) and benzyl-MgBr (50%), with *sec*-alkyl-MgBr giving lower yields (24-32%; two examples) and ar., methyl, or *tert*-alkyl analogs being unreactive. Additional alkylation at C was effected by *in situ* oxidation of the generated intermediate N,N-diethyl enolate to an iminium species and treatment with a second Grignard reagent in a one-pot procedure (six examples; Y 44-74%). Mixed N,N-dialkylation of the substrate was moderately successful starting from the (E)-*p*-nitrosulfonyloximo analog, which was mono-ethylated in toluene at -78°, and alkylated *in situ* with a second *prim-/sec*-alkyl-MgBr (crucially in dimethoxyethane) at the same temp. (four examples; Y 32-60%). Second alkylation with MeMgBr (0%) or *t*-BuMgBr (2%) was ineffective. F.e.s. S. Hata, T. Maeda, M. Shimizu, Bull. Chem. Soc. Jpn. *2012, 85* (11), 1203-5 [DOI: 10.1246/bcsj.20120195]; **α-*tert*-amino- from α-silox-, -alkox- or -sulfonylox-imino-carboxylic acid esters** and two Grignard compd. molecules cf. Y. Mizutani, H. Tanimoto, T. Morimoto, Y. Nishiyama, K. Kakiuchi, Tetrahedron Lett. *2012, 53* (44), 5903-6 [DOI: 10.1016/j.tetlet.2012.08.091].

Magnesium/iron(III) fluoride/1,3-bis(2,6-diisopropylphenyl)imidazol-2-ylidene *Mg/FeF$_3$/IPr*
Cross-coupling of Grignard compds. ArOTs → ArR
with phenolethers cf. *44,* 702; alkylarenes by coupling *prim-* or *sec*-alkylmagnesium chlorides **with aryl tosylates or sulfamates** with $FeF_3 \cdot 3H_2O$ and an N-heterocyclic carbene as ligand, e.g. IPr, s. T. Agrawal, S.P. Cook, Org. Lett. *2013, 15* (1), 96-9 [DOI: 10.1021/ol303130j].

Zinc oxide s. under Au *ZnO*

Magnesium bromide/ethyldiisopropylamine/4-dimethylaminopyridine *MgBr$_2$/i-Pr$_2$NEt/DMAP*
Carbalkoxylation with carbonic acid esters H → COOR
s. *28,* 732; β-ketocarboxylic acid esters from ketones with pentafluorophenyl methyl carbonate using $MgBr_2$-etherate/ethyldiisopropylamine/DMAP for enolate generation, also synthesis of β-keto-thiolic and -thionocarboxylic acid esters with pentafluorophenyl methyl thiol- or thionocarbonate, s. K.J. Hale, M. Grabski, J.T. Flasz, Org. Lett. *2013, 15* (2), 370-3 [DOI: 10.1021/ol303324a]; α-carboxylation of fatty acid methyl esters in dimethyl carbonate with NaH as base s. N. Kolb, M.A.R. Meier, Green Chem. *2012, 14* (9), 2429-35 [DOI: 10.1039/c2gc35793c].

Zinc bromide/copper(I) bromide/2(S)-[diphenyl(hydroxy)methyl]pyrrolidine ←

Asym. synthesis of allenes C≡CH → CH=C=CHR

from terminal acetylene derivs. and aldehydes

363.

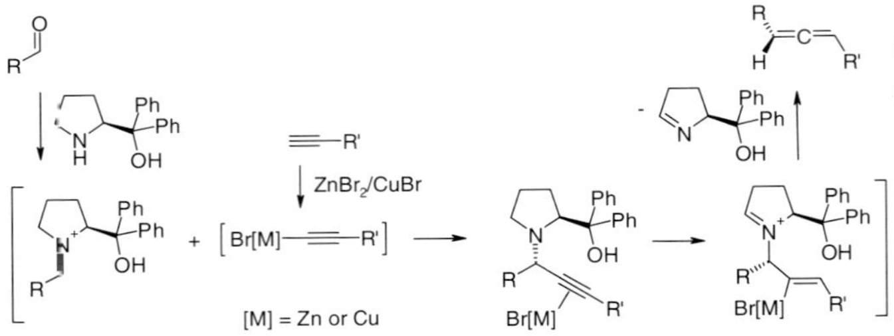

An asym. variant of *77*, 406 is reported with broad scope in terms of both alk-1-ynes (which are not limited to α-oxy-derivs.) and aldehydes (which are not limited to aliphatic ones) by performing the reaction **under bimetallic catalysis** in the presence of a cheap, readily-available chiral sec. amine. **E:** $ZnBr_2$ (0.5 eq.) and CuBr (0.2 eq.) added to an oven-dried Schlenk tube inside a glovebox, the tube then taken out of the glovebox and heated under vacuum with a heat gun until the inorganic salts molten, cooled to room temp., (S)-α,α-diphenylprolinol (1 mmol), startg. alkyne (1.5 eq.)/toluene (2 ml), and benzaldehyde (2 eq.)/toluene (1 ml) then added sequentially under argon, the tube heated in an oil bath at 110° with stirring for 13 h, cooled to room temp., the mixture filtered through a short pad of silica gel, eluting with ether, and after evaporation, the residue purified by chromatography on silica gel → (R)-product. Y 48% (e.e. 95%). The method is applicable to electron-rich or -poor ar. aldehydes or to cyclohexanal and to aryl- or alkyl-terminated alk-1-ynes (thirteen examples; Y 44-73%; e.e. 92-99%). Reaction may also be performed in dioxane at 120°, generally with slightly lower yield (40-63%) but higher enantioselectivity (96-99% e.e.; twelve examples). It is believed the alkynylmetal species, formed by deprotonation of the alk-1-yne in the presence of CuBr, $ZnBr_2$, and amine, adds via Re-face attack to the iminium ion generated *in situ* from the aldehyde and sec. amine, furnishing a highly enantio-enriched intermediate (S,S)-acetyleneamine, which then undergoes highly stereoselective intramolecular 1,5-hydride transfer and β-elimination to generate the axially-chiral disubst. allene under the cooperation of Cu(I) and $ZnBr_2$. F.e.s. R. Lü, J. Ye, T. Cao, B. Chen, W. Fan, W. Lin, J. Liu, H. Luo, B. Miao, S. Ni, X. Tang, N. Wang, Y. Wang, X. Xie, Q. Yu, W. Yuan, W. Zhang, C. Zhu, S. Ma, Org. Lett. *2013*, *15* (9), 2254-7 [DOI: 10.1021/ol400822m]; with CuBr/(R,R)-N-PINAP [N-PINAP = 4-[2-(diphenylphosphino)-1-naphthyl]-N-[1-phenylethyl]-1-phthalazinamine]/pyrrolidine then ZnI_2/NaI, especially for **axially-chiral 2-allenealcohols** from ethynylcarbinols, s. J. Ye, S. Li, B. Chen, W. Fan, J. Kuang, J. Liu, Y. Liu, B. Miao, B. Wan, Y. Wang, X. Xie, Q. Yu, W. Yuan, S. Ma, ibid. *2012*, *14* (5), 1346-9 [DOI: 10.1021/ol300296k]; with zinc halides/chiral sec. amines cf. M. Periasamy, N. Sanjeevakumar, M. Dalai, R. Gurubrahamam, P.O. Reddy, ibid. *14* (12), 2932-5 [DOI: 10.1021/ol300717e]; 2-allenealcohols from terminal siloxy-2-acetylenes and aldehydes with $ZnBr_2$/2(S)-[diphenyl(hydroxy)methyl]pyrrolidine then O-desilylation with Bu_4NF s. J. Ye, W. Fan, S. Ma, Chem. Eur. J. *2013*, *19* (2), 716-20 [DOI: 10.1002/chem.201201948].

Sodium hydrido(triacetoxo)borate s. under 2(S)-[Diphenyl(trimethylsiloxy)-methyl]pyrrolidine *$NaBH(OAc)_3$*

m-Chlorophenylboronic acid *$ArB(OH)_2$*

C_3-(α-Aminoalkyl)ation of indoles s. *73*, 369s*83* H → CH(NHR)R′

Mesoporous aluminosilicate s. under Copper(I) bis[2-(phenylthio)benzylidene]-1,2-ethylenediamine complexes ←

Indium(III) triflate *$In(OTf)_3$*

7-Alkylidene-7,11b-dihydroindeno[2,1-*c*]chromen-6(6a*H*)-one-6a-carboxylic acid esters ○
from α-(*o*-alk-1-ynylbenzylidene)malonic acid esters and phenols s. *83*, 368

Aluminum chloride *$AlCl_3$*
Friedel-Crafts acylation with carboxylic acid esters s. *10,* 594s*83* H → COR

Cerium dioxide s. under Au *CeO_2*

Scandium(III) triflate *$Sc(OTf)_3$*
Diastereoselective double Prins cyclization s. *43,* 706s*83* ○

Lanthanum(III) triflate *$La(OTf)_3$*
Benzo[1,4]heterocycle-condensed cyclopentanones from 2-furylcarbinols ←
and *o*-difunctionalized arenes via Piancatelli rearrangement – 3,3a,9,9a-Tetrahydrocyclopenta[*b*]-[1,4]benzoxazin-2(1*H*)-ones s. *83,* 131

Isopropanol s. under $Ni(cod)_2$ *i-PrOH*
Benzoquinone s. under $Pd(OAc)_2$ *BQ*

Per-6-amino-β-cyclodextrin *per-6-ABCD*
2,3-Dihydro-4-quinolones from *o*-aminoketones and aldehydes ○
Asym. cyclocondensation using per-6-amino-β-cyclodextrin
as supramolecular host and chiral base catalyst

364. 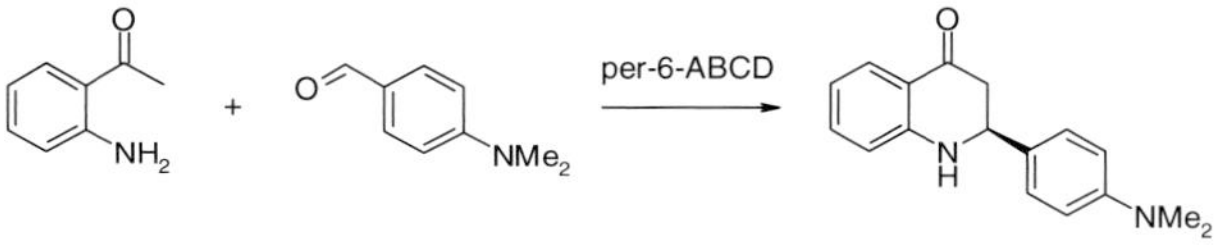

Use of a modified cyclodextrin [per-6-ABCD] with pendant amino groups at the primary face (providing hydrophobic and electrostatic binding of guest molecules) as a multifunctional catalyst for the novel preparation of the title compounds is reported. **E:** A soln. of *o*-aminoacetophenone (0.1 mmol) in ethanol (0.5 ml) added dropwise to a suspension of per-6-ABCD (1 eq.) in ethanol/water (1:1; 2 ml), the mixture stirred for 30 min, 4-dimethylaminobenzaldehyde (1 eq.) added to the orange suspension at -5°, the mixture stirred for 15 h, extracted with ethyl acetate, and purified by chromatography on silica → (S)-2-(4-dimethylaminophenyl)-2,3-dihydro-4-quinolone. Y 96% (e.e. 99%). This experimentally simple method appears general for electron-diverse (het)ar. and aliphatic (incl. benzyl and cinnamyl) aldehydes (e.e. 69-99%; twenty-six examples; Y 79-99%; *o-/p*-hydroxybenzaldehydes gave 18-20% and prim. aliphatic aldehydes gave 59-64%), with highest enantioselectivity (e.e. 88-99%) obtained for *m/p*-subst. benzaldehydes or heteroar. aldehydes. The catalyst was recycled seven times with only marginal reduction in yield and enantioselectivity. F.e. and optimization s. K. Kanagaraj, K. Pitchumani, J. Org. Chem. *2013*, *78* (2), 744-51 [DOI: 10.1021/jo302173a].

Ammonium formate s. under NH_4OH *$HCOONH_4$*
Ammonium acetate s. under Laccase *$AcONH_4$*
Chiral 2-[diphenyl(hydroxy)methyl]pyrrolidine s. under Zinc bromide ←

2(S)-[Diphenyl(trimethylsiloxy)methyl]pyrrolidine/sodium hydrido(triacetoxo)borate/ potassium tert-butoxide ←
Automated asym. 3-component synthesis of 2-pyrrolidon-3-ylacetic acid amides from maleimides, aldehydes and prim. amines in one pot

365. 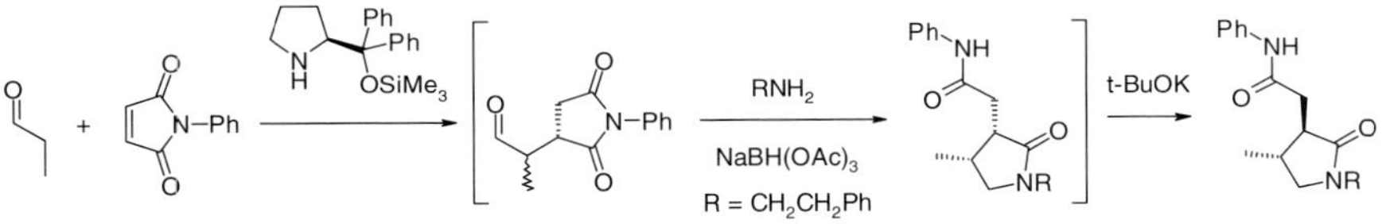

Optimization of a three-step procedure [s. J. Nöth et al., J. Comb. Chem. *2008*, *10* (3), 456-9 [DOI: 10.1021/cc700202c]] for use in an automated one-pot synthesis of a compound library is reported. **E:** As part of a 2x2x3 validation library using a Chemspeed Accelerator SLT-100 automated synthesizer, solns. of N-phenylmaleimide (0.4 mmol) in chloroform (1 ml), propanal (1.5 eq.) in the same solvent (1.5 ml) and chiral pyrrolidine catalyst (10 mol%) subjected to vortex mixing at 0° for 36 h, a soln. of 2-phenylethylamine (2 eq.) in the same solvent (0.8 ml) and $NaBH(OAc)_3$ (2.5 eq.) added, the mixture subjected to vortex mixing at room temp. for 6 h, concentrated *in vacuo* (30°; 6 mmHg) within the reactor vessel, methanol (10 ml) and K-*tert*-butoxide (10 eq.) added, the mixture subjected to vortex mixing for 12 h, concentrated *in vacuo* (40°; 6 mmHg) within the reactor vessel, quenched with 4 *M* aq. HCl, extracted with methylene chloride, filtered through silica, and purified chromatographically → 2-[(3R,4R)-4-methyl-2-oxo-1-(2-phenylethyl)pyrrolidin-3-yl]-N-phenylacetamide. Y 59% (purity 100%; e.e. 66%). The one-pot optimization also included management of by-products and excess reagents, with the validation set of 18 compounds affording yields of 4-59% (purity 83.1-100%; e.e. of the major *anti* enantiomers: 62-74%). The method was applied to a 4x4x8 matrix, affording 169 compounds in 15-70% yields, with the remaining compounds rejected on the basis of quantity (<20 mg) and purity (<90%) criteria. F.e.s. E. Fenster, D. Hill, O. Reiser, J. Aubé, Beilstein J. Org. Chem. *2012*, *8*, 1804-13 [DOI: 10.3762/bjoc.8.206].

1-Butyl-3-methylimidazolium hexafluorophosphate s. under $Bi(NO_3)_3$ *[bmim]PF_6*
1,3-Bis(2,6-diisopropylphenyl)imidazol-2-ylidene s. under Mg and $Ni(cod)_2$ *IPr*

1,3-Dimesitylimidazol-2-ylidene/1,8-diazabicyclo[5.4.0]undec-7-ene *NHC/DBU*
2-Aryl-2,3-dihydrocyclopenta[*c*]chromen-4(1*H*)-ones ○
from *o'*-hydroxychalcones and α,β-ethylenealdehydes
by stereoselective N-heterocyclic carbene-catalyzed double ring closure

366.

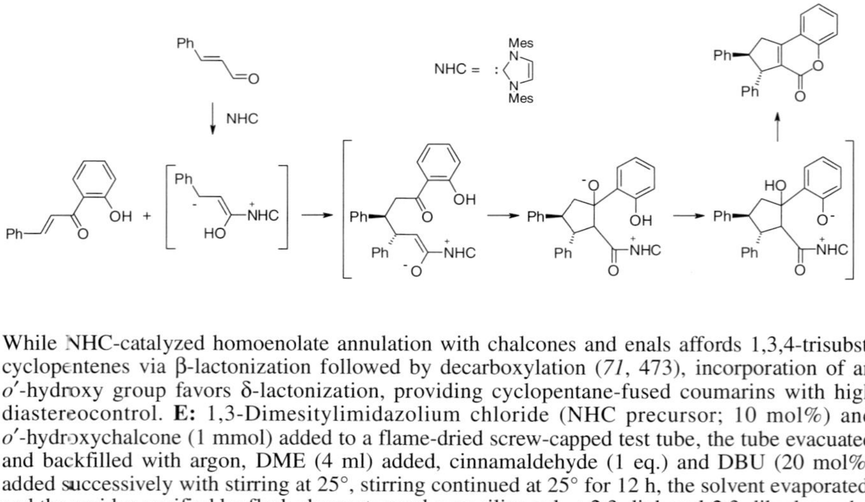

While NHC-catalyzed homoenolate annulation with chalcones and enals affords 1,3,4-trisubst. cyclopentenes via β-lactonization followed by decarboxylation (*71*, 473), incorporation of an *o'*-hydroxy group favors δ-lactonization, providing cyclopentane-fused coumarins with high diastereocontrol. **E:** 1,3-Dimesitylimidazolium chloride (NHC precursor; 10 mol%) and *o'*-hydroxychalcone (1 mmol) added to a flame-dried screw-capped test tube, the tube evacuated and backfilled with argon, DME (4 ml) added, cinnamaldehyde (1 eq.) and DBU (20 mol%) added successively with stirring at 25°, stirring continued at 25° for 12 h, the solvent evaporated, and the residue purified by flash chromatography on silica gel → 2,3-diphenyl-2,3-dihydrocyclopenta[*c*]chromen-4(1*H*)-one. Y 98% (d.r. >20:1). The reaction proceeds under mild conditions and gives high yields (74-97%; d.r. >20:1; twenty-four examples) from a broad range of substrates, tolerating ar. halo (F, Cl, Br), MeO, CF_3 or NO_2 groups; the 2-aryl group may also be replaced by styryl, furan-2-yl or ferrocenyl and the enal may bear a β-alkyl or aryl group. F.e.s. A. Bhunia, A. Patra, V.G. Puranik, A.T. Biju, Org. Lett. *2013*, *15* (7), 1756-9 [DOI: 10.1021/ol400562z].

Laccase/2,2'-azinobis(3-ethylbenzothiazoline-6-sulfonic acid) diammonium salt/air/ ammonium acetate ←
One pot enzymatic aromatizing Hantzsch pyridine synthesis s. *83,* 185

Lipase ←

Multicomponent syntheses with isonitriles ←

update on the Ugi condensation s. *17,* 809s*75*; *under enzymatic catalysis* with lipase [CALB] in toluene s. S. Klossowski, B. Wiraszka, S. Berlozecki, R. Ostaszewski, Org. Lett. *2013, 15* (3), 566-9 [DOI: 10.1021/ol3033829]; α-*sec*-acylaminocarboxylic acid amides *with ammonium carbonate* as amine component in trifluoroethanol for the synthesis of selenocysteine peptides from challenging (organoseleno)acetaldehydes s. M. Abbas, L.A. Wessjohann, Org. Biomol. Chem. *2012, 10* (27), 9330-3 [DOI: 10.1039/c2ob26552d]; *in situ*-conversion of the Ugi amides to the corresponding 2,2,2-trifluoroethyl esters with added H_2SO_4 as catalyst s. S. Balalaie, H. Motaghedi, D. Tahmassebi, M. Bararjanian, H.R. Bijanzadeh, Tetrahedron Lett. *2012, 53* (46), 6177-81 [DOI 10.1016/j.tetlet.2012.08.096]; Ugi condensation **with enols** (e.g. 3-hydroxy-Δ^3-2-pyrrolone-4-carboxylic acid esters) in place of carboxylic acids s. T.G. Castellano, A.G. Neo, S. Marcaccini, C.F. Marcos, Org. Lett. *2012, 14* (24), 6218-21 [DOI: 10.1021/ol302976g]; synthesis of sym. and unsym. bis(α-aminocarboxylic acid amides) by *double* 4-component condensation based on a sacrificial Mumm rearrangement s. F. La Spisa, A. Feo, R. Mossetti, G.C. Tron, Org. Lett. *2012, 14* (23), 6044-7 [DOI: 10.1021/ol302935y]; synthesis of N-(α-carbamyl)benzylcaprolactams by 3-component (4-point) condensation *in water* (cf. *52,* 351s*80*) s. M.A. Rasouli, M. Mahdavi, P.R. Ranjbar, M. Saeedi, A. Shafiee, A. Foroumadi, Tetrahedron Lett. *2012, 53* (52), 7088-92 [DOI: 10.1016/j.tetlet.2012.10.075]; solvent-free 4-component synthesis of 3-amino-1-pyrrolo-[1,2*a*][1,10]phenanthrolines s. M. Li, X.-L. Lv, L.-R. Wen, Z.-Q. Hu, Org. Lett. *2013, 15* (6), 1262-5 [DOI: 10.1021/ol400191b]; of 1,4-benzodiazepines s. Y. Huang, K. Khoury, T. Chanas, A. Dömling, Org. Lett. *2012, 14* (23), 5916-9 [DOI: 10.1021/ol302837h]; Ugi-type condensation in water *with trimethylsilyl cyanide as a functional isonitrile equivalent* for the KF-mediated synthesis of N-condensed 5-aminoimidazoles s. S.K. Guchhait, V. Chaudhary, C. Madaan, Org. Biomol. Chem. *2012, 10* (46), 9271-7 [DOI: 10.1039/c2ob26733k]; 3-component (4-molecule) synthesis of 4,5,6,7-tetrahydro-1*H*-1,4-diazepine-5-carboxylic acid amides from ene-1,2-diamines, ketones and isonitriles with readily recoverable silica-supported Fe_3O_4 as catalyst s. *83,* 314; isonitrile-based diastereoselective 3-component synthesis of polycyclic spiroindolines s. X. Wang, S.-Y. Wang, S.-J. Ji, Org. Lett. *2013, 15* (8), 1954-7 [DOI: 10.1021/ol400606c].

Synthase ←

Friedel-Crafts-type allylation with allyl esters H → C-C=C

with allyl phosphates using BF_3 cf. *29,* 747; regioselective prenylation of the indole ring of indolocarbazoles with prenyl diphosphate using a recombinant dimethylallyltryptophan synthase from an *Aspergillus* sp. (cf. *53,* 368) s. X. Yu, A. Yang, W. Lin, S.-M. Li, Tetrahedron Lett. *2012, 53* (50), 6861-4 [DOI: 10.1016/j.tetlet.2012.10.039]; regioselective C_2-allylation of 5-membered heteroarenes with α-subst. allyl acetates using $PdCl_2$ as the sole catalyst s. F.-Q. Yuan, F.-Y. Sun, F.-S. Han, Tetrahedron *2012, 68* (34), 6837-42 [DOI: 10.1016/j.tet.2012.06.038]; regioselective **C_3-allylation indoles** using a chiral palladium(II) N-[bis(*o*-sulfinylphenyl)phosphinyl]piperidine complex in a **dynamic kinetic asym. transformation** (cf. *79,* 384s*80*) s. L. Du, P. Cao, J. Xing, Y. Lou, L. Jiang, L. Li, J. Liao, Angew. Chem., Int. Ed. *2013, 52* (15), 4207-11 [DOI: 10.1002/anie.201209485]; asym. C_3-allylation of *3-subst.* indoles to give chiral 3-subst. 3-allylindolenines using a chiral palladium(II) 1,1′-binaphthyl-2,2′-diyl N-allylphosphoramidite as catalyst (and K_2CO_3 as base) s. Y. Liu, H. Du, Org. Lett. *2013, 15* (4), 740-3 [DOI: 10.1021/ol3032736].

tert-Butyl isocyanide s. under Palladium(II) chloride *t-BuNC*

Acetone cyanohydrin $Me_2C(OH)CN$

Strecker synthesis CO → C(N<)CN

of α-aminonitriles, update s. *52,* 449s*82*; *in water* using acetone cyanohydrin as cyanide source by coupling with enantio-pure [L-α-amino acid-derived] prim. amines **with asym. induction**, diastereoselectivity, s. M. Pori, P. Galletti, R. Soldati, D. Giacomini, Eur. J. Org. Chem. *2013* (9), 1683-95 [DOI: 10.1002/ejoc.201201533].

Carboxylic acid anhydrides *(RCO)₂O*

Stereoselective dehydrative double ring closure of tryptamines and *o*-hydroxyketones with N-acylation

367.

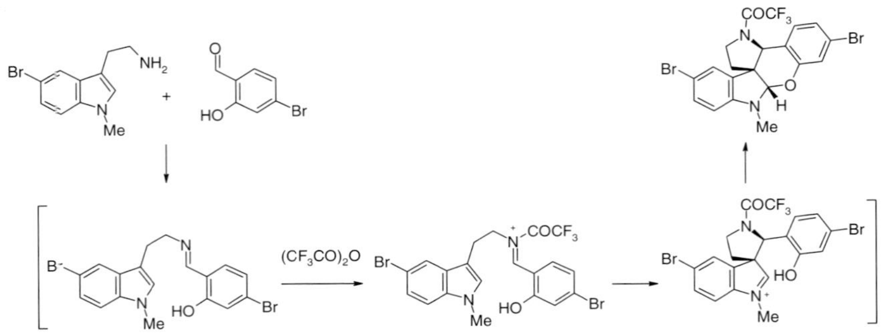

A novel one-pot synthesis of complex fused spiroindolines, bearing a highly congested contiguous spiro quaternary center and two tertiary stereocenters, via Mannich reaction of tryptamines with salicylaldehydes and subsequent intramolecular aminol ether formation, is reported. **E:** A mixture of 4 Å molecular sieves (30 mg), 5-bromo-1-methyltryptamine (0.2 mmol) and 4-bromosalicylaldehyde (1.2 eq.) in chloroform (0.44 ml) stirred at -40° for 8 h, a soln. of trifluoroacetic anhydride (1.2 eq.) in the same solvent (0.24 ml) added, stirring continued until reaction complete (TLC), quenched with satd. aq. $NaHCO_3$, and worked up with chromatographic purification → product. Y 86%. This experimentally simple and apparently general sequence uses readily available components (sixteen examples; Y 72-99%) and involves the use of an active anhydride for activation of imine/iminium ion intermediates (TFA, BF_3 and $TiCl_4$ were ineffective). F.e. and optimization s. J. Yang, X. Xie, Z. Wang, R. Mei, H. Zheng, X. Wang, L. Zhang, J. Qi, X. She, J. Org. Chem. *2013, 78* (3), 1230-5 [DOI: 10.1021/jo302404v].

Trifluoroacetic anhydride s. under Pd(OAc)₂ *(CF₃CO)₂O*
*Pivalic acid s. under [Cp*RhCl₂]₂* *t-BuCOOH*

Proline-functionalized polyacrylonitrile fiber ←
Heterogeneous Knoevenagel condensation s. *46,* 713s*83* CO → C=C

Trifluoroacetic acid s. under β-Isocupreidine *CF₃COOH*
Trichloroacetic acid s. under Chloro(1,5-cyclooctadiene)iridium(I) dimer *CCl₃COOH*

Cinchona-based 2-tert-aminothioureas ←
Asym. synthesis of β-tosylaminoketones CO → C(NHTs)C-CO
by 3-component Mannich reaction s. *60,* 336s*83*

1,3-Dibromo-5,5-dimethylhydantoin (s.a. under Magnesium) ←
Sym. 1,1-di[het]arylalkanes from oxo compds. CO → CAr_2
update s. *5,* 549s*82*; **1,1-bis(3-indolyl)alkanes** with 1,3-dibromo-5,5-dimethylhydantoin as catalyst under solvent-free conditions, also chemoselective conversion of ar. aldehydes, s. S.F. Hojati, T. Zeinali, Z. Nematdoust, Bull. Korean Chem. Soc. *2013, 34* (1), 117-20 [DOI: 10.5012/bkcs.2013.34.1.117]; 1,1-bis(5-hydroxypyrazol-4-yl)alkanes from ar. aldehydes using melamine trisulfonic acid as a recyclable solid acid catalyst s. N. Iravani, J. Albadi, H. Momtazan, M. Baghernejad, J. Chin. Chem. Soc. *2013, 60* (4), 418-24 [DOI: 10.1002/jccs.201200345].

Silica s. under Cu, Fe₃O₄ and Fe(HSO₄)₃ *SiO₂*
Titanium tetraethoxide s. under CuCl₂ *Ti(OEt)₄*

Titanium tetraisopropoxide *Ti(OPr-i)₄*
Polycarbosilane-supported siloxytitanium(IV) triisopropoxide ←
Knoevenagel condensation s. *46,* 713s*83* CO → C=C

Layered α-zirconium hydrogen phosphate/sodium calix[4]arenesulfonates ←

Hetero-Diels-Alder reaction with *in situ*-generated azomethines in water s. *52,* 363s*83* C

Trimethylsilyl chloride (s.a. under $FeCl_3$) *Me_3SiCl*

Pictet-Spengler cyclization

update s. *8,* 823s*82*; with trimethylsilyl chloride as condensing agent s. S.V. Ryabukhin, D.M. Panov, A.S. Plaskon, A.A. Tolmachev, R.V. Smaliy, Monatsh. Chem. *2012, 143* (11), 1507-17 [DOI: 10.1007/s00706-012-0804-7]; synthesis of 5,6-dihydro-6-spirophenanthridines from cyclic ketones, and 6-subst. phenanthridines from aldehydes, with propylphosphonic anhydride, s. J.K. Augustine, A. Bombrun, P. Alagarsamy, A. Jothi, Tetrahedron Lett. *2012, 53* (46), 6280-7 [DOI: 10.1016/j.tetlet.2012.09.034]; N-tosyl-1,2,3,4-tetrahydropyrido[4,3-*b*]indoles from 4-(*o*-aminoaryl)-3-acetylenetosylamines by **cycloisomerization-Pictet-Spengler cyclization** with [1,3-bis(2,6-diisopropylphenyl)imidazol-2-ylidene]gold(I) chloride/$AgSbF_6$ s. B.V.S. Reddy, M. Swain, S.M. Reddy, J.S. Yadav, B. Sridhar, J. Org. Chem. *2012, 77* (24), 11355-61 [DOI: 10.1021/jo302068e].

Zirconyl chloride *$ZrOCl_2$*

Solvent-free catalytic Biginelli synthesis s. *55,* 337s*83*

Tin(II) triflate *$Sn(OTf)_2$*

7-Alkylidene-7,11b-dihydroindeno[2,1-*c*]chromen-6(6a*H*)-one-6a-carboxylic acid esters from [*o*-(alk-1-ynyl)benzylidene]malonic acid esters and phenols

368. 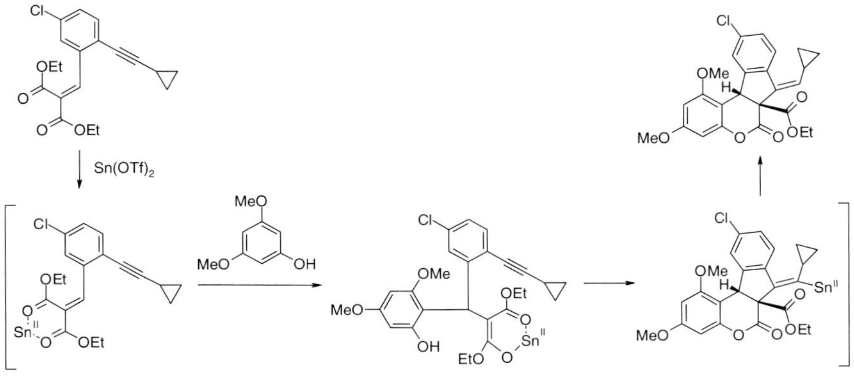

One pot double ring closure. $Sn(OTf)_2$ (5 mol%) added to a soln. of diethyl 5-chloro-2-cyclopropylethynylbenzylidenemalonate (1 eq.) and 3,5-dimethoxyphenol (1.2 eq.) in 1,2-dichloroethane (0.1 *M*), the sealed mixture stirred at 100° for 3 h, cooled to room temp., diluted with methylene chloride, washed with 1 *M* aq. NaOH, and purified by chromatography on silica → (E)-ethyl 10-chloro-7-cyclopropylmethylidene-1,3-dimethoxy-6-oxo-6,6a,7,11b-tetrahydroindeno[2,1-*c*]chromene-6a-carboxylate. Y 84% (E/Z 93:7). Both $In(OTf)_3$ and [optimally] $Sn(OTf)_2$ functioned as dual action catalysts, promoting the observed double cyclization (other catalysts produced variable yields of mono-cyclized products). The reaction was successful for electron-diverse di(m)ethyl 2-ethynylbenzylidenemalonates reacting with electron-rich phenols and 1-naphthol, generally with E-products predominating (often exclusively for ar. ethynes), with 3-methoxyphenols cyclizing exclusively at the least hindered position (thirty-two examples; Y 69-99%; E/Z 46:54 to 100:0). A bulky di-*tert*-butyl malonate analog underwent monocyclization to a 4-aryl-3,4-dihydrocoumarin (Y 60%), while replacing one ester group with acetyl afforded 4*H*-chromenes at room temp. (two examples; Y 25-38%), whereas at 100° only intramolecular cyclization of the ethyne to a naphthalene deriv. was observed (two examples; Y 16-25%). An α-cyanocinnamate analog, however, gave the expected tetracycle in low yield (25%), while a 2-ethynylcinnamate was unreactive. F.e. and optimization s. J.H. Joo, S.W. Youn, Adv. Synth. Catal. *2013, 355* (2-3), 559-68 [DOI: 10.1002/adsc.201200745].

Ethylammonium nitrate *$[EtNH_3]NO_3$*
3-Component Friedel-Crafts (α-aminoalkyl)ation H → CH(NH-)R′
o-(α-acylaminoalkyl)ation **of** 2-**naphthols**, update s. *73,* 369s*82*; solvent-free conversion with reusable ethylammonium nitrate as ionic liquid catalyst at ambient temperature, also reaction **of phenols**, s. S.A.R. Mulla, T.A. Salama, M.Y. Pathan, S.M. Inamdar, S.S. Chavan, Tetrahedron Lett. *2013, 54* (7), 672-5 [DOI: 10.1016/j.tetlet.2012.12.004]; green, solvent-free conversion of phenols with vine leaf-shaped Nano-S as catalyst s. V.K. Das, M. Borah, A.J. Thakur, J. Org. Chem. *2013, 78* (7), 3361-6 [DOI: 10.1021/jo302682k]; mild **C_3-(α-aminoalkyl)ation of indoles** with *m*-chlorophenylboronic acid as catalyst s. S.V. Goswami, P.B. Thorat, V.N. Kadam, S.A. Khiste, S.R. Bhusare, Chin. Chem. Lett. *2013, 24* (5), 422-4 [DOI: 10.1016/j.cclet.2013.03.024].

Tri-n-butylphosphine *Bu_3P*
(1,3-Dien)oxysilanes CHO → CH=C-C=C(OSi≤)
from α,β-ethylene-β-silylketones and aldehydes
via 1,4-Brook rearrangement-Wittig synthesis

369.

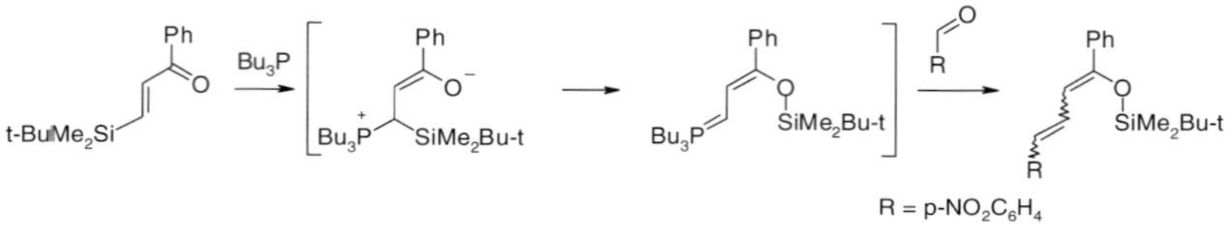

Tri-*n*-butylphosphine (1 eq.) added to a soln. of startg. β-*tert*-butyldimethylsilyl-enone and 4-nitrobenzaldehyde (1 eq.) in methylene chloride, the mixture stirred at room temp. for 1 h, and passed through a short silica gel column → product. Y ca. 100% (Y 96% in chloroform). The method is high yielding and operationally simple, avoiding the isolation of intermediate phosphonium salts/ alkylidenephosphoranes. Five further examples with 4-nitrobenzaldehyde and β-*tert*-butyldimethyl-, -triisopropyl- or -*tert*-butyldiphenyl-silyl-enones gave yields of 58-96% with some interconversion of geoisomers occurring (ratios of isomers not determined). Other aromatic (optionally bearing Cl, CF_3 or MeO), aliphatic or α,β-unsatd. aldehydes did not cause isomerization of the product, eleven examples giving yields of 39% (from hindered 2,6-dimethylbenzaldehyde) and 66-100% by reaction with the illustrated β-*tert*-butyldimethylsilyl-enone (Z,E/Z,Z 62:38 to 82:18). Crossover experiments indicated that the silyl migration is entirely intramolecular, producing the intermediate phosphorane with high stereoselectivity. F.e.s. Y. Matsuya, A. Koiwai, D. Minato, K. Sugimoto, N. Toyooka, Tetrahedron Lett. *2012, 53* (44), 5955-7 [DOI: 10.1016/ j.tetlet.2012.08.114].

2-Dicyclohexylphosphino-2′,6′-dimethoxybiphenyl s. under o-Pallad(II)acyclic complexes ←

Chiral 2-acetylamino-2′-(diphenylphosphino)-1,1′-binaphthyl ←
Regioselective organocatalyzed asym. α-allylation s. *82,* 366s*83* H → C-C=C

Tert. di(phosphines) s. under Palladium(II) acetate, Tris(dibenzylideneacetone)dipalladium, and Palladium(II) chloride ←

Propylphosphonic anhydride ←
Pictet-Spengler cyclization s. *8,* 823s*82* ○

Chiral 1,1′-binaphthyl-2,2′-diyl phosphoramidites s. under Chloro(1,5-cyclooctadiene)-iridium(I) dimer ←

1,1′-Binaphthyl-2,2′-diyl hydrogen phosphates ←
Organocatalyzed asym. Povarov reaction s. *75,* 403s*83*

Chiral 3,3′-diaryl-1,1′-binaphthyl-2,2′-diyl hydrogen phosphate s. under (Acetonitrile)-[2-(di-tert-butylphosphino)biphenyl]gold(I) hexafluoroantimonate ←

Chiral 3,3′-di(9-anthracenyl)-1,1′-binaphthyl-2,2′-diyl hydrogen phosphate ←
3-Component asym. [3+2]-cycloaddition s. *40,* 501s*83*

Chiral 3,3′-di(9-phenanthryl)-1,1′-binaphthyl-2,2′-diyl hydrogen phosphate ←
Organo-Brønsted acid asym. Friedel-Crafts alkylation of indoles H → R
with activated alcohols, e.g. with γ-hydroxy-γ-lactams, s. *43,* 703s*83*

(R)-6,6′-Bis(2,4,6-triisopropylphenyl)-1,1′-spirobiindan-7,7′-diyl hydrogen phosphate ←
Double ring closure via organocatalyzed asym. 3-component Povarov reaction with desymmetrizing intramolecular ring opening of an oxetane directing group C

370. 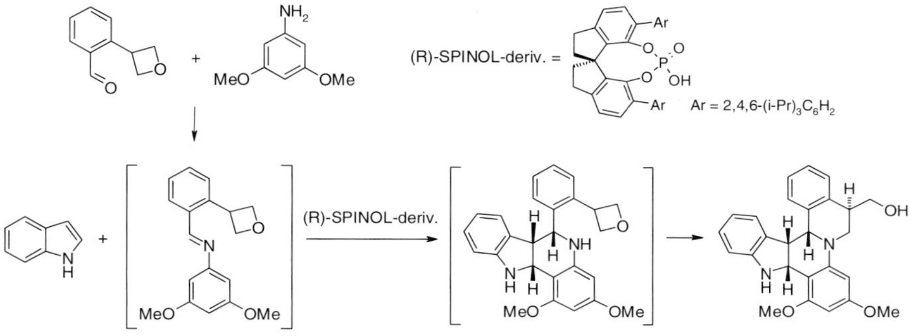

Chiral (5,6,11b,12,16b,16c-hexahydroindolo[3,2-*c*]isoquinolino[2,1-*a*]quinolin-5-yl)methanols. Anh. ether (2.5 ml) added to a centrifuge tube containing indole (0.25 mmol), 3,5-dimethoxyaniline (1 eq.) and the startg. oxetanylaldehyde (1 eq.), the mixture stirred for 10 min until the soln. became clear and homogeneous, (R)-6,6′-bis(2,4,6-triisopropylphenyl)-1,1′-spirodiindan-7,7′-diyl hydrogen phosphate (2.5 mol%) added, stirred at room temp. as the product precipitated out, diluted with hexanes after 12 h, stirring continued for 5 min, the mixture centrifuged, and the solid isolated after a second centrifugation and washing → ((5R,11bR,16bS,16cR)-9,11-dimethoxy-5,6,11b,12,16b,16c-hexahydroindolo[3,2-*c*]isoquinolino[2,1-*a*]quinolin-5-yl)methanol. Y 85% (d.r. >99:1; e.e. 98%). The procedure provides a rapid and simple entry into a pharmaceutically important complex polycyclic system, wherein four new bonds (two C-C and two C-N) are formed and four new stereogenic centers are created in one pot in high yield and with high diastereo- and enantio-selectivity. Although the yield and diastereoselectivity were still high using 2-methoxymethyl- or 2-(2-methoxyethyl)-benzaldehyde in place of the oxetane analog, the enantioselectivity was negligible (e.e. 2-5%), suggesting that the oxetane residue, apart from enabling closure of a second ring after the Povarov coupling, serves as a hydrogen-bonding acceptor stabilizing the transition state to effect efficient face selectivity. Electron-donating or -withdrawing groups are tolerated on the indole residue, while fluoro- or alkoxy-substitution are tolerated on the benzene ring of the benzaldehyde; a thienyl analog of the latter is also supported (twenty-one examples in all; Y 45-97%; d.r. generally >90:10 to >99:1; e.e. generally 75-98%). F.e.s. Z. Chen, B. Wang, Z. Wang, G. Zhu, J. Sun, Angew. Chem., Int. Ed. *2013, 52* (7), 2027-31 [DOI: 10.1002/anie.201206481]; asym. Povarov reactions s.a. *52,* 363s*83*.

Bismuth(III) nitrate/1-butyl-3-methylimidazolium hexafluorophosphate *$Bi(NO_3)_3$/[bmim]PF_6*
3,6-Dipropargylation of carbazoles with propargyl alcohols s. *43,* 703s*83* H → C-C≡C

Air s. under Laccase *O_2*

Tetra-n-butylammonium hydrogen sulfate *Bu_4NHSO_4*
Knoevenagel condensation-intramolecular hetero-Diels-Alder reaction ○
update s. *54,* 357s*82*; solvent-free synthesis of angular benzopyran-annulated pyrazoles from 2-(alkenyloxy)- or 2-(alkynyloxy)-acetophenones and Δ^2-5-pyrazolones with tetra-*n*-butylammonium hydrogen sulfate as catalyst s. N.J. Parmar, B.R. Pansuriya, B.M. Labana, T.R. Sutariya, R. Kant, V.K. Gupta, Eur. J. Org. Chem. *2012* (30), 5953-64 [DOI: 10.1002/ejoc.201200751].

Nafion-H ←
4-Component Hantzsch 1,4-dihydropyridine synthesis ○
update s. *68,* 368s*82*; *N-unsubst.* 1,4-dihydropyridine ring under *heterogeneous* conditions with recyclable Nafion-H in 60:40 PEG-400/water s. M. Kidwai, R. Chauhan, D. Bhatnagar, A.K. Singh, B. Mishra, S. Dey, Monatsh. Chem. *2012, 143* (12), 1675-80 [DOI: 10.1007/s00706-012-0742-4]; *N-subst.* 1,4-dihydropyridine ring with a nucleoside as the prim. amine component using 12-tungstophosphoric acid [$H_3PW_{12}O_{40}$] in ethanol s. A. Khalafi-Nezhad, M. Divar, F. Panahi, Tetrahedron Lett. *2013, 54* (3), 220-2 [DOI: 10.1016/j.tetlet.2012.11.003]; alternative 3-component route to N-unsubst. 1,4-dihydropyridines from enamines in water with readily recyclable HO_3SO-PEG-OSO_3H (cf. *47,* 727s*82*) s. S. Paul, A.R. Das, ibid. *54* (9), 1149-54 [DOI: 10.1016/j.tetlet.2012.12.079].

Camphorsulfonic acid *RSO_3H*
Friedel-Crafts alkylation with activated alcohols s. *43,* 703s*83* H → R

Pentafluorophenylammonium triflate *$[C_6F_5NH_3]OTf$*
Acid-catalyzed Biginelli synthesis s. *55,* 337s*83* ○

Melamine trisulfonic acid ←
Sym. 1,1-di[het]arylalkanes from oxo compds. s. *5,* 549s*83* CO → CAr_2

12-Tungstophosphoric acid *$H_3PW_{12}O_{40}$*
4-Component Hantzsch 1,4-dihydropyridine synthesis s. *68,* 368s*83* ○

Cesium 12-tungstophosphate nanoparticles ←
Claisen-Schmidt condensation CHO → CH=C-CO
updates s. *1,* 548s*79*; *47,* 710s*76*; solvent-free synthesis of chalcones with recyclable cesium 12-tungstophosphate nanoparticles s. E. Rafiee, F. Rahimi, Monatsh. Chem. *2013, 144* (3), 361-7 [DOI: 10.1007/s00706-012-0814-5].

N-(2-Pyridinioethyl)-o-hydroxyaldimine phosphotungstate ←
Knoevenagel condensation s. *46,* 713s*83* CO → C=C

Iodine/microwaves *I_2/[\\\\]*
Solvent-free catalytic Biginelli synthesis s. *55,* 337s*83* ○

Perchloric acid *$HClO_4$*
Friedel-Crafts alkylation with activated alcohols H → R
update s. *43,* 703s*82*; of electron-rich arenes with tertiary α-hydroxy-ketones or -esters using $HClO_4$ as catalyst in nitromethane s. L. Chen, J. Zhou, Chem. Asian J. *2012, 7* (11), 2510-5 [DOI: 10.1002/asia.201200693]; C_3-benzylation and C_3-allylation of indoles with $PdCl_2(MeCN)_2$ as lone catalyst (tolerating moist air) s. D. Das, S. Roy, Adv. Synth. Catal. *2013, 355* (7), 1308-14 [DOI: 10.1002/adsc.201300048]; with 4-hydroxychromans as the alcohol component using BF_3-etherate s. M. Deodhar, K. Wood, D.StC. Black, N. Kumar, Tetrahedron Lett. *2012, 53* (49), 6697-700 [DOI: 10.1016/j.tetlet.2012.09.117]; 3,6-dipropargylation of carbazoles (with the same or different propargyl alcohols sequentially) with $Bi(NO_3)_3 \cdot 5H_2O$ or TfOH in 1-butyl-3-methylimidazolium hexafluorophosphate as ionic liquid s. G.G.K.S.N. Kumar, K.K. Laali, ibid. *2013, 54* (8), 965-9 [DOI: 10.1016/j.tetlet.2012.12.023]; *o*-propargylation of phenols and subsequent cycloisomerization to benzofurans or 3-chromenes with $FeCl_3 \cdot 6H_2O$ s. F.-Q. Yuan, F.-S. Han, Adv. Synth. Catal. *2013, 355* (2-3), 537-47 [DOI: 10.1002/adsc.201200804]; regioselective synthesis of γ-aryl-α,β-ethylenephosphonic from β,γ-ethylene-α-hydroxyphosphonic acid esters and unactivated arenes with $FeCl_3$, also Friedel-Crafts reaction with α-phosphonylbenzyl alcohols, s. G. Pallikonda, M. Chakravarty, Eur. J. Org. Chem. *2013* (5), 944-51 [DOI: 10.1002/ejoc.201201352]; stereoselective C_8-glycosylation of flavones with aldoses as the alcohol component using $Pr(OTf)_3$ s. R.G. Santos, N.M. Xavier, J.C. Bordado, A.P. Rauter, ibid. *2013* (8), 1441-7 [DOI: 10.1002/ejoc.201201518]; coupling of indoles with 3-ethoxycarbonyl-3-hydroxyisoindolin-1-one as the alcohol component using camphorsulfonic acid s. A. Srivastava, S. Singh, S. Samanta, Tetrahedron

Lett. *2013, 54* (11), 1444-8 [DOI: 10.1016/j.tetlet.2013.01.010]; **organocatalyzed asym. Friedel-Crafts reaction of indoles** by coupling (at C_3) with γ-hydroxy-γ-lactams as the alcohol component using chiral 3,3′-di(9-phenanthryl)-1,1′-binaphthyl-2,2′-diyl hydrogen phosphate as organo-Brønsted acid catalyst s. T. Courant, S. Kumarn, L. He, P. Retailleau, G. Masson, Adv. Synth. Catal. *2013, 355* (5), 836-40 [DOI: 10.1002/adsc.201201008]; alternative *radical* C_3-*sec*-alkylation of pyrroles and indoles **with unactivated sec. alcohols** by a Brønsted acid-catalyzed redox chain reaction cf. X. Han, J. Wu, Angew. Chem., Int. Ed. *2013, 52* (17), 4637-40 [DOI: 10.1002/anie.201209810].

Hexadecyltrimethylammonium bromide s. under o-Pallad(II)acyclic complexes *REt_3NBr*

Silica-supported magnetite nanoparticles *Fe_3O_4-SiO_2*
4,5,6,7-Tetrahydro-1*H*-1,4-diazepine-5-carboxylic acid amides
from ene-1,2-diamines, ketones and isonitriles – 3-Component (4-molecule) synthesis s. *83,* 314

Iron(III) trifluoroacetate *$Fe(OCOCF_3)_3$*
Iron(III) hydrogen sulfate/silica *$Fe(HSO_4)_3/SiO_2$*
3-Component Mannich reaction CHO → CH(NHR)C-CC
update s. *68,* 361s*82*; simple, solvent-free synthesis of **β-aminoketones** with non-corrosive iron(III) trifluoroacetate s. M. Zhang, B. Xiong, W. Yang, L. Chen, F. Wu, Q. Wang, Y. Ding, Synth. Commun. *2012, 42* (19), 2831-43 [DOI: 10.1080/00397911.2011.569866]; *heterogeneous anti*-selective coupling of ketones with ar. aldehydes and ar. amines using silica-supported iron(III) hydrogen sulfate in ethanol s. H. Eshghi, M. Rahimizadeh, M. Hosseini, A. Javadian-Saraf, Monatsh. Chem. *2013, 144* (2), 197-203 [DOI: 10.1007/s00706-012-0800-y]; coupling of amines with acetophenone and ar. aldehydes using inexpensive K_3PO_4 as catalyst at room temperature s. B. Movassagh, S. Khosousi, ibid. *2012, 143* (11), 1503-6 [DOI: 10.1007/s00706-012-0729-1]; diastereoselective **asym. synthesis of β-tosylaminoketones** (cf. *60,* 336s*82*) by coupling tosylamine with unfunctionalized ketones and ar. aldehydes using a quinidine-based *tert*-aminothiourea as organocatalyst s. Q. Guo, J.C.-G. Zhao, Org. Lett. *2013, 15* (3), 508-11 [DOI: 10.1021/ol303315c].

Iron(III) tosylate *$Fe(OTs)_3$*
Catalytic Biginelli synthesis
update s. *55,* 337s*82*; with inexpensive, non-toxic iron(III) tosylate hexahydrate as catalyst, also from acetals in place of aldehydes, s. J.T. Starcevich, T.J. Laughlin, R.S. Mohan, Tetrahedron Lett. *2013, 54* (8), 983-5 [DOI: 10.1016/j.tetlet.2012.12.032]; rapid, simple, solvent-free conversion using dimedone as the β-dicarbonyl component with reusable $ZrOCl_2 \cdot 8H_2O$ as catalyst cf. S. Karami, B. Karami, S. Khodabakhshi, J. Chin. Chem. Soc. *2013, 60* (1), 22-6 [DOI: 10.1002/jccs. 201200145]; *heterogeneous* conversion with recyclable Cu/SiO_2 in ethanol s. M.M. Heravi, N. Karimi, H. Hamidi, H.A. Oskooie, Chin. Chem. Lett. *2013, 24* (2), 143-4 [10.1016/j.cclet.2013.01.003]; Biginelli synthesis *under acid catalysis* with air-stable, cost-effective pentafluorophenylammonium triflate as catalyst s. S. Khaksar, S.M. Vahdat, R.N. Moghaddamnejad, Monatsh. Chem. *2012, 143* (12), 1671-4 [DOI: 10.1007/s00706-012-0752-2]; with a β-keto-amide as the β-dicarbonyl component using TsOH, also 2-amino-3-cyano-4*H*-pyrans with malononitrile using piperidine as catalyst (cf. *61,* 340s*82*), s. R.K. Yadlapalli, O.P. Chourasia, R.S. Perali, Tetrahedron Lett. *2012, 53* (49), 6725-8 [DOI: 10.1016/j.tetlet.2012.10.020]; solvent-free conversion with iodine as catalyst under microwave enhancement s. Q. Liu, N. Pan, J. Xu, W. Zhang, F. Kong, Synth. Commun. *2013, 43* (1), 139-46 [DOI: 10.1080/00397911.2011.593289]; solvent-free *uncatalyzed* conversion under microwave enhancement for preparing sulfonylmethyl derivs. s. P.S. Harikrishnan, S.M. Rajesh, S. Perumal, A.I. Almansour, Tetrahedron Lett. *2013, 54* (9), 1076-9 [DOI: 10.1016/j.tetlet.2012.12.034]; bicyclic products by coupling of cyclopentanone with ar. aldehydes (excess) and [thio]urea in water **with simultaneous C-α-lateral benzylidenation** with added glycerol and boric acid as catalyst s. C. Mukhopadhyay, A. Datta, Synth. Commun. *2013, 43* (3), 438-49 [DOI: 10.1080/00397911.2011.602497].

Iron(III) fluoride s. under Mg *FeF_3*

Iron(III) chloride *$FeCl_3$*
Friedel-Crafts alkylation with activated alcohols s. *43,* 703s*83* H → R

Iron(III) chloride/trimethylsilyl chloride *$FeCl_3/Me_3SiCl$*

Prins cyclization

4-halogenotetrahydropyrans from 3-ethylenealcohols, update s. *43,* 706s*82*; *cis*-2,7-disubst. 4-chlorooxepanes from 4-ethylenealcohols and aldehydes using iron(III) chloride or acetoacetonate with added trimethylsilyl chloride s. M.A. Purino, M.A. Ramírez, A.H. Daranas, V.S. Martín, J.I. Padrón, Org. Lett. *2012, 14* (23), 5904-7 [DOI: 10.1021/ol3028016]; 3-(silylethynyl)-O-heterocyclics and -N-heterocyclics from hydroxy- and amino-functionalized, axially-chiral allenylsilanes, respectively, *with asym. induction* using trimethylsilyl triflate or trifluoroacetic acid as catalyst s. T. Okada, A. Shimoda, T. Shinada, K. Sakaguchi, Y. Ohfune, ibid. *14* (24), 6130-3 [DOI: 10.1021/ol302669q]; 4-chloro-1-tosyl-1,2,3,6-tetrahydropyridines or 3-chloromethyl-1-tosyl-Δ^3-pyrrolines from 2-allenetosylamines **by aza-Prins cyclization** with $FeCl_3$/trimethylsilyl chloride s. J. Cheng, X. Tang, Y. Yu, S. Ma, Chem. Commun. *2012, 48* (99), 12074-6 [DOI: 10.1039/c2cc36941a]; diastereoselective **Prins cyclization-intramolecular Friedel-Crafts reaction** with BF_3-etherate s. B.V.S. Reddy, P.S. Reddy, J.S. Yadav, B. Sridhar, Tetrahedron Lett. *2013, 54* (11), 1392-6 [DOI: 10.1016/j.tetlet.2012.12.113]; 1-(tetrahydrofuran-3-yl)-1,3-dihydroisobenzofurans by diastereoselective **double Prins cyclization** with $Sc(OTf)_3$ s. B.V.S. Reddy, S. Jalal, P. Borkar, J.S. Yadav, P.G. Reddy, A.V.S. Sarma, ibid. *54* (12), 1519-23 [DOI: 10.1016/j.tetlet.2013.01.006].

Bis(cyclooctadiene)nickel(0)/1,3-bis(2,6-diisopropylphenyl)-imidazol-2-ylidene/isopropanol *$Ni(cod)_2$/IPr/i-PrOH*

Pyridines from α,β-ethylenealkoximes and acetylene derivs. s. *83,* 371

Carbonyl(chloro)[6-(di-tert-butylphosphinomethyl)-2,2′-bipyridyl]hydridoruthenium(II)/potassium tert-butoxide ←

Pyrroles from 2-*prim*-aminoalcohols and sec. alcohols s. *83,* 379

Dichloro(pentamethylcyclopentadienyl)rhodium(III) dimer/silver acetate *$[Cp^*RhCl_2]_2$/AgOAc*

Pyridines from O-pivaloyl-α,β-ethyleneoximes and terminal ethylene derivs. via regioselective rhodium(III)-catalyzed ring closure

371. 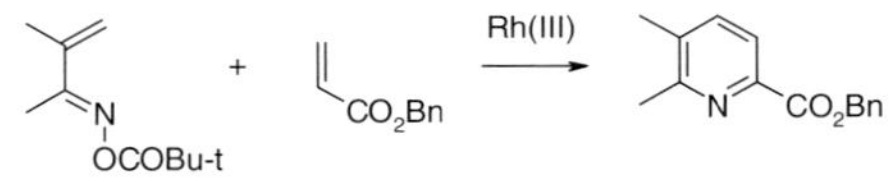

2,3,6-Trisubst. pyridines. A soln. of $[Cp^*RhCl_2]_2$ (2.5 mol%) and benzyl acrylate (1.2 eq.) in 1,2-dichloroethane/acetic acid (2:1; 0.7 ml) added to a mixture of (E)-3-methylbut-3-en-2-one O-pivaloyloxime (0.21 mmol) and AgOAc (2.1 eq.) under argon, the sealed mixture heated at 85° for 14 h, filtered, diluted with methylene chloride, washed with aq. Na_2CO_3, and purified by flash chromatography → benzyl 5,6-dimethylpicolinate. Y 94%. This novel pyridine synthesis gave best results for α-alkyl/aryl-oxime O-pivalates (oxime acetates and parent oxime gave isoxazoles only) with activated terminal alkenes (acrylates, acrylamide, vinyl ketones and styrenes), affording the 6-subst. pyridine regioisomers only (eighteen examples; Y 48-96%). α,β-Disubst. oxime pivalates gave moderate yields (44-55%; two examples;) as did the unsubst. analog (33%), while unactivated aliphatic alkenes gave mixtures of 5/6-subst. pyridines (6/5 1:1.3 to 7.2:1; five examples; 3,3-dimethylbut-1-ene gave a 17% yield of the 5-*tert*-butylpyridine (>20:1)). Internal activated alkenes, β-subst. acrylates, favored formation of 6-subst. pyridines, with regioselectivity (6/5 4.7:1 to >20:1) dependent on the substituent (five examples; Y 41-71%). F.e., optimization and substrate prepn. s. J.M. Neely, T. Rovis, J. Am. Chem. Soc. *2013, 135* (1), 66-9 [DOI: 10.1021/ja3104389]; **1,8-diazapyrenes** from 1,4-naphthoquinone bis(O-acetyloxime) and two [internal] acetylene deriv. molecules with $[Cp^*RhCl_2]_2/Cu(OAc)_2$ s. T. He, P.C. Too, R. Chen, S. Chiba, H. Sun, Chem. Asian J. *2012, 7* (9), 2090-5 [DOI: 10.1002/asia.201200192]; **pyridines from α,β-ethylenealkoximes and acetylene derivs.** with $Ni(cod)_2$/IPr (10 mol%) and *i*-PrOH (5 eq.), 2,3,4,6-tetrasubst. pyridines, s. Y. Yoshida, T. Kurahashi, S. Matsubara, Chem. Lett. *2012, 41* (11), 1498-9 [DOI: 10.1246/cl.2012.1498].

Dichloro(pentamethylcyclopentadienyl)rhodium(III) dimer/silver hexafluoroantimonate ←
2-Aryl-2*H*-indazoles from diar. azo compds. and aldehydes C

372.

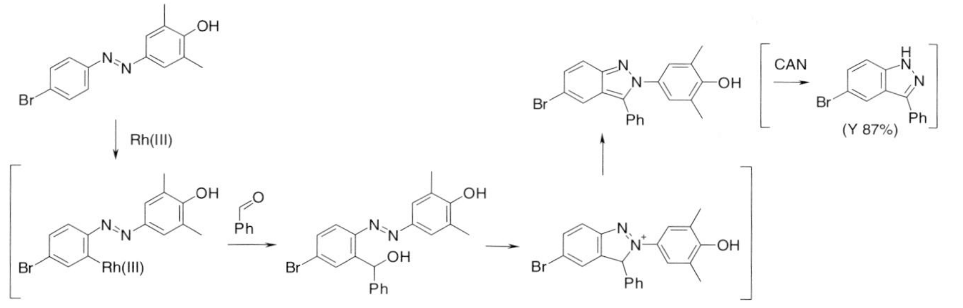

Regioselective rhodium-catalyzed ring closure. A sealed mixture of (E)-4-(4-bromophenylazo)-2,6-dimethylphenol (0.2 mmol), benzaldehyde (2 eq.), $[Cp^*RhCl_2]_2$ (5 mol%), $AgSbF_6$ (20 mol%) and dioxane (1 ml) stirred at 110° for 24 h, cooled to room temp., filtered through Celite, solvent removed *in vacuo*, and the residue purified by chromatography on silica → 4-(5-bromo-3-phenyl-2*H*-indazol-2-yl)-2,6-dimethylphenol. Y 80%. Rare direct conversion of azoarenes to 2-aryl-2*H*-indazoles, via *ortho* C-H activation and Rh(III)-promoted formal [4+1]-cycloaddition with ar. aldehydes, was successful for electron-diverse substrates. Regioselectivity in unsymmetrical azoarenes was controlled by electronic and steric factors, and for azoarenes containing one bulky 3,5-disubst. phenyl group, cyclization occurred near-exclusively on the other arene ring (9:1 to >19:1; twenty-three examples; Y 42-84%). 4-(Phenylazo)methoxybenzene gave a 1.8:1 mixture of regioisomers (Y 66%), via preferential cyclization in the unsubst. phenyl ring, and cyclization with cyclohexanecarboxaldehyde afforded a 36% yield of the indazole. A bulky 4-hydroxy-3,5-dimethylphenyl group was readily removed via oxidation (CAN), affording **N-unsubst. indazoles** in good yields (87-92%; six examples) in the presence of halo, ether and nitro functionality. F.e., optimization and substrate prepn. s. Y. Lian, R.G. Bergman, L.D. Lavis, J.A. Ellman, J. Am. Chem. Soc. *2013*, *135* (19), 7122-5 [DOI: 10.1021/ja402761p].

Dichloro(pentamethylcyclopentadienyl)rhodium(III) dimer/silver hexafluoroantimonate/ pivalic acid ←
Rhodium-catalyzed directed *o*-allylation with 2-ethylenecarbonic acid esters H → C-C=C

373.

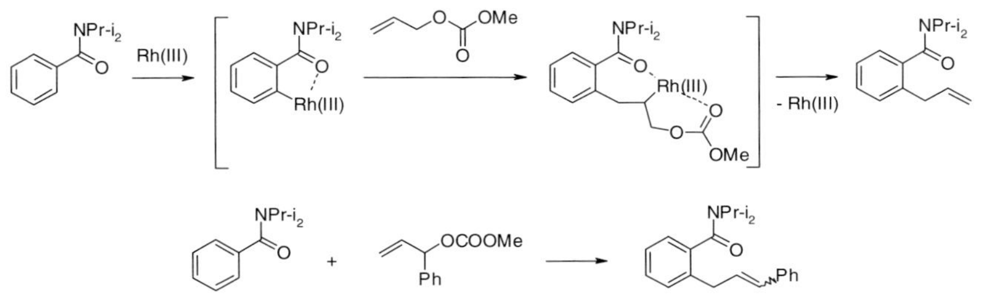

of N,N-disubst. arylcarboxylic acid amides. Dry chlorobenzene (2 ml), N,N-diisopropylbenzamide (0.4 mmol), pivalic acid (1 eq.) and allyl methyl carbonate (2 eq.) added under argon to a Schlenk tube containing $[Cp^*RhCl_2]_2$ (2.5 mol%) and $AgSbF_6$ (30 mol%), the mixture stirred at 35-50° for 18 h, and worked up with purification by flash chromatography on silica gel → 2-allyl-N,N-diisopropylbenzamide. Y 84% (*o*-allyl/*o*-vinyl isomer 37:1). On the 1 gram scale the yield was 86% (*o*-allyl/*o*-vinyl isomer 50:1). **Regioselective γ-allylation** was demonstrated by reaction with methyl 1-phenylallyl carbonate, affording *o*-cinnamyl-N,N-diisopropylbenzamide exclusively at 45° (Y 63%; E/Z 2:1). The procedure is mild, robust, high-yielding with excellent γ-regioselectivity, applicable on the gram scale, and a significant improvement on classical Friedel-Crafts allylation. It is also applicable to a wide array of electron-diverse benzamides, as well as naphthyl and heterocyclic analogs (fifteen examples; Y 48-87%; selectivity up to 50:1), while *N-directed o*-allylation took place with 2-phenylpyridine, 1-phenylpyrazole and 1-(2-pyrimidyl)-indole (nine examples; Y 42-82%). Interestingly, allyl bromide failed to react and allyl phosphates gave lower yields, due, it is supposed, to the undesirable generation of phosphoric acid. Reaction is thought to proceed most likely via migratory insertion of the allyl double bond into a chelated *o*-rhodacycle to give a 7-membered rhodium complex prior to β-oxygen elimination. F.e. and comparison of catalysts and solvents s. H. Wang, N. Schröder, F. Glorius, Angew. Chem., Int. Ed. *2013*, *52* (20), 5386-9 [DOI: 10.1002/anie.201301165].

Palladium- and amine-functionalized mesoporous silica nanoparticles ←
Palladium-catalyzed [Tsuji-Trost] α-allylation s. *38*, 772s*83* H → C-C=C

Palladium(II) acetate/potassium carbonate *Pd(OAc)₂/K₂CO₃*
***o,p*-Difunctionalized phenols from two α-functionalized ketone molecules** ○
Palladium(II)-catalyzed aerobic ring closure with carbon dioxide

374. 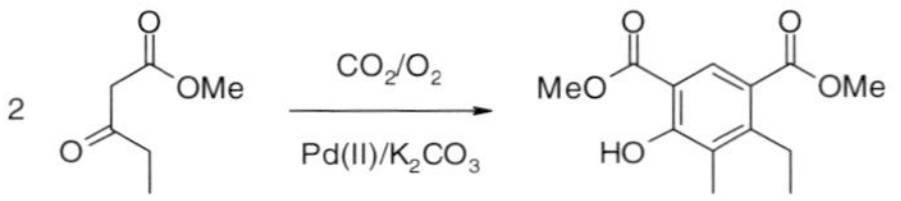

A mixture of methyl 3-oxopentanoate (0.6 mmol), $Pd(OAc)_2$ (5 mol%), K_2CO_3 (0.5 eq.) and DMF (1 ml) stirred at 100° *under air* for 2.5 h (TLC), diluted with water, extracted with ethyl acetate, and purified by flash chromatography → dimethyl 6-ethyl-4-hydroxy-5-methylisophthalate. Y 83%. This novel sequestration of CO_2 was successful for linear aliphatic ketones functionalized at the α-position (ester, ketone or ar. sulfone), affording the title phenols within 2-5 h at 100° (twelve examples; Y 63-92%). Control experiments showed that the reaction did not occur under either O_2 or CO_2 atmospheres, but was effective in air or (in marginally higher yield) in a 1:4 mixture of O_2/CO_2. F.e. and optimization s. Q. Gao, X.-c. Tan, Y.-m. Pan, H.-s. Wang, Y. Liang, Chem. Commun. *2012*, *48* (99), 12080-1 [DOI: 10.1039/c2cc37194d].

Palladium(II) acetate/1,10-phenanthroline/cesium carbonate *Pd(OAc)₂/phen/Cs₂CO₃*
Palladium-catalyzed C_3-arylation of pyridines with aryl tosylates H → Ar

375. 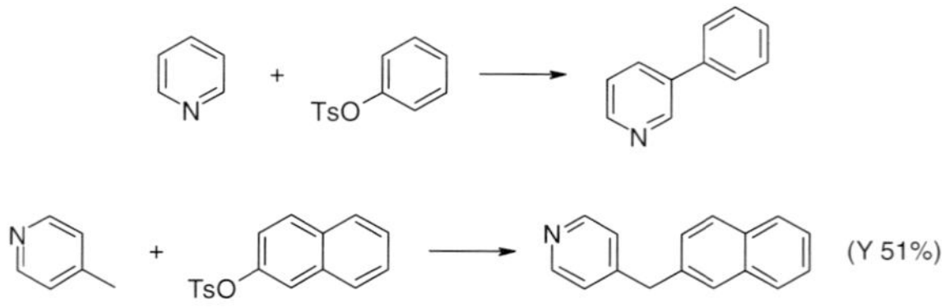

The first example of *undirected* C_3-arylation of pyridines is reported. **E:** A mixture of $Pd(OAc)_2$ (10 mol%), 1,10-phenanthroline (60 mol%), Cs_2CO_3 (3 eq.), phenyl tosylate (0.5 mmol) and pyridine (3 ml) under N_2 in a sealed tube stirred at 150° for 48 h, cooled to room temp., and worked up with purification by flash chromatography → 3-phenylpyridine. Y 65%. The procedure is applicable to the C_3-arylation of pyridine and certain alkylpyridines with electron-diverse aryl (incl. naphthyl) tosylates, notably tolerating methyl, fluoro, *tert*-butyl, methoxy and CF_3 groups (twenty-four examples; Y 47-67%). Among these, 3-methyl- and 3-ethyl-pyridine were arylated at the C_5-position, and 4-methylpyridine at the C_3-position by *phenyl* tosylates; more hindered naphthyl tosylates, however, *arylated the methyl group* of 4-methylpyridine instead (three examples; Y 47-59%). 2-Methylpyridines resisted arylation completely, and less stable aryl triflates were ineffective as arylating agents (being cleaved to the parent phenol). Reaction likely proceeds with formation of an intermediate aryl(3-pyridyl)palladium(II) complex, but a Pd(II)/Pd(IV) catalytic cycle cannot be excluded. F.e. and comparison of catalysts and bases s. F. Dai, Q. Gui, J. Liu, Z. Yang, X. Chen, R. Guo, Z. Tan, Chem. Commun. *2013*, *49* (41), 4634-6 [DOI: 10.1039/c3cc41066h].

Palladium(II) acetate/benzoquinone/lithium tert-butoxide *Pd(OAc)$_2$/BQ/t-BuOLi*
γ,δ-Ethyleneketones from 2-ethylenealcohols and ketones ←
Stereoselective palladium(II)-catalyzed oxidative cross-coupling via N-tosylhydrazones s. *83*, 399

Palladium(II) acetate/trifluoroacetic anhydride *Pd(OAc)$_2$/(CF$_3$CO)$_2$O*
Friedel-Crafts acylation H → COR
with carboxylic acids cf. *10*, 594s*80*; **N-directed *o*-acylation** with carboxylic acids under palladium catalysis with Pd(OAc)$_2$/trifluoroacetic anhydride s. J. Lu, H. Zhang, X. Chen, H. Liu, Y. Jiang, H. Fu, Adv. Synth. Catal. *2013*, *355* (2-3), 529-36 [DOI: 10.1002/adsc.201200743]; with aromatic or aliphatic carboxylic acid esters using $AlCl_3$ as Lewis acid, also intramolecular conversion, s. S.P. Chavan, S. Garai, A.K. Dutta, S. Pal, Eur. J. Org. Chem. *2012* (35), 6841-5 [DOI: 10.1002/ejoc.201201181].

Palladium(II) acetate/4,5-bis(diphenylphosphino)-9,9-dimethylxanthene/cesium carbonate ←
α-Arylation of ketones H → Ar
via enol acetates under palladium catalysis cf. *38*, 932; monoarylation of acetone with air- and moisture-resistant aryl imidazol-1-ylsulfonates using Pd(OAc)$_2$/XantPhos and Cs_2CO_3 as base s. L. Ackermann, V.P. Mehta, Chem. Eur. J. *2012*, *18* (33), 10230-3 [DOI: 10.1002/chem.201201394].

Tris(dibenzylideneacetone)dipalladium/1,3-bis(diphenylphosphino)propane or 2,2′-bis(diphenylphosphino)-1,1′-binaphthyl/potassium carbonate *[Pd(0)]/K$_2$CO$_3$*
N-Protected 2- or 3-vinyl-1,2,3,4-tetrahydroquinoline-4,4-dicarboxylic acid esters from α-(*o*-aminoaryl)malonic acid esters and 2-ene-1,4-diol esters ○
Effect of N-protective group on regioselectivity of palladium(0)-catalyzed ring closure

376. 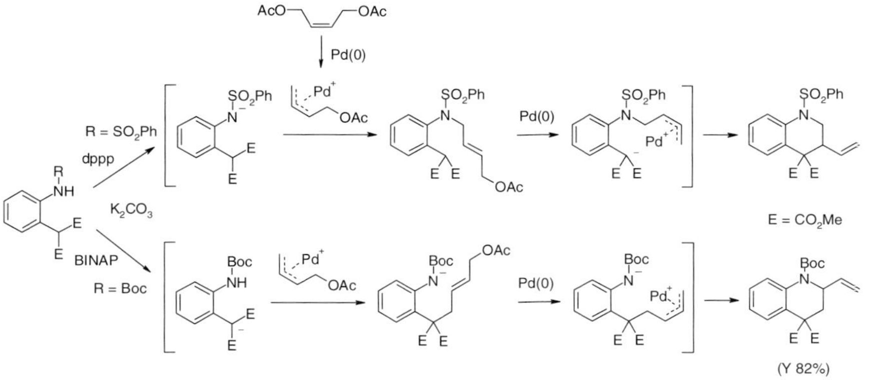

Palladium(0)-catalyzed cyclization of a series of N-protected *o*-aminophenylmalonates in the presence of base is reported to afford 2- *or* 3-vinyl-1,2,3,4-tetrahydroquinoline-4,4-dicarboxylic acid esters, depending on the nature of the N-protecting group (2-vinyl derivs. from carbamates and 3-vinyl derivs. from sulfonamides). The observed regioselectivity was rationalized in terms of preferential formation of either aza- or malonate anions. **E:** Pd$_2$(dba)$_3$·CHCl$_3$ (5 mol%), dppp (20 mol%) and K_2CO_3 (4 eq.) added to a stirred soln. of dimethyl 2-(2-benzenesulfonamidophenyl)-malonate (0.0795 mmol) and (Z)-1,4-diacetoxy-2-butene (1.2 eq.) in THF (2 ml) at room temp., the mixture stirred for 30 min, then at 80° for 2 h, filtered through silica gel, and purified by chromatography on silica → dimethyl 1-(benzenesulfonyl)-3-vinyl-2,3-dihydroquinoline-4,4(1*H*)-dicarboxylate. Y 89%. Cyclization of *o*-sulfonamidophenyl malonates with the (Z)-bis-allylic acetate afforded 3-vinyl derivs. (six examples; Y 80-92%; a bulky 2-naphthalenesulfonamide gave 55%), while carbamate analogs gave the best results in the presence of (±)-BINAP, affording 2-vinyl regioisomers (five examples; Y 66-85%). Replacing the (Z)-bis-allylic acetate with the (E)-stereoisomer or with the positional isomer, 3,4-diacetoxy-1-butene, afforded the same regioisomeric products (two examples of each regioisomer in variable yields) indicating the generation of a common π-allyl intermediate. Addition of a chiral ligand gave modest enantioselectivity (e.e. 55-76%) for substrates bearing a variety of electron-rich or -poor arenesulfonamide protecting groups; *p*-nitrobenzenesulfonamide derivs. were shown to be particularly effective (e.e. 99%), however, and, on optimization, it was even established that regioselectivity for these derivs. could be controlled to some extent by the choice of base, with K_2CO_3 promoting formation of a chiral 3-vinyl deriv. (Y 79%; e.e. 99%), and the 2-vinyl regioisomer being formed preferentially in the presence of K_3PO_4 (Y 46%; e.e. 92%, with 17% 3-vinyl isomer: e.e. 98%) or KOH (Y 52%; e.e. 83%, single regioisomer). F.e., optimization, reactions of products, and substrate prepn. s. M. Yoshida, Y. Maeyama, K. Shishido, Tetrahedron *2012*, *68* (48), 9962-72 [DOI: 10.1016/j.tet.2012.09.075].

Palladium phosphine complexes *[Pd]*
Palladium-catalyzed [Tsuji-Trost] α-allylation H → C-C=C
with allyl esters, update s. *38,* 772s*82*; of 2,2,2-trifluoroethyl phenyl sulfone with allyl carbonates using $Pd(PPh_3)_4$ as catalyst s. W. Zhang, Y. Zhao, C. Ni, T. Mathew, J. Hu, Tetrahedron Lett. *2012, 53* (48), 6565-8 [DOI: 10.1016/j.tetlet.2012.09.094]; of aldehydes and ketones with 4-acoxymethylcoumarins under palladium catalysis with added pyrrolidine (for *in situ*-enamine formation) s. K. Cattopadhyay, A. Recio III, J.A. Tunge, Org. Biomol. Chem. *2012, 10* (34), 6826-9 [DOI: 10.1039/c2ob25962a]; with palladium- and amine-functionalized mesoporous silica nanoparticles as co-operative catalyst for *heterogeneous* allylation/bis(allylation) of ethyl acetoacetate s. A.T. Dickschat, F. Behrends, S. Surmiak, M. Weiß, H. Eckert, A. Studer, Chem. Commun. *2013, 49* (22), 2195-7 [DOI: 10.1039/c3cc00235g]; palladium-catalyzed α-allylation of MIDA-type α-borylaldehydes **with 2-ethylenealcohols** (cf. *45,* 456) via a C,O-bis(enolate) s. J.D. St. Denis, Z. He, A.K. Yudin, Org. Biomol. Chem. *2012, 10* (39), 7900-2 [DOI: 10.1039/c2ob26503f].

Chiral palladium phosphine complexes *[Pd]**
Transition metal-catalyzed asym. α-allylation
update s. *48,* 772s*82*; regioselective asym. C_3-allylation of 2-benzofuranones under palladium catalysis with ion-paired chiral ligands, e.g. chiral [*o*-(diarylphosphino)benzyl]trimethylammonium 5,5′,6,6′,7,7′,8,8′-octahydro-1,1′-binaphthyl-2,2′-diyl phosphate, s. K. Ohmatsu, M. Ito, T. Kunieda, T. Ooi, J. Am. Chem. Soc. *2013, 135* (2), 590-3 [DOI: 10.1021/ja312125a]; asym. C_3-allylation of 3-aryloxindoles with allylidene dipivalate s. B.M. Trost, J.T. Masters, A.C. Burns, Angew. Chem., Int. Ed. *2013, 52* (8), 2260-4 [DOI: 10.1002/anie.201209783]; regio- and diastereo-selective asym. C_3-allylation of Δ^2-5-oxazolones with a chiral cationic iridium(I) 1,1′-binaphthyl-2,2′-diyl phosphoramidite complex and a racemic silver(I) biaryl-2,2′-diyl phosphate providing the essential phosphate counterion (cf. *80,* 84s*82*) s. W. Chen, J.F. Hartwig, J. Am. Chem. Soc. *2013, 135* (6), 2068-71 [DOI: 10.1021/ja311363a]; asym. synthesis of β-allenylmalonic from malonic acid esters and 1-subst. acoxy-2-allenes with $[Pd(\pi\text{-cinnamyl})Cl]_2$/(R)-DTBM-SegPhos/$K_2CO_3$ and added water (1 eq.) s. Q. Li, C. Fu, S. Ma, Angew. Chem., Int. Ed. *2012, 51* (47), 11783-6 [DOI: 10.1002/anie.201204346]; regioselective **organocatalyzed asym. α-allylation** (cf. *82,* 366) with Baylis-Hillman carbonates using chiral 2-acetylamino-2′-(diphenylphosphino)-1,1′-binaphthyl as catalyst, notably for asym. C_3-allylation of oxindoles and 2-benzofuranones, s. D. Wang, Y.-L. Yang, J.-J. Jiang, M. Shi, Org. Biomol. Chem. *2012, 10* (35), 7158-66 [DOI: 10.1039/c2ob25694k]; asym. C -allylation of 2-acyl-1-cyano-1,2-dihydroisoquinolines (Reissert compds.) with quinidine as catalyst s. T.-Y. Qin, W.-W. Liao, Y.-J. Zhang, S.X.-A. Zhang, ibid. *2013, 11* (6), 984-90 [DOI: 10.1039/c2ob27269e]; regioselective asym. allylation of N-subst. itaconimides with hydroquinidine s. W. Yang, D. Tan, L. Li, Z. Han, L. Yan, K.-W. Huang, C.-H. Tan, Z. Jiang, J. Org. Chem. *2012, 77* (15), 6600-7 [DOI: 10.1021/jo3012539].

Chiral palladium phosphine or phosphoromonoamidite complexes *[Pd]**
Regioselective asym. C_3-allylation of indoles
with acoxy-2-ethylenes s. *79,* 384s*83*

Palladium(II) chloride *$PdCl_2$*

Regioselective C-allylation of 5-membered heteroarenes $H \rightarrow C\text{-}C{=}C$

with acoxy-2-ethylenes s. *29*, 747s*83*

Palladium-catalyzed synthesis of 1,3-cyclopentadienes from (Z)-2-en-4-yn-1-yl acetates via 1,5- and 1,2-migration

377.

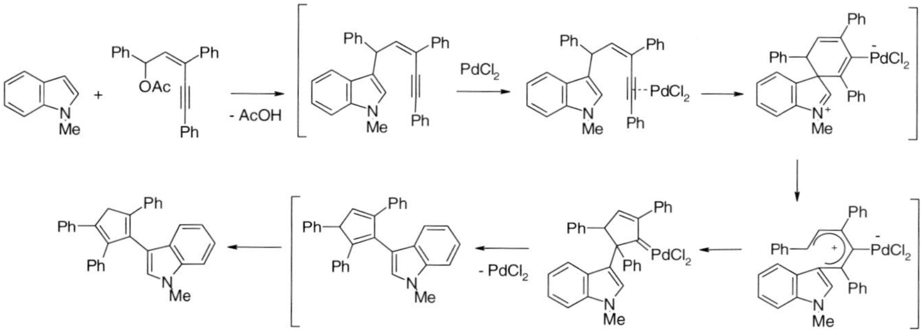

3-(3-Aryl-1,3-cyclopentadien-2-yl)indoles. A mixture of (Z)-1,3,5-triphenylpent-2-en-4-ynyl acetate (0.1 mmol), N-methylindole (1.1 eq.) and $PdCl_2$ (10 mol%) in acetonitrile stirred at 80° for 0.5 h (TLC), cooled to room temp., quenched with water, extracted with ethyl acetate, and purified chromatographically → 1-methyl-3-(2,4,5-triphenylcyclopenta-1,4-dienyl)-1*H*-indole. Y 86%. This cyclopentadiene formation is believed to proceed by Friedel-Crafts reaction, 1,5-indole migration via a spirocyclic intermediate, Nazarov-type electrocyclization, 1,2-indole migration and double bond migration. It was successful for electron-diverse triarylpent-2-en-4-ynyl acetates reacting with N-methylindoles, and tolerated 5/7-substituents (Br, OMe, Me, CN) on the indole ring (nineteen examples; Y 25%, 45-86%). A 5-propyl substituent was tolerated in the enyne substrate (Y 57%), but its 1-propyl isomer gave only an initial Friedel-Crafts arylation product, a 3-(pent-2-en-4-ynyl)indole, in 40% yield (two other examples gave 50% and 84% yields). Reaction of NH-indoles was effective for the parent indole (Y 68%) but 5/7-subst. 1*H*-indoles gave variable yields (36-50%; four examples). F.e.s. M.-J. Zhong, X.-C. Wang, H.-T. Zhu, J. Hu, L. Liu, X.-Y. Liu, Y.-M. Liang, J. Org. Chem. *2012*, *77* (23), 10955-61 [DOI: 10.1021/jo3021055].

Palladium(II) chloride/1,3-bis(diphenylphosphino)propane/cesium carbonate/ tert-butyl isocyanide ←

2-Aryl-Δ^2-imidazolines from 1,2-diamines and aryl triflates s. *83*, 424

Palladium(II) iodide/potassium iodide *PdI_2/KI*

Furan-3-carboxylic acid esters from 3-acetylene-1,2-diols via palladium(II)-catalyzed carbonylative ring closure

378.

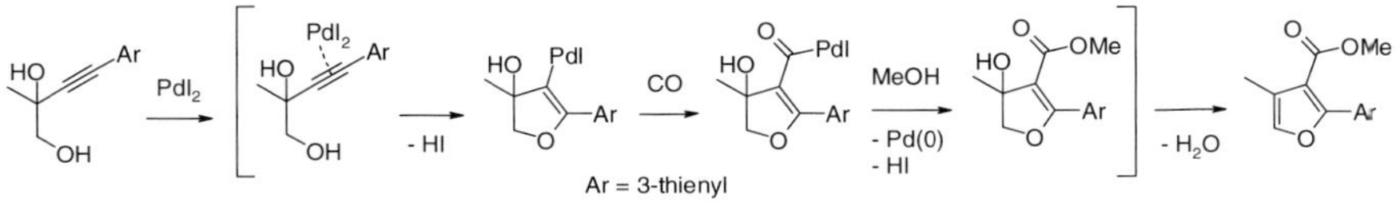

in one pot. A mixture of PdI_2 (2 mol%), KI (10 mol%), anh. methanol (28 ml) and 2-methyl-4-(thien-3-yl)but-3-yne-1,2-diol (1.39 mmol) stirred in a sealed autoclave under CO (32 atm.) and air (8 atm.) at 100° for 2 h, the autoclave cooled, degassed, solvent evaporated, and the residue purified by chromatography on silica → methyl 4-methyl-2-(thien-3-yl)furan-3-carboxylate. Y 88%. This general and atom-economical carbonylative 5-*endo-dig* cyclization was successful for but-3-yne-1,2-diols carrying (het)ar. or alkyl (incl. cyclohexenyl) at the alkyne terminus, alkyl, phenyl or *alkynyl* at C-2, and H or alkyl at C-1 (twenty-one examples; Y 56-93%). Cyclization of analogs containing a tert. alcohol at C-1 afforded the corresponding **4-methylene-4,5-dihydrofuran-3-carboxylic acid esters** (seven examples; Y 55-70%).

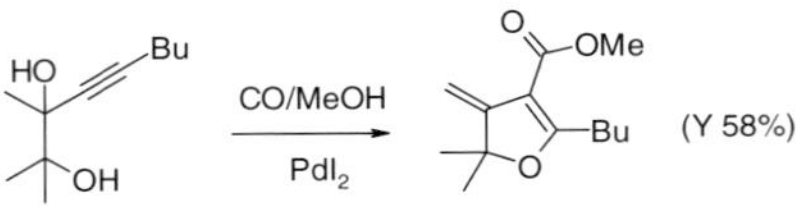

F.e., optimization and substrate prepn. s. B. Gabriele, R. Mancuso, V. Maltese, L. Veltri, G. Salerno, J. Org. Chem. *2012, 77* (19), 8657-68 [DOI: 10.1021/jo301628n].

Bis(acetonitrile)dichloropalladium(II) *$PdCl_2(MeCN)_2$*
C_3-Benzylation and C_3-allylation of indoles s. *43,* 703s*83* H → Bn or C-C=C

o-Pallad(II)acyclic complexes/2-dicyclohexylphosphino-2′,6′-dimethoxybiphenyl/ hexadecyltrimethylammonium bromide/triethylamine ←
Sonogashira coupling C≡CH → C≡CAr
with aryl sulfonates cf. *79,* 379; **with aryl imidazol-2-ylsulfonates** *in water* under microwave enhancement with a *p*-hydroxyacetophenone oxime-based *o*-pallad(II)acyclic complex and 2-dicyclohexylphosphino-2′,6′-dimethoxybiphenyl [S-Phos] as ligand in the presence of hexadecyltrimethylammonium bromide/Et_3N s. J.F. Cívicos, D.A. Alonso, C. Nájera, Adv. Synth. Catal. *2013, 355* (1), 203-8 [DOI: 10.1002/adsc.201200629].

Chiral iridium(III) di(phosphine) complexes/cesium carbonate *[Ir(III)]*/Cs_2CO_3*
Asym. synthesis of 3-ethylenealcohols H → C-C=C
from *in situ*-generated aldehydes and acoxy-2-ethylenes
by transfer-hydrogenative coupling with alcohols s. *79,* 385; diastereoselective (*anti* or *syn*) asym. synthesis of 2′-alkylated 3-ethylenealcohols with a cyclometalated π-allyliridium 3,4-dinitro-C,O-benzoate complex modified by (R)- or (S)-Cl,MeO-BIPHEP and Cs_2CO_3 as base s. D.C. Schmitt, A.-M.R. Dechert-Schmitt, M.J. Krische, Org. Lett. *2012, 14* (24), 6302-5 [DOI: 10.1021/ol3030692]; diastereoselective asym. coupling of partially protected polyols with an iridium(III) (S)- or (R)-SegPhos complex and an added arylcarboxylic acid as co-catalyst with the same base s. A.-M.R. Dechert-Schmitt, D.C. Schmitt, M.J. Krische, Angew. Chem., Int. Ed. *2013, 52* (11), 3195-8 [DOI: 10.1002/anie.201209863].

(1,5-Cyclooctadiene)[2,6-bis(diisopropylphosphinamino)-4-phenyl-1,3,5-triazine]iridium(I)/potassium tert-butoxide *[Ir(I)]/KOBu-t*
Pyrroles from 2-*prim*-aminoalcohols and sec. alcohols ○
Regioselective iridium-catalyzed dehydrogenative [3+2]-cyclocondensation

379.

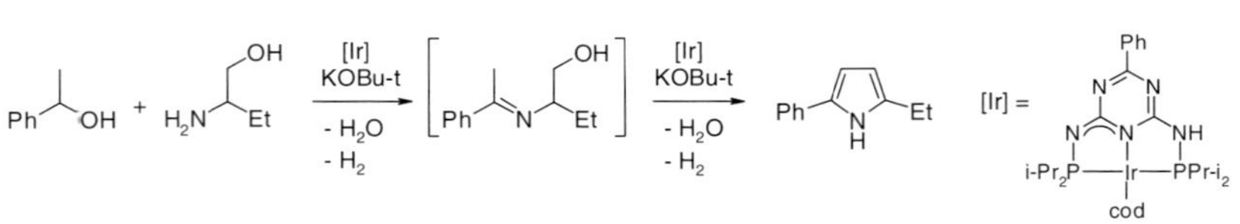

A sustainable iridium-catalyzed pyrrole synthesis is reported in which secondary alcohols and 2-*prim*-aminoalcohols, which may be obtained from renewable lignocellulosic materials, are deoxygenated and linked selectively via C-N and C-C bond formation. Two molecules of water as well as two molecules of hydrogen are produced in the course of the reaction but reduction (as in the 'borrowing hydrogen' methodology) is suppressed. **E:** Ir catalyst (0.05 mol%), THF (10 ml), 1-phenylethanol (2 eq.), 2-amino-1-butanol (10 mmol) and KOBu-*t* (1.1 eq.) combined in a pressure tube under dry N_2 using glovebox techniques, the tube closed with a silicone tube as a semi-permeable membrane, the mixture stirred at 90° for 24 h, cooled to room temp., quenched with water, the aq. layer extracted with ether, the combined organic layers dried (Na_2SO_4), concentrated *in vacuo*, and the residue purified by chromatography on silica gel → 2-ethyl-5-phenylpyrrole. Y 93%. The method tolerates a variety of functional groups, including olefins, chlorides, bromides, ferrocenyl, amines and hydroxyl groups, affording a range of 2,5-di- or 2,3,5-tri-subst. pyrroles in yields of 37-97% (thirty-two further examples, incl. several new compounds). Reaction of hexane-2,5-diol produced 2-γ-hydroxypyrroles which could undergo sequential reaction with a second aminoalcohol molecule to afford sym. (Y 42%) or unsym. (Y 83%) **1,2-di(pyrrol-2-yl)-ethanes**.

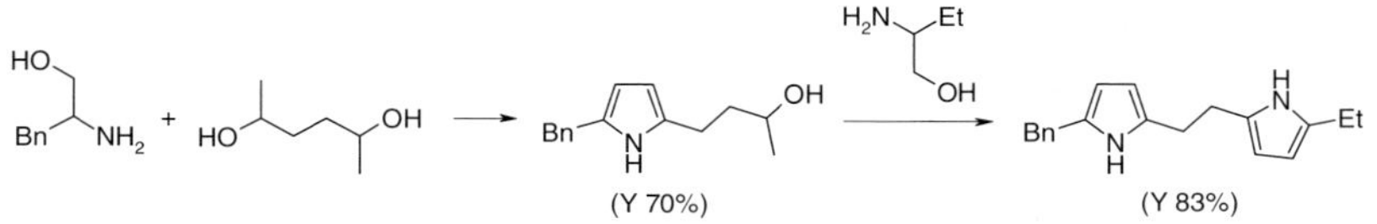

Furthermore, glycols could be used for N-alkylation of prim. ar. amines (or 2-aminopyridine) by the 'borrowing hydrogen' methodology with the same catalyst, then be linked selectively with a 2-*prim*-aminoalcohol molecule to afford **3-arylaminopyrroles** (three examples; Y 82-87%).

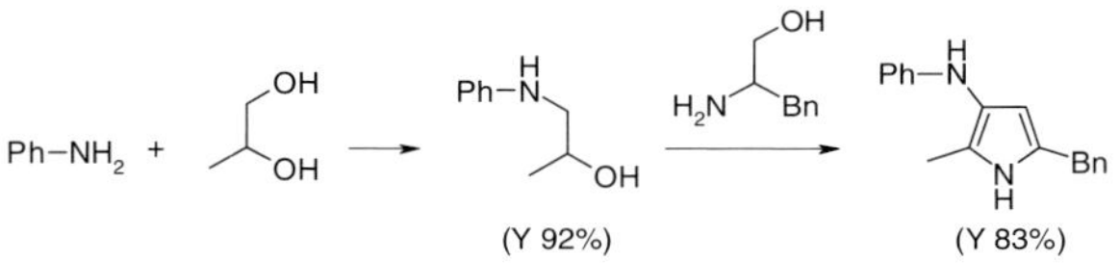

The catalyst is a crystalline material that can be handled in air, and in some cases loadings as low as 0.03 mol% sufficed; the catalyst resting state was found to be an iridium trihydride complex, however, formed by oxidation of the alcohol. It appears dehydrogenation of the secondary alcohol is significantly faster than that of the aminoalcohol, affording a ketone, which undergoes condensation with the aminoalcohol to form a Schiff base; subsequent intramolecular C-alkylation is fast in comparison to the oxidation of the secondary alcohol, and the role of the base is to mediate the C-C coupling/condensation step as well as accelerating the iridium-catalyzed alcohol oxidation. F.e.s. S. Michlik, R. Kempe, Nat. Chem. *2013*, *5* (2), 140-4 [DOI: 10.1038/nchem.1547]; with carbonyl(chloro)[6-(di-*tert*-butylphosphinomethyl)-2,2′-bipyridyl]hydridoruthenium(II)/KOBu-*t* cf. D. Srimani, Y. Ben-David, D. Milstein, Angew. Chem., Int. Ed. *2013*, *52* (14), 4012-5 [DOI: 10.1002/anie.201300574].

(1,5-Cyclooctadiene)[2,6-bis(diisopropylphosphinamino)-4-(p-trifluoromethyl-phenyl)-1,3,5-triazine]iridium(I)/sodium tert-butoxide ←

Pyridines from 3-*prim*-aminoalcohols and prim. or sec. alcohols ○

Regioselective iridium-catalyzed dehydrogenative [4+2]-cyclocondensation

380.

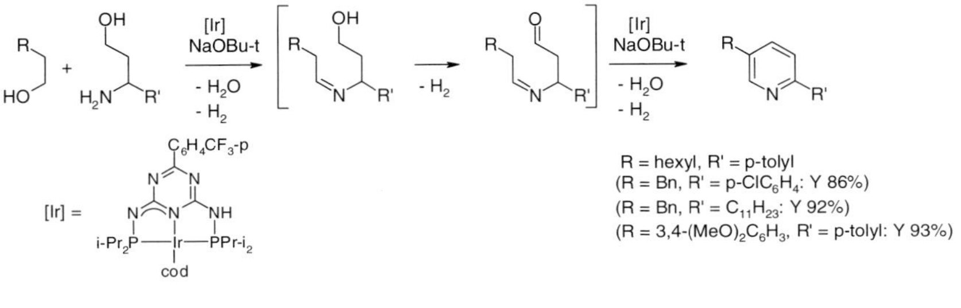

N-Heteroarene formation from aminoalcohols and alcohols via an iridium-catalyzed dehydrogenative Schiff-base reaction (cf. *83*, 379) has been extended to pyridine synthesis via loss of two water molecules and three hydrogen molecules. It is notably applicable to challenging **unsym. pyridines. E:** A mixture of Ir catalyst (0.01 *M* in THF; 0.5 mol%), THF (10 ml), 1-octanol (4 eq.), 3-amino-3-*p*-tolylpropan-1-ol (3 mmol) and NaOBu-*t* (1.1 eq.) in a pressure tube with a semi-permeable membrane stirred at 90° for 24 h then at 130° for 24 h under dry N_2 or argon, and worked up with purification by column chromatography → 5-hexyl-2-*p*-tolylpyridine. Y 94%. The method uses starting materials that may be obtained from renewable resources or waste feedstock. It has broad scope, being applicable to a variety of 2,4-, 2,5-, 2,6- and 3,5-disubst. pyridines (twenty-six further examples; Y 35-96%), including fused pyridines from cyclic sec. alcohols, and tolerating functional groups such as chlorides, amines, ethers, olefins, heteroarenes, and ferrocenyl moieties.

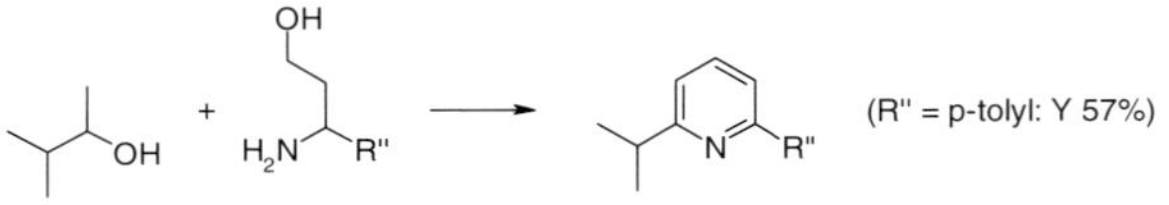

The catalyst resting state was found to be an [extremely air sensitive] iridium(III) trihydride complex. F.e.s. S. Michlik, R. Kempe, Angew. Chem., Int. Ed. *2013*, *52* (24), 6326-9 [DOI: 10.1002/anie.201301919].

Chloro(1,5-cyclooctadiene)iridium(I) dimer/chiral 1,1′-binaphthyl-2,2′-diyl phosphoramidites/cinchona-based prim. amines/trichloroacetic acid ←

Regio-, enantio- and diastereo-divergent α-allylation of aldehydes with 2-ethylenealcohols under dual catalysis H → C-C=C

381.

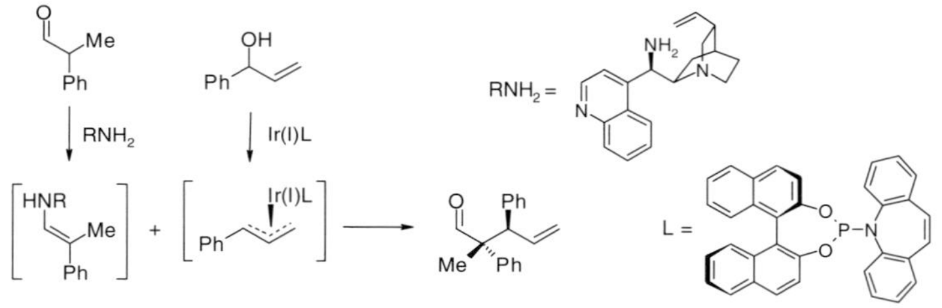

In the asym. α-allylation of α-subst. aldehydes with 1-subst. 2-ethylenealcohols each of the four possible diastereoisomers (possessing adjacent quaternary and tertiary stereocenters) can be secured regioselectively *with >99% enantioselectivity and very high diastereoselectivity* under dual catalysis simply by judicious choice of a chiral iridium(I) phosphoromonoamidite and a cinchona-based prim. amine as co-catalysts which *independently activate* the allyl alcohol and the aldehyde, respectively. **E:** A mixture of $[Ir(cod)Cl]_2$ (2 mol%) and the (R)-1,1′-binaphthyl-2,2′-diyl phosphoramidite (8 mol%) in 1,2-dichloroethane (0.5 ml) vigorously stirred under N_2 in a screw-capped glass vial for 15 min, the startg. allyl alcohol (0.25 mmol), aldehyde (1.1 eq.), cinchona-based prim. amine (10 mol%) and trichloroacetic acid (50 mol%) added sequentially to the resulting red soln., stirred at room temp. for 24 h, concentrated *in vacuo*, and worked up with purification by flash chromatography → (2R,3R)-2-methyl-2,3-diphenylpent-4-enal. Y 77% (d.r. >20:1; e.e. >99%). With the (S)-enantiomer of the phosphoromonoamidite (and all else the same) the (2R,3S)-product was obtained (Y 78%; d.r. 15:1; e.e. >99%); with the (R)-phosphoromonoamidite and the pseudo-enantiomeric prim. amine, however, the (2S,3R)-product was formed (Y 80%; d.r. 20:1; e.e. >99%) while the fourth diastereoisomeric product (the (2S,3S)-isomer) was obtained with the pseudo-enantiomeric prim. amine and the (S)-phosphoromonoamidite (Y 71%; d.r. >20:1; e.e. >99%). Each catalyst 'exerts high local stereocontrol irrespective of the other's inherent preference' in two distinct and highly face-selective catalytic cycles with minimal matched/mismatched interactions. The methodology is applicable to a wide range of α-arylallyl alcohols bearing electron-withdrawing or -donating groups on the benzene ring, while various functions and heterocyclic residues are tolerated on the aldehyde (twenty-four substrate pairs; Y 42-86%; d.r. 4:1 to >20:1; e.e. >99%). F.e.s. S. Krautwald, D. Sarlah, M.A. Schafroth, E.M. Carreira, Science *2013*, *340* (6136), 1065-8 [DOI: 10.1126/science.1237068].

Via intermediates *v.i.*

Asym. synthesis of 2-arylpiperidines ←
from pyridine N-oxides and arylmagnesium halides
via stereoselective reduction of 6-aryl-1-hydroxy-1,2,3,6-tetrahydropyridines

382. 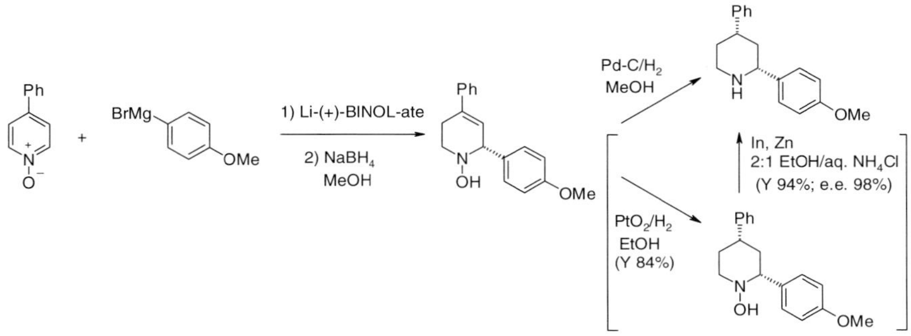

Chiral *cis*-2,4-diarylpiperidines. (+)-BINOL (1.2 eq.) dried *in vacuo* for 30 min then dissolved in THF, PhLi (1.8 *M*; 2.4 eq.) added dropwise at -78° under inert atmosphere, the mixture warmed to room temp. and stirred for 30 min, re-cooled to -78°, 4-phenylpyridine N-oxide (2 mmol) added, the mixture stirred for a further 60 min at room temp., re-cooled to -78°, startg. arylmagnesium bromide (3 eq.) added dropwise, stirred for 40 min (followed by LC-MS), a freshly-prepared suspension of $NaBH_4$ (3 eq.) in methanol (2 ml) added, the mixture allowed to reach room temp., methylene chloride (20 ml) and water (10 ml) added, the organic layer shaken with NaOH (10%; 3 x 10 ml) then washed with water, dried (Na_2SO_4), concentrated *in vacuo*, and the residue purified by flash silica column chromatography → (-)-2-(4-methoxyphenyl)-4-phenyl-5,6-dihydropyridin-1(2*H*)-ol (Y 94%; e.e. 74%, increased to 98% after recrystallization from ethanol), a portion of which (0.064 mmol) dissolved in methanol (2 ml) then added to a suspension of Pd-C (10% by weight) in methanol (1 ml), previously stirred for 30 min under H_2, the mixture stirred for 3 h under H_2, filtered (Celite), concentrated *in vacuo*, and purified by flash silica column chromatography → (-)-*cis*-2-(4-methoxyphenyl)-4-phenylpiperidine (Y 88%; e.e. 90%). Seven further examples with 4-phenylpyridine-N-oxide and arylmagnesium bromides (optionally bearing Me, Ph, Cl or MeO groups) gave the N-hydroxypyrideines in yields of 51-86% and e.e. of 46-80% (99% after recrystallization in one case), while two examples with pyridine N-oxide gave yields of 68-94% and e.e. of 55-80%. Results were improved on a 2 mmol scale compared to a 0.17 mmol scale. Results were poor with methyl-, phenylethynyl- or vinyl-Grignard compds. One further 4-phenylated example underwent hydrogenation over Pd-C to give the piperidine in 82% yield and 90% e.e. (from an intermediate with an e.e. of 99%), but a 4-unsubst. deriv. gave a racemic product, presumably due to double bond migration. Stepwise reduction under milder conditions with PtO_2/H_2 in ethanol followed by In/Zn in aq. NH_4Cl/ethanol was therefore examined, giving chiral 2-aryl- or 2,4-diaryl-1-hydroxypiperidines in yields of 84-96% (three examples) and the corresponding piperidines in yields of 71-94% (one 2,4-diaryl deriv.: e.e. 98%, two 4-unsubst. 2-aryl-derivs.: e.e. 90-98%). The BINOL may be recovered easily for reuse. F.e.s. M. Hussain, T. Sainte-Luce Banchelin, H. Andersson, R. Olsson, F. Almqvist, Org. Lett. *2013*, *15* (1), 54-7 [DOI: 10.1021/ol303085q]; regio- and stereo-selective synthesis of ***trans*-2,3-disubst. 2,3-dihydropyridine N-oxides** by sequential addition of Grignard compd. and an electrophile (such as an oxo compd.), also reduction to **piperidines**, s. H. Andersson, M. Gustafsson, D. Boström, R. Olsson, F. Almqvist, Angew. Chem., Int. Ed. *2009*, *48* (18), 3288-91 [DOI: 10.1002/anie.200900189]; **2-subst. pyridines** from pyridine N-oxides and organomagnesium chlorides followed by treatment with trifluoroacetic anhydride, **also 2,6-disubst. derivs.** by sequential addition of two different organomagnesium chloride molecules with addition of *o*-chloranil then trifluoroacetic anhydride, s. H. Andersson, T. Sainte-Luce Banchelin, S. Das, R. Olsson, F. Almqvist,

Chem. Commun. *2010*, *46* (19), 3384-6 [DOI: 10.1039/c000748j]; **2-subst. 4-pyridones or 4-aminopyridinium salts** from 4-benzyloxypyridine N-oxides via 4-benzyloxypyridinium salts s. H. Andersson, S. Das, M. Gustafsson, R. Olsson, F. Almqvist, Tetrahedron Lett. *2010*, *51* (31), 4218-20 [DOI: 10.1016/j.tetlet.2010.06.009]; **regioselective synthesis of 3-subst. 1-carbo-*tert*-butoxy-4-hydroxypiperazines** from pyrazine N-oxides, incl. one example of asym. synthesis in the presence of (-)-sparteine, cf. H. Andersson, T. Sainte-Luce Banchelin, S. Das, M. Gustafsson, R. Olsson, F. Almqvist, Org. Lett. *2010*, *12* (2), 284-6 [DOI: 10.1021/ol902619h].

Stereoselective synthesis of 2,6-*trans*-disubst. 4-acyltetrahydropyrans from 2-β-hydroxycyclopropanols and aldehydes via 2,4-dioxabicyclo[5.1.0]octanes ○

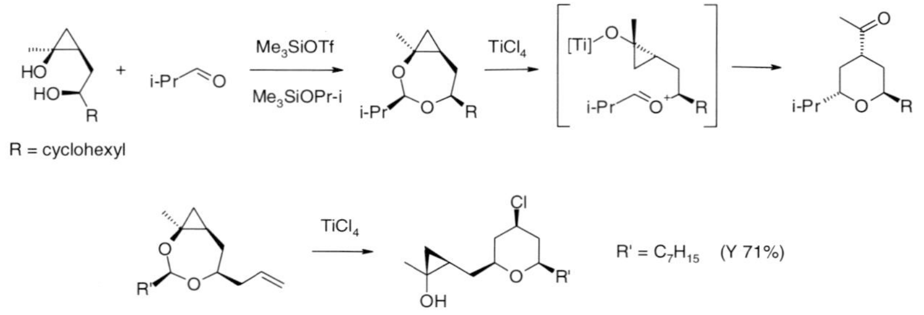

383.

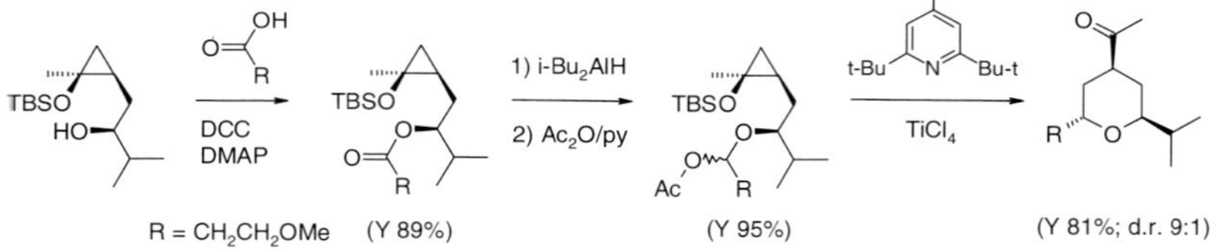

Isobutyraldehyde (1.2 eq.), trimethylsilyl isopropoxide (3.3 eq.) and a soln. of trimethylsilyl triflate (1 mol%) in methylene chloride added to a soln. of 2-(2-cyclohexyl-2-hydroxyethyl)-1-methylcyclopropanol (0.25 mmol) in the same solvent under argon at 0°, the mixture stirred for 3 h, quenched with pyridine, stirring continued for 15 min, concentrated *in vacuo*, purified by chromatography on silica, the cyclic acetal (0.85 mmol) dissolved in methylene chloride (6 ml) under argon at room temp., a soln. of 2,6-di-*tert*-butyl-4-methylpyridine (20 mol%) in the same solvent added, the mixture cooled to -78°, a soln. of $TiCl_4$ (1 eq.) in the same solvent added dropwise, the mixture stirred for 45 min, warmed to -25° over 1 h, stirring continued for 2 h, diluted with methylene chloride, quenched with pyridine (2 eq.), stirred at room temp. for 15 min, washed with satd. aq. $NaHCO_3$, extracted with methylene chloride, and purified by chromatography on silica → (2S*,4S*,6S*)-4-acetyl-2-cyclohexyl-6-isopropyl-tetrahydropyran. Y 72% (d.r. 1:0). Coupling of β-hydroxyethylcyclopropanols with ar. and aliphatic aldehydes (incl. β-alkoxy- or -siloxy-propanals) in the presence of Me_3SiOTf generated cyclic acetals that, in the presence of $TiCl_4$, underwent rearrangement to the energetically unfavorable title products (six examples; Y 62-90%; d.r. 6:1 to 1:0), with a γ-siloxybutanal proceeding to only 47% conversion (Y 22%; d.r. 1:0). A series of allyl-subst. intermediate acetals, with allyl introduced from the appropriate β-hydroxypent-4-en-1-ylcyclopropanols or from 4-pentenal derivs. gave the rearranged products near exclusively in most cases (six examples; Y 70-89%; d.r. 6:1 to 1:0), but a terminal allyl deriv. afforded the Prins product in 71% yield. In a final development, analogous products were obtained **from 2-β-hydroxy-1-siloxycyclopropanes and carboxylic acids**, via O-acylation (RCO_2H/DCC) and reductive O-acetylation (1) DIBAL, 2) Ac_2O/base) to afford α-acetoxyethers which underwent rearrangement under Lewis acid catalysis (optimally $TiCl_4$) (five examples; Y 71-81% from the acetoxyethers; d.r. 6:1 to 12:1). F.e. and optimization s. B.B. Parida, I.L. Lysenko, J.K. Cha, Org. Lett. *2012*, *14* (24), 6258-61 [DOI: 10.1021/ol3030204].

Pyrid[2,3-*b*]indoles from *o*-nitroarylacetonitriles and β-acoxy-α-methyleneketones via α-alkylidene-γ-(*o*-nitroaryl)-γ-cyanoketones ○

384.

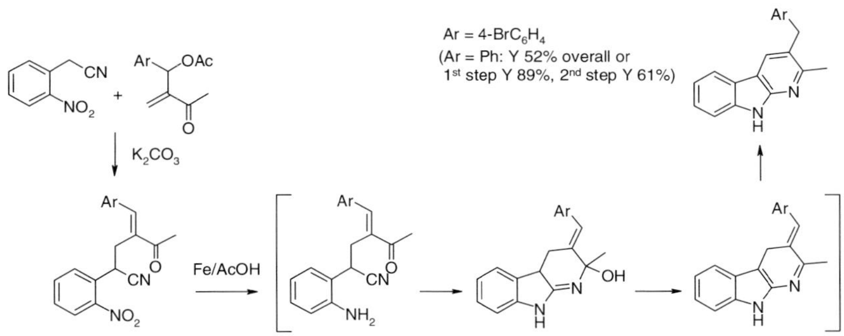

K_2CO_3 (1 eq.) added to a stirred mixture of *o*-nitrophenylacetonitrile (2 mmol) and 3-acetoxy-2-acetyl-3-(4-bromophenyl)prop-1-ene (1 eq.) in THF (6 ml) at room temp., the mixture stirred for 8 h, concentrated *in vacuo*, the crude intermediate keto(nitro)nitrile dissolved in acetic acid (5 ml), iron powder (6 eq.) added, the mixture heated under reflux for 1 h, cooled to room temp., acetic acid removed *in vacuo*, the residue diluted with ethyl acetate, filtered, solvent evaporated, and the residue purified by chromatography on silica → 3-(4-bromobenzyl)-2-methylpyrid[2,3-*b*]-indole. Y 59%. This experimentally simple procedure used readily available Baylis-Hillman acetates to monoalkylate *o*-nitroarylacetonitriles, with subsequent reduction of the nitro group resulting in spontaneous double cyclization to the products (seventeen examples; Y 48-67% overall). The method was tolerant of *ortho* substituents in the Baylis-Hillman acetates. F.e. and use of the method in a synthesis of neocryptolepine s. D. Basavaiah, D.M. Reddy, Org. Biomol. Chem. *2012*, *10* (44), 8774-7 [DOI: 10.1039/c2ob26339d].

Nitrogen ↑ CC ↓↑ N

Microwaves s. under Et_3N and Dodecacarbonyltriruthenium [\\\\]

Potassium iodide/tert-butyl peroxybenzoate/pivalic acid *$KI/PhCOO_2Bu$-t/t-BuCOOH*

C_3-Formylation of indoles with simultaneous N-aminomethylation H → CHO

385.

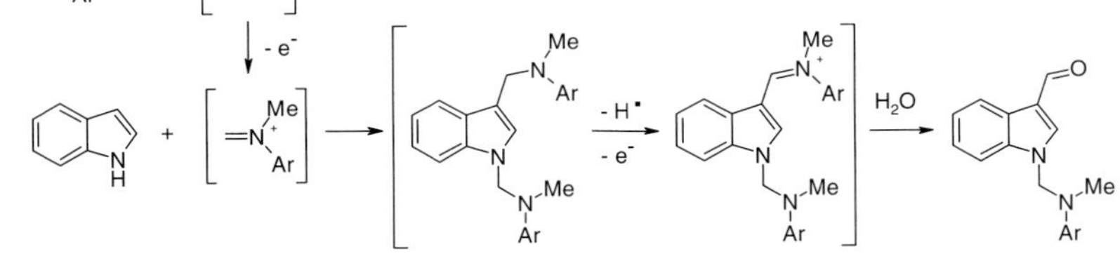

DMSO (1 ml) added under air to a mixture of indole (0.5 mmol), KI (20 mol%) and pivalic acid (5 eq.) in a reaction tube, N,N-dimethyl-*p*-methylaniline (4 eq.) and *tert*-butyl peroxybenzoate

(6 eq.) added separately, the mixture stirred at 60° for 16 h, quenched with satd. aq. $NaHCO_3$, extracted with ethyl acetate, and worked up with purification by chromatography on silica gel → product. Y 91%. This clean, mild and metal-free procedure, avoiding alkyl halides, toxic $POCl_3$ and stoichiometric amounts of base, is applicable to the coupling of electron-diverse indoles (bearing such functions as halide, ester, ether, nitro and cyano) with 4-methoxy-, 4-fluoro-, 4-chloro-, 4-bromo- and 4-*tert*-butyl-N,N-dimethylaniline (twenty examples; Y 32-91%). There was no reaction, however, with 7-methyl-, 7-ethyl- or 7-benzyloxy-indole, presumably for steric reasons. A radical mechanism is proposed via intermediate N,C_3-bis(N-methyl-N-arylamino-methyl)indoles. F.e. and comparison of oxidants s. L.-T. Li, H.-Y. Li, L.-J. Xing, L.-J. Wen, P. Wang, B. Wang, Org. Biomol. Chem. *2012*, *10* (48), 9519-22 [DOI: 10.1039/c2ob26636a].

Triethylamine *Et_3N*

Tetrasubst. *m*-amino-*p*-cyanocarboxylic acid esters from alkylidenemalononitriles and β-nitro-α,β-ethylenecarboxylic acid esters

386. 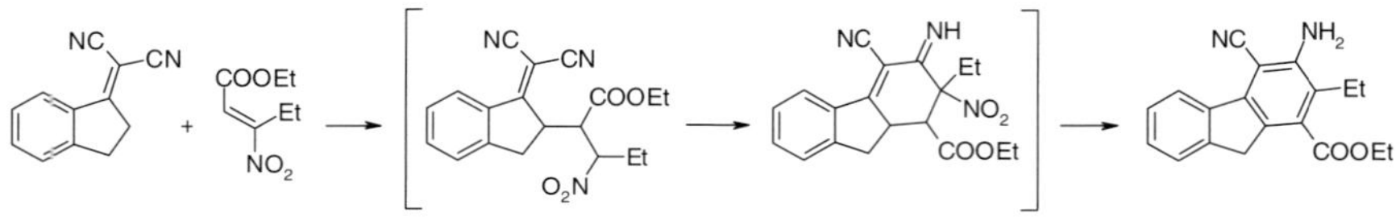

Triethylamine (0.15 eq.) added to a stirred, cooled (ice-water) soln. of the startg. alkylidenemalononitrile (1 mmol) and β-nitroacrylate (1 eq.) in acetonitrile (2.5 ml), the mixture allowed to reach room temp., left for 4 h, more acetonitrile (2.5 ml) and triethylamine (0.85 eq.) added, gradually heated to reflux temperature, refluxing continued for 18 h, cooled, quenched with 2 *M* HCl, extracted with ethyl acetate, and worked up with purification by flash chromatography → ethyl 3-amino-4-cyano-2-ethyl-9*H*-fluorene-1-carboxylate. Y 70%. Densely substituted anilines, notably *o*-alkyl derivs., are thus obtained in good yield (50-78%; twelve examples) from a wide range of substrates (incl. hindered, functionalized and cyclic derivs.). F.e. and comparison of promoters and solvents s. S. Gabrielli, A. Palmieri, D.S. Panmand, D. Lanari, L. Vaccaro, R. Ballini, Tetrahedron *2012*, *68* (39), 8231-5 [DOI: 10.1016/j.tet.2012.07.065].

Triethylamine/microwaves *Et_3N/[\\\\]*

1,4-Oxathiane-3-thiones from epoxides under microwave irradiation in water

387. 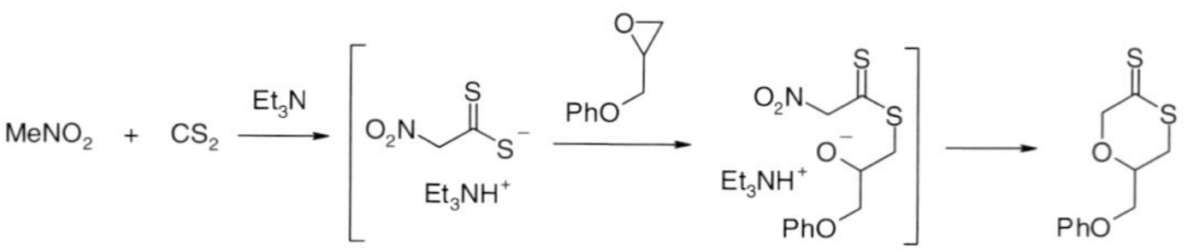

An efficient direct synthesis of the title products via a 3-component reaction of epoxides, nitromethane and carbon disulfide is reported. **E:** A soln. of nitromethane (1 eq.) and triethylamine (1 eq.) in water (10 ml) stirred at 70° for 30 min under microwave irradiation (800 W), 2-(phenoxymethyl)oxirane (2 mmol) and carbon disulfide (5 eq.) added, the mixture stirred for 2 h (TLC), filtered, and purified by recrystallization → 6-(phenoxymethyl)-1,4-oxathiane-3-thione. Y 90%. This method gave good yields from monosubst. oxiranes, with cyclohexene oxide affording a bicyclic analog (Y 78-90%; six examples). F.e.s. Z. Hossaini, F. Rostami-Charati, S. Seyfi, M. Ghambarian, Chin. Chem. Lett. *2013*, *24* (5), 376-8 [DOI: 10.1016/j.cclet.2013.03.003].

Pyridine *C_5H_5N*

5-Alkoxyindoles from *p*-quinone monoketals and hydrazines via Fischer-type ring closure

388.

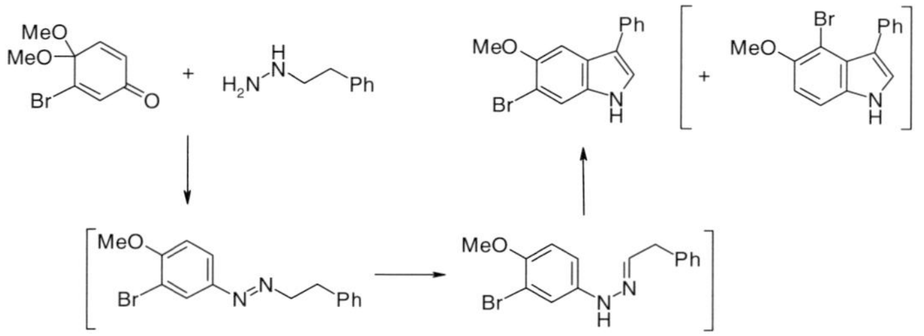

in one pot. Pyridine (20 eq.) added to a soln. of 3-bromo-4,4-dimethoxycyclohexa-2,5-dienone (0.3 mmol) and 2-phenylethylhydrazine hydrochloride (2 eq.) in dry chloroform (2 ml), the capped mixture stirred at 90° for 6 h (TLC), cooled to room temp., worked up (with 1 *M* aq. HCl and satd. aq. Na_2CO_3 washes), and purified by flash chromatography → 6-bromo-5-methoxy-3-phenylindole. Y 84% (as a mixture with the 4-bromo regioisomer; Y 13%). This novel, convenient and metal-free synthesis of 5-alkoxyindoles uses *p*-quinone monoketals (available in one step from phenols) and aliphatic hydrazines to generate azo compds. which are isomerized to hydrazones and cyclized *in situ*, affording the products in generally good yields (41%, 58-95%; twenty-six examples; butylhydrazine gave 37%). F.e. and optimization s. J. Zhang, Z. Yin, P. Leonard, J. Wu, K. Sioson, C. Liu, R. Lapo, S. Zheng, Angew. Chem., Int. Ed. *2013, 52* (6), 1753-7 [DOI: 10.1002/anie.201207533]; **oxidative Fischer indole synthesis with** prim. or sec. **alcohols** using $[Ru_3(CO)_{12}]$/2,2′-bis(diphenylphosphino)biphenyl/crotononitrile/$ZnCl_2$ in 2-methyl-2-butanol under microwave enhancement s. A. Porcheddu, M.G. Mura, L. De Luca, M. Pizzetti, M. Taddei, Org. Lett. *2012, 14* (23), 6112-5 [DOI: 10.1021/ol3030956]; **regioselective Fischer indole synthesis from N′-acyl-N-arylhydrazines and α,β-ethylenebromides** (CC↓↑C/v.i.) with CuI/K_2CO_3/DMEDA and subsequent ring closure with $ZnCl_2$ s. F. Zhan, G. Liang, Angew. Chem., Int. Ed. *2013, 52* (4), 1266-9 [DOI: 10.1002/anie.201207173].

Copper(I) thiophene-2-carboxylate/rhodium(II) pivalate *CuOCOR/[Rh(OCOBu-t)$_2$]$_2$*

C≡CH → COCH($NHSO_2R$)C-C=C

Regioselective 3-component synthesis of γ,δ-ethylene-α-(sulfonylamino)ketones from terminal acetylene derivs., 2-ethylenealcohols and sulfonic acid azides via 1-sulfonyl-1,2,3-triazoles and Claisen rearrangement of (Z)-2-(allyloxy)enesulfonylamines

389.

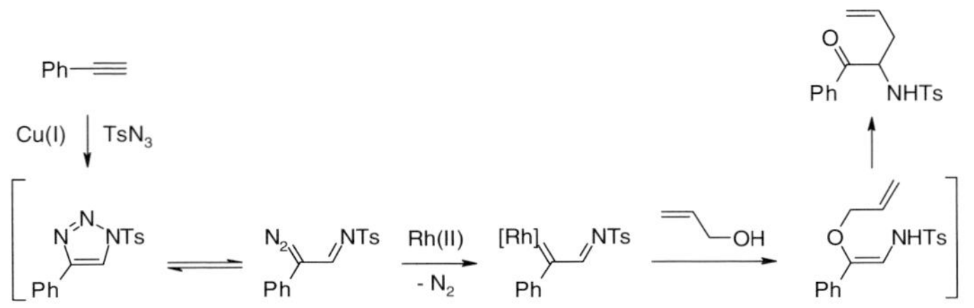

The triple bond of terminal acetylene derivs. can be simply, cleanly and regioselectively functionalized by C-O, C-N and C-C bonds **in one pot under sequential catalysis with two different metals** to provide γ,δ-ethylene-α-(sulfonylamino)ketones in good yield. **E:** A mixture of phenylacetylene (0.2 mmol), tosyl azide (1 eq.), allyl alcohol (1.5 eq.), Cu(I)-thiophene-

2-carboxylate (10 mol%), Rh(II)-pivalate (1 mol%), 3 Å molecular sieves (40 mg) and toluene (2 ml) stirred at room temp. in a sealed vial for 6 h, heated at 100° for 30 min, cooled, passed through a cotton stopper with methylene chloride, the filtrate concentrated under reduced pressure, and the residue purified by preparative thin-layer chromatography → product. Y 74%. Reaction is presumed to involve initial copper-catalyzed [3+2]-cycloaddition of the acetylene deriv. with tosyl azide, followed by ring-chain tautomerism of the resulting 1-sulfonyl-1,2,3-triazole; the generated diazo compd. is then converted to a rhodium carbenoid prior to reaction with the allyl alcohol to give a (Z)-2-(allyloxy)enesulfonylamine which undergoes Claisen rearrangement to give the product. Reaction of preformed 1-sulfonyl-1,2,3-triazoles (incl. 1-alkanesulfonyl derivs.) with 1- and 3-subst. allyl alcohols under rhodium(II) catalysis proceeded with double bond shift, confirming the intermediacy of the sigmatropic rearrangement; chiral α-subst. allyl alcohols also participated to give (E)-products *with* the anticipated *transfer of chirality*. The one-pot procedure is applicable to the coupling of electron-diverse arylacetylenes, as well as aliphatic terminal alkenes and functionalized derivs. (nine examples; Y 53-80%); furthermore, a wide range of products was also obtained from the preformed 1-sulfonyl-1,2,3-triazoles (fifteen examples; Y 62-94%), reaction failing in just one instance: with a 4-unsubst. deriv. (Y 13%). F.e.s. T. Miura, T. Tanaka, T. Biyajima, A. Yada, M. Murakami, Angew. Chem., Int. Ed. *2013*, *52* (14), 3883-6 [DOI: 10.1002/anie.201209603].

Copper(I) bromide/lithium tert-butoxide *CuBr/LiOBu-t*
Benzofurans from *o*-acetylenephenols and tosylhydrazones s. *44,* 781s*83* ○
Also N-protected indoles from *o*-acetyleneamines s. *44,* 815s*83*

Copper(I) iodide/lithium tert-butoxide *CuI/LiOBu-t*
Coupling with tosylhydrazones ←
copper-catalyzed synthesis of allenes from terminal acetylene derivs. with $Cu(MeCN)_4PF_6$/bis-(Δ^2-oxazolines)/Cs_2CO_3 cf. *79*, 390; with CuI/LiOBu-*t* s. M.L. Hossain, F. Ye, Y. Zhang, J. Wang, J. Org. Chem. *2013*, *78* (3), 1236-41 [DOI: 10.1021/jo3024686]; γ,δ-ethyleneketones from α,β-ethyleneketones or N,N-disubst. cyclopropanecarboxylic acid amides (cf. *23,* 819s*82*) from α,β-ethylenecarboxylic acid amides with $Pd(OAc)_2$/LiOBu-*t*, stereoselectivity, s. H. Jiang, W. Fu, H. Chen, Chem. Eur. J. *2012*, *18* (38), 11884-8 [DOI: 10.1002/chem.201202051]; 3-component synthesis of *o*-, *m*- or *p*-*tert*-amino-1,1-diarylethylenes from ar. dihalides and sec. amines (CC↓↑Hal, NC↓↑Hal) with $PdCl_2(MeCN)_2$/XPhos/NaOBu-*t* s. M. Roche, A. Hamze, J.-D. Brion, M. Alami, Org. Lett. *2013*, *15* (1), 148-51 [DOI: 10.1021/ol303194s]; triarylmethanes from benzophenone tosylhydrazones and ar. bromides under palladium catalysis in the presence of NH_4-formate/NH_4-acetate/Cs_2CO_3 s. Y. Xia, F. Hu, Z. Liu, P. Qu, R. Ge, C. Ma, Y. Zhang, J. Wang, ibid. *15* (7), 1784-7 [DOI: 10.1021/ol400690t].

Copper(II) bromide *$CuBr_2$*
Δ^2-4-Pyrrolone-3,5-dicarboxylic acid amides ○
from two α-diazo-β-ketocarboxylic acid amide molecules and prim. amines

390.
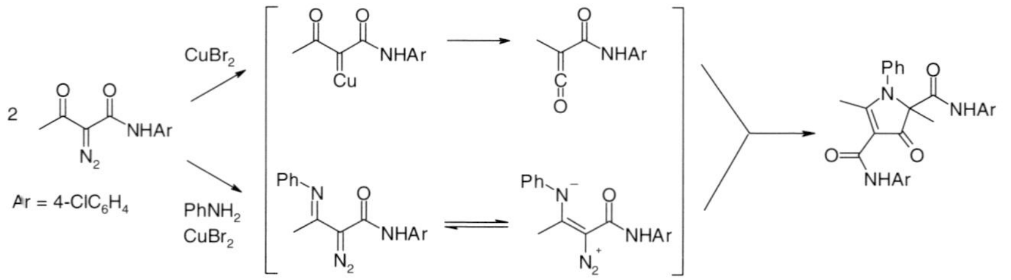

$CuBr_2$ (10 mol%) added to a soln. of N-(4-chlorophenyl)-2-diazo-3-oxobutanamide (2 mmol) and aniline (0.5 eq.) in DMF (6 ml), the mixture stirred at 90° for 4 h (TLC), quenched with brine, extracted with methylene chloride, and purified by chromatography on silica → N^2,N^4-bis(4-chlorophenyl)-2,5-dimethyl-3-oxo-1-phenyl-2,3-dihydro-1*H*-pyrrole-2,4-dicarboxamide. Y 77%. This novel and experimentally simple synthesis of the title products was successful for N-aryl-β-keto-amides reacting with electron-diverse ar., alkyl, allyl and propargyl amines (seventeen examples; Y 60-86%). In the presence of TsOH (0.5 eq.), NH_4OAc could be employed as the nitrogen source, affording the corresponding NH-pyrrolones (two examples; Y 60-76%). An α-diazo-β-keto-ester was unreactive under these conditions, while cyclic β-diketone and β-keto-amide analogs gave complex mixtures, and the tert. amide, N-methyl-N-phenyl-2-diazo-4-oxobutanamide, underwent intramolecular cyclization to an isatin deriv. (Y 61%). F.e. and optimization s. Z. Wang, X. Bi, P. Liao, X. Liu, D. Dong, Chem. Commun. *2013*, *49* (13), 1309-11 [DOI: 10.1039/c2cc38473f].

*Silver hexafluoroantimonate s. under [Cp*RhCl$_2$]$_2$* *$AgSbF_6$*
Zinc chloride s. under Dodecacarbonyltriruthenium *$ZnCl_2$*

Mercury(II) triflate or hexafluorophosphate/chiral 2,3;2′,3′-bis(difluoromethylenedioxy)-6,6′-bis(diphenylphosphino)biphenyl ←
Asym. cyclopropanation of ethylene derivs. with diazo compds. s. *23,* 819s*83*

Sodium trihydridocyanoborate s. under 2(S)-[Diphenyl(trimethylsiloxy)methyl]pyrrolidine ←

1,3,2-Oxazaborolidine·triflimide salts or chiral variants ←
Insertion of diazo compds. into carbon-hydrogen bonds H→R
under rhodium catalysis, update s. *23,* 816s*82*; cyclic β,γ-ethylene-δ-ketocarboxylic acid esters from cyclic enones and α-diazo-esters *by insertion into carbon(sp²)-hydrogen bonds* using a 1,3,2-oxazaborolidine·Tf_2NH salt or BF_3-etherate as Lewis acid s. S.I. Lee, B.C. Kang, G.-S. Hwang, D.H. Ryu, Org. Lett. *2013, 15* (7), 1428-31 [DOI: 10.1021/ol4000026]; **asym. insertion** using a chiral N-condensed 1,3,2-oxazaborolidine·Tf_2NH salt (cf. *54,* 390s*69*) s. S.I. Lee, G.-S. Hwang, D.H. Ryu, J. Am. Chem. Soc. *2013, 135* (19), 7126-9 [DOI: 10.1021/ja402873b].

Triphenyl borate/chiral 2,2′-bi-1-biphenanthrols or 2,2′-bi-1-naphthols ←
Asym. synthesis of aziridines from aldimines and diazo compds.
update s. *78,* 485s*82*; **asym. 3-component synthesis** of N-protected *cis*-aziridine-2-carboxylic acid esters from *in situ*-generated aldimines *under BORAX catalysis* using triphenyl borate and a chiral VAPOL- or VANOL-type [vaulted biaryl] diol as precursors of a self-assembled Brønsted acid s. A.K. Gupta, M. Mukherjee, G. Hu, W.D. Wulff, J. Org. Chem. *2012, 77* (18), 7932-44 [DOI: 10.1021/jo301064u].

(+)-Pinanediol s. under Titanium tetraisopropoxide ←
Benzoquinone s. under Pd(OAc)$_2$ *BQ*

2(S)-[Diphenyl(trimethylsiloxy)methyl]pyrrolidine/acetic acid/silica/sodium trihydridocyanoborate/acetic acid ←
2,4-Bis(hydroxymethyl)cyclopenten-1-ylacetic acid esters
from α-acoxy-β-nitro-β,γ-ethylenecarboxylic acid esters and glutaraldehyde
via formation of 4-carbalkoxymethyl-4-nitrocyclopentane-1,3-dicarboxaldehydes
by regioselective organocatalyzed asym. α-allylation-intramolecular Michael addition

391.

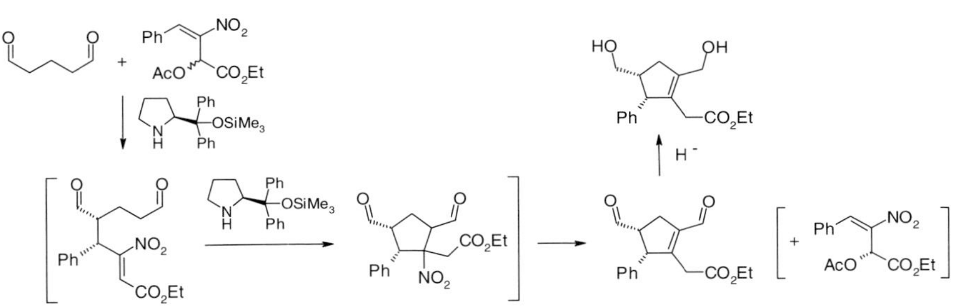

with kinetic resolution. 2(S)-[Diphenyl(trimethylsiloxy)methyl]pyrrolidine (20 mol%) added in one portion at 35° to a soln. of the startg. nitroallyl acetate (0.1 mmol), glutaraldehyde (4 eq.) and acetic acid (1.5 eq.) in toluene (0.5 ml), the mixture stirred, quenched after 6 h at ca. 50% conversion (TLC/NMR), subjected directly to flash chromatography on silica gel to separate the formed cyclopentanedicarboxaldehyde from the kinetically less reactive nitroallyl acetate enantiomer, and the resulting unstable dialdehyde isolated after conversion to the more stable 2,4-bis(hydroxymethyl)cyclopent-1-en-1-ylacetate [by reduction with $NaBH_3CN$ (4 eq.) on stirring with acetic acid (0.5 eq.) for 3 h in 1:1 methylene chloride/ethanol (2 ml), followed by purification by flash chromatography on silica gel] → ethyl [2,4-bis(hydroxymethyl)-5-phenylcyclopent-1-en-1-yl]-acetate. Y 45% (conversion 48%; d.r. 9:1; e.e. 96%). The slower reacting nitroallyl acetate enantiomer was also isolated (Y 33%; e.e. 88%). Reaction is applicable to variously ring-substituted α-acoxy-γ-aryl-β-nitro-β,γ-ethylenecarboxylic acid esters, with *m*-substituted derivs. generally outperforming *p*-substituted derivs.; heteroaryl-analogs also offered high yields and kinetic resolution, but aliphatic substrates reacted sluggishly (ten examples; Y of product: 32-45%, d.r. 12:1 to 3:1; e.e. 90-96%; Y of less reactive enantiomer: 31-46%, e.e. 76-99%; overall kinetic selectivity: S = 53-143]. Kinetic resolution is effected in the initial enamine-mediated α-allylation, which precedes the terminating intramolecular Michael addition, courtesy of the same organocatalyst, in a conjoined catalytic cycle. F.e. and comparison of chiral amines, solvents and acidic additives s. L.F. Yeh, S. Anwar, K. Chen, Tetrahedron *2012, 68* (36), 7317-21 [DOI: 10.1016/j.tet.2012.06.085].

Crotononitrile s. under Dodecacarbonyltriruthenium *MeCH═CHCN*
tert-Butyl peroxybenzoate s. under KI *$PhCOO_2Bu$-t*
Chiral 2,2′-bi-1-biphenanthrols or -1-naphthols s. under Triphenyl borate *VAPOL or VANOL*
Pivalic acid s. under KI *t-BuCOOH*
Trifluoroacetic acid s. under Palladium(II) acetate *CF_3COOH*

Titanium tetraisopropoxide/(+)-pinanediol ←
Epoxides from aldehydes and diazo compds.
under rhodium catalysis cf. *50,* 480; **asym. synthesis of *cis*-glycidic acid amides** with $Ti(OPr\text{-}i)_4$/(+)-pinanediol s. G. Liu, D. Zhang, J. Li, G. Xu, J. Sun, Org. Biomol. Chem. *2013, 11* (6), 900-4 [DOI: 10.1039/c2ob27179f].

2,2′-Bis(diphenylphosphino)biphenyl s. under Dodecacarbonyltriruthenium *BIPHEP*
Chiral 2,3;2′,3′-bis(difluoromethylenedioxy)-6,6′-bis(diphenylphosphino)biphenyl s. under Mercury(II) triflate or hexafluorophosphate ←

Phosphorus oxide chloride *$POCl_3$*
Oxazole-4-carboxaldehydes from azidomethyl ketones

392.
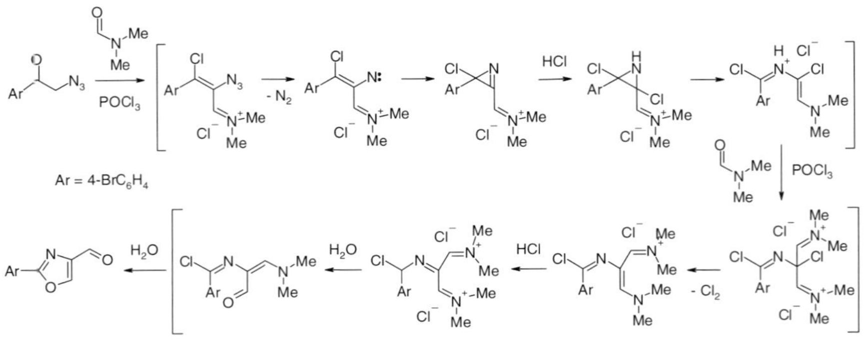

in one pot. $POCl_3$ (3 eq.) added dropwise over 30-45 min to a soln. of 4-bromophenacyl azide [generated *in situ* from NaN_3 (1.05 eq.) and 4-bromophenacyl bromide (50 mmol)] in DMF (50 ml) at 10-15°, the mixture stirred at room temp. for 3-4 h, then at 90-95° for 2-3 h, cooled to room temp., stirred with ice-water for 1 h, extracted with chloroform (with aq. $NaHCO_3$ wash), and purified chromatographically → 2-(4-bromophenyl)-1,3-oxazole-4-carboxaldehyde. Y 52%. Vilsmeier-Haack-type formylation of the substrates, expected to afford 5-aryloxazole-4-carboxaldehydes, gave the 2-aryl-isomers via an unprecedented rearrangement of alkenyl azide and azirine intermediates. Overall the reaction provides a general and direct route to **2-aryloxazole-4-carboxaldehydes from** readily available **phenacyl bromides** in moderate yields (29-52%; five examples). F.e.s. S.R. Shah, S.S. Navathe, A.G. Dikundwar, T.N. Guru Row, A.T. Vasella, Eur. J. Org. Chem. *2013* (2), 264-7 [DOI: 10.1002/ejoc.201201269].

Polyphosphoric or methanesulfonic acid ←
Isatins from alkoximinoacetic acid anilides s. *83*, 64 ○

Trifluoromethanesulfonic acid *TfOH*
Friedel-Crafts acylation with carboxylic acid amides s. *83*, 494 H → C(O)R

Quinoline-3-carboxylic acid esters from benzyl azides ○

393.

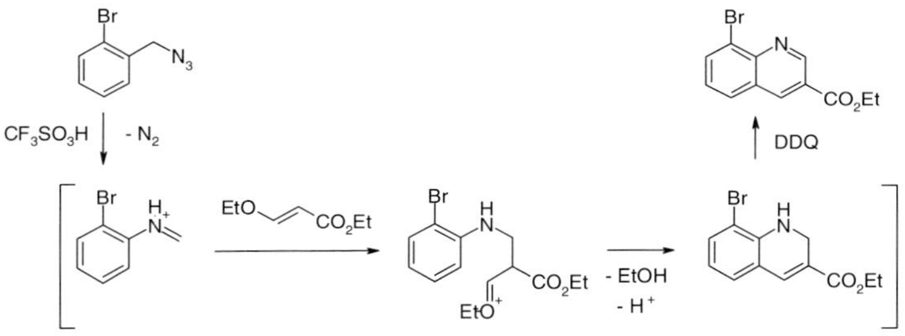

A novel method for synthesis of rare 4,5-unsubst. quinoline derivs., involving Aubé-Schmidt rearrangement of a benzylic azide to an N-aryliminium species [cf. J. Tummatorn et al., Tetrahedron *2012*, *68* (24), 4732-9 [DOI: 10.1016/j.tet.2012.04.014] for applications to N-benzyl-anilines and polycyclic heteroarenes incl. isocryptolepine], cyclization with *excess* 3-ethoxyacrylate, and dehydrogenation of an intermediate 1,2-dihydroquinoline, is reported. **E:** Triflic acid (1 eq.) added to a soln. of 2-bromobenzyl azide (0.6 mmol) in toluene (2.64 ml) at room temp., the mixture stirred until evolution of N_2 subsided, ethyl 3-ethoxyacrylate (2 eq.) added, the mixture stirred for 3 h, quenched with satd. aq. $NaHCO_3$, extracted with ethyl acetate, concentrated *in vacuo*, the residue dissolved in ethyl acetate (2.64 ml), stirred with DDQ (1 eq.) for 5 min, concentrated *in vacuo*, and purified by chromatography on silica → ethyl 8-bromoquinoline-3-carboxylate. Y 70%. The cyclization was successful for electron-diverse benzylic azides (incl. naphthyl analogs) (fourteen examples; Y 48-83%), with low yields observed for 4-nitro- (14%) and 4-methoxybenzyl azides (24%), and with 3-subst. benzyl azides affording mixtures of regioisomers. 4-Halo-(Cl, Br) benzyl azides afforded the expected quinolinecarboxylates as inseparable mixtures with an uncyclized by-product, while α-methylbenzyl azide gave only a trace of the expected quinoline. Addition of a Lewis acid, BF_3-etherate, allowed use of *stoichiometric* quantities of ethoxyacrylate, although generally at the expense of lower yields. F.e., optimization and substrate prepn. s. J. Tummatorn, C. Thongsornkleeb, S. Ruchirawat, T. Gettongsong, Org. Biomol. Chem. *2013*, *11* (9), 1463-7 [DOI: 10.1039/c3ob27493d].

Sulfuric acid *H_2SO_4*
Isatins from alkoximinoacetic acid anilides s. *83*, 64 ○

Manganese dioxide *MnO_2*
Radical *o*-arylation of anilines or phenols with arylhydrazines s. *38,* 793s*83* H → Ar

Cobalt(II) porphyrin complex *[Co(II)]*

Furans from α-diazocarbonyl compds. and terminal acetylene derivs.
Regioselective cobalt(II)-catalyzed ring closure under mild, neutral conditions

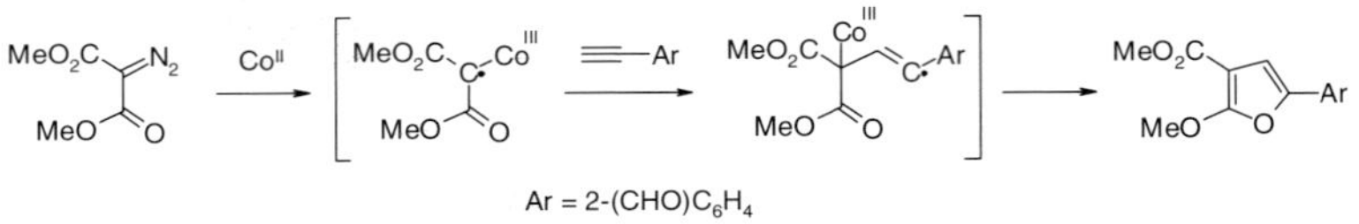

394.

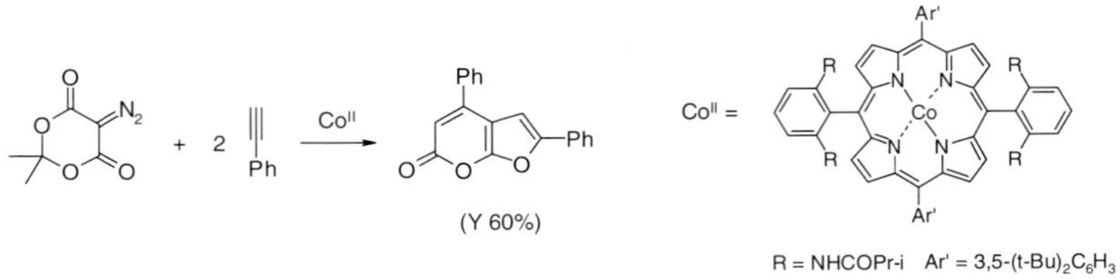

2-Alkoxyfurans. A soln. of 2-formylphenylacetylene (2 eq.) and dimethyl α-diazomalonate (0.1 mmol) in anh. 1,2-dichlorobenzene (0.5 ml) added to Co(II)-porphyrin catalyst (2 mol%) under N_2, the sealed mixture stirred at 80° for 24 h, concentrated, and purified by flash chromatography → methyl 5-(2-formylphenyl)-2-methoxyfuran-3-carboxylate. Y 96%. This novel method generates a Co(III)-carbene radical from reaction of the Co(II)-porphyrin deriv. with diactivated diazo compds. (activating groups included CN, COR, CO_2R, SO_2Ar and CHO), which at room temp. afforded cyclopropenes via reaction with terminal alkynes. At 80°, however, the major products were furan derivs., formed as single regioisomers. The reaction was successful with (het)ar., alkyl, vinyl, trimethylsilyl and keto/ester terminated alkynes, affording polyfunctionalized furans generally in good yield (34-98%; twenty-eight examples). Cyclization of 5-diazo-Meldrum's acid under these conditions utilized two molecules of alkyne to afford **6*H*-furo[2,3-*b*]pyran-6-ones** (two examples; Y 44-60%). An iterative process was also developed, using an α-diazo-propynoyl ketone, with temporary protection of the terminal alkyne as a trimethylsilane, to generate **oligofurans** (2-4 furan units; Y 75-89%). F.e., optimization and substrate prepn. s. X. Cui, X. Xu, L. Wojtas, M.M. Kim, X.P. Zhang, J. Am. Chem. Soc. *2012, 134* (49), 19981-4 [DOI: 10.1021/ja309446n]; **5-arylfuran-3-carboxylic acid esters** from α-diazo-β-ketocarboxylic acid esters and arylacetylenes with CuI in moderate yields s. A.K. Swenson, K.E. Higgins, M.G. Brewer, W.W. Brennessel, M.G. Coleman, Org. Biomol. Chem. *2012*, *10* (37), 7483-6 [DOI: 10.1039/c2ob26295a].

Dodecacarbonyltriruthenium/2,2′-bis(diphenylphosphino)biphenyl/crotononitrile/ zinc chloride/microwaves ←

Oxidative Fischer indole synthesis with alcohols s. *83*, 388

Chiral tetrakis(acetonitrile)[o-(4-phenyl-Δ²-oxazolin-2-yl)phenyl]ruthenium(II) hexafluorophosphate *[Ru(II)]**

Asym. cyclopropanation of ethylene derivs. with diazo compds. ▽

under rhodium(II) catalysis, update s. *23*, 819s*82*; **asym. synthesis of 2-aminocyclopropanecarboxylic acid esters** from N-(carbalkoxy)enamines with chiral tetrakis(acetonitrile)[*o*-(4-phenyl-Δ^2-oxazolin-2-yl)phenyl]ruthenium(II) hexafluorophosphate as catalyst s. S. Chanthamath, D.T. Nguyen, K. Shibatomi, S. Iwasa, Org. Lett. *2013, 15* (4), 772-5 [DOI: 10.1021/ol303404c]; *under mercury catalysis* with $Hg(OTf)_2$ or $Hg(PF_6)_2$ and chiral 2,3;2′,3′-bis(difluoromethylenedioxy)-6,6′-bis(diphenylphosphino)biphenyl as ligand for the diastereoselective synthesis of chiral 3-cyclopropanooxindoles s. Z.-Y. Cao, F. Zhou, Y.-H. Yu, J. Zhou, ibid. *15* (1), 42-5 [DOI: 10.1021/ol302998m].

Rhodium(II) acetate *$Rh_2(OAc)_4$*

α-Alkoxy-α-(2-aroyl-3-hydroxyindan-1-yl)carboxylic acid esters from *o*-formylchalcones, α-diazocarboxylic acid esters and alcohols Stereoselective rhodium(II)-catalyzed 3-component ring closure

395.

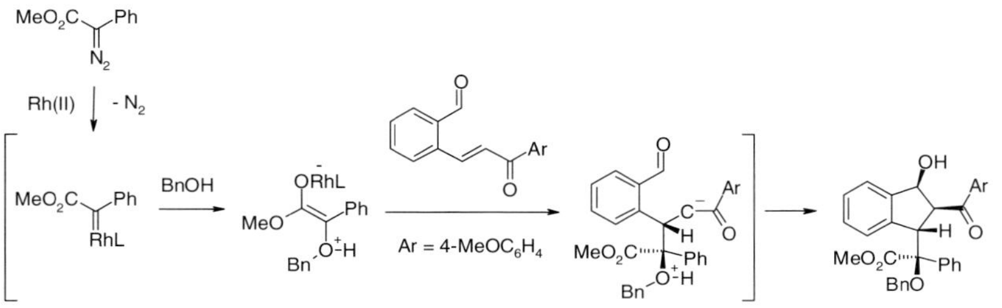

A novel rhodium(II)-catalyzed cascade process, wherein a generated oxonium moiety persists during addition to an enone before intramolecular trapping with an aldehyde moiety, is reported. **E:** A soln. of methyl α-diazophenylacetate (1.2 eq.) in 1,2-dichloroethane (1 ml) added by syringe pump over 1 h to a mixture of 2-(4-methoxybenzoylvinyl)benzaldehyde (0.2 mmol), benzyl alcohol (1.2 eq.), $Rh_2(OAc)_4$ (1 mol%) and 4 Å molecular sieves (0.1 g) in the same solvent (2 ml), followed by purification by flash chromatography on silica → methyl α-benzyloxy[2-(4-methoxybenzoyl)-1-hydroxyindan-3-yl]phenylacetate. Y 83%. Mild and rapid 3-component reaction of electron-diverse α-diazophenylacetates, aliphatic prim./sec. alcohols and electron-diverse *o*-formylchalcones afforded the title products with generation of four new stereogenic centers, with excellent diastereoselectivity (d.r. >20:1; seventeen examples; Y 60-83%). Use of *L*-menthyl α-diazophenylacetate gave the corresponding products **with asym. induction** (d.r. >20:1; e.e. 98-99%; two examples; Y 64-67%), with removal of the chiral auxiliary and concomitant reduction of the aroyl moiety [$LiAlH_4$] in one case affording the corresponding alcohol (five stereocenters) as a single enantiomer (Y 46%). Replacing *o*-formylchalcones with constituent aldehyde and enone molecules gave a mixture of products (75% combined yield), none of which was derived from all 4 components. F.e. and optimization s. J. Jiang, X. Guan, S. Liu, B. Ren, X. Ma, X. Guo, F. Lv, X. Wu, W. Hu, Angew. Chem., Int. Ed. *2013*, *52* (5), 1539-42 [DOI: 10.1002/anie.201208391].

Rhodium(II) pivalate s. under Copper(I) thiophene-2-carboxylate *$[Rh(OCOBu\text{-}t)_2]_2$*

Dirhodium(II) tetrakis[N-(p-dodecylbenzenesulfonyl)-2-pyrrolidone-5(S)-carboxylate] *[Rh(II)]*

3-β-Keto-1-sulfonylpyrroles from 1-sulfonyl-1,2,3-triazoles and furans ←

396.

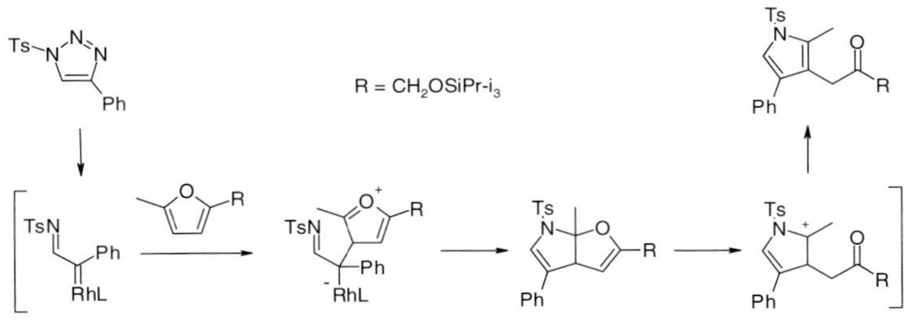

A mixture of 4-phenyl-1-tosyl-1,2,3-triazole (0.5 mmol), Rh_2(S-DOSP)$_4$ (10 mol%), 2-methyl-5-(triisopropylsiloxymethyl)furan (3 eq.) and freshly distilled 1,2-dichloroethane (2 ml) stirred under argon in a sealed vessel at 70° for 6 h (TLC), cooled to room temp., concentrated *in vacuo,* and the residue purified by flash chromatography → 1-(2-methyl-4-phenyl-1-tosyl-1*H*-pyrrol-

3-yl)-3-(triisopropylsiloxy)propan-2-one. Y 83% (regioselectivity >10:1). Electron-diverse N-sulfonyl [optimally tosyl] 4-aryl-1,2,3-triazoles reacted with 2,5-disubst. furans via formal [3+2]-cycloaddition and cleavage of the furan ring to afford trisubst. pyrroles. Unsymmetrically disubst. furans gave mixtures of regioisomers, generally with one isomer predominating (fourteen examples; Y 67-99%; regioselectivity 2:1 to >10:1). A 4-cyclohex-1-enyltriazole was also a good substrate (Y 70%), while the monosubst. furan, 2-methylfuran, gave a somewhat lower yield (41%). F.e. and optimization s. B.T. Parr, S.A. Green, H.M.L. Davies, J. Am. Chem. Soc. *2013*, *135* (12), 4716-8 [DOI: 10.1021/ja401386z].

Tris(acetonitrile)(pentamethylcyclopentadienyl)rhodium(III) bis(hexafluoroantimonate) ←
5-Alkylidene-Δ^3-pyrrol-2-ones from α,β-ethylenealkoximes and isocyanates ○

397.

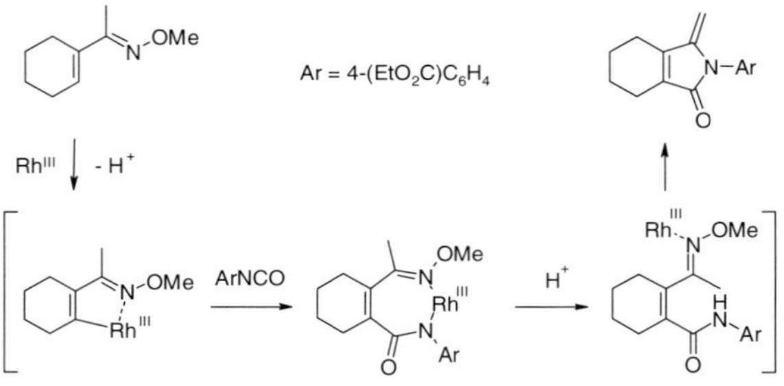

A mixture of $[Cp^*Rh(MeCN)_3](SbF_6)_2$ (5 mol%), 1,2-dichloroethane (1 ml), cyclohex-1-en-1-yl-(methyl)ketoxime O-methyl ether (0.2 mmol) and 4-ethoxycarbonylphenyl isocyanate (2 eq.) heated in a sealed vessel at 100° for 12 h, cooled to room temp., filtered through silica, and purified directly by chromatography on silica → 2-(4-ethoxycarbonylphenyl)-3-methylidene-4,5,6,7-tetrahydro-1*H*-isoindolin-1-one. Y 83%. This novel, experimentally simple and atom economical rhodium-catalyzed ring closure is scalable (to 2 mmol) and successful with electron-diverse ar. and alkyl isocyanates, affording best results with α,β-aliphatic disubst. ethylene-alkoximes (thirteen examples; Y 78-93%; bulky 1-naphthyl isocyanate gave 63%). α-Methyl-β-aryl-ethyleneoximes gave somewhat lower yields (55-65%; six examples), with an α-unsubst. β-phenyl deriv. giving 47% yield. F.e., optimization and substrate prepn. s. W. Hou, B. Zhou, Y. Yang, H. Feng, Y. Li, Org. Lett. *2013*, *15* (8), 1814-7 [DOI: 10.1021/ol4003674].

Dichloro(pentamethylcyclopentadienyl)rhodium(III) dimer/silver hexafluoroantimonate ←
Furans from α,β-ethylenealkoximes and aldehydes

398.

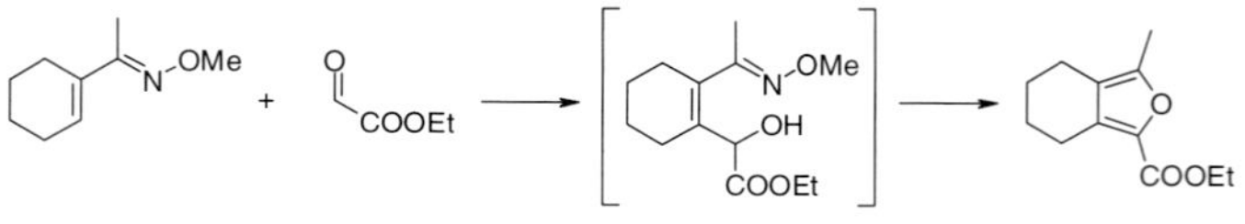

Dichloroethane (0.66 ml) added under N_2 to a screw-capped conical vial (kept in a glovebox) containing $[Cp^*RhCl_2]_2$ (2.5 mol%), $AgSbF_6$ (10 mol%), the startg. alkoxime (0.2 mmol) and ethyl glyoxylate (2 eq.), the vial sealed with a septum, removed from the glovebox, placed in an oil bath at 75°, the soln. stirred for 16 h, cooled to ambient temp., filtered through a pad of Celite, the solvent removed under reduced pressure, and the crude residue purified by chromatography on silica gel → product. Y 84%. The procedure is simple and applicable to the coupling of a wide range of cyclic or acyclic O-methyl α,β-ethylene-aldoximes or -ketoximes (excepting α-unsubst. derivs.), bearing alkyl or aryl substitution, with ethyl glyoxylate, electron-diverse aromatic aldehydes (substituted at the *o*-, *m*- or *p*-site) or [in moderate yield] with aliphatic aldehydes (twenty-two examples; Y 41-89%). **N-Tosylpyrrole-2-carboxylic acid esters** were obtained similarly with ethyl N-tosyliminoacetate in place of the aldehydes (six examples; Y 60-78%). Reaction is presumed to involve rhodium(III)-catalyzed addition of an alkenyl C-H bond across an aldehyde C=O or imine C=N bond, followed by cyclization and aromatization. F.e.s. Y. Lian, T. Huber, K.D. Hesp, R.G. Bergman, J.A. Ellman, Angew. Chem., Int. Ed. *2013*, *52* (2), 629-33 [DOI: 10.1002/anie.201207995].

Dichloro(pentamethylcyclopentadienyl)rhodium(III) dimer/acetic acid *[Cp*RhCl₂]₂/AcOH*
Benzofurans from acetylene derivs. and aryl hydroxamates s. *83,* 323 ○

Palladium nanoparticles-on-agarose ←
Palladium(II) acetate/1,10-phenanthroline/trifluoroacetic acid or calcium carbonate ←
Heck-Matsuda arylation of ethylene derivs. with diazonium salts H → Ar
update s. *38,* 793s*82*; *in aq. medium* with recyclable palladium nanoparticles-on-agarose at low catalyst loading, also Suzuki coupling (cf. *37,* 902s*83*), s. M. Gholinejad, Appl. Organomet. Chem. *2013, 27* (1), 19-22 [DOI: 10.1002/aoc.2932]; β-arylation of *o*-hydroxycinnamic acid esters and conversion to 4-arylcoumarins with $Pd(OAc)_2/CaCO_3$ s. D.A. Barancelli, A.G. Salles, Jr., J.G. Taylor, C.R.D. Correia, Org. Lett. *2012, 14* (23), 6036-9 [DOI: 10.1021/ol302923f]; **Heck-Matsuda-type arylation with acetanilides** under palladium catalysis (following N-deacetylation and diazotization), also Heck arylation (cf. *27,* 871s*83*), Suzuki coupling and further reactions, s. B. Schmidt, R. Berger, Adv. Synth. Catal. *2013, 355* (2-3), 463-76 [DOI: 10.1002/adsc.201200929]; **Heck-Matsuda biaryl synthesis** by coupling diazonium fluoroborates with polyfluoroarenes under copper catalysis, also one-pot conversion with anilines via diazotization, s. X. Zhu, F. Li, W. Su, Tetrahedron Lett. *2013, 54* (10), 1285-9 [DOI: 10.1016/j.tetlet.2012.12.110]; Heck-Matsuda-type arylation of indoles (at C_3) **with arylhydrazines** using $Pd(OAc)_2$/1,10-phen and trifluoroacetic acid under air s. Y. Chen, S. Guo, K. Li, J. Qu, H. Yuan, Q. Hua, B. Chen, Adv. Synth. Catal. *2013, 355* (4), 711-5 [DOI: 10.1002/adsc.201200997]; **directed radical *o*-arylation of anilines or phenols** with arylhydrazines using MnO_2 as oxidant cf. H. Jasch, J. Scheumann, M.R. Heinrich, J. Org. Chem. *2012, 77* (23), 10699-706 [DOI: 10.1021/jo301980j]; of anilines **with sodium aryldiazotates** ($ArN{=}N{-}O^-Na^+$) in aq. NaOH by a Gomberg-Bachmann-type reaction s. G. Pratsch, T. Wallaschkowski, M.R. Heinrich, Chem. Eur. J. *2012, 18* (37), 11555-9 [DOI: 10.1002/chem.201200430].

Palladium(II) acetate/benzoquinone/lithium tert-butoxide *Pd(OAc)₂/BQ/t-BuOLi*
γ,δ-Ethyleneketones from 2-ethylenealcohols and N-tosylhydrazones ←
Stereoselective palladium(II)-catalyzed oxidative cross-coupling

399. 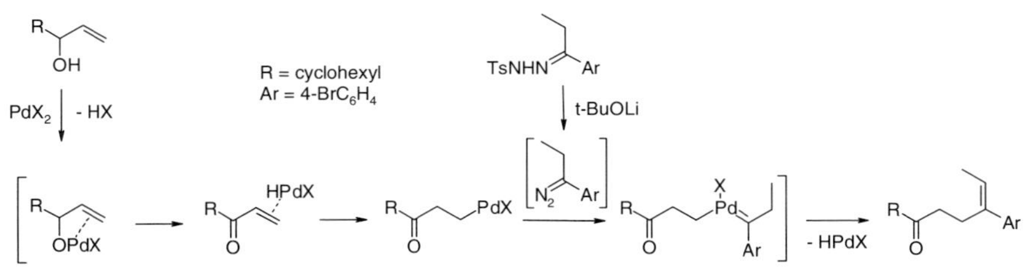

1-Cyclohexylprop-2-en-1-ol (0.3 mmol) added to a mixture of $Pd(OAc)_2$ (10 mol%), benzoquinone (2 eq.), Li-*tert*-butoxide (3 eq.), 4-bromopropiophenone N-tosylhydrazone (1 eq.) and acetonitrile (3 ml), the mixture stirred at 90° for 8 h, cooled to room temp., diluted with brine, extracted with ethyl acetate, and purified by flash chromatography → (E)-4-(4-bromophenyl)-1-cyclohexylhex-4-en-1-one. Y 77% (E/Z >10:1). This novel and scalable (to 10 mmol) coupling utilizes readily available compds. and is successful for allylic alcohols carrying electron-diverse aryl or alkyl groups at the α-position, reacting with tosylhydrazones derived from electron-diverse (het)ar. alkyl ketones, to afford 1,1-di- (thirteen examples; Y 53-84%) and (E)-trisubstituted pent-4-en-1-ones (sixteen examples; Y 52-82%; E/Z >10:1). The reaction was **also** carried out in one pot

from the corresponding **ketones**, generating the tosylhydrazone *in situ* before addition of the allylic alcohol (six examples; Y 51-80%). Other oxidants gave poor yields, and no reaction occurred using $Cu(OAc)_2$ as catalyst. F.e. and optimization s. H. Chen, L. Huang, W. Fu, X. Liu, H. Jiang, Chem. Eur. J. *2012, 18* (34), 10497-500 [DOI: 10.1002/chem.201200949].

Halogen ↑ CC ↓↑ Hal

Irradiation (s.a. under KOBu-t and KOBu-t/DMSO) ///
Benzofurans from *o*-chlorophenols and acetylene derivs. s. *44,* 781s*83* ○

Sodium hydroxide/tetra-n-butylammonium iodide *$NaOH/Bu_4NI$*
Potassium hydroxide or cesium carbonate/chiral quaternary ammonium bromides ←
Asym. α-alkylation H → R
under phase transfer catalysis, update s. *23,* 832s*82*; asym. α-benzylation and α-allylation of α-(carbo-*tert*-butoxy)lactones using a chiral fluorinated 1,1′-binaphthyl-2,2′-diyl-based spirocyclic quaternary ammonium bromide as catalyst and Cs_2CO_3 as base s. M.W. Ha, H. Lee, H.Y. Yi, Y. Park, S. Kim, S. Hong, M. Lee, M.-h. Kim, T.-S. Kim, H.-g. Park, Adv. Synth. Catal. *2013, 355* (4), 637-42 [DOI: 10.1002/adsc.201200976]; asym. α-alkylation of 2-arylcyclohexanones with a non-spirocyclic analog as catalyst in an aq. medium with KOH as base s. T. Kano, Y. Hayashi, K. Maruoka, J. Am. Chem. Soc. *2013, 135* (19), 7134-7 [DOI: 10.1021/ja403340r]; application of the methodology for the synthesis of chiral cyclic α-aminocarboxylic acid esters s. T. Kano, T. Kumano, R. Sakamoto, K. Maruoka, Org. Biomol. Chem. *2013, 11* (2), 271-8 [DOI: 10.1039/c2ob26778k]; phase transfer-catalyzed asym. α-alkylation of nickel(II) N-(*o*-acylaminobenzylidene)glycinates with a wide range of unactivated or activated halides *in 1,2-dichloroethane* containing 30% NaOH and Bu_4NI as catalyst (cf. *27,* 843s*79*) s. D. Houck, J.L. Aceña, V.A. Soloshonok, Helv. Chim. Acta *2012, 95* (12), 2672-9 [DOI: 10.1002/hlca.201200536].

Potassium hydroxide/chiral 1,1′-binaphthyl-based spirocyclic quaternary ammonium bromides ←
Regioselective asym. α-alkylation of allene-1,3-dicarboxylic acid esters ←
under phase transfer catalysis s. *83*, 274

Potassium tert-butoxide/irradiation *KOBu-t*////
Potassium tert-butoxide/alcohols or (S)-proline or macrocyclic aromatic 2-pyridone pentamers ←
Biaryls from arenes and ar. halides by metal-free coupling Ar-Ar′
update s. *77,* 458s*82*; inexpensive arylation of unactivated arenes with ar. halides using KOBu-*t* in *alcoholic* media s. W. Liu, F. Tian, X. Wang, H. Yu, Y. Bi, Chem. Commun. *2013, 49* (29), 2983-5 [DOI: 10.1039/c3cc40695d]; with electron-diverse ar. halides (and hetaryl halides) using KOBu-*t in DMSO under irradiation* at room temperature cf. M.E. Budén, J.F. Guastavino, R.A. Rossi, Org. Lett. *2013, 15* (6), 1174-7 [DOI: 10.1021/ol3034687]; of unactivated arenes using KOBu-*t* and a macrocyclic aromatic 2-pyridone pentamer as catalyst at 120° s. H. Zhao, J. Shen, J. Guo, R. Ye, H. Zeng, Chem. Commun. *2013, 49* (23), 2323-5 [DOI: 10.1039/c3cc00019b]; using KOBu-*t* and (S)-proline as catalyst s. K. Tanimoro, M. Ueno, K. Takeda, M. Kirihata, S. Tanimori, J. Org. Chem. *2012, 77* (18), 7844-9 [DOI: 10.1021/jo3008594]; of simple arenes **with diaryliodonium fluoroborates** under palladium catalysis with the Herrmann-Beller palladacyclic complex, diacetoxo[*o*-(di-*o*-tolylphosphino)benzyl]palladium(II) dimer, and added trifluoroacetic acid cf. T.E. Storr, M.F. Greaney, Org. Lett. *2013, 15* (6), 1410-3 [DOI: 10.1021/ol400412z].

Potassium tert-butoxide/1,10-phenanthroline *KOBu-t/phen*
Arylcarboxylic acid *tert*-butyl esters from ar. halides Hal → COOBu-*t*
by metal-free carbonylation s. *12,* 867s*83*

Potassium tert-butoxide/dimethyl sulfoxide/tris[2-phenylpyridinato-C2,N]iridium(III)/ irradiation ←

Biaryls from unactivated arenes and ar. halides Ar-Ar′

Regioselective radical arylation under visible-light photoredox catalysis

400.

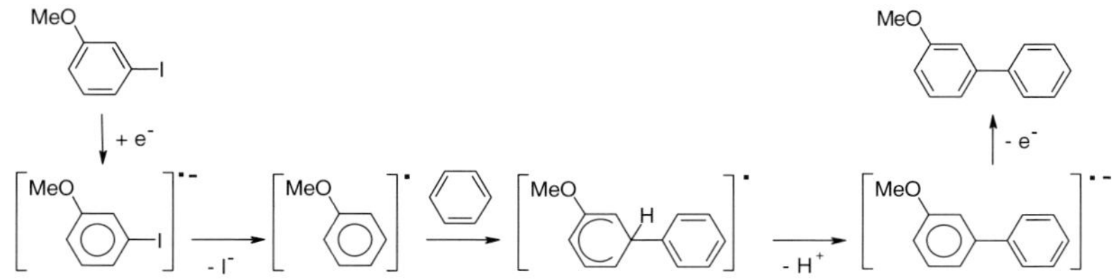

Arylation of unactivated arenes with ar. halides in basic media can now be performed *mildly* under visible-light photoredox catalysis, instead of under transition metal catalysis (cf. *77*, 458s*80,81*). **E:** Tris-[2-phenylpyridinato-C2,N]iridium(III) (0.5 mol%) and KOBu-*t* (3 eq.) introduced into a Schlenk tube under N_2 at room temp., benzene (3 ml), *m*-iodoanisole (0.5 mmol) and DMSO (0.5 ml) added via syringe, the mixture degassed by the freeze-pump-thaw procedure, placed at a distance of ca. 5 cm from a 14 W fluorescent lamp, stirred *at room temp.* for 24 h, and worked up with purification by chromatography on silica gel → 3-methoxybiphenyl. Y 89%. The procedure is notably applicable to the coupling of benzene, anisole, fluorobenzene and heteroarenes (pyridine, furan) with aryl iodides possessing electron-donating groups, yields being slightly lower with *o*-subst. iodobenzenes (thirteen examples; Y 62-89%). Less expensive aryl bromides, incl. naphthalene and phenanthrene derivs., also participated in the reaction but required heating at 70° (twenty-two examples; Y 35-79%), while less reactive chlorobenzene and benzene gave biphenyl in 35% yield at 90°. *p*-Chloroiodobenzene, *p*-bromoiodobenzene and 1,4-diiodobenzene, however, gave *p*-terphenyl on coupling with benzene (Y 46-61%). Reaction is presumed to involve initial generation of aryl halide radical anion via SET from photo-excited iridium(III), followed by cleavage to an aryl radical; the latter then adds to the arene to give a biaryl radical which is deprotonated to the corresponding biaryl radical anion prior to a second SET to give the product. The superbasic medium is essential for the conversion, yields being significantly lower with NaOBu-*t*, LiOBu-*t*, KOH and NaOH, while reaction failed with ethyldiisopropylamine and tributylamine. F.e., regioselectivity, and comparison of sensitizers s. Y. Cheng, X. Gu, P. Li, Org. Lett. *2013*, *15* (11), 2664-7 [DOI: 10.1021/ol400946k].

Organolithium compds. *RLi*

Diastereoselective synthesis of 1-amino-4-isochromanones from 3-chloroisoindole-1-carboxaldehydes

401.

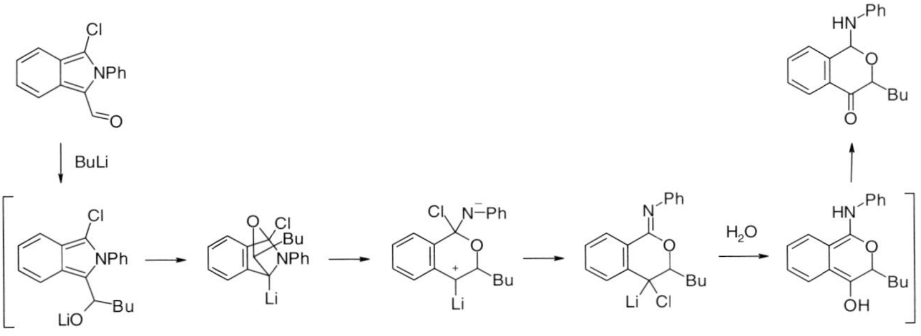

A soln. of *n*-butyllithium (1 eq.) in hexane (0.4 ml) added dropwise to a soln. of 3-chloro-2-phenylisoindole-1-carboxaldehyde (1 mmol) in THF (15 ml) at -78°, the mixture stirred for 2 h, quenched with satd. aq. NH_4Cl, and worked up with chromatographic purification → 3-butyl-1-(phenylamino)-3,4-dihydro-1*H*-2-benzopyran-4-one. Y 37% (as a ca. 2:1 mixture of diastereomers). The highly reactive but air stable 3-chloroisoindole-1-carboxaldehyde was prepared by Vilsmeier-Haack formylation of the corresponding isoindolinone. Attempted alkyl addition to the carbaldehyde group using BuLi or EtMgBr surprisingly resulted in formation of the title products in modest yields as 2:1 mixtures of diastereomers (two examples; Y 31-37%). The *trans* isomer was successfully isolated for X-ray crystallography but it was not confirmed as the major diastereomer. F.e. and additional reactions of the substrate s. I. Baglai, V. Maraval, Z.V. Voitenko, C. Duhayon, Y.M. Volovenko, R. Chauvin, Tetrahedron *2012*, *68* (34), 6908-13 [DOI: 10.1016/j.tet.2012.06.014].

Organolithium compds./copper(I) cyanide *RLi/CuCN*
Stereoselective 3-component synthesis of 3-siloxyenesilanes from 3,3-disilylacroleins via copper(I)-mediated 1,4-C→O-silyl group migration RCH(OSi≤)C=C(Si≤)R′

402.

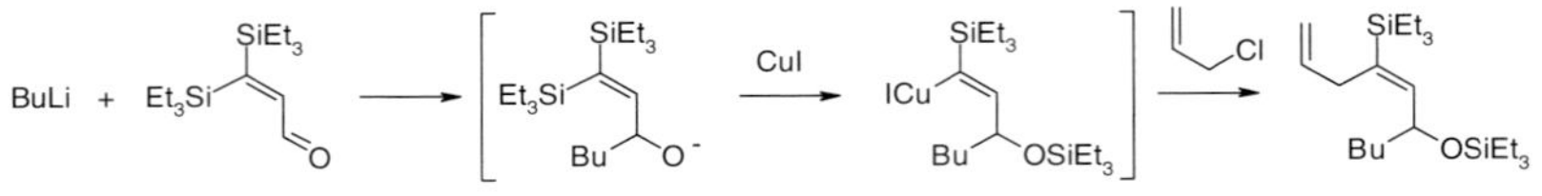

Butyllithium (2.5 *M* in hexane; 2 eq.) added dropwise to a soln. of 3,3-bis(triethylsilyl)acrolein (0.14 mmol) in dry THF (0.6 ml) under argon at -78°, stirred for 30 min, a soln. of CuCN (0.28 mmol) *in dry DMF* (1.8 ml) added via cannula at 0°, stirred for 30 min at room temp. followed by addition of allyl chloride (2 eq.), stirring continued for a further 2 h, the mixture quenched with 3.5% aq. NH_3, and worked up with purification by flash chromatography on silica gel → product. Y 85%. The procedure is applicable to the 3-component coupling of the disilylacrolein with various organolithium compds. (alkyl-, [hetero]aryl-, alkynyl- or vinyl-lithiums) and electrophiles, e.g. allyl or propargyl halides, allyl tosylates and diphenyl disulfide (twenty examples; Y 40-96%). A propargyl tosylate reacted regioselectively to give a 6-siloxy-4-silyl-1,2,4-triene (Y 72%). There was no reaction in the absence of copper(I) nor with CuI in place of CuCN. Reaction is presumed to involve initial 1,2-addition of the organolithium compd. to the aldehyde function, followed by copper-mediated 1,4-C→O-silyl group migration, thereby generating a siloxy-functionalized α-silylvinylcuprate prior to quenching with the electrophile. F.e. and conversion of the products to 1,1-disubst. (E)-crotylsilanes s. L. Yan, X. Sun, H. Li, Z. Song, Z. Liu, Org. Lett. *2013*, *15* (5), 1104-7 [DOI: 10.1021/ol400145z].

Lithium 2,2,6,6-tetramethylpiperidide *LiTMP*
Metal-free *o*-arylation of unprotected anilines with ar. halides [or triflates] H → Ar
s. *73*, 419s*83*

Bidentate or polymer-based lithium dialkylamides ←
α-Alkylation H → R
of esters with Li-amides s. *27*, 843s*32*; of ketones via amine-free lithium enolates using bidentate or recyclable polymer-based lithium dialkylamides as base s. R. Lazny, K. Wolosewicz, Tetrahedron Lett. *2013*, *54* (9), 1103-6 [DOI: 10.1016/j.tetlet.2012.12.067].

Potassium or cesium carbonate *K_2CO_3 or Cs_2CO_3*
Metal-free α-arylation of *o*-hydroxyacetophenones s. *61*, 392s*83* H → Ar

Sodium formate s. under Dichloro[1,2-bis(diphenylphosphino)ethane]palladium(II) *HCOONa*

Lithium or potassium iodide *LiI or KI*
Friedel-Crafts acylation H → Ac
update s. *5*, 603s*82*; of electron-rich arenes (incl. hindered phenols and N-methylpyrroles) with acyl chlorides using KI or LiI as promoter, also N-acylation of sulfonamides, s. R.J. Wakeham, J.E. Taylor, S.D. Bull, J.A. Morris, J.M.J. Williams, Org. Lett. *2013*, *15* (3), 702-5 [DOI: 10.1021/

ol400035f]; *heterogeneous* Friedel-Crafts acylation with moisture-insensitive, magnetically separable $CuFe_2O_4$ nanopowder s. R. Parella, Naveen, A. Kumar, S.A. Babu, Tetrahedron Lett. *2013, 54* (13), 1738-42 [DOI: 10.1016/j.tetlet.2013.01.081].

Macrocyclic aromatic 2-pyridone pentamers s. under Potassium tert-butoxide ←

Copper(0)/aluminum oxide *$Cu(0)/Al_2O_3$*
Heterogeneous copper-catalyzed Heck arylation with deactivated ar. chlorides H → Ar
s. *27,* 871s*83*

Copper(II) oxide s. under $Pd(OAc)_2$ *CuO*
Silver oxide s. under Copper(II) acetate and Bis(acetonitrile)dichloropalladium(II) *Ag_2O*

Copper ferrite nanoparticles *$CuFe_2O_4$*
Heterogeneous Friedel-Crafts acylation s. *5,* 603s*83* H → Ac

Chiral copper(I) bis(Δ^2-oxazoline) complexes *[Cu(I)]**
Cyclopropanation of ethylene derivs. with iodonium ylids ▽
cf. *75,* 440s*82*; cyclopropane-1,1-dicarboxylic acid esters from polysubst. alkenes by **asym. cyclopropanation** using a chiral copper(I) bis(Δ^2-oxazoline) complex s. C. Deng, L.-J. Wang, J. Zhu, Y. Tang, Angew. Chem., Int. Ed. *2012, 51* (46), 11620-3 [DOI: 10.1002/anie.201206376].

Copper(II) acetoacetonate/lithium tert-butoxide *$Cu(acac)_2$/LiOBu-t*
Copper-catalyzed C_5-arylation of thiazolo[3,2-*b*][1,2,4]triazoles s. *57,* 376s*83* H → Ar

Copper(II) acetate/triphenylphosphine oxide/silver oxide/acetonitrile ←
Ar. nitriles from halides s. *29,* 845s*83* Hal → CN

Copper(I) cyanide s.a. under Organolithium compds. *CuCN*

Copper(I) cyanide/sulfuric acid/dicyclohexylcarbodiimide/pyridine/triethylamine ←
Synthesis of α-arylnitriles C(Cl)CH=NOH → C(Ar)CN
from α-chloroaldoximes and lithium aryl(cyano)cuprates
via 1-nitrosoethylene derivs.

403. 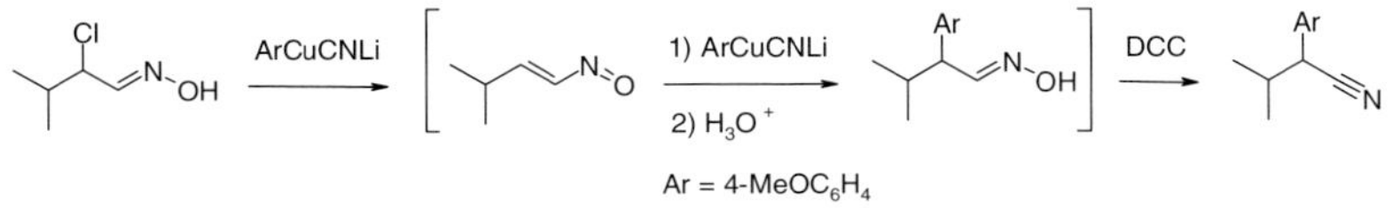

in one pot. A soln. of 4-methoxyphenyllithium (2.5 eq.) in hexane/ether (1:2 to 2:1; 0.5 to 1 *M*) added slowly via syringe to a mixture of CuCN (2.2 eq.) and anh. THF (1.5 ml) under argon at -10°, the colorless soln. stirred at -30° for 5 min, a soln. of 2-chloro-3-methylbutanaldoxime (0.15 mmol) in THF (1.5 ml) added dropwise, the yellow soln. stirred for 20 min, concd. H_2SO_4 (4.8 eq.) and methylene chloride (1.5 ml) added, the mixture warmed to room temp., anh. pyridine (15 eq.), anh. triethylamine (9.5 eq.) and solid DCC (8 eq.) added sequentially, the mixture stirred for 12 h, quenched with 2% aq. ethylenediamine, extracted with ether, and worked up with chromatographic purification → 2-(4-methoxyphenyl)-3-methylbutanenitrile. Y 98%. This experimentally simple dehydrochlorination-nucleophilic arylation-oxime dehydration was successful for α-chloroaldoximes (incl. 2-chloroacetaldoxime) reacting with electron-diverse lithium aryl(cyano)cuprates (eleven examples; Y 51-98%), via intermediate nitrosoethylenes. Reaction of a 1.5:1 diastereomeric mixture of 1-chloro-4-*tert*-butylcyclohexanecarbaldoximes apparently occurred via equatorial arylation of the intermediate nitrosoalkene to afford the product α-arylnitrile as a single diastereomer (Y 50%). F.e., optimization and substrate prepn. s. R. Sengupta, S.M. Weinreb, Synthesis *2012, 44* (18), 2933-7 [DOI: 10.1055/s-0032-1316984].

Copper(I) thiocyanate s. under Dichloro[1,2-bis(diphenylphosphino)ethane]palladium(II) ←

Copper(II) triflate *Cu(OTf)$_2$*

Regioselective copper-catalyzed 3-component synthesis of quinolines from diaryliodonium salts, acetylene derivs. and nitriles

404. 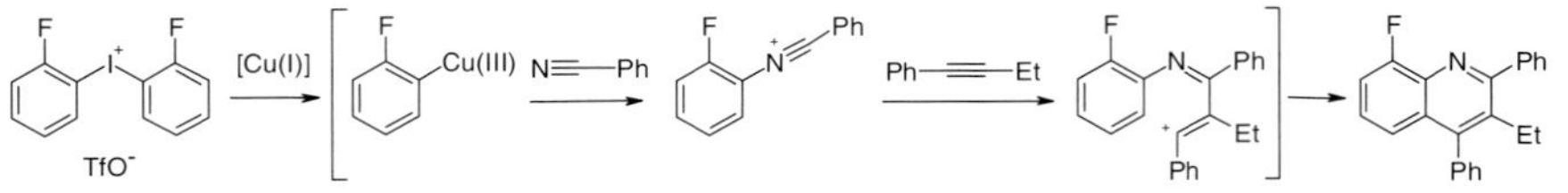

2,3,4-Trisubst. quinolines. The startg. alkyne (1.2 eq.), benzonitrile (1.2 eq.) and 1,2-dichloroethane (2 ml) added to a reaction tube charged with the startg. diaryliodonium salt (1 mmol) and $Cu(OTf)_2$ (10 mol%) under N_2, the tube sealed, the mixture cooled to room temp., aq. $NaHCO_3$ (5 ml) added, extracted with methylene chloride, and worked up with purification by chromatography on silica gel → product. Y 73%. The procedure is regioselective, economical, flexible and avoids handling aniline derivs. as commonly required for quinoline synthesis. It is applicable in high yield to the coupling of electron-diverse, symmetrical diaryliodonium salts possessing weakly coordinating anions with terminal or internal alkynes (possessing aryl, alkyl or ester groups) and electron-diverse [hetero]aromatic nitriles (fifteen examples; Y 52-95%). Aliphatic nitriles and phenylacetonitrile, however, gave low yields, while there was no reaction with ethyl cyanoformate and diethyl phosphorocyanidate. The electronic effect of substituents and the regioselectivity are in keeping with an electrophilic cascade process after initial formation of an arylcopper(III), following reduction or disproportionation of Cu(II) to Cu(I). F.e. incl. **2,4-disubst. quinolines** from terminal alkynes s. Y. Wang, C. Chen, J. Peng, M. Li, Angew. Chem., Int. Ed. *2013*, *52* (20), 5323-7 [DOI: 10.1002/anie.201300586].

Copper(I) chloride (s.a. under Pd-C) *CuCl*

Regio- and stereo-selective synthesis of enol triflates from acetylene derivs. and unsatd. iodonium triflates C≡C → C(R)=C(OTf)

405. 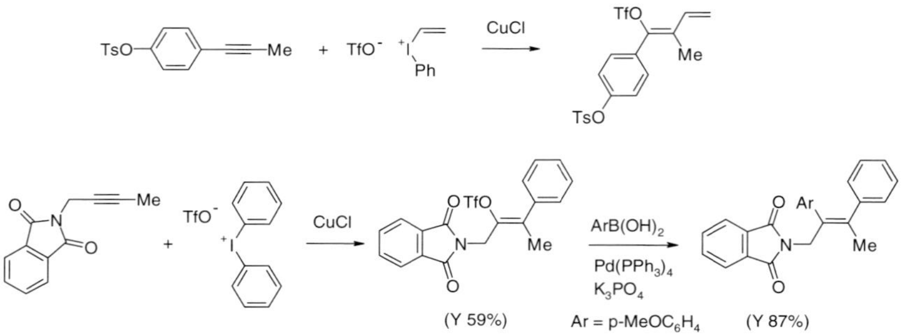

A 'polarity-reversed' [electrophilic] alkyne carbofunctionalization strategy has been reported, affording enol triflates which may then undergo cross-coupling to provide tetrasubst. alkenes (starting from internal alkynes). **E:** A reaction tube charged with ethenyl(phenyl)iodonium triflate (0.5 mmol) and CuCl (10 mol%), the tube sealed then evacuated and backfilled with N_2 three times, methylene chloride (3 ml) and 4-(prop-1-yn-1-yl)phenyl tosylate (1 eq.) added, the mixture stirred at 5° for 36 h, cooled to room temp., concentrated under reduced pressure, filtered through a short pad of silica gel, washed with ether, solvent removed under reduced pressure, and the residue purified by flash chromatography on silica gel → (Z)-4-[2-methyl-1-(trifluoromethylsulfonyloxy)buta-1,3-dien-1-yl]phenyl tosylate. Y 86% (isomeric ratio 3.1:1). The method is operationally simple, avoids handling organometallic reagents, and works with a range of alkynes and vinyl- or diaryl-iodonium triflates. Nine examples with sym. alkynes gave yields of 57-84% (Z/E 8:1 to >20:1 plus one Z/E 1:15). Six further examples with unsym. internal alkynes gave yields of 59-89% with good regio- and stereo-selectivity. Seven examples with terminal alkynes gave yields of 27% (with acetylene itself), 41-82%. Reaction is believed to take place via a copper(III) species. Elaboration to **tetrasubst. ethylene derivs.** via Suzuki coupling, Sonogashira coupling, Buchwald-Hartwig amination, Miyaura borylation or carbomethoxylation gave yields of 71-94% (six examples). Furthermore the enol triflates could be hydrolyzed to an α-arylketone (Y 90%), undergo asym. dihydroxylation with AD-mix β (Y 82%; e.e. 96%) or hydrogenation over Pd-C/$BaSO_4$ (Y 94%). F.e.s. M.G. Suero, E.D. Bayle, B.S.L. Collins, M.J. Gaunt, J. Am. Chem. Soc. *2013*, *135* (14), 5332-5 [DOI: 10.1021/ja401840j].

Copper(I) bromide s. under Magnesium *CuBr*

Copper(I) iodide [s.a. under $PdCl_2(PPh_3)_2$] *CuI*

Copper-catalyzed intramolecular carbo(trifluoromethyl)ation of ethylenearenes ○

406.

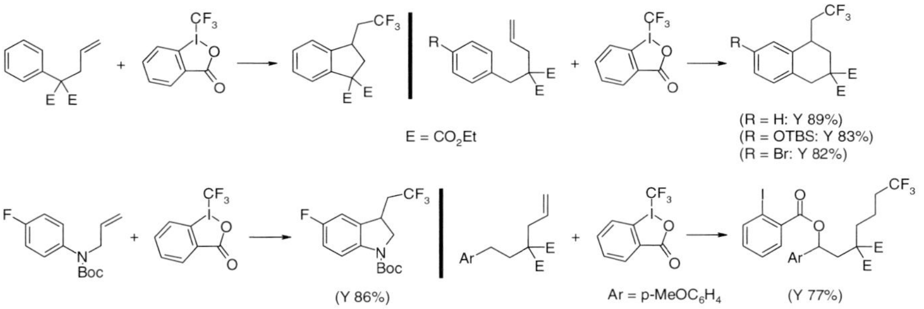

1-(2,2,2-Trifluoroethyl)indans. CuI (10 mol%) and Togni's reagent (1.2 eq.) added to a flame-dried Schlenk flask, the flask evacuated and back-filled with N_2, degassed 1,4-dioxane (1 ml) and startg. homoallylarene (0.2 mmol) added, the mixture stirred at 80° for 12 h, diluted with ethyl acetate (5 ml), washed with aq. $NaHCO_3$ and brine, the organic layer dried ($MgSO_4$), filtered, evaporated, and the residue subjected to column chromatography on silica gel → diethyl 3-(2,2,2-trifluoroethyl)indan-1,1-dicarboxylate. Y 78%. This copper-catalyzed intramolecular carbo-trifluoromethylation is applicable to **1-(2,2,2-trifluoroethyl)tetralins** (eleven examples; Y 61-98%), **N-Boc-3-(2,2,2-trifluoroethyl)indolines** (four examples; Y 82-86%; Y 67% for an N-TBS-deriv.) or N-Boc-4-(2,2,2-trifluoroethyl)-1,2,3,4-tetrahydroquinoline (Y 93%) as well as indans (three further examples; Y 59-77%). Functional groups such as ar. MeO, OTBS, OH, Cl or Br were tolerated. 6-Membered ring closure was generally faster than 5-membered, in line with acceleration by orbital interaction of aryl ring and alkene being crucial to promote the reaction. 7-Membered ring formation did not occur, however, **1-aryl-1-(*o*-iodoaroyloxy)-7,7,7-trifluorides** being formed instead (three examples; Y 45-77%), presumably via deprotonative trifluoromethylation, 1,5-hydride shift and trapping with *o*-iodobenzoate generated from the 1,2-benziodoxole. F.e.s. H. Egami, R. Shimizu, S. Kawamura, M. Sodeoka, Angew. Chem., Int. Ed. *2013*, *52* (14), 4000-3 [DOI: 10.1002/anie.201210250].

Copper(I) iodide/lithium carbonate *CuI/Li_2CO_3*

α-Arylketones from acetylene derivs. and diaryliodonium salts C≡C → C(O)CHAr

Regioselective copper(I)-catalyzed synthesis

407.

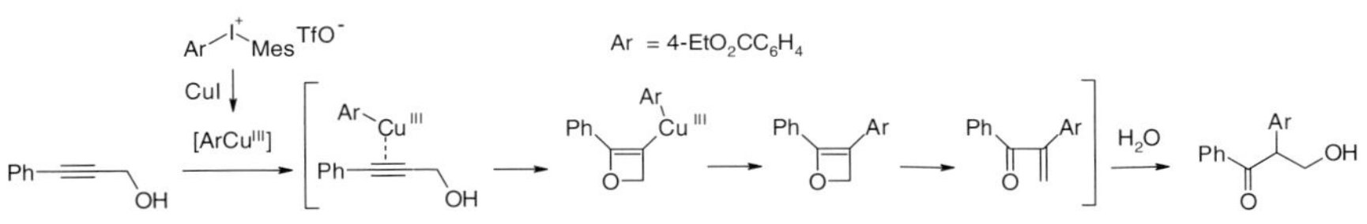

α-Aryl-β-hydroxyketones. Methylene chloride (5 ml) and 1-phenylprop-1-yn-3-ol (0.23 mmol) added to a mixture of 4-ethoxycarbonylphenyl(mesityl)iodonium triflate (1.1 eq.), CuI (20 mol%)

and Li_2CO_3 (1.8 eq.) under N_2, the mixture stirred at 70° for 48 h, quenched with satd. aq. NH_4Cl, extracted with ethyl acetate, and purified by flash chromatography → ethyl 4-(3-hydroxy-1-oxo-1-phenylpropan-2-yl)benzoate. Y 78%. This novel oxidative arylation of 1-phenylprop-1-yn-3-ols was shown to proceed via rearrangement to an intermediate α-alkylideneketone which was hydrated with adventitious water (use of additional water gave reduced yields), and was accomplished with electron-diverse symmetrical di(het)aryl iodonium triflates, or with unsymmetrical analogs containing a sterically hindered (and poorly transferred) mesityl group (thirteen examples; Y 63-85%). A *m*-nitrophenyliodonium salt gave only the intermediate α-methyleneketone (Y 46%), as did the α-subst. propargylic alcohol, 4,4-dimethyl-1-phenylpent-1-yn-3-ol (Y 48%), but α-isopropyl or α-4-chlorophenyl analogs gave complex mixtures, and tert. propargylic alcohols suffered dehydration to enynes only (two examples; Y 52-71%). The arylation was extended to use of 1-arylprop-1-ynes, using $Cu(OAc)_2/O_2$ as oxidant, and required the addition of 1 eq. of water to afford the product α-(het)aryl ketones (ten examples; Y 42-81%; a *m*-nitrophenyliodonium salt gave 24%), and was also successful with 1-phenylbut-1-yne (Y 42%) but gave complex mixtures with diarylacetylenes and 5-decyne.

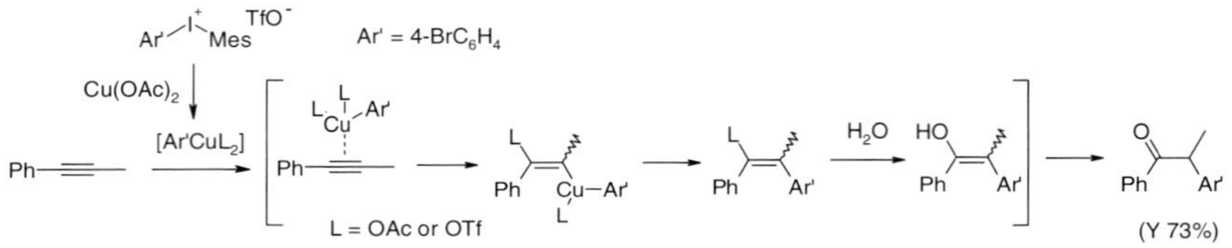

F.e., optimization and iodonium salt prepn. s. Z.-F. Xu, C.-X. Cai, J.-T. Liu, Org. Lett. *2013*, *15* (9), 2096-9 [DOI: 10.1021/ol4003543].

Copper(I) iodide/1,10-phenanthroline/cesium carbonate *CuI/phen/Cs$_2$CO$_3$*
α-Arylation of unactivated enolizable ketones s. *61,* 392s*83* H → Ar

Silver triflate *AgOTf*
Regioselective silver-promoted ar. formylation H → CHO
of phenolethers with dichloromethyl methyl ether s. *83,* 256

Gold(I) chloride *AuCl*
Regioselective gold-catalyzed direct C$_2$-(alk-1-ynyl)ation of furans H → C≡C

408. 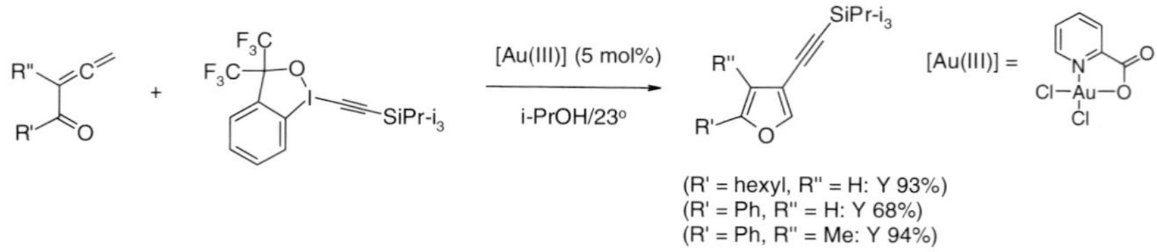

C$_2$-(Silylethynyl)ation. 1-[(Triisopropylsilyl)ethynyl]-1,2-benziodoxol-3(1*H*)-one [TIPS-EBX] (1.2 eq.), AuCl (0.05 eq.) and startg. furan (0.4 mmol) added to acetonitrile (1 ml) under air, the mixture stirred for 26 h at room temp., the organic layer washed with HCl (1 *M*; 20 ml), the aq. layer extracted with methylene chloride, the combined organic layers washed with satd. aq. Na_2CO_3 and brine, dried ($MgSO_4$), and concentrated under reduced pressure → product. Y 90%. Six further examples with 2-alkyl- or 2,3-dialkyl-derivs. or furan itself gave yields of 55-79%. 2-Aryl-derivs.

gave yields of 56-71% by reaction at 60° (four examples). 2,5-Dimethylfuran underwent C_3-(silylethynyl)ation in 45% yield. For alkynylation of pyrroles, indoles or thiophenes (with added trifluoroacetic acid) cf. *73*, 419s*82*. **3-(Silylethynyl)furans** could be obtained **from α-allene-ketones** by reaction with *3,3-bis(trifluoromethyl)*-1-[(triisopropylsilyl)ethynyl]-1,2-benziodoxole (2 eq.) in the presence of (2-picolinato)gold(III) dichloride (twelve examples; Y 53%, 68-97%). The clean interception of an Au-C intermediate with a well-defined alkynylation agent is believed to be an important breakthrough, success only being possible for a very narrow window of electronic density at the iodine atom. F.e. and optimization, also conversion to a sym. 2,5-bis(furan-2-yl)-thiophene (Y 51%) or a 2-(2-silylethyl)tetrahydrofuran (Y 74%), s. Y. Li, J.P. Brand, J. Waser, Angew. Chem., Int. Ed. *2013*, *52* (26), 6743-7 [DOI: 10.1002/anie.201302210]; **3-(silylethynyl)-indoles** from *o*-acetylene-*prim*-amines with TIPS-EBX under gold(III) catalysis s. J.P. Brand, C. Chevalley, J. Waser, Beilstein J. Org. Chem. *2011* (7), 565-9 [DOI: 10.3762/bjoc.7.65].

Magnesium (s.a. under $NiCl_2$/bpy) *Mg*

Metal-free coupling of 2-(*o*-halogenoaryl)pyridines with organomagnesium halides ArHal + RMgX → Ar-R

409.

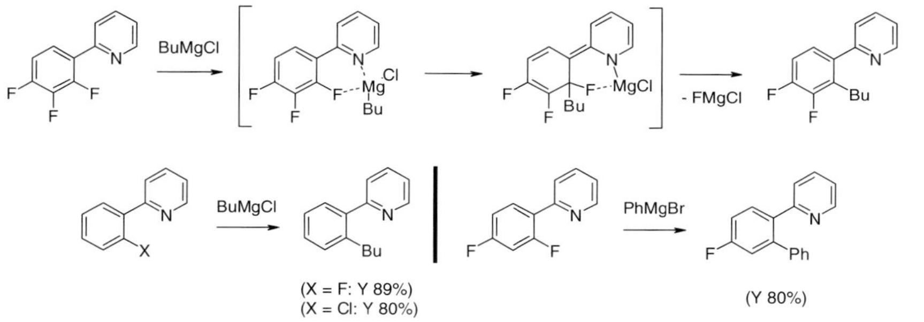

A soln. of *n*-butylmagnesium chloride (2.5 eq.) in THF (2 *M*) added dropwise to a stirred soln. of 2-(2,3,4-trifluorophenyl)pyridine (2 mmol) in THF (5 ml) at room temp., the mixture heated at 70° for 6-24 h (TLC), quenched with satd. aq. NH_4Cl and water, extracted with ethyl acetate, and purified by chromatography on silica gel → 2-(2-*n*-butyl-3,4-difluorophenyl)pyridine. Y 87%. High yields (82-90%; eight examples) were observed for nucleophilic alkylation of fluoro derivs. using BuMgCl, EtMgCl or MeMgCl (a single chloro analog afforded 80%). Yields were slightly depressed (80-83%; two examples) and reaction times prolonged for arylation with PhMgBr, affording **biaryls**. With the single exception of $MnCl_2$, yields were significantly lower using a wide range of metal catalysts. F.e.s. Y. Xiong, J. Wu, S. Xiao, J. Xiao, S. Cao, J. Org. Chem. *2013*, *78* (9), 4599-603 [DOI: 10.1021/jo400424d].

Magnesium/copper(I) bromide/chiral ferrocenyldi(phosphines) ←

Regioselective asym. synthesis of ethylene derivs. from β,γ-ethylenehalides C(Hal)C=C → C=C-C(R)

update s. *62*, 381s*82*; asym. synthesis of 3-methylated 1,4-dienes from 2,4-dien-1-yl bromides with CuBr·SMe_2 as catalyst and a chiral ferrocenyldi(phosphine) [(S,S)-TaniaPhos] as ligand s. Y. Huang, M. Fañanás-Mastral, A.J. Minnaard, B.L. Feringa, Chem. Commun. *2013*, *49* (32), 3309-11 [DOI: 10.1039/c3cc41021h].

Magnesium/dimethyl sulfoxide/trifluoroacetic anhydride *Mg/DMSO/$(CF_3CO)_2O$*

2-α-Substitution of indoles with Grignard compds. s. *83*, 50

Magnesium/transition metal salts or complexes ←

Cross-coupling with Grignard compds. (Kumada coupling) RMgHal + R′Hal → R-R′

update s. *26*, 875s*81*; Kumada **biaryl synthesis** of 2-aryl- from 2-chloro-N-heteroarenes with $FeBr_3$ or $CoCl_2$ as catalyst and quinoline or isoquinoline as accelerator s. O.M. Kuzmina, A.K. Steib, J.T. Markiewicz, D. Flubacher, P. Knochel, Angew. Chem., Int. Ed. *2013*, *52* (18), 4945-9 [DOI:

10.1002/anie.201210235]; Kumada synthesis **of diarylmethanes** from arylmagnesium bromides and benzyl chlorides with electronically-tuned dichlorobis[1,2-bis(diarylphosphino)benzene]-iron(II) complexes as catalyst s. S. Kawamura, M. Nakamura, Chem. Lett. *2013, 42* (2), 183-5 [DOI: 10.1246/cl.2013.183]; Kumada coupling of bromoporphyrins with α-silylalkylmagnesium halides using $Pd_2(dba)_3/Ph_2P(O)H$ as catalyst s. N. Sugita, S. Hayashi, F. Hino, T. Takanami, J. Org. Chem. *2012, 77* (23), 10488-97 [DOI: 10.1021/jo302122f]; asym. synthesis of 1,5-dienes from allyl bromides and allylmagnesium bromides **by asym. Kumada allyl-allyl coupling** with a chiral copper(I) phosphoromonoamidite as catalyst (cf. *62*, 381s*81*) s. V. Hornillos, M. Pérez, M. Fañanás-Mastral, B.L. Feringa, J. Am. Chem. Soc. *2013, 135* (6), 2140-3 [DOI: 10.1021/ja312487r].

Zinc s.a. under Pincer-type nickel(II) m-phenylenebis(phosphonite) complexes *Zn*

Zinc/copper *Zn/Cu*
Zinc/chiral 1,3,2-dioxaborolanes ←
Simmons-Smith cyclopropanation ▽
s. *26,* 876; accelerated conversion with methylene iodide and the Zn/Cu couple *in cyclopentyl methyl ether* s. K. Fujii, K. Shiine, T. Misaki, T. Sugimura, Appl. Organomet. Chem. *2013, 27* (12), 69-72 [DOI: 10.1002/aoc.2935]; chiral polysubst. *trans*-iodo- and *trans*-chloro-cyclopropanes from α-iodo- and α-chloro-zinc carbenoids, respectively, by **asym. cyclopropanation** in the presence of a chiral 1,3,2-dioxaborolane (cf. *47,* 806s*47-9*) s. L.-P.B. Beaulieu, L.E. Zimmer, A. Gagnon, A.B. Charette, Chem. Eur. J. *2012, 18* (46), 14784-91 [DOI: 10.1002/chem.201202528].

Zinc/nickel(II) acetoacetonate or dichlorobis(triphenylphosphine)palladium(II) ←
Ketones from carboxylic acid chlorides by Negishi-type coupling COCl → COR
update s. *41,* 796s*82*; synthesis of highly *hindered* diaryl ketones from *o*-subst. arylzinc halides and arylcarboxylic acid chlorides using $Ni(acac)_2$ or $PdCl_2(PPh_3)_2$ as catalyst s. H.-H. Cho, S.-H. Kim, Bull. Korean Chem. Soc. *2012, 33* (9), 3083-6 [DOI: 10.5012/bkcs.2012.33.9.3083].

Zinc/transition metal salts or complexes ←
Negishi coupling ←
update s. *38,* 836s*82*; synthesis of **alkylarenes** from ar. halides with the *bis(chelate)* complex, chlorobis[1,2-bis(diphenylphosphino)ethane]iron(II) or cationic bis(acetonitrile)bis[1,2-bis-(diphenylphosphino)ethane]iron(II) complexes, also application to the coupling of aryl-boron, -aluminum or -indium compds., s. R.B. Bedford, E. Carter, P.M. Cogswell, N.J. Gower, M.F. Haddow, J.N. Harvey, D.M. Murphy, E.C. Neeve, J. Nunn, Angew. Chem., Int. Ed. *2013, 52* (4), 1285-8 [DOI: 10.1002/anie.201207868]; with a cyclooctadienebis(silylene)nickel(II) complex as catalyst, also by Kumada coupling (cf. *26,* 875s*83*), s. C.I. Someya, M. Haberberger, W. Wang, S. Enthaler, S. Inoue, Chem. Lett. *2013, 42* (2), 286-8 [DOI: 10.1246/cl.2013.286]; *N-directed* coupling of ar. iodides with fluoro(iodo)methyl 2-pyridyl sulfone using Et_2Zn/CuI s. Y. Zhao, B. Gao, C. Ni, J. Hu, Org. Lett. *2012, 14* (23), 6080-3 [DOI: 10.1021/ol3029737]; mild and general **Negishi biaryl coupling** of [hetero]arylzinc chlorides with heteroaryl halides or triflates using the palladacyclic complex, mesyloxy[2-(dicyclohexylphosphino)-2′,4′,6′-triisopropylbiphenyl]-(2′-aminobiphenyl-2-yl)palladium(II), at low loading s. Y. Yang, N.J. Oldenhuis, S.L. Buchwald, Angew. Chem., Int. Ed. *2013, 52* (2), 615-9 [DOI: 10.1002/anie.201207750]; procedure for the site-selective oxidative addition of active zinc(0) to the C_2-Br bond of dibromothiophenes s. H.-S. Jung, H.-H. Cho, S.-H. Kim, Tetrahedron Lett. *2013, 54* (8), 960-4 [DOI: 10.1016/j.tetlet.2012.12.024]; an efficient and reliable procedure for the preparation of highly reactive Rieke zinc from $ZnCl_2$ and Li-naphthalenide (generated from naphthalene/Li *and added benzo[b]-thiophene*) s. S. Kudret, J. D′Haen, W.D. Oosterbaan, L. Lutsen, D. Vanderzande, W. Maes, Adv. Synth. Catal. *2013, 355* (2-3), 569-75 [DOI: 10.1002/adsc.201201077].

Diethylzinc s. under Nickel(II) acetoacetonate *Et_2Zn*

Magnesium bis(2,2,6,6-tetramethylpiperidide)·bis(lithium chloride)/boron fluoride ←
Chlorozinc 2,2,6,6-tetramethylpiperidide·lithium chloride/boron fluoride/ copper(I) cyanide·bis(lithium chloride) ←
Regioselective metalation of N-heteroarenes with boron fluoride/hindered metal amides as frustrated Lewis pairs ←

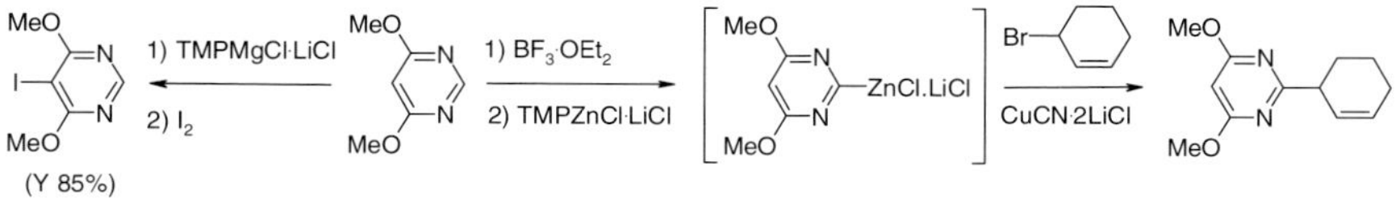

410. 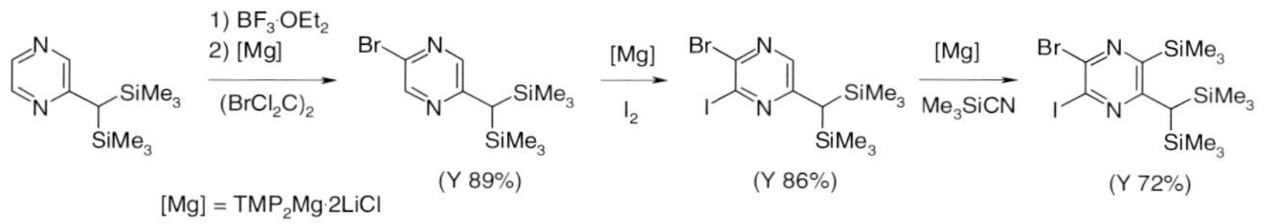

Metallic amide bases such as TMPZnCl·LiCl or TMP$_2$Mg·2LiCl (where TMP = 2,2,6,6-tetramethylpiperidide) have been found to be compatible with a strong Lewis acid, such as BF_3·OEt$_2$, a synergistic effect (dual activation) being observed, as a result of steric hindrance. This effect allows a switch in the regioselectivity of metalation of N-heteroarenes, such as pyrimidines (or thienopyrimidines), to be achieved by initial complexation with BF_3·OEt$_2$. **E: 2-Zincation of pyrimidines**. A dry argon-flushed Schlenk flask charged with a soln. of 4,6-dimethoxypyrimidine (2 mmol) in dry THF (2 ml), after cooling to 0°, BF_3·OEt$_2$ (1.1 eq.) added dropwise, the mixture stirred for 15 min at this temp., then cooled to -20°, followed by dropwise addition of TMPZnCl·LiCl (1.1 eq.) in THF (2.25 ml), stirred for 1 h (monitored by GC), the mixture cooled to -40°, CuCN·2LiCl (1.1 eq.) in THF (1.1 ml) added, after stirring for 15 min at this temp., 3-bromocyclohex-1-ene (1.5 eq.) added dropwise, the mixture slowly warmed to 25°, quenched with satd. aq. NH_4Cl (9 ml) and concd. NH_3 (1 ml), extracted with methylene chloride, dried ($MgSO_4$), solvents removed *in vacuo*, and the residue purified by flash chromatography on silica gel → product. Y 92%. In contrast, metalation in the absence of BF_3 followed by iodinolysis gave the 5-iodo-deriv. (Y 85%). Seven further examples gave 2-subst. pyrimidines in yields of 66-89% by Negishi cross-coupling, allylation or iodinolysis, while two examples with thienopyrimidines gave yields of 70% and 77%, and 2-bromination of a purine was achieved in 55% yield. The method was also adapted for the **sequential regioselective full functionalization of 2-[bis(trimethylsilyl)methyl]-pyrazine** via 5-subst. and 5,6-disubst. derivs. (1st step: five examples, Y 61-89%; 2nd step: six examples, Y 70-93%; 3rd step: six examples, Y 59-83%) by magnesiation using TMP$_2$Mg·2LiCl. Initial selective complexation at the least hindered heterocyclic N-atom directs magnesiation at position 5 as this is the most accessible activated proton and the second magnesiation occurs at position 6 due to steric factors. Further transformations of the bis(trimethylsilyl)methyl moiety to methyl, formyl or styryl groups are also described, and the group is compatible with palladium-catalyzed annelation to afford a imidazo[4,5-*b*]pyrazine deriv. The metalation procedure could not be extended to pyridazine derivs., however. F.e.s. K. Groll, S.M. Manolikakes, X.M. du Jourdin, M. Jaric, A. Bredihhin, K. Karaghiosoff, T. Carell, P. Knochel, Angew. Chem., Int. Ed. *2013*, *52* (26), 6776-80 [DOI: 10.1002/anie.201301694]; simple, mild prepn. of the zinc base from cheap, commercial zinc powder and 1-chloro-2,2,6,6-tetramethylpiperidine [TMPCl] and application to the regio- and chemo-selective functionalization of dibromo-pyridazines and -pyrazines cf. A. Unsinn, M.J. Ford, P. Knochel, Org. Lett. *2013*, *15* (5), 1128-31 [DOI: 10.1021/ol400248e]; for heteroarylcarbinols via generation of chloromagnesium 2-heteroaryl(trifluoro)borates with chloromagnesium 2,2,6,6-tetramethylpiperidide·BF_3 s. *1*, 680s*83*.

Zinc triflate/air/triethylamine *$Zn(OTf)_2/O_2/Et_3N$*

3-Acoxy-4-acylaminopyridines
from α-(N-acyl-N-vinylamino)aldehydes, isonitriles and carboxylic acid chlorides via N→O-acyl migration

411. 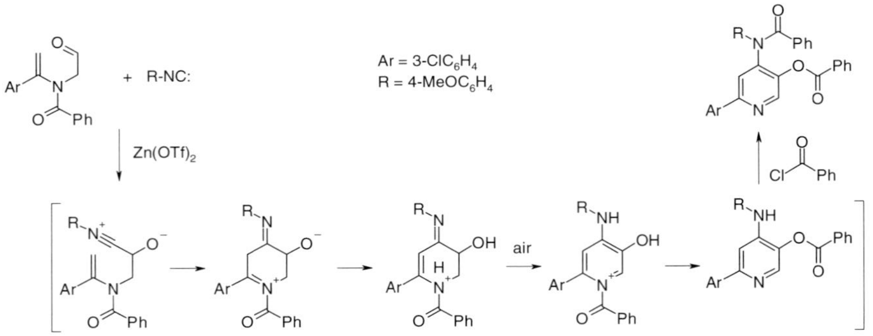

An unprecedented zinc triflate-mediated [5+1]-cycloaddition for the construction of pyridines **under mild conditions** is reported, the key step involving an α-addition of a tertiary enamide and an aldehyde to an isocyanide acting as a one-carbon component. **E:** N-[1-(3-Chlorophenyl)vinyl]-N-(2-oxoethyl)benzamide (0.3 mmol), 4 Å molecular sieves (300 mg) and $Zn(OTf)_2$ (1 eq.) added to a soln. of *p*-methoxyphenyl isocyanide (1.2 eq.) in acetonitrile (15 ml), the mixture stirred at room temp. for 7 h (TLC), then quenched with satd. aq. $NaHCO_3$, filtered through a pad of Celite, extracted with methylene chloride, the combined organic extracts washed with brine, dried ($MgSO_4$), the dried organic extracts left under atmospheric conditions overnight, Et_3N (4 eq.) and startg. acyl chloride (4 eq.) added, the resulting mixture stirred for 2 h, filtered, concentrated *in vacuo*, and the residue purified by flash chromatography on silica gel → 6-(3-chlorophenyl)-4-[N-(4-methoxyphenyl)benzamido]pyridin-3-yl benzoate. Y 90%. Fifteen further examples gave 2-aryl-5-aroyloxy-4-aroylaminopyridines starting from ar. or aliphatic isocyanides and benzoyl chloride in yields of 41-91%, while 2-(cyclohex-1-enyl)- and 2-cyclohexyl-analogs were obtained in yields of 60% and 70%, respectively. Three examples varying the acid chloride gave yields of 66% and 85% for aroyl chlorides and 30% with acetyl chloride with no scrambling of the acyl groups observed. This method provides a fast and straightforward route to analogs of acetylcholinesterase [AChE] inhibitors. F.e.s. C.-H. Lei, D.-X. Wang, L. Zhao, J. Zhu, M.-X. Wang, J. Am. Chem. Soc. *2013*, *135* (12), 4708-11 [DOI: 10.1021/ja401701x].

Borane·1,3-dimethylimidazol-2-ylidene/azodiisobutyronitrile ←
Benzyl alcohols from ar. iodides by reductive carbonylation s. *12*, 867s*83* I → CH_2OH

Chiral 1,3,2-dioxaborolanes s. under Zinc ←
Aluminum oxide s. under Cu(0) *Al_2O_3*
Boron fluoride s. under Chlorozinc 2,2,6,6-tetramethylpiperidide·lithium chloride *BF_3*
Norbornene s. under Bis(acetonitrile)dichloropalladium(II) ←
Alcohols s. under Potassium tert-butoxide *ROH*
Ammonium tartrate s. under Tris(dibenzylideneacetone)dipalladium ←
Dicyclohexylcarbodiimide s. under Copper(I) cyanide *DCC*
1,3-Di(adamantyl)imidazol-2-ylidene s. under [(allyl)PdCl]$_2$ *NHC*
Acetonitrile s. under Copper(II) acetate *MeCN*
Trifluoroacetic anhydride s. under Mg *$(CF_3CO)_2O$*
Formic acid s. under Palladium(II) chloride *HCOOH*
(S)-Proline s. under Potassium tert-butoxide *Pro-OH*
Trifluoroacetic acid s. under Bis(acetonitrile)dichloropalladium(II) *CF_3COOH*

Dimethyl sulfoxide s. under Potassium tert-butoxide and Mg *DMSO*
1,4-Bis(4-pyridyl)-2,3-diaza-1,3-butadiene s. under Bis(acetonitrile)dichloropalladium(II) ←
Tert. phosphines and di(phosphines) s. under $K_4Fe(CN)_6$, Nickel(II) acetoacetonate, Pd-C, Tris(dibenzylideneacetone)dipalladium, Palladium(II) chloride, Bis(acetonitrile)dichloropalladium(II) and $PdCl_2(PPh_3)_2$ *Ph_3P*
Triphenylphosphine oxide s. under Copper(II) acetate *Ph_3PO*
Air s. under $Zn(OTf)_2$ and $Pd(PPh_3)_4$ *O_2*
Trifluoromethanesulfonic acid s. under $Pd(OAc)_2$ *TfOH*
Sulfuric acid s. under Copper(I) cyanide and Palladium(II) chloride *H_2SO_4*
Tetra-n-butylammonium iodide s. under NaOH *Bu_4NI*
Hexadecyltrimethylammonium bromide s. under Bis(acetonitrile)dichloropalladium(II) ←
Chiral quaternary ammonium bromides s. under Potassium hydroxide ←
Ferrite s. Copper ferrite ←
Potassium hexacyanoferrate(II) s.a. under $Pd(OAc)_2$ *$K_4Fe(CN)_6$*
Potassium hexacyanoferrate(II)/triphenylphosphine/triethylamine *$K_4Fe(CN)_6/Ph_3P/Et_3N$*

2,3-Diaryl-2,3-dicyanooxiranes from two arylcarboxylic acid chloride molecules

412. 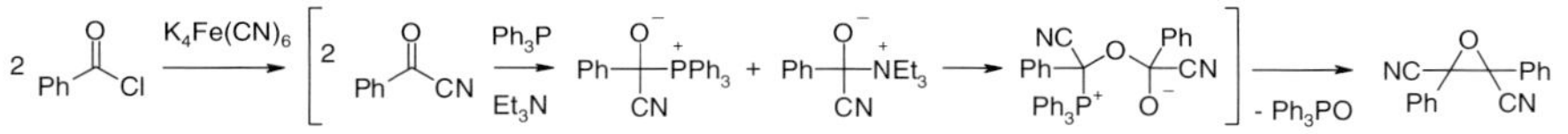

A mixture of $K_4Fe(CN)_6$ (0.2 mmol) and startg. aroyl chloride (1 mmol) heated at 160° for 3 h, then cooled to room temp., triphenylphosphine (0.5 mmol) and triethylamine (0.02 mmol) in methylene chloride (10 ml) added, the mixture stirred at room temp. until reaction complete (TLC), solvent evaporated, and the residue subjected to column chromatography → 2,3-diphenyl-oxirane-2,3-dicarbonitrile. Y 80%. This is a direct method using an eco-friendly cyanide source with simple work-up. Seven further examples gave yields of 65-77% with benzoyl chlorides bearing halo (Cl, F) or Me groups or with 2-furoyl chloride; the method failed with substrates bearing NO_2, Br, 2,4-Cl_2, or MeO substituents, however, or with aliphatic acid chlorides. F.e.s. Z. Li, J. Xu, P. Niu, C. Liu, J. Yang, Tetrahedron *2012*, *68* (43), 8880-3 [DOI: 10.1016/j.tet.2012.08.042].

Transition metal complexes (s.a. under Magnesium and Zinc) ←

N-Directed *o*-substitution H→R

updates s. *73,* 419s*82*; ***o*-arylation**, e.g. of 2-arylpyridines, with ar. chlorides under ruthenium catalysis using $RuCl_2(C_6H_6)$MOTPP [MOTPP = tris(4-methoxyphenyl)phosphine] and acetamide as cocatalyst with K_2CO_3 as base s. J. Zhang, Q. Yang, Z. Zhu, M.L. Yuan, H.Y. Fu, X.L. Zheng, H. Chen, R.X. Li, Eur. J. Org. Chem. *2012* (34), 6702-6 [DOI: 10.1002/ejoc.201201127]; with ar. halides using novel organoruthenium(II) β-diketonates [generated from [(η^6-*p*-cymene)Ru(μ-Cl)Cl]$_2$ and fluorinated β-diketones as ligand] cf. S. Seršen, J. Kljun, F. Pozgan, B. Štefane, I. Turel, Organometallics *2013, 32* (2), 609-16 [DOI: 10.1021/om3011189]; **with simultaneous C_2-prop-1-enylation** of the pyridine ring (with allyl bromide) in a one-pot conversion using [$RuCl_2$(*p*-cymene)]$_2$/K_2CO_3 and 1-adamantanecarboxylic acid as cocatalyst s. Y. Goriya, C.V. Ramana, Chem. Eur. J. *2012, 18* (42), 13288-92 [DOI: 10.1002/chem.201202379]; directed *o*-arylation of various functionalized arenes *in water* with ar. chlorides and heteroaryl halides using $RuCl_2(PPh_3)_2$(*p*-cymene), also sequential *o,o'*-diarylation with $Ru(OAc)_2$(*p*-cymene) for the second coupling, s. P.B. Arockiam, C. Fischmeister, C. Bruneau, P.H. Dixneuf, Green Chem. *2013, 15* (1), 67-71 [DOI: 10.1039/c2gc36222h]; *metal-free o*-arylation of unprotected anilines with ar. halides or triflates *via benzynes* using Li-2,2,6,6-tetramethylpiperidide as base s. T. Truong, O. Daugulis, Org. Lett. *2012, 14* (23), 5964-7 [DOI: 10.1021/ol302875x]; **N-directed β-arylation** of carboxylic acid N-(8-quinolyl)-amides with arylzinc compds. ($Ar_2Zn{\cdot}MgBr_2$/ArMgBr) using $Fe(acac)_3$/dppbz as catalyst and 1,2-dichloro-2-methylpropane as oxidant (cf. *82,* 418) s. R. Shang, L. Ilies, A. Matsumoto, E. Nakamura, J. Am. Chem. Soc. *2013, 135* (16), 6030-2 [DOI: 10.1021/ja402806f]; **N-directed**

γ-arylation of N-(2-pyridylsulfonyl)-α-aminocarboxylic acid esters (incl. dipeptides without racemization) with ar. iodides using $Pd(OAc)_2$/AgOAc s. N. Rodríguez, J.A. Romero-Revilla, M.Á. Fernández-Ibáñez, J.C. Carretero, Chem. Sci. *2013, 4* (1), 175-9 [DOI: 10.1039/c2sc21162a].

Transition metal salts s. under Magnesium and Zinc ←
1-(Di-tert-butylphosphino)-1′,2′,3′,4′,5′-pentaphenylferrocene s. under Palladium(II) acetate ←
1,1′-Bis(diphenylphosphino)ferrocene s. under $Pd(OAc)_2$ *dppf*
Chiral ferrocenyldi(phosphines) s. under Magnesium ←
Iron(III) chloride s. under $Pd(OAc)_2$ *$FeCl_3$*
Nickel(II) acetoacetonate s.a. under Zn *$Ni(acac)_2$*

Nickel(II) acetoacetonate/dicyclohexyl(phenyl)phosphine/diethylzinc *[Ni(II)]/Et_2Zn*
Regio- and stereo-selective nickel-catalyzed synthesis of 3-ethylene-3,1-fluorohydrins from aldehydes and 3-ethylene-2,2-difluorohalides CH(OH)C-C(F)=C-CH

413. 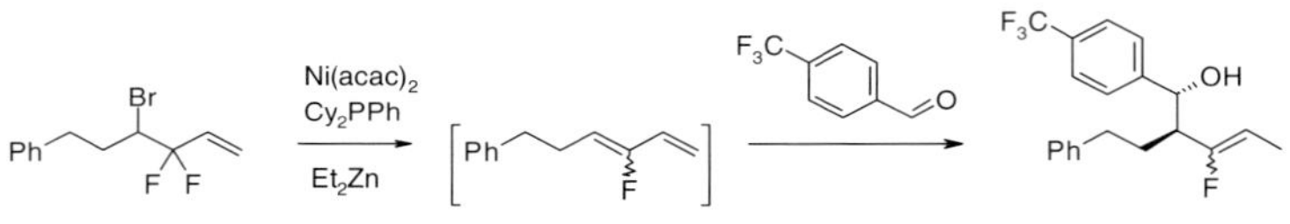

It is reported that generation of 2-fluoro-1,3-dienes from 3-ethylene-2,2-difluorohalides and reaction with aldehydes under nickel catalysis proceeds by unusual 1,4-adddition (to afford the allylation product) rather than 1,2-addition (affording the homoallyl deriv.; cf. *55*, 290), providing a useful route to the title compounds. **E:** (3-Bromo-4,4-difluorohex-5-en-1-yl)benzene (0.2 mmol) in THF (1 ml) and Et_2Zn (0.6 ml of 1 *M* hexane soln.; 3 eq.) added to a soln. of $Ni(acac)_2$ (10 mol%) and Cy_2PPh (20 mol%) in THF (2 ml) under argon, the mixture stirred for 1 h, then *p*-(trifluoromethyl)benzaldehyde (2 eq.) and Et_2Zn (0.9 ml of 1 *M* hexane soln.; 4.5 eq.) added dropwise, the mixture stirred at room temp. for an additional 1.5 h, quenched with 1 *M* HCl, the organic layer separated, the aq. layer extracted with ether, the combined organic extracts washed with brine, dried (Na_2SO_4), filtered, concentrated *in vacuo*, and the residue purified by column chromatography → *anti*-3-fluoro-2-phenethyl-1-[4-(trifluoromethyl)phenyl]pent-3-en-1-ol. Y 57% (Z/E 1.8:1). Reaction of methyl *p*-formylbenzoate under similar conditions gave the corresponding adduct in 54% yield (Z/E 3.5:1). In the case of electron-neutral or -rich ar. aldehydes the reaction was performed using the startg. ethylenetrihalide (0.2 mmol), aldehyde (2 eq.), Et_2Zn (*4.5 eq.*), $Ni(acac)_2$ (10 mol%), Cy_2PPh (20 mol%) *and $ZnCl_2$ (2 eq.)* in THF (1.5 ml) at room temp. for 1.5 h, without the need for preliminary generation of the diene, giving yields of 45-75% for sixteen examples (Z/E 5:1 to 11:1) with MeO, PhO, Br and Cl tolerated on the aromatic ring; under these conditions electron-poor aldehydes simply underwent addition of an ethyl group, however. The corresponding 2-fluorodiene was difficult to handle due to its ease of polymerization. F.e.s. X. Lin, F. Zheng, F.-L. Qing, J. Org. Chem. *2012, 77* (19), 8696-704 [DOI: 10.1021/jo301688d].

Nickel(II) chloride/2,2′-bipyridyl/magnesium *$NiCl_2$/bpy/Mg*
Pincer-type nickel(II) m-phenylenebis(phosphonite) complexes/zinc *[Ni(II)]/Zn*
Sym. biaryls from ar. halides 2 ArHal → Ar-Ar
update s. *34,* 825s*82*; by **Ullmann homocoupling** with $NiCl_2$/2,2′-bipyridyl/Mg, also polycondensation, s. T. Yamamoto, Chem. Lett. *2012, 41* (11), 1422-4 [DOI: 10.1246/cl.2012.1422]; **sym. 1,2-diarylalkanes from benzyl halides** with pincer-type nickel(II) *m*-phenylenebis-(phosphonite) complexes/zinc (cf. *65*, 394) s. T. Chen, L. Yang, L. Li, K.-W. Huang, Tetrahedron *2012, 68* (31), 6152-7 [DOI: 10.1016/j.tet.2012.05.075].

Dichloro(1,5-cyclooctadiene)ruthenium(II)/potassium acetate/sodium hydrogen carbonate ←
***o*-(*o*-Heteroaryl)benzophenones from *o*-aryl-N-heteroarenes and ar. halides** H → C(O)Ar
by ruthenium-catalyzed N-directed carbonylative *o*-aroylation in water

414.

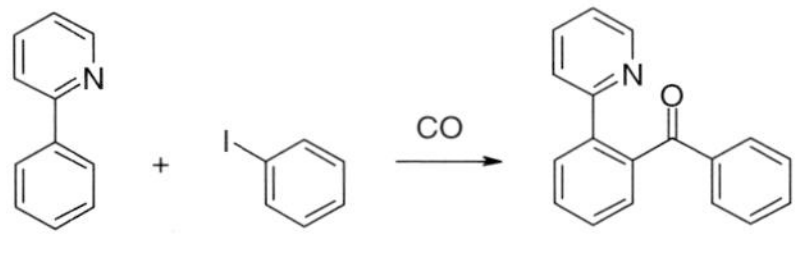

$[RuCl_2(cod)]_n$ (5 mol%), KOAc (20 mol%), $NaHCO_3$ (2 eq.), benzoxazole, and a magnetic stir bar placed in 6 Wheaton vials, which were then capped with a septum equipped with an inlet needle and flushed with argon, iodobenzene (2 eq.), 2-phenylpyridine (0.5 mmol), and water (1 ml) added via syringe, the vials placed in an alloy plate with 6 wells, then placed in a Parr autoclave, the latter sealed, purged several times with CO, then pressurized to 30 bar at room temp., heated at 120° for 20 h, cooled to room temp., vented to discharge excess CO, the mixture extracted with ethyl acetate, the organic layers washed with brine, dried (Na_2SO_4), evaporated, and the crude residue purified by flash chromatography on silica gel → phenyl[2-(pyridin-2-yl)phenyl]-methanone. Y 62%. This novel ruthenium-catalyzed carbonylative coupling, involving directed C-H bond activation, is applicable to electron-diverse ar. iodides (eleven examples; Y 40-74%) optionally bearing alkyl, MeO, F or CF_3 groups. Pyrazol-1-yl may also be used as directing group (Y 40%, 42%), while 3-(pyrid-2-yl)furan gave 50% and 3-(pyrid-2-yl)- or 3-(pyrimidin-2-yl)-thiophene gave 75% or 68% yield, respectively. Other ruthenium catalysts gave poor results and silver salts inhibited the reaction completely. F.e.s. A. Tlili, J. Schranck, J. Pospech, H. Neumann, M. Beller, Angew. Chem., Int. Ed. *2013, 52* (24), 6293-7 [DOI: 10.1002/anie.201301663].

Palladium nanoparticles or complexes or supported variants *[Pd]*
Heck arylation H → Ar
update s. *27,* 871s*82*; with palladium(0) di(phosphine) complexes generated from highly active and advantageous (η^5-cyclopentadienyl)(η^3-1-phenylallyl)palladium(II), also application to Suzuki and Sonogashira coupling, s. A.W. Fraser, B.E. Jaksic, R. Batcup, C.D. Sarsons, M. Woolman, M.C. Baird, Organometallics *2013, 32* (1), 9-11 [DOI: 10.1021/om301081r]; arylation of styrenes with activated or unactivated ar. chlorides using air-stable dichloro[1,3-bis(2,6-diisopropylphenyl)-imidazol-2-ylidene](1-methylimidazole)palladium(II) and Cs_2CO_3 as base in tetra-*n*-butyl-ammonium bromide as ionic liquid s. T.-T. Gao, A.-P. Jin, L.-X. Shao, Beilstein J. Org. Chem. *2012, 8,* 1916-9 [DOI: 10.3762/bjoc.8.222]; with air-stable [1,1′-(hexylene-1,6-diyl)bis(3-*n*-butyl-benzimidazol-2-ylidene)][$PdCl_2$(MeCN)]$_2$, also application to Suzuki coupling in water, s. Q.-X. Liu, W. Zhang, X.-J. Zhao, Z.-X. Zhao, M.-C. Shi, X.-G. Wang, Eur. J. Org. Chem. *2013* (7), 1253-61 [DOI: 10.1002/ejoc.201200954]; with air- and moisture-stable acyclic Chugaev-type bis(diamino-carbenes) as ligand for arylation with activated or deactivated ar. bromides and electron-deficient ar. chlorides s. M.O. Owusu, S. Handa, L.G.M. Slaughter, Appl. Organomet. Chem. *2012, 26* (12), 712-7 [DOI: 10.1002/aoc.2915]; of various olefins with electron-diverse ar. halides using a recyclable square-planar dinuclear palladium(II) terephthalaldehyde bis(thiosemicarbazone) complex s. R.N. Prabhu, R. Ramesh, Tetrahedron Lett. *2012, 53* (44), 5961-5 [DOI: 10.1016/j.tetlet.2012.08.120]; of crotonic or cinnamic derivs. with electron-rich ar. iodides using 1:2 $Pd(OAc)_2$/N-(4-carbethoxy-phenyl)urea as catalyst s. M.R. Smith, J.Y. Kim, M.A. Ciufolini, ibid. *2013, 54* (16), 2042-5 [DOI: 10.1016/j.tetlet.2013.01.077]; with $Pd(OAc)_2$ and a 1,3-diarylimidazolium salt as catalyst under aerobic conditions, also application to Suzuki coupling, s. M. Amini, M. Bagherzadeh, S. Rostamnia, Chin. Chem. Lett. *2013, 24* (5), 433-6 [DOI: 10.1016/j.cclet.2013.03.025]; with a phosphite-type dimeric P,C-palladacyclic complexes, $[Pd(\mu\text{-}Cl)[P(OPh)_2(OC_6H_4)]]_2$, also application to Suzuki coupling, s. K. Karami, M. Ghasemi, N.H. Naeini, Tetrahedron Lett. *2013, 54* (11), 1352-5 [DOI: 10.1016/j.tetlet.2012.12.071]; with inexpensive, robust 1-(2-pyridyl)-1*H*-benzotriazole as bidentate ligand, also application to Suzuki, Fujiwara-Moritani, Sonogashira, C-N and C-S coupling, s. A.K. Verma, R.R. Jha, R. Chaudhary, R.K. Tiwari, A.K. Danodia, Adv. Synth. Catal. *2013, 355* (2-3), 421-38 [DOI: 10.1002/adsc.201200583]; predominant (Z)-selective arylation under palladium catalysis with *sun-light irradiation*, also Suzuki and Sonogashira coupling, s. A.R. Chaudhary, A.V. Bedekar, Tetrahedron Lett. *2012, 53* (45), 6100-3 [DOI: 10.1016/j.tetlet.2012.08.139];

(E)-selective C_3-arylation of methyl 2-(acetoxymethyl)acrylate with $Pd(OAc)_2/AgOAc$, also diarylation, s. K.H. Kim, J.W. Lim, H.J. Lee, J.N. Kim, ibid. *2013, 54* (5), 419-23 [DOI: 10.1016/j.tetlet.2012.11.026]; ligand-free Heck coupling of 3-iodo-4-chromones with $Pd(OAc)_2/Et_3N$ *under microwave enhancement* s. Y. Zhang, Z. Lv, H. Zhong, M. Zhang, T. Zhang, W. Zhang, K. Li, Tetrahedron *2012, 68* (47), 9777-87 [DOI: 10.1016/j.tet.2012.09.017]; selective arylation at the *vinylic* site *of cyclic ethylene derivs.* with tri-*tert*-butylphosphine as ligand s. X. Wu, J. Zhou, Chem. Commun. *2013, 49* (42), 4794-6 [DOI: 10.1039/c3cc41722k]; *heterogeneous,* ligand-free Heck β-arylation and β,β-diarylation of electron-deficient ethylene derivs. with recyclable solid-supported palladium(0) nano- or micro-particles s. D. Sharma, S. Kumar, A.K. Shil, N.R. Guha, Bandna, P. Das, Tetrahedron Lett. *2012, 53* (52), 7044-51 [DOI: 10.1016/j.tetlet.2012.10.062]; with reusable palladium nanoparticles on a nano-silica-supported triazine-based dendritic polymer at low catalyst loading in water/DMF, also Suzuki coupling, s. A.L. Isfahani, I. Mohammadpoor-Baltork, V. Mirkhani, A.R. Khosropour, M. Moghadam, S. Tangestaninejad, R. Kia, Adv. Synth. Catal. *2013, 355* (5), 957-72 [DOI: 10.1002/adsc.201200707]; with palladium nanoparticles on silica-supported, DABCO-capped, amino-functionalized dendrimers s. Y. Xu, Z. Zhang, J. Zheng, Q. Du, Y. Li, Appl. Organomet. Chem. *2013, 27* (1), 13-8 [DOI: 10.1002/aoc.2930]; solvent-free synthesis of β-arylketones by arylation of Baylis-Hillman adducts using Pd-C as catalyst s. H.-S. Kim, S.-J. Lee, B. Choi, C.M. Yoon, Synthesis *2012, 44* (20), 3161-4 [DOI: 10.1055/s-0032-1316992]; arylation using palladium nanoparticles in hybrid materials formed from ionic liquid phases supported within macrocellular silica-based foams s. N. Brun, P. Hesemann, G. Laurent, C. Sanchez, M. Birot, H. Deleuze, R. Backov, New J. Chem. *2013, 37* (1), 157-68 [DOI: 10.1039/c2nj40527j]; arylation *by ball-milling* with palladium-PEG nanoparticles s. V. Declerck, E. Colacino, X. Bantreil, J. Martinez, F. Lamaty, Chem. Commun. *2012, 48* (96), 11778-80 [DOI: 10.1039/c2cc36286d]; Heck arylation and Suzuki coupling using porous (diimine)palladium(II) dichloride complexes embedded in heterometallic coordination polymers s. S.-L. Huang, A.-Q. Jia, G.-X. Jin, ibid. *2013, 49* (24), 2403-5 [DOI: 10.1039/c3cc38714c]; palladium-catalyzed Heck arylation within a supported aq. phase *under microflow* at low catalyst loading s. S.C. Stouten, Q. Wang, T. Noël, V. Hessel, Tetrahedron Lett. *2013, 54* (17), 2194-8 [DOI: 10.1016/j.tetlet.2013.02.064]; heterogeneous *copper-catalyzed* Heck arylation with deactivated ar. chlorides using a stable, reusable copper(0)/Al_2O_3 nanocomposite, also Suzuki and Sonogashira coupling, s. R. Arundhathi, D. Damodara, K.V. Mohan, M.L. Kantam, P.R. Likhar, Adv. Synth. Catal. *2013, 355* (4), 751-6 [DOI: 10.1002/adsc.201201007].

Palladium/copper/ethyldiisopropylamine *Pd/Cu/i-Pr_2NEt*
Sonogashira coupling C≡CH → C≡CR

update s. *27,* 851s*82*; *under continuous flow* using a palladium-coated tubular reactor in line with a copper tubing reactor with ethyldiisopropylamine as base (and the benefit of metal scavenging at the end of the flow) s. L.-M. Tan, Z.-Y. Sem, W.-Y. Chong, X. Liu, Hendra, W.L. Kwan, C.-L.K. Lee, Org. Lett. *2013, 15* (1), 65-7 [DOI: 10.1021/ol303046e]; for novel homogeneous and heterogeneous catalysts for Sonogashira coupling s.a. under *27,* 871s*83*.

Palladium-carbon/copper(I) chloride/cesium carbonate *Pd-C/CuCl/Cs_2CO_3*
Regioselective arylation of benzo[*b*]thiophenes with aryl chlorides H → Ar

415.

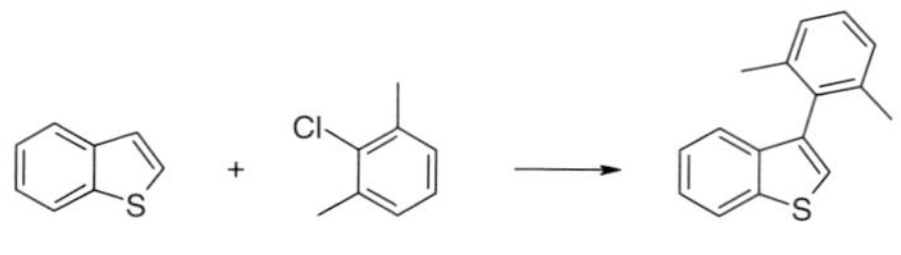

3-Arylbenzo[*b*]thiophenes. Challenging, direct C_3-arylation of benzo[*b*]thiophenes has been achieved using a heterogeneous Pd-C catalyst in the presence of CuCl, under ligand-free conditions. The method is scalable, operationally simple, and insensitive to air and moisture, affording the title compounds completely regioselectively. **E:** 2-Chloro-1,3-dimethylbenzene (2 eq.) added to a mixture of benzo[*b*]thiophene (0.3 mmol), Cs_2CO_3 (1.1 eq.), 5% Pd-C (type K-0219, containing <3% H_2O; 9.4 mol%) and CuCl (10 mol%) in 1,4-dioxane (1.5 ml), the resulting mixture stirred at 150° for 96 h, allowed to cool to room temp., filtered through a silica pad (with ethyl acetate), the filtrate concentrated *in vacuo*, and the residue purified by flash chromatography → 3-(2,6-dimethylphenyl)benzo[*b*]thiophene. Y 94% (C_3/C_2 selectivity >99:1). Nine further examples, using electron-neutral or -rich (Me, MeO subst.) ar. chlorides (incl. 1-naphthyl), and with the benzothiophene optionally substituted with phenyl or methoxy groups in the 5-position, afforded yields of 71-90%. Electron-deficient ar. chlorides reacted sluggishly, although yields were crucially determined by the nature of the electron-withdrawing substituent: *p*-CO_2Me (Y 24%), *m*-NO_2 (21%), *m*-F (73%), *m*-CF_3 (69%). Reaction between 1,4-dichlorobenzene and benzothiophene afforded a bis-derivative in 41% yield. C_3/C_2-Regioselectivity was >99:1 in all cases. Alternative heterogeneous (supported) Pd(0) catalysts were ineffective, but a variety of alternative Lewis additives were shown to enhance reactivity, of which CuCl was optimal; interestingly CuBr was almost as effective as CuCl, but the reaction was completely retarded using CuI. A series of mechanistic studies strongly suggested the active catalytic species to be *heterogeneous* in nature, the complete reversal of C_3/C_2-selectivity using an active, *homogeneous* Pd(0) catalyst being particularly noteworthy. The actual mechanistic pathway was not established, although a number of mechanisms traditionally associated with C-H activation under *homogeneous* catalysis were discounted. F.e.s. D.-T.D. Tang, K.D. Collins, F. Glorius, J. Am. Chem. Soc. *2013*, *135* (20), 7450-3 [DOI: 10.1021/ja403130g].

Palladium-carbon/copper iodide/triphenylphosphine/triethylamine *Pd-C/CuI/Ph_3P/Et_3N*
Ring closures by Sonogashira coupling-cycloisomerization s. *44*, 781s*83* ○

Palladium complexes *[Pd]*
Palladium-catalyzed α-arylation H → Ar
update s. *61*, 392s*82*; α-arylation **of ketones** and malonates with [hetero]ar. halides using $[Pd(cinnamyl)Cl]_2$/DalPhos and added base s. S.M. Crawford, P.G. Alsabeh, M. Stradiotto, Eur. J. Org. Chem. *2012* (30), 6042-50 [DOI: 10.1002/ejoc.201200951]; monoarylation of acetone with ar. chlorides using the same palladium catalyst and air-stable dicyclohexyl[2-methoxy-6-(N-methyl-N-phenylamino)phenyl]phosphine (Zheda-Phos) as ligand with Cs_2CO_3 as base s. P. Li, B. Lü, C. Fu, S. Ma, Adv. Synth. Catal. *2013, 355* (7), 1255-9 [DOI: 10.1002/adsc.201300207]; α-arylation **of methyl sulfoxides** with ar. bromides or chlorides using $Pd(OAc)_2$/N-(dicyclohexylphosphino)-2-(*o*-tolyl)indole and LiOBu-*t* as base s. T. Jia, A. Bellomo, K. El Baina, S.D. Dreher, P.J. Walsh, J. Am. Chem. Soc. *2013, 135* (10), 3740-3 [DOI: 10.1021/ja4009776]; **of methyl sulfones** with ar. bromides using the same system s. B. Zheng, T. Jia, P.J. Walsh, Org. Lett. *2013, 15* (7), 1690-3 [DOI: 10.1021/ol400472v]; α-arylation of unactivated *enolizable* ketones, e.g. aryl benzyl ketones, with ar. iodides *under copper catalysis* using CuI/1,10-phen and Cs_2CO_3 as base s. G. Danoun, A. Tlili, F. Monnier, M. Taillefer, Angew. Chem., Int. Ed. *2012, 51* (51), 12815-9 [DOI: 10.1002/anie.201206024]; microwave-assisted α-arylation of copper(I) amide enolates under palladium catalysis s. A. Verma, N. Prajapati, S. Salecha, R. Giridhar, M.R. Yadav, Tetrahedron Lett. *2013, 54* (15), 2029-32 [DOI: 10.1016/j.tetlet.2013.02.026]; *metal-free* α-arylation of *o*-hydroxyacetophenones with K_2CO_3 in DMSO or Cs_2CO_3 in acetonitrile s. Y. Liu, X. Zhang, Y. Ma, C. Ma, ibid. *54* (5), 402-5 [DOI: 10.1016/j.tetlet.2012.11.024].

Palladium complexes or salts ←
Transition metal-catalyzed carbonylation of halides Hal → C(O)X
update s. *12*, 867s*82*; arylcarboxylic acid esters and amides from ar. iodides with well-defined, ligand-free, ionic liquid-immobilized palladium s. M.V. Khedkar, T. Sasaki, B.M. Bhanage, ACS Catal. *2013, 3* (3), 287-93 [DOI: 10.1021/cs300719r]; preparation of *m*-carboranylcarboxylic acid amides s. K.B. Gona, V. Gómez-Vallejo, J. Llop, Tetrahedron Lett. *2013, 54* (8), 941-4 [DOI: 10.1016/j.tetlet.2012.12.019]; arylthiolic acid esters from electron-diverse ar. iodides with stoichiometric CO using $Pd(OAc)_2$/DPEPhos/NaOAc s. M.N. Burhardt, R.H. Taaning, T. Skrydstrup, Org. Lett. *2013, 15* (4), 948-51 [DOI: 10.1021/ol400138m]; arylcarboxylic acid vinyl esters by carbonylation of ar. halides and ketones using $[Pd(cinnamyl)Cl]_2$/diadamant-1-yl(*n*-butyl)-phosphine s. J. Schranck, A. Tlili, H. Neumann, P.G. Alsabeh, M. Stradiotto, M. Beller, Chem. Eur. J. *2012, 18* (49), 15592-7 [DOI: 10.1002/chem.201202895]; N′-Boc-protected N-aroylimino-sulfonamides from [hetero]ar. halides with $Mo(CO)_6$ and $Pd(OAc)_2$ or $PdCl_2$(dppf) using Et_3N/DBU as base s. S.R. Borhade, A. Sandström, P.I. Arvidsson, Org. Lett. *2013, 15* (5), 1056-9 [DOI: 10.1021/ol400049m]; carboxylic acid esters from alkyl iodides *by atom-transfer carbonylation*

under photoirradiation with the dimeric palladium isocyanide complex, $[Pd_2(CNMe)_6][PF_6]_2$, as catalyst, also 4-component synthesis of carboxylic acid derivs. by carbonylative *vic*-functionalization of alkenes, s. A. Fusano, S. Sumino, S. Nishitani, T. Inouye, K. Morimoto, T. Fukuyama, I. Ryu, Chem. Eur. J. *2012, 18* (30), 9415-22 [DOI: 10.1002/chem.201200752]; arylcarboxylic acid *tert*-butyl esters by *metal-free carbonylation* of ar. halides using KOBu-*t*/1,10-phenanthroline s. H. Zhang, R. Shi, A. Ding, L. Lu, B. Chen, A. Lei, Angew. Chem., Int. Ed. *2012, 51* (50), 12542-5 [DOI: 10.1002/anie.201206518]; **benzyl alcohols by reductive carbonylation** of ar. iodides with borane·1,3-dimethylimidazol-2-ylidene and AIBN as initiator s. T. Kawamoto, T. Okada, D.P. Curran, I. Ryu, Org. Lett. *2013, 15* (9), 2144-7 [DOI: 10.1021/ol4006294].

Copper-free Sonogashira coupling C≡CH → C≡CR

update s. *63,* 411s*82*; with a PNF pincer-type complex, [2-(di-*tert*-butylphosphinomethylene)-8-*fluoro*-1,2-dihydroquinolinato]arylpalladium(II), exhibiting a new *metal-fluorine coordination* mode s. A. Scharf, I. Goldberg, A. Vigalok, J. Am. Chem. Soc. *2013, 135* (3), 967-70 [DOI: 10.1021/ja310782k]; with (diaryl N-arylketimine)dichloro(isonitrile)palladium(II) complexes s. A.S. Kritchenkov, K.V. Luzyanin, N.A. Bokach, M.L. Kuznetsov, V.V. Gurzhiy, V.Y. Kukushkin, Organometallics *2013, 32* (6), 1979-87 [DOI: 10.1021/om4000665]; with $Pd(OAc)_2/Cs_2CO_3$ and tri-*tert*-butyl(decyl)phosphonium fluoroborate as phase transfer catalyst s. V. Ermolaev, V. Miluykov, D. Arkhipova, E. Zvereva, O. Sinyashin, Phosphorus, Sulfur Silicon Relat. Elem. *2012, 188* (1-3), 168-70 [DOI: 10.1080/10426507.2012.744013]; with an air- and moisture-insensitive bromo[*o*-(2-aminoethyl)aryl]palladium(II) dimer derived from homoveratrylamine s. A.R. Hajipour, H. Rahimi, F. Rafiee, Appl. Organomet. Chem. *2012, 26* (12), 727-30 [DOI: 10.1002/aoc.2920].

Regioselective transition metal-catalyzed coupling of [het]aryl halides with heteroarenes Ar-Ar′

update s. *57,* 376s*82*; C_5-arylation of 5-membered heteroarenes with bromopyridines using $PdCl(C_3H_5)$(dppb)/KOAc s. S. Bensaid, H. Doucet, Tetrahedron *2012, 68* (37), 7655-62 [DOI: 10.1016/j.tet.2012.06.034]; with bromopolyfluoroarenes using $Pd(OAc)_2$/KOAc, and subsequent biaryl coupling, s. T. Yan, L. Chen, C. Bruneau, P.H. Dixneuf, H. Doucet, J. Org. Chem. *2013, 78* (8), 4177-83 [DOI: 10.1021/jo400221x]; palladium-catalyzed C_2-arylation of [benz]azoles *with ar. fluorides* activated by *o*-directing groups s. D. Yu, L. Lu, Q. Shen, Org. Lett. *2013, 15* (4), 940-3 [DOI: 10.1021/ol303567t]; C_5-arylation of 2-vinylated 5-membered heteroarenes with ar. bromides *in cyclopentyl methyl ether* using $PdCl(C_3H_5)$(dppb)/KOAc s. Y. Bouazizi, K. Beydoun, A. Romdhane, H. Ben Jannet, H. Doucet, Tetrahedron Lett. *2012, 53* (50), 6801-5 [DOI: 10.1016/j.tetlet.2012.10.008]; sequential C_5- and C_2-arylation of 3-acylpyrroles with ar. bromides s. Y. Xu, L. Zhao, Y. Li, H. Doucet, Adv. Synth. Catal. *2013, 355* (1), 1423-32 [DOI: 10.1002/adsc.201300123]; 2-arylation of pyrroles with ar. chlorides using air-stable palladium(II) N-heterocyclic carbene complexes with KOAc s. I. Özdemir, N. Gürbüz, N. Kaloglu, Ö. Dogan, M. Kaloglu, C. Bruneau, H. Doucet, Beilstein J. Org. Chem. *2013, 9*, 303-12 [DOI: 10.3762/bjoc.9.35]; sequential regioselective C_3- and C_5-arylation of 1,2,4-triazoles s. J.M. Joo, P. Guo, D. Sames, J. Org. Chem. *2013, 78* (2), 738-43 [DOI: 10.1021/jo3021677]; C_3-arylation of indazoles with [hetero]ar. halides using $Pd(OAc)_2$/1,10-phen or bpy/K_2CO_3 s. A. Ben-Yahia, M. Naas, S. El Kazzouli, E.M. Essassi, G. Guillaumet, Eur. J. Org. Chem. *2012* (36), 7075-81 [DOI: 10.1002/ejoc.201200860]; C_3-arylation of imidazo[1,2-*b*]pyrazoles with [hetero]ar. bromides or chlorides using $Pd(OAc)_2$/Cy_3P s. S. Grosse, C. Pillard, S. Massip, J.M. Léger, C. Jarry, S. Bourg, P. Bernard, G. Guillaumet, Chem. Eur. J. *2012, 18* (47), 14943-7 [DOI: 10.1002/chem.201202593]; 5,7-diarylation of thieno-[3,4-*b*]pyrazines with 2-bromothiophenes using $Pd(OAc)_2/KOAc/Bu_4NBr$ s. N.I. Abdo, A.A. El-Shehawy, A.A. El-Barbary, J.-S. Lee, Eur. J. Org. Chem. *2012* (28), 5540-51 [DOI: 10.1002/ejoc.201200769]; *copper-catalyzed* C_5-arylation of thiazolo[3,2-*b*][1,2,4]triazoles using $Cu(acac)_2$/*t*-BuOLi s. Z. Xie, X. Zhu, Y. Guan, D. Zhu, H. Hu, C. Lin, Y. Pan, J. Jiang, L. Wang, Org. Biomol Chem. *2013, 11* (8), 1390-8 [DOI: 10.1039/c2ob27326h].

Carbonylative ring closures ○

flavones from *o*-halogenophenols and terminal acetylene derivs. by cyclocarbonylation s. *46,* 808s*50*; with $PdCl_2(PPh_3)_2$, an N-(1,2,3-triazolylmethyl)benzimidazole ligand and *n*-Pr_2NH, also aurones, s. J. Liu, M. Liu, Y. Yue, N. Zhang, Y. Zhang, K. Zhuo, Tetrahedron Lett. *2013, 54* (14), 1802-7 [DOI:

10.1016/j.tetlet.2013.01.043]; flavones from *o*-hydroxyacetophenones and ar. bromides with $Pd(OAc)_2$ s. X.-F. Wu, H. Neumann, M. Beller, Chem. Eur. J. *2012*, *18* (40), 12595-8 [DOI: 10.1002/chem.201202141]; 3-benzylidene-2(3*H*)-furanones from propargylarenes and ar. iodides with $Pd(OAc)_2$/tri-2-furylphosphine/ethyldiisopropylamine s. X.-F. Wu, H. Jiao, H. Neumann, M. Beller, ibid. *18* (50), 16177-85 [DOI: 10.1002/chem.201202138]; [2-alkylated] 3,1-benzoxazin-4-ones from *o*-bromamines and carboxylic acid anhydrides with K_2PdCl_4 s. X.-F. Wu, H. Neumann. M. Beller, ibid. *18* (40), 12599-602 [DOI: 10.1002/chem.201202142]; phthalazones from *o*-bromaldehydes and hydrazines with $Pd(OAc)_2$/dppf/DBU/$MgSO_4$ s. X.-F. Wu, H. Neumann, S. Neumann, M. Beller, ibid. *18* (28), 8596-9 [DOI: 10.1002/chem.201200267]; 2-aroylindolizines from 2-propargylpyridines and ar. iodides s. Z. Li, D. Chernyak, V. Gevorgyan, Org. Lett. *2012*, *14* (23), 6056-9 [DOI: 10.1021/ol302947r]; 2-aryl-Δ^2-oxazolines from 2-chloroethylamine and ar. bromides with $Pd(OAc)_2/Ad_2PBu/Et_3N$, also 2-aryl-5,6-dihydro-4*H*-1,3-oxazines, s. X.-F. Wu, H. Neumann, S. Neumann, M. Beller, Chem. Eur. J. *2012*, *18* (43), 13619-23 [DOI: 10.1002/chem.201202652].

Palladium(II) acetate/potassium carbonate *$Pd(OAc)_2/K_2CO_3$*
Biaryls from N-aryl-N′-tosylhydrazines and ar. iodides s. *83*, 460 ArNHNHTs → Ar-Ar′

Palladium(II) acetate/copper(II) oxide/trifluoromethanesulfonic acid *$Pd(OAc)_2$/CuO/TfOH*
Palladium-catalyzed directed *o*-arylation of N-arylphosphoromonoamidates with diaryliodonium triflates H → Ar

416. 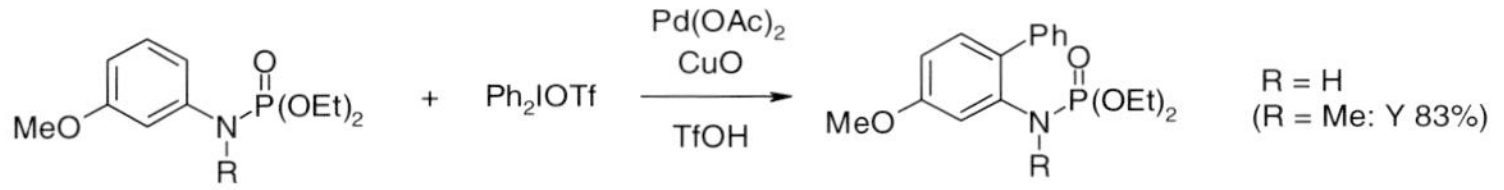

For the first time, phosphoramidates have been demonstrated to be synthetically useful as *o*-directing groups for transition metal-catalyzed C-H activation. **E:** Diethyl (3-methoxyphenyl)phosphoramidate (0.3 mmol) in 1,4-dioxane (1.2 ml) added to an oven-dried test tube charged with $Pd(OAc)_2$ (5 mol%), CuO (3 eq.) and diphenyliodonium triflate (1.2 eq.) under N_2, stirred for 5 min, trifluoromethanesulfonic acid (20 mol%) added, the resulting mixture stirred at room temp. for 3 h, quenched with water, extracted with methylene chloride, and purified by chromatography on silica gel → diethyl (4-methoxy-1,1′-biphenyl-2-yl)phosphoramidate. Y 84%. Fourteen examples, tolerating alkyl, *benzyl*, alkoxy, aryloxy and *bromo* substituents on the phosphoramidate N-aryl ring, were similarly phenylated in yields of 68-91%, unsymmetrical substrates affording single isomers in all cases, via arylation at the least hindered site. Similar results were obtained with N-methyl, -ethyl or -butyl analogs (ten examples; Y 68-86%). *p*-Methyl substitution was tolerated on the arylating moiety [Ar_2IOTf] (four examples; Y 78-81%), but a *p*-chloro analog was less successful (one example; Y 53%, with 32% rsm). Best results were obtained from *diethyl* phosphoramidates, in comparison with dimethyl, diisopropyl and cyclic analogs. The role of CuO is unclear, but appears to be as a base, promoting the formation of an intermediate palladacycle. F.e. and optimization s. B.C. Chary, S. Kim, Y. Park, J. Kim, P.H. Lee, Org. Lett. *2013*, *15* (11), 2692-5 [DOI: 10.1021/ol4009987].

Palladium(II) acetate/potassium hexacyanoferrate(II)/potassium carbonate ←
Dichloro[1,2-bis(diphenylphosphino)ethane]palladium(II)/sodium formate/copper(I) thiocyanate ←
Nitriles from halides Hal → CN
ar. nitriles from halides, update s. *29*, 845s*82*; with CuSCN as a safe cyanide source in the presence of $PdCl_2$(dppe)/Na-formate in an acidic aq. DMSO medium s. G.-Y. Zhang, J.-T. Yu, M.-L. Hu, J. Cheng, J. Org. Chem. *2013*, *78* (6), 2710-4 [DOI: 10.1021/jo3025829]; with commercially available aliphatic nitriles, e.g. acetonitrile, as cyanide source and added $Cu(OAc)_2$/triphenylphosphine oxide/Ag_2O under air s. R.-J. Song, J.-C. Wu, Y. Liu, G.-B. Deng, C.-Y. Wu, W.-T. Wei, J.-H. Li, Synlett *2012*, *23* (17), 2491-6 [DOI: 10.1055/s-0032-1317191]; ar. nitriles **from diaryliodonium triflates** using $K_4Fe(CN)_6$ with $Pd(OAc)_2/K_2CO_3$ cf. L. Liu, J. Li, J. Xu, J.-t. Sun, Tetrahedron Lett. *2012*, *53* (51), 6954-6 [DOI: 10.1016/j.tetlet.2012.10.036].

Palladium(II) acetate/1-(di-tert-butylphosphino)-1′,2′,3′,4′,5′-pentaphenylferrocene/triethylamine *[Pd(II)]/Et$_3$N*

β-Arylaldehydes from *trans*-2-subst. cyclopropanols and ar. bromides C
Regioselective palladium-catalyzed ring opening via aldehyde homoenolate equivalents

417.

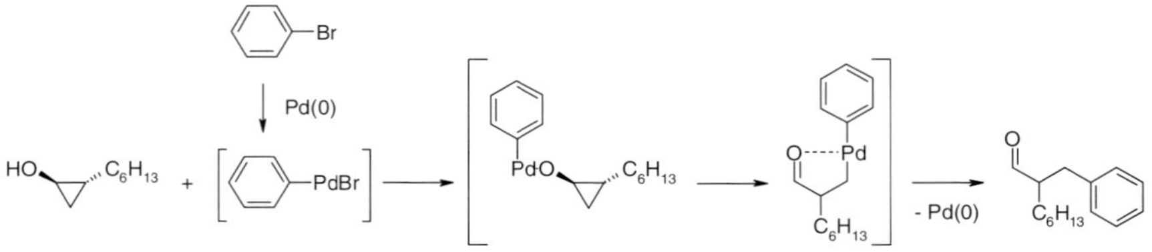

trans-2-Hexylcyclopropanol (2 eq.; 0.5 *M* soln. in THF), followed by bromobenzene (0.25 mmol) and triethylamine (2 eq.), added to an oven-dried reaction vial charged with Pd(OAc)$_2$ (5 mol%) and 1-(di-*tert*-butylphosphino)-1′,2′,3′,4′,5′-pentaphenylferrocene [QPhos] (10 mol%) under N$_2$, the mixture stirred for 12 h at room temp., quenched with satd. aq. NH$_4$Cl, and worked up with purification by chromatography on silica gel → 2-benzyloctanal. Y 86%. The procedure is mild, scalable (to the gram level), and applicable to the coupling of 2-[cyclo]alkyl-, 2-benzyl-, 2-indolylmethyl- and variously O-functionalized 2-alkyl-subst. *trans*-cyclopropanols with electron-diverse ar. bromides (twenty-two examples; Y 59-93%). Several monophosphines were assessed as ligand but only QPhos gave consistently high yields. There was no erosion of chirality on reaction with enantio-enriched *trans*-2-hexylcyclopropanol. Reaction is presumed to involve regioselective ring opening of an intermediate aryl(cyclopropyloxy)palladium(II) *by β-carbon elimination* to give a carbonyl-stabilized palladium homoenolate equivalent prior to reductive elimination to give the β-arylaldehyde *without* competing β-hydride elimination. F.e.s. K. Cheng, P.J. Walsh, Org. Lett. *2013*, *15* (9), 2298-301 [DOI: 10.1021/ol4008876].

Palladium(II) acetate/1,1′-bis(diphenylphosphino)ferrocene/iron(III) chloride/cesium carbonate ←

4-Imino-3,4-dihydroquinazolin-2-ylphosphonic acid esters ○
from *o*-iodocarbodiimides, isonitriles and phosphorous acid diesters

418.

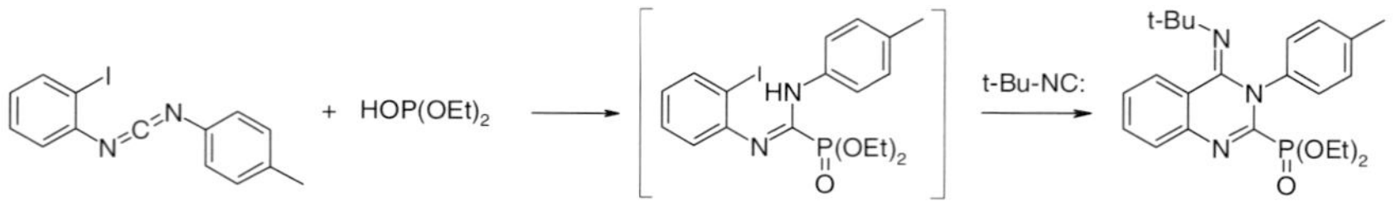

Pd(OAc)$_2$ (10 mol%), dppf (10 mol%) and FeCl$_3$ (10 mol%) added to a soln. of startg. *o*-iodocarbodiimide (0.2 mmol) in toluene (2 ml) at room temp. under N$_2$, then diethyl phosphite (1.5 eq.) added, the mixture stirred at room temp. for 1 h, *tert*-butyl isocyanide (2 eq.) and Cs$_2$CO$_3$ (2 eq.) added, the mixture stirred at reflux for 8-12 h (TLC), the solvent evaporated, and the residue purified on silica gel → diethyl (E)-4-(*tert*-butylimino)-3,4-dihydro-3-*p*-tolylquinazolin-2-ylphosphonate. Y 78%. This palladium-catalyzed 3-component ring closure involves formation of three bonds via nucleophilic attack of the phosphite diester, isocyanide insertion and C-N coupling in one pot. Eleven further examples starting from N′-aryl/alkyl/benzyl-N-*o*-iodoarylcarbodiimides, diethyl or di(iso)propyl phosphite and *tert*-butyl isocyanide gave yields of 47-76%, while cyclohexyl isocyanide gave 43% and *n*-butyl isocyanide gave 37%; an aryl [2,6-dimethylphenyl] isocyanide gave only a trace of product, however. F.e. and optimization s. G. Qiu, Y. Lu, J. Wu, Org. Biomol. Chem. *2013*, *11* (5), 798-802 [DOI: 10.1039/c2ob26979a].

Palladium(II) trifluoroacetate *$Pd(OCOCF_3)_2$*

Palladium-catalyzed O-directed *o*-arylation of aryl hydrogen phosphates with diaryliodonium triflates H → Ar

419. 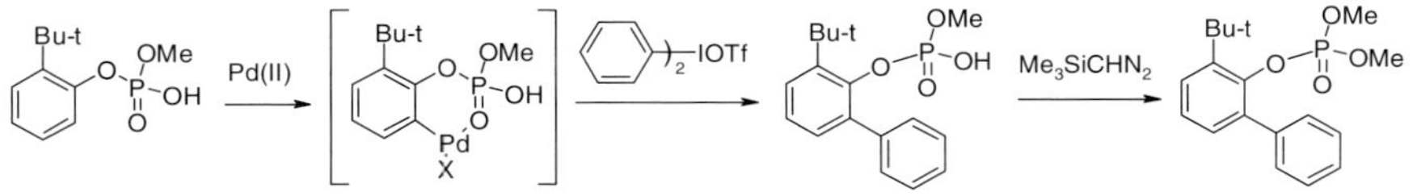

Weak coordination of the P=O bond of aryl hydrogen phosphates to palladium(II) is sufficient to direct arylation at the *ortho* site with diaryliodonium triflates, serving as both arylating agent and terminal oxidant in a Pd(II)/Pd(IV) catalytic cycle. **E:** $Pd(OAc)_2$ (10 mol%) and diphenyliodonium triflate (2 eq.) carefully weighed into a reaction vial, the startg. aryl methyl hydrogen phosphate (0.15 mmol) in 1,2-dichloroethane (1 ml) added, stirred at 80° for 15 h, cooled to room temp., the crude mixture filtered through a cotton plug, the filtrate concentrated *in vacuo*, worked up without chromatographic purification, and the product isolated as its dimethyl ester by treatment with trimethylsilyldiazomethane (5 eq.) in methanol at room temp. over 30 min → product. Y 83%. Reaction is applicable to *o*- and [regioselectively] to *m*-methylated substrates, but methoxyaryl derivs. gave low yields, while halogenated substrates reacted more slowly; the selectivity of methyl phenyl hydrogen phosphate itself was poor, but the diaryliodonium triflates supported various ring functions, e.g. ester, fluoride and amide (eighteen examples in all; Y 37-96%). Yields were generally lower with aryl dimethyl phosphates as substrates and regioselectivity unsatisfactory. F.e. and solvent effect s. L.Y. Chan, L. Cheong, S. Kim, Org. Lett. *2013*, *15* (9), 2186-9 [DOI: 10.1021/ol400732q].

Tris(dibenzylideneacetone)dipalladium/tri-2-furylphosphine/potassium carbonate/ammonium tartrate ←

Sym. sec. (Z,Z)-bis(2-arylallyl)amines from two allene and two ar. iodide molecules $[RCH{=}C(Ar)CH_2]_2NH$

420. 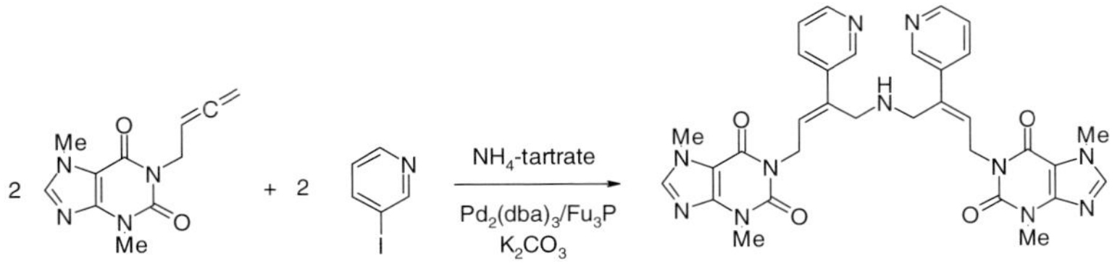

A facile and efficient **regio- and stereo-selective palladium-catalyzed multi-component** (five molecule) **synthesis** of multi-functional (Z,Z)-diallylamines is reported. It utilizes ammonium tartrate as a novel ammonia source, thus avoiding the need for ammonia gas or aqueous ammonia. **E: Sym. sec. (Z,Z)-bis(2-aryl-4-N-hetarylbut-2-enyl)amines**. A mixture of startg. allene (0.5 mmol), ar. iodide (1.2 eq.), ammonium tartrate (6 eq.), Pd_2dba_3 (2.5 mol%), TFP (0.1 mol%) and K_2CO_3 (2 eq.) in 5:1 dioxane/DMF (12 ml) stirred at 100° for 22 h (TLC), cooled to room temp., solvent evaporated, methylene chloride added, the soln. washed with 30% NH_4OH then water, the organic layer dried ($MgSO_4$), filtered, the filtrate concentrated *in vacuo*, and the residue purified by flash column chromatography → 1,1′-[iminobis[(2Z)-3-(pyridin-3-yl)but-2-ene-4,1-diyl]]bis(3,7-dimethyl-3,7-dihydro-1*H*-purine-2,6-dione). Y 80%. Thirty further examples starting from allenes based on purine, quinazolinone, thymidine or uridine cores and a variety of ar. iodides (including 2-iodothiophene as well as 3-iodopyridine) gave yields of 52-99%, with a broad range of functional groups tolerated (reactions with 4-iodonitrobenzene were generally the fastest). The proposed mechanism starts with oxidative addition of palladium into the carbon-iodine bond of the iodide followed by allene coordination and migratory insertion to furnish an arylated π-allyl complex which is attacked by the generated ammonia to afford an intermediate monoallylamine which then attacks a second molecule of the arylated π-allyl complex faster than ammonia to give the desired Z,Z-bisallylamine; the released H-Pd(II)-I species regenerates Pd(0) via reductive elimination in the presence of K_2CO_3. F.e.s. R. Grigg, S. Akkarasamiyo, N. Chuanopparat, E.E. Elboray, M.F. Aly, H.H. Abbas-Temirek, B. Kongkathip, N. Kongkathip, Chem. Commun. *2013*, *49* (20), 2007-9 [DOI: 10.1039/c3cc00067b].

Tetrakis(triphenylphosphine)palladium(0)/sodium acetate/air *$Pd(PPh_3)_4/NaOAc/O_2$*
Polysubst. 4-arylpyridines from N-alkylidene-2-alleneamines and ar. iodides ○
via 6π-electrocyclization of 5-aryl-2-aza-1,3,5-trienes

421. 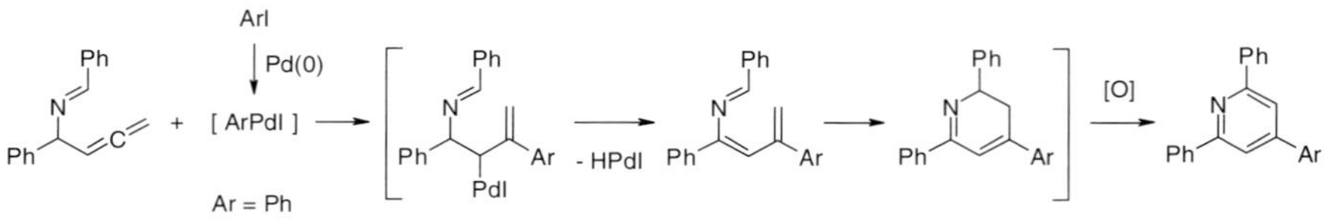

$Pd(PPh_3)_4$ (10 mol%) and NaOAc (2.5 eq.) added under N_2 to a mixture of the startg. N-alkylidene-2-alleneamine (0.689 mmol) [readily prepared by condensation of benzaldehyde with α-allenyl-benzylamine] and iodobenzene (1.1 eq.) in anh. DMF (2 ml), stirred at 80° for 12 h under N_2, exposed to air, stirring continued for another 12 h at 80°, cooled to room temp., water added, extracted with ether, and worked up with purification by flash chromatography on silica gel → 2,4,6-triphenylpyridine. The procedure is applicable to the coupling of electron-diverse iodoarenes, incl. heteroarene analogs, with N-[het]arylidene- and N-alkylidene-2-alleneamines, although yields from the latter were noticeably lower. An aryl or hetaryl group is also tolerated at the allene terminus to give the corresponding *2,3,4,6-tetrasubst.* pyridines (thirty-four examples in all; Y 7-90%). However, the more hindered N-benzylidene-2-phenyl-2-alleneamine gave a 3,4-diphenylpyrrole instead in low yield (34%). Reaction is presumed to involve regioselective arylpalladation of the allene, followed by β-elimination to the (Z)-5-aryl-2-aza-1,3,5-triene prior to 6π-electrocyclization and oxidative aromatization. Attempted one-pot conversion from the aldehyde, alleneamine and iodobenzene resulted in considerable decomposition and a low pyridine yield. F.e.s. Z. He, D. Dobrovolsky, P. Trinchera, A.K. Yudin, Org. Lett. *2013*, *15* (2), 334-7 [DOI: 10.1021/ol303246b].

Tetrakis(triphenylphosphine)palladium(0)/potassium phosphate *$Pd(PPh_3)_4/K_3PO_4$*
***o*-Hydroxybiaryls from 6-diazo-2-cyclohexenones and ar. iodides** ←
Palladium-catalyzed aromatizing coupling

422. 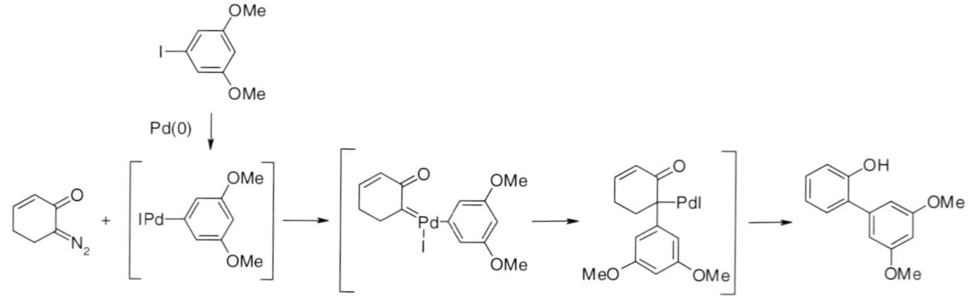

$Pd(PPh_3)_4$ (10 mol%) and K_3PO_4 (3 eq.) dissolved in dioxane (1 ml) under argon, the reaction flask placed in a pre-heated oil bath at 50° for ca. 5 min, a mixture of the startg. diazo compd. (2.1 eq.) and aryl iodide (0.25 mmol) in the same solvent (2.5 ml) added dropwise over 10 min, cooled to room temp. after 5 h, quenched with aq. HCl, stirred for 5 min, diluted with ethyl acetate and water, filtered through a short pad of Celite, the aq. layer of the filtrate extracted with ethyl acetate, and worked up with purification by chromatography on silica gel → 3′,5′-dimethoxybiphenyl-2-ol. Y 92%. The procedure is applicable to the coupling of electron-diverse ar. iodides (incl. *o*-subst. derivs. at a higher temperature and catalyst loading) with variously substituted 6-diazo-2-cyclohexenones, although reaction with 5-subst. derivs. required a longer time (seventeen examples; Y 54-95%). Chiral diazo compds. also participated **with central-to-axial asym. induction** (four examples; Y 54-75%; e.e. 37-72%). Reaction is presumed to involve initial oxidative addition of the aryl iodide to palladium(0), followed by coupling with the diazo compd. to give a palladium carbene complex prior to migratory insertion of the aryl group and β-elimination. F.e. and comparison of bases, solvent and catalysts s. K. Yang, J. Zhang, Y. Li, B. Cheng, L. Zhao, H. Zhai, Org. Lett. *2013*, *15* (4), 808-11 [DOI: 10.1021/ol303484f].

Diacetoxo[o-(di-o-tolylphosphino)benzyl]palladium(II) dimer *[Pd(II)]*
Biaryls from arenes and diaryliodonium fluoroborates s. *77*, 458s*83* Ar-Ar′

Palladium(II) chloride/triphenylphosphine/triethylamine/formic acid/sulfuric acid ←
Low-pressure carbonylation with formic acid/sulfuric acid ←

423.

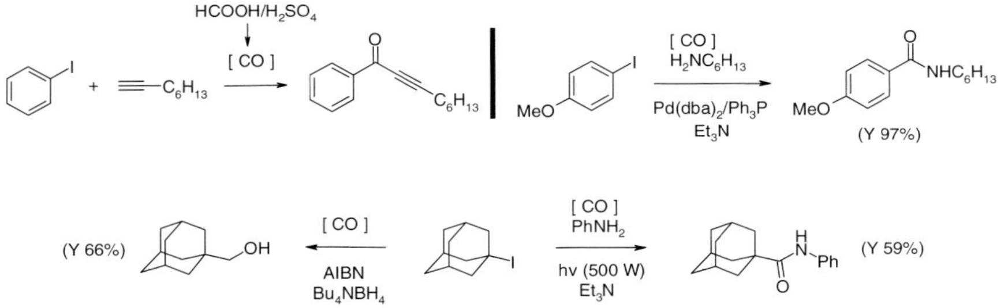

The Morgan reaction may be used to produce carbon monoxide from cheap, readily available reagents for application in carbonylations **in a two-chamber reactor** (COware). **E: Carbonylative Sonogashira coupling in water**. Iodobenzene (1 mmol), $PdCl_2$ (5 mol%), Ph_3P (10 mol%), triethylamine (3 eq.) and water (1.3 ml) added to the carbonylation chamber of a two-chamber reactor, the mixture stirred for 5 min before 1-octyne (1.4 eq.) added while the decarbonylation chamber filled with sulfuric acid (3 eq.), the reactor closed tightly, the decarbonylation chamber heated at 80° then formic acid (3 eq.) added, the acidic mixture stirred at this temp. for 15 min, then at room temp. for 24 h, evaporated, and the resulting brown oil purified by flash chromatography on silica gel → 1-phenylnon-2-yn-1-one. Y 91%. The method was also used for the palladium-catalyzed aminocarbonylation of 4-iodoanisole with *n*-hexylamine (Y 97%; pressure 2.6 atm.), radical hydroxymethylation of 1-iodoadamantane [(Y 66%); cf. S. Kobayashi, I. Ryu et al., Org. Lett. *2010*, *12* (7), 1548-51 [DOI: 10.1021/ol1002847]] and radical aminocarbonylation of 1-iodoadamantane (Y 59%). Aminocarbonylation of 4-iodoanisole was also examined in a continuous microflow 'tube in tube' reactor (Y 81%), unusually involving simultaneous use of a strong acid and base. F. details s. C. Brancour, T. Fukuyama, Y. Mukai, T. Skrydstrup, I. Ryu, Org. Lett. *2013*, *15* (11), 2794-7 [DOI: 10.1021/ol401092a]; **two-chamber process for** safe, laboratory-scale **palladium-catalyzed carbonylation** of ar. halides using *9-methylfluorene-9-carboxylic acid chloride as solid CO source* (cf. *12*, 867s*80*) s. P. Hermange, A.T. Lindhardt, R.H. Taaning, K. Bjerglund, D. Lupp, T. Skrydstrup, J. Am. Chem. Soc. *2011*, *133* (15), 6061-71 [DOI: 10.1021/ja200818w]; carbonylative Heck reaction cf. P. Hermange, T.M. Gøgsig, A.T. Lindhardt, R.H. Taaning, T. Skrydstrup, Org. Lett. *2011*, *13* (9), 2444-7 [DOI: 10.1021/ol200686h]; **aryl β-diketones from methyl ketones and ar. iodides** by palladium-catalyzed carbonylative arylation, incl. ^{13}C-labeled compds., also conversion to 4-chromones, 3-arylpyrazoles or 5-arylisoxazoles, cf. T.M. Gøgsig, R.H. Taaning, A.T. Lindhardt, T. Skrydstrup, Angew. Chem., Int. Ed. *2012*, *51* (3), 798-801 [DOI: 10.1002/anie.201107494]; arylcarboxylic acid amides under palladium catalysis using *$Mo(CO)_6$ as solid CO source* in a bridged 2-vial system, notably with retention of nitro groups, s. P. Nordeman, L.R. Odell, M. Larhed, J. Org. Chem. *2012*, *77* (24), 11393-8 [DOI: 10.1021/jo302322w].

Palladium(II) chloride/1,3-bis(diphenylphosphino)propane/ cesium carbonate/tert-butyl isocyanide *[Pd(II)]/Cs_2CO_3/t-BuNC*

2-Aryl-Δ²-imidazolines from 1,2-diamines and ar. halides or aryl triflates ○

424.

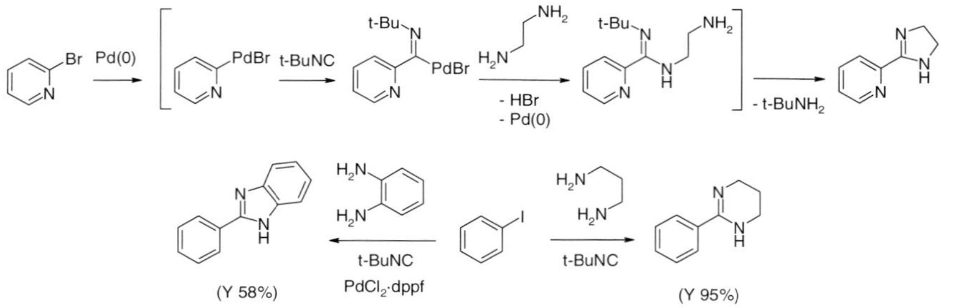

in one pot. A mixture of Cs_2CO_3 (1.3 eq.), anh. toluene (5 ml), 2-bromopyridine (1 mmol), *tert*-butyl isocyanide (1.5 eq.), ethylenediamine (5 eq.), $PdCl_2$ (5 mol%) and dppp (10 mol%) heated under reflux under N_2 for 16 h, cooled to room temp., filtered through Celite, concentrated *in vacuo,* and the residue purified by flash chromatography → 4,5-dihydro-2-(pyrid-2-yl)-1*H*-imidazole. Y 96%. Prepn. of the title products via Pd(II)-catalyzed 3-component reaction of *tert*-butyl isocyanide (optimally), ethylenediamine and (het)ar. halides (Cl, Br or I) or triflates (incl. bis-reaction with 2,6-dibromopyridine) gave highest yields with either dppp or 1,1′-bis(diphenylphosphino)ferrocene [dppf] ligands (fourteen examples; Y 51-96%; sterically-hindered 2-iodotoluene gave 29-32%), but failed with 2-chlorobenzene, 4-chloroanisole and 3-chloropyridine. Variation in the diamine component enabled synthesis of N-subst. 2-(het)aryl-Δ²-imidazolines (two examples; Y 60-82%), **2-arylbenzimidazoles** (three examples; Y 39-58%; 82% for a 4,5,6,7-tetrahydro deriv.) and a 2-aryl-1,4,5,6-tetrahydropyrimidine (Y 95%). F.e., optimization and use of the method for the one-pot synthesis of the Pybim ligand [2,6-bis[(4R,5R)-4,5-diphenyl-Δ²-imidazolin-2-yl]pyridine] in 51% yield s. J.V. Geden, A.K. Pancholi, M. Shipman, J. Org. Chem. *2013*, *78* (8), 4158-64 [DOI: 10.1021/jo400252n].

Bis(allylpalladium chloride)/1,3-di(adamantyl)imidazol-2-ylidene/cesium carbonate ←

Palladium-catalyzed β-(alk-l-ynyl)ation with α,β-acetylenebromides H→C≡C

425.

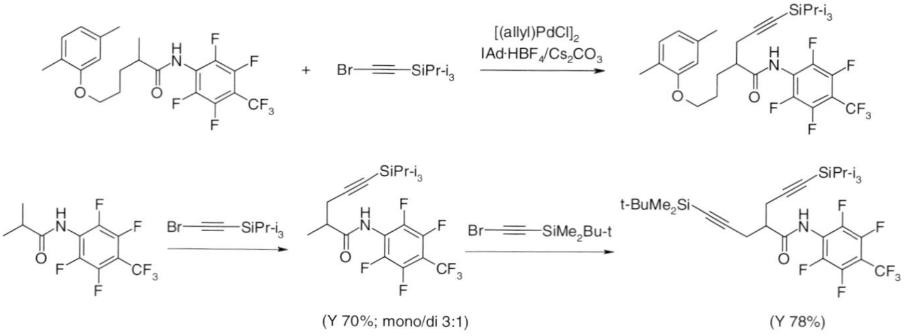

The first example of the application of [(alkynyl)Pd(II)L_n] complexes to activate C(sp^3)-H bonds is reported, β-(alk-l-ynyl)ation **of carboxylic acid anilides** [in particular 2,3,5,6-tetrafluoro-4-(trifluoromethyl)anilides] notably being achievable *without* the need for *co-oxidants* (cf. *80*, 447 via Pd(II)/Pd(IV) catalysis). **E: β-Silylethynylation**. Startg. anilide (0.1 mmol), [(allyl)$PdCl]_2$ (5 mol%),

1,3-di(adamantyl)imidazolium fluoroborate (NHC precursor; 20 mol%) and Cs_2CO_3 (2 eq.) added to a Schlenk tube under air, startg. alkynyl bromide (2 eq.) and ether (0.5 ml) added, the tube evacuated and backfilled with N_2 three times, the mixture stirred at room temp. for 5 min, then stirred vigorously at 85° for 8 h, cooled to room temp., solvent removed under reduced pressure, and the residue purified by flash chromatography on silica gel → 2-[3-(2,5-dimethylphenoxy)-propyl]-N-[2,3,5,6-tetrafluoro-4-(trifluoromethyl)phenyl]-5-(triisopropylsilyl)pent-4-ynamide. Y 82%. Ten further examples with propionic acid amides optionally bearing α-alkyl, -cyclohexyl, -phenethyl, -tetrahydropyran-4-yl, -benzyloxymethyl or -2,2,2-trifluoroethyl groups and TIPS- or TBS-ethynes gave yields of 61-81%, with reaction occurring selectively at β-methyl rather than β-methylene groups where appropriate. The method is also applicable to β-silylethynylation of the methylene groups of cyclohexane- and tetrahydropyran-2-yl-carboxamides (Y 70% (mono/di 6:1), 61%, respectively) with *cis*-selectivity. The product obtained by triisopropylsilyl-ethynylation of the dimethylacetic acid deriv. underwent subsequent β-*tert*-butyldimethylsilyl-ethynylation in 78% yield. A mechanism involving α-palladation and β-hydride elimination has been ruled out by deuterium labelling studies. F.e. and optimization s. J. He, M. Wasa, K.S.L. Chan, J.-Q. Yu, J. Am. Chem. Soc. *2013*, *135* (9), 3387-90 [DOI: 10.1021/ja400648w]; **N-directed γ-alkylation** of N-alkylpicolinamides using $Pd(OAc)_2/Ag_2CO_3$/dibenzyl hydrogen phosphate s. S.-Y. Zhang, G. He, W.A. Nack, Y. Zhao, Q. Li, G. Chen, ibid. *135* (6), 2124-7 [DOI: 10.1021/ja312277g].

Bis(acetonitrile)dichloropalladium(II)/silver oxide/trifluoroacetic acid *Pd(II)/Ag(I)/CF$_3$COOH*
9-Fluorenones from ar. nitriles and ar. iodides ○

426. 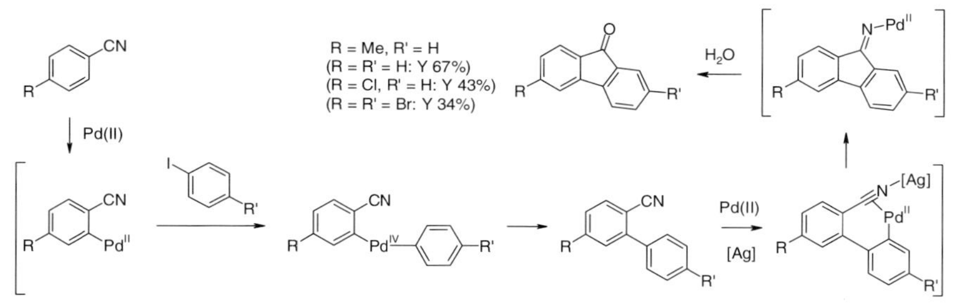

$PdCl_2(MeCN)_2$ (10 mol%), Ag_2O (1.1 eq.), startg. ar. nitrile (2 eq.) and startg. ar. iodide (0.5 mmol) in trifluoroacetic acid (99% with 0.2% water; 5 ml) and dimethylacetamide (0.25 ml) added to a screw-capped vial in a glovebox under N_2, stirred for a while, then water (1 eq.) slowly injected, the vial sealed with a cap, the mixture stirred at 140° for 96 h (method cf. *77*, 458s*79*), diluted with methylene chloride, filtered through a thin Celite pad, concentrated *in vacuo*, and the residue isolated by column chromatography → product. Y 60%. In this nitrile-directed fluorenone synthesis via dual C-H activation with intramolecular 1,2-addition to the cyano group, five further examples with benzo- or *p*-methylbenzo-nitrile and ar. iodides (optionally bearing a *p*-halo or *p*-methyl group) gave yields of 50-67%; yields were lower with *p*-chloro- (26-43%; four examples), *p*-bromo- (34%) or *p*-fluoro- (17%) nitriles. Twenty-two examples of fluorenone ring closure **from** the intermediate ***o*-cyanobiaryls** proceeded in 31-91% yield using Ag-trifluoroacetate in place of Ag_2O at 140° for 72 h. Use of a highly polar solvent was found to be essential to prevent formation of the corresponding *o*-carbamylbiaryl. F.e.s. J.-C. Wan, J.-M. Huang, Y.-H. Jhan, J.-C. Hsieh, Org. Lett. *2013*, *15* (11), 2742-5 [DOI: 10.1021/ol401063w]; activation of remote *meta* C-H bonds assisted by readily-removable nitrile-containing end-on templates s. D. Leow, G. Li, T.-S. Mei, J.-Q. Yu, Nature *2012*, *486* (7404), 518-22 [DOI: 10.1038/nature11158]; **aryl ketones from nitriles and ar. iodides** with $NiCl_2$(DME)/dppp/Zn in DME containing water at 100° s. J.-C. Hsieh, Y.-C. Chen, A.-Y. Cheng, H.-C. Tseng, Org. Lett. *2012*, *14* (5), 1282-5 [DOI: 10.1021/ol300153f].

Bis(acetonitrile)dichloropalladium(II)/norbornene/potassium hydrogen carbonate ←

Palladium-catalyzed direct C_2-alkylation of electron-deficient N-unsubst. pyrroles with prim. bromides H → R

427. 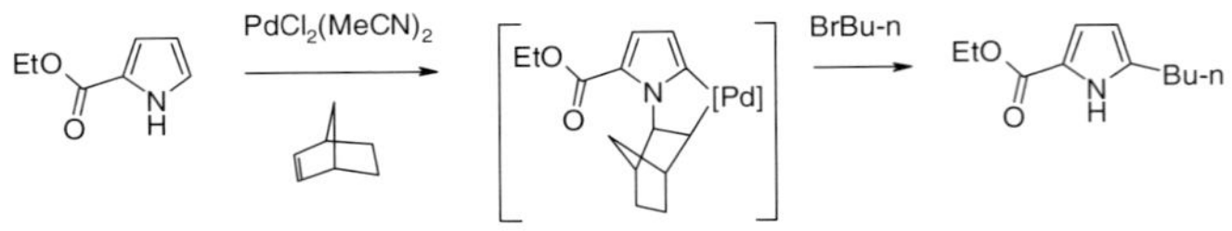

Regioselective direct alkylation via a norbornene-mediated cascade C-H activation process, previously described for N-unsubst. indoles (cf. *80*, 452), has been extended to pyrroles, for which the presence of an electron-withdrawing group was found to be necessary. **E: 5-Subst. pyrrole-2-carboxylic acid esters**. A reaction tube charged with ethyl pyrrole-2-carboxylate (1.01 mmol), norbornene (2.01 mmol), $KHCO_3$ (3.08 mmol), $PdCl_2(MeCN)_2$ (0.098 mmol) and *n*-butyl bromide (2.08 mmol), anh. DMA (1 ml) added, the tube heated in an aluminum block at 90° under air (balloon) for 22 h, the mixture cooled to room temp., diluted with ether, filtered, the filtrate washed with water, the aq. layer extracted with ether, the combined organic extracts washed with brine, dried (Na_2SO_4), filtered, solvent evaporated, and the residue purified by flash chromatography on silica gel → ethyl 5-butylpyrrole-2-carboxylate. Y 83%. Better results were obtained with esters than with 2-cyano, 2-carbamyl, 2-formyl or 2-acetyl derivs. so the scope of the reaction was explored with esters and a variety of prim. bromides, optionally bearing silyl ether, ester, acetal, cyano or alkenyl groups (eighteen further examples; Y 50-91%). In all cases the 5-alkylation product was the only regioisomer obtained. Five examples with [2-subst.] 3-acyl- or 3-carbalkoxy-pyrroles gave yields of 45-71%. F.e. and optimization, also application to the synthesis of mycalazal 14, a marine sponge metabolite, s. L. Jiao, T. Bach, Angew. Chem., Int. Ed. *2013*, *52* (23), 6080-3 [DOI: 10.1002/anie.201301154].

Bis(acetonitrile)dichloropalladium(II)/1,4-bis(4-pyridyl)-2,3-diaza-1,3-butadiene/ hexadecyltrimethylammonium bromide/1,8-diazabicyclo[5.4.0]undec-7-ene ←

Bis(acetonitrile)dichloropalladium(II)/2-dicyclohexylphosphino-2′,4′,6′-triisopropyl-biphenyl/cesium carbonate ←

Ring closures by Sonogashira coupling-cycloisomerization s. *44*, 781s*83* ○

Dichloro[N,N′-bis(2,6-diisopropylphenyl)imidazol-2-ylidene]-(N-methylimidazole)palladium(II)/potassium tert-butoxide *IPr(Im)PdCl$_2$/KOBu-t*

3-Aryl-3-hydroxyoxindoles from oxindoles and ar. chlorides via palladium(II)-catalyzed α-arylation-aerobic oxidation CH_2 → C(OH)Ar

428. 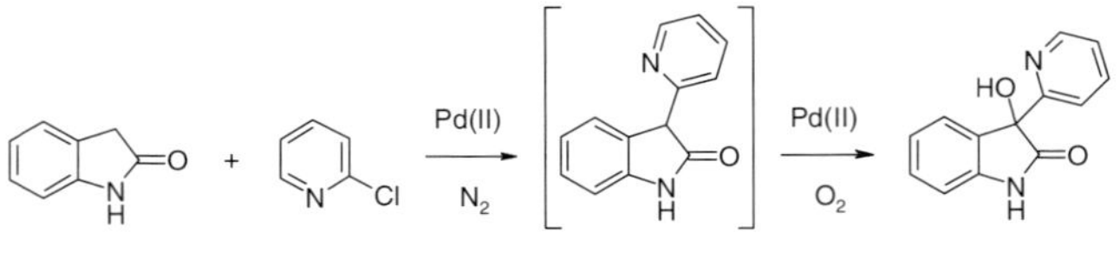

A mixture of oxindole (1.3 eq.), NHC-Pd(II)-Im complex (1 mol%), K-*tert*-butoxide (4 eq.), toluene (1 ml) and 2-chloropyridine (0.5 mmol) stirred at reflux under N_2 for 12 h, cooled to room temp., stirred under air for 3 h, solvent removed *in vacuo*, and the residue purified by flash chromatography → 3-hydroxy-3-pyrid-2-yloxindole. Y 72%. Novel, phosphine-free 3-arylation of NH-oxindoles was catalyzed by the palladium complex under N_2, affording 3-aryl-derivs. from 5-subst. (H, F, OMe) NH-oxindoles and electron-diverse (het)ar. chlorides in good yields (66-98%; twenty-five examples), with sterically-hindered 5,7-dimethyloxindole affording somewhat lower yields (44-55%; three examples). Further stirring of the reaction mixture under air at room temp. effected unprecedented oxidation of the 3-aryl-oxindoles to afford the title products in one pot in good yields (62-84%; eighteen examples; 2-chloroanisole gave 56%). F.e. and optimization s. Z.-K. Xiao, H.-Y. Yin, L.-X. Shao, Org. Lett. *2013*, *15* (6), 1254-7 [DOI: 10.1021/ol400186b].

Dichlorobis(triphenylphosphine)palladium(II) s.a. under Zn *$PdCl_2(PPh_3)_2$*

Dichlorobis(triphenylphosphine)palladium(II)/copper(I) iodide/triphenylphosphine/ triethylamine ←

Ring closures by Sonogashira coupling-cycloisomerization ○

benzofurans from *o*-halogenophenols and acetylene derivs. cf. *44,* 781; benzofuran-conjugated sugar and nucleoside derivs. with $PdCl_2(PPh_3)_2$/CuI/Ph_3P/Et_3N s. Y. Kommagalla, K. Srinivas, C.V. Ramana, Tetrahedron Lett. *2013, 54* (14), 1824-7 [DOI: 10.1016/j.tetlet.2013.01.089]; *copper-free* procedure in water *under visible-light irradiation* s. S. Ghosh, J. Das, F. Saikh, ibid. *2012, 53* (44), 5883-6 [DOI: 10.1016/j.tetlet.2012.08.078]; *metal-free* procedure from *o*-chlorophenols with acetone in aq. acetonitrile under photoirradiation cf. S. Protti, M. Fagnoni, A. Albini, J. Org. Chem. *2012, 77* (15), 6473-9 [DOI: 10.1021/jo3010183]; **2-indolyl-C-glycosides** (cf. *44,* 815) with $[Cu(phen)_2(PPh_3)_2]NO_3$/$K_3PO_4$ s. P. Subramanian, K.P. Kaliappan, Eur. J. Org. Chem. *2013* (3), 595-604 [DOI: 10.1002/ejoc.201201208]; **(Z)-3-alkylidenephthalimidines** from *o*-halogenocarboxylic acid amides with $PdCl_2(MeCN)_2$/1,4-bis(4-pyridyl)-2,3-diaza-1,3-butadiene *in aq. micelles* with added hexadecyltrimethylammonium bromide and DBU s. S. Sarkar, S. Dutta, R. Dey, S. Naskar, Tetrahedron Lett. *2012, 53* (50), 6789-92 [DOI: 10.1016/j.tetlet.2012.10.001]; (Z)-5-alkylidene-2(5*H*)-furanones from (Z)-3-iodoacrylic acid with Pd-C and CuI/Ph_3P/Et_3N s. D. Rambabu, S. Bhavani, K.S. Nalivela, S. Mukherjee, M.V.B. Rao, M. Pal, ibid. *2013, 54* (17), 2151-5 [DOI: 10.1016/j.tetlet.2013.02.040]; **4-alkylidene-3-azabicyclo[3.1.0]hexan-2-ones** from 2-iodocyclopropanecarboxamides with $PdCl_2(MeCN)_2$/X-Phos/Cs_2CO_3 s. B. de Carné-Carnavalet, A. Archambeau, C. Meyer, J. Cossy, B. Folléas, J.-L. Brayer, J.-P. Demoute, Chem. Eur. J. *2012, 18* (52), 16716-27 [DOI: 10.1002/chem.201203153]; alternative route to **benzofurans** from *o*-acetylenephenols and tosylhydrazones with CuBr/LiOBu-*t*, also N-protected **indoles** from *o*-acetyleneamines, s. T. Xiao, X. Dong, L. Zhou, Org. Biomol. Chem. *2013, 11* (9), 1490-7 [DOI: 10.1039/c2ob26867a].

Tris[2-phenylpyridinato-C2,N]iridium(III) s. under Potassium tert-butoxide *[Ir(III)]*

Via intermediates *v.i.*

3-Component synthesis of trisubst. ethylene from terminal acetylene derivs. via frustrated Lewis pair-mediated regio- and stereo-selective boroauration ←

429.

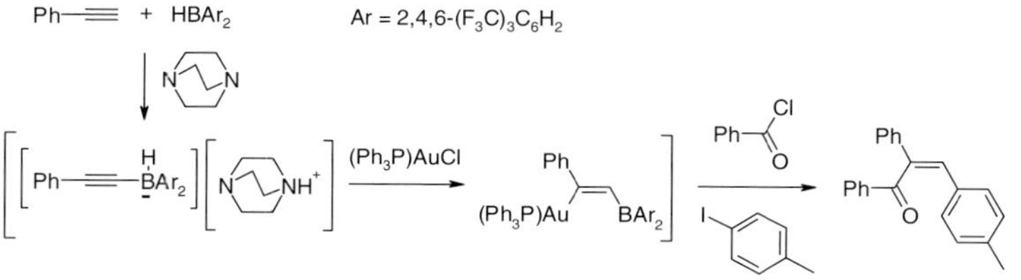

A soln. of bis[2,4,6-tris(trifluoromethyl)phenyl]borane (0.7 mmol) in toluene (3 ml) added slowly to a mixture of phenylacetylene (1.2 eq.), DABCO (1.5 eq.) and (triphenylphosphine)gold(I) chloride (1 eq.) in toluene/methylene chloride (1:1; 3 ml) at 0°, stirred for 3 h at the same temp., volatiles removed *in vacuo*, extracted with toluene at room temp., and worked up without chromatographic purification → intermediate boroauration adduct (Y 74%), a portion (0.26 mmol) of which dissolved in THF (1 ml), the soln. added to a mixture of benzoyl chloride (1.01 eq.) and $PdCl_2(PPh_3)_2$ (10 mol%) in THF (6 ml), stirred at 75° for 10 h, 4-iodotoluene (3.1 eq.) and NaOH (3 *M* in water, 6 ml) added, the two-phase mixture heated to 75° for 14 h, cooled to room temp., and worked up with chromatographic purification → product (Y 41%). Reaction is presumed to involve initial formation of a 'frustrated' triethylenediammonium alkynyl(diaryl)hydridoborate which undergoes auration regio- and stereo-selectively to give an (E)-2-borylvinylgold(I) compd. having differing reactivity at either end of the alkene residue; accordingly, under transition metal catalysis, the intermediate then couples in sequence with the two electrophiles with retention of configuration to give the trisubst. alkene to complete the one-pot synthesis. The initially-formed frustrated Lewis pair was isolable and characterized by ^{11}B-NMR spectroscopy and X-ray diffraction. F. details, also conversion of the Lewis pair to the corresponding (E)-eneborane on thermolysis (with liberation of DABCO), s. H. Ye, Z. Lu, D. You, Z. Chen, Z.H. Li, H. Wang, Angew. Chem., Int. Ed. *2012, 51* (48), 12047-50 [DOI: 10.1002/anie.201206927].

Isatins from prim. ar. amines via alkoximinoacetic acid anilides s. *83*, 64 ○

2-Aryloxazole-4-carboxaldehydes from phenacyl bromides s. *83*, 392

Sulfur ↾ CC ⇃↾ S

Without additional reagents *w.a.r.*
2-(Alkylthio)pyrroles from α-ketoketene mercaptals and 2-acetyleneamines ○
cf. *82*, 465; in water at 100°, also 2-aryl- or 2-acyl-pyrroles (with added K_2CO_3) from the corresponding 3-ketothioenolethers, s. C.-Q. Ren, C.-H. Di, Y.-L. Zhao, J.-P. Zhang, Tetrahedron Lett. *2013, 54* (11), 1478-81 [DOI: 10.1016/j.tetlet.2013.01.020].

Irradiation s. under [Ru(bpy)$_3$](PF$_6$)$_2$ and Ru(bpy)$_3$Cl$_2$ ///

Sodium hydride *NaH*
6-Acylbicyclo[3.1.0]hexan-2-one-3-carboxylic acid esters from 2-pyrone-3-carboxylic acid esters

430.
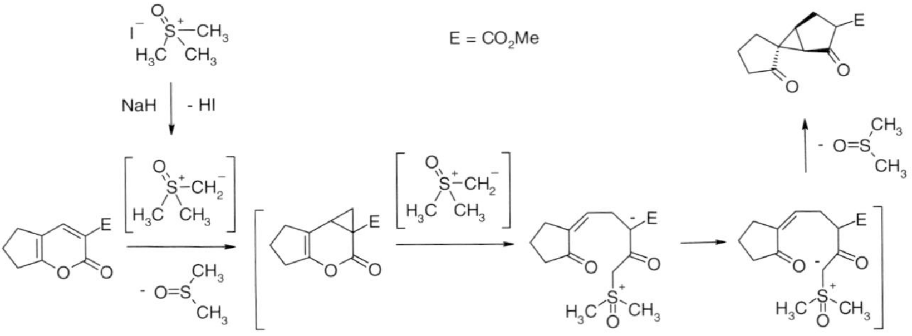

Trimethylsulfoxonium iodide (3.8 eq.) added in one portion to a suspension of NaH (3.2 eq.) in DMSO (2 ml) at room temp. under N_2, the mixture stirred for 30 min, methyl 2-oxo-2,5,6,7-tetrahydrocyclopentano[*b*]pyran-3-carboxylate (0.48 mmol) added, the mixture stirred for 4 h, quenched with 1 *M* aq. HCl at 0°, extracted with chloroform, and purified chromatographically → (1R*,1′R*,5S*)-2,2′-dioxospiro[bicyclo[3.1.0]hexane-6,1′-cyclopentane]-3-carboxylate. Y 75%. This novel skeletal rearrangement was applicable to 2-pyrones activated with a 3-carboxylate and carrying cycloalkano (5-12-membered ring; incl. oxa and aza analogs) at the 5,6-position (seven examples; Y 43-76%), with an acyclic 5-propyl-6-butyl deriv. giving a 42% yield. Methyl esters of the pyranone-3-carboxylates gave best results in comparison with benzyl, ethyl and *tert*-butyl analogs, and the products were characterized as their 2-methoxymethyl enol ethers (to suppress keto-enol tautomerism). Reaction appears to be initiated by Corey-Chaykovsky cyclopropanation, followed by double ring opening by attack of a second sulfoxonium methylid molecule, then subsequent intramolecular cyclopropanation. F.e. and optimization s. T. Miura, N.D. Yadav, H. Iwasaki, M. Ozeki, N. Kojima, M. Yamashita, Org. Lett. *2012, 14* (23), 6048-51 [DOI: 10.1021/ol302942m].

Potassium hydroxide/chiral 2-thioureidophosphonium halides ←
2-Nitro-N-carbo-*tert*-butoxyamines C(NHBoc)C(NO_2)
from α-(N-carbo-*tert*-butoxyamino)sulfones and aliphatic nitro compds.
via asym. aza-Henry [nitro-Mannich] reaction
using chiral 2-thioureidophosphonium halides as bifunctional phase transfer catalysts

431.

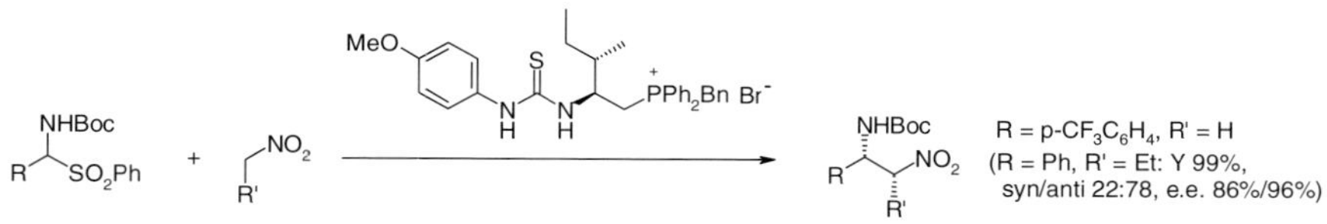

Nitromethane (5 eq.) and benzyl[(2S,3S)-2-[3-(4-methoxyphenyl)thioureido]-3-methylpentyl]-diphenylphosphonium bromide (5 mol%) added sequentially to a suspension of the startg. Boc-aminosulfone (0.1 mmol) in toluene (1 ml), the mixture stirred at -20° for 10 min, treated with freshly ground KOH (5 eq.), the resulting suspension stirred vigorously at -20° for 5 h, quenched with satd. aq. $NaHCO_3$, extracted with ethyl acetate, and purified by flash chromatography → *tert*-butyl [(1S)-2-nitro-1-[4-(trifluoromethyl)phenyl]ethyl]carbamate. Y 84% (e.e. 98%). The chiral bifunctional catalyst is easily tunable and was optimized in terms of: the amino acid skeleton from which it was derived; functionalization as an N-acyl-, ureido- or thioureido-deriv.; the N-aryl group on the thiourea; the groups on the phosphonium salt; and the use of a bromide or chloride counterion. The illustrated thioureidophosphonium salt was selected for optimization of reaction conditions, also taking factors such as cost and accessibility into consideration. The method is applicable to Boc-aminosulfones bearing electron-diverse α-aryl (optionally bearing F, Cl, Br, Me, MeO or NO_2 as well as CF_3) or 2-furyl or 2-thienyl groups, reaction with nitromethane giving high yields (68%, 77-95%) and enantioselectivity (91-98% e.e.) for twelve further examples; furthermore, α aliphatic (cyclohexyl, phenylethyl) groups resulted in only a slight drop in enantioselectivity (88% e.e.; Y 88% in both cases). Reactions with nitroethane or nitropropane gave quantitative yields of *syn/anti*-mixtures (two examples; *syn/anti* 37:63, 22:78) with high enantioselectivity (86-97% e.e.); ethyl 4-nitrobutanoate also gave an adduct (*syn/anti* 47:53; e.e. 89%/86%) which underwent reductive lactamization to the 5-α-Boc-amino-2-pyrrolidone with the stereochemical integrity basically retained (Y 65%; *syn/anti* 43:57; e.e. 88%/88%). F.e.s. D. Cao, Z. Chai, J. Zhang, Z. Ye, H. Xiao, H. Wang, J. Chen, X. Wu, G. Zhao, Chem. Commun. *2013*, *49* (53), 5972-4 [DOI: 10.1039/c3cc42864h].

Potassium tert-butoxide *t-BuOK*
Cyclopropanecarbonitriles from aldehydes and nitriles ▽
via diastereoselective Knoevenagel condensation-Corey-Chaykovsky reaction in one pot

432.

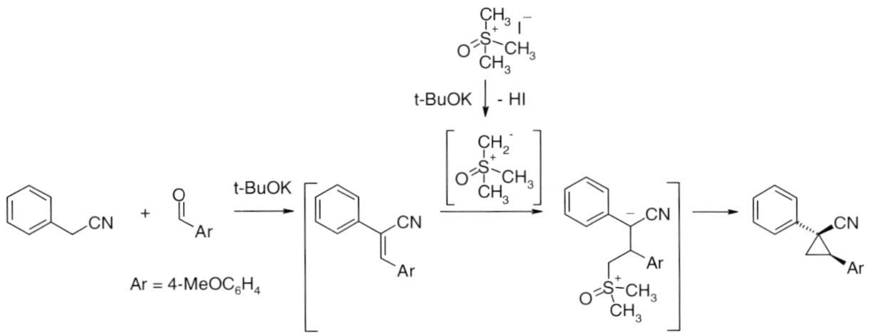

***cis*-2-Subst. 1-arylcyclopropanecarbonitriles**. Trimethylsulfoxonium iodide (2 eq.) and K-*tert*-butoxide (2 eq.) added to a stirred soln. of phenylacetonitrile (2 eq.) and 4-methoxybenzaldehyde

(1 mmol) in DMSO (1 ml), the suspension stirred at room temp. for 20 min, quenched with satd. aq. NH_4Cl, extracted with ether, and purified by chromatography on silica → *cis*-2-(4-methoxyphenyl)-1-phenylcyclopropanecarbonitrile. Y 93%. This novel, operationally simple sequence formally generates all-carbon quaternary centers from activated methylene compds. via cyclization of intermediate Corey ylids. The reaction was successful for electron-diverse (het)arylacetonitriles reacting with ar. aldehydes or benzyloxymethylacetaldehyde to afford the title products (sixteen examples; Y 45-74%, 93%). F.e., optimization, and application in a two step synthesis of (±)-bicafidine, s. J.J. Clemens, J.L. Asgian, B.B. Busch, T. Coon, J. Ernst, L. Kaljevic, P.J. Krenitsky, T.D. Neubert, E.J. Schweiger, A. Termin, D. Stamos, J. Org. Chem. *2013*, *78* (2), 780-5 [DOI: 10.1021/jo302443k].

n-Butyllithium/magnesium chloride *$BuLi/MgCl_2$*

Diastereoselective synthesis of γ-lactones from ketenes, aldehydes and S-aminosulfoxonium salts

433. 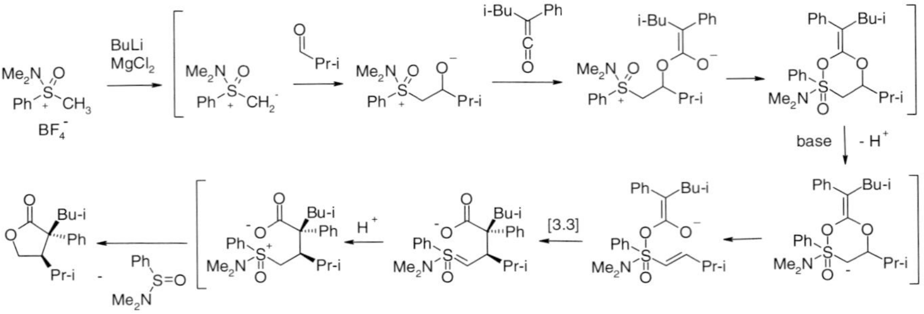

Direct access to ***trans*-α,α,β-trisubst. γ-lactones** via an unprecedented extension of the Johnson-Corey-Chaykovsky reaction is reported. **E:** A soln. of *n*-butyllithium (1 eq.) in hexane (0.1 ml) added dropwise to a suspension of anh. (dimethylamino)methyl(phenyl)oxosulfonium fluoroborate (1 eq.) in anh. THF (1.5 ml) at -78°, the mixture stirred for 45 min, $MgCl_2$ (2 eq.) added under N_2, the mixture stirred for 15 min, isobutyraldehyde (0.25 mmol) added dropwise, stirring continued for 1.5 h, a soln. of isobutyl(phenyl)ketene (1 eq.) in THF (0.5 ml) added over 1 h, stirring continued for 4 h, warmed to room temp. overnight, and worked up with chromatographic purification → *trans*-3-isobutyl-4-isopropyl-3-phenyldihydrofuran-2(3*H*)-one. Y 93% (d.r. 88:12). Addition of $MgCl_2$ in this case was assumed to activate the sulfoxonium ylid, via transmetalation, to reaction with the aliphatic aldehyde [lower yields were obtained for isobutyraldehyde in the presence of CuI (39%) or in the absence of additive (33%)]. Electron-diverse (het)ar. aldehydes generally required no additional transmetalation step, with the lithium ylids affording the *trans*-lactones in variable yields (40-82%; eleven examples; d.r. 75:25 to 92:8). F.e. and optimization s. M. Mondal, H.-J. Ho, N.J. Peraino, M.A. Gary, K.A. Wheeler, N.J. Kerrigan, J. Org. Chem. *2013*, *78* (9), 4587-93 [DOI: 10.1021/jo4003213].

n-Butyllithium/sodium tetrahydridoborate/sodium/methanol *$BuLi/NaBH_4/NaOMe$*

Regioselective synthesis of glycol monocarbamates from oxo compds. with addition of one C-atom C(OCON<)C(OH)

434. 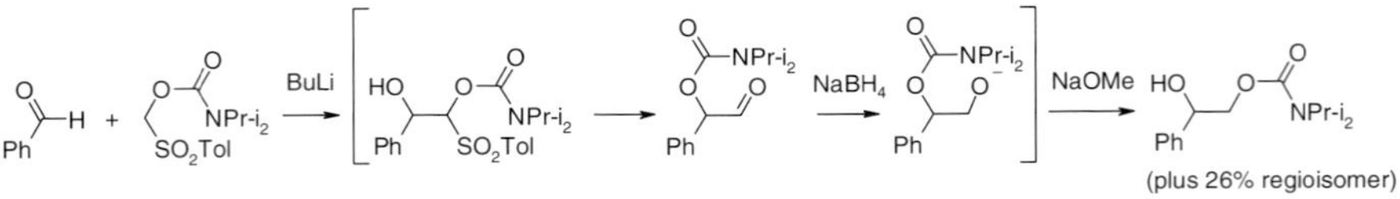

By judicious choice of reagents for each step it is possible to use the novel synthon, (N,N-diisopropylcarbamoyloxy)methyl *p*-tolyl sulfone, for the synthesis of either glycol monocarbamate regioisomer from oxo compds. **in one pot**, or for the synthesis of **glycols**. **E:** A soln. of (N,N-diisopropylcarbamoyloxy)methyl *p*-tolyl sulfone (0.32 mmol) in dry THF (2 ml) treated at -78° with *n*-BuLi (1.6 *M* in hexanes; 1.3 eq.) under N_2, the yellow soln. stirred at -78° for 45 min, treated with startg. aldehyde (neat; 1.1 eq.), stirred at -78° for a further 50 min, treated with $NaBH_4$ (10 eq.) and methanolic NaOMe (30% w/v; 0.34 ml; 6 eq.), the mixture gradually warmed to room temp., stirring continued for ca. 2 h (TLC), quenched with satd. aq. NH_4Cl, the organic layer separated, the aq. phase washed with ether, the combined organic phases washed with brine, concentrated, and purified by flash chromatography on silica gel → 1-phenyl-2-O-(N,N-diisopropylcarbamoyl)-1,2-ethanediol. Y 67% [plus 1-O-(N,N-diisopropylcarbamoyl)-1-phenyl-1,2-ethanediol: Y 26%]. The latter was the major product by treatment of the initial α-carbamyloxy-aldehyde with $NaBH_4$ in the absence of methanolic NaOMe (Y 49% after 2 h at 0° to room temp., with no 2-O-(N,N-diisopropylcarbamoyl)-deriv. detected). No glycol was detected under either set of conditions, but was the major product by deprotonation with *i*-PrMgCl and reduction with *i*-Bu_2AlH (Y 57%).

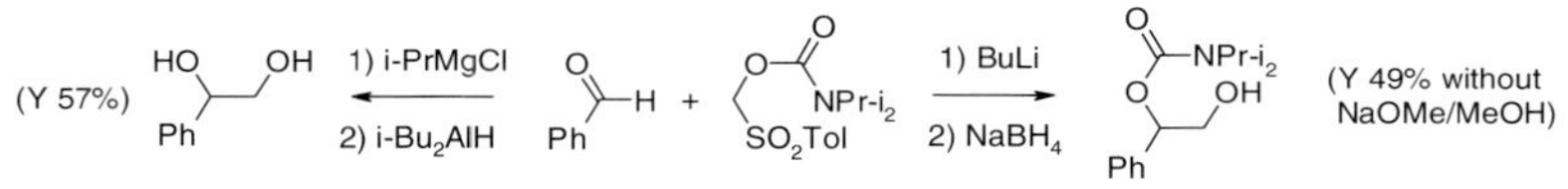

The methods are applicable to a variety of ar. aldehydes but failed with pyridine derivs. The first set of conditions gave yields of 54-80% with eight further ar. aldehydes (Y 0-27% for the 1-O-isomer), while *n*-hexanal gave only 12% yield and acetone gave 64%. The second set of conditions gave 1-O-isomers in yields of 49-70% (eight examples, all with ar. aldehydes). Glycols were obtained from ar. aldehydes in yields of 61-62% (three further examples), while benzophenone gave 57% yield. F.e. and prepn. of the startg. carbamyloxysulfone s. L. Ma, D. Zhao, L. Chen, X. Wang, Y.-L. Chen, J. Shen, Tetrahedron *2012*, *68* (42), 8704-11 [DOI: 10.1016/j.tet.2012.08.025].

Lithium acetylides *LiC≡CR*

(Z)-3-Aryl-3-ene-1,5-diynes C≡C-C(Ar)=CHC≡C

from β-chloro-β-styryl sulfoxides and two lithium acetylide molecules

435.

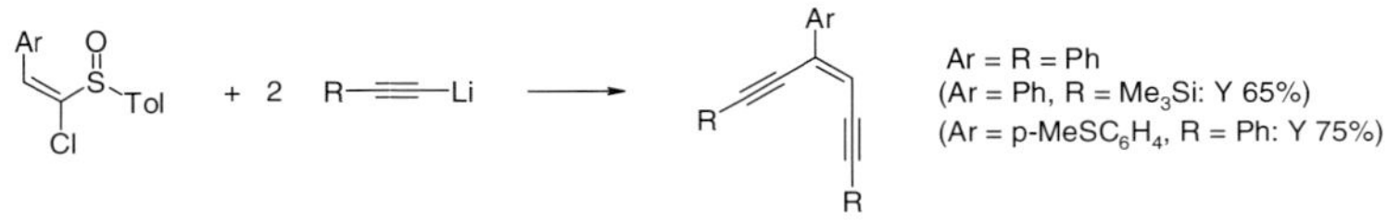

(Z)-1,3,6-Triaryl-3-ene-1,5-diynes. A soln. of phenylacetylene (5 eq.) in THF (1.2 ml) at room temp. treated with a soln. of BuLi in hexane (1.65 *M*; 5 eq.), the mixture stirred for 2 min, a soln. of (E)-1-chloro-2-phenylvinyl *p*-tolyl sulfoxide (0.1 mmol) in THF (0.8 ml) added, the mixture stirred for 30 min, quenched with satd. aq. NH_4Cl, extracted with chloroform, the organic layer dried ($MgSO_4$), concentrated under reduced pressure, and the residue purified by chromatography on silica gel → (Z)-product. Y 80%. The corresponding (Z)-1-chlorovinyl sulfoxide also gave the (Z)-product in 36% yield, while a Z/E mixture afforded the (Z)-enediyne in 49% yield as a single isomer. Ten further examples starting from 2-aryl-1-chlorovinyl sulfoxides optionally bearing MeS, methylenedioxy, CF_3 or Cl on the aromatic ring, or 1-naphthyl or 2-thienyl analogs, and Li-arylacetylides optionally bearing Me, MeO, F or Cl gave yields of 37-75%. An alkyl-terminated acetylide (phenethyl) gave a low yield (19%) but Li-trimethylsilylacetylide gave 65%. In all cases, the starting sulfoxides were completely consumed within 30 min, although low yields were obtained in some cases, with volatile by-products apparently lost during workup. F.e.s. T. Kimura, Y. Nishimura, N. Ishida, H. Momochi, H. Yamashita, T. Satoh, Tetrahedron Lett. *2013*, *54* (9), 1049-51 [DOI: 10.1016/j.tetlet.2012.11.152]; **1,2,4-trienes** from α-chloro-α,β-ethylenesulfoxides and α-lithiated β,γ-ethylenesulfoxides s. T. Kimura, G. Kobayashi, M. Ishigaki, M. Inumaru, J. Sakurada, T. Satoh, Synthesis *2012*, *44* (23), 3623-32 [DOI: 10.1055/s-0032-1317507]; **1,3-enynes** from β,β-dialkylated α-chloro-α,β-ethylenesulfoxides by 1,4-addition of lithium acetylides then organomagnesium chloride-induced 1,2-alkynyl migration, **also 1,3-enyn-1-yl sulfoxides** from β-monoalkylated α-chloro-α,β-ethylenesulfoxides by 1,4-addition-1,2-hydride shift (without added Grignard), s. N. Ishida, H. Saitoh, S. Sugiyama, T. Satoh, Tetrahedron *2011*, *67* (17), 3081-90 [DOI: 10.1016/j.tet.2011.02.079].

Lithium diisopropylamide *i-Pr₂NLi*

o-Allenesulfoxides $CH_2 \rightarrow C{=}C{=}CH$

from α,β-acetylenesulfones and o-subst. ar. sulfoxides with asym. induction

436.

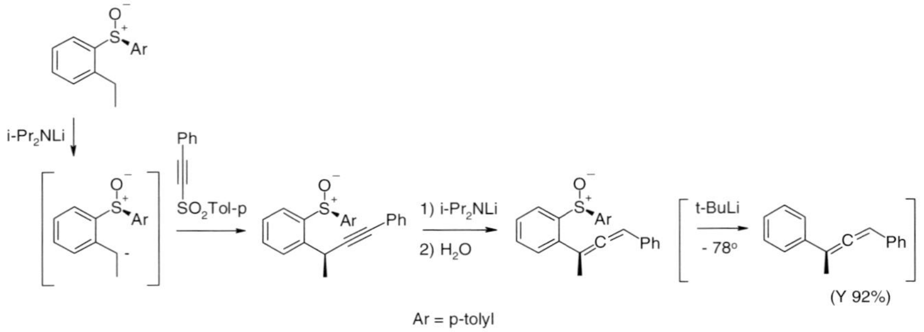

A novel two step approach to enantiomerically enriched allenes, involving reaction of chiral *o*-sulfinylbenzyllithiums with β-subst. sulfonylacetylenes, *in umpolung fashion*, and subsequent isomerization of the generated propargylic center with base, is reported. Selected examples were carried out as a **one-pot** procedure. **E:** A soln. of butyllithium (2.4 eq.) in hexanes (0.096 ml) added to a soln. of diisopropylamine (3.6 eq.) in THF (1 ml) at 0°, the mixture stirred for 10 min, cooled to -78°, a soln. of (S)-2-(*p*-toluenesulfinyl)ethylbenzene (0.1 mmol) in THF (1 ml) added, stirred for 10 min, cooled to -98°, a soln. of phenyl(*p*-toluenesulfonyl)acetylene (2.2 eq.) in THF (1 ml) added, stirred until reaction complete (20-30 min), quenched with water, stirred at room temp. for 15 min, extracted with ether, and purified by flash chromatography → (+)-(S*S*,a*R*)-1-(4-phenyl-buta-2,3-dien-2-yl)-2-(*p*-tolylsulfinyl)benzene. Y 69% (d.e. 90%). The one-pot procedure was carried out on a 0.1 mmol scale for a limited set of substrates (three examples; Y 62-69%; d.e. 80-90%). Stepwise generation of the propargylic intermediates (scalable to 2 mmol) was demonstrated for electron-diverse aryl-, silyl- and *tert*-butyl-(*p*-toluenesulfonyl)acetylenes in reaction with carbanions of chiral 2-alkyl- (incl. but-3-en-1-yl-) sulfinyltoluenes (twelve examples; Y 49-84%; d.e. 84-96%; an ethyl-terminated acetylene was unreactive). Simple isomerization of the products under basic conditions afforded the corresponding chiral allenes (eleven examples; Y 45-83%; d.e. 50%, 62 to >96%; isomerization failed for a silyl-terminated acetylene). Removal of the chiral sulfinyl group was achieved (*tert*-butyllithium/-78°) without loss of optical purity either from the allenes (two examples; Y 86-92%) or from the propargylic intermediates (two examples; Y 50-59%). F.e.s. J.L. Garcia Ruano, L. Marzo, V. Marcos, C. Alvarado, J. Alemán, Chem. Eur. J. *2012, 18* (32), 9775-9 [DOI: 10.1002/chem.201201607]; general, transition metal-free synthesis of **acetylene derivs.** from α,β-acetylenesulfones and organolithium compds. by anti-Michael addition s. J.L. Garcia Ruano, J. Alemán, L. Marzo, C. Alvarado, M. Tortosa, S. Díaz-Tendero, A. Fraile, ibid. *18* (27), 8414-22 [DOI: 10.1002/chem.201200939]; s.a. L. Marzo, J. Alemán, J.L. Garcia Ruano, Phosphorus, Sulfur Silicon Relat. Elem. *2013*, *188* (4), 403-7 [DOI: 10.1080/10426507.2012.748054].

Lithium bis(trimethylsilyl)amide/boron fluoride/1,8-diazabicyclo[5.4.0]undec-7-ene ←

Julia olefination with lactones and 4-ω-sulfonyl-1,3-dioxolanes C=C

en route to hydroxyspiroketals s. *83*, 93

Cesium fluoride/p-toluenesulfonic acid/B-[phenyl(dimethyl)silyl]-4,4,5,5-tetramethyl-1,3,2-dioxaborolane *CsF/TsOH/PhMe$_2$Si-B(pin)*

α-(Carbo-*tert*-butoxyamino)carboxylic acid esters from α-(carbo-*tert*-butoxyamino)sulfones via silylborane-mediated carboxylation C(NHBoc)SO$_2$R → C(NHBoc)COOR′

437. 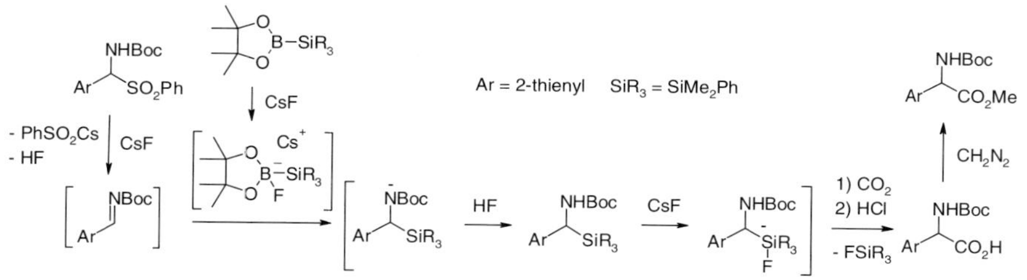

Use of silylboranes for preparation of the title compounds has been shown to offer a less toxic and more general alternative to the prior art using $Me_3SiSnBu_3$ and CsF [s. T. Mita et al., Angew. Chem., Int. Ed. *2011, 50* (6), 1393-6 [DOI: 10.1002/anie.201006422]], being applicable to both α-(het)aryl- and α-alkenyl-α-amidosulfones. **E:** *tert*-Butyl [2-thienyl(benzenesulfonyl)methyl]-carbamate (0.1 mmol), DMF (1.8 ml), a soln. of *p*-toluenesulfonic acid·H_2O (20 mol%) in the same solvent (0.2 ml) and PhMe$_2$Si-B(pin) (1.1 eq.) added to heat-gun-dried CsF (5 eq.), the mixture stirred in an autoclave under CO_2 (5 atm.) at room temp. for 3 h, diluted with water, acidified (pH 2) with 1 *M* aq. HCl, extracted with ether, concentrated *in vacuo*, the residue dissolved in ether (2 ml), cooled to 0°, a soln. of diazomethane (7 eq.) in the same solvent (4 ml) added, the mixture stirred for 10 min, quenched with acetic acid, concentrated *in vacuo*, and the residue purified by chromatography on silica-KF → methyl 2-(*tert*-butoxycarbonylamino)-2-(2-thienyl)-acetate. Y 75%. CsF was shown to activate the silylborane function as a base in the generation of an α-amidosilane intermediate, and activate the amidosilane to carboxy-desilylation. Reaction of electron-diverse α-(het)aryl-α-amidosulfones required the presence of a protic acid for efficient silylation, and afforded the title products in generally good yields (52%, 69-91%; eighteen examples) together with small amounts (1-6%) of N-benzylcarbamates (formed via proto-desilylation of intermediate α-amidosilanes). α-Alkenyl-α-amidosulfones gave somewhat lower yields (24-44%; five examples). F.e., optimization, substrate prepn., derivatization of products, and chromatography-free **resolution** of the products via precipitation of the crude amino acids with a chiral amine and subsequent recrystallization, s. T. Mita, J. Chen, M. Sugawara, Y. Sato, Org. Lett. *2012*, *14* (24), 6202-5 [DOI: 10.1021/ol302952r].

Triethylamine *Et$_3$N*

Diastereoselective double ring closure of α- or β-functionalized oxo compds. with [β-(trifluoromethyl)vinyl]diphenylsulfonium triflate ○

438. 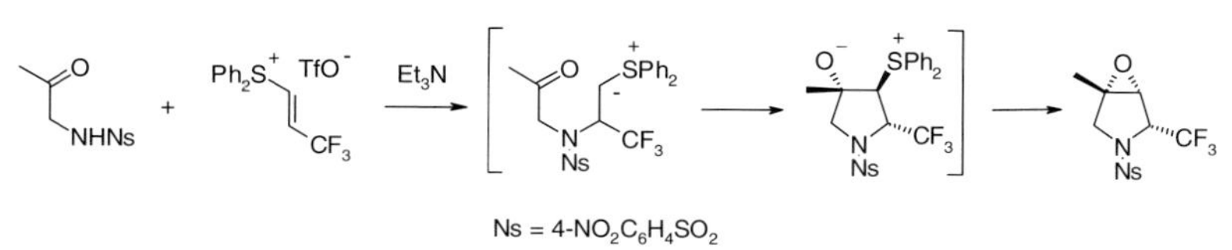

N-Protected 2-trifluoromethyl-6-oxa-3-azabicyclo[3.1.0]hexanes. Triethylamine (2 eq.) added to a soln. of 4-nitro-N-(2-oxopropyl)benzenesulfonamide (0.625 mmol) in methylene chloride (1.3 ml) under an inert atmosphere at room temp., the mixture stirred for 10 min, a soln. of [β-(trifluoromethyl)vinyl]diphenylsulfonium triflate (1.2 eq.) in the same solvent (0.3 ml) added dropwise, the mixture stirred for 24 h, quenched with 10% aq. citric acid, extracted with methylene chloride, washed with hexane (to remove Ph_2S), and concentrated → *cis*-1-methyl-3-(4-nitrophenylsulfonyl)-4-trifluoromethyl-6-oxa-3-azabicyclo[3.1.0]hexane. Y 90% (d.r. >20:1). This experimentally simple and efficient procedure effected cyclization of α-(t/n)osyl-amino-aldehydes and -ketones to afford the title compds. with good diastereocontrol (five examples; Y 49-90%; d.r. >20:1). An homologous β-tosylaminopropanal afforded a 7-oxa-3-azabicyclo[4.1.0]heptane after 36 h reaction (Y 72%; d.r. >20:1), and 2-tosylaminobenzaldehyde provided a benzo-fused deriv. (Y 45%; d.r. >20:1). Cyclization of α-hydroxyacetophenone afforded a 3,6-dioxabicyclo-[3.1.0]hexane (Y 88%; d.r. >20:1), but the corresponding thiol or homologous alcohol gave only the corresponding vinyl thioether (Y 91%) or ether (Y 63%) elimination products, respectively.

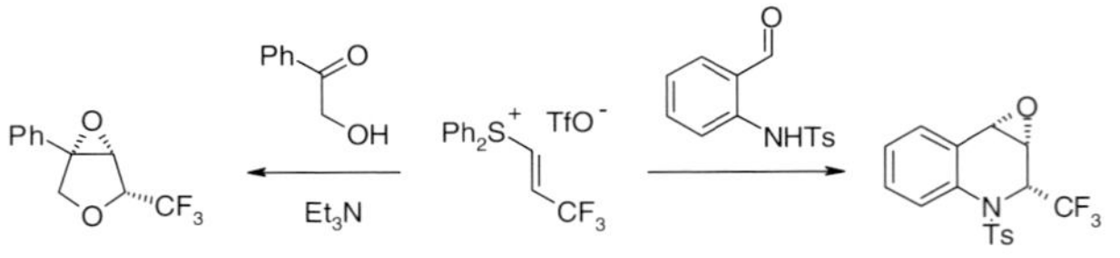

(Y 88%; d.r. >20:1) (Y 45%; d.r. >20:1)

Cyclization of chiral 3-(t/n)osylaminobutan-2-ones under these conditions gave 1:1 mixtures of diastereomers (Y 72-81%) and comparison with a similar cyclization using the des-trifluoromethylvinyl(diphenyl)sulfonium salt (cf. *57*, 390) suggested that CF_3 induced a diastereoselectivity of >92:1. F.e., optimization and reactions of products s. S.P. Fritz, T.H. West, E.M. McGarrigle, V.K. Aggarwal, Org. Lett. *2012*, *14* (24), 6370-3 [DOI: 10.1021/ol303200n]; analogous epoxide- or aziridine-fused heterocyclics with (2-bromoethyl)diphenylsulfonium triflate or with improved diastereoselectivity with a chiral sulfide-based analog s. S.P. Fritz, Z. Ali, M.G. Unthank, E.M. McGarrigle, V.K. Aggarwal, Helv. Chim. Acta *2012*, *95* (12), 2384-98 [DOI: 10.1002/hlca.201200455].

Copper(II) acetate s. under Pd(OAc)$_2$ *Cu(OAc)$_2$*

Gold(I) carbene complexes *[Au(I)]*
Synthesis of cyclopropanes from ethylene derivs. ▽
with sulfonium methylids s. *44*, 819s*48*; with a gold(I) carbene species generated in a catalytic cycle from an isolable [α-(imidazolio-2-ylsulfonyl)alkyl]gold(I) imidazol-2-ylidene salt s. D.H. Ringger, P. Chen, Angew. Chem., Int. Ed. *2013*, *52* (17), 4686-9 [DOI: 10.1002/anie.201209569].

Magnesium chloride s. under BuLi *MgCl$_2$*
Sodium tetrahydridoborate s. under BuLi *NaBH$_4$*
B-(Phenyldimethylsilyl)-4,4,5,5-tetramethyl-1,3,2-dioxaborolane s. under CsF *PhMe$_2$Si-B(pin)*
Boron fluoride s. under Lithium bis(trimethylsilyl)amide *BF$_3$*
Chiral quaternary 2-thioureidophosphonium salts s. under Potassium hydroxide ←
Air s. under Pd(OAc)$_2$ *O$_2$*
p-Toluenesulfonic acid s. under CsF *TsOH*

Tris(2,2′-bipyridyl)ruthenium(II) bis(hexafluorophosphate)/irradiation *[Ru(II)]/⫻*
3-Acylamino-1,1,1-trifluorides from ethylene derivs. and nitriles C=C → C(NHAc)C(CF_3)
via regioselective addition of trifluoromethyl radical under visible-light photoredox catalysis s. *83*, 318

Dichlorotris(2,2′-bipyridyl)ruthenium(II)/sodium carbonate/irradiation *[Ru(II)]/Na_2CO_3/⫻*
10a,11-Dihydro-10*H*-benzo[*b*]fluorenes ○
from *o*-allylarylacetylenes and arenesulfonic acid chlorides
Desulfitative double ring closure under visible-light photoredox catalysis

439. 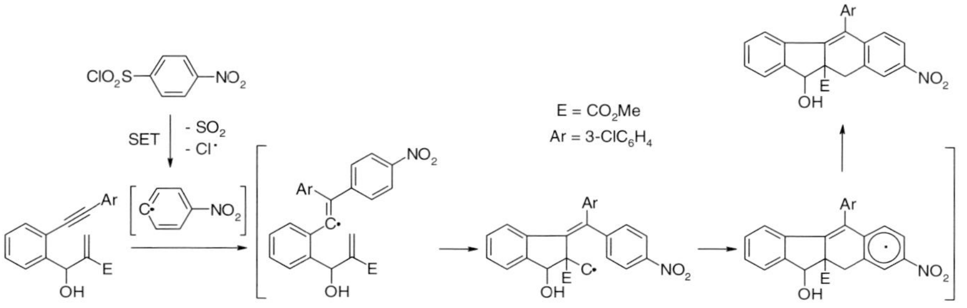

under mild conditions. A mixture of methyl 2-[[2-(3-chlorophenylethynyl)phenyl](hydroxy)-methyl]acrylate (0.3 mmol), *p*-nitrobenzenesulfonyl chloride (2 eq.), Ru(bpy)$_3$Cl$_2$ (5 mol%), Na_2CO_3 (1 eq.) and anh. acetonitrile (2 ml) stirred at 40° under argon for 18 h (TLC and GC-MS), cooled to room temp., and worked up with chromatographic purification → methyl 5-(3-chloro-phenyl)-11-hydroxy-8-nitro-10a,11-dihydro-10*H*-benzo[*b*]fluorene-10a-carboxylate. Y 88%. A series of *o*-2-prop-2-en-1-ylacetylenes, carrying H/OH at the 1- and H/CN/CO_2R/Ph at the 2-position of the prop-2-en-1-yl moiety, and terminated at the acetylene moiety with electron-diverse aryl, underwent double cyclization with electron-diverse arenesulfonyl chlorides (incl. 1-naphthalenesulfonyl chloride) to afford the title products in variable yields (40-88%; twenty-examples). Lower yields were obtained for 2-thienyl- (30%) and *n*-hexyl- (28%) terminated acetylenes. The method was compatible with nitrile, ester, alcohol, ketone, halo and ether functionality. F.e. and optimization s. G.-B. Deng, Z.-Q. Wang, J.-D. Xia, P.-C. Qian, R.-J. Song, M. Hu, L.-B. Gong, J.-H. Li, Angew. Chem., Int. Ed. *2013, 52* (5), 1535-8 [DOI: 10.1002/anie.201208380].

Palladium(II) acetate/copper(II) acetate/air *Pd(OAc)$_2$/Cu(OAc)$_2$/O_2*
Stereoselective Heck-type β-arylation of electron-deficient ethylene derivs. H → Ar
with sodium arenesulfinates s. *80,* 463s*83*

Palladium phosphine complexes ←
Transition metal-catalyzed desulfitative coupling ←
with alkali-metal arenesulfinates, update s. *80,* 463s*82*; **diarylmethanes** from Na-arenesulfinates and benzyl chlorides with Pd(OAc)$_2$/tri-*p*-tolylphosphine in heated cyclohexane s. F. Zhao, Q. Tan, F. Xiao, S. Zhang, G.-J. Deng, Org. Lett. *2013, 15* (7), 1520-3 [DOI: 10.1021/ol400295z]; **biaryls** by coupling with aryl triflates using Pd(OAc)$_2$/XPhos s. C. Zhou, Q. Liu, Y. Li, R. Zhang, X. Fu, C. Duan, J. Org. Chem. *2012, 77* (22), 10468-72 [DOI: 10.1021/jo302005s]; **3-arylquinolines** by coupling with 3-iodoquinolines using Pd$_2$(dba)$_3$/DavePhos/Bu$_4$NCl/K_2CO_3 s. J. Colomb, T. Billard, Tetrahedron Lett. *2013, 54* (11), 1471-4 [DOI: 10.1016/j.tetlet.2013.01.018]; **2-aryl-furans and -thiophenes** from ar. bromides with lithium 2-furyl- and 2-thienyl-sulfinates, respectively, *in water* or aq. DMF using PdCl$_2$/triphenylphosphine s. S. Sévigny, P. Forgione, New J. Chem. *2013, 37* (3), 589-92 [DOI: 10.1039/c3nj41153b]; (E)-selective Heck-type **oxidative β-arylation** of electron-deficient alkenes with Na-arenesulfinates using Pd(OAc)$_2$/Cu(OAc)$_2$ under air s. S. Hu, P. Xia, K. Cheng, C. Qi, Appl. Organomet. Chem. *2013, 27* (3), 188-90 [DOI: 10.1002/aoc.2970]; one-pot synthesis of **2-arylfurans** by coupling with *o*-hydroxy-β,β-dibromostyrenes and subsequent cyclodehydrobromination, also benzo[*b*]thiophene analogs from the corresponding thiophenols, using PdCl$_2$/Cu(OAc)$_2$/Bu$_4$NF/Et$_3$N s. W. Chen, P. Li, T. Miao, L.-G. Meng, L. Wang, Org. Biomol. Chem. *2013, 11* (3), 420-4 [DOI: 10.1039/c2ob27232f]; alternative ring closure with triaryl-bismuthines in place of arenesulfinates using Pd(PPh$_3$)$_4$/Cs_2CO_3 cf. M.L.N. Rao, D.N. Jadhav, P. Dasgupta, Eur. J. Org. Chem. *2013* (4), 781-8 [DOI: 10.1002/ejoc.201201314].

Remaining Elements ↑ CC ↓↑ Rem

Without additional reagents *w.a.r.*

Allylboration CO → C(OH)C-C=C

update s. *33,* 865s*82*; regio- and *anti*-selective crotylboration of the β-keto function of α,β-diketocarboxylic acid amides with (E)-crotylboronic acid pinacol ester, and *anti*-selective Lewis acid-catalyzed crotylboration of the α-keto function of α,β-diketocarboxylic acid esters or amides with BF_3-etherate, s. J. Roßbach, J. Baumeister, K. Harms, U. Koert, Eur. J. Org. Chem. *2013* (4), 662-5 [DOI: 10.1002/ejoc.201201517]; *anti-* and (E)-selective allylboration of aldehydes with chiral α-subst. allyl- and crotyl-boronic acid pinacol esters *with asym. induction* to give chiral (E)-*anti*-products via allyl(*n*-butyl)boronic acid esters [generated *in situ* with *n*-BuLi/trifluoroacetic anhydride], also chiral (Z)-*anti*-products directly from the pinacol esters, s. J.L.-Y. Chen, H.K. Scott, M.J. Hesse, C.L. Willis, V.K. Aggarwal, J. Am. Chem. Soc. *2013, 135* (14), 5316-9 [DOI: 10.1021/ja401564z]; synthesis of *anti*-3(E)-ethylene-4-silylalcohols by allylboration of aldehydes with β,γ-ethylene-α-silylboranes with asym. induction s. M. Chen, W.R. Roush, Org. Lett. *2013, 15* (7), 1662-5 [DOI: 10.1021/ol4004405]; allylboration of ketones, azomethines and hydrazones with allylboronates *under zinc catalysis* with Zn-bis(trimethylsilyl)amide via B→Zn-transmetalation s. Y. Cui, W. Li, T. Sato, Y. Yamashita, S. Kobayashi, Adv. Synth. Catal. *2013, 355* (6), 1193-205 [DOI: 10.1002/adsc.201201134]; regio- and diastereo-selective *uncatalyzed* allylboration of ketones **with β,γ-ethyleneboronic acids** s. M. Raducan, R. Alam, K.J. Szabó, Angew. Chem., Int. Ed. *2012, 51* (52), 13050-3 [DOI: 10.1002/anie.201207951]; 1,1-diallyllactones from dicarboxylic acid esters, and sym. 4-*prim*-amino-1,6-dienes from nitriles, with potassium allyl(trifluoro)borates by **double allylboration** using BF_3-etherate as Lewis acid s. T.R. Ramadhar, J. Bansagi, R.A. Batey, J. Org. Chem. *2013, 78* (3), 1216-21 [DOI: 10.1021/jo302194n]; allylboration of aldehydes with the same allylation agents using the recoverable, water-tolerant lanthanide(III)-based metal-organic framework, $[Eu_2(fum)_3(H_2O)_4]\cdot(3H_2O)$, as catalyst s. J.C.R. Freitas, C.K. de Oliveira, E.C. Cunha, I. Malvestiti, S. Alves Jr., R.L. Longo, P.H. Menezes, Tetrahedron Lett. *2013, 54* (12), 1558-61 [DOI: 10.1016/j.tetlet.2013.01.035]; **asym. allylboration** of ketones with 2-methylallylboronic acid esters using highly active and enantioselective (S)-3,3′-difluoro-1,1′-bi-2-naphthol as catalyst (cf. *33,* 865s*81*) s. Y. Zhang, N. Li, B. Qu, S. Ma, H. Lee, N.C. Gonnella, J. Gao, W. Li, Z. Tan, J.T. Reeves, J. Wang, J.C. Lorenz, G. Li, D.C. Reeves, A. Premasiri, N. Grinberg, N. Haddad, B.Z. Lu, J.J. Song, C.H. Senanayake, Org. Lett. *2013, 15* (7), 1710-3 [DOI: 10.1021/ol400498a].

Irradiation s. under In(Hal)₃ ⫻

Electrolysis/palladium(II) acetate/copper(I) iodide/p-benzoquinone/triethylamine/ tri-tert-butylphosphine·fluoroboric acid/cesium carbonate ←

1,4-Bis(biaryl)-1,3-butadiynes Ar′-Ar-C≡C-C≡C-Ar-Ar′

from two (bromoaryl)ethyne molecules and two arylboronic acid molecules

Site-selective sequential coupling controlled by ‘electrochemical reaction site switching’

440.

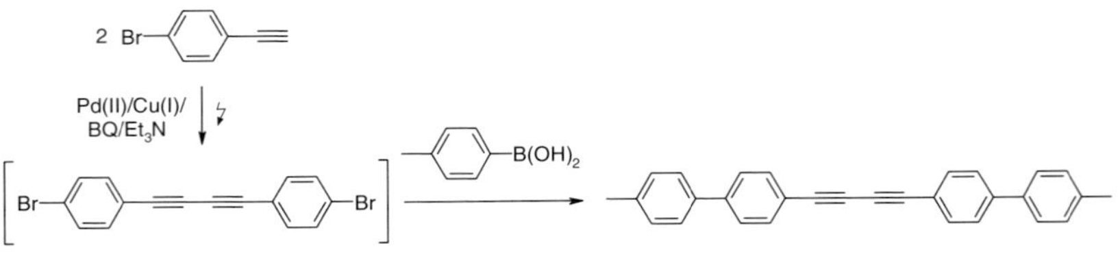

Site-selective sequential coupling reactions are reported in which the reaction site may be controlled completely by the on/off application of electricity. Thus, sym. 1,4-bis(biaryl)-1,3-butadiynes may be obtained from (bromoaryl)ethynes in high yields by a novel initial *electro-oxidative homo-coupling* of the ethynyl group mediated by *p*-benzoquinone, followed by *Suzuki-Miyaura coupling* (with electricity off). **E:** A soln. of 4-bromophenylacetylene (0.5 mmol), $Pd(OAc)_2$ (5 mol%), CuI (20 mol%), *p*-benzoquinone (15 mol%) and Et_3N (2 eq.) in a 0.1 *M* $KClO_4$ soln. in acetonitrile/water (7:1; 5 ml) placed in the anodic chamber of an H-type divided cell (glass filter) equipped with two Pt-electrodes using a 0.1 *M* $KClO_4$ soln. in acetonitrile/water (7:1; 5 ml) as catholyte, a constant current (10 mA; 2 F mol^{-1}) supplied at 30° with stirring under argon, the contents of the anodic chamber poured into a Schlenk flask, DME (5 ml), Cs_2CO_3 (3 mmol), *t*-$Bu_3P{\cdot}HBF_4$ (0.05 mmol), and *p*-tolylboronic acid (1.5 eq.) added, the mixture refluxed for 24 h, cooled to room temp., concentrated under reduced pressure, filtered, the filter cake washed with methanol, water and ethyl acetate, and the crude mixture purified by column chromatography on silica gel → 1,4-bis(4′-methyl-4-biphenyl)buta-1,3-diyne. Y 95%. Nine further examples gave yields of 47-86% from arylboronic acids bearing alkyl, MeO, Me_2N, F, MeCO, CN or styryl groups or a naphthyl analog. For sequential palladium-catalyzed electrochemical iodination-Suzuki biaryl coupling of 2-arylpyridines cf. *74,* 516s*82*. F.e. and optimization, also conversion to **2,5-bis(biaryl)-thiophenes** with $Na_2S{\cdot}9H_2O$ in DMF at 150° (five examples; Y 26%, 42-78%), s. K. Mitsudo, N. Kamimoto, H. Murakami, H. Mandai, A. Wakamiya, Y. Murata, S. Suga, Org. Biomol. Chem. *2012, 10* (48), 9562-9 [DOI: 10.1039/c2ob26567b].

sec- or tert-Butyllithium *s-BuLi or t-BuLi*
Generation of arynes from *o*-triflyloxyboronic acid esters ←

441.

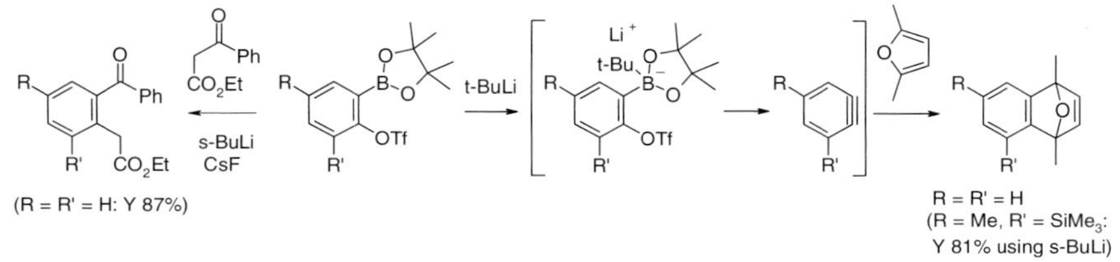

Arynes may be generated efficiently **via formation of boron-ate complexes** from *o*-triflyloxy-boronic acid esters with *sec-* or *tert*-butyllithium at low temperature followed by warming above 0°. The precursors are readily available, for example from phenols via iridium-catalyzed *o*-boryl-ation. **E:** A soln. of 2-(4,4,5,5-tetramethyl-1,3,2-dioxaborolan-2-yl)phenyl triflate (0.2 mmol) and 2,5-dimethylfuran (3 eq.) in ether (2 ml) treated with *t*-BuLi (1.54 *M* in pentane; 1.2 eq.) at -78°, the mixture allowed to warm to 25°, stirred for 30 min, treated with aq. phosphate buffer (pH 7), extracted with ethyl acetate, and purified by silica gel chromatography → 1,4-dihydro-1,4-dimethyl-1,4-epoxynaphthalene. Y 95%. Five further examples of Diels-Alder trapping with furans or pyrroles gave yields of 69-100%. Six examples with benzyl azide, diazomethyl(trimethyl)-silane, a nitrone, a ketene silyl acetal or a β-keto ester (with added CsF to facilitate enolization) as arynophile gave yields of 58-96%; arynophiles bearing base-sensitive groups may be used by adding them to the reaction mixture after formation of the boron-ate complex. Reaction of 2-(4,4,5,5-tetramethyl-1,3,2-dioxaborolan-2-yl)-6-trimethylsilyl-4-tolyl triflate selectively generated the 3-silylaryne on treatment with *s*-BuLi *or n-Bu_4NF* (Y 81% or 66% respectively for the 2,5-dimethylfuran adduct), while reaction of a 1:1 mixture of *o*-borylphenyl triflate and *o*-trimethylsilylphenyl triflate selectively generated the aryne from the former (Y of adduct 92%) with quantitative recovery of the latter with *t*-BuLi, or from the latter (Y of adduct 87%) with quantitative recovery of the former with CsF (4 eq.). F.e. and gram-scale reaction s. Y. Sumida, T. Kato, T. Hosoya, Org. Lett. *2013, 15* (11), 2806-9 [DOI: 10.1021/ol401140d].

Lithium diisopropylamide *i-Pr_2NLi*
Desilylative aldol condensation with α-diazo-α-(triethylsilyl)acetone ←
en route to α-diazo-β,β′-dihydroxyketones s. *83,* 259

Potassium carbonate/tetra-n-butylammonium bromide/palladium(II) acetate/ 1,3-bis(diphenylphosphino)propane ←
3-Component O-alkylative Wittig synthesis-Heck arylation s. *13,* 820s*83* ←

Alkali-metal fluoride *F⁻*
Annelation with arynes ○
generated from *o*-silyltriflates with fluoride ion, update s. *68,* 464s*82*; **Diels-Alder reaction** with *o*-benzoquinones using KF/18-crown-6 s. T. Kaicharla, S.S. Bhojgude, A.T. Biju, Org. Lett. *2012,*

14 (24), 6238-41 [DOI: 10.1021/ol302997q]; **naphthalene ring** by coupling with 1,3-diene-2,3-dicarboxylic acid esters under palladium catalysis using $Pd_2(dba)_3$/IMes·HCl/CsF s. R.M. Patel, N.P. Argade, ibid. *2013, 15* (1), 14-7 [DOI: 10.1021/ol3028658]; **1,3-dipolar cycloaddition** with carbohydrate-based Δ^1-pyrroline N-oxides and subsequent reductive N-O cleavage to give the corresponding 2-(*o*-hydroxyaryl)pyrrolidines s. R.K. Khangarot, K.P. Kaliappan, Eur. J. Org. Chem. *2012* (29), 5844-54 [DOI: 10.1002/ejoc.201200685]; one-pot synthesis of 2-(*o*-hydroxyaryl)-pyridines via cycloaddition of benzynes with pyridine N-oxides using excess of Bu_4NF s. B.S. Shaibu, R.K. Kawade, R.-S. Liu, Org. Biomol. Chem. *2012, 10* (34), 6834-9 [DOI: 10.1039/c2ob26130h]; bridged or angular tetracyclics from *o*-acetylene-N-tosylhydrazones **and pyridynes** with AgOTf/CsF/$BnEt_3NCl$ s. L. Jiang, X. Yu, B. Fang, J. Wu, ibid. *10* (40), 8102-7 [DOI: 10.1039/c2ob26379c].

Potassium fluoride s. under Copper(I) triflate, CuI, Dichlorobis(triphenylphosphine)-ruthenium(II), Bis(dibenzylideneacetone)palladium(0) and Tris(dibenzylideneacetone)dipalladium *KF*

Cesium fluoride [s.a. under Tris(dibenzylideneacetone)dipalladium] *CsF*
Carbazoles from ar. nitroso compds. and benzynes s. *83*, 450 ○

N-(4-Chloropicolinoyl)-9-amino-9-deoxy-epi-cinchonidine s. under Copper(II) acetate ←
Silver(I) oxide s. under $[Cp^*RhCl_2]_2$ Ag_2O

Chiral copper(II) bis(Δ²-oxazoline) complexes $[Cu(II)]^*$
Asym. Michael-type addition C=C → CHC(R)
to 2-(α,β-ethyleneacyl)pyridine N-oxides s. *41,* 111s*83*

Chiral copper(II) bis(Δ²-oxazoline) complexes/2,2,2-trifluoroethanol $[Cu(II)]^*/CF_3CH_2OH$
Asym. vinylogous aldol-type condensation of acylphosphonic acid esters s. *66,* 452s*83* ←

Copper(II) acetate s.a. under $Pd(OAc)_2$ $Cu(OAc)_2$

Copper(II) acetate/N-(4-chloropicolinoyl)-9-amino-9-deoxy-epi-cinchonidine/trimethylsilanol ←
Asym. vinylogous Mannich reaction of ketimines ←

442.

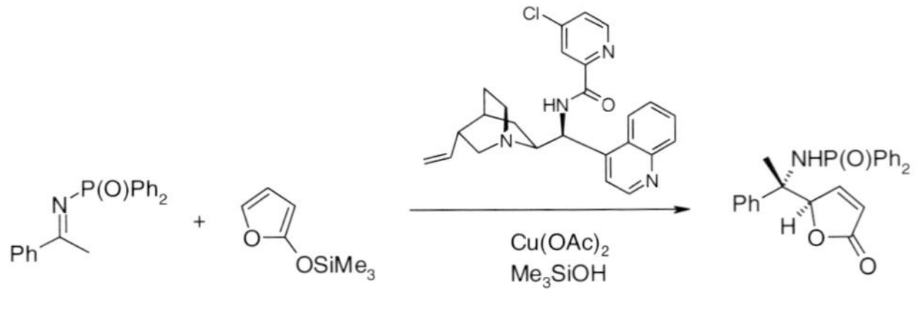

The first efficient diastereoselective asym. synthesis of the δ,δ-disubst. δ-amino-α,β-ethylene-γ-oxycarbonyl system has been reported by asym. vinylogous Mannich reaction of simple, unactivated N-phosphinylketimines **with 2-siloxyfurans** in the presence of a novel, readily accessible cinchona-based amide and $Cu(OAc)_2$ as cocatalyst. **E:** 2-(Trimethylsiloxy)furan (4 eq.) and trimethylsilanol (1.2 eq.) added at -20° to a soln. of the startg. N-phosphinylketimine (0.1 mmol), $Cu(OAc)_2$ (10 mol%) and N-(4-chloropicolinoyl)-9-amino-9-deoxy-*epi*-cinchonidine (10 mol%) in THF (0.2 ml), the mixture stirred for 48 h, diluted with ethyl acetate, quenched with silica gel, filtered through Celite, and worked up with purification by chromatography on silica gel → (R)-5-[(S)-1-(diphenylphosphinylamino)-1-phenylethyl]-2(5*H*)-furanone. Y 85% (*syn/anti* 1:99; e.e. 97%). The face-selectivity was reversed with the cinchonine analog of the organocatalyst. The procedure is applicable to the coupling of 2-(trimethylsiloxy)furan and its 3- and 4-methylated derivs. with a wide range of N-phosphinylimines derived from dialkyl ketones, electron-diverse aryl methyl ketones and heteroaromatic analogs (eighteen examples; Y 51-99%; *anti/syn* 96:4 to 99:1; e.e. 91-97%). The yield was low, however, with propiophenone imines and there was no reaction with hindered 5-methyl-2-(trimethylsiloxy)furan. Reaction is thought to involve initial desilylation of 2-siloxyfuran with a complex formed from the two reagents, which then bind the liberated 2-hydroxyfuran and the phosphinylimine by coordination with two oxygen atoms and three nitrogen atoms; the dienolate then attacks the imine by approaching the *Re* face, thereby minimizing steric repulsion. F.e. and comparison of organocatalysts, metal acetates and alcoholic additives s. M. Hayashi, M. Sano, Y. Funahashi, S. Nakamura, Angew. Chem., Int. Ed. *2013, 52* (21), 5557-60 [DOI: 10.1002/anie.201301917].

Copper(II) acetate/imidazole/2,4,6-collidine ←
Copper(II) trifluoroacetate *$Cu(OCOCF_3)_2$*
Introduction of the trifluoromethyl group CF_3
cf. *78,* 476s*82*; **trifluoromethylarenes** from arylboronic acids and Na-trifluoromethanesulfinate under copper catalysis with $Cu(OAc)_2$/imidazole/2,4,6-collidine and *tert*-butyl hydroperoxide as oxidant, also 2-ethylene-1,1,1-trifluorides from α,β-ethyleneboronic acids, s. Y. Li, L. Wu, H. Neumann, M. Beller, Chem. Commun. *2013, 49* (26), 2628-30 [DOI: 10.1039/c2cc36554e]; trifluoromethylarenes from potassium aryl(trifluoro)borates and 1-trifluoromethyl-3,3-dimethyl-1,2-benziodoxole [Togni′s reagent] with $Cu(OCOCF_3)_2$ as catalyst cf. Y. Huang, X. Fang, X. Lin, H. Li, W. He, K.-W. Huang, Y. Yuan, Z. Weng, Tetrahedron *2012, 68* (48), 9949-53 [DOI: 10.1016/j.tet.2012.09.083]; from 1-aryl-3,3-diisopropyltriazenes via iodination with methyl iodide and subsequent treatment with Me_3SiCF_3 and CuI/KF as promoter, also synthesis of other polyfluoroarenes, s. A. Hafner, S. Bräse, Adv. Synth. Catal. *2013, 355* (5), 996-1000 [DOI: 10.1002/adsc.201201040]; metal-free **oxidative ar. trifluoromethylation** of unactivated arenes with Me_3SiCF_3 in the presence of phenyl iodosoacetate/benzoquinone/K_3PO_4 s. X. Wu, L. Chu, F.-L. Qing, Tetrahedron Lett. *2013, 54* (3), 249-51 [DOI: 10.1016/j.tetlet.2012.11.011]; **directed *o*-trifluoromethylation** of anilides with 5-(trifluoromethyl)dibenzothiophenium fluoroborate [Umemoto′s reagent] and $Pd(OAc)_2$/$Cu(OAc)_2$/pivalic acid s. L.-S. Zhang, K. Chen, G. Chen, B.-J. Li, S. Luo, Q.-Y. Guo, J.-B. Wei, Z.-J. Shi, Org. Lett. *2013, 15* (1), 10-3 [DOI: 10.1021/ol302814x]; 2-acetylene-1,1,1-trifluorides from K-(alk-1-ynyl)trifluoroborates and Togni′s reagent using a ligated CuSCN complex as catalyst s. H. Zheng, Y. Huang, Z. Wang, H. Li, K.-W. Huang, Y. Yuan, Z. Weng, Tetrahedron Lett. *2012, 53* (49), 6646-9 [DOI: 10.1016/j.tetlet.2012.09.083].

Copper(II) isobutyrate/chiral 1,2-bis(2,5-dimethylphospholano)ethane/lithium tert-butoxide ←
Asym. propargylboration of oxo compds. CO → C(OH)C-C≡C
update s. *75,* 265s*82*; of aldehydes and ketones with propargylboronates using Cu(II)-isobutyrate/LiOBu-*t* and chiral 1,2-bis(2,5-dimethylphospholano)ethane as ligand, also regioselective asym. synthesis of the isomeric 2-allenealcohols with chiral 1,2-bis(2,5-diphenylphospholano)ethane as ligand, s. K.R. Fandrick, J. Ogikubo, D.R. Fandrick, N.D. Patel, J. Saha, H. Lee, S. Ma, N. Grinberg, C.A. Busacca, C.H. Senanayake, Org. Lett. *2013, 15* (6), 1214-7 [DOI: 10.1021/ol400124f]; asym. synthesis of *anti*-3-acetylenealcohols from aldehydes and α-alleneboronates *with kinetic resolution* using a chiral 3,3-diaryl-1,1′-binaphthyl-2,2′-diyl hydrogen phosphate as Brønsted acid s. A.S. Tsai, M. Chen, W.R. Roush, ibid. *15* (7), 1568-71 [DOI: 10.1021/ol4003459]; from diaryl ketones with AgF and a chiral ferrocenyldi(phosphine) [WalPhos-8] as ligand, also conversion to 2-allenealcohols without ligand, s. B.L. Kohn, N. Ichiishi, E.R. Jarvo, Angew. Chem., Int. Ed. *2013, 52* (16), 4414-7 [DOI: 10.1002/anie.201206971].

Copper(I) thiophene-2-carboxylate s. under $Pd(PPh_3)_4$ *CuTC*

Copper(I) triflate (or iodide)/3,4,7,8-tetramethyl-1,10-phenanthroline/silver fluoride/ silver tosylate (or fluoroborate)/potassium fluoride ←
1,1,1-Trifluorides from alkylboronic acids $B(OH)_2 \rightarrow CF_3$
Synthesis with addition of one C-atom

443. 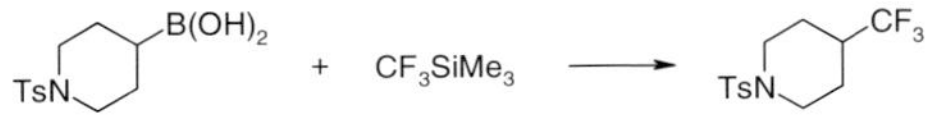

A general approach to the challenging formation of $C(sp^3)$-CF_3 bonds has been achieved by the novel copper(I)-catalyzed coupling of the Ruppert-Prakash reagent with *prim-* or *sec*-alkylboronic acids. **E:** Startg. *sec*-alkylboronic acid (0.2 mmol), AgOTs (1 eq.) and 3,4,7,8-tetramethyl-1,10-phenanthroline (40 mol%) added to a septum-sealed tube containing KF (3 eq.), AgF (3 eq.) and $CuOTf{\cdot}0.5C_6H_6$ (10 mol%), the tube evacuated and filled with argon (three times), DMF (1 ml) and (trifluoromethyl)trimethylsilane (3 eq.) added sequentially by syringe under argon, the mixture stirred for 24 h at 50°, water added at room temp., and the mixture worked up with purification by chromatography on silica gel → product. Y 54%. Conversion of marginally less challenging *prim*-alkylboronic acids was achieved with less expensive and more stable CuI in place of CuOTf, and with $AgBF_4$ in place of AgOTs. The procedure is generally applicable to the coupling of a wide range of *prim*-alkylboronic acids and (a)cyclic *sec*-alkylboronic acids, tolerating a number of common functional groups, incl. ether, ester, ketone, amide and protected amine (twenty-two examples in all; Y 35-78%). F.e. and comparison of copper and silver salts, ligands and bases s. J. Xu, B. Xiao, C.-Q. Xie, D.-F. Luo, L. Liu, Y. Fu, Angew. Chem., Int. Ed. *2012*, *51* (50), 12551-4 [DOI: 10.1002/anie.201206681].

Copper(II) triflate s. under $(MeCN)_2Pd(OTs)_2$ *$Cu(OTf)_2$*
Copper(I) chloride s. under Bis(dibenzylideneacetone)palladium(0) and Palladium(II) phosphine complexes *CuCl*
Copper(I) iodide s.a. under Electrolysis, Copper(I) triflate and Chloro-(1,5-cyclooctadiene)rhodium(I) dimer *CuI*

Copper(I) iodide/potassium fluoride *CuI/KF*
Introduction of the trifluoromethyl group s. *78*, 476s*83* CF_3

Silver fluoride/silver tosylate (or fluoroborate) s. under Copper(I) triflate *Ag^+*

Silver fluoride/chiral ferrocenyldi(phosphines) ←
Asym. synthesis of terminal 3-acetylene-*tert*-alcohols from ketones $CO \rightarrow C(OH)CH_2C{\equiv}CH$
and alleneboronic acid esters s. *75*, 265s*83*

Magnesium tert-butoxide/chiral 3,3′-bis(polyfluoroaryl)-1,1′-bi-2-naphthols ←
Asym. 1,4-addition of α,β-ethyleneboronic acids C=C → CHC-C=C
to heteroaryl α,β-ethyleneketones s. *55*, 452s*83*

Zinc bis(trimethylsilyl)amide *$Zn[N(SiMe_3)_2]_2$*
Zinc-catalyzed allylboration of C=O and C=N bonds C=X → C(XH)C=C=C
s. *33*, 865s*83*

Tetramethylammonium hydridotriacetoxoborate s. under Chlorobis(cyclooctene)rhodium(I) dimer *$Me_4NBH(OAc)_3$*
B-Methoxy-9-borabicyclo[3.3.1]nonane s. under Palladium(II) acetate *9-MeO-BBN*

Phenylboron dichloride/potassium carbonate *$PhBCl_2/K_2CO_3$*
4-Ethylenealcohols from aldehydes CHO → CH(OH)C-C-C=C
Synthesis with addition of five C-atoms with chirality transfer

444.

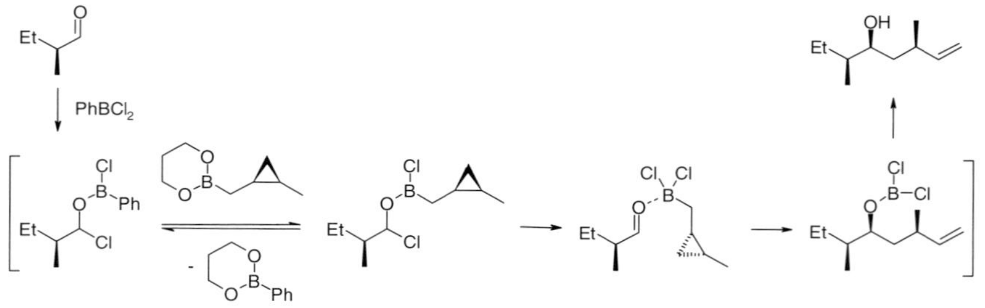

The first asym. *syn*-homocrotylation of aldehydes is reported, using a novel chiral 2-methylcyclopropylmethylboronate and based on a previously described diastereoselective homocrotylating (and homoallylating) procedure [s. W. Pei, I.J. Krauss, J. Am. Chem. Soc. *2011*, *133* (46), 18514-7

[DOI: 10.1021/ja2048122]]. **E:** $PhBCl_2$ (1.5 eq.) and K_2CO_3 (6 eq.) added to a soln. of propane-1,3-diyl (1S,2R)-2-methylcyclopropylmethylboronate (3 eq.) in methylene chloride (2 ml), (S)-2-methylbutanal (0.1 mmol) added at room temp. under N_2, the mixture stirred for 90 h, quenched with 3 *M* aq. NaOH, extracted with ethyl acetate, and purified by flash chromatography → (3R,5S,6S)-3,6-dimethyloct-1-en-5-ol. Y 76% (d.r. >99:1). The novel chiral homocrotylating agent afforded *syn*-adducts with non-chiral aliphatic aldehydes, with high enantioselectivity (e.e. 97-98%; seven examples; Y 62-89%). Asym. addition to chiral aliphatic aldehydes occurred with complete reagent control (cf. face control), affording the Felkin products with the illustrated chiral boronate (two examples; Y 76-83%; d.r. >20:1 to >99:1) and *anti*-Felkin products with its enantiomer (two examples; Y 78-81%; d.r. >20:1 to >99:1). F.e., optimization and multigram synthesis of the reagent s. H. Lin, W. Pei, H. Wang, K.N. Houk, I.J. Krauss, J. Am. Chem. Soc. *2013*, *135* (1), 82-5 [DOI: 10.1021/ja311061n].

Boron fluoride *BF_3*

o-Allyliodides from aryl iodosoacetates and allylsilanes via regioselective reductive iodonio-Claisen rearrangement ←

445.

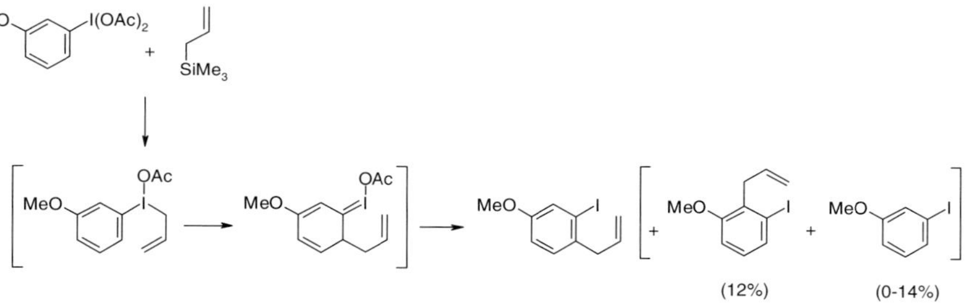

Allyltrimethylsilane (1.5 eq.) added to a soln. of *m*-methoxyphenyl iodosoacetate (0.6 mmol) in 1:1 acetonitrile/methylene chloride, cooled to -50°, BF_3-etherate (1.2 eq.) added dropwise, the mixture stirred for 2 h, quenched with satd. $NaHCO_3$, and worked up with purification by chromatography on silica gel → 2-iodo-4-methoxy-1-(prop-2-en-1-yl)benzene. Y 74% [plus 1-iodo-3-methoxy-2-(prop-2-en-1-yl)benzene: Y 12%]. Based on deuterium labelling studies, reaction is believed to involve initial Lewis acid-catalyzed generation of an allyl(aryl)iodine(III) species which undergoes reductive iodonio-Claisen rearrangement prior to re-aromatization, affording a mixture of regioisomeric *o*-allyliodides with predominant formation of the less hindered product. Depending largely on the electronic character and position of the substituent on the aromatic ring, reaction is accompanied to a greater or lesser extent by formation of aryl iodide via reductive elimination from the allyl(aryl)iodine species. This, however, is minimized with substrates bearing an electron-donating group at the *meta* position, e.g. alkoxy, acylamino or carbalkoxyamino (fourteen examples; Y 59-93%), the highest proportion of rearranged product being recorded with 3-methoxy-5-methylphenyl iodosoacetate. F.e. and comparison of Lewis acids, solvents and iodine(III) compds. s. H.R. Khatri, J. Zhu, Chem. Eur. J. *2012*, *18* (39), 12232-6 [DOI: 10.1002/chem.201202049].

Vinylogous aldol-type condensation ←

with asym. induction s. *53*, 453s*82*; by *syn*-selective Kobayashi coupling **of acetals** with chiral 1-(oxazolid-2-on-3-yl)-1-siloxy-1,3-dienes with *remote* asym. induction under Lewis acid catalysis with BF_3-etherate s. H. Tsukada, Y. Mukaeda, S. Hosokawa, Org. Lett. *2013*, *15* (3), 678-81 [DOI: 10.1021/ol303519y]; diastereoselective **asym. vinylogous aldol-type condensation** of acylphosphonic acid esters with 2-siloxyfurans using a chiral copper(II) bis(Δ^2-oxazoline) complex and 2,2,2-trifluoroethanol as additive (cf. *66*, 452s*79*) s. G. Hou, J. Yu, C. Yu, G. Wu, Z. Miao, Adv. Synth. Catal. *2013*, *355* (2-3), 589-93 [DOI: 10.1002/adsc.201200810].

Boron fluoride/s-collidine *BF_3/2,4,6-Me_3Py*

α,β-Ethylenelactones from lactolides and siloxyacetylenes with ring expansion by two C-atoms

446. 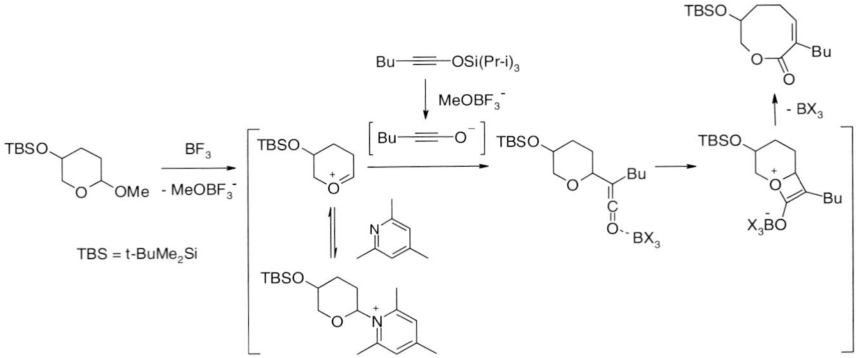

Two carbon ring expansion of cyclic α-methoxy/acetoxy-ethers via Lewis acid promoted generation of cyclic oxonium salts, and subsequent [2+2]-cycloaddition with electron-rich siloxyalkynes, is reported. This energetically favorable approach does not require high dilution or slow addition strategies, and provides access to 7- to 18-membered ring lactones. **E:** BF_3-etherate (2 eq.) added to a mixture of 5-*tert*-butyldimethylsiloxy-2-methoxytetrahydropyran (0.25 mmol), 1-triisopropyl-siloxyhex-1-yne (1.2 eq.), *s*-collidine (1 eq.) and anh. methylene chloride (2.5 ml) under N_2 at 0°, the mixture stirred at room temp. until reaction complete (TLC), quenched with satd. aq. $NaHCO_3$, extracted with methylene chloride, and purified by flash chromatography → (Z)-3-butyl-7-(*tert*-butyl-dimethylsiloxy)-5,6,7,8-tetrahydro-2*H*-oxocin-2-one. Y 75%. This mild and effective method provided 7- to 9-membered ring enelactones as single (Z)-isomers (Z/E >20:1; twenty-three examples; Y 56-92%) in the presence of silyl ether, ester, azide, allyl and propargyl ethers, and appears unaffected by sterically-hindering groups in the acetal substrates. A 10-membered ring was formed as a 1:1 mixture of stereoisomers (Y 88%) and an 18-membered ring as a single E-isomer (E/Z >20:1; Y 87%), but a highly oxygenated sugar deriv. was unreactive. An **iterative ring expansion** strategy was demonstrated starting from 2-acetoxy-tetrahydrofuran, via reduction and O-acylation [1) H_2/Pd-C; 2) *i*-Bu_2AlH/Ac_2O] of the first ethylenelactone, and a second ring-expansion affording the [5+2+2]-product in 41% overall yield for the four-steps. *s*-Collidine is thought to react with the initially generated oxonium ion to provide a stabilized source of the highly reactive intermediate. F.e., optimization and substrate prepn. s. W. Zhao, Z. Li, J. Sun, J. Am. Chem. Soc. *2013*, *135* (12), 4680-3 [DOI: 10.1021/ja400883q].

Indium(III) halides/irradiation *In(Hal)$_3$/⫻*

Radical Stille-type coupling with iodomethylphosphorus(V) compds. s. *34*, 862s*83* Sn⩽ → R

Chiral cationic bis(α-aminocarboxylate)lanthanide(III) complexes *[Ln(III)]**

Asym. aldol-type condensation CO → C(OH)C-CO

under lanthanide(III) catalysis cf. *47*, 885s*82*; of aliphatic or aromatic aldehydes with ketone-derived enoxysilanes *in aq. medium* with a chiral cationic bis(α-aminocarboxylate)lanthanide(III) complex s. Y. Mei, D.J. Averill, M.J. Allen, J. Org. Chem. *2012*, *77* (13), 5624-32 [DOI: 10.1021/jo300800b].

Lanthanide(III)-based metal-organic frameworks *Ln(III)-MOF*

Allylboration of aldehydes with potassium allyl(trifluoro)borates CHO → CH(OH)C-C=C

s. *33*, 865s*83*

p-Benzoquinone s. under Electrolysis and Phenyl iodosoacetate *BQ*

2(S)-[Diphenyl(trimethylsiloxy)methyl]pyrrolidine/p-nitrobenzoic acid ←
Organocatalyzed asym. vinylogous Michael-type addition of (1,3-dien)oxysilanes ←

447.
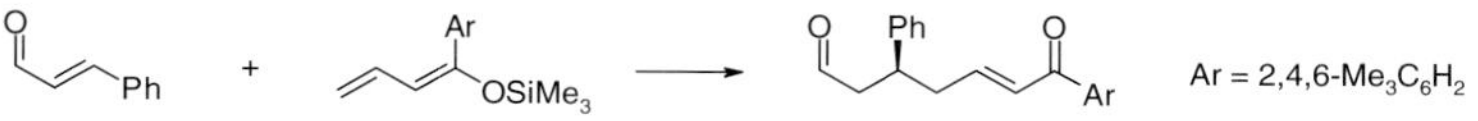

Chiral δ,ε-ethylene-ζ-ketoaldehydes. Startg. (1,3-dien)oxysilane (2 eq.) added to a soln. of 2(S)-[diphenyl(trimethylsiloxy)methyl]pyrrolidine (20 mol%), *p*-nitrobenzoic acid (20 mol%) and cinnamaldehyde (0.25 mmol) in toluene/water (10:1; 1 ml), stirred at room temp. for 48 h, solvent removed, and the residue worked up with purification by flash chromatography on silica gel → product. Y 74% (e.e. 99%). This is the first example of a catalytic asym. vinylogous Michael-type addition of *acyclic* (1,3-dien)oxysilanes, reaction proceeding in high yield and with high enantioselectivity (ten examples; Y 70-90%; e.e. 98 to >99%). Furthermore, 4-subst. (1,3(Z)-dien)oxysilanes reacted with high *anti*-diastereoselectivity and the same high yield and enantioselectivity (ten examples; Y 69-80%; *anti/syn* 82:18 to 92:8; e.e. 96-99%). The procedure is applicable to electron-diverse β-arylacroleins, β-(2-thienyl)acrolein and a β-silylacrolein, although the hindered mesityl group of the (1,3-dien)oxysilane appears critical in order to minimize undesirable 1,4-addition. The latter also complicated vinylogous addition of vinylketene S,O-silyl acetals, but enantioselectivity for the vinylogous addition was undiminished. Aliphatic acroleins also gave low yields but with significantly reduced enantioselectivity (one example; Y 50%; e.e. 71%); reaction was also lower yielding with (1,3(E)-dien)oxysilanes. Reaction is presumed to take place via classical iminium activation. F.e. and comparison of organocatalysts s. V. Gupta, S. Sudhir V., T. Mandal, C. Schneider, Angew. Chem., Int. Ed. *2012*, *51* (50), 12609-12 [DOI: 10.1002/anie.201207058].

1,3-Dimesitylimidazol-2-ylidene *NHC*
6-Oxabicyclo[3.2.0]heptan-7-one-3-carboxylic acid esters ←
from 2-siloxycyclopropanecarboxylic acid esters and α,β-ethylenecarboxylic acid fluorides
via diastereoselective N-heterocyclic carbene-catalyzed
Ireland-Coates Claisen rearrangement

448.
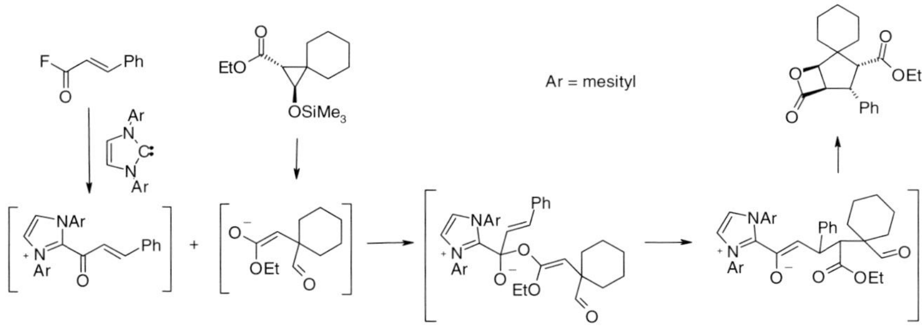

The startg. acid fluoride (0.2 mmol) added to a soln. of the startg. donor-acceptor cyclopropane (1 eq.) in THF (2 ml) containing powdered 4 Å molecular sieves, stirred for 1 h at room temp., cooled to -78°, a soln. of 1,3-dimesitylimidazol-2-ylidene (10 mol%) in THF (2 ml) [generated *in situ* by stirring $KN(SiMe_3)_2$ with 1,3-dimesitylimidazolium chloride in THF at room temp. for 1 h] added, the mixture stirred again for 1 h, the cooling bath removed, the mixture allowed to warm to room temp. slowly with stirring over 14 h, concentrated, and the crude residue purified via flash chromatography → ethyl 7-oxo-2-phenyl-6-oxaspiro[bicyclo[3.2.0]heptane-4,1′-cyclohexane]-3-carboxylate. Y 41% (d.r. >20:1). The procedure is high-yielding and highly diastereoselective, and applicable to electron-diverse cinnamyl fluorides and heterocyclic analogs, but yields were only modest with an aliphatic α,β-ethyleneacyl fluoride; the nature of the ester group of the siloxycyclopropanecarboxylate is critical: with the ethyl ester, as above, a significant amount of ethyl 1-formylcyclohex-1-ylacetate was formed as by-product, but this was minimized by using the more electron-rich *p*-methoxyphenyl ester when the yield was increased to 93% (eighteen examples in all; Y 41-93%). The yield was low (47%) when the cyclopropane pentamethylene substituent was replaced by *gem*-dimethyl; reaction was also sensitive to the bulk of the catalyst and diastereoselectivity lower with $LiN(SiMe_3)_2$ as base. Reaction is thought to be triggered by generation of an acylazolium ion by interaction of the acyl fluoride with the carbene, which couples with the γ-aldehydocarboxylic acid ester enolate formed from the cyclopropane by O-desilylation and retro-aldol scission; the resulting hemiacetal inner salt then undergoes the crucial Ireland-Coates Claisen rearrangement prior to intramolecular aldol condensation and β-lactonization. F.e.s. L. Candish, D.W. Lupton, J. Am. Chem. Soc. *2013*, *135* (1), 58-61 [DOI: 10.1021/ja310449k].

Chiral 2-pyridyl-Δ^2-oxazolines s. under $(MeCN)_2Pd(OTs)_2$ *L**

Microorganisms ←
Wittig synthesis of donor-acceptor stilbenes CO → C=C
under physiological conditions s. *13,* 820s*83*

2,2,2-Trifluoroethanol s. under Chiral copper(II) bis(Δ^2-oxazoline) complexes *CF_3CH_2OH*
Chiral 3,3′-bis(polyfluoroaryl)-1,1′-bi-2-naphthols s. under Magnesium tert-butoxide ←

(S)-3,3′-Difluoro-1,1′-bi-2-naphthol *3,3′-F_2-BINOL*
Asym. allylboration of ketones CO → C(OH)C-C=C
with β,γ-ethyleneboronic acid esters s. *33,* 865s*83*

Chiral 3,3′-dibromo-1,1′-bi-2-naphthol *3,3′-Br_2-BINOL*
***o*-Allylphenols from *o*-hydroxybenzyl ethers** OR → C=C
via organocatalyzed asym. addition of α,β-ethyleneboronic acid esters

449.

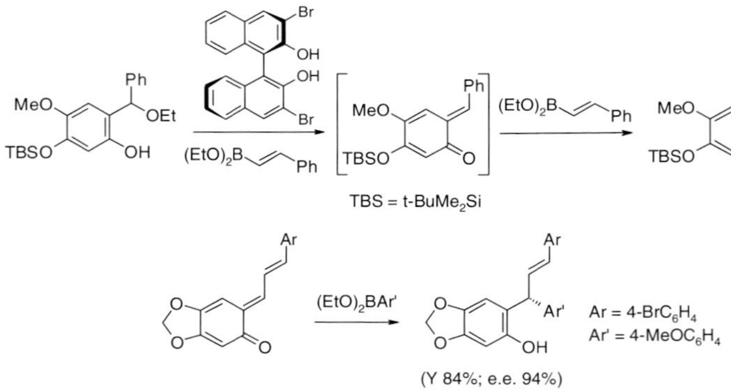

A novel enantioselective method for addition of aryl- or unsatd. boronates **to** stable or *in situ* generated ***o*-quinone methids** is reported. **E:** A soln. of diethyl styrylboronate (2 eq.) in toluene (0.25 ml) added slowly via syringe to a mixture of 5-*tert*-butyldimethylsiloxy-2-(α-ethoxybenzyl)-4-methoxyphenol (0.25 mmol) and chiral BINOL deriv. (10 mol%) in the same solvent (2.3 ml) at 0° under N_2, the mixture stirred vigorously for 6 h, and purified directly by flash chromatography → (S,E)-5-(*tert*-butyldimethylsiloxy)-2-(1,3-diphenylallyl)-4-methoxyphenol. Y 91% (e.e. 92%). In the presence of the chiral BINOL, alkenylboronates (incl. a benzothiazol-2-yl deriv.) were found to be sufficiently acidic to generate *o*-quinone methids under mild conditions from electron-diverse ethyl *o*-hydroxydiarylmethyl ethers and to trap them *in situ*, affording the title products with good enantioselectivity (e.e. 90-96%; nine examples; Y 84-95%). Stable isolable *o*-quinone methids were also vinylated (one example; Y 82%; e.e. 91%) or (het)arylated to afford **chiral *o*-hydroxy-1,1-diarylalkanes** (eleven examples; Y 46-85%; e.e. 80-96%) under similar conditions. F.e., optimization and use of the method in a short synthesis of the natural product, (S)-4-methoxydalbergione s. Y. Luan, S.E. Schaus, J. Am. Chem. Soc. *2012*, *134* (49), 19965-8 [DOI: 10.1021/ja309076g].

Pivalic acid s. under Pd(OAc)$_2$ *t-BuCOOH*
p-Nitrobenzoic acid s. under 2(S)-[Diphenyl(trimethylsiloxy)methyl]pyrrolidine *ArCOOH*

(S)-Proline [s.a. under Pd(OAc)$_2$] *Pro-OH*
Wittig synthesis CO → C=C
update s. *13,* 820s*82*; 'click' synthesis of donor-acceptor stilbenes under mild conditions *in water* with (S)-proline *or under physiological conditions* with *Calystegia sepium* s. D. McLeod, J. McNulty, Eur. J. Org. Chem. *2012* (31), 6127-31 [DOI: 10.1002/ejoc.201201042]; (Z)-enhanced synthesis of *o-hetero*-subst. β-styryl ketones (and trisubst. analogs) from *o*-hetero-subst. ar. aldehydes and (α-acylalkylidene)phosphoranes s. P.A. Byrne, L.J. Higham, P. McGovern, D.G. Gilheany, Tetrahedron Lett. *2012, 53* (49), 6701-4 [DOI: 10.1016/j.tetlet.2012.09.123]; Wittig synthesis of α,β-ethylene-α-fluoroaldehydes s. M. Kajjout, M. Smietana, J. Leroy, C. Rolando, ibid. *2013, 54* (13), 1658-60 [DOI: 10.1016/j.tetlet.2013.01.017]; one-pot **oxidative Wittig-type synthesis** of (Z)-α-bromo-α,β-ethylenecarboxylic acid esters from prim. alcohols and (carbethoxymethylene)triphenylphosphorane via α-bromination of the latter (with NBS) and subsequent oxidative coupling with the alcohol using SO_3·pyridine complex as oxidant and Et_3N as base s. U. Karama, R. Mahfouz, Z. Al-Othman, I. Warad, A. Almansour, Synth. Commun. *2013, 43* (6), 893-8 [DOI: 10.1080/00397911.2011.614712]; **Wittig synthesis-Heck arylation** under ligand-free conditions s. A.V. Bedekar, A.R. Chaudhary, M.S. Sundar, M. Rajappa, Tetrahedron Lett. *2013, 54* (5), 392-6 [DOI: 10.1016/j.tetlet.2012.11.022]; alkoxystilbenes by **3-component O-alkylative Wittig synthesis-Heck arylation** using K_2CO_3/Bu_4NBr/Pd(OAc)$_2$/dppp s. K.N. Patel, B.V. Kamath, A.V. Bedekar, ibid. *54* (1), 80-4 [DOI: 10.1016/j.tetlet.2012.10.102]; **Horner synthesis** of (Z)-α,β-ethylenenitriles from aromatic or aliphatic aldehydes (linear or branched) and bis(*o-tert*-butylphenoxy)phosphinylacetonitrile with *t*-BuOK, *t*-BuOK/18-crown-6 or NaH as base (cf. *39,* 854s*82*) s. K. Ando, M. Okumura, S. Nagaya, ibid. *54* (15), 2026-8 [DOI: 10.1016/j.tetlet.2013. 02.020].

Dimethyl sulfoxide s. under Tetrakis(acetonitrile)palladium(II) bis(fluoroborate) *DMSO*

S-Chiral sulfoxides *RS*(O)R′*
Asym. Hosomi-Sakurai reaction CO → C(OH)C-C=C
update s. *55,* 433s*82*; by coupling aldehydes with allyl(trichloro)silane *under organocatalysis* using an S-chiral sulfoxide as reagent s. G. Monaco, C. Vignes, F. De Piano, A. Bosco, A. Massa, Org. Biomol. Chem. *2012, 10* (48), 9650-9 [DOI: 10.1039/c2ob27022f]; coupling with [hetero]aromatic aldehydes by *counterion-directed* reaction using a chiral 1,1′-binaphthyl-2,2′-disulfonic acid imide as catalyst s. M. Mahlau, P. García-García, B. List, Chem. Eur. J. *2012, 18* (51), 16283-7 [DOI: 10.1002/chem.201203623].

Phenyl iodosoacetate/p-benzoquinone/potassium phosphate *PhI(OAc)$_2$/BQ/K$_3$PO$_4$*
Metal-free oxidative ar. trifluoromethylation s. *78,* 476s*83* H → CF_3

N-Bromosuccinimide/sulfur trioxide·pyridine/triethylamine *NBS/SO$_3$·py/Et$_3$N*
Oxidative Wittig-type synthesis of (Z)-α-bromo-α,β-ethylenecarboxylic acid esters ←
from prim. alcohols s. *13,* 820s*83*

Trimethylsilanol s. under Copper(II) acetate *Me$_3$SiOH*

Tetra-n-butylammonium difluorotriphenylsilicate *Bu$_4$NSiF$_2$Ph$_3$*
Carbazoles from ar. nitroso compds. and benzynes

450. 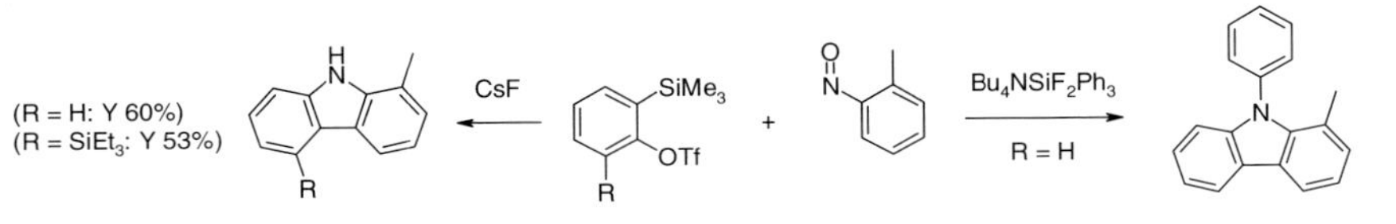

A transition metal-free 2-component approach to carbazoles is reported, by using benzynes generated from *o*-silyltriflates under mild conditions. **E: N-Arylcarbazoles from two benzyne molecules**. Tetra-*n*-butylammonium difluorotriphenylsilicate [TBAT] (2.6 eq.) and *o*-nitrosotoluene

(0.1 mmol) added under argon to a flame-dried Schlenk tube, followed by DME (2 ml), the soln. stirred for 5 min, *o*-(trimethylsilyl)phenyl triflate (2.2 eq.) added dropwise, the mixture stirred for 12-15 h (TLC) at room temp., solvent removed under reduced pressure, and the residue subjected to flash chromatography on silica gel → 1-methyl-9-phenyl-9*H*-carbazole. Y 76%. Seven further examples optionally bearing alkyl or OMe gave yields of 50-67%. In contrast, reaction with CsF as fluoride source in acetonitrile afforded **1-unsubst. carbazoles** (twenty-seven examples; Y 30-80%; optionally bearing alkyl, F, Br, I, OMe or $SiEt_3$ groups). For substituted arynes the regioselectivity is in accord with a mechanism involving initial [2+2]-cycloaddition with the nitroso compd., followed by cycloreversion to afford an N-aryl-*o*-quinone monoimine, which undergoes intramolecular addition to the carbonyl group and dehydration. F.e.s. S. Chakrabarty, I. Chatterjee, L. Tebben, A. Studer, Angew. Chem., Int. Ed. *2013*, *52* (10), 2968-71 [DOI: 10.1002/anie.201209447]; **1,5-iminobenzocycloheptadien-2-ones** by [3+2]-cycloaddition of benzynes with 3-oxidopyridinium ylids (with CsF) s. H. Ren, C. Wu, X. Ding, X. Chen, F. Shi, Org. Biomol. Chem. *2012*, *10* (45), 8975-84 [DOI: 10.1039/c2ob26519b]; **2-aryl-2*H*-indazole-3-carbonyl compds.** via a silver-catalyzed [3+2]-cycloaddition with α-diazocarbonyl compds. (with AgOTf/Bu_4NF) s. C.-D. Wang, R.-S. Liu, ibid. 8948-52 [DOI: 10.1039/c2ob26760h]; **3-carbalkoxy-1,2-benzisoxazolin-3-yl-acetic acid esters** from acetylenedicarboxylic acid esters and hydroxylamines via *in situ*-generation of the corresponding nitrones s. P. Li, C. Wu, J. Zhao, Y. Li, W. Xue, F. Shi, Can. J. Chem. *2013*, *91* (1), 43-50 [DOI: 10.1139/cjc-2012-0199]; **9-phenanthridones** from *o*-bromocarboxylic acid amides [with CsF/$Pd(OAc)_2$/Na_2CO_3] s. C. Lu, A.V. Dubrovskiy, R.C. Larock, J. Org. Chem. *2012*, *77* (19), 8648-56 [DOI: 10.1021/jo3016192]; f. annelations with arynes s. *68*, 464s*83*; N-aryl-acridones from three benzyne molecules and N-unsubst. 2-azetidinones or from preformed N-unsubst. 2,3-dihydro-4(1*H*)-quinolones via extrusion of ethylene s. *83*, 470.

Tri-tert-butylphosphine·fluoroboric acid s. under Electrolysis — *$t\text{-}Bu_3P{\cdot}HBF_4$*

Tert. phosphines and di(phosphines) s. under K_2CO_3, Chlorobis(cyclooctene)rhodium(I) dimer, Chloro(1,5-cyclooctadiene)rhodium(I) dimer, Palladium(II) acetate, Bis(dibenzylideneacetone)palladium(0) and Tris(dibenzylideneacetone)dipalladium — *⩾P*

2′,6′-Dimethyl-2′-(dicyclohexylphosphino)biphenyl-4-ylcarboxylic acid amide s. under Palladium(II) acetate ←

Chiral 1,2-bis(2,5-dimethylphospholano)ethane s. under Copper(II) isobutyrate ←

Chiral 1,3-bis(phosphorodiamidites) s. under π-Allyl(η^5-cyclopentadienyl)palladium(II) ←

Chiral 1,1′-binaphthyl-2,2′-diyl phosphoromonoamidites s. under Chloro(1,5-cyclooctadiene)iridium(I) dimer ←

Chiral bis(1,1′-binaphthyl-2,2′-diyl phosphoramidites) s. under Dichlorobis(triphenylphosphine)ruthenium(II) ←

Diphenyl phosphate s. under Chlorobis(cyclooctene)rhodium(I) dimer — *$(PhO)_2PO_2H$*

Chiral 3,3′-diaryl-1,1′-binaphthyl-2,2′-diyl hydrogen phosphates ←

Synthesis of *anti*-3-acetylenealcohols from aldehydes — CHO → CH(OH)C-C≡C

and α-alleneboronic acid esters with kinetic resolution s. *75*, 265s*83*

Oxygen (or Air) s. under Palladium(II) acetate, Tetrakis(acetonitrile)palladium(II) bis(fluoroborate), $(MeCN)_2Pd(OTs)_2$ and $PdCl_2$ — *O_2*

Chiral 1,1′-binaphthyl-2,2′-disulfonic acid imides ←

Counterion-directed asym. Hosomi-Sakurai reaction s. *55*, 433s*83* — CO → C(OH)C-C=C

3-Component organocatalyzed asym. aza-Hosomi-Sakurai reaction — CO → C(NHR)C-C=C

s. *72*, 590s*83*

Sulfur trioxide·pyridine s. under N-Bromosuccinimide — *$SO_3{\cdot}py$*

Tetra-n-butylammonium hydrogen sulfate s. under Chloro(1,5-cyclooctadiene)-iridium(I) dimer — *Bu_4NHSO_4*

Trimethylsilyl triflate — *Me_3SiOTf*

Michael-type addition — C=C → CHC(R)

with $Sc(OTf)_3$ cf. *48*, 661s*50*; synthesis of 2-β-keto-4-chromanones from 4-chromones with trimethylsilyl triflate as catalyst s. L.A. Stubbing, F.F. Li, D.P. Furkert, V.E. Caprio, M.A. Brimble,

Tetrahedron *2012, 68* (34), 6948-56 [DOI: 10.1016/j.tet.2012.05.115]; **asym. Michael-type addition** to 2-(α,β-ethyleneacyl)pyridine N-oxides with a chiral copper(II) bis(Δ^2-oxazoline) complex as catalyst (cf. *41,* 111s*80*) s. J. George, B.V.S. Reddy, Adv. Synth. Catal. *2013, 355* (2-3), 383-8 [DOI: 10.1002/adsc.201200631].

Sodium anthraquinone-2-sulfonate s. under Tetrakis(acetonitrile)palladium(II) bis(fluoroborate) ←

Fluoride ion *F^-*
Annelation with arynes s. *68,* 464s*83* ○

Tetra-n-butylammonium fluoride s. under $PdCl_2$ *Bu_4NF*
Tetra-n-butylammonium bromide s. under K_2CO_3 *Bu_4NBr*
Hydrogen fluoride s. under Chloro(1,5-cyclooctadiene)iridium(I) dimer *HF*

Iron, nickel or palladium salts/oxidant ←
Oxidative Heck-type arylation with arylboron compds. H → Ar
update s. *74,* 516s*82*; ligand- and base-free β-arylation of enones with $Pd(OAc)_2$/TfOH in *DMSO* [minimizing 1,4-addition] s. S.E. Walker, J. Boehnke, P.E. Glen, S. Levey, L. Patrick, J.A. Jordan-Hore, A.-L. Lee, Org. Lett. *2013, 15* (8), 1886-9 [DOI: 10.1021/ol400539h]; of *p*-benzoquinones with arylboronic acids or potassium aryl(trifluoro)borates with, preferably, $FeSO_4$ as catalyst and Na-persulfate as oxidant with added ascorbic acid s. K. Komeyama, T. Kashihara, K. Takaki, Tetrahedron Lett. *2013, 54* (9), 1084-6 [DOI: 10.1016/j.tetlet.2012.12.031]; with FeS/K-persulfate, also oxidative arylation of pyridines, cf. J. Wang, S. Wang, G. Wang, J. Zhang, X.-Q. Yu, Chem. Commun. *2012, 48* (96), 11769-71 [DOI: 10.1039/c2cc35468c]; oxidative arylation of planar-chiral aminomethylferrocenes with $Pd(OAc)_2$ and added Boc-L-valine s. D.-W. Gao, Y.-C. Shi, Q. Gu, Z.-L. Zhao, S.-L. You, J. Am. Chem. Soc. *2013, 135* (1), 86-9 [DOI: 10.1021/ja311082u]; **oxidative α-arylation** of cyclic ethers under nickel catalysis with $Ni(OAc)_2/Ph_3P/K_3PO_4$ and di-*tert*-butyl peroxide as oxidant s. D. Liu, C. Liu, H. Li, A. Lei, Angew. Chem., Int. Ed. *2013, 52* (16), 4453-6 [DOI: 10.1002/anie.201300459].

Chiral ferrocenyldi(phosphines) s. under Silver fluoride ←
Iron(III) triflate s. under Palladium(II) complexes *$Fe(OTf)_3$*

Dichloro[1,2-bis(diphenylphosphino)ethane]cobalt(II)/potassium phosphate ←
***cis*-1-*sec*-Aminoindan-2-carboxylic acid esters or amides** ○
from *o*-imidoylboronic acids and α,β-ethylenecarboxylic acid esters or amides

451. 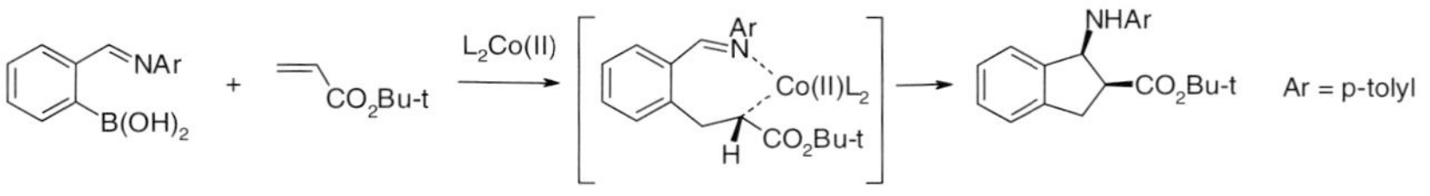

via 1,4-addition. A sealed tube containing $CoCl_2$(dppe) (10 mol%), the startg. *o*-imidoylboronic acid (1.5 eq.) and K_3PO_4 (1 eq.) evacuated and purged with N_2 three times, freshly distilled acetonitrile (2 ml) and the startg. activated alkene (0.2 mmol) added separately via syringe, the mixture stirred at 80° for 12 h, diluted with methylene chloride, filtered through a Celite and silica gel pad, the filtrate concentrated, and the residue purified on a silica gel column → product. Y 93%. The procedure is applicable in high yield and with high diastereo- and regio-selectivity to the coupling of acrylic acid esters or amides with *o*-borylated aldimines or ketimines possessing electron-diverse aryl or methyl substitution on nitrogen; a methylenedioxy group was also tolerated on the aromatic ring (fourteen examples; Y 52-96%). There was no reaction, however, with C-substituted acrylic acid derivs. or with cyclic enones as the ene component. Reaction is presumed to involve initial cobalt-mediated 1,4-addition of the arylboronic acids followed by formation of a conformationally favorable 7-membered cobaltacycle which undergoes intramolecular 1,2-addition to the C=N bond of the imine to give, after protonolysis, the 1,2-*cis*-disubst. indan preferentially. The 1,4-addition was studied preliminarily with isolation of the 1,4-adduct in high yield (65-93%; sixteen examples) from a wide range of electron-deficient alkenes (incl. methyl vinyl ketone and acrylonitrile) and electron-diverse arylboronic acids substituted at the *ortho*, *meta* or *para* site. F.e. and comparison of cobalt complexes s. M.-H. Chen, S. Mannathan, P.-S. Lin, C.-H. Cheng, Chem. Eur. J. *2012*, *18* (47), 14918-22 [DOI: 10.1002/chem.201202575].

Nickel(II) N-heterocyclic carbene or phosphine complexes *[Ni(II)]*
Nickel-catalyzed *sp²-sp³*-Suzuki coupling Ar-R
with inversion or retention of configuration s. *64,* 453s*83*

Nickel(II) bromide-diglyme/4,4′-di-tert-butyl-2,2′-bipyridyl/lithium tert-butoxide *[Ni(II)]/LiOBu-t*
***tert*-Alkylarenes from *tert*-alkyl halides and arylboronic acid esters** ⩾C-Hal → ⩾C-Ar

452.

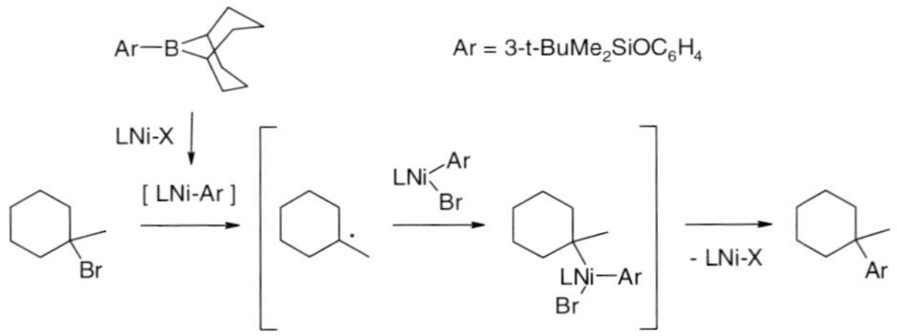

A novel Suzuki coupling of unactivated tert. alkyl halides using a commercially available nickel(II) catalyst, and free from alkyl isomerization, is reported. **E:** A sealed mixture of $NiBr_2$·diglyme (10 mol%), 4,4′-di-*tert*-butyl-2,2′-bipyridyl (11 mol%) and anh. benzene (24 ml) stirred vigorously in a glovebox for 2 h, a soln. of [3-(9-borabicyclo[3.3.1]nonan-9-yl)phenoxy]-*tert*-butyl(dimethyl)-silane (2.5 eq.) in the same solvent (1 ml) activated by stirring vigorously with Li-*tert*-butoxide (2.4 eq.) and isobutanol (2.4 eq.) for 20 min, added to the light green slurry, stirred for 20 min, 1-bromo-1-methylcyclohexane (1 mmol) added to the dark-green mixture via microsyringe, the sealed mixture stirred vigorously at 60° for 24 h, filtered through silica, and purified chromatographically → *tert*-butyldimethyl[3-(1-methylcyclohexyl)phenoxy]silane. Y 57%. This novel cross-coupling was effected by *tert*-alkyl chloride, bromide and iodide electrophiles but limited to *m*-subst. (H, alkyl, phenyl, ether, silyl ether) phenyl-BBN derivs. (fifteen examples; Y 53-86%). When the reaction was conducted in toluene, diphenylmethane was observed as a by-product, indicative of a radical mechanism. F.e., optimization and substrate prepn. s. S.L. Zultanski, G.C. Fu, J. Am. Chem. Soc. *2013*, *135* (2), 624-7 [DOI: 10.1021/ja311669p].

Chloro(cinnamyl)[1,1′-bis(diphenylphosphino)ferrocene]nickel(II)/potassium phosphate ←
Nickel-catalyzed Suzuki coupling s. *51,* 453s*83* Ar-Ar′

Dichloro[1,1′-bis(diphenylphosphino)ferrocene]nickel(II)/potassium phosphate *[Ni(II)]/K_3PO_4*
Suzuki coupling with aryl 4,6-dimethoxy-1,3,5-triazin-2-yl ethers

453.

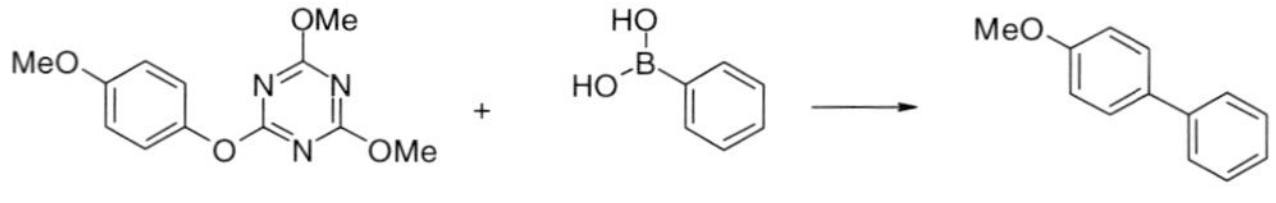

Phenol C-O bonds may be activated selectively for Suzuki-Miyaura coupling under nickel catalysis by a simple and efficient conversion to 4,6-dimethoxy-1,3,5-triazin-2-yl ethers with 2-chloro-4,6-dimethoxy-1,3,5-triazine in the presence of KOH. Such ethers are air- and moisture-stable crystalline solids that may be purified conveniently by recrystallization. Coupling with these ethers was tolerant of a wide range of functional groups, and was notably effective in the presence of electron-donating groups which are normally electronically deactivating. **E:** Startg. 1,3,5-triazin-2-yl ether (1 mmol), phenylboronic acid (4 eq.), $NiCl_2$(dppf) (5 mol%) and K_3PO_4 (7 eq.) added in turn to a Schlenk tube under argon, toluene (7 ml) injected by syringe, the mixture stirred in a preheated oil bath (110°) for 24 h, cooled to room temp., methylene chloride added, filtered through a short pad of Celite, the filtrate concentrated *in vacuo*, and the residue purified by flash column chromatography → 4-methoxybiphenyl. Y 90%. In contrast under the same conditions yields were lower with OMe (0%), OMs (21%), OTs (27%), OPiv (24%), $OCONMe_2$ (19%), OSO_2NMe_2 (68%) or $OPO(OEt)_2$ (46%) in place of the triazinyl ether. The method is compatible with functional groups such as amide, ester, lactone, nitrile, aldehyde or ketone, but failed with a nitro deriv., yields of 39-91% being obtained for thirty-six further examples. Reaction of pyrid-2-yl 4,6-dimethoxy-1,3,5-triazin-2-yl ether gave the corresponding 2-aryl-4,6-dimethoxy-1,3,5-triazines (Y 51%, 55%), however, as did N-(4,6-dimethoxy-1,3,5-triazin-2-yl)-4(1*H*)-pyridone (Y 64%, 67%). F.e. and optimization s. X.-J. Li, J.-L. Zhang, Y. Geng, Z. Jin, J. Org. Chem. *2013*, *78* (10), 5078-84 [DOI: 10.1021/jo4005537].

Dichloro(p-cymene)ruthenium(II) dimer *[RuCl$_2$(p-cymene)]$_2$*
1,4-Addition of arylboronic acids s. *62,* 449s*83* C=C → CHC(Ar)

Dichlorobis(triphenylphosphine)ruthenium(II)/chiral bis(1,1′-binaphthyl-2,2′-diyl phosphoramidites)/potassium fluoride *[Ru(II)]*/KF*
Asym. synthesis of benzyl alcohols from oxo compds. and arylboronic acids CO → C(OH)Ar
Chiral N-protected 3-aryl-3-hydroxyoxindoles s. *65,* 437s*83*

Cationic rhodium(I) complexes *[Rh(I)]*
1,4-Addition of organoboron compds. C=C → CHC(R)
of arylboronic acids *under rhodium catalysis*, update s. *62,* 449s*82*; to N-protected dehydroalaninates *en route* to fluorescent β-arylalaninates using ligated $[Rh(cod)_2]BF_4$ and CsF in aq. dioxane s. P.M.T. Ferreira, L.S. Monteiro, G. Pereira, E.M.S. Castanheira, C.G. Frost, Eur. J. Org. Chem. *2013* (3), 550-6 [DOI: 10.1002/ejoc.201201198]; to 4-chromones *under palladium(II) catalysis* with $Pd(OAc)_2$ with added $Fe(OTf)_3$ and pivalic acid, also oxidative 2-arylation with added DDQ/KNO_2 as catalyst (cf. *74,* 516s*83*), s. D. Kim, K. Ham, S. Hong, Org. Biomol. Chem. *2012, 10* (36), 7305-12 [DOI: 10.1039/c2ob26061a]; to enones *under ruthenium catalysis* with $[RuCl_2(p\text{-cymene})]_2$ in aq. dioxane *under neutral conditions* s. L. Zhang, X. Xie, L. Fu, Z. Zhang, J. Org. Chem. *2013, 78* (7), 3434-7 [DOI: 10.1021/jo4001367].

Chiral [bis(ferrocenyl)tetrafluorobenzobarrelene](hydroxo)rhodium(I) dimer *[Rh(I)]**
Rhodium-catalyzed asym. synthesis of 3,3-diarylphthalimidines [1(3*H*)-isoindolones] from 3-aryl-3-hydroxyphthalimidines and arylboroxines OH → Ar

454. 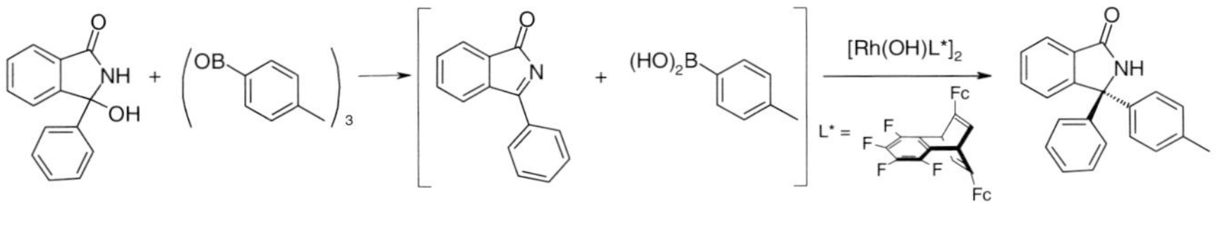

$[Rh(OH)((S,S)\text{-}Fc\text{-}tfb^*)]_2$ (5 mol% of Rh), 3-hydroxy-3-phenylisoindolin-1-one (0.2 mmol) and *p*-tolylboroxine (0.167 mmol) placed in a Schlenk tube under N_2, 1,4-dioxane (0.8 ml) added, the mixture stirred at 80° for 24 h, passed through a short column of alumina with ethyl acetate as eluent, solvent removed on a rotary evaporator, and the residue subjected to prep. TLC on silica gel → (R)-(+)-product. Y 98% (e.e. 95%). The method is applicable to arylboroxines bearing electron-rich or -poor substituents at the *ortho*, *meta* or *para* positions and a variety of 3-aryl-3-hydroxyphthalimidines also bearing electron-rich or -poor substituents and including 2-furyl or 2-thienyl analogs. The addition of phenylboroxine to 3-hydroxy-3-*p*-tolylisoindolin-1-one afforded the opposite enantiomer to the illustrated product (Y 95%; e.e. 94%). It is believed the arylboroxines act as dehydrating agents to generate intermediate cyclic N-acylketimines as well as arylating agents in a subsequent rhodium-catalyzed 1,2-arylation. F.e. (nineteen; Y 65%, 82-99%; e.e. 91-98%) s. T. Nishimura, A. Noishiki, Y. Ebe, T. Hayashi, Angew. Chem., Int. Ed. *2013*, *52* (6), 1777-80 [DOI: 10.1002/anie.201208593].

Chiral rhodium(I) diene or aminoethylenephosphite complexes *[Rh(I)]**
Asym. 1,4-Addition of organoboron compds. C=C → CHC(Ar)
under rhodium catalysis, update s. *55,* 452s*82*; of arylboronic acids to N-Boc-protected α,β-unsatd. δ-lactams with a chiral (bicyclo[3.3.0]octa-2,6-diene)hydroxorhodium(I) complex and KHF_2 as additive s. Z.-T. He, Y.-B. Wei, H.-J. Yu, C.-Y. Sun, C.-G. Feng, P. Tian, G.-Q. Lin, Tetrahedron *2012, 68* (45), 9186-91 [DOI: 10.1016/j.tet.2012.09.001]; of potassium trifluoro(vinyl)borates under the same conditions s. H.-J. Yu, C. Shao, Z. Cui, C.-G. Feng, G.-Q. Lin, Chem. Eur. J. *2012, 18* (42), 13274-8 [DOI: 10.1002/chem.201202660]; asym. 1,4-addition of arylboronic acids to acyclic or cyclic enones with a chiral tropane-based aminoethylenephosphite as ligand s. S. Vlahovic, N. Schädel, S. Tussetschläger, S. Laschat, Eur. J. Org. Chem. *2013* (8), 1580-90 [DOI: 10.1002/ejoc.201201539]; to α,β-ethylenealdehydes *under palladium catalysis* with $Pd(OAc)_2$ and a simple (S)-proline-based cyclic sec. amine as cocatalyst s. I. Ibrahem, G. Ma, S. Afewerki, A. Córdova, Angew. Chem., Int. Ed. *2013, 52* (3), 878-82 [DOI: 10.1002/anie.201208634]; of α,β-ethyleneboronic acids to heteroaryl α,β-ethyleneketones *under magnesium catalysis* with $Mg(OBu\text{-}t)_2$ and a chiral 3,3′-bis(polyfluoroaryl)-1,1′-bi-2-naphthol as ligand s. P.Q. Le, T.S. Nguyen, J.A. May, Org. Lett. *2012, 14* (23), 6104-7 [DOI: 10.1021/ol3030605]; asym. synthesis of *o*-allyl- or *o*-benzylphenols from *o*-hydroxybenzyl ethers via 1,4-addition of α,β-ethylene- or aryl-boronic acid esters **to *o*-quinone methids** *under organocatalysis* with chiral 3,3′-dibromo-1,1′-bi-2-naphthol s. Y. Luan, S.E. Schaus, J. Am. Chem. Soc. *2012, 134* (49), 19965-8 [DOI: 10.1021/ja309076g].

Rhodium(II) imidazol[idin]-2-ylidene complexes *[Rh(II)]*
Synthesis of alcohols from oxo compds. and boron compds. CO → C(OH)R
update s. *56,* 438s*82*; synthesis of α-hydroxycarboxylic acid esters from ethyl glyoxylate and aryl- or alkyl-boronic acids with rhodium(II) imidazol[idin]-2-ylidene complexes as catalyst s. C.S. Marques, A.J. Burke, Tetrahedron *2012, 68* (35), 7211-6 [DOI: 10.1016/j.tet.2012.05.129].

Chiral rhodium(I) N-heterocyclic carbene, phosphine or sulfoxide complexes *[Rh(I)]**
Asym. synthesis of benzyl alcohols CO → C(OH)Ar
from oxo compds. and arylboron compds.
under rhodium catalysis, update s. *65,* 437s*82*; asym. 1,2-addition of arylboronic acids to heteroaryl α-ketocarboxylic acid esters with $[Rh(coe)_2Cl]_2$/KOH and S-chiral N-(*tert*-butylsulfinyl)cinnamylamine as ligand (cf. *81*, 460) s. H. Wang, T.-S. Zhu, M.-H. Xu, Org. Biomol. Chem. *2012, 10* (46), 9158-64 [DOI: 10.1039/c2ob26316e]; to aryl α-ketocarboxylic acid esters *in water* at room temperature with the same catalyst, also asym. 1,4-addition (cf. *55,* 452s*83*), s. N. Khiar, V. Valdivia, Á. Salvador, A. Chelouan, A. Alcudia, I. Fernández, Adv. Synth. Catal. *2013, 355* (7), 1303-7 [DOI: 10.1002/adsc.201201065]; asym. 1,2-addition to trifluoromethyl ketones with a chiral 2,2′-bi-[3-*tert*-butyl-4-aryl-2,3-dihydro-1,3-benzoxaphospholyl] as ligand s. R. Luo, K. Li, Y. Hu, W. Tang, ibid. 1297-302 [DOI: 10.1002/adsc.201201125]; asym. synthesis of benzhydryls by 1,2-addition to ar. aldehydes with $Rh(OAc)_2$ and a chiral alkoxy- or sulfonato-subst. 1,3-bis-([2.2]paracyclophanyl)imidazol-2-ylidene as ligand s. W. Duan, Y. Ma, B. Qu, L. Zhao, J. Chen, C. Song, Tetrahedron: Asym. *2012, 23* (18-19), 1369-75 [DOI: 10.1016/j.tetasy.2012.09.001]; with chiral, [2.2]paracyclophanyl-subst. N-condensed imidazol-2-ylidenes as ligand s. D. Wang, Y. Ma, F. He, W. Duan, L. Zhao, C. Song, Synth. Commun. *2013, 43* (6), 810-25 [DOI: 10.1080/00397911.2011.610548]; asym. 1,2-addition to N-protected isatins *under ruthenium catalysis* with $RuCl_2(PPh_3)_2$/KF and the chiral, oxygen-bridged bis(1,1′-binaphthyl-2,2′-diyl phosphoramidite), [(R,R)-Me-BIPAM], as ligand in toluene/water s. Y. Yamamoto, M. Yohda, T. Shirai, H. Ito, N. Miyaura, Chem. Asian J. *2012, 7* (10), 2446-9 [DOI: 10.1002/asia.201200481]; **asym. 1,2-addition** of arylboronic acids **to the aldimine group** of N-aryliminoacetic acid esters *under palladium catalysis* with a chiral diacetoxo[2,2′-bi(Δ^2-oxazolinyl)]palladium(II) complex as catalyst (cf. *68,* 458s*82*) s. J. Chen, X. Lu, W. Lou, Y. Ye, H. Jiang, W. Zeng, J. Org. Chem. *2012, 77* (19), 8541-8 [DOI: 10.1021/jo301423e]; to the imino group of cyclic benzo-condensed N-sulfonylimines with $[Rh(coe)_2Cl]_2$/KF and S-chiral N-(*tert*-butylsulfinyl)-2-methylene-3-(1-naphthyl)propylamine as ligand s. H. Wang, T. Jiang, M.-H. Xu, J. Am. Chem. Soc. *2013, 135* (3), 971-4 [DOI: 10.1021/ja3110818].

Chlorobis(cyclooctene)rhodium(I) dimer/diethyl[4-(dimethylamino)phenyl]phosphine/ tetramethylammonium hydridotriacetoxoborate/diphenyl phosphate ←

Syntheses via *in situ* generation of 2-silyl-1,2-dihydropyridines from α,β-ethyleneazomethines and silylacetylenes ○

455.

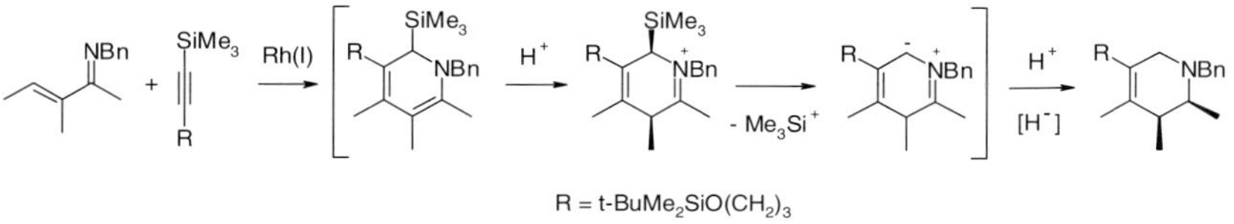

The generation of azomethine ylids via rhodium(I)-catalyzed cyclization of silylacetylenes with ethyleneimines, and treatment of the generated 2-silyl-1,2-dihydropyridines with acids, is reported. The ylids are versatile intermediates for the synthesis of N-heterocycles. **E: 1,2,3,6-Tetrahydropyridines**. A soln. of $[RhCl(coe)_2]_2$ (2.5 mol%) and diethyl[4-(dimethylamino)phenyl]phosphine (5 mol%) in toluene (0.25 ml) added to a mixture of N-benzyl-3-methylpent-3-en-2-one imine (0.23 mmol) and 5-*tert*-butyldimethylsiloxy-1-trimethylsilyl-1-pentyne (1.7 eq.) under N_2 in a glovebox, the sealed mixture stirred at 100° for 2 h, cooled to room temp., added via syringe to a soln. of $Me_4NBH(OAc)_3$ (3 eq.) in anh. methylene chloride (3 ml) under N_2, the homogeneous mixture stirred vigorously (>1000 rpm), a soln. of diphenyl phosphate (2.2 eq.) in the same solvent (2 ml) added dropwise via syringe over 2 min, the homogeneous mixture stirred for 3 h, quenched with 1 *M* aq. NaOH, stirred until gas evolution ceased (5-10 min), extracted with methylene chloride, and purified by flash chromatography → *cis*-5-(*tert*-butyldimethylsiloxyprop-3-yl)-2,3,4-trimethyl-1,2,3,6-tetrahydropyridine. Y 75% (d.r. >20:1). Generation of the azomethine ylids under reducing conditions resulted in protonation to afford tri-, tetra- and penta-subst. 1,2,3,6-tetrahydropyridines (as single diastereomers from polysubst. ethyleneimines) (twenty-four examples; Y 54-95%; d.r. >20:1) in the presence of silyl ether, carbamate, phthalimide, halo and tert. amine functionality. Generation of the ylid with diphenyl phosphate in the presence of acetylene mono-/di-carboxylates afforded **8-azabicyclo[3.2.1]octanes** [tropanes] (seven examples; Y 48-83%; d.r. >20:1; regioselectivity for the monoacetylenecarboxylates 3:1 to 6.5:1).

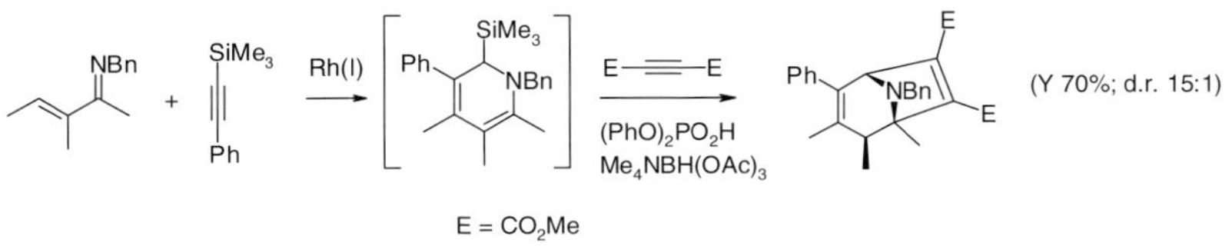

Interestingly, use of the stronger acid, benzenesulfonic acid, under reducing conditions at low temp (-78°) unexpectedly effected Nazarov-type 6π-electrocyclization to afford **2-azabicyclo-[3.1.0]hexanes** (two examples; Y 48-75%; d.r. 6:1 to 7:1).

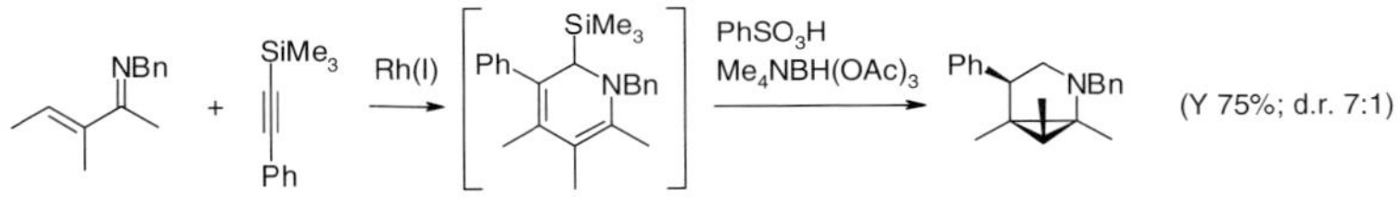

F.e., optimization and substrate prepn. s. M.A. Ischay, M.K. Takase, R.G. Bergman, J.A. Ellman, J. Am. Chem. Soc. *2013*, *135* (7), 2478-81 [DOI: 10.1021/ja312311k].

Chloro(1,5-cyclooctadiene)rhodium(I) dimer/bis(π-allylpalladium chloride)/potassium phosphate/2-dicyclohexylphosphino-2′,4′,6′-triisopropylbiphenyl/copper(I) iodide/ trans-cyclohexane-1,2-diamine ←

1-Aryl-3,4-dihydrocarbostyrils ○
from *o*-chloroboronic acids or esters, acrylamide and ar. iodides
Triple metal-catalyzed 1,4-addition-double N-arylation

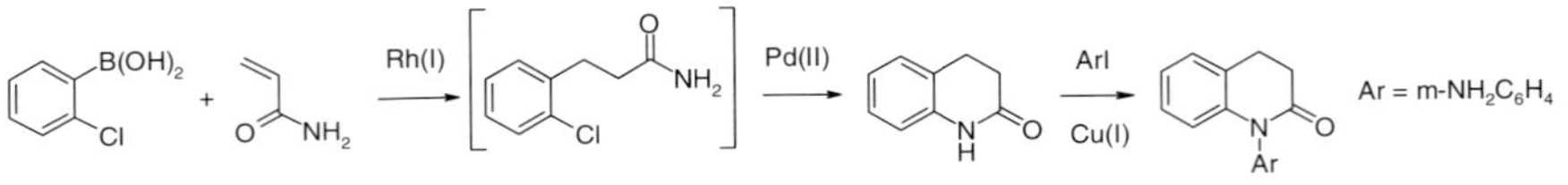

456. 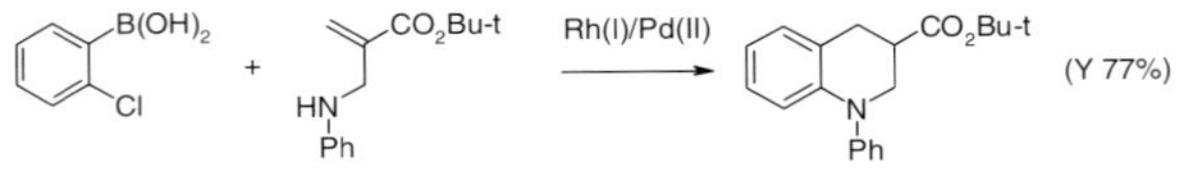

tert-Amyl alcohol (2 ml) and methanol (0.2 ml) added to a mixture of $[Rh(cod)Cl]_2$ (4 mol%), $[(allyl)PdCl]_2$ (5 mol%), XPhos (10 mol%), *o*-chlorobenzeneboronic acid (1.05 eq.), anh. ground K_3PO_4 (2.2 eq.) and acrylamide (0.4 mmol) under argon, the sealed mixture stirred vigorously at 100° for 16 h, cooled to room temp., CuI (10 mol%), 4 Å molecular sieves (50 mg), *m*-aminophenyl iodide (2.5 eq.), anh. K_3PO_4 (2.2 eq.) and *trans*-cyclohexane-1,2-diamine (20 mol%) added under argon, the sealed mixture stirred at room temp. for 10 min, then at 130° for 16 h, cooled to room temp., filtered through silica, concentrated *in vacuo*, and the residue purified by chromatography on silica → 1-(*m*-aminophenyl)-3,4-dihydroquinolin-2(1*H*)-one. Y 77%. Formation of the dihydroquinolone ring via rare Rh(I)-catalyzed conjugate addition to acrylamide, and Pd(II)-catalyzed intramolecular N-arylation was effective using *o*-chlorobenzeneboronic acids or their pinacolboronate esters (readily available from the corresponding phenols), and gave best yields with electron-poor boronic acids/esters (53-96%; ten examples) in the presence of ester, acetal, halo and ether functionality. Compatible Cu(I)-catalyzed N-arylation of the dihydroquinolones was achieved without isolation or solvent exchange, with best yields obtained from *o*-chlorobenzeneboronic acid (47-80%; thirteen examples). Variations in the acrylamide component were limited, with N-alkyl/aryl-derivs. affording complex mixtures and β-alkyl/aryl-derivs. being unreactive, but moderate yields of 3-methyl- (57%) and 3-phenyl-NH-dihydroquinolones (21%) were obtained from the corresponding α-subst. acrylamides. Interestingly, *tert*-butyl α-subst. (Ts or Ph) aminomethylacrylates provided a route to the corresponding **N-subst. 1,2,3,4-tetrahydroquinoline-3-carboxylic acid esters** (three examples; Y 69-71%). F.e., optimization and substrate prepn. s. L. Zhang, L. Sonaglia, J. Stacey, M. Lautens, Org. Lett. *2013*, *15* (9), 2128-31 [DOI: 10.1021/ ol4006008].

Dichloro(pentamethylcyclopentadienyl)rhodium(III) dimer/silver(I) oxide $[Cp^*RhCl_2]_2/Ag_2O$
10-Alkoxy-9(10*H*)-phenanthridones
from arylhydroxamic acid esters and arylboronic acids
Oxidative ring closure via *o*-rhodation under mild conditions

457. 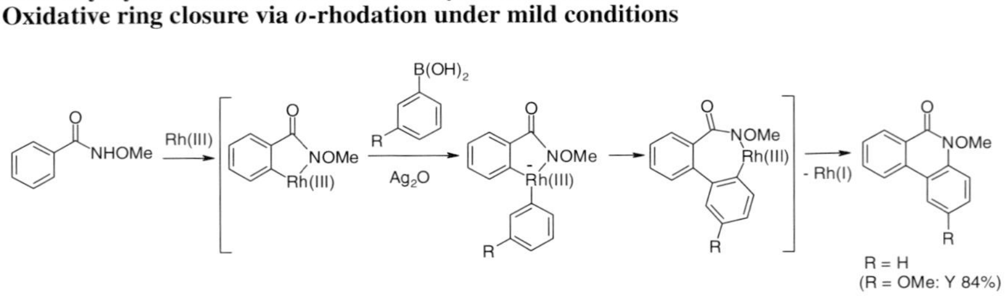

Methanol (3 ml) added under N_2 via syringe to a sealed tube containing $[Cp^*RhCl_2]_2$ (2 mol%), the startg. hydroxamate (0.6 mmol), phenylboronic acid (2 eq.) and Ag_2O (4 eq.), the mixture stirred at 60° for 3 h, diluted with methylene chloride, stirring continued in air for 10 min, filtered through a Celite and silica gel pad, and worked up with purification by chromatography on silica gel → 10-methoxy-9(10*H*)-phenanthridone. Y 94%. The procedure is widely applicable in high yield to the coupling of arylhydroxamates and arylboronic acids possessing alkyl, alkoxy, methylenedioxy, aryl or halogen substitution, **regioselective annelation** generally taking place with *m*-subst. arylboronic acids by coupling at the less hindered site (thirty-six examples; Y 72-94%). 2-Naphthylboronic acids, however, gave the more hindered product. Reaction involves a dual C-H activation, the catalytic cycle, it is thought, passing through 5- and 7-membered rhodium(III) metalacycles before the ultimate ring closing elimination with liberation of rhodium(I); Ag_2O then re-oxidizes the latter to rhodium(III) to continue the cycle. F.e. and comparison of solvents s. J. Karthikeyan, R. Haridharan, C.-H. Cheng, Angew. Chem., Int. Ed. *2012*, *51* (49), 12343-7 [DOI: 10.1002/anie.201206890].

Magnetically-recyclable, hollow nanocomposite palladium catalyst/potassium carbonate ←
Highly-active, magnetically-recyclable, self-assembled hollow nanocomposites as catalysts ←

458.

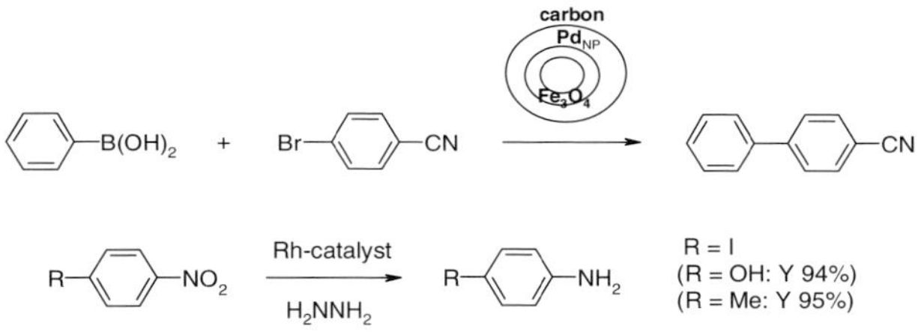

Heterogeneous Suzuki biaryl coupling in aq. medium. Startg. ar. halide (0.5 mmol), phenylboronic acid (1.2 eq.), Pd catalyst (1 mol%) and K_2CO_3 (4 eq.) in DMF/water (3:1) heated at 100° for 4 h → product. Y 95% (by GC-MS with decane as internal standard). Reaction of phenylboronic acid with five further ar. bromides (optionally bearing one or two Me groups) gave GC yields of 91-97% after 4 h, while six examples with ar. iodides (optionally bearing Me, MeO or *t*-Bu) gave similar yields after 1.5 h. The catalyst was prepared by adding $FeCl_3{\cdot}6H_2O$ (5.4 g) to sodium dodecyl sulfate [SDS] soln. (400 ml; 0.5 wt% in 1:1 water/ethanol), stirring for 1 h, heating to 80° for 5 h, stirring for another 12 h at room temp., adding $Pd(NO_3)_2{\cdot}H_2O$ (0.5 g) to the formed hollow iron hydroxide nanostructure [β-FeOOH], followed, with vigorous stirring, by pyrrole (as carbon source; 0.2 ml), stirring for 12 h at room temp., filtering, washing to yield a black solid, then heating at a rate of 0.9° min^{-1} to 300° under argon, maintaining that temp. for 5 h, then calcining at 200° for 2 h under H_2 gas flow (average size of Pd nanoparticles ~3 nm loaded uniformly on the carbon matrix; Pd loading ~1.7 wt%). This process is simple, economical and scalable and affords catalysts that are stable and readily recyclable after separation by applying a magnetic field. Such nanocomposites, which have increased surface area and low density, may be used as a general platform to load other noble metal catalyst nanoparticles such as rhodium, which can be utilized in reductions. **E: Prim. ar. amines from nitro compds.** A mixture of startg. nitrobenzene, hydrazine (2 eq.) and Rh catalyst (1 mol%) in ethanol heated at 80° for 2.5 h → *p*-iodoaniline. Y 99% by GC-MS. Six further examples gave yields of 91-99% by GC-MS. Both the palladium and rhodium catalysts may be reused for five consecutive cycles with little loss of activity. F.e.s. M. Shokouhimehr, J.E. Lee, S.I. Han, T. Hyeon, Chem. Commun. *2013*, *49* (42), 4779-81 [DOI: 10.1039/c3cc41034j]; palladium-catalyzed **Heck and Suzuki reactions** with a magnetic composite nanocatalyst [MP@NiSiO/Pd] having a rattle type nanostructure cf. S. Xuan, W. Jiang, X. Gong, Dalton Trans. *2011*, *40* (31), 7827-30 [DOI: 10.1039/c1dt10715a].

Palladium nanoparticles, complexes or supported variants ←
Suzuki coupling Ar-Ar′
update s. *37*, 902s*82*; *64*, 448s*82*; general and efficient synthesis of **biaryls** and heteroaromatic analogs from unactivated ar. chlorides or triflates or heteroar. halides with [hetero]arylboronic acids using *trifluoromethyl*-functionalized 1,3-diarylimidazolidin-2-ylidenes as ligand at low

catalyst loading s. T. Liu, X. Zhao, Q. Shen, L. Lu, Tetrahedron *2012, 68* (32), 6535-47 [DOI: 10.1016/j.tet.2012.05.068]; *in glycerol* under ultrasonic activation using N-heterocyclic carbene ligands possessing 3,4,5-trimethoxybenzyl, alkyl or sulfonate N-substituents s. A. Azua, J.A. Mata, P. Heymes, E. Peris, F. Lamaty, J. Martinez, E. Colacino, Adv. Synth. Catal. *2013, 355* (6), 1107-16 [DOI: 10.1002/adsc.201201047]; general procedure from ar. chlorides with dichloro-(3-chloropyridine)[1,4-bis(2,6-diisopropylphenyl)-3-methyl-1,2,3-triazolylidene]palladium(II) as catalyst in ethanol with KOBu-*t* as base s. J. Huang, J.-T. Hong, S.H. Hong, Eur. J. Org. Chem. *2012* (33), 6630-5 [DOI: 10.1002/ejoc.201201075]; with dichloro(isonitrile)palladium(II) diaminocarbene complexes as catalyst for Suzuki coupling with chlorine-subst. *meso*-tricarbo-indocyanine dyes s. S. Miltsov, V. Karavan, V. Boyarsky, S. Gómez-de Pedro, J. Alonso-Chamarro, M. Puyol, Tetrahedron Lett. *2013, 54* (10), 1202-4 [DOI: 10.1016/j.tetlet.2012.12.060]; coupling of bulky [*o*-subst.] arylboronic acids with bulky ar. bromides using $Pd(OAc)_2$ and 1,6-bis(di-*tert*-butylphosphino)hexane in aq. dioxane with Na_2CO_3 as base s. J.M. Mitchell, W.S. Brown, ibid. *2013, 54* (16), 2039-41 [DOI: 10.1016/j.tetlet.2013.01.060]; coupling with ar. chlorides using chloro(2′-aminobiphenyl-2-yl)palladium(II) 11-(dicyclohexylphosphino)-12-phenyl-9,10-dihydro-9,10-ethenoanthracene [KITPHOS] complexes at low loading s. S. Doherty, J.G. Knight, N.A.B. Ward, D.M. Bittner, C. Wills, W. McFarlane, W. Clegg, R.W. Harrington, Organometallics *2013, 32* (6), 1773-88 [DOI: 10.1021/om3011992]; coupling of heteroar. chlorides or tosylates at room temperature with $Pd(OAc)_2$/XPhos in aq. butanol with NaOH or CsOH as base s. J. Yang, S. Liu, J.-F. Zheng, J. Zhou, Eur. J. Org. Chem. *2012* (31), 6248-59 [DOI: 10.1002/ejoc.201200918]; coupling of heteroaryl halides with 3-pyridyl- or 3-thienyl-boronic acid using bis(η^3-allylpalladium chloride) and N,N,N′,N′-tetrakis(diphenylphosphinomethyl)ethylenediamine as ligand at low loading s. K. Wang, Q. Fu, R. Zhou, X. Zheng, H. Fu, H. Chen, R. Li, Appl. Organomet. Chem. *2013, 27* (4), 232-8 [DOI: 10.1002/aoc.2965]; coupling with O^6-alkyl-2-halogenoinosines (and with concomitant *in situ*-O-dealkylation) using dichloro[1,1′-bis(dicyclohexylphosphino)-ferrocene]palladium(II) as catalyst s. V. Gurram, N. Pottabathini, R. Garlapati, A.B. Chaudhary, B. Patro, M.K. Lakshman, Chem. Asian J. *2012, 7* (8), 1853-61 [DOI: 10.1002/asia.201200093]; coupling of *o*-subst. arylboronic acids with 10-halogenobenzo[*h*]quinolines in aq. media with triphenylphosphine oxide as ligand s. M. Weimar, M.J. Fuchter, Org. Biomol. Chem. *2013, 11* (1), 31-4 [DOI: 10.1039/c2ob26806j]; coupling of ar. chlorides using [2.2]paracyclophane-subst. tertiary 1-aryl-1,2,3-triazol-4-ylphosphines as ligand s. M. Austeri, M. Enders, M. Nieger, S. Bräse, Eur. J. Org. Chem. *2013* (9), 1667-70 [DOI: 10.1002/ejoc.201300031]; with planar-chiral analogs s. J.E. Glover, D.J. Martin, P.G. Plieger, G.J. Rowlands, ibid. *2013* (9), 1671-5 [DOI: 10.1002/ejoc.201201601]; Suzuki coupling with highly active calix[4]arene-derived triarylphosphine N-(*o*-methoxyphenyl)imines as ligand under palladium or nickel catalysis (TOFs up to ca. 4×10^5 h^{-1}) s. L. Monnereau, D. Sémeril, D. Matt, Adv. Synth. Catal. *2013, 355* (7), 1351-60 [DOI: 10.1002/adsc.201300091]; *phosphine-free coupling* with cyclopalladated thiosemicarbazone complexes cf. H. Yan, P. Chellan, T. Li, J. Mao, K. Chibale, G.S. Smith, Tetrahedron Lett. *2013, 54* (2), 154-7 [DOI: 10.1016/j.tetlet.2012.10.115]; with a recyclable alkoxopalladium(II) complex based on an N,O-ligand s. Y. Li, X. Mi, M. Huang, R. Cai, Y. Wu, Tetrahedron *2012, 68* (40), 8502-8 [DOI: 10.1016/j.tet.2012.07.026]; with a diacetatopalladium(II) biguanide complex (based on metformin) in aq. ethanol s. A. Alizadeh, M.M. Khodaei, D. Kordestani, M. Beygzadeh, Tetrahedron Lett. *2013, 54* (4), 291-4 [DOI: 10.1016/j.tetlet.2012.11.031]; with $PdCl_2$ and an N-(2-pyridyl)-N′-(2,6-diisopropylphenyl)benzamidine as ligand in aq. DMF s. Y.-T. Huang, X. Tang, Y. Yang, D.-S. Shen, C. Tan, F.-S. Liu, Appl. Organomet. Chem. *2012, 26* (12), 701-6 [DOI: 10.1002/aoc.2913]; with bis(N-benzyltriethylenediammonium) bis(trichloropalladate) as catalyst in ethanol s. A.R. Hajipour, I.M. Dehbane, F. Rafiee, ibid. *26* (12), 743-7 [DOI: 10.1002/aoc.2926]; study of various parameters affecting coupling *in water,* with particular reference to the water/organic interface s. C. Röhlich, A.S. Wirth, K. Köhler, Chem. Eur. J. *2012, 18* (48), 15485-94 [DOI: 10.1002/chem.201201266]; ligand-free Suzuki coupling with a wide range of ar. halides *in water* with $Pd(OAc)_2$ and diisopropylamine as base (in air) s. C. Liu, Y. Zhang, N. Liu, J. Qiu, Green Chem. *2012, 14* (11), 2999-3003 [DOI: 10.1039/c2gc36098e]; coupling of [hetero]aryl chlorides in water at *ultra-low* catalyst loading with $Pd(OAc)_2$/(2-mesitylinden-3-yl)dicyclohexylphosphine/methyltrioctylammonium chloride/$K_3PO_4\cdot3H_2O$ s. S.-L. Mao, Y. Sun, G.-A. Yu, C. Zhao, Z.-J. Han, J. Yuan, X. Zhu, Q. Yang, S.-H. Liu, Org. Biomol. Chem. *2012, 10* (47), 9410-7 [DOI: 10.1039/c2ob26463c]; *heterogeneous* coupling using robust, recyclable. ligand-free, polymer-based palladium *nanoparticles* s. H. Mahdavi, P. Hashemi, J. Chin. Chem. Soc. *2013, 60* (4), 425-36 [DOI: 10.1002/

jccs.201200257]; using recyclable palladium nanoparticles immobilized on amine-functionalized ionic liquid-modified magnetic nanoparticles at room temperature s. J. Wang, B. Xu, H. Sun, G. Song, Tetrahedron Lett. *2013, 54* (3), 238-41 [DOI: 10.1016/j.tetlet.2012.11.009]; using shell cross-linked palladium nanoparticles prepared from alkene-terminated phosphine-type dendron-stabilized palladium nanoparticles via olefin metathesis polymerization s. T. Shiomi, T. Nakahodo, H. Fujihara, Chem. Lett. *2012, 41* (12), 1700-2 [DOI: 10.1246/cl.2012.1700]; ligand-free heterogeneous coupling of 3-iodo-pyridines and -quinolines using Pd-C s. M. Lamblin, H. Bares, J. Dessolin, C. Marty, N. Bourgougnon, F.-X. Felpin, Eur. J. Org. Chem. *2012* (28), 5525-33 [DOI: 10.1002/ejoc.201200721]; coupling with palladium immobilized on amine-functionalized cellulose-aluminum oxide composite in water/DMF s. A. Kumbhar, S. Jadhav, S. Kamble, G. Rashinkar, R. Salunkhe, Tetrahedron Lett. *2013, 54* (11), 1331-7 [DOI: 10.1016/j.tetlet.2012.11.140]; in water with recyclable palladium (or nickel) nanocomposite carbon aerogels under aerobic and phosphine-free conditions s. L. Martín, E. Molins, A. Vallribera, Tetrahedron *2012, 68* (32), 6517-20 [DOI: 10.1016/j.tet.2012.05.099]; coupling with recyclable palladium ligated to phosphine-functionalized, pyrene-tagged dendrimers grafted onto magnetic Co-C nanoparticles s. M. Keller, V. Collière, O. Reiser, A.-M. Caminade, J.-P. Majoral, A. Ouali, Angew. Chem., Int. Ed. *2013, 52* (13), 3626-9 [DOI: 10.1002/anie.201209969]; with a palladium di(phosphine) complex covalently linked to mesoporous silica for coupling ar. chlorides s. S.-y. Onozawa, T. Miyaji, N. Fukaya, M. Yoshinaga, M. Ueda, Y. Takagi, H. Yasuda, Chem. Lett. *2013, 42* (3), 275-7 [DOI: 10.1246/cl.2013.275]; Suzuki coupling of ar. bromides or chlorides **with diarylborinic acids or anhydrides** using $Pd(OAc)_2$ ligated synergistically to 1,3-bis(2,6-diisopropylphenyl)imidazol-2-ylidene and triphenyl phosphite at low loading in *tert*-butanol with $K_3PO_4{\cdot}3H_2O$ as base s. X. Chen, H. Ke, Y. Chen, C. Guan, G. Zou, J. Org. Chem. *2012, 77* (17), 7572-8 [DOI: 10.1021/jo301335x]; **with triphenylborane·pyridine** using $Pd(OAc)_2/Cs_2CO_3$ s. M. Yang, Y. Gu, Y. Wang, X. Zhao, G. Yan, Chin. J. Org. Chem. *2012, 30* (10), 2581-6 [DOI: 10.1002/cjoc.201200775]; base-free coupling of arylboronic acid esters **with ar. fluorides** using palladium(0) phosphine complexes or nickel(0) NHC complexes s. M. Ohashi, H. Saijo, M. Shibata, S. Ogoshi, Eur. J. Org. Chem. *2013* (3), 443-7 [DOI: 10.1002/ejoc.201201405]; of arylboronic acids **with aryl sulfonates** using dichloro-[1,3-bis(2,6-diisopropylphenyl)imidazol-2-ylidene]-(N-methylimidazole)palladium(II)/morpholine/$K_3PO_4{\cdot}3H_2O$ s. Z.-Y. Wang, G.-Q. Chen, L.-X. Shao, J. Org. Chem. *2012, 77* (15), 6608-14 [DOI: 10.1021/jo301270t]; Suzuki coupling **in a flow microreactor** using a highly reactive palladium catalyst flash-generated from $Pd(OAc)_2$/*t*-Bu_3P s. A. Nagaki, N. Takabayashi, Y. Moriwaki, J.-i. Yoshida, Chem. Eur. J. *2012, 18* (38), 11871-5 [DOI: 10.1002/chem.201201579]; **nickel-catalyzed Suzuki coupling** of heteroarylboronic acids with heteroaryl halides using [(dppf)Ni(Cl)(cinnamyl)]/K_3PO_4 (cf. *51,* 453s*80*) s. S. Ge, J.F. Hartwig, Angew. Chem., Int. Ed. *2012, 51* (51), 12837-41 [DOI: 10.1002/anie.201207428].

Palladium nanoparticles *Pd*
Palladium(II) bis(hydrazone) or phosphine complexes *Pd(II)*
***sp²-sp³*-Suzuki coupling** Ar-R

of arylboronic acids or esters with alkyl halides, update s. *64,* 453s*82*; rapid synthesis of benzyl fluorides-^{18}F from arylboronic acid esters and fluoromethyl halides-^{18}F with $Pd_2(dba)_3$/tri-*o*-tolylphosphine/K_2CO_3 s. H. Doi, M. Goto, M. Suzuki, Bull. Chem. Soc. Jpn. *2012, 85* (11), 1233-8 [DOI: 10.1246/bcsj.20120151]; of *p*-aminotoluenes-^{11}C from arylboronic acid esters and methyl iodide-^{11}C s. V.L. Andersen, M.M. Herth, S. Lehel, G.M. Knudsen, J.L. Kristensen, Tetrahedron Lett. *2013, 54* (2), 213-6 [DOI: 10.1016/j.tetlet.2012.11.001]; of aryl(N-heteroaryl)methanes from arylboronic acids and N-heteroarylmethyl chlorides using a benzylamine-type pallad^{II}acyclic phosphine complex in aq. THF with K_3PO_4 as base s. J.R. Schmink, M.T. Tudge, ibid. *54* (1), 15-20 [DOI: 10.1016/j.tetlet.2012.09.112]; **allylarenes** from aryloxy-2-ethylenes with $Pd(OAc)_2$ and a glyoxal bis(hydrazone) as ligand, notably for preparing 1,3-diarylpropenes, s. T. Mino, T. Kogure, T. Abe, T. Koizumi, T. Fujita, M. Sakamoto, Eur. J. Org. Chem. *2013* (8), 1501-5 [DOI: 10.1002/ejoc.201201276]; from 2-ethylenecarbonic acid esters *with inversion of absolute configuration* and retention of geometry using highly active palladium nanoparticles *in water* s. J. Zhao, J. Ye, Y.J. Zhang, Adv. Synth. Catal. *2013, 355* (2-3), 491-8 [DOI: 10.1002/adsc.201200704]; **1,1-diarylalkanes** from α-subst. quaternary benzyl(trimethyl)ammonium salts with inversion of absolute configuration using $Ni(cod)_2$ with (*o*-Tol)$_3$P or *t*-Bu-XantPhos as ligand and K_3PO_4 or CsF as base s. P. Maity, D.M. Shacklady-McAtee, G.P.A. Yap, E.R. Sirianni,

M.P. Watson, J. Am. Chem. Soc. *2013, 135* (1), 280-5 [DOI: 10.1021/ja3089422]; from α-subst. benzyl carbamates or pivalates and arylboronic acid esters under $Ni(cod)_2$ catalysis with ligand-dependent [NHC cf. phosphine] inversion *or retention* of absolute configuration s. M.R. Harris, L.E. Hanna, M.A. Greene, C.E. Moore, E.R. Jarvo, ibid. *135* (9), 3303-6 [DOI: 10.1021/ja311783k]; one-pot **synthesis of ethylene derivs. from α,β-ethyleneiodides** and alkyl iodides **via potassium B-alkyl-B-methoxy-9-borabicyclo[3.3.1]nonanates** using $Pd(OAc)_2/K_3PO_4$ and a hemilabile 2′,6′-dimethyl-2′-(dicyclohexylphosphino)biphenyl-4-ylcarboxylic acid amide [Aphos-Y] as ligand s. N. Ye, W.-M. Dai, Eur. J. Org. Chem. *2013* (5), 831-5 [DOI: 10.1002/ejoc.201201602].

π-Allyl(η⁵-cyclopentadienyl)palladium(II)/chiral 1,3-bis(phosphorodiamidites) *[Pd(II)]**
2,3-Disubst. 3-cyano-4-methylene-N-tosylpyrrolidines
from 2-methylene-3-silylcyanohydrin acetates and N-tosylaldimines
via regioselective palladium(II)-catalyzed asym. [3+2]-annelation ○

459.

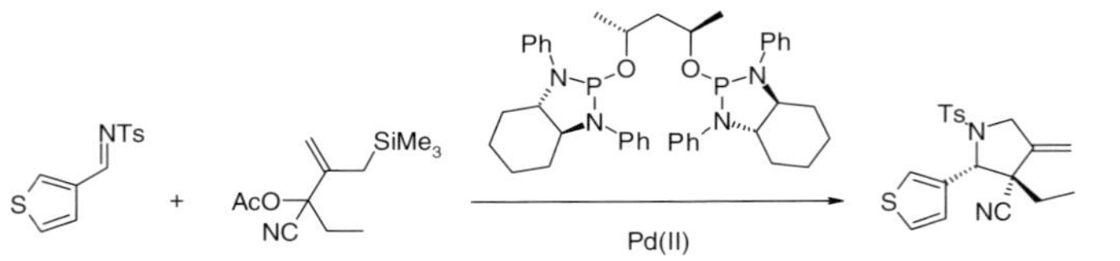

Novel use of disubst. cyanoalkyl donors in this trimethylenemethane cycloaddition, affording products with all-carbon quaternary centers, is reported. **E:** Chiral bis(phosphorodiamidite) ligand (6 mol%), $[PdCp(\eta^3\text{-}C_3H_5)]_2$ (5 mol%) and toluene (0.2 ml) added to 3-cyano-2-(trimethylsilylmethyl)pent-1-en-3-yl acetate (0.1 mmol) under argon, the soln. stirred for 5 min, N-tosylthien-3-ylcarbaldimine (1.3 eq.) added, the mixture stirred at 60° overnight, concentrated, and purified by flash chromatography → (2R,3S)-3-ethyl-4-methylene-2-(thien-3-yl)-1-tosylpyrrolidine-3-carbonitrile. Y 92% (d.r. 12:1; e.e. 91%). The cycloaddition appears general for electron-diverse N-tosyl (het)aryl and aliphatic aldimines reacting with 2-methylene-3-silylcyanohydrin acetates containing a linear α-alkyl group (incl. 4-trimethylsilylbut-3-yn-1-yl) (eighteen examples; Y 54-97%; d.r. 3:1 to >20:1; e.e. 74-93%; a cyanohydrin acetate containing α-isobutyl gave Y 32%). In general the illustrated ligand gave the best results, but in some cases, depending on the acceptor, an analogous ligand bearing monocyclic diazaphospholidine rings proved more effective. F.e., optimization, ligand/substrate prepn., and further transformations of the cycloadducts, s. B.M. Trost, T.M. Lam, M.A. Herbage, J. Am. Chem. Soc. *2013*, *135* (7), 2459-61 [DOI: 10.1021/ja312351s].

Palladium(II) acetate (s.a. under Electrolysis and K_2CO_3) or supported variant *[Pd(II)]*
Cross-coupling with triarylbismuthines ←
update s. *45*, 550s*82*; of tri[hetero]arylbismuthines with 3,6-dihalogenopyridazines using $Pd(OAc)_2$, chemoselectivity, s. K. Urgin, C. Aubé, M. Pipelier, V. Blot, C. Thobie-Gautier, S. Sengmany, J. Lebreton, E. Léonel, D. Dubreuil, S. Condon, Eur. J. Org. Chem. *2013* (1), 117-24 [DOI: 10.1002/ejoc.201200977]; **benzophenones** from arylcarboxylic acid chlorides and triarylbismuthines with recyclable, phosphine-free $Pd(OAc)_2$ immobilized through a 1,2-diamine linker to MCM-41 (in the presence of tributylamine) s. H. Zhao, L. Yin, M. Cai, Eur. J. Org. Chem. *2013* (7), 1337-45 [DOI: 10.1002/ejoc.201201432].

Palladium(II) acetate/potassium carbonate *$Pd(OAc)_2/K_2CO_3$*
Palladium-catalyzed coupling with N-aryl-N′-tosylhydrazines ArNHNHTs → Ar-Ar′

460.

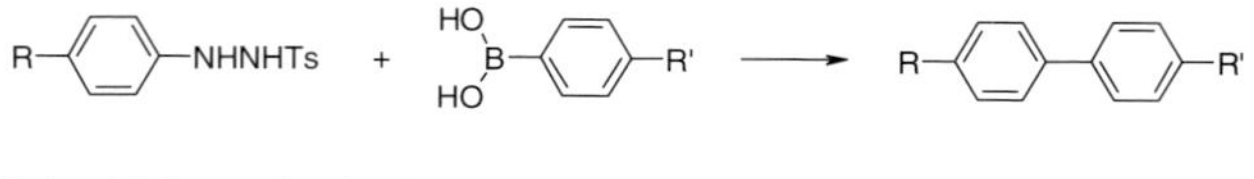

R = F_3CO, R' = OMe
(R = R' = H: Y 92%)
(R = NO_2, R' = OMe: Y 78%)
(R = Br, R' = OMe: Y 75%)

N-Aryl-N′-tosylhydrazines have been shown to be readily available and stable electrophiles which may be used in a variety of coupling reactions. **E: Suzuki biaryl coupling**. A mixture of startg. N-aryl-N′-tosylhydrazine (0.2 mmol), arylboronic acid (1.2 eq.), $Pd(OAc)_2$ (2 mol%) and K_2CO_3 (2 eq.) in methanol (2 ml) stirred at 60° for 2-12 h (TLC) under N_2, quenched with brine, extracted with ethyl acetate, and purified by flash column chromatography → 4-methoxy-4′-(trifluoromethoxy)biphenyl. Y 86%. Fifteen further examples gave yields of 70-93%, tolerating Cl, Br and NO_2 groups as well as ethers. Eight examples with diethyl(3-pyridyl)borane gave yields of 55-76% and two with pinacol 2-aminopyrimidin-2-ylboronate gave yields of 81% and 85% after re-optimization with DMSO as solvent. Extension to Hiyama coupling gave biphenyl in 88% yield [using $Pd(OAc)_2$/KF] while coupling with *p*-iodoanisole produced 4-methoxybiphenyl in 79% yield.

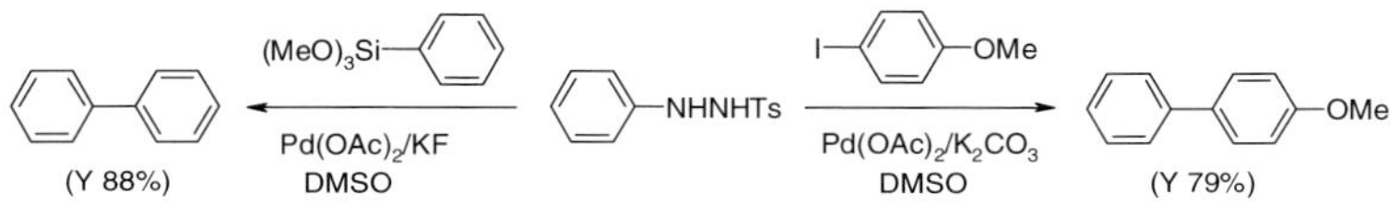

It is believed reaction proceeds via dehydrogenation of the hydrazine to form an N-aryl-N′-tosylazo compd. in equilibrium with the corresponding diazonium ion, which undergoes oxidative addition to the palladium catalyst; transmetalation of the arylboronic acid then affords a diarylpalladium species which undergoes reductive elimination. F.e.s. J.-B. Liu, H. Yan, H.-X. Chen, Y. Luo, J. Weng, G. Lu, Chem. Commun. *2013*, *49* (46), 5268-70 [DOI: 10.1039/c3cc41421c].

Palladium(II) acetate/copper(II) acetate/pivalic acid *Pd(OAc)$_2$/Cu(OAc)$_2$/t-BuCOOH*
Directed *o*-trifluoromethylation of anilides s. *78,* 476s*83* H → CF_3

Palladium(II) acetate/(S)-proline *Pd(OAc)$_2$/Pro-OH*
Asym. 1,4-addition of arylboronic acids to α,β-ethylenealdehydes C=C → CHC(Ar)
s. *55,* 452s*83*

Palladium(II) acetate/triphenylphosphine/potassium carbonate/air *[Pd]/K$_2$CO$_3$/O$_2$*
6*H*-Dibenzo[*b,d*]pyran-6-ones from *o*-hydroxyboronic acids and *o*-halogenaldehydes via Suzuki biaryl coupling-oxidative lactonization ○

461. 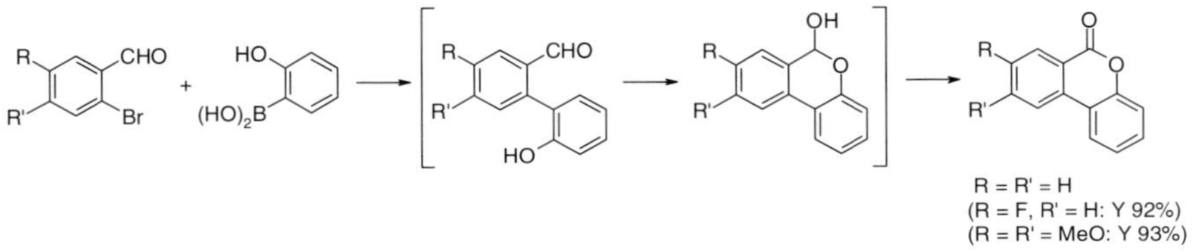

o-Bromobenzaldehyde (1 mmol), *o*-hydroxyphenylboronic acid (1 eq.), K_2CO_3 (1 eq.) and Ph_3P (25 mol%) added to a two-necked round bottom flask, flushed with N_2, DMF (3 ml) added and degassed with N_2, after addition of $Pd(OAc)_2$ (5 mol%) the mixture heated at 90° for 4 h, allowed to cool to room temp., diluted with water, extracted with ethyl acetate, the extracts dried (Na_2SO_4), evaporated under reduced pressure, and the residue purified by chromatography on silica gel → 6*H*-benzo[*c*]chromen-6-one. Y 90%. *o*-Chlorobenzaldehyde gave 87% yield after 8 h and the *o*-iodo-deriv. gave 91% after 3 h. This provides an efficient one pot route to 6*H*-benzo[*c*]chromen-6-ones, 5*H*-naphtho[1,2-*c*]chromen-5-ones and 6*H*-naphtho[2,1-*c*]chromen-6-ones, which serve as the core structures of a number of natural products and as important intermediates for several pharmaceutically important compounds, via Suzuki-Miyaura cross-coupling, hemiacetal formation and aerobic oxidative lactonization. F.e. (seven; Y 68-93%) s. R. Singha, S. Roy, S. Nandi, P. Ray, J.K. Ray, Tetrahedron Lett. *2013*, *54* (7), 657-60 [DOI: 10.1016/j.tetlet.2012.11.144]; **6-β-keto-6*H*-dibenzo[*b,d*]pyrans** from β-(*o*-bromoaryl)-α,β-ethyleneketones via aerobic, ligand-free Suzuki biaryl coupling-intramolecular oxa-Michael addition with $Pd(OAc)_2/K_3PO_4/Bu_4NBr$ in water (generating Pd nanoparticles) s. A. Ahmed, Y. Nuree, J.K. Ray, ibid. 665-8 [DOI: 10.1016/j.tetlet.2012.12.005].

Palladium(II) acetate/2′,6′-dimethyl-2′-(dicyclohexylphosphino)biphenyl-4-ylcarboxylic acid amides/potassium phosphate/B-methoxy-9-borabicyclo[3.3.1]nonane ←

Synthesis of ethylene derivs. from α,β-ethyleneiodides C=C(I) → C=C(R)

and alkyl iodides s. *64,* 453s*83*

Bis(dibenzylideneacetone)palladium(0)/2-[bis[3,5-bis(trifluoromethyl)phenyl]phosphino]-2′,4′,6′-triisopropyl-3,6-dimethoxybiphenyl/copper(I) chloride/potassium fluoride ←

***sec*-Alkylarenes** ArX + RSn⋞ → Ar-R

from Sn-*sec*-alkylazastannatranes and ar. halides or aryl triflates

Stille sp^2-sp^3-coupling with retention of absolute configuration

462.

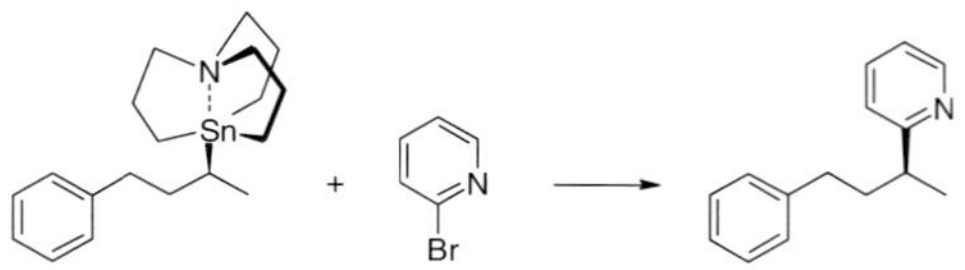

The authors report the first Stille coupling of a configurationally stable *sec*-alkyl nucleophile, specifically an Sn-*sec*-alkylazastannatrane, which proceeds cleanly *with retention* of absolute configuration and, additionally, *without isomerization* of the alkyl group. **E:** The startg. (S)-Sn-*sec*-alkylazastannatrane (1.5 eq.) and 2-pyridyl bromide (1 mmol), followed by degassed acetonitrile (3 ml), added via microsyringe to a Schlenk tube containing a mixture of $Pd(dba)_2$ (5 mol%), JackiePhos (10 mol%), CuCl (2 eq.) and KF (2 eq.) under argon, the tube sealed with a Teflon stopper, heated to 60° for 18 h (unoptimized reaction time), cooled to room temp., diluted with ether, the organic layer washed sequentially with satd. aq. KF and brine, dried, and worked up with purification by chromatography on silica gel → (S)-product. Y 65% (e.e. 91%). Notably, coupling does *not* require silyl activation by an α-heteroatom or a carbonyl group, and an extensive preliminary study of the coupling of racemic Sn-*sec*-alkylazastannatranes confirmed that the method is applicable to both electron-diverse nucleophiles and a wide range of [hetero]aryl chlorides, bromides and iodides or triflates (seventeen examples in all; Y 26%, 45-94%; with <1-4% alkyl isomerization, and three instances of retention of absolute configuration). F.e. and comparison of catalysts s. L. Li, C.-Y. Wang, R. Huang, M.R. Biscoe, Nat. Chem. *2013, 5* (7), 607-12 [DOI: 10.1038/nchem.1652].

Tris(dibenzylideneacetone)dipalladium/potassium fluoride *$Pd_2(dba)_3$/KF*

Regio- and stereo-selective synthesis of 1,4-dienes ←

from unsatd. boronic acids or esters, enol triflates and 1,3-butadiene

463.

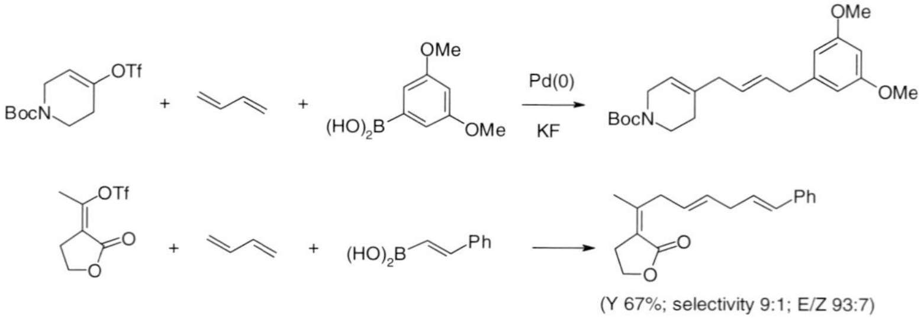

6-Aryl-1,4-dienes. A soln. of 1,3-butadiene (0.5 mmol) in *n*-hexane (0.36 ml) added to a mixture of KF (1.7 eq.), 3,5-dimethoxybenzeneboronic acid (1.5 eq.), $Pd_2(dba)_3$ (3 mol%) and *tert*-butyl 4-trifluoromethanesulfonyloxy-5,6-dihydropyridine-1(2*H*)-carboxylate (1.5 eq.) in DMA (2.5 ml) under N_2 in a glovebox, the mixture stirred at 55° for 16 h, cooled to room temp., diluted with ether, washed with water, and worked up with purification by flash chromatography → (E)-*tert*-butyl 4-[4-(3,5-dimethoxyphenyl)but-2-en-1-yl]-5,6-dihydropyridine-1(2*H*)-carboxylate. Y 66%

(selectivity >20:1; E/Z 93:7). This novel 3-component 1,4-addition of electron-diverse areneboronic acids and (a)cyclic enol triflates to 1,3-butadiene gave high ratios (>20:1) of 1,4:1,2-addition products due to removal of undesired 1,2-adducts by reaction with a stoichiometric excess of the enol triflates, with yield implications for the required 1,4-adducts (44-85%; thirteen examples; E/Z 89:11 to 94:6). The reaction was compatible with aldehyde, carbamate, lactone, ketone, halo, ether and amide functionality. The method was extended to the prepn. of skipped polyenes, **1,4,7-trienes**, using α,β-ethyleneboronic acids/esters (five examples; Y 43-71%; selectivity 2.9:1 to 9:1; E/Z 91:9 to 93:7), including a key intermediate containing the 1,4,7-triene core of the RNA-polymerase inhibitor, ripostatin A. F.e., optimization and substrate prepn. s. M.S. McCammant, L. Liao, M.S. Sigman, J. Am. Chem. Soc. *2013*, *135* (11), 4167-70 [DOI: 10.1021/ja3110544].

Tris(dibenzylideneacetone)dipalladium/1,2-bis(diphenylphosphino)benzene/cesium fluoride ←
Regio- and diastereo-selective synthesis of 4-allylcyclobut-2-enecarboxylic acids from 2-oxabicyclo[2.2.0]hex-5-en-3-ones and β,γ-ethyleneboronic acid esters C

464.

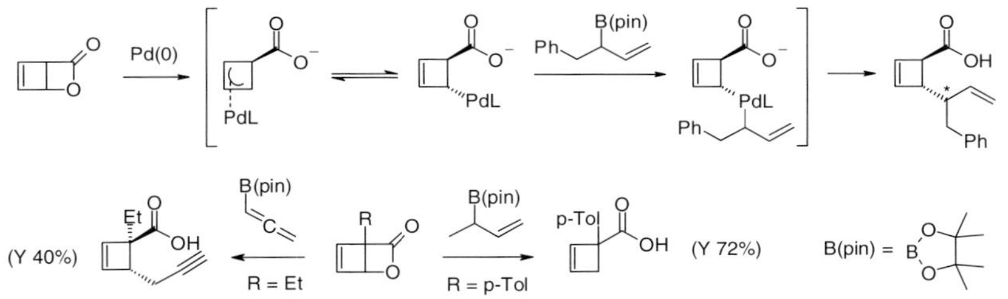

A soln. of $Pd_2(dba)_3$ (5 mol%), 1,2-bis(diphenylphosphino)benzene (10 mol%), anh. CsF (10 eq.) and THF (0.5 ml) stirred in a sealed vial under argon for 5 min, 4,4,5,5-tetramethyl-2-(1-phenylbut-3-en-2-yl)-1,3,2-dioxaborolane (1.2 eq.) and a soln. of 2-oxabicyclo[2.2.0]hex-5-en-3-one (0.2 mmol) in ether (0.95 ml) added, the mixture stirred at room temp. for 2 h, quenched with aq. $NaHCO_3$, washed with ether, acidified with 1.2 *M* aq. HCl (pH ~2), and extracted with ethyl acetate → crude *trans*-4-(1-phenylbut-3-en-2-yl)cyclobut-2-enecarboxylic acid (isolated as its N-benzylamide by treatment of the crude product with $BnNH_2$/EDCI/HOBt, followed by chromatographic purification on silica; Y 62%; d.r. 8:1). *trans*-Cleavage of the bicyclic lactones (carrying H, ethyl or *p*-tolyl adjacent to the carbonyl) with the allylboronate and α-subst. derivs., afforded branched products exclusively and diastereoselectively at the allylic carbon (five examples; Y 38-62%; d.r. 4:1 to 8:1). No reaction occurred with β-subst. allylboronates, while a γ-subst. analog gave a mixture of branched/linear regioisomers in 10% yield, and allenylboronates afforded **4-propargylcyclobut-2-encarboxylic acids** (two examples; Y 40-45%). An asym. variant [using (R)-4-*tert*-butyl-2-(2-diphenylphosphinophenyl)-Δ^2-oxazoline [*t*-BuPhox] as ligand] gave high enantioselectivity (absolute configuration not established) from the unsubst. bicyclic lactone (four examples; Y 56-75%; e.e. 89-98%; isolated as benzyl esters or N-benzylamides), with no diastereoselectivity observed for branched products. Ethyl- and *p*-tolyl-subst. lactones gave poor enantioselectivity with the unsubst. allylboronate (e.e. 44-69%; two examples; Y 34-51%), but with α-subst. allylboronates suffered β-elimination of the allylic moiety from the final intermediate to afford **1-subst. cyclobut-2-enecarboxylic acids** (three examples; Y 45-74%) with moderate (and undefined) enantioselectivity in the presence of the chiral ligand (e.e. 60-70%; three examples; Y 32-78%). F.e., optimization, substrate prepn. and reactions of the products s. S. Niyomchon, D. Audisio, M. Luparia, N. Maulide, Org. Lett. *2013*, *15* (9), 2318-21 [DOI: 10.1021/ol401033g].

Palladium complexes *[Pd]*
Hiyama coupling Si≡ → R
update s. *70*, 460s*82*; *fluorine-free* **biaryl synthesis** from trimethoxy(phenyl)silane and ar. bromides or chlorides with $Pd(OAc)_2$ (0.1 mol%) and *in situ*-generated 1-benzyl-3-(2-hydroxy-2-phenylethyl)imidazol-2-ylidene (0.5 mol%) as ligand under microwave enhancement s. I. Peñafiel, I.M. Pastor, M. Yus, Eur. J. Org. Chem. *2013* (8), 1479-84 [DOI: 10.1002/ejoc.201201066];

palladium-catalyzed **Stille *sp*²-*sp*³ coupling** of benzyl and allyl halides or sulfonates with Si,Si′-divinyldisiloxanes s. E.C. Frye, C.J. O′Connor, D.G. Twigg, B. Elbert, L. Laraia, D.G. Hulcoop, A.R. Venkitaraman, D.R. Spring, Chem. Eur. J. *2012, 18* (28), 8774-9 [DOI: 10.1002/chem.201200431].

Palladium complexes *[Pd]*
Stille coupling Sn≤ → R
updates s. *34,* 862s*82*; *39,* 887s*82*; widely applicable coupling of aryl-, alkenyl- or alkynyl-stannanes with aryl- or alkenyl-halides or triflates *in water* at room temperature under palladium catalysis with *nonionic amphiphile TPGS-750-M* as additive s. G.-p. Lu, C. Cai, B.H. Lipshutz, Green Chem. *2013, 15* (1), 105-9 [DOI: 10.1039/c2gc36042j]; coupling of bulky electrophiles *with gold(I) as cocatalyst* s. J. del Pozo, D. Carrasco, M.H. Pérez-Temprano, M. García-Melchor, R. Álvarez, J.A. Casares, P. Espinet, Angew. Chem., Int. Ed. *2013, 52* (8), 2189-93 [DOI: 10.1002/anie.201209262]; Stille **biaryl coupling** using bis(1-benzyl-4-aza-1-azoniabicyclo[2.2.2]octane) trichloropalladate dimer with added CsF under microwave irradiation s. A.R. Hajipour, N. Najafi, F. Rafiee, Appl. Organomet. Chem. *2013, 27* (4), 228-31 [DOI: 10.1002/aoc.2958]; chemoselective one-pot **Stille-Suzuki coupling** of 5-stannylthiophene-2-boronic acid esters with two different aryl halides s. J. Linshoeft, A.C.J. Heinrich, S.A.W. Segler, P.J. Gates, A. Staubitz, Org. Lett. *2012, 14* (22), 5644-7 [DOI: 10.1021/ol302571t]; alternative *radical coupling* of allyl-, alkynyl- or allenyl-stannanes with iodomethylphosphorus(V) compds. using an indium(III) halide under UV irradiation s. I. Suzuki, K. Kiyokawa, M. Yasuda, A. Baba, ibid. *2013, 15* (7), 1728-31 [DOI: 10.1021/ol4005257].

Palladium(II) complexes/iron(III) triflate *[Pd(II)]/Fe(OTf)₃*
1,4-Addition of arylboronic acids s. *62,* 449s*83* C=C → CHC(Ar)

Chiral diacetoxo[2,2′-bi(Δ²-oxazolinyl)]palladium(II) complexes *[Pd(II)]**
Asym. synthesis of sec. benzylamines from aldimines CH=NR → CH(NHR)Ar
and arylboronic acids s. *68,* 458s*83*

Tetrakis(acetonitrile)palladium(II) bis(fluoroborate)/6,6′-dimethyl-2,2′-bipyridyl/ sodium anthraquinone-2-sulfonate/oxygen/dimethyl sulfoxide ←
***m*-Hydroxybiaryls from 2-cyclohexenones and arylboronic acids** ←
via 3-aryl-2-cyclohexenones
via aerobic Heck-type arylation-dehydrogenative aromatization

465.
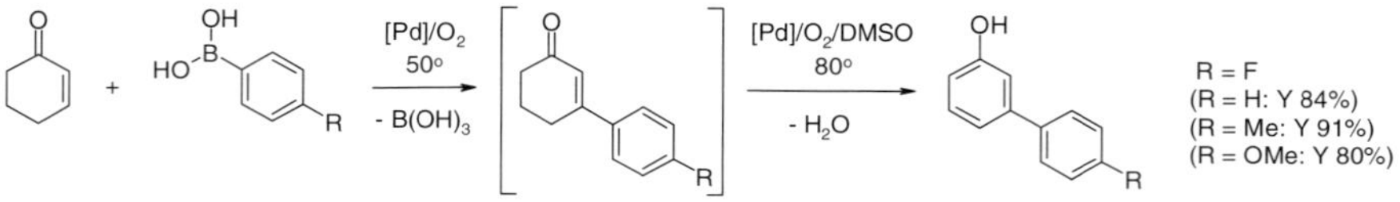

A novel palladium(II) catalyst system is reported to promote both the aerobic oxidative Heck coupling and dehydrogenation of cyclohexenones, and these reactions may be combined sequentially in one pot, providing a straightforward synthesis of *m*-subst. phenols under neutral conditions. **E:** $Pd(MeCN)_4(BF_4)_2$ (5 mol%), 6,6′-dimethyl-2,2′-bipyridyl (5 mol%), cyclohexenone (2 eq.), startg. arylboronic acid (0.25 mmol), anthraquinone-2-sulfonic acid sodium salt [AMS] (10 mol%) and N-methyl-2-pyrrolidone (0.08 ml) combined in a disposable heavy wall culture tube, the tubes placed in a 48-well aluminum block mounted on a Large Capacity Mixer [Glas-Col] [enabling several reactions to be performed simultaneously], the headspace above the tubes purged with O_2 for ca. 5 min, the mixture vortexed under 1 atm. O_2 at 50°, after 4 h, DMSO (0.2 ml) added to the formed soln. of arylcyclohexenone (isolable in 86% yield), the mixture heated at 80° for 32 h, then the O_2 inlet turned off, ethyl acetate added, followed by aq. NH_4Cl, and the mixture worked up with purification by flash chromatography on silica gel → product. Y 87%. Initially conditions for the formation of [*o*-, *m*- or *p*-subst.] **phenols from 2-cyclohexenones** were investigated and eleven examples gave yields of 46-96% in DMSO containing a little water, with functional groups such as CO_2Me, CHO, COMe, CF_3 or OH tolerated. $Pd(MeCN)_4(BF_4)_2$/6,6′-dimethyl-2,2′-bipyridyl was then found to be effective for oxidative Heck reactions, electron-rich arylboronic acids generally giving higher yields than electron-poor ones (which are more susceptible to homocoupling) in NMP (eighteen examples; Y 41-98%), with NO_2, CN, and halo (F, Cl, Br) also tolerated. One pot reactions gave 35-91% yields for seventeen further examples. F.e. incl. application to the synthesis of a potent fatty acid amide hydrolase inhibitor, URB597, s. Y. Izawa, C. Zheng, S.S. Stahl, Angew. Chem., Int. Ed. *2013*, *52* (13), 3672-5 [DOI: 10.1002/anie.201209457].

Bis(acetonitrile)palladium(II) ditosylate/chiral 2-(2-pyridyl)-Δ^2-oxazolines/ copper(II) triflate/oxygen *[Pd(II)]*/Cu(II)/O_2*

ω-Aryloxo compds. from ethylenealcohols and arylboronic acids by regioselective oxidative asym. isomerizing Heck-type arylation ←

466. 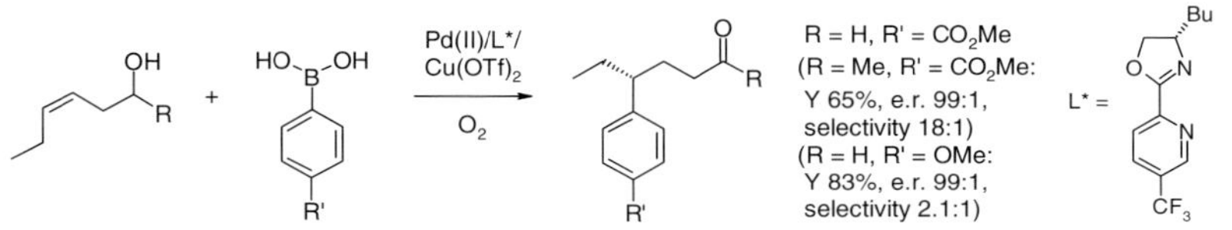

Palladium-catalyzed enantioselective redox-relay Heck-type reactions of acyclic alkenols, previously described with diazonium salts (cf. *82*, 393), have been developed in an oxidative variant with more practical arylboronic acids. **E: Chiral γ-arylaldehydes from (Z)-3-ethylene-*prim*-alcohols**. $Pd(MeCN)_2(OTs)_2$ (6 mol%), $Cu(OTf)_2$ (6 mol%), chiral oxazoline ligand (13 mol%), 3 Å molecular sieves (75 mg) and DMF (8 ml) added to a dry Schlenk flask, a three-way adapter fitted with a balloon of O_2 added, the flask evacuated and refilled with O_2 three times while stirring, the resulting mixture stirred for 10 min, a DMF soln. (2 ml) of *cis*-3-hexen-1-ol (0.5 mmol) and *p*-methoxycarbonylbenzeneboronic acid (3 eq.) added via syringe, the mixture stirred for 24 h at room temp., diluted with ether and water, and worked up with purification by flash chromatography on silica gel → product. Y 82% (e.r. 99:1; γ/β selectivity 19:1). The opposite enantiomer was obtained from *trans*-3-hexen-1-ol in good yield (72%) and high selectivity (e.r. 96:4; γ/β 6.2:1). Twelve further examples with *cis*-3-hexen-1-ol and arylboronic acids optionally bearing CN, SO_2Me, F, Cl, COMe, NO_2, OMe or alkyl groups gave e.r. 96:4 to 99:1 and yields of 35% (for the *p*-nitro-deriv.), 55-85%; site-selectivity was high for two examples of *ortho* subst. boronic acids (γ/β 30:1, 34:1), while for *meta* or *para* subst. derivs. electron-deficient examples gave higher site selectivity (γ/β 7.3:1, 14:1 to 49:1) than electron-rich ones (γ/β 2.1:1 to 5.1:1). Reaction of dibenzofuran-, indole- and 3-pyridyl-boronic acids showed high selectivity (e.r. 92:8 to 96:4; γ/β >20:1) but poor to moderate yields (16-53%; three examples), possibly due to higher levels of homocoupling or phenol formation. One example with a sec. alcohol gave the γ-arylketone in 99:1 e.r. (Y 65%; γ/β 18:1). Extension to other (Z)-alkenols to investigate alkene substitution and chain-length effects afforded β-, γ-, δ-, ε-, ζ- and η-aryloxo compds. in yields of 50-72%, e.r. 94:6 to 99:1 and site selectivity of 1.2:1 to 27:1 (ten examples). F.e.s. T.-S. Mei, E.W. Werner, A.J. Burckle, M.S. Sigman, J. Am. Chem. Soc. *2013*, *135* (18), 6830-3 [DOI: 10.1021/ja402916z].

Tetrakis(triphenylphosphine)palladium(0)/cesium carbonate *$Pd(PPh)_4/Cs_2CO_3$*

2-Arylbenzofurans from *o*-hydroxy-β,β-dibromostyrenes and triarylbismuthines ○

s. *80*, 463s*83*

Tetrakis(triphenylphosphine)palladium(0)/copper(I) thiophene-2-carboxylate *$Pd(PPh_3)_4$/CuTC*

N-(Carbomethoxy)iminoesters from boronic acids $RB(OH)_2$ → RC(OR)=NCOOMe

en route to carboxylic acid N-(carbomethoxy)amides or carboxylic acid esters s. *83*, 484

Palladium(II) phosphine complexes *[Pd(II)]*

Suzuki-type coupling with alkali-metal organoborates R-R′

under palladium catalysis, update s. *64*, 453s*82*; ***sp^2-sp^3* coupling** of potassium 2-alkoxyethyl-(trifluoro)borates with [hetero]aryl bromides s. N. Fleury-Brégeot, M. Presset, F. Beaumard, V. Colombel, D. Oehlrich, F. Rombouts, G.A. Molander, J. Org. Chem. *2012*, *77* (22), 10399-408 [DOI: 10.1021/jo3021665]; of potassium 2-(1,3-dioxolan-2-yl)ethyl(trifluoro)borates with [hetero]-aryl halides using dichlorobis[di-*tert*-butyl[4-(dimethylamino)phenyl]phosphine]palladium(II)/Cs_2CO_3 s. N. Fleury-Brégeot, D. Oehlrich, F. Rombouts, G.A. Molander, Org. Lett. *2013*, *15* (7),

1536-9 [DOI: 10.1021/ol400320q]; **N-subst. benzylamines** (and heterocyclic analogs) from potassium ammoniomethyl(trifluoro)borate inner salts and [hetero]aryl bromides s. N. Fleury-Brégeot, J. Raushel, D.L. Sandrock, S.D. Dreher, G.A. Molander, Chem. Eur. J. *2012, 18* (31), 9564-70 [DOI: 10.1002/chem.201200831]; ***sec*-benzyl ethers** (and heterocyclic analogs) from potassium α-alkoxyalkyl(trifluoro)borates and [hetero]aryl chlorides *with retention of absolute configuration* using a palladium(II) diadamant-1-yl(*n*-butyl)phosphine complex and CsOH·H_2O s. G.A. Molander, S.R. Wisniewski, J. Am. Chem. Soc. *2012, 134* (40), 16856-68 [DOI: 10.1021/ja307861n].

Palladium(II) phosphine complexes/copper(I) chloride *[Pd(II)]/Cu(I)*
Synthesis of acetylene derivs. by sila-Sonogashira coupling C≡C(Si≤) → C≡CR
cf. *54*, 469s*76* and under *70*, 460 in volume *82*; by coupling silylacetylenes with aryl tosylates or mesylates using $PdCl_2$(bpy)/BrettPhos and CuCl as cocatalyst for the synthesis of unsym. diarylacetylenes s. Y. Nishihara, D. Ogawa, S. Noyori, M. Iwasaki, Chem. Lett. *2012, 41* (11), 1503-5 [DOI: 10.1246/cl.2012.1503]; synthesis of hydroxyarylacetylenes from trimethylsilylalkynes and ar. iodides under standard Sonogashira conditions (Pd/Cu/base) *with endogenous phenoxide-assisted* cleavage of the silyl group s. M. Shigeta, J. Watanabe, G.-i. Konishi, Tetrahedron Lett. *2013, 54* (13), 1761-4 [DOI: 10.1016/j.tetlet.2013.01.091]; 1-aryl-4-silyl-1,3-butadiynes from 1,4-disilyl-1,3-butadiynes and ar. iodides via *exogenous* phenoxide-assisted monodesilylation with *p-tert*-butylphenoxide s. M. Shigeta, J. Watanabe, G.-i. Konishi, Asian J. Org. Chem. *2012, 1* (1), 43-6 [DOI: 10.1002/ajoc.201200018].

Palladium(II) chloride/tetra-n-butylammonium fluoride/air/sodium salt *$PdCl_2/Bu_4NF/O_2/Na^+$*
Biaryls from aryl(trialkoxy)silanes and sodium arenesulfinates Si≤ → Ar

467.

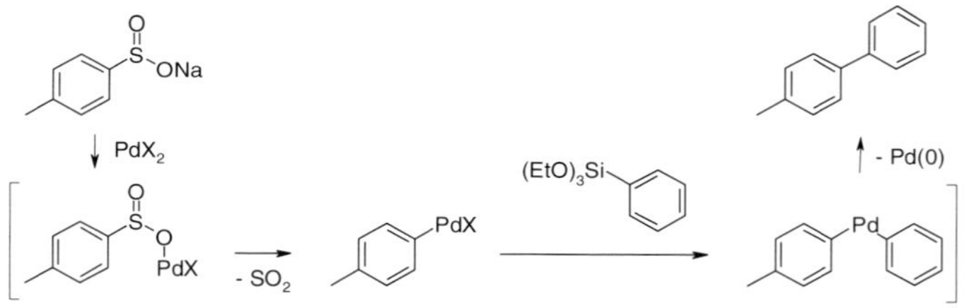

A mixture of sodium *p*-methylbenzenesulfinate (0.5 mmol), $PdCl_2$ (5 mol%) and phenyl(triethoxy)-silane (1.2 eq.) stirred uncapped in a soln. of Bu_4NF in THF (1 *M*; 1 ml) at 70° for 3 h, cooled to room temp., water (5 ml) added, the mixture filtered, extracted with ether, the extracts washed with water and brine, dried (Na_2SO_4), solvent removed under reduced pressure, and the residue purified by silica gel chromatography → 4-methylbiphenyl. Y 94%. Similar yields (90-91%) were obtained starting from phenyl(trimethoxy)silane or diphenyl(dimethoxy)silane. This **desulfitative Hiyama-type coupling** uses relatively non-toxic arylsilanes and widely available, stable, easy to handle sodium arenesulfinates **under mild, aerobic conditions in the absence of base and ligand**. Seventeen further examples gave yields of 73-92% with aryl(trialkoxy)silanes optionally bearing Me, F or CF_3 groups and sodium arenesulfinates optionally bearing Me, MeO, F, Cl, Br or NO_2 or with naphthyl analogs; *ortho* substituents are tolerated on the sulfinate. F.e. and optimization s. K. Cheng, S. Hu, B. Zhao, X.-M. Zhang, C. Qi, J. Org. Chem. *2013*, *78* (10), 5022-5 [DOI: 10.1021/jo302791q].

Bis(π-allylpalladium chloride) s. under Chloro(1,5-cyclooctadiene)rhodium(I) dimer *[Pd(II)]*

Chiral bis(π-allylpalladium chlorides) *[Pd(II)]**
Asym. aza-Hosomi-Sakurai synthesis of N-protected 3-ethyleneamines C=NR → C(NHR)C-C=C
update s. *72, 590s82*; with a chiral, (-)-β-pinene-based bis[π-allylpalladium chloride] as catalyst s. R.A. Fernandes, J.L. Nallasivam, Org. Biomol. Chem. *2012, 10* (38), 7789-800 [DOI: 10.1039/c2ob26188j]; **3-component organocatalyzed asym. aza-Hosomi-Sakurai reaction** using a chiral

1,1′-binaphthyl-2,2′-disulfonic acid imide as catalyst s. S. Gandhi, B. List, Angew. Chem., Int. Ed. *2013, 52* (9), 2573-6 [DOI: 10.1002/anie.201209776].

Chloro(1,5-cyclooctadiene)iridium(I) dimer/chiral 1,1′-binaphthyl-2,2′-diyl phosphoroamidites/tetra-n-butylammonium hydrogen sulfate/hydrogen fluoride ←
Regioselective iridium-catalyzed asym. synthesis of 1,4-dienes from potassium trifluoro(vinyl)borates and 2-ethylenealcohols OH → C=C

468.

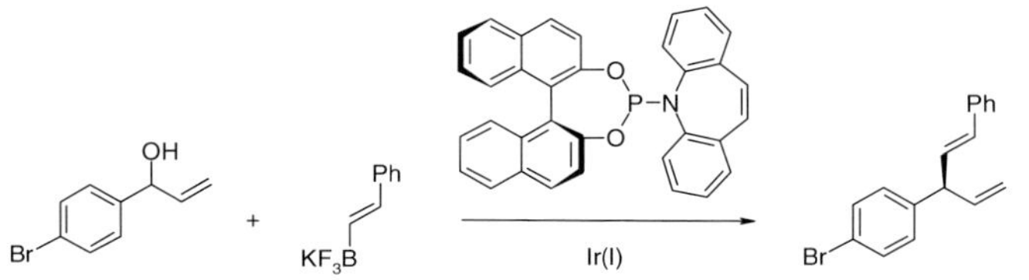

under mild conditions. A soln. of $[Ir(cod)Cl]_2$ (4 mol%) and chiral phosphoramidite ligand (16 mol%) in dioxane (0.5 ml) stirred vigorously for 15 min, 1-(4-bromophenyl)prop-2-en-1-ol (0.25 mmol), potassium (E)-styryltrifluoroborate (2 eq.), n-Bu_4NHSO_4 (10 mol%) and 50% aq. HF (2 eq.) added to the dark red soln., the resulting heterogeneous orange mixture stirred at room temp. for 4 h, concentrated *in vacuo*, and purified by flash chromatography → (R,E)-1-bromo-4-(1-phenylpenta-1,4-dien-3-yl)benzene. Y 78% (branched/linear >50:1; e.e. 98%). This novel, scalable (to 7 mmol) and experimentally simple vinylation was successful for mono-, di- and tri-substituted vinyltrifluoroborates reacting with electron-diverse sec. α-aryl-allylic alcohols in the presence of a phase transfer catalyst, generally with excellent branched/linear (b/l) and enantio-selectivity (b/l 3:1, 31:1 to >50:1; e.e. 89%, 94-99%; twenty-two examples; Y 55-99%). F.e., optimization, ligand prepn., and application to syntheses of the biologically active natural products, (-)-nyasol and (-)-hinokiresinol, s. J.Y. Hamilton, D. Sarlah, E.M. Carreira, J. Am. Chem. Soc. *2013, 135* (3), 994-7 [DOI: 10.1021/ja311422z].

Via intermediates *v.i.*
Carboxylic acid esters from boron compds. B< → COOR
via N-(carbomethoxy)iminoesters – Methyl N-[methoxy(methylthio)methylene]carbamate as carbonylating agent s. *83*, 484

Carbon ↑ CC ↓↑ C

Without additional reagents *w.a.r.*
Decarboxylative Mannich reaction C=N- → C(NH-)C-CO
update s. *77,* 514s*82*; mild, *uncatalyzed* reaction of β-ketocarboxylic acids with N-(*p*-methoxy-phenyl)aldimines derived from enolizable aldehydes, incl. reaction with chiral α-oxyaldimines to give *anti*-adducts *with asym. induction*, s. M. Böhm, K. Proksch, R. Mahrwald, Eur. J. Org. Chem. *2013* (6), 1046-9 [DOI: 10.1002/ejoc.201201644]; **organocatalyzed asym. decarboxylative Mannich reaction** (cf. *68,* 143s*82*) of malonic acid monothiolesters with N-protected isatin-based ketimines using a cinchona-based N-(heteroarenesulfonyl)amide as catalyst s. N. Hara, S. Nakamura, M. Sano, R. Tamura, Y. Funahashi, N. Shibata, Chem. Eur. J. *2012, 18* (30), 9276-80 [DOI: 10.1002/chem.201200367]; of β-ketocarboxylic acids with ketimines, notably 1-benzyl-4-(trifluoromethyl)-2(1*H*)-quinazolones, using a carbohydrate-based 2-*tert*-aminothiourea as catalyst s. H.-N. Yuan, S. Wang, J. Nie, W. Meng, Q. Yao, J.-A. Ma, Angew. Chem., Int. Ed. *2013, 52* (14), 3869-73 [DOI: 10.1002/anie.201210361].

Irradiation s. under N-Isopropylbenzylamine and fac-Ir(ppy)$_3$ ///
Microwaves s. under NaOMe and 3,4,5-Trimethylthiazol-2-ylidene [\\\\]

Sodium methoxide/microwaves *NaOMe/*

Fused pyridinium ring from α-diketones ○

via decarbalkoxylative Westphal condensation under microwave irradiation

469.

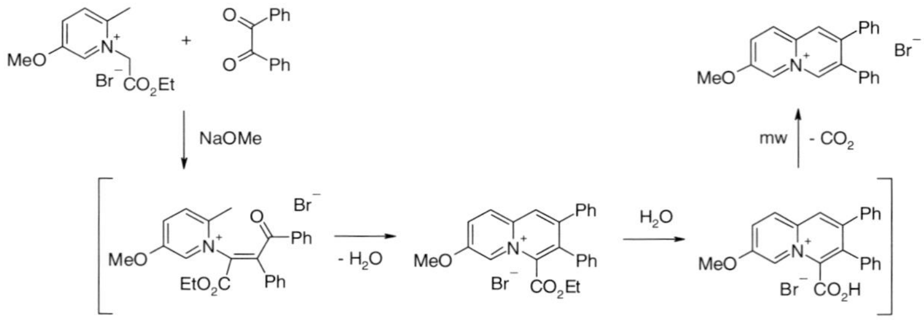

Quinolizinium salts. Benzil (1.2 eq.) and NaOMe (4 eq.) added to a soln. of N-ethoxycarbonylmethyl-5-methoxy-2-methylpyridinium bromide (1 mmol) in methanol at room temp., the mixture heated by microwaves (150 W) at 95-105° and 250 psi for 50 min, acidified with acetic acid, concentrated *in vacuo*, and purified by chromatography on silica → 7-methoxy-2,3-diphenylquinolizinium bromide. Y 58%. The use of microwaves effected rapid Westphal condensation of readily available N-ethoxycarbonylmethylpyridinium salts with (a)cyclic aliphatic/ar. α-diketones, with subsequent ester hydrolysis and decarboxylation, to afford pyridinium-fused heterocycles, some of them novel, in moderate yields (28-58%; nineteen examples), incl. analogs of semperivine. Subsequent experimentation suggested that tert. amines were more effective than NaOMe, affording improved yields and cleaner products but details are scant. F.e.s. T.S.C. Rao, S. Saha, G.B. Raolji, B. Patro, P. Risbood, M.J. Difilippantonio, J.E. Tomaszewski, S.V. Malhotra, Tetrahedron Lett. *2013*, *54* (6), 487-90 [DOI: 10.1016/j.tetlet.2012.11.059].

Lithium acetylides *LiC≡CR*

N-Protected α,β-acetylene-α′-aminoketones from 2,5-oxazolidiones C

en route to 4-iodo-Δ^4-3-pyrrolones with retention of chirality s. *83*, 199

Cesium fluoride *CsF*

N-Arylacridones from three benzyne molecules and N-unsubst. 2-azetidinones ○

470.

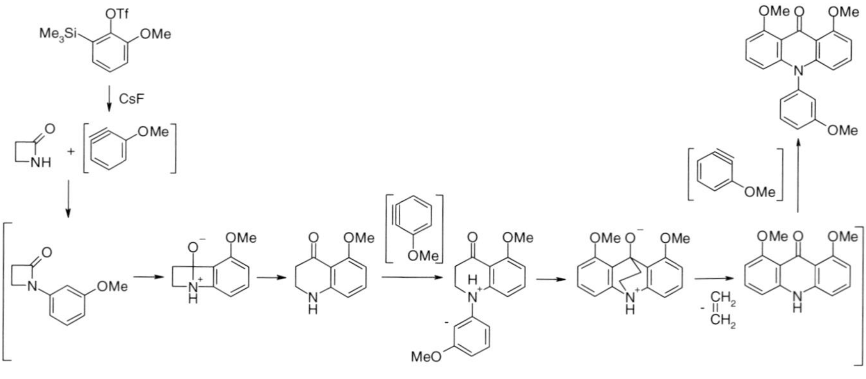

The reaction of N-unsubst. β-lactams with three aryne molecules to afford N-arylacridones **via** novel **extrusion of ethylene**, is reported. **E:** 2-Methoxy-6-trimethylsilylphenyl triflate (aryne precursor;

3.5 eq.), 2-azetidinone (0.25 mmol), acetonitrile (4 ml) and CsF (7 eq.) added sequentially to an oven-dried vial under a balloon of N_2 (for venting of ethylene), the mixture stirred for 24 h, quenched with aq. Na_2CO_3, extracted with ethyl acetate, and purified chromatographically → 1,8-dimethoxy-10-(3-methoxyphenyl)acridin-9(10*H*)-one. Y 83%. This unexpected transformation was shown to proceed via the expected 2,3-dihydro-4(1*H*)-quinolone, but attempts to modify the reaction for dihydroquinolone synthesis were unsuccessful. The reaction had limited success with azetidinone (four examples; Y 40-83%) and gave a moderate yield with 3,3-bis(allyl)azetidinone (30%), via extrusion of 4-methylhepta-1,6-diene, but N-benzyl (Y 14%), N-allyl (trace) and N-phenyl derivs. (no reaction) were poor substrates. The title products were **also** obtained **from** preformed electron-diverse N-unsubst. **2,3-dihydro-4(1*H*)-quinolones** and two molecules of aryne in good yields (69-90%; seven examples; a 6-fluoro deriv. was unreactive) under these conditions, and N-methyl/allyl derivs. gave the corresponding N-subst. acridones (seven examples; Y 48-63%; the N-phenyl deriv. was unreactive). Interestingly, N-methyl-2-oxazolidone reacted under these conditions, with extrusion of oxirane, to afford N-methylacridone (Y 33%). F.e.s. Y. Fang, D.C. Rogness, R.C. Larock, F. Shi, J. Org. Chem. *2012*, *77* (14), 6262-70 [DOI: 10.1021/jo3011073].

N-Isopropylbenzylamine/triethylenediamine/tris(2-phenylpyridinato-C2,N)iridium(III)/ acetic acid/irradiation ←

Organocatalyzed β-arylation of oxo compds. with ar. nitriles under visible-light photoredox catalysis H → Ar

471. 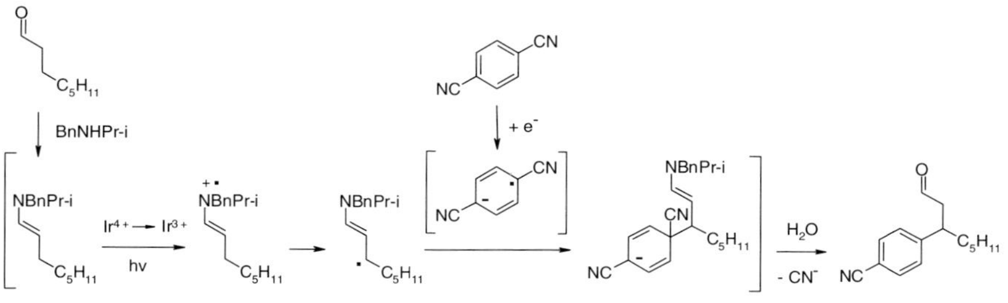

Organocatalysis has been combined with visible-light photoredox activation in a novel β-arylation of aldehydes or ketones with ar. nitriles. **E:** A dry vial containing terephthalonitrile (1 mmol), tris[2-phenylpyridinato-C2,N]iridium(III) (1 mol%) and DABCO (5 eq.) purged with a stream of N_2, DMPU (2 ml) added via syringe, followed by octanal (1.4 eq.), acetic acid (0.2 eq.), water (3 eq.) and N-isopropylbenzylamine (0.2 eq.), the mixture cooled to -78°, degassed via vacuum evacuation (5 min), backfilled with N_2, warmed to room temp., the process repeated three times, once thoroughly degassed the vial placed approximately 2 cm from a 26 W fluorescent lamp, irradiated for 26 h, diluted with ethyl acetate, added to 1 *M* aq. NaOH, and worked up with purification by flash chromatography on silica gel → 1-(oxooctan-3-yl)benzonitrile. Y 86%. The procedure is applicable to the coupling of a wide range of alkanals (notably possessing alkoxy, alkene, alkyne and carbamate groups, as well as electron-rich or heterocyclic aryl residues at the β-position) with cyano-arenes or -heteroarenes (incl. hindered substrates and notably those possessing electron-withdrawing groups at the *ortho, meta* or *para* site); cyclohexanones, incl. hindered derivs., and 4-chromanones also underwent β-arylation by using azepane as organocatalyst in place of N-iso-propylbenzylamine, notably with useful levels of diastereoselectivity in the case of 4-subst. cyclohexanones (twenty-eight examples in all; Y 44-88%). Promising levels of enantioselectivity were also recorded for the β-arylation of cyclohexanone in the presence of a cinchona-based prim. amine as catalyst (one example cited; Y 82%; e.e. 50%). Reaction is presumed to involve a dual catalytic cycle wherein a key 5π-electron β-enaminyl radical is generated from the intermediate enamine via SET to Ir^{4+} and subsequent deprotonation prior to reaction with the cyanoarene, with ultimate liberation of cyanide ion: the only apparent downside of the methodology. F.e.s. M.T. Pirnot, D.A. Rankic, D.B.C. Martin, D.W.C. MacMillan, Science *2013*, *339* (6127), 1593-6 [DOI: 10.1126/science.1232993].

N-Methylmorpholine N-oxide s. under Ozone *NMMO*
Copper(I) oxide s. under Bis(dibenzylideneacetone)palladium(0) *Cu_2O*
Silver carbonate s. under $Pd(OCOCF_3)_2$ *Ag_2CO_3*
Copper(I) chloride s. under $RhCl(PPh_3)_3$ *CuCl*

Copper(I) bromide/8-hydroxyquinoline/silver carbonate/potassium salt ←
β,γ-Acetylenenitriles from α,β-acetylenehalides C≡C(Hal) → C≡C-C(CN)
and potassium α-cyanocarboxylates s. *78,* 527s*83*

Copper(II) chloride/p-benzoquinone/triethylamine/pyridine *$CuCl_2$/BQ/Et_3N/py*
Decarboxylative Sonogashira-type coupling C≡C-COOH → C≡C(Ar)
with arylboronic acids s. *79,* 463s*83*

Zinc s. under Ozone *Zn*
Methylaluminum bis(2,6-di-tert-butyl-4-methylphenoxide) s. under $Ni(cod)_2$ *MAD*
p-Benzoquinone s. under $CuCl_2$ *BQ*
Ammonium acetate s. under AcOH *NH_4OAc*

Chiral 1,1′-binaphthyl-based 2-aminosquaramides ←
Organocatalyzed asym. decarboxylative Michael addition C=C → CHC(R)
update s. *68,* 143s*81*; of β-ketocarboxylic acids to nitroalkenes with a chiral, bifunctional 1,1′-binaphthyl-based 2-aminosquaramide as catalyst s. H.W. Moon, D.Y. Kim, Tetrahedron Lett. *2012, 53* (48), 6569-72 [DOI: 10.1016/j.tetlet.2012.09.100].

Malononitrile s. under AcOH *$CH_2(CN)_2$*
Pivalic anhydride s. under $[Rh(CO)_2Cl]_2$ *$(t\text{-}BuCO)_2O$*

Lauroyl peroxide *$(C_{11}H_{23}COO)_2$*
Ketones from 2-ethylene-*tert*-alcohols and xanthates via radical C-cleavage ←

472.

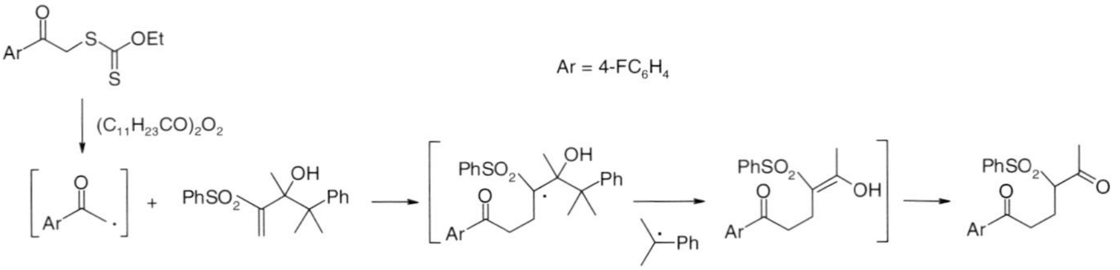

A soln. of 3,4-dimethyl-4-phenyl-2-(phenylsulfonyl)pent-1-en-3-ol (2 eq.) and O-ethyl S-4-fluorobenzoylmethyl xanthate (0.4 mmol) in ethyl acetate (0.4 ml) refluxed under N_2 for 15 min, lauroyl peroxide (30 mol%) then added every hour until xanthate consumed (2.1 eq.; 7 h), cooled to room temp., solvent removed *in vacuo*, and the residue purified by flash chromatography → 1-(4-fluorophenyl)-4-(phenylsulfonyl)hexane-1,5-dione. Y 70%. This versatile and effective radical alkylation involves rare cleavage of a C-C bond (extruding a cumyl radical), and affords synthetically useful 1,5-diketones, incl. trifluoromethyl ketones not easily available by other routes (fifteen examples; Y 55-84%), in the presence of silyl ether, phthalimide, sulfonamide, ester, tert. alcohol and sulfide functionality. The method was applicable to tert. propargylic-allylic alcohol analogs, affording the corresponding ynones (two examples; Y 57-60%; one product underwent partial lactonization), and its use in a sequential radical addition series to afford complex ketones is also described. F.e., substrate prepn. and cyclization of 1,5-diketones (with NH_4OAc) to **pyridines** (three examples; Y 53-84%) s. L. Debien, S.Z. Zard, J. Am. Chem. Soc. *2013*, *135* (10), 3808-11 [DOI: 10.1021/ja400831w].

Acetic acid/ammonium acetate/malononitrile *AcOH/NH$_4$OAc/CH$_2$(CN)$_2$*
Transition metal-free synthesis of styrenes from ar. aldehydes and aliphatic aldehydes ArCHO + OHCCH$_2$R → ArCH=CHR

473. 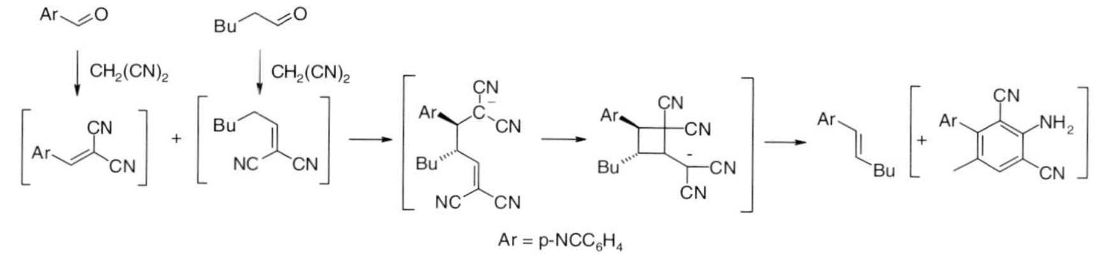

Acetic acid (1.5 eq.) added to a mixture of 4-cyanobenzaldehyde (4 mmol), hexanal (1.25 eq.), malononitrile (2.5 eq.) and acetonitrile (30 ml), the mixture stirred for 10 min, NH$_4$OAc (1.25 eq.) added, stirred at 80° for 8 h, cooled to room temp., filtered, concentrated *in vacuo*, dissolved in water, extracted with ethyl acetate, and worked up with chromatographic purification on silica → (E)-1-(hex-1-en-1-yl)-4-cyanobenzene. Y 56%. This novel but unoptimized method combined electron-diverse (het)ar. and linear aliphatic aldehydes via double Knoevenagel condensation with malononitrile in the presence of an acid (acetic acid, alanine or proline), with subsequent ammonium acetate or formate treatment promoting 1,4-addition-intramolecular 1,4-addition-retro-[2+2]-cycloaddition, **with** overall **loss of one C-atom**, to afford styrenes, only available by multistep routes previously in some cases (thirteen examples; Y 21-60%). The method was compatible with ether, nitro, cyano, halo, alkene and alkyne functionality and generated Knoevenagel [benzylidene-malononitrile] and *m*-amino-*o*,*p*-dicyanobiaryl by-products in variable and generally unspecified yields. With morpholine in DMF at 80°, the latter by-product was formed as major product (Y 77%) in one case. F.e.s. H.B. Borate, S.H. Gaikwad, A.S. Kudale, S.P. Chavan, S.G. Pharande, V.D. Wagh, V.S. Sawant, Tetrahedron Lett. *2013*, *54* (12), 1528-30 [DOI: 10.1016/j.tetlet.2013.01.008].

3,4,5-Trimethylthiazol-2-ylidene/microwaves *NHC/[\\\\]*
Stetter-type reaction with a formyl carbanion equivalent C=C-CO → C(CHO)CHCO
generated from aldoses by retro-benzoin condensation s. *83*, 305

N-[3,5-Bis(trifluoromethyl)phenyl]-N′-((R,R)-2-dimethylaminocyclohexyl)thiourea ←
α-Fluoro-γ-nitrocarbonyl compds. ←
from α-fluoro-α-(1,1-dihydroxy-2,2,2-trifluoroethyl)ketones and 1-nitroethylene derivs. via organocatalyzed asym. de(trifluoroacetyl)ative Michael addition

474. 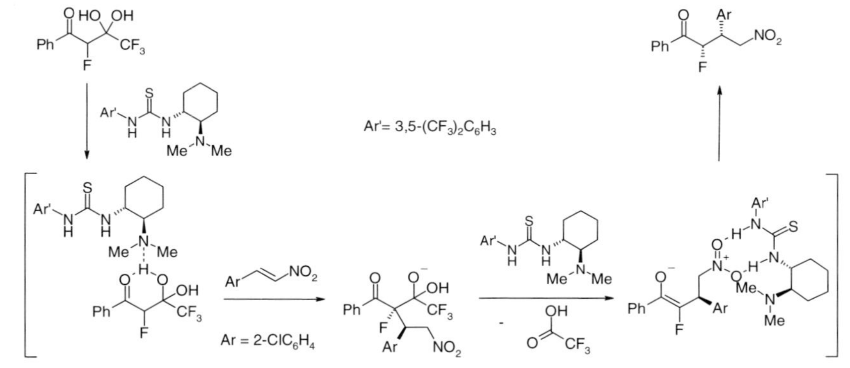

under mild conditions. Chiral thiourea catalyst (10 mol%) and 2,4,4,4-tetrafluoro-3,3-dihydroxy-1-phenylbutan-1-one (2.5 eq.) added to a soln. of 2-(2-chlorophenyl)-1-nitroethylene (0.2 mmol) in *tert*-butyl methyl ether (2 ml) at room temp., the mixture stirred for 3 d (TLC), concentrated, and purified chromatographically → (2S,3R)-3-(2-chlorophenyl)-2-fluoro-4-nitro-1-phenylbutan-1-one. Y 74% (d.r. 4:1; e.e. 98%). This novel, efficient and enantioselective procedure is driven by release of trifluoroacetic acid from the initial Michael adduct. The method was successful for electron-diverse (het)ar. ketones reacting with 2-subst. [alkyl, (het)ar.] 1-nitroethylenes (nineteen examples; Y 58-91%; d.r. 2:1 to 20:1; e.e. 78-98%), with an alkyl ketone - a 2-pentanone analog - affording a 28% yield (d.r. 3:1; e.e. 77%). F.e., optimization, prepn. of the substrates from the corresponding 1-subst. 3-hydroxy-4,4,4-trifluorobut-2-en-1-ones (ten examples; Y 62-96%), and reductive cyclization (H_2/Pd) to a chiral Δ^2-pyrroline (Y 73%) in one case, s. I. Saidalimu, X. Fang, W. Lv, X. Yang, X. He, J. Zhang, F. Wu, Adv. Synth. Catal. *2013, 355* (5), 857-63 [DOI: 10.1002/adsc.201200757].

Chiral 1,1′-bi-2-naphthol-based diorganotin dibromides/sodium methoxide/methanol ←
Aldol-type condensation with enolesters CO → C(OH)C-CO
diastereoselective aldol-type condensation of enol trichloroacetates using $Bu_2Sn(OMe)_2$/MeOH in THF cf. *58*, 468s*66*; **asym. condensation** of cyclic or acyclic enol trihalogenoacetates with isatins using an (S)-BINOL-based diorganotin dibromide as catalyst in toluene with added NaOMe and methanol s. A. Yanagisawa, N. Kushihara, T. Sugita, K. Yoshida, Synlett *2012, 23* (12), 1783-8 [DOI: 10.1055/s-0032-1316551].

Tert. phosphines s.a. under Ozone, Ni(cod)$_2$ and Bis(π-allylpalladium chloride) ⩾P
2-[o-(Dicyclohexylphosphino)phenyl]-5,6-dimethyl-1-octylbenzimidazole s. under Bis(dibenzylideneacetone)palladium(0) ←
Tris(2,4-di-tert-butylphenyl) phosphite s. under Chloro(1,5-cyclooctadiene)-rhodium(I) dimer *(ArO)$_3$P*

Ozone/N-methylmorpholine N-oxide/zinc/triphenylphosphine *O_3/NMMO/Zn/Ph_3P*
Dibromomethylene compds. from ethylene derivs. via ozonolytic C-cleavage-Corey-Fuchs reaction C=CRR′ → C=CBr_2

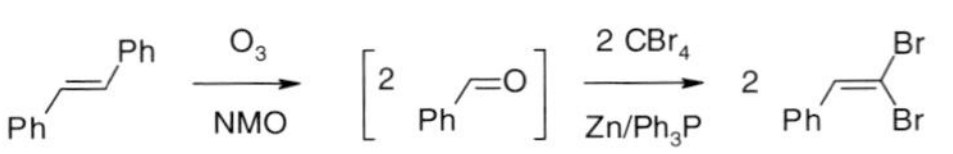

475. 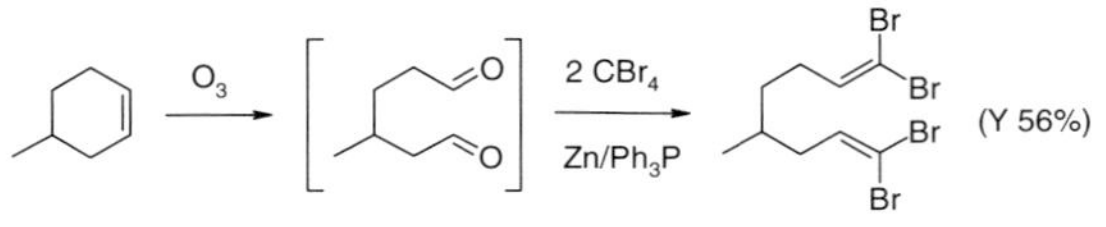

in one pot. A soln. of *trans*-stilbene (1 eq.) and N-methylmorpholine N-oxide (1 eq.) in methylene chloride stirred under ozone until ozonolysis complete, the soln. added via cannula to a previously stirred (23-48 h) suspension of Zn (6 eq.), carbon tetrabromide (6 eq.) and triphenylphosphine (6 eq.) in the same solvent (0.3 *M*) under argon, and the mixture stirred until reaction complete (TLC) → 1,1-dibromo-2-phenylethene. Y 85%. This rare tandem alkene cleavage with Corey-Fuchs dibromoolefination avoided the isolation of reactive intermediate aldehydes. Symmetrical (a)cyclic alkenes afforded single products in variable yields (24%, 56-64%; four further examples), and unsymmetrical methyl octadec-9-enoate afforded a mixture of 1,1-dibromodec-1-ene (Y 66%) and methyl 10,10-dibromodec-9-enoate (Y 54%). 4-Hydroxymethylstilbene gave a mixture of dibromomethylene compds. (in unspecified yields) formed by reaction of intermediate benzaldehyde and *p*-phthalaldehyde. F.e. and optimization s. B.A. Brown, J.G.C. Veinot, Tetrahedron Lett. *2013*, *54* (8), 792-5 [DOI: 10.1016/j.tetlet.2012.11.121].

Cinchona-based N-(heteroarenesulfonyl)amides ←
Organocatalyzed asym. decarboxylative Mannich reaction C=N- → C(NH-)C-CO
of malonic acid monothiolesters s. *68,* 143s*83*

Phenolato(pyrrolidinato)tungsten(VI) complex *[W(VI)]*
Alkylidyne-tungsten(VI) or -molybdenum(VI) complexes *[W(VI)] or [Mo(VI)]*
Cross-metathesis of ethylene derivs. C=C
under molybdenum or tungsten catalysis, update s. *30,* 467s*82*; synthesis of pinacol (Z)-α,β-ethylene- and β,γ-ethylene-boronic acid esters with a phenolato(pyrrolidinato)tungsten(VI) complex s. E.T. Kiesewetter, R.V. O'Brien, E.C. Yu, S.J. Meek, R.R. Schrock, A.H. Hoveyda, J. Am. Chem. Soc. *2013, 135* (16), 6026-9 [DOI: 10.1021/ja403188t]; **metathesis of acetylene derivs.** with finely-tunable, thermally stable, air- and moisture-tolerant tris(arylsiloxy)(1,10-phenanthroline or 2,2′-bipyridyl)(*p*-methoxybenzylidyne)molybdenum(VI) complexes (cf. *30,* 467s*79*) s. J. Heppekausen, R. Stade, A. Kondoh, G. Seidel, R. Goddard, A. Fürstner, Chem. Eur. J. *2012, 18* (33), 10281-99 [DOI: 10.1002/chem.201200621]; metathesis of terminal and internal acetylene derivs., and ring-closing metathesis (cf. *49,* 985s*83*) of terminal or internal diynes, with a tris(hexafluoro-*tert*-butanolato)(2,4,6-trimethylbenzylidyne)-molybdenum(VI) or -tungsten(VI) complex s. B. Haberlag, M. Freytag, C.G. Daniliuc, P.G. Jones, M. Tamm, Angew. Chem., Int. Ed. *2012, 51* (52), 13019-22 [DOI: 10.1002/anie.201207772]; homocoupling of acetylene derivs. and gram-scale formation of macrocyclic sym. triynes using silyl-bridged tris(phenolato)propylidynemolybdenum(VI) complexes at low catalyst loading s. H. Yang, Z. Liu, W. Zhang, Adv. Synth. Catal. *2013, 355* (5), 885-90 [DOI: 10.1002/adsc.201201105].

Anionic (ethylenediaminetetraacetic acid)(peroxo)iron(III) complexes s. under Palladium(II) acetate *[Fe(III)]*

Iron(III) chloride *$FeCl_3$*
Naphthalenes from *o*-β-ketoaldehydes and acetylene derivs. ○

476.

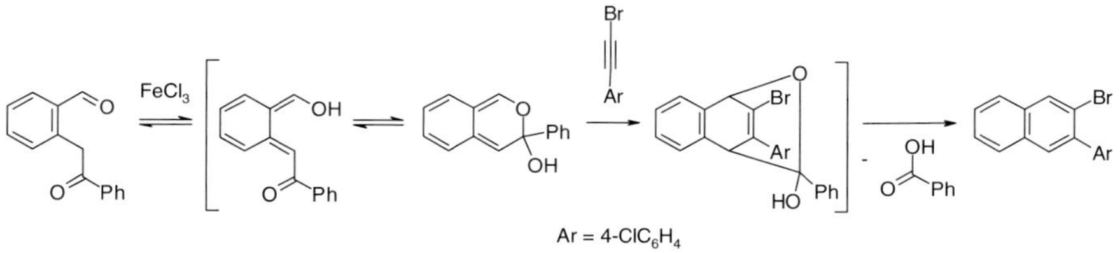

A mixture of 2-benzoylmethylbenzaldehyde (0.2 mmol), bromo(4-chlorophenyl)acetylene (1.1 eq.), $FeCl_3$ (5 mol%) and 1,2-dichloroethane (3 ml) stirred at 60° under argon until substrate consumed (TLC and GC-MS), cooled to room temp., extracted with ether (with aq. wash), and purified by chromatography on silica → 2-bromo-3-(4-chlorophenyl)naphthalene. Y 79%. This efficient benzannelation was catalyzed effectively by Zn(II), Fe(III) and Ag(I) salts (but not Cu(II)), and was successful for electron-diverse *o*-β-ketoaldehydes reacting with terminal and internal alkynes carrying electron-diverse aryl, linear/branched alkyl, bromo or trimethylsilyl substituents (twenty-two examples; Y 60-98%), with the aroyl moiety eliminated as an arylcarboxylic acid. The method was also applicable to diynes and triynes, affording the corresponding 1:2 (two examples; Y 51-75%) and 1:3 products (four examples; Y 56-87%). F.e., optimization and f. reactions of 2-bromonaphthalene products s. S. Zhu, Y. Xiao, Z. Guo, H. Jiang, Org. Lett. *2013, 15* (4), 898-901 [DOI: 10.1021/ol4000394].

Bis(1,5-cyclooctadiene)nickel(0)/methylaluminum bis(2,6-di-tert-butyl-4-methyl-phenoxide)/dimethyl(phenyl)phosphine ←
Carbostyrils from *o*-(aroylamino)nitriles and acetylene derivs. ○

477.

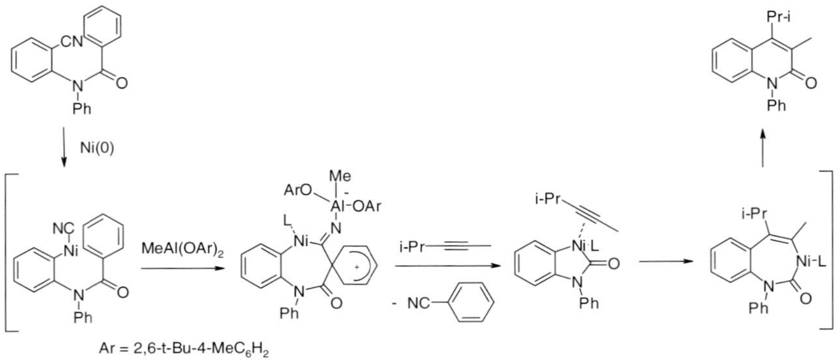

N-Phenyl-2-benzoylaminobenzonitrile (0.5 mmol) and 4-methylpent-2-yne (3 eq.) added to a soln. of Ni(cod)$_2$ (5 mol%), Me$_2$PPh (20 mol%) and MAD (20 mol%) in toluene (3 ml) in a dry box, the mixture heated under argon at 120° for 12 h, cooled to room temp., filtered through silica, concentrated *in vacuo*, and the residue purified by flash chromatography → 4-isopropyl-3-methyl-1-phenylquinolin-2(1*H*)-one. Y 96% (regioselectivity 3:1). This novel nickel-catalyzed ring closure was effective using Me$_3$P or Me$_2$PPh as ligands, and successful for N-subst. 2-benzoyl-aminobenzonitriles reacting with ar., alkyl, and silyl subst. internal alkynes, proceeding via loss of benzonitrile (sixteen examples; Y 59-99%). Unsymmetrical internal alkynes gave mixtures of regioisomers (1:1 to >20:1) dependent on the relative steric bulk of the two substituents. A 4-methoxybenzoylaminobenzonitrile gave a marginally improved yield compared with its benzoylamino analog (82% vs. 80%) but 4-trifluoromethylbenzoyl- (Y 39%) or acetyl-amino analogs (<1%) were inferior. F.e. and optimization s. K. Nakai, T. Kurahashi, S. Matsubara, Org. Lett. *2013*, *15* (4), 856-9 [DOI: 10.1021/ol303546p].

Ruthenium carbene complexes or supported variants *[Ru]*
Cross-metathesis of ethylene derivs. C═C
with ruthenium carbene complexes, update s. *49*, 932s*82*; (Z)-selective olefin metathesis, incl. ring-closing metathesis (cf. *49*, 985s*83*) with highly active (*o*-isopropoxybenzylidene)(nitrato)-[2-[3-(2,6-diisopropylphenyl)imidazolidin-2-yliden-1-yl]adamantyl]ruthenium (cf. *81*, 481) s. L.E. Rosebrugh, M.B. Herbert, V.M. Marx, B.K. Keitz, R.H. Grubbs, J. Am. Chem. Soc. *2013, 135* (4), 1276-9 [DOI: 10.1021/ja311916m]; application to pheromone synthesis s. M.B. Herbert, V.M. Marx, R.L. Pederson, R.H. Grubbs, Angew. Chem., Int. Ed. *2013, 52* (1), 310-4 [DOI: 10.1002/anie.201206079]; (Z)-selective cross-metathesis of terminal alkenes using robust, water- and air-tolerant chloro(1,3-dimesitylimidazolidin-2-ylidene)(2,4,6-triphenylbenzenethiolato)(*o*-iso-propoxybenzylidene)ruthenium s. G. Occhipinti, F.R. Hansen, K.W. Törnroos, V.R. Jensen, J. Am. Chem. Soc. *2013, 135* (9), 3331-4 [DOI: 10.1021/ja311505v]; olefin metathesis and other reactions with stable, readily tunable, isoxazolidine N-oxide-based chloro(2-nitronatoalkylidene)ruthenium NHC complexes s. T. Wdowik, C. Samojlowicz, M. Jawiczuk, M. Malinska, K. Wozniak, K. Grela, Chem. Commun. *2013, 49* (7), 674-6 [DOI: 10.1039/c2cc37385h]; rapid olefin metathesis with chloro[1,3-bis(2,6-diisopropylphenyl)imidazolidin-2-ylidene][*o*-(N-methylanilino)-benzylidene]ruthenium and related carbene complexes at low loading (0.0015-0.015 mol%) with TONs up to 58,000 s. L.H. Peeck, R.D. Savka, H. Plenio, Chem. Eur. J. *2012, 18* (40), 12845-53 [DOI: 10.1002/chem.201201010]; with a new family of dichloro(1,3-dimesitylimidazolidin-2-ylid-ene)(*o*-halogenobenzylidene)ruthenium complexes s. M. Barbasiewicz, M. Michalak, K. Grela, ibid. *18* (45), 14237-41 [DOI: 10.1002/chem.201202817]; with readily retrievable *trimethyl-ammoniomethyl-functionalized* dichloro(1,3-dimesitylimidazolidin-2-ylidene)(*o*-isopropoxyaryl)-

ruthenium complexes s. K. Skowerski, C. Wierzbicka, G. Szczepaniak, L. Gulajski, M. Bieniek, K. Grela, Green Chem. *2012, 14* (12), 3264-8 [DOI: 10.1039/c2gc36015b]; various types of olefin metathesis (incl. ROMP) with dichloro(1,3-dimesitylimidazolidin-2-ylidene)(4-isopropoxypyrimidin-5-ylmethylene)ruthenium complexes s. G.-L. Wu, S.-L. Cao, J. Chen, Z. Chen, Eur. J. Org. Chem. *2012* (34), 6777-84 [DOI: 10.1002/ejoc.201200989]; improved olefin metathesis with dichloro(tricyclohexylphosphine)(1,3-*di-o-tolyl*imidazolidin-2-ylidene)inden-1-ylidene-ruthenium complexes s. C. Torborg, G. Szczepaniak, A. Zielinski, M. Malinska, K. Wozniak, K. Grela, Chem. Commun. *2013, 49* (31), 3188-90 [DOI: 10.1039/c2cc37514a]; *heterogeneous* cross-metathesis and ring-opening-ring-closing metathesis with silica-*adsorbed*, linker-free dichloro[1,3-bis[2,6-dimethyl-4-(dimethylamino)phenyl]imidazolidin-2-ylidene](*o*-isopropoxyphenyl)ruthenium s. J. Cabrera, R. Padilla, M. Bru, R. Lindner, T. Kageyama, K. Wilckens, S.L. Balof, H.-J. Schanz, R. Dehn, J.H. Teles, S. Deuerlein, K. Müller, F. Rominger, M. Limbach, Chem. Eur. J. *2012, 18* (46), 14717-24 [DOI: 10.1002/chem.201202248]; with a robust pyridinium-tagged dichloro(3,4-dihydro-2*H*-1,4-benzoxazin-8-ylmethylene)ruthenium NHC complex anchored onto sulfonated silica fluoride for batch-mode metathesis *or under continuous flow* s. E. Borré, M. Rouen, I. Laurent, M. Magrez, F. Caijo, C. Crévisy, W. Solodenko, L. Toupet, R. Frankfurter, C. Vogt, A. Kirschning, M. Mauduit, Chem. Eur. J. *2012, 18* (51), 16369-82 [DOI: 10.1002/chem.201201589]; one-pot **O-desilylation-olefin metathesis** with added hexamethyldisiloxane s. S. Osman, K. Koide, Tetrahedron Lett. *2012, 53* (49), 6637-40 [DOI: 10.1016/j.tetlet.2012.09.035]; preparation of high-value cinnamate antioxidants by cross-metathesis s. J.A.M. Lummiss, K.C. Oliveira, A.M.T. Pranckevicius, A.G. Santos, E.N. dos Santos, D.E. Fogg, J. Am. Chem. Soc. *2012, 134* (46), 18889-91 [DOI: 10.1021/ja310054d]; synthesis of 6-subst. *o*-hydroxycarboxylic acid esters via olefin metathesis of 2,2-dimethyl-6-vinyl-1,3-dioxin-4-one with homoallyl alcohols s. P.S. Blencowe, A.G.M. Barrett, Can. J. Chem. *2012, 90* (11), 975-4 [DOI: 10.1139/v2012-071]; **ene-yne metathesis-Diels-Alder reaction** using 1,7-octadiene as an *in situ*-source of ethylene s. S. Fustero, P. Bello, J. Miró, A. Simón, C. del Pozo, Chem. Eur. J. *2012, 18* (35), 10991-7 [DOI: 10.1002/chem.201200835]; 2,6-*cis*-disubst. 2-β-ketotetrahydropyrans by **cross-metathesis-intramolecular Michael addition** s. H. Fuwa, T. Noguchi, K. Noto, M. Sasaki, Org. Biomol. Chem. *2012, 10* (40), 8108-12 [DOI: 10.1039/c2ob26189h]; one-pot **Friedel-Crafts allylation-cross-metathesis** with $InCl_3/NaHCO_3$ followed by Grubbs′ 2nd generation catalyst s. J.R. Henkie, S. Dhaliwal, J.R. Green, Synlett *2012, 23* (16), 2371-4 [DOI: 10.1055/s-0032-1317045].

Chloro(1,5-cyclooctadiene)rhodium(I) dimer/triethylenediamine *$[RhCl(cod)]_2$/DABCO*
Benzo[*b*]siloles from *o*-silylboronic acids and acetylene derivs. ○
cf. *76*, 522; s.a. M. Onoe, K. Baba, Y. Kim, Y. Kita, M. Tobisu, N. Chatani, J. Am. Chem. Soc. *2012, 134* (47), 19477-88 [DOI: 10.1021/ja3096174].

Chloro(1,5-cyclooctadiene)rhodium(I) dimer/tris(2,4-di-tert-butylphenyl) phosphite/ lithium bis(trimethylsilyl)amide ←
β,γ-Ethyleneketones OCOOR → C(O)R′
from α-siloxynitriles and 2-ethylenecarbonic acid esters

478.

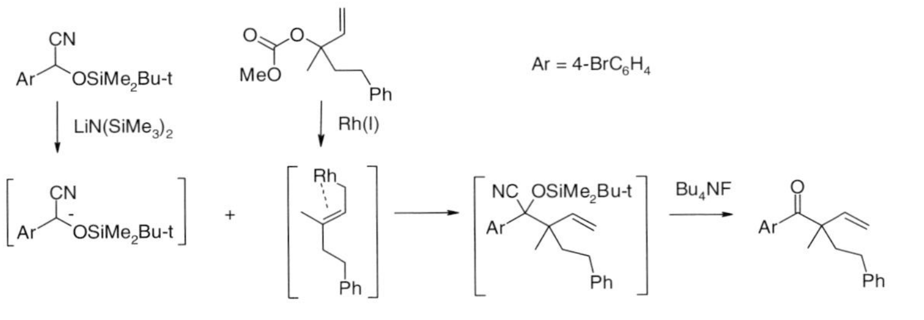

A novel and versatile alternative to conventional enolate alkylation, using an O-silyl cyanohydrin *as* masked *acyl anion equivalent* and affording α-quaternary subst. **ar. β,γ-ethyleneketones**, is reported. **E:** A soln. of $LiN(SiMe_3)_2$ (1.8 eq.) in anh. THF (0.9 ml) added dropwise to a soln. of

2-(*tert*-butyldimethylsiloxy)-2-(4-bromophenyl)acetonitrile (1.3 eq.) in the same solvent (3 ml) under argon at -10°, the mixture stirred for 30 min, a mixture of $[RhCl(cod)]_2$ (2.5 mol%) and tris(2,4-di-*tert*-butylphenyl) phosphite (10 mol%) in the same solvent (2 ml; previously stirred at room temp. under argon for 5 min), added via cannula, methyl (3-methyl-5-phenylpent-1-en-3-yl) carbonate (0.5 mmol) added via syringe, the mixture stirred for 16 h, a soln. of n-Bu_4NF (2.5 eq.) in the same solvent (1.25 ml) added, the mixture stirred at room temp. for 1 h, quenched with satd. aq. NH_4Cl, extracted with ether, and purified by chromatography on silica → 1-(4-bromophenyl)-2-methyl-2-phenethylbut-3-en-1-one. Y 88% (branched/linear >19:1). Metalation of allylic carbonate esters in the presence of the bulky phosphite ester, and treatment with lithiated siloxynitriles derived from electron-diverse benzaldehydes, proceeded with double inversion to afford the title products with high regioselectivity, following unmasking of the O-silyl cyanohydrin (fifteen examples; Y 70-91%; branched/linear >19:1). Reaction of a single chiral substrate (e.e. 98%) afforded the corresponding chiral α,α-disubst. aryl β,γ-ethyleneketone with good chirality transfer (e.e. 90%; Y 87% based on the alcohol precursor). F.e., optimization and substrate prepn. s. P.A. Evans, S. Oliver, J. Chae, J. Am. Chem. Soc. *2012, 134* (47), 19314-7 [DOI: 10.1021/ja306602g].

Chlorotris(triphenylphosphine)rhodium(I)/copper(I) chloride *$RhCl(PPh_3)_3/CuCl$*
Biaryls from arylcarboxylic acid esters and arylboronic acids ArCOOR → Ar-Ar′
N-Directed decarbalkoxylative C_{10}-arylation of ethyl benzo[*h*]quinoline-10-carboxylates s. *78,* 527s*83*

Palladium-carbon/sodium hydrogen carbonate *Pd-C/$NaHCO_3$*
Decarboxylative Heck arylation ←
update s. *77,* 42s*78*; β-arylketones from Baylis-Hillman acrylate adducts and ar. iodides in aq. ethanol with Pd-C and $NaHCO_3$ as base s. H.-S. Kim, S.-J. Lee, C.M. Yoon, Bull. Korean Chem. Soc. *2013, 34* (1), 325-7 [DOI: 10.5012/bkcs.2013.34.1.325].

Palladium(II) acetate/anionic (ethylenediaminetetraacetic acid)(peroxo)iron(III) complexes ←
N-Directed oxidative decarboxylative *o*-acylation H → COR
update s. *77,* 526s*78*; of acetanilides with α-ketocarboxylic acids in water at room temperature at biological pH with $Pd(OAc)_2$ as catalyst and the anionic iron peroxo complex, $[Fe(III)EDTA-(\eta^2-O_2)]^{3-}$, as oxidant s. S. Sharma, I.A. Khan, A.K. Saxena, Adv. Synth. Catal. *2013, 355* (4), 673-8 [DOI: 10.1002/adsc.201201085].

Palladium(II) trifluoroacetate/silver carbonate *$Pd(OCOCF_3)_2/Ag_2CO_3$*
β-Aryloxo compds. C=C-CHOH → C(Ar)CHCO
from arylcarboxylic acids and 2-ethylenealcohols
Palladium-catalyzed decarboxylative isomerizing Heck-type β-arylation

479.

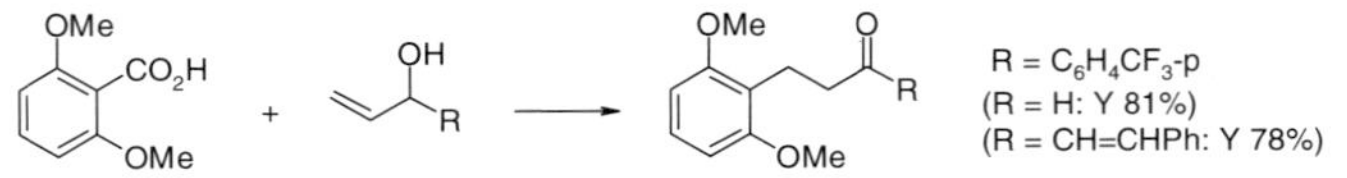

A mixture of 2,6-dimethoxybenzoic acid (0.5 mmol), startg. allylic alcohol (1.2 eq.), $Pd(OCOCF_3)_2$ (5 mol%), Ag_2CO_3 (1 eq.), and DMSO/toluene (1:20; 3 ml) stirred at 100° for 2-8 h under N_2, cooled to room temp., diluted with ether, washed with aq. HCl, the aq. phase re-extracted with ether, the combined organic extracts dried ($MgSO_4$), concentrated *in vacuo*, and the residue purified by silica gel column chromatography → product. Y 85%. The method is simple and efficient and uses readily available starting materials under mild conditions. It is applicable to allyl alcohol, 2-methylprop-2-en-1-ol and sec. allylic alcohols bearing [electron-rich or -poor] aryl, furyl, alkyl or alkenyl groups, dienols reacting selectively at the less substituted double bond. No product was obtained from *o*-methoxybenzoic acid but 2,4,5-trimethoxy-, 4-bromo-2,6-dimethoxy- or 2,6-difluoro-benzoic acids or 3-methylbenzofuran-2-yl- or 1,3-dimethylindol-2-yl-carboxylic acids may also be used. It is tentatively suggested that the hydrogen transfer may involve a Wacker-like oxidative process. F.e. (twenty-six; Y 41-84%) s. L. Huang, J. Qi, X. Wu, K. Huang, H. Jiang, Org. Lett. *2013, 15* (10), 2330-3 [DOI: 10.1021/ol400818v].

Palladium-catalyzed decarboxylative *o*-acylation of arylcarboxylic acids with α-ketocarboxylic acids H → C(O)R

480. 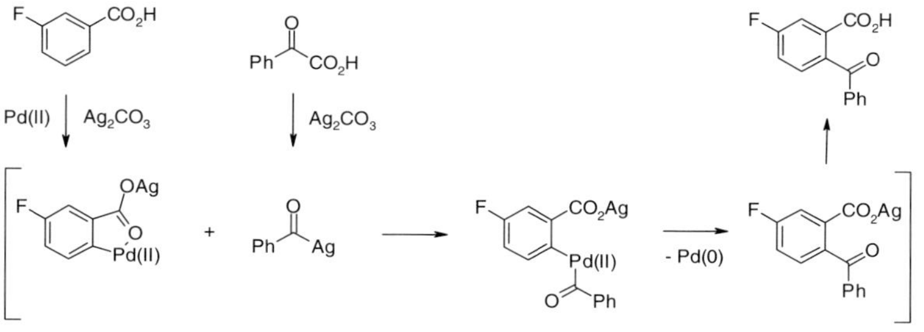

A 20 ml oven-dried pressure tube charged with 3-fluorobenzoic acid (0.2 mmol), startg. arylglyoxylic acid (3 eq.), $Pd(OCOCF_3)_2$ (10 mol%), Ag_2CO_3 (3 eq.), and DME (2 ml), the tube sealed, the mixture stirred vigorously at 150° for 24 h, cooled to room temp., diluted with ethyl acetate, filtered through a Celite pad, the filtrate concentrated *in vacuo*, and the residue purified by flash chromatography on silica gel → 2-benzoyl-5-fluorobenzoic acid. Y 80%. This unusual cross-coupling, involving two different carboxylic acids, was successful for benzoic acids and arylglyoxylic acids, optionally bearing electron-donating (Me, MeO) or mildly electron-withdrawing (F, Cl, Br) groups on either aromatic ring (twenty-four examples; Y 51-85%; a benzoic acid deriv. bearing a *para* CF_3 group required an elevated temp. - 165° - over a prolonged reaction time - 48 h - to afford 56%). With the exception of the illustrated example, electron-rich substrates generally afforded higher yields than their electron-deficient counterparts; *ortho* substitution was generally well-tolerated on either substrate. F.e. incl. application to the synthesis of pitofenone, an antispasmodic used in the combined drug Spasmalgon s. J. Miao, H. Ge, Org. Lett. *2013, 15* (12), 2930-3 [DOI: 10.1021/ol400919u]; **rhodium(II)-catalyzed *o*-acylation** of arylcarboxylic acids **with carboxylic acid anhydrides** (CC⇃↾O) using $[Rh(cod)Cl]_2/Cs_2CO_3$ or KF, complementing the *meta* selectivity of Friedel-Crafts reactions, s. P. Mamone, G. Danoun, L.J. Gooßen, Angew. Chem., Int. Ed. *2013, 52* (26), 6704-8 [DOI: 10.1002/anie.201301328].

Bis(dibenzylideneacetone)palladium(0)/2-[o-(dicyclohexylphosphino)phenyl]-5,6-dimethyl-1-octylbenzimidazole/copper(I) oxide/3,4,7,8-tetramethylphenanthroline ←

Biaryls by decarboxylative sp^2-sp^2 coupling of [hetero]arylcarboxylic acids (or esters) ArCOOH → Ar-Ar′

with ar. halides, update s. *78,* 527s*82*; **from** potassium arylcarboxylates and **aryl mesylates** with $Pd(dba)_2$/2-[*o*-(dicyclohexylphosphino)phenyl]-5,6-dimethyl-1-octylbenzimidazole and Cu_2O/3,4,7,8-tetramethylphenanthroline as cocatalyst, also styrenes with enol mesylates (with a slight variation in the cocatalysts), (cf. *74,* 54s*77*), s. B. Song, T. Knauber, L.J. Gooßen, Angew. Chem., Int. Ed. *2013, 52* (10), 2954-8 [DOI: 10.1002/anie.201208025]; **decarboxylative *o*-arylation** of arylcarboxylic acids with ar. iodides using $Pd(OAc)_2/Ag_2CO_3$, also *o*-arylation with added K_2CO_3 (cf. *73,* 419s*81*), s. C. Arroniz, A. Ironmonger, G. Rassias, I. Larrosa, Org. Lett. *2013, 15* (4), 910-3 [DOI: 10.1021/ol400065j]; *N-directed* [formal] decarboxylative C_{10}-arylation of ethyl benzo[*h*]quinoline-10-carboxylates **with arylboronic acids** using $RhCl(PPh_3)_3$/CuCl cf. J. Wang, B. Liu, H. Zhao, J. Wang, Organometallics *2012, 31* (12), 8598-607 [DOI: 10.1021/om300994j].

Palladium(II) bis(carbamyl)carbene complexes/cesium carbonate *[Pd(II)]/Cs_2CO_3*

Decarboxylative Sonogashira coupling C≡C-COOH → C≡C(Ar)

update s. *79,* 463s*81*; with [hetero]ar. chlorides or bromides with a *phosphine-free* palladium(II) bis(carbamyl)carbene complex and Cs_2CO_3 as base s. P.V. Reddy, P. Srinivas, M. Annapurna, S. Bhargava, J. Wagler, N. Mirzadeh, M.L. Kantam, Adv. Synth. Catal. *2013, 355* (4), 705-10 [DOI: 10.1002/adsc.201200876]; **by coupling with arylboronic acids** in place of ar. halides under copper catalysis with $CuCl_2·2H_2O$/*p*-benzoquinone/triethylamine/pyridine, also 1,3-diynes from α,β-acetyleneboronic acids, s. L. Shi, W. Jia, X. Li, N. Jiao, Tetrahedron Lett. *2013, 54* (15), 1951-5 [DOI: 10.1016/j.tetlet.2013.01.118].

Palladium phosphine complexes *[Pd]*

Decarboxylative coupling of halides C-COOH → C(R)

with functionalized [sp^3] carboxylic acids

decarboxylative sp^2-sp^3coupling with ar. halides cf. *78,* 527s*80*; arylacetic acid esters from ar. halides and malonic acid monoester potassium salts under palladium catalysis in mesitylene with added DMAP s. Y.-S. Feng, W. Wu, Z.-Q. Xu, Y. Li, M. Li, H.-J. Xu, Tetrahedron *2012, 68* (9), 2113-20 [DOI: 10.1016/j.tet.2012.01.032]; β-arylnitriles from alkali-metal α-cyanocarboxylates and benzylic electrophiles, e.g. benzyl chlorides, with $[Pd(allyl)Cl]_2$/2-(dicyclohexylphosphino)-2′,6′-dimethoxybiphenyl **by decarboxylative sp^3-sp^3 coupling** s. R. Shang, Z. Huang, X. Xiao, X. Lu, Y. Fu, L. Liu, Adv. Synth. Catal. *2012, 354* (13), 2465-72 [DOI: 10.1002/adsc.201200383]; synthesis of α,γ-peptide hybrids by decarboxylative C-alkylation of glutamic acid residues s. C.J. Saavedra, A. Boto, R. Hernández, Org. Lett. *2012, 14* (13), 3542-5 [DOI: 10.1021/ol301552c]; 3-β-keto- from 3-halogeno-oxindoles and β-ketocarboxylic acids by **organocatalyzed asym. decarboxylative sp^3-sp^3 coupling** s. J. Zuo, Y.-H. Liao, X.-M. Zhang, W.-C. Yuan, J. Org. Chem. *2012, 77* (24), 11325-32 [DOI: 10.1021/jo302048v]; α-quaternary β,γ-acetylenenitriles from α,β-acetylenehalides and potassium α-cyanocarboxylates by copper-catalyzed **decarboxylative sp^1-sp^3 coupling** with CuBr/8-hydroxyquinoline/Ag_2CO_3 s. Y.-S. Feng, Z.-Q. Xu, L. Mao, F.-F. Zhang, H.-J. Xu, Org. Lett. *2013, 15* (7), 1472-5 [DOI: 10.1021/ol400197y].

Bis(π-allylpalladium chloride)/triphenylphosphine/morpholine ←

4-Allyl-4-cyano-3-pyrrolidones ○

from α-(carballyloxyamino)-α′-cyanoketones and aldehydes

481.

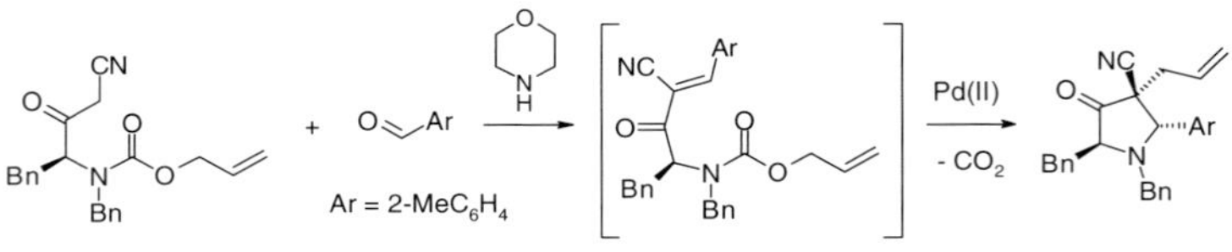

A novel **stereoselective** route to 3-pyrrolidones via **Knoevenagel condensation-decarboxylative intramolecular α-allylation-Michael addition** in one pot is reported, **with asym. induction** from chiral substrates. **E:** 2-Tolualdehyde (1.2 eq.) and morpholine (23 mol%) added to a mixture of (S)-N-allyloxycarbonyl-N-benzyl-4-amino-3-oxo-5-phenylpentanenitrile (1.19 mmol) and 4 Å molecular sieves (430 mg) in methylene chloride (0.5-0.6 *M*) under argon, the mixture stirred for 1-1.5 h (TLC), cooled to 0°, diluted with methylene chloride (0.11 *M*), $[(allyl)PdCl]_2$ (3 mol%) and triphenyl-phosphine (15 mol%) added, the mixture stirred for 0.5-16 h (TLC), and purified directly by chromatography on silica → (2S,4R,5S)-4-allyl-1,2-dibenzyl-4-cyano-5-(2-tolyl)-3-pyrrolidone. Y 83% (d.r. 11.1:1; inseparable mixture). This novel and convenient procedure optimally uses morpholine as base in the initial Knoevenagel condensation of cyanoketones (prepared from the corresponding amino acid esters) with electron-diverse (het)ar. aldehydes to minimize racemization (use of 4-nitrobenzaldehyde resulted in extensive decomposition). The condensates (isolated in four cases; Y 55-75%) cyclized in the presence of Pd(II) to afford inseparable mixtures of diastereomers (twenty-two examples; Y 51-99%). Generally, the illustrated *trans/trans* diastereomers predominated (d.r. 2.6:1 to >20:1; 3-cyanobenzaldehyde gave 1:1), but interestingly, 2,6-disubst. benzaldehydes gave the *cis/trans* diastereomers near exclusively (d.r. >20:1; two examples; Y 54-81%). In a final development, Luche reduction ($NaBH_4/CeCl_3$) of the products was incorporated into the one-pot procedure, affording the corresponding **chiral 3-allyl-3-cyano-4-hydroxypyrrolidines** with near complete control of the newly-formed hydroxyl stereocenter (e.e. 97 to >99%; three examples; Y 47-66%). F.e., optimization and substrate prepn. s. O.K. Karjalainen, M. Nieger, A.M.P. Koskinen, Angew. Chem., Int. Ed. *2013, 52* (9), 2551-4 [DOI: 10.1002/anie.201209443].

Tris(2-phenylpyridinato-C2,N)iridium(III) s.a. under N-Isopropylbenzylamine *[Ir(III)]*

fac-Tris(2-phenylpyridinato-C2,N)iridium(III)/irradiation *fac-Ir(ppy)$_3$/ꟿ*

Synthesis of oxindoles
from α-methylenecarboxylic acid anilides and iodosocarboxylates
via decarboxylative radical coupling under visible-light photoredox catalysis

482. 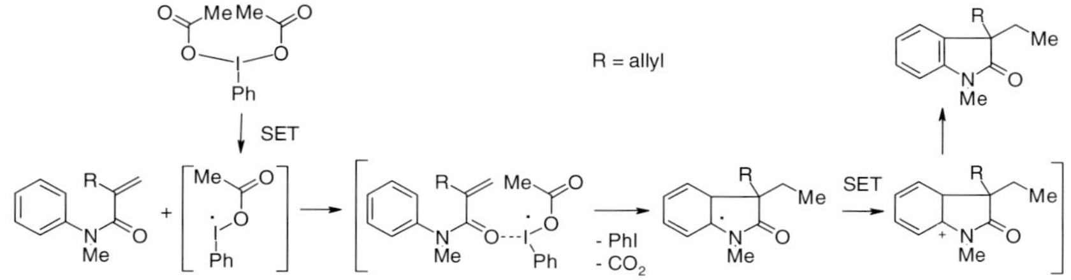

3,3-Disubst. oxindoles. Startg. α-methylenecarboxylic acid anilide (0.3 mmol), *fac*-Ir(ppy)$_3$ (1 mol%), phenyl iodosoacetate (3 eq.) and DMF (2 ml) added to an oven-dried Schlenk tube, the tube degassed three times by alternating vacuum evacuation (5 min) and argon backfill, placed ~5 cm from a 35 W fluorescent light bulb, the yellow soln. stirred at room temp. under visible-light irradiation for 24 h, diluted with ethyl acetate, washed with satd. K_2CO_3 soln., the aq. layer extracted with ethyl acetate, the combined organic layers washed with brine, dried (Na_2SO_4), filtered, the filtrate concentrated *in vacuo*, and the residue purified by chromatography on silica gel → 3-allyl-3-ethyl-1-methylindolin-2-one. Y 83%. Coupling of phenyl iodosoacetate with N-alkyl/benzyl/phenyl-α-methyl/ethyl/benzyl/allyl-subst. α-methylenecarboxylic acid anilides (optionally bearing Me, MeO, Cl or Br in the *para* position) or N-pyridyl analogs gave yields of 48-90% for fourteen further examples; reaction failed with a *p*-fluoroanilide, however. The method was extended to a variety of readily-available prim., sec. or tert. aliphatic carboxylic acids via initial ligand metathesis with phenyl iodosoacetate (twenty examples; Y 52-86%); ω,ω,ω-trifluoropropanoic acid or -butanoic acid could be used (but not trifluoroacetic acid), as could α-chloropropanoic acid and a lithocholic acid deriv. Benzoic acid was not suitable, however. F.e.s. J. Xie, P. Xu, H. Li, Q. Xue, H. Jin, Y. Cheng, C. Zhu, Chem. Commun. *2013*, *49* (50), 5672-4 [DOI: 10.1039/c3cc42672f].

Via intermediates *v.i.*

Synthesis of polysubst. biaryls via C_4-alkylation of 2,5-cyclohexadienecarboxylic acids and subsequent decarboxylative aromatizing C_3-arylation ←

483. 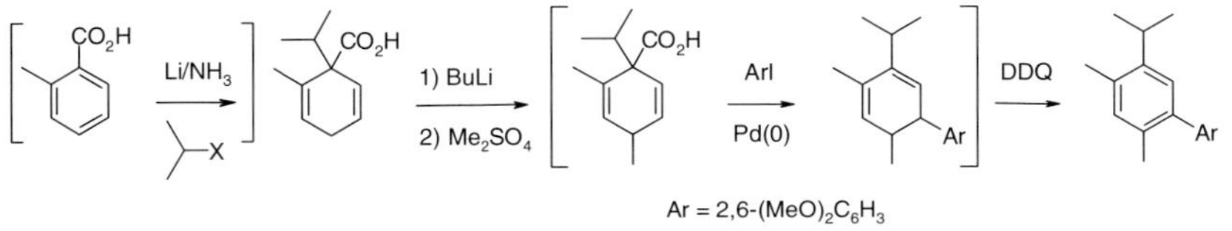

Preparation of polysubst. arenes via a novel method is reported, wherein pre-formed and storable 2,5-cyclohexadienecarboxylic acids, obtained by alkylative Birch reduction of arylcarboxylic acids, are subjected to a C_4-alkylation, decarboxylative C_3-arylation and aromatization sequence. **E:** A soln. of butyllithium (2.2 eq.) in hexane (1.6 *M*) and TMEDA (2.2 eq.) added slowly to a soln. of 1-isopropyl-2-methyl-2,5-cyclohexadiene-1-carboxylic acid (0.3 mmol) in anh. THF (0.3 *M*) at -78°, the yellow soln. stirred for 30 min, dimethyl sulfate (1.1-2 eq.) added, the mixture stirred at -78° or room temp. until soln. decolorized, stirring continued for a further 10 min, quenched with 2 *M* aq. HCl, extracted with methylene chloride, concentrated *in vacuo*, the residue dissolved in toluene (0.25 *M*), 2,6-dimethoxyphenyl iodide (1.1 eq.), Cs_2CO_3 (1.1 eq.), and $Pd(dba)_2$ (10 mol%) added under argon, the mixture stirred at 110° for 17-20 h, cooled to room temp., DDQ (1.5 eq.) added in one portion, the mixture stirred for 1 h, filtered, solvent removed *in vacuo*, and the residue purified by flash chromatography → 5′-isopropyl-2,6-dimethoxy-2′,4′-dimethylbiphenyl. Y 65%. 1,2-Di- and 1,2,3-tri-subst. 2,5-cyclohexadienylcarboxylic acids (incl. 1-isopropyl-1,2-dihydroxy-naphthalene-1-carboxylic acid) were subjected to C_4-alkylation (Me, Et, Bn), C_3-(het)arylation and re-aromatization (with DDQ) without purification of intermediates (twenty-examples; Y 38-91%; a *p*-isopropyl deriv. gave 10%). Results from individual C_3-arylation of the substrates and isolation of the dihydro products (seven examples; Y 54-95%), subsequent aromatization in some cases (four examples; Y 32-62%), and also C_4-alkylation of the substrates (nine examples; Y 55-95%) are also reported. F.e.s. E. Koch, A. Studer, Angew. Chem., Int. Ed. *2013*, *52* (18), 4933-6 [DOI: 10.1002/anie.201300259].

Carboxylic acid N-(carbomethoxy)amides $RB(OH)_2$ → RC(O)NHCOOMe
from boronic acids via N-(carbomethoxy)iminoesters
Methyl N-[methoxy(methylthio)methylene]carbamate as carbonylating agent

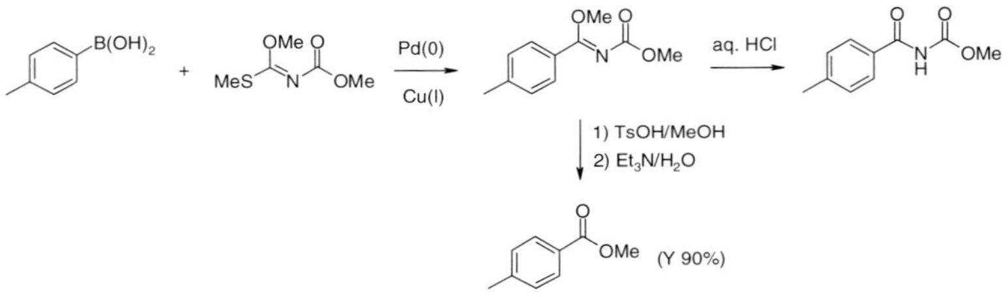

484.

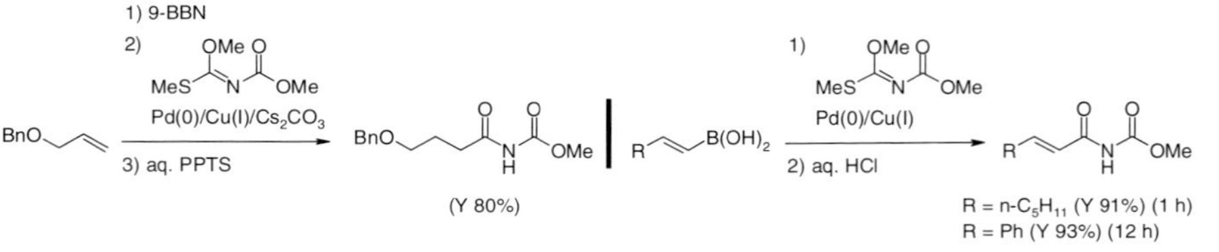

A soln. of $Pd(PPh_3)_4$ (10 mol%), Cu-thiophene-2-carboxylate (3 eq.), methyl N-[methoxy(methylthio)methylene]carbamate (1.1 eq.) and *p*-tolylboronic acid (0.168 mmol) in THF (2 ml) stirred at 50° for 12 h, filtered through a plug of Celite, the filtrate concentrated under reduced pressure, and the residue purified by flash chromatography on silica → methyl N-[methoxy(*p*-tolyl)-methylene]carbamate (Y 97%), a portion (0.514 mmol) of which dissolved in THF (2 ml), 1 *N* HCl (3 drops) added to the stirred soln. at room temp., stirring continued for 1 h, water (2 ml) added, extracted with methylene chloride, and purified by flash chromatography on silica → methyl 4-methylbenzoylcarbamate. Y 95%. Five further (het)arylboronic acids afforded yields of 60-97% for the coupling step (lowest for 2-thienyl- and electron-deficient aryl-boronic acids) and 91-97% for the hydrolysis. Alkenylboronic acids were similarly high-yielding (91-93%; two examples), and alkyl derivs. could be obtained **from** crude **B-alkyl-9-borabicyclo[3.3.1]nonanes**, obtained by treatment of a variety of alkenes (incl. styrene) with 9-BBN (five examples; overall Y 76-83% for the 3 steps), efficient coupling requiring Cs_2CO_3 as additive. Carboxylic acid esters could be obtained by treatment of the intermediate iminoethers with *p*-TsOH/MeOH (one example, illustrated; Y 90%), effecting a one-carbon coupling avoiding the use of toxic carbon monoxide. F.e., optimization, substrate preparation, and prospects for similar cross-coupling using organostannanes, s. T. Tomizawa, K. Orimoto, T. Niwa, M. Nakada, Org. Lett. *2012*, *14* (24), 6294-7 [DOI: 10.1021/ol303062a].

Elimination ⥣

Hydrogen ↑ CC ⥣ H

Microwaves s. under Copper(II) chloride [\\\\]
Potassium iodide s. under CuBr *KI*

Copper(II) acetate/sodium carbonate/pyridine/air $Cu(OAc)_2/Na_2CO_3/py/O_2$
Aromatization of 4,5-dihydro-3(2*H*)-pyridazones ←
with simultaneous replacement of ar. halogen by hydrogen s. *14*, 901s*83*

Silver trifluoroacetate s. under Bis(acetonitrile)dichloropalladium(II) $AgOCOCF_3$

Copper(I) bromide-dimethyl sulfide/potassium iodide/1,8-diazabicyclo[5.4.0]undec-7-ene/ cesium carbonate/oxygen ←
Pyrazoles from N,N-disubst. hydrazones ○
via copper(I)-catalyzed dehydrogenative ring closure

485.

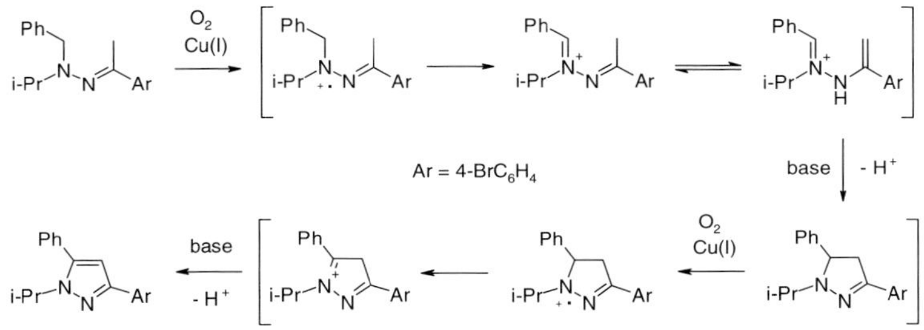

A mixture of 1-benzyl-1-isopropyl-2-[(1-(4-bromophenyl)ethylidene]hydrazine (0.3 mmol), $CuBr{\cdot}SMe_2$ (10 mol%), KI (0.5 eq.), DBU (30 mol%), Cs_2CO_3 (1.1 eq.) and 1,2-dichloroethane/ dimethyl sulfide (10:1 v/v; 3 ml) stirred vigorously under O_2 (1 atm.) at 135° for 5 h, solvent removed *in vacuo,* and the residue purified by flash chromatography → 3-(4-bromophenyl)-1-isopropyl-5-phenylpyrazole. Y 76%. This efficient method, which occurs via novel copper(I)-catalyzed intramolecular oxidative coupling of an iminium ion, was apparently general for 1-subst. (alkyl, benzyl, phenyl) 1-arylmethyl-2-(1-arylethylidene)hydrazines, wherein the aryl groups carried electron-donating or -withdrawing (incl. *ortho)* substituents (thirty-four examples; Y 45-92%). A 1-benzyl-1-methylhydrazine deriv. underwent competitive cyclization at benzyl or methyl carbons to afford an isomeric mixture of 1-methyl- (Y 39%) and 1-benzyl-pyrazoles (Y 22%). A 2-(1-phenylpropylidene)hydrazine gave a low yield of the corresponding tetrasubst. pyrazole (16%) while a cyclohexylidene analog gave 0%, attributed to competing decomposition of the substrates. F.e. and optimization s. G. Zhang, Y. Zhao, H. Ge, Angew. Chem., Int. Ed. *2013*, *52* (9), 2559-63 [DOI: 10.1002/anie.201209618].

Copper(II) chloride/nickel(II) nitrate/polyalkylene glycols/N,N,N',N'-tetramethylethylene-diamine/triethylamine/microwaves ←
Glaser 1,3-diyne coupling C≡C–C≡C
intermolecular coupling s. *16,* 780s*83*; **cyclic 1,3-diynes** by *intramolecular* aerobic coupling with $CuCl_2/Ni(NO_3)_2{\cdot}6H_2O$ in methanol containing a polyethylene or polypropylene glycol in the presence of TMEDA and triethylamine under microwave irradiation s. A.-C. Bédard, S.K. Collins, ACS Catal. *2013*, *3* (4), 773-82 [DOI: 10.1021/cs400050g].

Copper(II) salt s. under Iridium(I) complexes Cu^{2+}

Gold nanoparticles/poly(anilinesulfonic acid)/air ←
Aerobic aromatization of indolines and dihydropyridines in aq. media s. *14,* 901s*83* ←

Scandium(III) triflate/sodium sulfate *$Sc(OTf)_3/Na_2SO_4$*
Dehydrogenative intramolecular *o*-acylation with aldehydes ○
N-Methylacridones s. *82*, 485s*83*

Graphite oxide/molecular sieves ←
Dehydrogenative aromatization ←
update s. *14,* 901s*82*; of partially or fully satd. cycloalkanes and hydrogenated benzo-condensed heterocyclics with graphite oxide/4 Å molecular sieves s. X. Zhang, L. Xu, X. Wang, N. Ma, F. Sun, Chin. J. Org. Chem. *2012, 30* (7), 1525-30 [DOI: 10.1002/cjoc.201200174]; imidazoles from Δ^2-imidazolines by aerobic aromatization at room temperature with cationic chloro-(terpyridyl)-(*o*-pyrid-2-ylaryl)ruthenium(III) complexes as catalyst s. S. Aiki, Y. Kijima, J. Kuwabara, A. Taketoshi, T.-a. Koizumi, S. Akine, T. Kanbara, ACS Catal. *2013, 3* (5), 812-6 [DOI: 10.1021/cs300811w]; aerobic aromatization of 2-subst. indolines and dihydropyridines using gold nanoparticles as catalyst with redox-active poly(anilinesulfonic acid) in aq. media s. T. Amaya, T. Ito, Y. Inada, D. Saio, T. Hirao, Tetrahedron Lett. *2012, 53* (45), 6144-7 [DOI: 10.1016/j.tetlet.2012.09.002]; of 6-aryl-4,5-dihydro-3(2*H*)-pyridazones using $Cu(OAc)_2/Na_2CO_3$/pyridine with simultaneous replacement of ar. halogen by hydrogen s. L. Liang, G. Yang, W. Wang, F. Xu, Y. Niu, Q. Sun, P. Xu, Adv. Synth. Catal. *2013, 355* (7), 1284-90 [DOI: 10.1002/adsc.201300026]; base-catalyzed dehydrogenation of 2-aryl-3-hydroxy-4-chromanones [flavono-lignans] in (m)ethanol with, e.g. NH_4OH, amines or [bi]carbonates, also in the *absence* of oxygen, s. R. Gazák, P. Trouillas, D. Biedermann, K. Fuksová, P. Marhol, M. Kuzma, V. Kren, Tetrahedron Lett. *2013, 54* (4), 315-7 [DOI: 10.1016/j.tetlet.2012.11.049].

Polyalkylene glycols s. under Copper(II) chloride ←
p-Benzoquinone s. under Tetrakis(acetonitrile)palladium(II) bis(fluoroborate) *BQ*
tert-Butyl hydroperoxide s. under Bu_4NI and Cp_2Fe *t-BuOOH*
Acetic acid s. under Chloro(1,5-cyclooctadiene)iridium(I) dimer *AcOH*
Trifluoroacetic acid s. under $PhI(OCOCF_3)_2$ *CF_3COOH*
Dimethyl sulfoxide s. under Tetrakis(acetonitrile)palladium(II) bis(fluoroborate) *DMSO*
Phenyl iodosoacetate s. under Tetrakis(acetonitrile)palladium(II) bis(fluoroborate) *$PhI(OAc)_2$*

Phenyl iodosotrifluoroacetate/trifluoroacetic acid *$PhI(OCOCF_3)_2/CF_3COOH$*
2-Azaspiro[4.5]deca-3,6,9-triene-1,8-dione-3-carboxylic acid esters ○
from α-(*p*-hydroxyaroylamino)acrylic acid esters s. *83*, 92

2,3-Dichloro-5,6-dicyanoquinone *DDQ*
Dehydrogenative aromatization – Polysubst. biaryls s. *83*, 483 ←

1,2-Bis(diphenylphosphino)ethane s. under Chloro(1,5-cyclooctadiene)iridium(I) dimer *dppe*
Oxygen (or air) s. under CuBr, $Cu(OAc)_2$, Gold nanoparticles, Cationic chloro(terpyridyl)-(o-pyrid-2-ylaryl)ruthenium(III) complexes, $Pd(OAc)_2$ and Tetrakis(acetonitrile)-palladium(II) bis(fluoroborate) *O_2*
Sodium anthraquinone-2-sulfonate s. under Tetrakis(acetonitrile)palladium(II) bis(fluoroborate) ←
Poly(anilinesulfonic acid) s. under Gold nanoparticles ←
Tetra-n-butylammonium bromide s. under $Pd(OAc)_2$ *Bu_4NBr*

Tetra-n-butylammonium iodide/tert-butyl hydroperoxide *Bu_4NI/t-BuOOH*
3a-Cyano-1,2,3,3a-tetrahydrocyclopenta[*b*]naphthalen-4-ones from aryl ε,ζ-acetylene-α-cyanoketones ○

486.

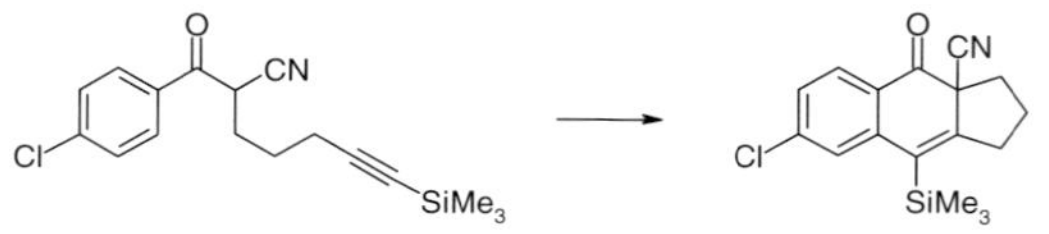

Oxidative double radical ring closure. A mixture of 2-(4-chlorobenzoyl)-7-(trimethylsilyl)hept-6-ynenitrile (0.32 mmol), *n*-Bu_4NI (20 mol%) and *t*-BuOOH (70% aq.; 1.1 eq.) in benzene (10 ml)

degassed with N_2 for 10 min, then heated at 80° for 1 h under N_2, quenched with satd. aq. Na_2SO_3, extracted with ethyl acetate, and purified by flash chromatography on silica gel → 7-chloro-4-oxo-9-(trimethylsilyl)-1,2,3,4-tetrahydro-3a*H*-cyclopenta[*b*]naphthalene-3a-carbonitrile. Y 95%. Mechanistically, reaction is thought to proceed via initial abstraction of the H-atom α to both cyano and keto groups, followed by 5-*exo-dig* cyclization with the alkyne moiety, and 6-*endo-trig* cyclization of the resulting vinyl radical with the aromatic ring. Nine further [6,6,5]-tricycles were similarly prepared, optionally bearing H, Me, MeO, CN, CF_3 or CO_2Me in place of Cl, and optionally bearing *gem*-dimethyl β to the keto group (Y 72%, 85-98%); corresponding [5,6,5]-tricycles could be prepared from 2-thienoyl or 2-furanoyl analogs (two examples; Y 85-95%). The presence of the terminal trimethylsilyl group was important in stabilizing the intermediate vinyl radical (Y 35% with an H-terminated substrate), but similar stabilization was also observed for analogs bearing a terminal aromatic group (three examples; Y 84-87%), which could be electron-rich or -deficient. Careful control of reaction conditions and rigorous exclusion of oxygen were critical in ensuring minimal by-product formation. F.e.s. Y.-C. Wong, C.-T. Tseng, T.-T. Kao, Y.-C. Yeh, K.-S. Shia, Org. Lett. *2012*, *14* (23), 6024-7 [DOI: 10.1021/ol3028972].

Ferrocene/tert-butyl hydroperoxide *Cp_2Fe/t-BuOOH*
Dehydrogenative intramolecular *o*-acylation with aldehydes ○
xanthones with $RhCl_3/Ph_3P/t$-BuOOH cf. *82*, 485; 9-fluorenones and xanthones with ferrocene as catalyst and *t*-BuOOH as oxidant s. S. Wertz, D. Leifert, A. Studer, Org. Lett. *2013, 15* (4), 928-31 [DOI: 10.1021/ol4000857]; synthesis of N-methylacridones with $Sc(OTf)_3$ and added Na_2SO_4 s. X.-A. Li, H.-L. Wang, S.-D. Yang, ibid. *15* (8), 1794-7 [DOI: 10.1021/ol400371h].

Nickel(II) nitrate s. under Copper(II) chloride *$Ni(NO_3)_2$*

Cationic chloro(terpyridyl)(o-pyrid-2-ylaryl)ruthenium(III) complexes/air *[Ru(III)]/O_2*
Imidazoles from Δ^2-imidazolines s. *14,* 901s*83* ←

Palladium(II) acetate/tetra-n-butylammonium bromide/oxygen *$Pd(OAc)_2/Bu_4NBr/O_2$*
Pyrroles from N-allylketimines ○
via palladium-catalyzed aerobic dehydrogenative intramolecular C-α-vinylation

487. 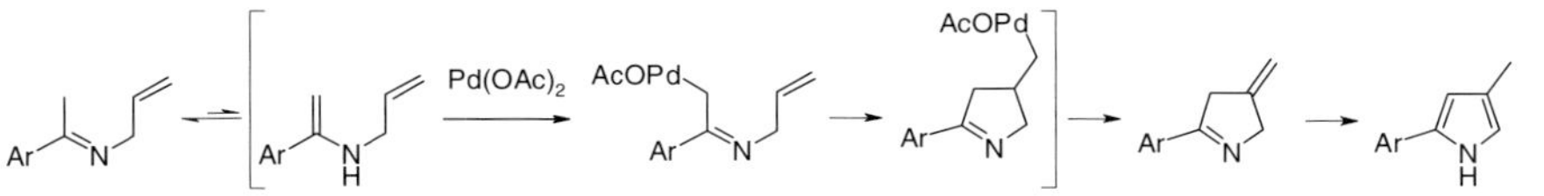

Ar = p-$CF_3C_6H_4$
(Ar = 3-pyridyl: Y 64%)

The first palladium-catalyzed aerobic oxidative intramolecular alkenylation reactions of C(*sp*³)-H bonds are reported for the construction of pyrrole derivatives. **E: 2-Aryl-4-methylpyrroles**. Toluene (2 ml), *DMSO* (0.2 ml) and startg. imine (0.5 mmol) added via syringe to a septum-sealed Schlenk tube containing $Pd(OAc)_2$ (10 mol%) and Bu_4NBr (1 eq.) under an O_2 balloon, the mixture stirred at 35° for 17 h, quenched with water, extracted with ether, and purified by flash chromatography on silica gel → 4-methyl-2-[(4-(trifluoromethyl)phenyl]-1*H*-pyrrole. Y 78%. Ten further examples with aryl N-allylketimines optionally bearing COOMe, CN, NO_2, OMe, F, Cl or Br on the aromatic ring gave yields of 46-71%. A 3-pyridyl analog gave a yield of 64% but reaction failed with a 2-pyridyl deriv., presumably due to its strong coordination to the palladium catalyst. Lower yields were obtained at higher temperatures and the method failed in initial experiments with nickel or copper catalysts. Kinetic experiments showed that the reaction exhibited a first-order dependence on the concentration of the catalyst, indicating that palladium nanoparticles are not involved, and suggested that C(*sp*³)-H metalation is a slow step. F.e.s. L. Meng, K. Wu,

C. Liu, A. Lei, Chem. Commun. *2013*, 49 (52), 5853-5 [DOI: 10.1039/c3cc42307g]; pyrroles or 4-vinyl-Δ^1-pyrrolines **from 2-ethylene-*prim*-amines and ketones** via palladium-catalyzed dehydrogenative ring closure of N-allylketimines under similar conditions s. Z. Shi, M. Suri, F. Glorius, Angew. Chem., Int. Ed. *2013*, *52* (18), 4892-6 [DOI: 10.1002/anie.201300477].

Tetrakis(acetonitrile)palladium(II) bis(fluoroborate)/6,6′-dimethyl-2,2′-bipyridyl/ sodium anthraquinone-2-sulfonate/oxygen/dimethyl sulfoxide ←
Phenols from 2-cyclohexenones ←
m-Hydroxybiaryls via aerobic Heck-type arylation-dehydrogenative aromatization s. *83*, 465

Tetrakis(acetonitrile)palladium(II) bis(fluoroborate)/phenyl iodosoacetate/p-benzoquinone ←
α,β-Ethyleneketones from ketones s. *83*, 57 CHCH → C=C

Bis(acetonitrile)dichloropalladium(II)/silver trifluoroacetate/trifluoroacetic acid ←
9-Fluorenones from *o*-cyanobiaryls s. *83*, 426 ○

Iridium(I) complexes/copper(II) salt *[Ir(I)]/Cu²⁺*
Oxidative intramolecular biaryl coupling
update s. *18*, 928s*82*; *82*, 484; 9-fluorenols from diarylcarbinols under iridium(I) catalysis with an added copper(II) salt as oxidant, also fluorenes from diarylacetic acids by oxidative *decarboxylative* intramolecular biaryl coupling under rhodium(I) catalysis with an added copper(II) salt, s. M. Itoh, K. Hirano, T. Satoh, Y. Shibata, K. Tanaka, M. Miura, J. Org. Chem. *2013*, *78* (4), 1365-70 [DOI: 10.1021/jo4000465].

Chloro(1,5-cyclooctadiene)iridium(I) dimer/1,2-bis(diphenylphosphino)ethane/acetic acid ←
4(1*H*)-Pyridone ring from *o-tert*-aminoketones
via iridium-catalyzed intramolecular dehydrogenative coupling

488.

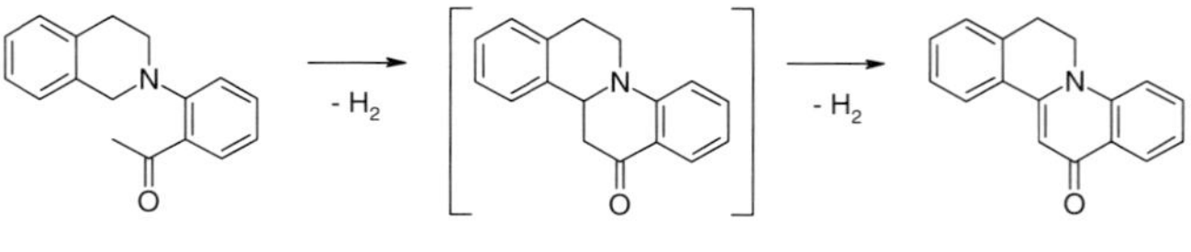

6*H*-Isoquinolino[2,1-*a*]quinolin-13(7*H*)-ones. $[Ir(cod)Cl]_2$ (2.5 mol%) and dppe (5 mol%) mixed in methylene chloride (1 ml) under argon at room temp., solvent removed *in vacuo* after 30 min, 1-[2-(3,4-dihydroisoquinolin-2(1*H*)-yl)phenyl]ethanone (0.2 mmol) in acetic acid (2 ml) added, the mixture stirred at 120° for 24 h under argon, solvent removed *in vacuo*, and the crude product purified by flash chromatography → 6*H*-isoquinolino[2,1-*a*]quinolin-13(7*H*)-one. Y 71%. This unprecedented, oxidant-free and acceptor-less intramolecular dehydrogenative coupling afforded the title compounds in yields of 63-72% (five examples; replacing acetyl with propionyl (one example) afforded the corresponding methyl deriv. in 48% yield), tolerating a variety of aryl substituents (e.g. Me, MeO, CF_3) on either the tetrahydroisoquinolinyl moiety or the aryl tether. A hetar. [1-(6,7-dihydrothieno[3,2-*c*]pyridin-5(4*H*)-yl)] analog (Y 51%) and a ketone O-methyl-oxime deriv. (Y 72%) were also suitable substrates, but non-fused piperidino or morpholino analogs gave none of the desired products (cyclization occurred via single C-C bond formation in the former case (Y 32%) but not in the latter). F.e., optimization, and preliminary studies into Ir-catalyzed *dehydrative* coupling to afford **5,6-dihydroindolo[2,1-*a*]isoquinolines**, s. S.-z. Nie, X. Sun, W.-t. Wei, X.-j. Zhang, M. Yan, J.-l. Xiao, Org. Lett. *2013*, *15* (10), 2394-7 [DOI: 10.1021/ol4008469].

Oxygen ↑

CC ⇑ O

Lithium tert-butoxide/tosylhydrazine *t-BuOLi/TsNHNH$_2$*

9-Phenanthrenols from 2′-acylbiaryl-2-carboxaldehydes
Transition metal-free ring closure via 1,2-migration

489.

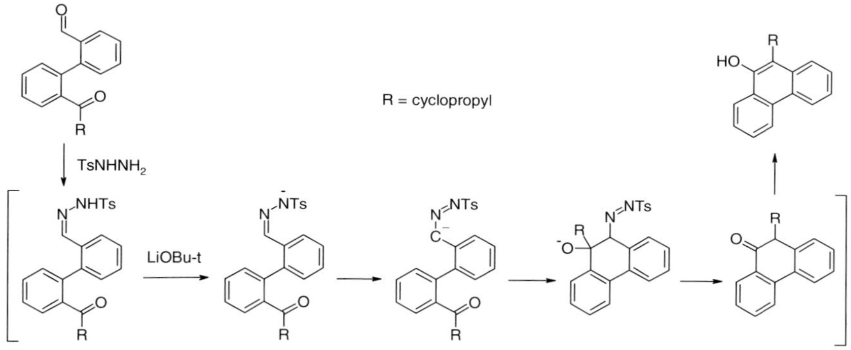

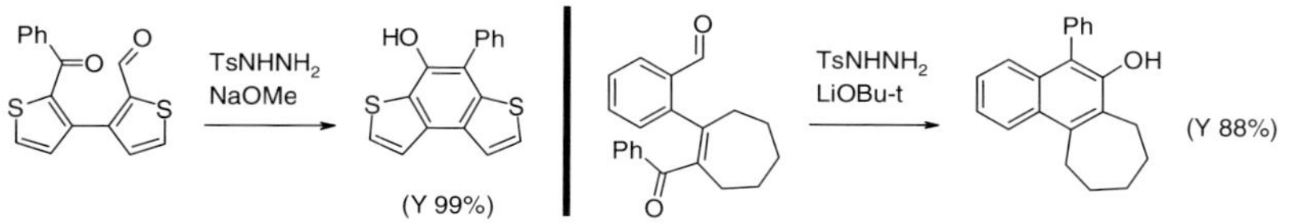

A mixture of 2-formyl-2′-cyclopropylcarbonylbiphenyl (0.3 mmol), tosylhydrazine (1.03 eq.) and toluene (6 ml) stirred in a flame-dried flask at 40° for 30 min (TLC), Li-*tert*-butoxide (2.5 eq.) added, the soln. stirred at 70° for 30 min, quenched with acetic acid (2.5 eq.), cooled to room temp., filtered through silica, solvent removed *in vacuo*, and the residue purified by chromatography on silica → 10-cyclopropylphenanthren-9-ol. Y 87%. This general and versatile reaction was successful for 2-formyl-2′-acyl/(het)aroyl-biaryls (incl. mono- and di-furyl/thienyl analogs), and involved initial chemoselective formation of the aldehyde tosylhydrazone which, on deprotonation (with NaOMe or LiOBu-*t*), effected intramolecular cyclization/1,2-migration to afford the air-sensitive products (twenty-seven examples; Y 62-99%) in the presence of nitro, halo, acetal, ether, alkyne and alkene functionality. Lower yields were obtained with 4-nitrobenzoyl (41%) and phenylpropioloyl derivs. (32%), and also from 3-(2-benzoylfur-3-yl)-2-formylthiophene (40%). The reaction was applicable to the cyclization of β-formyl-*o*-aroylstyrene derivs. to generate **1-naphthol rings** (thirteen examples; Y 64-99%), and also to analogous β-acyl/aroyl-*o*-formyl-styrenes, affording **2-naphthol rings** (seventeen examples; Y 53-96%). F.e., optimization, substrate prepn., and use of the method in a synthesis of a dibenzochrysene deriv. s. Y. Xia, P. Qu, Z. Liu, R. Ge, Q. Xiao, Y. Zhang, J. Wang, Angew. Chem., Int. Ed. *2013*, *52* (9), 2543-6 [DOI: 10.1002/anie.201209269].

Calcium bis(trimethylsilyl)amide/tetra-n-butylammonium hexafluorophosphate ←
Intramolecular Friedel-Crafts alkylation with alcohols
update s. 2, 794s*82*; indans and tetralins by regio- and diastereo-selective cyclodehydration with Ca(NTf$_2$)$_2$/[Bu$_4$N]PF$_6$ s. J.-M. Begouin, F. Capitta, X. Wu, M. Niggemann, Org. Lett. *2013, 15* (6), 1370-3 [DOI: 10.1021/ol400341p].

Samarium(II) iodide *SmI_2*

3-Ene-1,2-diols from 4-acyl-1,3-dioxolanes
via samarium diiodide-promoted eliminative intramolecular cross-pinacol coupling

490.

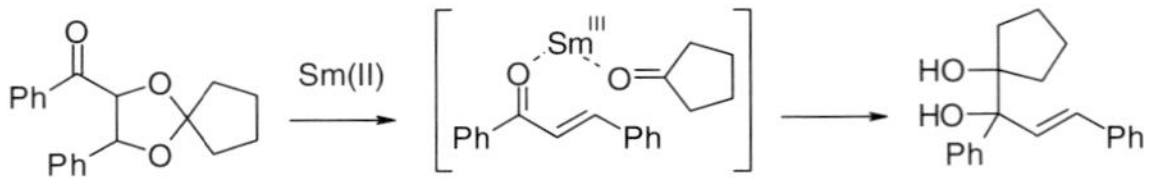

A novel method is reported for the preparation of rare 1,2-bis-*tert*-diols, via cross-pinacol coupling of ketones, themselves prepared *in situ* by reductive cleavage of 2,2-dialkyl-4-aroyl-5-aryl-1,3-dioxolanes. **E:** A soln. of 4-benzoyl-5-phenyl-2-spirocyclopentyl-1,3-dioxolane (0.3 mmol) in THF (2 ml) added to a stirred soln. of SmI_2 (4 eq.) in the same solvent (6 ml) at room temp. under N_2, the mixture stirred for 4 h, quenched with satd. aq. $Na_2S_2O_3$, extracted with ethyl acetate, concentrated *in vacuo*, and purified by flash chromatography → (E)-1-(1-hydroxy-1,3-diphenylallyl)cyclopentanol. Y 84%. Generation of the ketone components *in situ*, presumably as a Sm(III) complex, allowed 'pseudo-intramolecular' reaction of the component ketones, thereby competing effectively with homo-coupling (ten examples; Y 53-84%). Lower yields were obtained with 4-hetaroyl derivs. (21-35%; two examples), and also in the presence of additives (water, alcohols, HMPA). Comparable cross-couplings of dialkyl and α,β-unsaturated ketones under similar conditions gave significantly lower yields than the tandem process. F.e. and optimization s. Y.-H. Zhao, R.-L. Ma, X.-W. Sun, G.-Q. Lin, Synthesis *2012*, *44* (17), 2763-9 [DOI: 10.1055/s-0032-1316588].

Amberlyst-15 ←

Ethylene derivs. from alcohols CHC(OH) → C=C

update s. *5*, 627s*82*; heterogeneous dehydration of tert. alcohols with recyclable Amberlyst-15, notably tolerating N-protected amines and O-protected alcohols s. L.M.T. Frija, C.A.M. Afonso, Tetrahedron *2012*, *68* (36), 7414-21 [DOI: 10.1016/j.tet.2012.06.083]; (E)-2-aryl-1,3-enynes from 2-aryl-2-allenealcohols with acetyl chloride/NaOH in aq. medium via 2-aryl-1-acetoxy-2-allenes s. B. Alcaide, P. Almendros, T. Martínez del Campo, Org. Biomol. Chem. *2012*, *10* (37), 7603-9 [DOI: 10.1039/c2ob26085a]

Acetyl chloride/sodium hydroxide *AcCl/NaOH*

(E)-2-Aryl-1,3-enynes from 2-aryl-2-allenealcohols s. *5*, 627s*83* ←

Tosylhydrazine s. under t-BuOLi *$TsNHNH_2$*

(R)-3,3′-Di(anthracen-9-yl)-1,1′-binaphthyl-2,2′-diyl neomenthyl phosphite s. under Tris(dibenzylideneacetone)dipalladium ←

Tetra-n-butylammonium hexafluorophosphate s. under $Ca(NTf_2)_2$ *$[Bu_4N]PF_6$*

Sulfuric acid *H_2SO_4*

4-Thiochromones from (Z)-3-(arylthio)acrylic acids s. *83*, 235 ○

Tris(dibenzylideneacetone)dipalladium/(R)-3,3′-di(anthracen-9-yl)-1,1′-binaphthyl-2,2′-diyl neomenthyl phosphite/ethyldiisopropylamine ←

Allenes from (E)-enol triflates CHC(OTf)=C → C=C=C
by palladium-catalyzed asym. elimination

491.

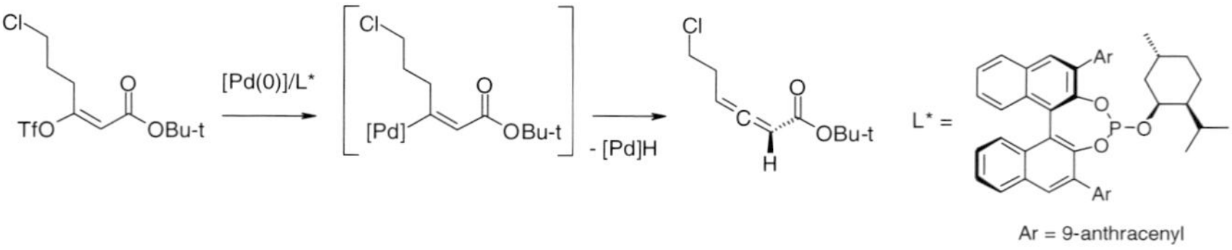

Isopropyl acetate (2.5 ml) added by syringe to a vial charged with $Pd_2(dba)_3$ (5 mol%) and chiral phosphite ligand (10 mol%) under N_2, the mixture stirred at 50° for 1 h, cooled to room temp., startg. (E)-enol triflate (0.5 mmol) and Hünig's base (4 eq.) added sequentially by syringe, the resulting mixture stirred for 8 h (TLC), concentrated to a volume of 0.8 ml under reduced pressure, and the resulting slurry purified by chromatography on silica gel → (R)-*tert*-butyl 6-chlorohexa-2,3-dienoate. Y 78% (e.e. 88%). Reaction was successful for a variety of β-subst. (E)-β-triflyloxy-α,β-ethylenecarboxylic acid esters wherein the ester group was bulky (benzyl, *tert*-butyl, adamantyl etc.) (eleven examples; Y 60%, 72-87%; e.e. 64%, 85-95%). Opposite enantiomers (twelve examples; Y 69-91%; e.e. 57%, 86-96%) were obtained using a *pseudo*-enantiomeric phosphite ligand. (Z)-Enol triflates were unreactive, even in the presence of added Lewis acids, and fully-subst. (E)-enol triflates were poor substrates, affording trisubst. allenes in low yield and low e.e. The choice of Pd-catalyst enables attenuation of the reaction at the allene stage [further isomerization to **1,3-dienes** occurs under catalysis with Pd(PBu-t_3)$_2$; s.a. I.T. Crouch, T. Dreier, D.E. Frantz, Angew. Chem., Int. Ed. *2011*, *50* (27), 6128-32 [DOI: 10.1002/anie.201101820]], and the use of novel phosphite, rather than phosphine, ligands helps to prevent erosion of enantioselectivity due to racemization of the product. F.e. and application to the formal synthesis of the natural product, (+)-epibatidine s. I.T. Crouch, R.K. Neff, D.E. Frantz, J. Am. Chem. Soc. *2013*, *135* (13), 4970-3 [DOI: 10.1021/ja401606e].

Chiral palladium ferrocenyldi(phosphine) complexes *[Pd]**
Asym. intramolecular Friedel-Crafts allylation ○
with 2-ethylenecarbonic acid esters using a chiral iridium catalysis cf. *77*, 531s*82*; asym. synthesis of large-ring *peri*-annulated indoles by intramolecular allylation at the C_3- or C_5-position of the indole ring (with double-bond shift), also at the C_3-position *without* double-bond shift using a chiral palladium ferrocenyldi(phosphine) complex, s. Q.-L. Xu, L.-X. Dai, S.-L. You, Chem. Sci. *2013*, *4* (1), 97-102 [DOI: 10.1039/c2sc21085a].

Nitrogen ↑ CC ⇑ N

Sodium methoxide or Lithium tert-butoxide *NaOMe or t-BuOLi*
9-Phenanthrenols from 2′-acylbiaryl-2-carboxaldehyde N-tosylhydrazones ○
Transition metal-free ring closure via 1,2-migration s. *83*, 489

(Triphenylphosphine)copper(I) iodide tetramer *[(Ph$_3$P)CuI]$_4$*
(E,E)-2-Sulfonyl-1,3-dienes from N-propargyl-N-sulfonylhydrazones ∩
via [3.3]-sigmatropic rearrangement-denitrogenative N→C-sulfonyl migration

492.

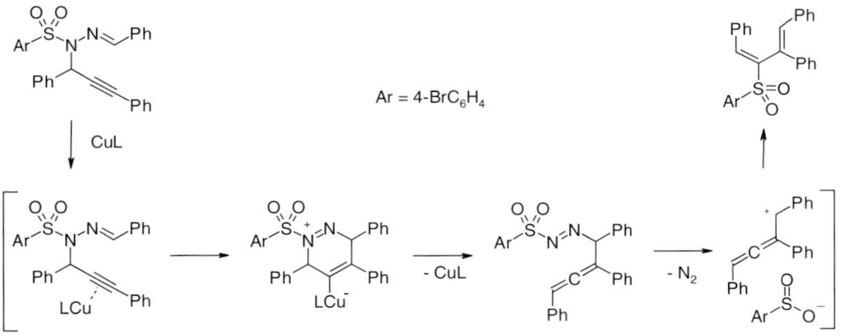

A suspension of [(Ph$_3$P)CuI]$_4$ (5 mol%) and N′-4-bromobenzenesulfonyl-N′-(1,3-diphenylprop-2-yn-1-yl)benzylidenehydrazine (0.5 mmol) in anh. toluene (2 ml) stirred at reflux under N_2 until reaction complete (TLC), cooled to room temp., solvent removed *in vacuo*, and the residue purified chromatographically → (E,E)-2-(4-bromobenzenesulfonyl)-1,3,4-triphenyl-1,3-butadiene. Y 91%. N-Arenesulfonyl-N-1-aryl-3-aryl/alkyl-prop-2-yn-1-yl (het)aryl aldehyde hydrazones, readily available from the corresponding hydrazones and propargylic acetates, were converted to the (E,E)-2-sulfonyl-1,3-butadienes in variable yields (48-92%; nineteen examples) under copper(I)

catalysis. Lowest yields were obtained with a thiophenecarboxaldehyde hydrazone (51%), a cyclopropyl-terminated propyne (55%) or a methanesulfonylhydrazone (48%), while a terminal propyne deriv. gave a complex mixture, and a primary propargylic hydrazone was unreactive. Reaction of two substrates in the same pot gave significant yields of crossover products, indicating *intermolecular migration of the sulfonyl moiety*. The proposed mechanism proceeds via intramolecular 6-*endo-dig* cyclization of the sulfonylhydrazone onto a copper-alkyne complex. F.e., optimization and substrate prepn. s. Y. Zhu, H.-T. Tang, Z.-P. Zhan, Adv. Synth. Catal. *2013*, *355* (7), 1291-6 [DOI: 10.1002/adsc.201300126].

Boron fluoride *BF_3*

Intramolecular Friedel-Crafts arylation with ar. triazenes s. *53*, 496s*83* ○

Oxalyl chloride/tin(IV) chloride/potassium hydroxide *$(COCl)_2/SnCl_4/KOH$*

2,3-Fused 3,4-dihydroisocarbostyrils from ω-azido-α-benzylcarboxylic acids

Double ring closure via intramolecular Schmidt reaction

493. 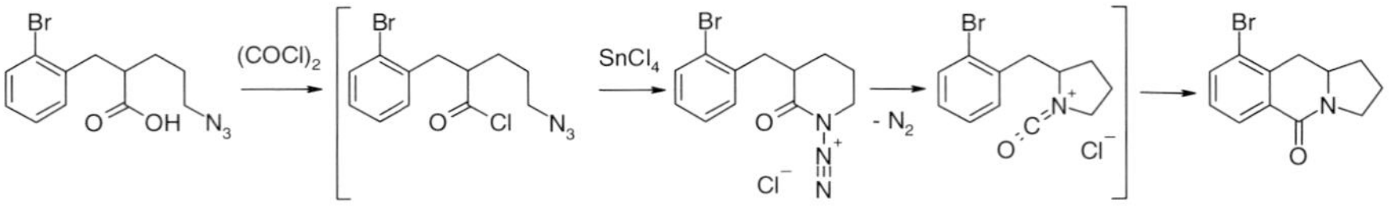

in one pot. Oxalyl chloride (3.5 eq.) added to a soln. of 5-azido-2-(2-bromobenzyl)pentanoic acid (0.86 mmol) in methylene chloride (14 ml) and DMF (one drop) at room temp., the mixture stirred for 1 h, $SnCl_4$ (4 eq.) added, the mixture heated under reflux for 4 h, cooled to room temp., quenched with water, 50% aq. KOH (3 ml) added, the mixture stirred for 30 min, extracted with methylene chloride, and purified by flash chromatography → 9-bromo-2,3,10,10a-tetrahydropyrrolo[1,2-*b*]isoquinolin-5(1*H*)-one. Y 78%. This practical and efficient three-step process created three new bonds and two rings, was effective for substrates carrying electron-donating or -withdrawing substituents on the aromatic ring, and appears unaffected by *ortho* substitution (eight examples; Y 61-85%). *m*-Subst. derivs. afforded mixtures of regioisomers, favoring cyclization at the least hindered *p*-position (two examples; Y 50-67%; 30:20 to 54:13). Homologous 2-benzyl-6-azidohexanoic acids cyclized to the 6,6,6-tricyclic lactams, **1,2,3,4,11,11a-hexahydropyrido[1,2-*b*]isoquinolin-6-ones** (two examples; Y 40-67%). The initial *in situ* generated acid chloride is presumed to undergo Schmidt reaction and cyclization of the generated isocyanate. Isolation of bicyclic azidoalkyl ketone by-products in some cases, presumably formed by intramolecular Friedel-Crafts acylation, suggested an alternative pathway but these intermediates did not cyclize to the products under the reaction conditions. F.e. and substrate prepn. s. P. Gu, X.-Y. Kang, J. Sun, B.-J. Wang, M. Yi, X.-Q. Li, P. Xue, R. Li, Org. Lett. *2012*, *14* (22), 5796-9 [DOI: 10.1021/ol302890a].

Trifluoromethanesulfonic acid *TfOH*

Friedel-Crafts acylation with carboxylic acid amides ←

494. 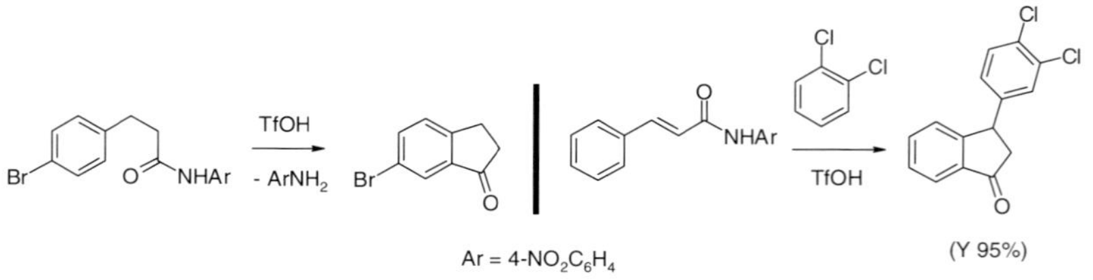

Inter- and intra-molecular Friedel-Crafts acylation using N-*p*-nitrophenylamides as acylating agents and triflic acid as catalyst are reported. Since by-product, 4-nitroaniline, and acid catalyst can be recycled in high yields the method produces minimal waste. **E: 1-Indanones**. Triflic acid (4 eq.) added to a soln. of N-4-nitrophenyl-3-(4-bromophenyl)propanamide (1 mmol) in anh. chloroform

(2 ml), the mixture stirred at 50° for 4 h, quenched with ice, extracted with chloroform, and purified chromatographically → 6-bromoindanone. Y 88%. Use of a strong electron-withdrawing group on nitrogen was essential for activation of the amide (2,4-dinitrophenyl and pyrazinyl gave comparable results), and triflic acid was superior to all other acid catalysts tested. A series of electron-diverse arylpropanamides were converted to the corresponding indanones (four examples; Y 68-96%). Interestingly, cyclization of a propenamide analog using arenes as solvent, effected both cyclization and electrophilic addition to the double bond, affording the corresponding **3-aryl-1-indanones** (two examples; Y 94-95%), and gave a 4-aryltetralone with the analogous butenamide (Y 76%). The method was also applicable to the intermolecular acylation of benzene, affording high yields of acetophenone (91%) and benzophenone derivs. (70%, 90-99%; four examples). In a final development, an amide, generated *in situ* under these conditions from 4-nitrophenyl isocyanate and benzene, reacted with a second molecule of benzene to afford benzophenone in 90% yield, with 4-nitrophenyl isocyanate functioning as a *novel phosgene equivalent.*

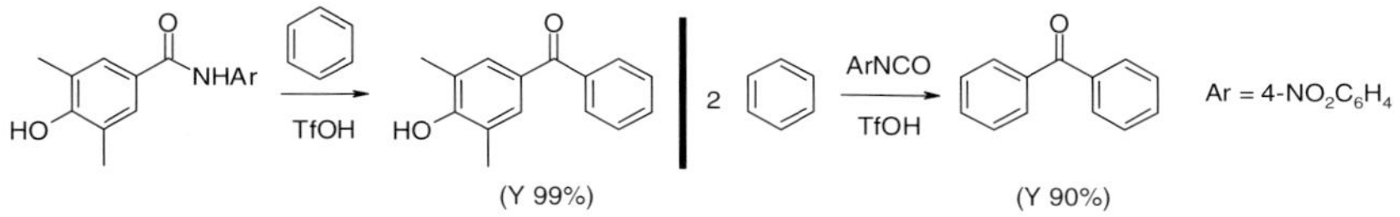

F.e., optimization, substrate prepn. and detailed mechanistic considerations s. E.K. Raja, D.J. DeSchepper, S.O. Nilsson Lill, D.A. Klumpp, J. Org. Chem. *2012*, *77* (13), 5788-93 [DOI: 10.1021/jo300922p].

Chiral rhodium(II) carboxylates *Rh(II)*
Regioselective asym. intramolecular insertion of diazo compds. into sp^3 carbon-hydrogen bonds ○
update s. *47*, 955s*82*; of alkanesulfonyl(diazo)acetic acid esters with dirhodium(II) tetrakis-[N-phthaloyl-(S)-*tert*-leucinate] especially for menthyl esters s. C.S. Jungong, A.V. Novikov, Tetrahedron: Asym. *2013*, *24* (2-3), 151-5 [DOI: 10.1016/j.tetasy.2012.12.002].

Palladium(II) acetate *$Pd(OAc)_2$*
Heck-type intramolecular biaryl synthesis with diazonium fluoroborates s. *53*, 496s*83*

Halogen ↑ CC ⇑ Hal

Irradiation s. under Ethyldiisopropylamine *///*

Sodium bis(trimethylsilyl)amide *$NaN(SiMe_3)_2$*
Intramolecular α-alkylation ○
of sulfones with K-bis(trimethylsilyl)amide/18-crown-6 cf. *32*, 961s*41*; of chiral α-subst. α-aryloxy-carboxylic acid esters *with retention of configuration* via generation of *short-lived* chiral enolates with Na-bis(trimethylsilyl)amide as base s. T. Yoshimura, K. Tomohara, T. Kawabata, J. Am. Chem. Soc. *2013*, *135* (19), 7102-5 [DOI: 10.1021/ja4018122].

Ethyldiisopropylamine/cationic iridium(III) complexes/irradiation *i-Pr_2NEt/[Ir(III)]////*
Regioselective radical ring closures of ethylenehalides
under visible-light photoredox catalysis s. *29*, 970s*83*

Titanocene dichloride/nickel(II) chloride/tricyclohexylphosphine/pyridine hydrochloride/manganese ←
Tri-n-butyltin hydride/azodiisobutyronitrile *Bu_3SnH/AIBN*
Regioselective radical ring closures of ethylenehalides
update s. *29*, 970s*82*; γ-lactones from α-bromoacetoxy-2-ethylenes *with asym. induction* using tri-*n*-butyltin hydride/AIBN at high temperature under high dilution, also low-temperature procedure (at -70°) with *i*-Bu_2AlH/tri-*n*-butyltin hydride/Et_3B/O_2 and subsequent Oppenauer-type treatment with oxo compds. as H-acceptor s. R. Bénéteau, J. Lebreton, F. Dénès, Chem. Asian J. *2012*, *7* (7), 1516-20 [DOI: 10.1002/asia.201200212]; ring closures under visible-light photoredox catalysis via SET with ethyldiisopropylamine and a cationic iridium(III) complex,

$[Ir(ppy)_2(dtbbpy)]PF_6$, as sensitizer, also direct replacement of halogen by hydrogen (HC⇓⇑Hal), s. H. Kim, C. Lee, Angew. Chem., Int. Ed. *2012, 51* (49), 12303-6 [DOI: 10.1002/anie.201203599]; catalytic radical ring closures of ethyleneiodides with $Cp_2TiCl_2/NiCl_2/Cy_3P$/pyridine hydrochloride and Mn as stoichiometric reductant in aq. THF, also intramolecular Heck reaction in anh. THF, s. A. Millán, L. Álvarez de Cienfuegos, D. Miguel, A.G. Campaña, J.M. Cuerva, Org. Lett. *2012, 14* (23), 5984-7 [DOI: 10.1021/ol3028913]; benzo-condensed N-heterocyclics by base-mediated ring closure of N-heterocyclic-bridged (iodoaryl)ethylenes **via intramolecular carbolithiation** with mesityl- or *n*-butyl-lithium (cf. *52*, 485s*68*) s. O. García-Calvo, E. Coya, S. Lage, I. Coldham, N. Sotomayor, E. Lete, Eur. J. Org. Chem. *2013* (8), 1460-70 [DOI: 10.1002/ejoc.201200994].

Tricyclohexylphosphine s. under Titanocene dichloride — *Cy_3P*
1,3-Bis(diphenylphosphino)propane s. under $Pd(OAc)_2$ — *dppp*

Pyridine hydrochloride s. under Titanocene dichloride — *Py·HCl*
Manganese s. under Titanocene dichloride — *Mn*
Manganese(0) s. under $NiBr_2(bpy)$ — *Mn(0)*
Nickel(II) chloride s. under Titanocene dichloride — *$NiCl_2$*

Dibromo(2,2′-bipyridyl)nickel(II)/manganese(0) — *$NiBr_2(bpy)/Mn(0)$*
Nickel-catalyzed intramolecular Heck-type arylation s. *36,* 957s*83* ○

Palladium(II) acetate/sodium acetate — *$Pd(OAc)_2/NaOAc$*
Intramolecular Heck-type biaryl synthesis
update s. *53,* 496s*82*; tetrahydroindeno[1,2,3-*ij*]isoquinoline [azafluoranthene] ring with $Pd(OAc)_2/K_2CO_3$ and a suitable ligand under microwave enhancement s. S. Ponnala, W.W. Harding, Eur. J. Org. Chem. *2013* (6), 1107-15 [DOI: 10.1002/ejoc.201201190]; 4-aza-9-fluorenones from aryl-(2-bromo-3-pyridyl)carbinols by *oxidative* intramolecular Heck-type biaryl synthesis with $Pd(OAc)_2$/NaOAc s. S. Dhara, A. Ahmed, S. Nandi, S. Baitalik, J.K. Ray, Tetrahedron Lett. *2013, 54* (1), 63-5 [DOI: 10.1016/j.tetlet.2012.10.085]; dibenzo[*a,c*]carbazoles s. S. Cacchi, G. Fabrizi, A. Goggiamani, A. Iazzetti, Org. Biomol. Chem. *2012, 10* (46), 9142-7 [DOI: 10.1039/c2ob26741a]; pyrazolo(benzo)thienoquinolines with $Pd(OAc)_2$, trifluoroacetoxo[2,6-bis(diphenylphosphinoxy)aryl]palladium(II) or related palladacyclic complexes s. F. Churruca, S. Hernández, M. Perea, R. SanMartin, E. Domínguez, Chem. Commun. *2013, 49* (14), 1413-5 [DOI: 10.1039/c2cc37905h]; 6*H*-benzo[*c*]chromenes by Heck-type intramolecular biaryl synthesis **with diazonium fluoroborates** using $Pd(OAc)_2$ as catalyst s. J. Zhou, L.-Z. Huang, Y.-Q. Li, Z.-T. Du, Tetrahedron Lett. *2012, 53* (52), 7036-9 [DOI: 10.1016/j.tetlet.2012.10.038]; polycyclic aromatics by **Friedel-Crafts intramolecular arylation with ar. triazenes** using BF_3-etherate cf. J. Zhou, W. Yang, B. Wang, H. Ren, Angew. Chem., Int. Ed. *2012, 51* (49), 12293-7 [DOI: 10.1002/anie.201206578].

Palladium(II) acetate/potassium acetate — *$Pd(OAc)_2/KOAc$*
Dibenzothiophene S-oxides from *o*-bromodiaryl sulfoxides

495.

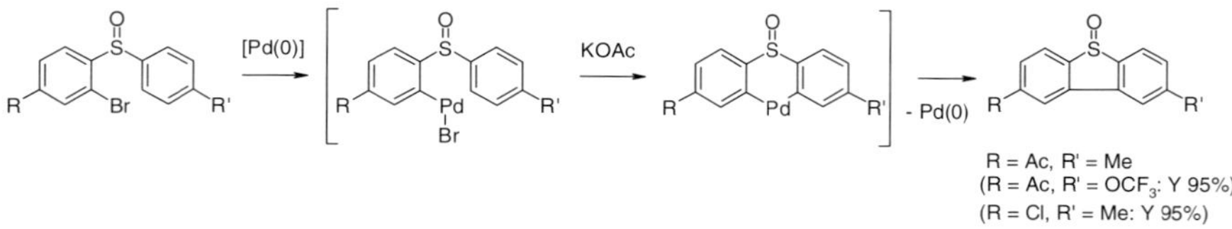

An efficient synthesis of dibenzothiophene S-oxides is reported **by direct palladium-catalyzed intramolecular biaryl coupling** in the absence of ligand, reaction of thioether analogs being precluded due to their high affinity for palladium. **E:** A mixture of 2-bromo-4-acetyl-1-(*p*-tolylsulfinyl)benzene (1.5 mmol), KOAc (2 eq.) and $Pd(OAc)_2$ (3 mol%) in dimethylacetamide [DMAc] (6 ml) heated at 130° for 8 h, cooled to room temp., diluted with methylene chloride, filtered through Celite, the filtrate washed with water and brine, dried ($MgSO_4$), concentrated *in vacuo*, and the residue purified by column chromatography → 2-acetyl-8-methyldibenzo[*b,d*]thiophene 5-oxide. Y 88%. The method is practical and versatile, using easily accessible starting materials.

It gave yields of 74-95% for ten further examples, tolerating acetyl, F, Cl, OMe, OCF_3, and CF_3 groups; 2-bromo-4-acetyl-1-(*m-tolyl*sulfinyl)benzene gave an inseparable mixture of regioisomers in only 30% yield, however. It is believed reaction proceeds via reduction of Pd(II) to Pd(0) by DMAc, insertion of Pd(0) into the C-Br bond, C-H activation to form a 6-membered palladacycle, and reductive elimination to afford the product with regeneration of Pd(0). The products may be reduced easily to dibenzothiophenes. F.e.s. T. Wesch, A. Berthelot-Bréhier, F.R. Leroux, F. Colobert, Org. Lett. *2013*, *15* (10), 2490-3 [DOI: 10.1021/ol400966q].

Palladium(II) acetate/1,3-bis(diphenylphosphino)propane/potassium carbonate ←
2,3-Dihydrospiro[indene-1,3′-indolin]-2′-ones ○
from *o*-subst. α-arylacrylic acid N-(*o*-halogenoaryl)amides

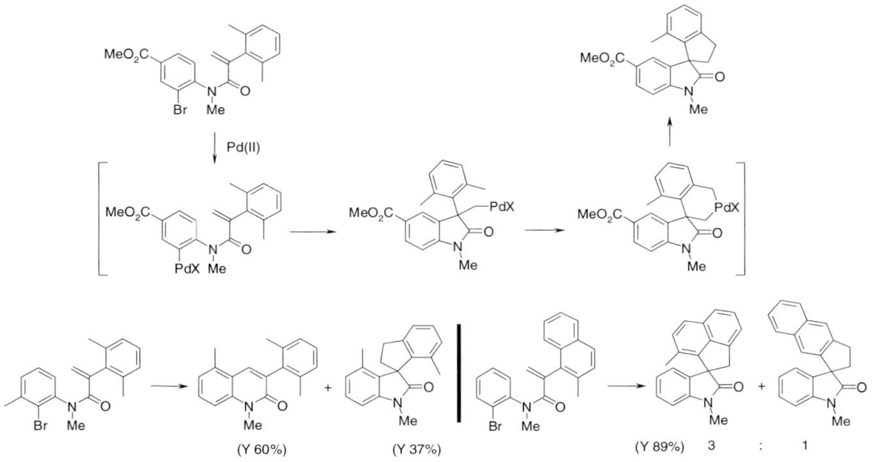

496.

in one pot. A mixture of methyl 3-bromo-4-[2-(2,6-dimethylphenyl)-N-methylacrylamido]-benzoate (0.1 mmol), $Pd(OAc)_2$ (10 mol%), dppp (20 mol%) and K_2CO_3 (1.3 eq.) in mesitylene (0.5 ml) stirred at 140° in a sealed tube under argon for 1-5 h (TLC), cooled to room temp., diluted with ethyl acetate, filtered through Celite, concentrated *in vacuo*, and purified by flash chromatography → methyl 1′,7-dimethyl-2′-oxo-2,3-dihydrospiro[indene-1,3′-indoline]-5′-carboxylate. Y 85%. The use of the bulky di(phosphine) ligand was key to the selectivity of this novel and efficient 5-*exo-trig* cyclization of *o*-halo- (Br, I) *tert*-acrylamides (other ligands gave significant amounts of the 6-*endo-trig* product, a quinolinone). The presence of a 2,6-dialkyl-(Me, Et) phenyl moiety was essential, with cyclization at methyl taking precedence over ethyl (eighteen examples; Y 78-99%; a 2-(2,6-diethylphenyl)acrylamide gave 68%). The method tolerates benzyl, 4-methoxybenzyl and 2-trimethylsilylethoxymethyl protection at nitrogen and was compatible with nitrile, ether and ester functionality. The reaction failed for a sec. acrylamide, and also for 2-*o*-tolyl- and N-*o*-chlorophenyl-acrylamides, while sterically-hindered N-*o*-bromo-*m*-methylphenylacrylamide favored formation of the carbostyril (60:37). A 2-(2-methyl-1-naphthyl)-acrylamide cyclized preferentially at the sp^2 carbon, affording a 3:1 mixture of spiroindoles in 89% yield. Use of chiral ligands failed to effect significant chiral induction. F.e., optimization and substrate prepn. s. T. Piou, L. Neuville, J. Zhu, Angew. Chem., Int. Ed. *2012*, *51* (46), 11561-5 [DOI: 10.1002/anie.201206267].

Palladium phosphine complexes *[Pd]*
Intramolecular Heck arylation
fused indoles s. *36*, 957s*47*,*50*; regioselective synthesis of 3-functionalized indoles from push-pull N-(*o*-bromoaryl)enamines with $Pd(PPh_3)_4/Et_3N$ under microwave enhancement s. H.H. Nguyen,

M.J. Kurth, Org. Lett. *2013, 15* (2), 362-5 [DOI: 10.1021/ol303314x]; N-protected 4,4-difluoro-1,2,3,4-tetrahydropyrid[4,3-*b*]indoles with $PdCl_2(PPh_3)_2$/base in pyridine s. M. Madaiah, M.K. Prashanth, H.D. Revanasiddappa, Tetrahedron Lett. *2013, 54* (11), 1424-7 [DOI: 10.1016/j.tetlet.2012.12.115]; tricyclic methylene-bridged arenesulfonamides by regioselective ring closure of partially hydrogenated N-(*o*-halogenosulfonyl)-N-heterocyclics under palladium(0) catalysis s. K. Geoghegan, P. Evans, I. Rozas, I. Alkorta, Chem. Eur. J. *2012, 18* (42), 13379-87 [DOI: 10.1002/chem.201201359]; ring closures of *o*-iodoaryloxy-2-ethylenes and arylamine analogs **via intramolecular carbonickelation** with $NiBr_2$(bpy)/Mn(0) s. R. Lhermet, M. Durandetti, J. Maddaluno, Beilstein J. Org. Chem. *2013, 9*, 710-6 [DOI: 10.3762/bjoc.9.81].

Bis[di-tert-butyl(methyl)phosphine]palladium(0)/dicyclohexylamine/potassium tert-butoxide ←
Dehydrobromination CHC(Br) → C=C
with AgF/py cf. *34*, 953; terminal ethylene derivs. from prim. bromides with bis[di-*tert*-butyl-(methyl)phosphine]palladium(0) in the presence of Cy_2NH/KOBu-*t* at room temperature s. A.C. Bissember, A. Levina, G.C. Fu, J. Am. Chem. Soc. *2012, 134* (34), 14232-7 [DOI: 10.1021/ja306323x].

Chiral palladium(0) cyclic phosphoromonoamidite complexes *[Pd(0)]**
Palladium-catalyzed remote asym. intramolecular arylation ○
asym. intramolecular β-arylation cf. *54,* 489s*82*; N-triflyl-1,2,3,4-tetrahydroquinolines by gram-scale ring closure of N-alkyl-*o*-triflylaminobromides via **asym. intramolecular γ-arylation** with a chiral TADDOL-based palladium(0) cyclic phosphoramidite as catalyst s. T. Saget, N. Cramer, Angew. Chem., Int. Ed. *2012, 51* (51), 12842-5 [DOI: 10.1002/anie.201207959].

Cationic iridium(III) complexes s. under Ethyldiisopropylamine *[Ir(III)]*

Sulfur ↑ CC ⇑ S

Dichlorobis(triphenylphosphine)palladium(II)/triphenylsilane *$PdCl_2(PPh_3)_2/Ph_3SiH$*
3-Organothio-2-cyclopentenones from β,γ-ethylene-α-ketoketene mercaptals via regio- and stereo-selective reductive intramolecular Heck-type reaction ○

497.

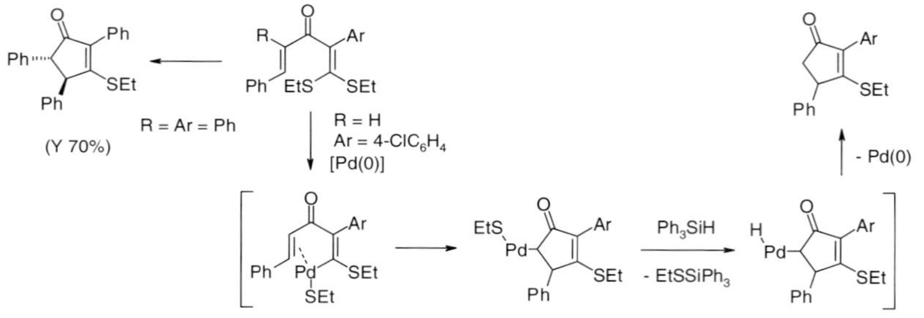

Triphenylsilane (2 eq.), $PdCl_2(PPh_3)_2$ (10 mol%) and anh. DMF (5 ml) added to (E)-2-(4-chloro-phenyl)-1,1-bis(ethylthio)-5-phenylpenta-1,4-dien-3-one (0.5 mmol) under N_2, the mixture stirred at room temp. for 0.5 h, then at 90° for 10 h (TLC), quenched with ice water, extracted with methylene chloride, and purified by chromatography on silica → 2-(4-chlorophenyl)-3-ethylthio-4-phenylcyclopent-2-enone. Y 91%. This novel Heck cyclization involved C-S activation, unexpectedly affording the title products only (cf. cyclopenta-2,4-dienones), and appears general for di(m)ethyl mercaptals carrying electron-diverse (het)aryl/alkyl at α and β′ positions (twenty-four examples; Y 56-94%). A low yield (25%) was obtained from an α-benzoyl deriv. (the yield was not improved at higher catalyst loading), and a complex mixture from an α-H deriv. Substrates carrying additional phenyl at the α′ position gave *trans*-tetrasubst. products exclusively (three examples; Y 66-72%). F.e., substrate prepn. and optimization s. B. Liu, G. Zheng, X. Liu, C. Xu, J. Liu, M. Wang, Chem. Commun. *2013*, *49* (22), 2201-3 [DOI: 10.1039/c3cc37571d].

Remaining Elements ↑ CC ⇑ Rem

Chlorobis(ethylene)rhodium(I) dimer/(R)-1,1′-binaphthyl-2,2′-diyl N,N-dimethyl-phosphoramidite *[Rh(I)]**

3-Aryl-1-indanones from *o*′-borylchalcones
via rhodium(I)-catalyzed asym. intramolecular 1,4-addition ○

498.

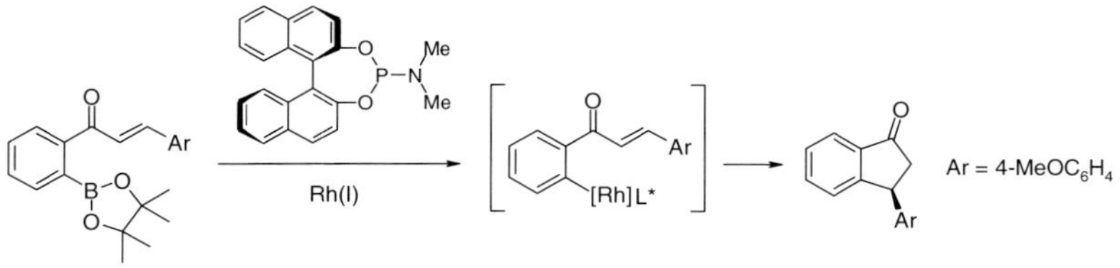

A mixture of 2′-pinacolboryl-4-methoxychalcone (0.2 mmol), $[RhCl(CH_2{=}CH_2)_2]_2$ (5 mol%) and (R)-MonoPhos (10 mol%) in toluene (2 ml) stirred under N_2 at 80° for 30 min, 1.5 *M* aq. K_3PO_4 (0.5 eq.) added, the mixture stirred at 40° for 3-5 h, concentrated *in vacuo*, and the residue purified by chromatography on silica → 3(S)-(4-methoxyphenyl)indan-1-one. Y 92% (e.e. 95%). The reaction appears general for *o*′-borylchalcones carrying electron-diverse substituents on either aromatic ring, with best results obtained using the simple chiral MonoPhos ligand (fifteen examples; Y 80-95%; e.e. 81-95%). F.e., optimization and substrate prepn. s. Y.-N. Yu, M.-H. Xu, J. Org. Chem. *2013*, *78* (6), 2736-41 [DOI: 10.1021/jo302656s].

Carbon ↑ CC ⇑ C

Copper(I) iodide/1,10-phenanthroline/potassium tert-butoxide *CuI/phen/KOBu-t*

Phthalimidines [1(3*H*)-isoindolones]
from *o*-halogenocarboxylic acids, aldehydes, isocyanates and prim. amines
via α-(*o*-halogenoaroylamino)carboxylic acid amides ○

499.

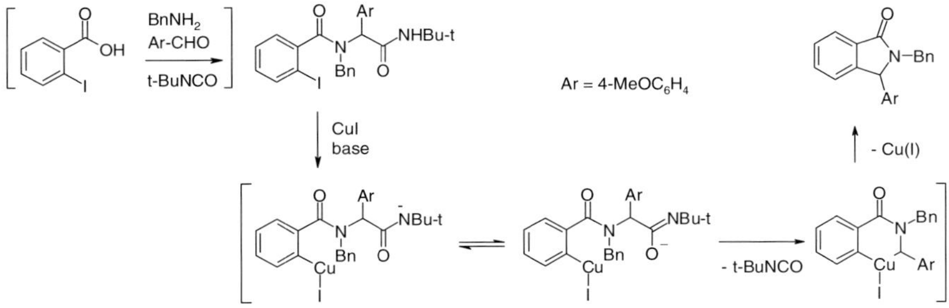

N-*tert*-Butyl-α-(N-benzyl-2-iodobenzoylamino)-4-methoxyphenylacetamide (1 mmol) [prepared in 85% yield by reaction of *o*-iodobenzoic acid, benzylamine, 4-methoxybenzaldehyde and *tert*-butyl isocyanide, according to Z. Xiang et al., Org. Lett. *2004*, *6* (18), 3155-8 [DOI: 10.1021/ol048791n]] mixed with CuI (10 mol%), 1,10-phenanthroline (10 mol%), K-*tert*-butoxide (4 eq.) and toluene (2 ml) in a dry Schlenk tube under N_2, the mixture stirred at 80° for 2-4 h (TLC), cooled to room temp., filtered through Celite, solvent removed *in vacuo*, and the residue purified by flash chromatography → 2-benzyl-3-(4-methoxyphenyl)-1(3*H*)-isoindolone. Y 72%. Eleven further 3-aryl-2-alkyl/benzyl-1(3*H*)-isoindolones were similarly prepared (Y 56-79% from *o*-iodo-derivs.; one example from an *o*-bromo-deriv., Y 68%) by this **copper(I)-catalyzed decarbamylative intramolecular α-arylation**, with elimination of the amide moiety as isocyanate or amine/

CO_2. F.e., optimization and preparation of the starting Ugi 4-component derivs. (Y 72-94%) s. V. Tyagi, S. Khan, P.M.S. Chauhan, Tetrahedron Lett. *2013*, *54* (10), 1279-84 [DOI: 10.1016/j.tetlet.2012.12.091].

Copper(I) iodide/air *CuI/O_2*
Acridones from *o*-(arylamino)ketones ○
Copper(I)-catalyzed aerobic ring closure via C-cleavage

500. 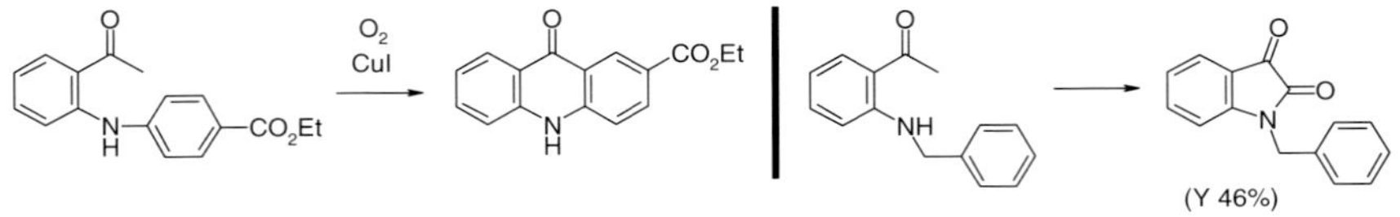

A mixture of N-(4-ethoxycarbonylphenyl)-2-aminoacetophenone (0.3 mmol) and CuI (20 mol%) in DMSO/chlorobenzene (1:1; 1.6 ml) stirred under air at 140° for 48 h (TLC), cooled to room temp., extracted with ethyl acetate (with brine wash), and purified by chromatography on silica → ethyl 10*H*-acridin-9-one-2-carboxylate. Y 90%. This novel oxidative C-C cleavage and intramolecular cyclization accommodated electron-donating and -withdrawing substituents on both aromatic rings, and an additional N-phenyl group (nineteen examples; Y 50-91%; a 3-methyl-2-aminoacetophenone gave 38%). Unsymmetrical N-(3-tolyl)-2-aminoacetophenone favored cyclization at the least hindered site (4:1; Y 84%). Replacing the methyl group of the methyl ketone moiety with ethyl, 4-methylbenzyl or 4-methoxybenzyl also gave good yields of the corresponding acridones (three examples; Y 60-80%; with benzaldehyde by-products detected from the benzyl ketones), but 2-N-benzylaminoacetophenone failed to give the homologous 6,7,6-system, instead affording N-benzyl-indole-2,3-dione in 46% yield. The mechanism was not elucidated, but a radical mechanism was discounted, and although 2-phenylaminobenzaldehyde also cyclized to the acridone in high yield (92%), it *was not* considered to be a key intermediate. F.e. and optimization s. W. Zhou, Y. Yang, Y. Liu, G.-J. Deng, Green Chem. *2013*, *15* (1), 76-80 [DOI: 10.1039/c2gc36502b].

(S)-4-tert-Butyl-2-[o-(diphenylphosphino)phenyl]-Δ^2-oxazoline s. under Tris(dibenzylideneacetone)dipalladium ←

Air s. under CuI *O_2*

Alkylidene- or alkylidyne-molybdenum(VI) complexes *[Mo(VI)]*
Ring-closing metathesis
of dienes under tungsten catalysis cf. *80,* 520; synthesis of macrocyclic trisubst. (Z)-ethylene derivs. with hindered bis(biphenylolato)(pentafluorophenylimido)alkylidenemolybdenum(VI) complexes s. C. Wang, F. Haeffner, R.R. Schrock, A.H. Hoveyda, Angew. Chem., Int. Ed. *2013, 52* (7), 1939-43 [DOI: 10.1002/anie.201209180]; **of terminal diynes** (cf. *49*, 985s*82*) with tris-(2-methylhexafluoroisopropanolato)(2,4,6-trimethylbenzylidyne)molybdenum(VI) or the tungsten analog s. B. Haberlag, M. Freytag, C.G. Daniliuc, P.G. Jones, M. Tamm, ibid. *2012, 51* (52), 13019-22 [DOI: 10.1002/anie.201207772].

Ruthenium carbene complexes or supported variants *[Ru]*
Ring-closing metathesis
of dienes s. *49,* 985s*82*; comparative study of the synthesis of 14-and 15-membered macrolactones and -lactams with various alkylideneruthenium phosphine or NHC complexes s. F. Grisi, C. Costabile, A. Grimaldi, C. Viscardi, C. Saturnino, P. Longo, Eur. J. Org. Chem. *2012* (30), 5928-34 [DOI: 10.1002/ejoc.201200960]; rapid synthesis of N-protected 2,5-dihydropyrroles and 1,2,3,6-tetrahydropyridines with highly active Hoveyda-type complexes formed by reaction of dichloro(1,3-diarylimidazolidin-2-ylidene)(pyridine)(3-phenyl-1-indenylidene)ruthenium complexes with 1-(aryloxy)-2-ethenylbenzenes s. P. Kos, R. Savka, H. Plenio, Adv. Synth. Catal. *2013, 355* (2-3), 439-47 [DOI: 10.1002/adsc.201200956]; synthesis of macrocyclic (Z)-ethylene derivs. with (*o*-isopropoxybenzylidene)(nitrato)[2-[3-(2,6-diisopropylphenyl)imidazolidin-2-yliden-1-yl]adamantyl]ruthenium, also conversion of (Z)/(E)-mixtures to (E)-isomers by ethenolysis, s. V.M. Marx, M.B. Herbert, B.K. Keitz, R.H. Grubbs, J. Am. Chem. Soc. *2013, 135* (1), 94-7 [DOI:

10.1021/ja311241q]; effect of Grubbs′ 1st and 2nd generation catalysts on the (Z)/(E)-selective synthesis of 13-membered unsatd. heterocyclics s. H.S.P. Rao, S. Rafi, P.R. Kumar, C. Guravaiah, N. Muthanna, Tetrahedron Lett. *2012, 53* (51), 6877-80 [DOI: 10.1016/j.tetlet.2012.09.065]; synthesis of macrocyclic (E)-ethylenesiloxanes and conversion to macrocyclic (E)- or (Z)-α,β-ethylenehalides (and thence macrocyclic trisubst. ethylene derivs.) s. Y. Wang, M. Jimenez, P. Sheehan, C. Zhong, A.W. Hung, C.P. Tam, D.W. Young, Org. Lett. *2013, 15* (6), 1218-21 [DOI: 10.1021/ol400134d]; reactivation of the Hoveyda-Grubbs 1st generation catalyst by designed decomposition in the presence of ethylene s. D.S. Tabari, D.R. Tolentino, Y. Schrodi, Organometallics *2013, 32* (1), 5-8 [DOI: 10.1021/om301042a]; synthesis of cyclic peptides by ring-closing metathesis of S,S′-diallylated dicysteine-containing peptides with added $MgCl_2$ in 7:3 water/*tert*-butanol s. S.A. Cochrane, Z. Huang, J.C. Vederas, Org. Biomol. Chem. *2013, 11* (4), 630-9 [DOI: 10.1039/c2ob26938d]; *heterogeneous* ring-closing metathesis with a recyclable dichloro(1,3-dimesitylimidazolidin-2-ylidene)(*o*-alkoxybenzylidene)ruthenium complex supported covalently through an imidazolium group to multi-walled carbon nanotubes s. S. Lee, J.Y. Shin, S.-g. Lee, Tetrahedron Lett. *2013, 54* (7), 684-7 [DOI: 10.1016/j.tetlet.2012.12.002]; synthesis of heteroarene-condensed 2,5-dihydro-1*H*-azepines with Grubbs′ 2nd generation catalyst s. T.A. Moss, ibid. *54* (8), 993-7 [DOI: 10.1016/j.tetlet.2012.12.042]; of chiral 1-azabicyclo[m.n.0]-alkenes s. T.J. Senter, M.L. Schulte, L.C. Konkol, T.E. Wadzinski, C.W. Lindsley, ibid. *54* (13), 1645-8 [DOI: 10.1016/j.tetlet.2013.01.041]; of P-chiral 2,3-dihydrophosphole·borane complexes s. X. Wu, P. O′Brien, S. Ellwood, F. Secci, B. Kelly, Org. Lett. *2013, 15* (1), 192-5 [DOI: 10.1021/ol303253h]; **ring-closing metathesis-1,3-dipolar cycloaddition** s. M. Arisawa, Y. Fujii, H. Kato, H. Fukuda, T. Matsumoto, M. Ito, H. Abe, Y. Ito, S. Shuto, Angew. Chem., Int. Ed. *2013, 52* (3), 1003-7 [DOI: 10.1002/anie.201206765]; synthesis of electron-diverse 1-arylpyrroles from N,N-diallylanilines by **dehydrogenative ring-closing metathesis** with Grubbs′ 1st or 2nd generation catalyst and added $FeCl_3$ or $CuCl_2$ as cocatalyst under oxygen s. W. Chen, J. Wang, Organometallics *2013, 32* (6), 1958-63 [DOI: 10.1021/om400046r]; one-pot asym. synthesis of benzo-condensed cyclic 3-ethylenealcohols by **asym. allylboration-ring-closing metathesis** with chiral 3,3′-bis-(2,4,6-triisopropylphenyl)-1,1′-binaphthyl-2,2′-diyl hydrogen phosphate for the initial allylboration prior to classical ruthenium carbene-catalyzed metathesis s. S. Fustero, E. Rodríguez, R. Lázaro, L. Herrera, S. Catalán, P. Barrio, Adv. Synth. Catal. *2013, 355* (6), 1058-64 [DOI: 10.1002/adsc.201201095].

Tris(dibenzylideneacetone)dipalladium/(S)-4-tert-butyl-2-[o-(diphenylphosphino)-phenyl]-Δ²-oxazoline *[Pd(0]**

Palladium-catalyzed asym. decarboxylative intramolecular α-allylation ←

update s. *71,* 499s*81*; synthesis of carbazolones and indolone derivs. with $Pd_2(dba)_3$ and (S)-4-*tert*-butyl-2-[*o*-(diphenylphosphino)phenyl]-Δ²-oxazoline as ligand s. C.J. Gartshore, D.W. Lupton, Angew. Chem., Int. Ed. *2013, 52* (15), 4113-6 [DOI: 10.1002/anie.201209069]; f. carbazolones s. Z. Li, S. Zhang, S. Wu, X. Shen, L. Zou, F. Wang, X. Li, F. Peng, H. Zhang, Z. Shao, ibid. 4117-21 [DOI: 10.1002/anie.201209878]; asym. synthesis of 3-allyl-2(3*H*)-furanones from allyl 2-furyl carbonates s. J. Fournier, O. Lozano, C. Menozzi, S. Arseniyadis, J. Cossy, ibid. *52* (4), 1257-61 [DOI: 10.1002/anie.201206368].

Formation of Electron Pair on Nitrogen

Elimination ⇑

Oxygen ↑ ElN ⇑ O

Cerium(III) chloride/zinc *$CeCl_3$/Zn*

Deoxygenation of cyclic N-oxides ⩾NO → ⩾N

of pyridine N-oxides with gold-on-hydroxyapatite cf. *37,* 994s*80*; N-deoxygenation of N-hetero-arene N-oxides with $CeCl_3{\cdot}7H_2O$/Zn s. B.W. Yoo, H.I. Jung, S.H. Kim, Y.S. Ahn, J.Y. Choi, Bull. Korean Chem. Soc. *2013, 34* (2), 359-60 [DOI: 10.5012/bkcs.2013.34.2.359].

Formation of Electron Pair on Sulfur

Elimination ⇑

Oxygen ↑ ElS ⇑ O

Trichloro(oxo)bis(triphenylphosphine)rhenium(V) *$ReOCl_3(PPh_3)_2$*
Thioethers from sulfoxides >SO → >S
update s. *52,* 495s*81*; deoxygenation of aliphatic or aromatic sulfoxides with $Re(O)Cl_3(PPh_3)_2$ as catalyst in the *absence* of a reductant s. S.C.A. Sousa, J.R. Bernardo, C.C. Romão, A.C. Fernandes, Tetrahedron *2012, 68* (39), 8194-7 [DOI: 10.1016/j.tet.2012.07.071].

Formation of Electron Pair on Remaining Elements

Elimination ⇑

Oxygen ↑ ElRem ⇑ O

Organosilicon hydrides/diaryl hydrogen phosphates *⩾SiH/$(ArO)_2P(O)H$*
Phosphines from phosphine oxides ⩾PO → ⩾P
update s. *44,* 993s*82*; from variously functionalized phosphine oxides with inexpensive polymethylhydrosiloxane or diethoxy(methyl)silane and a diaryl hydrogen phosphate as Brønsted acid catalyst s. Y. Li, L.-Q. Lu, S. Das, S. Pisiewicz, K. Junge, M. Beller, J. Am. Chem. Soc. *2012, 134* (44), 18325-9 [DOI: 10.1021/ja3069165].

Resolutions Res

(s.a. Subject Index under Resolution)

Without additional reagents *w.a.r.*
Separation of enantiomers by physical means ←
update s. *5,* 666s*82*; of 1-(benzofuran-2-yl)alkylamines on a chiral stationary phase based on (+)-(18-crown-6)-2,3,11,12-tetracarboxylic acid s. S. Park, S.J. Kim, M.H. Hyun, Bull. Korean Chem. Soc. *2012, 33* (10), 3497-500 [DOI: 10.5012/bkcs.2012.33.10.3497]; of α-amino acids and their esters using the same chiral tetracarboxylic acid as selector, comparing HPLC resolution with enantiomeric discrimination by NMR s. E. Bang, J.Y. Jin, J.H. Hong, J.S. Kang, W. Lee, W. Lee, ibid. 3481-4 [DOI: 10.5012/bkcs.2012.33.10.3481]; resolution of BINOLs via enantioselective inclusion complexation with (S)-proline and subsequent crystallization s. X. Hu, Z. Shan, Q. Chang, Tetrahedron: Asym. *2012, 23* (18-19), 1327-31 [DOI: 10.1016/j.tetasy.2012.08.014]; spontaneous resolution of N-subst. 3-aryl-3-hydroxyphthalimidines by *preferential dynamic crystallization* from toluene with added DBU s. F. Yagishita, H. Ishikawa, T. Onuki, S. Hachiya, T. Mino, M. Sakamoto, Angew. Chem., Int. Ed. *2012, 51* (52), 13023-5 [DOI: 10.1002/anie. 201205097]; of N-subst. *meso*-3,4-diphenylsuccinimides by the same technique s. S. Hachiya, Y. Kasashima, F. Yagishita, T. Mino, H. Masu, M. Sakamoto, Chem. Commun. *2013, 49* (42), 4776-8 [DOI: 10.1039/c3cc41678j]; pasteurian segregation of racemates by diastereoselective adsorption on an achiral surface by surface-mediated complex formation in a liquid s. H. Xu, W.J. Saletra, P. Iavicoli, B. Van Averbeke, E. Ghijsens, K.S. Mali, A.P.H.J. Schenning, D. Beljonne, R. Lazzaroni, D.B. Amabilino, S. De Feyter, Angew. Chem., Int. Ed. *2012, 51* (48), 11981-5 [DOI: 10.1002/anie.201202081]; resolution of α-(carbo-*tert*-butoxyamino)carboxylic acid esters via precipitation with a chiral amine and subsequent recrystallization s. *83*, 437.

Discrimination of enantiomers and absolute configuration determination ←
update s. *5,* 666s*82*; enantiodiscrimination (chiral recognition), e.g. of mandelic acid enantiomers, by UV absorption in macrocyclic hosts formed from resorcin[4]arenes possessing chiral urea or amide residues s. S. Wiegmann, G. Fukuhara, B. Neumann, H.-G. Stammler, Y. Inoue, J. Mattay, Eur. J. Org. Chem. *2013* (7), 1240-5 [DOI: 10.1002/ejoc.201201272]; chiral recognition of amines by ^{1}H NMR analysis by titration with chiral α-amino acid-functionalized resorcarene cavitands s. N. Li, F. Yang, H.A. Stock, D.V. Dearden, J.D. Lamb, R.G. Harrison, Org. Biomol. Chem. *2012, 10* (36), 7392-401 [DOI: 10.1039/c2ob25613d]; of α-aminocarboxylic acid anions by direct fluorescence on addition to chiral ionic polymers s. F. Song, N. Fei, F. Li, S. Zhang, Y. Cheng, C. Zhu, Chem. Commun. *2013, 49* (28), 2891-3 [DOI: 10.1039/c3cc40488a]; of α-aminocarboxylic acids with a chiral copper(I) N,N′-bis[2-hydroxy-1,1′-binaphthyl-3-ylmethyl]-1,2-diamine complex as fluorescent sensor s. F. Li, L. Li, W. Yang, L.-S. Zheng, Z.-J. Zheng, K. Jiang, Y. Lu, L.-W. Xu, Tetrahedron Lett. *2013, 54* (12), 1584-8 [DOI: 10.1016/j.tetlet.2013.01.047]; chiral recognition of sec. benzylamines using supramolecular chiral cyclodextrin-based nanocapsules in non-polar media, combined with kinetic resolution, s. T. Kida, T. Iwamoto, H. Asahara, T. Hinoue, M. Akashi, J. Am. Chem. Soc. *2013, 135* (9), 3371-4 [DOI: 10.1021/ja312367k]; determination of enantiopurity of carboxylic acids by NMR analysis in the presence of a chiral 3-(pyrrolidin-2-yl)-1,1′-bi-2-naphthol as shift reagent s. Q. Ma, M. Ma, H. Tian, X. Ye, H. Xiao, L.-h. Chen, X. Lei, Org. Lett. *2012, 14* (23), 5813-5 [DOI: 10.1021/ol3027686]; **absolute configuration determination** of sec. alcohols using a qualitative TLC analysis on kinetic resolution by O-acylation s. A.J. Wagner, S.D. Rychnovsky, J. Org. Chem. *2013, 78* (9), 4594-8 [DOI: 10.1021/jo400432q]; of carboxylic acids using a 1-amino-2-indanol-based chiral derivatizing agent for NMR analysis s. M. Park, S.-m. Kim, K. Choi, Org. Biomol. Chem. *2012, 10* (40), 8051-4 [DOI: 10.1039/c2ob26168e]; of carboxylate anions by CD analysis using a zinc porphyrin-based molecular tweezer as host in the presence of an achiral diamine s. H. Yoon, C.-H. Lee, W.-D. Jang, Chem. Eur. J. *2012, 18* (39), 12479-86 [DOI: 10.1002/chem.201200371]; of cyclic bis(bibenzyls) based on HPLC-CD and quantum-chemical CD calculations s. A. Schaumlöffel, M. Groh, M. Knauer, A. Speicher, G. Bringmann, Eur. J. Org. Chem. *2012* (35), 6878-87 [DOI: 10.1002/ejoc.201200873]; of subst. isoxazolidines based on chemical shift measurements of Mosher derivs. s. W. Wang, R.T. Cassell, K.S. Rein, Tetrahedron: Asym. *2013, 24* (4), 223-8 [DOI: 10.1016/j.tetasy.2013.01.016], of highly flexible macrocyclic natural products by comparing their solid-state electronic circular dichroism [ECD] spectra with calculations using time-dependent density functional theory [TDDFT] and X-ray geometry s. T. Kurtán, R. Jia, Y. Li, G. Pescitelli, Y.-W. Guo, Eur. J. Org. Chem. *2012* (34), 6722-8 [DOI: 10.1002/ejoc.201200862].

Chiral N-condensed 1,3,2-oxazaborolidines/borane-tetrahydrofuran/microwaves ←
***o*-Hydroxy-*o′*-(hydroxymethyl)biaryls from dibenzo[*b,d*]pyran-6-ones** ←
with dynamic kinetic resolution s. *83*, 7

2(S)-[Diphenyl(trimethylsiloxy)methyl]pyrrolidine/acetic acid/silica ←
4-(Carbalkoxymethyl)cyclopent-3-ene-1,3-dialdehydes ←
from α-acoxy-β-nitro-β,γ-ethylenecarboxylic acid esters with kinetic resolution s. *83*, 391

Horse liver alcohol dehydrogenase/nicotinamide adenine dinucleotide/1,4-butanediol ←
2-Aryl-*prim*-alcohols from α-arylaldehydes with dynamic kinetic resolution s. *83*, 8 ←

Chiral 1,1′-spirobiindan-based phosphate diesters ←
Kinetic resolution of γ-hydroxyacetals by intramolecular transacetalation s. *83*, 61 ←

Reviews

This is a collection of reviews gathered from literature
published mainly from June to December 2013,
arranged in the following sections:

1 Asymmetric synthesis - organocatalysis, immobilized catalysts, phase-transfer catalysis, dynamic kinetic resolution, desymmetrization, enantiomer separation

Molecular chirality: language, history, and significance, J. Gal, Top. Curr. Chem. *2013, 340,* 1-20; applications of **asymmetric organocatalysis** in medicinal chemistry, J. Alemán, S. Cabrera, Chem. Soc. Rev. *2013, 42* (1), 774-93; asymmetric organocatalysis mediated by α,α-(S)-diarylprolinols: recent advances, S. Meninno, A. Lattanzi, Chem. Commun. *2013, 49* (37), 3821-32; a snapshot of ionic-liquid-tagged proline-based organocatalysts, H.L. Brozinski, J.P. Delaney, L.C. Henderson, Austr. J. Chem. *2013,*

66 (8), 844-7; towards tartaric-acid-derived asymmetric organocatalysts, K. Gratzer, G.N. Gururaja, M. Waser, Eur. J. Org. Chem. *2013* (21), 4471-82; D-glucosamine-derived chiral catalysts for asymmetric reactions, C. Shen, P.-F. Zhang, Curr. Org. Chem. *2013, 17* (14), 1507-24; bifunctional *prim*-aminothioureas in asymmetric organocatalysis, O.V. Serdyuk, C.M. Heckel, S.B. Tsogoeva, Org. Biomol. Chem. *2013, 11* (41), 7051-71; terpene-derived bifunctional thioureas in asymmetric organocatalysis, S. Narayanaperumal, D.G. Rivera, R.C. Silva, M.W. Paixão, ChemCatChem *2013, 5* (10), 2756-73; powerful amino acid-derived bifunctional phosphine catalysts bearing a hydrogen bond donor in asymmetric synthesis, L.-W. Xu, ChemCatChem *2013, 5* (10), 2775-84; asymmetric organocatalysis and the nitro group functionality, L.S. Aitken, N.R. Arezki, A. Dell'Isola, A.J.A. Cobb, Synthesis *2013, 45* (19), 2627-48; when asymmetric aminocatalysis meets the **vinylogy** principle, I.D. Jurberg, I. Chatterjee, R. Tannert, P. Melchiorre, Chem. Commun. *2013, 49* (43), 4869-83; recent advances of **immobilized catalysts** for enantioselective processes, Q. Chen, Curr. Org. Chem. *2013, 17* (12), 1303-24; new advances in **dual stereocontrol** for asymmetric reactions, J. Escorihuela, M.I. Burguete, S.V. Luis, Chem. Soc. Rev. *2013, 42* (12), 5595-617; recent developments in **asymmetric phase-transfer reactions**, S. Shirakawa, K. Maruoka, Angew. Chem., Int. Ed. *2013, 52* (16), 4312-48; racemization of amines and related derivatives leading to their **dynamic kinetic resolution**, C.-J. Jiang, G.-L. Cheng, Curr. Org. Chem. *2013, 17* (11), 1225-34; organocatalyzed enantioselective **desymmetrization** of diols in the preparation of chiral building blocks, M.D. Díaz-de-Villegas, J.A. Gálvez, R. Badorrey, M.P. López-Ram-de-Víu, Chem. Eur. J. *2012, 18* (44), 13920-35; **recent developments in optical resolution**, R. Siedlecka, Tetrahedron *2013, 69* (31), 6331-63; chiral supramolecular selectors for enantiomer differentiation in liquid chromatography, A. Ciogli, D. Kotoni, F. Gasparrini, Top. Curr. Chem. *2013, 340,* 73-105; differentiation of enantiomers by capillary electrophoresis, G.K.E. Scriba, Top. Curr. Chem. *2013, 340,* 209-75; salient features of enantioselective gas chromatography: the enantiomeric differentiation of chiral inhalation anesthetics as a representative methodological case in point, V. Schurig, Top. Curr. Chem. *2013, 340,* 153-207; enantiomeric differentiation by synthetic helical polymers, E. Yashima, H. Iida, Y. Okamoto, Top. Curr. Chem. *2013, 340,* 41-72; chiroptical detectors for the study of unusual phenomena in chiral chromatography, N. Vanthuyne, C. Roussel, Top. Curr. Chem. *2013, 340,* 107-51; terms for the quantitation of a mixture of stereoisomers, V. Schurig, Top. Curr. Chem. *2013, 340,* 21-40.

2 Catalysis - carbene catalysis, nucleophilic catalysis, frustrated Lewis pairs, phase-transfer catalysis, supramolecular catalysis, photocatalysis

(for asym. catalysis s. under Section 1, heterogeneous catalysis s. under Section 3, for transition metal catalysis s. under Section 4, and biocatalysis s. under Section 5)

Acyl anion-free **N-heterocyclic carbene organocatalysis**, S.J. Ryan, L. Candish, D.W. Lupton, Chem. Soc. Rev. *2013, 42* (12), 4906-17; catalysis of organic reactions by carbenes and carbene complexes, N.I. Korotkikh, V.S. Saberov, N.V. Gliny-anaya, K.A. Marichev, A.V. Kiselyov, A.V. Knishevitsky, G.F. Rayenko, O.P. Shvaika, Chem. Heterocycl. Compd. *2013, 49* (1), 19-38; enantioselective cycloaddition reactions of ketenes catalyzed by N-heterocyclic carbenes, X.-Y. Chen, S. Ye, Synlett *2013, 24* (13), 1614-22; enantioselective cycloaddition reactions of ketenes catalyzed by N-heterocyclic carbenes, X.-Y. Chen, S. Ye, Synlett *2013, 24* (13), 1614-22; a quantitative approach to **nucleophilic organocatalysis**, H. Mayr, S. Lakhdar, B. Maji, A.R. Ofial, Beilstein J. Org. Chem. *2012, 8,* 1458-78; combining transition metal catalysis and organocatalysis - an update, Z. Du, Z. Shao, Chem. Soc. Rev. *2013, 42* (3), 1337-78; coexistence of Lewis acid and base functions: a generalized view of the **frustrated Lewis pair** concept with novel implications for reactivity, H. Berke, Y. Jiang, X. Yang, C. Jiang, Top. Curr. Chem. *2013, 334*, 27-57; frustrated Lewis pairs: some recent developments, G. Erker, Pure Appl. Chem. *2012, 84* (11), 2203-18; new organoboranes in frustrated Lewis pair chemistry, Z. Lu, H. Ye, H. Wang, Top. Curr. Chem. *2013, 334,* 59-80; N-heterocyclic carbenes in frustrated Lewis pair chemistry, E.L. Kolychev, E. Theuergarten, M. Tamm,

Top. Curr. Chem. *2013, 334,* 121-55; carbon-based frustrated Lewis pairs, S. Khan, M. Alcarazo, Top. Curr. Chem. *2013, 334,* 157-70; novel Al-based frustrated Lewis pair systems, W. Uhl, E.-U. Würthwein, Top. Curr. Chem. *2013, 334,* 101-19; paracyclophane derivatives in frustrated Lewis pair chemistry, L. Greb, J. Paradies, Top. Curr. Chem. *2013, 334,* 81-100; radical frustrated Lewis pairs, T.H. Warren, G. Erker, Top. Curr. Chem. *2013, 334,* 219-38; frustrated Lewis pairs beyond the main group: transition metal-containing systems, D.F. Wass, A.M. Chapman, Top. Curr. Chem. *2013, 334,* 261-80; reactions of phosphine-boranes and related frustrated Lewis pairs with transition metal complexes, A. Amgoune, G. Bouhadir, D. Bourissou, Top. Curr. Chem. *2013, 334,* 281-311; selective C-H activations using frustrated Lewis pairs: applications in organic synthesis, P. Knochel, K. Karaghiosoff, S. Manolikakes, Top. Curr. Chem. *2013, 334,* 171-90; metal-ligand multiple bonds as frustrated Lewis pairs for C-H functionalization, M.T. Whited, Beilstein J. Org. Chem. *2012, 8,* 1554-63; frustrated Lewis pair-mediated activations and reductions of CO_2 and CO, A.E. Ashley, D.O'Hare, Top. Curr. Chem. *2013, 334,* 191-217; polymerization by classical and frustrated Lewis pairs, E.Y.-X. Chen, Top. Curr. Chem. *2013, 334,* 239-60; recent developments in asymmetric **phase-transfer reactions**, S. Shirakawa, K. Maruoka, Angew. Chem., Int. Ed. *2013, 52* (16), 4312-48; microwave irradiation and phase transfer catalysis in C-, O- and N-alkylation reactions, G. Keglevich, A. Grün, E. Bálint, Curr. Org. Synth. *2013, 10* (5), 751-63; from containers to catalysts: **supramolecular catalysis** within cucurbiturils, B.C. Pemberton, R. Raghunathan, S. Volla, J. Sivaguru, Chem. Eur. J. *2012, 18* (39), 12178-90; **'release and catch' catalytic systems**, M. Gruttadauria, F. Giacalone, R. Noto, Green Chem. *2013, 15* (10), 2608-18; **visible-light photoredox catalysis** with transition metal complexes: applications in organic synthesis, C.K. Prier, D.A. Rankic, D.W.C. MacMillan, Chem. Rev. *2013, 113* (7), 5322-63; visible-light-driven photoredox catalysis in the construction of carbocyclic and heterocyclic ring systems, J. Xuan, L.-Q. Lu, J.-R. Chen, W.-J. Xiao, Eur. J. Org. Chem. *2013,* (30), 6755-70; **photoorganocatalysis**. what for? D. Ravelli, M. Fagnoni, A. Albini, Chem. Soc. Rev. *2013, 42* (1), 97-113; photocatalysis on supported gold and silver nanoparticles under ultraviolet and visible light irradiation, S. Sarina, E.R. Waclawik, H. Zhu, Green Chem. *2013, 15* (7), 1814-33; redox-active ligands in catalysis, O.R. Luca, R.H. Crabtree, Chem. Soc. Rev. *2013, 42* (3), 1440-59; semiconductor and **plasmonic photocatalysis** for selective organic transformations, X. Zou, Z. Tao, T. Asefa, Curr. Org. Chem. *2013, 17* (12), 1274-87.

3 Heterogeneous catalysis - nanostructured catalysts, nanoparticles, carbon catalysts, inorganic supports, polymeric and polymer-supported catalysts

(for copper, silver and gold nanocatalysts s. under Section 4, and applications of supported catalysts in continuous flow s. under Section 21)

Nanostructured catalysts for organic transformations, L.L. Chng, N. Erathodiyil, J.Y. Ying, Acc. Chem. Res. *2013, 46* (8), 1825-37; nanostructured materials for applications in heterogeneous catalysis, F. Zaera, Chem. Soc. Rev. *2013, 42* (3), 2746-62; tailoring nanohybrids and nanocomposites for catalytic applications, F. de Clippel, M. Dusselier, S. Van de Vyver, L. Peng, P.A. Jacobs, B.F. Sels, Green Chem. *2013, 15* (6), 1398-430; **metal nanoparticles** supported on solids as heterogeneous catalysts, Y. Wang, Z. Xiao, L. Wu, Curr. Org. Chem. *2013, 17* (12), 1325-33; advanced core-shell nanoparticle catalysts for efficient organic transformations, T. Mitsudomem, K. Kaneda, ChemCatChem *2013, 5* (7), 1681-91; fabrication of metal nanoparticles in metal-organic frameworks, H.R. Moon, D.-W. Lim, M.P. Suh, Chem. Soc. Rev. *2013, 42* (3), 1807-24; interface-confined **oxide nanostructures** for catalytic oxidation reactions, Q. Fu, F. Yang, X. Bao, Acc. Chem. Res. *2013, 46* (8), 1692-701; homogeneous recyclable catalysts based on metal nanoparticles for organic synthesis, L. Wu, Y. Zhang, Y.-G. Ji, Curr. Org. Chem. *2013, 17* (12), 1288-302; **nanocarbons** for the development of advanced catalysts, D.S. Su, S. Perathoner, G. Centi, Chem. Rev. *2013, 113* (8), 5782-816; an overview of **carbon nanotubes**: synthesis, purification and characterization, A. Manzano-Ramirez,

A. Moreno-Barcenas, M. Apatiga-Castro, E.M. Rivera-Munoz, R. Nava-Mendoza, R. Velazquez-Castillo, Curr. Org. Chem. *2013, 17* (17), 1858-66; oligoyne derivatives as reactive precursors for the preparation of carbon nanomaterials, J.-F. Morin, Synlett *2013, 24* (16), 2032-44; health and ecosystem risks of **graphene,** X. Hu, Q. Zhou, Chem. Rev. *2013, 113* (5), 3815-35; recent advances of immobilized catalysts for enantio-selective processes, Q. Chen, Curr. Org. Chem. *2013, 17* (12), 1303-24; catalysis using multifunctional **organosiliceous hybrid materials**, U. Díaz, D. Brunel, A. Corma, Chem. Soc. Rev. *2013, 42* (9), 4083-97; **porous materials** in catalysis: challenges for mesoporous materials, C. Perego, R. Millini, Chem. Soc. Rev. *2013, 42* (9), 3956-76; characterization, synthesis and catalysis of **hydrotalcite**-related materials for highly efficient materials transformations, S. Nishimura, A. Takagaki, K. Ebitani, Green Chem. *2013, 15* (8), 2026-42; multifunctional organic **polymeric catalysts** and reagents, J. Lu, P.H. Toy, Pure Appl. Chem. *2013, 85* (3), 543-56; **polymer-supported catalysts** obtained from metal-containing monomers, M.M. Dell'Anna, G. Romanazzi, P. Mastrorilli, Curr. Org. Chem. *2013, 17* (12), 1236-73; recyclable polymer-supported nanometal catalysts in water, A. Ohtaka, Chem. Record *2013, 13* (3), 274-85; soluble polymer-supported organocatalysts, Y.-C. Yang, D.E. Bergbreiter, Pure Appl. Chem. *2013, 85* (3), 493-510; the catalytic application of recoverable **magnetic** nanoparticle-supported organic compounds, H.-J. Xu, X. Wan, Y. Geng, X.-L. Xu, Curr. Org. Chem. *2013, 17* (10), 1034-50; **dendrimer**-supported catalysts for organic synthesis, Z. Li, J. Li, Curr. Org. Chem. *2013, 17* (12), 1334-49.

4 Transition metal catalysis - general aspects, ligands, Pd, Cu, Ag, Au, Fe, Mn

(for heterogeneous aspects s. under Section 3; for standard transition metal-catalyzed arylation, C-C and C-N cross-coupling, allylation, hydrogenation, hydro-functionalizations, olefin metathesis ... s. under Section 6, and applications in C-H activation s. under Section 7)

Biorelevant metals in sustainable metal catalysis - a survey, M.S. Holzwarth, B. Plietker, ChemCatChem *2013, 5* (7), 1650-79; **combining transition metal catalysis and organocatalysis** - an update, Z. Du, Z. Shao, Chem. Soc. Rev. *2013, 42* (3) 1337-78; reactions of phosphine-boranes and related frustrated Lewis pairs with transition metal complexes, A. Amgoune, G. Bouhadir, D. Bourissou, Top. Curr. Chem. *2013, 334,* 281-311; visible light photoredox catalysis with transition metal complexes: applications in organic synthesis, C.K. Prier, D.A. Rankic, D.W.C. MacMillan, Chem. Rev. *2013, 113* (7), 5322-63; **pincer ligands** as powerful tools for catalysis in organic synthesis, G. Van Koten, J. Organomet. Chem. *2013, 730*, 156-64; transition metal-catalyzed **C-C bond formation** *via* C-S bond cleavage: an overview, S.G. Modha, V.P. Mehta, E.V. Van der Eycken, Chem. Soc. Rev. *2013, 42* (12), 5042-55; transition-metal mediated **C-S bond activation** and transformations, L. Wang, W. He, Z. Yu, Chem. Soc. Rev. *2013, 42* (1), 599-621; **palladium-**mediated total synthesis of bioactive natural products, K.C. Majumdar, B. Sinha, Synthesis *2013, 45* (10), 1271-99; palladium-catalyzed oxidative carbocyclizations, Y. Deng, A.K.Å. Persson, J.-E. Bäckvall, Chem. Eur. J. *2012, 18* (37), 11498-523; an expedition through the last decade of heterocycle construction by using palladium, iron, copper, or iodine/*tert*-butyl hydroperoxide, L. Gao, S. Xiong, C. Wan, Z. Wang, Synlett *2013, 24* (11), 1322-39; synthesis of heterocycles via palladium-catalyzed Wacker-type oxidative cyclization reactions of hydroxy- and amino-alkenes, N.A. Butt, W. Zhang, Top. Heterocycl. Chem. *2013, 32,* 77-107; synthesis of functionalized cyclic molecules by palladium-catalyzed cyclization of propargylic esters with bis-nucleophiles, M. Yoshida, Heterocycles *2013, 87* (9), 1835-64; palladium-catalyzed oxidative carbonylation reactions, X.-F. Wu, H. Neuman, M. Beller, ChemSusChem *2013, 6* (2), 229-41; three carbons for complexity! recent developments of palladium-catalyzed reactions of allenes, T. Lechel, F. Pfrengle, H.-U. Reissig, R. Zimmer, ChemCatChem *2013, 5* (8), 2100-30; recent advances in transition metal-catalyzed trifluoromethylation and related transformations, P. Chen, G. Liu, Synthesis

2013, 45 (21), 2919-39; synthetic uses of ammonia in transition-metal catalysis, J. Kim, H.J. Kim, S. Chang, Eur. J. Org. Chem. *2013,* (16), 3201-13; unravelling the labyrinth of palladium-catalyzed reactions involving isocyanides, S. Lang, Chem. Soc. Rev. *2013, 42* (12), 4867-80; merging organocatalysis and **gold** catalysis - a critical evaluation of the underlying concepts, C.C.J. Loh, D. Enders, Chem. Eur. J. *2012, 18* (33), 10212-25; constructing molecular complexity from alkynol derivatives: a journey from Fischer carbene complexes to tandem catalysis with gold and other carbophilic Lewis acids, F. Rodriguez, F.J. Fañanás, Synlett *2013, 24* (14), 1757-71; combined coinage metal catalysis for the synthesis of bioactive molecules, N. Krause, O. Aksin-Artok, M. Asikainen, V. Breker, C. Deutsch, J. Erdsack, H.-T. Fan, B. Gockel, S. Minkler, M. Poonoth, Y. Sawama, Y. Sawama, T. Sun, F. Volz, C. Winter, J. Organomet. Chem. *2012* (704), 1-8; aerobic **copper**-catalyzed organic reactions, S.E. Allen, R.R. Walvoord, R. Padilla-Salinas, M.C. Kozlowski, Chem. Rev. *2013, 113* (8), 6234-458; copper-modified magnetic bimetallic nano-catalyst, ligand-regulated catalytic activity, N.B.R. Baig, R.S. Varma, Curr. Org. Chem. *2013, 17* (20), 2227-37; atomically precise gold nanoclusters as new model catalysts, G. Li, R. Jin, Acc. Chem. Res. *2013, 46* (8), 1749-58; model studies of heterogeneous catalytic hydrogenation reactions with gold, M. Pan, A.J. Brush, Z.D. Pozun, H.C. Ham, W.-Y. Yu, G. Henkelman, G.S. Hwang, C.B. Mullins, Chem. Soc. Rev. *2013, 42* (12), 5002-13; gold nanoparticle catalysts for selective hydrogenations, T. Mitsudome, K. Kaneda, Green Chem. *2013, 15* (10), 2636-54; gold-catalyzed direct arylation, L.T. Ball, G.C. Lloyd-Jones, C.A. Russell, Science *2012, 337* (6102), 1644-8; complexity-generating hydration reactions via gold-catalyzed addition of boronic acids to alkynes, T.D. Sheppard, Pure Appl. Chem. *2012, 84* (11), 2431-42; photocatalysis on supported gold and **silver** nanoparticles under ultraviolet and visible-light irradiation, S. Sarina, E.R. Waclawik, H. Zhu, Green Chem. *2013, 15* (7), 1814-33; C-C bond formation in di**iron** complexes, R. Mazzoni, M. Salmi, V. Zanotti, Chem. Eur. J. *2012, 18* (33), 10174-94; chiral iron catalysts for asymmetric synthesis, K. Gopalaiah, Chem. Rev. *2013, 113* (5), 3248-96; iron-catalyzed enantioselective reactions through the use of chiral bipyridyl-containing ligands, B. Plancq, T. Ollevier, Austr. J. Chem. *2012, 65* (12), 1564-72; **manganese** compounds in the catalysis of organic reactions, R.I. Khusnutdinov, A.R. Bayguzina, U.M. Dzhemilev, Russ. J. Org. Chem. *2012, 48* (3), 309-48.

5 Biocatalysis, enzymes ...

Biocatalytic methods for C-C bond formation, K. Fesko, M. Gruber-Khadjawi, ChemCatChem *2013, 5* (6), 1248-72; C-C hydrolases for biocatalysis, E. Siirola, A. Frank, G. Grogan, W. Kroutil, Eur. J. Org. Chem. *2013, 355* (9), 1677-91; unusual reactions mediated by FMN-dependent ene- and nitro-reductases, K. Durchschein, M. Hall, K. Faber, Green Chem. *2013, 15* (7), 1764-72; chemical and stereochemical actions of UDP-galactose 4-epimerase, P.A. Frey, A.D. Hegeman, Acc. Chem. Res. *2013, 46* (7), 1417-26; computational enzyme design, G. Kiss, N. Çelebi-Ölçüm, R. Moretti, D. Baker, K.N. Houk, Angew. Chem., Int. Ed. *2013, 52* (22), 5700-25; enzymatic glycosylation of multivalent scaffolds, P. Bojarová, R.R. Rosencrantz, L. Elling, V. Kren, Chem. Soc. Rev. *2013, 42* (11), 4774-97; enzymatic glycosylation of small molecules: challenging substrates require tailored catalysts, T. Desmet, W. Soetaert, P. Bojarová, V. Kren, L. Dijkhuizen, V. Eastwick-Field, A. Schiller, Chem. Eur. J. *2012, 18* (35), 10786-801; recent developments in design and synthesis of bicyclic azasugars, carbasugars and related molecules as glycosidase inhibitors, R. Lahiri, A.A. Ansari, Y.D. Vankar, Chem. Soc. Rev. *2013, 42* (12), 5102-18; biocatalysis in biomass-derived solvents: the quest for fully sustainable chemical processes, M. Perez-Sanchez, M. Sandoval, M.J. Hernaiz, P. Domínguez de María, Curr. Org. Chem. *2013, 17* (11), 1188-99.

6 Name reactions, standard transformations - arylation, transition metal-catalyzed cross-coupling, C-N coupling, allylation, hydrogenation, hydroformylation, hydrofunctionalization, olefin metathesis, aldol, Mannich, Henry, Reformatsky, Baylis-Hillman, McMurry, N-alkylation, diamination, halogenation, trifluoromethylation, borylation, cycloaddition, Nazarov rearrangement

(for the chemistry of specific functional groups s.a. under Section 17)

Development of recent total syntheses based on the **Heck** reaction, M.M. Heravi, E. Hashemi, N. Ghobadi, Curr. Org. Chem. *2013, 17* (19), 2192-224; enantioselective Heck arylation of acyclic alkenyl alcohols using a redox-relay strategy, E.W. Werner, T.-S. Mei, A.J. Burckle, M.S. Sigman, Science *2012, 338,* (6113), 1455-8; gold-catalyzed **direct arylation**, L.T. Ball, G.C. Lloyd-Jones, C.A. Russell, Science *2012, 337* (6102), 1644-8; the photocatalyzed Meerwein arylation: classic reaction of aryldiazonium salts in a new light, M.S. Durga, P. Hari, B. König, Angew. Chem., Int. Ed. *2013, 52* (18), 4734-43; transition-metal-catalyzed **Suzuki-Miyaura coupling** reactions: a remarkable advance from palladium to nickel catalysts, F.-S. Han, Chem. Soc. Rev. *2013, 42* (12), 5270-98; transmetalation in the Suzuki-Miyaura coupling: the fork in the trail, A.J.J. Lennox, G.C. Lloyd-Jones, Angew. Chem., Int. Ed. *2013, 52* (29), 7362-70; addressing challenges in **palladium-catalyzed cross-coupling** reactions through ligand design, R.J. Lundgren, M. Stradiotto, Chem. Eur. J. *2012, 18* (32), 9758-69; asymmetric transition metal-catalyzed cross-coupling reactions for the construction of tertiary stereocenters, E.C. Swift, E.R. Jarvo, Tetrahedron *2013, 69* (29), 5799-817; palladium-catalyzed migratory insertion of isocyanides: an emerging platform in cross-coupling chemistry, T. Vlaar, E. Ruijter, B.U.W. Maes, R.V.A. Orru, Angew. Chem., Int. Ed. *2013, 52* (28), 7084-97; synthesis of 4-[^{18}F]fluorohalobenzenes and palladium-mediated cross-coupling reactions for the synthesis of ^{18}F-labeled radiotracers, J. Way, V. Bouvet, F. Wuest, Curr. Org. Chem. *2013, 17* (19), 2138-52; boron-based pronucleophiles in catalytic, asymmetric C(sp^3)-allyl cross-couplings, U. Schneider, Y.-Y. Huang, A. Chakrabarti, H.T. Dao, N. Morita, S. Kobayashi, Pure Appl. Chem. *2012, 84* (11), 2417-30; recent approaches for C-C bond formation via direct dehydrative coupling strategies, R. Kumar, E.V. Van der Eycken, Chem. Soc. Rev. *2013, 42* (3), 1121-46; development and applications of highly selective palladium-catalyzed monocoupling reactions of (cyclo)-alkenes and 1,3-alkadienes bearing two or three electrophilic sites, and bis(enol triflates) with terminal alkynes, R. Rossi, F. Bellina, M. Lessi, C. Manzini, Tetrahedron *2013, 69* (37), 7869-909; photoinduced Ullmann **C-N coupling**: demonstrating the viability of a radical pathway, S.E. Creutz, K.J. Lotito, G.C. Fu, J.C. Peters, Science *2012, 338* (6107), 647-51; recent developments in recyclable copper catalyst systems for C-N bond forming cross-coupling reactions using aryl halides and arylboronic acids, M.L. Kantam, Top. Organomet. Chem. *2013, 46,* 119-71; metal-catalyzed C(sp^2)-N bond formation, A. Correa, C. Bolm, Top. Organomet. Chem. *2013, 46,* 55-85; palladium-catalyzed C(sp^2)-N bond forming reactions: recent developments and applications, G.S. Lemen, J.P. Wolfe, Top. Organomet. Chem. *2013, 46,* 1-53; the N-H functional group in organometallic catalysis, B. Zhao, Z. Han, K. Ding, Angew. Chem., Int. Ed. *2013, 52* (18), 4744-88; diastereodivergent processes in palladium-catalyzed **allylic alkylation**, M.T. Oliveira, D. Audisio, S. Niyomchon, N. Maulide, ChemCatChem *2013, 5* (6), 1239-47; the construction of *all*-carbon quaternary stereocenters by use of Pd-catalyzed asymmetric allylic alkylation reactions in total synthesis, A.Y. Hong, B.M. Stoltz, Eur. J. Org. Chem. *2013,* (14), 2745-59; a chronicle review: regioselective synthesis of trifluoromethyl group-containing allylic amines using palladium-catalyzed **N-allylation** pathway, T. Hirakawa, M. Kawatsura, T. Itoh, J. Fluorine Chem. *2013, 152,* 62-9; carbon-sulfur bond formation via metal-catalyzed **S-allylation** of sulfur nucleophiles, W. Liu, X. Zhao, Synthesis *2013, 45* (15), 2051-69; diastereoselective **allylation of carbonyl compounds** and imines: application to the synthesis of natural products, M. Yus, J.C. González-Gómez, F. Foubelo, Chem. Rev. *2013, 113* (7), 5595-698; allylic oxidations of olefins to enones, V. Weidmann, W. Maison, Synthesis *2013, 45* (16), 2201-21; homogeneous palladium-catalyzed asymmetric **hydrogenation**, Q.-A. Chen, Z.-S. Ye, Y. Duan, Y.-G.

Zhou, Chem. Soc. Rev. *2013, 42* (1), 497-511; development of iridium-catalyzed asymmetric hydrogenation: new catalysts, new substrate scope, A. Cadu, P.G. Andersson, J. Organomet. Chem. *2012* (714), 3-11; rhodium-catalyzed asymmetric hydrogenation as a valuable synthetic tool for the preparation of chiral drugs, P. Etayo, A. Vidal-Ferran, Chem. Soc. Rev. *2013, 42* (1), 728-54; model studies of heterogeneous catalytic hydrogenation reactions with gold, M. Pan, A.J. Brush, Z.D. Pozun, H.C. Ham, W.-Y. Yu, G. Henkelman, G.S. Hwang, C.B. Mullins, Chem. Soc. Rev. *2013, 42* (12), 5002-13; gold nanoparticle catalysts for selective hydrogenations, T. Mitsudome, K. Kaneda, Green Chem. *2013, 15* (10), 2636-54; outer sphere hydrogenation catalysis, O. Eisenstein, R.H. Crabtree, New J. Chem. *2013, 37* (1), 21-7; binuclear cooperative catalysts for the hydrogenation **and hydroformylation** of olefins, D.G.H. Hetterscheid, S.H. Chikkali, B. de Bruin, J.N.H. Reek, ChemCatChem *2013, 5* (10), 2785-93; amine-borane dehydrogenation chemistry: metal-free hydrogen transfer, new catalysts and mechanisms, and the synthesis of polyaminoboranes, N.E. Stubbs, A.P.M. Robertson, E.M. Leitao, I. Manners, J. Organomet. Chem. *2013, 730*, 84-9; applications of acceptorless **dehydrogenation** and related transformations in chemical synthesis, C. Gunanathan, D. Milstein, Science *2013, 341* (6143), 249; recent advances in catalytic sequential reactions involving hydroelement addition to [**hydrofunctionalization** of] carbon-carbon multiple bonds, X. Zeng, Chem. Rev. *2013, 113* (8), 6864-900; recent developments in Cope-type hydroamination reactions of hydroxylamine and hydrazine derivatives, A.M. Beauchemin, Org. Biomol. Chem. *2013, 11* (41), 7039-50; alkylboronate synthesis based on transition metal-catalyzed hydroboration, S. Huang, Y. Xie, S. Wu, M. Jia, J. Wang, W. Xu, H. Fang, Curr. Org. Synth. *2013, 10* (5), 683-96; platinum-based heterogeneously catalyzed hydrosilylation, M. Pagliaro, R. Ciriminna, V. Pandarus, F. Béland, Eur. J. Org. Chem. *2013,* (28), 6227-35; **olefin metathesis** in aqueous media, J. Tomasek, J. Schatz, Green Chem. *2013, 15* (9), 2317-38; a new tool in the toolbox: electron-withdrawing group-activated ruthenium catalysts for olefin metathesis, T.K. Olszewski, M. Bieniek, K. Skowerski, K. Grela, Synlett *2013, 24* (8), 903-19; removing ruthenium residues from olefin metathesis reaction products, G.C. Vougioukalakis, Chem. Eur. J. *2012, 18* (29), 8868-80; enantioselective aldol reactions catalyzed by chiral phosphine oxides, S. Kotani, M. Sugiura, M. Nakajima, Chem. Record *2013, 13* (4), 362-70; (S)-5-prolylamide-triazole as organocatalyst for direct asymmetric **aldol reactions**, L. Liu, M.-N. Gao, Y. Li, Z. Li, L. Song, Z.-W. Liu, D. Xue, Z.-T. Liu, Curr. Org. Chem. *2013, 17* (14), 1563-8; oxazolidinones as chiral auxiliaries in asymmetric aldol reactions applied to total synthesis, M.M. Heravi, V. Zadsirjan, Tetrahedron: Asym. *2013, 24* (19), 1149-88; the **Mukaiyama aldol reaction**: 40 years of continuous development, J.-I. Matsuo, M. Murakami, Angew. Chem., Int. Ed. *2013, 52* (35), 9109-18; Lewis base catalysis of the Mukaiyama aldol reaction: 40 years of inspiration and advances, G.L. Beutner, S.E. Denmark, Angew. Chem., Int. Ed. *2013, 52* (35), 9086-96; base-catalyzed Mukaiyama aldol reactions, a continued quest for stereoselectivity, R.L. Sutar, N.N. Joshi, Tetrahedron: Asym. *2013, 24* (21-22), 1345-63; the impact of the Mukaiyama aldol reactions in total synthesis, S.B.J. Kan, K.K.-H. Ng, I. Paterson, Angew. Chem., Int. Ed. *2013, 52* (35), 9097-108; recent advances in metal-catalyzed asymmetric **Mannich reaction,** B. Karimi, D. Enders, E. Jafari, Synthesis *2013, 45* (20), 2769-812; synthesis of β-nitroamines via classical Mannich and **aza-Henry** reactions, M. Telep El-Sayed, K. Mahmoud, A. Hilgeroth, Curr. Org. Chem. *2013, 17* (11), 1200-24; nitro-Mannich reaction, A. Noble, J.C. Anderson, Chem. Rev. *2013, 113* (5), 2887-9; recent advances in the diastereoselective **Reformatsky**-type reaction, S. Choppin, L. Ferreiro-Medeiros, M. Barbarotto, F. Colobert, Chem. Soc. Rev. *2013, 42* (3), 937-49; recent advances in organocatalytic asymmetric **Morita-Baylis-Hillman**/aza-Morita-Baylis-Hillman reactions, Y. Wei, M. Shi, Chem. Rev. *2013, 113* (8), 6659-90; the **McMurry reaction** in porphyrinoid chemistry, K.M. Smith, Top. Heterocycl. Chem. *2013, 33,* 1-34; microwave assisted **N-alkylation** of amines with alcohols: magnetically recoverable catalysis, S.R. Takkellapati, M.N. Nadagouda, Curr. Org. Chem. *2013, 17* (20), 2332-7; metal-free catalytic **vicinal diamination** of alkenes, K. Muñiz, Pure Appl. Chem. *2013, 85* (4), 755-62; protocols for the catalytic **oxidation of primary amines** to imines,

M. Largeron, Eur. J. Org. Chem. *2013*, (24), 5225-35; integrated reactions using addition to conjugated imines and iminium salts, T. Nishi, I. Mizota, M. Shimizu, Pure Appl. Chem. *2012, 84* (12), 2609-18; catalytic **Appel reactions**, H.A. van Kalkeren, F.L. van Delft, F.P.J.T. Rutjes, Pure Appl. Chem. *2013, 85* (4), 812-8; Yb(OTf)$_3$-catalyzed **bromination** reactions of natural product-like N-benzylcinnamamides: a facile route to diverse N-substituted amides of pharmacological interest, C.E. Puerto Galvis, J.G. Hernandez Barajas, V.V. Kouznetsov, Curr. Org. Chem. *2013, 17* (14), 1545-54; **introduction of fluorine** and fluorine-containing functional groups, T. Liang, C.N. Neumann, T. Ritter, Angew. Chem., Int. Ed. *2013, 52* (32), 8214-64; oxidative aliphatic C-H fluorination with fluoride ion catalyzed by a manganese porphyrin, W. Liu, X. Huang, M.-J. Cheng, R.J. Nielsen, W.A. Goddard III, J.T. Groves, Science *2012, 337* (6100), 1322-5; Mg-Cu bimetal system for selective C-F bond activation, S. Utsumi, T. Katagiri, K. Uneyama, J. Fluorine Chem. *2013, 152,* 84-9; organometallic aspects of **fluoroalkylation** reactions with copper and nickel, H. Wang, D.A. Vicic, Synlett *2013, 24* (15), 1887-98; fluoroalkylation of organic compounds, D.Y. Mikhaylov, Y.H. Budnikova, Russ. Chem. Rev. *2013, 82* (9), 835-64; taming of fluoroform: direct nucleophilic **trifluoromethylation** of Si, B, S, and C centers, G.K.S. Prakash, P.V. Jog, P.T.D. Batamack, G.A. Olah, Science *2012, 338* (6112), 1324-7; recent advances in transition-metal-catalyzed trifluoromethylation and related transformations, P. Chen, G. Liu, Synthesis *2013, 45* (21), 2919-39; recent progress in organocatalytic asymmetric **halocyclization**, K. Murai, H. Fujioka, Heterocycles *2013, 87* (4), 763-805; electrophilic activation of unsaturated systems: applications to selective organic synthesis, S. Suárez-Pantiga, J.M. González, Pure Appl. Chem. *2013, 85* (4), 721-40; new methods in organic synthesis through copper-catalyzed **borylation** reactions: stereoselective synthesis of 1,4-diols and vinylboronates, R. Alfaro, A. Parra, J. Alemán, M. Tortosa, Synlett *2013, 24* (7), 804-12; recent developments in catalytic asymmetric inverse-electron-demand Diels-Alder **cycloaddition,** X. Jiang, R. Wang, Chem. Rev. *2013, 113* (7), 5515-46; inverse electron demand Diels-Alder (IEDDA)-initiated conjugation: a (high) potential click chemistry scheme, A.-C. Knall, C. Slugovc, Chem. Soc. Rev. *2013, 42* (12), 5131-42; tandem inverse-electron-demand hetero-/retro-Diels-Alder reactions for aromatic nitrogen heterocycle synthesis, R.A.A. Foster, M.C. Willis, Chem. Soc. Rev. *2013, 42* (1), 63-76; bent bonds and the antiperiplanar hypothesis as a simple model to predict Diels-Alder reactivity: retrospective or perspective? G. Deslongchamps, P. Deslongchamps, Tetrahedron *2013, 69* (30), 6022-33; novel tetrahydropyridazines by unusual aza-Diels-Alder reaction of electron-poor 1,2-diaza-1,3-dienes with electron-poor alkenes under solvent free conditions, O.A. Attanasi, L. Bianchi, M.D'Auria, F. Mantellini, R. Racioppi, Curr. Org. Synth. *2013, 10* (4), 631-9; iminium ions as dienophiles in aza-Diels-Alder reactions: a closer look, M.G. Memeo, P. Quadrelli, Chem. Eur. J. *2012, 18* (40), 12554-82; catalytic enantioselective [4+2]-cycloaddition: a strategy to access aza-hexacycles, G. Masson, C. Lalli, M. Benohoud, G. Dagousset, Chem. Soc. Rev. *2013, 42* (3), 902-23; pericyclic [4+2]- and [3+2]-cycloaddition reactions of nitroarenes in heterocyclic synthesis, S.A. Shevelev, A.M. Starosotnikov, Chem. Heterocycl. Compd. *2013, 49* (1), 92-115; adventures of an occasional click chemist, U.H.F. Bunz, Synlett *2013, 24* (15), 1899-909; light-induced **click reactions**, M.A. Tasdelen, Y. Yagci, Angew. Chem., Int. Ed. *2013, 52* (23), 5930-8; graphene in copper-catalyzed azide-alkyne cycloaddition reactions: evolution from [60]fullerene and carbon nanotubes strategies, J.L. Segura, H.J. Salavagione, Curr. Org. Chem. *2013, 17* (15), 1680-93; enantioselective cycloaddition reactions of ketenes catalyzed by N-heterocyclic carbenes, X.-Y. Chen, S. Ye, Synlett *2013, 24* (13), 1614-22; from the development of catalysts for alkyne and alkyne-nitrile [2+2+2]-cycloaddition reactions to their use in polymerization reactions, S. Okamoto, Y.-K. Sugiyama, Synlett *2013, 24* (9), 1044-60; **Nazarov** reagents and their use in organic synthesis, G. Audran, P. Brémond, M. Feuerstein, S.R.A. Marque, M. Santelli, Tetrahedron *2013, 69* (39), 8325-48; beyond the divinyl ketone: innovations in the generation and Nazarov cyclization of pentadienyl cation intermediates, W.T. Spencer III, T. Vaidya, A.J. Frontier, Eur. J. Org. Chem. *2013,* (18), 3621-33; **Claisen** rearrangement of aliphatic allyl vinyl ethers from 1912 to 2012: 100 years of electrophilic catalysis, J. Rehbein, M. Hiersemann, Synthesis *2013, 45* (9), 1121-59; the Claisen rearrangement in the syntheses of bioactive natural products, K.C. Majumdar, R.K. Nandi, Tetrahedron *2013, 69* (34), 6921-57; propargyl Claisen rearrangement: allene synthesis and beyond, D. Tejedor, G. Méndez-Abt, L. Cotos, F. García-Tellado, Chem. Soc. Rev. *2013, 42* (1), 458-71; fluoro-**Pummerer** rearrangement and analogous reactions, V. Hugenberg, G. Haufe, J. Fluorine Chem. *2012, 143,* 238-62; advancements in the nascent **nitroso-ene** reaction, M. Baidya, H. Yamamoto, Synthesis *2013, 45* (14), 1931-8; synthesis of β-lactams by transition metal promoted **Staudinger** reactions: alternative synthetic approaches from transition metal enhanced organocatalysis to *in situ*, highly reactive intermediate synthesis and catalytic tandem reactions, R. Tuba, Org. Biomol. Chem. *2013, 11* (36), 5976-88; *tert*-amino effect: the **Meth-Cohn and Reinhoudt** reactions, A.Y. Platonova, T.V. Glukhareva, O.A. Zimovets, Y.Y. Morzherin, Chem. Heterocycl. Compd. *2013, 49* (3) 357-85.

7 Syntheses via C-H activation and functionalization ...

Palladium catalyzed C-H activation and its application to multi-bond forming reactions, K. Ramachandiran, T. Sreelatha, N.V. Lakshmi, T.H. Babu, D. Muralidharan, P.T. Perumal, Curr. Org. Chem. *2013, 17* (18), 2001-24; C-H activation: a complementary tool in the total synthesis of complex natural products, D.Y.-K. Chen, S.W. Youn, Chem. Eur. J. *2012, 18* (31), 9452-74; chiral cyclopentadienyl ligands as stereocontrolling element in asymmetric C-H functionalization, B. Ye, N. Cramer, Science *2012, 338* (6106), 504-6; organocatalytic C-H activation reactions, S.C. Pan, Beilstein J. Org. Chem. *2012, 8,* 1374-84; selective C-H activations using frustrated Lewis pairs: applications in organic synthesis, P. Knochel, K. Karaghiosoff, S. Manolikakes, Top. Curr. Chem. *2013, 334,* 171-90; direct nucleophilic functionalization of $C(sp^2)$-H bonds in arenes and hetarenes by electrochemical methods, A.V. Shchepochkin, O.N. Chupakhin, V.N. Charushin, V.A. Petrosyan, Russ. Chem. Rev. *2013, 82* (8), 747-71; transition metal-catalyzed C_{vinyl}-C_{vinyl} bond formation via double C_{vinyl}-H bond activation, X. Shang, Z.-Q. Liu, Chem. Soc. Rev. *2013, 42* (8), 3253-60; transition-metal-catalyzed additions of C-H bonds to C-X (X = N, O) multiple bonds via C-H bond activation, G. Yan, X. Wu, M. Yang, Org. Biomol. Chem. *2013, 11* (34), 5558-78; stereoselective cross-dehydrogenative coupling reactions based on $C(sp^3)$-H activation to give chiral heterocyclic compounds, M. Zhang, A. Zhang, J. Heterocycl. Chem. *2012, 49* (4), 721-5; metal-free oxidative carbon-heteroatom bond formation through C-H bond functionalization, R. Samanta, K. Matcha, A.P. Antonchick, Eur. J. Org. Chem. *2013,* (26), 5769-804; complex N-heterocycle synthesis via iron-catalyzed, direct C-H bond amination, E.T. Hennessy, T.A. Betley, Science *2013, 340* (6132), 591-5.

8 Reaction intermediates - radicals, carbenoids, zwitterions, carbocations, cumulenes

(for catalysis with carbenes s. under Section 2)

Diradicals, M. Abe, Chem. Rev. *2013, 113* (9), 7011-88; verdazyls: synthesis, properties, application, G.N. Lipunova, T.G. Fedorchenko, O.N. Chupakhin, Russ. Chem. Rev. *2013, 82* (8), 701-34; concerted reactions that produce diradicals and **zwitterions**: electronic, steric, conformational, and kinetic control of cycloaromatization processes, R.K. Mohamed, P.W. Peterson, I.V. Alabugin, Chem. Rev. *2013, 113* (9), 7089-129; radical reactions of fullerenes: from synthetic organic chemistry to materials science and biology, M.D. Tzirakis, M. Orfanopoulos, Chem. Rev. *2013, 113* (7), 5262-321; catalytic X-H insertion reactions based on **carbenoids**, D. Gillingham, N. Fei, Chem. Soc. Rev. *2013, 42* (12), 4918-31; carbene additions to fullerenes, M. Yamada, T. Akasaka, S. Nagase, Chem. Rev. *2013, 113* (9), 7209-64; heteroarylcarbenes, R.S. Sheridan, Chem. Rev. *2013, 113* (9), 7179-208; crystalline singlet phosphino**nitrene**: a nitrogen atom-transfer agent, F. Dielmann, O. Back, M. Henry-Ellinger, P. Jerabek, G. Frenking, G.A. Bertrand, Science *2012, 337* (6101), 1526-8; contemporary **carbocation** chemistry: applications in organic synthesis, R.R. Naredla, D.A. Klumpp, Chem. Rev. *2013, 113* (9), 6905-48;

ketenes and other cumulenes as reactive intermediates, A.D. Allen, T.T. Tidwell, Chem. Rev. *2013, 113* (9), 7287-342.

9 Multi-bond formation, multistep synthesis, multicomponent, tandem, sequential and domino reactions

Multi-bond forming processes in efficient synthesis, N.J. Green, M.S. Sherburn, Austr. J. Chem. *2013, 66* (3), 267-83; **multistep organic synthesis** of modular photosystems, N. Sakai, S. Matile, Beilstein J. Org. Chem. *2012, 8,* 897-904; higher-order **multi-component** reactions: beyond four reactants, S. Brauch, S.S. van Berkel, B. Westermann, Chem. Soc. Rev. *2013, 42* (12), 4948-62; synthesis of unnatural amino acids via microwave-assisted regioselective one-pot multicomponent reactions of sulfamidates, N.B.R. Baig, R.S. Varma, Curr. Org. Chem. *2013, 17* (20), 2323-31; multicomponent 1,3-dipolar cycloaddition reactions in the construction of hybrid spiroheterocycles, N. Arumugam, R. Suresh Kumar, A.I. Almansour, S. Perumal, Curr. Org. Chem. *2013, 17* (18), 1929-56; a new multicomponent protocol for the synthesis of pyridines and fused pyridines via a formal aza-[3+3]-cycloaddition, G. Tenti, M.T. Ramos, J.C. Menendez, Curr. Org. Synth. *2013, 10* (4), 645-54; recent application of 4-hydroxy-coumarin in multicomponent reactions, G. Mohammadi Ziarani, P. Hajibbasi, Heterocycles *2013, 87* (7), 1415-39; recent developments in enantioselective multi-catalyzed **tandem** reactions, H. Pellissier, Tetrahedron *2013, 69* (35), 7171-210; aziridinyl imines in organic synthesis: development of tandem reaction strategies and application to total synthesis of natural products, H.-Y. Lee, S.-B. Song, T. Kang, Y.J. Kim, S.J. Geum, Pure Appl. Chem. *2013, 85* (4), 741-54; ionic liquid-mediated tandem synthesis of bioactive quinoline-based thiophene/thiazole-linked multi-heterocomponent Ugi adducts, R.V. Patel, K.H. Chikhalia, S.H. Nile, S.W. Park, Curr. Org. Chem. *2013, 17* (10), 1125-9; tandem inverse electron-demand hetero-/retro-Diels-Alder reactions for aromatic N-heterocycle synthesis, R.A.A. Foster, M.C. Willis, Chem. Soc. Rev. *2013, 42* (1), 63-76; synthesis of β-lactams by transition-metal promoted Staudinger reactions: alternative synthetic approaches from transition metal-enhanced organocatalysis to *in situ*, highly reactive intermediate synthesis and catalytic tandem reactions, R. Tuba, Org. Biomol. Chem. *2013, 11* (36), 5976-88; Michael addition-initiated **sequential reactions** from 1,3-dicarbonyls for the synthesis of polycyclic heterocycles, X. Bugaut, D. Bonne, Y. Coquerel, J. Rodriguez, T. Constantieux, Curr. Org. Chem. *2013, 17* (18), 1920-8; (S)-proline-catalyzed **domino** reactions for the synthesis of heterocycles, S. Indumathi, J.C. Menendez, S. Perumal, Curr. Org. Chem. *2013, 17* (18), 2038-64; enantioselective organocatalyzed domino synthesis of six-membered carbocycles, S. Goudedranche, W. Raimondi, X. Bugaut, T. Constantieux, D. Bonne, J. Rodriguez, Synthesis *2013, 45* (14), 1909-30; synthesis of heterocycles via metal-catalyzed domino/one-pot reactions that generate a C-N or C-O bond, J. Keilitz, H.A. Malik, Top. Heterocycl. Chem. *2013, 32,* 187-223.

10 Heterocyclic chemistry - general methods; N-, O- and S-heterocyclics, condensed heterocyclics, macroheterocyclics

(s.a. under Section 6 for classical heterocyclic synthesis and cycloaddition, and under Section 11 for natural product synthesis)

Construction of **heterocyclic scaffolds** via transition metal-catalyzed $C(sp^3)$-H functionalization, M. Zhang, A.-Q. Zhang, Y. Peng, J. Organomet. Chem. *2013, 723,* 224-32; an expedition through the last decade of heterocycle construction by using palladium, iron, copper, or iodine/*tert*-butyl hydroperoxide, L. Gao, S. Xiong, C. Wan, Z. Wang, Synlett *2013, 24* (11), 1322-39; visible light-driven photoredox catalysis in the construction of carbocyclic and heterocyclic ring systems, J. Xuan, L.-Q. Lu, J.-R. Chen, W.-J. Xiao, Eur. J. Org. Chem. *2013,* (30), 6755-70; stereoselective cross-dehydrogenative coupling reactions based on $C(sp^3)$-H activation to give chiral heterocyclic compounds, M. Zhang, A. Zhang, J. Heterocycl. Chem. *2012, 49* (4), 721-5;

synthesis of heterocycles via palladium-catalyzed Wacker-type oxidative cyclization reactions of hydroxy- and amino-alkenes, N.A. Butt, W. Zhang, Top. Heterocycl. Chem. *2013, 32,* 77-107; the pseudo-intramolecular process: a novel synthetic method for functionalized heterocyclic compounds, N. Nishiwaki, Heterocycles *2013, 87* (5), 967-90; arylglyoxals in synthesis of heterocyclic compounds, B. Eftekhari-Sis, M. Zirak, A. Akbari, Chem. Rev. *2013, 113* (5), 2958-3043; **five-membered heterocycles** by reacting enolates with dipoles, P. Vitale, A. Scilimati, Curr. Org. Chem. *2013, 17* (18), 1986-2000; synthesis of **saturated heterocycles** via metal-catalyzed alkene hydroamination or hydroalkoxylation reactions, L.D. Julian, Top. Heterocycl. Chem. *2013, 32,* 109-55; synthesis of saturated heterocycles via metal-catalyzed alkene diamination, amino-alkoxylation, or dialkoxylation reactions, S.R. Chemler, D.A. Copeland, Top. Heterocycl. Chem. *2013, 32,* 39-75; synthesis of saturated heterocycles via metal-catalyzed alkene carboamination or carboalkoxylation reactions, J.P. Wolfe, Top. Heterocycl. Chem. *2013, 32,* 1-37; synthesis of saturated heterocycles via metal-catalyzed allylic alkylation reactions, J.M. Ketcham, A. Aponick, Top. Heterocycl. Chem. *2013, 32,* 157-86; synthesis of saturated heterocycles via metal-catalyzed formal cycloaddition reactions that generate a C-N or C-O bond, J. Waser, Top. Heterocycl. Chem. *2013, 32,* 225-69; azacycloalkanes from epoxides and aziridines, L.I. Kas′yan, V.A. Pal′chikov, Y.S. Bondarenko, Russ. J. Org. Chem. *2011, 47* (11), 1609-52; acetamidines and acetamidoximes containing an electron-withdrawing group at the α-carbon atom: their use in the synthesis of **N-heterocycles**, P.S. Lobanov, D.V. Dar′in, Chem. Heterocycl. Compd. *2013, 49* (4), 507-28; assembly of N-heterocycles via Pd- and Cu-catalyzed C-N bond formation reactions, Y. Jiang, D. Ma, Top. Organomet. Chem. *2013, 46,* 87-118; recent advances in the cyclization of N-heterocycles: the role of enabling techniques, G. Cravotto, S. Tagliapietra, M. Caporaso, D. Garella, E. Borretto, A. Di Stilo, Chem. Heterocycl. Compd. *2013, 49* (6), 811-26; use of sulfur fragments for the synthesis of N-heterocycles, A. Padwa, Pure Appl. Chem. *2013, 85* (4), 701-20; synthesis of N-heterocycles by ring enlargement reactions through bicyclic cyclopropane intermediates, A.A. Fesenko, A.D. Shutalev, Chem. Heterocycl. Compd. *2013, 49* (6), 827-48; catalytic ring expansion adventures, J.T. Njardarson, Synlett *2013, 24* (7), 787-803; transition metal-mediated synthesis of monocyclic heteroarenes, A.V. Gulevich, A.S. Dudnik, N. Chernyak, V. Gevorgyan, Chem. Rev. *2013, 113* (5), 3084-213; asymmetric reactions of a series of **aromatic azines** with nucleophiles, I.N. Egorov, T.A. Tseitler, G.V. Zyryanov, V.L. Rusinov, O.N. Chupakhin, Heterocycles *2012, 86* (2), 821-89; recent advances in the synthesis of regioselectively fluorinated homo- and hetero-cyclic compounds by complementary cyclization methods, R. Ruzziconi, F. Buonerba, J. Fluorine Chem. *2013, 152,* 12-28; **pyrrole**: chemical synthesis, microwave-assisted synthesis, reactions and applications, S.D. Joshi, U.A. More, V.H. Kulkarni, T.M. Aminabhavi, Curr. Org. Chem. *2013, 17* (20), 2279-304; synthesis of 3,4-disubstituted pyrroles, F.J. Leeper, J.M. Kelly, Org. Prep. Proced. Int. *2013, 45* (3), 171-210; [1,7]-electrocyclization reactions in the synthesis of **azepine** derivatives, N.A. Nedolya, B.A. Trofimov, Chem. Heterocycl. Compd. *2013, 49* (1), 152-76; synthesis of diversely fluorinated **pyrazoles** as novel active agrochemical ingredients, F. Giornal, S. Pazenok, L. Rodefeld, N. Lui, J.-P. Vors, F.R. Leroux, J. Fluorine Chem. *2013, 152,* 2-11; conjugates of **1,2,3-triazoles** and acyclic pyrimidine nucleoside analogues: syntheses and X-ray crystallographic studies, S. Korunda, S. Kristafor, M. Cetina, S. Raic-Malic, Curr. Org. Chem. *2013, 17* (10), 1114-24; properties, and structure of **tetrazoles**: certain achievements and prospects, S.V. Voitekhovich, Q.A. Ivashkevich, P.N. Gaponik, Russ. J. Org. Chem. *2013, 49* (5), 635-54; **N-aryl-C-nitroazoles**, M. Kurpet, R. Jedrysiak, J. Suwinski, Chem. Heterocycl. Compd. *2013, 48* (12), 1737-50; functionalized 3-aryl**isoxazoles** and 3-aryl-2-isoxazolines from reaction of aryl nitrile oxides and enolates: synthesis and reactivity, P. Vitale, A. Scilimati, Synthesis *2013, 45* (21), 2940-8; α-lithiated α-heterosubstituted 2-alkyloxazolines: key players in organic synthesis, S. Florio, Synlett *2013, 24* (9), 1061-85; construction of structurally inspiring spiroisoxazolines, S. Dadiboyena, Curr. Org. Synth. *2013, 10* (5), 661-82; **morpholines**: synthesis and biological activity, V.A. Pal′chikov, Russ. J. Org. Chem. *2013, 49* (6), 787-814; design, synthesis and molecular modeling studies of some heterocyclic compounds derived from the Suzuki coupling of 6-bromo-**1,3,4-oxadiazine** together with their antitumor and anti-leishmanial evaluations, R.M. Mohareb, S.M. Sherif, A.A. Elkair, Curr. Org. Chem. *2013, 17* (17), 1910-8; reactions of **pyridines**, benzopyridines and azapyridines with organomagnesiums and organolithiums, M. Schlosser, Top. Heterocycl. Chem. *2013, 31,* 171-222; metalation reactions of pyridines, quinolines, and isoquinolines with ate bases and their alkali metal salt-modified congeners, C. Metallinos, K. Stromski, Top. Heterocycl. Chem. *2013, 31,* 65-92; one-pot synthesis of 2,4,6-triarylpyridines by the oxidative cyclocondensation of benzaldehydes, aromatic alkynes and ammonium bifluoride, P. Wang, R. Hua, C.-J. Li, Curr. Org. Synth. *2013, 10* (4), 655-60; 2,2′:6′,2″-terpyridines functionalized with thienyl substituents: synthesis and applications, J. Husson, M. Knorr, J. Heterocycl. Chem. *2012, 49* (3), 453-78; dihydro**pyrimidine**, H. Cho, Heterocycles *2013, 87* (7), 1441-79; lithiations and Grignard reactions on pyrimidine and quinazoline, A. Kolarovic, Top. Heterocycl. Chem. *2013, 31,* 21-64; other stoichiometric metalation reactions on pyrimidine and quinazoline, P.C. Gros, Top. Heterocycl. Chem. *2013, 31,* 1-20; transition metal-catalyzed functionalization of **pyrazines**, N.I. Nikishkin, J. Huskens, W. Verboom, Org. Biomol. Chem. *2013, 11* (22), 3583-602; metalation of pyrazine and quinoxaline, N. Plé, C. Fruit, Top. Heterocycl. Chem. *2013, 31,* 131-70; synthesis of tetrahydro**pyridazine** amino acid derivatives by a formal [4+2]-cycloaddition reaction of 1,2-diaza-1,3-dienes with dehydroalanine esters, O.A. Attanasi, G. Favi, G. Giorgi, F. Mantellini, G. Moscatelli, G. Piersanti, Curr. Org. Synth. *2013, 10* (5), 803-11; metalation of pyridazine, cinnoline, and phthalazine, E. Horkel, Top. Heterocycl. Chem. *2013, 31,* 223-68; **1,2,4,5-tetrazines** and azolo[1,2,4,5]tetrazines: synthesis and reactions with nucleophiles, S.G. Tolshchina, G.L. Rusinov, V.N. Charushin, Chem. Heterocycl. Compd. *2013, 49* (1), 66-91; electrophilicity: the 'dark-side' of **indole chemistry**, M. Bandini, Org. Biomol. Chem. *2013, 11* (32), 5206-12; metalated indoles, indazoles, benzimidazoles, and azaindoles and their synthetic applications, H. Ila, J.T. Markiewicz, V. Malakhov, P. Knochel, Synthesis *2013, 45* (17), 2343-71; cyanoacetylindoles: versatile starting materials in organic synthesis, T. Govender, G.E.M. Maguire, H.G. Kruger, M. Shiri, Curr. Org. Synth. *2013, 10* (5), 737-50; chemistry of 3-acyl-2-haloindoles, S.V. Den′kina, K.F. Suzdalev, Chem. Heterocycl. Compd. *2013, 48* (11), 1581-612; synthesis of 2-(indol-3-yl)-ethanone-based arylhydrocarbon receptor agonist candidates via Weinreb amides of indole-3-acetic acid, M. Opietnik, A. Jungbauer, K. Mereiter. T. Rosenau, Curr. Org. Synth. *2013, 10* (5), 812-8; recent advances in asymmetric organocatalytic construction of 3,3′-spirocyclic **oxindoles**, L. Hong, R. Wang, Adv. Synth. Catal. *2013, 355* (6), 1023-52; advances in the synthesis of cycloalka[*b*]-indoles, R.R. Gataullin, Russ. J. Org. Chem. *2013, 49* (2), 151-85; recent advances in the synthesis and properties of 4-, 5-, 6- or 7-azaindoles, J.-Y. Mérour, S. Routier, F. Suzenet, B. Joseph, Tetrahedron *2013, 69* (24), 4767-834; enantioselective organocatalytic reactions with **isatin**, M. Flores, J. Pena, P. García-García, N.M. Garrido, D. Diez, Curr. Org. Chem. *2013, 17* (18), 1957-85; isatin as a strategic motif for asymmetric catalysis, S. Mohammadi, R. Heiran, R.P. Herrera, E. Marqués-López, ChemCatChem *2013, 5* (8), 2131-48; lithiations and magnesiations on **quinoline and isoquinoline**, F. Chevallier, F. Mongin, Top. Heterocycl. Chem. *2013, 31,* 93-129; recent developments in the synthesis of **quinoxaline** derivatives by green synthetic approaches, Y.V.D. Nageswar, K.H.V. Reddy, K. Ramesh, S.N. Murthy, *2013, 45* (1), 1-27; structure of NH-**benzazoles** (1*H*-benzimidazoles, 1*H*- and 2*H*-indazoles, 1*H*- and 2*H*-benzotriazoles), J. Elguero, I. Alkorta, R.M. Claramunt, P. Cabildo, P. Cornago, M.A. Farran, M.A. Garcia, C. Lopez, M. Perez-Torralba, D.S. Maria, D. Sanz, Chem. Heterocycl. Compd. *2013, 49* (1), 177-202; metal-catalyzed **furan synthesis**, W.J. Moran, A. Rodriguez, Org. Prep. Proced. Int. *2012, 44* (2), 103-30; the development of pot-, atom- and step-economic synthesis of functionalized tetrahydro**pyrans**, dihydropyrans and piperidines, P.A. Clarke, K. Ermanis, Curr. Org. Chem. *2013, 17* (18), 2025-37; synthesis of polyfluoroalkyl containing **thiopyran** derivatives and their applications in fluoroorganic chemistry, S.A. Siry, V.M. Timoshenko, J.-P. Bouillon, J. Fluorine Chem. *2012, 137*, 6-21; angular tricyclic **benzofurans** and related natural products of fungal origin: isolation, biological activity and synthesis, S.O. Simonetti, E.L. Larghi, A.B.J. Bracca, T.S. Kaufman, Nat. Prod. Rep. *2013, 30* (7), 941-69; asymmetric synthesis of chiral **flavanone and chromanone** derivatives, N.-X. Wang, Y. Xing, Y.-J. Wang, Curr. Org. Chem. *2013, 17* (14), 1555-62; syntheses and properties of **cyclosilazanes** and cyclocarbosilazanes, M. Kavala, A. Hawkins, P. Szolcsanyi, J. Organomet. Chem. *2013, 732*, 58-91; complex N-heterocycle synthesis via iron-catalyzed, direct C-H bond amination, E.T. Hennessy, T.A. Betley, Science *2013, 340* (6132), 591-5; 4-(2-R-aryl)-1,2,3-chalcogenadiazoles in the synthesis of fused heterocycles, M.L. Petrov, Russ. J. Org. Chem. *2013, 49* (4), 479-501; microwave-assisted chemistry of **purines and xanthines**: an overview, A. Manvar, A. Shah, Tetrahedron *2013, 69* (38), 8105-27; Michael addition-initiated sequential reactions from 1,3-dicarbonyls for the synthesis of **polycyclic heterocycles**, X. Bugaut, D. Bonne, Y. Coquerel, J. Rodriguez, T. Constantieux, Curr. Org. Chem. *2013, 17* (18), 1920-8; synthetic strategies to benzopyrrolo[1,2-*a*]azepines, S.A. El Bialy, B.F. Abdel-Wahab, Tetrahedron *2013, 69* (45), 9357-71; synthetic developments in functionalized pyrano[2,3-*c*]pyrazoles, B. Myrboh, H. Mecadon, M.R. Rohman, M. Rajbangshi, I. Kharkongor, B.M. Laloo, I. Kharbangar, B. Kshiar, Org. Prep. Proced. Int. *2013, 45* (4), 253-303; dithieno[3,2-*b*:2′,3′-*d*]phospholes: a look back at the first decade, C. Romero-Nieto, T. Baumgartner, Synlett *2013, 24* (8), 920-37; thieno[3,4-*b*]-pyrazines and its extended analogs: important building blocks for conjugated materials, S.C. Rasmussen, M.E. Mulholland, R.L. Schwiderski, C.A. Larsen, J. Heterocycl. Chem. *2012, 49* (3), 479-93; synthesis of new β-carboline derivatives fused with β-lactam rings: an experimental and theoretical study, M. Milen, P. Abranyi-Balogh, Z. Mucsi, A. Dancso, D. Frigyes, L. Pongo, G. Keglevich, Curr. Org. Chem. *2013, 17* (17), 1894-902; synthesis and characterization of poly([1,2,4]triazolyl- and [1,2,4]triazolo[3,4-*b*]-[1,3,4]thiadiazinylsulfanylmethyl)-arenes: novel multi-armed molecules, A.H.M. Elwahy, R.M. Sarhan, M.A. Badawy, Curr. Org. Synth. *2013, 10* (5), 786-90; hetero-cyclization of 1- and 2-alkynyl-9,10-anthraquinones: new synthetic and mechanistic aspects, S.F. Vasilevskiy, D.S. Baranov, Chem. Heterocycl. Compd. *2013, 49* (1), 140-51; synthesis of oxa-, aza- and thia-bowls and cages, S.J. Gharpure, S.K. Porwal, Org. Prep. Proced. Int. *2013, 45* (2), 81-153; synthesis of **phthalocyanines** with an extended system of π-electron conjugation, T.V. Dubinina, L.G. Tomilova, N.S. Zefirov, Russ. Chem. Rev. *2013, 82* (9), 865-95; *meso*-tetraaryl**porphyrins** bearing nitro or amino groups: synthetic strategies and reactivity profiles, V.I. Vaz Serra, S.M.G. Pires, Top. Heterocycl. Chem. *2013, 33,* 35-78; synthetic routes to unsymmetrical porphyrins, S. Nardis, Top. Heterocycl. Chem. *2013, 33,* 203-29; recent advancements in chiral porphyrin self-assembly, D. Monti, Top. Heterocycl. Chem. *2013, 33,* 231-91; functionalization of porphyrins through C-C bond formation reactions with functional group-bearing organometallic reagents, T. Takanami, Heterocycles *2013, 87* (8), 1659-89; degradation pathways for porphyrinoids, J. Wojaczynski, Top. Heterocycl. Chem. *2013, 33,* 143-202; supramolecular chemistry of protoporphyrin IX and its derivatives, S.V. Bhosale, S.V. Bhosale, G.V. Shitre, S.R. Bobe, A. Gupta, Eur. J. Org. Chem. *2013,* (19), 3939-54; tetrabenzoporphyrins: synthetic developments and applications, C.M.B. Carvalho, T.J. Brocksom, K.T. de Oliveira, Chem. Soc. Rev. *2013, 42* (8), 3302-17; synthesis and characterization of triphyrins: a new family of porphyrinoids, D. Kuzuhara, H. Yamada, Heterocycles *2013, 87* (6), 1209-40; the McMurry reaction in porphyrinoid chemistry, K.M. Smith, Top. Heterocycl. Chem. *2013, 33,* 1-34; functionalization of **corroles**, J.F.B. Barata, C.I.M. Santos, Top. Heterocycl. Chem. *2013, 33,* 79-141.

11 Natural product synthesis - general aspects, alkaloids, terpenes, drug development ...

Following the lead from nature: divergent pathways in **natural product synthesis** and diversity-oriented synthesis, C. Serba, N. Winssinger, Eur. J. Org. Chem. *2013* (20), 4195-214; C-H activation: a complementary tool in the total synthesis of complex natural products, D.Y.-K. Chen, S.W. Youn, Chem. Eur. J. *2012, 18* (31), 9452-74; aziridinyl imines in organic synthesis: development of tandem reaction strategies and application to total synthesis of natural products, H.-Y. Lee, S.-B. Song, T. Kang, Y.J. Kim, S.J.

Geum, Pure Appl. Chem. *2013, 85* (4), 741-54; palladium-mediated total synthesis of bioactive natural products, K.C. Majumdar, B. Sinha, Synthesis *2013, 45* (10), 1271-99; the Claisen rearrangement in the syntheses of bioactive natural products, K.C. Majumdar, R.K. Nandi, Tetrahedron *2013, 69* (34), 6921-57; phosphine organocatalysis in the synthesis of natural products and bioactive compounds, C. Gomez, J.-F. Betzer, A. Voituriez, A. Marinetti, ChemCatChem *2013, 5* (5), 1055-65; strained alkenes in natural product synthesis, M.R. Wilson, R.E. Taylor, Angew. Chem., Int. Ed. *2013, 52* (15), 4078-87; *ortho*-quinone methids in natural product synthesis, N.J. Willis, C.D. Bray, Chem. Eur. J. *2012, 18* (30), 9160-73; allylic oxidations in natural product synthesis, A. Nakamura, M. Nakada, Synthesis *2013, 45* (10), 1421-51; diastereoselective allylation of carbonyl compounds and imines: application to the synthesis of natural products, M. Yus, J.C. González-Gómez, F. Foubelo, Chem. Rev. *2013, 113* (7), 5595-698; recent contributions from the asymmetric aza-Michael reaction to alkaloids total synthesis, Z. Amara, J. Caron, D. Joseph, Nat. Prod. Rep. *2013, 30* (9), 1211-25; one-pot asymmetric 6π-azaelectrocyclization as a new strategy for **alkaloid synthesis**, T. Kobaysahi, T. Sakaguchi, S. Katsumura, Heterocycles *2013, 87* (4), 729-61; the akuammiline alkaloids: origin and synthesis, R. Eckermann, T. Gaich, Synthesis *2013, 45* (20), 2813-23; mitomycinoid alkaloids: mechanism of action, biosynthesis, total syntheses, and synthetic approaches, P.D. Bass, D.A. Gubler, T.C. Judd, R.M. Williams, Chem. Rev. *2013, 113* (8), 6816-63; simple indole alkaloids and those with a non-rearranged monoterpenoid unit, M. Ishikura, T. Abe, T. Choshi, S. Hibino, Nat. Prod. Rep. *2013, 30* (5), 694-752; muscarine, imidazole, oxazole and thiazole alkaloids, Z. Jin, Nat. Prod. Rep. *2013, 30* (6), 869-915; construction of decahydroquinoline ring systems: the case of lepadin alkaloids, A. Pelss, A.M.P. Koskinen, Chem. Heterocycl. Compd. *2013, 49* (2), 226-40; *amaryllidaceae* and *sceletium* alkaloids, Z. Jin, Nat. Prod. Rep. *2013, 30* (6), 849-68; the chemistry and biological activity of the *hyacinthaceae*, D.A. Mulholland, S.L. Schwikkard, N.R. Crouch, Nat. Prod. Rep. *2013, 30* (9), 1165-210; halogenated organic molecules of *rhodomelaceae* origin: chemistry and biology, B.-G. Wang, J.B. Gloer, N.-Y. Ji, J.-C. Zhao, Chem. Rev. *2013, 113* (5), 3632-85; upenamide: trials and tribulations, W.P. Unsworth, R.J.K. Taylor, Org. Biomol. Chem. *2013, 11* (42), 7250-61; **marine natural products**: synthetic aspects, J.C. Morris, Nat. Prod. Rep. *2013, 30* (6), 783-805; marine indole alkaloids containing an 1-(indol-3-yl)ethane-1,2-diamine fragment, N.E. Golantsov, A.A. Festa, A.V. Karchava, M.A. Yurovskaya, Chem. Heterocycl. Compd. *2013, 49* (2), 203-25; the marinopyrroles, D.L.J. Clive, P. Cheng, Tetrahedron *2013, 69* (25), 5067-78; tetrahydrofuran-containing macrolides: a fascinating gift from the deep sea, A. Lorente, J. Lamariano-Merketegi, F. Albericio, M Álvarez, Chem. Rev. *2013, 113* (7), 4567-610; diterpenoids of terrestrial origin, J.R. Hanson, Nat. Prod. Rep. *2013, 30* (10), 1346-56; natural **sesquiterpenoids**, B.M. Fraga, Nat. Prod. Rep. *2013, 30* (9), 1226-64; synthesis of **steroids** with nitrogen-containing substituents in ring, D.S.V. Stulov, A.Y. Misharin, Russ. J. Org. Chem. *2013, 48* (10), 1431-72; structure, bioactivity, and chemical synthesis of OSW-1 and other steroidal glycosides in the genus ornithogalum, Y. Tang, N. Li, J.-A. Duan, W. Tao, Chem. Rev. *2013, 113* (7), 5480-514; **(-)-isosteviol** as a versatile ex-chiral-pool building block for organic chemistry, C. Lohoelter, M. Weckbecker, S.R. Waldvogel, Eur. J. Org. Chem. *2013,* (25), 5539-54; recent advances in **cardanol** chemistry in a nutshell: from a nut to nanomaterials, V.S. Balachandran, S.R. Jadhav, P.K. Vemula, G. John, Chem. Soc. Rev. *2013, 42* (1), 427-38; recent advances in the syntheses of **decalin-type natural products**, H. Akita, Heterocycles *2013, 87* (8), 1625-58; distribution of cyclic and acyclic **bis-bibenzyls** in the marchantiophyta (liverworts), ferns and higher plants and their biological activities, biosynthesis, and total synthesis, Y. Asakawa, A. Ludwiczuk, Heterocycles *2012, 86* (2), 891-917; phytoalexins of the *pyrinae*: **biphenyls and dibenzofurans**, C. Chizzali, L. Beerhues, Beilstein J. Org. Chem. *2012, 8,* 613-20; toward the ideal synthesis and transformative therapies: the roles of step economy and function-oriented synthesis, P.A. Wender, Tetrahedron *2013, 69* (36), 7529-50; click chemistry for **drug development** and diverse chemical-biology applications, P. Thirumurugan, D. Matosiuk, K. Jozwiak, Chem. Rev. *2013, 113* (7), 4905-79;

bioorthogonal labelling of biomolecules: new functional handles and ligation methods, M.F. Debets, J.C.M. van Hest, F.P.J.T. Rutjes, Org. Biomol. Chem. *2013, 11* (38), 6439-55; biocatalyzed on-water synthesis of chiral building blocks for the preparation of **anti-cancer** drugs: a greener approach, P. Hoyos, V. Pace, A.R. Alcantara, Curr. Org. Chem. *2013, 17* (11), 1132-57; enantioselective synthesis of pactamycin, a complex antitumor antibiotic, J.T. Malinowski, R.J. Sharpe, J.S. Johnson, Science *2013, 340* (6129), 180-2; salient features of enantioselective gas chromatography: the enantiomeric differentiation of chiral **inhalation anesthetics** as a representative methodological case in point, V. Schurig, Top. Curr. Chem. *2013, 340,* 153-207; synthesis of two neurokinin NK1 receptor antagonists: (+)-l-733,060 and (-)-l-733,061, A. Cochi, D. Gomez Pardo, J. Cossy, Heterocycles *2012, 86* (1), 89-116; recent approaches for the synthesis of modified **cryptophycins**, C. Weiss, B. Sammet, N. Sewald, Nat. Prod. Rep. *2013, 30* (7), 924-40; advances in the total synthesis of biologically important **callipeltosides**, H. Hussain, I.R. Green, K. Krohn, I. Ahmed, Nat. Prod. Rep. *2013, 30* (5), 640-93; synthesis of **tetrodotoxin**, a classic but still fascinating natural product, T. Nishikawa, M. Isobe, Chem. Record *2013, 13* (3), 286-302; rational synthesis of **ibandronate and alendronate**, G. Keglevich, A. Grun, R. Kovacs, S. Garadnay, I. Greiner, Curr. Org. Synth. *2013, 10* (4), 640-4; **azaphilones:** chemistry and biology, J.-M. Gao, S.-X. Yang, J.-C. Qin, Chem. Rev. *2013, 113* (7), 4755-811; surveying the effects of **eldecalcitol** and related analogs from a biological perspective, N. Kubodera, Heterocycles *2012, 86* (1), 69-88; synthesis of **allocolchicinoids**: a 50 year journey, N.S. Sitnikov, A.Y. Fedorov, Russ. Chem. Rev. *2013, 82* (5), 393-411; the winding pathway to **erythropoietin** along the chemistry-biology frontier: a success at last, R.M. Wilson, S. Dong, P. Wang, S.J. Danishefsky, Angew. Chem., Int. Ed. *2013, 52* (30), 7646-65.

12 Carbocyclic chemistry - cyclopropanes, norbornadienes, fullerenes

(for carbocyclic synthesis by cycloaddition and Nazarov reaction s. under Section 6)

Palladium-catalyzed oxidative **carbocyclizations**, Y. Deng, A.K.Å. Persson, J.-E. Bäckvall, Chem. Eur. J. *2012, 18* (37), 11498-523; visible-light-driven photoredox catalysis in the construction of carbocyclic and heterocyclic ring systems, J. Xuan, L.-Q. Lu, J.-R. Chen, W.-J. Xiao, Eur. J. Org. Chem. *2013,* (30), 6755-70; syntheses and applications of monofluorinated **cyclopropanes**, E. David, G. Milanole, P. Ivashkin, S. Couve-Bonnaire, P. Jubault, X. Pannecoucke, Chem. Eur. J. *2012, 18* (47), 14904-17; aminocyclopropenium ions: synthesis, properties, and applications, J.S. Bandar, T.H. Lambert, Synthesis *2013, 45* (18), 2485-98; synthesis of substituted **norbornadienes**, K. Jack, B. Machin, A. Tigchelaar, W. Tam, Curr. Org. Synth. *2013, 10* (4), 584-630; end-group-differentiating ozonolysis of norbornene systems to afford highly substituted cyclopentane rings, S.A. Testero, M.I. Mangione, A.G. Suárez, R.A. Spanevello, Eur. J. Org. Chem. *2013,* (24), 5236-45; **fullerene** nanoarchitectonics: from zero to higher dimensions, L.K. Shrestha, Q. Ji, T. Mori, K. Miyazawa, Y. Yamauchi, J.P. Hill, K. Ariga, Chem. Asian J. *2013, 8* (8), 1662-79; synthesis and properties of [60]fullerene derivatives functionalized through copper catalyzed Huisgen cycloaddition reactions, J.L. Segura, M. Castelaín, H. Salavagione, Curr. Org. Synth. *2013, 10* (5), 724-36; carbene additions to fullerenes, M. Yamada, T. Akasaka, S. Nagase, Chem. Rev. *2013, 113* (9), 7209-64; radical reactions of fullerenes: from synthetic organic chemistry to materials science and biology, M.D. Tzirakis, M. Orfanopoulos, Chem. Rev. *2013, 113* (7), 5262-321; endohedral fullerenes, A.A. Popov, S. Yang, L. Dunsch, Chem. Rev. *2013, 113* (8), 5989-6113.

13 Aromatic chemistry - arynes, polyarenes, phenolic compds.

(for arylations, ar. cross-coupling and standard aromatic substitutions s. under Section 6; for aromatic C-H activation s. under Section 7)

Aromatic interactions as control elements in stereoselective organic reactions, E.H. Krenske, K.N. Houk, Acc. Chem. Res. *2013, 46* (4), 979-89; comparison of **oxidative**

aromatic coupling and the Scholl reaction, M. Grzybowski, K. Skonieczny, H. Butenschön, D.T. Gryko, Angew. Chem., Int. Ed. *2013, 52* (38), 9900-30; **aryne cycloaddition** reactions in the synthesis of large polycyclic aromatic compounds, D. Pérez, D. Peña, E. Guitián, Eur. J. Org. Chem. *2013,* (27), 5981-6013; concerted reactions that produce diradicals and zwitterions: electronic, steric, conformational, and kinetic control of **cycloaromatization** processes, R.K. Mohamed, P.W. Peterson, I.V. Alabugin, Chem. Rev. *2013, 113* (9), 7089-129; studies with **polyfunctional heteroaromatics and aromatics**: synthesis of tetra- and penta-substituted benzoic and phthalic acid derivatives through reactions of acetylenic esters or diesters with arylidenemalononitriles in the presence of (S)-proline or dabco, S.M. Al-Mousawi, M.S. Moustafa, N.M. Hilmy, M.H. Elnagdi, Curr. Org. Synth. *2013, 10* (5), 791-7; fluorescent photochromes of **diarylethene** series: synthesis and properties, V.Z. Shirinian, D.V. Lonshakov, A.G. Lvov, M.M. Krayushkin, Russ. Chem. Rev. *2013, 82* (6), 511-37; functionalization of **phenols** using a biomimetic couple of $NaNO_2$/hydrogen peroxide in an acidic medium, M. Jereb, Curr. Org. Chem. *2013, 17* (15), 1694-700; selective monoetherification of 1,4-hydroquinone promoted by $NaNO_2$, C. Gambarotti, L. Melone, C. Punta, S.U. Shisodia, Curr. Org. Chem. *2013, 17* (10), 1108-13; organic synthesis of C-prenylated phenolic compounds, O. Talhi, A.M.S. Silva, Curr. Org. Chem. *2013, 17* (10), 1067-102; syntheses of amido-, carbamido- and carbamato-alkylnaphthols, R. Csutortoki, I. Szatmari, F. Fulop, Curr. Org. Synth. *2013, 10* (4), 564-83; B-N as a C-C substitute in aromatic systems, M.J.D. Bosdet, W.E. Piers, Can. J. Chem. *2009, 87* (1), 8-29.

14 Amino acids, peptide and protein chemistry

Progress in glycine equivalent-based **α-amino acid synthesis**, A. Jakubowska, K. Kulig, Curr. Org. Synth. *2013, 10* (4), 547-63; synthesis of N-protected non-proteinogenic α-amino acid esters by using trichloroacetimidate- and acetate-coupling methods, I.A.I. Ali, W. Fathalla, Curr. Org. Chem. *2013, 17* (17), 1903-9; synthesis of unnatural amino acids via microwave-assisted regioselective, one-pot multicomponent reactions of sulfamidates, N.B.R. Baig, R.S. Varma, Curr. Org. Chem. *2013, 17* (20), 2323-31; strategies for asymmetric synthesis of amino acids with γ,δ-unsaturation, Z. Liu, S.J. Mehta, V.J. Hruby, Org. Prep. Proced. Int. *2012, 44* (3), 222-55; synthetic strategies for the development of fluorescent amino acids as optical probe, S.K. Maity, S. Bera, D. Haldar, Curr. Org. Synth. *2013, 10* (4), 525-46; organocatalytic asymmetric synthesis of **β-amino acid derivatives**, S.M. Kim, J.W. Yang, Org. Biomol. Chem. *2013, 11* (29), 4737-49; asymmetric synthesis of ^{11}C- labelled α-amino acids for PET, A. Popkov, P.H. Elsinga, Curr. Org. Chem. *2013, 17* (19), 2127-37; synthetic **glycopeptides and glycoproteins** with applications in biological research, U. Westerlind, Beilstein J. Org. Chem. *2012, 8,* 804-18; multivalent glyco(cyclo)peptides, M.C. Galan, P. Dumy, O. Renaudet, Chem. Soc. Rev. *2013, 42* (11), 4599-612; protein organic chemistry and applications for labeling and engineering in live-cell systems, Y. Takaoka, A. Ojida, I. Hamachi, Angew. Chem., Int. Ed. *2013, 52* (15), 4088-106.

15 Carbohydrate chemistry

Enzymatic **glycosylation** of multivalent scaffolds, P. Bojarová, R.R. Rosencrantz, L. Elling, V. Kren, Chem. Soc. Rev. *2013, 42* (11), 4774-97; enzymatic glycosylation of small molecules: challenging substrates require tailored catalysts, T. Desmet, W. Soetaert, P. Bojarová, V. Kren, L. Dijkhuizen, V. Eastwick-Field, A. Schiller, Chem. Eur. J. *2012, 18* (35), 10786-801; cross-coupling reactions for the synthesis of **C-glycosides** and related compounds, P. Merino, T. Tejero, E. Marca, F. Gomollón-Bel, I. Delso, R. Matute, Heterocycles *2012, 86* (2), 791-820; **glucuronides** from metabolites to medicines: a survey of the *in vivo* generation, chemical synthesis and properties of glucuronides, A.V. Stachulski, X. Meng, Nat. Prod. Rep. *2013, 30* (6), 806-48; **fluoro-C-glycosides** and fluoro-carbasugars, hydrolytically stable and synthetically challenging glycomimetics, E. Leclerc, X. Pannecoucke, M. Ethève-Quelquejeu, M. Sollogoub, Chem.

Soc. Rev. *2013, 42* (10), 4270-83; synthesis of *L*-**hexoses** and their related biomolecules, M.M.L. Zulueta, Y.-Q. Zhong, S.-C. Hung, Chem. Commun. *2013, 49* (32), 3275-87; glycal-derived δ-hydroxy-α,β-unsaturated aldehydes (perlin aldehydes): versatile building blocks in organic synthesis, L.V.R. Reddy, V. Kumar, R. Sagar, A.K. Shaw, Chem. Rev. *2013, 113* (5), 3605-31; recent developments in design and synthesis of **bicyclic azasugars, carbasugars** and related molecules as glycosidase inhibitors, R. Lahiri, A.A. Ansari, Y.D. Vankar, Chem. Soc. Rev. *2013, 42* (12), 5102-18; chemical approaches to study O-GlcNAcylation, P.S. Banerjee, G.W. Hart, J.W. Cho, Chem. Soc. Rev. *2013, 42* (10), 4345-57; hydrolysis and isomerization of **sugar phosphates** and carbohydrate phosphodiesters, S. Mikkola, Curr. Org. Chem. *2013, 17* (14), 1525-44; a general strategy for the chemo-enzymatic synthesis of asymmetrically branched **N-glycans**, Z. Wang, Z.S. Chinoy, S.G. Ambre, W. Peng, R. McBride, R.P. de Vries, J. Glushka, J.C. Paulson, G.-J. Boons, Science *2013, 341* (6144), 379-83; the use of glycoinformatics in glycochemistry, T. Lütteke, Beilstein J. Org. Chem. *2012, 8,* 915-29; new strategies for the synthesis of bio-medically relevant **oligosaccharides**: recent updates on 1,2-*cis*-O-glycosylation and α-O-sialylation, S. Crotti, R. Adamo, Curr. Org. Synth. *2013, 10* (4), 501-24.

16 Nucleosides, nucleotides

Advances in the enantioselective synthesis of carbocyclic nucleosides, O. Boutureira, M.I. Matheu, Y. Díaz, S. Castillón, Chem. Soc. Rev. *2013, 42* (12), 5056-72; click chemistry-based functionalization of nucleoside, nucleotide and nucleic acids, A.R. Kore, I. Charles, Curr. Org. Chem. *2013, 17* (19), 2164-91; synthesis and applications of fluorinated nucleoside analogs, H. Wojtowicz-Rajchel, J. Fluorine Chem. *2012, 143,* 11-48; conjugates of 1,2,3-triazoles and acyclic pyrimidine nucleoside analogs: syntheses and X-ray crystallographic studies, S. Korunda, S. Kristafor, M. Cetina, S. Raic-Malic, Curr. Org. Chem. *2013, 17* (10), 1114-24.

17 Functional groups - alkenes, allenes, alkynes, O-, N-, Hal-, S-, B-, Si-functions

(s.a. under Section 6 for classical functional group conversions)

Migratory insertion of **alkenes** into metal-oxygen and metal-nitrogen bonds, P.S. Hanley, J.F. Hartwig, Angew. Chem., Int. Ed. *2013, 52* (33), 8510-25; three carbons for complexity! recent developments of palladium-catalyzed reactions of **allenes**, T. Lechel, F. Pfrengle, H.-U. Reissig, R. Zimmer, ChemCatChem *2013, 5* (8), 2100-30; propargyl Claisen rearrangement: allene synthesis and beyond, D. Tejedor, G. Méndez-Abt, L. Cotos, F. García-Tellado, Chem. Soc. Rev. *2013, 42* (1), 458-71; alleneamides: powerful and versatile building blocks in organic synthesis, T. Lu, Z. Lu, Z.-X. Ma, Y. Zhang, R.P. Hsung, Chem. Rev. *2013, 113* (7), 4862-904; organic and organometallic derivatives of propargylallene: syntheses, structures, reactivity and rearrangements, M.J. McGlinchey, H. Hopf, Eur. J. Org. Chem. *2013,* (22), 4705-28; cyclizations of **enediynes** under the action of electrophiles, A.V. Gulevskaya, R.Y. Lazarevich, Chem. Heterocycl. Compd. *2013, 49* (1), 116-39; diverse transformations of chiral propargylic alcohols generated by BONOL-catalyzed **alkyne** addition to aldehydes, Q. Wang, L. Pu, Synlett *2013, 24* (11), 1340-63; organocatalyzed enantioselective desymmetrization of **diols** in the preparation of chiral building blocks, M.D. Díaz-de-Villegas, J.A. Gálvez, R. Badorrey, M.P. López-Ram-de-Víu, Chem. Eur. J. *2012, 18* (44), 13920-35; 1-(acyloxy)benzotriazoles: useful reagents for the regioselective acylation of diols, B.L. Wilkinson, Austr. J. Chem. *2013, 66* (8), 910-2; recent advances in the synthesis of **diphenylmethyl ethers**, M.T. Thornton, L.C. Henderson, Org. Prep. Proced. Int. *2013, 45* (5), 395-420; synthesis, reactions, applications, and biological activity of **diethanolamine** and its derivatives, M. Yar, N. Mushtaq, S. Afzal, Russ. J. Org. Chem. *2013, 49* (7), 949-67; entry to **tert. carbinamines** bearing a *gem*-diaryl group by copper-catalyzed Grignard addition to cyclic sulfamidate imines, A. Pitchaiah, M. Kim, I.T. Hwang, K.-I. Lee, Curr. Org. Synth.

2013, 10 (5), 798-802; direct α-functionalization of **saturated cyclic amines**, E.A. Mitchell, A. Peschiulli, N. Lefevre, L. Meerpoel, B.U.W. Maes, Chem. Eur. J. *2012, 18* (33), 10092-142; chemoselective activation strategies of **amidic carbonyls** towards nucleophilic reagents, V. Pace, W. Holzer, Austr. J. Chem. *2013, 66* (5), 507-10; substituted (E)-2-oxo-3-butenoates: reagents for every enantioselectively-catalyzed reaction, G. Desimoni, G. Faita, P. Quadrelli, Chem. Rev. *2013, 113* (8), 5924-88; recent advances in **isocyanide** insertion chemistry, G. Qiu, Q. Ding, J. Wu, Chem. Soc. Rev. *2013, 42* (12), 5257-69; palladium-catalyzed migratory insertion of isocyanides: an emerging platform in cross-coupling chemistry, T. Vlaar, E. Ruijter, B.U.W. Maes, R.V.A. Orru, Angew. Chem., Int. Ed. *2013, 52* (28), 7084-97; unravelling the labyrinth of palladium-catalyzed reactions involving isocyanides, S. Lang, Chem. Soc. Rev. *2013, 42* (12), 4867-80; highly selective introduction of heteroatom groups to isocyanides and its application to electrocyclic reactions, A. Ogawa, T. Tamai, T. Mitamura, A. Nomoto, Pure Appl. Chem. *2013, 85* (4), 785-800; **conjugated nitrodienes:** synthesis and reactivity, R. Ballini, N. Araújo, M.V. Gil, E. Román, J.A. Serrano, Chem. Rev. *2013. 113* (5), 3493-515; chemistry of bridged **lactams** and related heterocycles, M. Szostak, J. Aubé, Chem. Rev. *2013, 113* (8), 5701-65; 3,3-*gem*-difluorinated-**β-lactams**: synthesis pathways and applications, T. Orbegozo, F. Burel, P. Jubault, X. Pannecoucke, Tetrahedron *2013, 69* (20), 4015-39; methods of **sultam** synthesis, V.A. Rassadin, D.S. Grosheva, A.A. Tomashevskii, V.V. Sokolov, Chem. Heterocycl. Compd. *2013, 49* (1), 39-65; recent advances in enantioselective synthesis of γ-substituted **butenolides** via catalytic asymmetric vinylogous reactions, Q. Zhang, X. Liu, X. Feng, Curr. Org. Synth. *2013, 10* (5), 764-85; sustainable routes to **polyurethane precursors**, O. Kreye, H. Mutlu, M.A.R. Meier, Green Chem. *2013, 15* (6), 1431-55; 1-bromo-1-chloro-2,2,2-trifluoroethane (halothane) as a building block for **fluorine compounds**, W. Dmowski, J. Fluorine Chem. *2011, 132* (8), 504-11; 1,2-dibromotetrafluoroethane (Freon 114b2) as a building block for fluorine compounds, W. Dmowski, J. Fluorine Chem. *2012, 142*, 6-13; syntheses utilizing ***n*-perfluoroalkyl iodides** [R_FI, C_nF_{2n+1}-I] in 2000-10, P.M. Murphy, C.S. Baldwin, R.C. Buck, J. Fluorine Chem. *2012, 138*, 3-23; synthetic chemistry and biological activity of **pentafluorosulfenyl derivs.**, S.Altomonte, M. Zanda, J. Fluorine Chem. *2012, 143*, 57-93; recent developments of **ketene dithioacetal** chemistry, L. Pan, X. Bi, Q. Liu, Chem. Soc. Rev. *2013, 42* (3), 1251-86; **N-alkylsulfonamides** as useful carbon electrophiles, Y. Gu, S.-K. Tian, Synlett *2013, 24* (10), 1170-85; **sulfoxonium and sulfonium ylids** as diazocarbonyl equivalents in metal-catalyzed insertion reactions, A.C.B. Burtoloso, R.M.P. Dias, I.A. Leonarczyk, Eur. J. Org. Chem. *2013* (23), 5005-16; chemistry of **cyanoselenoacetamide**, V.V. Dotsenko, K.A. Frolov, S.G. Krivokolysko, Chem. Heterocycl. Compd. *2013, 49* (5), 657-75; catalytic enantioselective transformations of **borylated substrates**: preparation and synthetic applications of chiral alkylboronates, D.G. Hall, J.C.H. Lee, J. Ding, Pure Appl. Chem. *2012, 84* (11), 2263-78; alkynylboron compounds in organic synthesis, J. Jiao, Y. Nishihara, J. Organomet. Chem. *2012, 721-2*, 3-16; boroxine chemistry: from fundamental studies to applications in supramolecular and synthetic organic chemistry, Y. Tokunaga, Heterocycles *2013, 87* (5), 991-1021; isotope incorporation using organoboranes, G.W. Kabalka, M.-L. Yao, M. Akula, L. Yong, Pure Appl. Chem. *2012, 84* (11), 2309-16; chemical tricks to stabilize **silanones** and their heavier homologues with E=O bonds (E = Si-Pb): from elusive species to isolable building blocks, Y. Xiong, S. Yao, M. Driess, Angew. Chem., Int. Ed. *2013, 52* (16), 4302-11.

18 Protective group chemistry

N-Methyliminodiacetate (**MIDA**): a useful protecting group for the generation of complex organoboron compounds, Q.I. Churches, Austr. J. Chem. *2011, 64* (11), 1474.

19 Reagents and auxiliaries

(for organocatalysts and ligands s. under Sections 1 and 2; for nanocatalysts and supported catalysts s. under Section 3; for transition metal catalysts s. under Section 4; and for enzymes s. under Section 5)

A water-resistant solid superbase derived from **calcium**: synthesis, characterization and catalytic performance, L.-B. Sun, Y.-H. Li, Y. Yin, X.-L. Wu, X.-Q. Liu, Curr. Org. Chem. *2013, 17* (20), 2249-55; catalytic organic reactions on **ZnO**, M. Hosseini-Sarvari, Curr. Org. Synth. *2013, 10* (5), 697-723; **(2R,3R)-1,4-dimethoxy-1,1,4,4-tetraphenyl-butane-2,3-diol**: valuable reagent in the asymmetric synthesis of organoboronates, C.A. Berg, N.C. Eichenauer, J. Pietruszka, Pure Appl. Chem. *2012, 84* (11), 2339-416; **TEMPO** derivatives as alternative mild oxidants in carbon-carbon coupling reactions, O. García Mancheño, T. Stopka, Synthesis *2013, 45* (12), 1602-11; oxidative rearrangements with **hypervalent iodine** reagents, F.V. Singh, T. Wirth, Synthesis *2013, 45* (18), 2499-511; journey describing applications of **Oxone** in synthetic chemistry, H. Hussain, I.R. Green, I. Ahmed, Chem. Rev. *2013, 113* (5), 3329-71; **propanephosphonic anhydride** - a benign reagent for diverse applications inclusive of large-scale synthesis, Basavaprabhu, T.M. Vishwanatha, N.R. Panguluri, V.V. Sureshbabu, Synthesis *2013, 45* (12), 1569-601; catalytic enantioselective nucleophilic addition of **organolithium** derivatives: pitfalls and opportunities, A. Harrison-Marchand, H. Gerard, J. Maddaluno, New J. Chem. *2012, 36* (12), 2441-6; structurally defined **allyl compounds of main group metals**: coordination and reactivity, C. Lichtenberg, J. Okuda, Angew. Chem., Int. Ed. *2013, 52* (20), 5228-46; taming **sulfur dioxide**: a breakthrough for its wide utilization in chemistry and biology, P. Bisseret, N. Blanchard, Org. Biomol. Chem. *2013, 11* (33), 5393-8; synthetic uses of **ammonia** in transition-metal catalysis, J. Kim, H.J. Kim, S. Chang, Eur. J. Org. Chem. *2013,* (16), 3201-13; frontiers, opportunities, and challenges in biochemical and chemical catalysis of **CO_2** fixation, A.M. Appel, J.E. Bercaw, A.B. Bocarsly et al., Chem. Rev. *2013, 113* (8), 6621-58; catalyst development in the context of ring expansion-addition of carbon dioxide to epoxides to give organic carbonates, C.J. Whiteoak, A.W. Kleij, Synlett *2013, 24* (14), 1748-56.

20 Green chemistry - solvents, ionic liquids

Green solvents in organic synthesis: an overview, K. Shanab, C. Neudorfer, E. Schirmer, H. Spreitzer, Curr. Org. Chem. *2013, 17* (11), 1179-87; biocatalysis in biomass-derived solvents: the quest for fully sustainable chemical processes, M. Perez-Sanchez, M. Sandoval, M.J. Hernaiz, P. Domínguez de María, Curr. Org. Chem. *2013, 17* (11), 1188-99; **superheated water** as solvent in microwave-assisted organic synthesis of compounds of valuable pharmaceutical interest, F. Messina, O. Rosati, Curr. Org. Chem. *2013, 17* (11), 1158-78; recent development in functionalized **ionic liquids** as reaction media and promoters, W.-L. Wong, K.-Y. Wong, Austr. J. Chem. *2012*, *90* (1), 1-16; structural features of ionic liquids: consequences for material preparation and organic reactivity, C.C. Weber, A.F. Masters, T. Maschmeyer, Green Chem. *2013, 15* (10), 2655-79; **2-methyltetrahydrofuran**: a versatile eco-friendly alternative to THF in organometallic chemistry, V. Pace, Austr. J. Chem. *2012, 65* (3), 301-2.

21 Methodology - microwaves, ultrasound, flow chemistry, microreactors, electrochemical

Microwave-assisted chemical transformations, S.R. Takkellapati, Curr. Org. Chem. *2013, 17* (20), 2305-22; progress in microwave-aided chemical synthesis, H.M.A. Hassan, S. Harakeh, K.A. Sakkaf, I. Denetiu, Austr. J. Chem. *2012, 65* (12), 1647-54; microwave irradiation and phase transfer catalysis in C-, O- and N-alkylation reactions, G. Keglevich, A. Grün, E. Bálint, Curr. Org. Synth. *2013, 10* (5), 751-63; microwave-assisted alkylation of amines with alcohols: magnetically recoverable catalysis, S.R. Takkellapati, M.N. Nadagouda, Curr. Org. Chem. *2013, 17* (20), 2332-7; superheated water as solvent in microwave-assisted organic synthesis of compounds of valuable pharmaceutical interest,

F. Messina, O. Rosati, Curr. Org. Chem. *2013, 17* (11), 1158-78; microwave-assisted chemistry of purines and xanthines: an overview, A. Manvar, A. Shah, Tetrahedron *2013, 69* (38), 8105-27; synthesis of unnatural amino acids via microwave-assisted regioselective one-pot multicomponent reactions of sulfamidates, N.B.R. Baig, R.S. Varma, Curr. Org. Chem. *2013, 17* (20), 2323-31; applications of **ultrasound** in organic synthesis - a green approach, S. Puri, B. Kaur, A. Parmar, H. Kumar, Curr. Org. Chem. *2013, 17* (16), 1790-828; the role of **flow** in green chemistry and engineering, S.G. Newman, K.F. Jensen, Green Chem. *2013, 15* (6), 1456-72; asymmetric carbon-carbon bond formation under **continuous-flow** conditions with chiral heterogeneous catalysts, T. Tsubogo, T. Ishiwata, S. Kobayashi, Angew. Chem., Int. Ed. *2013, 52* (26), 6590-604; stereoselective organic reactions promoted by immobilized chiral catalysts in continuous flow systems, A. Puglisi, M. Benaglia, V. Chiroli, Green Chem. *2013, 15* (7), 1790-1813; supported ruthenium catalysts for sustainable flow chemistry, B.R. Vaddula, R.S. Varma, M.A.A. Gonzalez, Curr. Org. Chem. *2013, 17* (20), 2268-78; on-chip catalytic **microreactors** for modern catalysis research, B.-B. Xu, Y.-L. Zhang, S. Wei, H. Ding, H.-B. Sun, ChemCatChem *2013, 5* (8), 2091-9; direct nucleophilic functionalization of C(sp^2)-H-bonds in arenes and hetarenes by **electrochemical methods**, A.V. Shchepochkin, O.N. Chupakhin, V.N. Charushin, V.A. Petrosyan, Russ. Chem. Rev. *2013, 82* (8), 747-71.

Index to Volume 83

As in previous volumes, reactions are indexed from both the starting material and product aspects, e.g. '**Azides** startg. m.f. amines' and '**Amines** from azides'. Nomenclature for complex functions can be located under the 'special s.' sub-entry, e.g. '**Carboxylic acids** special s. aminocarboxylic acids' or by consulting the Formula Index of Complex Functional Groups (Volume *48*, p. 471).

Hydrogenated and functionalized ring systems are indexed by the conventional reversal, e.g. '**Pyridines, aryl-**', the only important exception to the rule being alkylideneisocyclics which are indexed as such, e.g. '**Alkylidenecyclopentanes**'.

As from Volume *51*, '**Epoxides**' has been used in place of 'Oxido compds.'; '**Thiiranes**' in place of 'Sulfido compds.'; '**Diels-Alder reaction**' in place of 'Diene synthesis'; and '**Benzo[*b*]thiophenes**' in place of 'Thianaphthenes'.

References to abstracts in this volume are in the format **83**, 234. An entry such as '**Suzuki coupling**, update **37,** 902s**83**' refers to the indexing of a supplementary reference, which may be followed up via the Supplementary References section (p. 519), from which the page number(s) on which the reference is located may be found.

Supplementary References in Volume 83

No.	Suppl. Vol. Ref. Page
257	*83*, 133
351	*83*, 179 (2)

Volume 32

29	*83*, 10
472	*83*, 172
716	*83*, 300
961	*83*, 437

Volume 33

47	*83*, 11
101	*83*, 33
108	*83*, 34
286	*83*, 94
501	*83*, 192
513	*83*, 207
546	*83*, 201
593	*83*, 219
618	*83*, 242
658	*83*, 278
865	*83*, 388

Volume 34

252	*83*, 30, 34, 45
528	*83*, 200
610	*83*, 233
668	*83*, 292
793	*83*, 247
825	*83*, 366
862	*83*, 412
953	*83*, 440

Volume 35

7	*83*, 1
85	*83*, 52
160	*83*, 78

Volume 36

117	*83*, 45
148	*83*, 49, 50
492	*83*, 201
665	*83*, 232
957	*83*, 439

Volume 37

236	*83*, 82
246	*83*, 160
657	*83*, 263
902	*83*, 4, 22, 353, 405
994	*83*, 443

Volume 38

140	*83*, 4
301	*83*, 103
375	*83*, 136
473	*83*, 175
584	*83*, 221
718	*83*, 297
772	*83*, 336
793	*83*, 353
836	*83*, 362
932	*83*, 335

Volume 39

83	*83*, 30, 44
189	*83*, 70
225	*83*, 79
593	*83*, 229
646	*83*, 263
701	*83*, 48, 166
854	*83*, 397
887	*83*, 412

Volume 40

1	*83*, 1
81	*83*, 46
109	*83*, 66
456	*83*, 246
501	*83*, 316
567	*83*, 235

Volume 41

111	*83*, 399
231	*83*, 76
325	*83*, 126
556	*83*, 211
671	*83*, 293
796	*83*, 362

Volume 42

45	*83*, 21
135	*83*, 48
354	*83*, 133
451	*83*, 35
654	*83*, 255, 263
682	*83*, 292

Volume 43

51	*83*, 12
200	*83*, 75
299	*83*, 125 (2)
319	*83*, 118
340	*83*, 153
568	*83*, 236
576	*83*, 234
703	*83*, 330
706	*83*, 332
710	*83*, 235, 236

No.	Suppl. Vol.	Ref. Page

Volume 54 continued

390	*83,* 347
469	*83,* 414
489	*83,* 440

Volume 55

146	*83,* 110, 127
149	*83,* 126
166	*83,* 155
290	*83,* 366
337	*83,* 331
433	*83,* 397
452	*83,* 402 (2)

Volume 56

129	*83,* 100
172	*83,* 167
438	*83,* 402

Volume 57

202	*83,* 199
376	*83,* 370
390	*83,* 386

Volume 58

78	*83,* 62
233	*83,* 229
236	*83,* 230
245	*83,* 227, 228
261	*83,* 245
468	*83,* 420

Volume 59

203	*83,* 195
311	*83,* 278

Volume 60

103	*83,* 76
135	*83,* 126
186	*83,* 183
201	*83,* 208
227	*83,* 227
307	*83,* 293
336	*83,* 331

Volume 61

71	*83,* 56
100	*83,* 72
195	*83,* 177
340	*83,* 331
392	*83,* 369

Volume 62

18	*83,* 7
39	*83,* 23
171	*83,* 146
282	*83,* 251
381	*83,* 361, 362
449	*83,* 401
650	*83,* 229

Volume 63

191	*83,* 169
253	*83,* 234
266	*83,* 241
411	*83,* 370

Volume 64

83	*83,* 48
141	*83,* 97
219	*83,* 219
249	*83,* 242
287	*83,* 279
448	*83,* 405
453	*83,* 407, 413

Volume 65

62	*83,* 185
83	*83,* 48
250	*83,* 246
266	*83,* 278
394	*83,* 366
437	*83,* 402

Volume 66

178	*83,* 133
290	*83,* 242
353	*83,* 319
452	*83,* 393

Volume 67

22	*83,* 12
290	*83,* 241
301	*83,* 259, 289
302	*83,* 266
310	*83,* 274
313	*83,* 267
336	*83,* 256
339	*83,* 293

Volume 68

143	*83,* 415, 418
184	*83,* 143, 207
230	*83,* 205

No.	Suppl. Vol.	Ref. Page

Volume 78 continued

48	*83,*	38
56	*83,*	47
88	*83,*	65
94	*83,*	69
128	*83,*	90
220	*83,*	179
229	*83,*	192
251	*83,*	213
285	*83,*	228
306	*83,*	270
319	*83,*	267
332	*83,*	276
476	*83,*	391
485	*83,*	347
527	*83,*	425, 426

Volume 79

69	*83,*	53, 54
81	*83,*	68
90	*83,*	72
103	*83,*	80
128	*83,*	102
145	*83,*	122
160	*83,*	6
206	*83,*	201
211	*83,*	205
225	*83,*	221
294	*83,*	276
319	*83,*	292
345	*83,*	307
346	*83,*	306
379	*83,*	338
384	*83,*	325
385	*83,*	338
390	*83,*	346
463	*83,*	425
470	*83,*	162

Volume 80

18	*83,*	12
23	*83,*	17
25	*83,*	18
29	*83,*	18
74	*83,*	54
84	*83,*	336
91	*83,*	68
97	*83,*	72
110	*83,*	54
117	*83,*	49
121	*83,*	92
128	*83,*	100
129	*83,*	99
139	*83,*	162
184	*83,*	162
189	*83,*	108
192	*83,*	164 (2)
214	*83,*	187
216	*83,*	195
221	*83,*	200
243	*83,*	213
284	*83,*	247
341	*83,*	285
363	*83,*	308
452	*83,*	378
463	*83,*	387
520	*83,*	442

Volume 81

25	*83,*	13
49	*83,*	63
160	*83,*	157
166	*83,*	162
170	*83,*	164
188	*83,*	189
206	*83,*	208
211	*83,*	208
220	*83,*	218
222	*83,*	180
306	*83,*	282
339	*83,*	299
460	*83,*	402
481	*83,*	422

Volume 82

30	*83,*	6
41	*83,*	172
116	*83,*	91
160	*83,*	141
166	*83,*	145
183	*83,*	164
189	*83,*	167
203	*83,*	184
262	*83,*	255
277	*83,*	249
301	*83,*	284
366	*83,*	336
376	*83,*	284
393	*83,*	413
400	*83,*	112
418	*83,*	365
465	*83,*	380
480	*83,*	97
484	*83,*	432
485	*83,*	431

Volume 83

181	*83,*	158
237	*83,*	98
314	*83,*	325
340	*83,*	304
379	*83,*	339
437	*83,*	444
470	*83,*	398